Encyclopedia of Microbiology

Second Edition

Volume 2 D–K

Editorial Board

Encyclopedia
of
MICROBIOLOGY

Second Edition

Volume 2 D–K

Editor-in-Chief

Joshua Lederberg

The Rockefeller University
New York, NY

ACADEMIC PRESS

A Harcourt Science and Technology Company

San Diego   San Francisco   New York   Boston   London   Sydney   Tokyo

Academic Press
A Harcourt Science and Technology Company
525 B Street, Suite 1900, San Diego, California 92101-4495, USA
http://www.apnet.com

Academic Press
24-28 Oval Road, London NW1 7DX, UK
http://www.hbuk.co.uk/ap/

Library of Congress Catalog Card Number: 99-65283

International Standard Book Number: 0-12-226800-8 (set)
International Standard Book Number: 0-12-226801-6 Volume 1
International Standard Book Number: 0-12-226802-4 Volume 2
International Standard Book Number: 0-12-226803-2 Volume 3
International Standard Book Number: 0-12-226804-0 Volume 4

PRINTED IN THE UNITED STATES OF AMERICA
00 01 02 03 04 05 MM 9 8 7 6 5 4 3 2 1

Contents

D

E

$$\boxed{G}$$

$$\boxed{F}$$

Contents of Other Volumes

O

P

V O L U M E   4

W

W

V

X

Y

Z

Zoonoses   **955**
   Bruno B. Chomel

Contents by Subject Area

INFECTIOUS AND NONINFECTIOUS DISEASE AND PATHOGENESIS: PLANT PATHOGENS, VIRUSES

INFECTIOUS AND NONINFECTIOUS DISEASE AND PATHOGENESIS: TREATMENT

PHYSIOLOGY, METABOLISM, AND GENE EXPRESSION

STRUCTURE AND MORPHOGENESIS

Cell Membrane: Structure and Function
Cell Walls, Bacterial
Crystalline Bacterial Cell Surface Layers
Developmental Processes in Bacteria
Fimbriae, Pili
Flagella
Outer Membrane, Gram-Negative Bacteria

SYSTEMATICS AND PHYLOGENY

Acetogenesis and Acetogenic Bacteria
Actinomycetes
Archaea
Azotobacter
Bacteriophages
Cyanobacteria
Dinoflagellates
Enteroviruses
Escherichia coli, General Biology

Extremophiles
Fungi, Filamentous
Heterotrophic Microorganisms
Retroviruses
Rhinoviruses
Spirochetes
Viruses
Yeasts

TECHNIQUES

Detection of Bacteria in Blood: Centrifugation and
    Filtration
Germfree Animal Techniques
Identification of Bacteria, Computerized
Microscopy, Confocal
Microscopy, Electron
Microscopy, Optical
Polymerase Chain Reaction (PCR)
Temperature Control
Transgenic Animal Technology

Preface

The scientific literature at large is believed to double about every 12 years. Though less than a decade has elapsed since the initiation of the first edition of this encyclopedia, it is a fair bet that the microbiology literature has more than doubled in the interval, though one might also say it has fissioned in the interval, with parasitology, virology, infectious disease, and immunology assuming more and more independent stature as disciplines.

According to the *Encyclopaedia Britannica*, the encyclopedias of classic and medieval times could be expected to contain "a compendium of all available knowledge." There is still an expectation of the "essence of all that is known." With the exponential growth and accumulation of scientific knowledge, this has become an elusive goal, hardly one that could be embraced in a mere two or three thousand pages of text. The encyclopedia's function has moved to becoming the first word, the initial introduction to knowledge of a comprehensive range of subjects, with pointers on where to find more as may be needed. One can hardly think of the last word, as this is an ever-moving target at the cutting edge of novel discovery, changing literally day by day.

For the renovation of an encyclopedia, these issues have then entailed a number of pragmatic compromises, designed to maximize its utility to an audience of initial look-uppers over a range of coherently linked interests. The core remains the biology of that group of organisms we think of as microbes. Though this constitutes a rather disparate set, crossing several taxonomic kingdoms, the more important principle is the unifying role of DNA and the genetic code and the shared ensemble of primary pathways of gene expression. Also shared is access to a "world wide web" of genetic information through the traffic of plasmids and other genetic elements right across the taxa. It is pathognomonic that the American Society for Microbiology has altered the name of *Microbiological Reviews* to *Microbiology and Molecular Biology Reviews*. At academic institutions, microbiology will be practiced in any or all of a dozen different departments, and these may be located at schools of arts and sciences, medicine, agriculture, engineering, marine sciences, and others.

Much of human physiology, pathology, or genetics is now practiced with cell culture, which involves a methodology indistinguishable from microbiology: it is hard to define a boundary that would demarcate microbiology from cell biology. Nor do we spend much energy on these niceties except when we have the burden of deciding the scope of an enterprise such as this one.

Probably more important has been the explosion of the Internet and the online availability of many sources of information. Whereas we spoke last decade of CDs, now the focus is the Web, and the anticipation is that we are not many years from the general availability of the entire scientific literature via this medium. The utility of the encyclopedia is no longer so much "how do I begin to get information on Topic X" as how to filter a surfeit of claimed information with some degree of dependability. The intervention of editors and of a peer-review process (in selection of authors even more important than in overseeing their papers) is the only foreseeable solution. We have then sought in each article to provide a digest of information with perspective and provided by responsible authors who can be proud of, and will then strive to maintain, reputations for knowledge and fairmindedness.

The further reach of more detailed information is endless. When available, many specific topics are elaborated in greater depth in the ASM (American Society of Microbiology) reviews and in *Annual Review of Microbiology*. These are indexed online. Medline, Biosis, and the Science Citation Index are further online bibliographic resources, which can be focused for the recovery of review articles.

The reputation of the authors and of the particular journals can further aid readers' assessments. Citation searches can be of further assistance in locating critical discussions, the dialectic which is far more important than "authority" in establishing authenticity in science.

Then there are the open-ended resources of the Web itself. It is not a fair test for recovery on a specialized topic, but my favorite browser, google.com, returned 15,000 hits for "microbiology"; netscape.com gave 46,000; excite.com a few score structured headings. These might be most useful in identifying other Web sites with specialized resources. Google's 641 hits for "luminescent bacteria" offer a more proximate indicator of the difficulty of coping with the massive returns of unfiltered verbiage that this wonderful new medium affords: how to extract the nuggets from the slag.

A great many academic libraries and departments of microbiology have posted extensive considered listings of secondary sources. One of my favorites is maintained at San Diego State University:

> http://libweb.sdsu.edu/scidiv/
> microbiologyblr.html

I am sure I have not begun to tap all that would be available.

The best strategy is a parallel attack: to use the encyclopedia and the major review journals as a secure starting point and then to try to filter Web-worked material for the most up-to-date or disparate detail. In many cases, direct enquiry to the experts, until they saturate, may be the best (or last) recourse. E-mail is best, and society or academic institutional directories can be found online. Some listservers will entertain questions from outsiders, if the questions are particularly difficult or challenging.

All publishers, Academic Press included, are updating their policies and practices by the week as to how they will integrate their traditional book offerings with new media. Updated information on electronic editions of this and cognate encyclopedias can be found by consulting www.academicpress.com/.

Joshua Lederberg

From the Preface to the First Edition

(Excerpted from the 1992 Edition)

For the purposes of this encyclopedia, microbiology has been understood to embrace the study of "microorganisms," including the basic science and the roles of these organisms in practical arts (agriculture and technology) and in disease (public health and medicine). Microorganisms do not constitute a well-defined taxonomic group; they include the two kingdoms of Archaebacteria and Eubacteria, as well as protozoa and those fungi and algae that are predominantly unicellular in their habit. Viruses are also an important constituent, albeit they are not quite "organisms." Whether to include the mitochondria and chloroplasts of higher eukaryotes is a matter of choice, since these organelles are believed to be descended from free-living bacteria. Cell biology is practiced extensively with tissue cells in culture, where the cells are manipulated very much as though they were autonomous microbes; however, we shall exclude this branch of research. Microbiology also is enmeshed thoroughly with biotechnology, biochemistry, and genetics, since microbes are the canonical substrates for many investigations of genes, enzymes, and metabolic pathways, as well as the technical vehicles for discovery and manufacture of new biological products, for example, recombinant human insulin. . . .

The *Encyclopedia of Microbiology* is intended to survey the entire field coherently, complementing material that would be included in an advanced undergraduate and graduate major course of university study. Particular topics should be accessible to talented high school and college students, as well as to graduates involved in teaching, research, and technical practice of microbiology.

Even these hefty volumes cannot embrace all current knowledge in the field. Each article does provide key references to the literature available at the time of writing. Acquisition of more detailed and up-to-date knowledge depends on (1) exploiting the review and monographic literature and (2) bibliographic retrieval of the preceding and current research literature. . . .

To access bibliographic materials in microbiology, the main retrieval resources are MEDLINE, sponsored by the U.S. National Library of Medicine, and the Science Citation Index of the ISI. With governmental subsidy, MEDLINE is widely available at modest cost: terminals are available at every medical school and at many other academic centers. MEDLINE provides searches of the recent literature by author, title, and key word and offers online displays of the relevant bibliographies and abstracts. Medical aspects of microbiology are covered exhaustively; general microbiology is covered in reasonable depth. The Science Citation Index must recover its costs from user fees, but is widely available at major research centers. It offers additional search capabilities, especially by citation linkage. Therefore, starting with the bibliography of a given encyclopedia article, one can quickly find (1) all articles more recently published that have cited those bibliographic reference starting points and (2) all other recent articles that share bibliographic information with the others. With luck, one of these articles may be identified as another comprehensive review that has digested more recent or broader primary material.

On a weekly basis, services such as Current Contents on Diskette (ISI) and Reference Update offer still more timely access to current literature as well as to abstracts with a variety of useful features. Under the impetus of intense competition, these services are evolving rapidly, to the great benefit of a user community desperate for electronic assistance in coping with the rapidly growing and intertwined networks of discovery. The bibliographic services of Chemical Abstracts and Biological Abstracts would also be potentially invaluable; however, their coverage of microbiology is rather limited.

In addition, major monographs have appeared from time to time—*The Bacteria, The Prokaryotes,* and many others. Your local reference library should be consulted for these volumes.

Valuable collections of reviews also include *Critical Reviews for Microbiology, Symposia of the Society for General Microbiology, Monographs of the American Society for Microbiology,* and *Proceedings of the International Congresses of Microbiology.*

The articles in this encyclopedia are intended to be accessible to a broader audience, not to take the place of review articles with comprehensive bibliographies. Citations should be sufficient to give the reader access to the latter, as may be required. We do apologize to many individuals whose contributions to the growth of microbiology could not be adequately embraced by the secondary bibliographies included here.

The organization of encyclopedic knowledge is a daunting task in any discipline; it is all the more complex in such a diversified and rapidly moving domain as microbiology. The best way to anticipate the rapid further growth that we can expect in the near future is unclear. Perhaps more specialized series in subfields of microbiology would be more appropriate. The publishers and editors would welcome readers' comments on these points, as well as on any deficiencies that may be perceived in the current effort.

My personal thanks are extended to my coeditors, Martin Alexander, David Hopwood, Barbara Iglewski, and Allen Laskin; and above all, to the many very busy scientists who took time to draft and review each of these articles.

Joshua Lederberg

Guide to the Encyclopedia

The *Encyclopedia of Microbiology, Second Edition* is a scholarly source of information on microorganisms, those life forms that are observable with a microscope rather than by the naked eye. The work consists of four volumes and includes 298 separate articles. Of these 298 articles, 171 are completely new topics commissioned for this edition, and 63 others are newly written articles on topics appearing in the first edition. In other words, approximately 80% of the content of the encyclopedia is entirely new to this edition. (The remaining 20% of the content has been carefully reviewed and revised to ensure currency.)

Each article in the encyclopedia provides a comprehensive overview of the selected topic to inform a broad spectrum of readers, from research professionals to students to the interested general public. In order that you, the reader, will derive the greatest possible benefit from your use of the *Encyclopedia of Microbiology*, we have provided this Guide. It explains how the encyclopedia is organized and how the information within it can be located.

ORGANIZATION

The *Encyclopedia of Microbiology* is organized to provide maximum ease of use. All of the articles are arranged in a single alphabetical sequence by title. Articles whose titles begin with the letters A to C are in Volume 1, articles with titles from D through K are in Volume 2, then L through P in Volume 3, and finally Q to Z in Volume 4. This last volume also includes a complete subject index for the entire

work, an alphabetical list of the contributors to the encyclopedia, and a glossary of key terms used in the articles.

Article titles generally begin with the key noun or noun phrase indicating the topic, with any descriptive terms following. For example, the article title is "Bioluminescence, Microbial" rather than "Microbial Bioluminescence," and "Foods, Quality Control" is the title rather than "Quality Control of Foods."

TABLE OF CONTENTS

A complete table of contents for the *Encyclopedia of Microbiology* appears at the front of each volume. This list of article titles represents topics that have been carefully selected by the Editor-in-Chief, Dr. Joshua Lederberg, and the nine Associate Editors. The Encyclopedia provides coverage of 20 different subject areas within the overall field of microbiology. Please see p. v for the alphabetical table of contents, and p. xix for a list of topics arranged by subject area.

INDEX

The Subject Index in Volume 4 indicates the volume and page number where information on a given topic can be found. In addition, the Table of Contents by Subject Area also functions as an index, since it lists all the topics within a given area; e.g., the encyclopedia includes eight different articles dealing with historic aspects of microbiology and nine dealing with techniques of microbiology.

ARTICLE FORMAT

In order to make information easy to locate, all of the articles in the *Encyclopedia of Microbiology* are arranged in a standard format, as follows:
- Title of Article
- Author's Name and Affiliation
- Outline
- Glossary
- Defining Statement
- Body of the Article
- Cross-References
- Bibliography

OUTLINE

Each entry in the Encyclopedia begins with a topical outline that indicates the general content of the article. This outline serves two functions. First, it provides a brief preview of the article, so that the reader can get a sense of what is contained there without having to leaf through the pages. Second, it serves to highlight important subtopics that will be discussed within the article. For example, the article "Biopesticides" includes subtopics such as "Selection of Biopesticides," "Production of Biopesticides," "Biopesticide Stabilization," and "Commercialization of Biopesticides."

The outline is intended as an overview and thus it lists only the major headings of the article. In addition, extensive second-level and third-level headings will be found within the article.

GLOSSARY

The Glossary contains terms that are important to an understanding of the article and that may be unfamiliar to the reader. Each term is defined in the context of the article in which it is used. Thus the same term may appear as a glossary entry in two or more articles, with the details of the definition varying slightly from one article to another. The encyclopedia has approximately 2500 glossary entries.

In addition, Volume 4 provides a comprehensive glossary that collects all the core vocabulary of microbiology in one A–Z list. This section can be consulted for definitions of terms not found in the individual glossary for a given article.

DEFINING STATEMENT

The text of each article in the encyclopedia begins with a single introductory paragraph that defines the topic under discussion and summarizes the content of the article. For example, the article "Eyespot" begins with the following statement:

> **EYESPOT** is a damaging stem base disease of cereal crops and other grasses caused by fungi of the genus *Tapsia*. It occurs in temperate regions world-wide including Europe, the USSR, Japan, South Africa, North America, and Australasia. In many of these countries eyespot can be found on the majority of autumn-sown barley and wheat crops and may cause an average of 5–10% loss in yield, although low rates of infection do not generally have a significant effect. . . .

CROSS-REFERENCES

Almost all of the articles in the Encyclopedia have cross-references to other articles. These cross-references appear at the conclusion of the article text. They indicate articles that can be consulted for further information on the same topic or for information on a related topic. For example, the article "Smallpox" has references to "Biological Warfare," "Polio," "Surveillance of Infectious Diseases," and "Vaccines, Viral."

BIBLIOGRAPHY

The Bibliography is the last element in an article. The reference sources listed there are the author's recommendations of the most appropriate materials for further research on the given topic. The bibliography entries are for the benefit of the reader and do not represent a complete listing of all materials consulted by the author in preparing the article.

COMPANION WORKS

The *Encyclopedia of Microbiology* is one of a series of multivolume reference works in the life sciences published by Academic Press. Other such titles include the *Encyclopedia of Human Biology, Encyclopedia of Reproduction, Encyclopedia of Toxicology, Encyclopedia of Immunology, Encyclopedia of Virology, Encyclopedia of Cancer,* and *Encyclopedia of Stress.*

Acknowledgments

The Editors and the Publisher wish to thank the following people who have generously provided their time, often at short notice, to review various articles in the *Encyclopedia of Microbiology* and in other ways to assist the Editors in their efforts to make this work as scientifically accurate and complete as possible. We gratefully acknowledge their assistance:

George A. M. Cross
Laboratory of Molecular Parasitology
The Rockefeller University
New York, NY, USA

Miklós Müller
Laboratory of Biochemical Parasitology
The Rockefeller University
New York, NY, USA

A. I. Scott
Department of Chemistry
Texas A&M University
College Station, Texas, USA

Robert W. Simons
Department of Microbiology and
    Molecular Genetics
University of California, Los Angeles
Los Angeles, California, USA

Peter H. A. Sneath
Department of Microbiology and Immunology
University of Leicester
Leicester, England, UK

John L. Spudich
Department of Microbiology and
    Molecular Genetics
University of Texas Medical School
Houston, Texas, USA

Pravod K. Srivastava
Center for Immunotherapy
University of Connecticut
Farmington, Connecticut, USA

Peter Staeheli
Department of Virology
University of Freiburg
Freiburg, Germany

Ralph M. Steinman
Laboratory of Cellular Physiology
    and Immunology
The Rockefeller University
New York, NY, USA

Sherri O. Stuver
Department of Epidemiology
Harvard School of Public Health
Boston, Massachusetts, USA

Alice Telesnitsky
Department of Microbiology and Immunology
University of Michigan Medical School
Ann Arbor, Michigan, USA

Robert G. Webster
Chairman and Professor
Rose Marie Thomas Chair
St. Jude Children's Research Hospital
Memphis, Tennessee, USA

Dairy Products

Mary Ellen Sanders
Dairy and Food Culture Technologies

GLOSSARY

bacteriophage A virus infecting a bacterium.

commercial sterility The result of processing (usually retort processing) of food to eliminate all pathogenic and spoilage microorganisms that can contribute to food spoilage under normal storage conditions. A commercially sterile product is not necessarily sterile. The only viable microbes, if any, remaining in a commercially sterile product are extremely heat-resistant bacterial spores, which can cause spoilage of product stored at unusually high storage temperatures.

fluid milk Milk that is prepared to be consumed as a natural liquid product, including raw or pasteurized milks with different fat contents, or milk solids or vitamin fortifications.

lactic acid bacteria The name of a group of bacteria belonging to a diversity of genera used to effect food fermentations. This group is composed chiefly of bacteria whose primary metabolic end product from carbohydrate metabolism is lactic acid, although poor lactate producers (such as leuconostocs and propionibacteria) are sometimes included due to their association with food fermentations.

milk products Products manufactured from fluid milk, including natural cheeses, processed cheeses, fermented milks, yogurts, butter, ice cream, sour cream, whipped cream, canned milks, and dried milk.

probiotic Living microorganisms, which upon ingestion in certain numbers exert health benefits beyond inherent basic nutrition.

starter culture A microbial strain or mixture of strains, species, or genera used to effect a fermentation and bring about functional changes in milk that lead to desirable characteristics in the fermented product.

THE MICROBIOLOGY OF DAIRY PRODUCTS is a field composed of both the positive and negative effects of microbes on milk and milk-based products. On the one hand, microbes are responsible for the transformation of milk into a wide array of fermented dairy products, such as cheeses, yogurts, and fermented milks produced worldwide. On the other hand, microbes can also cause food-borne disease and spoilage of dairy products. These two facets of dairy microbiology are intricately associated since, undoubtedly, the first fermented milk products made approximately 8000 years ago provided a means for preserving milk as a safe and wholesome food.

A diversity of microbes is associated with dairy products, including gram-positive and gram-negative bacteria, molds, yeasts, and bacteriophages. Spoilage and pathogenic microorganisms are chiefly controlled by pasteurization, refrigeration, fermentation, and by limiting post-process contamination. Reduced water activity, high salt content, and heat sterilization also contribute to the preservation of some dairy products. Great effort is expended to control contaminating microbes responsible for spoilage or pathogenicity. In contrast, microbes (dairy starter cultures) are intentional additives to milk destined for fermentation. These microbes serve to preserve milk primarily through the production of organic acids. The dairy microbiologist must balance microbial populations and activities in milk so that positive effects are enhanced and spoilage and pathogenesis are discouraged or eliminated.

I. NATURAL FLORA OF MILK

Milk, as it is produced by the mammal, is sterile. Bacteria inhabiting the teat or udder do, however, migrate up into the interior, causing even aseptically drawn milk to contain some bacteria, predominantly micrococci, streptococci, and *Corynebacterium bovis*. Milk taken from a mastitic animal (one with a teat infection) will show high levels of microbes, including streptococci, staphylococci, coliforms, *Pseudomonas aeruginosa*, and *Corynebacterium pyogenes*. Animals sick with other infections may also shed pathogenic microbes, including *Mycobacterium* species, *Brucella* species, mycoplasma, and *Coxiella burnetti*. Milk from a healthy animal develops a complex flora upon milking. Since milk is an animal product, microbes associated with mammals, farms, agricultural feedstuffs, and green plant material are often present in milk. Bacilli from the soil, clostridia from silage, coliforms from manure and bedding, and streptococci, lactococci, and lactobacilli from green plant material commonly contaminate milk. In addition, the storage and processing environment and equipment, including milking machines, farm storage tanks, transportation equipment, cooling tanks, and milk processing equipment, contribute greatly to the microbial flora of fluid milk.

II. MICROBIAL SPOILAGE

A. Psychrotrophs

By far the most significant group of microbes in the spoilage of high-moisture, refrigerated milk products (fluid milk, cottage cheese, and cream cheese) is the psychrotrophs. The term psychrotrophs is defined as microorganisms that are capable of growing at refrigeration temperatures, although their optimum growth temperature may be much higher. Psychrotrophs include species from at least 27 genera of bacteria, 4 genera of yeast, and 4 genera of molds. Proper refrigeration is of the utmost importance to the control of psychrotroph growth since a small increase in storage temperature can result in a large decrease in bacterial generation times. The production of lipases, proteases, exopolysaccharides, and

visible colonies of mold seriously affect the quality of milk products. Some psychrotrophic microbes produce heat-stable lipases and proteases which if produced prior to pasteurization can threaten product quality after pasteurization.

B. Fluid Milk

The spoilage of fluid milk is dictated by the effects of pasteurization, post-pasteurization contamination, and refrigeration. The pasteurization process for milk was originally designed to kill all pathogenic microbes. Modern pasteurization practices frequently exceed minimum pasteurization time/temperature requirements to provide extra safety margins and extended product shelf life. Along with the pathogens, yeasts, molds, and gram-negative and many gram-positive microbes are killed. A challenge to fluid milk processors is to limit the contamination that occurs after pasteurization during transport and packaging of pasteurized milk. The extent of this post-process contamination is directly related to the level of sanitation in the processing plant and effective refrigeration. Milk microbiologically stable at room-temperature is also produced using "ultra-high temperature" heat treatment and an aseptic packaging process.

C. Cheese

The final composition of cheese relative to moisture content, salt content, fat content, and pH can vary tremendously among different cheese varieties. Since all of these factors contribute to the microbial stability of cheese, the only general statement about microbial stability that can be made is that cheeses are more stable than the milk from which they were made. Moisture content can range from 80% in cottage and cream cheeses to 35% in hard grating cheeses. Salt ranges from 1.5 to 5%, although a more significant affect on water activity than might be expected is seen since the salt is concentrated in the aqueous phase. Final titratable acidities during fermentation also vary among different cheeses.

Molds, yeasts, and anaerobic spore-forming bacteria are involved most often in the spoilage of cheese, although psychrotrophic bacteria and molds spoil high-moisture cheeses, such as cottage cheese, and non-starter lactobacilli can lead to flavor and texture defects in cheese. Molds cause an unsightly appearance to cheese surfaces and can pose a health threat (see Section III). Anaerobic spore formers such as clostridia, coliforms and even unbalanced levels of gas-producing starter strains can cause abnormal gas formation. Spoilage during ripening of cheese can be controlled by maintaining low ripening temperatures and low humidity, factors that may inhibit microbial or enzymatic ripening processes. Spore formers have been successfully controlled using nisin, an antimicrobial peptide produced by *Lactococcus lactis* spp. *lactis,* or with the enzyme lysozyme.

D. Fermented Milk and Yogurt

Microbial growth is controlled in fermented milks by the low pH, high titratable acidity, and dominant starter culture numbers achieved during fermentation. The lactic cultures typically lower the pH of fermented milks to 3.5–4.5, depending on the product, and the pH may continue to decrease during refrigerated storage. At this level of acidity, pathogens are inhibited effectively and even killed upon storage. Spoilage is limited to yeasts and molds, which may accompany addition of flavorings and fruits that may be added to the product. As long as care is taken not to inhibit acid production during fermentation, a safe and long shelf life product results. Commercial yogurts frequently have a 5- to 7-week shelf life.

E. Dried Milk Products

Dried milk, skim milk, whey, buttermilk, cheese, and cream are popular for use as ingredients in other foods. The drying process, although conducted at elevated temperatures, is not a reliable method of microbial destruction. Proper pasteurization, sanitation and product handling are the only consistent controls over the safety of these products. Once dried, these products are microbiologically stable. However, any remaining pathogens or spoilage microbes are a threat in food subsequently formulated with the contaminated ingredient. Heat-resistant

spores can be especially problematic in the dried milk products.

F. Canned Milks

Canned evaporated milk is heated to achieve commercial sterility. Therefore, all pathogens are destroyed, although some extremely heat-resistant spores such as *Bacillus stearothermophilus* might survive. This product is both shelf stable under normal storage temperatures and pathogen-free. Sweetened condensed milks rely on pasteurization, low water activity and high sugar content for preservation. These products are not heat processed after canning and therefore can spoil due to contamination by molds or yeasts that enter during the fill operation or through can defects.

III. PATHOGENS OF CONCERN IN DAIRY PRODUCTS

Milk was once the vehicle of transmission of typhoid fever, scarlet fever, septic sore throat, diphtheria, tuberculosis, and shigellosis. The frequency and severity of these diseases prompted large-scale adoption of milk pasteurization by the end of World War II. This concern about milk-borne disease led the dairy industry to develop equipment and sanitation programs, which are still unsurpassed, decades before programs were implemented by other food industries. Milk-borne infections include listeriosis, salmonellosis, campylobacterosis, brucellosis and yersiniosis, although properly pasteurized milk which is free from post-process contaminants is not a vehicle for transmission of these or other diseases. Table I summarizes the cases in the United States of food poisoning associated with dairy products reported to the Center for Disease Control from 1988 to 1992. The pathogens involved in these cases were *Campylobacter, Escherichia coli,* and *Salmonella.*

A pathogen of great concern in recent years, enterohemorrhagic *E. coli,* including serotype O157:H7, is associated with cattle and has been isolated in raw milk. Its documented acid resistance and survival when inoculated into cottage and cheddar cheeses reinforces the importance of proper pasteurization

TABLE I
Cases of Food-Borne Disease Associated with Dairy Products in the United States from 1988 to 1992[a]

Year	Milk	Ice cream	Cheese	Other/unknown miscellaneous dairy
1988	120	163		
1989	48	16	164	22
1990	68	148	50	40
1991	37	62	25	23
1992	79	81		12

[a] From "Morbidity and Mortality Weekly Report" (1996).

and packaging of milk prior to consumption or use in the manufacture of other dairy foods. Another pathogen of concern in dairy products is the ubiquitous *Listeria monocytogenes.* In addition to milk and ice cream, cheeses such as Brie and Camembert can be vehicles for listeriosis transmission. Although fermentation reduces the pH to an inhibitory level, ripening of these cheeses involves the surface growth of mold. These molds produce alkaline metabolites and increase the surface pH of the cheese. The lack of refrigeration during ripening further encourages the growth of *Listeria,* although *Listeria* can also grow at refrigeration temperatures. These factors combine to make listeriosis a significant threat in these raw milk or contaminated cheeses. These same conditions can also encourage growth of other pathogens, emphasizing the importance of using pasteurized milk and eliminating post-process contamination during the manufacture of these cheeses.

Some microbial toxins can be found in dairy products or milk, including staphylococcal toxin, aflatoxins, other mycotoxins, and biogenic amines. *Staphylococcus aureus* can contaminate milk from mastitic cows (cows with an udder infection). If improperly pasteurized or raw, this milk can support the growth of *S. aureus,* leading to toxin production. Pasteurization will not inactivate this heat-stable toxin. Milk contaminated with *S. aureus* is especially dangerous if involved in a substandard fermentation. *S. aureus* can grow and produce toxin during the fermentation period if acid is not being generated. Some common *Aspergillus* spp. produce a family of toxic and carcinogenic aflatoxins during growth on damaged grains

and other substrates. Daily consumption of these grains by milking cows can lead to one type of aflatoxin, M_1, in milk. *Aspergillus* spp. producing aflatoxin can also grow on the surface of cheese, providing another means of contamination. Penicillic acid, patulin, and ochratoxin A are other mycotoxins that have been found in moldy cheese trimmings. The incidence of mycotoxin contamination of cheese is low; storage temperatures lower than 7°C greatly discourage toxin formation.

Biogenic amines, including tyramine and histamine, have been found in cheese. These vasoactive compounds can be toxic at high levels or in people with compromised metabolic ability to deaminate these amines. These amines are formed by the decarboxylation of tyrosine and histidine catalyzed by decarboxylases found in microbes (including nonstarter lactobacilli) commonly associated with cheese.

IV. FERMENTATION OF DAIRY PRODUCTS

A. Microbes Associated with Dairy Fermentations

Fermented dairy products derive their characteristic flavor and texture from the microbial action of starter cultures. There are hundreds of different varieties of cheeses and fermented milks produced worldwide. Although there is some difference in the source (cow, sheep, goat, and buffalo) and composition of the milk used as a raw material (e.g., for manufacture of reduced-fat cheeses), this great diversity of fermented foods stems primarily from the microbes used in the fermentation and the physical treatments of the fermented product.

Fermentation has been used to preserve milk for millennia. Before a technical understanding of the fermentation was available, microbes naturally present in the milk or in the containers used for milk storage were the agents of fermentation. In modern industrial fermentations, purified and characterized starter cultures are scaled-up and intentionally inoculated into the milk. This enables much more control over the fermentation and characteristics of the final product.

The microbes associated with fermented dairy products are listed in Table II. They are nonpathogenic and contribute specific attributes to the final product. They generally are present at levels of 10^6–10^9 per gram in freshly fermented products. During ripening or storage, the types and levels of microbes change, depending on the product.

B. Function of Microbes in Dairy Fermentations

The ability to ferment milk sugar, lactose, to lactic acid and to carry out the proteolytic degradation of milk protein, casein, to a useable nitrogen source are the primary requirements of a starter culture. However, these microbes contribute to flavor and texture development in many other ways. The production and degree of metabolic end products (lactic, propionic, and acetic acids and ethanol, CO_2, diacetyl, and dimethyl sulfide), other flavor compounds, proteases, and lipases all help to determine the unique attributes of a given fermented milk product. Lactic acid is important to all fermented dairy products. It provides the acidity necessary for a tart flavor and for changes in the structure of casein to achieve syneresis and desired functional characteristics. Propionic acid gives Swiss cheeses their characteristic nutty flavor. Acetic acid and ethanol must be balanced in yogurt to promote proper flavor. Ethanol, carbon dioxide, and lactic acid combine in mixed lactic bacteria/yeast fermented milks to provide the desired flavor. Carbon dioxide imparts effervescence to fermented milks, eye formation in some cheeses, and an open texture in others. Diacetyl provides the buttery flavor important in buttermilk, soft cheeses, and cottage cheese dressing. Starters which

TABLE II
Microbes Associated with Fermented Dairy Products

Microbes	Product association[a]	Function
Lactococcus lactis subspecies *cremoris* and *lactis*	American cheeses, buttermilk, cottage cheese, soft cheese	Lactic acid production at less than 40°C; some strains provide ropiness in fermented milk
Lactococcus lactis subspecies *lactis* var. *diacetylactis*	Buttermilk, soft cheese, sour cream, cottage cheese dressing	Lactic acid, CO_2, and diacetyl
Lactobacillus delbrueckii subspecies *bulgaricus*, *L. lactis*, *L. helveticus*, *L. casei*	Italian and Swiss cheeses, yogurt, fermented milks; often paired with *S. thermophilus*	Lactic acid production at less than 50°C; some ropy strains
Streptococcus thermophilus	Italian and Swiss cheeses, yogurt; often paired with the lactobacilli	Lactic acid production at less than 50°C; some ropy strains
Leuconostoc	Buttermilk, Roquefort cheese, cottage cheese	Ethanol, acetic acid, diacetyl, CO_2
Propionibacterium	Swiss cheeses	CO_2 causing eye formation; propionic and acetic acids from lactate
Lactobacillus acidophilus, *L. casei*, *L. reuteri*, *L. rhamnosus*, *Bifidobacterium*	Probiotic-containing dairy foods	Promotion of human health through modulation of undesirable intestinal microbiota activity, enhancement of immune function, or enzyme activity
Penicillium	Soft cheeses, blue-veined cheese	White surface mold; increased pH; blue vein production
Geotrichum	Soft cheeses	White surface mold
Lactose-fermenting yeast	Kefir and other mixed fermentation beverages	Ethanol and CO_2

[a] American cheeses: cheddar, brick, Monterey Jack, Muenster; soft cheeses: Camembert, Brie; Italian cheeses: parmesan, provolone, ricotta, mozzarella.

produce extracellular polysaccharides can improve the mouth-feel of some fermented milks and yogurts (or may cause a slime defect in products in which ropiness is not desirable). Starter proteases hydrolyze milk proteins to peptides, and starter and non-starter peptidases further their degradation. Some flavor defects, such as bitterness and brothiness, stem from the presence of certain amino acids or peptides. The action and control of starter proteases and peptidases are critical to proper ripening of most aged cheeses. Attempts to accelerate the ripening of aged cheeses have led to the development of culture and/or enzyme additives that promote proteolysis important in characteristic aged flavor and body. Lipases are very important to the proper flavor of some Italian cheeses. Frequently, starter lipases are not sufficient, and animal-derived lipases are added for proper ripening and flavor development. Non-starter bacteria can also contribute to flavor production during ripening (providing the justification for using raw milk in cheese manufacture). In addition to the contribution of microbial factors, there are numerous physical manipulations that help determine characteristics of the final product. The timing and extent of stirring, mixing, stretching, cheddaring, temperature and moisture control, and product formulation are important to the flavor, texture, and body of fermented dairy products.

C. Genetics of Dairy Starter Cultures

For years, observant dairy technologists noticed that lactic acid bacteria often did not retain some desirable traits when the cultures were held for long periods of time or sequentially transferred. Lactic cultures often lost their ability to rapidly ferment milk, to produce diacetyl, or to resist bacteriophage infection. Since the mid-1970's, researchers have focused on the genetic basis for this occurrence. Genes encoding lactose metabolism, protease production, diacetyl formation, bacteriophage resistance, and bacteriocin production have been found to be linked to plasmid DNA or, in a few cases, are flanked by insertion sequences. The linkage of these traits to naturally occurring unstable elements has provided a mechanism for rapid evolution and genetic shift in this group of bacteria.

This knowledge of the genetics of lactic acid bacteria led to efforts to apply directed genetic techniques to the improvement of these industrially important bacteria. Extensive worldwide research led to development of cloning vectors, integration vectors, gene sequencing, and genetic transfer systems, including conjugation, transformation, and transduction. The directed conjugal transfer of characterized phage-resistance plasmids into phage-sensitive recipient strains was the first example of use of this technology to genetically improve dairy strains. This field of research will continue to provide the tools for directed genetic improvement of starter cultures.

V. BACTERIOPHAGES IN DAIRY FERMENTATIONS

The failure of a milk fermentation due to poor starter culture growth results in significant product and efficiency loss to the manufacturer. Problems with starter culture performance can be caused by antibiotics in milk, natural inhibitors in milk, inactive starter inoculum, environmental conditions (e.g., temperature) during processing inhibitory to the culture, and bacteriophage. Bacteriophage, or phage, is the most significant of these factors.

Morphologically, there are many different types of lactic phages, and phages have been isolated for all genera and many species of lactic cultures. They are parasitic in nature, requiring host cell functions to replicate, and during a lytic cycle, they will lyse the host cell to release newly formed phage particles. The tremendous impact that a lytic phage infection can have on a fermentation is due to the potential of phage to rapidly infect, lyse, and release large numbers of progeny for subsequent infections. It may take less than 60 min for a complete replication cycle (latent period), and up to 200 viable phage progeny per cell can be released (burst size). This form of replication allows phage to out-pace the slower binary fission replication of bacteria.

Phage can exist in one of two life states: lytic or lysogenic. The lytic cycle results in rapid production of progeny phage, whereas the lysogenic cycle results in a latent infection mediated by incorporation of phage DNA into host DNA. Not all phages are capable of establishing a lysogenic state. Many lactic cultures have been shown to harbor lysogenic phage. However, phage species most commonly isolated from failed dairy fermentations do not appear to be homologous to lysogenic phage species, suggesting that lysogenic phage do not contribute greatly to the lytic phage problem.

Contributing factors to the impact of phage in dairy fermentations include: the non-aseptic nature of the fermentations; the desire for end-product consistency; the use of open vats in cheesemaking; the fluid nature of milk, which allows dissemination of phage particles; ongoing microbial growth in the fermentation environment providing potential hosts for the phage; the ability of phages to be transmitted as an aerosol; and the rapid replication rates of some lytic phages. Some strains, species, and genera seem more susceptible to phage infection than others. The lactococci appear to suffer more serious phage problems than the lactobacilli or *Streptococcus thermophilus*, and some strains with enhanced phage-resistance characteristics have been isolated from nature or developed in the laboratory.

The influence of phage on dairy fermentations has led to a variety of measures designed to control the effect of phage. The following are some control measures:

- *Sanitation:* A program of regular cleaning and sanitation (e.g., chlorine at 200 ppm) of equipment provides the most important means of reducing phage levels in the dairy plant.
- *Effective processing plant design:* A separate starter preparation room, control over air, personnel, and product flow in the plant to prevent downstream product from contaminating upstream processes, and the use of fermentation and starter vessels that can be easily and thoroughly cleaned all provide a manageable environment for effective phage control.
- *Use of phage-inhibitory media:* These media can be used during the scale-up of starter prior to inoculation of the product fermentation. Formulations generally involve chelation of Ca^{2+} ions with citrate or phosphate. Calcium ions are required for phage to infect the host cells. If phage levels are kept low in the starter tank, the

fermentation may proceed before phage levels build to a destructive level. The higher cost of these media, however, prevents their use in some production facilities.
- *Strain selection and systems for use:* After fermentative ability, the largest single factor for strain selection is resistance of strains to bacteriophage. Great emphasis by starter culture suppliers is placed on selecting and designing strains that will resist phage. Once effective strains are identified, their commercial performance is monitored through phage testing to determine the need for strain replacement. Strain replacement strategies vary, but at the heart of all is the availability for use as back-up of strains expressing strong phage resistance or at least expressing sensitivity to different types or ranges of phages. These phage-unrelated strains will not serve as hosts for the same phages, and they can function even in the presence of phages generated by a different strain.
- *Monitoring for phage:* Conducting phage tests on product or by-products (e.g., whey) of the fermentation informs the manufacturer of developing phage problems. This information can be used to make changes in sanitation schedules or strain usage.

VI. PROBIOTIC BACTERIA IN DAIRY PRODUCTS

Milk and milk products provide an excellent source of nutrition (protein, calories, vitamins, calcium, and other minerals). In addition to the nutritional benefits of milk consumption, certain bacteria, termed probiotic bacteria, associated with dairy products have been shown to promote health. The primary means for this health benefit is through the influence of the gastrointestinal tract physiology and microbiota. The gastrointestinal tract, composed of the stomach, small intestine, and colon, is colonized by populations of a diversity of microbes. The gastrointestinal tract is colonized, where physiological conditions permit, by these microbes soon after birth, and it continues to be exposed to ingested microbes. These endogenous and external microbes can engage in harmful activities, including acute intestinal pathogenesis, translocation through the gut into the blood-stream, and production of metabolic end products that can have procarcinogenic effects. The ingestion of beneficial bacteria can provide competition for these harmful bacteria, resulting in a reduction in their activities. Probiotic bacteria may attain competitive advantage in the gastrointestinal tract through the production of organic acids or bacteriocins or by competitively excluding harmful microbes from certain attachment sites.

Ingested probiotic bacteria, whether colonizing or transient, have been shown to mediate some beneficial effects in the gastrointestinal tract, including reduction of colon cancer-promoting activities (animal studies), improvement of lactose digestion, improvement of immune system function, and reduction of gastrointestinal infections. Further research is needed to determine the extent of the effect of probiotic bacteria in promoting human health. The most commonly cited cultures used for "therapeutic" purposes are *Lactobacillus* and *Bifidobacterium* species.

See Also the Following Articles

Food Spoilage and Preservation • Gastrointestinal Microbiology • Lactic Acid, Microbially Produced • Strain Improvement

Bibliography

Axelsson, L. (1998). Lactic acid bacteria: Classification and physiology. *In* "Lactic Acid Bacteria. Microbiology and Functional Aspects" (S. Salminen and A. von Wright, Eds.). Dekker, New York.

Bengmark, S. (1998). Ecological control of the gastrointestinal tract—The role of probiotic flora. *Gut* **42**, 2–7.

Brassart, D., and Schiffrin, E. J. (1997). The use of probiotics to reinforce mucosal defence mechanisms. *Trends Food Sci. Technol.* **8**, 321–326.

Champagne, C. P., Laing, R. R., Roy, D., and Mafu, A. A. (1994). Psychrotrophs in dairy products: Their effects and their control. *Crit. Rev. Food Sci. Nutr.* **34**, 1–30.

Dinsmore, P. K., and Klaenhammer, T. R. (1995). Bacteriophage resistance in *Lactococcus. Mol. Biotechnol.* **4**, 297–313.

Farkye, N. Y. (1999). Cheese—Microbiology of cheese-making and maturation. *In* "Encyclopedia of Food Microbiology" (R. K. Robinson, C. A. Batt, and P. D. Patel, Eds.). Academic Press, London.

Hansen, J. N. (1994). Nisin as a model food preservative. *Crit. Rev. Food Sci. Nutr.* **34**, 69–93.

International Dairy Federation Bulletin (1991). "Practical Phage Control." United States National Committee of the International Dairy Federation, Arlington, VA 703-5-28-3-049, (*http://www.fil-idf.org*).

Meng, J., and Doyle, M. P. (1997). Emerging issues in microbiological food safety. *Annu. Rev. Nutr.* **17**, 255–275.

Mital, B. K., and Garg, S. K. (1995). Anticarcinogenic, hypocholesterolemic, and antagonistic activities of lactobacillus acidophilus. *Crit. Rev. Microbiol.* **21**, 175–214.

"Morbidity and Mortality Weekly Report." Vol. 45, No. SS-5, (1996, October 25). Center for Disease Control, Atlanta, GA.

Muir, D. D. (1996). The shelf-life of dairy products. 1. Factors influencing raw milk and fresh products. *J. Soc. Dairy Technol.* **49**, 24–32.

Pouwels, P. H., and Leer, R. J. (1993). Genetics of lactobacilli—Plasmids and gene expression. *Antonie van Leeuwenhoek* **64**, 85–107.

Scheinbach, S. (1998). Probiotics—Functionality and commercial status. *Biotechnol. Adv.* **16**, 581–608.

Shah, N. P. (1994). Psychrotrophs in milk—A review. *Milchwissenschaft-Milk Sci. Int.* **49**, 432–437.

Sinell, H. J. (1995). Control of food-borne infections and intoxications. *Int. J. Food Microbiol.* **25**, 209–217.

Stiles, M. E., and Holzapfel, W. H. (1997). Lactic acid bacteria of foods and their current taxonomy. *Int. J. Food Microbiol.* **36**, 1–219.

Detection of Bacteria in Blood: Centrifugation and Filtration

Mathias Bernhardt, Laurel S. Almer, Erik L. Munson, and Ronald F. Schell

Wisconsin State Laboratory of Hygiene and University of Wisconsin, Madison

Steven M. Callister

Gundersen Lutheran Medical Center

GLOSSARY

aminoglycosides A class of antimicrobial agents that inhibit protein synthesis; used widely against gram-negative bacteria.

bacteremia The presence of bacteria in blood.

Ficoll–Hypaque A mixture of Ficoll, a polymer of sucrose, and Hypaque-M (diatrizoate meglumine), a contrast material used in angiography, which is used to separate bacteria from erythrocytes.

leukocytosis Increase in white blood cells.

septicemia The presence of bacteria in blood along with signs of clinical infection.

tachycardia Rapid heartbeat.

BACTEREMIA is the presence of bacteria in the blood. The presence of bacteria in itself is no cause for alarm because this is a daily occurrence. For instance, each time one brushes one's teeth or eats, microscopic abrasions in the gums caused by the toothbrush or the morning toast allow for the entry of oral flora. However, these bacteria are rapidly eliminated from the bloodstream via phagocytosis. Septicemia, on the other hand, is the presence of bacteria in the blood along with signs of clinical infection, such as fever, chills, tachycardia, hypotension, shock, or leukocytosis. In this scenario, the bacteria somehow evade the body's defense mechanisms and are able to colonize a normally sterile site. Septicemia is potentially life threatening. In the United States, approximately 400,000 cases of septicemia occur annually with a mortality rate of 40–50%.

I. OVERVIEW OF BLOOD CULTURING

A. Background of Blood Culturing

The basic procedure of performing a blood culture has changed little since its inception as a diagnostic tool. A blood sample is obtained from the patient via venipuncture, inoculated to the appropriate broth medium (a blood culture bottle), incubated, and repeatedly checked for growth of the organism(s). Once the organism is isolated on solid medium, drug susceptibilities are performed and the proper choice of antimicrobial agent can be administered to the patient. Although relatively simple, isolation of the causative organism can take from 24 hr to several days. Because of the high mortality rate, rapid detection and isolation of the infecting microorganism(s) are paramount for the administration of effective antimicrobial therapy and survival of the patient.

B. Current Technology

To better aid in the rapid detection of bacteremia, several rapid detection blood culturing devices have been developed, including the BACTEC 9000 series of blood culture systems, the Oxoid Signal blood bottle, the Roche Septi-Chek system, the Isolator, and other continuous monitoring systems. With the exception of the Isolator, these methods rely on the incubation of blood in a liquid medium but differ in the detection of bacterial growth. The BACTEC system uses infrared light and a computer to detect CO_2 production in the blood bottles. In the Oxoid bottle, CO_2 produced by the growing bacteria pushes the liquid medium containing the bacteria into a reservoir attached to the top of the bottle. The Septi-Chek system detects actual colonies on a slide paddle composed of chocolate, McConkey, and malt agars attached to the bottle. This paddle is coated with the blood–broth mixture immediately following inoculation of the bottle. The Isolator uses lysis of red blood cells and centrifugation for the recovery of bacteria. After centrifugation, the supernatant that contains the lysed blood cells is removed and the concentrate (containing the bacteria) is streaked out on several plates of different media. The plates are incubated and checked for the growth of isolated colonies. The continuous monitoring systems, the BacT/Alert, ESP, and the BACTEC 9240, check for elaboration of CO_2 with a colorimetric (BacT/Alert) or a fluorescent (BACTEC 9240) sensor. The ESP system monitors changes in gas pressure due to bacterial growth.

However, several days may be required for detection and isolation of bacteria using these new systems. In addition, although the BACTEC systems, Oxoid Signal, Roche Septi-Chek, and Isolator provide enhanced detection of bacteria, some drawbacks have been documented (such as contamination and different recovery rates for members of the family Enterobacteriaceae, staphylococci, yeasts, streptococci, and anaerobes).

II. LYSIS AND FILTRATION TECHNOLOGY: A BRIEF HISTORY

A. Lysis of Erythrocytes

The use of lysis technology in the detection of bacteremia is usually associated with the use of filtration technology. The original concept was to devise a detection method for bacteremia that would not necessarily involve the use of a liquid medium. Blood bottles, although widely accepted for the detection of microorganisms in blood, have disadvantages. The most critical is that blood bottles do not provide an optimum growth environment for microorganisms that may be present in the blood. This is because inhibitory agents, such as complement, antibodies, sodium polyanethol sulfonate (SPS) (an anticoagulant), and antimicrobial agents that may have been administered prior to culture, are not removed. This, in turn, leads to inhibited growth of microorganisms and hence longer detection times.

The challenge was to devise a system in which the bacteria could be cultured in the absence of these inhibitory agents. The simplest way to do this was to pass the blood through a membrane filter in which the bacteria would be trapped. Prior to this step, however, the blood cells had to be lysed. This was accomplished by several investigators through the use of lysing agents such as Triton X-100 and streptokinase–streptodornase. After filtration, the filters were either immersed in a broth or, more commonly, placed on an agar plate with the filtrate side up. Nutrients would diffuse through the filter, allowing the bacteria to grow as isolated colonies. However, the procedure has been considered slow, cumbersome, and impractical for clinical use. Furthermore, problems such as filter clogging and decreased recovery due to lysis of bacteria in addition to blood cells were also reported.

One system on the market today uses lysis centrifugation technology, which does not involve any filtration. This system is the Isolator, developed by Gordon Dorn. The Isolator uses a chemical cocktail that lyses blood cells in addition to preventing coagulation of the blood sample. The cocktail consists of Saponin, which lyses the red and white cells, SPS and ethylenediaminetetraacetic acid (EDTA), both of which are anticoagulants, and polypropylene glycerol, which inhibits the natural foaming tendency of Saponin. Some studies have shown that both EDTA and Saponin may be toxic toward *Streptococcus pneumoniae* and *Pseudomonas aeruginosa*, respectively. Another disadvantage is that although the Isolator may inhibit the action of aminoglycosides via SPS, it does not remove them entirely from the system.

B. Filtration

The use of membrane filter procedures for the isolation of microorganisms from blood has long been established. As already mentioned, these procedures were not very practical for clinical use. The early trials incorporated lysis technology as a vital component. However, filtration for the isolation of bacteria does not have to involve lysis. In recent years, a centrifugation–filtration method has been developed that employs the use of a density mixture for the separation of red blood cells from the microorganisms.

III. DETECTION OF BACTERIA BY CENTRIFUGATION PLUS FILTRATION

A. Background

Currently, the culture of blood in a liquid medium is the most common means of diagnosing bacteremia. However, this is a slow process. In addition, growth conditions in blood bottles are not necessarily conducive to rapid uninhibited growth. Therefore, a detection system is needed that allows the growth of bacteria on a filter membrane without use of lysing agents. The potential advantages of such a system include faster isolation of bacteremic agents, collection of leukocytes that may harbor phagocytized bacteria, no subculturing, removal of inhibitory agents (antibiotics) from blood, detection and faster identification of the causative organism, and detection of mixed infections or contaminants. Antibiotic susceptibility tests can be accelerated due to growth of distinct isolated colonies that do not need further purification. In addition, the efficacy of treatment (via the proper course of antibiotic therapy) can be determined by quantitation of bacteria on the filters.

B. Centrifugation and Filtration Procedure

The centrifugation and filtration procedure is shown in Fig. 1. Fresh human whole blood is first obtained from a healthy volunteer who has not received antibiotics during the preceding 2 weeks. This blood sample is then seeded with a known quantity of a single species of bacteria and mixed. Concurrently, sterile glass tubes are loaded with 8 ml of Ficoll–Hypaque (density mixture: $D = 1.149 \pm 0.002$ g/ml). Then, 5-ml aliquots of the seeded blood are added to the tubes containing the Ficoll–Hypaque. The tubes are stoppered, inverted five times to thoroughly mix the Ficoll–Hypaque with the seeded blood, and centrifuged ($386g$) for 30 min at room temperature.

During centrifugation, the red blood cells, which are denser than the bacteria, are driven to the bottom of the tube. This is further enhanced by the Ficoll, a polymer of sucrose. Ficoll acts by adhering to the red blood cells. This facilitates the sedimentation of the red blood cells during centrifugation. As the red blood cells are tumbling through the mixture, the bacteria are "washed free" from the erythrocytes and are retained throughout the Ficoll–Hypaque because they are less dense.

Centrifugation of the blood–density mixture results in the red blood cells being pelleted at the bottom of the tubes. Mononuclear and polymorphonuclear cells are distributed predominantly in the upper portion of the Ficoll–Hypaque. After centrifugation, the entire mixture, except for the erythrocytes, is removed and filtered through a 0.22-mm pore size filter under negative pressure with a single-place sterility test manifold attached to a vacuum pump.

The filter is then removed from the filtration apparatus, placed with the filtrate side up on a chocolate agar plate, and incubated at 35°C in a humidified atmosphere containing 5% CO_2. Isolated colonies of bacteria are detected within 18 hr.

C. Data

The recovery of bacteria from seeded whole blood is shown in Table I. In general, fewer microorganisms are recovered by the filters compared with the original inoculum. However, when 10 ml of Ficoll–Hypaque is used, improved recovery is observed (Table II) because the additional 2 ml of Ficoll–Hypaque provides an additional travel distance for the bacteria to be separated from the blood cells.

The centrifugation filtration procedure compares well to current blood culturing systems on the mar-

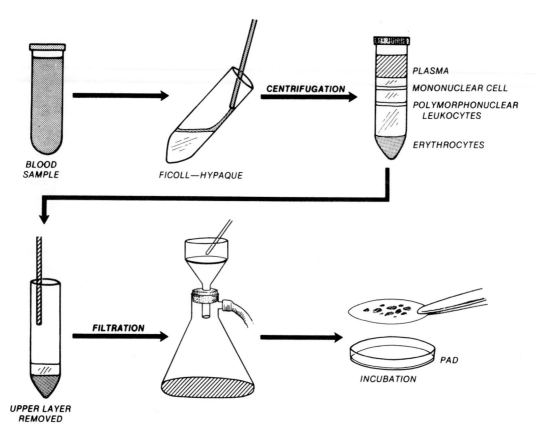

Fig. 1. Ficoll–Hypaque centrifugation and filtration procedure.

TABLE I
Microorganisms (Mean ± SD) Recovered from Blood by Centrifugation and Filtration[a]

Organism	Filtration	Inoculum	% Recovery
Staphylococcus aureus	15.3 ± 8.7	37.5 ± 4.9	40.8
Enterococcus faecalis	28.0 ± 4.5	32.0 ± 8.4	87.5
Streptococcus mitis	22.6 ± 6.6	65.0 ± 8.4	34.7
Streptococcus mutans	141.3 ± 10.0	172.6 ± 20.5	81.8
Streptococcus pneumoniae	52.3 ± 6.3	56.0 ± 1.4	93.3
Streptococcus salivarius	12.3 ± 8.0	15.5 ± 2.1	79.3
Streptococcus sanguis	40.0 ± 6.5	49.0 ± 4.2	81.6
Escherichia coli	38.6 ± 3.5	79.0 ± 9.8	48.8
Haemophilus influenzae	19.0 ± 6.2	46.3 ± 16.6	41.0
Klebsiella pneumoniae	35.3 ± 5.1	59.5 ± 7.7	59.3
Neisseria meningitidis	93.0 ± 7.8	131.0 ± 13.4	70.9
Pseudomonas aeruginosa	171.3 ± 12.0	159.5 ± 9.2	107.3

[a] Eight milliliters of Ficoll–Hypaque was used.

TABLE II
Microorganisms (Mean ± SD) Recovered from Blood by Centrifugation and Filtration[a]

Organism	Filtration	Inoculum	% Recovery
Staphylococcus aureus	45.5 ± 4.0	63.0 ± 6.0	72.0
Escherichia coli	18.0 ± 1.0	20.0 ± 2.0	90.0
Haemophilus influenzae	4.0 ± 1.0	7.0 ± 1.0	57.1
Klebsiella pneumoniae	49.0 ± 3.0	54.2 ± 7.0	90.4
Proteus mirabilis	5.0 ± 2.0	7.0 ± 1.0	71.4
Listeria monocytogenes	36.0 ± 4.0	47.0 ± 5.0	76.0

[a] Ten milliliters of Ficoll–Hypaque was used.

ket today, namely, the Isolator and the Septi-Chek. The centrifugation and filtration system has also proven to be very effective in the removal of antibiotics when used in conjunction with a nonionic polymeric adsorbent resin.

D. Concerns

Overall, the centrifugation and filtration procedure recovers less microorganisms than are present in the inoculum (Table I). Some loss of microorganisms is expected, mostly due to retention of small quantities of seeded blood in the pipette used to transfer blood to the gradient. In addition, the tubes retain a small portion of the Ficoll–Hypaque containing seeded blood. It is also possible that some bacteria adhered to the sides of the glass tubes and were missed by filtration. Some may argue that a 20% or 35% recovery of bacteria is poor; however, the sensitivity of the conventional blood culturing bottle is less. When 23 blood culture media were inoculated with 7–15 microorganisms per bottle, only 1 of the 23 different types of blood culture media supported the growth of all the bacteria tested. Furthermore, and most important, blood is normally a sterile body fluid. Any amount of bacteria isolated from blood is significant, be it 1 colony-forming unit (CFU)/ml or 50 CFU/ml of blood.

When the centrifugation and filtration system was used in clinical trials, the rate of isolation of bacteremic agents was significantly better than that of the conventional blood culture system. The main objection to centrifugation and filtration regards whether molecular diagnostic approaches will replace blood culturing, including the improved centrifugation and filtration system. Therefore, manufacturers have been reluctant to invest money in a new product they ultimately believe will be replaced by molecular detection systems. Culturing of blood is going to be a primary clinical sample for many years, even though molecular approaches are being developed. The centrifugation and filtration system offers an immediate improvement compared to the conventional blood culture systems.

The drawbacks of this system are minor compared to the advantages. Also, this system is in its infancy and has not been refined. Currently, centrifugation and filtration does not require lysing agents, multiple filters, dilutions, sophisticated equipment, or excessive centrifugation speeds. Most important, it offers faster isolation of bacteremic agents, which translates into better patient care.

IV. FUTURE DIRECTIONS

A. Refinement

The drawbacks mentioned earlier can be eliminated by making the centrifugation and filtration technique a closed system. This can be accomplished by manufacturing a double-ended blood collection tube that contains the Ficoll–Hypaque and an anticoagulant such as SPS. Blood would be drawn through one end of the tube. The other end of the tube would have a small rubber cone protruding into the tube. During centrifugation, the erythrocytes pellet around the cone but leave the tip exposed. The cone would then be punctured from the bottom of the tube during the filtration procedure. Ficoll–Hypaque containing bacteria would be drawn directly onto the filters. All this could be accomplished without ever opening the tube. Using this approach; loss of microorganisms should be minimal because there are no pipetting steps as in the current procedure.

Should there be any back-flushing of Ficoll–Hypaque into the vein during blood collection, this should cause no alarm. Ficoll is a polymer of sucrose and Hypaque is used commonly as a contrast medium in procedures such as angiograms.

B. Clinical Trials

Comparing a new technique, such as centrifugation and filtration, against established blood culturing devices in a laboratory setting is valuable in determining its usefulness. However, it needs to be evaluated in a clinical setting to determine its ultimate effectiveness.

See Also the Following Articles

Diagnostic Microbiology • Identification of Bacteria, Computerized

Bibliography

Bernhardt, M., Pennell, D. R., Almer, L. S., and Schell, R. F. (1991). Detection of bacteria in blood by centrifugation and filtration. *J. Clin. Microbiol.* **29**, 422–425.

Doern, G. V., Barton A., and Rao, S. (1998). Controlled comparative evaluation of BacT/Alert FAN and ESP 80A aerobic media as means for detecting bacteremia and fungemia. *J. Clin. Microbiol.* **36**, 2686–2689.

Dorn, G. L., Haynes, J. R., and Burson, G. (1976). Blood culture technique based on centrifugation: development phase. *J. Clin. Microbiol.* **3**, 251–257.

Herlich, M. B., Schell, R. F., Francisco, M., and LeFrock, J. L. (1982). Rapid detection of simulated bacteremia by centrifugation and filtration. *J. Clin. Microbiol.* **16**, 99–102.

Reimer, L. G., Wilson, M. L., and Weinstein, M. P. (1997). Update on detection of bacteremia and fungemia. *Clin. Microbiol. Rev.* **10**, 444–465.

Rohner, P., Pepey, B., and Auckenthaler, R. (1997). Advantage of combining resin with lytic BACTEC blood culture media. *J. Clin. Microbiol.* **35**, 2634–2638.

Sullivan, N. M., Sutter, V. L., and Finegold, S. M. (1975a). Practical aerobic membrane filtration blood culture technique: development of procedure. *J. Clin. Microbiol.* **1**, 30–36.

Sullivan, N. M., Sutter, V. L., and Finegold, S. M. (1975b). Practical aerobic membrane filtration blood culture technique: clinical blood culture trial. *J. Clin. Microbiol.* **1**, 37–43.

Weinstein, M. P. (1996). Current blood culture methods and systems: clinical concepts, technology and interpretation of results. *Clin. Infect. Dis.* **23**, 40–46.

Developmental Processes in Bacteria

Yves V. Brun

Indiana University

GLOSSARY

chemotaxis Movement toward or away from a chemical.
hypha (*pl.* hyphae) A single filament of a mycelium.
mycelium A network of cellular filaments formed by branching during the growth phase of fungi and actinomycetes.
phosphorelay A signal-transduction pathway in which a phosphate group is passed along a series of proteins.
regulon A group of genes controlled by the same regulatory molecule.
septum A partition that separates a cell into two compartments.
sigma factor The subunit of the RNA polymerase holoenzyme that confers promoter specificity.
surfactant A substance that reduces the surface tension of a liquid.
TCA cycle The cyclic pathway by which the two-carbon acetyl groups of acetyl-CoA are oxidized to carbon dioxide and water.
vegetative growth Exponential growth that usually occurs by simple binary cell division and produces two identical progeny cells.

BACTERIAL DEVELOPMENT generates specialized cell types that enhance the ability of bacteria to survive in their environment. In addition to changes in gene expression, developmental processes in bacteria involve changes in morphology and changes in function that play an important role in the life cycle of the organism. Eukaryotic organisms add sexual reproduction to the functions of development, but bacterial developmental processes are asexual.

Bacteria use two basic strategies to respond to changes in their environment. In the first and simplest strategy, they induce the expression of genes that enable them to deal with the environmental change. For example, starvation for inorganic phosphate, the preferred source of the essential element phosphorus, induces the Phosphate (Pho) regulon. The Pho regulon includes genes for the high-affinity phosphate-transport proteins that increase the ability of the bacterium to transport phosphate and genes that allow the bacterium to metabolize organic forms of phosphate. These responses are relatively simple in that they usually involve a two-component regulatory system that activates the transcription of a set of genes required for the response. At the other extreme of complexity are bacteria that undergo complex developmental transformations in response to stress or as part of their normal life cycle. These developmental responses involve not only changes in gene expression, but also changes in cellular morphology, metabolic chemistry, and association with cells of other species.

I. FUNCTION OF DEVELOPMENT

Bacterial development produces cells that have four basic types of functions. Representative examples of these functions are given in Table I. The most common product of bacterial development is

TABLE I
Examples of Prokaryotic Development[a]

Resting cells		
Resting cell	Representative genus	Group
Endospore	*Bacillus*	Gram positive
	Metabacterium	Gram positive
	Thermoactinomyces	Gram positive
Aerial spore	*Streptomyces*	Gram positive
Zoospore	*Dermatophilus*	Gram positive
Cyst	*Azotobacter*	Proteobacteria
	Methylomonas	Proteobacteria
	Bdellovibrio	Proteobacteria
Myxospore	*Myxococcus*	Proteobacteria
	Stigmatella	Proteobacteria
Exospore	*Methylosinus*	Proteobacteria
Small dense cell	*Coxiella*	Proteobacteria
Elementary body	*Chlamydia*	Chlamydia
Akinete	*Anabaena*	Cyanobacteria

Complementary cell types			
Cell	Function	Representative genus	Group
Heterocyst	Nitrogen fixation		
Vegetative cell	Oxygenic photo-synthesis	*Anabaena*	Cyanobacteria

Dispersal cells		
Cell	Representative genus	Group
Baeocyte	*Pleurocapsa*	Cyanobacteria
Elementary body	*Chylamidia*	Chlamydia
Gonidium	*Leucothrix*	Proteobacteria
Hormogonium	*Oscillatoria*	Cyanobacteria
Swarm cell	*Proteus*	Proteobacteria
Swarmer cell	*Caulobacter*	Proteobacteria
Zoospore	*Dermatophilus*	Gram positive

Symbiotic development			
Cell	Representative interaction	Function	Group
Bacteroid	*Rhizobium*–legume	N_2 fixation	Proteobacteria
	Frankia–alder	N_2 fixation	Gram positive

[a] From Shimkets, L., and Y. V. Brun (1999). Prokaryotic development: strategies to enhance survival. *In* "Prokaryotic Development." (Y. V. Brun and L. Shimkets, Eds.), pp. 1–7. American Society for Microbiology, ASM Press.

a resting cell with relatively low metabolic activity and a higher resistance to physical and chemical stress than the vegetative cell; the best-studied example is endospore formation in *Bacillus subtilis*. The second type of function of differentiated cells is dispersal. Dispersal can be propelled by flagella or can simply be aided by wind, water, or animals in the case of nonmotile cells; the dispersal swarmer cell of *Caulobacter crescentus* is the product of an asymmetric division that also produces a sessile stalked cell. The production of cells whose physiology is complementary represents the third type of function of bacterial development; this is best exemplified by formation of heterocysts that are specialized for nitrogen fixation in the Cyanobacterium *Anabaena*. Finally, bacterial development can lead to the establishment of a symbiotic relationship, as in the case of nodulation of legume roots by Rhizobium. Examples of the various developmental functions are presented here. In order to give a flavor of the research in this field, one example (*Caulobacter*) is described in slightly more detail.

II. ENDOSPORE FORMATION IN *BACILLUS SUBTILIS*

Endospore formation has been found exclusively in gram-positive bacteria and is best understood in *Bacillus subtilis*. The primary signal for the initiaton of sporulation is nutrient starvation. Cell density is also important for efficient sporulation, presumably to ensure that cells are sufficiently abundant. It may be that if starving cells are at a high density, it is better to sporulate rather than compete for nutrients. However, if cells are at a low density, the chances of finding additional nutrients is higher and sporulation less desirable. Before initiating sporulation, cells monitor many intracellular factors, such as DNA replication and the TCA cycle. The integration of the extracellular and intracellular signals is regulated through a multicomponent phosphorelay that controls the prosphorylation of the transcriptional regulator SpoOA. The initiation of sporulation by the formation of a polar septum instead of the vegetative midcell septum requires the accumulation of a sufficient concentration of SpoOA~P. The subsequent engulfment of the prespore by the mother cell compartmentalizes the prespore inside the mother cell (Fig. 1). The genome of the mother cell provides the components for constructing the spore exterior and the genome of the forespore provides the components for constructing the spore interior. The forespore ultimately becomes a metabolically quiescent and stress-resistant spore that can give rise to future progeny by germination when conditions improve. The mother cell is discarded by lysis after the completion of sporulation.

The regulation of events in the mother cell and the forespore is due to the presence of four different sigma factors, two in each compartment, which assures that each genome gives rise to a different set of products (Fig. 2). Activation of σ^F in the forespore depends on polar septation. σ^E is synthesized as an inactive precursor whose activation by proteolitic processing in the mother-cell compartment is dependent on activation of σ^F in the forespore. The transcription of the σ^G gene requires σ^F and thus only occurs in the forespore. σ^G activation depends on proteins made in the mother cell under the control of σ^E. Finally, σ^K is only synthesized in the mother cell under the control of σ^E and is activated in a σ^G-dependent manner.

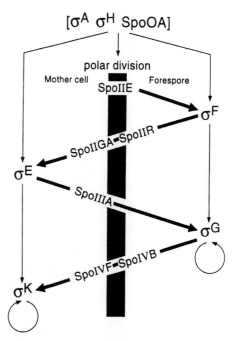

Fig. 2. Criss-cross regulation of cell type-specific sigma factors. [With permission, from P. Stragier and R. Losick (1996). Molecular genetics of sporulation in bacillus subtilis, *Annual Review of Genetics* 30: 297–341, © 1996 by Annual Reviews, www.annualreviews.org.]

III. SPORULATION IN *STREPTOMYCES COELICOLOR*

The aerial mycelium of *Streptomyces coelicolor* forms by directed cell growth and differentiates into a series of spores (Fig. 3). The vegetative mycelium grows in the nutrient substratum by the linear growth of cell wall close to the hyphal tip (Fig. 4). Branching of the vegetative mycelium allows close-to-exponential increase of the mycelial mass. Septation is infrequent in the vegetative mycelium and the vegetative septa do not allow cell separation. With time, the vegetative mycelium becomes more dense, producing aerial hyphae that grow quickly. Rapid growth occurs at the expense of nutrients derived from the substrate mycelium and aerial hyphae emerge from the surface of colonies. The formation of the aerial hyphae requires a set of genes called *bld* genes because mutants of these genes fail to develop a hairy surface layer (*bald*). Most of these mutants fail to produce a small extracellular surfactant protein SapB. SapB coats the surface of the aerial hyphae with a hy-

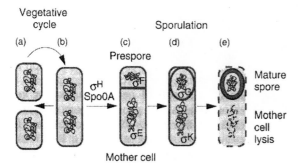

Fig. 1. Life cycle of *Bacillus subtilis*. (a) Vegetative growth occurs by binary fission when nutrients are plentiful. (b) and (c) When starved, the vegetative midcell division is replaced by a highly asymmetric polar division that compartmentalizes the cell into a prespore and a mother-cell compartment. (d) The prespore is engulfed by the mother cell. (e) After formation of the spore cortex (thick circle), the mother cell lyses and releases the mature spore. [From J. Errington (1996). Determination of cell fate in bacillus subtilis, *Trends in Genetics* 12: 31–34, Copyright (1996), with permission from Elsevier Science.]

Fig. 3. Scanning electron micrographs of spore chain in the aerial mycelium of *Streptomyces coelicolor.* Bar, 1 μm. [From K. Chater (1998). Taking a genetic scalpel to the Streptomyces colony. *Microbiology* 144: 1465–1478.]

drophobic outer surface. This may permit growth through the surface tension barrier at the air–colony interface. Antibiotics are produced, presumably to protect the nutrients released from lysing substrate mycelium from other bacteria. All antibiotic production, as well as aerial mycelium development, is prevented in *bldA* mutants. The *bldA* gene encodes the only tRNA that efficiently recognizes the rare leucine codon UUA. Genes required for vegetative growth do not contain TTA codons. TTA codons are found in regulatory genes involved in antibiotic production. Is it thought that an increased production of mature *bldA*-encoded tRNA during development allows the efficient translation of UUA codons in regulators of antibiotic production and in genes involved in development whose identify is still unknown. Growth of

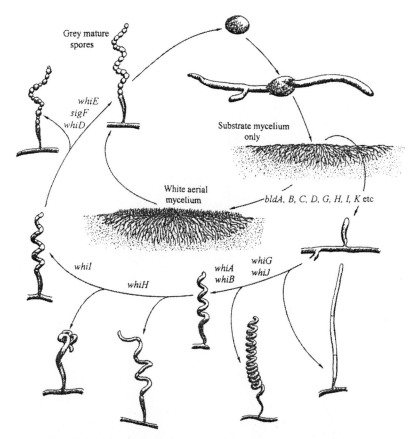

Fig. 4. Life cycle of *Streptomyces coelicolor.* Genes important in the different stages of the life cycle are shown. The phenotypes of various mutants are shown by arrows diverging from the normal life cycle immediately before the gene designation. [From K. Chater (1998). Taking a genetic scalpel to the Streptomyces colony. *Microbiology* 144: 1465–1478.]

the aerial hyphae eventually stops and regularly spaced sporulation septa are formed synchronously. Thus, the cell separation required for dispersal occurs by sporulation at the surface of colonies. Sporulation requires *whi* genes, identified because mutations in these genes prevent the formation of mature grey spores and the aerial mycelium remains white. Most of the early-acting *whi* genes appear to be regulatory. *whiG* encodes a sigma factor, while *whiI* and *whiH* encode transcriptional regulators.

IV. SWARMING BACTERIA

Swarming differentiation produces cells (swarm cell) capable of a specialized form of translocation

on a surface and occurs in a variety of bacteria, both gram-positive and gram-negative. These include *Proteus, Bacillus, Clostridium,* and *Vibrio* species, *Serratia marcescens, Rhodospirillum centenum, E. coli,* and *S. typhimurium.* The swarm cell differentiation is triggered by growth on an appropriate solid medium, for example a petri plate. Initially, cells grow vegetatively as short rods with a small number of flagella. Differentiated swarm cells are long (20–80 μm), multinucleate, non-dividing cells with up to 50-fold more flagella per unit cell-surface area than vegetative cells. Swarm cells migrate rapidly across the plate (Fig. 5). Swarming differentiation is not a starvation response and is not an obligatory stage in the life cycle of these bacteria. An important signal for swarm cell differentiation is bacterial contact with a solid

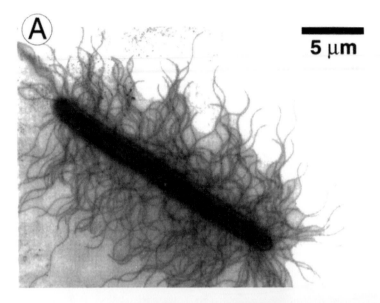

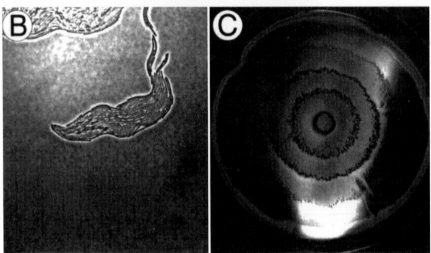

Fig. 5. Swarming in *Proteus mirabilis.* (A) Electron micrograph of a swarmer cell. (B) Movement of a mass of cells at the swarming periphery. (C) Characteristic colony morphology. The pattern is produced by alternating cycles of differentiation, movement, and consolidation. [From BACTERIA AS MULTICELLULAR ORGANISMS, edited by James A. Shapiro and M. Dworkin. Copyright © 1997 by Oxford University Press, Inc. Used by permission of Oxford University Press, Inc.]

surface. In *P. mirabilis* and *V. parahaemolyticus,* surface-sensing is mediated by the flagella. After a certain period, the migration of swarm cells slows down and the swarm cells divide (consolidation) and revert to the vegetative-swimmer cell type (Fig. 6). Vegetative growth continues until a second phase of swarm cell differentiation is initiated. This differentiation–consolidation cycle continues until the surface of the

plate is covered. Swarming can have a function in host–pathogen interactions. For example, *Proteus mirabilis* mutants that are deficient in swarming are unable to establish kidney infections.

The *flhDC* master regulatory operon is critical for the control of swarming differentiation. Artificial overexpression of FlhDC induces swarm cell differentiation without the need for contact with a solid sur-

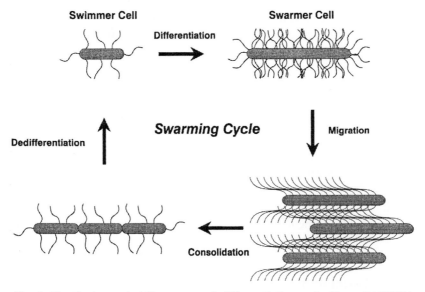

Fig. 6. The *Proteus mirabilis* swarm cell-differentiation cycle. [From BACTERIA AS MULTICELLULAR ORGANISMS, edited by James A. Shapiro and M. Dworkin. Copyright © 1997 by Oxford University Press, Inc. Used by permission of Oxford University Press, Inc.]

face. FlhD and FlhC form a complex that activates transcription of genes encoding flagellar export, structural, and regulatory proteins. In addition, FlhDC represses cell division.

V. DIMORPHIC LIFE CYCLE OF *CAULOBACTER CRESCENTUS*

A distinguishing feature of the development of stalked bacteria is that it is an integral part of the growth of the cell and not an alternative to it, as are the other bacterial developmental processes that occur in response to stress. The molecular mechanisms that control the developmental cycle of stalked bacteria have been studied most extensively in *Caulobacter crescentus*. Each division of *Caulobacter* cells gives rise to a swarmer cell and a stalked cell (Fig. 7). The swarmer cell is dedicated to dispersal and the stalked cell is dedicated to growth and the production of new swarmer cells. The obligatory time spent as a chemotactic swarmer cell presumably ensures that progeny cells will colonize a new environmental niche instead of competing with attached stalked cells. The swarmer cell has a single polar

flagellum and is chemotactically competent. During this dispersal stage of their life cycle, swarmer cells do not replicate DNA and do not divide. After approximately one-third of the cell cycle, in response to an unknown internal signal, the swarmer cell

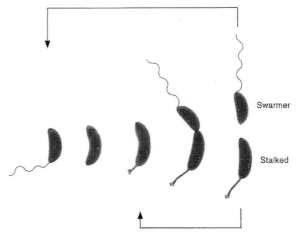

Fig. 7. The *Caulobacter crescentus* cell cycle. [From cover of *Genes and Development,* Vol. 12, No. 6. Kelly, A. J., M. Sackett, N. Din, E. Quardokus, and Y. V. Brun, Cell cycle dependent transcriptional and proteolytic regulation of FtsZ in *Caulobacter.* 1998. *Genes and Development,* 12: 880–893.]

sheds its flagellum, initiates DNA replication, and synthesizes a stalk at the pole that previously contained the flagellum. Located at the tip of the stalk is the holfast, the adhesion organelle that allows *Caulobacter* to attach to surfaces. The holdfast appears at the tip of nascent stalks during swarmer to stalked cell differentiation. Cell growth is accelerated at the time of swarmer-to-stalked cell differentiation and eventually leads to the formation of a predivisional cell in which a flagellum is synthesized *de novo* at the pole opposite the stalk. Unlike the swarmer cell, the progeny stalked cell is capable of initiating a new round of DNA replication immediately after cell division.

A. Cell-Cycle Regulation of Flagellum Synthesis

The best understood event in *Caulobacter* development is the biosynthesis of the flagellum (Fig. 8). A new flagellum is synthesized at every cell cycle and is localized at the pole opposite the stalk. The flagellum is, for the most part, similar to that of *E. coli* and its synthesis requires more than 50 genes. The expression of these flagellar genes is temporally ordered during the progression through the cell cycle; their order of transcription approximates the order of assembly of their protein products in the flagellum. Most flagellar genes can be grouped into four classes, forming a regulatory hierarchy that dictates their order of expression. First to be transcribed, immediately after the differentiation of the swarmer cell into a stalked cell, are the gene for the MS ring that anchors the flagellar basal body in the cytoplasmic membrane, genes that encode the proteins of the switch complex, and genes for the flagellar export apparatus. In addition, Class II genes encode the regulatory proteins FlbD and σ^{54} that are required for the transcription of most Class III and IV genes. These early-expressed genes make up Class II in

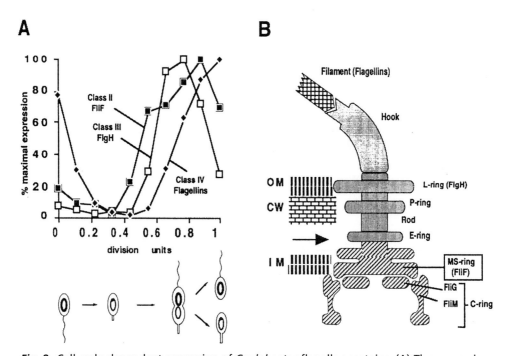

Fig. 8. Cell cycle-dependent expression of *Caulobacter* flagellar proteins. (A) The expression of a representative protein from each class is shown. (B) Diagram of the *Caulobacter* flagellum. [From U. Jenal and L. Shapiro (1996) Cell cycle-controlled proteolysis of a flagellar motor protein that is asymmetrically distributed in the *Caulobacter* predivisional cell. *The EMBO Journal* **15**: 2393–2406, by permission of Oxford University Press.]

the flagellar regulatory hierarchy. The expression of Class II genes is required for the transcription of the next set of flagellar genes, the Class III genes. Class III genes encode proteins that make up the rest of the basal body (the rings anchored in the peptidoglycan cell wall and in the outer membrane and the rod that traverses the rings) and the proteins that compose the hook structure. The expression of Class III genes is required for the transcription of the last flagellar genes to be expressed during the cell cycle, the Class IV genes that encode the flagellins that make up the helical filament. The flagellar regulatory cascade is triggered by the response regulator CtrA, which by definition occupies Class I of the regulatory hierarchy. CtrA activates the transcription of Class II genes by binding to a conserved sequence in their promoter region. *In vitro* experiments indicate that phosphorylated CtrA (CtrA~P) is required for the transcriptional activation of these genes. The *flbD* gene is the last gene of the Class II fliF operon whose

transcription depends on CtrA~P. The promoter of the *rpoN* gene contains a putative CtrA binding site. Consequently, the initiation of the transcription of Class II genes of the flagellar regulatory cascade by CtrA~P results in the synthesis of the regulatory proteins FlbD and σ^{54}, which are required for the transcription of Class III and IV genes. CtrA also regulates DNA replication and cell division, providing a mechanism to coordinate the expression of flagellar genes with those events (see next section).

B. Regulation of Cell Division and DNA Replication

In *Caulobacter,* different stages of development require the completion of specific stages of the replication and division cycles. The inhibition of DNA replication blocks flagellum synthesis by preventing the transcription of early flagellar genes that are at the

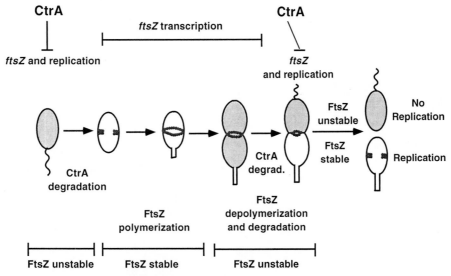

Fig. 9. Model of FtsZ regulation in *Caulobacter.* CtrA (gray shading inside cells) represses *ftsZ* transcription in swarmer cells. During swarm cell differentiation, CtrA is degraded and allows *ftsZ* transcription to be turned on. FtsZ concentration increases and FtsZ polymerizes and forms a ring at the site of cell division. During this time FtsZ is stable. The reappearance of CtrA inhibits *ftsZ* transcription. FtsZ depolymerizes as the cell and the FtsZ ring constrict. FtsZ is rapidly degraded, especially in the swarmer pole. [From Kelly, A. J., M. Sackett, N. Din, E. Quardokus, and Y.V. Brun. (1998). Cell cycle dependent transcriptional and proteolytic regulation of FtsZ in *Caulobacter. Genes and Development* **12**: 880–893.]

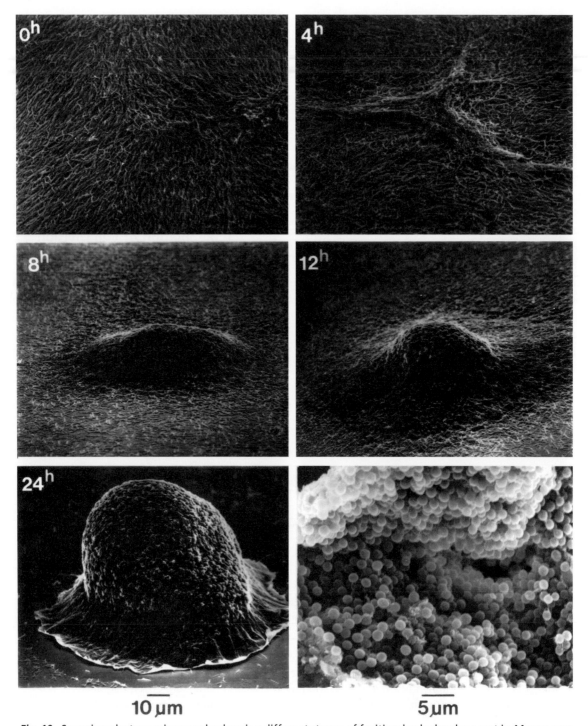

Fig. 10. Scanning electron micrographs showing different stages of fruiting-body development in *Myxococcus xanthus*. The lower right panel shows spores from an open fruiting body. [From Kaiser, D., L. Kroos, and A. Kuspa (1985). Cell interactions govern the temporal pattern of *Myxococcus* development. *Cold Spring Harbor Symposia on Quantitative Biology* **50**: 823–830.]

top of the flagellar regulatory hierarchy. Cells inhibited for DNA replication are also blocked for cell division and form long smooth filamentous cells with a stalk at one pole and flagella at the opposite pole. Cells that can replicate DNA, but that are blocked in cell division, are also affected in their progression through development. The initiation of cell division plays an essential role in the establishment of differential programs of gene expression that set up the fates of the progeny cells.

In all bacteria examined, the abundance and subcellular location of the tubulin-like GTPase, FtsZ, are critical factors in the initiation of cell division. FtsZ is a highly conserved protein that polymerizes into a ring structure associated with the cytoplasmic membrane at the site of cell division. FtsZ recruits other cell-division proteins to the site of cell division and may constrict, providing mechanical force for

division. In *Caulobacter*, FtsZ is subject to a tight developmental control. After cell division, only the stalked cell contains FtsZ. Transcriptional and proteolytic controls contribute to the cell cycle and developmental regulation of FtsZ (see Fig. 9).

The initiation of DNA replication and *ftsZ* transcription are controlled by the cell-cycle-response regulator CtrA. CtrA directly binds to five sites in the origin of replication and prevents the initiation of DNA replication. CtrA is present in swarmer cells, where it blocks DNA replication and represses *ftsZ* transcription. CtrA is degraded during swarmer cell differentiation, thus coordinating the onset of the replication and division cycles. The degradation of CtrA depends on the ClpXP protease. Late in the cell cycle, when DNA replication is complete and cell division has been initiated, CtrA is synthesized and represses *ftsZ* transcription and initiation of DNA

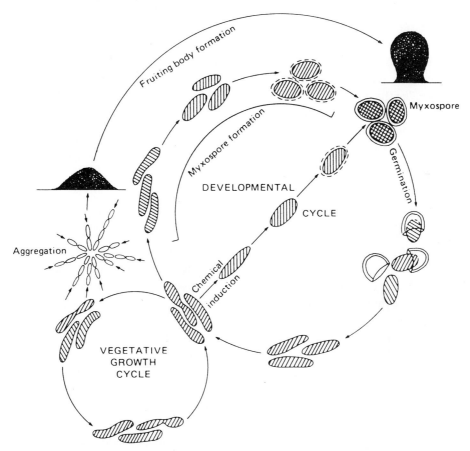

Fig. 11. Life cycle of *Myxococcus xanthus*. [From Dworkin, M. (1985). Developmental Biology of the Bacteria. Benjamin/Cummings, Menlo Park, CA.]

replication. Just before cell separation, CtrA is degraded in the stalked compartment. The absence of CtrA from stalked cells after cell division allows *ftsZ* transcription to resume and DNA replication to be initiated.

Proteolytic control of FtsZ is superimposed on the transcriptional control. FtsZ molecules are stable as they assemble into the FtsZ ring and are degraded rapidly once cells have begun to constrict. FtsZ is particularly unstable in the swarmer compartment of the predivisional cell, leading to its disappearance from swarmer cells after cell division. This two-tiered level of regulation ensures that FtsZ is only present in the cell that will initiate a new cell cycle immediately after cell division.

VI. FRUITING-BODY FORMATION IN MYXOBACTERIA

Fruiting-body formation in myxobacteria is only one example of bacterial social behavior. The entire life cycle of myxobacteria is pervaded by social behavior. Myxobacterial cells move together and feed cooperatively to maximize the efficiency of extracellular degradation. The enclosure of myxospores in the fruiting body allows then to be dispersed together and ensures that a sufficiently large population of cells will be present after germination to facilitate social interactions. The best-studied example of fruiting-body formation in myxobacteria is in *Myxococcus xanthus* (Fig. 10). When cells perceive a nutritional down shift, they enter the developmental pathway that leads to fruiting-body formation (Fig. 11). Fruiting-body formation can only occur if cells are on a solid surface, to allow gliding motility, and if the cell density is high. When these three conditions are met, cells move into aggregation centers and eventually form mounds containing approximately 10,000 cells. As many as 90% of cells lyse during aggregation. The surviving cells differentiate into resistant and metabolically quiescent myxospores during the last stages of fruiting-body formation. Fruiting-body development is regulated by a series of intercellular signals. Five categories of signals are involved: (1) the A signal is a mixture of amino acids and peptides that serves to monitor cell density; (2) the

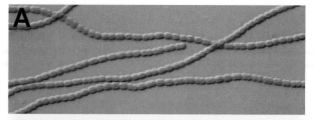
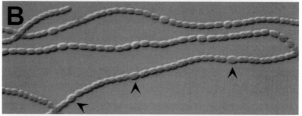
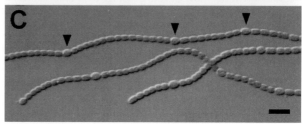

Fig. 12. *Anabaena* sp. strain PCC-7120 filaments grown in nitrate medium (A) Filaments were subjected to nitrogen step-down for 18 hr (B), and 24 hr (C) to induce heterocysts. Developing proheterocysts are indicated on one filament with arrowheads in (B), and mature heterocysts are indicated with triangles in (C). The strain used in these figures contains a reporter plasmid that does not affect wild-type development of heterocysts. Scale bar, 10 μm. Photo by Ho-Sung Yoon.

B signal acts early in development and its production depends on the Lon protease, but the nature of the signal has not been identified; (3) the C signal is associated with the cell surface and is the last of the five signals to act, controlling both aggregation and sporulation (the chemical nature of C signal is not known; all mutations that prevent C-signal formation map to the *csgA* gene and the CsgA protein itself could be the signal or it could produce the signal through an enzymatic activity); (4) the D signal requires the normal function of the *dsgA* gene that encodes the translation initiation factor 3 (IF3) but neither the manner by which *dsgA* functions in the production of the signal nor the identity of the signal are known; and (5) the E signal is thought to consist of branched-chain fatty acids liberated by a phospholipase and passed between cells to function as short-range signals.

VII. HETEROCYST DIFFERENTIATION IN CYANOBACTERIA

The purpose of heterocyst formation in cyanobacteria such as *Anabaena* is the production of a cell specialized for nitrogen fixation in order to separate two incompatible processes. The oxygen generated by the photosynthetic activity of vegetative cells is sufficient to inactivate the nitrogenase that is required to convert atmospheric N_2 to ammonium. When nitrogen fixation is required, the detrimental effect of oxygen is circumvented by sequestering nitrogenase in the anaerobic environment of the heterocyst, in an otherwise aerobic filament of vegetative cells. In the presence of ammonium or nitrate, cyanobacteria grow as undifferentiated vegetative filaments. When these cells are starved for nitrogen, heterocyst formation is induced (Fig. 12). The heterocyst is a terminally differentiated cell, but the differentiating cell passes through a proheterocyst stage that can go back to vegetative growth under appropriate conditions. Single heterocysts form at approximately every 10 cells in a filament. Heterocyst-pattern formation is controlled in part by a diffusable signal encoded by the *PatS* gene. The PatS peptide is produced by proheterocysts and inhibits development of neighboring cells by creating a gradient of inhibitory signal.

VIII. THE PREDATORY LIFESTYLE OF *BDELLOVIBRIO*

The predation of gram-negative bacteria by *Bdellovibrio* includes a dimorphic life cycle (Fig. 13). During the obligatory intraperiplasmic growth phase, *Bdellovibrio* use their prey's cytoplasmic contents as their growth substrate. During the attack phase, they search for a new prey but do not grow. No DNA replication occurs during the attack phase; however, RNA and protein are synthesized. Thus, in addition to their dispersal function, the attack cells have the ability to attach to and enter bacterial prey. The function of the intraperiplasmic cells is to grow and to produce more attack phase cells. Attack cells are highly motile (100 μm/s) by virtue of a single polar flagellum. The attack phase continues until a suitable prey is encountered or until energy is exhausted. The attack phase has been described as a "race against starvation to find a susceptible prey" (Diedrich, 1988). *Bdellovibrio* attack cells attach to prey cells and enter the prey 5–10 min after attachment. The flagellum is shed during the entry process. The prey cell is transformed into a bdelloplast by the action of

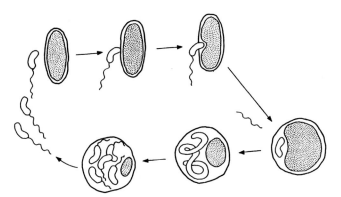

Fig. 13. Life cycle of *Bdellovibrio*. [From Thomashow, M. F. and T. W. Cotter. (1992). *Bdellovibrio* host dependence: the search for signal molecules and genes that regulate the intraperiplasmic growth cycle. *J. Bacteriol.* **174**: 5767–5771]; originally from Thomashow and Rittenberg (1979). *In* Developmental Biology of Prokaryotes, Blackwell Science Ltd.

a glycanase that solubilizes part of the peptidoglycan. The biochemical modification of the prey's peptidoglycan and lipopolysaccharide make the prey inaccessible to other *Bdellovibrio* cells. The *Bdellovibrio* cell then begins the systematic degradation of host macromolecules, which is complete in about 60 min. DNA replication begins during the intraperiplasmic growth phase and occurs without cell division to produce a multinucleate filament whose size is determined by the size of the prey, ranging from 4 to 100 times the length of an attack-phase cell. The inhibition of cell division during intraperiplasmic growth while DNA replication is occurring presents an interesting contrast to the usual coupling of replication and division in many bacteria. Swarming bacteria like *Proteus vulgaris* also inhibit cell division during growth as part of swarm cell differentiation. As part of the growth phase, *Bdellovibrio* incorporates some of the outer-mbrane proteins of the prey directly into its own membrane. Once growth becomes limited by the depletion of nutrients, elongation ceases and cell division is initiated simultaneously between the nucleoids. Flagella are synthesized *de novo* and the bdelloplast is lysed, releasing the attack-phase cells.

IX. CONCLUSION

The study of bacterial development has had a major impact on our understanding of the bacterial cell. In particular, it is now clear that bacterial cells, even those that do not differentiate, are not simply bags of enzymes. Bacterial cells are highly organized at the level of protein localization. The spatial constraints of differentiating cells are combined with temporal constraints that control the ordered progression through the developmental program. A major challenge will be to determine how temporal and spatial control are integrated during bacterial development.

See Also the Following Articles

CAULOBACTER, GENETICS • CELL DIVISION, PROKARYOTES • QUORUM SENSING IN GRAM-NEGATIVE BACTERIA • STARVATION, BACTERIAL

Bibliography

Belas, R. (1997). *Proteus mirabilis* and other swarming bacteria. *In* "Bacteria as Multicellular organisms" (J. A. Shapiro and M. Dworkin, eds.), pp. 183–219. Oxford University Press, Oxford.

Brun, Y. V., and L. Shimkets (eds.) (1999). "Prokaryotic Development." ASM Press, Washington, DC.

Chater, K. (1998). Taking a genetic scalpel to the Streptomyces colony. *Microbiology* **144**, 1465–1478.

Golden, J. W., and Yoon, H.-S. (1998). Heterocyst formation in Anabaena. *Curr. Opinion Microbiol.* **1**, 623–629.

Kim, S., Kaiser, D., et al. (1992). Control of cell density and pattern by intercellular signaling in *Myxococcus* development. *Annu. Rev. Microbiol.* **46**, 117–139.

Shapiro, J. A., and Dworkin, M., (eds.) (1997). "Bacteria as Multicellular Organisms." Oxford University Press, Oxford.

Stragier, P., and Losick, R. (1996). Molecular genetics of sporulation in bacillus subtilis. *Annu. Rev. Genet.* **30**, 297–341.

Thomashow, M. F., and Cotter, T. W. (1992). *Bdellovibrio* host dependence: The search for signal molecules and genes that regulate the intraperiplasmic growth cycle. *J. Bacteriol.* **174**, 5767–5771.

Wolk, C. P. (1996). Heterocyst formation. *Annu. Rev. Genet.* **30**, 59–78.

Wu, J., and Newton, A. (1997). Regulation of the Caulobacter flagellar gene hierarchy; Not just for motility. *Mol. Microbiol.* **24**, 233–239.

Diagnostic Microbiology

Vanderbilt University School of Medicine

David H. Persing

Corixa Corporation/Infectious Disease Research Institute

GLOSSARY

colony-forming unit Visible units counted in an agar plate which may be formed from a group of microorganisms rather than from one.

cytopathic effects (CPE) Tissue deterioration caused by viruses. CPE is widely used for the identification of virus isolates in the diagnostic virology laboratory.

dark-field microscope A microscope that has a device to scatter light from the illuminator so that the specimen appears white against a black background. It is widely used for direct examination of spirochete microorganisms.

fluorescent microscope A microscope that uses an ultraviolet light source to illuminate specimens that will fluoresce.

gold standard The best available approximation of the truth. It is a commonly used term, generally indicating a test method currently accepted as reasonably, but not necessarily 100%, accurate.

indigenous flora Microorganisms that colonize animals, humans, or plants without causing known disease.

nucleic acid probe A piece of labeled single-stranded nucleotide used to detect complementary DNA in clinical specimens or a culture and thus to specifically identify the presence of an organism identical to that used to make the probe.

probe amplification A nucleic acid amplification procedure in which many copies of the probe that hybridizes the target nucleic acid are made.

rapid plasma reagin (RPR) test Reagin is a substance made in response to a treponemal infection characterized by its ability to combine with lipids. The RPR test is a simple, rapid, non-treponemal card test for the diagnosis of syphilis.

serodiagnosis A high or rising titer of organism-specific IgG antibodies or the presence of organism-specific IgM antibodies may suggest or confirm a diagnosis.

shell vial culture A technique that combines cell culture and immunofluorescence assay for rapid detection of virus organisms.

signal amplification A nucleic acid amplification procedure in which a signal or reporter molecule attached to the probe is detected, and the signal is amplified enormously.

solid phase immunoassay An immunoassay in which the capture antigen or antibody is attached to the inside of a plastic tube, microwell, or the outside of a plastic bead, in a filter matrix, or some other solid support.

target amplification A nucleic acid amplification procedure in which many copies of the nucleic acid target are made.

Western blotting An immunologic technique for identification and characterization of protein antigen or antibody.

MICROBIOLOGISTS IN THE FIELD OF DIAGNOSTIC MICROBIOLOGY determine whether suspected pathogenic microorganisms are present in test specimens collected from human beings, animals, and the environment and, if they are present, identify them. In medical practice, a fundamental principle of diagnostic microbiology is to define infectious processes and elucidate treatment options through rapid detection and characterization of specific pathogens. Thus, beyond detection and identification of microorganisms in clinical specimens, diagnostic microbiology also provides physicians with antimicrobial susceptibility profiles of the identified microorganism. This article will provide a brief review of the fundamental principles of diagnostic microbiology. For greater detail, the reader is referred to several excellent textbooks listed in the bibliography.

I. DEVELOPMENT OF DIAGNOSTIC MICROBIOLOGY

The roots of diagnostic microbiology trace back more than three centuries when van Leeuwenhoek first observed bacteria and protozoa with his primitive microscope. However, it was not until the late 1800s that the work of Pasteur, Koch, and others ushered in the modern era of germ theory, as well as the use of isolation techniques. Since then, the capabilities of modern diagnostic microbiology have expanded and improved rapidly as a result of technological revolutions in microbiology, immunology, and molecular biology. A microorganism from a test sample can be detected and identified in any of four possible ways: (i) cultivation of microorganisms us-

ing artificial media or living hosts, (ii) direct microscopic examination, (iii) measurement of a microorganism-specific immune responses, and (iv) detection of microorganism-specific macromolecules, especially nucleic acids. These techniques are summarized in Table I, and the following sections will discuss them separately.

Significant changes have occurred in the field of diagnostic microbiology during the past 20 years. Until the early 1970s, definitive laboratory diagnoses of infectious diseases had been largely accomplished through the use of cumbersome, costly, time-consuming, and often subjective techniques. However, in the 1980s and 1990s, diagnostic technology evolved rapidly. For example, immunoassays came into wider usage with the development of radioimmunoassay and immunofluorescence and then enzyme immunoassay (EIA) and immunoblotting for routine diagnostic microbiology applications. These techniques, especially EIA, in many cases have supplanted labor-intensive and relatively insensitive and nonspecific procedures such as complement fixation and hemagglutination inhibition assays for viral se-

TABLE I
Methods Used for Microorganism Diagnosis

Test	Ease of performance	Turnaround time	Result interpretation	Advantages	Disadvantages
Direct examination	Could be performed in routine clinical lab and in nurse station	1–3 hr	Direct if correlated with symptoms	Rapid	Poor sensitivity and specificity; special skills are needed for interpretation
Culture	Could be performed in sophisticated clinical lab and in research lab	2–14 days	Definite	For phenotypic drug susceptibility testing	Time-consuming; poor sensitivity; limited microorganisms are culturable
Serology	Could be performed in larger and sophisticated clinical lab	4–6 hr	Indirect	Automation	Results are generally retrospective; immunosuppressed host may be unable to mount a response
Molecular diagnostics	Could be performed in only a few very sophisticated research and clinical labs	1–2 days	Direct without knowing microbial viability	High sensitivity and specificity	Facility requirement; false positive due to carryover contamination and false negative due to inhibitors in specimen

rology. Dependable commercially available direct immunoassay kits make detection of certain important groups of viruses economical, such as respiratory syncytial virus and rotavirus. Direct microorganism identification based on antigen or nucleic acid detection through immunologic and molecular techniques has significantly shortened the test turnaround time. Oligonucleotide probes have played a significant role in culture confirmation and differentiation of slowly growing microorganisms such as mycobacteria. At the same time, improvements in conventional methods have also kept pace; automated or semi-automated computerized commercial systems are available for microorganism identification and antimicrobial susceptibility determination. Bacteremia and fungemia are often detected by an automated instrument-assisted blood culture system, which has significantly improved the sensitivity and specificity for detecting all pathogens and, by virtue of automation, eliminated the hands-on time for detection of positive cultures.

Probably the greatest advance in the field of diagnostic microbiology has been in the area of nucleic acid probe detection. The polymerase chain reaction (PCR) and other recently developed amplification techniques have simplified and accelerated the *in vitro* process of nucleic acid amplification. Rapid techniques of nucleic acid amplification and characterization, along with the increased use of automation and user-friendly software, have significantly broadened microbiologists' diagnostic arsenal. The old diagnostic microbiology model, which was either labor-intensive or required days to months before test results became available, serviced largely only a hospitalized patient population. The modern diagnostic microbiology has gradually begun servicing an increasing number of outpatient populations, providing much more rapid results.

II. DIRECT IDENTIFICATION OF MICROORGANISMS

A. Microscopic Examination

Many test specimens may be examined in their native state under a microscope. A wet mount can be prepared by applying specimens directly onto the surface of a slide, which is often used to detect motile trophozoites of parasites such as *Giardia lamblia* in stools and *Trichomonas vaginalis* in vaginal discharges. Certain bacteria are so thin that they cannot be resolved in direct preparations. Their characteristic motility, however, is an important feature of presumptive identification. Dark-field examination, a method that allows light to be reflected or refracted off the surface of the object, is used to identify these bacteria, primarily spirochetes such as *Borrelia burgdorferi,* the microorganism causing Lyme disease. This method is also used for the demonstration of motile treponemes in exudates collected from a primary chancre of syphilis. Detection of viral inclusions in smears or tissues has been the traditional means of directly demonstrating virus infections. Electron microscopy has been used directly on stool specimens for detection of the viral causes of gastroenteritis such as rotaviruses.

Examination of stained material, either direct test specimens or samples of growth from cultures, is the most useful method for presumptive identification of several microorganisms. Microorganisms may be visualized and assigned to morphologic and functional groups using special stains. There are several staining methods that have been used in diagnostic microbiology (Table II). The most popular is the Gram stain, which is used to classify bacteria on the basis of their forms, sizes, cellular morphologies, and color reactions. Other commonly used staining methods include the acid-fast stain for mycobacteria, the acridine orange stain for cell wall-deficient bacteria (e.g., mycoplasma), the Giemsa stain for systemic protozoa (e.g., malaria), and iodine stains for intestinal helminthes. Calcofluor white, in place of 10% potassium hydroxide, binds to the cell walls of fungi; therefore, it is used for direct fungus detection, including detection of *Pneumocystis carinii*, a common opportunistic infection in AIDS and other immunocompromised hosts.

B. Microbial Antigen Testing

The method of using a specific antibody to detect microbial antigens is widely used in diagnostic microbiology. For immunologic detection of microbial antigens, latex particle agglutination, immunofluorescence assay, and EIA are the techniques most fre-

TABLE II
Stains Commonly Used for Detection of Microorganisms

Stain method	Organisms detected	Advantages	Disadvantages
Gram stain	Bacteria, yeast	Rapid; direct differentiation; assess specimen for culture	Cell wall-deficient bacteria stain unpredictably; pink background often masks gram-negative organisms
Acridine orange	Bacteria, mycoplasma	Good for organisms with damaged cell walls; background stain is relatively weak	Specific light source is needed; cannot differentiate bacterial gram reaction
Acid fast (Kinyoun or Ziehl–Neelsen) stain	Mycobacteria	Direct diagnosis of infection in untreated host	Cannot speciate mycobacteria; high background in tissue slide
Auramine-rhodamine	Mycobacteria	Lower power can be used for examining the slide	Fluorescence microscope is needed; artifact staining in tissue slide
Modified acid fast stain	Nocardia, cryptosporidia, isospora, cyclospora	Rapid and specific diagnosis	Tissue homogenates often mask the presence of the organism
Calcofluor white with potassium hydroxide	Pneumocystis, fungi	Rapid stain for fungi detection	Fluorescence microscope with specific filter is needed; species differentiation requires skills
India ink	*Cryptococcus neoformans*	Diagnosis of meningitis when positive in spinal fluid	Low sensitivity; a messy technique
Giemsa	Plasmodia, trypanosomes, leishmania, toxoplasma, histoplasma, pneumocystis	Detection of multiple organisms; shows the relationship between organisms and host cells	Not specific for viral inclusions. Cannot determine bacterial gram reaction

quently used. Antibody to a specific antigen is bound to latex particles to produce agglutination. This technique has been extremely useful in the detection of *Cryptococcus neoformans* in bodily fluid specimens, such as plasma and cerebrospinal fluid. Direct immunofluorescence assay has been included in the shell vial technique for rapid virus culture detection.

There are several approaches to enzyme antigen assays; the one most frequently designed for the detection of microbial antigens uses an antigen-specific antibody that is fixed to a solid phase well in a plastic tray. Antigen present in the specimen binds to the antibody. The test is then completed by adding a second antigen-specific antibody bound to an enzyme that can react with a substrate to produce a colored product. The use of microorganism-specific monoclonal antibodies has improved the reagent availability and reproducibility and enhanced test specificity. Examples include antigen detection kits developed by Abbott Diagnostics which have been used widely in the clinical virology laboratory for the rapid detection of respiratory syncytial virus and rotavirus in clinical specimens.

C. Genetic Probe Hybridization

Nucleic acid probes are based on the detection of unique nucleotide sequences within the DNA or RNA of a microorganism; these unique nucleotide "signatures" are surrogates for the presence of the organism.

This approach is used in diagnostic microbiology primarily for culture confirmation of organisms after a brief period of *in vitro* cultivation. Gen-Probe, Inc. has several culture identification nucleic acid probes available which have been approved by the U.S. Food and Drug Administration (FDA). Although these commercial products are more expensive than conventional approaches, the decrease in turnaround time has the potential to improve patient outcome and reduce overall health care costs. For example, these probes can be used for mycobacterial culture differentiation, in which different species are associated with different outcomes and required different treatment approaches.

III. MICROORGANISM CULTURE AND IDENTIFICATION

Isolation and cultivation of a microorganism, either in an artificial medium or in a living host, is definitive evidence for the presence of a microbe. In many cases, culture techniques remain the "gold standards" for diagnostic microbiology, even though lengthy incubation periods preclude the use of the test results as useful diagnostic procedures. Culture is usually the most specific method for establishing the presence of a particular pathogen in a suspected specimen. In addition, a pure isolation, either a virus or a bacterium, is essential for performing *in vitro* phenotypic antimicrobial susceptibility tests.

A. Culture Using Artificial Media

Bacteria, mycobacteria, mycoplasma, and fungi are cultured in either liquid or on a solid artificial media. Liquid media provide greater sensitivity for the isolation of small numbers of microorganisms; however, liquid media cannot be used for diagnosis of mixed infections and they preclude even the most rudimentary attempts at microorganism quantitation. Solid agar media, on the other hand, although less sensitive than liquid media, provide isolated colonies that can sometimes be identified based on their colony morphologies. Bacteria present in specimens can also be quantified by calculating the colony-forming units on an agar plate.

Culture media can be made selective by incorporating compounds such as antimicrobial agents that inhibit indigenous flora while permitting the growth of specific microorganisms resistant to these inhibitors. This is extremely important in the isolation and identification of diarrheaogenic pathogenic microorganisms in stool specimens. Sometimes, a growth indicator medium can be developed by incorporating one or more carbohydrates in the medium along with a suitable pH indicator. These differential media can be used to isolate and screen certain microorganisms.

B. Culture Using Living Cells

Chlamydiae and many viruses can be recovered and identified in a eukaryotic cell culture system. The virologic component of diagnostic microbiology has become increasingly more common with the advent of effective antiviral approaches as well as the need to identify and treat viral infections in the human immunodeficiency virus (HIV)-infected and other immunocompromised hosts. Isolation is a very sensitive method because, theoretically, a positive result can be obtained with a single infectious virion. However, not all viruses can be recovered *in vitro*. Several culture systems have been used for virus isolation, and their characteristics are listed in Table III. After inoculation, a virus can initially be identified and differentiated by (i) the pattern of the cytopathic effect (CPE), (ii) the specific cells in which CPE is induced, and (iii) the rapidity of the appearance of CPE. For example, the majority of enteroviruses cause CPE in rhesus monkey kidney cells. However, poliovirus-induced CPE appears within 1 day after inoculation, whereas it takes 4 days for echovirus to induce the same amount of cellular destruction. Typically, specimens are inoculated into as many different cell cultures as is reasonable to provide a susceptible host for each virus that may be present.

Shell vial isolation, which combines cell culture and antigen detection, has been adapted for the detection of several viruses as well as chlamydia. This method represents a relatively rapid means of identifying agents whose identification may ordinarily take as long as 3 weeks in a traditional tube culture. Vials

TABLE III
Cell Cultures Used in Diagnostic Microbiology

Culture type	Characteristic	Examples	Primary use	Remarks
Primary	Diploid	Primary monkey kidney	Influenza, parainfluenza, enteroviruses	Indigenous virus infection, e.g., SV-40; 1 or 2 passages only
Diploid cell lines	Diploid	Human diploid fibroblast MRC-5	Herpes simplex virus, cytomegalovirus, rhinovirus	Limited passage (50–70)
Established cell lines	Heteroploid	HEp-2, HeLa	Respiratory syncytial virus, adenovirus	Continuous passage; mycoplasma contamination

with coverslips are typically seeded with monolayer cells in growth medium. When the cells are nearly confluent, the growth medium is aspirated, and then a certain amount of specimen is added. The vials are centrifuged, after which growth medium is added. The vials are incubated for agent-specific times and temperatures, and the cells that have been grown on a coverslip are then incubated with fluorescein-conjugated virus-specific antibody directly in the shell vial. The stained coverslip is removed and placed on a slide for examination under the fluorescence microscope. This technique is used especially for the detection of cytomegalovirus (CMV) infections since the appearance of CMV-induced CPE occurs relatively late.

C. Culture Using Other Living Hosts

Microorganism isolation, especially for some viruses, occasionally requires the inoculation of test specimens into animals and embryonated eggs. Several arboviruses causing encephalitis can be isolated by inoculating the implicated specimen into suckling mice. Embryonated eggs are the best culture media for influenza virus proliferation. Susceptible animal inoculation, recently into SCID (severe combined immunodeficiency) mice and other rodents, has been a powerful tool to hunt for a variety of unknown pathogens. *Legionella pneumophila,* the bacterial organism causing Legionella disease, was first identified by inoculating the clinical materials into susceptible guinea pigs.

D. Phenotypic Identification of Microorganisms

After a bacterium or virus is isolated in a culture medium or a cell culture system, further identification of the microorganism is usually needed for clinical intervention and epidemiological investigation. Currently, identification of a bacterial microorganism in a clinical microbiology laboratory is mainly dependent on phenotypic characteristics, such as growth features, colony and microscopic morphology, physiologic composition, biochemical reactions, and antigenic characteristics. For example, determination of cellular fatty acid profiles is very useful for the identification of mycobacterial isolates. The identification of gram-negative bacilli is far more complex and often requires panels of tests for determining the biochemical and physiologic characteristics. Available commercial computer-assistant systems are very useful in the identification of such microorganisms. Identification of viruses, on the other hand, is usually based on the characteristic CPE in different cell cultures or on the detection of virus- or species-specific antigens.

IV. MEASUREMENT OF ANTIBODY RESPONSES TO INFECTION

Instead of detecting the presence of the microorganism, serologic tests measure the host's humoral immune response to a microorganism infection. Serologic tests are, in certain circumstances, the mainstay for the diagnosis of certain microorganism infections whenever recovery of those organisms in culture is difficult or impossible. However, there is usually a lag between the onset of microorganism infection and the development of antibody to the organism. Although IgM antibody appears relatively rapidly, IgM testing may be subject to interference by rheumatoid factors. For IgG testing, it is often necessary to obtain paired serum specimens, one taken during the acute phase of the disease and one taken during convalescence, to search for a rising titer of IgG antibody. Another limitation on the use of serology as a diagnostic tool is that immunosuppressed hosts may be unable to mount an antibody response.

A. Immunofluorescence Methods

Immunofluorescence techniques have been widely used in the past for detection of microorganism-specific antibodies and still have many applications. The common procedure used in a clinical serology laboratory is an indirect one. If present, the antibody in the serum specimen reacts with the microorganism-specific antigen which was prefixed on a glass slide. After a wash step, fluorescein-conjugated immunoglobulin directed against the test serum species is overlaid on the slide. After washing, the slide is examined for specific fluorescence under a fluorescence microscope. This technique is still the primary choice for diagnosis of several virus infections, including Epstein–Barr virus and measles virus, in the majority of clinical virology or serology laboratories. However, fluorescence techniques are quite labor-intensive and interpretation of the results is subjective; these techniques have therefore been gradually replaced by automatable and more objective serologic techniques such as enzyme immunoassays.

B. Enzyme Immunoassays

Enzyme immunoassays have replaced radioimmunoassays in most diagnostic applications and have become one of the most widely used diagnostic methods in the clinical serology laboratory. In its most used format, the solid phase EIA technique uses plastic microtiter plates or beads to which antigens are passively adsorbed. In a noncompetitive procedure, a test serum specimen is added. If present, the microorganism-specific antibody binds to the antigen. After washing, an enzyme-labeled immunoglobulin specific for the test serum species is added. After another wash step, chromogenic enzyme substrate is added, and the color that develops is proportional to the amount of specific antibody present. The application of EIA serologic techniques for diagnosis of viral and ricketsial infections has expanded significantly. This assay is most frequently used in screening for immune status, such as in rubella testing, but it is also used as a primary screening test for infections due to HIV, hepatitis C virus (HCV), hepatitis B virus, and human T cell lymphotrophic virus.

C. Immunoblotting

The specificity of a serologic test is determined largely by the antigen used to capture the antibody in the direct test formats. An immunoblot (or Western blot) is one of the most specific serologic methods available. The general components of immunoblotting as used for immunodiagnosis with human sera include (i) electrophoretic separation of protein antigen on sodium dodecyl sulfate polyacrylamide gels, (ii) electrophoretic transfer of the protein bands to a nitrocellulose or other support membrane, (iii) blocking of free-protein binding sites on the membrane, (iv) addition of the test serum, and (v) detection of the specifically bound serum antibodies. It is then possible to determine whether the patient's antibodies are directed against pathogen-specific or cross-reactive antigens. This additional level of analysis allows the discrimination of specific and nonspecific reactions. Immunoblotting has become a mainstay for confirming infection with HIV, HCV, and *B. burgdorferi* after initial positive EIA results.

D. Other Serologic Techniques

Other methods to detect immune humoral response to microorganism infections include agglutination and complement fixation assays. Agglutination reactions can be defined as the specific immunochemical aggregation of polystyrene (latex) particles coated with microorganism antigens that can be used to detect antigen-specific antibodies. Agglutination assays are simple to perform and require only 15–30 min for completion; thus, they are a useful alternative to EIA. This test has been used in the clinical serology laboratory for detection of rubella virus antibodies and quantitation of rapid plasma reagin, which can be used as an indication of the therapeutic efficacy of syphilis treatment. In a complement fixation test, an index system contains sheep erythrocytes, a hemolysin, and complement. If a test antibody is absent, an immune complex forms between erythrocytes and hemolysin and activates complement, leading to lysis of the erythrocytes. If the test serum contains antibody to the microbial antigen used, the formed antigen–antibody complex will preconsume (fix) the complement, thus preventing lysis of the erythrocytes. Although the complement fixation test is still used in the serodiagnosis of several fungal infections, the system is extremely labor-intensive and relatively insensitive.

V. MOLECULAR DETECTION AND IDENTIFICATION

Identification methods based on biochemical phenotypic parameters can, in some cases, result in confusing or misleading results, especially when the number of features is limited. During the past 10 years, nucleic acid amplification technology has opened new avenues for the detection, identification, and characterization of pathogenic organisms in diagnostic microbiology. Nucleic acid amplification techniques are classified into three general amplification categories, which all share certain advantages over traditional methods, particularly for the detection of fastidious, unculturable, and/or highly contagious organisms (Table VI). Molecular applications enhance the speed, sensitivity, and sometimes the specificity of an etiologic diagnosis. The promise of these techniques is the replacement of biological amplification—growth in culture—by enzymatic amplification of specific nucleic acid sequences.

A. Target Amplification Systems

Target amplification systems use PCR, transcription-based technologies, or strand displacement, in which many copies of the nucleic acid target are made. Among these, PCR and PCR-derived techniques are the best developed and most widely used methods of nucleic acid amplification. In 1992, Roche Diagnostics Systems, Inc. purchased the patent rights to the PCR technique with the goal of developing PCR-based kits for the diagnosis of genetic and infectious diseases. Semi-automated and automated systems have been manufactured by Roche for detection and/or quantitation of several organisms, such as HIV-1, HCV, CMV, and *Mycobacterium tuberculosis*. In addition, numerous user-developed PCR-based DNA amplification techniques have been applied to the detection of microbial pathogens, identification of clinical isolates, and strain subtyping. PCR-derived techniques, such as reverse-transcriptase PCR, nested PCR, multiplex PCR, arbitrary primed PCR, and broad-range PCR, have collectively expanded the flexibility and power of these methods in laboratories throughout the world.

Given the patent restrictions on PCR and the expanding interest in nucleic acid-based diagnosis, alternative amplification methods have been sought. Another target amplification system, transcription-mediated amplification (TMA) or nucleic acid sequence-based amplification (NASBA), begins with the synthesis of a DNA molecule complementary to the target nucleic acid (usually RNA). This technique involves several enzymes and a complex series of reactions which all occur simultaneously at the same temperature and in the same buffer. The advantages include very rapid kinetics and the lack of requirement for a thermocycler. Isothermal conditions in a single tube with a rapidly degradable product (RNA) help minimize (but may not eliminate) contamination risks. Amplification of RNA not only makes it possible to detect RNA viruses but also increases the sensitivity of detecting bacterial and fungal pathogens by targeting high copy number RNA templates. A TMA-based system manufactured by GenProbe, Inc. has been used to detect *M. tuberculosis* in smear-positive sputum specimens and to confirm *Chlamydia trachomatis* infection. A NASBA system is commercially available from Organon–Teknika Corporation for the detection and quantitation of HIV-1 infection. Recently, a TMA-based assay for detection of HIV and HCV RNA in donor blood specimens has received FDA clearance in the United States.

B. Probe Amplification Systems

In probe amplification systems, many copies of the probe that detects the target nucleic acid amplified after target-specific hybridization occurs. The ligase chain reaction (LCR) is the most successful application of probe amplification in diagnostic microbiology (Table IV). When used following a target amplification method, such as PCR, LCR can be sensitive and useful for the detection of point mutations. Although convenient and readily automated, a potential drawback of LCR is the difficult inactivation of post amplification products. The nature of the technique does not allow for the most widely used contamination control methods to be applied. The inclusion of a detection system within the same reaction tube (closed reaction systems) would significantly decrease the possibility of contamination which is associated with the opening of reaction tubes. A combination LCR kit for the detection of both *C. trachomatis* and *Neisseria gonorrhea* is commercially available from Abbott Laboratories.

C. Signal Amplification Systems

Signal amplification methods are designed to strengthen a signal by increasing the concentration of label attached to the target nucleic acid. Unlike procedures which increase the concentration of the probe or target, signal amplification increases the signal generated by a fixed amount of probe hybridized to a fixed amount of specific target. The fact that signal amplification procedures do not involve a nucleic acid target or probe amplification is a theoretical advantage, because of lower susceptibility to contamination problems inherent in enzyme-catalyzed nucleic acid amplification. Sensitivity, however, compared to that of target nucleic acid amplification techniques, may be a limiting factor. Another

TABLE IV
Nucleic Acid Amplification Methods

Amplification method	Amplification category	Manufacturer/license	Enzymes used	Temperature requirement	Nucleic acid target
Polymerase chain reaction (PCR)	Target	Roche Molecular System, Inc. (Branchburg, NJ)	*Taq* DNA polymerase	Thermal cycler	DNA (RNA)
Transcription-mediated amplification (TMA)	Target	Gen-Probe, Inc. (San Diego, CA)	Reverse transcriptase, RNA polymerase, RNase H	Isothermal	RNA or DNA
Nucleic acid sequence-based amplification (NASBA)	Target	Organon–Teknika Corporation (Durham, NC)	Reverse transcriptase, RNA polymerase, RNase H	Isothermal	RNA or DNA
Strand displacement amplification (SDA)	Target	Becton–Dickinson & Company (Rutherford, NJ)	Restrictive endonucleonase, DNA polymerase	Isothermal	DNA
Qβ replicase (QβR)	Probe	Gene-Trak Systems (Framingham, MA)	Qβ replicase	Isothermal	DNA or RNA
Cycling probe technology (CPT)	Probe	ID Biomedical (Vancouver, Canada)	RNase H	Isothermal	DNA (RNA)
Ligase chain reaction (LCR)	Probe	Abbott Laboratories (Abbott Park, IL)	DNA ligase	Thermal cycler	DNA
Hybrid capture system	Signal	DiGene Diagnostics, Inc. (Silver Spring, MD)	None	Isothermal	DNA
Branched DNA (bDNA)	Signal	Chiron Corporation (Emeryville, CA)	None	Isothermal	DNA or RNA

limitation of signal amplification is background noise due to the nonspecific binding of reporter probes. Currently, signal amplification products from two diagnostic companies are available for molecular diagnostic purposes. The Digene hybrid capture system is widely used to determine human papillomavirus infection and viral types in cervical swabs or fresh cervical biopsy specimens as well as other diagnostic targets. Another product is the branched DNA probe developed and manufactured by Chiron Corporation which represents an excellent method for quantitation and therapeutic response monitoring of HCV and HIV.

D. Molecular Identification of Microorganisms

During the past century, the identification of microorganisms has relied principally on phenotypic characteristics such as morphology and biochemical characterization. Genotypic identification is emerging as an alternative or complement to established phenotypic methods. Sequence analysis of certain microbial genes has been used for the identification of microorganisms. Broad-range PCR primers are designed to recognize conserved sequences in the phylogenetically informative gene of a variety of bacteria, and highly variable regions between the primer binding sites are amplified by PCR. The amplified segment is sequenced and compared with known databases to identify a close relative.

The most successful example is the small-subunit (16S) rRNA gene sequence-based microbial identification system, which includes the extraction of nucleic acids, PCR-mediated gene amplification, sequence determination, and computer-aided analysis (Fig. 1). Currently, the time and effort associated with data analysis and its cost are major limitations. The capital investment is high, particularly for automated analysis. Perkin–Elmer Biosystem has developed a 16S rRNA gene sequence database from more than 1200 American Type Culture Collection prototype bacterial strains. Their microbial identification system based on 16S sequence information is commercially available. With improved automation and decreased cost, both of which are likely in the next few years, such systems may become established in many diagnostic microbiology laboratories.

VI. ASSESSING THE PERFORMANCE OF DIAGNOSTIC TESTS

No diagnostic test is perfect. Therefore, several criteria should be established for the test evaluation, and a decision should be made regarding whether to do (i) a direct examination of the microorganism, (ii) a microorganism culture, (iii) a microorganism-specific antibody detection, or (iv) a microorganism nucleic acid amplification or a combination of these procedures for the evaluation. Although not all testing decisions are as complex as this, often a choice must be made among several methods. The most important factor is the validity of a test, which is evaluated by two parameters, i.e., accuracy and precision. Their relationship is demonstrated in Fig. 2. After test performance, other factors such as test turnaround time, technologist time, and test cost, are also important in assessing a diagnostic test for detection and identification of microorganisms.

A. Accuracy in Diagnostic Microbiology

The accuracy of a diagnostic test is its correspondence with the true value. In the clinical laboratory, accuracy of tests could be judged by the test performance such as sensitivity and specificity. The sensitivity of the test is the likelihood that it will be positive when the microorganism is present. The specificity of the test measures the likelihood that it will be negative if the microorganism is not present. To determine test sensitivity and specificity for a particular microorganism, the test must be compared against the gold standard, a procedure that defines the true state of the microorganism. The calculation of sensitivity and specificity is demonstrated in Table V.

Another parameter, the predictive values, can be calculated if the specimens collected represent the whole population. However, the predictive value of a diagnostic test is influenced by the frequency of infection prevalence in the population being tested.

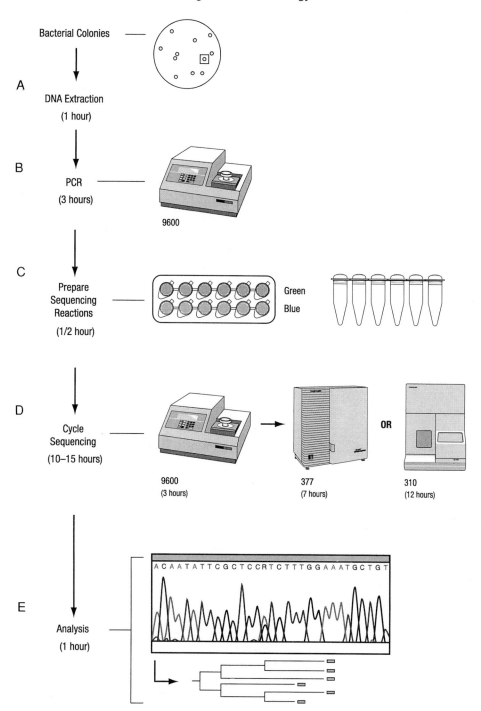

Fig. 1. Flowchart of microorganism identification based on 16S rRNA gene sequence analysis. The total identification time was 15.5–18.5 hr, comprising bacterial DNA extraction (A), PCR (B), sequencing reaction preparation (C), cycle sequencing (D), and analysis (E). The time for each step is shown in parentheses (adapted from Tang *et al.*, 1998).

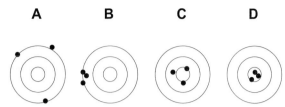

Fig. 2. Relationship between accuracy and precision in diagnostic tests. The center of the target represents the true value of the substance being tested. (A) A test which is imprecise and inaccurate. (B) A diagnostic technique which is precise but inaccurate. (C) A diagnostic technique which is accurate but imprecise. (D) An ideal test, one that is both precise and accurate.

Suppose we assess a population of 10,000 for a microorganism using a test with the 99% sensitivity and 95% specificity. The predictive value varies in proportion to the prevalence rate in the study population (Table VI). Therefore, the interpretation of a laboratory diagnostic result depends not only on the technical accuracy of the method used but also on the prevalence of the infection in the population to which the patient belongs.

B. Precision in Diagnostic Microbiology

Test precision is a measure of a test's reproducibility when repeated on the same sample, especially for quantitative testing. An imprecise test is one that yields widely varying results on repeated measurements (Fig. 2). The precision of microorganism diagnostic tests, which is monitored in clinical laboratories by repeatedly using a control material, must be good enough to distinguish clinically relevant changes in a patient's status from the analytic variability of the test. Precision can be assessed on both an intralaboratory (within the same laboratory) and interalaboratory (between different laboratories) basis.

C. Other Assessing Factors

Even when a test accurately analyzes a given specimen, there are other factors that need to be considered. The speed of microbial identification results, known as the test turnaround time, can have a major impact on clinical management. Generally, culture-based microorganism detection and identification

TABLE VI
Predictive Values Vary in the Test Population with Different Prevalence[a]

	Diagnostic test result	
	Positive	Negative
When the prevalence rate is 1%[b]		
Microorganism present	99	1
Microorganism absent	495	9405
When the prevalence rate is 10%[c]		
Microorganism present	990	10
Microorganism absent	450	8550

[a] Assume a sensitivity of 99% and a specificity of 95%.
[b] Positive predictive value = 99/(99 + 495) × 100% = 16.7; negative predictive value = 9405/(1 + 9405) × 100% = 99.9.
[c] Positive predictive value = 990/(990 + 450) × 100% = 68.8; negative predictive value = 8550/(10 + 8550) × 100% = 99.8.

TABLE V
Calculation of Sensitivity, Specificity, and Predictive Values

	Diagnostic test result[a]	
	Positive	Negative
Microorganism present	*a* (true positive)	*b* (false negative)
Microorganism absent	*c* (false positive)	*d* (true negative)

[a] Sensitivity = $a/(a + b) \times 100\%$; specificity = $d/(c + d) \times 100\%$; positive predictive value = $a/(a + c) \times 100\%$; negative predictive value = $d/(b + d) \times 100\%$.

have historically provided results with a long test turnaround time. Direct examination can be performed within hours, and microorganism-specific antibody and nucleic acid detection can be detected and identified in 1 day. The improved turnaround time provided by these techniques may translate into improved clinical outcomes. For example, a PCR test followed by a colorimetric microtiter plate identification for the detection of herpes simplex virus DNA in cerebrospinal fluids can be completed in 1 day if the specimen is processed in the morning. If the test results were positive, unnecessary antibacterial treatment could be eliminated and appropriate antiviral therapy instituted. Conversely, negative PCR results are being used to justify cessation of unnecessary intravenous acyclovir treatment ($180.00/day) and allow the diagnostic consideration of other etiologic causes of central nervous system infection.

Technologist time and cost of the test should also be considered when a diagnostic test is evaluated. We are in the middle of a health care revolution. While keeping in mind the goal of a high-quality diagnostic microbiology service, we are contemporaneously pressured to develop and maintain tests that require less technologist time and cost. Technologist time is calculated from two efforts: variable allied health effort, which is "hands-on" time, and fixed effort, which includes specimen processing, buffer preparation, maintenance, bench cleaning, and data entry. The hands-on time can be significantly improved by introducing an automatic procedure into the diagnostic test. One example is the serologic diagnosis of Epstein–Barr virus (EBV) infection. An EIA incorporated in an automated processing system has been replacing the manual immunofluorescence assay for the detection of EBV-specific antibody profiles.

VII. CONCLUDING REMARKS

Diagnostic microbiology rapidly detects and accurately identifies implicated microorganisms in test specimens through a variety of techniques. Technologic changes have made constant and enormous progress in various areas, including bacteriology, mycology, mycobacteriology, parasitology, and virology, during the past two decades in the field of diagnostic microbiology. The physical structure of laboratories, staffing patterns, work flow, and turnaround time have all been influenced profoundly by technical advances. These changes will continue and lead diagnostic microbiology inevitably to a modern discipline, which can face the challenges of the future.

See Also the Following Articles

DETECTION OF BACTERIA IN BLOOD: CENTRIFUGATION AND FILTRATION • IDENTIFICATION OF BACTERIA, COMPUTERIZED • MICROSCOPY, ELECTRON • POLYMERASE CHAIN REACTION

Bibliography

Henry, J. B. (Ed.) (1996). "Clinical Diagnosis and Management by Laboratory Methods," 19th ed. Saunders, Philadelphia.

Koneman, E. W., Allen, S. D., Janda, W. M., Schrechenberger, P. C., and Winn, W. C. (1997). "Color Atlas and Textbook of Diagnostic Microbiology," 5th ed. Lippincott-Raven, Philadelphia.

Lennette, E. H., Lennette, D. A., and Lennette, E. T. (Eds.) (1995). "Diagnostic Procedures for Viral, Rickettsial, and Chlamydial Infections," 7th ed. American Public Health Association Press, Washington, DC.

Murray, P. R., Baron, E. J., Pfaller, M. A., Tenover, F. C., and Yolken, R. H. (Eds.) (1999). "Manual of Clinical Microbiology," 7th ed. ASM, Washington, DC.

Persing, D. H., Smith, T. F., Tenover, F. C., and White, T. J. (Eds.) (1993). "Diagnostic Molecular Microbiology: Principles and Applications." ASM, Washington, DC.

Relman, D. A. (1999). The search for unrecognized pathogens. *Science* **284**, 1308–1310.

Rose, N. R., de Macario, E. C., Folds, J. D., Lane, H. C., and Nakamura, R. M. (Eds.) (1997). "Manual of Clinical Laboratory Immunology," 5th ed. ASM, Washington, DC.

Sahn, D. F., Forbes, B. A., and Weissfeld, A. S. (Eds.) (1998). "Bailey and Scott's Diagnostic Microbiology," 10th ed. Mosby-Year Book, St. Louis, Mo.

Tang, Y. W., Ellis, N. M., Hopkins, M. K., Smith, D. H., Dodge, D. E., and Persing, D. H. (1998). Comparison of phenotypic and genotypic techniques for identification of unusual pathogenic aerobic gram-negative bacilli. *J. Clin. Microbiol.* **36**, 3674–3679.

Dinoflagellates

Marie-Odile Soyer-Gobillard and Hervé Moreau

Centre National de Recherche Scientifique

GLOSSARY

basic nuclear proteins DNA-binding proteins of low molecular weight present in the chromatin of dinoflagellates which lacks nucleosomes.

cnidocyst Analogous organelle to coelenterate capsule containing a thread-like stinger used for defense or capturing prey found in some dinoflagellates.

coelome General cavity of coelomate animals (vertebrates and invertebrates).

cytostome Buccal aperture (mouth) that is permanently open but able to dilate. It allows the preys to be transported into the cell.

dinoflagellates (phylum Dinoflagellata) Protoctists that are essentially biflagellated and are characterized by a dinokaryon.

dinokaryon The nucleus of dinoflagellates, characterized by the presence of a permanent nuclear envelope and chromosomes quasi permanently condensed during the whole cell cycle.

dinomitosis Mitosis of the dinokaryon; characterized by the presence of an extranuclear mitotic spindle crossing the nucleus via cytoplasmic channels.

endosymbiosis The condition of one organism living inside a member of a different species.

myoneme Fibrillar striated contractile structure located in the cytoplasm of some protozoa (ciliates, sporozoa, and dinoflagellates). They are partially responsible for cell contraction.

nucleosomes Subunits present in the eukaryotic genomes except dinoflagellates; composed of octamers of histones and DNA.

pelagic Describing or referring to animals living in the marine domain, either swimming or passive; opposite of benthic (living on the bottom).

peridinin Carotenoid pigment specific to most autotrophic dinoflagellates.

protoctists Eukaryotic single-celled microorganisms; their kingdom is composed of 18 phyla.

taeniocyst Dinoflagellate (*Polykrikos*) extrusome; associated with nematocyst and forming a complex structure.

theca Total cell wall, composed of cellulose plates in dinoflagellates.

trichocyst Extrusome capable of sudden discharge outside of the cell; its exact role in dinoflagellates is not known.

DINOFLAGELLATES are a phylum of unicellular eukaryotic micro-organisms among the Protoctista. They show great diversity (Fig. 1), and can be autotrophic (Fig. 2B), heterotrophic (Fig. 2A), mixotrophic, parasitic, or symbiotic and are widely distributed throughout the seas and fresh waters throughout the world. They play a prominent role in the trophic chains. The diversity of this group is also displayed by their external morphology and by the organization of the thecal plates. The latter are the basis for the classification of the thecate members of approximately 2000 living species, 161 genera, 48 families, and 17 orders described to date.

I. GENERAL CHARACTERISTICS OF DINOFLAGELLATES AND THEIR CELL CYCLE

A. Features

In contrast to the great external diversity, several common features allow investigators to classify these organisms as a unique and homogeneous taxonomic group. First, dinoflagellates are typically biflagellated, with the two flagellae being oriented perpendicularly to one another: One transverse flagellum is wrapped around the cell, whereas the second is located longitudinally and beats posteriorly (Fig. 1A). This gives a characteristic spiral locomotion to dinoflagellate cells, from which the name of this group derives (*dino* in Greek means to turn). Another characteristic among most autotrophic dinoflagellate species is the presence of a typical carotenoid pigment, the peridine, in association with chlorophyll *a* and *c*. Several other accessory pigments are also present in these species: β carotene, diadinoxanthine, and dinoxanthine. Dinoflagellates have mitochondria with tubular cristae and possess ejectile proteinaceous organelles (extrusomes): trichocysts, nematocysts, or taeniocysts.

Furthermore, a few species show some unusual differentiations or adaptations which have been well described. For instance, *Noctiluca scintillans* (Macartney) Ehrenberg (Fig. 1B), which is heterotrophic and bioluminescent, has a cytostome and a tentacle with well-organized myonemes. It displays vegetative reproduction by binary fission of the trophonte (characterized by a nucleus with no condensed DNA), and it is also able to proliferate in seawater by fast gametogenesis creating red tides, with progressive condensation of the genome during the successive divisions. *Polykrikos kofoidii* Chatton possesses two to six nuclei, each being associated with a cytoplasmic territory and forming a syncytium, with the nuclei behaving in synchrony. *Syndinium turbo* Chatton, a coelomic plasmodial parasite of pelagic copepods (Fig. 1C), has free-living *Gymnodinium*-like sporocytes. All members of the genus *Blastodinium* are mixotrophic (except one, *Blastodinium contortum* Chatton var. *hyalinum*),

combining phagotrophy and phototrophy, and are located in the digestive tract of copepods. They reproduce by a complex iterative sporogenesis (Chatton, 1920) and also have free-living *Gymnodinium*-like sporocytes. *Erythropsis pavillardi* Hertwig possesses a sophisticated eyespot and a specific locomotor apparatus, the piston.

Generally haploid, cells display vegetative division and/or sexuality with a postzygotic meiosis, dependent on the prevailing environmental conditions. The most original and unifying features of dinoflagellates concern their nuclear organization and the course of their mitosis. Respectively called dinokaryon and dinomitosis by E. Chatton (1920), they have been well documented in several reviews (Raikov, 1982, 1995; Spector, 1984; Taylor, 1987; Margulis *et al.*, 1990, 1993). The current article will concentrate on data from the most recent studies within this area.

B. Life Cycle of Dinoflagellates

The complete life cycle of dinoflagellates has only been elucidated in four or five species. The vegetative cycle in *Prorocentrum micans* Ehrenberg seems quite simple, with division of swimming cells and alternative sexuality. In *Crypthecodinium cohnii* (Seligo) Chatton in Grassé, the cycle appears to be more complicated, with swimming cells corresponding to G_1 cells. During S phase, flagella drop off and the cell encysts, committing itself to one, two, or three successive mitoses, which give rise to two, four, or eight cells, respectively. *Pfiesteria piscicida* Steidinger and Burkholder, the dangerous "ambush-predator phantom" dinoflagellate toxic for fishes, displays a very complex life cycle, with transformations among an array of flagellated, amoeboid, and encysted stages—the amoeboid being quite undetectable. Symbiotic dinoflagellates also present complicated cycles. Furthermore, under conditions not well understood some species may proliferate, producing blooms or red tides, with thousands of cells per liter of seawater.

Chemical mutagenesis experiments performed in 1974 by Roberts *et al.* on vegetative cells of *C. cohnii* to select for non pigmented (carotenoids) mutants showed that they are functionally haploid, at least

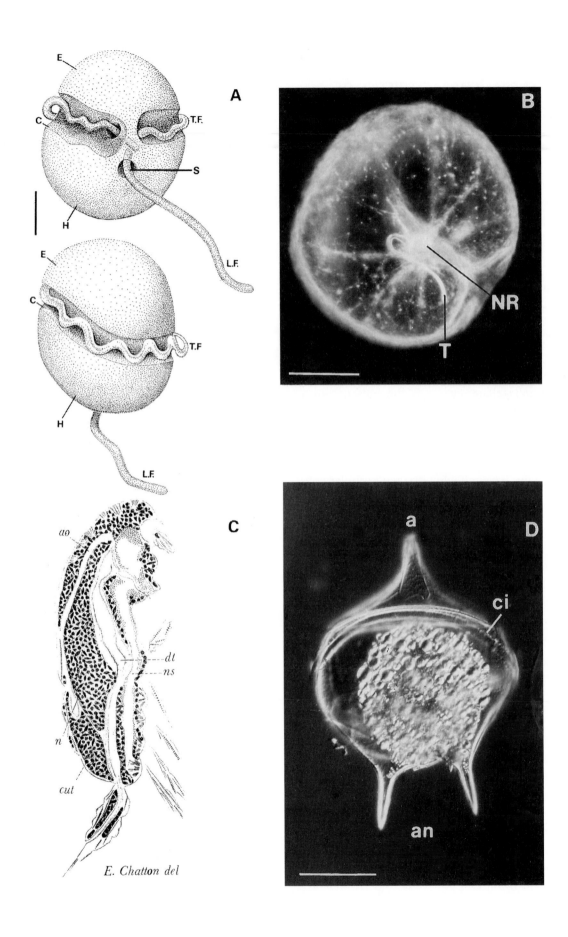

A

E

C

T.F.

S

H

L.F.

E

C

T.F.

H

L.F.

B

NR

T

C

ao

dt

ns

n

cut

E. Chatton del

D

a

ci

an

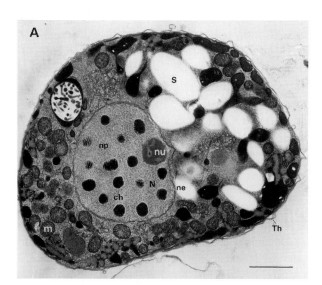

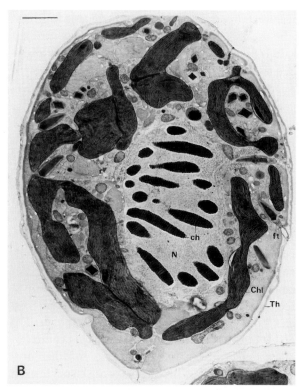

Fig. 2. Ultrastructural organization of two dinoflagellate species observed with the transmission electron microscope after helium fast-freeze preparation and OsO₄ cryosubstitution. (A) The heterotrophic *Crypthecodinium cohnii* cell. N, nucleus; nu, nucleolus; ch, chromosomes; np, nucleoplasm; ne, nuclear envelope; s, starch; m, mitochondria; Th, theca. Scale bar = 2 μm. (B) The autotrophic *Prorocentrum micans* cell. Chl, chloroplast; ft, fibrous trichocyst. Scale bar = 4.5 μm.

for this phenotypic trait. Evidence indicates that vegetative cells in other species of dinoflagellates are also haploid. When medium conditions become unfavorable in culture (sharp decrease in temperature or a depletion in nitrogen or phosphate), vegetative cells transform to morphologically identical gametes and can undergo sexuality to give a diploid zygote. Sexuality is seen either in the form of a fusion of cells and nuclei or by transmission of one nucleus

to the other gamete via a conjugation tube as is the case in *P. micans*.

Haploidy of the dinoflagellate genome is at odds with recent molecular genomic data showing that several genes (luciferase, luciferin-binding protein, Rubisco, and HCc) are tandemly repeated. This contradiction can be explained either by the particular nature of the genes studied (coding for proteins involved in bioluminescence or basic proteins), which

Fig. 1. General organization of dinoflagellates and various species. (A) Schematic of the heterotrophic dinoflagellate *Crypthecodinium cohnii*. E, episome; H, hyposome; S, sulcus; T.F., transverse flagellum; L.F., longitudinal flagellum; c, cingulum (reproduced from Perret *et al.,* 1993, by copyright permission of the Company of Biologists Ltd.). (B) *Noctiluca scintillans,* heterotrophic "naked" dinoflagellate provoking red tides. T, tentacle; NR, nuclear region. Scale bar = 240 mm. (C) Longitudinal section of the copepod *Paracalanus parvus* with the parasitic plasmodial dinoflagellate *Syndinium turbo* inside the coelomic cavity. ao, aorta; dt, digestive tract; n, nuclei of the plasmode; cut, cuticle of the copepod; ns, nervous system (reproduced from Chatton, 1920). (D) *Protoperidinium depressum,* an armoured heterotrophic dinoflagellate. a, apex; an, antapex; ci, cingulum. Scale bar = 25 μm.

are known in other organisms to be repetitive genes, or by the hypothesis that the dinoflagellate genome is duplicated with numerous nonexpressed pseudogenes and thus is functionally haploid.

II. STRUCTURAL ORIGINALITY OF THE NUCLEUS

The peculiar nature of the dinoflagellate nucleus has been documented for a long time and is one of the principal characters defining the group. Surrounded by a persistent nuclear envelope, the chromatin is quasi-permanently condensed in well-defined chromosomes, ranging from 4 to 200 depending on the species. Most of the species show a high content of DNA, which ranges from 7.0 pg/cell in *C. cohnii,* whose nucleus is haploid, to 200 pg/cell in *Gonyaulax polyedra* Stein. This compares with a range of 0.046–3 pg/nucleus in other unicellular eukaryotes. Chromosomes are characterized by a high G + C content and the presence of a rare base hydroxymethyl uracil in a high proportion (Soyer-Gobillard, 1996). They are devoid of longitudinal differentiation such as Q, G, or C banding, and all chromosomes in a nucleus appear to be morphologically identical, without any structural differentiation such as centromeric-associated structures (except for the presence of a kinetochore-like structure in rare species such as *Amphidinium carterae* Hulburt or *Syndinium* sp.). Dinoflagellates are the only eukaryotes in which the chromatin is totally devoid of histones and consequently lacks nucleosomes.

A. Chromosomes

At the electron microscopic level, and for most species, ultrathin sections of dinoflagellate chromosomes reveal a characteristic arch-shaped organization of the nucleofilaments (Fig. 3). No repeating subunit structures as nucleosomes could be detected on dispersed genomic DNA by electron microscopy, nuclease digestion, or a combination of these techniques. Furthermore, in all dinoflagellate species tested (except *Dinophysis* Ehrenberg), basic proteins

associated with genomic DNA represent only 10% of the chromosome mass, whereas in typical eukaryotes the histones : DNA ratio is close to 1. This is a unique characteristic among eukaryotic cells. In addition to the biochemical demonstration of several specific DNA-binding nuclear basic proteins (Rizzo, 1987), two variants of a basic protein (HCc = p14) have been cloned and sequenced in *C. cohnii.* Although they show some biochemical characteristics of histones (low molecular weight and basic charge), no convincing homology could be found between them and they did not appear to be quantitatively major nuclear proteins, as are histones in classical eukaryotes. HCc antigens were immunolocalized and mainly detected at the periphery of chromosomes and in the nucleolar organizing region; the putative function of HCc could be in the initiation of transcription.

Chromosome spreadings revealed double-twisted helix organization. The presence of divalent cations as Ca^{2+} and Mg^{2+} and structural RNAs has been demonstrated to be important for the maintenance of the architecture of the dinoflagellate chromosome (Soyer-Gobillard, 1996). Several models have been proposed to explain the right-handed superhelical organization of nucleofilaments and their replication and segregation processes. However, these models are speculative, and more data are needed before a definitive hypothesis can be made. Devoid of histones and nucleosomes, the very large genome of dinoflagellates is almost "naked" with chromosomes consisting of bare DNA. This intriguing characteristic is unique among eukaryotic cells, and it indicates that dinoflagellates could be a potentially very interesting model in cell biology.

Another characteristic of the dinoflagellate chromatin is the presence of an unusual base in a high proportion, the hydroxymethyl uracil, which can substitute between 10 and 60% of the thymines depending on the species. This unusual base is also found (albeit at lower proportions) in some bacteriophages. The function of this base is unclear, but it has been hypothesized that in bacteriophages it introduces a bending of DNA which is necessary for the recognition of DNA sequences by DNA-binding proteins.

Chromosomal nucleofilaments of dividing (Fig. 3C) or nondividing chromosomes, as seen in ultra-

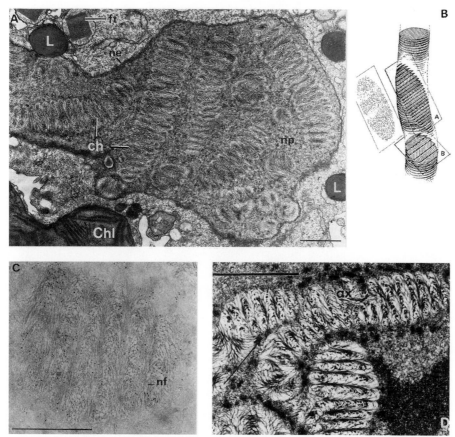

Fig. 3. Organization of dinokaryon and chromosomes in transmission electron microscopy. (A) Nucleus of *Prorocentrum micans*. Surrounded by the permanent nuclear envelope (ne), chromosomes composed of supertwisted nucleofilaments as seen in ultrathin sections show an arch-shaped organization. Scale bar = 1 μm (reproduced from *Biology of the Cell*, 1977, Vol. 30, pp. 297–300, by copyright permission of the Société Française des Microscopies). (B) Schematic interpretation (Bouligand's plywood model) of dinoflagellate chromosome ultrastructure as seen in ultrathin section. A, oblique; B, quite transverse plan section (reproduced from *The Journal of Cell Biology*, 1990, Vol. 111, pp. 293–308, by copyright permission of The Rockefeller University Press). (C) High magnification of an oblique chromosome section in which nucleofilaments (nf) are labeled with anti-B-DNA antibody. Gold particles are distributed in the whole chromosome. Scale bar = 1 μm (reproduced from *The Journal of Cell Biology*, 1990, Vol. 111, pp. 293–308, by copyright permission of The Rockefeller University Press). (D) Dividing chromosome of *Prorocentrum micans*. Observe the axis (ax) of segregation (arrow). The arch-shaped nucleofilament organization is not very modified. Scale bar = 1 μm (reproduced from *Vie et Milieu*, 1979, Vol. 28–29, pp. 461–472, by copyright permission of Laboratoire Arago, Université P. et M. Curie).

thin sections, can be labeled heavily with anti-B-DNA (= anti-right-handed DNA) antibody (Fig. 3C). Stretches of alternating purine–pyrimidine (GC rich) can form a left-handed helix (Z-DNA) and the B to Z transition is facilitated by the presence of divalent cations or constraints on DNA, such as superhelicity, as is the case in dinoflagellate chromosomes. In this material, Z-DNA was localized in limited areas inside the chromosomes, often at their periphery (Fig. 4A) (Soyer-Gobillard, 1996).

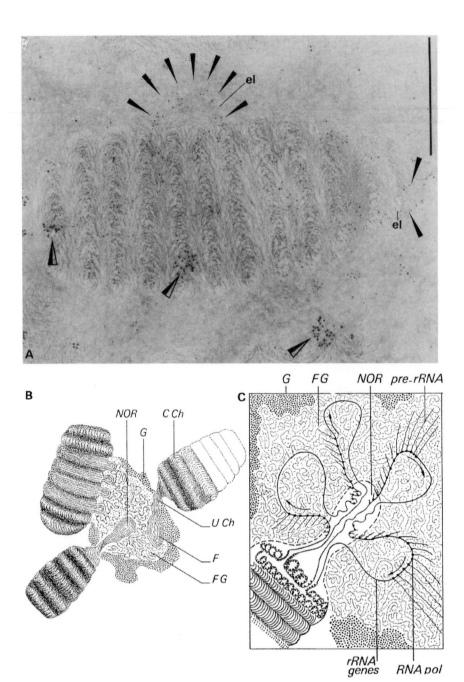

Fig. 4. How does the dinoflagellate chromosome function? (A) Double immuno-labeling with anti-Z and anti-B DNA antibodies. Clusters of Z (left-handed) DNA are located in the periphery of the chromosome (black and white arrows). Smaller gold particles labeling B-DNA are distributed in the whole chromosome and in their external periphery where opened extrachromosomal loops (el) are present (black arrows). Scale bar = 0.5 μm (reproduced from *The Journal of Cell Biology*, 1990, Vol. 111, pp. 293–308, by copyright permission of the Rockefeller University Press). (B) Schematic representation based on TEM observations of nucleolar chromosomes of *Prorocentrum micans*, showing the unwinding of nucleofilaments located in either telomeric or lateral regions. Several chromosomes are contributing to the formation of a new nucleolus. C Ch, condensed chromosome; U Ch, unwound chromosome region; NOR, nucleolar organizing region; F, fibrillar region;

B. Transcription/Replication

The presence of chromosomes in a quasi-permanently condensed state throughout the cell cycle raises the question of how such a structure can be transcribed and replicated. We have shown that a local microscale decondensation and recondensation occurs during transcription. Extrachromosomal loops on chromosomes and in the nucleolus have been described after *in situ* pronase treatment, and models propose that a local untwisting of DNA filaments allows transcription to occur (Figs. 4B and 4C). Extrachromosomal anti-B-DNA-labeling was also detected on the nucleoplasm that corresponds to DNA loops (Fig. 4A). The role of the Z-DNA conformation as a possible site for unwinding and DNA processing in chromosomes was strongly suggested. It was confirmed by Sigee (1984), using tritiated adenine incorporation, that transcription occurs in the same regions (i.e., in the periphery of chromosomes). More direct evidence of a different chromosomal localization of coding and noncoding sequences has been provided by restriction enzyme digestions of the genomic DNA. These experiments showed that bulk chromosomal DNA, which was inaccessible to these enzymes, contained few, if any, coding sequences (Soyer-Gobillard, 1996).

Renaturation kinetic studies demonstrated the presence of 55–60% repeated, interspersed DNA in the *C. cohnii* genome. This proportion of repeated sequences was confirmed later, and their organization in the genome was determined. Half of the genome is composed of unique sequences interspersed with repeated sequence elements of approximately 600 bp, representing approximately 95% of the total number of interspersed unique elements. In contrast to transcription, the process of DNA replication during S phase is understood less well, and contradictory observations have been reported.

C. Nucleolus

Observation of several dinoflagellates (*P. micans*, *C. cohnii*, and *A. carterae*) by light microscopy after silver staining of the nucleolar argyrophilic proteins has shown the presence of nucleolar material throughout the vegetative cell cycle and in particular during all stages of mitosis. Soyer-Gobillard and Géraud (1992) showed that during early prophase, when chromosomes begin to split, the nucleoli remain functional. This contrasts with most higher eukaryotes, in which nucleoli disappear at the end of the prophase and are reconstituted in daughter cells during the telophase. Three compartments are present in the dinoflagellate nucleolus: the nucleolar organizing regions (NORs) and the fibrillo granular and the (preribosomal) granular compartments. Several chromosomes can contribute to the formation of a single nucleolus (Fig. 4B). Coding sequences of ribosomal genes have been detected by *in situ* hybridization experiments both at the periphery of the nucleolar organizer region which corresponds to the unwound part of the nucleolar chromosomes and in the fibrillo granular region. These results suggest that rRNA gene transcription occurs predominantly at the periphery of the NOR where the coding sequences are located (Fig. 4C).

D. Nuclear Envelope and Lamins

As seen in ultrathin sections or on freeze-fractured *P. micans* cells, the nuclear envelope is similar to the classical eukaryotic model, and no distinctive

FG, fibrillogranular region; G, granular (preribosomal) region (reproduced from Soyer-Gobillard and Géraud, 1992, by copyright permission from the Company of Biologists Ltd.). (C) Predicted molecular organization of the dinoflagellate transcriptionally active nucleolus deduced from TEM observations after *in situ* hybridization with a ribosomal biotinylated probe. rDNA loops are represented unwound from the telomeric region of the otherwise condensed nucleolar chromosome. The rDNA transcription is initiated at the periphery of the NOR and carried on in the proximal part of the fibrillo granular (FG) compartment to generate the rRNA transcripts, whereas the distal FG region is devoted to rRNA processing and packaging of preribosomes of the granular (G) region (reprinted from *BioSystems* **26**, Géraud *et al.*, Nucleolar localization of rRNA coding sequences in *Prorocentrum micans* Ehr. (Dinomastigote, Kingdom Protoctist) by *in situ* hybridization, 61–74. Copyright 1991, with permission from Elsevier Science).

features have been observed except for the penetration of cytoplasmic channels during mitosis and the profound modifications in the pore distribution between interphase and mitosis.

The presence of lamin-like proteins, immunologically related to vertebrate lamins, has been demonstrated in the nuclear matrix, close to the nuclear envelope of *A. carterae*. These lamins probably play an important role during the chromosome division. A topoisomerase II homolog has also been identified in the nuclear matrix, suggesting that these enzymes could play a role in organizing the DNA in loop domains.

III. MITOTIC APPARATUS

There are several unusual features of mitosis of dinoflagellates. Principal among these are the following: (i) The nuclear envelope persists throughout mitosis; (ii) the chromosomes, which remain condensed throughout the cell cycle, attach to the inner part of the nuclear envelope before their segregation; and (iii) cytoplasmic channels containing the microtubular spindle traverse the nucleus during the mitosis.

A. Microtubules and Mitotic Spindle

During mitosis, the dinoflagellate nucleus becomes crossed by cytoplasmic channels made by invaginations of the cytoplasm and nuclear envelope (Fig. 5C). These channels have been described in all dinoflagellate species tested, and this feature is also a cellular characteristic of the group. Furthermore, they are only seen during mitosis, and not in G_1, S, or G_2 phases. In early and mid prophase, thick microtubular bundles pass through the nucleus in cytoplasmic channels and converge towards the two poles. Asters were never observed at the spindle poles, where centrosome regions are present but without the characteristic pair of centrioles. Among the proteins characterized by Perret *et al.*, (1993), in the centrosome regions are β-tubulin, α-actin, a homolog of a human protein CTR 210, and HSP 70 protein, which is conserved from dinoflagellates to humans. The microtubular bundles of the spindle split in two during late prophase and lengthen in early anaphase. The spindle bundles diverge in late anaphase, extend almost to the plasma membrane, and depolymerize during telophase. The cleavage furrow, in which tubulin and actin were characterized, appears in anaphase and is formed by an invagination of the plasma membrane in the kinetosome region. As shown by using anti-β-tubulin antibody on whole cells and confocal scanning laser microscope observations, the cortical tubular network remains polymerized during mitosis, contrary to what occurs in other eukaryotic cells (Figs. 5A and 5B).

B. Chromosome Segregation

The system of cytoplasmic channels traversing the intact nucleus indicates that the microtubular mitotic spindle is not in direct contact with the chromosomes but is separated from them by the persistent nuclear envelope. Highly condensed chromosomes do not form a typical metaphase plate and appear in contact with the nuclear envelope around the cytoplasmic channels and/or the surrounding nuclear envelope depending on species. At the electron microscope level, no dense material or overt differentiated structures could be found on the chromosomes or on the nuclear membranes, except in one species studied by Oakley and Dodge (1974)—*A. carterae* Hulburt. In this species, kinetochore-like structures have been described, but these are located on the cytoplasmic side of the channels. Due to the low number of dinoflagellate species studied regarding this aspect, it is difficult to determine whether *A. carterae* is exceptional in this regard. In contrast to prokaryotes, the presence of a microtubular spindle in dinoflagellates suggests a different segregation mechanism, resembling higher eukaryotes, via the driving force of β-tubulin bundles. The possible effects of microtubule polymerization or depolymerization inhibitors on dinoflagellates are not known, and it remains to be determined whether there is a physical link between the chromosomes and the spindle via the nuclear envelope.

IV. PHYLOGENY

A. Evolution of Dinoflagellates

The absence of histones and nucleosomes, the permanently condensed and highly ordered chromo-

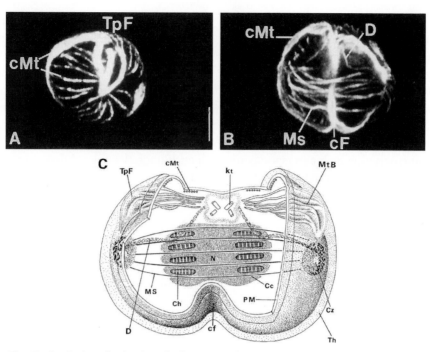

Fig. 5. Cortical and mitotic spindle microtubular organization as observed with confocal laser scanning microscope (CLSM) after incorporation of anti-β-tubulin antibody (A and B). (A) Cortex microtubule organization of *Crypthecodinium cohnii*. Note the rows of tubules (cMt) and the particular microtubular organization as a three-pronged fork (TpF). (B) In this *C. cohnii* dividing cell, mitotic spindle (Ms) and two microtubular desmoses (D) binding the kinetosome region to the mitotic spindle poles coexist with the cortical microtubules (cMt). The cleavage furrow (cF) is also present and labeled. (A and B) Scale bar = 10 μm. (C) Schematic representation based on CLSM and TEM observations of a dividing cell of *C. cohnii* showing the 2 × 2 kinetosomes (kt), the split microtubular desmose (D) linking the kinetosomes region to the centrosome-like zones (Cz), which are closely connected with the microtubular mitotic spindle (MS) passing through the nucleus by cytoplasmic channels (Cc). (MtB, microtubular basket; N, nucleus; Th, theca; cMt, cortical microtubules; PM, plasma membrane; Ch, chromosomes (reproduced from Perret *et al.*, 1993, by copyright permission of the Company of Biologists Ltd.).

somes bound to the nuclear envelope during their segregation, led to the Dodge mesokaryote concept being postulated for dinoflagellates in the 1970s and 1980s. This concept considered dinoflagellates as having prokaryotic traits conserved along with eukaryotic features, the idea being that dinoflagellates are very primitive eukaryotes, intermediate between prokaryotes and eukaryotes. However, later studies showed that dinoflagellates are true eukaryotes, having cellular compartments, distinct cell cycle phases, and a typical eukaryotic genomic organization (high percentage of noncoding repeated sequences and the

presence of intragenic introns). Furthermore, different phylogenetic studies based on ribosomal gene sequences showed that dinoflagellates emerged late in evolution and have a common ancestor with apicomplexa and ciliates (Cavalier-Smith, 1993; Van de Peer *et al.*, 1996). These studies suggested that dinoflagellates were indeed true eukaryotes which could have lost their histones and consequently their nucleosomes, leading to the peculiar highly condensed DNA structure as hypothesized for the first time by Cavalier-Smith (1981). According to this concept, similarities between prokaryotes and dinoflagellates reflect convergence rather than a special evolutionary relationship (Raikov, 1995).

This view is not contradictory with geological analysis based on the examination of fossilized theca. The first unambiguous dinoflagellate fossils occurring in the Triassic belong to Gymnodiniales but biogeochemical analysis of early Cambrian (approximately 520 millions years ago) sediments detected specific dinosterols. This time period, which corresponds to the earlier trace of dinoflagellates found, is later than the time period in which the appearance of the first photosynthetic eukaryotes occurred (750 millions years ago). Very old (proterozoic) fossils (acritarchs) are ambiguous, and currently there is no unequivocal fossil data that allows us to identify dinoflagellates as being more primitive than other protists.

B. Origin of Plastids in Dinoflagellates

The presence of heterotrophic and autotrophic species within the same phylum has attracted the attention of botanists and zoologists. Today, the dinoflagellates are viewed as protoctists (Margulis and Schwartz, 1998) and their phototrophic members as having acquired their plastids by cellular secondary endosymbiosis from eukaryotic microorganisms that already acquired photosynthesis.

The classical dinoflagellate plastid which contains peridinin, the specific pigment of autotrophic dinoflagellates, is surrounded by three membranes. Two membranes are common around chloroplasts in green and red algae, but protoctists having acquired their plastids by endosymbiosis may have more than two. As models for organelle acquisition by secondary endosymbiosis, the dinoflagellates are parti-cularly well suited since two species, *Peridinium foliaceum* Biecheler and *Peridinium balticum* Lemmermann, have two nuclei: that of the "host dinoflagellate" and that of the endosymbiont (e.g., diatom endosymbiont). Furthermore, authors showed that they are members of phototrophic dinoflagellates with various degrees of membrane reduction around chloroplasts and pigments other than peridinin, reflecting a different origin of their chloroplasts.

V. TOXIC DINOFLAGELLATES

Many dinoflagellate species are able to proliferate and produce blooms (or red tides for species having red pigments such as peridinin). Among these, approximately 200 toxic dinoflagellate species have been described and classified as a function of the symptoms they produce after ingestion of their toxins (Table I). Most important toxic species belong to a few genus and are briefly described as follows (Andersen, 1996).

Several species of the genus *Dinophysis* can cause intoxications of the diarrhetic shellfish poisoning type by production of different chemical forms of the okadaic acid. This molecule has been shown *in vitro* to be a potent protein phosphatase inhibitor and a skin tumor promoter in mice. Only a few tens of cells per liter of these species, which only exceptionally form blooms, can be toxic for humans, and they exert their damage after consumption of shellfish.

Alexandrium species (*tamarense, fundyense,* and *minutum*) and *Gymnodinium breve* produce a family of toxins causing paralytic shellfish poisoning by selec-

TABLE I

Families of Toxins Involved in Human Food Poisoning with the Indication of Syndromes, Solubility of Toxins, and the Target of the Toxins[a]

Toxin family	Syndrome	Solubility	Action on
Brevetoxin	Neurotoxic shellfish poisoning	Fat	Nerve, muscle, lung, brain
Ciguatoxin	Ciguatera fish poisoning	Fat	Nerve, muscle, heart, brain
Domoic acid	Amnesic shellfish poisoning	Water	Brain
Okadaic acid	Diarrhetic shellfish poisoning	Fat	Enzymes
Saxitoxin	Paralytic shellfish poisoning	Water	Nerve, brain

[a] Adapted from Anderson *et al.* (1993).

tively blocking the influx of sodium ions through excitable membranes, thus interrupting the formation of an action potential. Human intoxication also occurs by consumption of contaminated shellfish able to accumulate toxins.

Other intoxications occur via *G. breve* for neurotoxic shellfish poisoning syndromes and *Gambierdiscus toxicus* for ciguatera. This latter intoxication is endemic to a few tropical islands, such as Tahiti and the Gambier Islands, and it occurs after consumption of contaminated fish.

Pfisteria-like dinoflagellates cause major fish kills in estuaries and coastal waters of the mid-Atlantic and southeastern United States and have also been associated with serious human health impacts. The substance(s) responsible for this toxicity has not been identified and appears to be linked to excretions/secretions of contaminated fish.

Species producing harmless proliferations (e.g., *N. scintillans*) must also be mentioned for their environmental pollution in tourist areas due to low visibility in seawaters and/or oxygen depletion, inducing escape and/or kills of local organisms.

Toxic incidents resulting from different dinoflagellate species are reported increasingly more frequently throughout the world near the coasts. This increase of toxic blooms seems to be linked to the eutrophization of coasts and the spreading of more or less endemic species due to an increase in commercial exchanges. However, an artificial bias due to better detection of these blooms for health and economic reasons in many countries cannot be excluded.

VI. CONCLUSIONS

Dinoflagellates present many unusual (and sometimes unique) characteristics among eukaryotic cells. These characteristics explain the interest in these organisms as particularly fruitful models in cellular and molecular biology. The mechanisms which have been used during adaptive evolution are often more or less conserved in other eukaryotic cells but used in other metabolic routes. Knowledge of these mechanisms in dinoflagellates could aid in elucidating poorly understood phenomena in higher eukaryotes.

Furthermore, the global increase of toxic blooms creates more economic and health problems. External factors controlling these blooms are difficult to understand, and a better knowledge of the molecular and cellular characteristics controlling the cell cycle of dinoflagellates could lead to better prevention of these problems.

Acknowledgments

We thank Professor Dag Klaveness (University of Oslo) for helpful comments and critical reading of the manuscript. We gratefully acknowledge Professor Terry Preston (University of London) for suggestions and critical reading of the manuscript.

See Also the Following Articles

DNA Replication • Flagella • Origin of Life • Ribosome Synthesis and Regulation

Bibliography

Andersen, P. (1996). Design and implementation of some harmful algal monitoring systems. UNESCO Intergovernmental Oceanographic Commission Technical Series No. 44, pp. 1–26.

Anderson, D. M., Galloway, S. B., and Joseph, J. D. (1993). "Marine Biotoxins and Harmful Algae: A National Plan" pp. 3–13. Woods Hole Analytical Chemists, Washington, DC.

Bhaud, Y., Soyer-Gobillard, M. O., and Salmon, J. M. (1988). Transmission of gametic nuclei through a fertilization tube during mating in a primitive dinoflagellate, *Prorocentrum micans* Ehr. *J. Cell Sci.* **89**, 197–206.

Bhaud, Y., Barbier, M., and Soyer-Gobillard, M. O. (1994). A detailed study of the complex cell cycle of the dinoflagellate *Crypthecodinium cohnii* Biecheler and evidence for variation in histone H1 kinase activity. *J. Euk. Microbiol.* **41**, 519–526.

Burkholder, J. M., and Glasgow, H. B. (1997). *Pfiesteria piscicida* and other *Pfiesteria*-like dinoflagellates: Behavior, impacts and environmental controls. *Limnol. Oceanogr.* **42**, 1052–1075.

Cavalier-Smith, T. (1981). The origin and early evolution of the eukaryotic cell. *In* "Molecular and Cellular Aspects of Microbial Evolution" (Society for General Microbiology Ltd., Symposium 32) (Carlile, Collins, and Mosely, eds.), pp. 33–84. Cambridge Univ. Press.

Cavalier-Smith, T. (1993). Kingdom Protozoa and its 18 phyla. *Microbial Rev.* **57**(4), 953–994.

Chatton, E. (1920). Les Péridiniens parasites. Morphologie, reproduction, ethologie. *Arch. Zool. Exp. Gén.* **59**, 1–75.

Dodge, J. D. (1965). Chromosome structure in the dinoflagellates and the problem of the mesocaryotic cell. *In* "Progress in Protozoology" (Abstr. II Internat. Conf. Protozool.), pp. 264–265. London Excerpta Medica.

Dodge, J. D. (1985). "Atlas of Dinoflagellates. A Scanning Electron Microscope Survey." Farrand, London.

Géraud, M. L., Herzog, M., and Soyer-Gobillard, M. L. (1991). Nucleolar localization of rRNA coding sequences in *Prorocentrum micans* Ehr. (Dinomastigote, Kingdom Protoctist) by *in situ* hybridization. *BioSystems* 26, 61–74.

Margulis, L., Corliss, J. O., Melkonian, M., and Chapman, D. J. (1990). "Handbook of Protoctista," pp. 419–437. Jones & Bartlett, Boston.

Margulis, L., McKahann, H., and Olendzenski, L. (1993). "Illustrated Glossary of Protoctista." Jones and Bartlett, Boston.

Margulis, L., and Schwartz, K. V. (1998). "Five Kingdoms: An illustrated Guide to the Phyla of Life on Earth." Freeman, San Francisco.

Minguez, A., Franca, S., and Moreno Diaz de la Espina, S. (1994). Dinoflagellates have a eukaryotic nuclear matrix with lamin-like proteins and topoisomerase II. *J. Cell Sci.* 107, 2861–2873.

Moldowan, J. M., and Talyzina, N. (1998). Biogeochemical evidence for dinoflagellate ancestors in the early cambrian. *Science,* 281, 1168–1170.

Oakley, B. R., and Dodge, J. D. (1974). Kinetochores associated with the nuclear envelope in the mitosis of a Dinoflagellate. *J. Cell Biol.* 63, 322–325.

Palmer, J. D., and Delwiche, C. F. (1996). Second-hand chloroplast and the case of disappearing nucleus. *Proc. Natl. Acad. Sci. USA* 93, 7432–7435.

Perret, E., Davoust, J., Albert, M., Besseau, L., and Soyer-Gobillard, M. O. (1993). Microtubule organization during the cell cycle of the primitive eukaryote dinoflagellate *Crypthecodinium cohnii. J. Cell. Sci.* 104, 639–651.

Raikov, I. B. (1982). "The Protozoan Nucleus. Morphology and Evolution," Cell Biology Monographs Vol. 9. Springer-Verlag, New York.

Raikov, I. B. (1995). The dinoflagellate nucleus and chromosomes: Mesokaryote concept reconsidered. *Phycologia* 34, 239–247.

Rizzo, P. J. (1987). Biochemistry of the dinoflagellate nucleus. *In* "The Biology of Dinoflagellates" (F. J. R. Taylor, Ed.), pp. 143–173. Blackwell, Oxford.

Roberts, T. M., Tuttle, R. C., Allen, J. R., Loeblich, A. R., III, and Klotz, L. C. (1974). New genetic and physicochemical data on the structure of dinoflagellate chromosomes. *Nature (London)* 248, 446–447.

Rowan, R., Whitney, S. M., Fowler, A., and Yellowlees, D. (1996). Rubisco in marine symbiotic dinoflagellates: Form II enzymes in eukaryotic oxygenic phototrophs encoded by a nuclear multigen family. *Plant Cell* 8, 539–553.

Saunders, G. W., Hill, D. R. A., Sexton, J. P., and Andersen, R. A. (1997). Small-subunit ribosomal RNA sequences from selected dinoflagellates: Testing classical evolutionary hypotheses with molecular systematic methods. *Plant Syst. Evol.* 11(Suppl.), 237–259.

Schnepf, E., and Drebes, G. (1993). Anisogamie in the dinoflagellate *Noctiluca. Helgoländer Meeresuntersuchungen* 47, 265–273.

Schnepf, E., and Elbrächter, M. (1988). Crypthophycean-like double membrane-bound chloroplasts in the dinoflagellate *Dinophysis* Ehrenb.: Evolutionary, phylogenetic and toxicological implications. *Bot. Acta* 101, 196–203.

Soyer, M. O. (1970a). Les ultrastructures liées aux fonctions de relation chez *Noctiluca miliaris* S. (*Dinoflagellata*). *Z. Zellforsch* 104, 29–55.

Soyer, M. O. (1970b). Etude ultrastructurale de l'endoplasme et des vacuoles chez deux types de Dinoflagellés appartenant aux genres *Noctiluca* (Suriray) and *Blastodinium* (Chatton). *Z. Zellforsch.* 105, 350–388.

Soyer, M. O. (1971). Structure du noyau des *Blastodinium* (dinoflagellés parasites). Division et condensation chromatique. *Chromosoma Berlin* 33, 70–114.

Soyer, M. O. (1972). Les ultrastructures nucléaires de la Noctiluque (Dinoflagellé libre) au cours de la sporogenèse. *Chromosoma Berlin* 39, 419–441.

Soyer, M. O. (1977). Préservation des structures nucléaires dinoflagellés. *Biol. Cell* 30, 297–300.

Soyer, M. O. (1979). Axe chromosomique et division chez *Prorocentrum micans* E., Dinoflagellé libre. *Vie Milieu* 28/29, 461–472.

Soyer-Gobillard, M. O., Géraud, M. L., Coulaud, D., Barray, M., Théveny, B., and Delain, E. (1990). Location of B- and Z-DNA in the chromosomes of a primitive eukaryote Dinoflagellate. *J. Cell Biol.* 111, 293–308.

Soyer-Gobillard, M. O. (1996). The genome of the primitive eukaryote Dinoflagellates: Organization and functioning. *Zool. Stud.* 35, 78–84.

Soyer-Gobillard, M. O., and Géraud, M. L. (1992). Nucleolus behaviour during the cell cycle of a primitive dinoflagellate eukaryote, *Prorocentrum micans* Ehr., seen by light microscopy and electron microscopy. *J. Cell Sci.* 102(3), 475–485.

Spector, D. L. (1984). Dinoflagellate nuclei. *In* "Dinoflagellates" (D. L. Spector, Ed.), pp. 107–147. Academic Press, New York.

Taylor, F. J. R. (1987). The biology of dinoflagellates. *In* "Botany Monographs," Vol. 21. Blackwell, London.

Van de Peer, Y., Van der Auweva, G., and De Wachter, R. (1996). The evolution of stramenopiles and alveolates as derived by "Substitution rate calibration" of small ribosomal subunit RNA. *J. Mol. Evol.* 42, 201–210.

Diversity, Microbial

Charles R. Lovell
University of South Carolina

GLOSSARY

division The largest phylogenetic grouping within a domain. This grouping can be considered analogous to the kingdom and includes one or more phyla.

domain The largest phylogenetically coherent grouping of organisms. The three domains of living organisms are the Bacteria, the Archaea, and the Eucarya.

monophyletic A group of organisms descending from a common ancestor having the significant traits that define the group.

species (bacterial or archaeal) The smallest functional unit of organismal diversity consisting of a lineage evolving separately from others and with its own specific ecological niche.

subdivision The largest phylogenetic grouping within a division. This grouping can be considered analogous to a phylum, and includes one or more orders.

THE MICROORGANISMS include all organism taxa containing a preponderance of species that are not readily visible to the naked eye. As a consequence of this anthropocentric definition, the microbiota have historically included numerous types of organisms differing from each other at the most fundamental levels of cellular organization and evolutionary history. We typically consider the major groups of microorganisms to be the bacteria, archaea, fungi, algae, protozoa, and viruses. This collection spans the breadth of organismal evolutionary history and ecological function, and represents the great majority of the major phylogenetic divisions of living organisms. Although these microbial groups differ from each other in many important ways, they have in common enormous diversity and great importance in the biosphere.

I. THE SIGNIFICANCE OF MICROBIAL DIVERSITY

We are only beginning to appreciate the true extent of microbial diversity and have only recently developed methods suitable for its exploration. Interest in this area has been spurred by the crucial roles microorganisms play in global ecology and in numerous human endeavors. Major ecological roles of microorganisms and applications of microbial products and processes in human enterprises are still being uncovered, but what we currently know provides an imposing picture of the importance of the microbiota in global and human ecology.

Microorganisms are prime movers in all ecosystems. They mobilize some elements by converting them from nonvolatile to volatile forms, permitting large-scale mass transport of these elements among different geographic locations. They capture carbon dioxide (and other one-carbon compounds) through several known fixation pathways, accounting for approximately 50% of global primary productivity. Microorganisms participate in and frequently dominate the interconversions of chemical compounds, producing and regulating the great variety of materials necessary to the biosphere. Effectively all decomposition processes, including spoilage of foods, are dominated by the microbiota. Microorganisms are found in all environments in which liquid water and an energy source can be found. These include such extreme environments as polar sea ice (in fluid-filled channels and pockets), hydrothermal vents, hot and cold deserts, brines approaching salt saturation, extremely acidic and alkaline waters, and deep (>1 km) subsurface solid rock formations. It can be accurately stated that the Earth is a microbial world.

Microorganisms also participate in a plethora of symbiotic (literally "living together") interactions with each other and with higher organisms. Some are common commensals using the products of their hosts without causing any discernible harm. Some form mutualistic interactions in which they and their hosts both actively benefit from the symbiosis. These mutualisms range from production of essential vitamins and facilitation of digestive processes by the microflora found in the digestive tracts of animals to the elegant and species- or strain-specific nitrogen-fixing interactions that occur between many plant species and bacteria. Finally, microorganisms are infamous for their parasitic interactions with higher organisms. These range from subtle, such as the influence of *Wolbachia* infections on sex determination in arthropods, to catastrophic. The impacts of pathogenic microorganisms on human populations have on occasion been devastating and the plagues that ravaged Europe and Asia throughout the Middle Ages are no more notable in this regard than the AIDS pandemic of the present. The impacts of parasites on the structuring of higher organism communities can obviously be very significant and, as Dutch elm disease demonstrates, even a single pathogen can drive a host species to local extinction.

The exploitation of microorganisms having useful properties is central to many human enterprises and has fostered substantial bioprospecting efforts. Numerous microbial products ranging from antibiotics to foods are important commodities, but perhaps the most significant impact of beneficial microorganisms lies in agriculture. Many microorganisms that grow in mutualistic interactions with plants have been used to increase crop yields and to control pests.

Nitrogen-fixing mutualisms between plants and bacteria form the basis for crop rotation and can facilitate reclamation of overexploited or marginal lands. The mycorrhizal fungi, which grow in intimate association with plant roots, enhance the ability of plants to take up mineral nutrients and water. This type of symbiosis permits plant growth in many marginal environments and most plants participate in it. Numerous bacteria and non-mycorrhizal fungi colonize the surfaces of plants and can successfully outcompete potentially pathogenic organisms, preventing infections. In addition, the highly specific insecticidal toxins produced by *Bacillus thuringiensis* have been harnessed in insect pest control strategies and offer the major advantage of not harming beneficial insect species. Agriculture is, and always has been, highly dependent on microorganisms.

The vast array of microbial activities and their importance to the biosphere and to human economies provide strong rationales for examining their diversity. Loss of biodiversity is an ongoing global crisis and loss of the microbiota indigenous to human-impacted environments or associated with higher organisms facing extinction is a very real concern. As a consequence of the growing recognition of the importance of species loss, achieving a meaningful assessment of microbial diversity has been the topic of several major workshops and meetings. This is a period of very active inquiry into microbial diversity.

A. Approaches to the Examination of Microbial Diversity

Past studies of microbial diversity were highly dependent on our ability to isolate these organisms into pure (i.e., single species) laboratory cultures for characterization. However, in most environments only 1–5% of the microscopically visible Bacteria and Archaea can be successfully cultivated, and obtaining pure cultures of many algae, protozoa, and fungi is problematic. In addition, detailed morphological, physiological, and behavioral characterization of any microorganism can be extremely time-consuming. As a consequence, many pure culture isolates have only been partially characterized. Recent applications of molecular biological tools, chiefly the polymerase chain reaction (PCR), DNA–DNA hybridization, DNA cloning and sequencing, and efficient analysis of DNA sequences, have provided a means to catalog microorganisms without the necessity for isolation and offer useful methods for streamlining the characterization of pure cultures.

Commonly used strategies for assessing microbial diversity employ PCR to amplify specific bacterial genes extracted from samples from the environment of interest. Nucleotide sequences of the genes encoding the small subunit ribosomal RNA (16S rRNA for Bacteria and Archaea and 18S rRNA for the Eucarya), when properly aligned, provide substantial information on the relatedness of unknown organisms to known species and support the construction of phylogenetic trees for microorganisms (Fig. 1). Other genes of interest for characterizing microbial diversity encode key enzymes involved in microbially mediated processes [i.e., functional group genes; *nifH* (nitrogenase iron protein) for nitrogen fixation, *amoA* (ammonium monooxygenase) for ammonium oxidation, *rbcL* (ribulose bisphosphate carboxylase/ oxygenase) for CO_2 fixation, etc.]. The amplified gene sequences can be resolved using various gel electrophoresis methods to provide a fingerprint of the microbial community present in the sample, or they can be cloned into an appropriate vector for propagation and nucleotide sequence analysis. The sequences of these genes are then available for phylogenetic analysis and species-specific domains within them can be used in DNA–DNA hybridization studies to quantify organisms of interest in environmental samples. Similar phylogenetic analyses of uncharacterized pure cultures allow the relatedness of these organisms to known species to be efficiently determined. Knowledge of the type of organism studied greatly facilitates selection of appropriate testing criteria for its complete identification. These technologies, although not perfect, have opened the door to the vast unknown world of microorganisms, clarifying the relationships of microorganism groups to each other and to higher organisms and facilitating the cataloging of species that have thus far resisted isolation and cultivation efforts. The picture that has emerged shows that of the three known domains of living organisms two, the Bacteria and the Archaea, consist entirely of microorganisms, and the third, the Eucarya, contains many major microbial taxonomic divisions. It is also clear from recent and ongoing studies that many new taxonomic lineages of microorganisms await discovery. The following sections summarize what is known, or at least widely accepted, about the diversities of the major groups of microorganisms and indicate some interesting areas in which information is mostly lacking. It should be noted that most current estimates of the diversity of all major microbial groups (Table I) are quite

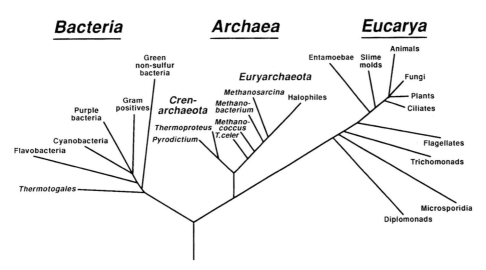

Fig. 1. The universal phylogenetic tree showing the three domains of living organisms—the Archaea, the Bacteria, and the Eucarya (reproduced with permission from Woese, 1994).

TABLE I
Estimated Numbers of Described and Undescribed Species in the Major Groups of Microorganisms According to Recent Surveys[a]

Group	Estimated described species	Estimated undescribed species
Bacteria	4,000	>1,000,000
Archaea	175	?
Fungi	72,000	>1,500,000
Algae	40,000	>200,000
Protozoa	40,000	>200,000
Viruses	6,500	>1,000,000

[a] See Bibliography for sources.

conservative and may be significant underestimates of the true numbers.

II. BACTERIAL DIVERSITY

The Bacteria constitute an extremely diverse domain of prokaryotic microorganisms. They have a very simple cellular architecture, lacking the membrane-bound organelles characteristic of eukaryotic cells. They reproduce asexually, typically by simple fission, and are genetically haploid. Bacteria acquire soluble nutrients by transport across the plasma membrane but are unable to engulf colloids or particles through any endocytosis mechanisms. Virtually all known bacterial taxa have cell walls composed of an aminosugar polymer called peptidoglycan and this material is unique to this domain. Numerous distinctions can also be found between the Bacteria and other types of organisms in the details of their information processing machinery (i.e., DNA replication, RNA synthesis, and protein synthesis systems), chromosome structure, flagellar structure, and many physiological features.

Although unified by many features of cell structure and function and clearly forming a monophyletic domain, the Bacteria encompass enormous physiological diversity and can be found in almost every conceivable ecological niche. These organisms are found in extreme environments ranging from Antarctic sea ice to superheated marine hydrothermal vents,

from extremely dilute aqueous systems to solutions approaching salt saturation, from highly acidic waters approaching a pH value of 0 to alkaline waters with pHs in excess of 12, and from the near vacuum of the upper atmosphere to the crushing pressures of the Challenger Deep. Bacteria can even be found in the highly radioactive environment of nuclear power plant reactor chambers. Although a few very hot environments appear to be dominated by the Archaea, the bacteria are ubiquitous inhabitants of all other environments and are important (and sometimes the only) participants in numerous ecological processes. Bacteria dominate mineral cycling and degradative processes in most environments, are important primary producers, and participate in all known types of symbiotic interactions, including parasitism on higher organisms, mutualistic interactions with other microorganisms and with higher organisms, and even active predation by some bacteria (such as *Bdellovibrio* species) on other bacterial species. The Bacteria dominate the biosphere like no other type of organism on Earth.

The bacterial domain encompasses an imposing diversity of physiological properties and ecological functions. The bacteria also clearly exemplify many of the major problems frequently encountered in assessing microbial diversity. Simply stated, nothing concerning the proper identification, classification, or phylogenetic analysis of bacteria is simple. Detailed observation of natural bacteria is complicated by their small size and their frequent growth on surfaces and in multispecies associations (biofilms, microcolonies, etc.). Even when they can be observed with a suitable degree of precision, bacteria display very few taxonomically useful morphological features. Most are simply spherical or rod shaped and a given species cannot be differentiated from organisms having similar shapes, but quite different activities, on the basis of appearance. As a consequence, bacterial identification is highly dependent on laboratory cultivation of species of interest in order to perform detailed physiological testing. However, the overwhelming majority of bacteria in nature cannot be readily cultivated on artificial laboratory growth media. Whether this is due to some incapacity of some observable cells or to our failure to formulate suitable growth media is not known, but the consequences for systematic appraisal of bacterial diversity are extremely important.

The lack of information obtainable through microscopic observations and the difficulties encountered in isolating even numerically dominant species into pure culture make bacterial diversity studies highly dependent on molecular biological approaches. Through such methods it is possible to determine phylogenetic relatedness of unknown bacterial species, including those that resist cultivation, to known species. However, complete characterization of new bacterial species still requires extensive physiological characterization of laboratory cultures. In addition, since bacteria reproduce asexually, the exact definition of a species is not completely clear. The general trend has been that intensively studied taxa are highly speciose, whereas less studied taxa tend to have few and sometimes genetically heterogeneous species. Finally, the species is not always the finest phylogenetically or ecologically significant subdivision of the Bacteria. Many (all?) species contain numerous strains that can be differentiated only through detailed molecular, physiological, and/or immunological testing. Strains are particularly important to us when they differ in their interactions with higher organism hosts. For example, different strains of a given *Rhizobium* species are nitrogen-fixing mutualists of different leguminous plant hosts. Each bacterial species, whether it associates with higher organisms or not, appears to be a radiation of closely related strains which may differ significantly in the ecological niches they exploit and which contribute to even greater bacterial diversity. Deciding the level of detail desirable in diversity studies and evaluating the feasibility of obtaining a complete catalog of bacterial diversity in all but the most species-poor environments are ongoing concerns.

Currently, there are 36 monophyletic divisions of bacteria (Fig. 2). These range from the *Proteobacteria,* a division which includes a great many well-characterized species, to the *Acidobacterium* division, which has only three cultivated species, and the 13 "candidate" divisions (Hugenholtz *et al.,* 1998) which currently lack characterized species altogether. Several additional division-level lineages are only represented by unique environmental 16S rRNA sequences. The total number of bacterial divisions is

unknown but almost certainly in excess of 40. In addition to their phylogenetic coherence, some divisions are also unified in ecophysiological function (the Cyanobacteria, the Green Sulfur Bacteria, etc.), but it is important to note that this is the exception rather than the rule. For example, the *Proteobacteria* division includes numerous different physiological types ranging from aerobic heterotrophs to strictly anaerobic photoautotrophs and obligate intracellular parasites. The degree of physiological diversity within many divisions is unknown and is particularly problematic for divisions having no characterized species.

The combination of molecular phylogenetic analysis and classical physiological characterization provides a powerful approach for identifying new bacterial species and placing them into a phylogenetically coherent framework, but bacterial diversity presents an imposing challenge at all levels. Currently, there are only about 4000 validly published bacterial species representing 728 genera. A total of only about 8000 bacterial 16S rRNA gene sequences are currently available. It is clear that only a small fraction of bacterial diversity has been cataloged, even for most well-studied environments. The vast majority of bacteria have not been cultivated and unknown 16S rRNA gene sequences are routinely recovered. In addition, many ecosystems, particularly in the tropics, await detailed study. Even common and easily accessed environments, such as temperate terrestrial soils and freshwater sediments, have been the subjects of few systematic surveys. Currently, it seems unlikely that an accurate estimate of bacterial diversity can be presented or defended.

III. ARCHAEAL DIVERSITY

The Archaea, like the Bacteria, are prokaryotic in terms of their cellular architecture. They reproduce asexually and depend on dissolved substances for carbon and energy. There, however, the similarity between the two domains ends. The Archaea are quite different from the Bacteria in the cell wall polymers they produce, their unusual lipids, resistance to broad spectrum antibiotics that inhibit most bacteria, and numerous details of their information processing

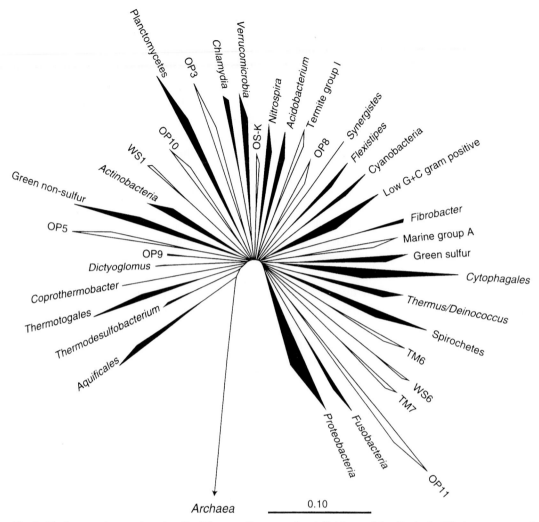

Fig. 2. Phylogenetic tree showing the 36 currently recognized divisions of the Bacteria. Wedges represent division level groups of two or more sequences. Divisions containing characterized species are shown in black, and those containing only environmental sequences are shown in outline (reproduced with permission from Hugenholtz et al., 1998). Scale bar represents 0.10 base changes per nucleotide in the 16S rRNA gene sequences examined.

machinery. The discovery by Carl Woese that the Archaea constitute a distinct monophyletic domain divergent from both the Bacteria and the Eucarya (Fig. 1) not only made sense of the disparate characteristics of these organisms but also it revolutionized our understanding of the evolution and diversity of life on Earth.

Most of the Archaea that have been validly described and published are extremophiles falling into four major categories. These include the high salt-requiring extreme halophiles, the anaerobic

methane-producing methanogens, various types of Archaea that grow optimally at extremely high temperatures (thermophiles or hyperthermophiles), and the cell wall-less archaeon *Thermoplasma*. The United States National Center for Biological Information GenBank database in 1999 listed only 63 genera and 175 described species of Archaea. Several hundred sequences from unknown Archaea are also available, but we cannot make predictions about the characteristics of these organisms. Some general features of the known archaeal groups are given here.

The extreme halophiles inhabit highly saline environments and generally require oxygen and at least 1.5 M NaCl for growth. Virtually all these organisms can grow at saturating levels of salt. Although such highly saline environments are relatively rare, the extremely halophilic Archaea display a surprising diversity of cell morphologies, from rods and cocci to flattened disks, triangles, and rectangles. Given the unusual environment in which these organisms grow, the currently described nine genera may include most of the diversity of extreme halophiles.

The methanogens are restricted to anoxic habitats and all produce methane. Suitable habitats for methanogens are much more common than the high-salt environments required by the extreme halophiles and consequently the known diversity of methanogens is greater. There are seven major groups of methanogens containing at least 25 genera, including extremely halophilic methanogens, some methanogens that require highly alkaline growth conditions, and several thermophilic types. Major methanogenic habitats include soils ranging from pH neutral to acidic peat-rich soils, all types of freshwater sediments, both geothermally heated and unheated marine sediments, and the digestive tracts of higher animals, such as ruminants, and those of arthropods, including insects and crustaceans. Given this broad range of methanogenic environments, it is clear that the diversity of these organisms could greatly exceed the current rather sparse collection of species.

The hyperthermophilic Archaea include all nonmethanogenic organisms that require temperatures in excess of 80°C for optimum growth. These Archaea are the most extremely thermophilic of all known organisms and several can grow at temperatures in excess of 100°C. Most hyperthermophiles are obligate anaerobes and most use reduced sulfur compounds (mineral sulfur or H_2S) in their metabolism. Ideal conditions for these organisms can be found in sulfur and sulfide-rich geothermally heated waters, such as those of hot springs and deep-sea hydrothermal vents. Although such habitats can have extremely low pH values, most hyperthermophiles have been recovered from neutral pH to mildly acidic locations. The widespread occurrence of geothermally heated waters, soils, and sediments, and the range of pH values and nutrient chemistries occurring within

them, tends to support the idea of substantial diversity of hyperthermophilic Archaea. Certainly, the current catalog contains a broad range of organisms differing in physiology and growth requirements.

Thermoplasma, the cell wall-less thermophilic Archaeon, appears to be very restricted in its habitat range. All but one strain of this organism has been isolated from self-heating coal refuse piles. Although other, less transient habitats for this organism have been sought, only one strain of *Thermoplasma* has been recovered from geothermal environments. With such a limited distribution and narrow habitat range, this archaeal subdivision would be predicted to have very limited diversity.

Based on 16S rRNA gene sequence analysis, these diverse organisms can be readily grouped into two major divisions (kingdoms; Fig. 3). The *Euryarchaeota* include the halophilic and methanogenic Archaea growing at moderate temperatures and *Thermoplasma*. The *Crenarchaeota* include all the hyperthermophilic organisms. It is easy to see why most researchers have long considered Archaea to be confined to extreme environments and thus likely to be restricted to very limited diversity. However, new findings from moderate- to low-temperature environments show that unknown Archaea are prominent members of the microbial communities found there. DNA extracted from samples of soils, marine and freshwater sediments, marine plankton, sponges, and the gut contents of holothurians (sea cucumbers) all contained 16S rRNA sequences specific to the Archaea. Archaea from these environments include a crenarchaeotal group that represents as much as 20% of the planktonic marine prokaryote assemblage (Fuhrman and Davis, 1997) and appears to occur in both surface and deep marine waters worldwide. Sequences related to this marine group have also been recovered from freshwater sediments and terrestrial soils. Two newly recognized marine euryarchaeotal groups distantly related to *Thermoplasma* have also been discovered (Fuhrman and Davis, 1997). None of these organisms have been isolated into pure culture and little can be deduced about their physiologies and growth requirements. All that we currently know is that the Archaea are more diverse and far more abundant in nonextreme environments than was previously thought. It is worth-

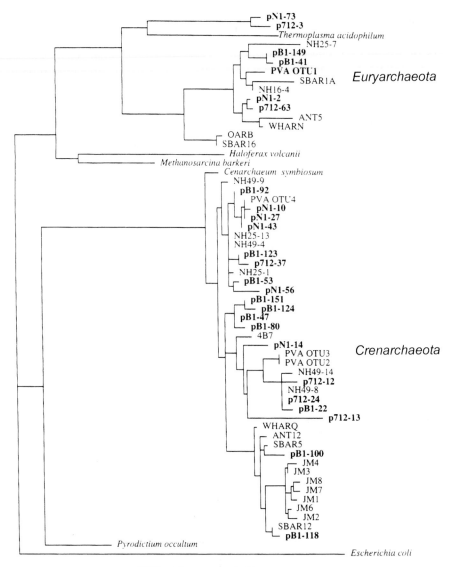

0.10 base changes per nucleotide

Fig. 3. Phylogenetic tree for the Archaea showing the major groups of currently undescribed Archaea. Abbreviations represent cloned environmental 16S rRNA gene sequences (for published sources, see Fuhrman and Davis, 1997). The following clones were obtained from water samples: NH clones, pN1 clones, p712 clones, and 4B7 from the Pacific Ocean; WHAR clones from Woods Hole, MA; SBAR from Santa Barbara, CA; OAR from Oregon; and ANT clones from Antarctica. pGrfA4 is from freshwater lake sediment and PVA clones are from a volcanic seamount near Hawaii. JM clones are from an abyssal holothurian and *Crenarchaeum symbiosum* is from an uncharacterized marine sponge symbiont. Scale bar represents 0.10 base changes per nucleotide in the 16S rRNA gene sequences examined (adapted with permission from Fuhrman and Davis, 1997).

while to consider that we have had the molecular biological tools required for discovery of these unknown Archaea for only a short time and that the Archaea have only been considered a separate domain for slightly longer than 20 years. We need to know a great deal more than we currently know in order to develop any meaningful estimates of archaeal diversity or any clear realization of their significance in most environments.

IV. FUNGAL DIVERSITY

The fungi are an extremely diverse division of the Eucarya that includes both unicellular species and species that grow as finely divided networks of filaments called hyphae. Unicellular species are all microscopic, but hyphal networks can support fruiting bodies that are readily visible without magnification. All are included among the microorganisms on the basis of their lack of differentiated tissues. Primary features that distinguish the fungi from other Eucarya are (i) use of chitin as a cell wall polymer (but not in all fungi) and (ii) the absence of photopigments. Fungal nutritional requirements and core metabolic processes are neither unusual for eukaryotes nor particularly diverse. However, their morphologies and life cycles display considerable diversity, providing the basis for classical fungal taxonomy. This scheme has a serious limitation in that mature reproductive structures are required for correct identification of some organisms and these structures can be difficult to find or are produced haphazardly by some fungi. The operational solution to this problem has been inclusion of all fungi for which sexual reproductive structures have not been seen in a single polyphyletic subdivision. The seven classically defined fungal taxonomic subdivisions are given here.

The Ascomycota (Ascomycetes or sac fungi) include about 35,000 described species. Many species of the red, brown, and blue-green molds that cause food spoilage, as well as the powdery mildews that cause Dutch elm disease and chestnut blight, are Ascomycetes. The Basidiomycota (Basidiomycetes or club fungi) include about 30,000 known species.

Among these are the smuts, rusts, shelf fungi, stinkhorns, puffballs, toadstools, mushrooms, and bird's nest fungi. The human pathogen *Cryptococcus neoformans,* which causes cryptococcosis, is also a Basidiomycete, as are the important plant pathogens, the smuts and rusts. The Zygomycota (Zygomycetes) include about 600 known species. Among these are the common bread molds and a few species parasitic on plants and animals. The Deuteromycota (Deuteromycetes or fungi imperfecti) include all species for which a sexual reproductive phase has not been observed—about 30,000 known species. Several human pathogens, including the organisms causing ringworm, athlete's foot, and histoplasmosis, are included in this group. The aflatoxin-producing species *Aspergillis flavus* and *A. parasiticus,* important organisms in fungal food poisonings, are also in this group. Other Deuteromycetes are responsible for production of antibiotics or are used in production of foods, such as cheeses and soy sauce. When the reproductive (perfect) stage of a Deuteromycete is characterized, it is transferred from the fungi imperfecti to the appropriate group. These four subdivisions of fungi form a monophyletic cluster of true fungi. Molecular phylogenetic analysis has revealed that the remaining three classically defined fungal groups are more closely related to other types of organisms than to the four true fungal subdivisions. The *Oomycota* (Oomycetes or water molds) resemble fungi in that they grow in a finely branched network of hyphal filaments. However, the Oomycetes have cell walls composed of cellulose, produce motile asexual zoospores, and are closely related to the algal Phaeophyta and Chrysophyta. The Oomycetes are saprotrophs and several are important plant pathogens, the most famous of which is *Phytophthora infestans,* the organism responsible for the Irish potato famine. The *Myxomycota* (plasmodial slime molds; about 700 known species) and *Acrasiomycota* (cellular slime molds; about 50 known species) lack cell walls of any kind during vegetative growth, display amoeboid motility, and are related to the protozoa. Both types are saprotrophic and feed by phagocytosis. When food resources become limiting, both types of slime molds will form fruiting structures. None of the slime molds are known to be important plant or animal parasites or to produce commercially valuable products. However, slime molds are significant participants in organic matter turnover in soils.

The major ecological functions of fungi are well understood. Soil fungi are ubiquitous saprotrophs and responsible for much of the decay of soil organic matter. These organisms are highly adapted to the saprotroph function, producing a variety of important extracellular enzymes for degradation of insoluble substrates and infiltrating the soil matrix and decaying organic matter through hyphal growth. Although primarily terrestrial, fungi are also found in aquatic ecosystems, including highly saline intertidal soils, and are particularly significant in the decay of plant materials in these environments. Fungi are also important members of the digestive tract flora of many animals and actively participate in the breakdown of plant materials, particularly in the ruminant animals and many insects.

Fungal growth at the expense of plant biomass is not confined to dead biomass. Some fungi are also very significant plant parasites responsible for economically important damage to crop plants. The complex relationship of the mycorrhizal fungi with their plant hosts, which can be characterized as mutualistic, parasitic, or fluctuating between these roles, should not be overlooked in this regard. The mycorrhizae promote plant growth in many cases and are very important agents in the structuring of plant communities, but they can also damage their hosts under stressful environmental conditions. Fungal parasitism in some cases results in spoilage of food products, particularly the seed heads of maturing grain crops. An interesting fungus important in the spoilage of grains is the Ascomycete *Claviceps purpurea,* which produces the alkaloid mycotoxin ergot. Ergot is a potent hallucinogen and consumption of ergot-contaminated rye is thought to have been the cause of medieval dancing fits, in which the populations of whole villages danced wildly until exhausted. Accusations of witchcraft and the resulting Salem witch trials and executions may also have been due to ergotism.

Although they are less frequently cited as inhabitants of extreme environments than the extremist species among the Bacteria or Archaea, the fungi can grow across a broad range of environmental conditions and include species that could certainly be considered extremophiles. In general, fungi can tolerate greater extremes of salinity, pH, and desiccation than non-extremophilic species of bacteria. The diverse array of fungi participating in the lichen symbiosis can be found from temperate to very cold and dry environments worldwide. The microbiota of the rock varnishes common in hot deserts and the endolithic microflora growing within rocks in the Antarctic dry valleys both include fungi. Fungi are found in anaerobic environments including subsurface marine sediments, water-saturated decaying organic matter, the rumens of ruminant mammals, and the hindguts of numerous insect species. Fungi inhabit soils having temperatures from below freezing up to about 60°C. Any effort to catalog total fungal diversity should certainly include such extreme environments, which are clearly more common than is usually realized.

There are between 72,000 and 100,000 known species of fungi, but total species numbers have been estimated at approximately 1.5 million, approximately six times the estimated number of vascular plant species (Hawksworth and Rossman, 1997). This estimate should certainly be considered conservative since it does not include the mostly undescribed fungi growing commensalistically, mutualistically, or parasitically in or on animals. The estimated millions of undescribed insect species in the tropics are particularly noteworthy in this regard. Since mutualistic and parasitic fungi are typically quite host specific, many of these insect species may harbor unknown fungi. Given the important ecological functions of fungi as saprotrophs in organic matter decomposition and symbionts of terrestrial plants, it is sensible to consider plants first in estimates of fungal diversity. However, the many interactions of fungi with animals are easily overlooked and a great many undescribed fungal species are likely to be involved in them. The tropics, with their abundance of plant and insect species that are found nowhere else, likely represent the largest global reservoir of undescribed fungi, and characterization of the diversity of tropical fungi has only started.

The introduction of molecular biological methods for phylogenetic characterization of fungi should greatly facilitate identification of unknown species, particularly those that do not reliably produce fruiting bodies. Classical description of new species on the basis of morphological and life cycle characteristics continues at a respectable rate, but most fungal species, including many already available in culture, herbarium, and private collections, have yet to be validly described. It is certain that discovery of new fungal species will be limited primarily by the resources available for this effort for many years to come.

V. ALGAL DIVERSITY

The algae are a diverse assemblage of photosynthetic Eucarya. Most species are unicellular and thus counted among microorganisms. Some are much larger and these macroalgae will not be discussed here. The algae all contain chlorophyll *a* as well as a variety of additional pigments, giving them colorations ranging from green to yellow, red, and brown. All algae carry out oxygenic photosynthesis and most are essentially aquatic in character, including those found at the surface of saturated soils. The algae inhabit all moderate (i.e., nonextreme) marine and freshwater environments that receive sufficient light to support photosynthesis. They are also found in a variety of unusual or extreme environments, including Antarctic sea ice, mountain snow fields, hot springs (up to maximum temperatures of about 60°C), and acidic waters (pH 4 or 5). A few species are able to tolerate very dry environments, such as dry soils, and the endolithic algae grow within rocks in the extremely cold, dry environment of Antarctic dry valleys. The halophilic *Dunaliella* can be found in many salt lakes and is often the only oxygenic phototroph present. The algae are ubiquitous, although sometimes unnoticed, members of virtually all microbial communities wherever light is available.

Algal diversity is impressive. There are at least 40,000 known algal species, with an extrapolated total of more than 200,000 species. These organisms are not monophyletic and their classical taxonomy is largely based on morphology and a few key phenotypic traits. The most important of these is their performance of oxygenic photosynthesis, their panoply of photopigments, their cell wall structure and chemistry, and the carbon reserve materials they synthesize when light is not a limiting resource. Consideration of these traits lead to the division of the algae into six major groups; the Chlorophytes, the Euglenophytes, the Crysophytes, the Rhodophytes, the Phaeophytes, and the Pyrrophytes (Table II). This phenotypic classification system, although internally consistent, does not reflect the evolutionary history or the true phylogeny of the algae. The cyanobacteria, once known as the Cyanophytes, are clearly bacteria and should not be included among the algae. The Chlorophytes and Rhodophytes are closely related

TABLE II
Characteristics of the Major Groups of Algae

Group	Common name	Morphology	Photopigments	Carbon reserves	Cell wall
Chlorophytes	Green algae	Unicellular or macroscopic	Chlorophyll *a* and *b*	Starch (α-1,4-glucan), sucrose	Cellulose
Euglenophytes	Euglenoids	Unicellular	Chlorophyll *a* and *b*	Paramylon (β-1,2-glucan)	None
Chysophytes	Golden-brown algae	Unicellular	Chlorophyll *a*, *c*, and *e*	Lipids	Silicate
Rhodophytes	Red algae	Unicellular or macroscopic	Chlorophyll *a* and *d*, phycocyanin, phycoeurythrin	Floridean starch (α-1,4-glucan and α-1,6-glucan), fluoridoside (glycerol-galactoside)	Cellulose
Phaeophytes	Brown algae	Macroscopic	Chlorophyll *a* and *c*, xanthophylls	Laminarin (β-1,3-glucan), mannitol	Cellulose
Pyrrophytes	Dinoflagellates	Unicellular	Chlorophyll *a* and *c*	Starch (α-1,4-glucan)	Cellulose

to the higher plants, but the Crysophytes and Phaeophytes are more closely related to the oomycetes, and the Pyrrophytes are more closely related to the ciliated protozoa.

The major ecological functions of the microalgae are reasonably well understood. Algae are important primary producers in most aquatic environments, including sediments exposed by tidal action, where they can contribute to the formation of elaborate and highly active microbial mats. The algae contribute an estimated 30–50% of global primary production and are perhaps most significant in offshore marine waters in which fixed carbon is in short supply. In coastal marine systems, a quite different role of algae is also observed. Many marine microalgae produce potent toxins. Toxin-producing dinoflagellates commonly belonging to the genera *Gymnodinium* and *Gonyaulax* can form massive blooms (red tides) resulting in severe losses to fisheries. Recently, the toxin-producing "ambush predator" dinoflagellate, *Pfiesteria piscicida,* was identified as an important pathogen responsible for fish kills along the Atlantic and Gulf coasts of North America. This organism has a very complex life cycle, including 24 distinct stages, and its toxin has potent activity against a variety of animals, including humans. Toxic algal blooms are increasing in frequency and have been linked to sewage and agricultural runoff into coastal marine waters and to transport of dinoflagellate cysts in ship ballast water. Such environmental impacts may acquaint us with other, once obscure taxa and reveal additional capacities in better known organisms.

The majority of algal diversity remains to be detected and described, and it is very likely that numerous new organisms will be recognized in the near future. This may be particularly true of bloom-producing species. However, the major algal phylogenetic lineages have probably been identified and important improvements in methods for analyzing photopigment profiles and algal morphologies should facilitate recognition of new microalgae in field samples. Much work remains, particularly in terms of detailed phylogenetic analyses, but it is fair to say that we understand the range of important algal niches in the environment and that the major ecotypes of algae have been identified.

VI. PROTOZOAL DIVERSITY

Like the algae, the protozoa are a diverse collection of eukaryotic microorganisms defined on the basis of a small set of phenotypic traits. The protozoa lack cell walls and pigmentation, and most are motile. The absence of photopigments separates these organisms taxonomically from the algae (Euglenophytes and Pyrrophytes are currently placed in the algae) and motility distinguishes them from the true fungi. The slime molds are motile, phylogenetically related to the protozoa, and probably should be counted among them. The protozoa are very widely distributed, inhabiting all aquatic environments in which temperatures are above freezing but below about 60°C as well as soils and the digestive tracts of many animals. The major ecological function of the protozoa is as primary consumers of other microorganisms, and they are important predators on bacteria, small microalgae, and each other. Protozoa can also take up high-molecular-weight organic solutes and colloids by means of pinocytosis and low-molecular-weight organic molecules by simple diffusion. Some protozoa grow primarily as saptrotrophs, whereas others are important parasites of higher organisms. About 40,000 known species of protozoa have been documented, but this may be less than 25% of the total.

The major taxonomic groups of protozoa are the Mastigophora, the Sarcodina, the Ciliophora, and the Sporozoa. The Mastigophora are motile by means of flagella and are commonly known as flagellates. They are closely allied with the Euglenophytes, which are capable of purely heterotrophic growth and can lose their chloroplasts if maintained in the dark. Important flagellate parasites of higher organisms include the trypanosomes, such as *Trypanosoma gambiense,* the organism responsible for African sleeping sickness. The Sarcodina include the amoebas, which lack shells, and the foraminifera, which produce calcium carbonate shells during active growth. The Ciliophora are motile by means of cilia and feed primarily on particulate materials and microbial cells. Few ciliates are parasitic on higher organisms. In contrast, the Sporozoa are all obligate parasites. This large group is characterized by the lack of motile adult stages and by absorption of organic solutes as their primary means of obtaining nutrition. The Sporozoa include the plasmodia, which are the pathogens responsible for malaria.

As is the case for the algae, the protozoa are not monophyletic. Phylogenetic analysis has revealed that the Ciliophora and Pyrrophytes are more closely related to each other than to the other protozoal or algal groups. Mastigophora and Euglenophytes are also more closely related to each other than to other groups. The Sporozoa are very diverse and include some lineages that are highly divergent from all other known eukaryotic lines of descent. Data from more protozoa, particularly among the Sporozoa, will be required to fully evaluate the several known phylogenetic lineages and to properly place newly discovered lineages. Such data will also be very helpful in eliminating synonymous species from the published literature.

Considerations of the ecological functions of the protozoa, as well as the absence of solid evidence for specific geographic distributions of protozoal species, imply fairly restricted species diversity for these organisms (Finlay *et al.,* 1998). The best studied large protozoal group in terms of diversity is the Ciliophora native to marine and freshwater sediments. Only about 3000 species are known, but based on the appearance of certain species in all suitable habitats these may represent the majority of extant species. In other environments even ciliate diversity is poorly characterized. Foissner (1997) estimated that 70–80% of soil ciliates are unknown and that global diversity of these organisms is in the range of 1300–2000 species. Ciliates in the digestive tracts of animals are also very poorly characterized and may represent a substantial pool of undocumented protozoal diversity. Certainly the gut flora of an animal species would be considered to reflect the habitat, population structure, food source(s), and digestive tract architecture of that animal and consequently be somewhat characteristic of that species. The diversity of all other groups of protozoa is much more poorly characterized than that of the ciliates, and estimates of 200,000 protozoal species may not be excessive.

Although the taxonomic characterization of protozoa remains difficult and time-consuming, molecular biological methods for phylogenetic analysis should greatly facilitate identification and description of new species. Given the ubiquity of some protozoal species across suitable habitats worldwide (Finlay *et al.,* 1998), characterization of protozoal diversity in certain habitats may not be as difficult as currently thought. The largest underexplored reservoir of protozoal diversity appears to be animals, particularly arthropods, in which important protozoal digestive tract flora can be abundant and are poorly characterized. This presents a particular problem in the case of endangered animal species, whose flora may also be lost in the event of extinction.

VII. VIRAL DIVERSITY

No matter whether it is more correct to consider viruses as "living organisms" or as renegade genome fragments, the viruses have very important impacts on other types of organisms and should not be neglected in discussions of microbial diversity. These obligately parasitic entities have no independent physiological activities and require a suitable host for reproduction, which damages or destroys host cells. The importance and success of viruses as parasites in higher organisms is well-known. Numerous illnesses in animals and plants have viral origins and the rapidity with which viruses spread through a susceptible host population can be alarming. The Influenza pandemic of 1918 was particularly noteworthy for its rapid propagation and lethality. Identification of such pathogenic viral strains and development of vaccines against them are foci of major national and international efforts.

In addition to the damage done by out-of-control viral reproduction, many viruses can affect their hosts more subtly through genetic modification. Temperate viruses can insert their genome into that of an appropriate host and be propagated as a stable provirus within the host for many generations. This interaction, called lysogeny in bacteria–virus systems, is a form of recombination and can confer new properties to the host. Important examples of this type of recombination-driven change in host phenotype are provided by the pathogenicity islands of *Vibrio cholerae,* enterohemorrhagic *Escherichia coli* strains, and some other enteric pathogens. These bacterial species and strains display pathogenic traits, such as adhesion to specific host receptors and toxin production, only if the correct provirus (prophage) is present. The genetic modifications brought about by lysogeny can not only dictate whether or not a host bacterium is pathogenic but also control the severity of infection by the recombinant pathogen. Many organisms, particularly microorganisms, harbor proviruses or provirus-like sequences in their genomes and the importance of most of these to the host organisms is unknown. Although we have been chiefly interested in viruses that infect humans, our domesticated animals, and crop plants, all types of organisms have viral parasites. The impacts of these less studied viruses on their host populations in nature are only beginning to be unraveled. Recent findings indicate that viruses are active participants in microbial food web processes and may exert some control over microbial population dynamics in nature.

Perhaps the diversity of no other microbial group is as poorly characterized and as widely underestimated as that of the viruses. This is partly due to difficulties in identifying and cultivating the host organisms necessary for propagating the viruses found in the environment. Viral taxonomy, which was long dependent on phenotypic characteristics, such as host range, symptoms of infection, morphology of the viral particle, and the type of nucleic acid (single-stranded DNA or RNA or double-stranded DNA or RNA) composing the genome, now also employs nucleotide sequence analysis of viral genomes. However, most types of viruses remain to be identified and no hosts for these organisms are known. There are about 6500 described species of viruses, including 2500 animal viruses, 2000 plant viruses, and 2000 bacterial viruses. However, the total number of virus species has been estimated to be 500,000, and this may be a significant underestimate of true viral diversity.

Recent findings, particularly in aquatic systems, show that viruses are extremely numerous in nature (Paul *et al.*, 1996). Viral numbers are typically on the order of 10^9 to 10^{10} virions per liter of water and estimates of viral production in nature are very high. From 1 to 4% of bacteria in freshwater and marine systems contain mature viruses and, since complete viruses are only visible in cells during the final 10%

of the viral replication cycle, actual frequencies of infected cells could be much higher. In addition, as many as 4% of bacteria in coastal marine waters may contain stable proviruses and much higher proportions of culturable aquatic species are known to be lysogenic. Viruses are present in all the microbial environments that have been examined to date and with numbers as high as those reported for aquatic systems, and because of the diversity of morphologies observed it is clear that viruses represent a vastly underestimated pool of microbial diversity. This is particularly apparent when we consider the fact that more than 70 different viruses, representing six major viral groups, have been isolated using a single marine bacterium as the host strain (Paul *et al.*, 1996). If this pattern holds true across all types of organisms, and there is no compelling reason to think that it may not, viral diversity is much greater than currently appreciated. In addition, many viruses are quite mutable, adding genetic variation among related strains to the already large task of cataloging viral diversity. It is likely that each viral isolate represents only one of a radiation of closely related but distinguishable genotypes. This additional diversity within strains is not inconsequential, as is clear from the differences in pathogenicity among different variants of HIV and influenza A.

Although the occasional human pathogen, such as the Ebola virus, introduces itself through grim displays of lethality, most viruses will only be discovered and described through painstaking, systematic efforts. It is clear that this will require a major commitment of time and resources, but the foundations for describing natural viral diversity have been laid and this effort can be now undertaken with some expectation of long-term success. A key consideration is the types of microbial, plant, and animal hosts that should be emphasized in these studies although, as Ebola illustrates, the host range for a given virus may prove very elusive.

VIII. TOWARD A FUNCTIONAL SURVEY OF MICROBIAL DIVERSITY

Currently, most of the earth's microbial diversity is completely unknown and current estimates of the numbers of species extant are widely considered to be significant underestimates. The microbiota likely dominate global biodiversity, and efforts to characterize microbial diversity and to determine the interplay between this diversity and ecosystem function have recently accelerated. Given the ever-expanding rate of global change brought about by human activities, two extremely important organizing foci for diversity assessment can be identified.

First, habitats that are currently threatened by human activities are clearly of immediate interest to diversity survey efforts. Most tropical regions of the world are either severely human impacted or threatened. This is of particular interest since these regions are home to an enormous diversity of plant and animal species. Each of these higher organism species is host to numerous commensal, mutualist, and parasitic microorganisms, and when a host organism faces extinction so too may some members of its microflora. Bioprospecting in the tropics for microorganisms that produce useful products has begun, but many novel species may be lost before they can be documented and preserved. Consideration of habitat loss and the potential for host organism extinction provide spatial and temporal frames of reference for microbial diversity survey efforts.

Second, some microbial activities, such as nitrogen fixation, primary production, and methane oxidation, are clearly essential to proper ecosystem function and are restricted to specific functional groups of microorganisms. The diversity of a given functional group may be limited, as seems to be the case for methane-oxidizing bacteria and for nitrogen-fixing bacteria in some habitats. Low diversity may also be an important consideration for microorganisms that grow in symbiotic relationships with higher organisms in which the degree of specificity of the microorganisms for their hosts can be great. If the active species within an essential microbial functional group are lost due to habitat modification or local extinction of a higher organism host species, key local ecosystem functions could deteriorate. Evaluation of environmentally sensitive, low-diversity functional groups provides a rationale for exploring connections between microbial diversity and ecosystem function and an ecological focus for diversity survey efforts.

TABLE III
Web Sites Providing Taxonomic or Phylogenetic Information on Microorganisms[a]

Web sites for bacterial nomenclature
  http://www-sv.cict.fr/bacterio/ (J. P. Euzeby)
  http://www.dsmz.de/bactnom/bactname.htm (Web site of the DSMZ, M. Kracht, database administrator)
Index of fungi (not free of charge)
  http://www.cabi.org/catalog/taxonomy/indfungi.htm (CABI)
Index of viruses
  http://life.anu.edu.au/viruses/lctv/index.html
Other sites with links to culture collections, databases, and phylogenetic analysis
  http://www.cme.msu.edu/RDP (Ribosome Database Project, B. L. Maidak, curator)
  http://wdcm.nig.ac.jp/ (WDCM, World Data Centre for Microorganisms)
  http://ftp.ccug.gu.se/ (CCUG)

[a] Information provided by Dr. Manfred Kracht, Deutsche Sammlung von Mikroorganismen und Zellkulturen GmbH, Braunschweig, Germany.

Clearly, much work will have to be done to support a meaningful assessment of microbial diversity. However, the technology necessary to pursue this effort exists, and some important work has already been done. Compiling this information and systematic elimination of synonymous listings will be greatly facilitated by use of the World Wide Web, and several relevant websites have already been constructed (Table III). The future will certainly produce a much greater understanding of the extent and importance of microbial diversity and most likely even more questions.

See Also the Following Articles

Archaea • Ecology, Microbial • Fungi, Filamentous • Origin of Life • Viruses, Overview

Bibliography

Amann, R. I., Ludwig, W., and Schleifer, K.-H. (1995). Phylogenetic identification and *in situ* detection of individual microbial cells without cultivation. *Microbiol. Rev.* **59**, 143–169.

Finlay, B. J., Esteban, G. F., and Fenchel, T. (1998). Protozoan diversity; Converging estimates of the global number of free-living ciliate species. *Protist* **149**, 29–37.

Foissner, W. (1997). Global soil ciliate (Protozoa, ciliophora) diversity. A probability-based approach using large sample collections from Africa, Australia, and Antarctica. *Biodiversity Conservation* 6, 1627–1638.

Fuhrman, J. A., and Davis, A. A. (1997). Widespread Archaea and novel Bacteria from the deep sea as shown by 16S rRNA gene sequences. *Marine Ecol. Progr. Ser.* 150, 275–285.

Hammond, P. M. (1995). Described and estimated species numbers: An objective assessment of current knowledge. *In* "Microbial Diversity and Ecosystem Function" (D. Allsopp, R. R. Colwell, and D. L. Hawksworth, Eds.), pp. 29–71. Cambridge Univ. Press, Cambridge, UK.

Hawksworth, D. B., and Rossman, A. Y. (1997). Where are all the undescribed fungi? *Phytopathology* 87, 888–891

Hugenholz, P. B., Goebel, B. M., and Pace, N. R. (1998). Impact of culture-independent studies on the emerging phylogenetic view of bacterial diversity. *J. Bacteriol.* 180, 4765–4774.

Pace, N. R. (1997). A molecular view of microbial diversity and the biosphere. *Science* 276, 734–740.

Paul, J. H., Kellogg, C. A., and Jiang, S. C. (1996). Viruses and DNA in marine environments. *In* "Microbial Diversity in Time and Space" (R. R. Colwell, U. Simidu, and K. Ohwada, Eds.), pp. 115–124. Plenum, New York.

Williams, D. M., and Embley, T. M. (1996). Microbial diversity: Domains and kingdoms. *Annu. Rev. Ecol. Systematics* 27, 569–595.

Woese, C. R. (1994). There must be a prokaryote somewhere: Microbiology's search for itself. *Microbiol. Rev.* 58, 1–9.

DNA Repair

Lawrence Grossman

The Johns Hopkins University

GLOSSARY

endonuclease   Nuclease that hydrolyzes internal phosphodiester bonds.

excision   Removal of damaged nucleotides from incised nucleic acids.

exonuclease   Nuclease that hydrolyzes terminal phosphodiester bonds.

glycosylase   Enzymes that hydrolyze N-glycosyl bonds linking purines and pyrimidines to carbohydrate components of nucleic acids.

incision   Endonucleolytic break in damaged nucleic acids.

ligation   Phosphodiester bond formation as the final stage in repair.

nuclease   Enzyme that hydrolyzes in the internucleotide phosphodiester bonds in nucleic acids.

resynthesis   Polymerization of nucleotides into excised regions of damaged nucleic acids.

transcription   The synthesis of messenger and RNA from template DNA.

THE ABILITY OF CELLS   to survive hostile environments is due, in part, to surveillance systems which recognize damaged sites in DNA and are capable of either reversing the damage or of removing damaged bases or nucleotides, generating sites which lead to a cascade of events restoring DNA to its original structural and biological integrity.

Both endogenous and exogenous environmental agents can damage DNA. Many repair systems are regulated by the stressful effects of such damage, affecting the levels of responsible enzymes, or by modifying their specificity. Repair enzymes appear to be the most highly conserved proteins showing their important role throughout evolution. The enzyme systems can directly reverse the damage to form the normal purine or pyrimidine bases or the modified bases can be removed together with surrounding bases through a succession of events involving nucleases, DNA-polymerizing enzymes, and polynucleotide ligases which assist in restoring the biological and genetic integrity to DNA.

I. DAMAGE

As a target for damage, DNA possesses a multitude of sites which differ in their receptiveness to modification. On a stereochemical level, nucleotides in the major groove are more receptive to modification than those in the minor groove, the termini of DNA chains expose reactive groups, and some atoms of a purine or pyrimidine are more susceptible than others. As a consequence, the structure of DNA represents a heterogeneous target in which certain nucleotide sequences also contribute to the susceptibility of DNA to genotoxic agents.

A. Endogenous Damage

Even at physiological pH's and temperatures in the absence of extraneous agents, the primary structure of DNA undergoes alterations (Table I). Many specific reactions directly influence the informational

TABLE I
Hours Needed for a Single Event at pH 7.4 at 37°C

Event	Single-stranded DNA (2×10^6 base pairs)	Double-stranded DNA (2×10^6 base pairs)	Molecular events per genome per day
Depurination	2.5	10	24×10^3
Depyrimidination	50.8	200	12×10^2
Deamination of C	2.8	700	3.4×10^2
Deamination of A	140.0	?	?

content as well as the integrity of DNA. Although the rate constants for many reactions are inherently low, it is because of the enormous size of DNA and its persistence in cellular life cycles that the accumulation of these changes can have significant long-term effects.

1. Deamination

The hydrolytic conversion of adenine to hypoxanthine, guanine to xanthine, and cytosine to uracil-containing nucleotides is of sufficient magnitude to affect the informational content of DNA (Fig. 1).

2. Depurination

The glycosylic bonds linking guanine in nucleotides are especially sensitive to hydrolysis—more than the adenine and pyrimidine glycosylic links. The result is that apurinic (or apyrimidinic sites) are

recognized by surveillance systems and as a consequence are repaired.

3. Mismatched Bases

During the course of DNA replication there are non-complementary nucleotides which are incorrectly incorporated into DNA and manage to escape the editing functions of the DNA polymerases. The proper strand and the mismatched base are recognized and repaired.

4. Metabolic Damage

When thymine incorporation into DNA is limited either through restricted precursor UTP availability or through inhibition of the thymidylate synthetase system, dUTP is utilized as a substitute for thymidine triphosphate. The presence of uracil is identified as a damaged site and acted on by repair processes.

5. Oxygen Damage

The production of oxygen, superoxide, or hydroxyl radicals as a metabolic consequence, as well as the oxidative reactions at inflammatory sites, causes sugar destruction which eventually leads to strand breakage.

B. Exogenous Damage

The concept of DNA repair in biological systems arose from studies by photobiologists and radiobiologists who studied the viability and mutagenicity in biological systems exposed to either ionizing or ultraviolet (UV) irradiation. Target theories, derived from

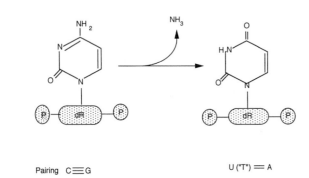

Fig. 1. Deamination reactions have mutagenic consequences because the deaminated bases cause false recognition.

the random statistical nature of photon bombardment, led to the identification of DNA as the primary target for the cytotoxicity and mutagenicity of ultraviolet light. In addition, most of the structural and regulatory genes controlling DNA repair in *Escherichia coli* were identified, facilitating the isolation and molecular characterization of the relevant enzymes and proteins.

1. Ionizing Radiation

The primary cellular effect of ionizing radiation (Fig. 2) is the radiolysis of water which mainly generates hydroxyl radicals (HO $\cdot$). The hydroxyl radical is capable of abstracting protons from the C4′ position of the deoxyribose moiety of DNA, thereby labilizing the phosphodiester bonds and generating single- and double-strand breaks. The pyrimidine bases are also subject to HO $\cdot$ addition reactions.

2. Alkylation Damage

This modification occurs on purine ring nitrogens (cytotoxic adducts), at the O^6 position on guanine,

and at the O4 positions of the pyrimidines (mutagenic lesions) and the oxygen residues of the phosphodiester bonds of the DNA backbone (biologically silent) (Fig. 3). Alkylating agents are environmentally pervasive, arising indirectly from many foodstuffs and from automobile exhaust in which internal combustion of atmospheric nitrogen results in the formation of nitrate and nitrites.

3. Bulky Adducts

Large, bulky polycyclic aromatic hydrocarbon modification occurs primarily on the N^2, N^7, and C^8 positions of guanine, invariably from the metabolic activation of these large hydrophobic uncharged macromolecules to their epoxide analogs (Fig. 4). The major source of these substances is from the combustion of tobacco, petroleum products, and foodstuffs.

4. Ultraviolet Irradiation

Most of the UV photoproducts are chemically stable; their recognition provided direct biochemical

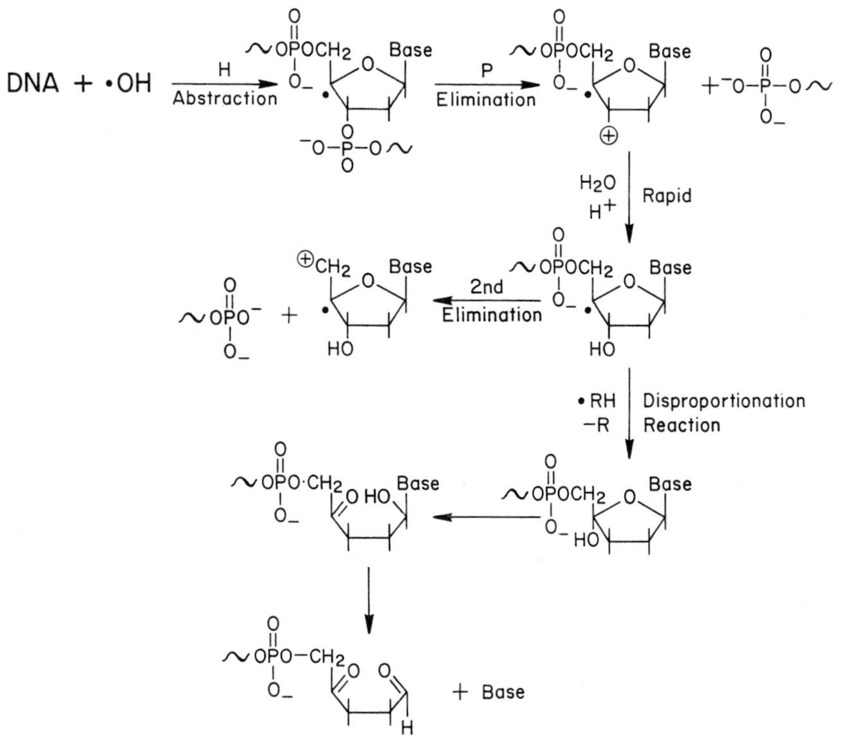

Fig. 2. DNA backbone breakage by ionizing radiation.

$$NO_2^- + H^+ \rightleftharpoons HNO_2 \qquad \text{(Gastric Reaction)}$$

$$HNO_2 + R_1 \cdot R_2 - NH \cdot HCl \underset{HCl}{\overset{H_2O}{\rightleftharpoons}} R_1 \cdot R_2 - N - N = O \qquad \text{(Fischer–Hepp)}$$

$$N^I - \text{Nitrosamine}$$

DISTRIBUTION OF ALKYL PRODUCTS FROM
IN VIVO METHYLATION OF MAMMALIAN DNA'S

ALKYLATING AGENT	GUANINE			ADENINE		
	3	O⁶	7	I	3	7
$\overset{O}{\underset{O}{R}} \cdot O - \overset{O}{\underset{O}{S}} - R$ (MMS, EtMs)	0.6	< 0.3	82	0.5	7	0.8
$O = N - \overset{R}{\underset{O=C-NH_2}{N}}$	1.5	5	64	0.1	5	1.3

Fig. 3. The formation of alkylation sites in DNA exposed to nitrites.

evidence for DNA repair. The major photoproducts are 5,6-cyclobutane dimers of neighboring pyrimidines (intrastrand dimers), 6,4-pyrimidine-pyrimidone dimers (6-4 adducts), and 5,6-water-addition products of cytosine (cytosine hydrates).

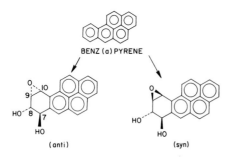

BENZ (a) PYRENE

(anti)   (syn)

BENZ (a) PYRENE DIOLEPOXIDES

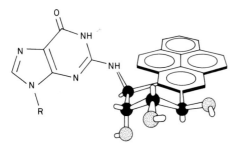

Fig. 4. Bulky adducts formed in DNA exposed to benz(a)-pyrene.

II. DIRECT REMOVAL MECHANISMS

The simplest repair mechanisms involve the direct photoreversal of pyrimidine dimers to their normal homologs and the removal of O-alkyl groups from the O^6-methylguanine and from the phosphotriester backbone as a consequence of alkylation damage to DNA.

A. Photolyases (Photoreversal)

The direct reversal of pyrimidine dimers to the monomeric pyrimidines is the simplest mechanism, and it is chronologically the first mechanism described for the repair of photochemically damaged DNA (Fig. 5). It is a unique mechanism characterized by a requirement for visible light as the sole source of energy for breaking two carbon–carbon bonds.

The enzyme protein has two associated light-absorbing molecules (chromophores) which can form an active light-dependent enzyme. One of the chromophores is $FADH_2$ and the other is either a pterin or a deazaflavin able to absorb the effective wavelengths of 365–400 nm required for photoreactivation of pyrimidine dimers. It is suggested that photoreversal involves energy transfer from the pterin molecule to $FADH_2$ with electron transfer to the pyrimidine dimer resulting in nonsynchronous cleavage of the C^5 and C^6 cyclobutane bonds. Enzymes that carry out photoreactivation have been identified in both prokaryotes and eukaryotes.

B. Alkyl Group Removal (Methyl Transferases)

Bacterial cells pretreated with less than cytotoxic or genotoxic levels of alkylating agents before lethal or mutagenic doses are more resistant (Fig. 6). This is an adaptive phenomenon with anti-mutagenic and anti-cytotoxic significance. During this adaptive period, a 39-kDa Ada protein is synthesized that specifically removes a methyl group from a phosphotriester bond and from an O^6-methyl group of guanine (or from O^4-methyl thymine). The O^6-methyl group of guanine is not liberated as free O^6-methyl guanine during this process, but it is transferred directly from the alkylated DNA to this

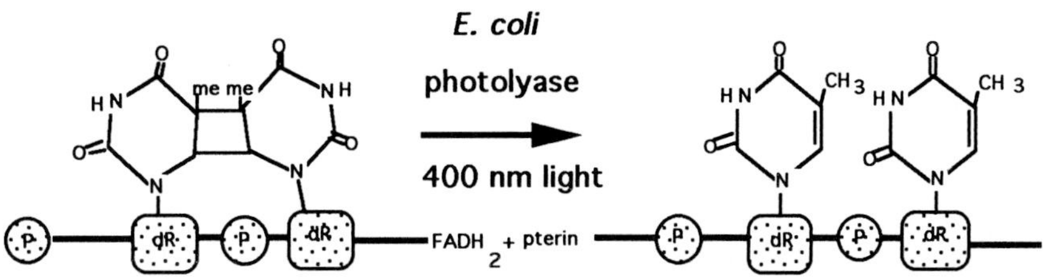

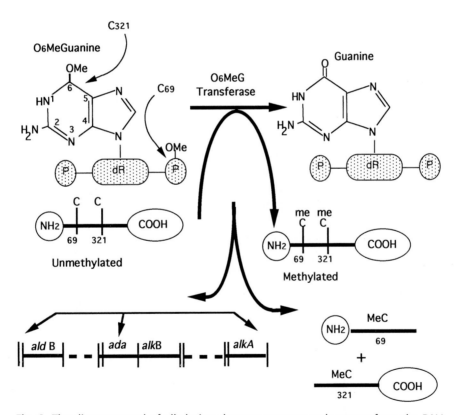

Fig. 5. The direct photoreversal of pyrimidine dimers in the presence of visible light.

protein; the Ada protein (methyl transferase) and an unmodified guanine are simultaneously generated. These alkyl groups specifically methylate cysteine 69 and cysteine 321, respectively, in the protein.

The methyl transferase is used stoichiometrically in the process (does not turnover) and is permanently inactivated in the process. Nascent enzyme, however, is generated because the mono- or dimethylated transferase activates transcription of its own "regulon" which includes, in addition to the *ada* gene, the *alk* B gene of undefined activity and the *alk* A gene which sponsors a DNA glycosylase. The latter enzyme acts on 3-methyl adenine, 3-methyl guanine, O^2-methyl cytosine, and O^2-methyl thymine. The

Fig. 6. The direct reversal of alkylation damage removes such groups from the DNA backbone and the O^6 position of guanine. Such alkyl groups are transferred directly to specific cysteine residues on the transferase, the levels of which are influenced adaptively by the levels of the alkylating agents. The methylation of the transferase inactivates the enzyme which is used up stoichiometrically in the reaction. The alkylated transferase acts as a positive transcriptive signal turning on the synthesis of unique mRNA. Regulation of transferase levels may be influenced by a unique protease.

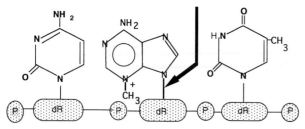

Fig. 7. DNA glycosylases hydrolyze the N-glycosyl bond between damaged bases and deoxyribose generating an AP (apyrimidinic or apurinic) site (arrow).

methylated Ada protein can specifically bind to the operator of the *ada* gene acting as a positive regulator. Down regulation may be controlled by proteases acting at two hinge sites in the Ada protein.

III. BASE EXCISION REPAIR

A. Base Excision Repair by Glycosylases and Apyrimidinic or Apurinic Endonucleases

Bases modified by deamination can be repaired by a group of enzymes called DNA glycosylases, which specifically hydrolyze the N-glycosyl bond of that base and the deoxyribose of the DNA backbone generating an apyrimidinic or an apurinic site (AP site) (Fig. 7). These are small, highly specific enzymes which require no cofactor for functioning. They are the most highly conserved proteins, attesting to the evolutionary unity both structurally and mechanistically from bacteria to man. As a consequence of DNA glycosylase action, the AP sites generated in the DNA are acted on by a phosphodiesterase (Fig. 8) specific

for such sites which can nick the DNA 5′ and/or 3′ to such damaged sites. If there is a sequential action of a 5′ acting and a 3′ acting AP endonuclease, the AP site is excised generating a gap in the DNA strand.

B. Glycosylase-Associated AP Endonucleases

An enzyme from bacteria and phage-infected bacteria, encoded in the latter case by a single gene (*den*), hydrolyzes the N-glycosyl bond of the 5′ thymine moiety of a pyrimidine dimer followed by hydrolysis of the phosphodiester bond between the two thymine residues of the dimer (Fig. 9). This enzyme,

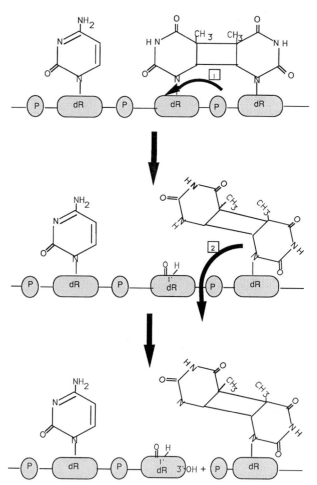

Fig. 9. The same enzyme that can hydrolyze the N-glycosyl residue of a damaged nucleotide also hydrolyzes the phosphodiester bond linking the AP site generated in the first N-glycosylase reaction.

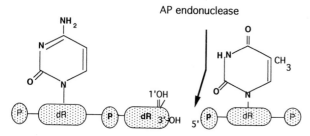

Fig. 8. Endonucleases recognize AP sites and hydrolyze the phosphodiester bonds 3′, 5′, or both sides of the deoxyribose moiety in damaged DNA.

referred to as the pyrimidine dimer DNA glycosylase, is found in *Micrococcus luteus* and phage T4-infected *E. coli*. This small uncomplicated enzyme does not require cofactors and is presumed to act by a series of linked β elimination reactions. An enzyme behaving in a similar glycosylase–endonuclease fashion but acting on the radiolysis product of thymine is thymine glycol, which has been isolated from *E. coli* and is referred to as endonuclease III.

IV. NUCLEOTIDE EXCISION REPAIR

The ideal repair system is one that is somewhat indiscriminate and which can respond to virtually any kind of damage. Such a repair system has been characterized in *E. coli* in which it consists of at least six gene products of the *uvr* system. This ensemble of proteins consists of the UvrA protein that binds as a dimer to DNA in the presence of ATP, followed by the UvrB protein which cannot bind DNA by itself. Translocation of the $UvrA_2B$ complex from initial undamaged DNA sites to damaged sites is driven by a cryptic ATPase associated with UvrB which is activated by the formation of the $UvrA_2B$-undamaged DNA complex. This complex is now poised for endonucleolytic activity catalyzed by the interaction of the $UvrA_2B$-damaged DNA complex with UvrC to generate two nicks in the DNA seven nucleotides 5′ to the damaged site and the three or four nucleotides 3′ to the same site. These sites of breakage are invariant regardless of the nature of the damage. In the presence of the UvrD (helicase III), DNA polymerase I, and substrate deoxynucleoside triphosphates, the damaged fragment is released and this is accompanied by the turnover of the UvrA, UvrB, and UvrC proteins. The continuity of the DNA helix is maintained based on the sequence of the opposite strand. The final integrity of the interrupted strands is restored by the action of DNA polymerase I, which copies the other strand, and by polynucleotide ligase, which seals the gap (Fig. 10). The levels of the Uvr proteins are regulated in *E. coli* by an "SOS" regulon monitoring many genes, including *uvr*A, *uvr*B, possibly *uvr*C, and *uvr*D, as part of the excision repair system. It also includes the regulators of the SOS system, the uvrD and recA proteins; cell division

genes *sul*A and *sul*B; recombination genes *rec*A, *rec*N, *rec*Q *uvr*D, and *ruv*; mutagenic by-pass mechanisms (*umu*DC and *rec*A); damage-inducible genes; and the lysogenic phage λ. The LexA protein negatively regulates these genes as a repressor by binding to unique operator regions. When the DNA is damaged (e.g., by UV light), a signal in the form of a DNA repair intermediate induces the synthesis of the RecA protein. When induced, the RecA protein acts as a protease assisting the LexA protein to degrade itself, activating its own synthesis and that of the RecA protein as well as approximately 20 other genes. These genes permit the survival of the cell in the face of life-threatening environmental damages. Upon repair of the damaged DNA, the level of the signal subsides, reducing the level of RecA and thus stabilizing the integrity of the intact LexA protein and its repressive properties on all the other genes (Fig. 11). Then the cell returns to its normal state.

V. TRANSCRIPTION-COUPLED NUCLEOTIDE EXCISION REPAIR

It appears that the structural and biological specificty associated with transcriptional processes limit DNA repair to those damaged regions of the chromosome undergoing transcription and that damage in those quiescent regions is persistent. Within expressed genes the repair process is selective for the transcribed DNA strand for damage such as pyrimidine dimers, and this "coupling" to transcription has been shown in *E. coli*. As a consequence, DNA repair occurs preferentially in active transcribed genes. Preferential repair occurs in the transcribed strand in actively expressed genes. Nucleotide excision repair (NER) differs in the two separate DNA strands of the lactose operon of UV-irradiated *E. coli*. The level of repair examined in the uninduced condition is about 50% after 20 min in both strands. As a consequence of a 436-fold induction of β-galactosidase, most of the dimers (70%) are removed from the transcribed strand of the induced operon within 5 min, whereas the extent of repair in nontranscribed strands is similar to that of the uninduced condition. This selective removal of pyrimidine dimers from

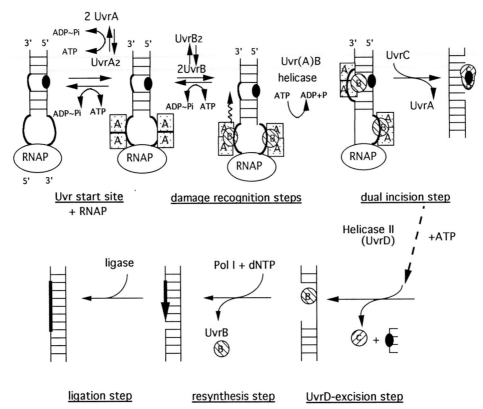

Fig. 10. Nucleotide excision reactions. In this multiprotein enzyme system the UvrABC proteins catalyze a dual-incision reaction seven nucleotides 5′ and three or four nucleotides 3′ to a damaged site. The UvrA protein, as a dimer, binds to undamaged sites initially and in the presence of UvrB, whose cryptic ATPase is manifested in the presence of UvrA providing the energy necessary for translocation to a damaged site. This pre-incision complex interacts with UvrC, leading to the dual-incision reaction. The incised DNA–UvrABC does not turnover and requires the coordinated participation of the UvrD and DNA polymerase reactions for damaged fragment release and turnover of the UvrABC proteins. Ligation, the final reaction, restores to integrity the DNA stands.

the transcribed strand of a gene is abolished in the absence of significant levels of transcription.

As shown in Figs. 12 and 13, the RNAP when binding to its promoter site generates a defined distortion 3′ (downstream) to the RNAP binding site which provides a "landing site" for the UvrA$_2$B complex. Strand specificity is dictated by the 5′ → 3′ directionality of the UvrA$_2$B helicase which can translocate only on the non-transcribed strand because RNAP interferes with the directionality on the lower strand. Nicking occurs only on the strand opposite to the strand which the UvrA$_2$B endonuclease binds; hence, it is the transcribed strand which is initially repaired in this model.

A. Effect of Pyrimidine Dimers on Transcription and Effect of RNAP on Repair

T–T photodimers in the template strand constitute an absolute block for transcription, whereas those in the complementary strand have no effect. Irradiation of cells with UV results in truncated transcripts because the pyrimidine dimers become a "stop" site. In the absence of ribonucleoside triphosphates, promoter-bound RNAP does not translocate and, hence, has no effect on repair on a T–T photo-dimer downstream from the transcriptional initiation site no matter whether the photodimer

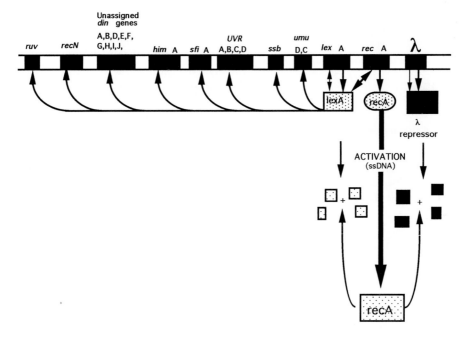

Fig. 11. Regulation of the nucleotide excision pathway by the "SOS" system. The *lex* A and phage λ repressors negatively control a multitude of genes which are turned on when bacterial cells are damaged. This leads to the over-production of the recA protein which assists in the proteolysis of the LexA and phage λ proteins, thereby derepressing the controlled gene systems. When DNA is fully repaired the level of recA declines, restoring the "SOS" system to negative control.

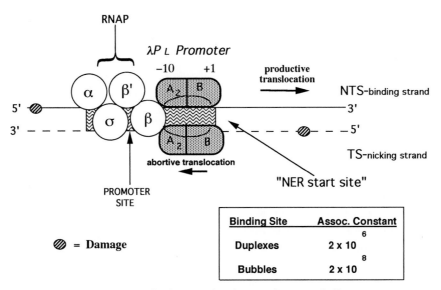

Fig. 12. Strand selectivity by the *E. coli* UvrA$_2$B helicase.

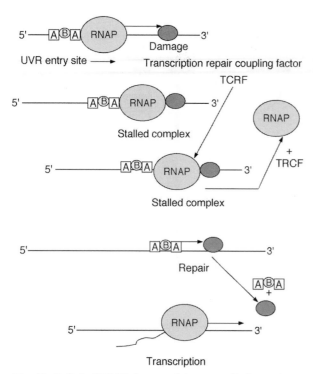

Fig. 13. Relief of RNAP-damaged sites as stalled complexes by the TRCF when RNA precedes the Uvr complex.

is in the transcribed or in the nontranscribed strand.

B. Transcription Repair Coupling Factor

Transcription repair coupling (TRC) is achieved through the action of the transcription repair coupling factor (TRCF). TRCF is the product of the *mfd* gene (*mutation frequency decline*), which maps at 25 min on the *E. coli* chromosome. The cloned *mfd* gene is translated into a 1148-amino acid protein of ~130 kDa. The Mfd protein can nonspecifically bind dsDNA (and less efficiently ssDNA) in an ATP-binding-dependent manner, with the ATP hydrolysis promoting its dissociation. The amino acid sequence of Mfd reveals motifs which are characteristic of many DNA and RNA helicases. However, *in vitro* purified Mfd does not show either DNA or RNA helicase activity. N-terminal 1–378 residues of Mfd, have a 140-amino acid region of homology with UvrB and bind UvrA protein.

In vitro transcription is inhibited by NER damage in a transcribed strand, whereas it has no effect on the noncoding strand. This inhibition is thought to result from a stalled RNAP at the site of damage. TCRF is able to release the stalled RNAP in an ATP hydrolysis-dependent manner. Moreover, it actually stimulates NER of the transcribed strand, so that it becomes faster than the nontranscribed one. Based on all these observations, it is concluded that TRCF-Mfd carries out preferential repair of the transcribed strand by (i) releasing RNAP stalled at damaged sites

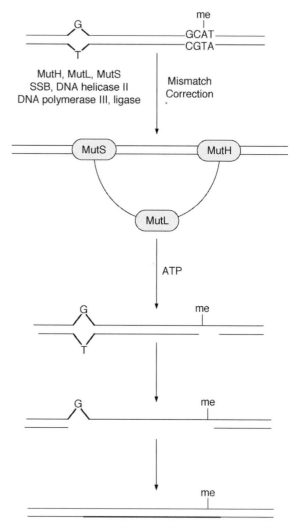

Fig. 14. In the repair of mismatched bases strand distinction can be achieved by the delay in adenine methylation during replication. It is the nascent unmethylated strand which serves as a template for the incision reactions catalyzed by many proteins specifically engaged in mismatch repair processes.

and (ii) recruiting the UvrA$_2$B complex to damaged sites through the high-affinity interaction with UvrA.

VI. MISMATCH CORRECTION

Many mechanisms do not recognize damage but do recognize mispairing errors that occur in all biological systems (Fig. 14). In *E. coli* mismatch correction is controlled by seven mutator genes; *dam* (methyl directed), *mutD*, *mutH mutL*, *mutS*, *mutU*, *uvrD*, and *mutY*. In mismatch correction one of the two strands of the mismatches is corrected to conform with the other strand. Strand selection is one of the intrinsic problems in mismatch repair and the selection is achieved in bacterial systems by adenine methylation, which occurs at d(GATC) sequences. Since such methylation occurs after DNA has replicated, only the template strand of the nascent duplex is methylated. In mismatch repair only the unmethylated strand is repaired, thus retaining the original nucleotide sequence. The MutH, MutL, and MutS proteins appear to be involved in the incision reac-

tion on this strand, with the remainder of the proteins plus DNA polymerase III and polynucleotide ligase participating in the excision–resynthesis reactions.

See Also the Following Articles

Oxidative Stress • *RecA* • SOS Response

Bibliography

Friedberg, E. C. (1984). "DNA Repair." Freeman, New York.

Grossman, L., Caron, P. R., Mazur, S. J., and Oh, E. Y. (1988). "Repair of DNA containing pyrimidine dimers." *FASEB J.* **2**, 2696–2701.

Lindahl, T., Sedgwick, B., Sekiguchi, M., and Nakabeppu, Y. (1988). "Regulation and expression of the adaptive response to alkylating agents". *Annu. Rev. Biochem.* **57**, 133–157.

Modrich, P. (1989). "Methyl-directed DNA mismatch correction." *J. Biol. Chem.* **264**, 6597–6600.

Sancar, A., and Sancar, G. B. (1988). "DNA repair enzymes." *Annu. Rev. Biochem.* **57**, 29–67.

Walker, G. C. (1985). "Inducible DNA repair systems". *Annu. Rev. Biochem.* **54**, 457.

Weiss, B., and Grossman, L. (1987). "Phosphodiesterases involved in DNA repair". *Adv. Enzymol.* **60**, 1–34.

DNA Replication

James A. Hejna and Robb E. Moses

Oregon Health Sciences University

structure by both DNA sequence and topology and requires specific protein–DNA interactions. DNA replication in *E. coli* is bidirectional and symmetrical.

GLOSSARY

chromosome   Package of genes representing part or all of the inherited information of the organism.

exonuclease   Enzyme that degrades DNA from a terminus.

gyrase   Enzyme that introduces supercoiling into the DNA duplex in an adenosine triphosphate-dependent reaction.

helicase   Enzyme that unwinds duplex DNA and requires adenosine triphosphate.

polymerase   Enzyme that synthesizes a nucleic acid polymer.

replication   Act of duplicating the genome of a cell.

replicon   Replicative unit, either part or the whole of the genome; in *Escherichia coli,*the entire genome is considered a replicon.

topoisomerase   An enzyme that alters the topology of DNA, either one strand at a time (type I) or two strands at a time (type II). Gyrase and Topoisomerase IV are type II enzymes.

DNA REPLICATION   in *Escherichia coli* is a carefully regulated process involving multiple components representing more than 20 genes participating in duplication of the genome. The process is divided into distinct phases: initiation, elongation, and termination. The synthesis of a new chromosome involves an array of complex protein assemblies acting in sequential fashion in a carefully regulated and reiterated overall pattern. The scheme for DNA replication is under careful genetic control. The process is localized on the DNA

I. DEVELOPMENT OF THE FIELD

This article will focus primarily on *Escherichia coli* as a model organism, with the assumption that what is true for *E. coli* is generally true for other prokaryotes. In gram-positive bacteria such as *Bacillus subtilis,* this assumption has been largely substantiated. Sequencing of several prokaryotic genomes has revealed homology to many of the genes involved in *E. coli* DNA replication. The development of our understanding of DNA replication in prokaryotes depends on a combination of biochemical and genetic approaches. Using several selection techniques, many laboratories isolated *E. coli* mutants that were conditionally defective (usually temperature sensitive) in DNA replication. This method of identifying genes involved in DNA replication assumed that defects of such genes resulted in the death of the cell. When a large series of mutants was assembled, they fell clearly into two broad categories: those in which DNA replication ceased abruptly following a shift to restrictive conditions and those in which DNA replication ceased slowly. The former class is called fast stop and the latter slow stop. The first category represents cells containing mutations in gene products that are required for the elongation phase of DNA replication, and the latter category contains cells with defects in gene products that are required for the initiation of new rounds of DNA replication.

The identification of temperature-sensitive, *dnats,*

mutants was one requirement for understanding DNA replication. The second requirement was the development of systems that could be biochemically manipulated but that represented all or part of the authentic DNA replication process in *E. coli*. Several successive systems offered increasing advantages. Systems in which permeable cells allowed free access of small molecules to minimally disturbed chromosomes were the earliest. These systems allowed definition of the energy and cofactor requirements for the elongation phase of DNA replication. The limitation was that they did not permit access by macromolecules to the replication apparatus and therefore did not allow complementation of defects in proteins required for DNA replication. Also, such systems did not allow the initiation of new rounds of replication.

The development of lysate replication systems rested on the recognition that the failure to maintain the complex process of DNA replication in early studies was due to dilution of the components, resulting in disassembly of the replication structure and loss of functions required for DNA replication. The concentrated lysate systems depended on the bacterial chromosome, but it was quickly recognized that small bacterial phage chromosomes could be utilized as exogenous templates for DNA replication because they permitted the addition of proteins to allow complementation of defects in DNA replication. Lysates made from mutants defective in a specific step of DNA replication could be used to define the step at which the defect occurred and to identify the protein product complementing the defect. This allowed assignment of protein products to genes. Such systems, however, did not allow the study of initiation of new rounds of DNA synthesis on the host chromosome.

Two theoretical shortcomings of such systems are (i) that such systems might not define all the proteins required for DNA replication by the host and (ii) that such systems might require a protein for replication of the phage DNA not ordinarily required by the host chromosome. The following general point derived from these studies is worth remembering: Although *E. coli* contains numerous proteins that have overlapping or similar enzymatic function, the participation of a protein in the replication process is carefully regulated, reflecting a specific role. The

DNA polymerases are the best example. Each of the three recognized DNA polymerases of *E. coli* (Table I) has similar enzymatic capabilities, but ordinarily only DNA polymerase III catalyzes replication. This restriction of activity can be partially explained on the basis of protein–protein interactions. The complete basis for the regulation is not understood, and the reasons for its being advantageous to the cell are not clear.

II. CONTROL OF DNA REPLICATION

Control of DNA replication relies on the regulation of new rounds of replication. In *E. coli*, there are two components of control: the DnaA protein and the structure of the origin of DNA replication (*oriC*). The region of the *E. coli* origin of DNA replication is at 85 min on the genetic map (based on a total of 100 min). On the sequenced *E. coli* genome, the origin is located between nucleotides 3923371 and 3923602. Thus, there is a fixed site on the *E. coli* genome that represents the appropriate place for the initiation of DNA replication. This DNA initiation is referred to as "macroinitiation" as opposed to repetitive initiation, which must occur multiple times during the "elongation" phase of DNA replication. The latter is referred to as "microinitiation." It seems that the requirements for macroinitiation at the *oriC* region include those needed for microinitiation plus additional requirements. There is a region of approximately 250 bp that must be present for DNA replication to initiate.

TABLE I
DNA Polymerases of *E. coli*

	I	*II*	*III*
Molecular mass (kDa)	103	88	130
Synthesis	$5' \rightarrow 3'$	$5' \rightarrow 3'$	$5' \rightarrow 3'$
Initiation	No	No	No
5'-Exonuclease	Yes	No	No
3'-Exonuclease	Yes	Yes	Yes[a]
Gene	*polA*	*polB*	*polC* (*dnaE*)

[a] In a separate protein.

A. Macroinitiation

Several features regarding this region are notable. There are multiple binding sites for the DnaA protein (the DnaA boxes) (Fig. 1). This is a nine-nucleotide sequence that has been shown to bind the DnaA protein. There are also multiple promoter elements, suggesting the involvement of RNA polymerase in macroinitiation. Possible DNA gyrase binding sites are also present. Another notable feature is the presence of multiple Dam-methylase sites (GATC sequence).

The definition of the *oriC* region depends on cloning of this region into plasmids constructed so that replication of the plasmid depends on the function of the *oriC* sequence. This has allowed development of an *in vitro* assay system for macroinitiation of DNA replication. The cloning of *oriC* confirmed that the specificity of macroinitiation in *E. coli* resides in the origin.

The primary protein actor in macroinitiation is the DnaA protein. This protein binds at multiple sites within the *oriC* structure as noted. In addition to binding at the DnaA box consensus sequence, the DnaA protein displays a DNA-dependent adenosine triphosphatase (ATPase) activity and appears to display cooperative binding properties. This suggests that the possible role in initiation is a change of conformation of DNA by DnaA protein interactions. The DnaA protein also binds in the promoter region of the DnaA gene, suggesting autoregulation, which is supported by genetic studies. It appears that the DnaA protein must act positively to initiate DNA replication in *E. coli*.

Both protein and RNA synthesis are required for macroinitiation to occur in *E. coli*. The macroinitia-

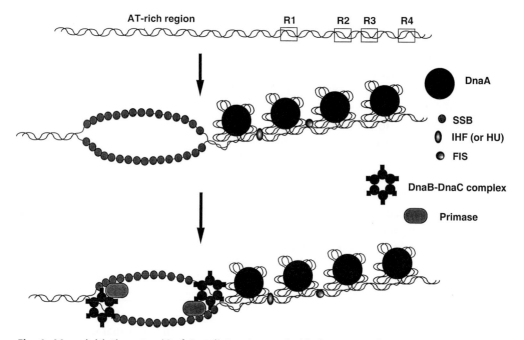

Fig. 1. Macroinitiation at *oriC* of *E. coli*. DnaA protein binds to DnaA boxes R1–R4. The DNA is probably wrapped around the DnaA proteins, and a higher order nucleoprotein complex involving as many as 30–40 DnaA polypeptides is generated with the assistance of DNA-binding proteins, such as IHF, HU, and FIS. The winding of DNA into this complex leads to the compensatory unwinding of an AT-rich region adjacent to the DnaA boxes. Single-stranded DNA is then coated by Ssb protein. DnaC recruits DnaB, the helicase which drives the DNA unwinding at the replication fork, to the junction between single-stranded DNA and double-stranded DNA. Primase then associates with DnaB and synthesizes the RNA primer. The RNA primer is subsequently extended by DNA polymerase III holoenzyme as the DnaB helicase unwinds the chromosome ahead of the polymerase.

tion phase may be further subdivided into stages. The earliest step involves the binding of the DnaA protein to the *oriC* structure. This results in a conformational change of the origin. This complex then binds the DnaA to DnaB and DnaC proteins (which are also required for the elongation phase of DNA replication). DnaC plays a unique role in the delivery of the DnaB protein to the replication structure. The resulting complex appears to unwind the DNA strands since the DnaB protein functions as a helicase, an ATP-dependent unwinding activity. This allows the binding of single-strand binding (Ssb) protein, which allows priming such as that which occurs in the elongation of DNA replication. This stage is followed by the propagation of microinitiation and elongation phases of replication.

Thus, the proteins required for the macroinitiation of *E. coli* DNA replication appear to include DnaA and DnaC, which have specific roles, as well as DnaB, Ssb protein, gyrase, the DnaG primase protein, and the replicative apparatus of DNA polymerase III holoenzyme complex (see Section III,B). Studies also suggest a direct role for RNA polymerase in the macroinitiation of DNA replication.

The DnaA protein offers important support of the replicon hypothesis. Mutations in the DnaA protein demonstrate that all of the *E. coli* chromosome is under a unit control mechanism which defines it as a single replicon. Integration of certain low-copy number plasmids into the chromosome suppresses the phenotype in DnaA mutants that were defective in macroinitiation. This "integrative suppression" shows a general control of macroinitiation and supports the replicon hypothesis.

B. Microinitiation

Microinitiation is the hallmark of the elongation phase of DNA replication. During this phase, repeated initiation occurs along the DNA. The microinitiation step appears to be analogous to the initiation step studied in the *in vitro* lysate systems using small circular phage genomes. In the prokaryotic cell, the requirements for microinitiation appear to mimic those of the phage systems G4 and φX174, which do not display a requirement for the DnaA protein or the features of the *oriC* region.

Cell proteins required for microinitiation include the DnaB protein. The DnaB protein contains a nucleoside triphosphate activity that is stimulated by single-stranded DNA. It also displays DNA helicase activity. In addition, it appears to undergo protein–protein interactions with the DnaC protein. The DnaB mutants are notable for a rapid cessation of DNA synthesis at restrictive conditions. It appears that the DnaB protein is one of the "motors" that moves the replication complex along the DNA (or moves the DNA through the replication complex). The DnaB protein is typical of the proteins involved in DNA replication in that it may have more than one role.

The DnaG protein of *E. coli* is the primase. This protein is capable of synthesizing oligonucleotides utilizing nucleoside (or deoxynucleoside) triphosphates. It appears that physiologically its role is to synthesize RNA primers, which can be utilized by the DNA polymerase III holoenzyme complex to initiate DNA synthesis. As shown in Fig. 2, at least some portion of DNA replication in most organisms is discontinuous. That is, part of the DNA is synthesized in short pieces (termed Okazaki pieces). This is the result of the restriction for DNA synthesis in the $5' \rightarrow 3'$ direction. Since the replication fork requires apparent growth of the nascent strands in both the $5' \rightarrow 3'$ direction and the $3' \rightarrow 5'$ direction, studies were initiated that searched for precursors

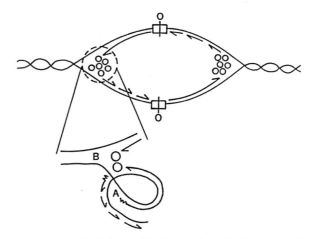

Fig. 2. Model of the replication region. O, the origin; A, primed, elongating nascent strand; B, the point at which the replicative enzyme will release from the A strand and reinitiate a new discontinuous strand.

or enzymes that would allow growth in the $5' \rightarrow 5'$ direction. None were found. The hypothesis of discontinuous synthesis states that, on a microscopic scale, DNA is synthesized discontinuously in a $5' \rightarrow 3'$ direction in small pieces to allow an overall growth in the $3' \rightarrow 5'$ direction on one strand (the lagging strand). This hypothesis predicts the existence of a relatively uniform class of small nascent DNA strands prior to joining, and it also predicts joining activity for such DNA strands. Both of these predictions are fulfilled.

It appears that in *E. coli* DNA is synthesized on one lagging strand in pieces of approximately 1000 nucleotides, which are then covalently linked via the action of DNA ligase following synthesis. Because none of the DNA polymerases in prokaryotes have been found to initiate synthesis *de novo*, this hypothesis leads to the prediction that RNA synthesis, which can be demonstrated to initiate *de novo*, forms primers that are utilized for DNA strand synthesis. Identification of the DnaG primase activity satisfies this prediction.

The polarity restriction of DNA synthesis by DNA polymerases permits one strand to be made continuously, as indicated in the model. It appears that relatively few initiations are made in this (leading) strand and, in fact, that macroinitiation may serve to prime the whole length of the strand.

Ssb protein has analogs throughout nature. In *E. coli*, this protein is relatively small (approximately 19 kDa) and functions as a tetramer. This protein is required for DNA replication. It appears to have several roles. In the single-strand phage systems *in vitro* it confers specificity on the origin of DNA replication and there is no reason to doubt that it performs a similar role in *E. coli*. It probably maintains DNA in a more open state under physiologic conditions during DNA replication in the cell. Evidence suggests that the Ssb protein may participate in a nucleosome-like structure (a nucleoprotein complex that compacts the chromosome, analogous to eukaryotic chromatin), perhaps with *E. coli* HU protein. HU, like IHF and FIS, is a small DNA-binding protein that bends the double helix, thereby facilitating the action of other DNA-binding proteins such as DnaA. It is possible that the coating of DNA strands by Ssb protein protects against nucleolytic degradation during replication. Lastly, Ssb protein appears to

stimulate the rate of synthesis of DNA polymerases under particular conditions. Whether or not this is the case during DNA replication is not clear.

The $5' \rightarrow 3'$ exonuclease of polymerase I is essential in *E. coli*, and such an enzymatic activity meets the requirement for the elimination of leftover RNA primers on the lagging strand. After the replication fork has moved on, leaving a $3'$ terminus of the newly synthesized Okazaki fragment adjacent to an RNA primer, polymerase I extends the $3'$ terminus of the nascent DNA strand while digesting the RNA primer in a process called "nick translation." DNA ligase is required for joining the Okazaki pieces made during discontinuous DNA synthesis. This enzyme is also required for DNA replication because mutants conditionally defective in ligase are also conditionally defective in DNA synthesis.

In addition to the previously mentioned proteins, proteins such as DNA gyrase and DNA topoisomerase I (ω protein) may play a critical role during the microinitiation phase of DNA replication.

Genes *priA*, *-B*, and *-C* (for primosome), *dnaT*, and their products play a role in the assembly of the primosome structure and are required for replication of at least some single-stranded phages and plasmids. However, the effect on the cell of a deficiency is modest.

III. DNA STRAND SYNTHESIS

The genetics of DNA strand synthesis are reflected in the DNA polymerase. *Escherichia coli* is known to contain at least three distinct and separate DNA polymerases, all possessing the following enzymatic activities: synthesis exclusively in the $5' \rightarrow 3'$ direction, utilization of $5'$ deoxynucleoside triphosphates for substrates, the copying of single-stranded DNA template, incorporation of base analogs or ribonucleoside triphosphates at low efficiency under altered conditions (such as in the presence of manganese), a $3'$ editorial exonuclease that preferentially removes mismatched $3'$ termini, and a rate of synthesis that does not approach that of DNA replication in the cell (Table I). Despite these similarities, distinct physical differences exist, and the cell uses exclusively DNA polymerase III for DNA replication. The synthesis subunit for DNA polymerase III, the α subunit, is encoded by the *polC* (*dnaE*) gene.

A. DNA Polymerases

DNA polymerase I, encoded by the *polA* gene, appears to be an auxiliary protein for DNA replication. Cells lacking this enzyme demonstrate viability, although those lacking the notable $5' \rightarrow 3'$ exonuclease activity of this enzyme are only partially viable unless grown in high salt. DNA polymerase I is very important for survival of the cell following many types of DNA damage, and in its absence the cell has persistent single-stranded breaks that promote DNA recombination. DNA polymerase I appears to be a particularly potent effector with DNA ligase in sealing single-stranded nicks, perhaps because of its ability to catalyze nick translation in which the $5'$ exonucleolytic removal of bases is coupled to the synthesis activity. Neither of the other DNA polymerases appear to possess this property.

DNA polymerase II is an enzyme without a defined role in the cell. It has been cloned and overproduced and has been found to bear a closer relationship to T4 DNA polymerase and human polymerase alpha than to either of the other two *E. coli* DNA polymerases. Nevertheless, it is capable of interacting with the subunits of the DNA polymerase III complex. Cells that completely lack the structural gene for DNA polymerase II (the *polB* gene) show normal viability and normal repair after DNA damage in many circumstances. Among suggested roles is synthesis to bypass DNA damage.

DNA polymerase III is the required replicase of *E. coli*. The fact that it plays a significant role in DNA replication is demonstrated because *dnaEts* mutants contain a temperature-sensitive DNA polymerase III. Despite having properties similar to those of DNA polymerase I and II, DNA polymerase III is specifically required for DNA replication. This is a reflection of its ability to interact with a set of subunits that confer particular properties on the complex. In the complex (termed the holoenzyme) DNA polymerase III takes on the properties of a high rate of synthesis and great processivity. Intuitively, processivity may be thought of as the ability of an enzyme catalyzing

the synthesis of DNA to remain tracking on one template for a long period of time before disassociating and initiating synthesis on another template. Highly processive enzymes are capable of synthesizing thousands of nucleotides at a single stretch before releasing the template. DNA polymerase III appears to be uniquely processive among the *E. coli* DNA polymerases.

B. Holoenzyme DNA Polymerase III

In addition to the α-synthesis subunit, there are at least nine constituents of the DNA polymerase III holoenzyme complex (Table II). Most of them have been shown to be the products of required genes as demonstrated by the fact that mutations in that gene produce conditional cessation of DNA replication (*dnaE*, *dnaQ*, *dnaN*, and *dnaX*) or that "knockout" mutants are inviable (*holA*, encoding the δ subunit, and *holB*, encoding the δ' subunit). A knockout mutation of *holE* did not impair cell viability, implying that the θ subunit is dispensable for normal growth. The γ protein and the τ protin appear to be products of the same *dnaX* gene. This is a case in which frame-shifting termination of protein synthesis plays a role in producing different proteins from the same gene. Mutants constructed with a frameshift in the *dnaX* gene that abolish production of γ but do not affect τ are viable; however, τ has been shown to be essential. The β subunit is the product of the *dnaN* gene. This subunit appears to confer specificity for primer utilization upon the complex and to increase the processivity. The ε protein of the holoenzyme complex is known to provide a powerful 3 editorial exonuclease activity. This is manifest by the fact that in addition to lethal mutants in this gene (*mutD*), mutants that show increased error rates in DNA replication (mutators) can be isolated.

Physical studies as well as genetic studies indicate that the DNA polymerase III holoenzyme complex exists in a dimer form. The stoichiometry of the various subunits suggests that the dimer is not exactly symmetrical, but it does appear to be symmetrical for the α β, and ε subunits. The holoenzyme comprises two dimerized β subunits (β_4), a dimeric core pol III', ($\alpha_2\varepsilon_2\theta_2\tau_2$), and a single γ complex ($\gamma_2\delta_1\delta_1'\chi_1\psi_1$) that appears to be involved in loading the β processivity clamp onto the DNA template.

TABLE II
Proteins Involved in *E. coli* DNA Replication

Protein	Subunit	Gene	Size	Function
DnaA		*dnaA*	58	Conformational change of DNA at *oriC*, macroinitiation
DnaB		*dnaB*	52	ATP-dependent DNA helicase, unwinding DNA at the replication fork
DnaC		*dnaC*	27	Recruitment of DnaB to *oriC*–DnaA complex
Gyrase	A	*gyrA*	96	GyrA52–GyrB$_2$ tetramer maintains the chromosome in a negatively supercoiled state, affecting global regulation of replication, and possibly local regulation of initiation at *oriC*
	B	*gyrB*	88	ATPase subunit of gyrase
Topoisomerase IV	A	*parC*	83	Decatenation of replicated chromosomes
	B	*parE*	70	Subunit of topo IV
Topoisomerase I		*topA*	110	Relaxes supercoils, affecting global regulation of replication
Ssb		*ssb*	19	Protects single-stranded DNA from nucleolytic degradation
DNA polymerase III	α	*dnaE*	129.9	DNA synthesis
	ε	*dnaQ*	27.5	$3' \rightarrow 5'$ proofreading exonuclease
	θ	*holE*	8.6	Stimulates ε
	τ	*dnaX*	71.1	Coordinates both halves of pol III holoenzyme by linking DnaB with pol III, interacts with primase
	γ	*dnaX*	47.5	Subunit of γ complex, β clamp loading; binds ATP
	δ	*holA*	38.7	Subunit of γ complex, interacts with β
	δ'	*holB*	36.9	Subunit of γ complex, cofactor of γ ATPase
	χ	*holC*	16.6	Subunit of γ complex, interacts with Ssb
	ψ	*holD*	15.2	Subunit of γ complex; links χ and γ
	β	*dnaN*	40.6	Processivity, sliding clamp
Primosome	Pri A (N′)	*priA*	80	Recognition and binding to primosome assembly site (PAS), ATPase, helicase
	Pri B (N)	*priB*	11.4	stabilization of Pri A–PAS interaction
	Pri C (N″)	*priC*	20.3	Primosome assembly
	Dna T (I)	*dnaT*	20	Primosome assembly
	Dna G (primase)	*dnaG* (*parB*)	64	Primase, synthesizes RNA primer, interacts with Dna B
Ter (Tus, Tau)		*tus*	34	Contrahelicase, blocks Dna B by binding to TER sites
HU	HU-α	*HupA*	10	Histone-like protein, condenses DNA
	HU-β	*HupB*	10	Forms heterodimer with HU-α
FIS		*fis*	11	Regulates initiation at *oriC*, small DNA-binding protein, modulates transcription factors and inversion
IHF	IHF-α	*himA*	11	Modulates initiation at *oriC*, involved in site-specific recombination
	IHF-β	*himD*	10	Forms heterodimer with IHF-α
Polymerase I		*polA*	103	Eliminates Okazaki primers, DNA repair polymerase
DNA Ligase		*lig*	74	Joins DNA fragments during replication

The physical and genetic evidence supporting dimerization of DNA polymerase III is in accordance with a structural model for replication. This is a so-called inch-worm, or trombone, model of DNA replication (Fig. 3). As indicated in the model, a dimer at the growing fork would allow coupling of rates of synthesis on the leading and lagging strands, i.e., the strand made continuously and the strand made discontinuously. Because the strand made discontinuously may require frequent initiation, one might expect synthesis of this nascent DNA strand to be slower. To prevent a discrepancy in growth rate between the strands, dimerization of the synthesis units for the two strands is a method for locking the rates in step. It is possible to do this by assuming that the DNA template for the lagging strand loops out in such a way as to provide permissible polarity for the nascent strand.

A key player in organizing the replisome (that is, the multi-protein complex that replicates the chromosome) appears to be the τ subunit, which has been shown to interact with both DnaB helicase and primase. The coupling of pol III to DnaB explains the high level of processivity on the leading strand while allowing the other half of the pol III holoenzyme to cycle on and off the lagging strand during microinitiation.

IV. TERMINATION OF REPLICATION

The termination of DNA replication is complicated by topological problems created by the circular na-

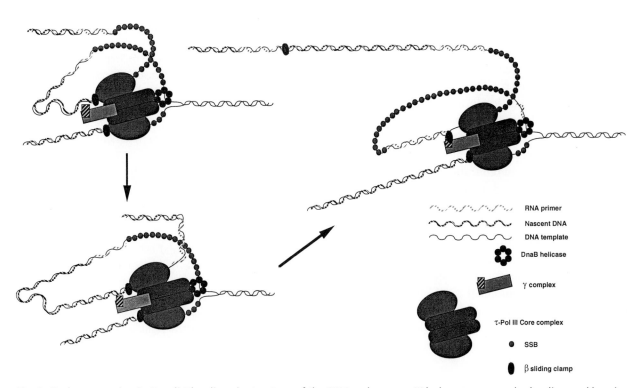

Fig. 3. Fork progression in *E. coli*. The dimeric structure of the DNA polymerase III holoenzyme couples leading and lagging strand DNA synthesis during replication. Leading strand synthesis (bottom strand) is continuous and processive, but lagging strand synthesis must reinitiate many times during replication of the chromosome. This constraint is due to the antiparallel nature of DNA and the ability of DNA polymerases to synthesize DNA only in the 5′ → 3′ direction. As the replication fork progresses, new primers are synthesized by primase on the lagging strand. The asymmetrical γ complex loads a new β dimer onto the primed DNA template, and then the β complex associates with the core polymerase to extend DNA synthesis from the 3′ end of the primer. DNA synthesis on the lagging strand proceeds only as far as the previously replicated Okazaki fragment, at which point the β subunit, along with the nascent DNA, is released from the core polymerase, allowing the next cycle of synthesis to initiate at the next primer.

ture of the bacterial chromosome. The double-helical structure of the template DNA, given a semiconservative mode of DNA replication, results in two interwound chromosomes which must be unlinked by topoisomerases. Decatenation of the newly replicated chromosomes is accomplished primarily by topoisomerase IV, a double-stranded DNA topoisomerase that closely resembles DNA gyrase (topoisomerase II). Another problem is that bidirectional replication on a circular template must be coordinated so that the opposing replication forks meet and terminate DNA synthesis at a defined location. Otherwise, part of the genome might be overreplicated.

A specific region of the chromosome, approximately directly opposite the origin on the *E. coli* map at the 30- to 32-min region, represents the termination region. This region is particularly sparce in genetic markers. A specific terminator protein, ter (or Tus), binds to sequences called Ter (or τ). Ter sequences are arrayed in the termination region in an inverted repeat configuration such that binding of Tus protein to Ter sequences confers a polarity to blockage of the replication fork. Tus protein has been described as a polar contrahelicase because its ability to inhibit DnaB, the replication fork-specific helicase, depends on the polarity of the Ter sequence with respect to the origin of replication. A replication fork is unimpeded by Tus bound to a ter sequence in the forward orientation, but it is blocked by Tus bound to a ter sequence in the reverse orientation. Because of the inverted configuration of Ter sites in the termination region, a replication fork that progresses through a forward-oriented Ter site will be stopped at the next reverse-oriented site. The termination region thus "traps" replication forks, ensuring that part of the genome is not overreplicated. *Bacillus subtilis* has evolved a similar mechanism, with a ter-

mination protein, RTP, that binds to specific inversely-oriented repeats, also called Ter sequences. Of note is the fact that the *E. coli* Ter sites are widely separated by approximately 350 kb, whereas the innermost oppositely oriented Ter sites in *B. subtilis* are only 59 bp apart. The basic strategy for termination in the two organisms is quite similar, but the molecular mechanisms appear to be different. DNA replication termination may also be a control for cell division.

Bibliography

Bramhill, D., and Kornberg, A. (1988). A model for initiation at origins of DNA replication. *Cell* **54**, 915–918.

Echols, H. (1986). Multiple DNA-protein interactions governing high-precision DNA transactions. *Science* **233**, 1050–1056.

Kelman, Z., and O'Donnell, M. (1995). DNA polymerase III holoenzyme: structure and function of a chromosomal replicating machine. *Annu. Rev. Biochem.* **64**, 171–200.

Manna, A. C., Karnire, S. P., Bussiere, D. E., Davies, C., White, S. W., and Bastia, D. (1996). Helicase-contrahelicase interaction and the mechanism of termination of DNA replication. *Cell* **87**, 881–891.

Marians, K. J. (1996). Replication for propagation. *In* "*Escherichia coli* and *Salmonella*: Cellular and Molecular Biology" (R. Curtis, III, *et al.*, Eds.), Second Edition, pp. 749–763. ASM, Washington, DC.

McHenry, C. S. (1988). DNA polymerase III holoenzyme of *Escherichia coli*. *Annu. Rev. Biochem.* **57**, 519–550.

Peng, H., and Marians, K. J. (1993). Decatenation activity of topoisomerase IV during oriC and pBR322 DNA replication *in vitro*. *Proc. Natl. Acad. Sci. USA* **90**, 8571–8575.

Yuzhakov, A., Turner, J., and O'Donnell, M. (1996). Replisome assembly reveals the basis for asymmetric function in leading and lagging strand replication. *Cell* **86**, 877–886.

Zavitz, K. H., and Marians, K. J., (1991). Dissecting the functional role of PriA protein-catalysed primosome assembly in *Escherichia coli* DNA replication. *Mol. Microbiol.* **5**, 2869–2873.

DNA Restriction and Modification

Noreen E. Murray

University of Edinburgh, Institute of Cell and Molecular Biology

I. Detection of Restriction Systems
II. Nomenclature and Classification
III. R–M Enzymes as Model Systems
IV. Control and Alleviation of Restriction
V. Distribution, Diversity, and Evolution
VI. Biological Significance
VII. Applications and Commercial Relevance

GLOSSARY

ATP and ATP hydrolysis Adenosine triphosphate is a primary repository of energy that is released for other catalytic activities when ATP is hydrolyzed (split) to yield adenosine diphosphate.

bacteriophages (lambda and T-even) Bacterial viruses. Phage lambda (λ) is a temperate phage and therefore on infection of a bacterial cell one of two alternative pathways may result; either the lytic pathway in which the bacterium is sacrificed and progeny phages are produced or the temperate (lysogenic) pathway in which the phage genome is repressed and, if it integrates into the host chromosome, will be stably maintained in the progeny of the surviving bacterium. Phage λ was isolated from *Escherichia coli* K-12 in which it resided in its temperate (prophage) state. T-even phages (T2, T4, and T6) are virulent coliphages, i.e., infection of a sensitive strain of *E. coli* leads to the production of phages at the inevitable expense of the host. T-even phages share the unusual characteristic that their DNA includes hydroxymethylcytosine rather than cytosine.

conjugation, conjugational transfer Gene transfer by conjugation requires cell-to-cell contact. Conjugative, or self-transmissible, plasmids such as the F factor of *E. coli* encode the necessary functions to mobilize one strand of their DNA with a defined polarity from an origin of transfer determined by a specific nick. The complementary strand is then made in the recipient cell. Some plasmids are trans-

missible but only on provision *in trans* of the necessary functions by a conjugative plasmid. A conjugative plasmid, such as the F factor, can mobilize transfer of the bacterial genome following integration of the plasmid into the bacterial chromosome.

DNA methyltransferases Enzymes (MTases) that catalyze the transfer of a methyl group from the donor *S*-adenosylmethionine to adenine or cytosine residues in the DNA.

efficiency of plating This usually refers to the ratio of the plaque count on a test strain relative to that obtained on a standard, or reference, strain.

endonucleases Enzymes that can fragment polynucleotides by the hydrolysis of internal phosphodiester bonds.

Escherichia coli strain K-12 The strain used by Lederberg and Tatum in their discovery of recombination in *E. coli*.

glucosylation of DNA The DNA of T-even phages in addition to the pentose sugar, deoxyribose, contains glucose attached to the hydroxymethyl group of hydroxymethylcytosine. Glucosylation of the DNA is mediated by phage-encoded enzymes, but the host provides the glucose donor.

helicases Enzymes that separate paired strands of polynucleotides.

recombination pathway The process by which new combinations of DNA sequences are generated. The general recombination process relies on enzymes that use DNA sequence homology for the recognition of the recombining partner. In the major pathway in *E. coli*, RecA protein promotes synapsis and RecBCD generates the DNA strands for transfer. The RecBCD enzyme, also recognized as exonuclease V, enters DNA via a double-strand end. It tracks along the DNA, promoting unwinding of the strands and degradation of the strand with a 3′ end. The degradation is halted by special sequences termed Chi, following which strand separation continues and the single-stranded DNA with a 3′ end becomes available for synapsis with homologous DNA.

SOS response DNA damage induces expression of a set of

91

genes, the SOS genes, involved in the repair of DNA damage.

Southern transfer   The transfer of denatured DNA from a gel to a solid matrix, such as a nitrocellulose filter, within which the denatured DNA can be maintained and hybridized to labeled probes (single-stranded DNA or RNA molecules). Fragments previously separated by electrophoresis through a gel may be identified by hybridization to a specific probe.

transformation   The direct assimilation of DNA by a cell, which results in the recipient being changed genetically.

AWARENESS OF THE BIOLOGICAL PHENOMENON OF RESTRICTION AND MODIFICATION (R–M)

grew from the observations of microbiologists that the host range of a bacterial virus (phage) was influenced by the bacterial strain in which the phage was last propagated. Although phages produced in one strain of *Escherichia coli* would readily infect a culture of the same strain, they might only rarely achieve the successful infection of cells from a different strain of *E. coli*. This finding implied that the phages carried an ''imprint'' that identified their immediate provenance. Simple biological tests showed that the occasional successful infection of a different strain resulted in the production of phages that had lost their previous imprint and had acquired a new one, i.e., they acquired a new host range.

In the 1960s, elegant molecular experiments showed the "imprint" to be a DNA modification that was lost when the phage DNA replicated within a different bacterial strain; those phages that conserved one of their original DNA strands retained the imprint, or modification, whereas phages containing two strands of newly synthesized DNA did not. The modification was shown to provide protection against an endonuclease, the barrier that prevented the replication of incoming phage genomes. Later it was proven that the modification and restriction enzymes both recognized the same target, a specific nucleotide sequence. The modification enzyme was a DNA methyltransferase that methylated specific bases within the target sequence, and in the absence of the specific methylation the target sequence rendered the DNA sensitive to the restriction enzyme.

When DNA lacking the appropriate modification imprint enters a restriction-proficient cell it is recognized as foreign and degraded by the endonuclease. The host-controlled barrier to successful infection by phages that lacked the correct modification was referred to as "restriction" and the relevant endonucleases have acquired the colloquial name of restriction enzymes. Similarly, the methyltransferases are more commonly termed modification enzymes. Classically, a restriction enzyme is accompanied by its cognate modification enzyme and the two comprise a R–M system. Most restriction systems conform to this classical pattern. There are, however, some restriction endonucleases that attack DNA only when their target sequence is modified. A restriction system that responds to its target sequence only when it is identified by modified bases does not, therefore, coexist with a cognate modification enzyme.

Two early papers documented the phenomenon of restriction. In one, Bertani and Weigle (1953), using temperate phages (λ and P2), identified the classical restriction and modification systems characteristic of *E. coli* K-12 and *E. coli* B. In the other, Luria and Human (1952) identified a restriction system of the second, nonclassical kind. In the experiments of Luria and Human, T-even phages were used as test phages, and after their growth in a mutant *E. coli* host they were found to be restricted by wild-type *E. coli* K-12 but not by *Shigella dysenteriae*. An understanding of the restriction phenomenon observed by Luria and Human requires knowledge of the special nature of the DNA of T-even phages. During replication of T-even phages the unusual base 5-hydroxymethylcytosine (HMC) completely substitutes for cytosine in the T-phage DNA, and the hydroxymethyl group is subsequently glucosylated in a phage-specific pattern at the polynucleotide level. In the mutant strain of *E. coli* used by Luria and Human as host for the T-even phages, glucosylation fails and, in its absence, the nonglucosylated phage DNA becomes sensitive to an endonuclease present in *E. coli* K-12 but not in *S. dysenteriae*. Particular nucleotide sequences normally protected by glucosylation are recognized by an endonuclease in *E. coli* K-12 when they include the modified base, HMC, rather than cytosine residues. In the T-phage experiments the modified base is hydroxymethylcytosine, but much later it was discovered that methylated cytosine residues can also evoke restriction by the same endonucleases.

The classical (R–M) systems and the modification-dependent restriction enzymes share the potential to attack DNA derived from different strains and thereby "restrict" DNA transfer. They differ in that in one case an associated modification enzyme is required to protect DNA from attack by the cognate restriction enzyme and in the other modification enzymes specified by different strains impart signals that provoke the degradative activity of restriction endonucleases.

I. DETECTION OF RESTRICTION SYSTEMS

A. As a Barrier to Gene Transfer

This is exemplified by the original detection of the R–M systems of *E. coli* K-12 and *E. coli* B by Bertani and Weigle in 1953. Phage λ grown on *E. coli* strain C (λC), where *E. coli* C is a strain that apparently lacks an R–M system, forms plaques with poor efficiency [efficiency of plating (EOP) of 2×10^{-4}] on *E. coli* K-12 because the phage DNA is attacked by a restriction endonuclease (Fig. 1). Phage λ grown on *E. coli* K-12 (λK) forms plaques with equal efficiency on *E. coli* K-12 and *E. coli* C since it has the modification required to protect against the restriction system of *E. coli* K-12 and *E. coli* C has no restriction system (Fig. 1). In contrast, λK will form plaques with very low efficiency on a third strain, *E. coli* B, since *E. coli* B has an R–M system with different sequence specificity from that of *E. coli* K-12.

Phages often provide a useful and sensitive test for the presence of R–M systems in laboratory strains of bacteria, but they are not a suitable vehicle for the general detection of barriers to gene transfer. Many bacterial strains, even within the same species and particularly when isolated from natural habitats, are unable to support the propagation of the available test phages, and some phages (e.g., P1) have the means to antagonize at least some restriction systems (see Section IV.C). Gene transfer by conjugation can

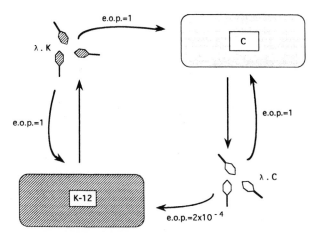

Fig. 1. Host-controlled restriction of bacteriophage λ. *Escherichia coli* K-12 possesses, whereas *E. coli* C lacks, a type I R–M system. Phage λ propagated in *E. coli* C (λC) is not protected from restriction by *Eco*KI and thus forms plaques with reduced efficiency of plating (EOP) on *E. coli* K-12 as compared to *E. coli* C. Phages escaping restriction are modified by the *Eco*KI methyltransferase (λK) and consequently form plaques with the same efficiency on *E. coli* K-12 and C. Modified DNA is indicated by hatch marks (reproduced with permission from Barcus and Murray, 1995).

monitor restriction although some natural plasmids, but probably not the F factor of *E. coli*, are equipped with antirestriction systems. The single-stranded DNA that enters a recipient cell by conjugation, or following infection by a phage such as M13, becomes sensitive to restriction only after the synthesis of its complementary second strand, whereas the single-stranded DNA that transforms naturally competent bacteria may not become a target for restriction because it forms heteroduplex DNA with resident (and therefore modified) DNA and one modified strand is sufficient to endow protection. Transformation can be used to detect restriction systems, but only when the target DNA is the double-stranded DNA of a plasmid.

B. *In Vitro* Assays for DNA Fragmentation

Endonuclease activities yielding discrete fragments of DNA are commonly detected in crude extracts of bacterial cells. More than one DNA may be used to increase the chance of providing a substrate that includes target sequences. DNA fragments diagnostic of endonuclease activity are separated according to their size by electrophoresis through a matrix, usually an agarose gel, and are visualized by the use of a fluorescent dye, ethidium bromide, that intercalates between stacked base pairs.

Extensive screening of many bacteria, often obscure species for which there is no genetic test, has produced a wealth of endonucleases with different target sequence specificities. These endonucleases are referred to as restriction enzymes, even in the absence of biological experiments to indicate their role as a barrier to the transfer of DNA. Many of these enzymes are among the commercially available endonucleases that serve molecular biologists in the analysis of DNA (Table I: see Section VII). *In vitro* screens are applicable to all organisms, but to date restriction and modification systems have not been found in eukaryotes, although some algal viruses encode them.

C. Sequence-Specific Screens

The identification of new R–M genes via sequence similarities is sometimes possible. Only occasionally are gene sequences sufficiently conserved that the presence of related systems can be detected by probing Southern transfers of bacterial DNA. Generally, screening databases of predicted polypeptide se-

TABLE I
Some Type II Restriction Endonucleases and Their Cleavage Sites[a]

Bacterial source	Enzyme abbreviation	Sequences $5' \to 3'$ $3' \leftarrow 5'$	Note[b]
Haemophilus influenzae Rd	HindII	GTPy ↓ PuAC CAPu ↑ PyTG	1, 5
	HindIII	↓ AAGCTT TTCGAA ↑	2
Haemophilus aegyptius	HaeIII	GG ↓ CC CC ↑ GG	1
Staphylococcus aureus 3A	Sau3AI	↓ GATC CTAG ↑	2, 3
Bacillus amyloliquefaciens II	BamHI	↓ GGATCC CCTAGG ↑	2, 3
Escherichia coli RY13	EcoRI	↓ GAATTC CTTAAG ↑	2
Providencia stuartii	PstI	↑ ↓ CTGCAG GACGTC ↑	4

[a] The cleavage site for each enzyme is shown by the arrows within the target sequence.

[b] 1, produces blunt ends; 2, produces cohesive ends with 5′ single-stranded overhangs; 3, cohesive ends of Sau3AI and BamHI are identical; 4, produces cohesive ends with 3′ single-stranded overhangs; 5, Pu is any purine (A or G), and Py is any pyrimidine (C or T).

quences for relevant motifs has identified putative R–M systems in the rapidly growing list of bacteria for which the genomic sequence is available. Currently, this approach is more dependable for modification methyltransferases than for restriction endonucleases, but the genes encoding the modification and restriction enzymes are usually adjacent. Many putative R–M systems have been identified in bacterial genomic sequences.

II. NOMENCLATURE AND CLASSIFICATION

A. Nomenclature

R–M systems are designated by a three-letter acronym derived from the name of the organism in which they occur. The first letter is derived from the genus and the second and third letters from the species. The strain designation, if any, follows the acronym. Different systems in the same organism are distinguished by Roman numerals. Thus, *Hind*II and *Hind*III are two enzymes from *Hemophilus influenzae* strain Rd. Restriction endonuclease and modification methyltransferases (ENases and MTases) are sometimes distinguished by the prefixes R.*Eco*RI and M.*Eco*RI, but the prefix is commonly omitted if the context is unambiguous.

B. Classification of R–M Systems

R–M systems are classified according to the composition and cofactor requirements of the enzymes, the nature of the target sequence, and the position of the site of DNA cleavage with respect to the target sequence. Currently, three distinct, well-characterized types of classical R–M systems are known (I–III), although a few do not share all the characteristics of any of these three types. In addition, there are modification-dependent systems. Type I systems were identified first, but the type II systems are the simplest and for this reason will be described first. A summary of the properties of different types of R–M systems is given in Fig. 2.

C. Type II R–M Systems

A type II R–M system comprises two separate enzymes; one is the restriction ENase and the other the modification MTase. The nuclease activity requires Mg^{2+}, and DNA methylation requires S-adenosylmethionine (AdoMet) as methyl donor. The target sequence of both enzymes is the same; the modification enzyme ensures that a specific base within the target sequence, one on each strand of the duplex, is methylated and the restriction endonuclease cleaves unmodified substrates within, or close to, the target sequence. The target sequences are usually rotationally symmetrical sequences of from 4 to 8 bp; for example, a duplex of the sequence 5'-GAA*TTC is recognized by *Eco*RI. The modification enzyme methylates the adenine residue identified by the asterisk, but in the absence of methylated adenine residues on both strands of the target sequence the restriction endonuclease breaks the phosphodiester backbones of the DNA duplex to generate ends with 3' hydroxyl and 5' phosphate groups. The type II systems can be subdivided according to the nature of the modification introduced by the MTase: N6-methyladenine (m6A) and N5 and N4 methylcytosine (m5C and m4C). Irrespective of the target sequence or the nature of the modification, ENases differ in that some cut the DNA to generate ends with 5' overhangs, some generate 3' overhangs, and others produce ends which are "blunt" or "flush" (Table I).

Type II restriction enzymes are generally active as symmetrically arranged homodimers, an association that facilitates the coordinated cleavage of both strands of the DNA. In contrast, type II modification enzymes act as monomers, an organization consistent with their normal role in the methylation of newly replicated DNA in which one strand is already methylated.

The genes encoding type II R–M systems derive from the name of the system. The genes specifying R.*Bam*HI and M.*Bam*HI, for example, are designated *bamHIR* and *bamHIM*. Transfer of the gene encoding a restriction enzyme in the absence of the transfer of the partner encoding the protective MTase is likely to be lethal if the recipient cell does not provide the relevant protection. Experimental evidence supports

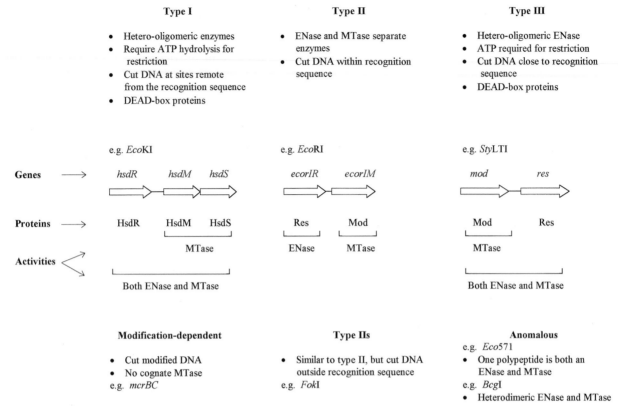

Fig. 2. The distinguishing characteristics and organization of the genetic determinants and subunits of the different types of restriction and modification (R–M) systems. R–M systems are classified on the basis of their complexity, cofactor requirements, and position of DNA cleavage with respect to their DNA target sequence. The types I–III systems are the classical R–M systems. The restriction enzymes of the types I and III systems contain motifs characteristic of DEAD-box proteins and may therefore be helicases. Helicase activity could be associated with DNA translocation. The type IIS systems are a subgroup of the type II systems that cleave DNA outside their recognition sequence. Some systems do not fit readily into the current classification and *Eco*571 has been tentatively termed a type IV system. The modification-dependent systems are not classical R–M systems, because they have no cognate methylase and only cut DNA that contains certain modifications. ENase, restriction endonuclease; MTase, methyltransferase (reprinted from *Trends in Microbiology* **2**(12), G. King and N. E. Murray, pp. 461–501. Copyright 1995, with permission from Elsevier Science).

the expectation that the genes encoding the two components of R–M systems are usually closely linked so that cotransfer will be efficient.

A subgroup of type II systems, type IIS, recognizes asymmetric DNA sequences of 4 to 7 bp in length. These ENases cleave the DNA at a precise but short distance outside their recognition sequence; their name is derived from their shifted (S) position of cutting. Type IIS systems have simple cofactor requirements and comprise two separate enzymes, but they differ from type II systems in the recognition of an asymmetric DNA sequence by a monomeric ENase, and require two MTase activities.

D. Type I R–M Systems

Type I R–M systems are multifunctional enzymes comprising three subunits that catalyze both restriction and modification. In addition to Mg^{2+}, endonucleolytic activity requires both AdoMet and adenosine triphosphate (ATP). The restriction activity of type I enzymes is associated with the hydrolysis of ATP, an activity that may correlate with the peculiar characteristic of these enzymes—that of cutting DNA at nonspecific nucleotide sequences at considerable distances from their target sequences. The type I R–M enzyme binds to its target sequence and its activity as an ENase or a MTase is determined by the methylation state of the target sequence. If the target sequence is unmodified, the enzyme, while bound to its target site, is believed to translocate (move) the DNA from both sides toward itself in an ATP-dependent manner. This translocation process brings the bound enzymes closer to each other and experimental evidence suggests that DNA cleavage occurs when translocation is impeded, either by collision with another translocating complex or by the topology of the DNA substrate.

The nucleotide sequences recognized by type I enzymes are asymmetric and comprise two components, one of 3 or 4 bp and the other of 4 or 5 bp, separated by a non-specific spacer of 6–8 bp. All known type I enzymes methylate adenine residues, one on each strand of the target sequence.

The three subunits of a type I R–M enzyme are encoded by three contiguous genes: *hsdR*, *hsdM*, and *hsdS*. The acronym *hsd* was chosen at a time when R–M systems were referred to as host specificity systems, and *hsd* denotes *host* specificity of *DNA*. *hsdM* and *hsdS* are transcribed from the same promoter, but *hsdR* is from a separate one. The two subunits encoded by *hsdM* and *hsdS*, sometimes referred to as M and S, are both necessary and sufficient for MTase activity. The third subunit (R) is essential only for restriction. The S (specificity) subunit includes two target recognition domains (TRDs) that impart target sequence specificity to both restriction and modification activities of the complex; the M subunits include the binding site for AdoMet and the active site for DNA methylation. Two complexes of Hsd subunits are functional in bacterial cells: one comprises all three subunits ($R_2M_2S_1$) and is an R–M system, and the other lacks R (M_2S_1) and has only MTase activity.

E. Type III R–M Systems

Type III R–M systems are less complex than type I systems but nevertheless share some similarities with them. A single heterooligomeric complex catalyzes both restriction and modification activities. Modification requires the cofactor AdoMet and is stimulated by Mg^{2+} and ATP. Restriction requires Mg^{2+} and ATP and is stimulated by AdoMet. The

recognition sequences of type III enzymes are asymmetric sequences of 5 or 6 bp. Restriction requires two unmodified sequences in inverse orientation (Fig. 3a). Recent evidence indicates that type III R–M enzymes, like those of type I, can translocate DNA in a process dependent on ATP hydrolysis, but they hydrolyze less ATP than do type I systems and probably only translocate DNA for a relatively short distance. Cleavage is stimulated by collision of the translocating complexes and occurs on the 3′ side of the recognition sequence at a distance of approximately 25–27 bp. This contrasts with cleavage by type I enzymes in which cutting occurs at sites remote from the recognition sequence. Because only one strand of the recognition sequence of a type III R–M system is a substrate for methylation, it might be anticipated that the immediate product of replication would be sensitive to restriction. It is necessary to distinguish the target for modification from that needed for restriction in order to understand why this is not so. Restriction is only elicited when two unmethylated target sequences are in inverse orientation with respect to each other and, as shown in Fig. 3b, replication of modified DNA leaves all unmodified targets in the same orientation.

The bifunctional R–M complex is made up of two subunits, the products of the *mod* and *res* genes. The Mod subunit is sufficient for modification, whereas the Res and Mod subunits together form a complex with both activities (Fig. 2). The Mod subunit is functionally equivalent to the MTase (M_2S) of type I systems and, as in type I R–M systems, imparts sequence specificity to both activities.

F. Modification-Dependent Restriction Systems

These systems only cut modified DNA. They are variable in their complexity and requirements. The simplest is *DpnI* from *Streptococcus pneumoniae* (previously called *Diplococcus pneumoniae*). The ENase is encoded by one gene, and the protein looks and behaves like a type II enzyme except that it only cuts its target sequence when it includes methylated adenine residues.

Escherichia coli K-12 encodes three other distinct, sequence specific, modification-dependent systems.

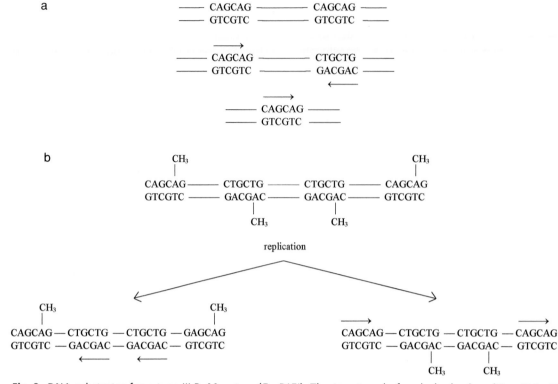

Fig. 3. DNA substrates for a type III R–M system (*Eco*P15I). The top strand of each duplex is written 5′ to 3′ the arrows identify the orientation of the target sequences. Solid lines indicate polynucleotide chains of undefined sequence. (a) Only pairs of target sequences shown in inverse orientation (line 2) are substrates for restriction. A single site in any orientation is a target for modification. (b) Replication of modified DNA leaves all unmodified target sites in the daughter molecules in the same orientation, and therefore insensitive to restriction.

Two, Mrr and McrA, are specified by single genes. Mrr is distinguished by its ability to recognize DNA containing either methylated adenine or 5-methylcytosine in the context of particular, but undefined, sequences. McrA and McrBC both restrict DNA containing modified cytosines (hydroxymethylcytosine or methylcytosine). The *mcr* systems (*modified cytosine restriction*) are those first recognized by Luria and Human by their ability to restrict nonglucosylated T-even phages (*rglA* and *B*; restricts glucoseless phage). McrBC is a complex enzyme with a requirement for GTP rather than ATP.

G. Other Systems

As more R–M systems are identified, new enzymes with novel properties continue to be found. *Eco*571 and *Bcg*I, for example, are most similar to type II systems, but are not ideally suited to this classification (Fig. 2). *Eco*571 comprises a joint ENase–MTase and a separate MTase. The former behaves as if it were a type IIS endonuclease fused to a MTase; it cleaves to one side of the target sequence and methylates the sequence on one strand. Cleavage is stimulated by AdoMet but not by ATP. The separate MTase behaves like a type II modification enzyme.

III. R–M ENZYMES AS MODEL SYSTEMS

A. Sequence Recognition, Including Base Flipping

Structures of the crystals of several restriction ENases have been determined, some in both the presence and the absence of DNA. The symmetrically arranged dimers of the type II enzymes interact with their specific target sequences by the combined effects of different types of interactions including hydrogen bonding and electrostatic interactions of amino acid residues with the bases and the phosphate backbone of the DNA. No general structure, such as a helix–turn–helix or zinc finger (often found in proteins that interact with DNA) is characteristic of the protein–DNA interface, and amino acids that are widely separated in the primary sequence may be involved in interactions with the target nucleotide sequence. Comparisons of the active sites of *Eco*RV, *Eco*RI, and *Pvu*II identify a conserved tripeptide sequence close to the target phosphodiester group and a conserved acidic dipeptide that may represent the ligands for the catalytic cofactor Mg^{2+} essential for ENase activity.

The structure of a monomeric MTase interacting with its target sequence provided an important solution to the question of how enzymes that modify a base within a DNA molecule can reach their substrate. The cocrystal structure of M.*Hha*I bound to its substrate showed that the target cytidine rotates on its sugar–phosphate bonds such that it projects out of the DNA and fits into the catalytic pocket of the enzyme. Such base flipping was confirmed for a second enzyme, M.*Hae*III, which also modifies cytosine, and circumstantial evidence supports the notion that this mechanism may be true for all MTases regardless of whether they methylate cytosine or adenine residues.

Comparative analyses of the amino acid sequences of many MTases identified a series of motifs, many of which are common to MTases irrespective of whether the target base is cytosine or adenine. These motifs enable structural predictions to be made about the catalytic site for DNA methylation in complex enzymes for which crystals are not available.

B. DNA Translocation

Specific interactions of large R–M enzymes with their DNA substrates are not readily amenable to structural analysis. The relative molecular weight of *Eco*KI is in excess of 400,000 and useful crystals have not been obtained. Nevertheless, these complex enzymes have other features of mechanistic interest.

Much evidence supports models in which DNA restriction involves the translocation of DNA in an ATP-dependent process prior to the cutting of the substrate. In the case of type I R–M enzymes, the breaks in the DNA may be many kilobases remote from the target sequence. Molineux and colleagues (1999), using assays with phages, have shown that *Eco*KI can transfer (translocate) the entire genome (39 kb) of phage T7 from its capsid to the bacterial cell. For linear DNA, the evidence supports the idea that cutting by type I R–M systems occurs preferentially midway between two target sequences. For type III enzymes the breaks are close to the target sequence, but in both cases the endonuclease activity may be stimulated by the collision of two translocating protein complexes.

The most conserved features of the polypeptide sequences of type I and type III R–M systems are the motifs characteristic of adenine MTases and the so-called DEAD-box motifs found in RNA and DNA helicases. The latter motifs acquired their collective name because a common variant of one element is Asp-Glu-Ala-Asp, or DEAD when written in a single-letter code. The DEAD-box motifs, which include sequences diagnostic of ATP binding, are found in the subunit that is essential for restriction (HsdR or Res) but not for modification. It is not known whether an ATP-dependent helicase activity drives the translocation of DNA, although circumstantial evidence correlates ATPase activity with DNA translocation. Mutations in each DEAD-box motif have been shown to impair the ATPase and endonuclease activities of a type I ENase.

IV. CONTROL AND ALLEVIATION OF RESTRICTION

A. Control of Gene Expression

The expression of genes coding for R–M systems requires careful regulation. Not only is this essential to maintain the protection of host DNA in restriction-proficient cells but also it is especially important when R–M genes enter a new host. Experiments show that many R–M genes are readily transferred from one laboratory strain to another. The protection of host DNA against the endonucleolytic activity of a newly acquired restriction system can be achieved if the functional cognate MTase is produced before the restriction enzyme. Representatives of all three types of classical R–M systems have been shown to be equipped with promoters that might permit transcriptional regulation of the two activities.

Transcriptional regulation of some of the genes encoding type II systems has been demonstrated. Genes encoding repressor-like proteins, referred to as C-proteins for controlling proteins, have been identified in some instances. The C-protein for the *Bam*HI system has been shown to activate efficient expression of the restriction gene and modulate the expression of the modification gene. When the R–M genes are transferred to a new environment, in the absence of C-protein there will be preferential expression of the modification gene, and only after the production of the C-protein will the cells become restriction proficient.

For complex R–M systems, despite the presence of two promoters, there is no evidence for transcriptional regulation of gene expression. The heterooligomeric nature of these systems presents the opportunity for the regulation of the restriction and modification activities by the intracellular concentrations of the subunits and the affinities with which different subunits bind to each other. Nevertheless, efficient transmission of the functional R–M genes of some type I systems requires ClpX and ClpP in the recipient cell. Together these proteins comprise a protease, but ClpX has chaperone activity (i.e., an ability to help other proteins to fold correctly). These proteins function to permit the acquisition of new R–M systems; they degrade the HsR subunit of active R–M complexes before the endonuclease activity has the opportunity to cleave unmodified DNA.

B. Restriction Alleviation

The efficiency with which a bacterial cell restricts unmodified DNA is influenced by a number of stimuli, all of which share the ability to damage DNA. Restriction provokes the induction of the SOS response to DNA damage, and a consequence of this is a marked reduction in the efficiency with which *E. coli* K-12 restricts incoming DNA. This alleviation

of restriction is usually monitored by following the EOP of phages—unmodified in the case of classical systems or modified in the case of modification-dependent restriction systems. Alleviation of restriction is characteristic of complex systems and can be induced by ultraviolet light, nalidixic-acid, 2-aminopurine, and the absence of Dam-mediated methylation. The effect can be appreciable and a variety of host systems contribute to more than one pathway of alleviation. Recent experiments have shown that ClpXP is necessary for restriction alleviation; therefore there appears to be a connection between the complex mechanisms by which restriction activity is normally controlled and its alleviation in response to DNA damage.

C. Antirestriction Systems

Many phages and some conjugative plasmids specify functions that antagonize restriction. An apparent bias of functions that inhibit restriction by type I R–M systems may reflect the genotype of the classical laboratory strain *E. coli* K-12, which is a strain with a type I but no type II R–M system.

The coliphages T3 and T7 include an "early" gene, $O \cdot 3$, the product of which binds type I R–M enzymes and abolishes both restriction and modification activities. $O \cdot 3$ protein does not affect type II systems. The $O \cdot 3$ gene is expressed before targets in the phage genome are accessible to host restriction enzymes so that $O \cdot 3^+$ phages are protected from restriction and modification by type I systems. Phage T3 $O \cdot 3$ protein has an additional activity; it hydrolyzes AdoMet, the cofactor essential for both restriction and modification by *Eco*KI and its relatives. Bacteriophage P1 also protects its DNA from type I restriction, but the antirestriction function Dar does not interfere with modification. The Dar proteins are coinjected with encapsidated DNA so that any DNA packaged in a P1 head is protected. This allows efficient generalized transduction to occur between strains with different type I R–M systems.

Coliphage T5 has a well-documented system for protection against the type II system *Eco*RI. As with the $O \cdot 3$ systems of T3 and T7, the gene is expressed early when the first part of the phage genome enters the bacterium. This first segment lacks *Eco*RI targets, whereas the rest of the genome, which enters later, has targets that would be susceptible in the absence of the antirestriction protein.

Some conjugative plasmids of *E. coli*, members of the incompatibility groups I and N, also encode antirestriction functions. They are specified by the *ard* genes located close to the origin of DNA transfer by conjugation so that they are among the first genes to be expressed following DNA transfer. Like the *ocr* proteins of T3 and T7, the protein encoded by *ard* is active against type I R–M systems.

Bacteriophage λ encodes a very specialized antirestriction function, RaI, which modulates the *in vivo* activity of some type I R–M systems by enhancing modification and alleviating restriction. The systems influenced by RaI are those that have a modification enzyme with a strong preference for hemimethylated DNA. RaI may act by changing the MTase activity of the R–M system to one that is efficient on unmethylated target sequences. Unmodified *raI*⁺ λ DNA is restricted on infection of a restriction-proficient bacterium, because *raI* is not normally expressed before the genome is attacked by the host R–M system, but phages that escape restriction and express *raI* are more likely to become modified if RaI serves to enhance the efficiency with which the modification enzyme methylates unmodified DNA.

Some phages are made resistant to many types of R–M systems by the presence of glucosylated hydroxymethylcytosine (HMC) in their DNA (e.g., the *E. coli* T-even phages and the *Shigella* phage DDVI). The glucosylation also identifies phage DNA and allows selective degradation of host DNA by phage-encoded nucleases. Nonglucosylated T-even phages are resistant to some classical R–M systems because their DNA contains the modified base HMC, but they are sensitive to modification-dependent systems, although T-even phages encode a protein (Arn) that protects superinfecting phages from McrBC restriction. It has been suggested that phages have evolved DNA containing HMC to counteract classical R–M systems, and that host-encoded modification-dependent endonucleases are a response to this phage adaptation. In this evolutionary story, the glucosylation of HMC would be the latest mechanism that renders T-even phages totally resistant to most R–M systems.

In some cases, a phage genome can tolerate a few targets for certain restriction enzymes. The few *Eco*RII sites in T3 and T7 DNA are not sensitive to restriction because this unusual enzyme requires at least two targets in close proximity and the targets in these genomes are not sufficiently close. For the type III enzymes, the orientation of the target sequences is also relevant. Since the target for restriction requires two inversely oriented recognition sequences, the T7 genome remains refractory to *Eco*P15I because all 36 recognition sequences are in the same orientation. The unidirectional orientation of the target sequences is consistent with selection for a genome that will avoid restriction. Considerable evidence supports the significance of counterselection of target sequences in phage genomes, in some cases correlating the lack of target sequences for enzymes found in those hosts in which the phages can propagate.

V. DISTRIBUTION, DIVERSITY, AND EVOLUTION

A. Distribution and Diversity

R–M systems are probably ubiquitous among prokaryotes. The most complete documentation is that for type II systems. This is a result of the many *in vitro* screens for sequence-specific endonucleases put to effective use in the search for enzymes with different specificities. The endonucleases identified include more than 200 different sequence specificities. Type II restriction enzymes have been detected in 11 of the 13 phyla of Bacteria and Archaea, and the lack of representation in two phyla, Chlamydia and Spirochaetes, could result from a sampling bias since representatives of these two phyla are commonly difficult to culture.

There is no reliable screen for type I systems and there has been no practical incentive to search for enzymes that cut DNA at variable distances from their recognition sequence. The apparent prevalence of type I systems in enteric bacteria may simply reflect the common use of *E. coli* and its relatives in genetic studies. Genetic tests currently provide evidence for type I enzymes with approximately 20 different specificities, and recent analyses of genomic sequences from diverse species support the idea that type I systems are widely distributed throughout the prokaryotic kingdom.

Despite the identification of many restriction enzymes, little has been reported about the relative frequency, distribution, and diversity of R–M systems within natural populations of bacteria. Janulaitis and coworkers (1988) screened natural isolates of *E. coli* for sequence-specific ENases and detected activity in 25% of nearly 1000 strains tested. Another screening experiment searched for restriction activity encoded by transmissible resistance plasmids in *E. coli*. The plasmids were transferred to *E. coli* and the EOP of phage λ was determined on the exconjugants. Approximately 10% of the transmissible antibiotic resistance plasmids were correlated with the restriction of λ and the ENases responsible were shown to be type II. However, plasmid-borne type I, type II, and type III systems are known in *E. coli*. Because of the transmissible nature of many plasmids, the frequency with which R–M systems are transferred between strains could be high and their maintenance subject to a variety of selection pressures not necessarily associated with the restriction phenotype.

The evidence, where available (i.e., for *E. coli* and *Salmonella enterica*), is consistent with intraspecific diversity irrespective of the level at which the diversity is examined. In *E. coli*, there are at least six distinct mechanistic classes of restriction enzyme (types I–III, and three modification-dependent types) and the anomalous *Eco*571 (Fig. 2). The type II systems in *E. coli* currently include approximately 30 specificities, and at least 14 type I specificities have been identified. Bacterial strains frequently have more than one active restriction system. Four systems are present in *E. coli* K-12, and *H. influenzae* Rd has at least three systems that are known to be biologically active, whereas three more are indicated in the genomic sequence.

B. Evolution

R–M enzymes may be dissected into modules. A type II MTase comprises a TRD and a module that is responsible for catalyzing the transfer of the methyl group from AdoMet to the defined position on the relevant base. The catalytic domains share sequence similarities, and these are most similar when the catalytic reaction is the same, i.e., yields the same product (e.g., 5mC). Given the matching specificities of cognate ENase and MTase, it might be expected that their TRDs would be of similar amino acid sequence. This is not the case; it seems likely that the two enzymes use different strategies to recognize their target sequence. Each subunit of the dimeric ENase needs to recognize one-half of the rotationally symmetrical sequence, whereas the monomeric MTase must recognize the entire sequence. The absence of similarity between the TRDs of the ENase and its cognate MTase suggests that they may have evolved from different origins.

Restriction enzymes that recognize the same target sequence are referred to as isoschizomers. A simple expectation is that the TRDs of two such enzymes would be very similar. This is not necessarily true. Furthermore, the similarities observed do not appear to correlate with taxonomic distance. The amino acid sequences of the isoschizomers *Hae*III and *Ngo*PII, which are from bacteria in the same phylum, show little if any similarity, whereas the isoschizomers *Fnu*DI and *Ngo*PII, which are isolated from bacteria in different phyla, are very similar (59% identity).

Type I R–M systems are complex in composition and cumbersome in their mode of action, but they are well suited for the diversification of sequence specificity. A single subunit (HsdS or S) confers specificity to the entire R–M complex and to the additional smaller complex that is an MTase. Any change in specificity affects restriction and modification concomitantly. Consistent with their potential to evolve new specificities, type I systems exist as families, within which members (e.g., *Eco*KI and *Eco*BI) are distinguished only by their S subunits. Currently, allelic genes have been identified for at least seven members of one family (IA); each member has a different specificity. It is more surprising that allelic genes in *E. coli,* and its relatives, also specify at least two more families of type I enzymes. Although members of a family include only major sequence differences in their S polypeptides, those in different families share very limited sequence identities (usually 18–30%). Clearly, the differences between gene sequences for type I R–M systems are no indication of the phylogenetic relatedness of the strains that encode them. Note that despite the general absence of sequence similarities between members of different families of type I enzymes, pronounced similarities have been identified for TRDs from different families when they confer the same sequence specificity.

The information from gene sequences for both type I and type II systems, as stated by Raleigh and Brooks (1998), "yields a picture of a pool of genes that have circulated with few taxonomic limitations for a very long time."

Allelic variability is one of the most striking features of type I R–M systems. Both the bipartite and asymmetrical natures of the target sequence confer more scope for diversity of sequence specificity than the symmetrical recognition sequences of type II systems. The S subunit of type I enzymes includes two TRDs, each specifying one component of the target sequence. This organization of domains makes the subunit well suited to the generation of new specificities as the consequences of either new combinations of TRDs or minor changes in the spacing between TRDs. In the first case, recombination merely reassorts the regions specifying the TRDs, and in the second case unequal crossing over within a short duplicated sequence leads to a change in the spacing between the TRDs. Both of these processes have occurred in the laboratory by chance and by design.

For type I R–M systems the swapping or repositioning of domains can create enzymes with novel specificities, but the evolution of new TRDs with different specificities has not been witnessed. In one experiment, strong selection for a change that permitted a degeneracy at one of the seven positions within the target sequence failed to yield mutants with a relaxed specificity.

VI. BIOLOGICAL SIGNIFICANCE

The wide distribution and extraordinary diversity of R–M systems, particularly the allelic diversity documented in enteric bacteria, suggest that R–M systems have an important role in bacterial communities. This role has traditionally been considered to be protection against phage. Laboratory studies following bacterial populations under conditions of phage infection indicate that R–M systems provide only a transitory advantage to bacteria. Essentially, an R–M system with a different specificity could assist bacteria in the colonization of a new habitat in which phage are present, but this advantage would be short-lived as phages that escape restriction acquire the new protective modification and bacteria acquire mutations conferring resistance to the infecting phages. It can be argued that one R–M system protects against a variety of phages, and the maintenance of one R–M system may compromise the fitness of the bacterium less than the multiple mutations required to confer resistance to a variety of phages. No direct evidence supports this expectation. It is relevant to remember that the restriction barrier is generally incomplete, irrespective of the mechanism of DNA transfer, and that the fate of phage and bacterial DNA fragmented by ENases may differ. A single cut in a phage genome is sufficient to prevent infectivity. Fragments generated from bacterial DNA will generally share homology with the host chromosome and could therefore be rescued by recombination. The rescue of viable phages by homologous recombination requires infection by more than one phage or recombination with phage genomes that reside within the host chromosome. A protective role for R–M systems in no way excludes an additional role that influences genetic recombination.

In *E. coli,* and probably bacteria in general, linear DNA fragments are vulnerable to degradation by exonucleases, particularly ExoV (RecBCD). Therefore, the products of restriction are substrates for degradation by the same enzyme that is an essential component of the major recombination pathway in *E. coli.* However, degradation by RecBCD is impeded by the special sequences, designated Chi, that stimulate recombination. It has been shown that a Chi sequence can stimulate recombination when RecBCD enters a DNA molecule at the site generated by cutting with *Eco*RI. It seems inevitable that fragmentation of DNA by restriction would reduce the opportunity for recombination to incorporate long stretches of DNA; however, given that DNA ends are recombinogenic, restriction could promote the acquisition of short segments of DNA.

Radman and colleagues (1989) suggested that R–M systems are not required as interspecific barriers to recombination since the DNA sequence differences between *E. coli* and *Salmonella* are sufficient to hinder recombination. It is evident, however, that selection has maintained a diversity of restriction specificities within one species, and consequently restriction is presumed to play a significant role within a species, in which DNA sequence differences are less likely to affect recombination. Detailed analyses of the effects of restriction on the transfer of DNA between strains of *E. coli* are currently under way and the molecular techniques are available to monitor the sizes and distributions of the DNA fragments transferred between strains.

Recently, Kobayashi and colleagues viewed R–M genes as "selfish" entities on the grounds that loss of the plasmid that encodes them leads to cell death. The experimental evidence for some type II R–M systems implies that the cells die because residual ENase activity cuts incompletely modified chromosomal DNA. The behavior of type I systems, on the other hand, is different and is consistent with their ability to diversify sequence specificity; when new specificities are generated by recombination, old ones are readily lost without impairing cell viability.

VII. APPLICATIONS AND COMMERCIAL RELEVANCE

Initially, the opportunity to use enzymes that cut DNA molecules within specific nucleotide sequences added a new dimension to the physical analysis of small genomes. In the early 1970s, maps (restriction maps) could be made in which restriction targets were charted within viral genomes and their mutant derivatives. Within a few years the same approach was generally applicable to larger genomes. The general extension of molecular methods to eukaryotic genomes depended on the technology that enabled the cloning of DNA fragments, i.e., the generation of a population of identical copies of a DNA fragment. In short, DNA from any source could be broken into discrete fragments by restriction ENases, the fragments could be linked together covalently by the enzymatic activity of DNA ligase, and the resulting

new combinations of DNA could be amplified following their recovery in *E. coli*. Of course, to achieve amplification of a DNA fragment, and hence a molecular clone, it was necessary to link the DNA fragment to a special DNA molecule capable of autonomous replication in a bacterial cell. This molecule, the vector, may be a plasmid or a virus. Importantly, it is usual for only one recombinant molecule to be amplified within a single bacterial cell. In principle, therefore, one gene can be separated from the many thousands of other genes present in a eukaryotic cell, and this gene can be isolated, amplified, and purified for analysis. The efficiency and power of molecular cloning have evolved quickly, and the new opportunities have catalyzed the rapid development of associated techniques, most notably those for determining the nucleotide sequences of DNA molecules, the chemical synthesis of DNA, and recently, the extraordinarily efficient amplification of gene sequences *in vitro* by the polymerase chain reaction (PCR). In some cases, amplification *in vitro* obviates the need for amplification *in vivo* since the nucleotide sequence of PCR products can be obtained directly.

The bacterium *E. coli* remains the usual host for the recovery, manipulation, and amplification of recombinant DNA molecules. However, for many of the commonly used experimental organisms the consequence of a mutation can be determined by returning a manipulated gene to the chromosome of the species of origin.

The recombinant DNA technology, including screens based on the detection of DNA by hybridization to a specific probe and the analysis of DNA sequence, is now basic to all fields of biology, biochemistry, and medical research as well as the "biotech industry." Tests dependent on DNA are used to identify contaminants in food, parents of children, persons at the scene of a crime, and the putative position of a specimen in a phylogenetic tree. Mutations in specific genes may be made, their nature confirmed, and their effects monitored. Gene products may be amplified for study and use as experimental or medical reagents. Hormones, cytokines, blood-clotting factors, and vaccines are amongst the medically relevant proteins that have been produced in microorganisms, obviating the need to isolate them from animal tissues.

Most of the enzymes used as reagents in the laboratory are readily available because the genes specifying them have been cloned in vectors designed to increase gene expression. This is true for the ENases used to cut DNA. It is amusing that in the 1980s the generally forgotten, nonclassical, restriction systems identified by Luria and Human (1952) were rediscovered when difficulties were encountered in cloning type II R–M genes. It was soon appreciated that cloning the genes for particular MTases was a problem in "wild-type" *E. coli* K-12; the transformed bacteria were killed when modification of their DNA made this DNA a target for the resident Mcr restriction systems. Rare survivors were *mcr* mutants, ideal strains for recovering clones of foreign DNA rich in 5mC, as well as genes encoding MTases.

See Also the Following Articles

BACTERIOPHAGES • CONJUGATION, BACTERIAL • RECOMBINANT DNA, BASIC PROCEDURES • TRANSFORMATION, GENETIC

Bibliography

Barcus, V. A., and Murray, N. E. (1995). Barriers to recombination: Restriction. *In* "Population Genetics of Bacteria" (S. Baumberg, J. P. W. Young, S. R. Saunders, and E. M. H. Wellington, Eds.), Society for General Microbiology Symposium No. 52, pp. 31–58. Cambridge Univ. Press, Cambridge, UK.

Bickle, T. A., and Krüger, D. H. (1993). Biology of DNA restriction. *Microbiol. Rev.* **57**, 434–450.

Cheng, X., and Blumenthal, R. M. (1996). Finding a basis for flipping bases. *Structure* **4**, 639–645.

Raleigh, E. A., and Brooks, J. E. (1998). Restriction modification systems: Where they are and what they do. *In* "Bacterial Genomes: Physical Structure and Analysis" (F. J. de Bruijn, J. R. Lupski, and G. M. Weinstock, Eds.), pp. 78–92. Chapman & Hall, New York.

Wilson, G. G., and Murray, N. E. (1991). Restriction and modification systems. *Annu. Rev. Genet.* **25**, 585–627.

DNA Sequencing and Genomics

Brian A. Dougherty

Bristol–Myers Squibb Pharmaceutical Research Institute

GLOSSARY

bioinformatics Biological informatics, specifically the computer-assisted analysis of genome sequence and experimental data.

DNA microarrays Miniaturized arrays containing thousands of DNA fragments representing genes in an area of a few square centimeters. These "chips" are used mostly to monitor the expression of genomes at the level of messenger RNA.

functional genomics The study of gene function at the genome level.

genomics The study of the genome, or all of the genes in an organism.

proteomics The study of the protein complement, or proteome, of an organism.

whole genome shotgun sequencing A random approach to genome sequencing based on shearing of genomic DNA, followed by cloning, sequencing, and assembly of the entire genome, and ultimately leading to finished, annotated genomic sequence. Sequencing of entire genomes can also be achieved by a sequential shotgun sequencing of a set of overlapping, large insert clones.

ADVANCES IN DNA SEQUENCING, computing, and automation technologies have allowed the sequencing of entire genomes from living organisms. Bacterial genomes range in size from 0.5 to at least 10 Mb (1×10^6 base pairs) and are composed of approximately 90% coding regions; therefore, high-throughput sequencing of randomly cloned DNA fragments represents the most cost-effective way to rapidly obtain genome information for these organisms.

Since the publication in 1995 of the first whole genome sequence for a bacterium, a complete microbial genome sequence has been published every several months (Table I), and high coverage sequence data have been added to the public domain throughout this period for more than 50 other genomes. Sequencing of a microbial genome requires logistical planning for the challenges of high-throughput sequencing, bioinformatic analysis, and presentation of the data in a concise, user-friendly format, usually in the form of a World Wide Web site to supplement a journal publication; this is an intensive process that is often underappreciated. Following the sequencing, analysis, and initial assignment of function based on similarity to previously identified genes in sequence databases, analysis of gene function at the laboratory bench is required to confirm gene function. These functional genomics studies will also require a substantial scale-up in order to keep up with the high-throughput pace of genome sequencing. Simply stated, in the pregenomic era genes were usually identified based on a selection for function, followed by the laborious tasks of cloning, subcloning, and sequencing; genome sequencing has reversed this process today, and the challenge of the postgenomic era will be to determine how all the gene products of a genome function, interact, and allow the organism to live.

TABLE I
Published Microbial Genomes[a]

Genome	Size (Mb)	% G + C	Institution	Reference
Aquifex aeolicus	1.55	43	Diversa	Deckert et al., Nature 392, 353 (1998)
Archaeoglobus fulgidus	2.18	48	TIGR	Klenk et al., Nature 390, 364 (1997)
Bacillus subtilis	4.21	43	International Consortium	Kunst et al., Nature 390, 249 (1997)
Borrelia burgdorferi	1.52	28	TIGR	Fraser et al., Nature 390, 580 (1997)
Chlamydia trachomatis	1.04	41	Stanford University	Stephens et al., Science 282, 754 (1998)
Escherichia coli	4.64	50	University of Wisconsin	Blattner et al., Science 277, 1453 (1997)
Haemophilus influenzae Rd	1.83	39	TIGR	Fleischmann et al., Science 269, 496 (1995)
Helicobacter pylori	1.66	39	TIGR	Tomb et al., Nature 388, 539 (1997)
Methannococcus jannaschii	1.74	31	TIGR	Bult et al., Science 273, 1058 (1996)
Methanobacterium thermo-autotrophicum	1.75	49	GTC/Ohio State University	Smith et al., J. Bacteriol. 179, 7135 (1997)
Mycobacterium tuberculosis	4.41	65	Sanger Centre	Cole et al., Nature 393, 537 (1998)
Mycoplasma genitalium	0.58	31	TIGR	Fraser et al., Science 270, 397 (1995)
Mycoplasma pneumoniae	0.82	40	University of Heidelberg	Himmeireich et al., Nucleic Acid Res. 24, 4420 (1996)
Pyrococcus horikoshii	1.74	42	NITE	Kawarabayasi et al., DNA Res. 5, 55 (1998)
Rickettsia prowazekii	1.11	29	University of Uppsala	Andersson et al., Nature 396, 133 (1998)
Saccharomyces cerevisiae	12.07	38	International Consortium	Goffeau et al., Nature 387 (Suppl.), 5–105 (1997)
Synechocystis sp.	3.57	47	Kazusa DNA Research Institute	Kaneko et al., DNA Res. 3, 109 (1996)
Treponema pallidum	1.14	52	TIGR/University of Texas	Fraser et al., Science 281, 375 (1998)

[a] Current as of October 1998.

I. INTRODUCTION: DNA SEQUENCING

DNA sequencing has seen many technological changes during the past two decades, evolving from slab polyacrylamide gels that could resolve 20–50 bp at a time to the 700-bp reads of today. In the late 1970s, DNA sequences were manually read from x-ray film using chemical and enzymatic methods to incorporate radiolabel; for the past 10 years, computers have been making automated base calls after detecting fluorescently labeled DNA fragments resolved on sequencers such as the Applied BioSystems 377 model. The current state-of-the-art capillary sequencers should deliver higher throughput with a greater degree of automation. Due to the recent availability of these instruments, a greatly accelerated schedule for completing the Human Genome Project has been proposed by multiple groups of researchers. Two of these 96-capillary sequencers are the Mega-BACE sequencer (Molecular Dynamics/Amersham-Pharmacia Biotech) and the much anticipated Applied BioSystems 3700 model, which promises about 300–400 kb of sequence per day with nearly unattended operation (a 1-hr manual-loading step for 1 day's worth of automated sequencing).

Upon publication of the 1.8-Mb chromosome of *Haemophilus influenzae* in 1995, it became apparent to the scientific community that sequencing of bacterial and even eukaryotic genomes was possible by scaling-up existing sequencing technologies. Although the utility of the whole-genome shotgun approach (Fig. 1) for sequencing microbial genomes is unquestionable today, the original *H. influenzae* project was in fact deemed an unproven, high-risk venture and was not funded because such a large segment of DNA had not been sequenced by a random approach. The standard paradigm for any large-scale sequencing projects was based on strategies used for sequencing the human genome and related model organisms. These "top-down" techniques consisted of two phases, an up-front mapping and cos-mid-ordering phase, followed by subcloning and sequencing of a minimal tiling path (a complete set of cloned genome fragments with minimal overlap among the cosmids). Sequencing of larger segments of a microbial genome was not attempted due to a computational limitation at that time—existing software packages for the assembly of random DNA fragments were not sufficiently robust to assemble a DNA segment much larger than the size of a cosmid insert (~40 kbp). However, advances based on an unconventional approach to sequencing the human genome set the stage for whole-genome sequencing of smaller genomes. The Venter laboratory at the National Institutes of Health (NIH) and later at The Institute for Genomic Research (TIGR) developed the expressed sequence tag (EST) method for streamlining gene discovery in human genome sequencing by enriching for and sequencing messenger RNAs. By optimizing the conditions for high-throughput fluorescent sequencing and developing the computational software to assemble hundreds of thousands of random sequences into contiguous sequences (contigs), the stage was set to attempt a shotgun approach to sequencing an entire genome, which had last been done for bacteriophage λ in 1977.

The whole genome shotgun sequencing technique is described in the following section, and it consists of four phases: library construction, random se-

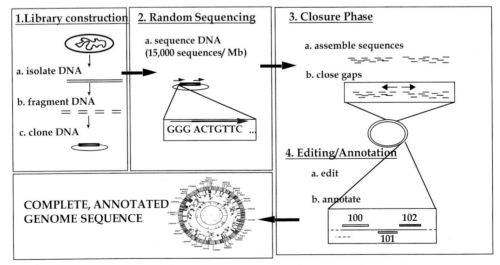

Fig. 1. Strategy for a whole genome random sequencing project.

quencing, gap closure, and editing/annotation (Fig. 1).

II. MICROBIAL GENOME SEQUENCING PROJECT

A. Library Construction Phase

Library construction is the most critical phase of a microbial genome sequencing project. Assuming a completely random library, Lander and Waterman calculated that the sequencing statisitics would follow a Poisson distribution (i.e., sequencing enough random reads to add up to one genome equivalent would leave 37% of the genome unsequenced; see Table II). Sequencing of a nonrandom library will result in a significant deviation from the Lander–Waterman model for random sequencing, and the generated assemblies will have too many gaps to be efficiently closed. A library construction procedure has been developed (Fleischmann *et al.*, 1995) to achieve a high order of randomness, and it consists of the following steps:

1. Random shearing of genomic DNA and purification of ~2 kbp fragments
2. Fragment end repair and ligation to blunt-ended, dephosphorylated vector
3. Gel purification of the "v+i" (vector plus single insert) band from other forms

4. Final polishing of free ends and intramolecular ligation of v+i DNA
5. High-efficiency transformation of *Escherichia coli* strains (DH10B, SURE, etc.)
6. Direct plating onto two-layer antibiotic diffusion plates

This optimized method results in efficient sequencing because the library is a collection of highly random, single-insert fragments that can be sequenced from both ends using universal primers. Key factors for success include random shearing of the DNA (usually by nebulization), purification of a narrow size range of insert (using minimal amounts of longwave UV light), a second gel purification of v+i to minimize plasmid clones lacking insert or containing chimeric inserts, propagation of clones in a highly restriction-deficient *E. coli* background, and outgrowth of transformed cells as individual colony-forming units on an antibiotic diffusion plate (rather than standard outgrowth as a mixture of clones in liquid medium, which may select against certain slower growing clones in the transformed population). Even with all these safeguards in place, however, gaps in the genome do occur. This is due to both the statistics of random sequencing and the inability to clone certain DNA fragments, such as those containing strong promoters (e.g., the 16S rRNA promoter in *H. influenzae*) or "toxic" genes (e.g., the complete *Hind*III restriction enzyme gene

TABLE II
**Lander–Waterman Calculation for Random Sequencing of a Microbial Genome
(2-Mb Size, 500-bp Average Sequence Read Length)**

No. sequences	% genome sequenced	No. gaps	Average gap size	Fold coverage	Total bp sequenced
4,000	63.21	1472	500	1×	2,000,000
8,000	86.47	1083	250	2×	4,000,000
12,000	95.02	597	167	3×	6,000,000
16,000	98.17	293	125	4×	8,000,000
20,000	99.33	135	100	5×	10,000,000
24,000	99.75	59	83	6×	12,000,000
28,000	99.91	26	71	7×	14,000,000
32,000	99.97	11	63	8×	16,000,000

without the cognate methylase gene present in the transformed *E. coli* strain).

B. Random Sequencing Phase

After the library is constructed, random sequencing of the individual clones is performed. This involves (i) the production of tens of thousands of templates in 96-well blocks, (ii) sequencing of the purified plasmid template DNA in both directions, (iii) transfer of the edited sequence reads to a database, and (iv) assembly of the sequence data into contigs.

To ensure that the expected Poisson distribution of random sequences is obtained, the assembled sequence data from the first several thousand sequences are plotted relative to the Lander–Waterman equation (Fig. 2). The sequencing and assembly of a verified random library continues until approximately 7- or 8 × genome coverage is achieved, which should result in a manageable number of gaps to close (Table II). Based on diminishing return, after sequencing to 7- or 8× random genome coverage with the 2-kbp insert library, directed sequencing strategies are employed for gap closure; for genome sequencing projects in which gap closure will not be pursued, the project goes directly to the annotation phase.

C. Closure Phase

Of the three phases of a whole-genome sequencing project discussed so far, this is the most time-consuming; however, closure of a genome results in a complete list of the set of genes. If a certain gene or metabolic pathway is missing from a closed genome, its absence is not due to it being located in an unsequenced gap. Moreover, a complete genome is a linear string of nucleotides, rather than a collection of contiguous sequences of uncertain orientation relative to one another, and contains genes in linear progression with exact 5′ and 3′ coordinates.

The forward and reverse reads of the plasmid clones are an important tool used for genome closure. Once the assembled sequences are obtained, any template with a forward sequence read in one assembly and a reverse in the other represents a link between assemblies and is used as a template for primer walking to close the gap. Clone linkage for closure of potentially larger gaps is provided by end sequencing a large insert lambda library and mapping positions of the forward and reverse reads (separated by about 20 kbp) on the assembled contigs.

The previously mentioned gaps are called "sequencing" gaps since a DNA template is available; closure of "physical" gaps is more challenging because no template is immediately available. De-

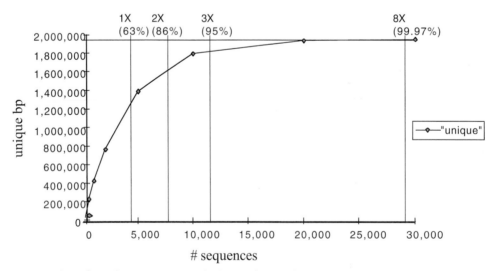

Fig. 2. Plot of Lander–Waterman calculation for random sequencing for a 2-Mb genome with 500-bp average sequence read lengths.

pending on the number of gaps to close and tools at one's disposal, a template is generated for a physical gap by several means, including combinatorial polymerase chain reaction (PCR), screening for large insert clones that span the gap, Southern hybridization fingerprinting, and cycle sequencing using genomic DNA as a template. Once a template is obtained, it is sequenced until the physical gap is closed. The gap closure data are then incorporated into the assembly data, the assembled contigs are joined, and the process is repeated until closure of the entire genome is achieved.

D. Editing and Annotation Phase

Following genome closure the sequence is edited and annotated. Editing consists of proofreading the sequence and resolving ambiguities or regions of low coverage by dye terminator sequencing. In addition, possible frameshifts are identified, PCR amplified from genomic DNA, and resequenced. It should be noted, however, that some apparent base insertions, deletions, and frameshifts are authentic. Examples include polymorphisms in repetitive sequences of the *Campylobacter jejuni* genome and more than 20 authentic frameshifts in protein coding regions of the lab-propagated *H. influenzae* Rd organism (several of the counterparts of these genes in other *H. influenzae* strains have been demonstrated to be full-length, functional genes.)

Annotation begins with gene finding, which is not an insignificant task in prokaryotes; although stop sites of open reading frames (ORFs) are straightforward, identifying the proper translation initiation codon can be challenging. Annotation also involves characterization of other features of the genome, including repeated nucleotide sequences, gene families, and variations in nucleotide composition. Furthermore, all genes are classified into functional role categories and analyzed as a whole for the presence of various metabolic pathways. In this way, whole-genome sequencing provides a complete picture of the metabolic potential of an organism and lays the groundwork for many follow-up studies once the genetic complement of the cell has been defined.

III. BIOINFORMATICS AND GENOME DATABASES

The need for computational tools to handle genome project data is increasing, and the URLs listed in Table III represent a fraction of the genomic websites that continue to proliferate. The genome project monitoring sites given are among the most comprehensive and are continually updated as new genome projects are both initiated and completed. In fact, it is estimated that in addition to the 18 microbial genomes currently completed (Table I), approximately 50 more should be done by the Year 2000. The genome project websites listed provide extensive annotation, role categorization, search capabilities, and genome segment display and retrieval. In the near future, experimental data from functional genomics work (expression of gene under various conditions, phentoype of knockout mutant, protein–protein interaction information, etc.) will be incorporated into these existing structures and provide much added value to the basic research community.

During the course of a genome sequencing project, a variety of tools are required. At the DNA sequencing and assembly stage, software for high-throughput sequencing is needed for automated base calling (e.g., phred and ABI base caller) and data tracking/management. An assembly program (e.g., phrap and TIGR Assembler) is then used to find overlaps among the random sequences and build large sets of contiguous sequence. Gaps in the sequence are then closed by directed sequencing; programs to clearly display, track, and update this information are crucial. Gene finding is then performed using programs that search for ORFs among the assembled sequence data (e.g., Genemark, Critica, and Glimmer) and that use computational techniques to help discriminate the signal of real genes from the noise of potential ORFs found in a microbial genome. The annotation stage involves human inspection of the list of candidate genes and other data to help generate a catalog of genes from the organism (more like a rough draft rather than a final list). Procedures used to determine if a gene is authentic include examining similarity scores, multiple sequence alignments, and published data when similar to that of known genes and examining the

TABLE III
Selected Genomic Resources from the World Wide Web

Complete and in-progress genome projects

http://www.tigr.org/tdb/mdb/mdb.html	TIGR Microbial Database
http://geta.life.uiuc.edu/~nikos/genomes.html	Nikos' Genome Sequencing Projects List
http://www-c.mcs.anl.gov/home/gaasterl/genomes.html	MAGPIE Genome Sequencing Projects List
http://www.ncbi.nlm.nih.gov/BLAST/unfin_databases.html	Entrez Microbial Genomes Listing

Genome sequence data

http://www.ncbi.nim.nih.gov/Entrez/Genome/org.html	Entrez Genome Browser
http://www.tigr.org/tdb/mdb/mdb.html	TIGR Microbial Database; links to projects
http://www.pasteur.fr/Bio/SubtiList.html	Subtilist server for *B. subtilis*
http://www.pasteur.fr/Bio/Colibri.html	Colibri server for *E. coli*
http://www.ncbi.nim.nih.gov/PMGifs/Genomes/yc4.html	NCBI Yeast Genome; links to YPD, MIPS, SGD

Analysis of Genome Sequence Data

http://mol.genes.nig.ac.jp/gib/	Genome Information Broker
http://motif.stanford.edu/	The Brutlag Bioinformatics Group
http://www.expasy.ch/	Expasy Molecular Biology Server
http://www.ncbi.nlm.nih.gov/COG/	Clusters of Orthologous Groups (COGs) site
http://www.sanger.ac.uk/Software/Pfam/	P fam (Protein families database, HMMs)
http://lion.cabm.rutgers.edu/~bruc/microbes/index.html	SEEBUGS Microbial Genome Analysis
http://www.genome.ad.jp/kegg/	KEGG: Kyoto Encyclopedie of Genes and Genomes
http://ecocyc.PangeaSystems.com/ecocyc/ecocyc.html	EcoCyc: Encyclopedia of *E. coli* Genes/Metabolism

Compilations of genome-related sites/links

http://www.hgmp.mrc.ac.uk/GenomeWeb/	GenomeWeb List of Other Genome Sites
http://www.genome.ad.jp/	GenomeNet WWW Server
http://www.public.iastate.edu/~pedro/research_tools.html	Pedro's Biomolecular Research Tools

sequence for gene-like characteristics (e.g., the presence and proper spacing of a promoter, ribosome binding site, and initiation codon; the presence of a similar gene in the database; the position of the gene in an operon; or nucleotide or codon usage similar to that of the rest of the genome). Functional role category can also be assigned based on similarity to genes of known function, and a picture of the metabolic capacity of the organism emerges. Finally, it should be emphasized that quality annotation is based on bioinformatic analysis beyond assigning the top BLAST similarity score. Motif searches (e.g., Prosite, Blocks, and e-motif) reveal local similarity that may be missed at the global level of a BLAST search. Also, multiple sequence alignments and phylogenetic relationships between members of a gene family can provide insight into possible function. Searches using hidden Markov models are currently providing information about relationships between

predicted proteins that may be missed by BLAST searches, and gene families are being constructed using hidden Markov modeling (Pfam) and other techniques (e.g., COGs). Finally, molecular modeling comparisons such as threading are being used to gain insight into relationships between proteins seen only at the three-dimensional level.

IV. FUNCTIONAL GENOMICS APPROACHES

A. Comparative Genomics

The resources required for whole-genome sequencing of a 2-Mb genome are not insignificant, with a cost of approximately $1 million and requiring 1 or 2 years to complete. However, high-throughput sequencing of microbial DNA is the most cost-effective method to obtain genomic information: given a coding density of 85–90%, a 2-Mb genome would give 1800 gene sequences at a cost of ~$500/gene (significantly less than the per gene cost of the pregenomic era). This cost is reduced even further by sequencing at a lower average redundancy (3–6× coverage), providing a catalog of the majority of the genes in an organism of interest. These data are sufficient for the purpose of rapidly acquiring a proprietary database of genomic data, often from multiple organisms, and performing functional genomics work with the ultimate goal of producing vaccines or antibiotics. However, the goal of the complete genome project, published in a peer-reviewed journal, is to sequence every base pair of a genome and become a foundation for downstream research by providing a contiguous list of genes, exact 5′ and 3′ coordinates for each gene, reconstruction of metabolic pathways while accounting for all genes in an organism, and other qualities not available from collections of large contigs.

There are some cases in which genome sequences from two isolates of the same species exist, but often one genome is held as proprietary information (this happens regularly for important microbial pathogens). Genome projects in which data are publicly available for two species of bacteria include those for *Neisseria meningitidis* (one serotype A and the other serotype B) and for *Mycobacterium tuberculosis* (H37Rv is completed, and CSU-93 is in progress). The latter case will be of special interest because the CSU-93 is a highly infectious strain of *M. tuberculosis,* and it is possible that "*in silico*" comparisons of the genome sequences relative to H37Rv may point to the mutation(s) or gene(s) responsible for this phenotype. Finally, comparisons by the Blattner laboratory of the completed *E. coli* K-12 strain to the enterohemorrhagic *E. coli* O157:H7 have provided some interesting findings. Although conventional wisdom would have suggested that the pathogenic O157 was a K-12 backbone plus some large segments of additional DNA, such as the 92-kb virulence plasmid or the 43-kb LEE pathogenicity island, it was determined from shotgun sequencing that the O157 genome is actually a fine mosaic of O157 and K-12 DNA and includes 1.2 Mb of DNA unique to O157 while lacking ~0.15 Mb of DNA known to be unique

to the completed K-12 genome. These results provide a glimpse into the importance of comparative genomics work and the complex ways in which genomes/microbes evolve and adapt.

B. RNA-Level Differential Gene Expression

Researchers are interested in differential gene expression; that is, what genes are up- and down-regulated under a specific set of conditions (wild-type vs isogenic mutant, growth with and without an antibiotic present, and growth in the lab vs in a host for pathogens). Prior to genome sequences being widely available, differential display PCR was used to identify differentially expressed genes. Random primers are used for reverse transcription of message followed by amplification and agarose electrophoresis. Differentially labeled bands are isolated, cloned, and sequenced to determine the expressed genes. With eukarotic cells, quantitative EST sequencing or SAGE (serial anaylsis of gene expression) can also be used for differential gene expression without *a priori* knowledge of genome sequences, but the density of genes in bacteria coupled with the inability to apply mRNA-enrichment procedures based on polyadenylation has limited the use of these techniques in bacteria.

Molecular techniques previously used to study a single gene are now being applied to entire genomes. In RNA-level gene expression, researchers have applied DNA-blot hybridization and RT-PCR techniques to monitor the expression of thousands of genes at once. DNA microarrays can be thought of as high-throughput dot blots using DNA on a solid "chip" surface to report expression patterns for thousands of genes. Several different microarraying technologies are being used, but costs are sufficiently high that the equipment is currently beyond the reach of the average lab. For synthetic arrays, such as the Affymetrix chip (*http://www.affymetrix.com*), photolithographic techniques are used to synthesize oligonucleotides on a 1.2-cm^2 silicon chip (~64,000 addressable positions per chip). For spotted arrays, such as those made by Stanford University (*http://cmgm.stanford.edu/pbrown*) or Molecular Dynamics (*http://www.mdyn.com*), an x-y-z-stage robot is used to deposit DNA spots, usually PCR products, onto a treated glass microscope slide. In either case, DNA representing thousands of genes are linked to a solid surface, then hybridized with fluorescently labeled RNA from different conditions, followed by confocal scanning at different wavelengths to quantitate hybridized probe. The use of different fluorescent dyes for each experimental condition allows for greater sensitivity and, in the case of the spotted arrays, simultaneous hybridization of two or more samples on the same chip. Thus, expression data from thousands of genes (even whole genomes in the case of microbes) are then deconvoluted and analyzed.

C. Protein-Level Differential Gene Expression

Just as gene expression can be monitored at the transcription level, it can also be measured at the translation level by analyzing the entire protein complement of the genome, or proteome. At the core of these proteomics experiments is a separation technology that was developed more than two decades ago—two-dimensional polyacrylamide gel electrophoresis (2D-PAGE). Today, however, a new generation of instruments using mass spectroscopy (MS; *http://www.asms.org*) techniques afford highly accurate mass determination, giving a rapid, high-throughput path from a spot on a 2D-PAGE gel to an identified protein. Translation of whole-genome DNA sequence data gives the theoretical mass and isoelectric point for all protein gene products, and studies indicate that proteins identified from 2D-PAGE gels by electrospray MS or amino acid sequencing techniques correlate closely to these theoretical values (although approximately 20% of the spots represent isoforms, which are protein modifications that alter predicted migration positions of proteins). Variations on these techniques are being used, such as MALDI-TOF (matrix-assisted laser desorption/ionization time-of-flight) and SELDI (surface-enhanced laser desorption/ionization). The SELDI technique is an interesting twist on chip technology, giving a "protein chip" (Ciphergen, *http://www.ciphergen.com*) capable of on-chip protein enrichment and real-time estimates of protein mass.

One question that needs to be resolved in the postgenomic era is whether differential gene expression is best measured at the RNA or protein level; it is most likely, however, that researchers doing whole-genome expression studies will want to use both microarraying and proteomics technologies to assess gene expression levels.

D. Mutagenesis

With microbes, one of the classic methods for determining gene function is via mutagenesis. For many large-scale studies, transposons or antibiotic cassettes are used to disrupt genes, providing an isogenic pair of strains, a wild type, and a null mutant. By sequencing out from transposon or cassette, one can determine a "genome sequence tag" for the disrupted gene. In the pregenomic era, identification of the entire gene and surrounding genes/operons would have involved a substantial effort of cloning and sequencing, but because of the availability of genomic sequence data, the tag and sequence data allow one to rapidly map large numbers of mutants. The recent development of *in vitro* transposition systems has simplified the transposon-hopping step to a microfuge tube reaction and has great potential for use with naturally transformable organisms. Genome footprinting is another technique based on saturation mutagenesis that allows PCR-based scanning of a whole genome for gene essentiality under different experimental conditions. Finally, rather than being restricted to a random approach for mutagenesis, the availability of genome sequence data allows the directed knockout of genes in genetically amenable systems, and publicly announced programs to knockout every gene in the genome are proceeding for *Bacillus subtilis* (*http://locus.jouy.inra.fr/cgi-bin/genmic/madbase/progs/madbase.operl*) and *Saccharomyces cerevisiae* (*http://sequence-www.stanford.edu/group/yeast_deletion_project deletions3.html*).

One advance that is invaluable for studies involving pools of mutants is oligonucleotide tag (or "bar-code") mutagenesis. In this technique, a unique oligonucleotide tag is included in the transposon/cassette and then introduced into each mutant. In the case of "signature-tagged mutagenesis" of pathogenic bacteria, developed in the Holden laboratory, pools of 96 bar-coded mutants are used to infect a host organism, the bacteria are collected, the DNA are isolated, and the bar code is amplified and then hybridized to a 96-spot array of DNA from each mutant. By comparing hybridization of the bar-code tag from *in vivo*-grown pools of mutants with that from *in vitro*-grown mutants, one can identify those mutants less successful at causing infection in a host—potential new virulence factors. Infection of mammals, as opposed to growing mutants in laboratory media, is a much more realistic environment for learning about microbial pathogenesis, and the bar-coding procedure has been very useful because it minimizes the number of experimental animals needed for infection. The whole-genome knockout program for *S. cerevisiae* incorporates bar codes into the cassettes, and preliminary experiments for monitoring pools of mutants have successfully employed microarray chips for the DNA–DNA hybridization step. Thus, the use of DNA microarrays for parallel processing of both mutant growth rates and global transcription patterns has enormous implications for changing molecular microbiology.

E. Other Functional Studies

Numerous functional studies in addition to those mentioned previously are being performed for sequenced bacteria. This list is not inclusive, and it should again be mentioned that the most well-coordinated functional analysis programs currently being performed are those for *B. subtilis* and *S. cerevisiae*. Methods currently being used include (i) reporter fusion analyses, constructed for entire genomes, to provide gene expression data using whole cells; (ii) protein–protein interactions to determine what proteins interact with each other in a cell are providing important information to be used in conjunction with gene function studies; (iii) protein overexpression, using a high-throughput, brute-force protein overexpression methodology involving designing primers to the 5′ and 3′ ends of sequenced genes, cloning into T7 polymerase-driven vectors with affinity tags, and expressing and purifying the gene products using *E. coli,* with purified protein being used for numerous applications including *in vitro* screening for novel inhibitors, use as a component for

vaccines, and for determining the three-dimensional structure of the expressed protein; (iv) structural genomics—in addition to modeling the theoretical structures of all the predicted proteins in a genome, determining the actual structure of expressed proteins of interest is a focal point for structural genomics, for use in rational drug design and for adding to the accumulating structure databases and helping to refine predictive programs; and (v) metabolic reconstruction studies, with initial efforts focusing on the well-studied *E. coli,* are being used as a framework for other microbes. Complete sequence data and excellent annotation are critical for all of these types of studies.

The trend for functional analysis work is increased throughput, which usually implies both miniaturization and automation. Besides the microarrays mentioned previously, miniaturization technology is progressing to the point of "lab-on-a-chip" design. These nanoreactors are capable of performing routine molecular biological techniques in tandem; for example, add a sample to the nanoreactor and DNA is extracted, purified, and separated by size for such applications as Southern hybridization or nucleotide sequencing. Automation is also apparent in any high-throughput sequencing lab, which incorporates robots for every step possible, including clone picking and arraying, plasmid purification, and numerous pipeting and reaction steps. Clearly, the automation that has provided an explosion of genome sequences will be applied to functional studies so that laboratory experimentation can keep pace with the high-throughput of genome sequencing.

V. CONCLUSIONS

The applications of high-throughput sequencing and computational methods that led to the first whole genome sequence for a living organism is revolutionizing microbiology. The present golden era of genome sequencing will soon lead to a challenging postgenomic era, in which laboratory experiments must be designed in order to determine how the thousands of genes in each microbe's genome function, interact, and allow these organisms to survive and evolve. To meet this challenge, researchers will become even more dependent on automation and computers to allow them to continually push the limits of what is achievable in genomic sequencing and functional genomics.

See Also the Following Articles

HAEMOPHILUS INFLUENZAE, GENETICS • MAPPING BACTERIAL GENOMES • MUTAGENESIS

Bibliography

Anderson, N. L., and Anderson, N. G. (1998). Proteome and proteomics: New technologies, new concepts, and new words. *Electrophoresis* **19**, 1853–1861.

Ash, C. (1997). Year of the genome. *Trends Microbiol.* **5**, 135–139.

Benton, D. (1996). Bioinformatics—Principles and potential of a new multidisciplinary tool. *Trends Biotechnol.* **14**, 261–272.

Clayton, R. A., White, O., and Fraser, C. M. (1998). Findings emerging from complete microbial genome sequences. *Curr. Opin. Microbiol.* **1**, 562–566.

Dujon, B. (1998). European Functional Analysis Network (EUROFAN) and the functional analysis of the *Saccharomyces cerevisiae* genome. *Electrophoresis* **19**, 617–624.

Fleischmann, R. D., *et al.* (1995). Whole-genome random sequencing and assembly of *Haemophilus influenzae* Rd. *Science* **269**, 496–512.

Fraser, C. M., and Fleischmann, R. D. (1997). Strategies for whole microbial genome sequencing and analysis. *Electrophoresis* **18**, 1207–1216.

Moszer, I. (1998). The complete genome of *Bacillus subtilis*: From sequence annotation to data management and analysis. *FEBS Lett.* **430**, 28–36.

Schena, M., Heller, R. A., Theriault, T. P., Konrad, K., Lachenmeier, E., and Davis, R. W. (1998). Microarrays: Biotechnology's discovery platform for functional genomics. *Trends Biotechnol.* **16**, 301–306.

Tang, C. M., Hood, D. W., and Moxon, E. R. (1997). *Haemophilus* influence: The impact of whole genome sequencing on microbiology. *Trends Genet.* **13**, 399–404.

Downy Mildews

Jeremy S. C. Clark and Peter T. N. Spencer-Phillips

University of the West of England

I. Taxonomy and Host Range
II. The Infection Cycle
III. Physiological and Biochemical Interactions
IV. Manipulation and Disease Assessment
V. Control of Downy Mildew Diseases
VI. The Future

GLOSSARY

biotrophy Symbiosis whereby a parasite extracts nutrients from living host tissues.

haustoria Intracellular hyphae of determinate growth thought to have an important role in nutrient uptake.

hypersensitive response A host defense response resulting in rapid, localized death of plant cells and restricted growth of the invading pathogen.

mycelium A network of hyphae that comprises the vegetative body of a fungus or oomycete.

oomycetes Organisms with a growth form similar to that of fungi; now classified as related to brown algae.

pathogen A parasite that causes disease symptoms.

race-specific resistance Plant resistance to pathogen invasion conferred by a single gene held by a particular cultivar of a plant, corresponding to an avirulence gene held by a particular race of the pathogen.

recovery resistance The ability of plant tissues to grow away from an infection to produce a healthy shoot.

systemic acquired resistance Plant resistance to pathogen invasion conferred by a previous infection.

systemic infections Infections which pervade the host tissue and sometimes the whole plant.

DOWNY MILDEWS are devastating pathogens of plants. In a world with a population approaching 6 billion, twice what it was less than 40 years ago, and still showing near-exponential growth, crop devastation by fungal pathogens and other pests and diseases is increasingly important. This is as true in developed countries, in which people are demanding improved quality but reduced environmental damage by pesticides, as it is in developing countries, which sometimes struggle to provide funding for basic control measures such as fungicides.

Downy mildews, so named because sporulating structures typically appear as a white, gray, brown, or mauve "down" on infected leaves, are important plant pathogens causing considerable losses of staple crops in Africa, India, and elsewhere in the tropics and are therefore contributing causes of malnutrition and starvation.

Maize is the most important cereal in Africa and South America (38 and 56 million metric tons produced in 1997, respectively), and pearl millet and sorghum are also important staple crops in Africa, especially in the sub-Sahara zone. These two latter crops are grown for their grain for brewing, flour, and malted and fermented foods; stems for fencing; and total biomass for stock feed. In Africa the downy mildews which attack pearl millet, *Sclerospora graminicola*, and sorghum, *Peronosclerospora sorghi*, have a variable incidence from 0 to 50%, leading to 0–20% crop loss. In Zaire, maize losses due to various downy mildew species have been reported to be between 10 and 100%.

In Southeast Asia, maize is the third most important cereal, after rice and wheat, in terms of area and production, and it is generally acknowledged that increases in area are no longer possible. Only increased productivity will be able to meet the increasing demand for maize as a food commodity, animal feed, and industrial crop. Maize diseases are therefore of increasing importance. Since the introduction of maize in the 1800s, this crop has proved vulnerable to the five downy mildew diseases in this region: *Peronosclerospora philippinensis, P. sorghi, P. maydis, P. sacchari,* and *Sclerophthora rayssiae* f.sp. *zeae.* Losses due to *P. philippinensis* in Northern India are frequently 40–60%, and a disease incidence of 80–100% is not uncommon. In Indonesia, losses due to *P. maydis* average 40%.

In the United Kingdom, pea downy mildew (*Peronospora viciae*) can still cause up to 55% yield loss in fields in which plant resistance has not been effective. In 1989, a disease survey by the Agricultural Development and Advisory Service estimated that lettuce downy mildew (*Bremia lactucae*) caused approximately £1.2 million loss in the United Kingdom. In Italy, lettuce downy mildew can lead to complete loss of marketable yield due to its effect on crop quality. In the late 1980s in the former Czechoslovakia, cucumber yields decreased by 80–85% due to downy mildew. Additionally, downy mildews seriously affect the yield or quality of a wide range of other cereals, fruits, vegetables, and ornamentals.

Thus, the economic importance of downy mildew diseases is enormous, reflected by the world fungicide market ($4.7 billion), from which 20% (i.e., $1 billion) was used for the control of downy mildews in 1994. Downy mildews of grapes, vegetables, tobacco, and hops are currently the main targets for control, but there are numerous potential crop targets in developing countries. Fungicide insensitivity against the most widely used fungicides (e.g., metalaxyl) is prevalent in many downy mildews, including those which attack lettuce (*B. lactucae*), cucumbers (*Pseudoperonospora cubensis*), grapes (*Plasmopara viticola*), peas (*P. viciae*), and sunflowers (*Plasmopara halstedii*), and is likely to become of increasing importance (see Section V.B).

I. TAXONOMY AND HOST RANGE

A. The Kingdom Straminipila

Originally, the downy mildews and other oomycetes were classified as fungi (i.e., belonging to the kingdom Mycota). Although superficially the growth form of the downy mildews is very similar to that of powdery mildew (ascomycete) and rust (basidiomycete) fungi, the fact that this is due to convergent evolution has been known for many years.

Recent advances in systematics, molecular biology, and the study of ultrastructure have led to a new classification in which the oomycetes belong to a group with an equidistant relationship between plants and fungi, and this group has been assigned kingdom status (Fig. 1). All classes of this new kingdom sometimes display a very advanced type of flagellum, the straminipilous flagellum, and the kingdom is therefore named Straminipila. The kin of the oomycetes are therefore the photosynthetic brown algae and some heterotrophic protoctists.

That downy mildews are equally related to plants and to fungi is seen, for example, in the deduced amino acid sequences for a heat shock protein isolated from *B. lactucae* that is equally similar to yeast and maize heat shock proteins. Other indicators include oomycetes having cellulose rather than chitin within their cell walls and using the α,ε-diaminopimelic acid lysine synthesis pathway, as in plants. Differences compared to fungal groups include those in ultrastructure, amino acid and lipid metabolism, the absence of metachromatic storage granules, and diversity in nuclear division (Lebeda and Schwinn, 1994).

Little is known about the genomes of downy mildews, except that they have variable numbers of chromosomes: *B. lactucae* has 7 or 8 and *Peronospora parasitica* has 18–20. Most downy mildews are probably diploid, although polyploid or heterokaryotic isolates of *B. lactucae* have also been reported.

B. The Class Peronosporomycetes

According to Dick (in Spencer-Phillips, 2000), the 11 downy mildew genera belong to the kingdom Straminipila, division Oomycota, class Peronosporomycetes in the orders Peronosporales (genera *Albugo, Basidiophora, Bremia, Bremiella, Paraperonospora, Peronospora, Plasmopara,* and *Pseudoperonospora*) or Sclerosporales (genera *Peronosclerospora, Sclerospora,* and *Sclerophthora*). The related genera *Pythium* and *Phytophthora,* from which

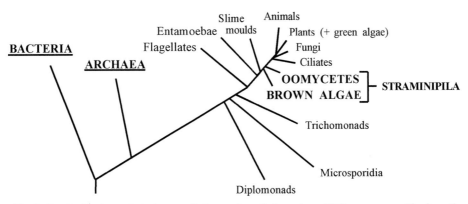

Fig. 1. Rooted universal phylogenetic tree primarily based on rRNA sequences. The length of each line represents genetic distance. The Bacteria, Archaea, and Eucarya comprise the three domains of living organisms; within the Eucarya, the oomycetes (including the downy mildews) are placed along with brown algae in the kingdom Straminipila (adapted with permission from Madigan *et al.*, 1997).

downy mildews probably evolved, belong to the order Pythiales.

Downy mildews attack an estimated 15% of the families of flowering plants, including a vast range of nonwoody and woody dicots as well as monocots, but they do not attack gymnosperms or any lower plants. They have a worldwide distribution, with the order Peronoporales tending to be found in temperate zones and the Sclerosporales in tropical regions.

C. Species and Host Range

Each downy mildew species has a very specific and usually narrow host range, sometimes restricted to one plant species (e.g., *P. viticola, P. halstedii,* and *Plasmopara manshurica*). Typically, a few host species in several genera are attacked, for example, *P. viciae* attacks peas (*Pisum* spp.), beans (*Vicia* spp.), and sweet peas (*Lathyrus* spp.). Some, however, attack a broad range of host species, sometimes covering a whole family; for example, *Sclerophthora macrospora* attacks at least 140 graminaceous species, *P. parasitica* attacks more than 80 cruciferous species, and *P. cubensis* attacks all cucurbitaceae. Within a species of downy mildew pathogen, isolates usually only attack plants within one genus, therefore delineating *formae speciales* or "pathotypes" (although notable exceptions are isolates of *P. cubensis*

that can attack five genera of cucurbitaceae). Lastly, not all isolates of a downy mildew species or *forma specialis* are virulent on all cultivars of a particular plant species. This race-specific resistance is discussed in Section V.B. In fact, DNA fingerprinting has demonstrated a very high level of genetic variation in the natural populations of several downy mildews (e.g., millet downy mildew, *S. graminicola*).

The broad host range of some downy mildews means that collateral weed species can be a potential source of inoculum for the infection of crops. In Southeast Asia the wild grass *Saccharum spontaneum*, which is also planted as fencing around fields, is known to harbor *P. philippinensis*, which attacks neighboring maize, and eradication of the grass leads to reduced incidence of downy mildew.

II. THE INFECTION CYCLE

A. Oospores

Oospores are large (25–50 mm diameter), thick-walled sexual spores which form within plant tissue. In *P. viciae,* for example, oospores form within the leaves, stems, and seed coats and along the internal pod walls of pea plants. They result from the fusion of oogonia and antheridia from two different mating types in heterothallic species (e.g., *S. graminicola*) or from the same mycelium in homothallic species (e.g., *P. sorghi*). Some species (e.g., *P. parasitica*) comprise both heterothallic and homothallic isolates. A mature oospore is assumed to contain only one diploid nucleus.

Oospores are the survival propagules and cause the primary infections. Within dried plant material and in soil oospores can survive at least 10 years; oospores of *Peronospora destructor* in onion debris are known to germinate well after 25 years of outdoor storage. Oospores either infect directly using a germ tube or, in some species (e.g., *P. viticola*), the germ tube produces a sporangium and consequent zoospores. Oospores germinate sporadically or in response to root exudates, and they infect germinating seedlings or sometimes older plants.

Mycelia (and perhaps oospores) within seed coats can form another source of inoculum. Mycelium within cabbage seeds has been shown to infect developing seedlings and is thought to be a major source of inoculum in South Africa. In contrast, mycelium within pea seed coats has never been found to infect seedlings growing from these seeds, although it does affect their quality and germinability.

B. Sporangia/Conidia

Sporulation results in the production of asexual spores 15–30 mm in diameter which contain up to 30 nuclei. These spores are termed either sporangia if they produce zoospores internally which are then released and encyst to germinate via a germ tube (e.g., *S. graminicola*) or conidia if they germinate directly via a germ tube (e.g., *P. viciae;* Fig. 2).

Sporangia and conidia are produced on the ends of stalked sporangiophores or conidiophores, which usually protrude through the open stomata of leaves or stems (only two species are reported to sporulate on roots—*P. halstedii* and *Plasmopara lactucae-radicis*). The spores are generally released by a twisting motion of the sporangiophore or conidiophore during changes in humidity. They are ephemeral with a life span of only a few hours or days, depending on the humidity once released, and they will not survive lack of liquid water after germination. It is not surprising, therefore, that the downy mildew mycelium, which is protected within the host plant, is programmed not to sporulate unless humid conditions prevail. In some species, for example, millet downy mildew (*S. graminicola*), leaf wetness is an absolute requirement for sporulation.

The spores are transferred from plant to plant to produce secondary infections by rain splash and can be dispersed hundreds of kilometers by wind. Epidemics therefore tend to occur with cool or moderate temperatures and a humid environment, typically in the spring and autumn in temperate countries or during the rainy season in the tropics.

C. Symptoms

The biotrophic symbiosis of downy mildews with their hosts means that, under certain conditions, infected plant tissue can appear to be perfectly healthy. In the laboratory, pea leaves with localized secondary lesions can remain green even during sporulation, and they may only senesce at the same rate as uninfected leaves. In compatible interactions between arabidopsis and *P. parasitica*, no symptoms are visible before sporulation. However, other downy mildews (and pea downy mildew under field conditions) usually give rise to varying amounts of chlorotic and necrotic patches of shoot tissue, especially in the later stages of infection, although it has been suggested that necrosis is the result of secondary infecting microorganisms.

Oospores and sporangia or conidia can lead to either localized lesions or systemic infections, and all types of attack can result in death of a proportion of infected plants. Systemic infections, however, are more commonly found after oospore infection and occur when the whole of the growing shoot becomes infected via the apical meristem. Profuse sporulation results, with stunted growth or distortion and early death of the plant. Infected pearl millet plants show signs of hormonal imbalance, with leafy growths produced in place of floral parts, to give the so-called "crazy top" or "green ear" syndrome.

In sorghum infected by *P. sorghi*, whitish stripes can appear on the leaves where rows of oospores have been produced between the vascular strands. These later necrose and shred, releasing oospores into the air. Alternatively, oospores can remain in

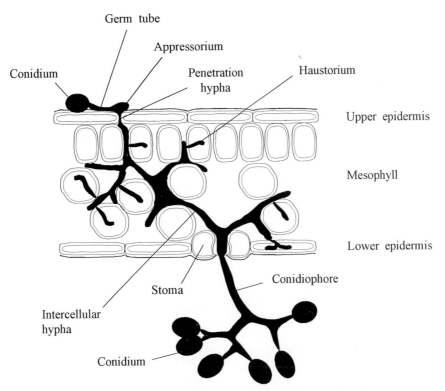

Fig. 2. Diagrammatic representation of the asexual infection cycle of *Peronospora viciae* in a pea leaf. The conidium germinates to form a germ tube; a penetration hypha develops from under an appressorium and penetrates either via epidermal cell walls or through a stoma. Intercellular hyphae can vary considerably in width, sometimes filling intercellular spaces. Haustoria are intracellular structures formed within host cells. Asexual conidia form mostly on the lower surface of the leaf on conidiophores which emerge through stomata.

necrotic plant tissue to infect via the soil in a future season.

Most plant species become more resistant with age. Sorghum, for example, becomes completely resistant to systemic infection a few weeks after seed germination.

D. The Endophytic Mycelium

A conidium (or zoospore released from a sporangium) germinates to give rise to a germ tube which then forms an appressorium over the intended point of entry into a healthy plant leaf or stem (Fig. 2). Depending on the species, the entry point could be either a stoma or an anticlinal wall between two epidermal cells (Fig. 2) or the upper periclinal wall of an epidermal cell. A single conidium can be suffi-

cient to initiate an infection (e.g., of Kohlrabi by *P. parasitica*).

A penetration hypha passes through the stoma or epidermal cell wall and gives rise to an intercellular hypha. In some species (e.g., *B. lactucae*), secondary spore-like structures termed infection vesicles are produced within epidermal cells, and a hypha grows out from these. The hyphae grow by branching that has been described as dichotomous, to form a coenocytic mycelium with apparently indeterminate growth. At intervals along the hyphae of most species, specialized intracellular hyphae called haustoria are formed within adjacent plant cells. These are finger-like, branched, or globose projections of determinate growth and are formed as follows. A penetration peg penetrates a plant cell wall and invaginates the plant plasma membrane, which enlarges to encompass the growing haustorium. This new membrane is called the extrahaustorial membrane, and the space between this and the haustorial wall is called the extrahaustorial matrix. Haustorial cytoplasm is therefore surrounded by the haustorial plasma membrane, haustorial wall, extrahaustorial matrix, and extrahaustorial membrane.

A biotrophic relationship with the plant tissue, whereby nutrients are obtained from the living host cells which are kept alive at least for a short time and often until sporulation, has been adopted by all downy mildew pathogens. Most species produce haustoria which do not penetrate or destroy the plant cell plasma membrane or cytoplasm and presumably allow the transfer of biomolecules which are not able to traverse between intercellular hyphae and plant cells. It should be noted, however, that the details of nutrient transfer, and the functions of downy mildew haustoria, are not known.

One theory, that haustoria provide a surface area for nutrient absorption, has arisen by direct comparison with haustoria of powdery mildew and rust fungi. In powdery mildews, elegant experiments have shown that electrochemical gradients set up by ATPase proton pumps across the various membranes drive transport of nutrients from epidermal cells to the fungus via haustoria (Spencer-Phillips, 1997). With rusts, recent evidence has shown the presence of amino acid transporters only on the haustorial membranes, suggesting that these haustoria are at least important for uptake of amino acids.

However, as mentioned previously, the taxonomic relationships between these groups and downy mildews mean that assumptions regarding parallel functions should be treated with caution, especially because it has already been shown that carbon from sucrose can be transferred directly from the apoplast to intercellular hyphae of *P. viciae* without the need for haustoria. Additionally, some downy mildew species either have no haustoria or form haustoria only rarely.

Other roles for haustoria can also be envisaged. For example, the mechanism by which downy mildews influence the physiology of the plant is not known and may require an interface across which large signaling molecules are able to pass directly into the plant cells in order to affect DNA transcription or other metabolic processes.

III. PHYSIOLOGICAL AND BIOCHEMICAL INTERACTIONS

A. The Compatible Interaction

In the compatible interaction, host cell damage is minimal, with no evidence of toxin production and very localized production of cell wall-degrading enzymes during the penetration of plant cells walls (e.g., by *B. lactucae*). One well-publicized reaction of the plant to the formation of haustoria (which must, in any case, include a large number of complex biochemical responses) is the formation of a callose ring round the point of entry into the plant cell.

As the infection progresses, changes in translocation patterns, hormonal levels, and host–cell permeability typify a biotrophic interaction. Sucrose and other nutrients are diverted from source leaves to the new infection sink created by a localized lesion. Other changes include increases in the activity of several enzymes, including invertases, α-glucosidase, ribonuclease, and peroxidase (see Section III.B.2).

In many species of downy mildew, severe damage to the host coincides with sporulation, mostly because this is when the pathogen's demand on the host for nutrients is maximal. In onion, up to 10% of stomata can be blocked by conidiophores bearing conidia, the production of which can amount to 5% of the dry weight of a leaf in a single night.

As mentioned in Section II.C systemic infections can lead to changes in plant morphology. However, the causal details of many of these processes are unknown.

B. The Incompatible Interaction

1. The Hypersensitive Response

In an incompatible interaction, for example, between a downy mildew isolate and a plant cultivar expressing race-specific resistance (see Section V.B), recognition of the pathogen by the host usually leads to a hypersensitive response (HR). This HR is the rapid and localized death of plant cells at the site of infection, preventing the establishment of a biotrophic symbiosis. Other responses include the activation of defense genes resulting in reactions such as an oxidative burst, deposition of cell wall reinforcements such as lignin, and the accumulation of pathogenesis-related proteins and the antimicrobial phytoalexins (Table I). Callose plaques at the sites where penetration pegs have failed to produce haustoria and callose sheaths surrounding haustoria have also been inferred or demonstrated to be part of the resistance response in *Peronospora* or *P. lactucae-radicis* (but not *Bremia*) pathosystems.

The determination of specificity and the mechanisms of recognition are highly complex phenomena which can occur rapidly, resulting in an HR within 20 hr in an incompatible interaction. The details of each interaction may vary, especially in relation to the speed and infection stage at which a resistant response occurs. Lebeda (in Spencer-Phillips, 2000) concluded that there are at least three different biochemical mechanisms underlying the race-specific resistance of different lettuce species to *B. lactucae*. It is still unclear whether the HR alone is sufficient to prevent infection, and a mutant of arabidopsis has recently been identified which expresses race-specific resistance to *P. parasitica* with no hypersensitive response. Other attributes of incompatible interactions are therefore also of importance.

TABLE I
Predicted Sequence of Events Leading to a Hypersensitive Response[a]

Cells responsible	Sequence of events
Germ tube or hypha	Infection, differentiation in response to plant signals, and production of cultivar-specific elicitors
Penetrated plant cell	Primary recognition, activation of resistance genes, and *de novo* protein synthesis in penetrated cell
Penetrated plant cell	Irreversible membrane damage and release of phenolics
Penetrated plant cell	Release of endogenous elicitors and secondary signals and accumulation of wall-bound phenolics
Surrounding plant cells	Secondary recognition, and transcription of mRNAs controlling biosynthesis of defense precursors by surrounding cells
Surrounding plant cells	Deposition of lignin, phytoalexins, and other defense chemicals and structures in and around the infection site

[a] Adapted from Mansfield (1990).

2. Other Biochemical Aspects of Resistance

One biochemical marker for resistance seems to be the level of peroxidase activity in cultivars of muskmelon (*Cucumis melo*) and lettuce. Pre-infection levels are higher in resistant cultivars, but peroxidase activity increases after infection in both resistant and susceptible cultivars. There is evidence that the level of peroxidase is linked to age-related resistance in many host–pathogen interactions including *P. viticola* on grape vines. Here, older leaves are found to have a partial resistance with reduced lesion areas and sporulation and a higher level of peroxidase activity than that of younger leaves. Significantly, a reverse age effect has been found with the related oomycete pathogen *Phytophthora infestans*, which causes late blight on potato. Here, peroxidase activity is higher in the younger leaves, which are more resistant than in mature, more susceptible leaves.

Another chemical essential for the primary resistance of arabidopsis to *P. parasitica,* and for systemic acquired resistance (see Section V.C.3.e), is salicylic acid. This indicates the importance of the enzyme phenylammonia lyase, which provides the precursors for lignin production and other phenolics, including salicylic acid. Lastly, lipoxygenase activity in pearl millet also appears to be a useful marker for resistance.

The biochemical details of resistance responses, for example, signal transduction pathways, are being studied at the molecular level, especially using the arabidopsis–*P. parasitica* interaction. Arabidopsis is currently the focus of a worldwide sequencing project to characterize the first complete plant genome, which in this species is exceptionally small. Together with the lettuce–*B. lactucae* system, these interactions are amongst the most important model systems for the study of the genetics of resistance in plants (see Section V.B).

IV. MANIPULATION AND DISEASE ASSESSMENT

A. Culture

Downy mildews must be grown on a living host and attempts to culture axenically have been unsuccessful. They can be grown and maintained on living cotyledons and leaf disks, provided that species specificity and closely controlled environmental conditions are maintained. Dual cultures have also been produced. For example, cultivation on host callus tissue has been successful for some species (e.g., *P. halstedii*, *P. sorghi*, and *S. graminicola*), and *P. halstedii* has also been grown on "hairy" roots (roots stimulated to grow with *Agrobacterium rhizogenes*).

Methods to store conidia and sporangia in liquid nitrogen with cryoprotectants (e.g., 10% glycerine) and at $-80°C$ have been developed, and oospores can remain viable in soil or an air-dried state at room temperature for many years. However, oospores are remarkably fastidious with regard to germination and infection. For example, *P. viticola* oospores only germinated after 5 months of weekly alternating temperatures of 10 and $-5°C$. The species-specific conditions required for these processes have not been described for many species.

The study of downy mildews is therefore particularly difficult, and the problems with germination, infection, and storage, as well as the fact that there are many homothallic isolates which do not interbreed, have severely hampered genetic studies. To date, only one microscopically visible genetic marker has been found (large lipid droplets in isolates of *B. lactucae* insensitive to Metalaxyl). Recently, however, two homothallic isolates of *P. parasitica* grown on the same arabidopsis plant were found to out-cross, which will initiate rapid progress in the study of this interaction.

B. Assessment of Disease

Incidence and severity of downy mildew are measured quantitatively using simple rating scales. For example, pearl millet downy mildew incidence is measured as the percentage of diseased seedlings 30 days after emergence in the field or 14 days after inoculation in a greenhouse. Severity is measured by assessing the tillers at the soft-dough stage on a scale devised by the International Crops Research Institute for the Semi-Arid Tropics (ICRISAT), where 1 = no symptoms, 3 = only nodal tillers diseased, 5 = <50% of the basal tillers diseased, 7 = >50%

of the basal tillers diseased, and 9 = all tillers and main shoot diseased or plant killed. A severity index is calculated as the sum of (no. of plants at one severity rating × severity rating)/total no. of plants. In the future, measurements from bulked samples will be possible using antisera to downy mildews and ELISA, especially because few crops harbor more than one downy mildew species (note that maize is an exception).

C. Screening for Different Pathotypes

In order to discern the presence of different pathotypes, an isolate can be grown on a set of differential cultivars (preferably isogenic or near isogenic) on adult plants, on seedlings, or in dual culture as single-spore solates or as native mixtures. For example, in 1992 and 1993 the International Virulence Nursery for Pearl Millet Downy Mildew (at ICRISAT) used a set of 12 differential cultivars to differentiate pathotypes of *S. graminicola* in various parts of the tropics by assessing mean disease severity. These methods can allow monitoring for pathogen variation and identification of new virulence genes in the population.

Care must be taken if deductions are to be made concerning resistance reactions with dual cultures. Incongruous results have been found between responses to the infection of some callus cultures and whole plants in several downy mildews, and between inoculated leaf discs and whole seedlings of sunflower. Differences have even been found between the relative aggressiveness of isolates of *P. halstedii* on sunflower seedlings in the laboratory compared with that found in the field and of *P. destructor* on onion seedlings and field-grown onions.

Two chemical methods to screen pearl millet cultivars for resistance to *S. graminicola* have been developed recently. The first uses a soluble protein, produced by *S. graminicola* mycelia, which only binds to suspension cells of susceptible cultivars. This has been used to develop an ELISA to detect susceptible suspension cultures, eliminating the need to use living downy mildew. The second method uses a known elicitor, arachidonic acid, which is used to induce the hypersensitive response in seedlings of pearl millet cultivars. The speed at which the hypersensitive response occurs correlated with disease incidence in the field.

V. CONTROL OF DOWNY MILDEW DISEASES

A. Fungicides

Several broad-spectrum, protectant chemicals are used against downy mildews. The original Bordeaux mixture of copper sulfate and calcium hydroxide was first used against grape downy mildew (*P. viticola*) in 1885, and mixtures utilizing the Cu^{2+} ion are still of great economic significance today. However, copper fungicides do not achieve complete disease control.

In the past 30 years more advanced systemic fungicides have been developed. The vast majority of fungicides effective on ascomycetes, basidiomycetes, and deuteromycetes, however, are completely ineffective against oomycetes because of either different pathways of sterol biosynthesis or cell wall composition. This again demonstrates the distinction between these groups of organisms and the strength of the new taxonomic classification.

The major "oomyceticide" which can give complete control and which has a long track record of effectiveness against downy mildews is the systemic phenylamide fungicide metalaxyl [methyl *N*-(2,6-dimethylphenyl)-*N*-(methoxyacetyl)-DL-alaninate], which affects oomycete rRNA synthesis. This can be applied to seeds to prevent primary infection (effective up to 1 month after sowing) or as a foliar spray to prevent secondary infections. The dithiocarbamates (e.g., mancozeb) are almost as important.

Unfortunately, control is minimal if infections have already begun and, in peas grown for freezing, late pod infections can occur for which there is no effective fungicide treatment. An additional problem is the increasing concern regarding fungicide insensitivity. Fungicides which attack a particular metabolic step usually give very good or complete control of downy mildew for a short time but, as discussed later, are usually overcome quite rapidly, sometimes after only a few years, by mutations conferring fungicide insensitivity.

Other fungicides recently recommended for use against pea downy mildew include cymoxanil (2-cyano-*N*-[(ethylamino)carbonyl]-2-(methoxyimino)-acetamide) and fosetyl-Al (aluminum tris-*o*-ethyl phosphonate), which also show good activity against grape, hop, and lettuce downy mildews. In addition, the carbamates show some activity. New fungicides discovered recently (e.g., the *β*-methoxyacrylate strobilurins) seem to have broad specificity against all groups of fungi and the oomycetes and may be very effective.

B. Race-Specific (or R Gene) Resistance and Fungicide Insensitivity

Plant resistance can be classified according to whether it is monogenic, polygenic, race specific or race nonspecific, or non-host resistant or partial field resistant. All types have been found against lettuce downy mildew (see Section V.C.3), but perhaps the best studied is race-specific resistance.

Race-specific resistance is conferred by a single resistance (R) gene, which corresponds to a single avirulence (Avr) gene in the pathogen. Plant resistance or susceptibility therefore depends on which alleles of these R and Avr genes are present within the two populations. This gene-for-gene theory, first proposed by Flor (1955), is remarkably useful for many oomycete–host interactions. A classic study on gene-for-gene interactions and the relationship between these and fungicide sensitivity was initiated by Crute (United Kingdom) in the 1980s with lettuce downy mildew. Although the following discussion concentrates on lettuce downy mildew, race-specific resistance has been found against all crop downy mildew species except grape and sugarcane. In only a few species, however, have the pathotype–cultivar relationships been studied in detail.

At least 23 dominant plant resistance R genes against downy mildew (*Dm* genes) have been found in the lettuce population, each with a corresponding downy mildew avirulence gene. Additionally, in the United Kingdom in the 1980s a downy mildew pathotype with avirulence gene 11 was found to be insensitive to the fungicide metalaxyl. Consequently, lettuce cultivars were developed which contained *Dm*11 (plus other *Dm* genes) and these were used effectively, together with metalaxyl, to keep the various pathotypes in check. However, pathogen populations can change remarkably quickly, and both resistance genes and fungicides seem to be equally vulnerable. In cucumber crops in Israel, metalaxyl-insensitive isolates were first reported in 1979 and increased from 30 to 100% between 1982 and 1985.

Metalaxyl was only introduced in northwest Italy in 1990 and found to give effective control of lettuce downy mildew with two applications per year. However, in 1993 reduced efficacy was observed, and in 1994 metalaxyl failed to give control in some regions. Isolates from these regions were insensitive to 100× normal inhibitory concentrations of metalaxyl. Even worse, some prevalent isolates of lettuce downy mildew were found to have virulence against all *Dm* genes except *Dm*18 as well as insensitivity to metalaxyl. Although theoretically it would be possible to use *Dm*18 plus metalaxyl, this strategy is considered too risky and likely to result in a superbug (i.e., a pathogen insensitive to metalaxyl and virulent on all cultivars). Because other effective, registered systemic fungicides are currently not available, the current recommendation for control of lettuce downy mildew in Italy is to use *Dm*18 plus copper protectant fungicides in order to try to preserve the effectiveness of the resistance gene for as long as possible.

Similar situations exist in France with grape and sunflower downy mildews, which developed metalaxyl insensitivity within 2 and 6 years, respectively, after first fungicide use, and pathotypes with many virulence genes were also found. In the United Kingdom, the resistant pea cultivar Carrera first introduced in 1995 was recently found to have become completely susceptible in some areas (Jane Thomas, personal communication, National Institute of Agricultural Botany, UK).

C. Integrated Control

The design of integrated strategies in which two or more control measures are used simultaneously is becoming more urgent, as the previously discussed examples suggest. Long-term concerns regarding the threats to health and the environment have meant that usage of fungicides is increasingly restricted by legislation. A good example of the problems that can

occur is given by the constraints on the control of grape diseases, including downy mildew, that pertained in New York in the late 1980s. Captan ($C_9H_8Cl_3NO_2S$) and the dithiocarbamate protectant fungicide mancozeb were and still are severely restricted by grape processors. Metalaxyl was not registered for use on grapevine in the late 1980s and severe insensitivity to metalaxyl had occurred elsewhere. Also, the copper fungicides were phytotoxic to important grape cultivars. At this stage, research into numerous unorthodox treatments was initiated, including the use of mycoparasites, triazoles with vapor action, calcium polysulfide treatment against oospores, and heat or UV light treatments of the vines.

It appears that due to resistance breakdown, fungicide insensitivity, and the necessity of reducing the environmental burden of toxic chemicals, integrated control is absolutely essential if the threat of future epidemics is to be averted in developed countries. The first step toward integrated control could be the rational use of fungicides based on weather forecasting and disease monitoring.

In developing countries, integrated control measures provide the only hope of sustainable control because of the high cost of fungicides and the cost and availability of new resistant cultivars. Each crop species and location have to be assessed independently to take into account economic and cultural factors as well as the environmental and biological factors. In addition to race-specific plant resistance and fungicides, cultural or biological control and nonspecific resistance are also being considered.

1. Cultural Control

Possibilities for the cultural control of downy mildews of millet, maize, and sorghum in the tropics include (i) crop rotation, especially with bait plants to remove oospores from the soil; (ii) deep tillage to reduce oospores in the upper soil; (iii) oversowing, where uninfected plants compensate for those infected; (iv) roguing, the time-consuming removal of infected plants which also relies on compensation; (v) early sowing to avoid sporangial infections in places where these are more important than oospores; and (vi) removal of adjacent wild plant hosts, as mentioned in Section I.

Crop rotation sometimes has the effect of reducing the number of fungicide-insensitive isolates of a pathogen in a population because these isolates typically have a reduced fitness compared with isolates which do not carry the genes for insensitivity. Unfortunately, metalaxyl-insensitive downy mildew strains can sometimes have a higher degree of fitness and competitiveness.

2. Biological Control

There have been many examples of small-scale research studies into the possibility of using biological control agents against downy mildews, but to date no one agent has gained commercial acceptability. The bacterium *Pseudomonas fluorescens* sprayed either as an aqueous suspension or in talc onto pearl millet in India has reduced heavy infestations dramatically from 90% to approximately 20% incidence in the field. The fungus *Fusarium proliferatum* has been sprayed at weekly intervals onto grapevines, and it reduced the incidence of downy mildew. It also reduced severity by 50–99% depending on the year and cultivar, and it was proposed that this fungus could be applied in areas where downy mildew is not severe or in combination with metalaxyl or copper- or sulfur-based fungicides (to which *F. proliferatum* is insensitive). A chytrid fungus *Gaertneriomyces* sp. has been found to be an effective parasite of *P. sorghi* oospores, but field applications have not been tested.

3. Nonspecific Resistance

Resistance based on cohorts of minor nonspecific genes could contribute to a durable, broad-based resistance. Initially, it was thought that a broad-based genetic control of resistance would be predictable by studying complex, partial, or late-onset resistance. However, it has been found that the underlying genetics cannot be predicted from the structural or biochemical responses from resistant plants since even the most complex responses can be caused by single genes triggering a cascade of biochemical or physiological events.

There are, however, some types of resistance which may be conferred by more than a few genes and which therefore might contribute to a more durable type of resistance than the *Dm* genes described in

Section V.B. For example, the pea cultivar Dark Skin Perfection was used for more than 35 years and, although it is affected by downy mildew under conditions favourable to the pathogen, usually exhibits partial resistance.

a. Landraces

One advantage of African small-scale agriculture is that cereal landraces are usually grown which are heterogeneous and well adapted to the local environment. This sometimes results in a stable and acceptable level of disease incidence because of the continuous evolution and coevolution of host and pathogen. Germplasm from these landraces may provide useful sources for resistance breeding programs.

In India in 1993, 60% of the pearl millet varieties grown were F1 hybrids, which give uniformity of growth, early maturity, and a relatively high yield. However, downy mildew incidence varied from 27 to 53%, which was higher than that for the genetically heterogeneous, open-pollinated cultivars that have shown durable resistance for many years. Clearly, losses from downy mildew must be balanced against the other traits which lead to marketable yield.

b. Field Resistance

One component of the resistance of landraces is probably similar to the field resistance studied in lettuce and also found in brassicas, cucumbers, and peas. With cultivars expressing field resistance the downy mildews were found to have a low infection efficiency and a low reproductive output in the field, but this epidemiology was not reflected accurately in laboratory tests using the same cultivars.

Contrary to expectations, the expression of field resistance was not found to imply anything about its inheritance, durability, or race specificity. This type of resistance can be either race nonspecific or race-specific, and it is probably controlled by few genes. It is characterized by part of the response being triggered very early following infection, and it results in the delayed appearance of symptoms, reduced severity, a reduced number of diseased plants and infected leaves, and a slow epidemic.

c. Resistance Transferred from Wild Hosts

In the past it has been found that resistance can be transferred to a crop from related wild species.

For example, at least five race-specific (*Dm*) genes have been introgressed from the wild lettuce *Lactuca serriola* into cultivated lettuce (*Lactuca sativa*). As with all race-specific resistance, however, it is not expected to last long.

In contrast, some accessions of *Lactuca saligna* are completely resistant to all isolates of *B. lactucae* isolated from *L. sativa* and are therefore classified as nonhosts to these isolates of *B. lactucae*. There is evidence that the basis of this resistance is not due to genes similar to the *Dm* genes used in lettuce breeding, and that this resistance is effective at a later stage of mycelial development than that for the *Dm* genes. Additionally, it seems that this type of resistance may have a multigenetic basis and therefore durability, although transfer of this from one species to another may prove difficult and one attempt has already failed (in which resistance introgressed into lettuce was overcome quickly by isolates which remain avirulent on *L. saligna*). This type of non-*Dm* gene-mediated resistance has been patented by Sandoz.

d. Recovery Resistance

In millet, sorghum, and maize another type of resistance has been found. This is where uninfected parts of a plant can compensate for early infection by continued growth to produce perfectly healthy leaves, tillers, and seed sets. Surprisingly, this can occur even after systemic infection, and it is particularly interesting because the plant allows the fungus to complete its asexual life cycle (apparently no oospores are produced). This presumably means that the selection pressure for a change in virulence is either reduced or eliminated. Pearl millet cultivars selected for this trait are being used for the production of commercial hybrids in India and show recoveries of between 90 and 98% in areas heavily infested (70–98% incidence) with downy mildew.

The mechanism of recovery resistance is unknown, but it might be related to systemic acquired resistance (discussed below) and perhaps to the phenomenon whereby seedlings from systemically infected pea plants rarely become systemically infected themselves.

e. Systemic Acquired Resistance

Systemic acquired resistance (SAR) is a plant immune system that is activated in many dicots and monocots following an initial infection. For example, local inoculations of tobacco with *Peronospora tabacini* or tobacco mosaic virus will induce resistance to many fungi (including *P. tabacini*), a bacterium and tobacco mosaic virus. One SAR mechanism involves salicyclic acid production, expression of SAR genes and pathogenesis-related proteins PR1 and PR5, and appears to be very similar throughout the plant kingdom. Interestingly, there are several chemical inducers which can be used instead of an initial infection, such as salicylic acid, 2,6-dichloroisonicotinic acid, and benzo(1,2,3)thiadiazole-7-carbothioic acid S-methyl ester (BTH). BTH is as effective as the fungicide cymoxanil against downy mildew in maize, and it can induce disease resistance in wheat for the entire growing season. It is also effective in *Brassica* species against *P. parasitica*, even when given as a seed treatment. Another chemical, BABA (DL-3-aminobutyric acid), probably acts via a different SAR mechanism and has been reported to protect tomato against *P. infestans*, tobacco against *P. tabacini*, peppers against *Phytophthora capsici*, and grapes against *P. viticola*.

Perhaps the most exciting possibility is given by the discovery of lesion mimic mutants which carry mutations in the SAR gene pathway over-expressing these genes. Genetic engineering leading to crops with constitutively expressed systemic acquired resistance may provide the most effective means of controlling downy mildews as well as other diseases. It should be noted, however, that infections by *Albugo candida* can predispose horseradish leaves to infection by *P. parasitica*.

VI. THE FUTURE

A key driving force in downy mildew research is the fact that "the enormously dynamic nature of these pathogen populations dictates an equally dynamic approach for their control" (Crute, 1989). The instabilities that characterize the use of resistance genes and fungicides require that, even in developed countries such as the United Kingdom and the United States, continued monitoring of the downy mildew pathogen and introduction of new cultivars are essential to avoid disaster. Research into the fundamental basis of the various types of resistance, the various alternatives to fungicides, and the basic biology of these versatile pathogens is of utmost importance.

See Also the Following Articles

Plant Disease Resistance • Plant Pathogens • Sporulation

Bibliography

Crute, I. R. (1989). Lettuce downy mildew: A case study in integrated control. *In* "Plant Disease Epidemiology, Vol. 2. Genetics, Resistance and Management" (K. J. Leonard and W. E. Fry, Eds.), pp. 30–53. McGraw-Hill, New York.

Hall, G. S. (1996). Modern approaches to species concepts in downy mildews. *Plant Pathol.* **45**, 1009–1026.

Jeger, M. J., Gilijamse, E., Bock, C. H., and Frinking, H. D. (1998). The epidemiology, variability and control of the downy mildews of pearl millet and sorghum, with particular reference to Africa. *Plant Pathol.* **47**, 544–569.

Lebeda, A., and Schwinn, F. J. (1994). The downy mildews—An overview of recent research progress. *J. Plant Dis. Protection* **101**, 225–254.

Lucas, J. A., and Sherriff, C. (1988). Pathogenesis and host specificity in downy mildew fungi. *In* "Experimental and Conceptual Plant Pathology. Vol. 2. Pathogenesis and Host–Parasite Specificity" (R. S. Singh, W. M. Hess, and D. J. Weber, Eds.), pp. 321–349. Gordon & Breach, London.

Madigan, M. T., Martinko, J. M., and Parker, J. (1997). "Biology of Microorganisms," 8th ed. Prentice Hall, Englewood Cliffs, NJ.

Mansfield, J. W. (1990). Recognition and response in plant/fungus interactions. *In* "Recognition and Response in Plant/Virus Interactions" (R. S. S. Fraser, Ed.), NATO ASI Series, Vol. H41. pp. 31–52. Springer-Verlag, Berlin.

Saharan, G. S., Verma, P. R., and Nashaat, N. I. (1997). "Monograph on Downy Mildew of Crucifers," Saskatoon Research Centre Technical Bulletin 1997–01. Minister of Supply & Services, Canada.

Sharma, R. C., De Leon, C., and Payak, M. M. (1993). Diseases of maize in South and South-East Asia: Problems and progress. *Crop Protection* **12**, 414–422.

Spencer, D. M. (1981). "The Downy Mildews." Academic Press, London.

Spencer-Phillips, P. T. N. (1997). Function of fungal haustoria in epiphytic and endophytic infections. *Adv. Bot. Res.* **24**, 309–333.

Spencer-Phillips, P. T. N., Ed. (2000). Advances in Downy Mildew Research. Manuscript in preparation.

Stegmark, R. (1994). Downy mildew on peas (*Peronospora viciae* f. sp. *pisi*). *Agronomie* **14**, 641–647.

Thakur, R. P. (1995). Status of international sorghum anthracnose and pearl millet downy mildew virulence nurseries. *In* "Disease Analysis through Genetics and Biotechnology" (J. F. Leslie and R. A. Frederiksen, Eds.), Iowa State Univ. Press, Ames.

Ecology, Microbial

Michael J. Klug

Michigan State University

David A. Odelson

Eco Soil Systems, Inc., San Diego, CA

GLOSSARY

biofilm Matrix-enclosed bacterial populations adherent to each other and/or to surfaces or interfaces.

cell sorter An instrument that uses optical or mechanical technologies which allow the separation of cells on the basis of size or cellular properties.

chemotaxis A movement response by microorganisms in which the stimulus is a chemical concentration gradient.

community An assemblage of populations of microorganisms which occur and interact within a given habitat.

ecosystem In terms of microorganisms, the totality of biotic and abiotic interactions to which the organism is exposed.

food web The interaction of communities of organisms with varying functional capabilities.

habitat A location where microorganisms occur.

interspecies hydrogen transfer The coupling or syntrophic relationship of hydrogen producers and hydrogen consumers.

Koch's postulates A concept embodied in these postulates defines the fundamental questions needed to address the function of a microbial population within a given habitat or its role in a function.

laser scanning confocal microscope An approach to microscopy which uses intense laser light beams, optics which exclude light from parts other than the specimen, and computer-assisted image enhancement to provide a nearly three-dimensional image without sample destruction or fixation.

microelectrode A micro version of pH, O_2, and specific ion electrodes which allow the exploration of microhabitats.

PCR Polymerase chain reaction; a method for increasing the number of copies of a target nucleic-acid sequence without having to culture the organism.

phylloplane The aboveground exposed surfaces of plants that are available for the colonization of microorganisms.

population A group of individuals of one species within a defined area or space.

probe A chemical or molecular technique which allows the detection or quantitation of a population or activity of microorganisms.

rhizosphere The region of soil which adheres to plant roots or is influenced by the activities of plant roots.

rumen A chamber anterior to the digestive tract of animals which harbors microorganisms that metabolize ingesta and provide metabolic intermediates to the animal, which serve as energy and biochemical precursors.

syntrophic An interpopulation interaction that involves two or more populations which provide nutritional requirements for each other.

MICROORGANISMS are often considered to be among the first life forms on earth. Fascination with these beings was initially predisposed to their small size and simple forms. Early observations of microorganisms (in the seventeenth century) by Leeuwenhoek, with the aid of the first microscope, were followed by the demonstration of the role of microorganisms in the process of fermentation and spoilage by Pasteur and, eventually, the development of a means to prevent the growth of microorganisms, i.e., pasteurization.

I. EVOLUTION OF MICROBIAL ECOLOGY AS A DISCIPLINE

Isolation of causative microorganisms of disease, as well as pure culture techniques, evolved in the late 1800s; Koch's postulates provided a solid basis for studying microorganisms and their roles. The era that followed provided for the continued isolation of microorganisms, the definition of their metabolic capabilities, and, in turn, their implied roles in important biogeochemical processes, i.e., in the nitrogen and sulfur cycle. Attention was given to organisms from specific habitats, i.e., soil, water, animal, and plants. The findings demonstrated the vast numbers of diverse microbiological forms and functions found in nearly every location that was sampled. Eventually, microbiological subdisciplines dealing with microbial associations of natural (soil, water) and manmade (food, industrial, and other) environments were established.

Within the last 30 years, it has been recognized that common microorganisms are observed in many habitats and common principles are involved in the mechanisms describing the associations of those microbes in varying habitats. The observation that individual populations of microorganisms are rarely found alone suggests potential interactions between populations and with their surrounding physicochemical environments. Additionally, the observation that they associate or align themselves within specific "strata" or gradients of physiochemical parameters points to the large number of metabolic functions of these organisms. They also have been shown to have the ability to sense (e.g., chemotaxis) and move by various means of locomotion within these gradients to maintain selected conditions for their growth.

The field of ecology is defined as a discipline of biology which deals with organisms' interactions with each other and surrounding environments. As one might expect, these aforementioned observations of microorganisms and their habitats led to the development of the subdiscipline known as microbial ecology. This development further emphasizes the need to establish a union between the examinations of the physicochemical nature of a habitat and microbiological investigations. An equally important emphasis in microbial ecology is the structure and activities associated with microbial communities rather than the earlier emphasis in causative populations of disease or specific processes.

II. MICROBIAL COMMUNITIES

Although observations of microorganisms and their associations have been made in numerous habitats, a few of these observations are felt to highlight the importance of the interactions between microorganisms which leads to community vs population responses.

The rumen ecosystem has received considerable attention, which has been driven principally by economic considerations. Research in this system has, however, defined the syntrophic relationship between members of the microbial fermentative community and the animal's growth and survival. Technically, approaches to this ecological niche present somewhat greater difficulties than those of soil or natural waters. The ecosystem is internal to the animal and the microbes are strict anaerobes. Mechanistically, however, rumen microbial ecology is relatively easier to discern, inasmuch as both input and output to and from the system are clearly defined, in much a similar relationship as an industrial bioreactor. It is interesting to note that syntrophic community-level interactions are frequently illustrated with the rumen system, in regard to hydrogen transfer and methane production.

The classic anaerobic food web first described by Hungate, and later by Wolf, which occurs in the rumen, involves the interpopulation interactions among bacterial communities capable of plant polysaccharide hydrolysis (e.g., cellulose), monomer fermenting communities, fatty acid-oxidizing communities, and, finally, terminal communities (e.g., methanogenesis), which oxidize fatty acids and reduce CO_2 to CH_4. The hydrogen is derived from previous oxidative steps. The coupling, or syntrophic, relationship between hydrogen producers and hydrogen consumers (interspecies hydrogen transfer) is now recognized as a fundamental relationship in other systems dominated by anaerobic microorganisms. In the rumen, rates and extent of metabolism are controlled by the interdependence of one community on another. The rumen system has also served as an example of a strategy for, an approach to, and methods to conduct similar investigations of animal–microbe associations, be it the crop of tropical birds or the intestinal tract of termites. Similar interactions have been observed in other anaerobic systems (such as sewage sludge, lake and marine sediments), which suggest common controls and mechanisms associated with metabolism of complex organic compounds in all of these systems. These relationships have also been shown to be involved in the metabolism of naturally occurring and manmade halogenated compounds. One such reaction involved dehalogenation by replacement of a halogen substituent of a molecule with a hydrogen atom. The hydrogen is derived from hydrogen-producing fermentative microorganisms.

Questions related to the response of these communities and the interactions among populations to disturbance of their physical–chemical environment will rely on tools which allow analyses of change in both the nature of the populations involved and changes in their functions. Do populations adapt and respond phenotypically or do they replace each other?

Traditional food web descriptions in aquatic and terrestrial ecosystems have often failed to consider the role of microbial interactions as a contribution to carbon and nitrogen cycling in these systems. Early recognition of these contributions was blurred by the examination of individual populations of organisms and by inadequately examining the interactions and controls of population size and distribution in the surrounding communities. Figure 1 illustrates, in its simplest form, the relationship among primary producers in aquatic/marine or terrestrial environments, heterotrophic bacteria, and phagotrophic protozoans or zooplankton.

Primary producers release soluble organic com-

pounds (root exudates, algal metabolites), which are consumed by heterotrophic bacteria, which are subsequently grazed by phagotrophic protozoans or zooplankton. The grazers excrete nitrogen and phosphorus, which is used by the primary producers.

In sum, these observations have strikingly modified the contemporary view of the structure and mechanistic controls and regulation of growth of higher plant and animal forms in aquatic and terrestrial systems. They also point to the importance of a thorough understanding of these interactions if one is to consider management of these associations in applied applications, e.g., sustainable agriculture or aquaculture.

Microbial biofilm communities were first described by Zobell and Anderson. Over the past two decades, their significance and ubiquity have been documented. It is now recognized that biofilm communities predominate numerically and metabolically in most ecosystems. Further, it is clear that biofilm cells are fundamentally phenotypically distinct from nonadhering cells. It now appears that the ability to form surface-associated structured biofilms is a common characteristic of, at least, bacteria. Remarkable intercellular and interspecies interactions, facilitated by chemicals released by these organisms, leads to complex structured communities, made up of both prokaryotic and eukaryotic organisms.

These examples, however, primarily involve interactions based on an exchange of chemical metabolites. The fact that these described interactions involve high densities of cells in close proximity to each other, e.g., biofilm, can also lead to intercellular exchange of genetic information, which can lead to adaptive change in function within communities. Knowledge of naturally occurring phenotypic and genotypic viability of microorganisms has been hampered because the majority of our knowledge comes from laboratory-selected strains, grown under controlled conditions, and on medium which has no resemblance to the environment from which they were selected. It is important to realize that approaches to microbial ecology must recognize not only isolated individual populations but, more directly, the interactions of populations within a community and the resultant effect on the overall functions of the community.

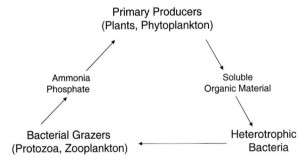

Fig. 1. Relationship of primary producers, heterotrophic bacteria, and bacterial consumers in aquatic/marine and terrestrial environments.

III. APPROACHES TO MICROBIAL ECOLOGY

The fundamental approach to higher plant and animal ecological studies uses quantitative observations of specific populations in various environments. For over 100 years, plant and animal ecologists have observed and detailed the frequency of occurrence of specific plant and animal populations under various environmental conditions. These observations have led to the development of models for predicting the occurrence of, as well as relationship between, the organisms and the associated environments in which they were observed. Recently, these observations have been complemented with physiological and genetic approaches that suggest specific mechanisms of selection which have led to the observed frequency or distribution of organisms. Unfortunately, our understanding of microbial systems is still embryonic in both description and prediction.

Kluyver, in 1956, estimated that about one half the "living protoplasm" on earth is microbial. This estimate is now considered conservative, and the sheer numbers represent a daunting challenge to the microbial ecologist. Even the most modern techniques do not allow routine quantification of microbial numbers and definition of specific populations. It is principally the size of the microorganism, and, equally important, the size of the local habitat, that creates a considerable constraint on the observation of microorganisms in natural surroundings. Numbers of organisms ranging from 1 million cells per milliliter of water to 10 billion cells per gram of human fecal material. Further, the spatial distribution of these organisms within these habitats makes it difficult to recover a representative sample to examine. *In situ* observations are further complicated by our inability to observe this environment without disturbing the microorganism's natural habitat. The various surfaces and potential differences in chemistries of the habitat provide varying degrees of carbon and nitrogen resources and physicochemical environments (e.g., gaseous exchange) for colonizing microorganisms. Additionally, other organisms, micro- and macro- alike, have a potential impact on the native organisms. Grazing of microorganisms by other microbes or faunal components of a habitat

have a significant impact on numbers and types of microorganisms present.

As noted, microbial ecology evolved as a cross-discipline of standard microbiology and environmental analytical analysis. The activity of a specific population or community of microorganisms was inferred by estimating their relative numbers by direct microscope counts, viable colonial or turbidimetric determinations, or specific chemical analyses (e.g., chlorophyll for algae). Unfortunately, limitations by both the microbial and the environmental analyses have, in most cases, failed to accurately define the ecology of microbial habitats. Recently, novel analytical and microbiological methodologies have provided tools for expanding our view of microbial habitats, in a fashion that is not constrained by the microscale of the environment, nor limited by our ability to selectively cultivate members of a community.

In situ observation of microorganisms has been expanded through use of laser confocal microscopy, an advancement which provides for a kind of "x-ray" imaging of material requiring neither disturbance nor fixation. Similarly, laser optical trapping allows the removal of individual cells from a habitat. These techniques provide new insights to our understanding of population components of communities. These increases in optical resolution also provide a means of sorting communities either by size or chemical factors through the use of cell sorters.

Advancements have also been made in developing methodology to simulate the microhabitats and physiochemical gradients which occur therein. It is interesting to note that one of the pioneering environmental microbiologists, Winogradsky, simulated gradients of light, water, sulfide, and oxygen to describe relationships between sulfide-oxidizing photoautotrophic bacteria, sulfate-reducing, and chemoautotrophic sulfur-oxidizing organisms in his Winogradsky column. Advances using gels and gradostats to simulate diffusion barriers allow us to expand our knowledge concerning relationships between the spatial heterogeneity of the physicochemical environment and the distribution of diverse groups of microorganisms. Combined with microelectrode techniques, precise analytical measurements will complement these microbial investigations. Although our abilities to observe organisms and to better understand the physical and chemical nature of their habits have improved, we still are unable to isolate a high percentage of these observable organisms. Nevertheless, within the last 30 years, the number of previously underscribed microorganisms has increased significantly.

In an effort to forego the inherent problems of the microscopic and numerical diversity of the microbial world, recent method development has approached the microbial component of the unseen habitats at the macromolecular level. The recognition that diversity is intrinsically related to the organism's genetics has provided a sound basis for separation and characterization of both microscopic and macroscopic life.

Examination of the heterogenity of DNA in soil suggests that as many as 10,000 bacterial species can be harbored in 100 g of soil. This estimate only includes the bacterial fraction and, again, reinforces the challenge presented to understanding the diversity of microorganisms in natural systems. From an ecological perspective, it also presents a challenge to understanding how so many species coexist in such small areas.

Recombinant DNA technology has allowed the identification and determination of the specific nucleotide sequence of genes. It has provided ecologists with novel methods to pursue community structure. In practice, microbial community samples can be analyzed without cultivation or microscopic observation of microorganisms. In brief, nucleic acid is isolated from the sample and utilized for hybridization studies with the corresponding gene of interest (i.e., the probe). This probe can represent either a metabolic gene or a systematic determinant, such as the 16S rRNA gene sequence. In turn, as one may infer, the sample can be evaluated in terms of population diversity at the species level or community diversity at the kingdom level (Eukaryotic vs Prokaryotic). All of these analyses have been expanded by inclusion of the polymerase chain reaction (PCR), a method allowing amplification and subsequent detection of as few as 10 microorganisms. Interestingly, these methods have, in fact, relied on previous isolation and characterization of specific microbial populations prior to isolation of a gene for use as a probe. These investigations have, however, illustrated the universal nature of certain sequences, such as from

16S rRNA genes, for random use in community analyses. An example of this technology is the use of various profiling techniques. Separate PCR-amplified 16S rRNA gene-fragment sequences are separated, and the resulting numbers provide an estimate of diversity within the community.

Areas that still remain to be explored in microbial ecology are the relationship of microbial diversity to the response of microbial communities to varying degrees of disturbance and the relationship between diversity and the function of the community. In higher plant and animal systems, the relationship of diversity within communities to their resistance to change or recovery after disturbance (e.g., fire, tillage) has been discussed and debated for decades. Various indices have been used to calculate diversity within communities. In their simplest form, these indices represent the number of species found within the community; therefore, communities with many species are described as having high diversity. Other indices relate the dominance of specific species to the total diversity, such that even communities with high diversity can have a few populations which dominate activity within the community. Our current inabilities to adequately isolate all microorganisms and the lack of distinctive morphological characteristics fail to provide us with accurate methods for measuring indices of diversity. Continued improvements in our analytical skills in identifying the macromolecular characteristics of microorganisms will provide a basis for estimating diversity at the phylogenetic level.

Profiles of cellular or phospholipid fatty acids from lipids extracted from communities or specific chemical or antigenic determinants may provide a snapshot of changes within microbial communities after disturbance. Community-level assays of the functional diversity of populations based on carbon source utilization also provide indications of changes following disturbance.

IV. FUTURE DIRECTIONS IN MICROBIAL ECOLOGY

It has often been said that advancements in microbial ecology are limited by the methods which are available to analyze microbial systems. Some of the advances made in the analytical and molecular methods over the past decade have been illustrated in this overview. Although continued advancements in the use of microelectrodes and optical methods increase our abilities to measure and observe microorganisms emphasis must increase on the refinement of techniques to discern changes in microbial community structure and the impact of these changes on the function of the community. A significant area of contemporary concern is the remediation of habitats contaminated with anthropogenic sources of organic and inorganic compounds.

The applied area of bioremediation is, and will continue to be, a timely subject in the decades to come. The use and management of the intrinsic properties of microbial communities will provide stimulus for applied microbial ecology. Although this area is yet to be accurately defined, in principle, the directive is to promote microbial dissimilation of anthropogenic compounds in a controlled manner.

In a similar manner, a reduced input of anthropogenic chemicals in agricultural systems to reduce environmental contamination will result in greater reliance on activities of the soil microbial communities.

These and other applications require further understanding of the structure and activities of microbial communities. Also required is an understanding of changes in the structure and function of these communities in relation to changes in the physiochemical environments associated with them. This, in fact, is microbial ecology.

See Also the Following Articles

BIOFILMS AND BIOFOULING • BIOREMEDIATION • CHEMOTAXIS • DIVERSITY, MICROBIAL • RUMEN FERMENTATION

Bibliography

Amann, R. I., Binder, B. J., Olson, R. J., Chisholm, S. W., Devereux, R., and Stahl, D. A. (1990). Combination of 16S rRNA-targeted oligonucleotide probes with flow cytometry for analyzing mixed microbial populations. *Appl. Environ. Microbiol.* 56(6) 1919–1925.

Arndt-Jovin, D. J., Robert-Nicoud, M., Kaufman, S. J., and Jovin, T. M. (1985). Fluorescence digital imaging microscopy in cell biology. *Science* 235, 247–256.

Atlas, R. M., and Bartha, R. (1998). In "Microbial Ecology." Benjamin and Cummings, Menlo Park, CA.

Clarholm, M. (1994). The microbial loop in soil. In "Beyond the Biomass Compositional and Functional Analysis of Soil Microbial Communities" (K. Ritz, J. Dighton, and K. E. Giller, eds.), pp. 221–230. John Wiley and Sons.

Costerton, J. W., Lewandowski, Z., Caldwell, D. E., Korber, D. R., and Lappin-Scott, H. M. (1995). Microbial Biofilms. *Annu. Rev. Microbiol.* 49, 711–745.

Garland, J. L. (1996). Patterns of potential C source utilization by rhizosphere communities. *Soil Biol. Biochem.* 28, 223–230.

Hedrick, D. B., Richards, B., Jewell, W., Guckert, J. B., and White, D. C. (1991). Disturbance, starvation, and overfeeding stresses detected by microbial lipid biomarkers in high-solids high-yield methanogenic reactors. *J. Indust. Microbiol.* 9, 91–98.

Kluyver, A. J., and van Neil, C. B. (1956). "The Microbe's Contribution to Biology." Harvard Univ. Press, Cambridge, MA.

Mohn, W. W., and Tiedje, J. M. (1992). Microbial reductive dechlorination. *Microbiol. Rev.* 56, 482–507.

Revsbech, N. P., and Jorgensen, B. B. (1986). Microelectrodes: Their use in microbial ecology. In "Advances in Microbial Ecology" (K. C. Marshall, ed.), pp. 293–352. Plenum Press, New York.

Santegoeds, C. M., Nold, S. C., and Ward, D. M. (1996). Denaturing gradient gel electrophoresis used to monitor the enrichment culture of aerobic chemoorganotrophic bacteria from a hot spring cyanobacterial mat. *Appl. Environ. Microbiol.* 62, 3922–3928.

Shapiro, J. A. (1998). Thinking about bacterial populations as multicellular organisms. *Annu. Rev. Microbiol.* 52, 81–104.

Steffan, R. J., and Atlas, E. J. (1988). DNA amplification to enhance the detection of genetically engineered bacteria in environmental samples. *Appl. Environ. Microbiol.* 54, 2185–2191.

Torsvik, V., Goksøyr, J., and Daae, F. L. (1990). High diversity in DNA of soil bacteria. *Appl. Environ. Microbiol.* 56, 782–787.

Whitman, W. B., Coleman, D. C., and Wiebe, W. J. (1998). Prokaryotes: The unseen majority. *Proc. Natl. Acad. Sci. USA* 95, 6578–6583.

Wolin, M. J., and Miller, T. L. (1982). Interspecies hydrogen transfer: 15 years later. *ASM News* 48, 561–565.

Economic Consequences of Infectious Diseases

Martin I. Meltzer

Centers for Disease Control and Prevention

GLOSSARY

categories of costs Costs can be categorized as direct medical, direct nonmedical, or indirect costs of lost productivity. Costs can also be categorized as either fixed or variable.

cost–benefit analysis (CBA) A comparison of all the costs and benefits that might occur due to an intervention to control an infectious disease over a prespecified analytic time horizon. These costs and benefits are discounted to the year zero.

cost-effectiveness analysis (CEA) The total net costs of the intervention divided by the number of health outcomes averted (the denominator). The result is the total net cost per unit health outcome averted (e.g., total net cost per death averted).

cost–utility analysis (CUA) A specialized form of CEA in which the health outcomes used for the denominator are valued in terms of utility or quality. An example of a utility

or quality measure of valuation of a health outcome is the quality-adjusted life year.

discounting A fundamental economic concept is that communities and individuals have a time-based preference for goods and services that are available now. Resources spent and benefits gained in the future are therefore discounted to present values to allow for time preferences. Discounting allows for the direct comparison of costs and benefits that occur during different time periods.

effectiveness The reduction in incidence of a disease due to routinely applying an intervention to a population. The level of effectiveness is often less than the efficacy.

efficacy The maximum possible reduction in incidence of a disease attributable to an intervention designed to prevent and control a disease. Often measured using random controlled trials.

fixed costs Costs that do not vary in the short to medium term and are unlikely to vary even if there are fluctuations in the number of cases (e.g., cost of a building).

incidence The number of new cases (infection or disease) that occur during a defined time period.

opportunity costs The value of the resources lost due to, or used in treating and preventing, infectious diseases, valued in terms of foregone alternative uses. The use of opportunity costs is a core concept of economic analyses.

perspective The viewpoint of who suffers the consequences of an infectious disease as well as who benefits from an intervention to prevent or treat an infectious disease.

prevalence The proportion of a defined population that has a given disease or condition at one specific point in time or over a defined time period.

societal perspective Includes all the costs caused by an infectious disease and all the benefits of treating and preventing the disease, regardless of who pays and who benefits.

variable costs Costs that vary depending on the numbers of cases treated (e.g., physician time and drugs administered).

INFECTIOUS DISEASES can impact societies on such a vast scale that disease-induced changes cannot readily be reduced to a single estimate of economic impact. Furthermore, because the time scales associated with these impacts can vary from 12 to 24 months for pandemic influenza to persistent annual growth of cases of dengue fever, it is essential to carefully define the time period covered by an assessment of impact.

Assessments of the economic impact of infectious diseases are probably most relevant when there are definite needs for such information, such as when evaluating potential interventions. When conducting an assessment of economic impact, the analyst has to decide on a variety of factors that influence the final result, including choosing the perspective of the study (e.g., patient, health care provider, or societal) and what categories of costs to include. When conducting an economic analysis, as opposed to a financial analysis, it is also essential to use opportunity or "true" costs as well as to discount all future costs (and benefits) at some predetermined rate. Costs associated with the impact of an infectious disease can be categorized as direct medical (e.g., physician time and drugs), direct nonmedical (e.g., administration and patient's travel costs), indirect costs (e.g., time lost from work because of illness), and intangible costs (e.g., pain and suffering). There are three main methods of assessing economic costs and benefits of interventions designed to control and prevent infectious diseases: cost–benefit analysis, cost-effectiveness analysis, and cost–utility analysis. When using any of these methods, it is essential to distinguish between the efficacy and effectiveness of the intervention(s) being studied and to include estimates of the harm associated with an intervention (e.g., side effects from a drug) as well as the benefits.

I. THE HISTORY OF THE IMPACT OF INFECTIOUS DISEASES: TWO THEMES

When considering the economic impact of infectious diseases there are two themes that emerge that are discussed in the following sections.

A. Infectious Diseases as Agents of Change

Throughout recorded history, infectious diseases have demonstrated the ability to notably impact the course of human affairs. Thucydides, for example, records that a plague spread through Athens when it was besieged by the Peloponnesians in 430 BC, and that the plague likely influenced the Peloponnesians to withdraw earlier than originally planned. Alexander the Great, who conquered enough territories to assemble one of the largest empires ever, died from what some have speculated was a case of malaria. His death hastened the shrinkage of the Macedonian empire. The "Black Death" that swept through Europe in the 1340s and early 1350s essentially altered the very fabric of society by causing the demise of the feudal system (see Box 1). Such large-scale devastations caused by infectious diseases can and do continue to occur. Acquired immunodeficiency syndrome (AIDS), caused by the human immunodeficiency virus (HIV), is currently wrecking havoc on the African continent, causing an enormous death rate among some of the most productive members of society (Box 1).

It is a principle of military history that "diseases destroy more soldiers than do powder and the sword" (Maj. W. S. King, U.S. Army surgeon, after the first battle of Bull Run, U.S. Civil War, 1861). In earlier centuries, soldiers died from plagues that swept through encampments as well as from infections that took hold after battle-related injuries, no matter how minor such injuries may have initially been. Even the advent of antibiotic drugs and vaccines against infectious diseases has not removed the problem of infectious diseases among late twentieth-century armies. For example, the official history of the United States Medical Corp in the Vietnam conflict records that approximately 71% of all hospital admissions of active-duty personnel were due to diseases (with malaria accounting for a large portion of those admitted with disease). It is pure speculation to discuss how society and economies would have been changed if infectious diseases had not been such a constant harvester of troops on almost every military campaign of note.

The impacts of infectious diseases on history are

BOX 1

The Impact of the Black Death and AIDS: Defying Simple Economic Analysis?

It has been estimated that the Black Death that ravaged Europe from the late 1340s through the early 1350s killed between 25% and 50% of Europe's population. Such a large and rapid depopulation had direct consequences on societies and their economics. For example, labor on feudal manors in England had become so scarce that villeins often left their fiefs (manors) to which they were indentured, and went to work for another lord for a wage. With labor scarce, such wages quickly increased, and there were several attempts by king and parliament to pass ordinances, including the famous Statute of Labourers of 1351, attempting to regulate wages and to stop serfs from running away. Such attempts essentially failed, and the shortage of labor so empowered the laborers that they rebelled in 1381, causing a formal abolition of villenage (although traces of such practices continued for some time afterward). In France, in which the king was trying to raise money and troops to reverse the defeat by the English at Crecy (1346), the Black Death so ravaged the taxpayers of Montpellier that the king abolished their tax assessment. The high death rate caused by the Black Death in Cairo, Egypt, caused the prices of manufactured goods, such as linens, to decrease to one-fifth of their preplague prices. Furthermore, the reduction in available labor caused rents from land to decline while the price of agricultural goods such as grains increased.

Although not as rapid or as severe as the Black Death, globally AIDS accounted for approximately 1% of all deaths in 1990, and this percentage is estimated to increase to 2% by 2020. The World Health Organization estimated that, at the end of 1997, there were almost 31 million people living with HIV/AIDS (approximately 90% of those infected live in Africa). This is three to six times higher than the estimate made for 1987. It has been difficult for experts to agree on whether or not AIDS will have a net negative economic impact, as measured by per capita gross national product (GNP), on those countries most affected. However, it is also true that GNP does not adequately measure all aspects of human welfare and, at this point in history, predicting the long-term impact of AIDS on a nation's GNP is more art than science. Some examples of the economic impact of AIDS include the World Bank estimate that the annual treatment cost of an AIDS patient in any country is equal to approximately 2.7 times the per capita GNP of that country. As an example of the opportunity cost, this amount could, on average, pay for a year of primary education for 10 students. Even more striking is the fact that AIDS has markedly reduced live expectancy in Africa. For example, in Burkina Faso, AIDS has reduced average life expectancy by 11 years, whereas in Zimbabwe the reduction is approximately 22 years.

---

not limited to military history. Trade routes, commerce, and new enterprises can be prevented and disrupted by infectious diseases. Yellow fever, for example, defeated the initial attempts to build the Panama Canal. The discoveries that malaria and yellow fever are spread by mosquitoes allowed sanitation engineers to target their disease control efforts on eliminating mosquito breeding grounds. The resultant reduction in the number of disease vectors (insects or animals that spread an infectious disease) allowed the eventual completion of the monumental engineering feat.

From this discussion, it is clear that infectious diseases can often cause large, long-lasting, and wide-ranging economic impacts. The effects can be so great and diverse that it can be difficult to calculate, in monetary units, reliable estimates of their economic impact.

B. Time and the Impact of Infectious Diseases

Time is the second theme to consider when appraising the economic impact of infectious diseases.

Epidemics of infectious diseases can come and go, in varying lengths of time, and the duration of an epidemic is not a good measure of total impact. The Black Death, for example, was not a single epidemic that swept through Europe in one single cycle of events. Rather, after the initial onslaught in the 1340s and 1350s, it remained a threat that appeared and reappeared on numerous occasions over a time scale that spanned several centuries. In contrast, the influenza pandemic of 1918 lasted just 12–24 months but caused an estimated 20–25 million deaths worldwide. In the twentieth century, the influenza pandemic was unique in the number of deaths caused by a single infectious agent during such a relatively short period of time (Fig. 1).

Not all infectious diseases cause significant economic impact through epidemics. Some diseases have exhibited a persistent, and increasing, burden. An example of such a disease is the mosquito-borne virus that causes dengue fever. Although classic dengue is rarely fatal, with symptoms such as fever, headaches, and malaise, it can lead to dengue hemorrhagic fever (DHF) which can result in death due to internal bleeding and circulatory failure. Those patients who receive adequate care, often in a hospi-

tal setting, usually recover. It has been estimated that, although there were probably less than 1 million cases of dengue per year in the early 1970s, by the late 1990s there are approximately 20–30 million annual cases worldwide of dengue and DHF. Less than 10% of these were reported to public health authorities such as the World Health Organization. There are probably many reasons for underreporting, including the lack of a cure (although symptoms can be treated) and the absence of definitive public health interventions (vaccines are being developed).

Figure 1 can also give the impression that, in developed countries such as the United States, infectious diseases no longer pose the threat that they used to. The reduction in deaths due to infectious diseases has been due to a multitude of factors, such as improved sanitation, less crowded housing, an explosion of knowledge concerning the transmission of infectious diseases, and a new "arsenal" of antibiotic drugs and vaccines. However, the insert in Fig. 1 shows a distinct increase in deaths due to infectious diseases in the United States since the early 1980s. Much of this increase is due to HIV/AIDS and AIDS-related complications. Although the number of deaths in the United States due to HIV/AIDS in a single year has

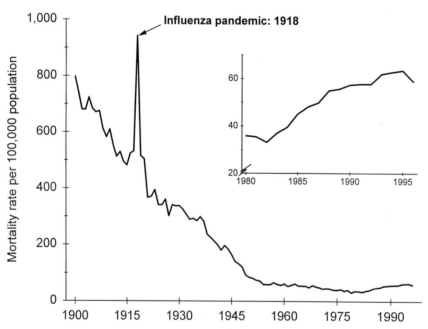

Fig. 1. Crude infectious disease mortality rate in the United States from 1900 through 1996 (adapted from Armstrong *et al.*, 1999).

not reached the proportions of deaths due to the 1918 influenza pandemic, the longevity of the disease in a patient and the cost of treatments (but not cures) makes HIV/AIDS one of the most important infectious diseases in terms of economic impact (see Box 1).

Another time-related element is the discovery that many chronic diseases may initially be caused by an infectious pathogen. For example, it has been estimated that 82% of cervical cancers are caused by human papillomavirus, and that many stomach ulcers are caused by the bacteria *Helicobactor pylori*. Some of these chronic diseases can take years, if not decades, to manifest themselves in a patient, and then they can take a similar time scale to reach a point of resolution, be it cure or death.

This second theme of time emphasizes that any assessment of the economic impact of an infectious disease has to carefully define the time period covered by the assessment. Such careful definition will inevitably mean that some important economic impacts are likely to occur outside the period studied.

II. INCIDENCE AND PREVALENCE

The first step in determining the economic impact of an infectious disease is to calculate the number of persons in a given population who succumb to the disease within a defined time period. Prevalence describes what proportion of a defined population has a given disease or condition at one specific point in time or over a defined time period (a "snapshot"). Incidence describes the frequency of occurrence of new cases over a defined time period.

A. The Connection between Prevalence and Incidence

Most burden of disease estimates are prevalence estimates, often the number of cases (new and existing) that exist in a year. To calculate the economic impact of a disease, and the benefits of an intervention, an analyst often needs to know the incidence of disease. Assume, for example, that an intervention has been proposed that can prevent infection from a defined pathogen but cannot cure those who are

already infected (e.g., polio vaccination). Assessing the benefits of such an intervention will require measuring the incidence of the disease before and after the intervention. Unfortunately, it is often expensive and difficult to organize the collection of data needed to determine disease incidence. It is possible, however, to calculate estimates of incidence from prevalence data using the following standard equation:

$$\text{Incidence} = P/(1 - P)D$$

where P is the prevalence and D is the average duration of the disease.

Cumulative incidence is the number of cases of a disease that can occur over a predefined time period. The average time during which an infected person can readily transmit the infection to others (i.e., be infectious) is often used as the time period for calculating cumulative incidence:

$$\text{Cumulative incidence} = 1 - e^{(-I \times t)}$$

where t is the total time period in which transmission/infection can occur, I is the incidence, e is the constant for natural logarithms, which is approximately 2.71828.

III. ASSESSING THE ECONOMIC IMPACT OF AN INFECTIOUS DISEASE

Although infectious diseases can cause momentous impacts that appear to defy being quantified in economic terms (see Section I), there are many instances in which the impact of infectious diseases can be at least partially assessed. Table I provides examples of estimates of costs incurred while treating some infectious diseases in the United States in the early and mid-1990s. No matter how interesting the data in Table I, such numbers probably have little impact or value unless the estimates are applied in some manner. For example, an estimate of the economic impact of a disease could be used to evaluate the costs and benefits of an intervention designed to prevent and control the disease. The data in Table I were used to help explain to policymakers and the public the need to invest government funds into programs focusing on the prevention and control

Disease	Annual cost ($)[b]	Type of cost[c]
AIDS	5.8 billion	Direct medical charges (1993 dollars)
Tuberculosis	703 million	Direct medical charges (1991 dollars)
Nosocomial infections (acquired in hospital)	4.5 billion	Hospital charges (1992 dollars)
Food-borne bacteria	2.9–6.7 billion[d]	Direct and indirect costs (1993 dollars)
Human papillomavirus	1.23 billion[e]	Direct medical charges (1991 dollars)
Neonatal group B streptococcal infections	294 million	Direct medical charges (1993 dollars)
Bacterial vaginosis	1.0 billion	Direct medical charges (1993 dollars)

[a] Adapted from Centers for Disease Control and Prevention (1998).

[b] The sources of these estimates are cited in the Centers for Disease Control and Prevention (1998).

[c] Cost can be categorized into direct medical costs, direct nonmedical costs, indirect costs, and intangible costs. See Section III for further details.

[d] Range of direct + indirect costs due to food-borne illnesses caused by six pathogens: *Campylobacter jejuni* or *coli, Clostridium perfringens, Escherichia coli* O157:H7, *Listeria monocytogenes, Salmonella* (nontyphoid), and *Staphylococcus aureus.*

[e] Preliminary estimates are based on the fact that human papillomavirus causes 82% of all cervical cancers; thus, annual charges were calculated by assuming that 82% of the following treatment charges are attributable to human papillomavirus: 1 million follow-up visits due to precancerous lesions identified by Pap smear costing $1,100 each for a subtotal of $1.1 billion, 55,000 treatments of carcinoma *in situ* at $4,360 each for a subtotal of $0.2 billion, and 15,800 treatments of cervical cancer costing $11,300 each for the subtotal of $0.2 billion. These estimates do not include indirect costs attributable to lost productivity or the cost of screening for cervical cancer.

of emerging and reemerging infectious disease. The following sections present concepts and methods to measure and evaluate the economic impact of infectious diseases.

A. Infection versus Clinical Disease: An Economic Perspective

In many infectious diseases, a portion of those who become infected do not show any clinical symptoms as a result of the infection (i.e., they become subclinically ill). For epidemiologists and others studying how diseases are spread, those with subclinical illnesses can be very important because they can transmit the infectious agent to susceptible persons. By definition, subclinical cases do not consume resources and thus do not cause a negative economic impact. Indeed, subclinical cases may represent a positive economic impact because the infection could naturally immunize the person against future illness. Valuing natural immunization as an economic benefit is rarely, if ever, done. Thus, an economic analysis of an infectious disease usually focuses on clinical cases that cause some verifiable economic impact,

such as a half day lost from work, a visit to a physician, or the cost of a drug taken as treatment.

B. Opportunity Costs: Financial versus Economic Analyses

There are at least two methods that can be used to assess the economic impact of an infectious disease—financial and economic. A financial analysis of the impact of an infectious disease is essentially an accounting approach wherein the analysis is restricted to actual cash-based costs, such as the cost of any drugs used or the physician's fee. An economic analysis, however, will use the "true," or opportunity, costs. Opportunity cost is defined as the value of the resources used to treat or prevent an infectious disease, where the resources are valued in terms of foregone alternative uses. An example of an opportunity cost is the time lost from work by a patient suffering from a "mild" case of influenza. The time lost has a value to both the patient and society. Even if the person is unemployed, the illness may prevent household and community chores from being done, which are valued. Perhaps the greatest opportunity costs exerted by infectious diseases is that of premature death. From an economics point of view, every person who dies early represents a loss of potential return on resources invested up to the point of death in the feeding, clothing, housing, and education of that person.[1]

C. Perspective of the Analysis

The choice of perspective of the analysis is as important as the choice between conducting an economic or financial analysis. The perspective is the viewpoint of who suffers the consequences of an infectious disease, as well as who benefits from an intervention to prevent or treat an infectious disease. Typically, textbooks on health economics will list at least four different perspectives: that of (i) the individual patient, (ii) the provider of health care services (e.g., physician, clinic, and hospital), (iii) the third-party payers of health care services (e.g., private and national health insurance schemes), and (iv) society. The societal perspective includes all the costs due to an infectious disease and all the benefits of treating and preventing the disease, regardlesss of who pays and who benefits. Thus, any other perspectives is a subset of the societal perspective (see Section VII.B).

---

1. Some have argued that, while premature death represents a loss of future productivity, it also represents a savings in terms of future consumption avoided. For example, although tobacco smokers often die earlier than non-smokers (and often require costly medical care to treat smoking related diseases such as lung cancer), it has been calculated that such early death results in a net savings to U.S. society. The net savings is due to the reduced payments from the social security and Medicare trust funds. What the argument for including avoided consumption lacks is a clear definition of society's objective. Inclusion of such savings implies that the "efficient" consumption of resources is a primary objective of a society. We are then in danger of routinely condemning to death all those with long-term illnesses, physical or mental handicaps, or even infirmity as a result of old age. When evaluating the economic impact of infectious diseases, we should not ignore the implicit social contract that human life itself has value greater than the sum of resources required to sustain it (see Section XI.B for further comments on the valuation of life).

D. Approaches to Estimating the Value of Health and Nonhealth Outcomes

There are two standard methods of measuring the opportunity costs of a disease: the cost-of-illness (COI) method and the willingness-to-pay (WTP) approach.

1. COI

The COI methodology collects and uses data concerning direct medical, direct nonmedical costs and productivity losses, or indirect nonmedical costs (see Section III.E for additional definitions and examples).

2. WTP

The WTP approach estimates what an individual or society would be willing to pay to avoid contracting a given disease or condition. The WTP approach goes beyond the COI methodology in that it implicitly provides a valuation on some of the intangible costs associated with a case of an infectious disease, such as fear and pain. WTP estimates are often obtained directly through surveys, although indirect methods are available. In a typical WTP survey, a respondent is given a hypothetical scenario regarding an intervention (e.g., suppose a vaccine were available to protect against disease X). The respondent is then asked what is the maximum that they would be willing to pay for the intervention. This maximum amount could be solicited as either a single, open-ended question (e.g., "What is the maximum that you would be willing to pay?") or the respondent could be given a set amount and asked if he or she would pay that amount (e.g., "Would you be willing to pay $Y?"). The latter technique can be made more sophisticated by randomly assigning each respondent to one of several different set amounts. An analysis of the distribution of percentage of positive responses versus the amount "offered" can provide a decision maker with a view of how people value an intervention over a wide range of values. WTP surveys also usually collect some sociodemographic information (education, income, age, sex, etc.) and data regarding the respondent's knowledge and attitude toward the problem. By using statistical models, the analysts can obtain an understanding of how valuations are influenced by sociodemographics and knowledge and attitudes.

E. Categorizing Costs

In the COI methodology, costs can be categorized as direct medical, direct nonmedical and indirect costs of lost productivity. Examples of costs in each category are given in Table II, which also includes an illustration of how the perspective of the study (see Section III.A) can determine whether such costs are included in an economic analysis of the impact of a disease.

1. Fixed and Variable Costs

Costs can be categorized as either variable or fixed, where variable costs are those costs that vary depending on the numbers of cases treated (e.g., physi-

TABLE II
Categorizing Costs: Examples of Costs and When to Include in, or Exclude from, a Study

		Include (+) in or exclude (−) from study?[a]			
		Perspective of study			
Cost category	*Example of costs*	*Patient with insurance[b]*	*Physician (private practice)[c]*	*Payer[d]*	*Society[e]*
Direct medical	Physician time	Copayment	+	+	+
	Medical personnel time (e.g., nurse and technician)	−	+	+	+
	Drugs	Copayment	+	+	+
	Medical devices (e.g., syringes and ultrasound)	−	+	+	+
	Expandables used in laboratory tests (e.g., x-ray film and reagents)	−	+	+	+
Direct nonmedical	Administration[f]	−	+	+	+
	Physical facility (e.g., clinic and office)	−	+	−	+
	Utilities (e.g., telephone and electricity)	−	+	−	+
	Patient's travel costs	+	−	−	+
	Temporary hired caregiver[g]	+	−	−	+
Indirect costs	Time off from work to visit physician	+	−	−	+
	Time off work while ill and recuperating	+	−	−	+
	Hire temporary household help while ill[h]	+	−	−	+

[a] Inclusion of a cost item in a study will depend on the chosen perspective. The four perspectives shown in the table do not cover all possible perspectives. For example, a researcher may wish to study the cost of an infectious disease from the perspective of a hospital (another form of health care provider). In such a case, the analysis would most likely include the cost of the time of the physicians hired by the hospital.

[b] Assumed patient is covered by some form of health care insurance.

[c] Physician perspective assumed to be that of a physician who has his/her own private practice. Many of the costs listed, such as laboratory-based costs, are likely to be reimbursed by a health insurance company (assuming the patient has insurance that covers the treatment).

[d] Payer represents the perspective of the third-party health insurance payer, who reimburses the physician for services rendered to a patient covered by an insurance scheme (private or public).

[e] Societal perspective is the sum of all perspectives. Care must be taken, however, not to double account, such as including laboratory fees from both physicians and payer perspectives.

[f] Physician's practice and health insurance may each have separate administration costs.

[g] Temporary caregiver may be hired to look after family members while adult visits physician.

[h] Temporary help may be hired to do household chores and look after family while an adult is ill or to allow an adult to concentrate on nursing a sick child. These costs may or may not be reimbursed by an insurance scheme.

cian time and drugs administered). Fixed costs are those costs that do not vary in the short to medium term and are unlikely to vary if there are fluctuations in the number of cases (e.g., cost of a building). It is recommended that one set of cost terminology, appropriate for the intended audience, be used throughout a single study.

2. Intangible Costs

Another category of costs, labeled intangible costs, includes items such as pain, suffering, and fear. It is often difficult to find widely accepted values for such items. The WTP approach of costing health outcomes (see Section III.D.2) can be used to obtain values for such items. However, if a COI approach is used, then researchers may resort to simply listing the number of readily identifiable intangible costs that may be associated with the impact of an infectious disease. These non-dollar-valued costs and benefits may become crucial in the public debate concerning the adoption of an intervention designed to prevent or control an infectious disease.

3. Sources for Obtaining Economic Costs

It was recommended in Section III.B that an economic analysis of the impact of an infectious disease use opportunity costs. However, financial charge data, such as the price of drugs bought at a for-profit pharmacy, are often more readily available than opportunity costs. The analyst can obtain estimates of the economic costs in a variety of ways. Large-volume discount prices, for example, can be used as proxies for opportunity costs. An example of large volume discount purchases is the Medicare health insurance scheme in the United States for those 65 years of age and older. Under the Medicare system, the U.S. federal government sets the amount of money that it will reimburse hospitals for inpatient care, reported by Diagnostic-Related Groups. These amounts are usually published annually in the *Federal Register* by the Health Care Financing system of the Department of Health and Human Services (e.g., *Federal Register* **61**,(170) 46166–46328, August 30, 1996). Textbooks on health economics, such as that by Haddix *et al.* (1996), typically provide a list of resources that a researcher can use to start collecting financial and economic data associated with health care systems.

4. Data Limitations

Many sources of financial and economic data related to health care have limitations. The most common limitation is that many report average cost per procedure but not the cost per case of disease. A researcher, for example, may readily find the Medicare national average reimbursement for an office visit or an x-ray. However, it is far more difficult to find data concerning the "average" cost of treating, for example, an uncomplicated case of influenza. There are even fewer data concerning the indirect costs borne by the patient (Table II), such as time lost from work due to a specified disease.

Another important limitation of many cost databases is that they only report average costs. Cost data are most likely not normally distributed (i.e., they do not have a "bell-shaped" probability distribution curve), and median costs are often much lower than average costs. Thus, average costs may overstate the costs accrued by a patient with a "typical" case of a given disease.

5. Overcoming Data Limitations

The researcher can overcome the lack of data concerning the cost of treating a given disease by making some assumptions regarding the number of physician visits required, the number and type of diagnostic tests that may be ordered, and the type of antibiotics prescribed. In effect, the researcher is modeling the impact of the infectious disease in terms of health care resources used to treat the disease. Each of these components can then be priced using average or median costs, such as average Medicare allowable rates. Sometimes even average cost data are not readily available, and a researcher will have to conduct original research, such as examining medical charts and conducting interviews, to obtain relevant cost data. If the researcher is able to model the costs of an infectious disease using average costs, there is no assurance that such a model reflects actual average costs. The researcher, however, knowing the elements used to construct such an estimate, can conduct sensitivity analyses (see Section VIII.C.1) to determine the impact of increases or decreases in the costs of various components.

IV. DISCOUNTING COSTS OVER TIME: A KEY ECONOMIC CONCEPT

When measuring the impact of infectious diseases, analysts will frequently consider the impact over time periods of more than 1 year. A key economic concept is that society places a premium on benefits gained in the present when compared to those that may be gained in the future. To reflect this preference for goods and services that are delivered now, both resources spent and benefits gained in the future are discounted when being compared to resources spent and benefits gained in the present.

To illustrate the concept of discounting, imagine a stockpile of drugs established, using public funds, to treat and prevent a sudden outbreak of an infectious disease. This stockpile might be considered as an investment made by society. If an outbreak occurred this year, and the stockpile was used to treat and prevent disease and reduce the number of adverse outcomes, then those saved from adverse health outcomes related to the disease (including death) will be able to continue contributing to society. Such contributions represent the "return" to the investment made in the resources expended to build the stockpile of drugs. However, if the stockpile remains unused for 10 years, then the resources invested in the stockpile are "idle" and cannot be invested in other opportunities such as public-financed education. Discounting, therefore, allows for the direct comparison of costs and benefits that occur during different time periods.

A. Formula for Discounting

To calculate the present value of a "stream" of costs or benefits that extend into the future, the following formula is applied (see Section VI.A for an example of the formula applied to data):

$$PV = \sum_{t=0}^{N} \frac{\$_t}{(1+r)^t}$$

where PV is the present value, $\$_t$ is the dollar value of cost or benefit in year t, r is the discount rate

expressed in decimals (e.g., 3% = 0.03), t is the time period ranging from 0 to N, and N is the maximum time period being examined.

For example, suppose that a proposed infectious disease control program will save $15,000 in direct medical costs every year for 5 years (first year = Year 0). The PV of this stream of savings is as follows:

$$\frac{\$15,000_{year0}}{(1+0.03)^0} + \frac{\$15,000_{year1}}{(1+0.03)^1} + \frac{\$15,000_{year2}}{(1+0.03)^2}$$
$$+ \frac{\$15,000_{year3}}{(1+0.03)^3} + \frac{\$15,000_{year4}}{(1+0.03)^4} = \$70,756.58$$

B. Choosing a Discount Rate

The present value of future costs and benefits is dependent on the discount rate r. A common question is: What is the "correct" discount rate to use? The answer depends partly on the perspective of the analysis (see Section III.C). If the perspective is that of a provider (e.g., a single physician's private practice), then an appropriate discount rate for the cost of time and other inputs may be the rate on some instrument of savings or the rate on an investment that could be a potential alternative. If, however, the analysis has a societal perspective, then a much lower discount may be applied. The exact rate used in an analysis with a societal perspective is usually based on recommendations from expert panels and government agencies such as the U.S. Federal Office of Management and Budget (OMB). In the past decade, the OMB has recommended that a discount rate of 3–5% be used for analyses with a U.S. societal perspective. Given that there are no exact methods for choosing the appropriate discount rate, analysts should also consider conducting sensitivity analyses (see Section VIII). The goal of such sensitivity analyses would be to determine if altering the discount rate would alter the overall conclusions. Note that, when doing a sensitivity analysis, a researcher might include a scenario which has a 0% discount rate.

C. Discounting Nonmonetary Costs and Benefits

It is important to emphasize that all future nonmonetary costs associated with an infectious disease, such as future deaths directly due to infection in the present, should also be discounted. Society also has a time preference for such nonmonetary costs and routinely values the life of somebody living now compared to a birth or death in some future time.

V. CONCEPTS NEEDED TO ASSESS THE ECONOMICS OF AN INTERVENTION

The science and technology developed in the industrial and postindustrial ages have produced a variety of technologies and concepts that have been applied both to treat clinical cases due to an infectious pathogen and to systematically prevent and control the spread of infectious diseases (e.g., antibiotics, vaccines, and improved sanitation). Thus, the costs associated with an infectious disease often extend beyond the cost per patient treated (Table I) and include the costs of resources expended in attempts to interrupt the transmission of infectious diseases. In addition to the economic concepts that must be employed to measure the economic impact of an infectious disease, there are at least three other concepts that must be employed when measuring the economic costs and benefits associated with an intervention.

A. The Need to Assess the Economic Costs and Benefits of an Intervention

At any given point in time a society has a limited set of resources for communitywide activities. Usually, there are more opportunities to use those resources than there are actual resources. Members of a community have a need for, among other items, education, housing, security, food, and health care. Therefore, given the variety of demands on resources, decision makers often ask: Does it make economic "sense" to invest in the control and prevention of a specified infectious disease? That is, will the benefits associated with the intervention be greater than the costs of implementing the intervention? To answer such questions, the analyst has to consider both the costs of an intervention and the benefits associated with a potential reduction in the number of cases that will need treatment.

B. Efficacy versus Effectiveness

It is rare for an intervention to totally eliminate a disease. Randomized controlled trials (RCTs) are often conducted to provide estimates of the maximum reduction in the number of cases of a disease that can be achieved by using a specified intervention. This maximum possible reduction in incidence of a disease is termed the efficacy of an intervention. The maximum impact (efficacy) is achieved because well-done RCTs carefully select patients who will comply with the protocol and often do not have other diseases or conditions that may "interfere" with the intervention being studied. Furthermore, clinicians participating in RCTs are usually carefully selected for their interest and expertise in the problem, and they work with carefully trained staff who have time for follow-up and record keeping.

It is not surprising that in most cases the benefit obtained from an intervention applied in routine clinical or public health settings is less than that achieved by the RCTs. The benefit that accrues from an intervention routinely applied to a population larger than that used in the RCTs is called effectiveness. The difference between efficacy and effectiveness can be significant, and it is a major challenge to obtain realistic measures of effectiveness. Unless data can be obtained from an observational study, an analyst will probably have to resort to modeling to estimate the level of effectiveness, conducting many sensitivity analyses (see Section VIII.C) to determine the economic impact of various levels of effectiveness.

C. Recognizing Both Benefits and Harms Resulting from an Intervention

Although we often focus on the benefits of an intervention, the associated harms must also be recognized. The problem is that many costs and harms may be hidden, take some time to become apparent, or be of such low probability of occurrence that they do not show up in the initial trials and thus are excluded from an initial evaluation of an intervention. For example, Guillain-Barré syndrome (GBS) has been associated with viral, bacterial, and other infections as well as vaccinations. Although GBS is

TABLE III
Three Methods of Conducting an Economic Analysis of an Intervention to Control and Prevent and Infectious Disease[a]

| | Costs included?[b] | | Outcome measure |
Method of analysis	*Direct*	*Indirect*	*(benefit)*[c]
Cost–benefit	Yes	Yes	Dollars
Cost-effectiveness	Yes	Often	Health outcome[b]
Cost–utility	Yes	Occasionally	Utility measure[c]

[a] Adapted from Meltzer and Teutsch (1998).

[b] All future costs and benefits, monetary and nonmonetary, should be discounted to year zero (see Section IV).

[c] An example of a health outcome is cases averted.

[d] An example of a utility measure is a QALY.

associated with a wide variety of outcomes, including partial paralysis from which a victim often gradually recovers, the risk associated with contracting GBS from vaccination is approximately 1 per 1 million vaccinees. Thus, although the risk of contracting a case of GBS from certain vaccinations is extremely small, the potentially high cost associated with the harmful outcomes suggests that this harmful side effect should be included as a cost associated with the appropriate intervention.

Many interventions, particularly for infectious diseases, are large-scale population-based programs in which the harms and costs are accrued by large groups of subjects, most of whom do not have, nor will they contract, the disease that is the target of the program. The benefits, however, may only accrue to a few who will avoid contracting the disease as a result of the program. It is therefore important to fully assess the magnitude and severity of both the benefits and the harms and who accrues these benefits and harms and then compare the overall "balance" (i.e., do the benefits outweigh the harms for society?).

VI. METHODS OF ASSESSING THE ECONOMIC COSTS AND BENEFITS OF DISEASE CONTROL AND PREVENTION

There are three main methods used to assess the economics of an intervention designed to control

and prevent a disease: cost–benefit analysis (CBA), cost-effectiveness analysis (CEA), and cost–utility analysis (CUA) (Table III). Some texts, in addition to the three methods listed here, include cost analysis as a fourth method. However, because cost analysis does not include estimates of the benefits associated with an intervention, it will not be considered here as a distinct method of evaluating the economics of a disease control intervention.

A. Cost–Benefit Analysis (CBA)

For many applied economists, CBA is considered the "gold standard" by which all other methods are judged. In its simplest form, a CBA lists all the costs and benefits that might occur due to an intervention over a prespecified analytic time horizon. These costs and benefits are discounted (see Section IV.A) to the year zero. If the total discounted benefits are greater than the total discounted costs, then the intervention is said to have a positive net present value (NPV). The formula for calculating NPV is as follows:

$$\text{NPV} = \sum_{t=0}^{N} \frac{(benefits - costs)_t}{(1 + r)^t}$$

where t is the year from 0 to N, N is the number of years being evaluated, and r is the discount rate.

CBA is most useful in three circumstances. First, it is useful when a choice has to be made between two or more options. In such a case, the logical action is to give top priority to those options that give the highest positive NPVs. Second, the results of a CBA analysis can also indicate the economic impact of a single intervention. Third, CBA is useful because it can include an array of important benefits or costs not directly associated with a health outcome, such as time off from work taken by family members to care for sick relatives. One of the most important quantitative issues related to CBA is the fact that all costs and benefits must be expressed in monetary terms, including the value of human lives lost or saved as a result of the intervention. Quantifying all benefits and costs is a nontrivial task.

B. Cost-Effectiveness Analysis (CEA)

A CEA expresses the net direct and indirect costs and costs savings in terms of a predefined unit of health outcome (e.g., lives saved and cases of illness avoided). The total net costs of the intervention are calculated, and then this is divided by the number of health outcomes averted (the denominator). The result is the total net cost per unit health outcome (e.g., net $ cost or savings per death averted). Many of the data required for an economic CEA, with a societal perspective, are the same as those needed for a CBA, with the most important exception being the avoidance of valuing a human life. Explicitly or implicitly, the value of a human life is part of the health outcome used (e.g., lives saved).

A distinct limitation of CEA is that there is no numerical valuation of the actual health outcome. For example, CEA can only provide estimates of the net costs of averting a case of polio. CEA cannot provide any information regarding how society might value each averted case, even in a seemingly similar outcome. For example, how might a community value averting a case of influenza in a 75-year-old person versus a case in an infant? Thus, CEA is best used when comparing two or more strategies or interventions that have the same health outcome. For example, is vaccination more cost-effective than chemoprophylaxis in preventing a case of influenza?

C. Cost–Utility Analysis (CUA)

CUA is a specialized form of CEA in which the health outcomes used for the denominator are valued in terms of utility or quality. A CUA, for example, may attempt to differentiate between the quality associated with an averted case of polio and an averted case of influenza. Examples of the nonmonetary units of valuation include the quality-adjusted life year (QALY; see Box 2) (Patrick and Erickson, 1993) and the disability-adjusted life year (DALY) (Murray, 1994). The result of a CUA is usually expressed as the total net cost per unit of utility or measure of quality (e.g., net $ cost or savings per QALY gained).

Similar to CEA, in CUA the value of life is implicit in the denominator (e.g., calculation of a QALY requires the valuation of life). An unresolved quantitative issue, however, is how to deal with time costs, such as time lost from work due to illness. Some analysts maintain that indirect productivity losses associated with a disease should not be incorporated in the numerator because utility measures used in the denominator (e.g., QALYs) implicitly incorporate a value for lost productivity. Other analysts merely state that care should be taken to avoid having productivity losses in both the numerator and the denominator (i.e., avoid double accounting). Thus, when conducting a CUA, the analyst should explicitly state if morbidity costs, such as lost productivity, are included in the numerator.

There are other, more fundamental problems with using CUA. One key problem is that the techniques used to measure quality of life lost due to a disease (see Box 2) often focus on long-term disabilities. Therefore, are QALYs an appropriate tool to measure the impact of diseases, such as influenza and dengue fever, that cause short-duration morbidity among large numbers of a population? A related problem occurs when attempting to use CUA to compare between noticeably different diseases and health states. Is it feasible to compare, for example, the loss of utility due to diabetes with the loss of utility due to influenza? The danger is that "league tables," in which interventions are ranked according to their cost–utilities, and decisions made on such rankings, may be constructed without regard to differences

---

BOX 2

Utility and QALYs: A Brief Explanation

QALYs are a measurement of the "usefulness" or utility of a given health state and the length of life lived under that health state. For example, is the value or quality of living a year with both legs paralyzed due to polio equal to 0.65 of a year without the polio-induced paralysis? There are three basic methodologies for obtaining values of the utility of a defined health state: expert opinion, values used in previous studies, and conducting surveys. The latter can be divided into two categories—direct and indirect.

Direct Valuation of Utilities

Techniques for directly soliciting utility evaluations via surveys include the standard gamble, time trade-off, and the use of a rating scale. Typically, in the standard gamble method, an individual is asked to choose between two options. One option is a gamble. For example, assume that there is a probability (p) of dying due to a medical intervention to alleviate polio-induced paralysis and, for the same intervention, a probability ($1 - p$) of healthy life for 29 years. The other option, with perfect certainty, is to live, for example, 30 years with polio-induced paralysis of the legs. During an interview, the probabilities are systematically changed until the respondent is indifferent between the two options (the gamble and the perfect certainty). The

probability that generated the point of indifference (e.g., 0.65) is then equivalent to the quality-adjusted life year of a person living with polio-induced paralysis of the legs. Problems with these methods include the fact that each method is a simplification of the number of options/outcomes facing an individual, and that respondents may have difficulty evaluating the probabilities assigned to the gamble.

Indirect Valuation of Utilities

Indirect valuation of utilities is often attempted using questionnaires that break down a health state into sub-groups, or domains. Examples of these domains are opportunity (e.g., social and cultural), health perceptions (self-satisfaction with health state), and physical function (e.g., mobility and self-care). Within each domain, a respondent must rate the impact of the disease from a set of descriptions. For example, in the domain of mobility, a respondent might state that he or she is able to walk around his or her house and neighborhood without help but with some limitations. This response is assigned a preference weight, such as 0.9, on a scale of 0 to 1. The preference-weighted responses from all the other domains are then used to construct a single utility index using a pre-determined equation. The key to such a system, however, is determining the appropriate set of preference weights for each response within each domain.

---

due to estimation techniques and natural differences in disease states.

VII. CONDUCTING AN ECONOMIC ANALYSIS: FIVE ESSENTIAL STEPS

Each of the following steps should be fully addressed at the start of an economic evaluation of an intervention to control and prevent an infectious disease.

A. Define the Question and the Intended Audience for the Answer

This step should define both the type of study (e.g., financial and economic; see Section III) and the methodology (CBA, CEA, or CUA; see Section VI). Many studies suffer from attempting to answer too many questions at once, confusing readers. "Stakeholders" are usually the most interested in the answer, where a stakeholder is defined as a person or group who may have a vested interest in seeing the intervention deployed (or even preventing the intervention from being deployed). Examples of stakeholders for public health interventions include government public health officials, politicians, those suffering from the disease that the intervention aims to alleviate, private firms producing the intervention (e.g., pharmaceutical companies), nongovernment aid agencies, providers, health insurance firms, and advocacy groups.

B. Decide on the Perspective

The perspective defines those costs and benefits that are included in a study (see Section III.C). To help decide the appropriate perspective, ask who will pay for the costs and consequences of the intervention. The choice of perspective will often depend on the question and the intended audience. Whenever interventions will use societal resources such as tax revenues, it is best to use a societal perspective.

C. List Other Possible Interventions

Although the potential audience may focus on the economics of a particular intervention, including other practical interventions in a study will allow a direct comparison of the costs and benefits associated with other options. The "do-nothing" option should also be considered for inclusion because it can be used to illustrate both the burden imposed by the disease and the magnitude of the benefits from deploying an intervention(s).

D. Decide on the Health Outcome of Interest

Potential health outcomes include cases averted, lives saved, years of healthy life saved, treatments prevented, QALYs gained, and DALYs gained. The choice will depend, to some extent, on the question, audience, perspective, and analytic method. For example, a hospital administrator may be most interested in treatments prevented because this would represent potential financial losses. Conversely, a health insurance company may consider reduced number of treatments a savings.

E. Determine the Time Frame and Discount Rate

Although the choice of the time frame for the intervention is often obvious, the choice of the analytic time frame is dependent on the analyst's interpretation of how long the costs and benefits will "linger." For example, in a vaccination campaign of a cohort of children, the intervention costs may be spread over a relatively short period (e.g., 2 years). Some other costs may be generated over time, such as the costs of care associated with those children that develop long-term sequelae in reaction to either the vaccine or the disease. The benefits, however, may accrue over many years in the form of reduced mortality and morbidity. Thus, the time frame should be sufficiently long to capture all the significant costs and benefits. The absolute necessity of discounting future costs and benefits is discussed in Section IV along with a description of methods.

F. Reporting Outcomes: Cost-Effectiveness Ratios

In addition to the basic results of an economic analysis, such as the NPV of a CBA, ratios can be calculated for each of the three methods of conducting a economic analysis (see Section VI). For CBA, the present value of the benefits can be divided by the present value of the costs to give the benefit:cost ratio (BCR). The problem with the BCR is that, although it provides an estimate of the return (benefit) obtained for each dollar of cost, it does not provide a sense of perspective. Thus, although a BCR of 6:1 indicates that there are $6 of benefits for each $1 of costs, there is no indication of how much the intervention actually costs. Thus, when reporting a BCR, it is essential to also report the present value of both the costs and benefits as well as the resultant NPV.

For CEA, one can calculate at least two different ratios: average cost per unit of health outcome and the marginal cost of an additional unit of health outcome. The average cost is simply the total costs of an intervention divided by the total number of health outcomes provided by the intervention. The marginal cost is the cost of obtaining one extra unit of health outcome. For example, if a planned intervention is estimated to save 150 lives, at an average cost of $X/life saved, what would be the cost of saving the 151st life?

The cost-effective ratios for CUA are similar to those used in CEA, with the exception that the outcome is always indicated in a utility measure. For example, one can calculate average cost per QALY saved or estimate the marginal cost of expanding an intervention so that one additional QALY is saved.

VIII. TOOLS FOR CONDUCTING AN ECONOMIC ANALYSIS: DECISION TREES

Sections VI and VII outlined concepts and methods for conducting an economic analysis of an intervention to control and prevent an infectious disease. There is often a need, however, to combine these concepts and methods in a manner that allows the intended audience to readily appreciate the methods and data used. Decision trees are one commonly used methodology that can be used to compare the economic benefits and harms of one or more interventions versus no intervention.

A. Outline for Building a Decision Tree

1. List all appropriate options for intervention. Two or more interventions can be evaluated in a side-by-side comparison. It is generally appropriate to consider including the do-nothing, or no intervention, option. Indeed, the do-nothing option must be included if only one intervention is being evaluated.

2. Decide on the unit of outcome. The appropriate unit is dependent on the perspective (see Section VII.D). For example, clinicians may primarily be interested in clinical and epidemiological parameters (e.g., cases averted and reduction in mortality), whereas policymakers and analysts may emphasize the need to have the benefits and harms measured in economic terms (e.g., dollars). The "construction rules" listed later contain an important rule related to outcomes.

3. Construct a decision tree. A decision tree is a "skeleton," or schematic diagram, that presents all

the options being studied and all the different paths that a patient(s) or a population may follow to end-up at one of the defined outcomes. Figure 2 is an example of a decision tree examining the economic costs and benefits of vaccinating an individual adult against Lyme disease. Reading from the left- to the right-hand side of the figure, the first set of "branches" are the interventions being studied: vaccination versus no vaccination.

4. Attach probabilities to each branch. After the initial split that differentiates the interventions being studied, each time a branch splits into two or more possible occurrences the point of splitting is termed a "probability node." In Fig. 2, for the strategy labeled "Vaccinate? YES," the first probability node describes the probability of a patient who is vaccinated of contracting a case of Lyme disease (probability = 0.00075). This probability is equivalent to the probability of vaccine failure, which is a harm associated with vaccination. This probability can be compared in Fig. 2 with the probability of 0.005 of contracting Lyme disease under the strategy labeled "Vaccinate? NO." If a vaccinated person does contract a case of Lyme disease, the next probability node describes the probabilities of being diagnosed with an early case of Lyme disease (there is a probability of 0.80 of being diagnosed with early Lyme disease). If diagnosed with early Lyme disease, then the patient faces one of four health outcomes: cardiac (probability = 0.01), neurologic (probability = 0.01), arthritic (probability = 0.05), and case resolved (probability = 0.93). In Fig. 2, the probabilities of each of these outcomes are different than those for a patient diagnosed with a late-stage case of Lyme disease (from the branch labeled "Recognized early LD? No"). Note that the final outcomes are clinical outcomes and do not indicate any valuation of each outcome.

5. Calculate the expected values for each option. This is done by multiplying and adding, or "folding back," the values of the outcomes by the probabilities along the branches. This procedure provides a single estimate of the expected value associated with each option. For example, in Fig. 2 the expected probability of successfully identifying and treating an early case of Lyme disease under the "Vaccinate? YES" strategy is determined as follows: probability of

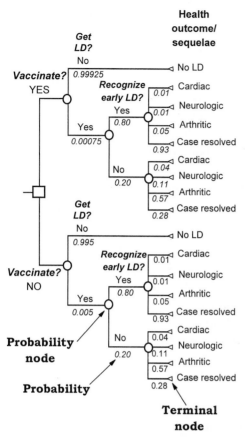

Fig. 2. A decision tree: determining the costs and benefits of vaccinating an individual adult to prevent Lyme disease (LD). Health outcomes and sequelae have been simplified into five categories: No LD, no Lyme disease; cardiac, cardiovascular sequelae (e.g., high-grade atrioventricular blocks); neurologic, neurological sequelae (e.g., isolated nerve palsy); arthritic, arthritic or rheumatological/musculoskeletal sequelae (e.g., episodic oligoarticular arthritis); and case resolved, case resolved after a course of an oral antibiotic, with no further complications. Probabilities are only for illustration of the techniques of building and analyzing data using a decision tree. The actual risk of contracting Lyme disease, the probability of being diagnosed with early Lyme disease, probabilities associated with the various health outcomes, and the reduction in probability of contracting Lyme disease after vaccination may vary by individual and by locale (reproduced with permission from Meltzer *et al.*, 1999).

"arthritic" (0.05); multiplied by the probability of "Recognize early LD? Yes" (0.80) multiplied by the probability of "Get LD? Yes" (0.00075), which equals a final expected probability of 0.00003 (0.05 × 0.80 × 0.00075 = 0.00003). This figure should then be multiplied by the pre-determined cost of treating and resolving a case of Lyme disease-related arthritis plus the original cost of vaccination (which, in this branch, failed to protect against disease). For example, assume that the total direct and indirect costs from a societal perspective (Table II) for treating Lyme disease-related arthritis equal $1,000 and the total direct and indirect costs of vaccination equal $150, for a total value of $1,150. Then the expected value of that particular outcome would be $0.0345 ($1,150 × 0.00003). The same methodology would be repeated for each arm of the "Vaccinate? YES" strategy, and the results would be added to give a total expected value for the vaccination strategy. Note that the total would include the cost of vaccinating under the "Get LD? No" branch (0.99925 × assumed $150 per person vaccinated = $149.8875). The total expected value for the "Vaccinate? YES" strategy would then be compared to a similarly calculated total expected value for the "Vaccinate? NO" strategy. In this particular example, the strategy with the lowest expected value would represent the lowest cost to society.

B. Two "Construction Rules" for Decision Trees

First, the probabilities for all the branches that split from a given node must sum to 1. For example, in Fig. 2 the probabilities of "Recognize early LD? Yes" (0.80) and "Recognize early LD? No" (0.20) sum to a total of 1.0. Second, the outcomes must be both exhaustive and mutually exclusive. That is, all possible options must be listed, and each individual can only reach one outcome.

C. The Importance of Sensitivity Analyses

Decision trees use quantitative data, but estimates of probabilities and the values of outcomes are often uncertain. It is therefore essential that sensitivity analyses be conducted in order to evaluate the robustness of the initial results.

1. Univariate and Multivariate Sensitivity Analyses

Sensitivity analyses can be univariate (changing one variable, such as altering just the probability of getting Lyme disease after vaccination in Fig. 2) or multivariate (two or more variables altered simultaneously). Within these broad categories, there are many types of sensitivity analyses. An analysis can be "worst case" or "best case," or it can consist of altering predefined parameters by given amounts or percentages. The amount by which a given parameter is altered may be based on some known data or may be some arbitrary amount set by the analyst. More sophisticated methodologies for sensitivity analyses include Monte Carlo simulations, in which the amount by which a parameter is increased or decreased is dependent on some previously defined probability distribution.

2. Threshold Analysis as a Form of Sensitivity Analysis

Univariate and multivariate sensitivity analyses depend on a combination of data and the judgment of the analyst as to how much a value might differ. Threshold analysis is a form of sensitivity analysis that removes the subjective element of the analyst's judgment. The goal of threshold analysis is to find the value of a key parameter which will cause the conclusion to change.

To illustrate threshold analysis, consider the following hypothetical example: For a given infectious disease, an analyst may compare a proposed large-scale screen-and-treat program to a treat-all program, with the measure of outcome being the cost per actual case successfully treated (note that the do-nothing option is not valid because one program will definitely be implemented). Based on some expert opinion, the analyst assumed that the test will have a sensitivity (probability of giving a true positive result) of 95%. Assume that the initial results of the decision tree analysis determined that the screen-and-treat option provides the highest expected value (i.e., had the lowest cost per case successfully treated). Threshold analysis, however, determines that the screen-and-treat option only provides the lowest cost per case successfully treated if the sensitivity of the screening test is 90% or greater. That is, with a sensitivity of less than 90%, the treat-all option would be preferred. Therefore, how realistic is it to expect the test to be 90% sensitive "in the field"? This is an example of how results from threshold analyses can help define critical gaps in a database and thus help prioritize applied research agendas.

D. Other Forms of Decision Analysis

Decision trees are not the only methodology for conducting decision analysis. For example, screening for cervical cancer may be done annually or less frequently, but the probability associated with each test is altered based on the results of previous tests. In such a case, the process may be best analyzed using Markov models. It is important to note that modeling is as much of an art as a science, and there is often more than one method that can be used to model a particular problem. The choice of model depends on a variety of factors, one of the most important of which is the ability of the intended audience to both understand and accept the results.

IX. RESOURCE ALLOCATION: WHICH INTERVENTION TO CHOOSE

Resource allocations can be made on the basis of the results of a CBA or CEA study. In the case of CBA, resources should logically be allocated to those interventions that will provide the largest NPV, assuming that the resources needed are available. The strength of CBA is that such allocations can be made across several different types of interventions, even for very different diseases or conditions.

In the case of CEA-based studies, resources should be allocated to the intervention that provides the lowest cost per unit of health outcome. A problem occurs when trying to decide resource allocation between two projects with very different health outcomes. For example, how would you choose between Project A, which costs $13 million and provides a CE ratio of $1,000/case of pneumoccal disease in a person over 65 years of age, and Project B, which costs $7 million and provides a cost-effectiveness ratio of $950/case of childhood measles?

CUA attempts to avoid some of these problems by reducing all interventions to a common health-utility outcome such as QALYs. With CUA data, resources are allocated to those projects that cost the least per unit of health-utility saved (or save the most per unit of health-utility saved). As mentioned earlier, the problem is that this method of resource allocation implies that a QALY saved by preventing heart attacks is equal to a QALY saved by preventing childhood diabetes.

A. Cost-Effectiveness Does Not Equal Cost Savings

An important point to note regarding the comments made on resource allocation is that cost-effectiveness does not equal cost savings. Indeed, many prevention-oriented interventions result in a net cost for society rather than a net savings. Tengs *et al.* (1995) reviewed published literature and evaluated the cost per life saved for 587 health interventions. Approximately 10% of the interventions saved more money than they cost. The median cost per life saved for all interventions was $42,000. Most childhood immunizations and some drug treatment and some prenatal care programs save society money.

B. Valuing Outcomes: Quality and Other Adjustments

Allocating resources to interventions based solely on economic criteria is fraught with many problems. There are obvious limits to an economic analysis of health interventions. For example, what is the cut-off point or threshold value for cost-effectiveness in terms of dollars spent per life saved? Is $50,000/life saved too much? What about $1.5 million/life saved? Also, does it matter if the life saved or QALY gained would belong to somebody aged 3 years versus someone 63 years old? None of these questions can be answered simply by an economic analysis. In the end, an economic evaluation can only provide data for a societal debate as to the value of a given intervention or set of interventions.

See Also the Following Articles

AIDS, HISTORICAL • GLOBAL BURDEN OF INFECTIOUS DISEASES • SURVEILLANCE OF INFECTIOUS DISEASES

Bibliography

Armstrong, G. L., Conn, L. A., and Pinner, R. W. (1999). Trends in infectious disease mortality in the United States during the 20th century, *J. Am. Med. Assoc.* **281**, 61–66.

Centers for Disease Control and Prevention (1998). Preventing emerging infectious diseases: A strategy for the 21st century. *Morb. Mort. Weekly Rep.* **47** (No. RR-15), 1–14.

Creighton, C. (1891). "A History of Epidemics in Britain from A.D. 664 to the Extinction of the Plague." Cambridge Univ. Press, Cambridge, UK.

Drummond, M. F., Stoddart, G. L., and Torrance, G. W. (1997). "Methods for the Economic Evaluation of Health Care Programmes," 2nd ed. Oxford Univ. Press, New York.

Gold, M. R., Siegel, J. E., Russell, L. B., and Weinstein, M. C. (Eds.) (1996). "Cost-Effectiveness in Health and Medicine." Oxford Univ. Press, New York.

Gottfried, R. S. (1983). "The Black Death." Free Press, New York.

Haddix, A. C., Teutsch, S. M., Shaffer, P. A., and Dunet, D. O. (Eds.) (1996). "Prevention Effectiveness: A Guide to Decision Analysis and Economic Evaluation." Oxford Univ. Press, New York.

Meltzer, M. I., and Teutsch, S. M. (1998). Setting priorities for health needs and managing resources. *In* "Statistics in Public Health: Quantitative Approaches to Public Health Problems" (D. F. Stroup and S. M. Teutsch, Eds.), pp. 123–150. Oxford Univ. Press, New York.

Meltzer, M. I., Dennis, D. T., and Orloski, K. (1999). Cost effectiveness of Lyme disease vaccine. *Emerging Infect. Dis.* **5**, 321–328.

Murray, C. J. L. (1994). Quantifying the burden of disease: The technical basis for disability-adjusted life years. *Bull. World Health Org.* **72**, 429–445.

Patrick, D. L., and Erickson, P. (1993). "Health Status and Health Policy: Allocating Resources to Health Care." Oxford Univ. Press, New York.

Tengs, T. O., Adams, M. E., Pliskin, J. S., *et al.* (1995). Five-hundred life-saving interventions and their cost-effectiveness. *Risk Anal.* **15**, 369–389.

The World Bank (1997). "Confronting AIDS: Public Priorities in a Global Epidemic." Oxford Univ. Press, New York.

Education in Microbiology

Ronald H. Bishop
University of Ulster

GLOSSARY

active learning   A process in which students participate in classroom activities to develop and reinforce understanding of key concepts rather than simply receive information passively.

curriculum   The totality of the aims, processes, and desired outcomes of a course of education.

deep learning   A process leading to thorough and long-lasting understanding of concepts and factual information.

independent (open) learning   A process in which individuals study at their own pace using materials designed to instruct and assess, with little if any direct contact between the students and the educator.

surface learning   A process resulting in fragmented factual knowledge and limited, short-term understanding.

syllabus   The specification of course objectives, indicating the knowledge and skills to be developed during it.

MICROBES   affect almost every aspect of our lives. To influence their activities to our advantage, we must understand them. Without education, our understanding will be little better than supposition or even myth. An effective education must go far beyond the knowledge of the interested layperson; science is both technically and intellectually a difficult discipline, often appearing counterintuitive to the general public, and its application to microbiological activities requires much specialized knowledge and experience. Hence, the education of members of the public to help them act to reduce microbiological hazards and contribute sensibly to the political debate over contentious issues in microbiology is a particularly difficult problem. A review of education necessarily differs from one of a scientific topic, for two reasons—the national goals and mechanisms of education differ widely across the world, whereas a scientific study attempts to reflect universal truths; and scientific reviews are based on the critical evaluation of reproducible evidence rather than on personal opinion, whereas reproducible evidence is rare in discussions of education, which are much more opinion-based. Hence, this article is inevitably more influenced by the personal opinion and national experience (in this case, mainly in the UK) of the author. However, it will survey principles rather than attempt to make a comprehensive review of practice because the latter differs so much from country to country. The article, in effect, addresses four key questions about education in microbiology: Why? Who? What? and How?

I. WHY EDUCATE IN MICROBIOLOGY?

A. Scientific Reasons

Many of the world's eminent microbiologists were trained in other scientific disciplines. The big advances in microbiology of the second half of twentieth century were in our molecular understanding of microbes. The skills of chemists and biochemists in particular were the tools that enabled microbes to be exploited to answer fundamental questions in molecular biology and disease. The underlying philosophy of using prokaryotes as simple models of eukaryotes, especially of human cells, was spectacularly successful for several decades. It is still essential that students of related disciplines be educated in microbiology to appreciate the opportunities that these organisms offer, and, conversely, that microbiologists become familiar with and unafraid of chemical and mathematical approaches.

Since the 1980s, however, there has been increasing recognition that microbes are useful and interesting in themselves—as industrial biocatalysts, as nutrient cyclers and environmental bioremediators, and as extremely successful parasites. It is now clear that microbes, as any other successful organisms, are extremely sensitive and responsive to their environments, be these human tissues, chilled foods or deep subsurface rock formations. Thus, their molecular mechanisms can only be interpreted and understood, let alone be influenced by humans, in the context of their environments. Medical microbiology and food microbiology, for example, are increasingly seen as studies of ecological systems. Microbiology now demands a more integrated or holistic view of the organism and its environment, a "feel for the organism," which comes only from a prolonged study of the biology and ecology of microorganisms and which cannot be developed simply as an adjunct to another discipline.

B. Political Reasons

Because microbes affect so many human activities, national political goals in agriculture, health, wealth generation, and many other areas are underpinned by microbiology. Without an adequately educated and trained workforce, as well as the necessary funding, countries will fall far short of achieving these goals. Biotechnology, for example, has been identified by the European Union (EU) as one of the key growth areas of the economy in the twenty-first century. Recognizing the importance of education, the EU launched the European Initiative for Biotechnology Education (EIBE), which provides supporting materials and encourages good practice in biotechnology education (Grainger, 1996). Significantly, this support is targeted particularly at secondary schools. Interaction between microbiologists and politicians is necessary for the latter to have a realistic view of the time scales and benefits of investment in the science. Informed political debate is necessary on many contentious microbiological issues, ranging from environmental management to the sale of genetically manipulated foods and the treatment of HIV infection. Although these debates need specialist input from microbiologists, all sectors of society are entitled to have their views considered and the more informed and educated those views are, the higher will be the quality of the debate and the more likely that a consensus view that is to the best advantage of society will emerge.

C. Personal Reasons

The decline of infectious disease as a major factor in human illness is directly proportional to a society's ability to implement public health measures. Although many of these are unseen and perhaps unknown to the general public, it is still a fundamental responsibility of every individual in society to maintain an adequate level of hygiene for the protection of him- or herself and those around them. Because microbes are often undetected by human senses, unlike many other dangerous stimuli, avoidance must be based on anticipation rather than direct warning. This requires an understanding of cause and effect, which has to result from an educational process of some kind.

II. WHO SHOULD BE EDUCATED IN MICROBIOLOGY?

There are four categories of people who require or desire some level of education in microbiology.

1. Experienced microbiologists undertaking continuing professional development by updating or widening their skills and knowledge.
2. Students studying microbiological theory and techniques for their later professional use.
3. Those wishing to study and understand microbiology for their own interest.
4. Those working in other areas in which microbiological safety is an important aspect.

Within each category, the range of requirement may be very wide. Professionals include environmental health officers, food scientists, physicians, and a wide variety of others, in addition to the technicians and researchers who are the first to come to mind when the term "microbiologist" is used. The continual informal updating of its practitioners has always been a central feature of any science, but continuing professional development is now becoming much more formalized as certificated programs. Those with an interest in microbiology, but who would be unlikely to use their knowledge in a practical sense, range from the interested layperson, who realizes that no world picture is complete without some understanding of the role of microbes, to nonmajor students and professionals such as geologists, economists, and historians who cannot properly understand the systems they work with unless they appreciate the roles of microorganisms underlying them. Category 4 in effect includes everybody—not just food handlers in the catering and retail sectors but homemakers of every kind who can use a basic knowledge of microbiology to reduce the hazards associated with food, water, and illness in the domestic situation.

Microbiological education should begin as early as possible, preferably in elementary school, where children can be intrigued by the concept of invisible organisms having such far-reaching effects on their lives. This early acquaintance may be of great importance to those who stop secondary education early, but who still need to know about everyday microbial hazards (category 4). High school science is greatly enriched by microbiological topics, although the very cautious approach to the use of pure cultures on the grounds of safety has limited the opportunity for laboratory or project work at this level.

III. WHAT SHOULD BE STUDIED?

Careful comparison of the knowledge and skills of the individual with the prerequisites for any level of microbiology education must be made before embarking on a course of study.

A. Who Chooses the Topics?

The choice of topics or subject matter to be included in the syllabus of an educational programme in microbiology may be determined directly by the individual educator, imposed by an authority outside the classroom (either from within or outside the education or microbiology professions), or set largely by a major textbook teaching package (but still selected by the individual educator).

1. *Internal versus External*

At the pre-university level, the educator often has little if any direct experience as a microbiologist and it is usual for the syllabus, and often the entire curriculum, to be drawn up by an outside group of specialists (often at national level) who have both the academic and educational knowledge to determine what is important and practicable for students to learn at each particular stage in their education.

At the university level, the individual educator has much more influence over the syllabus for a microbiology course. At this level, it is usual for students to learn directly from practitioners of whatever academic discipline they are studying. The latter are assumed to be more expert and thus to know better than anyone else what students need to know to become competent and independent practitioners of the discipline themselves. University educators are expected to be sufficiently active in research to be able to keep abreast of and interpret the relevant research literature. Their knowledge, experience, and teaching skill help them to choose the most appropriate topics for their students to study. This approach to choice of syllabus content is widespread, particularly in countries where there is a strong tradition of academic freedom.

Increasingly, however, the academic choice of the individual educator is becoming more limited, even at university level. Professional bodies may instead specify certain areas of expertise as necessary for certification or registration. In countries where the state provides significant resources for university-level education, there may be pressure to teach areas defined by national political and economic objectives. This may also be a means whereby the state tries to ensure value for their taxpayers' money.

2. *Textbook Packages*

In many institutions throughout the world, university educators choose to base their course content on the major textbooks of microbiology. This may be because they are not always specialists or experienced microbiologists, as would ideally be the case, and they feel that the experience of the authors of the major textbooks is a better guide to the needs of students than their own. The authors of the major texts marketed by big publishing companies have international reputations both as microbiologists and as educators, and can make authoritative choices of what to emphasize in a microbiology course and how to present it. Even experienced educators in microbiology, however, make increasing use of such textbook packages because they no longer consist of just a book but come with student and tutor guides, multimedia teaching and review aids, laboratory guides, banks of test questions for assessment, and a supporting website, all of which have been tested and developed much more thoroughly than could any individual educator. The production costs of a major textbook package aimed at the first-year undergraduate market now approach $1 million, and few if any institutions could fund the development of teaching packages of similar quality just for their own courses. Which of several competing textbook packages is chosen still depends on the view of the individual educator after a careful consideration of the content. Inevitably, the content of general textbooks is oriented toward more popular areas of microbiology, so that medical aspects, for example, may be emphasized more strongly than some educators would like for their own courses.

More advanced courses have specific requirements that are much less easy to satisfy from a general source like a textbook, and the expertise of the educator concentrating on the aims and objectives of the particular program may be the only way of satisfactorily deciding on the content.

B. Core Content

Many attempts have been made to define a core content for microbiology courses—the topics and skills that students in particular circumstances must know and understand. A core content could be imposed directly by the state or other bureaucracies, but preferably and more commonly a core content is agreed on by consensus among the educators. A defined core content or often an entire syllabus is more likely at the pre-university level, where educa-

tors are less experienced and where students from many institutions are preparing to take common national examinations. At the university level, the concept of a defined core content common to a number of microbiology courses has often been opposed on the grounds of academic freedom or held to be unnecessary because educators consider it inevitable that the same basic topics will be taught in similar courses across the country. However, the current political emphasis toward public accountability and appraisal make it more likely that microbiology teaching in many countries will be based on a national core syllabus, at least at introductory or general level.

The biggest consultation exercise for the development of consensus core themes for introductory or general microbiology courses was mounted by the American Society for Microbiology (ASM) through its undergraduate education conferences. Several hundred microbiology educators from community colleges, undergraduate universities, and research universities throughout the United States engaged in lengthy discussions over several years to develop and revise a list of core themes and concepts for an introductory microbiology course (Table I).

Subsequently, the conference devised a core content for an introductory microbiology laboratory course to be part of the overall course. This defined guidelines for laboratory thinking, skills, safety, and content, identifying only those items considered essential in any introductory microbiology course, regardless of content or specific purpose (Table II).

Teaching laboratories across the world will vary greatly in their level of facilities, but it should be possible for any course anywhere to meet at least these minimum requirements for introductory training in safe microbiological procedure. If not, it is unlikely that an educational program in practical microbiology can safely be undertaken.

C. Customized Content

1. *Advanced Study*

For advanced students, educators who select topics at the cutting edge of the science with which they are directly familiar and who teach by personal example should always be a major part of the educational experience. A recent survey of 100 UK undergraduate

TABLE I
**Core Themes and Concepts for an
Introductory Microbiology Course**[a,b]

Theme 1: Microbial cell biology*

Information flow within a cell
Regulation of cellular activities
Cellular structure and function*
Growth and division*
Cell energy metabolism*

Theme 2: Microbial genetics*

Inheritance of genetic information
Causes, consequences, and uses of mutations*
Exchange and acquisition of genetic information

*Theme 3: Interaction and impact of
microorganisms and humans**

Host defense mechanisms
Microbial pathogenicity mechanisms*
Disease transmission
Antibiotics and chemotherapy*
Genetic engineering
Biotechnology

*Theme 4: Interactions and impact of
microorganisms in the environment**

Environmental pressure for survival
Adaptation and natural selection
Symbiosis
Microbial recycling of resources
Microbes transforming environment
Harnessing of microbes for productive uses

Theme 5: Integrating theses*

Microbial evolution
Microbial diversity

[a] Reproduced with permission from ASM News, Vol. 64, No. 1, p. 12, 1998.

[b] The terms "microbes," "microbial," and "microorganisms" refer to all microorganisms from subcellular viruses and other infectious agents to all prokaryotic and eukaryotic microorganisms. Asterisks denote themes and concepts that constitute the Lab Core for Content.

degree courses with a high microbiology content (C.R.A.C., 1999) showed a range of 20 topic areas being offered in the final year of study. Syllabuses will change annually as topics develop and are often revised completely at intervals of about 5 years. More

than half of the courses required a compulsory study of molecular biology and genetics, biochemistry and physiology, and industrial microbiology in the final year. The emphasis on molecular genetics is understandable, given its contribution to our current understanding of cellular activities, but there is much concern that students are seeing it as an end in itself and are not developing the breadth of microbiological expertise to place it in context. This difficulty is made worse by the trend to absorb departments of microbiology into larger academic units of cell and molecular biology, and the resulting dilution of studies specifically in microbiology (Woese, 1994).

2. Nonmajors

The content of courses in microbiology for nonprofessionals or nonmajors will obviously depend on their interests, their requirements, and the extent of their prior understanding, particularly of biochemical concepts. If it is required for academic study credit at the undergraduate level, something close to the ASM scheme of core themes and concepts in Table I would be an appropriate start. Alternatively, students may want or need some familiarization with one or more of the advanced aspects of microbiology—medical bacteriology or microbial molecular genetics, for example. As the demand is usually very occasional, it may not be cost-effective to create a short course tailored to the specific requirements of these students and usually they simply join an existing course for microbiology majors. Care must obviously be taken by the academic adviser to ensure that the student is sufficiently familiar with the necessary core concepts to ensure a good understanding of the advanced material.

3. Noncollege Education

Food handlers in catering and retail outlets usually have a much more limited experience and interest than those studying microbiology at university level. In many countries, such workers are required to undergo training courses to learn and show competence in safe practices. The essential content is usually dictated by the regulatory requirements. These training courses will be more effective when they result in understanding rather than in rote compliance with rules and procedures. However, both the

Core Themes and Concepts for the Introductory Microbiology Laboratory[a]

Laboratory content
All of the Core Themes and Concepts for an Introductory Microbiology Course (see Table I) taught through laboratory exercises.

Laboratory skills
Use of a bright-field light microscope to view and interpret slides, including correctly setting up and focusing the microscope, proper handling, cleaning, and storage of the microscope, correct use of all lenses, recording microscopic observations
Proper preparation of slides for microbiological examination, including cleaning and disposing of slides, preparing smears from solid and liquid cultures, performing wet mount and/or hanging drop preparations, performing Gram stains
Proper aseptic techniques for the transfer and handling of microorganisms and instruments, including sterilizing and maintaining sterility of transfer instruments, performing aseptic transfer, obtaining microbial samples
Use of appropriate microbiological media and test systetms, including isolating colonies and/or plaques, maintaining pure cultures, using biochemical test media, accurately recording macroscopic observations
Estimation of the number of microbes in a sample using serial dilution techniques, including correctly choosing and using pipettes and pipetting devices, correctly spreading diluted samples for counting, estimating appropriate dilutions, extrapolating plate counts to obtain the correct CFU or PFU in the starting sample
Correct use of standard microbiology laboratory equipment, including using the standard metric system for weights, lengths, diameters and volumes, lighting and adjusting a laboratory burner, using an incubator

Laboratory thinking skills
Demonstration of an increased skill level in formulating and testing a hypothesis, collecting and analyzing data, discussing and presenting lab results or findings in the laboratory, and working effectively in teams.

Laboratory safety
Demonstration of the ability to explain and practice safe microbiological procedures, procedures for protecting oneself, and emergency procedures. In addition, institutions where microbiology laboratories are taught will train faculty and staff in proper waste stream management; provide and maintain all necessary safety equipment and information resources; train faculty, staff and students in the use of safety equipment and procedures; and train faculty and staff in use of materials safety data sheets (MSDS).

[a] Reproduced with permission from ASM News, Vol. 64, No. 1, p. 13, 1998.

content and delivery necessary to promote understanding requires more thought and time, running counter to the desire of employers to reduce costs. Courses in food hygiene are certificated by national bodies responsible for the maintenance of food safety (e.g., the UK's Chartered Institute of Environmental Health), and may be available worldwide in partnership with local providers. Staff training in large food retailers is often based on these courses.

Informing and educating members of the public who have little microbiological interest or knowledge is a major challenge for the communicator. There are very few examples of sustained and effective programs that manage to reach all sectors of the public.

One such program is "Operation Clean Hands," mounted by the ASM in response to a survey on public hand-washing habits. Tracking surveys showed that half the population of the United States was exposed to the resulting media program outlining the microbiological and health consequences. Prime-time television programs can reach a large proportion of the public and can be very successful in highlighting an occasional individual issue. However, the links between microbiologists and program planners and makers are not strong, and science takes little advantage of the medium to promote, educate, and entertain in microbiology. Again, the ASM has recognized this and formed a coalition, the "Microbial Literacy Collaborative," to prepare a public television series on microbiology. An important component is a preliminary 4-week intensive immersion of TV production staff in microbes and how they are studied. Improving the microbial awareness of media personnel in this way should have an influence on them and on their future broadcasting activities that goes far beyond the television series itself.

IV. HOW SHOULD WE EDUCATE IN MICROBIOLOGY?

Education is a professional activity because its practice is underpinned by a substantial body of theoretical concepts and supporting evidence. To be professional, educators need to base their actions on that body of understanding. Teaching while being unaware of it is comparable to researching in microbiology without knowing the relevant literature and with just a few old techniques learned through direct observation. Despite this, many microbiology educators have received no formal training in education. For this reason, some of current thinking in education will be summarized and related to problems in microbiology teaching.

A. How Students Learn

A paradigm change in our understanding of how students actually learn has taken place over the last two decades, leading to a shift in emphasis from "How should students be taught?" to "How do students learn?" Earlier, learning was often interpreted in behaviorist terms. Students were thought to learn by a stimulus–response process, learning appropriate responses through conditioning. This encouraged a didactic approach by teachers, effectively telling students what they needed to know and rewarding the correct responses.

The phenomenographic approach pioneered in the 1970s by Marton in Sweden and Pask and Entwistle in the UK has given rise to a new and now widely accepted paradigm. This distinguishes two fundamentally distinct approaches by students to learning, the deep and surface approaches. The deep approach involves an active search for meaning, with a conscious intention by the student to understand the concepts and to relate them to previous knowledge and experience. The surface approach focuses on specific facts, which students attempt to memorize without emphasizing the connections among them and with no real intention of understanding. Table III summarizes the characteristic features of these two approaches to learning.

Many studies have suggested that more effective learning and greater student satisfaction derive from a deep approach to learning. While surface approaches may be effective in passing examinations, understanding is poor and short-lived and is often recognized to be so by the student, who gains little personal satisfaction from the process. It is important to recognize that:

1. The surface approach is not always intrinsically wrong and may usefully contribute to learning, particularly in disciplines like microbiology with a high factual content, but a full understanding cannot follow from a surface approach alone.

2. A particular approach to learning is not a fixed feature of a student's personality or academic ability. Most students are capable of both approaches, and a fundamental role of the teaching process is to encourage the student to employ the most appropriate approach to the task in hand. Indeed, some students clearly use different approaches at different times in an attempt to achieve the highest marks possible, leading to very variable levels of understanding (sometimes referred to as the strategic approach to learning).

TABLE III
Different Approaches to Learning[a]

Deep approach

Intention to understand; student maintains structure of task

Focus on what is signified (e.g., the author's argument, or the concepts applicable to solving the problem)

Relate previous knowledge to new knowledge

Relate knowledge from different courses

Relate theoretical ideas to everyday experience

Relate and distinguish evidence and argument

Organize and structure content into a coherent whole

Internal emphasis: "A window through which aspects of reality become visible, and more intelligible"

Surface approach

Intention only to complete task requirements; student distorts structure of task

Focus on "the signs" (e.g., the words and sentences of the text, or unthinkingly on the formula needed to solve the problem)

Focus on unrelated parts of the task

Memorize information for assessments

Associate facts and concepts unreflectively

Fail to distinguish principles from examples

Treat the task as an external imposition

External emphasis: demands of assessments, knowledge cut off from everyday reality

[a] Reproduced with permission from Ramsden (1992).

TABLE IV
Characteristics of the Context of Learning Associated with Deep and Surface Approaches[a]

Surface approaches

Assessment methods emphasizing recall or the application of trivial procedural knowledge

Assessment methods that create anxiety

Cynical or conflicting messages about rewards

An excessive amount of material in the curriculum

Poor or absent feedback on progress

Lack of independence in studying

Lack of interest in and background knowledge of the subject matter

Previous experiences of educational settings that encourage these approaches

Deep approaches

Teaching and assessment methods that foster active and long-term engagement with teaching tasks

Stimulating and considerate teaching, especially teaching that demonstrates the lecturer's personal commitment to the subject matter and stresses its meaning and relevance to students

Clearly stated academic expectations

Opportunities to exercise responsible choice in the method and content of study

Interest in and background knowledge of the subject matter

Previous experiences of educational settings that encourage these approaches

[a] Reproduced with permission from Ramsden (1992).

3. There is much evidence that student workloads and the teaching styles and attitudes of both individual educators and whole academic departments strongly influence the learning strategy that students will adopt (Table IV).

Point 3 has fundamental implications for the whole curriculum, particularly in the sciences where new knowledge and interpretations are added to old, rather than partly replacing them, as in some other disciplines. More and more topics being introduced into the microbiology syllabus when student workloads are already so high means that only a surface approach will allow most students to cope. A deep understanding becomes progressively more difficult to achieve.

B. Teaching and Learning Methods

There is an extensive and readily accessible literature on teaching and learning methodology (see the Bibliography). This section will emphasize features that the author has found to be particularly relevant to his teaching of microbiologists. It will consider only university-level teaching.

1. Lectures

The traditional lecture format allows a large body of factual information concerning microbiology to be selected, defined, and transmitted to a large number of students at one time. A good lecturer can give this an enthusiastic and personal flavor, bringing the process of doing microbiology to life in a way that other information sources, particularly textbooks, cannot. Within limits, the lecturer can also vary the pace of learning to suit the student group. To be a useful learning experience, however, students must be explicitly aware of the aims and objectives of each lecture.

Written objectives and summaries are useful, but there is less agreement over the value of giving students copies of lecture notes. Students themselves always value these, but there is little evidence that providing them improves either the knowledge or understanding gained, as assessed by examinations, and they tend to encourage a surface approach to learning. Partial notes, where gaps are left for students to add their own during the lecture, may be a slight improvement. Students should always be given their own copies of diagrams and complex tables or figures that are used during a lecture.

a. Active Learning

It is important to break up the lecture period (typically 50–60 min) into shorter spells of 15–20 min, as student concentration drops markedly after this time. These breaks can usefully be employed to engage the students more actively in learning by setting small problems or discussion topics to reinforce the key points in the lecture material, to be considered by small groups of students sitting near each other to talk over in "buzz groups." Students may be very apprehensive of active learning, especially if they feel they may be singled out to make a response in front of a large group of their peers. Tactful handling of these sessions is thus necessary, but students usually lose their fears with experience. Frequent active-learning interludes should be a general feature of lecture and other sessions from the beginning; once students are allowed to get used to teaching as a passive experience that is done to them, they are seldom happy to take an active part in any learning exercise. This requires a concerted strategy by all the lecturers involved with the course.

b. Prior Preparation

An alternative approach is the prior-preparation lecture, in which students are directed to read relevant sections of a course text before coming in to the lecture (and, if necessary, encouraged to do so by a short test at the start of the session). The lecturer then reviews and expands on the material, putting it into context and providing a number of "buzz group" exercises to make students think about the key points and their application. This approach is possible with quite large classes and, although it is much harder work for both the students and teacher than a traditional lecture, students consistently rate the approach as at least no worse and usually better than the traditional lecture in maintaining their interest and promoting understanding.

2. Small-Group Sessions

Learning sessions in which small groups of (5–20) students interact with educators and each other have long been a feature of university-level education, though increased student–staff ratios in many courses have resulted in their becoming less common. Tutorials, problem-solving sessions, and seminars are common examples of small-group teaching. The main aims are to develop the thinking and discussion skills of students. They do require a particularly wide range of teaching and personal skills from the educator, however, and many studies have shown that the majority of the talking is done by the teacher, suggesting that the necessary skills, particularly in listening and questioning, are often not well developed. It is particularly important that students do not feel they will be exposed to ridicule or outright criticism for their views or mistakes. Even an incorrect response often contains points that can be praised and used to lead students toward a better understanding. Allowing students to try their ideas on their colleagues in subgroups of two or three before exposing them to the teacher and the whole group will also improve their confidence in the process. The traditional practice of selecting one student to prepare and read a paper, which is then criticized by the teaching and (though not often) other student members of the group is, for these reasons, one of the least effective methods in small-group learning. Students are seldom assessed on their performance in small-group sessions, perhaps because the main benefit is felt to be in the general skills of thinking and discussion, which are not easy to quantify.

3. Laboratory Classes

Laboratory work is always considered essential to a satisfactory education in microbiology, even though the undergraduate practical work tradition in science dates only from the mid-1800s. Students typically spend one-half to two-thirds of their total class time in the laboratory, although the laboratory work's contribution to their overall assessment mark is usually much smaller. It is also usually the most costly aspect of microbiology education.

a. Aims

Laboratory work has many different aims, and students and faculty may well perceive these differently. A survey of students, graduates, and practicing scientists found that the only three aims that were rated highly by all three groups were teaching basic practical skills, familiarization with important measurement techniques and apparatus, and data analysis and interpretation. (It is notable that the last of these need not involve the student in laboratory work at all.) Thus it is important that the aims and objectives of laboratory work be clearly articulated and understood by both the educator and students, and that assessment methods (see Section IV.B.6) address all of them. Frequently, laboratory assessments are based on written reports and do not assess skills used in the laboratory itself or carry any guarantee that the student concerned actually did any of the work.

Sterile transfer techniques in particular are fundamental to safe working in almost every aspect of microbiology. Thorough training and practice in sterile transfer should be the earliest practical experience of any microbiology student. Competence in this cannot be assessed from written reports and needs to be observed in the laboratory. A satisfactory grade in sterile techniques should be an absolute requirement for success in any introductory microbiology course. This training is very labor-intensive, however, and a useful approach is for more senior students to teach and guide beginners in the lab on a one-to-one basis.

b. Structured Exercises

Demonstrations and tightly structured exercises predominate in the earlier stages of study, introducing fundamental practices such as sterile transfer technique, cell enumeration methods, and staining procedures. However, students must be convinced that the experiment is a meaningful exercise or they will have little interest in it. Many laboratory manuals have been published with programs of basic exercises covering all the skills in Table II, and they can be very useful to both educators and students. It is important at this stage to have experiments that reliably work, in the sense of yielding data that the students can usefully discuss. It is also useful to have the steps of the experimental procedure laid out in detail for the beginning student, who already has many things to think about at the same time in an unfamiliar environment.

Written instructions for experimental procedures are usually provided as separate handouts or incorporated into a laboratory manual. For the operation of instruments such as a spectrophotometer, however, these may be more useful in the form of a wall-chart, a short video, or a computer program placed beside the instrument, rather than in written words. It is quite common for students to come into the laboratory without having read through their instructions and thus to be unaware of the aims of the exercise or of the procedures to be followed. Clearly, they will gain less from the laboratory experience. A simple prelab test administered by computer or multiple-choice test will take only a few minutes at the start of the laboratory session and often acts as an effective stimulus for prior preparation.

Tightly structured exercises foster a "cookbook" mentality, in which the student sees an experiment simply as following a prescribed series of steps. There is little development of the important skill of experimental design, other than making suggestions in the written report for improvements or remedies if the experiment has not worked. Many laboratory classes are scheduled for too short a time for the student to be able to propose and implement an improvement to the experimental design. For this, a hybrid of actual laboratory work and computer simulation may be useful. For example, students can make total and viable counts of a microbial culture in the laboratory to experience all the real-life difficulties that are never apparent from a textbook description of the procedure (e.g., overlapping colonies, ill-defined lumps in the agar, clumps of cells, and dilution errors). Then, rather than using the counting methods to follow growth of a culture in real time, which may take all day to do just once, they can move to a simple computer simulation of batch growth that gives count data as its output almost instantaneously.

Here, they can repeatedly test the effects of various parameters on growth and continuously refine the design of their simulated experiment until satisfactory results are obtained (see Section IV.B.3.d). Mild failure, in terms of imperfect data, often allows a student to learn more than does a completely successful experiment, but failure to gain any useful data at all is more common than most educators would like to admit and probably does more harm than good.

c. Open Investigations

More open investigations, in which not only the data but also the methods, materials, and even aims are selected and obtained by the student, offer greater potential benefits. With greater ownership of the work, student motivation, understanding, and reward is much higher, but the costs, in materials and in staff time for more individual attention, and in the risks of failure are also higher. Thus, these investigations are usually restricted to advanced students working in smaller groups. The topics are usually selected from the research specializations of the department concerned, and thus vary widely from institution to institution. Again, published manuals of advanced laboratory exercises are available, but, by the nature of the work, the exercises are less likely to be successful unless the educators using them are already well practiced in the techniques. The open investigation is often taken to its extreme in final-year undergraduate courses in the form of a small-scale research project. As an exposure to front-line microbiology, this is valuable as an experience and as an indicator of the student's interest and aptitude for professional development in this direction. However, its effectiveness depends very much on the individual project and on the people and facilities associated with it. When it is reasonably successful, students value the experience highly; when it is not, the negative influence on attitudes and confidence can be damaging out of all proportion. Miniprojects at earlier stages of study have been attempted and are often considered to be well worth the extra effort they require.

d. Simulations and Data-Handling Exercises

The constraints of time, safety, cost, and availability of equipment or expertise limit the experimental work that students can actually do. There is no substitute for hands-on experience at the lab bench, but simulated experiments or data-handling exercises can greatly extend students' understanding. Simulations allow students to design experimental approaches, select parameter values, and obtain reasonably realistic data for analysis and interpretation. With improvements in computer graphics, many simulations now provide realistic visual presentations of results rather than simple numerical data. Excellent computer simulation exercises are now available, ranging from the identification of pathogens in food-poisoning outbreaks to complex protein-purification protocols (see Section IV.C). Data-handling exercises, either on paper or computer-based, can be used to simulate the final analysis stage of almost any kind of investigation. They are limited only by the ingenuity of the educator (particularly in the popular studies of extraterrestrial microorganisms). Real data taken from research papers (copyright restrictions permitting) or data prepared by the educator can be used. The context must be sufficiently convincing for the student to take the exercise seriously, and enough variation should be introduced into the data to give a flavor of real-life ambiguity. These exercises are used widely particularly in molecular biology.

4. Work Experience

In some countries, there is a tradition of students spending a substantial period working in practicing microbiological laboratories in the public or private sector, just before they start their final year of study. Usually, this is equal to about half or one year's study, though periods as short as a few weeks can be of value to both students and employers. The range of work experience positions in which students may be placed is very large, covering almost the whole spectrum of microbiology. The selection procedures and conditions of employment (including a stipend or the lack of one) are also very varied. With a proper program of training and management in the laboratory to ensure that the student is not used simply as an extra pair of hands, the benefits to the student include a great boost to the students' self-confidence and self-esteem when they are taken seriously by a major employer, the experience of the discipline of a daily work regime, the experience of the commercial- or public-sector social and economic culture, the development of interpersonal relationships and other work-based personal skills, an appreciation of the application of microbiology to real-life situations, the experience of using techniques that may be beyond those practicable in a teaching laboratory, and help toward a career choice. There are also many benefits to employers and to the academic institution from these kind of links with each other.

Work experience opportunities are extremely time-consuming to set up, even in a heavily industrialized country (perhaps consuming as much as one worker-year to set up 30–40 placements). They also require substantial time commitment to visit and monitor the students' activity and progress. Nevertheless, the benefits to all three participants, and particularly to the students, are very large.

5. Independent Study

Independent study, outside the timetable of class sessions, may be carried out individually or in small groups. Apart from review for examinations or the occasional piece of personal reading, this typically involves directed coursework exercises—researching case studies, writing term papers, problem-solving exercises, lab-report writing and so on. Many skills, both personal and academic, are developed.

Although cooperation between individual students should be tolerated or even encouraged, as this is the normal process of science, there is concern that outright copying (plagiarism) during these exercises is hard to prevent and distorts the assessment process. A careful definition of topics and a requirement that students submit rough working drafts with the final product can make undetected copying more difficult. Assessing individual contributions to group exercises can also be difficult.

Independent (or open) learning takes this process much further. Self-learning packages allow the study of a topic or even a whole discipline at a pace and flexibility suited to the individual student. They form the basis of much professional development work in many disciplines, as well as degree-level study with institutions like the Open Universities of the UK and The Netherlands. In these cases, their use allows distance learning with little or no attendance at the institution and limited contact with the educator.

Open-learning packages are increasingly useful in conventional higher education, both for remedial work in basic chemistry, mathematics, and other skills in which microbiology students are commonly weak, and in mainstream-course delivery. They have many advantages over traditional lectures in transmitting information and developing ideas, and the greater face-to-face availability of the educator on campus to help with problems means that the supporting material need not take as long to develop as it does for full distance learning.

Both distance and on-campus open learning will undoubtedly be the major growth area in higher education and continuing professional development in the early twenty-first century, particularly when based on material posted on the World Wide Web, with its wide accessibility.

6. Assessment (Evaluation) of Students

For almost all students, it is unfortunately the case that assessment effectively defines the curriculum. The learning styles they adopt will be those they perceive will gain them the highest marks in their course assessments (see Section IV.A). Hence, methods of assessment must be an integral part of the curriculum and be considered to be part of the teaching and learning process. They should not, as is so often the case, be added on after the content and organization of the course has been decided. The importance of assessment as a learning tool cannot be overestimated.

Student assessment has many functions, but they can be categorized and summarized as measuring and reporting on the progress of students (summative assessment), giving students feedback on their progress and understanding as a method of learning (formative assessment), and giving educators feedback on the effectiveness of their teaching. Any one item of assessment should fulfil at least two and preferably all three functions at the same time.

Assessment is time-consuming, as each student has to be considered individually. However, students rate rapid and constructive feedback on their work more highly than almost any other aspect of good teaching. If there is a large number of students, detailed and rapid feedback is obviously difficult to provide. Experience or a quick preliminary look through the essays or practical reports will often identify common errors or misinterpretations that can be listed, with explanatory comments, on a feedback sheet and supplied to the individual student with the appropriate points checked.

Traditional written examinations are still widely used as the major part of a course assessment, despite decades of studies showing that they largely test rote learning and that students get no useful feedback from them. These are exactly the circumstances that encourage and reward surface rather than deep learning.

In general, it is essential to use a variety of assessment methods to give a more accurate indication of student achievement, to make the criteria and methods of assessment explicit to both the student and the assessor, and where possible to involve the student actively in the assessment—both self and peer assessment can be extremely effective learning experiences, even if concerns over reliability mean that the grades are not in the end recorded. None of these points about assessment is specific to education in microbiology, but are emphasized because assessment is probably the least adequately considered aspect of the curriculum regardless of discipline.

7. Education in Microbiological Research

The supervision of graduate research students is probably the most personal activity in microbiology education and, in consequence, often the most variable in quality. Serious attention is now being paid to issues of effective supervision, completion rates, and the integration of the research student into a collaborative rather than individual venture. The personal and professional skills required of the supervisor are many and varied, as are the styles that can be employed successfully. This aspect of microbiology education is too complex to be usefully considered here; the reader is referred to items in the Bibliography, particularly to Brown and Atkins (1988). The supervision of undergraduate research projects requires similar skills and planning, over a shorter time scale.

C. Resources

Traditional sources of information and teaching materials include textbooks and published laboratory manuals. Many excellent examples are available and are featured in publishers' catalogues. These also list a number of good computer simulations for microbiology teaching. Many other simulations are not available commercially but can be obtained free or at small cost from their authors. Several countries have national programs for the development of science teaching resources, particularly computer-based ones.

The Internet, particularly through the World Wide Web, is a huge resource of information, ideas, and experience. Tens of thousands of microbiology-related pages are registered with the major search engines. Teaching materials, including simulations, educational experience, and even entire online courses of microbiology instruction may be found with a carefully devised search strategy. A good way to start searching is through the websites of the large microbiological societies, which are usually easy to find and contain many links to other useful educational sites.

Learned societies of microbiology exist in most countries and often have a division or interest group concerned with education. These organize meetings and workshops on many aspects of microbiology education and provide unique opportunities to exchange experiences with educators across the country and beyond. They may carry features on education in their newsletters to members, and sometimes can supply teaching materials or the funding to help develop them. The ASM has set up an International Collection for Microbiology Teaching and Learning, consisting of a large variety of peer-reviewed teaching materials available from the society's website, along with many other teaching resources and activities.

D. Educating the Educators

University educators are seldom appointed for their teaching ability, but instead for their competence in their discipline itself. Most have never received formal training in education. New faculty increasingly have an opportunity to pursue this training, while in-service and reflective courses for more experienced educators are becoming more widely available. Nevertheless, the most experienced can learn much from reading about teaching and reflecting, with the help of colleagues, on his or her own practice. Good teachers constantly evaluate their performance, but formal teaching-evaluation schemes are used increasingly for quality assurance. Well-designed evaluation schemes can provide feedback that helps even good teachers to improve further, but most are far too simplistic to serve any really useful purpose.

Wide use is also made in university microbiology teaching of graduate students as instructors and lab demonstrators. Although they are usually very competent in the science, they too need an awareness of the extra skills of observation, intervention, and guidance that they will require and some help in developing them.

See Also the Following Article

CAREERS IN MICROBIOLOGY

Bibliography

American Society for Microbiology Web site: *http://www.asmusa.org*

Boud, D., Dunn, J., and Hegarty-Hazel, E. (1986). "Teaching in Laboratories." Open University Press, London.

Brown, G., and Atkins, M. (1988). "Effective Teaching in Higher Education." Routledge, London.

C.R.A.C. (1999). "Degree Course Guide to Microbiology, Immunology and Biotechnology." Hobsons, London.

Entwistle, N., and Tait, H. (1990). Approaches to learning, evaluation of teaching and preferences for contrasting academic environments. *Higher Education* **19**, 169–194.

Grainger, J. M. (1996). Needs and means for education and training in biotechnology: Perspectives from developing countries and Europe. *World J. Microbiol. Biotechnol.* **12**, 451–456.

McKeachie, W. J. (1999). "McKeachie's Teaching Tips: Strategies, Research and Theory for College and University Teachers," 10th ed. Houghton Miflin, New York.

Microbiology Network Web site: *http://microbiol.org*

Race, P. (1994). "The Open Learning Handbook," 2nd ed. Kogan Page, London.

Ramsden, P. (1992). "Learning to Teach in Higher Education." Routledge, London.

Rowntree, D. (1987). "Assessing Students: How Shall We Know Them?" 2nd ed. Kogan Page, London.

Society for General Microbiology Web site: *http://www.socgenmicrobiol.org.uk*

Woese, C. L. (1994). There must be a prokaryote somewhere: Microbiology's search for itself. *Microbiol. Rev.* **58**, 1–9.

Emerging Infections

David L. Heymann

World Health Organization

I. A 20-Year Perspective
II. Misplaced Optimism
III. Weaknesses Facilitating Emergence and Re-emergence
IV. Further Amplification
V. Solutions

GLOSSARY

amplification of transmission The increased spread of infectious disease that occurs naturally or because of facilitating factors, such as nonsterilized needles and syringes, that can result in an increase in transmission of infections such as hepatitis.

anti-infective (drug) resistance The ability of a virus, bacterium, or parasite to defend itself against a drug that was previously effective. Drug resistance is occurring for bacterial infections such as tuberculosis and gonorrhoea, for parasitic infections such as malaria, and for the human immunodeficiency virus (HIV).

eradication The complete interruption of transmission of an infectious disease and the disappearance of the virus, bacterium, or parasite that caused that infection. The only infectious disease that has been eradicated is smallpox, which was declared eradicated in 1980.

International Health Regulations Principles for protection against infectious diseases aimed at ensuring maximum security against the international spread of infectious disease. The Regulations provide public health norms and standards for air- and seaports to prevent the entry of infectious diseases, and require reporting to the World Health Organization (WHO) the occurrence of three infectious diseases: cholera, plague, and yellow fever.

reemerging infection A known infectious disease that had fallen to such low prevalence or incidence that it was no longer considered a public health problem, but that is presently increasing in prevalence or incidence. Reemerging infections include tuberculosis, which has increased worldwide since the early 1980s, dengue in tropical regions, and diphtheria in eastern Europe.

EMERGING INFECTIONS are newly identified and previously unknown infectious diseases. Since 1970 there have been over 30 emerging infections identified, causing diseases ranging from diarrheal disease among children, hepatitis, and AIDS to Ebola hemorrhagic fever.

I. A 20-YEAR PERSPECTIVE

In the Democratic Republic of the Congo (DRC, formerly-Zaire), the decrease in smallpox vaccination coverage, poverty, and civil unrest causing humans to penetrate deep into the tropical rain forest in search of food may have resulted in breeches in the species barrier between humans and animals, causing an extended and continuing outbreak of human monkeypox. During the 1970s and 1980s, when this zoonotic disease was the subject of extensive studies, it was shown that the monkeypox-virus-infected humans, but that person-to-person transmission beyond three generations was rare. The outbreak of human monkeypox in 1996–1997 is a clear example of the ability of infectious diseases to exploit weaknesses in our defenses against them.

Numerous infectious diseases have found weakened entry points into human populations and emerged or re-emerged since the 1970s (see Table I). In the early to mid-1970s, for example, classic dengue fever had just begun to reappear in Latin America after it had been almost eliminated as a result of mosquito control efforts in the 1950s and

TABLE I
Principal Newly Identified Infectious Organisms Associated with Diseases[a]

Year	Newly identified organism	Disease (year and place of first recognized or documented case)
Diseases primarily transmitted by food and drinking water		
1973	Rotavirus	Infantile diarrhoea
1974	Parvovirus B19	Fifth disease
1976	*Cryptosporidium parvum*	Acute enterocolitis
1977	*Campylobacter jejuni*	Enteric pathogens
1982	*Escherichia coli* 0157:H7	Haemorrhagic colitis with haemolytic uremic syndrome
1983	*Helicobacter pylori*	Gastric ulcers
1986	*Cyclospora cayatanensis*	Persistent diarrhoea
1989	Hepatitis E virus	Enterically transmitted non-A and non-B hepatitis (1979, India)
1992	*Vibrio cholerae 0139*	New strain of epidemic cholera (1992, India)
Unclear modes of transmission, thought to be primarily transmitted by drinking water		
1985	*Enterocytozoon bieneusi*	Diarrhea
1991	*Encephalitozoon hellem*	Systemic disease with conjunctivitis, in AIDS patients
1993	*Encephalitozoon cunicali*	Parasitic disseminated disease, seizures (1959, Japan)
1993	*Septata intestinalis*	Persistent diarrhea in AIDS patients
Diseases primarily transmitted by close contact with infectious individuals, excluding sexually transmitted diseases, nosocomial infections and viral haemorrhagic fevers		
1980	HTLV-1	T-cell lymphoma leukemia
1982	HTLV II	Hairy cell leukemia
1988	HHV-6	Rosela subitum
1993	*Influenza* A/Beijing/32 virus	Influenza
1995	HHV-8	Associated with Kaposi sarcoma in AIDS patients
1995	*Influenza* A/Wuhan/359/95 virus	Influenza
Sexually transmitted diseases		
1983	HIV-1	AIDS (1981)
1986	HIV-2	Less pathogenic than HIV-1 infection
Nosocomial and related infections		
1981	Staphylococcus toxin	Toxic shock syndrome
1988	Hepatitis C	Parenterally transmitted non-A non-B hepatitis
1995	Hepatitis G viruses	Parenterally transmitted non A non B hepatitis
Human zoonoses and vector-borne diseases, including viral haemorrhagic fevers		
Transmitted by close contact with animals or animal products, excluding food-borne diseases		
1977	Hantaan virus	Hemorrhagic fever with renal syndrome (1951)
1990	Reston strain of Ebola virus	Human infection documented but without symptoms (1990)
1991	Guanarito virus	Venezuelan hemorrhagic fever (1989)
1992	*Bartonella henselae*	Cat-scratch disease (1950s)
1993	Sin nombre virus	Hantavirus pulmonary syndrome (1993)
1994	Sabià virus	Brazilian hemorrhagic fever (1955)
Tick-borne		
1982	*Borrelia burgdorferi*	Lyme disease (1975)
1989	*Ehrlichia chaffeensis*	Human ehrlichiosis
1991	New species of *Babesia*	Atypical babesiosis
Unknown animal vector		
1977	Ebola virus	Ebola hemorrhagic fever (1976, Zaire and Sudan)
1994	Ebola virus, Ivory Coast strain	Ebola hemorrhagic fever
Soil-borne diseases, airborne diseases and diseases associated with recreational water with no evidence of direct person-to-person transmission		
1977	*Legionella pneumophilia*	Legionnaires' disease (1974)

[a] From G. Rodier, WHO.

1960s. Twenty years later, dengue has become hyper-epidemic in most of Latin America, with over 500,000 cases reported in 1995–1996 of which over 13,000 were the hemorrhagic form diagnosed in 25 countries (WHO, unpublished). In 1991, cholera, which had not been reported in Latin America for over 100 years, re-emerged in Peru with over 320,000 cases and nearly 3000 deaths, and rapidly spread throughout the continent to cause well over 1 million cases in a continuing and widespread epidemic.

In North America, *Legionella* infection was first identified in 1976 in an outbreak among war veterans. Legionellosis is now known to occur worldwide and poses a threat to travellers exposed to poorly maintained air conditioning systems. During 1995 in Europe, 172 people, thought to have been infected in hotels at which they stayed during travel, were identified with legionellosis through the European Working Group for Legionella Detection.

During this same 20-year period a new disease in cattle, bovine spongiform encephalopathy, was identified in Europe and became associated in time and place with a previously unknown variant of Creutzfeldt–Jakob disease; and diphtheria outbreaks occurred in the newly independent states of the former USSR, with over 50,000 cases reported in 1995. By 1996, food-borne infection by *E. coli* 0157, unknown in 1976, had become a food-safety concern in Japan, Europe, and in the Americas. Hepatitis C was first identified in 1989 and is now thought to be present in at least 3% of the world's population, while hepatitis B has reached levels exceeding 90% in populations at high risk from the tropics to eastern Europe.

In 1976 the Ebola virus was identified for the first time as causing a disease that has come to symbolize emerging diseases and their potential impact on populations without previous immunological experience. Ebola has caused at least four severe epidemics and numerous smaller outbreaks since its identification in simultaneous outbreaks in Zaire and Sudan. In 1976, at the time of the first Ebola outbreak in Zaire, HIV seroprevalence was already almost 1% in some rural parts of Zaire, as shown retrospectively in blood that had been drawn from persons living in communities around the site of the 1976 outbreak,

and HIV has since become a preoccupying problem in public health worldwide.

II. MISPLACED OPTIMISM

In this same 20-year period the eradication of smallpox was achieved. This unparalleled public health accomplishment resulted in immeasurable savings in human suffering, death, and money, and stimulated other eradication initiatives. The transmission of poliomyelitis has been interrupted in the Americas and the disease is expected to be eradicated from the world during the first decade of the twenty-first century. During 1996, 2090 cases of polio were reported to WHO, a decrease from 32,251 cases reported in 1988. Reported cases of dracunculiasis have decreased from over 900,000 in 1989 to less than 200,000 in 1996, with the majority of cases in one endemic country. Leprosy and Chagas disease likewise continue their downward trends towards elimination.

The eradication of smallpox boosted an already growing optimism that infectious diseases were no longer a threat, at least to industrialized countries. This optimism had prevailed in many industrialized countries since the 1950s, a period that saw an unprecedented development of new vaccines and antimicrobial agents and encouraged a transfer of resources and public health specialists away from infectious disease control. Optimism is now being replaced by an understanding that the infrastructure for infectious disease surveillance and control has suffered and in some cases become ineffective. A combination of population shifts and movements with changes in environment and human behavior has created weaknesses in the defense systems against infectious diseases in both industrialized and developing countries.

III. WEAKNESSES FACILITATING EMERGENCE AND RE-EMERGENCE

The weakening of the public health infrastructure for infectious disease control is evidenced by failures such as in mosquito control in Latin America and Asia with the re-emergence of dengue now causing major epidemics; in the vaccination programs in eastern Europe, which contributed to the re-emergence of epidemic diphtheria and polio; and in yellow fever vaccination, facilitating yellow fever outbreaks in Latin America and sub-Saharan Africa. It is also clearly demonstrated by the high levels of hepatitis B and the nosocornial transmission of other pathogens such as HIV in the former USSR and Romania, and the nosocomial amplification of outbreaks of Ebola in Zaire, where syringes and failed barrier nursing drove outbreaks into major epidemics.

Population increases and rapid urbanization during this 20-year period have resulted in a breakdown of sanitation and water systems in large coastal cities in Latin America, Asia, and Africa, promoting the transmission of cholera and shigellosis. In 1950, there were only two urban areas in the world with populations greater than 7 million, but by 1990 this number had risen to 23, with increasing populations in and around all major cities, challenging the capacity of existing sanitary systems.

Anthropogenic or natural effects on the environment also contribute to the emergence and re-emergence of infectious diseases. The effects range from global warning and the consequent extension of vector-borne diseases, to ecological changes due to deforestation that increase contact between humans and animals, and also the possibility that microorganisms will breach the species barrier. These changes have occurred on almost every continent. They are exemplified by zoonotic diseases such as Lassa fever first identified in West Africa 1969 and now known to be transmitted to humans from human food supplies contaminated with the urine of rats that were in search of food, as their natural habitat could no longer support their needs. Other zoonotic diseases include Lyme borreliosis in Europe and North America, transmitted to humans who come into contact with ticks that normally feed on rodents and deer, the reservoir of *Borrelia burgdorferi* in nature; and the Hantavirus pulmonary syndrome in North America. The narrow band of desert in sub-Saharan Africa, in which epidemic *Neisseria meningitidis* infections traditionally occur, has enlarged as drought spreads south, so that Uganda and Tanzania experience epidemic meningitis, while outbreaks of malaria and other vector-borne diseases have been linked to the cutting of the rainforests.

And finally human behavior has played a role in the emergence and re-emergence of infectious diseases, best exemplified by the increase in gonorrhea and syphilis during the late 1970s, and the emergence and amplification of HIV worldwide, which are directly linked to unsafe sexual practices.

IV. FURTHER AMPLIFICATION

The emergence and re-emergence of infectious diseases are amplified by two major factors—the continuing and increasing evolution of anti-infective (drug) resistance (see Table II), and dramatic increases in international travel. Anti-infective agents are the basis for the management of important public health problems such as tuberculosis, malaria, sexually transmitted diseases, and lower-respiratory infections. Shortly after penicillin became widely available in 1942, Fleming sounded the first warning of the potential importance of the development of resistance. In 1946, a hospital in the U.K. reported that 14% of all *Staphylococcus aureus* infections were resistant to penicillin, and by 1950 this had increased to 59%. In the 1990s, penicillin-resistant *S. aureus* had attained levels greater than 80% in both hospitals

TABLE II
Resistance of Common Infectious Diseases to Anti-Infective Drugs, 1998

Disease	Anti-infective drug	Range (%)
Acute respiratory infection (*S. pneumoniae*)	Penicillin	12–55
Diarrhea (*Shigella*)	Ampicilline Trimethoprim	10–90
	Sulfamethoxazole	9–95
Gonorrhea (*N. gonorrhoeal*)	Penicillin	5–98
Malaria	Chloroquine	4–97
Tuberculosis	Rifampicin Isonizid	2–40

and the community. Levels of resistance of *S. aureus* to other anti-infectives, and among other bacteria increased with great rapidity. By 1976, chloroquine resistant *Plasmodium falclparum* malaria was highly prevalent in southeastern Asia and 20 years later was found worldwide, as was high-level resistance to two back-up drugs, sulfadoxine-pyrimethamine and mefloquine. In the early 1970s, *Neisseria gonorrhoeae* that was resistant to usual doses of penicillin was just being introduced into Europe and the United States from Southeast Asia, where it is thought to have first emerged. By 1996, *N. gonorrhoeae* resistance to penicillin had become worldwide, and strains resistant to all major families of antibiotics had been identified wherever these antibiotics had been widely used. Countries in the western Pacific, for example, have registered quinolone resistance levels up to 69%.

The mechanisms of resistance, a natural defense of microorganisms exposed to anti-infectives, include both spontaneous mutation and genetic transfer. The selection and spread of resistant strains are facilitated by many factors, including human behavior in over-prescribing drugs, in poor compliance, and in the unregulated sale by nonhealth workers. In Thailand, among 307 hospitalized patients, 36% who were treated with anti-infective drugs did not have an infectious disease. The over-prescribing of anti-infectives occurs in most other countries as well. In Canada, it has been estimated of the more than 26 million people treated with anti-infective drugs, 50% were treated inappropriately. Findings from community surveys of *Escherichia coli* in the stool samples of healthy children in China, Venezuela, and the United States suggest that although multiresistant strains were present in each country, they were more widespread in Venezuela and China, countries where less control is maintained over antibiotic prescribing. Animal husbandry and agriculture use large amounts of anti-infectives, and the selection of resistant strains in animals, which then genetically transfer the resistance factors to human pathogens or infect humans as zoonotic diseases, is a confounding factor that requires better understanding. Direct evidence exists that four multiresistant bacteria infecting humans, *Salmonella, Campylobacter, Enterococci,* and *Escherichia coli*, are directly linked to resistant organisms in

animals (WHO Conference on the Medical Impact of the Use of Antimicrobial Drugs in Food Animals, Berlin, 13-17 October 1997).

Infections with resistant organisms require increased length of treatment with more expensive anti-infective drugs or drug combinations; and a doubling of mortality has been observed in some resistant infections. At the same time, fewer new antibiotics reach the market, possibly in part due to the financial risk of developing a new anti-infective drug that may itself become ineffective before the investment is recovered. There is no new class of broad-spectrum antibiotic currently on the horizon.

The role of travel in the spread of infectious diseases has been known for centuries. Because a traveller can be in a European or Latin American capital one day and the next day be in the center of Africa or Asia, humans, like mosquitoes, have become important vectors of disease. During 1995, over 500 million people travelled by air (World Tourism Organization), and contributed to the growing risk of exporting or importing infection or drug resistance. In 1988, a clone of multiresistant *Streptococcus pneumoniae* first isolated in Spain was later identified in Iceland. Another clone of multiresistant *S. pneumoniae*, also first identified in Spain, was subsequently found in the United States, Mexico, Portugal, France, Croatia, Republic of Korea, and South Africa. A study conducted by the Ministry of Health of Thailand on 411 exiting tourists showed that 11% had an acute infectious disease, mostly diarrheal, but also respiratory infections, malaria, hepatitis, and gonorrhea (B. Natth, personal communication). Forced migration such as by refugees is also associated with the risk re-emergence and spread of infectious diseases. By January 1, 1996, there were over 26 million refugees in the world (UNHCR, 1996). In a refugee population estimated to be between 500,000 and 800,000 in one African country in 1994, an estimated 60,000 developed cholera in the first month after the influx, and an estimated 33,000 died.

V. SOLUTIONS

Eradication and regulation may contribute to the containment of infectious diseases, but do not replace sound public health practices that prevent the weaknesses through which infectious diseases penetrate. Eradication was successful for smallpox and is advancing for poliomyelitis with virus transmission interrupted in the Americas. Eradication or elimination applies to very few infectious diseases—those that have no reservoir other than humans, that trigger solid immunity after infection, and for which there exists an affordable and effective intervention.

Attempts at regulation to prevent the spread of infectious diseases were first recorded in 1377 in quarantine legislation to protect the city of Venice from plague-carrying rats on ships from foreign ports. Similar legislation in Europe, and later the Americas and other regions, led to the first international sanitary conference in 1851, which laid down a principle for protection against the international spread of infectious diseases—maximum protection with minimum restriction. Uniform quarantine measures were determined at that time, but a full century elapsed, with multiple regional and inter-regional initiatives, before the International Sanitary Rules were adopted in 1951. These were amended in 1969 to become the International Health Regulations (IHR), which are implemented by the World Health Organization (WHO).

The IHR provide a universal code of practice, which ranges from strong national disease detection systems and measures of prevention and control including vaccination to disinfection and de-ratting. Currently the IHR require the reporting of three infectious diseases—cholera, plague, and yellow fever. But when these diseases are reported, regulations are often misapplied, resulting in the disruption of international travel and trade, and huge economic losses. For example, when the cholera pandemic reached Peru in 1991, it was immediately reported to WHO. In addition to its enormous public health impact, however, misapplication of the regulations caused a severe loss in trade (due to concerns for food safety) and travel, which has been estimated as high as $770 million.

In 1994, an outbreak of plague occurred in India with approximately 1000 presumptive cases. The appearance of pneumonic plague resulted in thousands of Indians fleeing from the outbreak area, risking spread of the disease to new areas. Plague did not

spread, but the outbreak led to tremendous economic disruption and concern worldwide, compounded by misinterpretation and misapplication of the IHR. Airports were closed to airplanes arriving from India, exports of foodstuffs were blocked, and in some countries Indian guest workers were forced to return to India even though they had not been in India for several years before the plague epidemic occurred. Estimates of the cost of lost trade and travel are as high as $1700 million. Again, the country suffered negative consequences from reporting an IHR-mandated diseases due to the misapplication of the IHR.

A further problem with the IHR is that many infectious diseases, including those which are new or reemerging, are not covered even though they have great potential for international spread. These range from relatively infrequent diseases such as viral hemorrhagic fevers to the more common threat of meningococcal meningitis.

Because of the problematic application and disease coverage of the IHR, WHO has undertaken a revision and updating of the IHR to make them more applicable to infection control in the twenty-first century. The revised regulations will replace reporting of specific diseases, such as cholera, with reporting of disease syndromes, such as epidemic diarrheal disease with high mortality. They will have a broader scope to include all infectious diseases of international importance and will clearly indicate what measures are appropriate internationally, as well as those are inappropriate. It is envisaged that the revised IHR will become a true global alert-and-that response system to ensure maximum protection with minimum restriction.

But eradication and elimination cannot substitute for good public health—rebuilding of the weakened public health infrastructure and strengthening water and sanitary systems; minimizing the impact of natural and anthropogenic changes in the environment; effectively communicating information about the prevention of infectious diseases; and using antibiotics appropriately. The challenge in the twenty-first century will be to continue to provide resources to strengthen and ensure more cost-effective infectious disease control while also providing additional resources for other emerging public health problems, such as those related to smoking and aging.

See Also the Following Articles

Bibliography

(1969). "The International Health Regulations," 3rd ed.

(1996). "Fighting Disease, Fostering Development," World Health Report.

Fenner, F., Henderson, D. A., *et al.* (1988). "Smallpox and Its Eradication." World Health Organization, Geneva.

Garrett, L. (1995). "The Coming Plague: Newly Emerging Disease in a World out of Balance." Penguin Books.

Levy, S. B. (1992). "The Antibiotic Paradox: How Miracle Drugs Are Destroying the Miracle." Plenum Press, New York.

Energy Transduction Processes: From Respiration to Photosynthesis

Stuart J. Ferguson
University of Oxford

GLOSSARY

aerobic respiration   The energetically downhill electron transfer from a donor molecule or ion to oxygen, which is reduced to water, with concomitant coupled ion translocation and thus generation of an electrochemical gradient.

anaerobic respiration   The energetically downhill electron transfer, from a donor molecule or ion to a molecule other than oxygen, or to an ionic species, with concomitant coupled ion translocation and thus generation of an electrochemical gradient. The reduction products of the acceptors can either be released from the cell or, sometimes, used as further electron acceptors.

antiport   The transport of a molecule or ion up its chemical or electrochemical gradient with the concomitant movement in the opposite direction, but down its electrochemical gradient, of one or more protons or sodium ions.

bacteriorhodopsin   A protein of the cytoplasmic membrane of the halophilic archaebacterium *Halobacterium salinarum* (formerly *halobium*) that has a covalently attached retinal molecule. Absorption of light by the latter pigment results in proton translocation across the membrane.

chemiosmotic mechanism   The transduction of energy between two forms via an ion electrochemical gradient (q.v.) (usually of protons but sometimes of sodium) across a membrane. Examples of such membranes are the cytoplasmic membranes of bacteria, the inner mitochondrial membranes of eukaryotes, and the thylakoid membranes of algae.

cytochrome   Hemoprotein in which one or more hemes is alternately oxidized and reduced in electron-transfer processes.

electrochemical gradient   The sum of the electrical gradient or membrane potential ($\Delta\psi$) and the ion concentration gradient across a membrane (the latter is often defined as ΔpH for protons).

electron acceptor   Low-molecular-weight inorganic or organic species (compound or ion) that is reduced in the final step of an electron-transfer process.

electron donor   Low-molecular-weight inorganic or organic species (compound or ion) that is oxidized in the first step of an electron-transfer process.

electron transport   The transfer of electrons from a donor molecule (or ion) to an acceptor molecule (or ion) via a series of components (a respiratory chain, q.v.), each capable of undergoing alternate oxidation and reduction. The electron transfer can either be energetically downhill, in which case it is often called respiration (q.v.), or energetically uphill when it is called reversed electron transfer (q.v.).

F_oF_1 ATP synthase   The enzyme that converts the protonmotive force into the synthesis of ATP. Protons (or more rarely sodium ions) flow through the membrane sector of the enzyme, known for historical reasons as F_o, and thereby cause conformational changes, with concomitant ATP synthesis, in the globular F_1 part of the molecule. In some circumstances the enzyme can generate the proton-motive force at the expense of ATP hydrolysis.

oxidase   The hemoprotein that binds and reduces oxygen, generally water. Oxygen reductase is the function.

oxidative phosphorylation   Adenosine triphosphate

(ATP) synthesis coupled to a proton or sodium electrochemical gradient (q.v.), generated by electron transport, across an energy transducing membrane.

P/O (P/2e) ratio   The number of molecules of ATP synthesized per pair of electrons reaching oxygen, or more generally any electron acceptor.

photophosphorylation   Adenosine triphosphate (ATP) synthesis coupled to a proton or sodium electrochemical gradient generated by light-driven electron transport, which is often cyclic in bacteria.

proton-motive force   The proton electrochemical gradient (q.v.) across an energy-transducing membrane in units of volts or millivolts.

quinone   Lipid-soluble hydrogen (i.e., proton plus electron) carrier that mediates electron transfer between respiratory chain components.

respiration   The sum of electron transfer reactions resulting in reduction of oxygen (aerobically) or other electron acceptor (anaerobically) and generation of proton-motive force.

respiratory chain   Set of electron-transfer components, which may be arranged in a linear or branched fashion, that mediate electron transfer from a donor to an acceptor in aerobic or anaerobic respiration (q.v.).

redox potential   A measure of the thermodynamic tendency of an ion or molecule to accept or donate one or more electrons. By convention, the more negative the redox potential the greater is the propensity for donating electrons and vice versa.

reversed electron transport   The transfer of electrons energetically uphill toward the components of an electron-transfer chain that have more negative redox potentials. Such electron transfer can be regarded as the opposite of the respiration (q.v.) and is driven by the proton-motive force.

symport   The transport of a molecule up its chemical or electrochemical gradient with the concomitant movement, in the same direction but down its electrochemical gradient, of one or more protons or sodium ions.

uniport   The transport of an ionic species in direct response to the membrane potential across a membrane.

THE INNER MITOCHONDRIAL MEMBRANE of the microbial eukaryote and the cytoplasmic membrane of the prokaryote are the key sites where energy available from processes such as the oxidation of nutrients or from light is converted into other forms that the cell needs. Most prominent among these other forms is ATP and thus these types of membrane are concerned with oxidative phosphorylation (or photophosphorylation).

I. INTRODUCTION

Energy-transducing membranes share many common components, but most importantly they operate according to the same fundamental chemiosmotic principle. This states that energetically downhill reactions that are catalyzed by the components of these membranes are coupled to the translocation of protons (or more rarely sodium ions) across the membranes. The direction of movement is outward from the matrix of the mitochondria or the cytoplasm of bacteria. The consequence of this translocation is the establishment of a proton electrochemical gradient. This means that the matrix of the mitochondria or cytoplasm of the bacteria tends to become both relatively negatively charged (thus, called the N side) and alkaline relative to the other side of the membranes, the intermembrane space in the mitochondria and the periplasm in gram-negative bacteria (and equivalent zone in gram-positive bacteria and archae), which is thus called the P side (Fig. 1). This electrochemical gradient is in most circumstances dominated by the charge term, which means that there is often a substantial membrane potential across the membranes, frequently estimated to be on the order of 150–200 mV. In most circumstances, the pH gradient generated by the proton translocation is small, 0.5 unit would be an approximate average value. The membrane potential is added to the pH gradient to give the total gradient, which is usually called the proton-motive force if it is given in millivolts. The conversion factor is such that 0.5 pH unit is approximately equivalent to 30 mV. Strictly speaking, the expression of the gradient as an electrochemical potential requires that units of kJ/mol be used; in practice this is rarely done, which sometimes causes confusion. I use the term proton-motive force in this article.

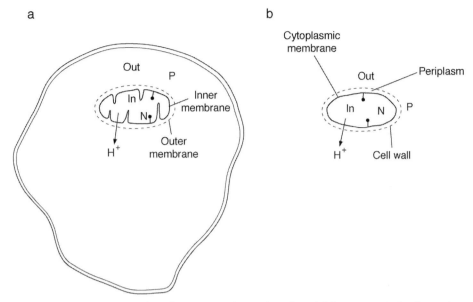

Fig. 1. (a) Idealized mitochondrion in a eukaryotic cell and (b) gram-negative bacterium showing the direction of proton translocation linked to an exergonic (energetically downhill) reaction. P means a relatively positive aqueous phase (i.e., outside the inner mitochondrial membrane or the bacterial cytoplasmic membrane); N means a relatively negative aqueous phase (i.e., inside the mitochondrion (the matrix) or the cytoplasm of a bacterium. ᶦ represents the ATP synthase enzyme.

II. MITOCHONDRIAL ENERGETICS

The best-known machinery for generating the proton-motive force is the mitochondrial respiratory chain. The standard mitochondrial respiratory chain is found, at least under some growth conditions, in eukaryotic microbes. The key point is that as a pair of electrons traverses the chain from NADH to oxygen there are three segments (formerly called sites, but this term is inappropriate because it implies equivalence and relates to a very old idea that ATP is made at three sites within the electron-transport chain) where protons can be translocated across the membrane. The first and last of these segments move four protons per two electrons, while the middle segment moves only two (Fig. 2) (consideration of the mechanisms of these proton translocations is beyond the scope of this article). Thus, 10 protons are moved per two electrons moving along the chain from NADH to O_2. Electrons may enter the chain such that they miss the first proton-translocating segment of the chain; succinate and the intermediates generated during fatty-acid oxidation are the most prominent examples of electron sources for this. In these cases, six protons are translocated per two electrons. The entry of electrons at the third segment would obviously give a translocation stoichiometry of four.

The proton-motive force generated can then be used to drive various uphill reactions. Most prominent is ATP synthesis. This is achieved by protons flowing back across the membranes and through the ATP synthase enzyme, often called FoF1 ATP synthase. There is increasing insight into the mechanism of this enzyme; it appears to function akin to a rotary motor in which the flow of protons through the F_0 is coupled to rotation and structural changes in the F_1 part of the molecule, events that are somehow linked to ATP synthesis. It is not settled how many protons must pass through the ATP synthase to make one ATP molecule; a consensus value adopted here, even though it is not fully confirmed, is three. On the basis of "what goes one way across the membrane must come back the other," it might therefore be thought that the stoichiometry of ATP production per pair of electrons (called the P/O or P/2e ratio) flowing from NADH to oxygen would be 10/3 (i.e.,

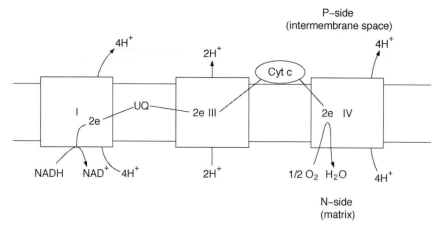

Fig. 2. Considerably simplified representation of the proton-translocation stoichiometry, per two electrons, of the mitochondrial respiratory chain. I indicates complex I (otherwise known as NADH dehydrogenase); III, complex III (otherwise known as the ubiquinol-cytochrome c oxidoreductase or cytochrome bc_1 complex); IV, complex IV (otherwise known as cytochrome C oxidase or cytochrome aa_3 oxidase); P, positive aqueous phase; and N, negative aqueous phase.

3.3) for NADH and 6/3 (i.e., 2) for succinate. However, matters are a little more complicated. The combined process of entry of ADP and Pi (phosphate) into mitochondria and the export of ATP to the cytoplasm involves the movement of one proton into the matrix (Fig. 3). Thus for each ATP made, the expected stoichiometry is 10/(3 + 1) = 2.5 for NADH and 6/(3 + 1) = 1.5 for succinate. These values differ from the classic textbook values of 3 and 2, respectively, but they are rapidly becoming accepted.

It was generally thought that eukaryotes were only capable of aerobic respiration. However, there is now evidence for a form of mitochondrial anaerobic respiration in which nitrate is reduced to nitrous oxide (more typically a prokaryotic characteristic, see the following) and for a novel type of mitochondrion from the ciliate protist *Nyctotherus ovalis* that reduces protons to hydrogen. In both these examples, electrons are derived from NADH.

III. BACTERIAL ENERGETICS

Many species of bacteria employ a respiratory chain similar to that found in mitochondria in order to generate a proton-motive force. However, there are many more types of electron donor and acceptor

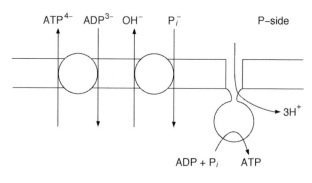

Fig. 3. Charge movement, associated with ATP synthesis and translocation of adenine nucleotides and phosphate across the inner mitochondrial membrane. The stoichiometry of proton translocation through the ATP synthase is commonly taken to be 3 but this is not a definite value. The translocated protons are not believed to pass through the active site of the ATP synthase enzyme. Note that the adenine nucleotide exchange moves one positive charge into the matrix per nucleotide exchanged and the operation of the phosphate transporter effectively moves the chemical part of the proton (but not the charge) into the matrix. Thus, in combination, the two transporters move one positive charge into the mitochondrion per ATP synthesized and returned to the P phase. Note that these transporters do not operate in bacterial ATP synthesis.

species that can be used by bacteria (eukaryotes are restricted to the oxidative breakdown of reduced carbon compounds), and various forms of anaerobic respiration are widespread. A further general difference between bacteria and microbial eukaryotes is that in the former the protonmotive force can drive a wider range of functions and be generated in more diverse ways than in the latter. Thus, functions alongside ATP synthesis (for which the enzyme is very similar to that found in mitochondria), such as driving of many active transport processes and the motion of the flagella, are important processes that depend on the protonmotive force in many organisms (Fig. 4).

A common mode of active transport is known as the symport; the classic example of this is the lactose–proton symporter coded for by the *lacY* gene of the *lac* operon of *E. coli*. In this case, a transmembrane protein translocates together a proton down its electrochemical gradient and a lactose molecule up its concentration gradient (Fig. 5); the exact mechanism is presently unknown. There are cases known where Na is the translocated ion. A second type of transport system is the antiport (Fig. 5). Here the movement via a protein of the proton down its electrochemical gradient is obligatorily linked to the movement of another species, typically an ion, in the opposite direction and up its electrochemical gradient. The third type, uniport, is the case where an ion moves in direct response to the membrane potential and is probably rarer than the other two exmaples in the prokaryotic world.

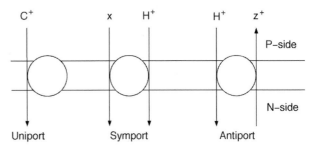

Fig. 5. The three common modes of substrate transport across the bacterial cytoplasmic membrane.

It is important to appreciate that not all transport processes across the bacterial cytoplasmic membrane are directly driven by the proton-motive force. Some transport reactions are driven directly by ATP. Notable among such systems are ABC (ATP binding cassette) transporters.

A more subtle aspect of prokaryotic energetics is that in some species of bacteria the proton-motive force must drive reversed electron transport under some circumstances (see later).

A common misconception when multiple functions of the proton-motive force are discussed is that this force can be divided (e.g., 100 mV for ATP synthesis and 50 mV for flagella motion). This notion is incorrect because the proton-motive force across a membrane has a single value at any one time and it is the magnitude of this force that is acting simultaneously on all energy-transducing units, be they ATP synthases, active transporters, or flagella.

IV. PRINCIPLES OF RESPIRATORY ELECTRON-TRANSPORT LINKED ATP SYNTHESIS IN BACTERIA

In principle, energy transduction on the cytoplasmic membrane is possible if any downhill reaction is coupled to proton translocation. The most familiar examples are probably those that also occur in mitochondria, for example, electron transfer from NADH to oxygen or from succinate to oxygen. In these cases, the electrons pass over a sizeable redox drop (Table I). In contrast to mitochondria, various species of bacteria can use a wide variety of electron donors and acceptors. The fundamental principle is that the redox drop should be sufficient for the elec-

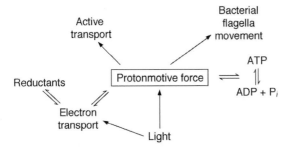

Fig. 4. The central role of the proton-motive force in linking diverse reactions. Note that the direct generation of proton-motive force from light is unusual, but is exemplified by the proton-pumping bacteriorhodospin protein found in halobacteria. Normally light drives electron-transport processes, which in turn generate the proton-motive force.

TABLE I
Approximate Standard Redox Potentials of Some Electron-Donor and -Acceptor Couples Used in Respiratory Processes[a]

Couple	$E^{\circ\prime}$ (mV)
N_2O/N_2	+1360
NO/N_2O	+1180
O_2/H_2O	+820
NO_3^-/NO_2^-	+430
NO_2^-/NH_4^+	+360
NO_2^-/NO	+350
fumarate/succinate	−30
methanol/formaldehyde	−180
$NAD^+/NADH$	−320
$CO_2/formate$	−430
CO_2/CO	−540

[a] Redox potential refers to the standard state (1 M concentrations for solutes and 1 atm pressure for gases). Conditions experienced by cells may vary significantly from these and thus the actual redox potentials of the couples should be calculated from the Nernst equation and may differ substantially from those in this table.

tron transfer to be coupled to the translocation of protons across the cytoplasmic membrane. Table I shows that such sizeable drops are associated with the aerobic oxidation of hydrogen, sulfide, carbon monoxide, and methanol, to cite just a few electron donors. Anaerobic respiration is also common with many suitable pairings of reductants and oxidants (e.g., Table I). Thus NADH can be oxidized by nitrate, nitrite, nitric oxide, or nitrous oxide. The flow of electrons to these acceptors, each of which (other than nitrate) is generated by the reduction of the preceding ion or molecule, is the process known as denitrification. In *E. coli* under anaerobic conditions, formate is frequently an electron donor, and nitrate and nitrite are the acceptors, with the latter being reduced to ammonia (Table I) rather than to nitric oxide, as occurs in denitrifying bacteria. A wide variety of electron-transport components, including many different types of cytochrome are involved in catalyzing these reactions. The mechanisms whereby electron transport is linked to the generation of the proton-motive force are frequently complex. How-

ever, nitrate respiration (Fig. 6) provides an example of one of the simplest mechanisms that corresponds to Mitchell's original redox loop mechanism.

An important point is that the consideration of the energy drop between the donor and acceptor (Table I) is only a guide as to whether proton translocation, and thus ATP synthesis, can occur and, if so, with what stoichiometry. Thus while many bacterial species can form a respiratory chain with considerable similarity to that found in mitochondria, others vary from this pattern. Notable here is *Escherichia coli*, which always lacks the cytochrome bc1 complex, and which, following some growth conditions, has cytochrome bo as the terminal oxidase, but which under others has cytochrome bd. The consequence is that when the former proton-pumping oxidase is operating only eight protons are translocated per pair of electrons flowing from NADH to oxygen, while with the latter oxidase the stoichiometry would be six. The corresponding stoichiometry for mitochondria is ten. This example illustrates the important

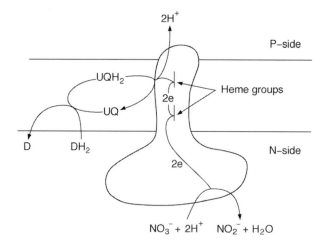

Fig. 6. A simple mechanism for generating the proton-motive force, the bacterial nitrate reductase system. Oxidation of ubiquniol (UQH_2) to ubiquinone (UQ) at one side of the membrane is accompanied by release of protons to the P side and translocation of the electrons towards the N side, where they combine with protons and nitrate to produce nitrite. Overall, the process effectively translocates two protons per two electrons across the membrane. The mitochondrial electron-transport chain (see Fig. 2) involves more complex mechanisms for proton translocation. DH_2 is an unspecified donor to the ubiquinone, and D is the product of oxidizing DH_2.

point that it is not just the energy drop between a donor and an acceptor that is important, but also the details of the components (or molecular machinery) in between. Another example is methanol to oxygen. Periplasmic oxidation of methanol feeds electrons into the electron-transport chain close to the terminal oxidase, yet energetic considerations alone would indicate that electrons could span more proton-translocating sites just as they do when succinate is the electron donor (compare the redox potentials for fumarate–succinate and methanol–formaldehyde Table I). A final example to consider is the case in which both the electron donor and acceptor are in the periplasm and they are connected purely by periplasmic components. In such a case, which applies to methanol (as donor) and nitrous oxide (N_2O as acceptor), the electrons do not pass through any proton-translocating complex. Thus, no proton translocation would occur no matter what the redox drop between the two components.

It is not necessary for electrons to flow over such a large energy drop as they do when they pass from NADH to oxygen (Table I) in order to generate a proton-motive force. Thus, if the driving force associated with a reaction was very small, it might still be energetically be possible for the passage of two electrons from a donor to an acceptor to cause the translocation of just one proton. If three protons are required for the synthesis of ATP, then the ATP yield stoichiometry would be 0.166 per electron flowing from electron donor to acceptor. This seemingly bizarre stoichiometry is not only energetically possible but also mechanistically possible because the chemiosmotic principle involves the delocalized proton-motive force that is generated by all the enzymes of the membrane and also consumed by them all. There is no case known that matches this extreme; nevertheless, there may well be organisms yet to be discovered that have such low stoichiometries of ATP synthesis.

One example of the lowest known stoichiometries of ATP synthesis per pair of electrons reaching the terminal electron acceptor (oxygen) occurs in *Nitrobacter*. Table I shows that the redox drop is small between nitrite and oxygen. This organism also illustrates the versatility and subtlety of the chemiosmotic mode of energy transduction. *Nitrobacter* species oxi-

dize nitrite to nitrate at the expense of the reduction of oxygen to water in order to sustain growth. The energy available as a pair of electrons flows from nitrite to oxygen is sufficient to translocate two protons (a more detailed consideration of how this is done is outside the scope of this chapter). This means, recalling the current consensus that three protons are needed for the synthesis of one ATP molecule, that the ATP yield stoichiometry would be 0.66/2e. Nitrobacter also illustrates another important facet of energy transduction in the bacterial world. The organism is chemolithotrophic, which means that it grows on nitrite as the source not only of ATP but also of reductant (NADPH), which is required for reducing CO_2 into cellular material. Energetic considerations immediately show that nitrite cannot reduce NADP directly. What happens in the cell is that a minority of the electrons originating from nitrite are driven backward up the electron-transfer system to reduce NAD(P) to NAD(P)H. This is achieved by the inward movement of protons reversing the usual direction of proton movement (Fig. 7). This reversed electron-transport process is an important phenomenon in a variety of bacteria, especially those growing in the chemolithotrophic mode.

Most studies of electron transport-linked ion trans-

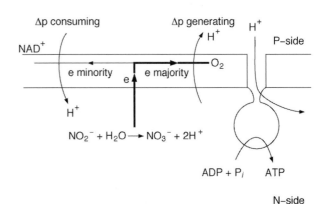

Fig. 7. Reversed electron transport illustrated by the example of *Nitrobacter*. The majority of eletrons derived from nitrite flow energetically downhill to oxygen via a cytochrome oxidase, which generates a protonmotive force. A minority of electrons is driven energetically uphill by the protonmotive force so as to reduce NAD^+ to NADH. Note that in this diagram no proton stoichiometry values are implied.

location have been done with species of eubacteria. However, the same fundamental process also occurs in archaebacteria, although with some novel features that reflect some of the extreme growth modes tolerated by these organisms. For example, a key step in methane formation by methanogenic bacteria is electron transfer from hydrogen or other reductant to a small molecule contains a disulfide bond. The latter is reduced to two sulfides and the overall process is coupled to the translocation of protons across the cytoplasmic membrane. The proton-motive force thus set up can be used to drive ATP synthesis. Interestingly the ATP synthase in archaebacteria shows significant molecular differences from its counterpart in eubacteria and mitochondria, but is believed to function according to the same principle.

V. GENERATION OF THE ION ELECTROCHEMICAL GRADIENT OTHER THAN BY ELECTRON TRANSPORT

A. ATP Hydrolysis

Organisms that are incapable of any form of respiration still require an ion electrochemical gradient across the cytoplasmic membrane for purposes such as nutrient uptake. One way in which this requirement can be met is for some of the ATP synthesized by fermentation to be used for ATP hydrolysis by the F_oF_1 ATPase. This means that this enzyme works in the reverse of its usual direction and pumps protons out of the cell. Thus there are many organisms that can prosper in the absence of any electron-transport process, either as an option or as an obligatory aspect of their growth physiology.

B. Bacteriorhodopsin

A specialized form of light-driven generation of proton-motive force, and hence of ATP, occurs in halobacteria; these organisms are archaebacteria. The key protein is bacteriorhodopsin, which is a transmembrane protein with seven α-helices that has a covalently bound retinal. The absorption of light by this pigment initiates a complex photocycle that is linked to the translocation of one proton across the cytoplasmic membrane for each quantum absorbed. Bacteriorhodopsin is one of a family of related molecules. Another, halorhodospin, is structurally very similar and yet catalyzes the inward movement of chloride ions driven by light.

C. Methyl Transferase

One step of energy transduction in methanogenic bacteria involves an electron-transfer process (see earlier). Another important process in methanogens is the transfer of a methyl group from a pterin to a thiol compound. This exergonic (energetically downhill) reaction is coupled to ion, in this case sodium, translocation across the cytoplasmic membrane.

D. Decarboxylation Linked to Ion Translocation

In the bacterial world, the electrochemical gradients can be generated by diverse processes other than electron transport or ATP hydrolysis. For example, *Propionegenium modestum* grows on the basis of catalyzing the conversion of succinate to propionate and carbon dioxide. One of the steps in this conversion is decarboxylation of methyl-malonyl coenzyme A (CoA) to propionyl CoA. This reaction is catalyzed by a membrane-bound enzyme that pumps sodium out of the cells, thus setting up a sodium electrochemical gradient (or sodium-motive force). This gradient in turn drives the synthesis of ATP as a consequence of sodium ions reentering the cells through a sodium-translocating ATP synthase enzyme. Apart from illustrating that sodium, instead of proton circuits, can be used for energy transduction in association with the bacterial cytoplasmic membrane, this organism also illustrates that the stoichiometry of ATP synthesis can be less than one per CO_2 formed. It is believed that each decarboxylation event is associated with the translocation of two sodium ions and the synthesis of ATP with three. Thus non-integral stoichiometry is consistent with the energetics of decarboxylation and ATP synthesis. This is an important paradigm to appreciate; the underpinning growth reaction for an organism does not have to be capable of supporting the synthesis of one or more integral numbers of ATP molecules.

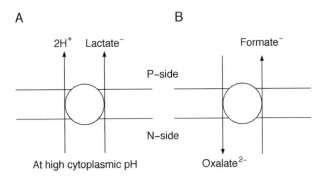

A
B

2H⁺ Lactate⁻

P–side

N–side

At high cytoplasmic pH

Formate⁻

Oxalate²⁻

Fig. 8. Two examples of generation of protonmotive force by end-product extrusion from fermenting bacteria.

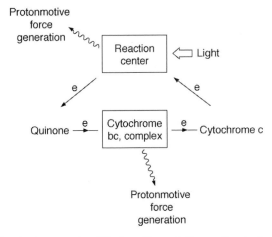

Protonmotive force generation

Reaction center

Light

e

e

Quinone → e → Cytochrome bc, complex → e → Cytochrome c

Protonmotive force generation

Fig. 9. An oversimplified outline of the cyclic electron-transport process of photosynthetic bacteria. There are two types of reaction centers, depending on the organism. The molecular composition of the system depends on the organism. Two components contribute to the generation of the protonmotive force.

E. Metabolite Ion-Exchange Mechanisms

Another example of the generation of a proton-motive force is ion exchange across the membrane. For example, in fermenting bacteria there is evidence that under some conditions an end-product of metabolism, lactic acid, leaves the cell together (i.e., in symport) with than one proton; this results in the generation of a protonmotive force (Fig. 8). A second example is provided by *Oxalobacter formigenes*, in which the entry of the bivalent anion oxalate is in exchange for the exit of the monovalent formate ion generated by decarboxylation of the oxalate, leading to the net generation of membrane potential (Fig. 8). This seems to be the principal mode of generating membrane potential in this organism.

VI. PHOTOSYNTHETIC ELECTRON TRANSPORT

Prokaryotic photosynthesis involves a cyclic electron-transport process in which a single photosystem captures light energy and uses it to drive electrons around the cycle (Fig. 9). The consequence of this cyclic electron flow is the generation of the proton-motive force. There are two types of photosystem found in prokaryotes. One is related to the water-splitting photosystem that is found is plants; typically this bacterial photosystem is found in organisms such as *Rhodobacter sphaeroides*. The second type of photosystem is closely related to the second photosystem of plants, the one that is concerned with the genera-

tion of NADPH. *Heliobacter* is an example of an organism carrying this type of center. Some microorganisms have both of these photosystems, arranged to operate in series as in plants. In this group are the prokaryotic blue-green algae and the eukaryotic algae.

VII. ALKALIPHILES

An interesting unresolved problem relates to energy transduction in the alkaliphilic bacteria. The problem is straightforward. These organisms can grow in an environment with a pH as high as 11 or 12. A cytoplasmic pH even as high as 9 means that the pH gradient could be as much as 3 units (equivalent to 180 mV) the wrong way around in the context of the chemiosmotic mechanism. The membrane potential always seems to be larger than 180 mV, but the total protonmotive force can be very low (e.g., around 50 mV). For some organisms that use a conventional proton-translocating respiratory chain and ATP synthase, it is not understood how they survive energetically. In other organisms, there is evidence for the role of a sodiummotive force. This would sidestep the problem of the adverse proton concentration gradient.

See Also the Following Articles

ABC TRANSPORT • AEROBIC RESPIRATION • FLAGELLA • METHANO-
GENESIS

Bibliography

Brock, T., Madigan, M. T. (1997). "Brock Biology of Microorganisms," 8th ed. Prentice-Hall, London.

Embley, T. M., and Martin, W. (1998). A hydrogen-producing mitochondrion. *Nature* **396**, 517–519.

Ferguson, S. J. (1998). Nitrogen cycle enzymology. *Curr. Opinion Chem. Biol.* **2**, 182–193.

Harris, D. A. (1995). "Bioenergetics at a Glance." Blackwells, Oxford.

Konings, W. N., Lolkema, J. S., and Poolman, B. (1995). The generation of metabolic energy by solute transport. *Arch. Microbiol.* **164**, 235–242.

Nicholls, D. G., and Ferguson, S. J. (1992). "Bioenergetics 2." Academic Press, London.

Unden, G., and Bongaerts, J. (1997). Alternative respiratory pathways of *Escherichia coli*: energetics and transcriptional regulation in response to electron acceptors. *Biochim. Biophys. Acta* **1320**, 217–234.

Walker, J. E. (1998). ATP synthesis by rotary catalysis (Nobel lecture). *Angew Chem. Intl. Ed.* **37**, 2308–2319.

White, D. (1995). "The Physiology and Biochemistry of Prokaryotes." Oxford University Press, Oxford.

Enteropathogenic Bacteria

Farah K. Bahrani-Mougeot and Michael S. Donnenberg

University of Maryland, Baltimore

GLOSSARY

bacteriophage A virus that infects bacteria.

colitis Inflammation of the large intestine (colon).

cytotoxins Bacterial products that damage cells.

diarrhea Increase in the frequency of bowel movement and decrease in the consistency of stool.

dysentery Inflammatory disease of the large bowel with severe abdominal cramps, rectal urgency, and pain during stool passage and the presence of blood, pus, and mucus in stool.

enteritis Inflammation of the small intestine.

enterotoxins Bacterial products that cause fluid secretion from intestinal cells.

fimbriae (or pili) Rigid rod surface organelles with diameters of about 2 to 7 nm in gram-negative bacteria that often mediate bacterial adherence to host cells.

flagellae Ropelike surface organelles of 15 to 20 nm in diameter that provide bacteria with motility and the ability to move toward nutrients and away from toxic substances (chemotaxis).

gastritis Inflammation of the stomach.

pathogenicity island Segment of DNA that is foreign to the bacterial host and carries virulence genes.

plasmid Extrachromosomal self-replicating DNA element.

type III secretion pathway A specialized protein secretion system, responsible for export of virulence determinants by some gram-negative bacterial pathogens and symbionts of animals and plants. Some of the proteins secreted by this pathway are translocated by the bacteria into host cells.

type IV fimbriae Special type of fimbriae produced by certain pathogenic gram-negative bacteria. These fimbriae have subunits with different primary structures and often different morphologies from common fimbriae.

ENTERIC INFECTIONS are caused by a variety of microorganisms, including bacterial pathogens. Among these infections, diarrheal diseases are a major cause of mortality in the children of third-world countries, due to malnutrition, poor personal hygiene, and insufficient environmental sanitation. In industrialized countries, diarrhea may result from food-borne outbreaks and is common in day care centers, hospitals, and chronic care institutions, among homosexual men and immunocompromised patients. Diarrhea can result from inflammatory infections in the colon and/or small intestine caused by pathogens, such as *Shigella, Salmonella,* and *Campylobacter,* or noninflammatory infections in the small intestine by pathogens such as *Vibrio* and enterotoxigenic *Escherichia coli.* Diarrhea can also result from ingestion of preformed toxins, produced by bacteria such as *Clostridium perfringens* and *Bacillus cereus.* In addition, infections by enteropathogenic bacteria may cause systemic syndromes, such as typhoid fever caused by *Salmonella typhi.* Another form of enteric infection, gastritis, results exclusively from infections with *Helicobacter pylori.* We will briefly review the epidemiology, pathogenesis, and clinical features of the most important bacterial enteric pathogens of humans.

I. BACTERIAL AGENTS OF INFLAMMATORY DIARRHEA

In the majority of cases, inflammatory diarrhea in the distal small bowel and colon occurs in response

TABLE I
Epidemiology and Clinical Characteristics of Bacterial Enteric Pathogens

Pathogen	Route of transmission	Site of infection	Clinical syndrome
B. cereus	Foods such as fried rice and vanilla sauce	Small intestine	Watery diarrhea Vomiting Abdominal cramps
C. jejuni	Contaminated food and water Fecal–oral	Colon Small intestine	Watery diarrhea Dysentery
C. difficile	Environmental contamination with spores Fecal–oral	Colon	Watery diarrhea Pseudomembranous colitis
C. perfringens	Foods such as meat, turkey and chicken	Small intestine	Watery diarrhea Necrotic enteritis
EAEC[a]	?	Small intestine	Watery and mucoid diarrhea
EHEC[b]	Food contaminated with cattle feces Person–person contact	Colon	Diarrhea Hemorrhagic colitis Hemolytic uremic syndrome
EIEC[c]	Contaminated food and water Fecal–oral	Colon	Watery diarrhea Dysentery
EPEC[d]	Person–person contact	Small intestine Colon	Watery diarrhea
ETEC[e]	Contaminated food and water	Small intestine	Watery diarrhea
H. pylori	Fecal–oral (?) Person–person contact (?)	Stomach	Gastritis Peptic ulcer Gastric cancer
Salmonella (nontyphi)	Contaminated food and water Animal–person contact	Small intestine Colon	Watery diarrhea Dysentery
S. typhi	Contaminated food and water	Systemic	Typhoid fever
Shigella	Fecal–oral Contaminated food and water Person–person contact	Colon	Watery diarrhea Dysentery
S. aureus	High salt- or high sugar-containing foods	Small intestine	Watery diarrhea Vomiting
V. cholerae	Contaminated food and water Shellfish	Small intestine	Rice-water diarrhea
Y. enterocolitica	Contaminated food Person–pig contact	Small intestine Systemic	Acute diarrhea Enterocolitis Mesenteric adenitis

[a] Enteroaggregative *E. coli.*
[b] Enterohemorrhagic *E. coli.*
[c] Enteroinvasive *E. coli.*
[d] Enteropathogenic *E. coli.*
[e] Enterotoxigenic *E. coli.*

to bacterial invasion of the intestinal tissues. However, in some infections, enterocolitis can be an outcome of bacterial toxicity without invasion. Causative agents of these infections include:

A. *Shigella* spp.

Shigella, the major etiologic agent of bacillary dysentery, is traditionally divided into four species based on biochemical and serological characteristics. These species include *S. dysenteriae*, *S. flexneri*, *S. sonnei*, and *S. boydii*. However, techniques such as multilocus enzyme electrophoresis have revealed that all *Shigella* strains are actually encompassed within the species *E. coli*. Clinical syndromes of shigellosis include a mild watery diarrhea, which is often followed by severe dysentery with blood, mucus, and inflammatory cells in feces. The incubation period ranges from 6 hr to 5 days. Epidemiological studies show that *Shigella* is transmitted by the fecal–oral route or by contaminated food and water. As few as 100 organisms can cause infection in an adult.

Shigella invades the colonic mucosa and this invasion involves entry and intercellular dissemination. Bacteria enter the cells by a micropinocytic process, which requires polymerization of actin at the site of entry. Shortly after entry, bacteria lyse the phagocytic vacuole and move into the cytoplasm where they multiply. Within the cytoplasm, *Shigella* recruits actin microfilaments at one pole of the bacterium, which leads to the formation of a polymerized actin tail behind the bacterium and, consequently, movement of the bacteria (Fig. 1). The movement of the bacteria leads to the formation of cell membrane protrusions that extend from one cell into the adjacent cell and allow dissemination of bacteria without their release into the extracellular environment. The ability of *Shigella* to spread from cell to cell is measured *in vitro* by the formation of plaques on a confluent cell monolayer (plaque assay) and *in vivo* by formation of keratoconjunctivitis in guinea pigs (Sereny test).

All of the genes required for *Shigella* invasion are carried on a 200-kb virulence plasmid. A 30-kb fragment contains the genes that encode secreted proteins called IpaA, B, C, and D, the chaperones for these proteins called Ipgs, and a specialized type III

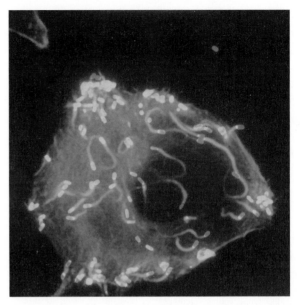

Fig. 1. Actin tail formation by *Shigella flexneri*. Actin microfilaments are polymerized at one end of the bacterium and help it to move within the host cell cytoplasm. Bacteria are stained in red and actin in green. The areas where bacteria and actin colocalize appear in yellow (courtesy of Coumarin Egile and Philippe J. Sansonetti). See color insert.

secretion system for these proteins called Mxi–Spa. Several other proteins that might be important in the entry process are also secreted by the Mxi–Spa machinery. This machinery becomes activated upon contact of bacteria with the host epithelial cells. The *icsA* (*virG*) gene, which confers the ability of *Shigella* to spread from cell to cell, is located approximately 40 kb away from the *ipa–ipg–mxi–spa* region.

Following secretion, IpaB and IpaC form a complex, which triggers recruitment of actin at the site of bacterial entry by a mechanism that involves the small host G protein, Rho. IpaA contributes to the entry process by interacting with other cytoskeletal proteins, such as vinculin and α-actinin. Contact with macrophages results in delivery of IpaB into the cytoplasm of macrophages where IpaB induces programmed cell death and release of interleukin-1. Release of this cytokine triggers a cascade of proinflammatory responses, which opens intercellular junctions and destabilizes the epithelia, thus facilitating bacterial invasion and ulceration of the colon.

Infections due to *S. dysenteriae* are more severe and more likely to lead to complications than are infections with other *Shigella* species. *S. dysenteriae* produces a lethal cytotoxin, called Shiga toxin. This toxin is composed of an enzymatically active A subunit and five B subunits that mediate binding of the toxin to target cell receptors. The toxin functions by cleaving 28S rRNA of eukaryotic cells and inhibiting protein synthesis. Shiga toxin may play a role in manifestations of hemolytic uremic syndrome, an occasional consequence of shigellosis.

B. Enteroinvasive *E. coli*

E. coli is commonly regarded as a harmless commensal of the intestines of humans. However, at least six varieties have been identified that possess specific pathogenic mechanisms allowing them to cause diarrhea. These pathogens include Enteroinvasive *E. coli* (EIEC). EIEC is closely related to *Shigella* spp. in biochemical and serological characteristics, pathogenic mechanisms, and virulence determinants (see preceding). EIEC is identified as *E. coli* based on its biochemical profile but is distinguished from other strains of *E. coli* based on genotypic and phenotypic characteristics of *Shigella* spp. Therefore, the Sereny test and plaque assay are appropriate tests for identification of EIEC.

Like *Shigella* spp., EIEC is transmitted by contaminated food and water. The infectious dose, however, is 2 to 3 logs higher than that for *Shigella* infection.

Therefore, transmission from person to person is likely less than is the case for *Shigella*. Infection with EIEC leads to watery diarrhea, with dysentery syndrome in only some of the patients.

C. Nontyphoidal *Salmonella*

Salmonella spp. other than *Salmonella typhi* are the cause of salmonellosis in humans. There are over 2000 serotypes of *Salmonella* that infect a wide range of hosts, from humans to domestic animals, birds, reptiles, and insects. Most of the *Salmonella* serotypes associated with human infections belong to subgroup 1 of the 6 subgroups of *Salmonella enterica*. Many serotypes are species-specific; thus, a particular serotype may be nonpathogenic in one host species and cause severe infection in another.

Gastroenteritis is the most common manifestation of *Salmonella* infections. Diarrhea begins 8 to 48 hr after ingestion of contaminated food and lasts for 3 to 7 days. Infection is often food-borne and associated with consumption of foods of animal origin, such as chicken, raw milk, and undercooked eggs. The infectious dose ranges from 10^5 to 10^{10} organisms and depends on serotype, source of infection, and host factors. Food-borne salmonellosis seems to be predominantly a disease of industrialized countries. Infections can also be transmitted by the fecal–oral route, particularly among homosexual men. Immu-

TABLE II
Mechanisms of Pathogenicity of Enteropathogenic Bacteria

Pathogenicity mechanism	*Organism*
1. Invasion of epithelial cells	*Shigella*, EIEC[a], *Salmonella*, *Yersinia*, *C. jejuni*, *L. monocytogenes*
2. Colonization of epithelial cells	EPEC[c], EHEC[d]
A/E[b] of intestinal epithelial cells	*C. difficile*, EHEC[d] EAEC[e], non-Cholerae Vibrios, *A. hydrophilia*, *P. shigelloides*, *H. pylori*, *V. cholera*, ETEC[f]
toxin production with overt damage to epithelial cells	
production of an enterotoxin without overt damage to epithelial cells	
3. Release of toxins in the absence of colonization	*C. perfringens*, *B. cereus*, *S. aureus*

[a] Enteroinvasive *E. coli*.
[b] Attaching and effacing.
[c] Enteropathogenic *E. coli*.
[d] Enterohemorrhagic *E. coli*.
[e] Enteroaggregative *E. coli*.
[f] Enterotoxigenic *E. coli*.

TABLE III
Mechanisms of Action of Toxins Produced by Enteropathogenic Bacteria

Mode of Action	Toxin	Organism	Type of toxin
Inhibition of protein synthesis by cleaving 28S rRNA	Shiga toxin	*Shigella*, EHEC[a]	Cytotoxin
ADP-ribosylation of G_{s_α}	Cholera toxin	*V. cholerae*	Enterotoxin
	Heat-labile toxin	ETEC[b]	Enterotoxin
Activation of guanylate cyclase	Heat-stable toxin	ETEC[b]	Enterotoxin
	EAST1[c]	EAEC[d], EPEC[e], EHEC[a]	Enterotoxin
Glucosylation of Rho	Toxin B	*C. difficile*	Cytotoxin
Pore formation	CPE[f]	*C. perfringens*	Cytotoxin
	Ace[g]	*V. cholerae*	Cytotoxin
	Aerolysin	*A. hydrophila*	Cytotoxin

[a] Enterohemorrhagic *E. coli*.
[b] Enterotoxigenic *E. coli*.
[c] Enteroaggregative heat-stable toxin.
[d] Enteroaggregative *E. coli*.
[e] Enteropathogenic *E. coli*.
[f] *C. perfringens* enterotoxin.
[g] Accessory cholerae enterotoxin.

nocompromised hosts, such as HIV-infected individuals, are more prone to *Salmonella* infections. In these patients, nontyphoidal *Salmonella* usually causes bacteremia.

Salmonella spp. invade mucosal cells of the small intestine by a bacterial-mediated endocytosis process similar to *Shigella* entry. Bacteria stimulate signal transduction pathways in epithelial cells that lead to cytoskeletal rearrangements and membrane ruffling, similar to those induced by growth factors on mammalian cells. Membrane ruffling results in uptake of bacteria into cells. Unlike *Shigella*, *Salmonella* remains in membrane-bound vacuoles, modifies the pH of the phagolysosome, and multiplies. The organisms can also be taken up by macrophages, where they multiply and penetrate into deeper tissues (see the section following on *S. typhi*).

Most of the research on the molecular genetic basis of *Salmonella* invasion has been done on *S. typhimurium*. The genes that confer the ability of *Salmonella* to invade are located on a 40-kb pathogenicity island, called SPI-1, at the 63 minutes of *S. typhimurium* chromosome. SPI-1 encodes the components of a type III secretion pathway, called Inv–Spa, and the Sip proteins secreted via this machinery. Sip proteins have functions similar to *Shigella* Ipa proteins in the process of entry and induction of

proinflammatory response and tissue destruction (see preceding section on *Shigella*).

Lipopolysaccharide (LPS) is another factor involved in invasiveness of *Salmonella*. Rough mutants with short O-side chains are less virulent. *Salmonella* also produces several adhesins, which might facilitate attachment to epithelial cells prior to penetration. In addition, nontyphi *Salmonella* strains produce an enterotoxin that has cytoskeleton-altering activity.

Salmonella strains induce secretion of cytokines such as IL-8 from epithelial cells. IL-8 is a chemoattractant for polymorphonuclear leukocytes (PMNs) and stimulates transmigration of PMNs through epithelial cell tight junctions into the intestinal lumen. The passage of PMNs through cellular junctions can lead to fluid leakage and consequent diarrhea.

D. *Campylobacter* spp.

Campylobacters are slender, spirally curved gram-negative rods, carrying a relatively small (1.6×10^6 bp) AT-rich genome that has been already sequenced. These organisms are microaerophilic, thermophilic, and require complex media for growth. Campylobacters are one of the most commonly reported causes of diarrhea worldwide. Infections with these organisms usually result in inflammatory dysentery-like diarrhea in adults of industrialized nations and watery diarrhea in children of developing countries. Campylobacters can also cause systemic disease. *C. jejuni* and *C. fetus* are the prototypes for diarrheal and systemic infections, respectively.

Campylobacter spp. are found in the gastrointestinal tracts of most domesticated mammals and fowls. Transmission occurs by contaminated food or water or by oral contact with feces of infected animals or humans. The organism cannot tolerate drying or freezing, thus being limited in transmission. Infections occur all year long, with a sharp peak in summer. The infectious dose can be as small as 500 or as high as 10^9 organisms, depending on the source of infection. The incubation period ranges from one to seven days and the duration of the illness is usually one week, with occasional relapses in untreated patients.

C. jejuni causes inflammation in both the colon and the small intestine. The resulting nonspecific colitis can be mistaken for acute ulcerative colitis. The inflammatory process can be extended to the appendix, mesenteric lymph nodes, and gall bladder. Bacteremia can also occur in some cases. Infections with *C. jejuni* may be followed by noninfectious complications, such as Guillain–Barre syndrome (GBS), an acute disease of peripheral nerves, and reactive arthritis.

Inflammation and bacteremia caused by *C. jejuni* suggest tissue invasion by this organism. Invasion seems to be linked to the presence of flagellae. Flagellae enable the organisms to move along the viscous environments and penetrate the intestinal mucosa. Binding of *C. jejuni* to host epithelial cells leads to bacterial uptake by a complex mechanism that remains controversial. Bacteria also penetrate the underlying lymphoid tissue and survive within the macrophages.

C. jejuni strains also produce several toxins. These toxins include a heat-labile cholera-like enterotoxin (CLT), which correlates with watery diarrhea, a cytolethal distending cytotoxin (CLDT), which alters the host cytoskeleton, and a hemolysin(s).

C. fetus causes febrile systemic illness more often than diarrheal infections. This organism has a tropism for vascular sites, thus causing bacteremia. The organism uses lipopolysaccharide (LPS) and a sur-

face (S) layer protein, which functions as a capsule, to resist phagocytosis and serum-killing. *C. fetus* can disseminate to cause meningoencephalitis, lung abscess, septic arthritis, and urinary tract infections.

E. *Clostridium difficile*

C. difficile is a gram-positive anaerobic spore-forming bacillus. This organism is widespread in the environment and is found in the intestines of several mammals, including humans. *C. difficile* is the most common recognized cause of diarrhea in hospitals and chronic care facilities in developed countries. Antibiotic therapy disrupts the normal flora of the colon and allows colonization or proliferation of *C. difficile*. Infection occurs by ingestion of spores from the environment. The spores resist the gastric acid and germinate into the vegetative form in the colon. The symptoms of disease range from mild diarrhea to severe pseudomembranous colitis. Infection with *C. difficile* is more common in the elderly, whereas neonates are resistant to colonization by this organism.

Pathogenic strains of *C. difficile* produce two very large toxins: toxin A and toxin B. Toxin A is a cytotoxin with a molecular weight of 308 kDa. This toxin stimulates cytokine production by macrophages and infiltration of PMNs, which results in the inflammation seen in pseudomembranous colitis. Toxin B is a protein of 207 kDa that also possesses cytotoxic activity. This toxin has glucosyltransferase activity that catalyzes transfer of glucose to the small GTP-binding protein, Rho. Modification of Rho leads to actin cytoskeletal disruption, which results in rounding of the cells. *C. difficile* also produces hydrolytic enzymes, such as hyaluronidase, gelatinase, and collagenase, which might contribute to destruction of connective tissue and subsequent fluid accumulation.

II. BACTERIAL AGENTS OF NONINFLAMMATORY DIARRHEA

The noninflammatory watery diarrhea caused by bacteria is usually associated with the production of an enterotoxin after bacterial colonization of the small intestine or with the presence of preformed enterotoxins in food. Occasionally, bacteria can cause a drastic effect on intestinal epithelial cells in the absence of an enterotoxin. This category of diarrhea is caused by:

A. *Vibrio cholerae*

V. cholerae belongs to the family *Vibrionaceae* and has been the cause of seven cholera pandemics since 1817. It is transmitted by contaminated food and water. Food-borne transmission often occurs by ingestion of raw or undercooked shellfish. Since the acid-sensitive bacteria must pass through the stomach to colonize the small intestine, a high inoculum of 10^9 organisms is required to cause disease. The diarrhea can be extremely severe, with characteristic "rice water" stools, which can lead to rapid dehydration, circulatory collapse, and death.

Cholera is caused by toxigenic strains of *V. cholerae* O1 and O139 Bengal. The O1 strains can be divided into El Tor and classical biotypes that are epidemiologically distinct. *V. cholerae* O139 is a new strain that caused a major epidemic in 1992 in India in a population which was already immune to *V. cholerae* O1 strains. The non-O1 serogroups of *V. cholerae* cause cholera and dysentery but have not been linked to cholera epidemics.

V. cholerae secretes cholera toxin (CT), which is responsible for the characteristic secretory diarrhea. CT is an enterotoxin that binds to enterocytes via five B subunits that facilitate the entry of the enzymatically active A subunit. The A subunit then catalyzes the ADP-ribosylation of the GTP binding protein, $G_{s\alpha}$, which results in activation of adenylate cyclase, accumulation of cAMP in enterocytes, and increase in secretion of chloride and water. Increased release of water into the intestinal lumen leads to secretory diarrhea. CT is encoded by the *ctx*AB genes carried on a filamentous bacteriophage. The receptor for the phage is a type IV fimbria, called the toxin-coregulated pilus (TCP), which is an essential colonization factor of *V. cholerae*. The gene encoding TCP is located on a 40-Kb pathogenicity island (PI). This PI is associated with pandemic and epidemic strains of *V. cholerae*.

V. cholerae O1 also produces two other toxins, called the zonula occludens toxin (Zot) and the accessory cholera enterotoxin (Ace). Zot affects the structure of the intercellular tight junction, zonula occludens. Ace is postulated to form ion-permeable channels in the host cellular membrane. The role of these toxins in pathogenesis is unknown.

B. Enteropathogenic *E. coli*

Enteropathogenic *E. coli* (EPEC) is an important cause of diarrhea in infants less than 2 years of age. EPEC is transmitted by the fecal–oral route by person-to-person contact. The infection occurs more frequently during the warm seasons. Infection with EPEC is often severe and leads to a high mortality rate in developing countries. The symptoms of the disease include watery diarrhea, vomiting, and fever. All EPEC strains induce a characteristic attaching and effacing (A/E) lesion on the brush border of the intestine which can be mimicked in tissue culture (Fig. 2). Pedestal-like structures form beneath the intimately adhering bacteria, due to the polymerization of actin. The accumulation of actin beneath the bacteria can be detected by a fluorescent-actin staining (FAS) assay. In addition, EPEC adheres to epithelial cells in tissue culture in a localized pattern, which can be detected by light microscopy.

Intimate attachment of EPEC to epithelial cells is mediated by an adhesin called intimin, which is encoded by the *eae* gene located on a 35-kb pathogenicity island, called the locus of enterocyte effacement (LEE). The LEE also encodes a type III secretion system called Esc, several proteins secreted by this secretion system called EspA, B, D, and F, and Tir. Tir becomes localized to the host cell membrane, where it serves as a receptor for intimin. The formation of the A/E lesions requires the Esc proteins, EspA, B, D, Tir, and intimin. The mechanisms that lead to diarrhea are unknown but may be related to changes in ion secretion and intestinal barrier function that have been detected *in vitro* and/or to loss of microvilli.

Localized adherence of EPEC to epithelial cells is dependent on the presence of a 90-kb plasmid, which carries the genes required for the biogenesis of type IV fimbriae, called bundle-forming pili (BFP). BFP

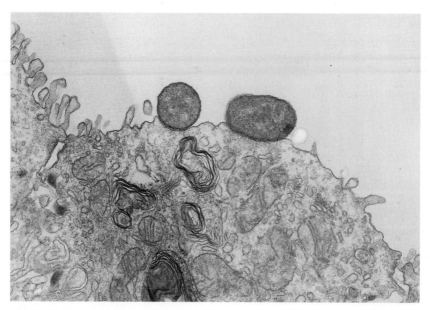

Fig. 2. Typical attaching/effacing (A/E) lesions caused by enteropathogenic *E. coli* (EPEC) in an intestinal epithelial tissue culture model. Bacteria adhere intimately to the epithelial cells and induce formation of pedestals in the cell membrane (courtesy of Barry McNamara).

form ropelike structures (Fig. 3) that are responsible for the aggregation of EPEC bacteria to each other and for the localized adherence of EPEC to host epithelial cells. In addition to BFP, some EPEC strains produce other types of fimbriae that may also contribute to localized adherence. Some EPEC strains also produce a low molecular heat stable toxin called EAST1, similar to EAST1 of enteroaggregative *E. coli,* but the importance of this toxin in disease is unknown.

C. Enterotoxigenic *E. coli*

Enterotoxigenic *E. coli* (ETEC) is a cause of infantile and childhood diarrhea in developing countries and travelers' diarrhea in adults of industrialized countries visiting ETEC-endemic areas. Infection in travelers results in mild watery diarrhea in the majority of cases, whereas in infants from endemic areas, it can cause more severe diarrhea. ETEC has a short incubation period of 14 to 50 hr. Epidemiological studies indicate that ETEC infection is more common in warm seasons and is transmitted through fecally contaminated food and water. The infectious dose for ETEC is approximately 10^8.

ETEC colonizes the mucosal epithelial cells of the small intestine and produces at least one of the two enterotoxins known as heat-labile toxin (LT) and heat-stable toxin (ST); both are encoded on a plasmid. LT is similar to cholera toxin (CT) in structure, function, and mode of action (see preceding). ST is a small polypeptide that activates intestinal guanylate cyclase, leading to accumulation of GMP, secretion of chloride and water, and, thus, diarrhea. ST resembles a peptide, guanilyn, normally found in the intestinal epithelium.

ETEC strains produce multiple fimbriae that are host species-specific. Human ETEC fimbriae, called colonization factor antigens (CFAs), are associated with specific O serogroups. These fimbriae exhibit different morphological features, such as rigid rods similar to common fimbriae, bundle-forming flexible rods, and thin wavy filaments. In addition, many human ETEC strains produce a type IV fimbria, called Longus.

D. Enterohemorrhagic *E. coli*

Enterohemorrhagic *E. coli* (EHEC) is an emerging enteropathogen, which can cause watery diarrhea

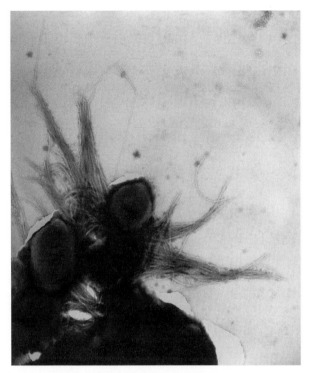

Fig. 3. Bundle-forming pili produced by Enteropathogenic *E. coli* (EPEC). Each bundle consists of several individual fimbrial filaments about 7 nm in diameter (reproduced with permission from Anantha *et al.*, (1998). *Infect. Immun.* **66**, 122–131).

followed by bloody diarrhea, an illness designated hemorrhagic colitis (HC). EHEC is also associated with severe cases of hemolytic uremic syndrome (HUS), with a mortality rate of 5–10% The reservoir for this organism is the intestinal tract of cattle and, therefore, undercooked contaminated beef is the major source of infection. Contaminated milk, juice, lettuce, sprouts, and fast food have also caused outbreaks. Since the infectious dose is as low as 50 to 200 organisms, it is not surprising that EHEC is also spread by direct person-to-person contact.

The most important EHEC serotype, O157 : H7, has been the cause of several food-borne outbreaks in the United States, Canada, Japan, and Europe since the 1980s. The mortality related to HUS has brought public attention to EHEC infections and has been the impetus for new regulations for handling and cooking of beef products.

The major pathogenic feature of EHEC is production of the bacteriophage-encoded Shiga toxin (or verotoxin), that is closely related to the Shiga toxin of *S. dysenteriae*. HUS is thought to be the result of hematogenous dissemination of Shiga toxin, cytotoxicity to endothelial cells, and microscopic thrombosis in the kidneys and elsewhere. Shiga toxins are also essential for development of bloody diarrhea and hemorrhagic colitis. Many EHEC strains also produce EAST1 toxin, similar to the toxin of enteroaggregative *E. coli*.

Similar to EPEC, EHEC strains also possess the LEE pathogenicity island on the chromosome (see preceding) and exhibit classic A/E histopathology. The A/E phenomenon is thought to be responsible for development of watery diarrhea, in a manner similar to EPEC diarrhea.

E. Enteroaggregative *E. coli*

Enteroaggregative *E. coli* (EAEC) is a cause of persistent diarrhea in children of developing and developed countries. EAEC infects the small bowel and causes diarrhea in less than 8 hr, which may persist for >14 days. The diarrhea is usually mucoid and may be watery, with low-grade fever. EAEC has also been associated with diarrhea in human immunodeficiency virus-infected patients. More importantly, colonization with EAEC is linked to growth retardation in children, independent of the symptoms of diarrhea.

EAEC induces the formation of a mucous biofilm, which can trap and may protect the bacteria, leading to persistent colonization and diarrhea. Bacteria also elicit a cytotoxic effect on the intestinal mucosa. The cytotoxic effect is mediated by genes present on a 100-kb plasmid.

EAEC exhibits an aggregative adherence phenotype, mediated in some strains by a flexible bundle-forming fimbria called Aggregative Adherence Fimbriae I (AAF/I). AAF/I is a member of the Dr family of adhesins, present in uropathogenic *E. coli*.

EAEC strains also produce a ST-like toxin, EAST1, which is linked to the AAF/1 gene cluster on the 100-kb plasmid. The role of EAST1 in diarrhea is unknown, In addition, EAEC strains produce a 108-kDa cytotoxin that belongs to the autotransporter family of proteins. This protein also exhibits enterotoxin activity.

F. *Bacillus cereus*

B. cereus is a gram-positive spore-forming rod that resides in water, soil, and as part of the normal flora in humans. This organism produces several toxins and causes two forms of toxin-mediated food poisoning, one characterized by emesis and the other by diarrhea. The production of either the emesis or diarrheal toxin is dependent on the type of the food on which the bacteria grow. The emetic toxin is a small toxin that is resistant to heat, extreme pH, and proteolytic enzymes. The toxin acts on the enteric nervous system through unknown mechanisms. The emetic toxin is associated with fried rice in the majority of cases. The emetic form of the disease has an incubation period of 2 to 3 hr and elicits symptoms of vomiting and abdominal cramps that last 8 to 10 hr.

The diarrheal toxin is a secretory cytotoxin consisting of a two- to three-component protein complex. This toxin induces secretion in the rabbit-ligated ileal loop and is cytotoxic in tissue culture. The mechanism of action of this toxin is unknown. The diarrheal toxin is associated with a variety of foods, such as sausage, vanilla sauce, and puddings. The incubation period for the diarrheal form ranges from 6 to 14 hr. The illness is characterized by diarrhea and abdominal cramps, which may last 20 to 36 hr.

G. *Staphylococcus aureus*

S. aureus is a gram-positive coccus that is among the most common causes of bacterial food-borne disease. Food poisoning usually occurs by contamination of food with infected wounds on the hands, from the normal flora of skin or from the respiratory tract of food handlers. Foods such as ham and custard, which have a high concentration of salt or sugar, provide a good growth medium for *S. aureus*. One to 6 hr after ingestion of the contaminated food, the symptoms of food poisoning begin, with severe vomiting and abdominal pain followed by diarrhea, which may last 24 to 48 hr.

Food poisoning with *S. aureus* results from ingestion of the small enterotoxins A, B, C, D, or E, which are superantigens. Enterotoxin A is the most common one associated with food poisoning. These toxins act on enteric nervous system through unknown mechanisms. As little as 100 to 200 nanogram of the toxins can cause food poisoning. The *S. aureus* toxins are resistant to heat, irradiation, pH extremes, and proteolytic enzymes. Therefore, even overcooking of the contaminated food does not prevent the food poisoning. Involvement of *S. aureus* in outbreaks of food-borne disease is confirmed by detection of enterotoxins in food and by phage typing.

H. *Clostridium perfringens*

C. perfringens is a gram-positive spore-forming organism that can tolerate aerobic conditions, unlike other members of Clostridia. *C. perfringens* type A exists in soil and in the intestinal tracts of most animals. This organism can cause a relatively mild food poisoning, more frequently in winter. Diarrhea and abdominal cramps develop 6 to 24 hr following ingestion of preformed toxin and last up to 24 hr. Toxigenic strains of *C. perfringens* usually grow on foods such as meat and poultry at temperatures between 15 to 50°C, with a doubling time as short as 10 min. The spores are heat-resistant and can survive cooking and germinate after cooling. *C. perfringens* type A produces a 35-kDa cytotoxin called *C. perfringens* enterotoxin (CPE) during sporulation. CPE binds irreversibly to cells and forms ion-permeable channels in intestinal epithelial cells and acts as a superantigen that reacts with human T cells.

C. perfringens can also cause non-food-poisoning diarrhea. The diarrhea is more severe, with blood and mucus in feces, and lasts longer. The disease occurs predominantly in the elderly or results from antibiotic therapy, similar to the cases with *C. difficile*.

Infections with *C. perfringens* type C can lead to necrotizing enteritis known as "pig-bel," a syndrome described in New Guinea related to consumption of large undercooked pork meals in native feasts. The symptoms of pig-bel include severe abdominal pain, bloody diarrhea, vomiting, and death due to intestinal perforation. The symptoms of pig-bel are associated with a toxin called β toxin. β toxin is a cytotoxin with an unknown mechanism of action. This toxin is usually inactivated by proteolytic enzymes in the intestine. However, lack of proteolytic enzymes in malnourished hosts or inhibition of these enzymes by certain foods, such as sweet potato, allows the activity of β toxin and subsequent necrotizing enteritis.

III. BACTERIAL AGENTS OF ENTERIC FEVER

A. *Salmonella typhi*

Typhoid fever is a severe systemic disease caused by *S. typhi*. The disease is characterized by fever and abdominal symptoms. Infection can be transmitted by consumption of water or food contaminated with the feces of a patient or a chronic carrier. Humans are the only known reservoir for this organism, making the studies of typhoid infection difficult. However, *S. typhimurium,* which normally causes gastroenteritis in humans, causes a disease similar to typhoid in mice. Therefore, most of the studies of typhoid have focused on infection of mice and mouse macrophages with *S. typhimurium*. However, it is not clear that all of the conclusions from studies of *S. typhimurium* in mice apply to *S. typhi* in humans.

Following oral inoculation in mice, *S. typhimurium* survives the gastric acid barrier and reaches M cells, specialized epithelial cells that cover lymphoid tissues of the small intestine. Bacteria use M cells to penetrate the intestinal mucosa, whereupon they are engulfed by macrophages. Within macrophages, *Salmonella* attenuates the acidification process and multiplies. Survival within macrophages results in spread of the organisms and systemic infection. Bacteria enter the blood through the thoracic duct. Finally, bacteria are taken up by tissue macrophages in the bone marrow, liver and spleen.

The ability to multiply in macrophages and cause systemic infections is encoded by genes that reside on a second pathogenicity island, SPI-2, at 30 min on the *S. typhimurium* chromosome. (See preceding section on nontyphoidal *Salmonella* about SPI-I.) In addition to the products of these genes, intramacrophage survival is also modulated by the PhoP/PhoQ two-component regulatory system. These proteins regulate acid phosphatase synthesis and unknown genes, which are essential for survival in the acidic environment of the macrophage.

B. *Yersinia* spp.

Within the genus *Yersinia, Y. pestis, Y. pseudotuberculosis,* and *Y. enterocolitica* are pathogenic for humans. Based on DNA hybridization techniques, *Y. enterocolitica* has recently been subdivided into eight other species. *Y. pestis* is the cause of bubonic plague, the "Black Death," which claimed one-fourth of Europe's population in the fourteenth century. *Y. pseudotuberculosis* and *Y. enterocolitica* are the causes of yersiniosis, a disease more prevalent in developed countries. Yersiniosis is characterized by an enteric feverlike illness, which is accompanied by acute diarrhea. Mesenteric adenitis is a common manifestation of the disease, which causes an acute appendicitis-like syndrome, with fever and abdominal pain. Extraintestinal manifestations can also include septicemia and nonpurulent arthritis.

Infections with *Y. pseudotuberculosis* are more common in animals and less frequent in humans. *Y. enterocolitica* is carried by healthy pigs but is pathogenic for humans. It is transmitted by ingestion of contaminated water or food, more commonly, contaminated milk. The organisms can multiply at low temperatures, such as those of refrigerated food. The infectious inoculum may be 10^9 organisms and the incubation period may last 4 to 7 days.

Following ingestion of *Yersinia,* bacteria adhere to and enter the intestinal epithelial cells. Infection then spreads to the mesenteric lymph nodes, where abscesses develop. Adherence of *Yersinia* to host cells is mediated by a plasmid-encoded adhesin, called YadA, and a chromosomally encoded protein, called Invasin. Invasin mediates entry into cells by interacting with $\beta1$ integrin receptors. This interaction leads to the extension of a pseudopod that forms a "zipper" around the bacterium, resulting in internalization.

At 37°C under low calcium conditions *in vitro, Yersinia* spp. secrete a set of proteins called Yops. Yops are virulence factors that enable bacteria to survive and multiply within lymphoid tissues of the host. The genes encoding Yops reside on a 70-kb virulence plasmid. Yops are secreted by a type III secretion system called Ysc and the secretion is regulated by temperature and contact with eukaryotic cells. Upon contact of bacteria with host cells, Yops are synthesized, secreted, and some are injected into the cytoplasm of host cells. Following injection, YopE and YopT act as cytotoxins that disrupt the actin microfilament structure. YopH is a protein tyrosine phosphatase that dephosphorylates certain proteins of macrophages. These Yops together inhibit phagocytosis by macrophages. YopP (YopJ) induces apoptosis of macrophages, which results in release of proinflammatory cytokines and subsequent inflammatory responses to infection.

Y. enterocolitica also produces an enterotoxin similar to heat-stable toxin of *E. coli,* called Yst. This enterotoxin might be responsible for cases of food poisoning caused by this organism.

IV. BACTERIAL AGENTS OF GASTRITIS

A. *Helicobacter pylori*

H. pylori is a spiral, microaerophilic gram-negative bacterium with two to six polar sheathed flagellae that endow the bacterium with a corkscrew mode of motility. It has a relatively small genome of 1.7×10^6 bp which is highly AT-rich, similar to that of Campylobacters. The complete sequence of the genome has been determined for two strains. This organism is extremely prevalent, residing in large numbers (10^8–10^{10} organisms per stomach) in the stomachs of at least half of the human population. *H. pylori* probably is not found in the environment or in animals and, therefore, person-to-person contact and the fecal–oral route are the likely means of transmission. Once inoculated, the incubation period is estimated to be 3 to 7 days and infection can last for the lifetime of the host. *H. pylori* exhibits tissue specificity exclusively for gastric mucosal epithelial cells and does not invade beyond these tissues. Detection of the organisms is best accomplished by biopsy of stomach tissue and subsequent testing for urease activity, by serologic testing, or by culture.

Infections with *H. pylori* result in acute and chronic gastric inflammation, which, when untreated, can lead to peptic ulcers or stomach carcinoma. The majority of cases of gastric and duodenal ulcers are caused by *H. pylori.* The outcome of infection with *H. pylori* depends on a variety of bacterial, host, and environmental factors.

Gastric inflammation by *H. pylori* is mediated by several virulence factors. All *H. pylori* strains produce urease in very high amounts. Urease is a nickel-containing hexameric enzyme, which catalyzes hydrolysis of urea to ammonia. Ammonia neutralizes gastric acid of the stomach, allowing the organism to colonize. Flagellae are another important colonization factor of this organism, allowing the bacteria to move along the mucous layer of the stomach. Most *H. pylori* strains produce a vacuolating cytotoxin, called VacA, which is an autotransporter protein. This cytotoxin induces acidic vacuoles in the cytoplasm of eukaryotic cells, Also, the *H. pylori* strains that are more associated with duodenal ulcer and stomach cancer carry the cytotoxin-associated gene (*cag*) pathogenicity island. Genes on the *cag* PI are required for secretion of IL-8 and tyrosine phophorylation of host proteins.

V. MISCELLANEOUS BACTERIAL ENTERIC PATHOGENS

Thorough coverage of all enteric bacterial pathogens is beyond the scope of this article. However, we briefly describe a few other common enteropathogenic bacteria in the following section:

A. Non-Cholerae Vibrios

V. parahaemolyticus, V. vulnificus, V. mimicus, V. fluvialis, V. furnissii, and *V. hollisae* reside in aquatic environments and prefer high salt and warm temperature habitats. These organisms have been implicated in intestinal infections following consumption of contaminated raw or undercooked shellfish. *V. parahaemolyticus* and *V. vulnificus* can also cause wound infections and septicemia. *V. parahaemolyticus* produces the thermostable direct hemolysin (TDH), which acts as an enterotoxin to stimulate intestinal secretion and subsequent diarrhea. Hemolysin may also be involved in tissue damage observed in wound infection. This hemolysin elicits the Kanagawa phenomenon on Wagatsuma agar, a phenotype specific to pathogenic strains.

V. vulnificus is an invasive pathogen. Eating contaminated raw oysters leads to septicemia 24 to 48 hr later, particularly in people with underlying liver disease. This organism produces several proteolytic enzymes and a hemolysin, which contribute to overt damage to epithelial cells in wound infections. *V. vulnificus* also produces a capsule, which has been associated with virulence.

B. *Aeromonas* spp.

Aeromonas spp. are widely distributed in marine environments and can be transmitted to humans via contaminated food, especially during summer. Three species, *A. hydrophila*, *A. sobria*, and *A. caviae*, cause diarrhea in humans. *A. hydrophila* produces a hemolysin called aerolysin or β-hemolysin, which, in addition to cytolytic activity, acts as an enterotoxin and induces diarrhea. It also produces other enterotoxins and cytoskeleton-altering toxins. In addition, *A. hydrophila* produces a type IV fimbria that might contribute to virulence.

C. *Plesiomonas shigelloides*

P. shigelloides has biochemical similarities to *Aeromonas* and antigenic similarities to *Shigella* spp. It is primarily a marine microorganism, which can be transmitted to humans by raw or undercooked seafood. Diarrhea, accompanied with relatively severe abdominal cramps, occurs 24 hr after ingestion of the organism. The stool may contain mucus, blood, and pus, suggesting an invasive mechanism of the disease. *P. shigelloides* produces a heat-labile enterotoxin with cytoskeleton-altering activity and a heat-stable enterotoxin with unknown mechanism of action.

P. shigelloides also causes extraintestinal infections, such as meningitis in neonates, septicemia in immunocompromised hosts, and septic arthritis.

D. *Listeria monocytogenes*

L. monocytogenes is the only species of *Listeria* that is pathogenic to humans. This organism is a gram-positive rod, which is present in soil, as part of the fecal flora of many animals, and on many foods, such as raw vegetables, raw milk, cheese, fish, meat, and poultry. Diarrhea can occur following ingestion of 10^9 organisms via contaminated food. The incubation period is long and can last between 11 to 70 days. Infections with *L. monocytogenes* are uncommon in the normal population, but in immunocompromised patients, neonates, the elderly, and pregnant women, infection can lead to encephalitis, meningitis, and stillbirth.

L. monocytogenes invades both epithelial cells and phagocytes. Entry involves a "zippering" mechanism, similar to that described for *Yersinia* entry (see preceding). Interaction between Internalin, a bacterial protein, and E-cadherin, a receptor on epithelial cells, induces phagocytosis. Once inside the phagocytic vacuole, bacteria lyse the vacuolar membrane by a hemolysin called listeriolysin O and spread within the cytoplasm. In the cytoplasm, a bacterial protein called ActA induces assembly of actin microfilaments behind the bacterium, which results in movement of bacteria in a manner similar to movement of *Shigella* inside the cytoplasm (see preceding). Bacteria move to adjacent cells and spread. Genes involved in escape from the vacuole and in intra/intercelullar spread are carried on a pathogenicity island in the *L. monocytogenes* chromosome.

E. Enterotoxigenic *Bacteroides fragilis*

B. fragilis is a gram-negative non-spore-forming anaerobic rod that comprises a part of the normal intestinal flora of nearly all humans. Enterotoxigenic *B. fragilis* (ETBG) strains produce a toxin that stimulates fluid secretion in the intestinal lumen and causes rounding of epithelial cells and loss of intestinal microvilli in an *in vivo* intestinal model. These strains have been linked to diarrhea in animals and in a small number of studies of humans.

F. *Clostridium botulinum*

C. botulinum is the cause of food poisoning, which can occur by ingestion of preformed toxins in inadequately processed food, such as home-canned vegetables and fish. Vomiting and diarrhea occur before neurological symptoms begin. The disease is caused by neurotoxins A, B, or E. Infant botulinum is another manifestation, which is different from food poisoning in that toxins are produced after germination of spores in gut. *C. botulinum* also produces a cytotoxin called C2 toxin, which can alter the cell cytoskeleton by ADP-ribosylation of G actin and preventing polymerization of G to F actin. The role of this toxin in disease is unknown.

Acknowledgments

We acknowledge Rick Blank and David McGee for careful reading of this chapter and Philippe Sansonetti, Coumarin Egile, Barry McNamara, and Ravi Anantha for providing the figures. This work was supported by Public Health Services Awards AI37606, AI32074, and DK49720 from the National Institutes of Health.

See Also the Following Articles

Cholera • Clostridia • Fimbriae, Pili • Food-borne Illnesses • Gastrointestinal Microbiology

Bibliography

Armstrong, G. L., Hollingsworth, J., and Morris, J. G. (1998). Bacterial food-borne disease. *In* "Bacterial Infections of Humans, Epidemiology and Control" (3rd ed.) (A. S. Evans and P. S. Brachman, eds.), pp. 109–138. Plenum Publishing Co., New York.

Blaser, M. J. (1995). Campylobacter and related species. *In* "Principles and Practice of Infectious Diseases," Vol. 2 (4th ed.) (G. L. Mandel, J. E. Bennett, and R. Dolin, eds.), pp. 1948–1956. Churchill Livingstone, New York.

Borriello, S. P. (1998). Pathogenesis of *Clostridium difficile* infection. *J. Antimicro. Chemo.* **41**, 13–19.

Cornelis, R. G. (1998). The *Yersinia* deadly kiss. *J. Bacteriol.* **180**, 5495–5504.

Donnenberg, M. S., Zhang, H. Z., and Stone, K. D. (1997). Biogenesis of the bundle-forming pilus of enteropathogenic *Eschericia coli*: Reconstruction of fimbriae in recombinant *E. coli* and role of DsbA in pilin stability—A Review. *Gene.* **192** (1), 33–38.

Dunn, B. E., Cohen, H., and Blaser, M. J. (1997). *Helicobacter pylori. Clin. Microbiol. Rev.* **10**, 720–741.

Guerrant, R. L. (1995). Principles and syndromes of enteric infection. *In* "Principles and Practice of Infectious Diseases," Vol. 1 (4th ed.) (G. L. Mandel, J. E. Bennett, and R. Dolin, eds), pp. 945–962. Churchill Livingstone, New York.

Lorber, B. (1997). Listeriosis. *Clin. Infect. Dis.* **24**, 1–11.

Miller, S. I., Hohmann, E. L., and Pegus, D. A. (1995). *Salmonella* (including *Salmonella typhi*). *In* "Principles and Practice of Infectious Diseases," Vol. 2 (4th ed.) (G. L. Mandel, J. E. Bennett, and R. Dolin, eds.), pp. 2013–2033. Churchill Livingstone, New York.

Nataro, J. P., and Kaper, J. B. (1998). Diarrheagenic *Escherichia coli. Clin. Microbiol. Rev.* **11**, 142–201.

Parsot, C., and Sansonetti, P. J. (1996). Invasion and the pathogenesis of Shigella infections. *Curr. Top. Microbiol. Immunol.* **209**, 25–42.

Sears, C. L., Guerrant, R. L., and Kaper, J. B. (1995). Enteric bacterial toxins. *In* "Infections of the Gastrointestinal Tract" (M. J. Blaser, P. D. Smith, J. I. Radvin, H. B. Greenberg, and R. L. Guerrant, eds.), pp. 617–634. Raven Press, Ltd., New York.

Tauxe, R. V. (1998). Cholera. *In* "Bacterial Infections of Humans, Epidemiology and Control" (3rd ed.) (A. S. Evans and P. S. Brachman, eds.), pp. 223–242. Plenum Publishing Co., New York.

Taux, R. V., and Hughes, J. M. (1995). Food-borne disease. *In* "Principles and Practice of Infectious Diseases." Vol. 1 (4th ed.) (G. L. Mandel, J. E. Bennett, and R Dolin, eds.), pp.1012–1024. Churchill Livingstone, New York.

Enteroviruses

Nora M. Chapman and Steven Tracy

University of Nebraska Medical Center

Charles J. Gauntt

University of Texas Health Sciences Center

I. Classification
II. Replication
III. Epidemiology
IV. Pathogenesis
V. Control

GLOSSARY

enterovirus A nonenveloped icosahedral virus with a single-stranded, positive-sense RNA genome; the name suggests the enteric tract as the primary site of replication.

host A cell or an organism, simple or complex, in which a virus may replicate.

positive strand polarity An RNA molecule with a sequence that can be translated by the cell's translational machinery, similar to a cellular mRNA.

RNA-dependent RNA polymerase An enzyme encoded by all picornaviruses for the synthesis of all viral RNA species.

virus receptor A structure, usually a protein, on the outer surface of the cell, that serves as a site for the adsorption of the virus to the cell and facilitates its entry into the cytoplasm.

THE ENTEROVIRUSES are a large group of RNA-genome viruses that infect animals and are classified as a genus in the family Picornaviridae (literally "small RNA viruses"). Enteroviruses are etiologic agents of numerous diseases in humans and have been isolated from a variety of animal species. The best-known enterovirus is poliovirus, the cause of poliomyelitis. Although poliovirus and its disease will soon be con- quered worldwide by the successful administration of poliovirus vaccines, there are no vaccines against any of the other 63 human enteroviruses. Among the diseases caused by nonpolio enteroviruses are myocarditis, encephalitis, meningitis, hand-foot-and-mouth disease, and pancreatitis. The three-dimensional shape of an enterovirus is icosahedral, a structural design commonly found in the virus world. Four viral proteins, each in 60 copies, make up the capsid. The structures of several enteroviruses (as well as enterovirus relatives in the Picornaviruses) have been solved at the near-atomic level, revealing intricate protein–protein interactions. Enteroviruses replicate in the infected cell cytoplasm via a negative-strand RNA intermediate and exit by lysis of the cell. Although relatively small, enteroviruses can be used as cloning vectors to express antigenic or biologically active foreign protein sequences.

I. CLASSIFICATION

The human enteroviruses make up one genus in the virus family Picornaviridae (Table I); enteroviruses have also been isolated from animals, including cows, pigs, and sheep. Enteroviruses cause or are etiologically linked to a wide variety of disease (Table II). The polioviruses (PV1-3) are expected to be completely eradicated worldwide well before the year 2010, and it is expected that further research on these viruses, as well as use of the live poliovirus vaccine, will then be proscribed. The approach of typing the enteroviruses by comparison of numerous enteroviral nucleotide sequences has suggested that

Encyclopedia of Microbiology, Volume 2
SECOND EDITION

TABLE I
Human Enteroviruses[a]

	Number of Serotypes
Polioviruses	3
Coxsackieviruses[b] A1–A22, 24	23
Coxsackieviruses B1–6	6
Echoviruses[c] 1–7, 9, 11–27, 29–34	30
Enteroviruses 68–71	4

[a] Enteroviruses belong to the family Picornavirus (Picornaviridae), members of which are characterized by icosahedral capsids, comprising four capsid proteins, that contain small single-stranded, positive-sense RNA genomes. Realignment to contain two new groups of viruses that contain echovirus 22 and Aichi virus may be forthcoming.

[b] Coxsackieviruses were named after Coxsackie, NY, the town in which the child lived from whom the first such virus was recovered.

[c] ECHO stands for enteric cytopathic human orphan. The designation as "orphan" reflects the initial belief that these were agents without an associated disease.

TABLE II
Common Human Diseases Caused by Enteroviruses[a]

Poliomyelitis
Encephalitis
Aseptic meningitis
Hand-foot-and-mouth disease
Herpangina
Exanthema
Febrile illness
Acute hemorrhagic conjunctivitis
Meningoencephalitis
Inflammatory heart disease (peri-, myo-, and endocarditis)
Pancreatitis
Myositis
Pleurodynia (Bornholm disease)
Pharyngitis
Common cold-like symptoms

[a] Some viruses cause clearly defined diseases (such as poliovirus and poliomyelitis), whereas the majority of enteroviruses are implicated or known to cause several types of diseases (such as the CVBs and heart disease, pancreatitis, and meningitis).

the enteroviruses might best be grouped according to their genetic and inferred evolutionary relationships.

A. Isolation and Propagation

Enteroviruses are frequently isolated from fecal samples for clinical identification, although they also can be obtained from throat swabs, as well as from various biopsy or autopsy tissue samples. The growth of enteroviruses in established or primary cell cultures is not guaranteed; many enteroviruses serotypes from the CVA and EV do not replicate well, or at all, in a cell culture. Several cell lines can be used for the detection of enteroviral cytopathic effects (cpe) or the propagation of most enteroviruses; these are human foreskin WI-38, HeLa, the human colon cancer line CaCo-2, and the buffalo green monkey kidney (BGMK) cell line or primary monkey kidney cells. Human RD (rhabdosarcoma) cells are often the host of choice for several group A coxsackieviruses and echoviruses. Some enteroviruses multiply in cell lines and do not induce cpe that is observable using light microscopy. Coxsackie A viruses can be propagated in suckling mice. Because

of these limitations on the certainty of replication in cell cultures and because of the cost of the assay in terms of money and time required, molecular approaches to the detection and identification of enteroviruses have been and are being developed.

B. Identification

The serologic typing of enteroviruses is carried out using pools of polyclonal serotype-specific anti-virus neutralizing antibodies; based on a specific pattern of neutralization using such pools, the serologic identity of the virus can be ascertained. Each distinct enterovirus (for example, coxsackievirus B3 or echovirus 22) is characterized as a serotype; an antiserum raised against one serotype of an echovirus, for example echovirus 22, in a naive animal does not neutralize other enteroviruses. Powerful, rapid, and relatively inexpensive molecular approaches such as reverse transcriptase (RT)-mediated amplification (generally using the polymerase chain reaction, PCR) of a specific genomic sequence are increasingly being linked with nucleotide-sequence determination to genetically type viruses. Involvement of enteroviruses in specific disease can best be assessed by assaying tissue samples obtained by biopsy, from transplanted tissue, or at autopsy; in this way, for example, the group B coxsackieviruses have been linked clearly to the causation of acute inflammatory heart disease. The interpretation of titers of circulating anti-viral antibodies against a specific virus can be complex because of frequent heterotypic responses generated within a group (e.g., group B coxsackieviruses) and are at best inferential.

C. Physical and Structural Characteristics

1. *General Characteristics*

Enteroviruses are approximately spherical, 28–30 nm in diameter. One milligram of virus contains about 7×10^{13} virions. Enteroviruses are stable to pH 3, a factor that permits enteroviruses to pass from the stomach into the intestine following oral infection via contaminated hands or objects, and aerosol, food, or water intake. Enteroviruses are not inactivated in the presence of ether or 70% ethanol, a common disinfectant, but are readily inactivated by chlorine (bleach) or iodine-containing solutions, UV irradiation, or heat (50°C or more). However, organic matter (such as in sewage) can protect enteroviruses from chlorine.

2. *Structure of Enteroviruses*

When studied at the near-atomic level, enteroviral capsids show an icosahedral symmetry. Enteroviruses encode four capsid proteins, termed proteins 1A–D (formerly called VP4, VP2, VP3, and VP1, respectively). Capsid proteins are arranged in protomers, each protomer containing one each of the four capsid proteins. Five protomers in turn form a pentamer, 12 of which form the icosahedral capsid structure; thus, each enterovirus capsid contains 60 copies of each of the four capsid proteins. The proteins 1B–D are prominent on the exterior surface of the capsid, while protein 1A is highly conserved and is entirely internal in the capsid structure. The icosahedral structure involves twofold, threefold, and fivefold axes of rotation. The enterovirus pentamer is structured about the fivefold axis of symmetry, and pentamers abut each other at twofold and threefold axes. Around the apex of each pentamer can be found a depression that has been termed "the canyon." This structural element was first proposed by Michael Rossmann at Purdue University to be the site of cell-membrane receptor molecule interaction with the virus particle, subsequent work based on the mutational analysis of the amino acids lining the canyon, as well as structural studies, has provided strong supportive data for this canyon hypothesis. The derivation of the near-atomic structure for coxsackievirus B3 described a significant planar region astride the twofold axes of symmetry that was more prominent than in other enterovirus structures, this planar region has also been postulated as a potential binding site for a host-cell receptor protein. Recent research identifying the proteins that are involved as cellular receptors, as well as their function and interaction with enterovirus virions, suggest that viruses may will need more than a single cellular molecule in order to efficiently enter cells (Table III).

Neutralizing anti-enterovirus antibodies bind to specific sites on the exterior of the virus capsid. Although the sites (epitopes) to which neutralizing antibodies bind are found on the external enterovirus capsid surface, they are not always on the most prominent protrusions. The immune systems of different animals also identify different sites against which to direct immune responses; for example, human and murine anti-poliovirus neutralizing antibodies are directed against differing epitopes on the poliovirus exterior. As for many amino acid sequences recognized as antigenic by B-cells, some T-cell epitopes may often be found located close to or overlapping the same epitopes. In addition, specific amino acid sequences that are recognized by the T-cell responses are also be found in many other (nonstructural) enterovirus proteins.

II. REPLICATION

A. Binding of the Virus to a Cell-Surface Receptor

Enteroviruses initiate the process of entry to a host cell by binding to one or more specific host-cell membrane proteins (called virus receptors; Table

Polioviruses	Human poliovirus receptor
Group B coxsackievirus	CAR
Enterovirus 70	DAF (CD55)
Echovirus 7	DAF (CD55)
Echovirus 1	VLA-2
Group A coxsackievirus type 6	Alpha v beta 6 integrin

[a] Abbreviations: CAR (coxsackievirus adenovirus receptor); DAF (decay accelerating factor); VLA (very late antigen).

III). It is intuitively obvious that the specific expression of receptors by cells is a primary determinant of virus species restriction to a cell type and also of host tropism. The six stereotypes of the group B coxsackieviruses may be studied in mouse and in humans because the human and mouse receptors for these viruses have been recently shown to be nearly identical; however, the closely related polioviruses cannot replicate naturally in mice due to the lack of a recognizable receptor. But there is no block to poliovirus replication in mouse cells if the correct poliovirus receptor is supplied: when the cDNA encoding the human poliovirus receptor is expressed in mouse cells or introduced into the murine genome in transgenic mice, such cells or mice can be infected productively with human polioviruses. While several receptors for enteroviruses, such as that for the polioviruses and the group B coxsackieviruses, cluster in the immunoglobulin superfamily of proteins, other proteins that serve as receptors have been characterized (Table III). The productive binding to the cell or binding to the receptor changes the conformation of the enterovirus capsid, making the virion no longer infectious if it dissociates from the receptor. The amino terminus of the capsid proteins 1-D and 1-A are externalized on binding to the receptor protein. The precise mechanism by which the viral RNA leaves the capsid and enters the cell cytoplasm is not understood.

B. Translation of the Viral Genome

The first major event in the replication cycle of positive-strand RNA viruses such as the enterovir-

uses is to translate the viral RNA genome into viral proteins. The 11 enterovirus proteins (four structural, seven nonstructural) are translated from a single open reading frame (ORF). Preceding the ORF is a 5′ nontranslated region (NTR) that represents about 10% of the genome length. Picornaviruses (the family to which the enteroviruses belong) share an unusual form of initiation of translation. The normal mode of translation initiation in eukaryotes is that a "cap structure" (m^7Gppp(5′)N) interacts with eIF-4E (a cap-binding protein found in cells) to form the 43S pre-initiation complex; the ribosome then scans to the first initiation codon in an acceptable context. However, within the enteroviral 5′ NTR, there is a region termed the internal ribosome entry site (IRES). Enteroviruses form a translation complex within the IRES, as opposed to the 5′ terminus of the RNA, and translation of the viral ORF begins at an initiation codon approximately 740 nucleotides downstream from the 5′ terminus of the uncapped viral genomic RNA. Several AUG codons are found upstream of this authentic start site of translation, but none of them function to initiate translation. The IRES region (found in the enteroviral 5′ NTR approximately from nucleotides 120–600) contains conserved nucleotide sequences and secondary (RNA base-pairing) structures to which a variety of host proteins have been shown to bind; these proteins include the La autoantigen, the polyprimidine tract-binding protein, and binding protein 2. Host proteins such as these are, very likely, crucial for the correct initiation of translation of the enteroviral genome at the outset of replication.

As the ribosome proceeds along a positive strand of viral RNA, the nascent polyprotein is processed by two self-cleaving viral proteases to produce the structural and nonstructural proteins. The four capsid proteins (making up the P1 region of the genome) are cleaved from the nonstructural protein regions P2 and P3 by the action of protease 2A, which cleaves at its own amino terminus. The proteolytic activity of 3C or 3CD cleaves all other processing sites in the viral polyprotein with the exception of an autocatalytic cleavage of the final capsid precursor, VP0, to VP4 and VP2 that occurs within the newly assembled virus particles.

C. Transcription of the Viral RNA

Once translation of the viral genome has begun, the incoming (viral) positive strand of RNA must be transcribed into negative-strand copies. From these negative-strand copies, multiple complete positive-strand RNA molecules will be transcribed for translational purposes and for encapsidation as genomes in new virions of the next virus generation.

Enterovirus RNA transcription occurs at membranous vesicles in replication complexes containing several viral proteins. The positive-strand RNA serves not only as the mRNA for the viral proteins, but also for the synthesis of the (negative-strand) antigenome. Positive-strand RNA is normally produced greatly in excess of negative-strand RNA. RNA structures (secondary and tertiary) in the short are important for the transcription of negative-strand RNA. The first 100 nucleotides of the 5′ NTR form distinct secondary structures (commonly termed the cloverleaf) and contain primary sequences important for the initiation of transcription of the positive strand from the negative-strand template. Viral protein 3D is the RNA-dependent RNA polymerase. The data support a model of enterovirus RNA replication that requires several viral proteins for successful transcription. The viral protein 3CD forms a complex with viral protein 3AB and the 3′-end of viral RNAs to initiate transcription. This intermediate is then processed by the 3C protease to yield a 3D–3AB polymerase complex. Progeny viral RNAs have a covalently linked viral protein (VPg or 3B) at the 5′-end. Viral proteins 2C (which contains helicase motifs), 2BC, and 2A also have vital functions in RNA replication. Host-cell factors are also involved. The poly (rC) binding protein 2 is a host protein of the RNA-replication complex bound to the 5′ NTR cloverleaf and is also an essential host factor for enterovirus IRES-dependent translation.

Positive-strand genomic RNA becomes associated with the newly synthesized capsid proteins in the membrane-bound replication complexes the infected cell cytoplasm. This requires processing of the P1 polyprotein (produced by cleavage at the VP1-2A site by protease 2A) to the capsid proteins 1AB, 1C, and 1D by the 3CD protease; these three proteins form the 5S protomer. Five such protomers interact to form the 14S pentamer, which then associates with other pentamers and with positive-strand viral RNA to form capsids containing 60 copies of each viral protein. Upon assembly with viral RNA, an autocatalytic cleavage processes most of 1AB to 1AB to 1A and 1B to yield infectious particles. The 1AB cleavage is not completely efficient; some 1AB can usually be found among the proteins extracted frompurified enterovirus particles. The interior of the capsid can accommodate a range of sizes of RNA molecules. Using recombinant-DNA engineering techniues, enterovirus RNA molecules that range in size from 80–110% of the wild-type genome have been encapsidated in infectious virions. Indeed, the entire capsid protein (P1) region can be excised and substituted with RNA that encodes one or more foreign proteins; such RNAs are encpsidated using a packaging system that supplies the viral capsid proteins in *trans*. A genomic RNA packaging signal has yet to be identified. Recent work suggests, however, that some recognition exists as RNA molecules from one enterovirus are packaged only inefficiently by capsid proteins of other enteroviruses.

III. EPIDEMIOLOGY

A. Enteroviral Diseases

The 66 serotypes of enteroviruses cause or are implicated as causative agents in numerous human diseases (Table II). Infections with group B coxsackieviruses are often associated with cariovascular symptoms; of the cases of acute inflammatory cardiomyopathy (myocarditis) or dilated cardiomyopathy investigated for the presence of enteroviral RNA, approximately 20–25% are positive (this translates to about 5000–6000 cases per year in a population the size of the United States). Roughly two-thirds of aseptic meningitis cases (>7000 cases annually in the United States) are also positive for enteroviruses. Enteroviral disease outbreaks, such as pleurodynia, hemorrhagic conjuncts, and hand-foot-and-mouth disease, are occasionally seen. Nosocomial infections, particularly in neonates, occur frequently and can be fatal.

The predominant fecal–oral mode of transmission increases the chance that children will be infected and that household members of the infected child will also be infected. Enteroviruses are commonly transmitted through the contamination of hands and objects. Enteroviruses can also be transmitted by the contamination of food or water supplies, and the presence of enteroviruses in sewage and in shellfish exposed to sewage is well documented. Neonatal enteroviral infection may be transmitted *in utero* or at birth by blood-borne contact with the infected mother and can be the cause of severe illness in the child. Infected asymptomatic individuals are often sources of an enterovirus infection. Although transmission of enteroviral diseases from animals is exceedingly rare, swine vesicular disease virus (SVDV) and coxsackievirus B5 (CVB5) are closely related viruses, and humans can be infected with SVDV. It is likely that SVDV evolved from CVB5.

Enteroviral diseases tend to occur seasonally in temperat climates. In the United States, the majority of enteroviral illness tends to occur between June and October. In tropical countries, there is a greater incidence during the rainy season. The relative abundance of nonpoliovirus enteroviruses in sewage or the number of clinical isolates correlates with increased relative humidity.

IV. PATHOGENESIS

A. Diseases Caused by Enteroviruses

Most (perhaps ~90%) enterovirus infections in humans are asymptomatic and do not lead to disease. Enteroviral disease in humans is generally not the enteric tract, as the virus's classification (entero) would imply, but rather are transmitted in a fecal–oral transmission route, from feces-contaminated water, food, and inanimate objects (fomites: e.g., parents changing an infant's diaper might be exposed to the virus in the urine and feces) to other humans. Enteroviral diseases range from mild illness in the upper respiratory tract, pharyngitis (sore throat), conjunctivitis, and rashes (herpangina and hand-foot-and-mouth disease) to severe diseases, particularly in infants and small children, such as aseptic meningitis, inflammatory heart disease, and pancreatitis. There is a reasonable inferential body of evidence from serological, epidemiological and animal studies to suspect enteroviruses, most likely the group B coxsackieviruses, as possible etiologic agents of type 1 (insulin-dependent) diabetes; this etiologic link has not been confirmed, however.

B. Pathogenesis of Infection

An enterovirus infection that results in any pathological changes in the host, whether or not clinically recognized disease occurs, requires at minimum a susceptible host and an enterovirus strain whose genotype (the total genetic information of the virus) is capable of setting in motion the molecular changes in the infected cell that ultimately lead to pathologic changes in tissues. The infection of an individual in one animal species, including humans, by one of many strains of these viruses may not result in disease, whereas infection of another individual within that species or individuals in a different species by the same viral strain might produce serious disease and death. Susceptibility or resistance to enteroviral induction of acute disease is ultimately a consequence of innate and specific immune responses to the infection. These characteristics reflect the genetic background of the individual or strain of animal infected, and the sex and age of the individual infected. For example, reproducible experimental induction of acute inflammatory heart disease (myocarditis) in inbred mice by CVB3 generally requires mice between 3 weeks (just weaned) to about 6 weeks (young adult) of age for most murine strains challenged with cardiovirulent strains of CVB3; some strains are fairly resistant to acute disease, while others become extremely ill. In the outbred human population, however, the results of any infection are far more difficult to predict. The transition from acute to chronic myocarditis in CVB3–mouse models correlates significantly with the genetics of the mouse. There may also be a genetic component in humans, although it is not understood. Most murine strains and the majority of humans that develop CVB-induced acute inflammatory cardiomyopathy resolve the disease, whereas relatively few murine strains and a small proportion of humans (estimated at 10%) make the transition to chronic disease. CVB3 infections of immunosuppressed or immunodificient mice (and humans; e.g., echovirus infection of the central nervous system in agammaglobulinemic patients) can result in persistent infections during which virus can be excreted for weeks to months.

The molecular basis in the host for these significantly different outcomes is not known, but the host's response to infection with the production of soluble regulatory molecules such as cytokines, chemokines, prostaglandins, and sex hormones certainly plays a major role in suppressing (and resolving) the acute disease or, in some cases, in participating in continued (or chronic) inflammation. For example, female mice are generally more resistant than males to CVB3-induced cardiopathology. Another contributing factor in sustaining disease in some strains or individuals may be the persistence of the CVB3 genome in infected cells that results in the production of a few viral proteins that chronically induce pro-inflammatory mediators. Molecular mimicry, that is, the sharing of similar antigenic sequences (epitopes) between viral and cellular proteins, can contribute to the induction of chronic disease via autoimmune cell-mediated and antibody responses.

The infection of cells in an organ that leads to the death of the cells or the dysregulation of the critical functions of the cells in that organ begins with tropism (the predilection of a virus for infecting that particular cell). In most cases, tropism reflects the capacity of a virus for binding (absorbing) to a specific receptor on the cell's plasma membrane. These cellular receptors normally have interactive functions for binding soluble mediators or providing contacts with other cells to maintain homeostasis in the organ or tissue. Viruses have evolved to bind these receptors and gain entrance into the cell, where they carry out replication of their component parts or assembly into new virus particles that continue the infection. In cell infections that lead to minor acute disease that the host resolves by innate and specific immune responses, most often it is the superficial epithelial cells on mucous membranes of the respiratory or gastrointestinal tracts, the endothelial cells lining blood vessels, or the cells of the reticuloendothelial system that are destroyed by the viral infection. Recovery occurs because the body can rapidly replace such cells. In serious enterovirus disease involving, for example, the heart, pancreas, or central

nervous system, the infected cells that die may not be able to regenerate or regenerate sufficiently rapidly; for example, it is generally believed that humans are born with their total number of myocytes (beating cells) in the heart. Data also suggest that some regeneration may occur in the heart, but the extent of such regeneration is unclear. Infections of different cells may result in a quick (hours to a day) death of the cells or prolonged infections that may last for days to weeks without apparent cytopathology. The infection of some remodeling cells within these organs, such as fibroblasts or macrophages, may not result in any obvious cytopathology or cell death for weeks, and thus serve as a focus for re-infection and inflammatory responses for months. Most textbooks state that enteroviruses are highly cytopathic viruses, replicating quickly in host cells and destroying these cells within a few to a dozen hours. While in this statement is true for some cell cultures used in the laboratory for virological research and for clinical diagnosis of enteroviruses infections and in the host, such as infections of superficial cells described here, it is not true for all cells of the body or in culture. There are numerous examples of enterovirus infections of cells that result in long-term infections that permit the release of viruses for days to months without cell destruction. A detailed study of infections of long-lived cells in culture is difficult because most such cells are difficult to propagate for any significant length of time in the laboratory.

C. Enteroviral Genetics of Virulence

Enteroviruses are predominantly benign viruses, which is to say the likelihood of a serious disease resulting from an enteroviral infection is low. However, some strains of enteroviruses are indeed more virulent (able to cause serious disease) than other strains of the same serotype; this difference in phenotype relates directly to the viral genotype. RNA viruses exist as mixed populations and are easily able to adapt to new environments due to the RNA polymerase's inherently poor ability to correct errors during the transcription of the viral RNA. These mixed populations of virus strains have been termed quasispecies, in that they are all closely related but are all different from each other. It is incorrect to think of any virus population as "a virus," for indeed the population is not generally homogenous.

The viral genetics that determine whether a poliovirus isolate, for example, is naturally highly neurovirulent (has a high probability of causing poliomyelitis) or much less so (effectively a naturally occurring attenuated or vaccine strain) have not been determined, nor are the genetics of naturally occurring virulence well understood for any of the other enteroviruses. The primary molecular basis for the artificial attenuation of neurovirulence in the Sabin vaccine strains of polioviruses was elucidated in the 1980s. For each of the three poliovirus serotypes, a single nucleotide transition within the same oligomeric region of the 5 NTR has been demonstrated to be primary site that determines whether the strain is attenuated or neurovirulent. A change in two amino acids in two of the capsid proteins determines virulence in a strain of CVB4 that was repeatedly passaged in mice until it was virulent for the pancreatic acinar tissue. An artificial but strongly attenuating mutation in the 5′NTR of CVB3 has been identified as well. Recent data suggest the 5′ NTR may also play a significant role in determining the natural genetics of cardiovirulence in the CVBs. The study of the genetics of viral virulence remains a complex molecular problem.

V. CONTROL

A. Environmental Factors

Sanitation is important in controlling enterovirus spread. Contamination of hands or objects with virus from feces followed by ingestion (fecal–oral) is the primary mode of transmission. Consequently, disinfection and hand washing are critical to limiting enterovirus infections. Hand washing can decrease enterovirus levels by 100- to 10,000-fold. The treatment of contaminated objects with 5% sodium hypochlorite (bleach) or 1% glutaraldehyde disinfects effectively; however, the use of alcohol as a disinfectant is of limited benefit. Treatment of sewage and sludge to inactivate enteroviruses is necessary when this material is to be released into aquatic enterovironments or used in agriculture. Lack of treatment can result in the sewage contamination of waters in which the number of enteroviruses per milliliter can be as high as 1,000,000, approaching the titers of viruses obtained in laboratory cell cultures. Because considerable variation is seen in the genomic sequences of isolates of a serotype from different times and locations, genotyping can be used to track circulating strains of an enterovirus. The use of RT-PCR has increased the utility of this type of analysis. Because enteroviruses are excreted in feces, the examination of sewage or water supplies for enteroviruses can be used as an indicator of their circulation within a population.

B. Vaccines

Poliovirus infection has been nearly eradicated worldwide through the use of inactivated Salk and attenuated Sabin poliovirus vaccines. The great success of both types of poliovirus vaccines strongly suggests that other enteroviral diseases may be successfully prevented by vaccines for nonpoliovirus enteroviruses. Inactivated and live attenuated vaccines against CVBs have successfully generated protective immunity to CVB-induced disease in animal models and support the notion that such enteroviruses vaccines will also allow the elimination of significant enteroviral diseases. Because there are 63 nonpolio enteroviruses, vaccines against all human enteroviruses will be much more difficult to administer. However, one approach might be to target those enterovirus serotypes commonly associated with clinically serious diseases. For example, vaccines against the CVBs (likely to cause much of the enterovirus-induced myocarditis and dilated cardiomyopathy), against those, echovirus serotypes that are most often isolated from cases of aseptic meningitis, and against enterovirus 70 and coxsackievirus A24 (causes of hemorrhagic conjunctivitis) would eliminate a large number of deaths, serious or lingering illness, and medical expense due to these illnesses (e.g., inappropriate treatment, loss of working days, and secondary infections). Another approach may be to genetically engineer chimeric enteroviruses that can express antigenic epitopes of other enteroviruses. Because enteroviral diseases often have a rapid progression, vaccination is more likely to be effective in reducing the incidence of disease than is the treatment of diagnosed disease with antiviral compounds. It is clear that vaccination against enteroviruses is an extremely effective approach to preventing diseases caused by the targeted virus.

C. Antiviral Drugs

Anti-enterovirus compounds have not enjoyed great success. A number of different antivirals are being and have been tested in animal models for effectiveness in reducing enterovirus (and rhinovirus) replication. One well-studied group of antivirals bind in a pocket-like structure the enteroviral capsid beneath the canyons in these viruses, thereby inhibiting conformational changes subsequent to binding the receptor, which are thought to be necessary for virus uncoating. The use of these specific compounds has been limited due to a significant variation in the natural susceptibility of different viral strains and serotypes to the compounds, as well as the viruses ability to adapt to the presence of antiviral drugs by selection of dominant viral quasi-species that are resistant to the drugs. Intravenous immuoglobulin treatment can also reduce or prevent enteroviral disease in adults and children, and it has been particularly helpful in treating chronic enterovirus infection in agammaglobulinemic patients.

See Also the Following Articles

Food-borne Illnesses • Gastrointestinal Microbiology • Vaccines, Viral • Water, Drinking

Bibliography

Ansardi, D. C., Porter, D. C., Anderson, M. J., and Marrow, C. D. (1996). Poliovirus assembly and encapsidation RNA. *Adv. Vir. Res.* **46**, 1–68.

Belsham, G., and Sonenberg, N. (1996). RNA-protein interactions in regulation of picornavirus RNA translation. *Microbiol. Rev.* **60**, 499–511.

Minor, P., Macadam, A., Stone, D., and Almond, J. (1993). Genetic basis of attenuation of the Sabin oral poliovirus vaccines. *Biologicals* **21**, 357–363.

Palmenberg, A. C., and Sgro, J.-Y. (1998). Topological organization of picornaviral genomes: Statistical prediction of RNA structural signals. *Sem. Virol.* **8**, 231–241.

Racaniello, V. (1996a). Early events in poliovirus infection: Virus-receptor interactions. *Proc. Natl. Acad. Sci. U.S.A.* **93**, 11378–11381.

Rancaniello, V. (1996b). The poliovirus receptor: A hook, or an unzipper? *Structure* **4**, 769–773.

Robart, H. A. (ed.) (1995). "Human Enteroviral Infections." ASM Press, Washington, D.C.

Tracy, S., Chapman, N. M., and Mahy, B. W. J. (eds.) (1997). Current Topics in Microbiology and Immunology, Vol. 223, "The coxsackie B Viruses." Springer-Verlag, Heidelberg.

Xiang, W., Paul, A. V., and Wimmer, E. (1998). RNA signals in entero- and rhinovirus genome replication. *Sem. Virol.* **8**, 256–273.

Enzymes, Extracellular

Fergus G. Priest

Heriot Watt University

GLOSSARY

chaperone A cytoplasmic protein that maintains a protein destined for secretion in a secretion-compatible configuration.

periplasm The space between the cytoplasmic and outer membranes in gram-negative bacteria and between the cytoplasmic membrane and cell wall in gram-positive bacteria.

preprotein A protein bearing a signal peptide prior to maturation by signal peptidase.

signal peptide An extension to a protein, normally N-terminal and 15–40 amino acids long, that is responsible for secretion of the protein across the membrane.

Sec protein A protein involved in the secretory apparatus of an organism.

temporal regulation The derepression of enzyme synthesis that occurs as a batch culture enters stationary phase.

EXTRACELLULAR ENZYMES are generally secreted by bacteria and molds in order to hydrolyze high-molecular-weight molecules that are in the environment and that, in their native form, would be too large to enter the cell. The resultant products are then assimilated by the microorganism as sources of nutrients. Consequently, extracellular enzymes are commonly secreted by soil microorganisms such as filamentous fungi and bacteria of the genera *Bacillus* and *Streptomyces*. The enzymes are responsible for the hydrolysis of plant polysaccharides, including cellulose, lipids, nucleic acids, starch, and plant cell-wall components. The high yields of these enzymes in culture fluids and their robust properties, especially their tolerance to extremes of pH and temperature, led to their application in various industries, particularly in starch- and food-processing industries, where they replaced several unsatisfactory chemical processes, and in household laundry detergents.

I. LOCALIZATION OF ENZYMES

Extracellular enzymes are generally defined as enzymes that have crossed the cytoplasmic membrane of the cell. This definition includes proteins attached to the outer surface of the membrane because such molecules have at least initiated the export process. The final destination of an extracellular enzyme, therefore, depends on the cell structure. Gram-positive bacteria and fungi are surrounded by a thick cell wall comprising peptidoglycan or chitin-glucan, respectively. There is no apparent compartmentation in these walls, so enzymes either are released from the outer surface of the membrane and subsequently move through the wall to accumulate in the surrounding environment (truly extracellular enzymes) or are anchored to the membrane (membrane-bound enzymes). However, evidence from electron microscopy and physiological studies points to a region external to the membrane in gram-positive cells that is the equivalent of the gram-negative periplasm.

Encyclopedia of Microbiology, Volume 2
SECOND EDITION

Gram-negative bacteria are surrounded by two hydrophobic barriers, the cytoplasmic and the outer membranes, between which lies the periplasm. Enzymes may therefore be located on or in the cytoplasmic membrane, in the periplasm, or in the outer membrane of these bacteria. The periplasmic enzymes are commonly considered to be the equivalent of the membrane-bound enzymes of the gram-positive cell, which are often released by protoplasting, and, indeed, many similar enzymes such as alkaline phosphatase and maltodextrin-hydrolyzing enzymes are found in these locations. Although gram-negative bacteria are often considered to seldom secrete enzymes, bacteria such as *Erwinia, Pseudomonas,* and *Serratia* secrete large amounts of extracellular enzymes to the external environment.

Studies have revealed some interesting variations on the localization of extracellular hydrolases. Some gram-positive (e.g., *Clostridium thermocellum*) and gram-negative (e.g., *Bacteroides* species) celluloytic bacteria synthesize large numbers of protuberances on the cell surface called cellulosomes. The cellulosome comprises at least 14 enzymes (mostly endocellulases; no exohydrolases), assuming a total molecular weight of about 2 MDa. An attachment factor in the cellulosome is responsible for the attachment of the bacterium to the insoluble cellulose fiber and, in providing close contact, enables cellulose hydrolysis by the endocellulases.

II. EXTRACELLULAR ENZYMES: THEIR CHARACTERISTICS AND APPLICATIONS

A. Starch-Hydrolyzing Enzymes

Starch comprises a combination of two polysaccharides: amylose, which is a linear chain of 1,4-α-linked glucose residues, and amylopectin, which is a branched molecule of amylose chains linked by 1,6-α branch points. On average, amylose comprises chains of about 10^3 glucose residues, whereas amylopectin contains about 10^4–10^5 residues in chains of about 20–25 residues in length. The proportion of the two molecules in starches varies depending on the source, but is generally in the region of 70% amylopectin. Related α-glucans are glycogen, which

is a highly branched form of amylopectin, and pullulan, which is a linear molecule of maltotriose units joined by 1,6-α linkages and is derived from the mold *Aureobasidium pullulans.*

1. α-Amylase

α-Amylase is the most common of microbial starch-hydrolyzing enzymes. It is an endo-acting enzyme that hydrolyzes the internal 1,4-α bonds in amylose, amylopectin, and glycogen. This results in the rapid reduction in viscosity and iodine-staining power of starch and the slow release of reducing sugars. Most α-amylases can be grouped into two broad classes, the liquefying enzymes and the saccharifying enzymes. The former hydrolyze amylose to oligosaccharides of five or six glucose residues with small amounts of maltose and glucose. The best studied examples derive from *Bacillus*, in particular *Bacillus amyloliquefaciens* (a close relative of *Bacillus subtilis*), *Bacillus licheniformis,* and the thermophile *Bacillus stearothermophilus.* Saccharifying enzymes conduct a more extensive depolymerization of amylose, resulting in large amounts of glucose, maltose, and maltotriose. *B. subtilis* secretes a saccharifying α-amylase, but the industrial enzymes are produced from fungi such as *Aspergillus oryzae.*

α-Amylases are calcium metalloenzymes, generally requiring at least one atom of this metal per molecule. In the presence of calcium, they are stable to extremes pH values with optimal activity occurring between pH 4.8 and 6.5 (exceptional enzymes from acidophilic or alkaliphilic bacteria have more extreme pH optima). The fungal enzymes usually have the lower pH optima. Most have molecular masses of about 50 kDa, the fungal (but not the bacterial) enzymes being glycoproteins. Although many α-amylases are rapidly denatured above 50°C in their pure state, the liquefying enzymes from some bacilli are exceptionally thermostable. For example, the optimum operating temperature of the α-amylase from *B. licheniformis* is 90°C, and it can be used industrially at temperatures in excess of 100°C.

2. Glucoamylase

Glucoamylase (also called amyloglucosidase) also hydrolyzes the 1,4-α linkages in amylose chains, but it does so by attacking consecutive bonds starting at the nonreducing chain end. It is therefore an exo-acting enzyme, and it releases glucose residues in their α-anomeric form. Like other exo-attacking enzymes, it reduces the viscosity and iodine-staining power of starch slowly, but produces a rapid release of reducing power. Glucoamylase will also hydrolyze the 1,6-α branch points in amylopectin, although it does this inefficiently. Some bacteria secrete glucoamylases but the industrial enzymes are produced from *Aspergillus* and *Rhizopus* species. Like other fungal amylases, the enzymes are thermolabile (optimum temperature about 60°C), have low pH optima (about 4.5–5), and are glycoproteins containing about 5–20% carbohydrate. Molecular masses range from 27–112 kDa.

3. β-Amylase

Several bacteria such as *Bacillus cereus, Bacillus megaterium,* and *Bacillus polymyxa,* as well as some clostridia, secrete β-amylase. This enzyme attacks amylose in an exo-fashion, hydrolyzing alternate 1,4-α linkages starting at the nonreducing end of the chain and releasing maltose in the β-configuration. It does not hydrolyze the 1,6-α bond; therefore, the end-product from amylopectin hydrolysis is a large limit β-dextrin. These enzymes are rather thermolabile and are not produced on an industrial scale. However, it is required for maltose production, and this has prompted the search for alternative enzymes. This has resulted in the discovery of maltogenic α-amylases from some bacteria, that is, enzymes that are exo-acting, but that produce maltose in the α-configuration from starch.

4. Debranching Enzymes

Enzymes that hydrolyze the 1,6-α bonds in amylopectin, glycogen, and pullulan are called debranching enzymes. The two major classes are isoamylase, which hydrolyzes glycogen but not pullulan, and pullulanase, which hydrolyzes the 1,6-α linkages in both amylopectin and pullulan, but which has low activity on glycogen. The hydrolysis of amylopectin by pullulanase yields amylose, and the hydrolysis of pullulan produces maltotriose. Pullulanases are common in bacteria and are produced commercially from *Bacillus acidopullulyticus.*

5. Applications of Starch-Hydrolyzing Enzymes

A major use of these enzymes is in the starch-processing industry, where they have now replaced the traditional acid hydrolysis of starch. The basic process begins with liquefaction. A starch–water slurry (30–35% dry solids) is heated at 105–110°C to burst the granules and release the starch into solution. Thermostable α-amylase (from *B. licheniformis*) is added both before (to reduce viscosity) and after this gelatinization step during the liquefaction stage in which the partial hydrolysis of the highly viscous starch solution to maltodextrins of about 40 glucose units is achieved. Liquefaction at the high temperatures that are used (about 85°C) is rapid (generally within 2 hr) and the degree of hydrolysis can be varied according to requirements by adjusting the period of hydrolysis or amount of enzyme used. The dextrins may be dried and used at this stage, but usually the material is processed further.

Saccharifying α-amylases and glucoamylase are used to produce syrups from the liquefied starch in a process termed saccharification. These fungal enzymes are relatively thermolabile, and so the process is conducted at lower temperatures and for longer periods (e.g., 55°C for about 40–96 hr). Moreover, the pH has to be reduced to the more acid optima of these enzymes. High-maltose syrups are produced using fungal α-amylase and contain about 50% maltose, the residue being some glucose, maltotriose, and α-limit dextrins. The highly maltogenic α-amylases have been introduced (in conjunction with a debranching enzyme) for the production of syrups comprising maltose almost entirely. These have application in the brewing industry because they resemble malt hydrolysates more closely than do other syrups, and their low hygroscopicity and resistance to crystallization makes them attractive to the confectionery and food industry.

High-conversion syrups are produced from liquefied starch by hydrolysis with fungal α-amylase and glucoamylase. These syrups comprise about 40% glucose, 45% maltose, and the remainder maltotriose. They are used extensively in the brewing, baking, and soft drink industries. Varying the ratios of amylase and glucoamylase gives different proportions of end-products, but the maximum glucose concentration is about 43%—above this crystallization becomes a problem.

A major development in the starch-processing industry in recent years has been the introduction of high-fructose syrups. Because fructose tastes about twice as sweet as glucose, it can be used to replace sucrose in foods and beverages, and provides the same sweetness and calorific value. High-fructose corn syrups (HFCS) are made from liquefied starch by exhaustive hydrolysis with glucoamylase in combination with pullulanase. This yields about 96% glucose. The syrup is then isomerized using the enzyme glucose isomerase. This intracellular enzyme derived from various bacteria catalyzes the reversible isomerization of xylose to xyulose. It also converts glucose to fructose. The enzyme is used in immobilized form. First, the pH of the glucose syrup is raised and the material deionized (calcium inhibits glucose isomerase). After isomerization, the syrup contains about 42% fructose and 54% glucose. Higher levels of fructose can only be obtained by nonenzymatic treatment because the reversible reaction reaches an equilibrium. In the United States, HFCS have replaced sucrose in many important applications, particularly the sweetening of beverages, but in the European community, strict production quotas have been imposed.

Other applications for α-amylases are various. In the textile industry, cotton is soaked in starch size before weaving to provide tensile strength. After weaving, the size is removed with α-amylase before the fabric is dyed. The distilling and brewing industries use amylases extensively to aid the conversion of starch to fermentable sugars during mashing. Similarly, in bread production, the flour used may contain insufficient endogenous amylase. Wheat raised in the hot dry climate of North America, for example, is often deficient in amylase. Thus, when the yeast begins to raise the dough, it is restrained by the lack of fermentable sugar and the process halts. The supplementation of the flour with fungal amylase ensures adequate maltose levels and consistent leavening of the dough. Crust color, which is brought about by chemical browning of free glucose can be enhanced by the addition of glucoamylase. There are numerous other minor applications of amylases that together make amylases and related enzymes an im-

portant product in the enzyme market, with a global annual sales in 1996 of around $156 million.

B. Proteases

From an economic point of view, the proteases are the most important industrial enzymes and together make up some 40% of the enzyme market. Proteases can be classified on the basis of their catalytic properties into four groups—the serine proteases, metalloproteases, cysteine proteases, and aspartic proteases. This correlates well with the pH optima; thus, the serine proteases have alkaline optima, the metalloproteases are optimally active around neutrality, and the cysteine and aspartic proteases have acidic optima. All of these proteases are of commercial interest.

1. Serine Proteases

Serine proteases are secreted by many bacteria, the enzymes from *B. licheniformis* and *B. amyloliquefaciens*, and various alkaliphilic bacilli being produced for industrial use. These small enzymes (25–30 kDa) exhibit maximal activity at pH 9–11, have no metal-ion requirement, and resemble the animal enzyme trypsin. The most famous alkaline serine proteases are the subtilisins. Subtilisin Carlsberg was prepared and crystallized in 1952. It is secreted by *B. licheniformis* (then confused with *B. subtilis*), whereas Subtilisin Novo, sometimes called Subtilisin BPN, is produced by *B. amyloliquefaciens*.

The enzymes are very similar, but *Subtilisin Carlsberg* is the enzyme made famous by its inclusion in detergents. The concept of improving laundering efficiency by the inclusion of protease was developed in 1913 by Rohm, who used pancreatic enzyme preparations. These were not very effective, but the properties of the alkaline protease from *B. licheniformis*, notably the high pH optimum (pH 8–9), reasonable resistance to oxidizing agents (now improved by protein engineering), lack of metal-ion requirements (detergents contain metal chelators), and reasonable temperature tolerance made it ideal as a laundering aid. Enzyme-containing detergents quickly gained a large market volume, particularly in Europe. Dirt often adheres to fabrics by being bound by protein, so protease-containing detergents not only remove obvious proteinaceous stains such as sweat and blood, but also help overall laundering efficiency.

The search for improved alkali stability encouraged the isolation and screening of alkaliphilic bacilli for proteases. These bacteria grow at high pH and not at neutrality, so it was considered likely that their extracellular enzymes would be well suited to an alkaline environment. Indeed, serine proteases from these bacteria have exceptionally high pH optima (up to ~12), and they are now preferred for use in high-pH detergent formulations.

These highly alkali-tolerant serine proteases also opened up new markets in the leather industry. The traditional way to remove hair from cowhides is to use sodium sulfide and slaked lime. These chemicals dissolve the hair and open up the fiber structure. Enzyme-assisted dehairing uses alkaline proteases in combination with lime, thus reducing the toxic sulfide requirement. One of the oldest applications of industrially made enzymes is in the bating process, in which dehaired skins are soaked in enzyme to make them pliable. Originally this used pancreatic extracts, but alkaline proteases quickly replaced the animal extracts.

2. Metalloproteases

These enzymes are widely distributed in microorganisms. They contain an essential metal atom, usually zinc, have a molecular weight of 35–40 kDa, and a pH optimum near neutrality. They are generally less thermostable than the serine proteases. They are produced commercially from *B. amyloliquefaciens* and *B. stearothermophilus* ("*Thermolysin*") and have application in the food industry. For example, flour for biscuits and cookies should ideally be low in gluten to produce the desired dough, and metalloproteases are well suited to this. They also have uses in brewing to increase the amount of nitrogen available to fermentation, and in fish-meal processing.

3. Cysteine Proteases

The thiol proteases have a cysteine residue at their active site and are typified by the plant enzyme papain. There are few microbial sources of these enzymes, and the industrial products are still manufactured from plant sources. The enzymes have optimal activity just below pH neutrality, and they have applications in brewing for removal of protein hazes, which occur when the beer is chilled, and for tenderizing meat. Both applications require an enzyme with a high specificity to avoid spoilage of the product.

4. Acid Proteases

These enzymes are widely distributed in molds and yeast, but are seldom found in bacteria. They have pH optima at 3–4 and, as do the animal enzymes pepsin and rennin, their active sites contain an aspartate residue. The most important use of these enzymes is in the dairy industry as rennin substitutes. The coagulation of milk to produce cheese is traditionally brought about using extracts (rennet) from the stomachs of young calves. This largely consists of the enzyme chymosin with some pepsin. As the amount of calf slaughter has declined, cheese manufacture and consumption has increased, providing a requirement for alternative sources of milk-coagulating enzymes. It is necessary that the enzyme has the correct specificity for the coagulation of milk and does not lead to proteolytic degradation of the casein. Rennets from *Mucor pusillus* and *Rhizomucor miehi* contain acid proteases with pH optima around 4–4.5 and produce a pattern of hydrolysis products from casein similar to calf rennet. These microbial rennets are now used extensively for the production of hard cheddar-style cheeses, but for specialist cheeses animal rennets are still used. The cloning of the gene for calf chymosin has led to the introduction of genetically engineered rennets. Several products are now available and legislation in countries that until recently have restricted use of these coagulants is expected to change.

C. Other Enzymes

Together, proteases and starch-hydrolyzing enzymes constitute about 70% of the world market; however, some other important extracellular enzymes are also manufactured on a commerical scale.

1. Cellulase

The hydrolysis of crystalline cellulose to sugar is an attractive proposition, but with the current enzyme preparations it is uneconomic. Cellulose comprises linear chains of 1,4-β-linked glucose residues held in a crystalline form by hydrogen bonds, and it is very resistant to enzymatic hydrolysis. Fungal enzymes, such as those from *Trichoderma*, are mixtures of exo-acting (cellobiohydrolases) and endo-acting (endo-cellulases) enzymes together with β-glucosidase. These are among the most effective at attacking crystalline cellulose. Some bacteria such as *Clostridium thermocellum* produce endoglucanases in the form of cellulosomes (see Section I) that are effective at hydrolyzing native cellulose, apparently in the absence of exo-enzymes. Many bacilli produce endocellulases that hydrolyze soluble cellulose derivatives such as carboxymethyl cellulose, but are inactive on the crystalline substrate. Some of these enzymes are now used in the detergent industry as color "brighteners." Tiny fibrils are generated by wear on cotton fabrics. These reflect light and give a worn, bleached appearance to the garment. Removal of these fibrils by presoaking or washing in cellulase-supplemented detergents restores the original color and softens the fabric. They are also used to obtain the "stone-washed" appearance of denim jeans without the need for bleach.

2. Pectinase

Molds are primarily used for the production of pectinases on a commercial scale. These enzymes from aspergilli are mixtures of pectin esterases and depolymerizing enzymes and are particularly well suited to their principal application, which is the clarification of and reduction in viscosity of fruit juices by the removal of pectin. The application of pectinase-mash enzymes in the processing of apples, berries, grapes, and pears is standard practice, and these enzymes are also used in the citrus industry for juice processing.

3. β-Glucanase

Mixed linkage 1,3–1,4-β-glucans are present in barley cell walls and can cause viscosity and turbidity problems in brewing. These difficulties can be alleviated by adding β-glucanase from *Bacillus* species to the mash. Barley β-glucans can also cause problems in poultry feeds when the animal is fed on cereal-rich diets. Poultry cannot digest β-glucans, resulting in wet, sticky droppings and a reduction in the use of the carbohydrate content of the feed. The addition of β-glucanase to the feed not only cures this problem, but also provides more metabollically useful sugars.

4. Lipase

The natural substrates of lipases are triglycerides of long-chain fatty acids. The substrate is therefore insoluble in water and the enzymes characteristically catalyze the hydrolysis of the ester bonds at the interface between the aqueous phase and the solid substrate. Lipases are included in washing detergents for the removal of fatty stains, but major developments are expected in the oils and fats industry, in which the specificity of microbial lipases can be harnessed for the modification of plant oils for their inclusion in margarines and cocoa.

These major enzymes and some others of minor or developing importance are listed in Table I.

III. REGULATION OF SYNTHESIS

Extracellular enzymes are secreted by microorganisms to provide assimilable sources of low-molecular-weight nutrients. It is particularly important for the organism to exercise some control over the synthesis of these enzymes for two reasons. First, the external conditions may not be conducive to enzymatic activity due to extremes of pH or the presence of inhibitory or inactivating ions. It would therefore be wasteful to continue to secrete enzymes. Second, even if the conditions are appropriate for the enzyme, the products from the enzymatic degradation of the macromolecular substrate might be rapidly assimilated by a large number of competing microorganisms. It might therefore be expected that sophisticated regulatory systems have evolved for the control of synthesis of these enzymes. Such systems do indeed exist and can be grouped into three classes.

A. Induction

Extracellular enzymes are commonly secreted at a low basal level. If they encounter substrate, the

TABLE I
Some Common Extracellular Enzymes and Their Sources and Uses

Enzyme	Source	Principal uses
α-Amylase	*Aspergillus oryzae*	Starch hydrolysis for sugar syrups, brewing, textiles, and paper processing
	Bacillus amyloliquefaciens	
	Bacillus licheniformis	
β-Glucanase	*Aspergillus niger*	β-glucan hydrolysis in beer brewing
	Bacillus subtilis	
Cellulase	*Aspergillus* sp.	Fruit and vegetable processing
	Penicillium sp.	Detergents
	Trichoderma reesei	
Glucoamylase	*Aspergillus niger*	Glucose syrup production
	Rhizopus sp.	
Glucose isomerase[a]	*Actinoplanes missouriensis*	Isomerization of glucose into high-fructose syrups
	Bacillus coagulans	
	Streptomyces sp.	
Lactase	*Kluyveromyces marxianus*	Hydrolysis of lactose in milk and whey
	Saccharamyces sp.	
Lipase	*Aspergillus* sp.	Cheese and butter flavor modification; fat and oil processing
	Mucor sp.	
	Rhizopus sp.	
Pectinase	*Aspergillus niger*	Extraction and clarification of fruit juices
Penicillin acylase	*Bacillus megaterium*	Synthesis of 6-aminopenicillanic acid for manufacture of semisynthetic antibiotics
	Escherichia coli	
Protease (acid)	*Mucor pusillus*	Cheese manufacture
	Rhizomucor miehi	
Protease (neutral)	*Bacillus amyloliquefaciens*	Baking and brewing
Protease (alkaline)	*Bacillus licheniformis*	Detergent and leather industries
	Alkaliphilic bacilli	
Pullulanase	*Bacillus acidopullulyticus*	Debranching starch in sugar syrup manufacture
Xylanase	*Aspergillus niger*	Paper-pulp bleaching
	Aspergillus oryzae	

[a] Glucose isomerase is an intracellular enzyme.

material is hydrolyzed and the low-molecular-weight products are released and diffuse. In the absence of competing microorganisms, the product will accumulate to a threshold concentration and will be transported into the cell. The intracellular product then induces further synthesis of the enzyme. Thus, the microorganism can monitor the level of product in the environment and is able to regulate synthesis of the enzyme accordingly, and the external macromolecule is hydrolyzed and catabolized in the most efficient manner. Many amylases, cellulases, β-glucanases, and other extracellular enzymes are regulated in this way. Inducing sugars generally range from disaccharides to tetrasaccharides and sometimes higher oligosaccharides. In the case of endocellulase (carboxymethyl cellulase) from several bacilli, glucose is often the inducer, which is unusual given its role in catabolite repression (see Section III.B). Not all extracellular enzymes are inducible in this way, and the high levels of amylase and protease are synthesized by *B. amyloliquefaciens* and *B. licheniformis* in a constitutive fashion. The mechanisms of induction are now being elucidated in several bacteria, in particular *Eschericichia coli*, and are generally based on standard inducible operon systems (see Boos *et al.*, 1996 for regulation of maltodextrin catabolism in *E. coli*).

B. Catabolite Repression

Virtually all extracellular enzymes are controlled by catabolite repression. That is, in the presence of glucose or some other rapidly metabolized carbon source, there is repression of enzyme synthesis. It must be remembered that this control is not exclusively operated by sugars—in pseudomonads, for example, succinate is a strong catabolite repressor. Catabolite repression provides energy efficiency. It is wasteful for a microorganism to synthesize a range of enzymes in response to a range of inducers—better to use the carbon sources in turn, starting with the most easily metabolized. Catabolite repression provides this coordination of carbon metabolism and to exert its influence must override induction control. Although virtually all microorganisms display catabolite control of their metabolism, the molecular mechanisms vary. Here we consider only how catabolite repression operates in the case of the extracellular enzymes of *Bacillus*.

The gram-positive bacteria of the genus *Bacillus* do not contain cyclic adenosine monophosphate (AMP) or its receptor protein CRP, which together are responsible for the operation of most aspects of catabolite repression in *E. coli*, a different system operates. Catabolite control protein (CcpA) was recognized as a repressor of amylase synthesis in *B. subtilis* through the isolation of mutant strains. Under catabolite-repressing conditions (i.e., in the presence of glucose), CcpA binds to the promoter of the amylase (and other extracellular enzyme) structural gene and inhibits transcription. But how does CcpA sense catabolite repressing conditions? A protein component of the phosphoenolpyruvate-dependent phosphotransferase (PTS) sugar-transport system, called Hpr, is involved, possibly in the following way. It has been established that in the presence of glucose, Hpr becomes phosporylated at a specific serine residue. Hpr(serP) interacts with CcpA (perhaps in the presence of a metabolite such as fructose 1,6-biphosphate) and the complex binds to the target regions in catabolite repressible operons to inhibit transcription. There is substantial evidence for this process.

C. Temporal Regulation

In bacilli and streptomycetes, the synthesis of many inducible and constitutive extracellular enzymes is repressed during exponential growth and derepressed during the early stationary phase. This is distinct from catabolite repression and is more closely related to differentiation. These bacteria form endospores (*Bacillus*) or conidiospores (*Streptomyces*) in response to starvation. But differentiation into a spore is a decision not to be taken lightly. As a spore, the bacterium does not divide, it does not metabolize, and it is no longer contributing effectively to the success of the species. Sporulation is an absolute last resort; the bacterium will try anything to remain in its vegetative state. Between exponential growth and entry into the sporulation process is a transition phase during which *B. subtilis*, and presumably other spore-forming bacteria, attempt to discover new sources of nutrients. They become particularly motile, enabling them to swim to new microenvironments, they secrete antibiotics to reduce competition from other microorganisms, some such as *B. subtilis* become competent for DNA-mediated transformation and thus incorporate new genetic material, and they secrete extracellular enzymes. This is all done in a last desperate search for food, and if this fails they form spores. Although the transition phase is of short duration in flask cultures grown in the laboratory, in the soil where most bacilli live, starvation is the norm and the transition phase may predominate. Bacilli live in this poised state, ready to exploit any opportunity for nutrients and rarely do they grow exponentially, as they would in a flask of nutrient broth. It is therefore predictable that comprehensive and sophisticated mechanisms for sensing and responding to starvation in all its ramifications have evolved.

All bacteria constantly sense their environment and respond to it by regulating protein synthesis in a coordinated way. One mechanism for achieving this involves two-component signal-transduction systems. Briefly, a sensor protein, often but not invariably located spanning the cytoplasmic membrane, responds to changes in a specific factor (such as pH, osmolarity, or oxygen) by phosphorylating an internal portion of the protein at a conserved histidine residue using ATP. This histidine protein kinase sensor then transfers the phosphate to a second protein in the chain, a response regulator. In its phosphorylated form, the response regulator usually activates (but may repress) the transcription of certain operons under its control. In this way, sets of genes (regulons) respond to certain environmental stimuli. Extracellular enzymes in *Bacillus* (and other bacteria such as *Staphylococcus aureus* and *Pseudomonas aeruginosa*) are controlled by such two-component systems.

Early genetic studies revealed several mutations (e.g., *pap*, *amyB*, and *sacU*) that caused hyperproduction of various extracellular enzymes, including amylase, β-glucanase, levansucrase, and proteases, but that also had confusing (at that time) pleiotropic effects on motility and DNA-mediated transformation. These mutations have now been attributed to changes in the DegS–DegU two-component regulatory system. DegS refers to a histidine protein kinase sensor protein (although it is not located in the membrane) that, under certain conditions (perhaps associated with entry into stationary phase), phosphorylates DegU. The latter is a response regulator that, when phosphorylated, activates the operons under its control, including the various extracellular enzymes mentioned previously. It also represses flagella synthesis when phosphorylated, thus explaining the early mutations affected in motility.

A second regulatory system that is more closely involved with sporulation controls serine protease synthesis in *B. subtilis*. The AbrB protein is a repressor of several transition- and stationary-phase genes including some sporulation genes and *aprE,* the gene for extracellular serine protease. During rapid growth, AbrB ensures the repression of these important genes. As the cell enters the stationary phase, SpoOA protein, which is a response regulator in a signal-transduction pathway associated with the initiation of sporulation, is phosphorylated and represses *abrB* expression. This, in turn, alleviates the repression of *aprE,* and serine protease is synthesized and secreted.

Numerous other genes that affect extracellular en-

zyme synthesis in *B. subtilis* have been cloned and characterized. Serine protease is influenced to a great extent by these regulators and its synthesis is controlled by at least nine regulatory genes.

IV. SECRETION

The process of the translocation of proteins across membranes in both eukaryotic and prokaryotic cells is largely in accord with the signal model. In this scheme, proteins destined for the membrane are synthesized with an N-terminal extension of some 15–40 amino acids, the signal peptide. In the original model based on eukaryotic cells, this peptide interacted with membrane proteins that formed a pore through which the protein was transported as it was translated. This co-translational secretion resulted in a nascent protein on the outside of the membrane that was processed by the removal of the signal peptide by signal peptidase. The protein then adopted its native configuration.

The signal hypothesis was so influential that it was adopted as a universal scheme for the secretion of proteins in eukaryotes and prokaryotes. Indeed, the process has been conserved to a remarkable extent throughout evolution. The analysis of signal peptide sequences from a variety of sources reveals a limited homology of primary sequences but a conserved structure. Signal peptides comprise three characteristic regions. First, a positively charged N-terminal end. This seems to be important in targeting the peptide to the membrane. It is followed by a hydrophobic region of 10–15 amino acids, which is important for the translocation of the protein across the membrane. The inclusion of charged amino acid residues in this region interrupts secretion, as do changes to its length. The C-terminal end of the signal peptide includes the cleavage site for signal peptidase. The universal structure of signal peptides indicates that they could function in heterologous hosts, and this is indeed sometimes the case, particularly for closely related species.

Just as the signal sequence is firmly established as having a central role in secretion, all classes of organisms contain a structurally similar pore or heterotrimeric complex of membrane proteins, called

Sec61p in eukaryotes and making up the SecYEG proteins in *E. coli*. SecY is a 10-transmembrane helix protein that has been conserved in all organisms, and SecE and G are small membrane proteins found in bacterial translocases. However, the mechanics of secretion differ between eukaryotic and prokaryotic cells. Co-translational secretion is the norm for eukaryotic cells, but in bacteria posttranslational secretion predominates. Morover, bacteria contain SecA, a central component of the secretion pathway that is unique to bacteria. SecA provides translation-independent transfer of large hydrophilic proteins through the translocase channel driven by ATP. SecA not only acts as the physical link among all components of the secretion pathway, the preprotein, SecYEG, lipids, and ATP, but it also acts as a translocase and is responsible for the export of the protein through this general secretory pathway (GSP).

The protein destined for secretion is translated by the ribosome complete with signal peptide as a preprotein. This preprotein must be maintained in a suitable conformation for export; it must not be allowed to fold inappropriately in the cytoplasm. SecB is a chaperone protein that interacts with the preprotein and maintains its secretion-compatible form. SecB accompanies the preprotein to the membrane, where SecB interacts with SecA thus delivering the preprotein to the translocase. SecA recognizes the signal peptide, but also interacts with mature parts of the protein. ATP-binding by SecA results in a conformational change that inserts SecA more deeply into the membrane, taking about 20 amino acid residues of the bound protein with it. On ATP-hydrolysis, SecA releases the peptide chain, comes out of the membrane, and binds a new portion of the protein. ATP is bound again by SecA, which inserts again into the membrane, and so this cycle repeats itself with 20–30 amino acid residues of the protein crossing the membrane at each stage.

The final stage of secretion is processing to remove the signal peptide. In *E. coli,* there are two signal peptidases. Signal peptidase I is the general enzyme that processes several periplasmic and outer membrane proteins, whereas signal peptidase II, or prolipoprotein signal peptidase, seems to be specific for processing of prolipoproteins, such as the outer-membrane protein OmpA.

In gram-positive bacteria, the exported protein is often thought to diffuse through the cell wall to become completely extracellular, but it has been pointed out that the barrier imposed by the peptidoglycan should not be so readily dismissed and that some form of translocation might operate to mobilize the proteins. In support of this, pulse-labeling studies have shown that some proteins are externalized by *B. subtilis* more quickly than others.

In gram-negative bacteria, the protein must pass through the outer membrane to become completely external. The main terminal branch (MTB) of the general secretory pathway is used by most of these bacteria to transport proteins from the periplasm to the environment. It is complex with at least 14 components, some of which may form a pore through which the protein is exported; however, progress in understanding the mechanisms of the process has been slow.

V. INDUSTRIAL PRODUCTION

The manufacture of extracellular enzymes is done on an industrial scale and so the processes involved must be kept relatively simple. For example, in most cases no attempt is made to purify the enzymes; they are sold as enzyme mixtures, and in many applications this is desirable. The "contaminating" amylase in a β-glucanase preparation destined for the mash tun of a brewery will usefully catalyze hydrolysis of starch. Similarly, enzyme mixtures in a washing detergent will be beneficial. Where purity is necessary, additional steps are taken to purify the enzyme, and this, obviously, adds to the cost.

The main problems for production are to devise an economic process that complies with the strict codes for safety and production laid down by the regulatory authorities, such as the Food and Drug Administration in the United States or the Health and Safety Inspectorate in the United Kingdom. In this respect, the producer organism is obviously of crucial importance. Some bacteria such as *B. amyloliquefaciens* or *B. licheniformis* automatically have clearance for production of food-grade enzymes, but other products from unlisted sources must be subject to extensive safety testing. Many enzymes are now manufactured from genetically engineered strains (over 60%) and this introduces additional legislative problems, although these can be fewer if the host already has clearance. Fermentation media are generally based on agricultural by-products such as corn and barley starches, soybean extracts, fish meal, corn steep liquor, and other similar materials, depending on local availability and world markets. Medium composition is a closely guarded secret, but usually the fermentation is balanced empirically to provide for a rapid period of exponential growth (e.g., in an amylase fermentation, 10–20 hr) followed by a prolonged transitional phase and extensive stationary phase, lasting perhaps 100 hr, during which the enzyme accumulates. Virtually all fermentations now use submerged culture, although some fungal enzymes are manufactured as solid-substrate fermentations on moistened wheat bran in open trays.

Asporogenous mutants of bacilli are usually employed in production to give higher yields and to prevent spores of the producer strain being present in the finished product. At the end of the fermentation, the broth is cooled and centrifuged to remove the cells or mycelium, perhaps with some form of flocculating pretreatment. The enzyme is then concentrated by ultrafiltration or vacuum drying. The enzyme can then be filtered, preservatives can be added, and the material can be standardized prior to being packaged as a liquid concentrate. Alternatively, after filtration, extending agents are added, it is spray-dried, and it is packaged as a powder after standardization. In the case of proteases, the powder is pearled, or marumized, into a dust-free form by covering it in a waxy coating in the form of small beads.

VI. THE FUTURE

Developments in biotechnology have encouraged new enzyme technologies, but in many respects this is an industry in which the available products are far more numerous than the market can bear. Nevertheless, new technologies based on enzymes are the subjects of intensive research in biotechnology companies and several trends are evident.

Most applications involve the hydrolysis of macro-molecules, but the industrial use of enzymes for biosynthetic and speciality purposes holds great promise. Such enzymes now make up 10% of the market and include oxido-reductases for the synthesis of gluconic acid, fumarase for the synthesis of L-malic acid, and glycosyltransferases for the manufacture of isomaltooligosaccharides, which have potential in the food and health industries. Enzyme applications in medicine include the production of several drugs—for example, penicillins and cortisone for arthritis treatment. The cost of the latter has decreased 200-fold to less than $1/g using an enzyme-based manufacturing process.

Enzymes will probably play a key role in the introduction of environmentally friendly processes that form cleaner products in a much milder, efficient, and economical way than traditional chemical processes based on large amounts of noxious solvents, acids, alkali, or other corrosive agents. For example, chlorine gas is used to pretreat Kraft wood pulp before alakli extraction and bleaching with chlorine dioxide to make quality white paper. The bleaching sequence removes the lignin, which imparts the "brown bag" color to the paper, from the wood, but the liquid waste from pulping mills contains highly toxic chlorinated organic compounds. Xylanase, which solubilizes the lignin so that it can be drawn off in alkaline fluids prior to "brightening," was predicted throughout the 1980s as being the next major development in bulk enzyme application and it is heartening that it is being used in numerous pulp mills with an associated reduction in chlorine wastes.

Enzymes from thermophilic microorganisms, sometimes called thermozymes, can operate at 60–125°C. These offer new processes and much improved existing processes, such as single-step conversion of starch to a glucose–maltose syrup using the α-glucosidase from *Thermococcus hydrothermalis*. The enzyme is optimally active at 110°C and, together with the α-amylase from *B. licheniformis*, produces sugar syrups from starch in a single gelatinization, liquefaction, and saccharification process, all operating at 90°C. With developments in microbiology and genetic and protein engineering, new enzymes and enzyme processes are being introduced, which suggests a strong future for extracellular enzymes.

See Also the Following Articles

CELLULASES • LIPASES, INDUSTRIAL USES • PECTINASES

Bibliography

Beveridge, T. J. (1995). The periplasmic space and the periplasm of gram-positive and gram-negative bacteria. *Am. Soc. Microbiol. News* **61**, 125–130.

Boos, W., Peist, R., Decker, K., and Zdych, E. (1996). The maltose system. *In* "Regulation of Gene Expression in *Escherichia coli*" (E. C. C. Lin and A. S. Lynch, eds.), pp. 201–230. Chapman & Hall, London.

Crabb, W. D., and Mitchinson, C. (1997). Enzymes involved in processing starch to sugars. *Trends Biotechnol.* **15**, 349–352.

Economou, A. (1998). Bacterial preprotein translocase: Mechanism and conformational dynamics of a processive enzyme. *Mol. Microbiol.* **27**, 511–518.

Godfrey, T., and West, S. I. (eds.) (1996). "Industrial Enzymology," 2nd. ed. Macmillan, London.

Guzmá-Maldonado, H., and Paredes-López, O. (1995). Amylolytic enzymes and products derived from starch: A review. *CRC Crit. Rev. Food Sci. Technol.* **35**, 373–403.

Harwood, C. R. (1992). *Bacillus subtilis* and its relatives: Molecular biological and industrial workhorses. *Trends Biotechnol.* **10**, 247–256.

Pero, J., and Sloma, A. (1993). Proteases. *In* "*Bacillus subtilis* and Other Gram-Positive Bacteria" (A. L. Sonenshein, J. A. Hoch, and R. Losick, eds.), pp. 713–726. American Society for Microbiology, Washington, D.C.

Rapoport, T. A., Jungnickel, B., and Kutay, U. (1996). Protein transport across the eukaryotic endoplasmic reticulum and bacterial inner membrane. *Ann. Rev. Biochem.* **65**, 271–303.

Strauch, M. A., and Hoch, I. A. (1993). Transition-state regulators: Sentinels of *Bacillus subtilis* post-exponential gene expression. *Mol. Microbiol.* **7**, 337–342.

Wong, S.-L. (1995). Advances in the use of *Bacillus subtilis* for the expression and secretion of heterologous proteins. *Curr. Opinion Biotechnol.* **6**, 517–522.

Enzymes in Biotechnology

Badal C. Saha and Rodney J. Bothast

U.S. Department of Agriculture

GLOSSARY

biotechnology The use of living organisms or their products to make or modify a product for commercial purposes.

chiral A geometric attribute of a compound, describing a molecule that cannot be superimposed on its mirror image.

dextrose equivalent (DE) An indication of total reducing sugars as a percentage of glucose; unhydrolyzed starch has a DE of zero and glucose has a DE of 100.

enantioselective reaction A chemical reaction or synthesis that produces the two enantiomers of a chiral product in unequal amounts.

liquefaction A process in which starch granules are dispersed or gelatinized in aqueous solution and then partially hydrolyzed by a thermostable α-amylase.

pitch A mixture of lipids, resins, and other extraneous compounds found in wood.

prebiotics Nondigestive food ingredients that beneficially affect the host by selectively stimulating the growth or activity of a limited number of bacterial species already in the colon.

racemate An equimolar mixture of two enantiometric species.

recombinant DNA DNA that is formed by joining pieces of DNA from two or more organisms.

reversion The process of forming di- and higher oligosaccharides through the condensation of glucose, maltose, or other reducing sugars.

saccharification The process in which a substrate is converted into low-molecular-weight saccharides by enzymes.

tannin Water-soluble phenolic compounds (molecular weight 500–3000) that have the property of combining with proteins, cellulose, gelatin, and pectin to form an insoluble complex.

ENZYMES play important roles in various biotechnology products and processes in the food and beverage industry and have already been recognized as valuable catalysts for various organic transformations and production of fine chemicals and pharmaceuticals. At present, the most commonly used enzymes in biotechnology are hydrolytic enzymes, which catalyze the breakdown of larger biopolymers into smaller units. Enzymes catalyze reactions in a selective manner, not only regio- but also stereoselectively, and have been used both for asymmetric synthesis and racemic resolution. The chiral selectivity of enzymes has been employed to prepare enantiomerically pure pharmaceuticals, agrochemicals, and food additives. Enzymatic methods have already replaced some conventional chemical processes. Microbial enzymes have largely replaced the traditional plant and animal enzymes used in industry with the industrial enzyme market grown at the level of about 10% annually. About 50 enzymes are used in industry, most of them (~90%) produced by submerged or solid-state fermentation by microorganisms. Enzymes work under moderate conditions of temperature, pH, and pressure. They are also highly biodegradable and generally pose no threat to the environment.

I. CLASSIFICATION

Enzymes, produced by living systems, are proteineous in nature with catalytic properties. As catalysts, enzymes are both efficient and highly specific for a particular chemical reaction involving the synthetic, degradative, or alteration of a compound. Cofactors are involved in reactions in which molecules are oxidized, reduced, rearranged, or connected. Enzymes have been divided into six major classes based on the types of reactions they catalyze.

1. *Hydrolases* catalyze the hydrolytic reactions (e.g., glycosidases, peptidases, and esterases). Water is the acceptor of the transferred group.
2. *Oxidoreductases* catalyze oxidation or reduction reactions (e.g., dehydrogenases, oxidases, and peroxidases).
3. *Isomerases* catalyze isomerization and racemization reactions (e.g., racemases and epimerases).
4. *Transferases* catalyze the transfer of a group from one molecule to another one (e.g., glycosyl transferases and acetyl transferases).
5. *Lyases* catalyze elimination reactions in which a bond is broken without oxidoreduction or hydrolysis (e.g., decarboxylases and hydrolyases).
6. *Ligases* catalyze the joining of two molecules with ATP or other nucleoside triphosphate cleavage (e.g., DNA ligases).

Biotechnology is broadly defined as a technique that involves the use of living organisms or their products to make or modify a product for commercial purposes. This article aims to familiarize the general reader with the use of enzymes in several specific areas in biotechnology.

II. STARCH CONVERSION

Foodstocks containing starch include most of the cereal grains (corn, sorghum, barley) and tuberous crops (potatoes). Starch contains about 15–30% amylose and 70–85% amylopectin. Amylose (molecular weight about 300,000) is a long linear chain polymer of α-1,4–linked glucose residues, whereas, amylopectin (molecular weight about 1.5–3×10^9) is a branched polymer having both α-1,4 and α-1,6 linkages. The branched chains may contain from 20–30 glucose units. Enzymes have already replaced the use of strong acid and high temperature to break down starchy materials.

Three types of enzymes are involved in starch bioconversion—endo-amylase (α-amylase, EC 3.2.1.1), exo-amylase (glucoamylase, EC 3.2.1.3; β-amylase, EC 3.2.1.2), and debranching enzyme (pullulanase, EC 3.2.1. 41; isoamylase, EC 3.2.1.68). α-Amylase hydrolyzes the internal α-1,4 glycosidic bonds of starch at random in an endo-fashion, producing maltooligosaccharides of varying chain lengths. α-Amylase cannot act on α-1,6 linkages. The enzyme is produced by bacteria, such as *Bacillus lichiniformis, B. subtilis,* and *B. amyloliquefaciens;* and fungi, such as *Aspergillus oryzae.* Glucoamylase cleaves glucose units from the nonreducing end of starch and it can hydrolyze both α-1,4 and α-1,6 linkages of starch. It is, however, slower in hydrolyzing α-1,6 linkages. Glucoamylase is produced by various genera of fungi such as *Endomycopsis, Aspergillus, Penicillium, Rhizopus,* and *Mucor.* β-Amylase hydrolyzes the α-1,4-glycosidic bonds in starch from the nonreducing ends, generating maltose. The enzyme is unable to bypass the α-1,6 linkages and leaves dextrins, known as β-limit dextrins. β-Amylase is produced by a number of microorganisms, including *B. megaterium, B. cereus, B. polymyxa, Thermoanaerobacter thermosulfurogenes,* and *Pseudomonas* sp. Pullulanase (pullulan 6-glucanohydrolase) or isoamylase (glycogen 6-glucanohydrolase) cleaves the α-1,6-linked branch points of starch and produces linear-chain amylosaccharides of varying lengths. Pullulanase is produced by *Aerobacter aerogenes* and isoamylase is produced by *Pseudomonas amyloderamosa.* A new enzyme, amylopullulanase, hydrolyzes both α-1,4 and α-1,6 linkages of starch and generates DP_2–DP_4 (DP, degree of polymerization) as products. The enzyme has the potential to be used in both the liquefaction and saccharification of starch. Amylopullulanase is produced by various anaerobic as well as aerobic bacterial species.

A. Production of Glucose

Glucose (dextrose) can easily be produced from starch by an acid hydrolysis process. However, this process has many disadvantages, such as a low yield of glucose (~85%), formation of undesirable bitter sugar (gentiobiose), and the inevitable formation of salt and a large amount of coloring materials. With the discovery of thermostable α-amylase from *B. licheniformis,* an enzymatic process has replaced the acid hydrolysis process.

The enzymatic production of glucose from starch usually involves two essential and basic steps, liquefaction and saccharification. First, an aqueous slurry of corn starch (30–35% dry substance basis, DS) is gelatinized (105°C for 5 min) and partially hydrolyzed (95°C for 2 hr) by a highly thermostable α-amylase to a dextrose equivalent (DE) of 10–15. The optimal pH for the reaction is 6.0–6.5 and Ca^{2+} (50 ppm) is required. During liquefaction, the α-amylase hydrolyzes the α-1,4 linkages at random, reducing the viscosity of the gelatinized starch. The liquefied and partially hydrolyzed starches are generally known as maltodextrins and are widely used in the food industry as thickeners. In the saccharification step, the temperature of the reaction is brought to 55–60°C, the pH is lowered to 4.0–5.0, and glucoamylase is added. A long reaction time of 24–72 hr is required, depending on the enzyme dose and the percentage of glucose desired in the product. The efficiency of the saccharification reaction with glucoamylase can be improved by adding pullulanase or isoamylase. This enzyme addition increases the glucose yield (about 2%), reduces the saccharification time from 72 hr to 48 hr, allows an increase in substrate concentration to 40% DS, and allows a reduction in the use of glucoamylase up to 50%. The glucose yield is about 95–97.5% by using a starch-debranching enzyme with glucoamylase.

At high-glucose concentrations, glucose molecules polymerize in a reaction called reversion, forming unwanted by-products, such as maltose, isomaltose, and higher saccharides, and decreasing the glucose yield. The polymerization reaction is generally catalyzed by glucoamylase or by another enzyme called transglucosidase (EC 2.4.1.24), which is often present in crude glucoamylase preparations. Typically, glucose syrups (DE 97–98) having 96% glucose contain 2–3% disaccharides (maltose and isomaltose) and 1–2% higher saccharides.

B. Production of High-Fructose Corn Syrups

Glucose isomerase (EC 5.3.1.5) is another example of the successful use of enzymes in biotechnology. It is produced intracellularly by *Streptomyces, Bacillus, Arthobacter,* and *Actinoplanes.* The enzyme is used to convert glucose into fructose, which is significantly sweeter than glucose. High-fructose corn syrups (HFCS) are prepared by enzymatic isomerization of glucose syrups (DE 95–98, 40–50% DS) in a column reactor containing immobilized glucose isomerase. The pH of the reaction is 7.5–8.0, temperature 55–60°C, and 5 mM Mg^{2+} is usually added as an activator and stabilizer of glucose isomerase, which also compensates for the inhibitory action of Ca^{2+} (used in starch liquefaction with α-amylase) on the enzyme. The most stable commercial glucose isomerase has a half-life of around 200 days in industrial practice.

The conversion of glucose to fructose by glucose isomerase is an equilibrium reaction; therefore, the continuous removal of fructose is essential to achieve a high-fructose content in the syrups. Two grades of HFCS are available on the market—HFCS-42 and HFCS-55, which contain 42 and 55% DS fructose, respectively. Glucose isomerase can only produce HFCS containing 42% fructose from glucose syrups (~95% glucose) under normal operating conditions. HFCS-42 syrup is then fractionated in a chromatographic column to yield 90% fructose syrup, which can be blended back to make 55% fructose syrup for use in soft drinks. Fig. 1 summarizes the steps involved for the conversion of starch to high fructose corn syrups.

C. Production of High-Maltose Conversion Syrups

Various maltose-containing syrups are used in the brewing, baking, soft drink, canning, confectionery,

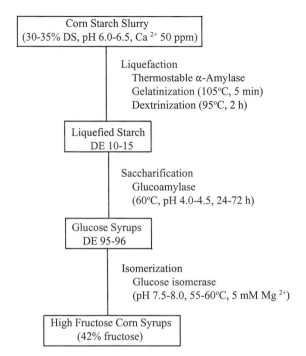

Fig. 1. Production of high-fructose corn syrups from starch.

and other food industries. There are three types of maltose-containing syrups—high-maltose syrup (DE 35–50, 45–60% maltose, 10–25% maltotriose, 0.5–3% glucose), extra-high-maltose syrup (DE 45–60, 70–85% maltose, 8–21% maltotriose, 1.5–2% glucose), and high-conversion syrup (DE 60–70, 30–47% maltose, 35–43% glucose, 8–15% maltotriose). The production of these syrups from starch generally involves liquefaction and saccharification, as in the production of glucose. The liquefaction reaction is terminated when the DE is about 5–10 becuase a low DE value increases the final maltose content. In this case, depending on the maltose content of the syrup, saccharification is generally performed by using a maltogenic amylase such as β-amylase, β-amylase with pullulanase or isoamylase, or a fungal α-amylase at pH 5.0–5.5 and 50–55°C. High-conversion syrups are produced from liquefied starch (DE ~40) by saccharifying it with a carefully balanced mixture of β-amylase or fungal α-amylase and glucoamylase. After the completion of saccharification, the syrup is heated to destroy the enzyme action and to stop further glucose formation.

D. Production of Cyclodextrins

Cyclodextrins (CDs) are cyclic oligosaccharides composed of six or more α-1,4-linked glucose units. CDs of six, seven, and eight α-D-glucose residues are called α-, β-, and γ-CD, respectively. CDs possess the unique ability to form inclusion complexes with various chemicals and are useful in the food, cosmetic, pharmaceutical, and plastic industries as emulsifiers, antioxidants, and stabilizing agents. The CDs are produced from starch or dextrins by cyclomaltodextrin glucanotransferase (EC 2.4.1.19, CGTase) by intramolecular transglycosylation (cyclization reaction). The ratio of α-, β-, and γ-CDs produced from soluble starch depends on the source of CGTase (*B. megaratium, B. macerans, B. circulans, B. ohbensis,* or *Klebsiella pneumoniae*). CDs are industrially produced from amylose by using CGTase. CDs are not highly soluble in water. Branched CDs such as maltosyl CD can be prepared by using the reverse reaction of pullulanase with maltose and CDs as substrates. Branched CDs are highly soluble in water and organic solvents.

E. Production of Alcohol

The process of making alcohol from starch involves three steps: (1) preparation of the feedstock, (2) fermentation of simple sugars to ethanol, and (3) recovery of the alcohol. Enzymes play an important role in preparing the feedstock, that is, in converting starch to fermentable sugars, mainly glucose. Corn contains about 60–70% starch. The wet-milling process obtains starch by successively removing other fractions (oil, protein, and fiber) of the corn kernel. The liquefaction and saccharification of starch are then carried out as described for the preparation of glucose. In the alcohol-beverage industry, β-amylase is also used. Microbial enzymes have replaced the traditional enzymes supplied by adding malt. Glucose is then fermented by the traditional yeast *Saccharomyces cerevisiae* to ethanol, which can be recovered by distillation. The saccharification and fermentation can also be carried out in a process known as simultaneous saccharification and fermentation (SSF).

Alcohol is also produced from corn starch by the dry-milling of corn. The addition of a protein-splitting enzyme (protease) releases the soluble nitrogen compounds from the fermentation mash, which helps the growth of the yeast and decreases fermentation time. The residue left after fermenting the sugars is known as distiller's dried grain (DDG) and is used as animal feed. Typical large-scale industrial fermentation processes make 10–13% (v/v) ethanol with an ethanol yield as high as 95% of theoretical, on the basis of starch. At present, ethanol is the most widely used renewable fuel in the transportation sector.

III. LIGNOCELLULOSIC BIOMASS CONVERSION

The lignocellulosic biomass includes various agricultural residues (straws, hulls, stems, and stalks), deciduous and coniferous woods, municipal solid wastes (paper, cardboard, yard trash, and wood products), waste from pulp and paper industry, and herbaceous energy crops. The composition of these materials varies, but the major component is cellulose (35–50%), followed by hemicellulose (20–35%), and lignin (10–25%). Proteins, oils, and ash make up the remaining fraction of the lignocellulosic biomass.

The structure of these materials is very complex and native lignocellulosic biomass is resistant to an enzymatic hydrolysis. The pretreatment of lignocellulosic biomass is crucial before enzymatic hydrolysis. Various pretreatment options are available to fractionate, solubilize, hydrolyze, and separate the cellulose, hemicellulose, and lignin components. These include steam explosion, dilute-acid treatment, concentrated-acid treatment, alkaline treatment, treatment with SO_2, treatment with hydrogen peroxide, ammonia–fiber explosion, and organic-solvent treatments. In each option, the biomass is treated to reduce its size and to open its structure. Acid pretreatment at high temperature usually hydrolyzes hemicellulose to its sugars (xylose, arabinose, and other sugars), which are water soluble. The residue contains cellulose and lignin. The lignin can be extracted with solvents such as ethanol, butanol, or formic acid. Alternatively, the hydrolysis of cellulose with lignin present makes water-soluble sugars and the residues are lignin plus any unreacted materials.

No commercial process exists for the conversion of lignocellulosic biomass to fermentable sugars, which can then be fermented by microorganisms into a variety of value-added products.

A. Cellulose Conversion

Cellulose is a linear polymer of 8000–12,000 glucose units linked by 1,4-β-D-glucosidic bonds. The enzyme system for the conversion of cellulose to glucose comprises endo-1,4-β-glucanase (EC 3.2.1.4), exo-1,4-β-glucanase (EC 3.2.1.91), and β-glucosidase (EC 3.2.1.21). These cellulolytic enzymes with β-glucosidase act sequentially and cooperatively to degrade crystalline cellulose to glucose. Endoglucanase acts in a random fashion on the regions of low crystallinity of the cellulosic fiber, whereas exoglucanase removes cellobiose (β-1,4 glucose dimer) units from the nonreducing ends of cellulose chains. Synergism between these two enzymes is attributed to the endo–exo form of cooperativity and has been studied extensively in the degradation of cellulose in *Trichoderma reesei*. β-Glucosidase hydrolyzes cellobiose and in some cases cellooligosaccharides to glucose. The enzyme is generally responsible for the regulation of the whole cellulolytic process and is a rate-limiting factor during enzymatic hydrolysis of cellulose, as both endoglucanase and cellobiohydrolase activities are often inhibited by cellobiose. Thus, β-glucosidase not only produces glucose from cellobiose, but also reduces cellobiose inhibition, allowing the celluloytic enzymes to function more efficiently. However, as are β-glucanases, most β-glucosidases are subject to end-product (glucose) inhibition. For a complete hydrolysis of cellulose to glucose, the enzyme system must contain the three enzymes in right proportions. Many microorganisms are cellulolytic, however, only two microorganisms (*Trichoderma* and *Aspergillus*) have been studied extensively for cellulase. The saccharification of cellulosic materials at commercial scale is hindered by the cost of the cellulase enzymes and also by the cost of pretreatment of the substrate.

B. Hemicellulose Conversion

Hemicelluloses are heterogeneous polymers of pentoses (xylose and L-arabinose), hexoses (mannose), and sugar acid (glucuronic acid). Xylans, which are major hemicelluloses of many plant materials, are heteropolysaccharides with a homopolymeric backbone chain of 1,4-linked β-D-xylopyranose units. In addition to xylose, xylans may contain arabinose, glucuronic acid, or its 4-*O*-methyl ether, and acetic, *p*-coumaric, and ferulic acids.

Total hydrolysis of xylan requires endo β-1,4-xylanase (EC 3.2.1.8), β-xylosidase (EC 3.2.1.37), and several accessory enzyme activities, which are necessary for hydrolyzing various substituted xylans. These accessory enzymes include α-L-arabinofuranosidase (EC 3.2.1.55), α-glucuronidase (EC 3.2.1.1), acetyl xylan esterase (EC 3.2.1.6), feruloyl esterase, and *p*-coumaroyl esterase. The endoxylanse randomly attacks the main chains of xylans, and the β-xylosidase hydrolyzes xylobiose and xylooligosaccharides to xylose. α-L-Arabinofuranosidase and α-glucuronidase remove the arabinose and 4-*O*-methyl glucuronic acid substituents, respectively, from the xylan backbone. The esterases hydrolyze the ester linkages between xylose units of the xylan and acetic acid (acetyl xylan esterase) or between arabinose side-chain residues and phenolic acids such as ferulic acid (feruloyl esterase) and *p*-coumaric acid (*p*-coumaroyl esterase).

Depolymerizing and side-group-cleaving enzymes work synergistically to break down hemicellulose to simple sugars. Some xylanases do not cleave glycosidic bonds between xylose units that are substituted; thus the side-chains must be cleaved before the xylan backbone can be completely hydrolyzed. On the other hand, several accessory enzymes only remove side-chains from xylooligosaccharides. These enzymes require xylanases to partially hydrolyze hemicellulose before side-chains can be cleaved. Although the structure of xylan is more complex than cellulose and requires several different enzymes with different specificities for a complete hydrolysis, the polysaccharide does not form tightly packed crystalline structures and is thus more accessible to enzymatic hydrolysis. Xylanolytic enzymes are produced by numerous microorganisms such as *Trichoderma* sp., *Aspergillus* sp., *Fusarium* sp., and *Bacillus* sp.

C. Lignin Conversion

Lignin is a long-chain heterogeneous polymer composed largely of phenylpropane units that are most commonly linked by ether bonds. The conversion of cellulose and hemicellulose to fuels and chemicals generates lignin as a by-product. In recent years, removal of lignin from lignin–carbohydrate complex (LCC) has received much attention because of potential application in the pulp and paper industry. The lignin barrier can be disrupted by a variety of pretreatments, rendering the cellulose and hemicellulose more susceptible to enzymatic attack. The basidiomycete *Phanerochaete chrysosporium* is able to degrade lignin in a H_2O_2-dependent process catalyzed by extracellular peroxidases (lignin peroxidase, Lip; and managanese peroxidase, MnP). Due to the extreme complexity of the problem, a great deal of research needs to be done to understand all the factors involved in lignin-biodegradation process.

IV. ENZYMATIC PRODUCTION OF FUNCTIONAL OLIGOSACCHARIDES

Various oligosaccharides, prepared by using enzymes, are the relatively new functional food ingredients used to promote the growth of bifidobacteria in the human intestine. They provide a good example of the successful use of reverse reactions of glycosidases and transglycosylation reactions catalyzed by glycosyltransferases in biotechnology. Isomaltooligosaccharides are prepared from maltose by the transglucosidase activity of α-glucosidase (EC 3.2.1.20). These oligosaccharides contain glucose residues linked by α-1,6 glycosidic linkages and induce a bifidogenic response. Glycosyl sucrose (coupling sugar) is prepared from starch and sucrose by using CGTase. CGTase catalyzes a transglycosylation reaction whereby glycosyl moieties are transferred from starch to the glucose residues of sucrose by α-1,4-linkage. Gentiooligosaccharides are prepared from glucose by enzymatic transglucosylation. They contain glucose residues linked by β-1,6 glycosidic bonds and promote the growth of bifidobacteria. Xylooligosaccharides are produced from xylan substrates by the action of end 0-1,4-β-xylanase. These oligosaccharides promote the growth of bifidobacteria and are used in prebiotic drinks. Fructooligosaccharides (Neosugars, mixtures of Glu-Fru$_2$, Glu-Fru$_3$, Glu-Fru$_4$) are produced from sucrose using the transfructosylation activity of the enzyme β-

fructofuranosidase (EC 3.2.1.26) from *Aspergillus niger*. These are also produced from inulin, a linear, β-2,1-linked fructose polymer initiated by a glucose unit, through controlled enzymatic hydrolysis using inulinase. Palatinose oligosaccharides are prepared from sucrose by first synthesizing palatinose using the enzyme palatinose synthase and then intermolecular condensation. Lactosucrose is prepared from lactose and sucrose by transfructosylation with β-fructofuranosidase. It consists of a lactose molecule to which a fructose moiety is joined at the glucose residue by a β-2,1 glycosidic bond. Galactooligosaccharides are produced from lactose using the galactosyl transferase activity of β-galactosidase from *Kluyveromyces fragilis*, *K. lactis*, *B. circulans*, and *A. oryzae*.

V. ENZYMES IN THE MODIFICATION OF FATS AND OILS

Lipases (EC 3.1.1.3, glycerol ester hydrolases) are a ubiquitous class of enzymes that catalyze hydrolysis, esterification (synthesis), and transesterification (group exchange of esters). These enzymes are used for diverse purposes, such as fat hydrolysis, flavor development in dairy products, ester synthesis, transesterification of fats and oils, production of chiral organic compounds, production of washing and cleaning products, and treatment of domestic and industrial products. Lipolytic reactions occur at the lipid–water interface. There are two broad types of lipases based on their positional specificity. Nonspecific lipase releases fatty acids from all three positions of the glycerol moiety and is used to hydrolyze fats and oils completely to free fatty acids and glycerol. These are produced by *Candida* sp., *Staphylococcus* sp., and *Geotrichum* sp. The other type of the enzyme is 1,3-specific lipase, which releases fatty acids from 1,3 positions and preferentially frees fatty acids and di- and monoglycerides as the reaction products. This type of lipase is produced by *Aspergillus*, *Mucor*, *Rhizopus*, and *Pseudomonas* sp.

Conventional fat hydrolysis uses high-temperature steam-splitting (250–260°C at 50 bar), which requires high energy and creates environmental concerns. Lipases can be used to hydrolyze fats and oils with excellent yields, but the high cost of the enzyme makes the process uneconomical. Edible fats and oils with improved or new properties are produced by the action of regiospecific lipases. The enrichment of oils with highly unsaturated fatty acids can be made by lipase-mediated transesterification reactions. Enzymatic transesterification reactions are usually used to incorporate a new fatty acid into a triglyceride. Lipases are also used to assist in the extraction of fats and oils and for the development of cheese flavor. There is an increasing use of the versatile microbial lipases in the large-scale synthesis of fine chemicals, in various organic synthesis, and for the preparation of chiral compounds in enantiometrically pure form. The Unilever company has patented a process using a fixed-bed reactor containing immobilized lipase to convert palm oil and stearic acid to a cocoa-butter substitute.

VI. ENZYMES IN THE ANIMAL FEED INDUSTRY

Animal feed is composed mainly of plant materials, such as cereals, agricultural and milling by-products, and waste agricultural residues. These contain non-starch polysaccharides (such as β-glucan, cellulose, hemicellulose, and lignin), protein, and phytic acid. Monogastric animals (those with one stomach) generally cannot fully digest and use the fiber-rich feedstuff. Due to the complex nature of the feed materials, the starch sequestered by β-glucans and pentosans is also not degradable in the digestive process. The β-Glucan, arabinoxylan, and phytic acid, present in many feedstuffs, act as anti-nutritional factors, as these materials interfere with the digestibility, absorption, and use of nutrients, affecting animal production. β-Glucan contains glucose residues linked via β-1,3 and β-1,4 glycosidic linkages. The use of β-glucanases in poultry feed helps the incorporation of large quantities of barley in the diet. Arabinoxylans (pentosans) are heteropolymers containing mainly xylose and arabinose. Arabinose side-chains in arabinoxylans are often associated with ferulic acids, which form linkages between the arabinoxylans and lignin molecules in the plant cell walls. The addition of arabinofuranosidase to wheat-based diets can facilitate the removal of anti-nutritional characteristics of these diets.

Phytic acid (myo-inositol hexaphosphate) represents the major storage form of phosphates in plants. Up to 80% of the grain phosphorus is bound in the form of phytic acid. Phytic acid forms complexes with metal ions such as calcium, magnesium, iron, and zinc, thus preventing their assimilation by the animal. The addition of suitable microbial phytase (EC 3.1.3.16, phytate-6-phosphatase) liberates part of the bound phosphorus and makes it possible to reduce the phosphorus content of the feed by 30%. Using phytase, which is now commercially available, in animal feed can also alleviate environmental pollution and control certain diseases. The most common source of microbial phytase is *Aspergillus ficuum*.

Vegetable protein use can be enhanced by using microbial proteases. Thus, feed digestion and use by animals can be enhanced by adding the proper enzymes as feed additives. This also reduces overall waste by enhancing the efficiency of the digestion process. Various microbial enzymes are now used as feed enhancers and hold the prospect of playing an indispensible role in animal and poultry production.

VII. ENZYMES IN THE PULP AND PAPER INDUSTRY

In recent years, tremendous research efforts have been made to reduce the amount of chlorine used for bleaching of kraft pulp after the pulping processes. Chlorine-based (chlorine, chlorine dioxide, and hypochlorite) bleaching can result in the discharge of chloroorganics such as chlorinated phenols and chlorinated dioxins into the environment. Environmental pressures are forcing the pulp and paper industry to develop new technology to reduce or eliminate the presence of various contaminants in the bleaching-plant effluents.

The main constituents of wood are cellulose, hemicellulose, and lignin. Research on the use of enzymes in pulp manufacture involves the degradation or modification of hemicellulose and lignin without disturbing the cellulose fibers. Pretreatment of kraft pulp with xylanase promotes a decrease in lignin content (kappa number) and a brightness increase

of the treated pulp. Xylanase facilitates the lignin removal with high specificity from lignin–carbohydrate complexes. This enzymatic pretreatment reduces the amount of chlorine needed to reach a target brightness in pulp by chemical bleaching by 20–30%.

For high-quality pulps, extensive debarking is essential, but consumes a substantial amount of energy. The border between wood and bark is the cambium, which consists of only one layer of cells. This living cell layer produces xylem cells toward the inside of the stem and phloem cells toward the outside. Both cambium and phloem layers have a high content of pectins. Enzymes (mainly polygalacturonase) specific for the hydrolysis of the cambium and phloem layers facilitate bark removal. This new and attractive approach of enzymatic debarking shows great potential for saving both energy and raw material.

Pitch is the sticky resinous material in wood. The removal of pitch by chemical pulping and bleaching is not particularly efficient. Treatment with lipases has been found to be useful in reducing pitch deposits because lipases hydrolyze the triglycerides in the wood resin to fatty acids and glycerol. The enzyme does not affect the quality of and has a good effect on the product.

The removal of ink is an important part of waste-paper processing. Conventional de-linking involves the pulping of the paper in highly alkaline solution. It has been reported that cellulase enzymes can increase the efficiency of the de-linking process. But there is a concern that treatment of the secondary fibers by cellulases may decrease the fiber strength. More research needs to be done.

VIII. ENZYMES IN THE FRUIT-JUICE PROCESSING INDUSTRY

Enzymes can play an important role in the preparation and processing of various fruit and vegetable juices such as apple, orange, grapefruit, cranberry, pineapple, grape, carrot, and lemon. Fruits and vegetables are particularly rich in pectic substances. Pectin, a hydrocolloid, has a great affinity for water and can form gels under certain conditions. The addition of exogenous enzymes, such as pectinases, pectin lyase, pectin esterase, and polygalacturanase, allows a reduction in viscosity by pectin hydrolysis and improves the pressibility as the pectin gel collapses. For the complete liquefaction of fruits and vegetables, hemicellulases, cellulases, and amylases can be used in addition to pectinases. Arabinan, a polymer of arabinose, is an important component of fruit cell walls and may cause haze in fruit-juice concentrate. Haze formation can be prevented by using endoarabinases to break down arabinan into low-molecular-weight degradation products that do not precipitate. Various commercial enzyme preparations are now available for use in fruit juice industry.

Enzymatic de-bittering of grapefruit can be achieved through the application of fungal naringinase preparations. The enzyme preparation contains both α-rhamnosidase (EC 3.2.1.40) and β-glucosidase activities. α-Rhamnosidase first breaks down naringin, 7-(2-rhamnosido-β-glucoside), (an extremely bitter flavanoid) to rhamnose and prunin and then β-glucosidase hydrolyzes prunin to glucose and naringenin. Prunin bitterness is less than one-third of that of naringin. However, α-rhamnosidase is competitively inhibited by rhamnose, and β-glucosidase is inhibited by glucose. Immobilized enzymes are used to solve the inhibition problems. Another enzyme, glucose oxidase (EC 1.1.3.4), is used to scavenge oxygen in fruit juice and beverages to prevent color and taste changes. Glucose oxidase is produced by various fungi such as *Aspergillus niger* and *Penicillium purpurogenum*.

by the indigenous enzymes (cathepsins and the Ca^{2+}-dependent neutral protease). Current practice in the United States favors immersion of meat cuts in concentrated enzyme solutions, followed by vacuum packaging and refrigeration for up to 3 weeks. Enzymes are also used for the separation of hemoglobin from blood proteins and the removal of meat from bones. For the preparation of pet food, minced meat or meat by-products are hydrolyzed by a protease to produce a liquid meat digest or a slurry with a much lower viscosity. Pets like the savory flavor generated by peptides and amino acids produced during the enzymatic hydrolysis of meat.

Fish protein concentrates (FPC) are generally prepared by treating ground fish parts with a protease. After hydrolysis, the bones and scales are removed by screening, and the mixture of solubles and undigested fish solids is dried or separated by centrifugation. The yield of solubles, mainly amino acids and peptides, is generally 60–70% of the initial fish solids. Enzymatic treatment of fish stickwater (press water obtained in fish-meal manufacture) is performed with a microbial protease (from *B. lichiniformis*) prior to concentration by evaporation. This reduces its viscosity to 20–50%, which helps to increase the final solid contents by 55–73% under industrial conditions. Thus, the use of enzymes substantially saves in drying costs. Enzymes (xanthin oxidase, catalase, nucleoside phosphorylase, and nucleotidase) can also be used for testing the quality and freshness of fish.

IX. ENZYMES IN THE MEAT- AND FISH-PROCESSING INDUSTRY

Proteinases, either indigenous (cathepsin) or obtained from plants and microorganisms, are used in the meat and fish industries to tenderize meat and solubilize fish products. Plant proteinases include papain (EC 3.4.4.12) from papaya, ficin (EC 3.4.4.24) from figs, and bromelain (EC 3.4.4.24) from pineapples. Microbial enzymes include fungal proteases (from *A. oryzae* and *A. niger*) and bacterial proteases (from *B. subtilus* and *B. lichiniformis*). Tenderization of meat can be achieved by keeping the rapidly chilled meat at 1–2°C to allow proteolysis

X. ENZYMES IN THE DAIRY INDUSTRY

The application of enzymes (proteases, lipases/esterases, lactase, and catalase) in dairy technology is well established. Rennets (renin, a mixture of chymosin and pepsin obtained mainly from animal and microbial sources) are used for the coagulation of milk during cheese production. Proteases of various kinds are used for the acceleration of cheese ripening, for modification of functional properties, and for modification of milk proteins to reduce the allergenic properties of cow milk products for infants. Lipases are used mainly in cheese ripening for development of lipolytic flavors. Lactase (β-galactosidase, EC

3.2.1.23) is used to hydrolyze lactose to glucose and galactose in order to increase digestion or to improve the solubility and sweetness of various dairy products. Many people do not have sufficient lactase to digest milk sugar. Lactose hydrolysis helps lactose-intolerant people to drink milk and eat various dairy products. β-Galactosidase from *Kluyveromyces fragilis*, *A. niger*, or *A. oryzae* are inhibited by galactose. Immobilized enzyme systems are used to overcome this inhibition problem and to lower the cost of lactase use. The cheese-manufacturing industry produces large quantities of whey as a by-product, of which lactose represents 70–75% of the whey solids. The hydrolysis of lactose by lactase converts whey into more useful food ingredients. Lactases have also been used in the processing of dairy wastes.

Hydrogen peroxide is used as an effective chemical sterilant for the treatment of raw milk as an alternative to pasteurization with heat. Catalase (EC 1.11.1.6), which catalyzes the decomposition of hydrogen peroxide, is then used to hydrolyze the residual peroxide. The bitter off-flavors that develop in ripened cheese as it matures are due to the formation of bitter-flavored peptides from milk proteins. Peptidases can break down the bitter peptides as they are formed and thus help to maintain the traditional flavor of cheese.

XI. ENZYMES IN DETERGENTS

Currently, the detergent industry occupies about 25–30% of the entire industrial enzyme market. Over half of all laundry detergents contain enzymes such as protease, amylase, lipase, and cellulase. In order to perform well in laundry-detergent environments, these enzymes must be very efficient; work at highly alkaline pH conditions (pH 9–11) and high temperatures; be stable in the presence of chelating agents, perborates, and surfactants; and possess long storage stability (>1 year). The use of enzymes allows the launderer to use lower temperatures and a shorter period of agitation after a preliminary period of soaking. Proteases are the most widely used enzymes in the detergent industry. They remove protein-based stains such as blood, egg, grass, meat sauce, and body secretions. These detergent enzymes (serine

proteases) are produced by the fermentation of *B. lichiniformis*, *B. amyloliquefaciens*, or *Bacillus* sp. Amylases remove the residues of starchy foods such as mashed potato, spaghetti, and oatmeal. Lipases facilitate the removal of fatty stains such as lipsticks, frying fats, butter, salad oil, sauces, and tough stains on collars and cuffs. The lipase from *Humicola lanuginosa*, produced by recombinant DNA technology, is effective at pH ≤ 12 and temperatures ≤ 60°C. Recently, an alkali-stable fungal cellulase preparation has been introduced for use in washing cotton fabrics. Treatment with these cellulase enzymes removes the small fibers without apparently damaging the major fibers, and restore the fabric to its "as new" condition by improving the color brightness, enhancing the softness, and removing particulate soiling. Bleach-stable enzymes (amylase and protease) are now available for use in automatic dishwashing detergents.

XII. ENZYMES IN THE LEATHER INDUSTRY

Animal skin is composed of 60–65% water, 30–32% protein, approximately 10% fat, and 0.5–1% minerals. Skins are soaked initially to clean them and to allow rehydration. Proteolytic enzymes effectively facilitate the soaking process. In addition, lipases have also been used to dissolve and remove fat. Dehairing is then carried out using an alkaline protease, such as subtilisin, in a very alkaline bath. The alkaline condition tends to swell the hair roots, easing the removal of the hair by allowing proteases to selectively attack the protein in the hair follicle. The conventional dehairing process required harsh chemicals such as slaked lime and sodium sulfide, which essentially swells the hide and loosens and damages the hair. The use of enzyme-based dehairing process has lead to much less pollution from tanneries.

Tanning wastes, consisting primarily of glue stock, cannot be discarded at waste dump sites and must be completely processed. All proteins are solubilized at a high alkaline pH with bacterial proteases in order to partition the fat from the hydrolyzate. After being concentrated, the protein is dried and used as a nutritional supplement in animal feed.

TABLE I
Microbial Enzymes and Their Applications in Biotechnology

Enzyme	Producer microorganism	Use
Acetolactate decarboxylase	*Lactobacillius* sp.	Acceleration of maturation process in beer production
α-Amylase	*Bacillus amyloliquefacians*	Starch liquefaction, viscosity reduction, brewing, baking, confectionery, textile, detergents
	Bacillus licheniformis	
	Bacillus stearothermophilus	
β-Amylase	*Bacillus polymyxa*	Starch saccharification to maltose, brewing, baking
Aminopeptidase	*Lactococcus lactis*	Protein break down to peptides, cheese, soy sauce, removes bitter taste, tenderize meat, remove haze from beer
Anthocyanase	*Aspergillus niger*	Decolorizing fruit juices and wines
Arabinofuranosidase	*Aspergillus niger*	Liberates arabinose, animal feed, wine making, lignocellulosic biomass conversion, delignification of pulp
	Aureobasidium pullulans	
Catalase	*Aspergillus niger*	Oxygen removal, lemonades, egg whites, contact lens care
Cellulases	*Trichoderma reesei*	Fruit and vegetable processing, juice clarification, oil extraction, biomass conversion, textile industries, detergents, deinking, animal feeds, digestive aid
	Penicillum purpurogenum	
	Clostridium thermocellum	
Chitinase	*Trichoderma harzianum*	Antimicrobial activity
	Streptomyces sp.	
Cyclomaltodextrin glucano-transferase	*Bacillus* sp.	Cyclodextrin production
Fumarase	*Brevibacterium flavum*	Conversion of fumaric acid to malic acid
α-Galactosidase	*Aspergillus niger*	Raffinose hydrolysis to sucrose and galactose in sugar beet, flatulence reduction, processed legume-based foods
	Aspergillus pheonics	
	Saccharomyces cerevisiae	
β-Galactosidase (lactase)	*Aspergillus niger*	Milk sugar hydrolysis to glucose and galactose; cheese whey hydrolysis
	Kluyveromyces fragilis	
	Candida pseudotropicalis	
β-Glucanase	*Aspergillus niger*	Improved feed utilization, brewing, baked goods
	Bacillus subtilis	
Glucoamylase	*Aspergillus niger*	Starch saccharification to glucose, brewing, baking, alcohol production
	Endomycopsis fibuligera	
	Rhizopus niveus	
Glucose isomerase	*Bacillus coagulans*	Conversion of glucose to fructose, high fructose corn syrups
	Actinoplanes missouriensis	
	Streptomyces olivochromogenes	
Glucose oxidase	*Aspergillus niger*	Oxygen scavenger, beverages, eggs, fruit juices, wine
	Penicillium amagasakiense	
Invertase	*Aspergillus* sp.	Invert sugar production, baking, confectionery, artificial honey manufacture
	Saccharomyces sp.	
Isoamylase	*Pseudomonas amylodermosa*	Starch debranching, brewing, maltose production
Laccase	*Corioltus versicolor*	Oxygen removal

continues

Continued

Enzyme	Producer microorganism	Use
Lipase	*Aspergillus niger* *Candida rugosa* *Geotrichum candidum* *Humicola lanuginosa*	Cheese ripening, flavor development, fat and oil modification, transesterification of fatty acids, stereoselective transformation, pitch control in pulp and paper industry, detergents, organic synthesis
Naringinase	*Aspergillus niger*	Citrus fruit juice debittering
Nitrile hydratase	*Rhodococcus rhodochrous*	Acrylamide production from acrylonitrile, nicotinamide production
Pectinases	*Aspergillus niger* *Aspergillus oryzae*	Fruit juice processing, coffee processing, wine
Penicillin acylase	*Bacillus megaterium*	Antibiotic production
Phytase	*Aspergillus ficuum* *Aspergillus niger*	Liberation of phosphate groups, animal feed
Proteinases (serine proteinase, carboxylproteinase, metalloproteinase)	*Bacillus subtilis* *Aspergillus*, sp.	Protein hydrolysis, cheese, meat, fish, cereal, fruit, beverage, baking, leather, laundry detergenets, gelatin hydrolysis, peptide synthesis
Pullulanase	*Aerobacter aerogenes* *Bacillus acidopullulyticus*	Starch debranching, glucose and maltose production, brewing
Rennet	*Mucor* sp.	Milk coagulation, dairy industry
Tannase	*Aspergillus oryzae*	Improved solubility of instant tea, reduces chill haze formation in beer, wine making, animal feed additive
Thermolysin	*Bacillus thermoproteolyticus*	Synthesis of aspartame
Xylanases	*Trichoderma* sp. *Aspergillus* sp.	Pulp and paper making, fruit and vegetable processing, juice clarification, extraction processes, biomass conversion, textile industries, detergents, de-inking, animal feeds, digestive aid

XIII. ENZYMES IN THE PRODUCTION OF FINE CHEMICALS

Enzymes or whole cells have been extensively used as catalysts in organic solvents for production of a variety of fine chemicals and pharmaceuticals. One such chemical is indigo dye, which has a market size of $200 million. The manufacture of indigo dye used for the dying of clothes, particularly denims, requires a harsh chemical process and generates carcinogenes and toxic wastes. The company Amegen has developed a biocatalytic process for the production of indigo. The pathway forming indigo involves converting tryptophan to indole by the enzyme tryptophanase, hydroxylating indole to indoxyl by naphthalene dioxygenase, and a nonenzymic step to form indigo from indoxyl (spontaneous oxidation).

Aspartame (L-aspartyl-L-phenylalanine methyl ester) is an artificial dipeptide sweetener about 200 times sweeter than sucrose, which has a market of more than $1 billion. Aspartame can now be made by using an enzymatic process. The enzyme thermolysin (a metaloprotease produced by *B. thermoproteolyticus*) catalyzes the amide bond formation between L-aspartic acid and phenylalanine methyl ester. It is enantioselective and forms the peptide bond only with L-phenylalanine methyl ester. The enzyme is also regioselective and does not react with the β-carboxy of the aspartic acid, so no bitter-tasting β-aspartame is formed. L-Aspartic acid can also be prepared enzymatically from ammonium fumarate substrate by a single enzyme L-aspartate ammonia lyase obtained from *Escherichia coli*.

The β-lactam antibiotics, the penicillins and cephalosporins, are the most widely used antibiotics. 6-Aminopenicillanic acid (6-APA) can be produced enzymatically (pH 7.0, 35°C) from fermentation-derived penicillins (Penicillin G) by deacylation using immobilized penicillin amidase (EC 3.5.1.11), frequently called penicillin acylases, from *E. coli* or *B. megaterium.* 6-APA is an important intermediate compound for the manufacture about 20 semisynthetic penicillins.

Other fine chemicals produced by enzymes include acrylamide and catechol. Acrylamide, an industrially important chemical, can be prepared by transforming acrylonitrile into acrylamide by using the enzyme nitrile hydratase from *Rhodococcus rhodochrous* in immobilized form. A biocatalytic alternative to the usually employed industrial synthesis of catechol has been developed using glucose as substrate.

XIV. ANALYTICAL APPLICATIONS OF ENZYMES

Enzymes are used in various analytical methods, both for medical and nonmedical purposes. Immobilized enzymes, for example, are used as biosensors for the analysis of organic and inorganic compounds in biological fluids. Biosensors have three major components—a biological component (e.g., enzyme and whole cell), an interface (e.g., polymeric thick or thin film), and a transducing element that converts the biochemical interaction into a quantifiable electrical or optical signal. A glucose biosensor consists of a glucose oxidase membrane and an oxygen electrode, whereas a biosensor for lactate consists of immobilized lactate oxidase and an oxygen electrode. The lactate sensor functions by monitoring the decrease in dissolved oxygen that results from the oxidation of lactate in the presence of lactate oxidase. The amperometric determination of pyruvate can be carried out with the pyruvate oxidase sensor, which consists of a pyruvate oxidase membrane and an oxygen electrode. For the determination of ethanol, the biochemical reaction cell using an alcohol dehydrogenase (ADH, EC 1.1.1.1) membrane anode is used.

A bioelectrochemical system for total cholesterol estimation was developed, based on a double-enzymatic method. In this system, an immobilized enzyme reactor containing cholesterol esterase (EC 3.1.1.13) and cholesterol oxidase (EC 1.1.3.6) is coupled with an amperometric detector system. An amino acid electrode for the determination of total amino acids has also been developed using the enzymes L-glutamate oxidase, L-lysine oxidase, and tyrosinase. Enzyme electrodes are used for the continuous control of fermentation processes. Table I summarizes some microbial enzymes and their applications in biotechnology.

XV. CONCLUSION

This article gives a general overview of the use of enzymes in biotechnology. The use of enzymes in various fields including agrochemicals, pharmaceutical, and organic synthesis is increasing rapidly. In the twenty-first century, we will observe tremendous progress being made in the use of microbial enzymes in biotechnology. Some industrial enzymes are already produced in high yields by using genetically engineered microorganisms. New and improved enzymes, with respect to stability, activity, and specificity, will be made possible by the use of site-directed mutagenesis. Thermophilic microorganisms, especially extremophiles, hold great potential for developing thermostable enzymes for biotechnology. Psychrophilic microorganisms, on the other hand, will be useful in the development of cold-active enzymes. There is an increasing demand to replace some of the traditional chemical processes with environmentally benign chemistry due to the growing awareness of the toxic effects of certain chemicals; this will also help to create a market for new enzymes.

See Also the Following Articles

DAIRY PRODUCTS • LIGNOCELLULOSE, LIGNIN, LIGNINASES • PULP AND PAPER

Bibliography

Cabral, J. M. S., Best, D., Boross, L., and Tramper, J. (eds.) (1994). "Applied Biocatalysis." Harwood Academic, Chur, Switzerland.

Crittenden, R. G., and Playne, M. J. (1996). Production, properties and applications of food-grade oligosaccharides. *Trends in Food Sci. Technol.* 7, 353–361.

Dordick, J. S. (ed.) (1991). "Biocatalysts for Industry." Plenum Press, New York.

"Enzymes at Work." (1995). Novo Nordisk, Bagsvaerd, Denmark.

Godfrey, T., and West, S. (eds.) (1996). "Industrial Enzymology," 2nd ed. Macmillan Press, London.

Kennedy, J. F. (ed.) (1987). "Biotechnology, Vol. 7a, Enzyme Technology." H.-J. Rehm, and G. Reed (series eds.). VCH Verlagsgesellschaft, Weinheim, Germany.

Laskin, A. I. (ed.) (1985). "Enzymes and Immobilized Cells in Biotechnology." Benjamin/Cumming, Menlo Park, CA.

Saha, B. C., and Woodward, J. (eds.) (1997). "Fuels and Chemicals from Biomass." American Chemical Society, Washington, D.C.

Tucker, G. A., and Woods, L. F. J. (eds.) (1995). "Enzymes in Food Processing," 2nd ed. Blackie Academic, Bishopbriggs, UK.

Uhlig, H. (1998). "Industrial Enzymes and Their Applications." John Wiley & Sons, New York.

Erwinia: Genetics of Pathogenicity Factors

Arun K. Chatterjee, C. Korsi Dumenyo, Yang Liu, and Asita Chatterjee

University of Missouri-Columbia

GLOSSARY

bacteriocin An antibacterial compound or protein produced by some bacterial strains that is active against certain strains of the same or closely related species. Bacteriocin-producing strains usually carry gene(s) that confer resistance to the bacteriocin.

blackleg A bacterial disease of potato caused by *Erwinia carotovora* subspecies *atroseptica* that is characterized by black shriveled lower stems and rotting of the tuber.

chaperone A protein that is required for the proper folding, secretion, or assembly of another protein and that protects the target protein(s) from proteolytic attack or premature processing.

global regulator A protein or RNA that functions directly or indirectly to increase or decrease the expression of specific sets of genes or phenotypes.

hypersensitive reaction (HR) Localized and rapid physiological reaction in a nonhost plant in response to bacterial, fungal, or viral plant pathogens or their elicitors. The manifestation of the HR is due to the death of the plant cells that have come in contact with the pathogen or the elicitor.

metalloprotease An enzyme that breaks down other proteins by the cleavage of the peptide bonds and in which heavy metal ions are bound directly to some of the amino acid side-chains.

pathogenicity The ability of a pathogen to infect the host and cause disease.

peritrichous Describing flagella that are uniformly distributed over the bacterial body.

quorum sensing A phenomenon by which bacteria monitor their population via the levels of signal molecules (e.g., acyl homoserine lactones) present in the culture. The signal molecules are usually diffusible and their production is cell density dependent. The accumulation of such molecules to the threshold concentration signals the bacteria that their density has reached a quorum. The bacteria then begin expressing the genes or phenotypes that are controlled in a cell density-dependent manner.

regulon A group of genes whose expression is regulated by a common regulatory factor or signal.

secondary metabolite A molecule or substance not usually required for sustenance, whose accumulation occurs at the later stages of growth, regulated differently than primary metabolism.

secretion systems Pathways by which bacteria target molecules, such as proteins, across the cell envelope to the periplasm, extracellular medium, or directly into the host cells.

sigma (σ) factor A component of RNA polymerase holoenzyme (RNA polymerase + sigma factor), that determines the specificity of the polymerase. Each sigma factor recognizes specific sequences in the target promoters and directs the holoenzyme to initiate transcription. Sigma factors other than the predominant (σ^{70}) are referred to as alternate.

THE GENUS *ERWINIA* of the family Enterobacteriaceae is an assemblage of gram-negative, nonsporulating, facultatively anaerobic, and peritrichously flagellated bacteria isolated from plants, soil, and animals, including humans. Disease-causing organisms including *Erwinia* produce substances that enable them to survive within and outside the host, to colonize the host, and to ultimately inflict harm. Such substances, also referred to as pathogenicity factors, include enzymes, nonenzymatic proteins, peptides, hormones, toxins, and polysaccharides. These factors may alter host physiology or degrade preformed structures leading to plant cell death, cause cell enlargement, or induce uncontrolled cell multiplication. The ensuing symptoms to plant hosts may range from wilts to maceration to necrosis to galls, depending on the host plant and the pathogenicity factor. In addition to such factors, many *Erwinia* species produce secondary metabolites including antimicrobial substances such as antibiotics, pigments, and homoserine lactone derivatives, the quorum-sensing signals. Many *Erwinia* species normally produce most of these substances at low basal levels, but these are induced *in planta* or in culture in response to specific environmental conditions as well as plant and bacterial signals.

I. INTRODUCTION

The lumping of strains colonizing diverse habitats into the genus *Erwinia* has brought together bacteria that are metabolically and genetically diverse. Recognition of this diversity has from time to time prompted proposals for taxonomic revision by reshuffling the individual members into different genera of the family Enterobacteriaceae. A proposal recommends the reorganization of the members of *Erwinia* into four genera (Table I) based on ribosomal DNA sequence homology. Accordingly, the genus *Erwinia* would mostly include members of the Amylovora group; the genus *Pectobacterium* would comprise strains belonging to the Cartovora or the soft rot group; the genus *Pantoea* would comprise members of the Herbicola group as well as some enterobacterial species previously classified into the genus *Enterobacter*; and the new genus, *Brenneria*, would house such species as *alni, nigrifluens, paradisiaca, quercina, rubrifaciens, and salicis*. With the exception of *Pantoea*, this and other proposals for splitting the genus *Erwinia* have not been widely accepted. Therefore, for the sake of clarity, this article follows the nomenclature in vogue, such as *E. amylovora, E. rhapontici, E. carotovora* subsp. *atroseptica* [*E. c. atro-*

septica], *E. carotovora* subsp. *betavasculorum* [*E. c. betavasculorum*], *E. carotovora* subsp. *carotovora* [*E. c. carotovora*], *E. chrysanthemi, E. herbicola, E. herbicola* pv. *gypsophilae* [*E. h. gypsophilae*], *E. milletiae, E. stewartii,* and *E. uredovora*. Some salient characteristics of these bacteria and other members of the genus are given in Table I.

This article focuses on the genetics of pathogenicity factors and secondary metabolites and their regulation in *Erwinia* species. The intent is to provide a general but brief overview of these topics. Space limitation does not permit an in-depth analysis of the deficiencies that remain in knowledge of pathogenicity mechanisms or the controversies that surround some of the prevailing notions regarding regulatory systems. The reader should consult comprehensive review articles that deal with specific aspects of plant–bacteria interactions.

II. EXOENZYMES

Soft rot *Erwinia* species including *E. c. atroseptica, E. c. betavasculorum, E. c. carotovora,* and *E. chrysanthemi* produce an assortment of exoenzymes (Table II), which cause degradation of plant cell-wall components, resulting in tissue maceration and cell death. Pectate lyase (Pel) and polygalacturonase (Peh) act on pectate (polygalacturonic acid) as well as on pectin that is partially methyl esterified. Pectin lyase (Pnl) and pectin methyl esterase (Pem), on the other hand, act on the methyl esterified derivatives of pectate. In fact, for the Pnls of soft rot *Erwinia*, the preferred substrate is highly methyl esterified pectins including Link pectin, which is >95% methyl esterified. Oligogacturonate lyase (Ogl) acts on oligouronides, the products resulting from depolymerization of pectin or pectate. In addition, soft rot *Erwinia* produce extracellular cellulase (Cel), which cleaves carboxymethyl cellulose, as well as extracellular protease (Prt), which acts on a variety of substrates including plant cell-wall proteins. *E. chrysanthemi* also produces other exoenzymes such as phospholipase C (Plc), pectin acetyl esterase (Pae) and xylanase (Xyn). Because mutants of *E. chrysanthemi* strains deficient in the production of Xyn or Plc are unaffected in pathogenicity, it is doubtful

TABLE I
Taxonomic Status of Plant Pathogenic *Erwinia* Species

Erwinia *species*	*Groups* (new genera)	*Host plants*	*Disease symptoms*
E. amylovora	Amylovora (*Erwinia*)	Rosaceous species	Fire blight
E. rhapontici		Rhubarb, wheat, pea, onion	Rotting, pink seed
E. persicinus[a]		Tomato, banana, cucumber	Rotting
E. mallotivora		Oak	Leaf spot
E. tracheiphila		Melon, cucumber	Wilt
E. psidii		Guava	Die back, vascular tissue collapse
E. alni[a]	(*Brenneria*)	*Alder* species	Bark canker
E. salicis		Willow (*Salix* species)	Watermark disease
E. nigrifluens		Walnut	Bark canker
E. paradisiaca		Banana	Pseudostem pit
E. quercina		Acorn	Rotting
E. rubrifaciens		Walnut	Bark canker
E. herbicola	Herbicola (*Pantoea*)	Beet, gypsophila	Gall
E. milletiae		*Milletia japonica*	Gall
E. ananas		Pineapple, sugarcane	Rotting
E. uredovora		*Puccinia* uredia	
E. stewartii		Maize	Vascular wilt, leaf blight
E. aphidicola[a]		Pea aphids	Motality
E. carotovora subsp.	Carotovora (soft rot) (*Pectobacterium*)	Potato, tomato, sugar beet	Soft rot, blackleg
E. chrysanthemi		Potato	Soft rot, blackleg
E. cypripedii		Orchids	Brown rot
E. cacticida[a]		Cacti	Rotting

[a] These species designations were not included in the Approved Lists of Bacterial Names.

that these enzymes play significant roles in host–pathogen interactions.

A. Protease (Prt)

Most soft rotting *Erwinia* produce extracellular Prts, many of which are metalloproteases. In *E. chrysanthemi*, four Prt isoenzymes, PrtA, PrtB, PrtC, and PrtG, are encoded by the *prtA*, *prtB*, *prtC*, and *prtG* genes, respectively. The DNA cluster containing *prtA*, *prtB*, and *prtC* genes is closely linked with the *prtG-inh-prtDEF* operon. *prtD*, *prtE*, and *prtF* specify cell envelope-localized components of the Prt secretion machinery. The *inh* gene encodes a protease inhibitor that could inhibit the processing of the Prt precursor in the periplasm. In *E. c. carotovora*, one protease, Prt1 and the corresponding gene *prt1* have been characterized. The residual Prt activity in Prt1-deficient mutants of *E. c. carotovora* is apparently due to other Prt isoenzymes. Prt1 has less sequence similarity with the *E. chrysanthemi* Prts than with the Prts produced by other bacteria. However, *Erwinia* Prts have conserved amino acid sequences involved in Zn^{2+} binding and catalytic activity.

B. Pectate Lyase (Pel)

Multiple molecular forms (isoenzymes) of Pel occur in *E. carotovora* subspecies and *E. chrysanthemi*. Isoenzymes are distinguished by their molecular masses, isoelectric points, and kinetic properties. Pels cleave polygalacturonic acid (pectate) or partially methyl esterified pectate by a trans-eliminative reaction with the introduction of double bonds between C4 and C5 of the uronic acid residues (Table II). The catalytic activity of Pel is calcium-dependent and is optimal at relatively high pH values. In *E. chrysanthemi*, its cognate *pel* gene encodes each of the nine Pel isoenzymes: *pelA*, *pelB*, *pelC*, *pelD*, *pelE*, *pelI*, *pelL*, *pelX*, and *pelZ*. Of these, the products of

TABLE II
Major Exoenzymes Produced by the Soft Rot *Erwinia* Species

Enzyme	Producing species	Isoenzymes[a]	Mode of action[b]	Preferred substrate
Pectate lyase (Pel)	*E. carotovora* subsp., *E. chrysanthemi*	Yes	Trans-elimination (endo and exo)	Polygalacturonic acid
Polygalacturonase (Peh)	*E. carotovora* subsp., *E. chrysanthemi*	No	Hydrolysis (endo and exo)	Polygalacturonic acid
Cellulase (Cel)	*E. carotovora* subsp., *E. chrysanthemi*	Yes	Hydrolysis	Carboxy-methyl cellulose
Protease (Prt)	*E. carotovora* subsp., *E. chrysanthemi*	Yes	Hydrolysis	Gelatin, Azocasein
Pectin lyase (Pnl)	*E. c. atroseptica, E. c. carotovora, E. chrysanthemi, E. herbicola, E. rhapontici, E. milletiae*	No	Trans-elimination	Pectin
Pectin methyl esterase (Pem)	*E. chrysanthemi*	Yes	Hydrolysis	Pectin
Phospholipase (Plc)	*E. chrysanthemi*	No	Hydrolysis	Lecithin
Pectin acetyl esterase (Pae)	*E. chrysanthemi*	No	Hydrolysis	Demethylated acetyl-esterified pectin oligomers
Oligo galacturonate lyase (Ogl)	*E. c. carotovora, E. chrysanthemi*	No	Trans-elimination (endo and exo)	Oligogalacturonate
Xylanase (Xyn)	*E. chrysanthemi*	No	Hydrolysis	RBB-Xylan[c]

[a] Multiple isoenzymes are present in at least one strain.

[b] Endo refers to random cleavage of the substrate yielding oligomers and exo refers to cleavage from the ends of the polymeric substrate.

[c] RBB = remazol brilliant blue.

the first five genes are considered the primary Pels, whereas the rest are considered the secondary Pels. In *E. c. carotovora*, up to three *pel* genes have been characterized in various strains. Pels have been classified into six families based on their primary amino acid sequence. The five primary Pels of *E. chrysanthemi* (PelA, PelB, PelC, PelD and PelE) as well as several bacterial and fungal Pels and Pel-like plant proteins make up family I. Family II comprises the *E. carotovora* periplasmic Pels, PelB and Pel153. PelI of *E. chrysanthemi* and Pel3 of *E. c. carotovora*, which share homology with the Hrp W proteins of *E. amylovora* and *Pseudomonas syringae* (see Section IV.B for HrpW proteins), belong to family III. PelX and PelL of *E. chrysanthemi* make up family IV, while pelZ of *E. chrysanthemi* belongs to family V. The product of *pelA* of *Azospirillum irakense* is the sole member of family VI.

The genes encoding Pels in *Erwinia* are chromosomally located as single transcriptional units and

may be scattered or clustered on the genome. In *E. chrysanthemi*, *pelADE* and *pelBC* are two distinct clusters. The clustering of highly related *pel* genes has prompted the notion that some of these genes may have evolved by gene duplication. Each Pel has a signal peptide that is cleaved during Sec-dependent (type II) secretion, resulting in a mature secreted product.

C. Polygalacturonase (Peh)

Strains of *E. chrysanthemi* produce exo-Pehs, whereas *E. c. carotovora* strains produce endo-Pehs. The Pehs from *E. c. carotovora* strains are highly similar to each other and with numerous fungal and bacterial enzymes. PehX of *E. chrysanthemi* has higher homology with exo-Pehs from other bacteria, such as exo-Peh from *Yersinia enterocolitica*, PehB from *Ralstonia solanacearum*, and PglA from *Thermoanaerobacterium thermosulfurigenes*, than with endo-

Pehs from *E. c. carotovora*. Pehs act on pectate and pectin (partially methyl esterified pectate), but cleave the substrate by hydrolysis (Table II). Pehs are usually differentiated from Pels by their catalytic properties, especially their relatively low pH optima and reaction products. Although every strain of soft rot *Erwinia* produces multiple Pel isoenzymes, each of these bacteria generally produces a single Peh species. This contrasts with the production by several fungi of a single Pel but of many Peh isoenzymes, which are usually encoded by gene families.

D. Cellulase (Cel)

E. chrysanthemi and most *E. c. carotovora* strains produce at least two isoenzymes of Cel. In *E. c. carotovora* strain LY34, up to five Cel isoenzymes were detected by gel activity staining, although only two *cel* genes have been isolated from this and other strains. *celV* and *celS* (or their homologs) encode the two Cels of *E. c. carotovora*; and in *E. chrysanthemi*, the two Cel isoenzymes, EGZ (endoglucanase Z) and EGY (endoglucanase Y) are encoded by *celZ* and *celY*, respectively. The two pairs of genes do not share significant homology with each other. EGZ in *E. chrysanthemi* and CelV in *E. c. carotovora* are considered the major cellulases, as these account for at least 95% of the total Cel activity in each species. Strains deficient in EGZ or CelV are reduced in pathogenicity. Cels have N-terminal signal peptides, the secretion signals found in proteins secreted by the type II system (see Section II.H). In addition, Cels possess discrete functional domains—the catalytic domain; the cellulose-binding domain, which has also been implicated in secretion; and a linker region connecting the two domains. All glycosyl hydrolases (i.e., enzymes that catalyze the hydrolytic cleavage of bonds between two monomers in a polysaccharide) have been classified into families based on the amino acid sequences of their catalytic domains. Seventy families of these enzymes are recognized, of which 11 (families 5, 6, 7, 8, 9, 10, 11, 12, 26, 44, and 48) make up the cellulases.

E. Pectin Lyase (Pnl)

Pnl production is induced upon DNA damage in several *Erwinia* species, including *E. chrysanthemi*,

E. c. atroseptica, *E. c. carotovora*, *E. herbicola*, *E. rhapontici*, and *E. milletiae*. Immunological studies with antibodies raised against an *E. chrysanthemi* pnl have indicated that the *E. chrysanthemi* Pnl is more closely related to Pnl from *E. c. atroseptica* than to Pnls produced by *E. c. carotovora*, *E. rhapontici*, and *E. milletiae*. The main distinguishing characteristics of the *Erwinia* Pnls are their preference for highly methyl esterified pectin, calcium-independent catalytic activities, and induction upon DNA damage. Like Pel, Pnl cleaves the substrate by trans-elimination. The pectin lyase structural genes of strains 71 and Er of *E. c. carotovora* have been characterized. The enzymes from these two strains are very similar in structural and kinetic properties and these have a high sequence homology with Pnls from *Pseudomonas marginalis* and *Bacillus subtilis* as well as Pels from *P. marginalis* and various fungi.

F. Pectin Methyl Esterase (Pem)

Pems act by hydrolysis, cleaving the methyl group from the esterified pectin, have a pH optimum around 7.5, and do not require metal ions as cofactors. Pem isoenzymes, PemA and PemB are encoded in *E. chrysanthemi* by *pemA* and *pemB* genes, respectively. The two isoenzymes, however, differ in their substrate preference. PemA prefers highly polymerized pectin and is secreted to the extracellular space by the type II system (see Section II.H). PemB on the other hand is an outer-membrane-localized lipoprotein that prefers oligomers as substrates. PemA and PemB share only 20% identity. In fact, PemA is more closely related to plant Pems, whereas PemB is related to a Pem of *Ralstonia solanacearum*. A PemA-deficient mutant is reduced in virulence as it produces 83% of the wild-type levels of systemic infection in *Saintpaulia ionantha* plants.

G. Pectin Acetyl Esterase (Pae)

In *E. chrysanthemi* strain 3937, one pectin acetyl esterase, PaeY, which is encoded by *paeY*, has been characterized. However, the genetic basis for the occurrence of three additional isoenzymes in the PaeY mutant is not known. *paeY* is located between *pelD* and *pemA* on the chromosome, and this genetic organization is conserved in other *E. chrysanthemi* strains. The enzyme has low homology with rhamnogalacturonan esterase from *Aspergillus aceleatus*. As does Pem, this enzyme acts by hydrolysis, cleaving the acetyl group from the polymeric substrate and releasing acetate (Table II). PaeY possesses an N-terminal signal sequence typical of the proteins secreted by the type II secretion pathway. The protein is secreted to the periplasm in recombinant *E. coli* and to the extracellular milieu by *E. chrysanthemi*. Pae has a pH optimum of about 8.0, prefers de-esterified oligomers as substrates, and can also act on synthetic substrates such as X-acetate and triacetin. The enzyme does not require a metal ion cofactor, but $HgCl_2$ and $CuCl_2$ inhibit the activity. PaeY mutants of *E. chrysanthemi* are attenuated in their virulence as they cause only about 53% of the wild-type level of systemic infection in *Saintpaulia ionantha*.

H. Protein Secretion Systems

Secretion of pathogenicity factors and pathogenicity-associated proteins by *Erwinia* occurs through three distinct pathways, generally referred to as type I, type II, and type III secretion systems. The salient features of the type I and type II systems are given here; the characteristics of the type III system, which is used for the secretion of Avr (Avirulence) proteins, Dsp (Disease-specific) proteins, and Harpins, are presented in Section IV.C.

Prts of *E. chrysanthemi* and most likely of *E. c. carotovora* and other soft rot *Erwinia* are secreted in a signal peptide-independent manner by the type I secretion system, which recognizes the secretion signal located at the C-terminal segment of the protein. The Prt secretion system in *E. chrysanthemi* is composed of three membrane proteins, PrtD, PrtE, and PrtF, the respective homologs of HlyB, HlyD, and TolC of the hemolysin secretion system of *E. coli*. PrtD is inner-membrane localized and has an ATP binding cassette (ABC). PrtE connects PrtD and the outer-membrane-localized PrtF. Prts are synthesized as zymogens and secreted directly to the extracellular space without periplasmic intermediates. Once secreted, the enzyme is autocatalyzed into an active form by cleavage.

The type II secretion system in *Erwinia*, like the pullulanase secretion (Pul) system of *Klebsiella oxytoca*, involves a two-step process in which Pel, Peh, and Cel are first secreted to the periplasmic space by the Sec-dependent secretion system. The Sec secretion system, as mostly defined in *E. coli*, consists of a cytosolic chaperone, SecB, and a complex multicomponent system called the translocase. The translocase complex comprises an ATP-hydrolyzing protein, SecA, and several other inner membrane proteins. The secreted proteins are processed by a signal peptidase in the periplasm. In the second step, Out proteins direct the translocation of the secreted proteins across the outer membrane to the extracellular milieu. In *E. c. carotovora*, at least 13 Out proteins are involved in this step. The specific locations of most Out proteins have been predicted based on their homology with Pul proteins. Most of the *out* gene products, OutC, OutE, OutF, OutG, OutH, OutI, OutJ, OutK, OutL, OutM, OutN, and OutO, most likely are located in the inner membrane. OutS, an outer-membrane lipoprotein appears to facilitate the correct localization of OutD, which is presumed to form a channel in the outer membrane. The genetic arrangement of the *out* clusters of *E. c. carotovora* and *E. chrysanthemi* is very similar, with the notable exception of the absence of *outN* in *E. chrysanthemi*.

I. Regulation of Exoenzyme Production

Based on the regulation of their production, exoenzymes of *E. c. carotovora* have been divided into two main classes; Pel, Peh, Prt, and Cel are grouped in Class I, whereas Pnl belongs to Class II. Class I enzymes are co-regulated and induced by plant components and intermediates of the pectate catabolic pathway. The production of Pel, Peh, Cel, and Prt also is growth phase-dependent, as most of the enzymatic activities are produced in late log phase or early stationary phase. Some Class I enzymes are susceptible to catabolite repression or glucose effect. The synthesis of Pnl, the Class II enzyme, is induced in most soft rot *Erwinia* by DNA-damaging agents such as mitomycin C, and not by catabolic products of pectate or non-DNA damaging plant components.

Studies in both *E. c. carotovora* and *E. chrysanthemi* have revealed that, in addition to N-acyl homoserine

TABLE III
Regulatory Factors Controlling Exoenzyme Production and Pathogenicity in Soft Rot *Erwinia*

Regulatory factor	Species or subspecies	Effect on exoenzymes[a]	Comments
AepA	E. c. carotovora	+	Transmembrane protein
AepB	E. c. carotovora	+	Not characterized
RsmA	E. carotovora subspecies, E. chrysanthemi	−	RNA-binding protein, acts by lowering mRNA stability
rsmB RNA	E. carotovora subspecies, E. chrysanthemi	+	Regulatory RNA molecule, counteracts RsmA effect
RsmC	E. c. carotovora	−	A regulator of RsmA and rsmB
PecT/HexA	E. c. carotovora, E. c. atroseptica, E. chrysanthemi	−	LysR type transcriptional regulator
Hor	E. c. carotovora	+	Positive regulator of antibiotics and exoenzymes
ExpS/RpfA	E. c. carotovora	+	Two-component sensor kinase
ExpA	E. c. carotovora	+	Two-component response regulator
RpoS	E. c. carotovora	−	Stationary phase-specific sigma factor
KdgR	E. c. carotovora, E. chrysanthemi	−	IclR type regulator, acts by binding to KdgR box
PecS	E. chrysanthemi	−	A cytoplasmic regulator of pectate lyase and cellulase
PecM	E. chrysanthemi	−	A transmembrane protein
Crp	E. chrysanthemi	+	cAMP receptor (catabolite activator) protein, complexes with cAMP to activate pel genes
Pir	E. chrysanthemi	+	Activates pel genes by modulating plant signals
PehR	E. c. carotovora	+	Specific regulator of polygalacturonase production
RecA	E. c. carotovora	+	Inactivates RdgA by cleavage
RdgA	E. c. carotovora	−	Repressor of rdgA and rdgB
RdgB	E. c. carotovora	+	Activator of pnlA, ctv and cellular lysis

[a] (+) = positive effect on gene expression, (−) = negative effect on gene expression.

lactones (AHLs) and other signals, a number of negative and positive regulatory factors (Table III) control the production of exoenzymes. The negative regulatory factors include KdgR, PecS, PecM, PecT/HexA, RsmA, RsmC, RpoS, and RdgA. The positive regulatory factors are AepA, AepB, Crp, Hor, ExpS (RpfA), ExpA, Pir, CRP, RecA, RdgB, and *rsmB* RNA regulator. The characteristics of the signals and the regulators are summarized here.

1. Regulation by Quorum Sensing

Many bacteria, including the *Erwinia* species, produce AHLs, also known as autoinducers. The basic structure of such molecules consists of a homoserine lactone ring and an *N*-acylated side-chain of variable length. The side-chains, with different degrees of saturation and substitution, determine the diffusibility of the molecules as well as their species specificity.

The synthesis of these metabolites is generally autoinduced in a growth phase-dependent fashion. The accumulation of AHL in culture is accompanied by enhanced expression of the AHL synthase gene, and this process usually occurs during late log phase and early stationary phase when bacterial cells are present at relatively high densities. The AHLs are also referred to as quorum-sensing signals because the expression of AHL-regulated genes is initiated only when the concentration of AHL molecules has reached a threshold value, which, in turn, is controlled by cell density. Thus, the bacteria use AHL concentration to determine if the population has reached a quorum.

AHLs are typified by *N*-(3-oxohexanoyl)-L-homoserine lactone (OHHL), the autoinducer produced by *Vibrio fischeri*, *E. carotovora* subspecies, *E. chrysanthemi*, *E. stewartii*, and *E. herbicola*, as well as by other bacteria. AHLs are synthesized by AHL synthases, encoded by *luxI* homologs. In most bacteria, there are cognate regulatory factors, usually transcriptional activators called the LuxR homologs. Generally, LuxR homologs bind AHL, and such complexes activate the AHL-regulated genes as well as *luxI* homologs. Homologs of the LuxR and LuxI pair control many phenotypes in diverse bacteria.

The production of AHL by *E. carotovora* subspecies is required for pathogenicity and the synthesis of exoenzymes and antibiotics. The AHL synthases from different *E. c. carotovora* strains; the products of *expI*, *carI*, and *hslI* genes, share 70–98% sequence homology. *E. c. carotovora* mutants defective in these genes are devoid of AHL activity, indicating that OHHL probably is the only AHL analog produced by these bacteria. Similar to the *esaI/esaR* regulatory system in *E. stewartii* (see Section III.B), the biosynthesis of AHL by HslI in *E. c. carotovora* strain 71 appears to be constitutive and not autoinducible via a cognate LuxR-type activator. In *E. c. carotovora* strain GS101, the CarR protein, the LuxR homolog that functions as the AHL-dependent regulator of antibiotic (carbapenem) production, is not involved in the regulation of exoenzyme production. In fact, the identity of the putative LuxR-type regulator for AHL-mediated regulation of the genes for exoenzymes in *E. c. carotovora* has remained elusive. These observations suggest that multiple LuxR homologs may be acting through a single AHL signal molecule to control different phenotypes in *E. c. carotovora*.

E. chrysanthemi strain 3937 produces three analogs of AHL; two, OHHL and *N*-(hexanoyl)-L-homoserine lactone, synthesized by the *expI* gene product, and a third, *N*-(decanoyl)-L-homoserine lactone synthesized by a unidentified gene product. Multiple AHLs are also produced by the related strain EC16. ExpR, a LuxR homolog of *E. chrysanthemi* strain 3937, binds *pel* promoters in the presence of AHL; but in the absence of AHL, ExpR binds its own promoter presumably to repress its expression. The ExpR–AHL complex, as are those of LuxR of *V. fischeri*, LasR of *P. aeruginosa*, and TraR of *A. tumefaciens*, is thought to activate the expression of *pel* genes. However, *E. chrysanthemi* 3937 derivatives deficient in either ExpR or the two AHLs synthesized by the *expI* product are unaffected in enzyme synthesis and pathoge-

nicity. The possibility remains that another putative LuxR homolog acting in conjunction with *N*-(decanoyl)-L-homoserine lactone may function as a transcriptional factor.

Bioluminescence in *Vibrio harveyi* is controlled by two other autoinducer (AI)-like systems—AI-1, which is *N*-(3-hydroxy) butanoyl-L-homoserine lactone; and AI-2, which awaits chemical characterization. Strains of *E. coli* and *Salmonella typhimurium*, two organisms that for a long time were thought to be AHL-deficient, were recently reported to produce AI-2. Subsequent studies disclosed that some strains of *E. amylovora*, but not the highly virulent strain E9, produce AI-2-like molecules. Also, *E. chrysanthemi* strain EC16 and its Ahl⁻ derivative, and *E. c. carotovora* strain 71 and its nonpathogenic AHL⁻ derivative produce similar compounds. These observations, albeit preliminary, suggest that pathogenicity of *E. amylovora* strain E9 and *E. c. carotovora* strain 71 is not significantly affected by AI-2-like molecules.

2. Regulation by Other Signals

Exoenzyme production in soft rot *Erwinia* occurs in response to other signals in addition to AHLs. For example, the synthesis of Class I enzymes in *E. carotovora* subspecies is affected by calcium, pectate catabolic intermediates, plant components, and temperature.

The extracellular concentration of calcium affects Peh production and Pel activity in *E. c. atroseptica* and *E. c. carotovora*. High (millimolar) concentrations of calcium specifically inhibit the activity of *pehA* promoter of *E. c. carotovora* and the ability of the pathogen to cause disease. These observations also indicate a relationship between the resistance of plants containing high calcium to *E. c. carotovora* infection and inhibition of Peh production *in planta*. Calcium, released from plant cell wall by Peh, may play opposing regulatory roles by repressing *pehA* expression and stimulating Pel activity. These observations support the hypothesis that Peh is required during the early stages of infection, whereas Pels probably are responsible for continued plant-tissue maceration in environments containing high levels of calcium.

The synthesis of Class I exoenzymes is also highly dependent on cultural conditions. Generally, the production of Class I enzymes is negligible in rich medium, low in minimal medium, and high in medium containing plant extracts or pectate degradation products. The exact nature of plant components is unknown, but pectate catabolic products induce exoenzyme production by inactivating the global repressor, KdgR (see Section II.J.3). The induction of exoenzymes in *E. carotovora* subspecies is higher with plant components than with pectate and its catabolic intermediates. In *E. chrysanthemi,* the gene (*pir*) that controls Pel production in response to plant signals has been characterized.

Several soft rot *Erwinia* grow at elevated temperatures, for example, *E. c. atroseptica* at 31°C and *E. c. betavasculorum, E. c. carotovora,* and *E. chrysanthemi* at 35°C. However, the levels of the exoenzymes and transcripts of some of the cognate genes produced at these temperatures are very low in *E. c. atroseptica, E. c. betavasculorum,* and *E. c. carotovora.* Other factors that positively affect exoenzyme production are low osmolarity, low oxygen availability, low iron, high nitrogen, and cAMP. Some of these conditions probably mimic the environment within the plant tissues under which the bacteria must sustain physiological activities to cause diseases.

3. Negative Regulators

KdgR, a member of IclR family of transcriptional regulators, is a global negative regulator. In *E. chrysanthemi,* the KdgR-regulated genes are involved in the degradation of pectate as well as in the Out (type II) secretion system. In addition to Pel, *E. c. carotovora* KdgR controls virulence; the production of Peh, Cel, Prt, and harpin (the elicitor of the hypersensitive reaction; and the expression of *rsmB,* which specifies a regulatory RNA. As would be expected, the $kdgR_{Ecc}$ gene of *E. c. carotovora* strain 71 has high homology with $kdgR_{Ech}$. KdgR is a DNA-binding protein with a helix-turn-helix (HTH) motif. The recognition sequences (KdgR box) to which KdgR binds are located in the 5′ regulatory regions of most of the target genes. The three compounds that induce pectinolysis in *E. chrysanthemi* are KDG (2-keto-3-deoxy-D-gluconate), DKI (DTH, 4-deoxy-L-threo-5-hexoseulose uronate), and DKII (DGH, 3-deoxy-D-glycero-2,5-hexodiulosonate). These molecules bind the KdgR

protein and dissociate it from the operators of the target genes, thereby allowing their expression. In addition to promoter occlusion, KdgR inhibits transcription elongation by a roadblock mechanism. In *rsmB,* three KdgR-binding sequences are present in tandem in the transcriptional unit. KdgR binds these sequences to create a roadblock, which stops transcription elongation.

PectT, a LysR type regulator, controls the production of Pel and extracellular polysaccharide in *E. chrysanthemi.* PecT was found to repress the expression of *pelC, pelD, pelI, pelL* and *kdgC* (*pelW*), to have no effect on *pelA, pemA* and *pemB,* and to activate the expression of *pelB.* The purified PecT protein binds the promoters of *pelA, pelB, pelC, pelD, pelE, pelI, pelL, pelZ,* and *kdgC.* It is curious that PecT binds the *pelA* promoter, but does not affect its expression. Also, how this protein apparently acts both as an activator and a repressor is not clear.

HexA⁻ mutants of *E. c. carotovora* overproduce Pel, Prt, Cel, and AHL, and are hypervirulent. In addition, a HexA⁻ mutant of *E. c. atroseptica* has a spreading motility phenotype and overexpresses *fliA* and *fliC,* two genes specifying flagellum formation. HexA of *E. c. carotovora* shares 80% sequence identity with PecT of *E. chrysanthemi* and is also a member of the LysR type transcriptional regulators. Unlike other members of the LysR family, *hexA* of Ecc strain SCRI193 is autoinduced.

An RsmC⁻ mutant of *E. c. carotovora* strain 71 was identified for its hypervirulence and high basal levels of exoenzymes. Exoenzyme production by this mutant is still affected by plant signals and is partially AHL-dependent. The 16-kDa RsmC protein has only a limited homology to a segment of a predicted eukaryotic transcriptional adaptor. RsmC positively regulates RsmA production, but negatively regulates *rsmB* transcription (see Section II.I.5 for RsmA and *rsmB* regulators). Thus, a plausible explanation for the RsmC effect is that it modifies the production of global regulators such as RsmA and *rsmB* RNA and these, in turn, affect various phenotypes. The findings with a RsmA⁻ and RsmC⁻ double mutant are certainly consistent with this hypothesis.

RpoS (σ^s), the stationary-phase-specific sigma factor, also affects exoenzyme production in *E. c. carotovora* strain 71, most likely through its positive effect on *rsmA*. Because the production of Class I exoenzymes is growth phase-dependent with more being produced at the later growth stages, it was initially thought that RpoS probably is a positive regulator of exoenzyme genes. Instead, RpoS⁻ strains of *E. c. carotovora* strain 71 compared to their RpoS⁺ parent strains produce higher levels of exoenzymes, are hypervirulent, and poorly express *rsmA* (see Section II.I.5). Subsequent studies established that RpoS activates transcription of *rsmA*, which encodes a negative regulator. In *E. c. carotovora* strain SCC3193, another virulence factor, ExpM has been identified that negatively controls RpoS. The *rpoS* homologs of other soft rot *Erwinia* are presumed to perform similar regulatory roles. Note that although RpoS controls various stress responses in *E. amylovora*, it does not affect virulence.

Mutants of the divergently transcribed *pecS* and *pecM* genes of *E. chrysanthemi* are derepressed in the production of Pel, Cel, and the blue pigment, indigoidine. However, exoenzyme production in PecS⁻ and PecM⁻ mutants is still affected by other signals such as pectate, plant signals, and growth phase. PecS, a member of MarR family of transcriptional regulators, represses its own expression and that of *pecM*. PecS also binds the regulatory regions of several *pel* and *celZ* genes of *E. chrysanthemi*. PecM potentiates the DNA-binding ability of PecS. The possibility remains that as an inner-membrane protein, PecM also is involved in the transduction of extracellular signal.

4. Positive Regulators

Exoenzyme production in *Erwinia* is also positively controlled. One of the first such regulators was identified as AepA (*aep*, activator of *e*xtracellular *p*rotein production). An AepA⁻ mutant of *E. c. carotovora* strain 71 is deficient in exoenzyme synthesis and nonpathogenic. *aepA* is predicted to encode a protein with a prokaryotic signal peptide of 21 amino acid residues and also to have three hydrophobic regions, suggesting membrane localization. Homologs of *aepA* are present in strains of *E. c. atroseptica* and *E. c. carotovora*, and *aepA* of *E. c. carotovora* is functional in these strains, suggesting that this gene may be a conserved regulator of exoenzymes in these bacteria.

Yet another positive regulator of exoenzymes in *E. c. carotovora* is the protein Hor. Mutants deficient in Hor are defective in the synthesis of Pel, Cel, Prt, and carbapenem antibiotic. Further details of this protein are given in Section VII.A.

Regulators belonging to the bacterial two-component regulatory system also control virulence and exoenzyme synthesis in *E. c. carotovora*. *E. c. carotovora* strain SCC3193 contains the putative response regulator, ExpA, and a sensor kinase, ExpS. A homolog of ExpS, designated as RpfA, has been characterized in *E. c. carotovora* strain AH2 as well. An ExpS⁻ mutant and a RpfA⁻ mutant are reduced in virulence, whereas an ExpA⁻ mutant is nonpathogenic. The Exp⁻ mutants are also reduced in the production of exoenzymes and the expression of *celV*, *pelC*, and *pehA*.

The cyclic AMP (cAMP) receptor protein (CRP or CAP) positively controls virulence and expression of some of the pectinase-encoding genes involved in the catabolism of pectin in *E. chrysanthemi* and *E. c. carotovora*. Extensive studies of *E. coli* and several other bacteria have established that the cAMP–CRP complex activates the transcription of genes subject to cAMP catabolite repression. The synthesis of cAMP, which is suppressed by glucose, is catalyzed by adenylate cyclase, the product of the *cya* gene. The *crp* gene of *E. chrysanthemi* strain 3937 consists of a 630-bp open reading frame, which encodes a protein product of 23.6 kDa. CRP of *E. chrysanthemi* shares high homology with CRP of *E. coli* and other enterobacteria. The expression of *pemA*, *pelB*, *pelC*, *pelD*, *pelE*, *kduI*, *kduD*, *kdgT*, and *kdgK* is reduced in a Crp⁻ mutant compared to the Crp⁺ *E. chrysanthemi* under inducing conditions. Crp protein binds the promoter regions of *pelB*, *pelC*, *pelD*, *pelE*, *ogl*, *kduI*, and *kdgT*, the genes that are positively controlled by the cAMP–CRP complex.

In addition to global regulators, gene-specific regulators also control Peh and Pel production. Transposon mutagenesis of *E. c. carotovora* strain SCC3193 produced mutants that are specifically affected in the expression of a *pehA-bla* fusion (*pehA*, Peh structural gene). Transcription of *pehA* is severely reduced in the mutants that carry insertions in *pehR* (Peh regulator gene). PehR⁻ mutants are, however, only slightly affected in pathogenicity. Additional work is needed to fully characterize PehR and to understand its mechanism of action.

The *pir* locus of *E. chrysanthemi* encodes a 30-kDa protein of the IclR family, which positively controls its own expression and the expression of the *pelE* in strain EC16 in response to plant signals. In a Pir⁻ mutant, *pelE* expression is not fully induced by plant extracts. Pir binds its own promoter region as well as the promoter region of *pelE*, and interferes with KdgR binding (see Section II.I.3 for KdgR effects on gene expression). The Pir⁻ mutant produces less total Pel activity, suggesting that Pir action may be required for the expression of other *pel* genes. However, the mutant is not affected in Prt and Cel production, and Pir does not bind the promoter regions of *prtC* and *celY* genes. Thus Pir is a specific regulator of *pel* genes and apparently does not participate in induction of the genes for other exoenzymes by plant signals in *E. chrysanthemi*.

5. Posttranscriptional Regulation

The RsmA–*rsmB* pair posttranscriptionally regulates pathogenicity factor and secondary metabolite production in *Erwinia* and several other enterobacteria. RsmA is a small 6.8-kDa RNA-binding protein. RsmA⁻ mutants of *E. c. carotovora* strain 71 are hypervirulent, derepressed in exoenzyme and harpin$_{Ecc}$ production, and elicit the nonhost defense response, HR, in tobacco leaves. Also, the RsmA⁻ mutants of *E. c. carotovora* are Ahl-independent as Ahl⁻ RsmA⁻ double mutants have the same phenotypes as do their Ahl⁺ RsmA⁻ parents. This observation suggests that Ahl may be acting through a regulatory pathway that is controlled by RsmA in *E. c. carotovora*.

rsmB, which was identified as a positive regulator of exoenzyme production in *E. carotovora* subspecies, specifies an untranslated regulatory RNA. RsmA and *rsmB* RNA control many phenotypes in soft rot *Erwinia*, including the production of Pel, Peh, Cel, and Prt; motility; flagellum formation; pigment synthesis; elicitation of the HR; and pathogenicity. In addition, the RsmA–*rsmB* system controls pathogenicity, elicitation of the HR, extracellular polysaccharide (EPS), motility, flagella, and Prt in *E. amylovora*; mucoid phenotype and phytohormone in *E. h. gypsophilae*; pigment and EPS in *E. stewartii*, and pigment and extracellular Prt in *Serratia marcescens*.

RsmA acts by lowering the half-life of mRNA species. The mechanism by which RsmA affects transcript stability is not understood. *rsmB* is believed to neutralize the RsmA effect, but the underlying mechanism remains to be clarified. However, two key observations that provide some insights into the possible regulatory mechanisms are that RsmA binds *rsmB* RNA, and *rsmB* RNA is then processed, yielding a relatively stable and regulatory *rsmB′* RNA. Thus, *rsmB* RNA may act by depleting the pool of free RsmA molecules, that bind mRNAs of the RsmA-regulated genes to promote their decay. In addition, the *rsmB′* RNA may repress transcription or translation of *rsmA* or promote RsmA degradation.

Homologs of *rsmA* and *rsmB* exist in all the *Erwinia* species tested and in other enterobacteria, such as *E. coli*, *Salmonella typhimurium*, *Shigella flexneri*, *Yersinia psudotuberculosis*, and *S. marcescens*. In *E. coli*, CsrA–*csrB*, the homologs of RsmA–*rsmB*, control glycogen biosynthesis and other phenotypes in much the same way that RsmA–*rsmB* system controls exoenzymes in *Erwinia*. This suggests that RsmA and *rsmB* homologs could similarly function to control virulence and pathogenicity in other enterobacteria.

6. Regulation of Pectin Lyase (Pnl) by RecA–Rdg Regulon

DNA-damaging agents induce Pnl in most soft rot *Erwinia* (see Section II.E). In *E. c. carotovora* strain 71, Pnl is co-induced with the bacteriocin, carotovoriocin (Ctv), and cellular lysis. Thus, DNA damage produces signal(s) required for this induction. The putative signal is then channeled via the RecA–Rdg regulatory pathway to induce gene expression.

The induction of Pnl upon DNA damage is reminiscent of the activation of an SOS response. This apparent similarity prompted the notion that RecA protein effecting DNA repair and recombination might be involved in Pnl induction. This was confirmed by the finding that Pnl and Ctv are noninducible with mitomycin C in a RecA⁻ mutant of *E. c. carotovora* strain 71.

The isolation of noninducible mutants of *E. c. carotovora* in which *recA⁺* DNA could not restore the induction of Pnl and Ctv upon DNA damage led to the discovery of the *rdg* locus (*rdg*, regulator of damage inducible genes). The locus comprises two genes, *rdgA* and *rdgB,* as separate transcriptional units. RdgA is an HTH-type DNA-binding protein and has high homology with bacterial and bacteriophage repressors such as Φ80 cI and 434 cI proteins. RdgA represses its own expression and that of the *rdgB* gene. The cleavage of RdgA by the proteolytic RecA* inactivates the RdgA repressor.

rdgB encodes a 13.5-kDa product, which shares high homology with bacteriophage Mu activators C and Mor. RdgB contains a HTH motif in the C-terminal region and activates transcription of *pnlA* and the *ctv* operon by specifically binding the cognate promoters. Also, a DNA segment of *Haemophilus influenzae* is predicted to encode a RdgB homolog.

The regulatory events leading to the activation of Pnl, Ctv and cellular lysis can be summarized as follows. The signal(s) generated upon DNA damage activates RecA into a proteolytically active RecA*, which then cleaves RdgA. This cleavage inactivates the RdgA repressor, resulting in the initiation of *rdgB* transcription. The consequent increase in RdgB pool leads to activation of the expression of the *ctv* operon, *pnlA,* and the genes specifying cellular lysis. The significance of this regulation in ecology and plant virulence of *E. c. carotovora* or other soft rot *Erwinia* awaits elucidation.

III. EXTRACELLULAR POLYSACCHARIDE

Many *Erwinia* species produce EPS, and the genetics of EPS production have been studied in some detail in *E. amylovora, E. chrysanthemi,* and *E. stewartii.* In *E. amylovora* and *E. stewartii,* EPS is required for pathogenicity. On the other hand, in *E. chrysanthemi,* EPS appears to function as an ancillary virulence factor because EPS-deficient strains are less effective in the initiation of soft rot than are their wild-type parents. Aside from its role as pathogenicity factor, EPS is believed to protect bacteria against host defense and desiccation, to help in vector-mediated transmission of bacterial pathogens, and to sequester nutrients released from host cells. Some EPS-producing plant pathogenic bacteria cause wilting symptoms. In pathogenesis of *E. amylovora,* resulting in fire-blight, the EPS protects the bacterium from host defenses; enables it to move through

the cortex; clogs the plant's vascular tissues, preventing water flow; and ultimately leading to wilting. In addition, the polysaccharide material is frequently discharged onto the plant surfaces as milky-white and sticky droplets, generally referred to as "the ooze."

A. EPS Composition and EPS Biosynthetic Genes

E. amylovora produces three types of EPS, referred to as levan, amylovoran, and glucan. Levan, a high-molecular-weight and low-viscosity homopolymer of fructose, is synthesized extracellularly from sucrose by levansucrase, an enzyme encoded by the *lsc* gene. Amylovoran, the main component of the bacterial ooze, is an acidic heteropolymer containing branched repeating units of one glucuronic acid and four galactose residues. About 10% of the subunits also contain a single branched glucose residue. The repeating unit of amylovoran is variably decorated with pyruvate and acetate groups. Levan-deficient mutants are reduced in virulence, whereas amylovoran-deficient mutants are nonpathogenic. The genetics and biosynthesis of glucan, a low-molecular-weight homopolymer of glucose, have not been studied in *Erwinia.*

The EPS produced by *E. stewartii,* known as stewartan, is an acidic heteropolymer made up of branched chain repeating units of seven monosaccharides, comprising galactose, glucose, and glucuronic acid in the ratio of 3:3:1. It is similar in structure to amylovoran with the addition of a terminal glucose on the major side-chain and the substitution of a glucose residue in the backbone. Stewartan-deficient mutants of *E. stewartii* are weakly pathogenic on corn. The *cps* (capsular polysaccharide synthesis) genes specifying stewartan production are partially allelic with the *ams* (amylovoran synthesis) genes of *E. amylovora,* permitting interspecific complementation between *cps* and *ams* mutations using the cloned gene clusters.

Amylovoran biosynthesis is specified by the *ams* locus, a large DNA segment spanning more than a 16-kb region of the *E. amylovora* genome. Downstream of the large *amsGHIABCDEFJKL* operon are the *amsM* and *galE* genes, which probably also belong to this operon. There is direct evidence that *amsGHI-*

ABCDEFJKL and *galE* are required for EPS production, and indirect evidence (i.e., sequence homology of AmsM with GalU and GalF of *E. coli*,) suggests that AmsM also could be involved in EPS synthesis. The *amsGHIABCDEFJKL* operon has an *rcs*-controlled promoter about 500 bp upstream from *amsG*, the first gene of the operon, and the 39-bp JUMPstart sequence, which is found in the 5′ regions of EPS biosynthetic operons. The *ams* gene products range in size from 15.7–82.2 kDa with some of the proteins having high homology with the products of EPS biosynthetic genes from other bacteria. For example, AmsG, H, L, and M of *E. amylovora* are 75–100% homologous with CpsA, B, L, and M of *E. stewartii*, respectively. Also, AmsI, AmsA, AmsB, AmsF, AmsJ, and AmsK of *E. amylovora* share 50–75% amino acid identity with CpsI, CpsC, CpsE, CpsH, CpsJ, and CpsK of *E. stewartii*, respectively. Some of the *ams* gene products, based on sequence homology, have been predicted to function as glucosyl and galactosyl transferases, acid phosphatase, and ABC transporters. It is noteworthy that three genes differ in each species (*amsC*, *amsE*, *amsD* of *E. amylovora*; and *cpsD*, *cpsF*, and *cpsG* of *E. stewartii*), and these differences are sufficient to account for the variations in amylovoran and stewartan structures. As a prelude to the analysis of functions of the various *ams* genes, the product of *amsI* has initially been characterized. The gene encodes a small-molecular-weight acid phosphatase. Its overexpression leads to a strong phosphatase activity and a severe reduction in EPS production in *E. amylovora*, resulting in a phenotype similar to other *ams* mutants. AmsI has been postulated to participate in various steps in EPS production, including dephosphorylation of the lipid carrier diphosphate.

Unlike *E. amylovora* and *E. stewartii*, soft rot *Erwinia* species, including *E. chrysanthemi*, do not produce EPS in carbohydrate-rich media. However, a PecT⁻ mutant of *E. chrysanthemi* was found to have, in addition to elevated levels of pectolytic enzymes, a mucoid phenotype. The EPS produced by *E. chrysanthemi* is composed of L-rhamnose, D-galactose, and galacturonic acid in the ratio of 4:1:1. Mutants with insertions in the *eps* locus, which specifies EPS biosynthesis, are reduced in virulence on *Saintpaulia ionantha*. The analysis of the *eps* locus of *E. chrysan-*themi strain 3937 indicates that, as in other EPS producing *Erwinia*, the transposon insertions map to a large operon with the promoter region containing the JUMPstart sequences.

B. Regulation of EPS Biosynthesis

The regulation of EPS biosynthesis in *E. amylovora* and *E. stewartii* appears to be generally similar to the regulation of capsule biosynthesis in *E. coli*. In contrast, the findings in *E. chrysanthemi* indicate that EPS in this strain may be regulated differently from *E. amylovora*, *E. stewartii*, and *E. coli*. For example, while PecT tightly regulates EPS synthesis in the wild-type *E. chrysanthemi*, there is no indication that the homologs of this regulator, if present in *E. amylovora* and *E. stewartii*, play similar roles. Moreover, the expression of *eps* genes in *E. chrysanthemi* is repressed by the cAMP receptor protein and induced by low osmolarity. It also remains to be determined if the *rcs* genes and AHLs (see later) function in EPS regulation in *E. chrysanthemi*, as they do in *E. amylovora* or *E. stewartii*.

In both *E. amylovora* and *E. stewartii*, the *rcs* genes (regulator of capsule synthesis), *rcsA*, *rcsB*, and *rcsC*, as well as Lon protease, are the key regulators of EPS biosynthesis. RcsA, RcsB, and RcsC positively regulate EPS synthesis, whereas the Lon protease has a negative effect on EPS production. *rcsB* and *rcsC* are linked and divergently transcribed, and belong to the classical family of bacterial two-component regulatory systems. RcsC, the sensor kinase, is a transmembrane protein, which is presumed to phosphorylate RcsB, the transcriptional activator, on sensing an unknown signal. However, in *E. coli* and *Salmonella typhi*, RcsC senses high osmotic conditions and desiccation. The sequences of these regulatory factors are highly conserved and the proteins are functionally interchangeable between *E. coli* and *E. amylovora* or *E. stewartii*. Interestingly, RcsA, an HTH-type regulator, by itself does not bind DNA, but forms a complex with RcsB, which activates transcription of the *ams* (and *cps*) genes. The recognition sequence for the RcsA–RcsB complex has been determined to be a 23-bp region spanning nt −555 to −533 upstream of *amsG*, the first gene of the *ams* operon. A homologous sequence was also found in the *cpsA* promoter of *E. stewartii*.

RcsA, but not the RcsA–RcsB complex, is susceptible to Lon proteases; hence the cytoplasmic pool of free RcsA is generally limiting and overproduction of RcsA results in excess EPS production. In *E. amylovora*, DNA binding by RcsB alone occurs only at high protein concentrations. The efficiency of interaction of the RcsA–RcsB complex of *E. amylovora* with DNA is higher at 28°C than at 37°C, and this probably contributes to temperature-dependent synthesis of EPS.

A study in *E. coli* has documented that *rcsA* in *E. coli* is self-inducible. Moreover, a 25-bp putative RcsA box has been identified in the upstream sequences of the promoters of *rcsA* and *cps* genes of *E. coli*. These observations imply that RcsA may bind the RcsA box upstream of *rcsA* to activate its transcription. A putative RcsA box is also present in the RcsA–RcsB binding site in the vicinity of the *ams* promoter region of *E. amylovora*, but the sequences of the RcsA boxes are not well conserved. This sequence divergence may explain the apparent lack of RcsA binding to the RcsA–RcsB binding region of *E. amylovora*.

In *E. coli*, *rcsA* expression is increased by RcsB and negatively controlled by H-NS, a modulator of environmentally regulated genes. The effect of H-NS, in turn, is counteracted by DsrA, a small regulatory RNA, by RNA–RNA base-pairing interactions at the *hns* locus to inhibit H-NS translation. Although a *dsrA* homolog is not present in *E. stewartii*, it remains to be determined if in other EPS-producing *Erwinia* species *rcsA* expression is regulated by an RNA regulator. In this context, it is perhaps significant that the H-NS gene of *E. chrysanthemi* is structurally and functionally related to the *E. coli* gene.

Another positive regulatory gene, *rcsV*, controls EPS production in *E. amylovora*. The product of *rcsV* suppresses an *rcsA* mutation in *E. stewartii*, as well as activating EPS production in *E. ananas* and *E. herbicola*. The latter observation provides indirect evidence for the presence of EPS biosynthetic genes in these *Erwinia* species. The transcription of *rcsV* is very weak in culture and RcsV⁻ mutants of *E. amylovora* have no detectable phenotype. However, the finding that overexpression of *rcsV* gene in *E.*

amylovora by translational fusions with *lacZ′* or *malE′* alleviated the effects of RcsA deficiency, implicates RcsV as a nonallelic RcsA analog affecting EPS regulation. Unlike RcsA, the presence of both a phosphorylation site and the HTH motif in RcsV indicates that it is most likely a typical response regulator of bacterial two-component regulatory system. The possibility remains that a plant signal may alter the level of *rcsV* expression, thereby activating the *rcsV* pathway during pathogenesis.

In addition to the *rcs* genes, the *esaR–esaI* regulatory system controls EPS synthesis in *E. stewartii*. In fact, in the regulatory hierarchy, *esaR–esaI* has been placed above the *rcs* pathway. OHHL is synthesized by the EsaI protein (OHHL synthase), encoded by the *esaI* gene of *E. stewartii*. The OHHL signal is required for the growth phase-dependent production of stewartan by *E. stewartii* strain DC283. *esaI* and *esaR*, the latter encoding a transcriptional factor, are closely linked and are homologs of the *luxR–luxI* family. In contrast to the typical LuxR-type regulators, EsaR functions as a repressor for EPS production. In an EsaR⁻ mutant, excess EPS is produced independently of cell density. The EPS-overproducing EsaR⁻ mutant is less virulent than the wild-type parent. The latter observation suggests that AHL tightly controls EPS production during the early stages of infection and that uncontrolled EPS production interferes with other steps in pathogenesis of *E. stewartii*. EsaR is believed to negatively control the expression of one of the *rcs* genes or to interfere with the formation of RcsA–RcsB, and the binding of AHL to EsaR could abolish these activities.

Several bioassay systems have revealed that *E. amylovora* strains do not produce OHHL. In light of the findings that EPS production in *E. amylovora* is growth phase- (cell density-) dependent and OHHL controls EPS in *E. stewartii*, and given the similaries in structure, function and Rcs-mediated regulation of *cps* and *ams* genes, it is surprising that an *esaR–esaI*-like system apparently does not control EPS production in *E. amylovora*. However, several *E. amylovora* strains, with the notable exception of the highly virulent and EPS⁺ wild-type strain E9, produce AI-2, which is postulated to function as a quorum-sensing signal in *E. coli* and *Salmonella typhimurium* (see Section II.I.1 for details). Despite the finding that

EPS production and virulence in strain E9 is AI-2 independent, additional studies are needed to rule out a role of a quorum-sensing system in EPS production in *E. amylovora*.

IV. ELICITORS OF HYPERSENSITIVE REACTION

A. Organization of the *hrp* Genes in *Erwinia*

The *hrp* (hypersensitive reaction [HR] and pathogenicity) genes in *E. amylovora*, *E. chrysanthemi*, and *E. h. gypsophilae*, as in *Pseudomonas* species, specify harpin production, regulation of *hrp* (and *avr*) genes, and the type III pathway for the secretion of Harpin and other proteins affecting plant–bacteria interactions. The *hrp* locus of *E. stewartii*, originally designated *wts* (water soaking), was initially characterized as responsible for water-soaking symptoms produced by the bacterium on the maize host. The locus spans a 28-kb stretch of DNA, which resolved into eight complementation group—*wtsE*, *hrpN*, *hrpV*, *hrpG*, *hrpS*, *hrpXY*, *hrpL*, and *hrpJ*. Some genes in the *wts* loci were subsequently found to be homologs of *hrp* genes of *P. syringae* pv. *phaseolicola* and to complement Hrp⁻ mutants of *E. amylovora*. Although *E. c. carotovora* carries *hrp* genes such as *hrpX*, *hrpY*, *hrpS*, *hrpL*, and *hrpN*, the organization of these *hrp* genes and their structure have not been elucidated. However, some segments of the *hrp* clusters are quite similar in other *Erwinia* species and many Hrp proteins from different bacteria share homology. For example, *hrpFG*, *hrcC*, *hrpTV*, and *hrpN* genes of *E. amylovora* and *E. chrysanthemi* are arranged in the same order and are oriented similarly, and the genes are also of similar sizes. Sequence analysis has also confirmed a similar organization of the conserved regulatory and secretion genes in the *E. stewartii* and *E. amylovora hrp* clusters. Homologs of many *hrp* genes are widespread in gram-negative bacteria, including human pathogens such as the *Yersinia* and *Salmonella* species, where they are involved in the secretion of various pathogenicassociated factors. The *hrc* (*hrp* and conserved) designation has been adopted for the *hrp* genes that are conserved between

plant and animal pathogens. The last letter designations of *hrc* genes correspond to those of *Yersinia ysc* (yersinia secretion) genes.

B. Harpins

The *hrpN* genes of *E. amylovora*, *E. c. carotovora*, *E. chrysanthemi*, and *E. stewartii* make up single gene transcriptional units and encode approximately similar-size products, designated as Harpin$_{Ea}$, Harpin$_{Ecc}$, Harpin$_{Ech}$, and Harpin$_{Es}$, respectively. *hrpN$_{Ecc}$* does not possess a typical sigma-70 promoter; instead, it possesses the consensus sequence typical of σ^{54} promoters, but the physiological relevance of these sequences is not known. However, a more significant observation in the context of *hrpN* expression in *Erwinia* species is the presence of promoter sequences typical of HrpL- controlled genes in *hrpN$_{Ea}$*, hrpN$_{Ecc}$, hrpN$_{Es}$, and possibly *hrpN$_{Ech}$* (see Section IV.D). As expected, *hrpN* genes of these *Erwinia* species share high homology.

Harpins are glycin-rich, acidic, heat-stable proteins without cysteine residues and apparently devoid of enzymatic activity. Many plant pathogenic bacteria, especially those capable of eliciting the HR, synthesize harpins and secrete them into the milieu. When infiltrated into nonhost tissues, harpins trigger incompatible interactions, which result in the typical HR symptoms. This was demonstrated for the first time with Harpin$_{Ea}$. Subsequent studies have shown that harpins produced by *E. amylovora*, *E. chrysanthemi*, *E. c. carotovora*, and *E. stewartii* are structurally and functionally quite similar. The *Erwinia* harpins do not have significant sequence homology with the harpins or HrpZ proteins of plant pathogenic pseudomonads. Harpins also appear to function as pathogenicity factors in susceptible interactions, although their effects can be variable. For example, while *hrpN* mutants of *E. amylovora* strain 321 are drastically reduced in pathogenicity, similar mutants of strain CFBP1430 are only slightly reduced in virulence, and HrpN⁻ mutant of *E. stewartii* are fully pathogenic. Moreover, HrpN⁻ mutants of *E. chrysanthemi* are affected in their pathogenicity toward some hosts but not others.

HrpW$_{Ea}$ of *E. amylovora* strains Ea321 and CFBP1430 and HrpW$_{Pst}$ of *P. syringae* pathovar *to-*

mato (and pathovar *syringae*) represent new or secondary harpins. These proteins have the features of harpin; that is, they are acidic, glycine, and serine-rich; devoid of cysteine residues; and capable of eliciting the HR. The N termini of HrpW$_{Ea}$ and HrpW$_{Pst}$ share homologies with *Erwinia* and *Pseudomonas* harpins, respectively, and are sufficient for the HR elicitor activity. The C termini of HrpW proteins have homology with Pels. HrpW proteins can bind pectate, although they lack pectolytic activity. The homology of HrpW with Pels and, more significantly, the pectate-binding ability of HrpW raise the intriguing possibility of a relationship between Pel structure and the elicitation of the HR. Indeed, this idea is supported by the finding that PelI of *E. chrysanthemi* is extracellularly processed proteolytically by PrtA and PrtC, and that the processed product is capable of eliciting a necrotic response in tobacco.

Multiple copies of *hrpW* in *trans* enhance the efficiency of induction of the HR, although a HrpW⁻ HrpN$_{Ea}$⁺ mutant is unaffected in elicitation of the HR. These observations indicate that the HrpW proteins are accessory harpins. Furthermore, the presence of *hrpW*$_{Ea}$ homologs in other *Erwinia* species and in possible *hrpW*$_{Pst}$ homologs in other plant pathogenic *Pseudomonas*, *Ralstonia*, and *Xanthomonas* species suggests that HrpW may be an important component of the Hrp machinery.

C. Type III Secretion Pathway

Harpins are secreted by the type III system, which is fundamentally different from the types I and II systems used for the secretion of exoenzymes by soft rot *Erwinia*. In addition to harpins, other proteins required for pathogenesis or HR, including DspE (see Section IV.E) and *avr* gene products, are secreted via the type III secretion system. Several animal-pathogenic bacteria, including *Yersinia*, *Salmonella*, and *Shigella* species, also use similar secretion systems to deliver virulence or pathogenicity-associated factors. In fact, most of our current understanding of the Harpin and Avr protein secretion system comes from studies of these bacteria, especially the *Yersinia* species. The type III system, like the type I system, is Sec-independent (see Section II.H). At least 20 and possibly more proteins, many of which are localized

in the cytoplasmic membrane, constitute the type III secretion apparatus. A cytoplasmic, probably membrane-associated ATPase is required for type III secretion. In addition, the secreted proteins sometimes require small cytoplasmic proteins, which function as chaperones and direct proper translocation of the proteins through the secretion apparatus. Most of the inner-membrane proteins share homology with flagellar assembly proteins. In contrast, an outer-membrane protein (YscC of *Yersinia* and possibly HrcC of *E. amylovora*) of the type III system is homologous with PulD (OutD of *Erwinia*), the outer-membrane secretin (multimeric outer-membrane component) of the type II system. However, unlike the type II secretion system, periplasmic intermediates of the secreted proteins generally do not occur. Nevertheless, mutant studies in *P. syringae* have shown that secretion through the inner membrane and the outer membrane occurs in distinct steps that are genetically separable.

As would be expected, most of the *hrp* gene products are dedicated towards the synthesis and assembly of the type III secretion apparatus. In *E. amylovora*, at least 15 *hrc* and *hrp* genes specify the secretion pathway. These include *hrcV, hrcN, hrcQ, hrcR, hrcS, hrcT, hrcU, hrcJ, hrcC, hrpE, hrpJ, hrpQ, hrpB, hrpO,* and *hrpD*. Although the functions of many *Erwinia hrp* and *hrc* genes are not yet known, several predictions can be made based on genetic homology and the studies of a few specific genes. It has become apparent that, unlike the general secretion (type II) pathway, the recognition signals for the type III system are not in the N-terminal regions of the proteins. Instead, these may reside within the 5′ regions of the mRNAs of secreted proteins (i.e., Harpins, Avr proteins, and DspE). The type III system may also be induced on contact with the host. In some instances, Hrp pili probably are integral parts of the secretion apparatus. These pili are believed to act either by promoting bacterial attachment to the host cell or by serving as conduits for the delivery of proteins into the host cells.

D. Regulation of *hrp* Genes

Although some wild-type strains of *Erwinia* possess *hrpN* and most likely the full complement of accessory *hrp* genes, they differ with respect to their ability to elicit the HR under a standard set of experimental conditions. For example, although wild-type strains of *E. amylovora* elicit the HR, *E. stewartii* and *E. h. gypsophilae* require preinduction, and the wild-type *E. c. carotovora* and *E. chrysanthemi* strains do not elicit HR. This inability of *E. crysanthemi* and *E. c. carotovora* strains may be attributed to low levels of *hrp* expression or to a suppressive effect of a HR-inhibitory factor. *E. chrysanthemi* elicits HR only when all the primary *pel* genes are deleted or Pel secretion is prevented by *out* mutations. High levels of Pel activity were, therefore, thought to interfere in some manner with the elicitation of HR by *E. chrysanthemi*. However, in *E. c. carotovora*, it is the RsmA⁻ mutants and not the RsmA⁺ parent that elicit the HR, although the RsmA⁻ strains overproduce exoenzymes, including Pels, along with Harpin$_{Ecc}$. Thus, a low level of harpin production and possibly poor expression of other *hrp* genes may be responsible for the HR⁻ phenotype of the wild-type *E. c. carotovora* strains. The finding that various cultural conditions drastically affect *hrp* expression in *Erwinia* species also suggests a tight regulation of *hrp* genes. For example, the expression of *hrp* genes is repressed in rich media, but is stimulated in minimal media whose compositions apparently are similar to those of the intercellular fluids of host plants. In *E. c. carotovora* strain 71, *hrpN* is expressed at a basal level in minimal sucrose medium but is induced by plant extracts, as is the case with *Xanthomonas campestris* pv. *vesicatoria*. In addition, in *E. c. carotovora*, *hrpN* expression is AHL-dependent and co-regulated with class I enzymes by KdgR, HexA, RsmA, RsmC, and *rsmB* RNA that mediate global regulation (see Sections II.I.3 and II.I.4).

Extensive studies in *P. syringae* pathovars and *X. c. vesicatoria* have disclosed that the expression of *hrp* genes is controlled by intricate regulatory cascades comprising an array of regulatory factors. In *P. s. phaseolicola*, σ^{70}-coupled RNA polymerase activates the expression of *hrpR*, and HrpR, in turn, activates *hrpS* transcription. *hrpR* and *hrpS* are tandemly arranged and apparently expressed as separate transcriptional units. HrpR and HrpS share about 57% sequence identity with each other and with response regulators of two component regulatory systems.

HrpS, together with an unknown component and in cooperation with σ^{54}-RNA polymerase holoenzyme, activates the expression of *hrpL*, which encodes a member of the extra-cytoplasmic function (ECF) subfamily of sigma factors. The regulatory hierarchy may be somewhat different in *P. s. syringae*, in which the products of *hrpR* and *hrpS* are proposed to form a heterodimmer to activate *hrpL* expression. HrpL (σ^L)-RNA polymerase holoenzyme probably is responsible for the activation of transcription of the remaining *hrp* genes through its recognition of specific promoter sequences, called the *hrp* box. Recently, HrpV has also been identified as a negative regulator of *hrp* gene expression in *P. s. syringae*, acting above *hrpRS* in the regulatory hierarchy. A HrpV⁻ mutant overproduced HrpZ, an elicitor of HR, as well as HrcJ, HrcC, and HrcQ(B), all components of the type III secretion apparatus. It has not been determined if the *hrpV* gene of *E. amylovora* also has such a regulatory role. The available evidence suggests that at least some components of *hrp* regulation in *E. amylovora* and *E. chrysanthemi* may be similar to those of *P. s. syringae*. For example, HrpL positively regulates *hrp* gene expression in *E. amylovora*. HrpL$_{Ea}$ is a 21.7-kDa protein and the product of a single-gene operon in the *E. amylovora hrp* cluster; this alternative sigma factor is required for both pathogenicity and HR. The expression of *hrpL* is also regulated by environmental conditions and by *hrpS* as in *P. s. syringae*, but a homolog of *hrpR* has not yet been identified in *E. amylovora*. The lack of cross-complementation by *hrpL* genes of *P. s. syringae* and *E. amylovora* raises the possibility that despite the overall similarity in regulatory pathways, some key regulatory events may have diverged in these plant-pathogenic bacteria to sustain species-specific interactions with different sets of host and nonhost plants.

E. Disease-Specific (*dsp*) and Host-Specific Virulence (*hsvG*) Loci Associated with the *hrp* Regulon

The disease-specific (*dsp*) locus of *E. amylovora* is closely linked with the *hrp* cluster. The locus contains two open reading frames, *dspE* and *dspF* in *E. amylovora* strain 321 (*dspA* and *dspB* in strain CFBP1430). The expression of *dsp* genes is controlled by HrpL (see Section IV.D). DspE is a hydrophilic protein of 198 kDa that has 30% sequence identity with AvrE of *P. syringae* pv. *tomato;* DspE is also a functional homolog of AvrE. The *dspF* gene is located downstream of *dspE*, encodes a 16-kDa acidic protein with 43% homology with AvrE of *P. s. tomato,* and is structurally similar to chaperones of virulence factors (Syc proteins) of *Yersinia.* This similarity and the findings that a functional DspF is required for the secretion of DspE by the *hrp* secretion system strongly suggest that DspF functions as a chaperone for the secretion of DspE. A DspE homolog (WtsE) and associated chaperone are found adjacent to the *E. stewartii hrp* cluster and are likewise required for pathogenicity, but not for the induction of the HR.

The *hsvG* gene specifies a host-specific virulence factor in *E. h. gypsophilae,* which infects only *Gypsophila,* and in *E. herbicola* pv. *betae,* which infects both *Gypsophila* and beet. In both *E. h. gypsophilae* and *E. h. betae, hsvG* is required for infection of *Gypsophila* because HsvG⁻ mutants of both pathovars are nonpathogenic on this host. The gene is located on the pathogenicity-associated pPATH plasmid together with *hrp* and phytohormone biosynthetic genes. HsvG probably is regulated by the *hrp* regulon as its regulatory sequences contain a putative *hrp* box. *hsvG* is predicted to encode a predominantly hydrophilic product of 71.3 kDa without a signal peptide or membrane-spanning domains. HsvG has no homology with any known protein in the databases. The presence of *hsvG* homologs in only pathogenic *E. h. gypsophilae* strains, together with the observations that an HsvG⁻ mutant of *E. h. betae* is unaffected in its ability to infect sugar beet, strongly indicate that HsvG is a *Gypsophila*-specific virulence factor.

V. PHYTOHORMONE

E. h. gypsophilae induces gall formation in *Gypsophila* plants. The galls result from cell enlargement, as well as increased cell division, primarily due to plant growth-regulating substances (auxins and cytokinins) produced by this bacterium. The cytokinins produced by this bacterium have been identified as zeatin, zeatin riboside, iso-pentenyladenine, and two other immunoreactive zeatin type compounds, whereas the auxin is indole-3-acetic acid (IAA). In *E. h. gypsophilae,* the genes for the synthesis of the cytokinins and indole-3-acetamide (IAM) pathway of IAA biosynthesis are located on a mega-plasmid, designated pPATH. Along with the phytohormone genes, this plasmid also carries the *hrp* cluster and a host-specificity gene. In a nonpathogenic epiphytic strain of *E. herbicola,* genes located on the chromosome specify IAA production through the indole-3-pyruvate (IpyA) pathway. Analysis of the *etz* (erwinia *trans* zeatin) locus for the synthesis of cytokinin in *E. h. gypsophilae* revealed two open reading frames (ORFs) with low G + C content. The first ORF (*pre-etz*) encodes a 169-amino-acid putative product that does not share homology with any known cytokinin gene. In contrast, the second ORF (*etz*) encodes a 237-amino-acid product and has high homology with cytokinin biosynthetic genes from other bacteria. Northern analysis has revealed the presence of a common transcript for both ORFs, as well as an *etz*-specific transcript. The genes are differentially expressed, with the *etz* transcript dominating during late log phase and the *pre-etz* transcripts accumulating during the stationary phase. Mutations in *pre-etz* and *etz* caused 42% and 30% reductions in gall formation, respectively.

IAA is produced by the epiphytic *E. herbicola* through the IpyA pathway, while the pathogenic *E. h. gypsophilae* uses both the IAM and the IpyA pathways. The regulation of synthesis of IAA has been studied *in vitro* and *in planta* in the nonpathogenic *E. herbicola* strain 299R through the use of an *ipdC::inaZ* fusion (*inaZ* specifies ice nucleation activity and is used as a reporter); *ipdC* encodes the enzyme responsible for the formation of indole-3-acetaldehyde, the immediate precursor of IAA in the IpyA pathway. The expression of *ipdC::inaZ* is generally low in liquid medium at all growth stages, irrespective of medium richness, pH, or nitrogen availability. In contrast, *ipdC::inaZ* expression is induced 18-fold, 32-fold, and 1000-fold on bean-leaf surfaces, on tobacco-leaf surfaces, and in pear flowers, respectively. Thus, IAA production by the IpyA pathway appears to be linked to the epiphytic lifestyle. A similar study has been conducted to determine the relative contributions of the two IAA pathways and cytokinin production on pathogenic states of *E. h. gypsophilae*. Mutants defective in IAM or the cytokinin pathway induce galls which are 30–42% smaller than those induced by the wild-type strain, while the IpyA pathway mutants produce galls similar in size to the wild-type. This observation seems to suggest that the chromosomally encoded IpyA pathway does not contribute to pathogenicity in *E. h. gypsophilae*. The inactivation of the two auxin pathways or both auxin and cytokinin pathways does not cause any further reduction in gall size. This study, therefore, suggests that, although the hormones play a major role in gall formation, other factors such as the products of *hrp* and *avr*-like genes may also be involved. In a recent study of the plant growth-promoting bacterium *Azospirillum brasilense*, expression of an *ipdC::gusA* fusion was induced by a structurally related group of auxins including IAA itself. This induction is cell density-(growth phase)-dependent, as higher expression was detected in cells during the stationary phase than in exponential growth stage. It remains to be determined if auxin production in *E. h. gypsophilae* similarly responds to these regulatory parameters.

VI. OTHER PATHOGENICITY FACTORS

Several factors in addition to those described are required for full virulence in various *Erwinia* species. These include the peptide methionine sulfoxide reductase (MsrA), the *sap* (*s*ensitivity to *a*ntimicrobial *p*eptides) gene products, the proteins involved in iron acquisition and transport, and flavohemoglobin.

A. Methionine Sulfoxide Reductase

The plant-inducible gene, *msrA*, which is predicted to encode *m*ethionine *s*ulfoxide *r*eductase, is a virulence determinant in *E. chrysanthemi*. In *E. coli*, MsrA has been implicated in protecting the cell from oxidative stress. An MsrA⁻ mutant of *E. chrysanthemi* is less efficient in the maceration of chicory leaves and causes no systemic symptoms in *Saintpaulia ionantha* plants. The sensitivity of the *E. chrysanthemi* mutant to oxidative agents, such as hydrogen peroxide and the herbicide paraquat, suggests that MsrA plays a

similar role in *E. chrysanthemi* as has been implicated in *E. coli*.

B. The *sap* Locus

The *sap* locus of *E. chrysanthemi* strain AC4150 confers the ability to resist the plant host's antimicrobial peptides. The predicted products of the *sapABCDF* operon of *E. chrysanthemi* exhibit similarity with ABC transporters, and have 70, 69, 63, 78, and 75% similarities with the corresponding genes from *S. typhimurium*. In *S. typhimurium*, the periplasmic SapA is proposed to bind antimicrobial peptides. Subsequently, the peptide is transported into the cytoplasm for degradation or activation of resistance determinants. An *E. chrysanthemi* Sap⁻ mutant is less virulent in potato and more susceptible to the antimicrobial peptides snakin-1 and α-thionin produced by potato and wheat endosperm, respectively. The specificity of Sap proteins for certain but not all peptides and the conservation of *sap* locus in several *Erwinia* species suggest that this locus confers resistance against specific antimicrobial peptides in *Erwinia* species.

C. Iron-Acquisition Systems

Many microorganisms, including some *Erwinia* species, require iron for pathogenicity. However, because of its high reactivity and the ability to exist in multiple oxidation states, iron is not always available to the invading bacteria in the host tissues. Microorganisms, therefore, use multiple systems, including siderophores, to acquire iron for use in biological reactions. Siderophores are low-molecular-weight iron chelators, which act as iron scavengers. The siderophores secreted into the milieu bind Fe(III) with high affinity and make it available to bacteria through internalization of the complex into the bacterial cells. The genetics of iron acquisition in *Erwinia* have been studied the most in *E. chrysanthemi* and to some extent in *E. amylovora* and *E. c. carotovora*.

E. chrysanthemi synthesizes chrysobactin, a catechol-type siderophore, as well as achromobactin. The *fct-cbsCBEA* operon specifies iron uptake and its subsequent transport into the cell via chrysobactin. The operon encodes the ferrichrysobactin receptor Fct (*f*errichrysobactin *t*ransport), localized in the outer membrane, and enzymes for the first four steps of chrysobactin biosynthesis (*cbs*). The biosynthesis of achromobactin is specified by the *acs* (achromobactin synthesis) locus, whereas the internalization of ferriachromobactin is performed by the ferriachromobactin receptor (Acr) and the ferriachromobactin permease encoded by the *cbrABCD* operon. The *ferric uptake repressor*, Fur, in conjunction with the Fe(II) as a co-repressor represses the expression of the *fct-cbsCBEA* operon, most likely by binding to the Fur box in the promoter region of the operon. *E. chrysanthemi* strains deficient in iron transport produce only localized symptoms, suggesting that iron is required for full virulence. Also, the expression of *pelD* and *pelE* is lower in a Cbr⁻ mutant and induced in the presence of iron chelators, suggesting that iron deficiency has a positive regulatory effect on *pelD* and *pelE* expression.

E. c. carotovora also makes a catechol-type siderophore, probably also chrysobactin, and several strains in addition synthesize the hydroxamate siderophore, aerobactin. *E. c. carotovora* mutants, deficient in the production of either or both siderophores, do not differ from the wild-type strain in the capacity to macerate potato tuber tissue or to cause aerial stem rot of potato.

E. amylovora synthesizes cyclic hydroxamate-type siderophores mostly in the form of desferrioxamine (DFO), whose biosynthesis is controlled by a gene cluster. A transposon insertion in *dfoA* disrupts DFO biosynthesis. DfoA shares 58% homology with the product of the *alcA* gene, which specifies alcaligin siderophore biosynthesis in *Bordetella* spp. The DFO biosynthetic mutant induces less necrosis than the wild-type on apple flowers, but is unaffected in its ability to cause necrosis in seedlings. In contrast, the *E. amylovora* strain deficient in FoxR, the receptor for desferrioxamine, induces less necrosis in both apple seedlings and flowers. In addition to iron acquisition, desferrioxamine may also protect bacteria against oxidative damage.

D. Flavohemoglobin (HmpX)

The flavohaemoglobin HmpX, encoded by the *hmpX* gene, is a pathogenicity factor in *E. chrysanthemi* strain 3937. *E. chrysanthemi* HmpX, having 64% amino acid identity with *E. coli* Hmp, has homol-

ogies at its N-terminal end with hemoglobins and at its C-terminal with a family of reductases. Pel activity of the parent *E. chrysanthemi* is induced 50- and 300-fold over the basal level under microaerobic conditions and in *Saintpaulia ionantha* leaves, respectively, whereas Pel activity in a HmpX⁻ mutant is induced only fourfold under microaerobic conditions and the mutant does not grow in *Saintpaulia ionantha* leaves. Consistent with those observations is the finding that the mutant produces only necrotic spots or no symptoms when inoculated into *Saintpaulia ionantha* plants. There is a 60-fold increase in *hmpX::gus* expression when the bacterium is co-cultured with tobacco cells compared to the levels in the bacterium grown in artificial medium. This observation suggests that *hmpX* expression is activated by an unidentified plant signal. The presence of both the hemoglobin domain and the reductase domain coupled with the phenotypes of an HmpX⁻ strain *in planta* and under limiting oxygen conditions suggest a dual role for HmpX during pathogenesis. Accordingly, HmpX could act as a scavenger of the plant-produced reactive oxygen species and subsequently reduce these species by its reductase activity.

E. Pathogenicity Islands in *Erwinia*?

The genes specifying pathogenicity in some bacterial pathogens of humans are clustered together on large stretches of DNA and these are referred to as pathogenicity islands (PAI). The criteria generally used to define PAI are large, compact and distinct genetic units (often >30-kb DNA stretch) containing several virulence genes; the presence of the genetic units specifically in pathogenic strains; different G + C composition compared to the remainder of the genomic DNA; the association of the genetic units with tRNA genes or IS (insertion sequence) elements as their boundaries; the presence of mobility genes; and instability. In *E. amylovora*, *E. chrysanthemi*, and *E. h. gypsophilae*, the *hrp* clusters are linked with some other pathogenicity-associated loci, and the genetic organizations share some features with PAIs. In *E. h. gypsophilae*, the phytohormone genes, as well as the gene for host specificity (*hsvG*), are close to the *hrp* genes on the pPATH megaplasmid, which is present only in pathogenic strains. This region also contains insertion sequences with direct repeats.

Closely linked to the *hrp* cluster in *E. amylovora* is the disease specific (*dsp*) locus for virulence in host plants. In *E. chrysanthemi*, the *hrpC* and *hrpN* operons at the ends of the *hrp* cluster are flanked by *plcA*, a gene encoding phospholipase C, and *hecA* and *hecB* genes, which encode homologs of hemolysin and a hemolysin activator–transporter, respectively. Phospholipases and hemolysins are virulence factors of several animal and human pathogens, but the roles of these proteins in interactions of *E. chrysanthemi* with host plants await clarification. It also remains to be determined if, in addition to genes for pathogenicity, *hrp* clusters possess other features that are found in PAIs.

VII. ANTIMICROBIAL SUBSTANCES

A. Antibiotics

E. c. carotovora, *E. c. betavasculorum*, and *E. herbicola* produce antibacterial antibiotics. *E. c. carotovora* strain GS101 makes a carbapenem antibiotic, 1-carbapen-2-em-3-carboxylic acid. The antibiotics produced by *E. c. betavasculorum* and *E. herbicola* strain Eh1087 appear to be β-lactams, while that of *E. herbicola* strain Eh252 is a histidine analog that interferes with histidine biosynthesis. Genetic analysis of antibiotic production in *E. c. betavasculorum* has revealed a class of Ant⁻ mutants that neither produce exoenzymes nor antibiotics. Such mutants have now been found to be AHL-deficient (see Section II.I.1).

The genetics of antibiotic production have been examined in *E. herbicola* strains EH1087 and EH252. Through random mutagenesis of EH1087, Ant⁻ mutants were isolated, and subsequently an antibiotic locus complementing these mutants was identified. The *ant* locus that specifies antibiotic biosynthesis in this strain has six ORFs oriented in the same direction. The putative gene products of these ORFs have homology with aldehyde dehydrogenase, RNAse E, ornithine amino transferases, and drug-resistance translocase. The *ant* locus of EH252 has also been cloned, mapped, and expressed in *E. coli*.

Genetics of carbapenem production have been extensively studied in *E. c. carotovora* strain GS101, leading to the identification of the regulatory and biosynthetic genes. The eight-gene *carA-H* operon specifies carbapenem antibiotic biosynthesis and resistance. The gene products range in size from 9.8 kDa (CarE) to 55.9 kDa (CarA). The products of the first five genes are responsible for the synthesis of the carbapenem molecule. The predicted sequences of CarA, CarB, CarC, CarD, and CarE have homologies with the *orf1* gene product of the clavulanic acid biosynthetic gene cluster from *S. clavuligerus*, enoyl CoA hydratase from *Rhizobium meliloti*, isoenzymes of clavalaminate synthase of *S. clavuligerus*, proline oxidase precursor from *Drosophilla melanogaster*, and ferredoxin from *Synechococcus*, respectively. CarFGH products, localized in the periplasmic fraction, specify carbapenem resistance. The CarF and CarG proteins presumably bind carbapenem and thereby block the access of the antibiotic to the target site. The lack of similarity to any known β-lactam resistance mechanisms, including that due to β-lactamases, suggests that carbapenem resistance in *E. c. carotovora* may be mediated by a novel mechanism.

Closely linked but not part of the *carA-H* operon is *carR*, the gene encoding the transcriptional activator of carbapenem biosynthesis. *carR* and *carI*, members of *luxR–luxI* family of regulatory genes, control antibiotic production in *E. c. carotovora*. *carI* encodes an AHL synthase that synthesizes OHHL (see Section II.I.1). CarR, the transcriptional activator of the *carA-H* operon, presumably binds OHHL to activate *carA-H* genes. Although mutations in *carI* affect both antibiotic and exoenzymes, the deficiency of CarR only affects antibiotic production. The *carA-H* operon together with *carR* and OHHL are sufficient for carbapenem biosynthesis in *E. coli*.

Another positive regulator, Hor (homologue of rap), which controls the production of Pels, Cel Prt, and carbapenem antibiotic and pathogenicity, has been identified in *E. c. carotovora* strain GS101. The 16-kDa Hor protein belongs to a novel group of regulators in plant, human, and animal pathogens that control multiple phenotypes. For example, Rap of *Serratia marcescens* positively controls the production of carbapenem antibiotic and pigment. SlyA proteins of *S. typhimurium* and *E. coli* function as positive-acting regulatory factors for the induction of hemolysins. In fact, Hor of *E. c. carotovora* and Rap of *S. marcescens* are functional in heterologous systems.

Hor also is distantly related to PecS of *E. chrysanthemi,* a repressor controlling the production of Pel, Cel, and the blue pigment indigoidine (see Section II.I.3), and *E. chrysanthemi* possesses a *hor* homolog as well as *pecS.* Hor affects carbapenem antibiotic production via transcription of the *car* biosynthetic genes, rather than through the expression of the regulatory genes, *carI* and *carR.*

B. Bacteriocins

The virulent *E. c. carotovora* strain 71 and other *E. c. carotovora* strains produce bacteriocins, including carotovorocin (Ctv). Also, a nonvirulent biocontrol strain of *E. c. carotovora,* CGE234-M403, produces a low-molecular-weight bacteriocin as well as a high-molecular-weight bacteriocin. The relationship between Ctv of strain 71 and the high-molecular-weight bacteriocin of CGE234-M403 has not been examined.

The production of Ctv is induced by DNA-damaging agents such as mitomycin C and UV light, and regulated by the same factors (i.e., RecA, RdgA, and RdgB) that control Pnl and cellular lysis (see Section II.I.6 for the details of these regulators). The genetic locus *ctv* comprises a two-gene operon in *E. c. carotovora* strain 71; the promoter proximal gene encodes a protein (CtvA) having the bacteriocin function, whereas the other encodes an accessory protein. In *E. c. carotovora* strain 379, Ctv production is specified by an indigenous plasmid, but the details of genetic organization are not known. A positive regulatory gene *brg* (bacteriocin regulator gene) controls the synthesis of the low-molecular-weight bacteriocin in *E. c. carotovora* strain CGE234-M403. *brg* is predicted to encode a 99-amino-acid protein that has high homology with several regulatory proteins including Yrp, a positive regulator of enterotoxin production in *Yersinia enterocolitica.* Thus, the regulatory systems controlling Ctv in *E. c. carotovora* strain 71 and the low-molecular-weight bacteriocin in *E. c. carotovora* strain CGE234-M403 are quite different.

Bacteriocin production has also been detected in other *Erwinia* species. For example, *E. herbicola* produces a bacteriocin known as herbicolacin. The genes for herbicolacin production are located on a plasmid.

The finding that herbicolacin is active against *E. amylovora,* the bacterium that causes fire-blight in apples and pears (Table I), raises the possibility of using this system for biological control.

VIII. PIGMENTS

Several *Erwinia* species synthesize pigments. For example, *E. rhapontici* and *E. rubrifaciens* produce pink pigments, *E. herbicola* and *E. uredovora* produce carotenoid pigments, and *E. chrysanthemi* produces the blue pigment indigoidine, especially when grown on YGC (yeast extract–glucose–calcium carbonate) medium. Also, *E. h. gypsophilae, E. stewartii,* and *E. ananas* produce yellow pigments that probably are carotenoids as well. Carotenoids are a class of yellow, orange, and red pigments that differ greatly with respect to structure and function. Both prokaryotes and eukaryotes produce these compounds. Most carotenoids are made up of a linear C_{40} hydrocarbon with 3–15 conjugated double bonds and absorb light in the range of 400–500 nm. Carotenoids provide protection against the effects of reactive oxygen species and other radicals generated in the presence of visible or UV light. Thus, these pigments may protect the invading bacteria from the host-produced reactive oxygen species (ROS) or ROS-induced defense responses.

The genetics of carotenoid biosynthesis have been studied in *E. herbicola* and *E. uredovora,* as well as in other prokaryotes and eukaryotes. The *crt* gene clusters dedicated to the synthesis of carotenoids in these bacteria have been cloned. These genes may be chromosomally located or plasmid-borne. Sequences of *crt* and their putative products have revealed that the genes are homologous in *E. herbicola* and *E. uredovora* irrespective of their plasmid or chromosomal location. In *E. herbicola* strain Eho13, the biosynthetic genes include *crtE, crtX, crtY, crtI, crtB,* and *crtZ,* which encode GGPP synthase, zeaxanthine glycosylase, lycopene cyclase, phytoene desaturase, phytoene synthase, and β-carotene hydroxylase, respectively. cAMP and RpoS positively regulate the biosynthesis of carotenoids in some strains of *E. herbicola.*

IX. CONCLUSION

During the 1990s, three classes of pathogenic *Erwinia* have received the most attention—those that cause soft rot or tissue maceration, represented by *E. chrysanthemi* and *E. c. carotovora*; those responsible for wilt, represented by *E. amylovora* and *E. stewartii*; and those that cause hypertrophy, represented by *E. h. gypsophilae*. Studies of these and related bacteria have produced a better overall understanding of the chemical nature of virulence factors and their modes of action, genetic organization, regulation, and secretion. Harpins, the elicitors of nonhost resistance have also attracted considerable attention. Thus, researchers are able to make few generalizations and predictions: (1) pathogenicity genes in several of these bacteria are clustered and some such clusters may represent pathogenicity islands; (2) pathogenicity factors are secreted by at least three distinct pathways whose genetic components have been identified; (3) intricate regulatory systems control the production of transcripts of pathogenicity genes; (4) gene expression not only requires plant signals and appropriate environmental conditions, but also bacterial metabolites that sense cell density and metabolic state; and (5) posttranscriptional regulation plays a crucial role in the expression of pathogenicity genes.

Despite impressive progress, much remains to be learned about the pathogenicity and ecology of these bacteria. Several immediate questions and issues are as follows. What is the role of Harpins in susceptible interactions and how do their actions interface with those of the known virulence factors, say plant cell-wall degrading enzymes, hormones, or EPS? Is there a temporal sequence in the production and actions of those factors and, if so, is it responsible for the switch from endophytic or epiphytic states to the pathogenic phase? What are the functions of the Hrp proteins in type III secretion pathway? How many proteins are secreted by each of the three secretion pathways and what roles do the secreted proteins play in host interaction and pathogen fitness? What is the biological significance of co-regulation of Harpins and virulence factors? How do various signals interact with regulatory proteins to control expression of genes determining plant interaction? How do the regulatory RNAs and RNA-binding proteins control gene expression? It is expected that most of these issues and perhaps others not listed here will be resolved in the near future. This optimism rests on the fact that the *Erwinia* species have become powerful model systems for the analysis of various facets of host–pathogen interactions. Aside from fundamental interests, *Erwinia* species are also attractive from the applied and biotechnological perspectives, for example, for hyperproduction of degradative enzymes, antibiotics and bacteriocins, pigments, polysaccharides, and other useful metabolites. Paradoxical though it may sound, genetic manipulation of some *Erwinia* species may also lead to the development of effective biological control systems. If the past is any guide, there is ample reason to remain optimistic about developments along these lines.

Acknowledgments

We are very grateful to Thomas Burr; Alan Collmer; David Coplin, Jeanne Erickson, Sheng Yang He, Joyce Loper, and Shula Manulis for critically reviewing the manuscript. We also thank David Coplin for sharing unpublished data with us. Research in our laboratory is supported by National Science Foundation (grant MCB-9728505) and Food for the 21st Century Program of University of Missouri.

See Also the Following Articles

PIGMENTS, MICROBIALLY PRODUCED • PLANT PATHOGENS • QUORUM SENSING IN GRAM-NEGATIVE BACTERIA

Bibliography

Andersson, R. A., Palva, E. T., and Pirhonen, M. (1999). The response regulator ExpM is essential for the virulence of *Erwinia carotovora* subsp. *carotovora* and acts negatively on the sigma factor RpoS (σ^s). *Mol. Plant-Microbe Interact.* **12**, 575–584.

Armstrong, G. A. (1997). Genetics of eubacterial carotenoid biosynthesis: a colorful tale. *Annu. Rev. Microbiol.* **51**, 629–659.

Broek, A. V., Lambrecht, M., Eggermont, K., and Vanderleyden, J. (1999). Auxins upregulate expression of the indole-3-pyruvate decarboxylase gene in *Azospirillum brasilense*. *J. Bacteriol.* **181**, 1338–1342.

Chuang, D., Kyeremeh, A. G., Gunji, Y., Takahara, Y., Ehara, Y., and Kikumoto, T. (1999). Identification and cloning of an *Erwinia carotovora* subsp. *carotovora* bacteriocin regulator gene by insertional mutagenesis. *J. Bacteriol.* **181**, 1953–1957.

Condemine, G., Castillo, A., Passeri, F., and Enard, C. (1999). The PecT repressor coregulates synthesis of exopolysaccharides and virulence factors in *Erwinia chrysanthemi*. *Mol. Plant-Microbe Interact.* **12**, 45–52.

Hassouni, M. E., Chambost, J. P., Expert, D., Van Gijsegem, F., and Barras, F. (1999). The minimal gene set member *msrA*, encoding peptide methionine sulfoxide reductase, is a virulence determinant of the plant pathogen *Erwinia chrysanthemi*. *Proc. Natl. Acad. Sci. U.S.A.* **96**, 887–892.

Franza, T., Sauvage, C., and Expert, D. (1999). Iron regulation and pathogenicity in *Erwinia chrysanthemi* 3937: Role of the fur repressor protein. *Mol. Plant-Microbe Interact.* **12**, 119–128.

Galan, J. E., and Collmer, A. (1999). Type III secretion machines: Bacterial devices for protein delivery into host cells. *Science* **284**, 1322–1328.

Harris, S. J., Shih, Y., Bentley, S. D., and Salmond, G. P. C. (1998). The *hexA* gene of *Erwinia carotovora* encodes a LysR homologue and regulates motility and the expression of multiple virulence determinants. *Mol. Microbiol.* **28**, 705–717.

Hauben, L., Moore, E. R. B., Vauterin, L., Steenackers, M., Mergaert, J., Verdonck, L., and Swings, J. (1998). Phylogenetic position of phytopathogens within the *Enterobacteriaceae*. *Syst. Appl. Microbiol.* **21**, 384–397.

Liu, Y., Cui, Y., Mukherjee, A., and Chatterjee, A. K. (1998). Characterization of a novel RNA regulator of *Erwinia carotovora* ssp. *carotovora* that controls production of extracellular enzymes and secondary metabolites. *Mol. Microbiol.* **29**, 219–234.

Lopez-Solanilla, E., Garcia-Olmedo, F., and Rodriguez-Palenzuela, P. (1998). Inactivation of the *sapA* to *sapF* locus of *Erwinia chrysanthemi* reveals common features in plant and animal bacterial pathogenesis. *Plant Cell* **10**, 917–924.

Lory, S. (1998). Secretion of proteins and assembly of bacterial surface organelles: Shared pathways of extracellular protein targeting. *Curr. Opinion Microbiol.* **1**, 27–35.

Manulis, S., Havivchesner, A., Brandl, M. T., Lindow, S. E., and Barash, I. (1998). Differential involvement of indole-3-acetic acid biosynthetic pathways in pathogenicity and epiphytic fitness of *Erwinia herbicola* pv. *gypsophilae*. *Mol. Plant-Microbe Interact.* **11**, 634–642.

McGowan, S. J., Bycroft, B. Y., and Salmond, G. P. C. (1998). Bacterial production of carbapenems and clavums: evolution of β-lactam antibiotic pathways. *Trends Microbiol.* **6**, 203–208.

Nomura, K., Nasser, W., Kawagishi, H., and Tsuyumu, S. (1998). The *pir* gene of *Erwinia chrysanthemi* EC16 regulates hyperinduction of pectate lyase virulence genes in response to plant signals. *Proc. Natl. Acad. Sci. U.S.A.* **95**, 14034–14039.

Reverchon, S., Bouillant, M. L., Salmond, G., and Nasser, W. (1998). Integration of the quorum-sensing system in the regulatory networks controlling virulence factor synthesis in *Erwinia chrysanthemi*. *Mol. Microbiol.* **29**, 1407–1418.

Shevchik, V., Condemine, G., Robert-Baudouy, J., and Hugouvieux-Cotte-Pattat, N. (1999). The exopolygalacturonate lyase PelW and the oligogalacturonate lyase Ogl, two cytoplasmic enzymes of pectin catabolism in *Erwinia chrysanthemi* 3937. *J. Bacteriol.* **181**, 3912–3919.

von Bodman, S. B., Majerczak, D. R., and Coplin, D. L. (1998). A negative regulator mediates quorum-sensing control of expolysaccharide production in *Pantoea stewartii* subsp. *stewartii*. *Proc. Natl. Acad. Sci. U.S.A.* **95**, 7687–7692.

Wehland, M., Kiecker, C., Coplin, D. L., Kelm, O., Saenger, W., and Bernhard, F. (1999). Identification of an RcsA/RcsB recognition motif in the promoters of exopolysaccharide biosynthetic operons from *Erwinia amylovora* and *Pantoea stewartii* subspecies *stewartii*. *J. Biol. Chem.* **274**, 3300–3307.

Escherichia coli, General Biology

Moselio Schaechter

San Diego State University

I. Taxonomy
II. Ecology
III. Structure and Function of Cell Parts
IV. Metabolism and Growth
V. Pathogenesis
VI. Principles of Diagnosis Using Clinical Specimens

GLOSSARY

enterobacteriaceae A family of the γ-Proteobacteria that includes *Escherichia coli* and related gram-negative bacteria.

enterohemorrhagic Escherichia coli (EHEC) A bacterium that causes hemorrhagic colitis.

enteropathogenic Escherichia coli (EPEC) A bacterium that causes diarrhea after colonization of the middistal small intestine.

enterotoxigenic Escherichia coli (ETEC) The bacterium that is the most common agent of watery "tourist's" diarrhea.

hemorrhagic colitis Bloody diarrhea caused by enterohemorrhagic *Escherichia coli* (EHEC).

phase variation A phenomenon occurring in several strains of the *Escherichia coli* in which the bacteria alternate between the fimbriated and nonfimbriated conditions.

ESCHERICHIA COLI is a gram-negative facultative anaerobic nonspore-forming motile rod. The species belongs to the Enterobacteriaceae family of the γ-Proteobacteria and includes a large number of strains that differ in pathogenic potential. Certain strains are common innocuous residents of the intestine of mammals; others cause human and animal infections of the digestive and urinary tracts, blood, and central nervous system. The structure, biochemical functions, and genetics of this organism are well studied, making it the best known of all cellular forms of life. This organism has occupied center stage in the development of molecular biology.

I. TAXONOMY

E. coli is one of five recognized species of the genus *Escherichia*. The genus is named after Theodor Escherich, who first isolated *E. coli* in 1884. The species is defined on the basis of certain readily measurable biochemical activities shared by most strains. Thus, *E. coli* generally ferments lactose, possesses lysine decarboxylase, produces indole, does not grow on citrate, does not produce H_2S, and is Voges-Proskauer–negative (does not produce acetoin). Its chromosomal DNA is 49–52% G + C.

The limits of the taxonomic definition are under scrutiny because DNA hybridization data suggest that, contrary to tradition, *Escherichia* and *Shigella* belong to the same genus. *E. coli* comprises a number of strains that share the same basic taxonomic features, with about 70% DNA homology at the extremes.

E. coli strains are defined mainly by their antigenic composition. Of taxonomic relevance are over 170 different serological types of lipopolysaccharide antigens (O antigens) and 80 types of capsular (K antigens). Other properties that are used to define individual strains are H antigens (flagellar proteins), F antigens (fimbrial proteins), and phage and colicin sensitivity.

Recent quantitative approaches to defining taxonomic relationships are based on patterns of isozymes of metabolically important enzymes, restriction fragment-length polymorphism, and protein composition of the outer membrane. Although there is a huge number of combinations of these properties, the variety of strains that have been isolated is circumscribed (perhaps in the thousands). By the criteria used, most of the strains in today's world appear to be the clonal descendants of relatively few ancestors. Genetic exchange leading to recombinational events thus seems to be an infrequent event in the environment. It has been estimated that, for this species, major episodes of selection have occurred once in 30,000 years.

II. ECOLOGY

E. coli is the most abundant facultative anaerobe in the feces, and therefore the colon, of normal humans and many mammals. It is commonly present in concentrations of $\sim 10^7 - 10^8$ live organisms per gram of feces. Thus, the total number of individual *E. coli* cells present on Earth at any one time exceeds 10^{20}.

E. coli is far from the most abundant organism in the colon and is outnumbered 100-fold or more by strict anaerobes. *E. coli* and other intestinal facultative anaerobes colonize not only in the large intestine of vertebrates but also the ileum, the distal segment of the small intestine. In the ileum, *E. coli* is present in numbers that approximate those of the organisms in feces. The ileal population is transient, being rapidly propelled into the cecum by peristalsis. These organisms are apparently derived via a reflux mechanism from the cecum and are rarely acquired anew by ingestion. A better understanding of what shaped *E. coli* during evolution will require a systematic study of the distinct selective pressures it may have faced in these two very different regions of the intestine, the ileum and the colon. In the ileum, *E. coli* may have been selected for rapid growth under aerobic conditions; in the colon, in competition for limited nutrients and in the presence of noxious chemicals under anaerobic conditions.

Most fecal *E. coli* isolates are well adapted to colonizing the mammalian intestine and seldom cause disease. When human strains are cultivated in the laboratory, they tend to lose the ability to colonize.

Included among these is K12, the most widely used strain in the molecular microbiology.

E. coli cells are periodically deposited from their intestinal residence into soils and waters. It has been thought that they do not survive for extended periods of time in such environments and could be cultured only for a few days (seldom weeks) after their introduction. For this reason, their presence has been taken as a measure of recent fecal contamination, and the coliform count of the drinking water supply or of swimming facilities is still a common measure of microbiological water purity. The notion that *E. coli* has a short survival time in the environment has been challenged, and new work suggests that the presence of these organisms may not be a reliable indication of recent fecal pollution.

Mammals become colonized with *E. coli* within a few days of birth, possibly from the mother or other attendants. How the organisms are transmitted to the neonate is not known with certainty; this may occur during passage through the birth canal or, shortly after birth, via the fecal–oral route.

III. STRUCTURE AND FUNCTION OF CELL PARTS

In both structure and function, *E. coli* serves as the prototype for members of the Enterobacteriaceae. An example of the overall composition of this organism in its growth phase is shown in Table I.

A. Fimbriae (Pili)

E. coli strains carry one or two kinds of fimbriae, common and conjugative (sex pili). Common fimbriae are usually found in numbers of 100–1000/cell and consist mainly of an acidic hydrophobic protein called fimbrin. The common fimbriae fall into seven groups according to the amino acid sequence of their major fimbrin. Fimbriae are highly antigenic, comprising many F antigens.

E. coli strain K12 possesses only type 1 common fimbriae. This strain and others alternate between the fimbriated and nonfimbriated condition, a phenomenon known as phase variation. It is thought that the presence of common fimbriae allows organ-

TABLE I
Overall Macromolecular Composition of an Average *E. coli* Cell[a]

Macromolecule	Percentage of total dry weight	Weight per cell ($10^{-15} \times g$)	Molecular weight	Number of molecules per cell	Different kinds of molecule
Protein	55.0	155.0	4.0×10^4	2,360,000	1050
RNA	20.5	59.0			
23S ribosomal RNA		31.0	1.0×10^6	18,700	
16S ribosomal RNA		16.0	5.0×10^5	18,700	
5S ribosomal RNA		1.0	3.9×10^4	18,700	
Transfer		8.6	2.5×10^4	205,000	60
Messenger		2.4	1.0×10^9	1380	400
DNA	3.1	9.0	2.5×10^9	2.13	
Lipid	9.1	26.0	705	22,000,000	
Lipopolysaccharide	3.4	10.0	4346	1,200,000	
Murein	2.5	7.0			
Glycogen	2.5	7.0	1.0×10^6	4360	
Total macromolecules	96.1	273.0			
Soluble pool	2.9	8.0			
Building blocks		7.0			
Metabolites, vitamins		1.0			
Inorganic ions	1.0	3.0			
Total dry weight	100.0	284.00			
Total dry weight/cell		2.8×10^{-13} g			
Water (at 70% of cell)		6.7×10^{-13} g			

[a] These values are for *E. coli* in balanced growth in a glucose-minimal medium at 37°C.

From Neidhardt, F. C., Ingraham, J. L., and Schaechter, M. (1990). "Physiology of the Bacterial Cell." Sunderland, MA: Sinauer Associates.

isms in their first efforts to colonize their host by attaching to epithelial cells. Inside the body, turning off the synthesis of common fimbriae may lessen the chances that the organisms will be phagocytized by white blood cells.

The sex fimbriae (usually called pili) are encoded by plasmids such as F or R and are usually present in one or a few copies per cell. These structures cause the donor and recipient bacteria to make contact, allowing the transfer of DNA during conjugation.

B. Flagella

E. coli is usually endowed with only 5–10 flagella/cell. However, this complement suffices to endow the cells with brisk motility. The flagella are typically 5–10 μm long and are arranged randomly around the cell surface, a pattern called peritrichous flagellation. As is typical of bacterial flagella, those of *E. coli* are composed of a long filament, a hook, and a basal body. The principal component of *E. coli* flagella is an *N*-methyl-lysine-rich protein known as flagellin. Its size, usually around 55 kDa, varies among strains. Around 20,000 subunits of this protein assemble to make the flagellar filament. *In vitro,* flagellin self-assembles into flagella-like filamentous cylindrical lattices with hexagonal packing.

The *E. coli* flagellar genetic system consists of about 40 genes arranged in five regions. These genes are involved in structure, function, assembly, and regulation. Flagella are highly antigenic, comprising a large number of H antigens. The N and C termini of various H antigens are highly conserved, the major antigenic divergence being found in the central region of the molecule.

C. Capsule and Outer Membrane

In some strains, the outer membrane of *E. coli* covered by a polysaccharide capsule composed of K antigens. Other polysaccharides, the M antigens (colanic acids, which are polymers of glucose, galactose, fucose, and galacturonic acid), are synthesized under conditions of high osmolarity, low temperature, and low humidity, suggesting that normally these compounds may be made in response to stressful conditions in the external environment. In addition, *E. coli* and other enteric bacteria possess a glycolipid anchored in the outer leaflet of their outer membrane, called the enterobacterial common antigen (ECA).

The outer membrane of *E. coli* is typical of that of gram-negative bacteria, consisting of a lipid bilayer whose inner leaflet is made up largely of phospholipids and whose outer leaflet is made of lipopolysaccharide (LPS). Interspersed are several kinds of membrane proteins. One, the murein lipoprotein, is small (7.2 kDa) and exists in 7×10^5 copies per cell. This protein contains three fatty acid residues that help anchor it into the inner leaflet of the outer membrane, whereas the rest of the molecule is located in the periplasm. About one-third of the molecules are covalently linked to the cell-wall peptidoglycan. The major outer-membrane proteins include the pore-forming proteins, called porins Omp C, Omp F, and Pho E. Together, these porins are present in about 10^5 copies per cell. Their sizes vary from 36.7–38.3 kDa. The diameters of the pores are 1.16 nm for Omp F and Pho E and 1.08 nm for Omp C. The synthesis of these porins is regulated by environmental conditions; thus, Omp F is repressed by high osmotic conditions or high temperature, Omp C is derepressed by high osmotic conditions, and Pho E is synthesized when cells are starved for phosphate. These findings suggest that the organisms use narrower porin channels in the animal host than in the outside environment, which invites speculation about the nature of chemical and metabolic challenges faced under the two conditions.

Certain compounds that are too large to diffuse through *E. coli* porin channels are carried across the outer membrane by special transport proteins. These compounds include maltose oligosaccharides, nucle-osides, various iron chelates, and vitamin B_{12}. The proteins involved, as well as the porins, also act as receptors for the attachment of bacteriophages and colicins.

D. Periplasm and Cell Wall

By functional tests of solute partition and electron microscopy, the periplasm of *E. coli* makes up 20–40% of the cell volume. This compartment is osmotically active, in part because it contains large amounts of membrane-derived oligosaccharides (molecules of 8–10 linked glucose residues substituted with 1-phosphoglycerol and O-succinyl esters). There is evidence that the contents of the periplasmic space form a gel. The *E. coli* periplasm contains over 60 known proteins, including binding proteins for amino acids, sugars, vitamins, and ions; degradative enzymes (phosphatases, proteases, and endonucleases); and antibiotic detoxifying enzymes (β-lactamases, alkyl sulfodehydrases, and aminoglycoside phosphorylating enzymes). The periplasmic environment is oxidizing, whereas that of the cytoplasm is reducing. This explains why certain secretory proteins that require disulfide bonds for activity are inactive in the cytoplasm.

As in most bacteria, the cell wall of *E. coli* consists of a peptidoglycan layer responsible for cell shape and rigidity. In this organism, peptidoglycan is one or, at most, a few molecules thick. It is anchored to the outer membrane at some 400,000 sites via covalent links to the major membrane lipoprotein and noncovalent links to porins.

Evidence indicates that the periplasm is spanned by 200–400 adhesion zones between the outer membrane and the cytoplasmic membrane. These appear to be the sites of attachment of certain bacteriophages and of export of outer-membrane proteins and lipopolysaccharide. In addition to these apparently scattered junctions, the two membranes appear joined at defined periseptal annuli, ring-shaped adhesion zones that are formed near the septum.

E. Cytoplasmic Membrane

The cytoplasmic membrane of *E. coli* is made up of about 200 distinct proteins and four kinds of phospholipids. Proteins make up about 70% of the weight of the structure. Under aerobic conditions, the *E. coli* cytoplasmic membrane contains a number of dehydrogenases (e.g., NADH-, D- and L-lactate, and succinate dehydrogenases), pyruvate oxidase, cytochromes (of the o and d complexes), and quinones (mainly 8-ubiquinone). Anaerobically grown *E. coli* may contain other dehydrogenases (e.g., formate- and glycerol-3-phosphate dehydrogenases) and enzymes involved in anaerobic respiration (nitrate and fumarate reductases). The cytoplasmic membrane is the site of adenosine triphosphate synthesis. The cytoplasmic membrane systems involved in the transport of solutes are highly efficient and permit this species to grow in relatively dilute nutrient solutions. An example of the variety of transport systems is shown for amino acids in Table II.

The cytoplasmic membrane of *E. coli* contains over 20 proteins involved in various aspects of peptidoglycan biosynthesis, cell wall elongation, and cell division. About 10 of these proteins have been identified by their ability to covalently bind β-lactam antibiotics. They are known as the penicillin-binding proteins, and some have been shown to be involved directly in cell-wall synthesis.

TABLE II
Transport Systems for Amino Acids in *E. coli*[a]

Glycine–alanine
Threonine–serine
Leucine–isoleucine–valine
Phenylalanine–tyrosine–tryptophan
Methionine
Proline
Lysine–ornithine–arginine
Cystine
Asparagine
Glutamine
Aspartate
Glutamate
Histidine
Cysteine (probably)

---

[a] From Neidhardt, F. C., Ingraham, J. L., and Schaechter, M. (1990). "Physiology of the Bacterial Cell." Sunderland, MA: Sinauer Associates.

F. Cytoplasm

Most of the 1000 or so biochemical reactions necessary for the growth of *E. coli* take place in the cytoplasm. These activities are divided into those concerned with metabolic fueling (production of energy, reducing power, and precursor metabolites), biosynthesis of building blocks, polymerization into macromolecules, and assembly of cell structures. For *E. coli*, each of these activities is generally the same as for all microorganisms, but with certain species-specific characteristics.

In fast-growing *E. coli*, much of the cytoplasmic space is taken up by ribosomes. The number of ribosomes per cell is proportional to the growth rate, ranging from about 2000 in cells growing at 37°C at doubling rates of 0.2 hr^{-1} to >70,000 at a doubling rate of 2.5 h^{-1}, where they make up about 40% of the cell mass. In *E. coli*, the genes for the four ribosomal RNAs (16*S*, 23*S*, 5*S*, transfer) are arranged in seven operons located at different sites on the chromosome. Most of these operons are found near the origin of chromosome replication and, thus, are replicated early. This arrangement ensures that ribosomal RNAs are made in large amounts during rapid growth. Each operon encodes one, two, or three transfer RNAs at sites between the 16*S* and 23*S* RNA genes or at the end of the operon. The 52 ribosomal proteins are encoded by 21 transcriptional units. Ribosomal RNAs and proteins assemble into particles via a precise sequence of reactions that has been well studied *in vitro*.

G. DNA and Nucleoid

The entire sequence of the *E. coli* genome has been determined. It consists of 4,639,221 bp and codes for an estimated 4288 proteins. Slightly fewer than 2000 of these proteins and about one-third of all the genes have been characterized. The number of predicted operons is 2192, of which 75% have only one gene, 16.6% two genes, and the rest three or more genes. Protein-coding sequences account for 87% of the genome, stable RNAs for 0.8%, and noncoding repeats for 0.7%, leaving about 11% of the genome for regulatory and other functions. The genome contains a number of repeated sequences, most of unknown function, as well as several dozen insertion sequences, cryptic prophages, and phage remnants.

The nucleoid of *E. coli* is a highly lobular intracytoplasmic region, generally located toward the center of the cell. In this region, the DNA is found at a local concentration of 2–5% (w/v). The reason why this long molecule is folded and physically limited to the nucleoid region is not well understood. It is known that *in vivo* the DNA is negatively supercoiled into some 50 individual domains. Nucleoids of superhelicity and dimension similar to those seen intracellularly can be isolated by the gentle breakage of cells in the presence of divalent cations.

Transcription takes place at the nucleoid–cytoplasm interface, as the nucleoid is thought to form a significant barrier to the diffusion of many macromolecules. The reason for the highly irregular shape of the nucleoid may be to contribute to the availability of genes for transcription. At least four small-molecular-weight proteins that bind to DNA are known to play a role in transcription, recombination, and replication. These nucleoid-associated proteins range in molecular weight from 9,200 to 15,400. Two, HU and IHF, are among the abundant *E. coli* proteins and are present in 20–50,000 monomers per cell.

The initiation of DNA replication takes place at a specific origin site, *oriC*, and is under the influence of a protein that is highly conserved among many bacteria, DnaA. Once initiated, DNA replication takes place at a nearly constant rate in moderately fast and fast-growing *E. coli,* until it reaches a terminus. The doubling of the chromosome takes 40 mm at 37°C, which requires that the time of initiation of chromosome replication takes place before the end of the previous round in cultures growing faster than this.

Little is known about the mode of segregation of the nucleoids. The process takes place with considerable fidelity and, thus, cannot result from partitioning into progeny cells by chance alone. A widespread view is that the chromosome is attached to the cell membrane and that movement of the membrane serves as a primitive mitotic apparatus. However, this view is based largely on cell fractionation studies and is not supported by functional tests. At least *in vitro,*

recently replicated (hemimethylated) origin DNA binds to the membrane with great specificity.

IV. METABOLISM AND GROWTH

A. Biosynthetic and Fueling Reactions

The central metabolism of *E. coli* is carried out via the Embden–Meyerhof–Parnas pathway, the pentose pathway, and the tricarboxylic acid cycle plus, for the metabolism of gluconate, the Entner–Doudoroff pathway. As a facultative anaerobe, *E. coli* meets its energy needs either by respiratory or fermentative pathways. *E. coli* carries out a mixed acid fermentation of glucose that results in the formation of a large number of products. Under anaerobic conditions, the main products (and the moles formed per 100 moles of glucose used) are formate (2.4), acetate (37), lactate (80), succinate (12), ethanol (50), 2,3-butanediol (0.3), CO_2 (88), and H_2 (75).

The need for biosynthetic building blocks is met in *E. coli* by the production of 12 precursor metabolites common to all bacteria. The energy requirements for the manufacture of the major building blocks are shown in Table III.

The precursor metabolites do not contain nitrogen or sulfur, which must enter the metabolic circuit independently. *E. coli* does not fix dinitrogen gas, but can use a number of compounds as a source of nitrogen, including ammonium ions and various amino acids. It can use nitrate and nitrite as terminal electron acceptors during anaerobic respiration by activating nitrate or nitrite reductases. However, under anaerobic conditions, no energy is generated by this process, and a source of reduced nitrogen is necessary for the anaerobic growth of *E. coli*. The incorporation of ammonium ion into organic compounds is catalyzed either by L-glutamate dehydrogenase when ammonia is abundant or by glutamine synthetase and glutamate synthase, acting together, when ammonia is limiting.

The common sources of sulfur for *E. coli* are sulfate and sulfur-containing amino acids. Sulfate is transported into the cell after being reduced to H_2S by a sulfite reductase. Sulfur is then assimilated from H_2S using O-acetylserine sulfohydrolase to

TABLE III
Energy Requirements for Polymerization of the Macromolecules in 1 g of Cells[a]

Macromolecule	Amount of energy required (μmol P)
From activated building blocks	
DNA from dNTPs	136
RNA from NTPs	236
Protein from aminoacyltransfer RNAs	11,808
Murein, in part from activated building blocks	138
Phospholipids, in part from activated building blocks	258
Lipopolysaccharide	0
Polysaccharide (glycogen)	0
Total energy	12,576
From unactivated building blocks	
DNA from dNMPs	336
RNA from NMPs	1516
Protein from amino acids	21,970
Murein, in part from activated building blocks	138
Phospholipids, in part from activated building blocks	258
Lipopolysaccharide	0
Polysaccharide (glycogen)	0
Total energy	24,218

[a] From Neidhardt, F. C., Ingraham, J. L., and Schaechter, M. (1990). "Physiology of the Bacterial Cell." Sunderland, MA: Sinauer Associates.

produce L-cysteine. *E. coli* has a comple system of transporting and using organic phosphates, including an inducible alkaline phosphatase in its periplasm.

B. Nutrition and Growth

E. coli is a chemoheterotroph capable of growing on any of a large number of sugars or amino acids provided individually or in mixtures. Some strains found in nature have single auxotrophic requirements, among which thiamin deficiency is common.

The growth of many strains is inhibited by the presence of single amino acids, such as serine, valine, or cysteine. *E. coli* grows faster with glucose than with any other single carbon and energy source and reaches a doubling time of 50 min under well-oxygenated conditions at 37°C. Doubling times with less favored substrates may be hours in length. Slow rates of growth can also be achieved by using an externally controlled continuous culture device or by adding a metabolic analog and its antagonist to the culture at proper ratios. When the medium is supplemented with building blocks such as amino acids, nucleosides, sugars, and vitamin precursors, *E. coli* grows more rapidly, reaching doubling times of 20 minutes at 37°C in rich-nutrient broths.

E. coli can grow at temperatures between 8°C and 48°C, depending on the strain and the nutrient medium. Its optimum growth temperature is 39°C. *E. coli* does not grow in media containing a NaCl concentration greater than about 0.65 M. In response to changes in the osmotic pressure of the medium, *E. coli* increases its concentration of ions, especially K^+ and glutamate. The pH range for growth is between pH 6.0 and 8.0, although some growth is possible at values approximately 1 pH unit above and below this range.

V. PATHOGENESIS

Strains of *E. coli* are responsible for a large number of clinical diseases. In their manifestation, some of these diseases overlap with those caused by other species (e.g., *Shigella* and *Salmonella*). The most common infections caused by *E. coli* involve the intestinal and urinary tracts of humans and other mammals, where they produce simple watery diarrhea or locally invasive forms of infection (e.g., dysentery). *E. coli* infects deeper tissues, including the blood (septicemia) in patients with compromised defense mechanisms and, additionally, the meninges in infants. The organism also causes mastitis in cattle. Strains of *E. coli* that produce intestinal infections are divided into groups according to the clinical picture they produce and

TABLE IV
Classification of Pathogenic *E. coli*

Group	Symptoms
Enterotoxigenic (ETEC)	Watery diarrhea ("travelers' disease")
Enteropathogenic (EPEC)	Watery diarrhea
Enterohemorrhagic (EHEC)	Bloody diarrhea, hemorrhagic colitis, hemolytic-uremic syndrome, thrombocytopenic purpura
Enteroinvasive (EIEC)	Bloody diarrhea
Enteroaggregative (EAggEC)	Watery diarrhea, persistent diarrhea

their known virulence factors (Table IV). These strains are denoted by abbreviations (e.g., ETEC and EPEC, where the terminal EC stands for *E. coli*) and are proliferating.

A. Enterotoxigenic Strains

Enterotoxigenic *E. coli* (ETEC) strains acquired from food or water contaminated with human or animal feces are the most common bacterial agents of diarrhea in the United States and Europe. These strains circulate among a population, but the majority of people (especially adults) usually remain asymptomatic, most likely due to immunity afforded by previous exposure. ETEC strains are responsible for the "tourist's diarrhea" that frequently affects persons traveling to countries with a low level of sanitation. Watery diarrhea due to *E. coli* resembles that seen in mild cases of cholera.

ETEC strains colonize the small intestine, where they produce one or both of two enterotoxins called heat labile (LT) and heat stable (ST). Both toxins act by changing the net fluid transport activity in the gut from absorption to secretion. LT is structurally similar to cholera toxin and activates the adenylate cyclase–cyclic adenosine monophosphate system, whereas ST works on guanylate cyclase. The intestinal mucosa is not visibly damaged, the watery stool does not contain white or red blood cells, and no inflammatory process occurs in the gut wall. Gut cells activated by LT or cholera toxin remain in that state until they die, whereas the effects of ST on guanylate cyclase are turned off when the toxin is washed away from the cell.

B. Enteropathogenic Strains

Enteropathogenic *E. coli* (EPEC) strains cause diarrhea after the colonization of the mid-distal small intestine (ileum). They recognize their preferred hosts and tissues by means of plasmid-encoded surface adhesins specific for receptors on the intestinal brush-border membranes, called the bundle-forming pili (or fimbriae). Characteristic of EPEC infection is an attachment–effacement (A/E) lesion on the surface of enteric epithelial cells. The affected cells form a broad flat pedestal (effacement) beneath the attached microorganism, which, by damaging the absorptive surface, contributes to the diarrhea. The genes required for the formation of A/E lesions are located on a 35-kb pathogenicity island called the locus of enterocyte effacement (LEE). Once bound to the epithelial cells, EPEC export critical virulence factors by a type III secretion apparatus, causing several host signals to be activated.

C. Enterohemorrhagic Strains

The enterohemorrhagic *E. coli* (EHEC) comprise a limited number of serotypes that cause a characteristic nonfebrile bloody diarrhea known as hemorrhagic colitis. The most common of these serotypes in the United States is O157:H7, whereas others, particularly O26, are found with greater frequency elsewhere in the world. *E. coli* O157:H7 causes both outbreaks and sporadic disease. The organism is commonly isolated from cattle, and several outbreaks have been traced to undercooked hamburger meat.

EHEC strains have two special characteristics of pathogenic importance. First, they produce high levels of two related cytotoxins that resemble toxins of *Shigella*, with the same protein synthesis-inhibitory action and binding specificity. These toxins are therefore called Shiga-like toxins (SLT) I and II. The SLT are cytotoxic for endothelial cells in culture. Second, they possess a gene highly homologous to the EPEC attaching and effacing gene. In combination, the protein encoded by this gene and the SLT presumably damage the gut mucosa in a manner characteristic of hemorrhagic colitis.

EHEC strains cause systemic manifestations (hemolytic–uremic syndrome or thrombotic thrombocytopenic purpura) that are believed to be related to systemic absorption of SLT, possibly in combination with endotoxin. These syndromes represent the clinical response to endothelial damage of glomeruli and the central nervous system.

D. Other Strains That Cause Intestinal Infections

Enteroinvasive *E. coli* (EIEC) strains cause dysentery, resembling that due to *Shigella*. Unlike the strains already described, which are noninvasive, EIEC strains are selectively taken up into epithelial cells of the colon, requiring for this process a specific outer-membrane protein. EIEC strains also make Shiga-like toxins. Cell damage by these strains triggers an intense inflammatory response.

Other strains, called enteroaggregative *E. coli*, (EAggEC), are associated with diarrhea in infants under 6 months of age, often persisting for weeks with marked nutritional consequences. EAggEC strains spontaneously agglutinate (aggregate) in tissue culture.

E. Strains That Infect the Genitourinary Tract

E. coli strains are the most common cause of genitourinary tract infections in humans, including cystitis, pyelonephritis, and prostatitis. Many of the strains that cause pyelonephritis possess fimbriae (pili) that bind specifically to a glycolipid constituent of kidney tissue. These structures are called P pili because the receptor is a complex galactose-containing molecule that is part of the P blood group antigen. About 1% of humans are P antigen-negative and, not carrying the P pilus receptor, are not suscep-

tible to colonization by P pili carrying strains. These people do not suffer from urinary infections mediated by the usual route (i.e., mucosal colonization followed by ascending invasion of the bladder). Such individuals may, however, become infected when the normal route is bypassed (e.g., by the use of an indwelling urinary catheter).

F. Other Invasive Strains

Strains that possess a sialic acid-containing capsular polysaccharide, called K1 antigen, cause invasive disease, such as septicemia and meningitis in infants. *E. coli* is also a common cause of septicemia in adults, especially in patients that are immunocompromised. Many patients with this manifestation acquire it as a consequence of infections of the urinary tract, often following manipulations such as urinary catheterizations.

VI. PRINCIPLES OF DIAGNOSIS USING CLINICAL SPECIMENS

Naturally occurring strains of *E. coli* are generally similar with respect to both their colonial morphology on agar plates and their shape under the microscope. *E. coli* can be distinguished from other enteric bacteria on the basis of biochemical and nutritional properties. Most strains of *E. coli* differ from some of the classic intestinal pathogens, such as *Salmonella* and *Shigella*, in that they ferment lactose. For this reason, lactose is included as the sole added sugar together with a pH indicator in agar media. Lactose-fermenting colonies (presumptively those of *E. coli*) turn a distinctive color due to the production of acid.

With the help of an ingenious array of differential and selective media, it is usually simple to isolate *E. coli* from samples, such as feces, that contain a preponderance of other bacteria. These media and other special tests permit laboratories to narrow down the identification to the main genera of the Enterobacteriaceae.

Classifying *E. coli* into serological subgroups is not a task that most clinical laboratories are prepared to carry out. The serological reagents most readily available commercially are those directed against EPEC strains. EHEC strains of the serotype O157:H7 are nearly unique in their inability to ferment sorbitol. Nonfermenting colonies are detected on sorbitol-containing MacConkey agar and confirmed with a sensitive and specific latex agglutination test.

See Also the Following Articles

Enteropathogenic Bacteria • Flagella • Gastrointestinal Microbiology • Outer Membrane, Gram-Negative Bacteria • Ribosome Synthesis and Regulation

Bibliography

Levi, Primo (1989). An interview with *Escherichia coli* (one of five intimate interviews). *In* "The Mirror Maker," pp. 42–46. Schoecken Books, New York.

Neidhardt, F. C., Ingraham, J. L., and Schaechter, M. (1990). "Physiology of the Bacterial Cell." Sinauer Associates, Sunderland, MA.

Neidhardt, F. C., Curtiss, R., III, Ingraham, J. L., Lin, E. C. C., Low, K. B., Magasanik, B., Reznikoff, W. S., Riley, M., Schaechter, M., and Umbarger, H. E. (1996). "*Escherichia coli* and *Salmonella,* Cellular and Molecular Biology," 2nd ed. ASM Press, Washington, D.C.

Escherichia coli and Salmonella, Genetics

K. Brooks Low
Yale University

GLOSSARY

bacteriophage (*phage*) A virus that infects bacteria. Lysoganic bacteriophages sometimes insert stably into the chromosome without replicating. Lytic bacteriophages always replicate following infection, and usually kill their host bacterial cell as they produce a burst of progeny phage particles.

conjugation A process of DNA transfer between bacterial cells involving cell-to-cell contact and requiring the functions of fertility factor genes in the donor cell.

curing The elimination of a replicon such as a plasmid or prophage from a bacterial strain.

F-prime A plasmid derived by excision from a chromosomally integrated fertility factor (F), so that it carries a portion of the main chromosome, thus enabling a partial diploid (merodiploid) state.

fimbriae Hairlike proteinaceous appendages (100–1000 per cell), 2–8 nm in diameter, on the surface of bacteria that are often involved in adhesion to specific types of eukaryotic cells.

flagella Thin helical appendages on the surface of bacteria (5–10 per cell; originating at random points on the surface—peritrichous), about 20 nm in diameter, that can rotate and cause the cell to swim.

genome The total stably inherited genetic content of a cell, including chromosome and plasmids.

genotype The particular nucleotide sequence of the genome of a particular strain. A genotypic symbol (e.g., *leuB44*, a mutation in a leucine bioynthetic gene) is used to notate any change from the wild type (*leu*$^+$).

horizontal gene transfer The introduction of new genetic material into a cell's genome as a result of uptake from an outside source such as by mating (conjugation), transduction or transformation (i.e., not simply by mutation and vertical inheritance into vegetative descendants); usually used to denote gene transfer between somewhat unrelated species.

lysogenic conversion (*phage conversion*) A change in bacterial phenotype due to either phage infection or stable establishment (lysogeny) of a bacteriophage in the genome, whereby new (bacteriophage) genes are expressed, such as genes for toxin production or a new cell surface antigen.

mutation Any inherited change in nucleotide sequence in the genome.

operon A set of contiguous genes that are transcribed together into a single mRNA, together with any *cis*-acting regulatory sequences. The gene products are thus controlled coordinately. If two or more genes are controlled coordinately, but are at separate sites on the genome, they constitute a regulon.

palindrome An inverted repeated sequence. In double-stranded DNA, the ends of a palindrome are self-complementary, and can melt and reanneal to form hairpin-like structures.

pathogenicity island A cluster of contiguous genes in a genome that appear to originate from a distant evolutionary source and that is involved in conferring a pathogenic phenotype to a strain.

phase variation A reversible change in cell surface antigens, controlled by a genetic switch in the genome.

phenotype The observable characteristics of an organism based on its cellular or colonial morphology, biochemical structure, growth characteristics, or serological specificity. A phenotypic symbol (e.g., Leu$^-$, denoting a growth requirement for leucine) is used to indicate any change from the wild type (Leu$^+$).

plasmid An independently replicating component of the genome, usually much smaller than the main chromosome and usually not essential for cell survival.

recombination, genetic The interaction of nucleic acid elements of the genome to produce a new arrangement of base pairs; sometimes also including the addition from an outside source of new independently replicating elements such as plasmids, thus enlarging the genome.

R-prime A plasmid derived by excision from a chromosomally integated R factor, so that it carries a portion of the main chromosome, thus enabling a partial diploid (merodiploid) state.

serotype (*serovar; bioserotype; ser*) A strain-specific immunological reactivity due to particular antigenic components on or near the cell surface of a particular strain.

transduction A gene transfer involving the injection of genetic material carried in a bacteriophage capsid.

transformation, bacterial A gene transfer by uptake of naked DNA into a cell.

OVER 2300 WILD-TYPE (naturally occurring) varieties of *Salmonella* are known to inhabit Earth and an estimated vast excess of over 10,000 (perhaps more than 100,000) wild-type varieties of *Escherichia coli* exist as well. To help explain this striking diversity, there is overwhelming evidence that a complex spectrum of vertically inherited mutational events and horizontal gene transfer events has taken place over the course of evolution, since the time these two species diverged from a common ancestor approximately 150 million year ago. The resulting population of genetic variants includes mostly innocuous inhabitants of avian, human, and other animal intestinal tracts (*E. coli* cells normally constitute about 1% of the bacterial flora in the human intestine), and also a large number of highly pathogenic variants that are able to invade the intestine, blood, and other organs, exacting a huge toll in terms of disease and death (there are 3.6 million deaths annually from *Salmonella* infections worldwide) and the associated economic burden (over a billion dollars annually in the United States alone). As is well known, certain strains of *E. coli* and *Salmonella* have been used since the 1940s to serve as laboratory workhorses for studies of mutation, physiology, gene transfer, and fundamental life processes such as mechanisms of genetic inheritance, macromolecular synthesis, biosynthetic pathways, and gene regulation. The huge families of laboratory-derived strains and seminal findings from these experiments is a crucial counterpart to the lives of *E. coli* and *Salmonella* in the wild. The genetics of these organisms is based on the study of their genomic structure and mechanisms of variation observed both in the laboratory and in nature.

I. TAXONOMIC NICHE OF *ESCHERICHIA COLI* AND *SALMONELLA*

The description of the phylogeny of *Escherichia coli* and *Salmonella* has had a particularly tortuous and confusing history. This is in part because of the medical relevance of the many pathogenic varieties of these strains and the ensuing detail of study, and in part due to the intrinsically vast spectrum of subtle antigenic variations that have evolved on the surface of these cells, and, furthermore, the ability of the cells of even one strain to alternate between antigenic types in their normal course of growth.

The larger family of bacterial genera to which *E. coli* and *Salmonella* belong is the Enterobacteriaceae (or Enterobacteria), defined in 1937 by Rahn, which has diverse natural habitats including animal intestines, as well as plants, soil, and water. The number of genera, let alone species, defined to be in this family has varied repeatedly for almost 100 years, depending on available diagnostic measures and criteria of relatedness agreed to and disagreed to by many investigators. For example, the number of genera of Enterobacteriaceae listed in Bergey's *Manual of Determinative Bacteriology*, 5th edition (1939), a widely used standard, is 9, with 67 species (not including *Salmonella*, whose taxonomic history has varied widely depending on the concept of species), whereas the 8th edition (1974) lists 12 genera and 37 species. As of the 1990s, the use of methods of DNA–DNA hybridization to determine relatedness has led to the clear resolution of over 30 genera and 100 species within the "extended" family of Enterobacteriaceae.

Within this family, certain genera such as *E. coli* and *Shigella* are very closely related (70–100% homologous); yet they are maintained in separate genera because of the practical difference in clinical diseases they cause. The next closest relatives of *E. coli* are the *Salmonella* and *Citrobacter* genera, then *Klebsiella* and *Enterobacter*. More distantly related genera include *Erwinia*, *Hafnia*, *Serratia*, *Morganella*,

Edwardsiella, Yersinia, Providencia and, most distantly, *Proteus*.

The above genera were all known before 1965 (starting with *Serratia* in 1823), mostly as a result of infection in and the fecal content of humans or animals. Since the late 1970s, 17 more genera in this family have been discovered, some from clinical origin and some from water, plants, or insects.

The major characteristics that serve to identify the Enterobacteriaceae are that they are gram-negative rod-shaped bacteria that, with minor exceptions, can grow aerobically and anaerobically, produce a catalase but not oxidase, ferment glucose and convert nitrates to nitrates, contain a common enterobacterial antigen, are usually motile by means of flagella, do not require sodium for growth, and are not spore-forming. Further subclassification depends on other subtleties in metabolic capacities, such as the ability to use various sugars. For practical reasons, many medically important strains are diagnosed also with antisera specific to components of the outer membrane. Horizontal gene transfer is believed to occur between virtually all members of the family.

E. coli was discovered in feces from breast fed infants by Theodor Escherich, in 1885, and was first named *Bacterium coli*. By the early 1900s it became clear that it differed from another similar organism known primarily as a pathogen, *Bacillus typhi* (known since 1930 as *Salmonella typhi*), and hence was renamed *Escherichia* in honor of its discoverer. *E. coli* was found to ferment lactose, whereas the (*Salmonella*) *typhi* strains could not. In hindsight, this ability to ferment lactose and the divergence of *E. coli* and *Salmonella* roughly 150 million years ago coincides rather strikingly with the assumed appearance of mammals, and the assumed first synthesis of lactose in nature, at about the same evolutionary time. The differing abilities of *E. coli* and *Salmonella* to ferment lactose, and invade mammals in a pathogenic fashion suggests a rapid adaptation to the newly emerging families of mammals on Earth.

In contrast to the discovery of *E. coli* as a common and compatible inhabitant of the human intestine, *Salmonella*, named in 1900 after Salmon, was first defined as the causative agent for hog cholera (hence the initial strain name, "choleraesuis"). Investigations of the varied and varying antigenic properties of many similar strains from diseased animals and humans led to a classification scheme (Kauffmann–White) based on three types of antigens. Further classification was based on differences in sensitivity to various bacteriophages.

The three broad classes of antigens used to categorize both *E. coli* and *Salmonella* isolates are based on three types of surface structures. The outermost are the flagella that are present in motile strains. The antigens of flagella are termed H antigens (Hauch, meaning "cloud" or "film," describing colonies of motile bacteria). Next is the group of outermost surface layers including the capsule (in some strains), envelope and fimbriae. The antigens of this group are called K (for Kapsule). Internal to the K group is a group of antigens associated with the phospholipids in the outer membrane—termed O antigens because strains that lose their motility (and are thus "ohne Hauch," i.e., without cloud or film as colonies) through loss of flagella can expose the lower-lying (O) antigens. What makes this overall system feasible is that by selective treatment the outer antigens can be removed or inactivated stepwise by alcohol (for the H antigen, i.e., the flagella) and by heating to 100°C for 2 hr (which inactivates most or all of the K antigens), which thus leaves material containing only the O antigens. By preparing antibodies to the bacteria in all three states, the antisera can be subtractively purified to obtain fractions specific for each of the three antigen classes.

The results of decades of effort in this direction in the early 1900s (and continuing into the present, for more specific isolates) resulted in a well-defined spectrum of antigenic determinants that define a multitude of *E. coli* and *Salmonella* strains. In *E. coli* at least 173 distinct O antigens, 80 K antigens, and 56 H antigens are known. Although it is not likely that all conceivable combinations exist in nature, it is thought that well over 10,000 variants exist, more than for any other species. Note that the common laboratory strain, *E. coli* K-12 (also written K12 or K_{12}), does not show any K or O antigenic determinants. The name K-12 was applied in a Stanford clinic where it was isolated in 1922, long before the K, O, H system was developed.

As for *E. coli*, *Salmonella* strains from natural sources display a huge variety in O and H (and sometimes K) antigenic determinants, and over 2300 combinations have now been registered, second only to *E. coli* in complexity. Some of this complexity is due to the ability of *Salmonella* to present more than one form of H antigen (phase variation). Additional variation is introduced by the phenomenon of lysogenic conversion, whereby certain bacteriophages infect a strain, become lysogenized as a stable prophage, and express one or more new genes that alter the O surface antigens. Until the early 1940s, the names used for different isolates were taken to mean separate species, thus hundreds of *Salmonella* species are described in the early literature. In 1973 a major simplification occurred as a result of DNA-relatedness studies of Crosa and colleagues, who found that virtually all *Salmonella* strains are closely related and can be considered a single species, divided into six subgenera (subspecies). Officially, this species is named *Salmonella choleraesuis*; however a subsequent proposal is pending to name the species *Salmonella enterica* and thus use a name not associated with any of the earlier isolates. Many investigators are already using this terminology, wherein the particular strain is defined with a serovar name that corresponds in most cases with the older "species" name, that is, *Salmonella enterica* serovar Typhimurium, or simply *Salmonella* Typhimurium.

II. THE GENOME

A. The Chromosome

The genomes of *E. coli* and *Salmonella* strains consist of a single major circular chromosome of over 4 Mb of DNA (the genomes from natural isolates of *E. coli* range in size from 4.5–5.5 Mb), and commonly one or more small independently replicating (extrachromosomal) DNA plasmids, usually 100 kb or less. Thus, these organisms are haploid, although a second copy of a portion of the chromosome can be carried on a plasmid, thus creating a partial diploid (merodiploid) state. Strains of this type have been used extensively in the laboratory in genetic studies of dominance and regulation.

Though the genome is haploid, the number of copies of the main chromosome varies, ranging between approximately one per cell in the resting (stationary phase) state to sometimes four per cell in rapidly dividing cells in rich growth media. In this state there are more than one set of replication growth forks along the chromosome, which replicates divergently beginning at a site called the origin (of replication) and ending in a region called the terminus (see Fig. 1). Thus, in rapidly growing cultures there are at any given time more copies of genes located near the origin than near the terminus. As a result of this multiplicity of growth forks, rapidly growing cells divide and produce two daughter cells as often as every 20 min, even though the time required for any one growth fork to move from the origin to the terminus is 40 min. Genes whose products are used in high concentration during rapid growth (such as genes for components of the ribosome) tend to be located nearer the origin than the terminus, thus allowing relatively high amounts of product formation.

The chromosomes of *E. coli* and *Salmonella* are clearly related, as can be seen from Fig. 1, which shows the general congruence of the sequence of genes around the chromosomes for the two most commonly used laboratory strains (originally isolated from the wild), *E. coli* K-12 and *Salmonella* Typhimurium LT2. The vast majority of the genes in *E. coli* K-12 have counterparts at the corresponding regions of the *Salmonella* Typhimurium chromosome, and vice versa. At the base-sequence level, the DNA homology varies from approximately 75–99% for various protein-coding genes. The G + C to A + T DNA base ratios are also similar (50.8% G + C content for *E. coli* K-12; 52% for *Salmonella* Typhimurium). Recombination can occur between the *E. coli* and *Salmonella* chromosomes, albeit rarely (see later discussion), to form viable hybrid strains that carry portions from each chromosome, and most of the gene functions from one organism can substitute for the analogous function in the other.

Superimposed on the general congruence of these two related genomes are a number of sites in which an insertion or deletion shows a distinct difference between the two, as indicated in Fig. 1 (and see later discussion). Figure 1 also shows a region of the chromosome, in the 26–40 min region, where a large inversion has reversed the gene sequence in *E. coli* K-

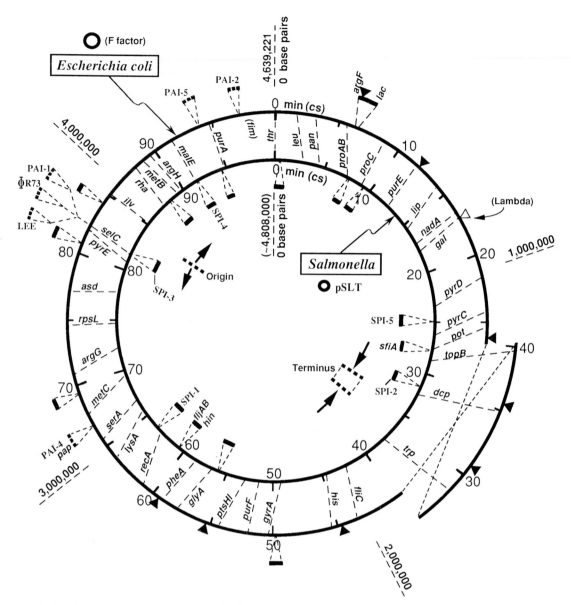

Fig. 1. The two circular genetic maps of the chromosomes of representative strains of *Escherichia coli* (strain K-12 MG1655) and *Salmonella* (serotype Typhimurium, strain LT2), drawn concentrically for comparison. The maps are oriented relative to position 0 at the top, historically chosen to be at the threonine (*thr*) locus in both organisms. The outermost (radially extending) numbers (1,000,000, etc.) indicate the approximate number of nucleotide base pairs in distance clockwise along the chromosome from position 0. The numbers written along the circles (10, 20, etc.) indicate the corresponding percentage distance around the chromosome, also denoted in units of minutes (min) or, equivalently, centisomes (cs). Between the two map circles, the relative positions of representative genes or gene clusters common to both chromosomes are indicated by the gene names (*thr, leu,* etc.). Origin and Terminus indicate the positions of initiation and termination, respectively, of normal bidirectional chromosome replication during vegetative growth. The region between approximately 27 and 40 min on the *E. coli* map has an inverted gene order relative to *Salmonella*, as indicated. Representative short extra arcs of chromosome are indicated as either present on the *E. coli* chromosome but not *Salmonella* (drawn exterior to the *E. coli* map circle), or present on the *Salmonella* chromosome but not *E. coli* (drawn interior to the *Salmonella* map circle). The arcs labeled SPI-1, SPI-2,

12 relative to *Salmonella* Typhimurium. It is believed that the *Salmonella* Typhimurium configuration of this region is the more distantly ancestral one because certain other types of Enterobacteria related to *E. coli* and *Salmonella typhimurium* carry inversions analogous to the 26–40 min inversion but differing greatly in extent, ranging from a much smaller inversion (in *Klebsiella aerogenes*) to a much larger inversion (in *Salmonella* Enteriditis). Thus, it is believed that over the course of evolution a number of independent inversion events in this region occurred, starting from a configuration similar to that of *Salmonella* Typhimurium.

A summary of types of genes and other genetic elements on the *E. coli* K-12 chromosome is listed in Table I. For structural genes that encode proteins, Table II indicates the major classes based on broad functional role in the cell.

B. Extrachromosomal Elements

Most natural isolates of *Salmonella,* and many of *E. coli,* carry extrachromosomal plasmids. These are broadly classified as either self-transmissable (conjugative) or not (nonconjugative). Conjugative plasmids carry a number of genes needed for conjugative DNA transfer between cells (see later discussion). Some nonconjugative plasmids can be conjugationally transferred (mobilized) when a conjugative plasmid is also present to provide the necessary gene functions. Plasmids can be classified into groups based on whether they can co-exist stably in the same cell. (Two different plasmids that can coexist are defined to different incompatibility, Inc, groups.) There are more than 30 Inc groups among the Enterobacterial plasmids.

About two-thirds of natural *Salmonella* Typhimurium strains carry a conjugative plasmid roughly 90 kb in size, similar to pSLT, which is present in strain LT2 as indicated in Fig. 1. These plasmids usually carry some of the genes that contribute to the virulence of *Salmonella* as a pathogen. Some wild-type *E. coli* strains (roughly 10% or more) also carry conjugative plasmids. The one denoted F from *E. coli* K-12 is about 100 kb in size and is unusual in that it is self-transmissable at very high frequency (see later discussion). The F factor and other F-like factors from *E. coli* strains have a number of genes involved in conjugation that are homologous to pSLT and similar plasmids from *Salmonella* Typhimurium strains.

Another type of plasmid, found in about 30% of wild-type *E. coli* strains, is the colicin factor. Colicins are toxins produced by one bacterium to kill or inhibit another bacterium. At least 20 distinct types of colicin factor, some of which are self-transmissable, have been found in *E. coli*.

III. WAYS IN WHICH THE GENOME CAN CHANGE

A. Mutation

Various changes in base sequence occur spontaneously at the rate of approximately 6×10^{-10} per base pair per generation (i.e., about 0.003 per genome) during vegetative growth. Much higher rates of mutagenesis are caused by mutagenic agents such as irradiation or exposure to certain chemicals, or even by simple prolonged starvation, as may occur in nature. The consequences of mutation can range from almost no effect (silent mutations, e.g., a base substitution in an amino acid-coding triplet in which the encoded amino acid does not change; however the change in tRNA used can lead to subtle changes in translation

---

and so on denote Salmonella pathogenicity islands. Analogous islands of genes indicated by dashed arcs (PAI-1, PAI-2, etc., pathogenicity island, or LEE, locus of enterocyte effacement) on the *E. coli* map are sometimes present in certain pathogenic *E. coli* strains. The small circles represent examples of plasmids that are typically present in *E. coli* K-12 (F factor, fertility factor) or *Salmonella* Typhimurium LT2 (pSLT). The small open triangle at 17 min on the *E. coli* map indicates the chromosomal attachment site of bacteriophage lambda, which is usually present as a lysogenic prophage in *E. coli* K-12 strains. Other similar sites where other lysogenic bacteriophages sometimes are found integrated are not shown, except for ΦR73, a lysogenic phage whose chromosomal insertion site is the same one (near 82 min) used by PAI-1 and by LEE. The small filled-in triangles on the *E. coli* map indicate the positions of cryptic prophages.

TABLE I

Escherichia coli K-12 Genome, Major Features[a]

Elements	Number	Characteristics
Base pairs	4,639,221	Single circular chromosome
Replichores	2	Divergently replicating halves, from single origin to terminus region
G + C content	50.8%	Overabundance of G in leading replicative strands (G ~26.2%; C, A, T ~24.6%)
Protein-codable genes	4288	88% of genome; 55% transcribed in direction of replication
average size (amino acids)	317	
number with 1000–2383 amino acids	56	
number with <100 amino acids	381	
operons (known and predicted)	2584	73% single gene (predicted), 17% two genes, 10% three or more genes; 68% single promoter (predicted), 20% two promoters, 12% three or more promoters; >10% regulated by more than one protein
Stable RNA genes		
rRNA	7 operons	All operons transcribed in replication direction; 5 CW, 2 CCW; location biased toward origin of replication.
tRNA	86	43 operons of various sizes widely dispersed in location and orientation.
Cryptic prophages	8	All located in late-replicating halves of replichores
other phage-like remnants	33	
Integration sites (att) for various lysogenic phages	>15	Widely dispersed around genome; sites for lambda-related phages located in late replicating portions of replichores
Insertion sequences	42	Number of repeats: 1 to 11; 4 classes (gene arrangements); size range 768–1426 bp
Insertion sites for fertility factors (F, Col V, etc.)	>20	Some are at IS sequences
Repeated sequences (or, pseudo-repeated)		
Chi (GCTGGTGG)	761	RecBCD-mediated recombination initiator; highly skewed (~2:1) to leading replicative strand; widely distributed
Rhs	5	5.7–9.6 kb; components and G + C content variable; function unknown; not found in most other *E. coli* strains or in *Salmonella*
REP	697	~40-bp palindromes; sometimes tandemly repeated (1–12 copies) in 355 BIME sites; possible role in mRNA stabilization, gene expression and/or replication
IRU	19	Imperfect palindromes; size ~125 bp; function unknown
RSA	6	Imperfect palindromes; size ~152 bp; function unknown
Box C	33	High G + C; size ~56 bp; function unknown
iap	23	29-bp imperfect palindromes clustered near *iap* gene on the chromosome; function unknown.

[a] Strain K-12 MG1655. Abbreviations: BIME, bacterial interdispersed mosaic elements; CCW, counter-clockwise; CW, clockwise; IRU, intergenic repeat units (ERIC, enterobacterial repetitive intergenic consensus); REP, repetitive extragenic palindromic unit (PU, palindromic unit); Rhs, rearrangement hot spot.

TABLE II
Distribution of Major Functional Classes of *E. coli* Protein-Coding Genes[a]

Functional class	Number of genes	Percent of total
DNA replication, repair, recombination	115	2.7
Nucleotide metabolism	58	1.4
RNA metabolism, transcription	55	1.3
Protein synthesis	182	4.2
Amino acid metabolism	131	3.0
Cell membranes and structure	237	5.5
Fatty-acid and lipid metabolism	48	1.1
Binding and transport	427	10.0
Carbon-compound catabolism	130	3.0
Intermediary metabolism	188	4.4
Energy metabolism	243	5.7
Regulation	178	4.2
Phage, transposons	87	2.0
Miscellaneous	577	13.5
Unknown	<u>1632</u>	38.0
Total	4288	

[a] Strain K-12 MG1655. Values estimated.

efficiency) to mild phenotypic changes (e.g., temperature sensitivity or leaky requirement for a growth factor; i.e., a leaky auxotroph or bradytroph, due to point mutations that alter just one or a few amino acids) to severe phenotypic changes (e.g., the inactivation of one or more genes due to deletions, frameshift mutations, or nonsense mutations). Sometimes a single mutation can result in a change in more than one phenotype or a change in production of a number of gene products, in which case the mutation is called pleiotropic. A broad range of phenotypes can be observed in *E. coli* or *Salmonella* mutants, listed in Table III.

B. Recombination

Three distinct types of rearrangements of genetic elements (i.e., recombination) are observed throughout the living world: homologous, site-specific, and transpositional. Figure 2 shows diagrammatically the main topological features of these classes.

1. Homologous Recombination

This refers to the interaction of two very similar DNA molecules (i.e., having almost the same base sequence for many hundreds of continuous bases) at equivalent sites along their DNA chains, which can result in crossover or DNA-repair events almost randomly at any site along the pair of parental DNA molecules. This can thereby result in a change in linkage, or proximity on the same chromosome, for various small (but nevertheless significant) chromosomal differences situated along the lengths of the two recombining DNA molecules. If the two recombining DNA chains are generally homologous, but contain scattered differences from one to the other (e.g., a few percent difference or more for the *E. coli* vs. *Salmonella* chromosomes) recombination can still occur, but is more infrequent with increasing differences in sequences. This nearly homologous configuration is known as homeologous recombination.

Homologous recombination can thus occur following gene transfer between closely related organisms (see later discussion). However, homologous recombination also occurs, very frequently, even between elements within one cell. For instance, crossovers between some of the seven homologous copies of rRNA operons can occur during vegetative growth to produce large genetic duplications or inversions. These rearrangements are unstable and readily recombine back to the normal haploid state. Neverthe-

TABLE III
Classes of Phenotypic Characteristics of Mutant Strains

Phenotypic Class	Examples
Morphological	Colony size, texture, color; cell shape, motility, staining properties
Physiological	Dependence on temperature, oxygen, pH
Nutritional	Requirements for growth; possible carbon and energy sources
Biochemical	Breakdown products (e.g., acid) from carbohydrates, nitrates
Inhibition	Sensitivity to antibiotics, dyes, etc.
Serological	Agglutination by specific antisera
Pathogenic	Invasion capacity; virulence in variety of hosts

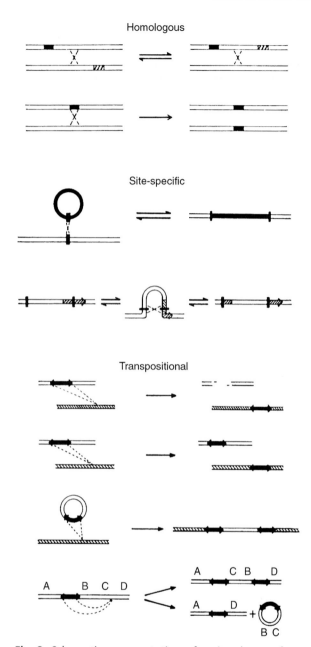

Homologous

Site-specific

Transpositional

Fig. 2. Schematic representation of major classes of genetic recombination. Adapted from Low, B. In Low, B. (ed.) (1988) *The Recombination of Genetic Material.* Academic Press, pp. 6–7, with permission from the author and publisher.

less their mere existence (in up to one-third of a growing population of cells) makes possible various changes in gene dosage or gene arrangement, which likely contributes to the evolutionary flux over long periods of time due to rare accompanying recombina-

tion events that are sometimes nonhomologous (or illegitimate).

2. Site-Specific Recombination

As the name implies, this form of recombination is a crossover event between two DNA double helices, which is catalyzed to occur at unique sites on the DNA strands, defined by the local DNA sequence involving usually 20–30 bases. A specialized enzyme recognizes these sequences, on both parental molecules, and causes a double-strand breakage and rejoining event that exchanges the partners of flanking DNA. The crossover is completely conservative, that is, no bases are added or lost at the crossover site.

One configuration of site-specific recombination enables the circular form of a lysogenic phage, such as lambda (found in *E. coli* K-12) or P22 (one of a large family of related bacteriophages found in most natural *Salmonella* isolates), to integrate into the chromosome (and thus become a stable lysogen), or to be excised from it (and thus begin to replicate vegetatively). See Fig. 2 (Site-specific, upper).

Another DNA configuration involved in site-specific recombination involves a hairpin-like intermediate such that the end result of the crossover is an inversion of the stretch of DNA located between the two crossover sites (Fig. 2, Site-specific, lower). An example of this occurs in *Salmonella* and results in a change in gene expression for two different genes for the production of flagellin, which is assembled into filaments on the surface of the cell (see Fig. 3). By switching the gene expression in this way, *Salmonella* is able to change its filament antigenic structure (i.e., undergo phase variation) and thus reduce attack by host immune systems. Another mechanism for phase variation also exists in some strains, in which methylation of certain critical adenine residues will activate or deactivate gene expression, and this methylation pattern can be quasi-stably inherited for many generations, until it flip-flops back to a nonmethylated state, which reverses the antigenic phase.

3. Transpositional Recombination

As in the case of site-specific recombination, certain specialized short DNA sequences are involved in recombination involving DNA elements called

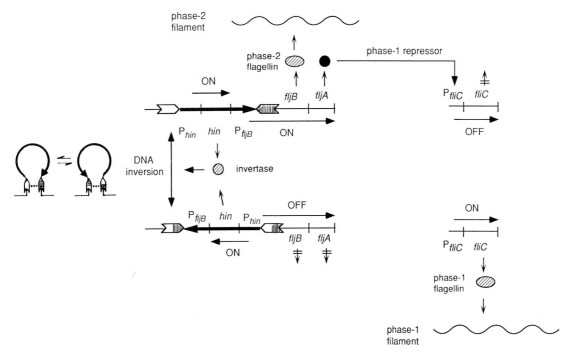

Fig. 3. Mechanism for flagellar phase variation in *S. typhimurium* (i.e., the alternate expression of two structural genes for flagellin, an H antigen). The two genes, *fliC* (formerly named H1) and *fljB* (formerly named H2), are distant from each other on the chromosome at 43 and 61 min, respectively (see Fig. 1). The promoter for the *fljB* operon is part of a region of the chromosome that is bounded by an inverted repeat (white or shaded arrow bars), which can be inverted by site-specific recombination mediated by the Hin invertase protein, a product of gene *hin* in the invertable region. In the orientation shown at the top of the figure (phase 2), the promoter for the *fljB* operon (P$_{fljB}$) is correct and FljB or phase 2 flagellin is synthesized, together with the repressor of the phase 1 *fliC* operon, the *fljA* gene product (formerly named rh1). In the other orientation (phase 1), shown at the bottom of the figure, P$_{fljB}$ is separated from the operon it promotes and neither the phase 2 flagellin nor the phase 1 repressor is synthesized. The *fliC* operon is therefore expressed, and FliC or phase 1 flagellin is synthesized. Adapted from Macnab, R. *In* Neidhardt, F., *et al.* (1996), p. 137, with permission from the author and the American Society for Microbiology.

transposons. Transposons are sequences of DNA, usually in the size range of 700–10,000 bases, which can move by recombination from one location in the genome to any of a multitude of other sites (targets). The specialized sequences that promote this recombination are usually 15–30 bases long and located at the two ends of the transposon as an inverted repeat. The DNA between the two ends can encode a number of genes whose functions are either required for the transpositional recombination (e.g., transposase, resolvase) or other functions that alter cell phenotype, such as antibiotic resistance. Transposons that do not encode any internal genes except those needed for transposition are called in-

sertion sequences (IS). The *E. coli* chromosome contains several copies of five IS, and approximately 15% of the spontaneous mutations in *E. coli* occur due to the movement of these IS to new sites. None of these particular IS are normally found in *Salmonella*, but a different IS specific to *Salmonella* strains is normally present somewhere on its chromosome.

Transposition events can either involve the movement of the entire original transposon DNA sequence to a new site (conservative transposition) or a replicative process in which the transposon is replicated to form two copies, one of which is inserted at the new (target) site and the other of which is retained at the original site (replicative transposition) (see Fig. 2).

Even in the case of conservative transposition, a small amount of DNA synthesis is involved, in which a few of bases at the target site are copied to create a small duplication at each end of the newly inserted transposon.

IV. GENE TRANSFER BETWEEN CELLS

A. Conjugation

In Section II it was noted that most *Salmonella* strains and many *E. coli* strains carry fertility factors (conjugative plasmids) that can promote gene transfer from cell to cell by conjugation. Furthermore, the F-like plasmids in *E. coli* and the pSLT-like plasmids in *Salmonella* have a number of genes in common, and considerable homology between them. These facts strongly imply that over evolutionary history there has been considerable cross-talk and gene transfer among these strains. The particular F factor in *E. coli* strain K-12 is (fortunately for J. Lederberg and other early investigators in the discovery and study of conjugation) highly efficient for promoting conjugation between F⁺ cells (those that carry the F factor as a plasmid) and F⁻ cells (those which have no F factor in any form). This is due to a mutation on the K-12 F factor in the regulatory system that, for the vast majority of fertility factors, represses the genes for conjugation. With these "normal" conjugative plasmids, this repression keeps all the cells except about 1/1000 from acting as donors. At any given time any one of the cells can, with a probability of 1/1000, become temporarily derepressed for conjugative functions and cause its own transfer into an F⁻ cell (thus converting it into an F⁺ cell). Immediately following this transfer into the F⁻ cell, there is a delay before the repression of the conjugative functions builds up. Thus, such a newly formed F⁺ cell is temporarily able to conjugate (act as a donor) efficiently again, with another F⁻ cell. In this way a rapid epidemic spread of conjugative plasmids can move into a new population of F⁻ cells when the opportunity arises.

On rare occasions an F-like plasmid can recombine with the chromosome to form an Hfr (high frequency of recombination) cell (see Fig. 4). In the process of conjugation from such a cell, the transfer of the normal leading region (next to a specific site called the origin of transfer, indicated by the small arrowhead in Fig. 4), leads to the transfer of chromosomal genes because in the Hfr state the F factor and chromosome form one continuous DNA molecule. Considerable amounts of chromosomal genetic information can be transferred in this way, although the transfer process tends to be interrupted spontaneously so that the genes furthest around the chromosome have the least chance of being transferred.

B. Transduction

Recall that a large family of lysogenic bacteriophages related to P22 exist among wild-type *Salmonella* strains. It so happens that when phage P22 is replicating lytically in a cell and being packaged into new phage particles, occasionally (about once in every 30 times) the packaging process makes a mistake and packages a segment (about 1%) of the bacterial genome instead of the newly replicated P22 DNA. These aberrant particles are called transducing particles and can come out of the cell along with the burst of normal new phage particles. The transducing particles can travel to another cell and inject in the genomic DNA, followed in some cases by recombination into the new cell's genome. The analogous processes occur in *E. coli* strains with a phage known as P1, which can package about 2% of the genome into transducing particles. The process, discovered in 1952 by Zinder and Lederberg, is called generalized transduction and it most likely has had a major role in the horizontal transfer of genetic information among enteric bacteria. In some cases, the transduced chromosomal DNA fragment does not recombine into the chromosome but forms a stable (nonreplicating) structure that continues to be unilaterally inherited by only one of the daughter cells at each division. If the transducing fragment carries a functional gene that allows the transduced cell to grow on selective medium, only a tiny colony will grow because only one of the two daughter cells after each division will divide again. This is termed abortive transduction and has been useful in determining functional complementation between mutated genes.

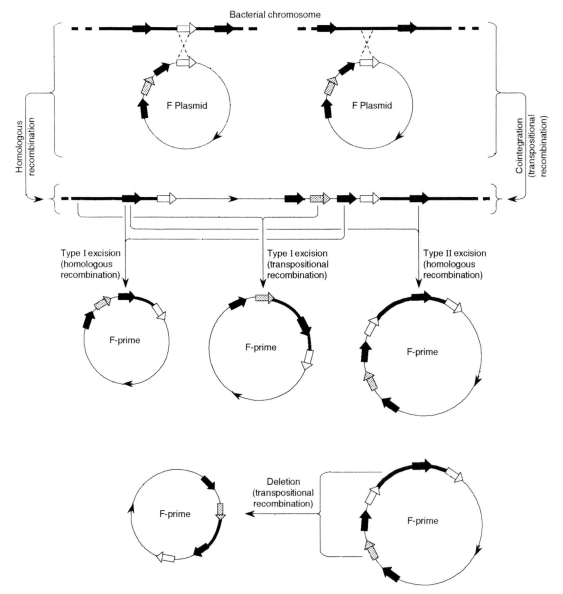

Fig. 4. Interconversion between various states of the F factor in *E. coli*. The top heavy horizontal lines represent parts of the chromosome, with arrows that represent the various positions of IS. IS that are also located on the F factor (thin circular line) can recombine with the chromosome either by homologous recombination with a chromosomal IS of the same type (top left), or by transpositional recombination at a chromosomal site where there is not normally an IS (top right). The resulting chromosome with the integrated F factor (center horizontal line) is called Hfr. The Hfr chromosome can undergo subsequent recombination events, and they produce plasmids (F-primes) that carry most or all of the F factor plus a section of the chromosome. From Holloway, B. and Low, B. *In* Neidhardt, F., *et al.* (1996), p. 2414, with permission from the author and the American Society for Microbiology.

Another mode of transduction, termed specialized transduction, is distinctly different from generalized transduction. Specialized transduction involves the excision of a chromosomally located lysogenic bacteriophage (i.e., in the prophage state) to form a circular replicating form. However, rarely, this excision event is imprecise and sometimes involves a crossover within the adjacent bacterial chromosomal DNA (analogous to F-prime formation, Fig. 4), such that the replicating phage and subsequent phage particles produced contain a small amount (a few genes) of chromosomal DNA. This DNA can thus be carried by the progeny phage particles and injected into new host cells, followed by possible recombination events. Thus, in specialized transduction only genes close to the prophage integration site can be transduced. Moreover, the altered phage particles carrying the chromosomal fragment can be grown and amplified to produce large quantities of the specific chromosomal genes for experimental purposes.

C. Transformation

The uptake of naked DNA by bacteria and the subsequent inheritance of its genetic information in the genome occur naturally in some gram-positive and gram-negative bacterial species such as *Streptococcus pneumoniae, Bacillus subtilus, Haemophilis,* and *Acinetobacter,* but only under certain growth conditions that induce competence, or the ability to take up DNA. In contrast, *E. coli* and *Salmonella* are not known to have this natural competence for DNA uptake, and it is unlikely that transformation has contributed significantly to their evolutionary genomic development. However, laboratory methods have been devised that promote the uptake of DNA into *E. coli* and *Salmonella,* thus greatly facilitating their use in molecular biological studies. These methods for artificially induced transformation competence generally involve either treatment with di- or multivalent cations (e.g., Ca^{2+}, Mg^{2+}) and heat shock, which is believed to partially disrupt the lipopolysaccharide in the outer membrane (chemical competence), or treatment with a brief pulse of high-voltage electricity (electroshock), which renders the membrane temporarily permeable to a variety of macromolecules including DNA.

See Also the Following Articles

Conjugation, Bacterial • DNA Replication • Mapping Bacterial Genomes • Transduction • Transposable Elements

Bibliography

Brock, T. D. (1990). "The Emergence of Bacterial Genetics." Cold Spring Harbor Laboratory Press, Cold Spring Harbor, NY.

Clewell, D. B. (ed.) (1993). "Bacterial Conjugation." Plenum Press, New York.

Goodfellow, M., and O'Donnell, A. G. (eds.) (1993). "Handbook of New Bacterial Systematics." Academic Press, San Diego, CA.

Janda, J. M., and Abbott, S. L. (1998). "The Enterobacteria." Lippincott-Raven, Philadelphia.

Logan, N. A. (1994). "Bacterial Systematics." Blackwell Scientific, London.

Lund, B. M., Sussman, M., Jones, D., and Stringer, M. F. (eds.) (1988). "Enterobacteriaceae in the Environment and as Pathogens." Blackwell Scientific, London.

Neidhardt, F. C., Curtiss, R., III, Ingraham, J. L., Lin, E. C. C., Low, K. B., Magasanik, B., Reznikoff, W. S., Riley, M., Schaecter, M., and Umbarger, H. E. (eds.) (1996). "*Escherichia coli* and *Salmonella,* Cellular and Molecular Biology," 2nd ed. ASM Press, Washington, D.C.

Riley, M., and Sanderson, K. E. (1990). Comparative genetics of *Escherichia coli* and *Salmonella typhimurium. In* "The Bacterial Chromosome" (K. Drlica and M. Riley, eds.), pp. 85–95. ASM Press, Washington, D.C.

Sussman, M. (ed.) (1997). "*Escherichia coli*: Mechanisms of Virulence." Cambridge University Press, Cambridge.

Evolution, Theory and Experiments

Richard E. Lenski
Michigan State University

GLOSSARY

adaptation The match between a particular feature of an organism and its environment that results from natural selection.

evolution The change in the genetic properties of populations and species over generations, which requires the origin of variation (by mutation and mixis) as well as the subsequent spread or extinction of variants (by natural selection and genetic drift).

fitness The average reproductive success of a genotype in a particular environment, usually expressed relative to another genotype.

genetic drift Changes in gene frequency caused by the random sampling of genes during transmission across generations (rather than by natural selection).

mixis The production of a new genotype by recombination of genes from two sources.

natural selection Changes in gene frequency caused by the specific detrimental or beneficial effects of those genes.

population A group of individuals belonging to the same species and living in close proximity, so that individuals may potentially recombine, compete for limiting resources, or otherwise interact.

EXPERIMENTAL EVOLUTION is the laboratory study of the fundamental processes of evolutionary change. These processes include mutation and adaptation by natural selection, and they give rise to patterns of genetic diversity within and among populations. Microorganisms have proven to be useful subjects for this research as a consequence of their large population sizes and short generation times, as well as the ease with which their environments and genetic systems can be manipulated. Experimental studies of microbial evolution have generally confirmed the basic principles of modern evolutionary theory, while also providing new insights into the genetics, physiology, and ecology of microorganisms.

I. REVIEW OF EVOLUTIONARY THEORY

Evolutionary theory seeks to explain patterns of biological diversity in terms of a few fundamental evolutionary processes. These processes are presumed not only to have operated in the past, but also to continue to operate today. Thus, they can be studied experimentally in the laboratory. Before discussing experiments that have used microorganisms to examine evolutionary processes, the major elements of evolutionary theory will be reviewed.

A. Evolutionary Patterns

The three most conspicuous products of organic evolution are the wealth of genetic variation that exists within almost every species; the divergence of populations and species from one another and from their common ancestors, and the manifest adaptation, or fit, or organisms to the environments in which they live.

1. Genetic Variation

The existence of extensive genetic variation within species has been demonstrated by a variety of means. Variation in certain traits, such as seed shape in pea plants and blood type in humans, can be shown to have a genetic basis by a careful examination of pedigrees. For many other traits, such as milk production in cows or body weight in humans, quantitative genetic analyses are required to partition the phenotypic variation that is due to genetic versus environmental influences. Biochemical and molecular techniques have also revealed extensive variation in DNA sequences and the proteins they encode.

2. Divergence and Speciation

All biological species differ from one another in some respects. It is generally possible to arrange species hierarchically, depending on the extent and nature of their similarities and differences. This hierarchy is reflected in the classification scheme of Linnaeus (species, genus, family, and so on). This hierarchical arrangement also suggests a sort of "tree of life," in which the degree of taxonomic relatedness between any two species reflects descent with modification from some common ancestor in the more or less distant past. Investigating the origins of groups and their relationships requires an historical approach, which is not amenable to direct experimentation. Even so, historically based hypotheses can be tested by phylogenetic and comparative methods, which use data on the distribution of character states across various groups and environments, sometimes supplemented with information from the fossil record.

The extent of evolutionary divergence that is necessary for two groups of organisms to be regarded as distinct species is embodied in the concept of biological species, according to which "species are groups of actually or potentially interbreeding populations, which are reproductively isolated from other such groups" (E. Mayr, 1942. "Systematics and the Origin of Species," Columbia University Press). Speciation thus refers to the historical process by which groups of organisms become so different from one another that they no longer can interbreed. However, many organisms (including most microorganisms) reproduce primarily or exclusively asexually, and this definition is not applicable. For such organisms,

the extent of evolutionary divergence that corresponds to distinct species is somewhat arbitrary and often more a matter of convenience than of scientific principle.

3. Adaptation

The various features of organisms often exhibit an exquisite match to their environments. For example, the bacteria that live in hot springs have special physiological and biochemical properties that allow them to survive and grow at very high temperatures that would kill most other bacteria; often these thermophiles cannot grow at all in the much more benign conditions in which most other bacteria thrive. Nevertheless, organisms are by no means always perfectly adapted to the environments in which they live. Evidence for the imperfection of organisms is seen whenever species become extinct, usually as a consequence of some change in the environment to which they cannot quickly adapt.

B. Evolutionary Processes

Biological evolution occurs whenever the genetic composition of a population or species changes over a period of generations. Four basic processes contribute to such change: mutation, mixis, natural selection, and genetic drift. Selection and drift cannot act unless genetic variation among individuals exists.

1. Sources of Genetic Variation

Genetic variation among individuals is generated by two distinct processes, mutation and mixis. In terms of evolutionary theory, these processes are usually distinguished as follows. Mutation refers to a change at a single gene locus from one allelic state to another (e.g., *abcd* → *Abcd*, where each letter indicates a locus), whereas mixis refers to the production of some new multilocus genotype by the recombination of two different genotypes (e.g., *abcd* × *ABCD* → *aBcD*).

a. Mutation

There are many types of mutations, including point mutations, rearrangements, and transposition of mobile elements from one site in the genome to another. Some mutations cause major changes in an organism's phenotype; for example, a bacterium may become resistant to attack by a virus (bacteriophage) as the result of a mutation that alters a receptor on the cell surface. Other mutations have little or even no effect on an organism's phenotype; many point mutations have absolutely no effect on amino acid sequence (and hence protein structure and function) because of the redundancy in the genetic code.

Any number of factors may affect mutation rates, including environmental agents (e.g., UV irradiation) and the organism's own genetic constitution (e.g., defective DNA-repair genes). Evolutionary theory makes no assumptions about the rates of mutations or their biophysical bases, with one exception: Mutations are assumed to occur randomly, that is irrespective of their beneficial or harmful effects on the organism.

b. Mixis

Recombination among genomes can occur by a number of mechanisms. The most familiar mechanism is eukaryotic sex, which occurs by Mendelian segregation (meiosis) and reassortment of chromosomes (fertilization). Many eukaryotic microorganisms, including fungi and protozoa, engage in sexual mixis. Bacteria generally reproduce asexually, but may undergo mixis via conjugation (plasmid-mediated), transduction (virus-mediated), or transformation. Even viruses may recombine when two or more coinfect a single cell.

Unlike mutation, these various mechanisms do not necessarily produce organisms with new genes; instead, they may produce organisms that possess new combinations of genes. This can have very important evolutionary consequences. In the absence of mixis, two or more beneficial mutations can be incorporated into an evolving population only if they occur sequentially in a single lineage. Mixis allows beneficial mutations that occur in separate lineages to be combined and thereby incorporated simultaneously by an evolving population. Thus, mixis may accelerate the rate of adaptive evolution by bringing together favorable combinations of alleles.

2. Natural Selection

One of the most conspicuous features of biological evolution is the evident fit (adaptation) of organisms to the environments in which they live. For centuries, this match between organism and environment was taken as evidence for the design of a Creator. But in 1859, Charles Darwin published *The Origin of Species,* in which he set forth the principle of adaptation by natural selection. This principle follows logically from three simple premises. First, variation among individuals exists for many phenotypic traits. Second, these phenotypic traits influence survival and reproductive success. Third, phenotypic variation in those characters that affect survival and reproductive success is heritable, at least in part. (Many phenotypic traits are subject to both genetic and environmental influences.) Hence, individuals in later generations will tend to be better adapted to their environment than were individuals in earlier generations, provided that there is heritable phenotypic variation and the environment has not changed too much in the intervening time. (Environments do sometimes change, of course, and when this happens a population or species may become extinct if it cannot adapt to these changes.)

Darwin himself did not know about the material basis of heredity (DNA and chromosomes), nor did he even understand the precise causes of heritable variation among individuals (mutation and mixis). What he clearly understood, however, was that this heritable variation did exist and its causes could be logically separated from its consequences for the reproductive success of individuals and the resulting adaptation of species to their environments.

Darwin's theories were influenced, in part, by his observations of the practices of breeders of domesticated animals and plants. These practices are now commonly referred to as artificial selection. It is useful to distinguish between artificial and natural selection, and to relate this distinction to experimental evolution in the laboratory. Under artificial selection, organisms are chosen by a breeder, who allows some but not all individuals to survive and reproduce. Individuals are thus selected on the basis of particular traits that are deemed desirable to the breeder. By contrast, under natural selection, no one consciously chooses which individuals within a population will survive and reproduce and which will not. Instead, the match between organismal traits and environmental factors determines whether or not any given individual will survive and reproduce.

At first glance, one might regard laboratory studies as examples of artificial selection. This, however, would not reflect the critical distinction between artificial and natural selection, that is, whether a breeder or the environment determines which individuals survive and reproduce. In experimental evolution, an investigator manipulates environmental factors, such as temperature and resource concentration, but he or she does not directly choose which individuals within an experimental population will survive and reproduce. Instead, natural selection in the laboratory, like natural selection in the wild, depends on the match between organismal traits and environmental factors.

3. *Genetic Drift*

The frequency of genes in populations, and hence also the distribution of phenotypic traits, may change not only as a consequence of natural selection, but also as a consequence of the random sampling of genes during transmission across generations. This random sampling is called genetic drift. In practice, it can be difficult to distinguish between natural selection and genetic drift. This difficulty is especially evident when the only available data consist of static distributions of gene frequencies or phenotypic traits. What is needed to resolve this problem is some independent method of determining the effects of particular genes or phenotypic traits on survival and reproductive success. By using microorganisms to study evolution experimentally, it is possible to compare the survival and reproductive success of different genotypes that are placed in direct competition with one another. With proper replication of such experiments, one can distinguish systematic differences in survival and reproductive success from chance deviations due to drift.

II. EXPERIMENTAL TESTS OF FUNDAMENTAL PRINCIPLES

Two important principles of modern evolutionary theory are the randomness of mutation and adaptation by natural selection. According to the former, mutations occur irrespective of any beneficial or harmful effects they have on the individual. According to the latter, individuals in later generations tend to be better adapted to their environment than were individuals in earlier generations, provided that the necessary genetic variation exists and the environment itself does not change.

A. Random Mutation

For many years, it was known that bacteria could adapt to various environmental challenges. For example, the introduction of bacteriophage into a population of susceptible bacteria often caused the bacterial population to become resistant to viral infection. It was unclear, however, whether mutations responsible for bacterial adaptation were caused directly by exposure to the selective agent, or whether this adaptation was the result of random mutation and subsequent natural selection. Two elegant experiments were performed in the 1940s and 1950s that demonstrated that mutations existed prior to exposure to the selective agent, so that these mutations could not logically have been caused by that exposure.

1. *Fluctuation Test*

The first of these experiments was published by Salvador Luria and Max Delbrück in 1943, and relies on subtle mathematical reasoning. Imagine a set of bacterial populations, each of which grows from a single cell to some large number (N) of cells; the founding cells are identical in all of the populations. If exposure to the selective agent causes a bacterial cell to mutate with some low probability (p), then the number of mutants in a population is expected to be, on average, pN. Although this probability is the same for each of the replicate populations, the exact number of mutants in each population may vary somewhat due to chance (just as the number of heads and tails in 20 flips of a coin will not always equal exactly 10). Mathematical theory shows that the expected variance in the number of mutants among the set of replicate populations is equal to the average number of mutants under this hypothesis.

Now imagine this same set of populations, but assume that mutations occur randomly, that is, independent of exposure to the selective agent. During each cell generation, there is a certain probability that one of the daughter cells is a mutant. A mutant cell's progeny are also mutants, and so on. According to the mathematical theory for this hypothesis, the variance in the number of mutants among the replicate populations should be much greater than the average number of mutants. This large variance comes about because mutations will, by chance, occur earlier in some replicate populations than in others, and each of the early ("jackpot") mutations will leave many descendant mutants as a consequence of the subsequent population growth.

Luria and Delbrück designed experiments that allowed them to measure both the average and the variance of the number of mutants in a set of populations of the bacterium *Escherichia coli*. The observed variance was much greater than expected if exposure to the selective agent had caused the mutations. Hence, their results supported the hypothesis of random mutation.

2. Replica-Plating Experiment

Joshua and Esther Lederberg devised a more direct demonstration of the random origin of bacterial mutations, which they published in 1952. In their experiment, thousands of cells are spread on an agar plate that does not contain the selective agent; each cell grows until it makes a small colony, and the many colonies together form a lawn of bacteria (master plate). By making an impression of this plate using a pad of velvet, cells from all of the colonies are then transferred to several other agar plates (replica plates) that contain the selective agent, which prevents the growth of colonies except from those cells having the appropriate mutation. If mutations are caused by exposure to the selective agent, then there should be no tendency for mutant colonies found on the replica plates to be derived from the same subset of colonies on the master plate. But if mutations occur during the growth of the colony on the master plate (i.e., prior to exposure to the selective agent), then those master colonies that give rise to mutant colonies on one replica plate should also give rise to mutant colonies on the other replica plates (Fig. 1). Indeed, Lederberg and Lederberg observed that master colonies giving rise to mutants on one replica plate gave rise to mutants on the other replica plates. Moreover, they could isolate resistant mutants from those master colonies without the cells having ever been exposed to the selective agent. This experi-

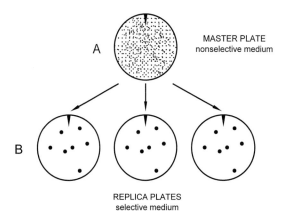

Fig. 1. Schematic illustration of the replica-plating experiment, devised by Lederberg and Lederberg to demonstrate the random occurrence of mutations. (A) Thousands of bacteria are spread on a master plate which does not contain any selective agent. Each cell grows until it makes a small colony, and the many colonies together form a dense lawn. (B) A piece of velvet is used to make an impression of the colonies on the master plate, and this impression is then transferred to several replica plates that contain a selective agent, such as bacteriophage or antibiotic. The selective agent kills all the cells except those with an appropriate resistance mutation, which can grow and form colonies. Notice the correspondence in the location of the mutant colonies on the replica plates. This spatial correspondence implies that these mutations occurred randomly during growth of the colonies on the master plate, that is, prior to the cells' exposure to the selective agent.

ment thus demonstrates that the mutations had occurred randomly during the growth of the colony on the master plate.

B. Adaptation by Natural Selection

In addition to demonstrating the random occurrence of mutations, the fluctuation test and the replica-plating experiment both demonstrate adaptation by natural selection. Two other types of experiments also demonstrate adaptation by natural selection. One type is very complicated and involves monitoring the dynamics of mutation accumulation at one genetic locus, which is not under selection, in order to study indirectly what is happening at some other loci, which are under selection.

The other type of experiment that demonstrates adaptation by natural selection involves direct estimation of an evolving population's fitness relative to its ancestor, and it is conceptually very simple. A population is founded by an ancestral clone, which is also stored in a dormant state (usually at very low temperature). The population is then propagated in a defined environment, and one or more samples are obtained after many generations have elapsed. These derived organisms are placed in direct competition with the ancestral clone under the same defined environmental conditions (after both types have acclimated physiologically to these conditions). If the derived organisms increase in number relative to the ancestral clone, in a systematic and reproducible fashion, then the evolving population has evidently become better adapted to the environment as the result of mutation and natural selection (Fig. 2).

To distinguish the derived and ancestral types from one another in a competition experiment, it is usually necessary to introduce a genetic marker that can be scored into one of them. This genetic manipulation

necessitates an appropriate control experiment to estimate the effect of the genetic marker on fitness.

III. GENETIC AND PHYSIOLOGICAL BASES OF FITNESS

The fact that one strain may be more fit than another in a particular environment usually says little or nothing about the causes of that difference. It is interesting to know why one strain is more fit than another in terms of their genotypes and their physiological properties. By using both classic and molecular genetic methods, one can construct various genotypes of interest and then determine the effects of their differences on physiological performance and fitness.

A. Effects of Single and Multiple Mutations

A study by Santiago Elena and Richard Lenski was the first to examine systematically the fitness effects of a large set of random mutations. Using transposon mutagenesis, they made 225 genotypes of *E. coli* that carried either one, two, or three insertion mutations. Each genotype was placed in competition with an unmutated strain in order to measure its relative fitness. Single-insertion mutations reduced fitness by about 2.5%, on average; there was tremendous variability among the mutational effects, which ranged from approximately neutral to effectively lethal.

The relationship between average fitness and mutation number was approximately log-linear; that is, subsequent mutations were neither more nor less harmful, on average, than was the first mutation. However, many pairs of mutations interacted strongly with one another; that is, their combined effect on fitness was different from what was expected, given their separate effects. (The relationship between average fitness and mutation number was approximately log-linear because some interactions were synergistic, whereas others were antagonistic.) These data imply that a satisfactory understanding of the genetic basis of any organism's functional capacity cannot depend entirely on the step-by-step

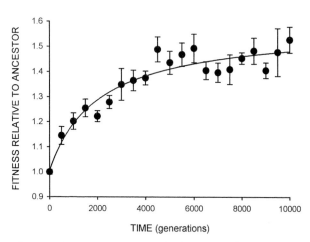

Fig. 2. Adaptation by natural selection is demonstrated by the improvement in mean fitness of 12 populations of *E. coli* during their serial propagation for 10,000 generations in a glucose-limited minimal medium. Fitness is expressed relative to the ancestral strain, which was stored at −80°C; relative fitness was measured by competing the ancestral and derived strains in this same medium. Error bars show 95% confidence intervals around the sample means. The mean fitness of the derived strains relative to their ancestor improved by ~50% during the 10,000 generations of experimental evolution. [Adapted from Lenski, R. E., and Travisano, M. (1994). *Proc. Natl. Acad. Sci. USA* **91**, 6808–6814.]

analysis of individual genes and pathways, but that instead a more integrative approach is necessary.

B. Fitness Effects Due to Possession of Unused Functions

A number of studies have used well-characterized bacterial genotypes to examine the effects on fitness caused by the carriage and expression of superfluous gene functions. These studies have measured the relative fitnesses of (1) bacteria with constitutive (high-level) and repressed (low-level) expression of enzymes for catabolism of carbon sources in media where those resources are not available; (2) prototrophic bacteria (which produce an amino acid or other required nutrient) and auxotrophic mutants (which cannot produce the required nutrient) in media where the required nutrients are supplied; (3) phage-sensitive bacteria and resistant mutants in environments where the phages are absent; and (4) antibiotic-sensitive bacteria and resistant genotypes in media that contain no antibiotics.

These studies have often, but not always, demonstrated substantial fitness disadvantages due to the possession of unnecessary gene functions. In some cases in which such disadvantages have been detected, they are much greater than can be explained by the energetic costs of synthesizing the extra proteins and other metabolites. For example, one study found that the fitness disadvantage associated with the synthesis of the amino acid tryptophan, when it was supplied in the medium, was much greater than could be explained on the basis of energetic costs. Evidently, the expression of superfluous functions can sometimes have strong indirect effects, which presumably arise through the disruption of other physiological processes.

The idea that microorganisms may have reduced fitness due to possession of unnecessary functions has two important practical implications. First, in many bioengineering applications, microorganisms are constructed that constitutively express high levels of some compound (e.g., a pharmaceutical) that one wishes to harvest for its commercial value. However, the producing microorganisms themselves do not benefit from that compound, and so any mutant that arises that no longer produces the compound

may have a selective advantage. Such a mutant would thus spread through the population and thereby reduce the efficiency of the production process. Second, the spread of antibiotic-resistant bacteria has become a serious concern in public health. It has been proposed that the prudent use of antibiotics, including the elimination of their use in certain environments (e.g., animal feeds), might favor the return of antibiotic-sensitive bacteria, thereby restoring the efficacy of antibiotics. This proposal rests, in part, on the presumption that resistant bacteria are less fit than their sensitive counterparts, in the absence of antibiotic, due to their possession of the superfluous resistance function. This often appears to be the case, but sometimes antibiotic-resistant bacteria evolve solutions that minimize or even eliminate the cost of resistance, thus complicating efforts to contain their spread.

C. Effects Due to Variation in Essential Metabolic Activities

It is clear that the expression of unnecessary metabolic functions is often disadvantageous to a microorganism. An equally important issue concerns the relationship between fitness and the level of expression of functions that are required for growth in a particular environment. This latter issue is more difficult to address experimentally, because it demands careful analyses of subtle differences between strains in biochemical activities rather than merely manipulating the presence or absence of some function.

Daniel Dykhuizen, Antony Dean, and Daniel Hartl performed a pioneering study to examine the relationships between genotype, biochemical activities in a required metabolic pathway, and fitness. Their study examined growth on lactose by genotypes of *E. coli* that varied in levels of expression of the permease used for active uptake of lactose and the β-galactosidase required for hydrolysis of the lactose. (The genotypes were otherwise essentially identical.) Given that both enzymes are necessary for growth on lactose, how do the activities at each step affect the net flux through this metabolic pathway? And how does net flux affect fitness?

Metabolic control theory consists of mathematical models to describe the dynamics across multiple

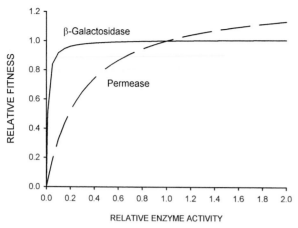

Fig. 3. Relative fitness of *E. coli* expressing different levels of permease and β-galactosidase activities in the lactose operon. The fitness curves were predicted from metabolic control theory using estimates of the biochemical activities of the two enzymes. Estimates of relative fitness were then obtained from competing strains with different enzyme activities in chemostats in which lactose was the sole source of energy. The resulting fitness values were very close to those predicted from biochemical data and metabolic theory. [Adapted from Dykhuizen, D. E., Dean, A. M., and Hartl, D. L. (1987). *Genetics* **115**, 25–31.]

steps in a biochemical pathway. Using this theory, Dykhuizen and his colleagues predicted how flux through this pathway would depend on the activities of the permease and β-galactosidase enzymes (Fig. 3). They estimated these activities for a number of genotypes using biochemical methods. They then predicted that the relative fitness of two genotypes would be directly proportional to their relative fluxes if lactose were provided as the sole energy source. In order to test the theory and its predictions, they estimated the relative fitnesses of the various genotypes in a medium in which lactose was the sole source of energy for growth. The observed fitnesses were very close to the predicted values.

Dykhuizen and Dean also successfully extended this mechanistic approach to predicting fitness in competition for mixtures of lactose and glucose. However, the results (for both single and mixed sugars) were obtained with genotypes in which gene regulation was eliminated in order to simplify the analysis. An important challenge for the future is to include the complex dynamical effects of gene regulation in the models and experiments.

D. Effects of Genetic Background

It is obvious that the fitness effects caused by particular genetic differences depend strongly on the environment. For example, an antibiotic-resistance gene function that is essential for the survival and replication of a bacterium in the presence of antibiotic may hinder growth in an antibiotic-free environment. Similarly, the fitness effect that is due to a particular gene function may often depend on the genetic background in which that gene is found.

For example, one study found that different alleles at the 6-phosphogluconate dehydrogenase (6PGD) locus in *E. coli* had similar fitnesses in a gluconate-limited medium, provided that these alleles were present in a genetic background that also encoded an alternative metabolic pathway for 6-phosphogluconate use. In a genetic background where this alternative pathway was defective, however, these alleles had quite variable fitnesses in that same medium.

In another study with *E. coli*, it was observed that the selective disadvantage associated with bacteriophage-resistance mutations in a virus-free environment was substantially reduced during several hundred generations of experimental evolution. This fitness improvement resulted from secondary mutations in the genetic background that compensated for the maladaptive side effects of the resistance mutations, but which did not diminish the expression of resistance.

Other studies have demonstrated similar compensatory evolution among antibiotic-resistant bacteria. When bacteria resistant to a particular antibiotic first arise, they are typically less fit in the absence of that antibiotic, thus helping to control their proliferation. Over time, however, the evolving bacteria become very good competitors in the absence of antibiotic while retaining their resistance to antibiotic. That is, the bacteria find ways to "have their cake and eat it, too." Such compensatory evolution unfortunately makes it more difficult to devise strategies to manage the spread of antibiotic-resistant pathogens.

IV. GENETIC VARIATION WITHIN POPULATIONS

In nature, abundant genetic variation exists in most species, including microorganisms. Some of this variation exists within local populations, while other variation may distinguish one population from another. This section describes some of the dynamic processes that influence genetic variation within populations.

A. Transient Polymorphisms

A population is polymorphic whenever two or more genotypes are present above some defined frequency (e.g., 1%). A polymorphism arises whenever an advantageous mutation is increasing in frequency relative to the ancestral allele. This type of polymorphism is called transient, because eventually the favored allele will exclude the ancestral allele by natural selection.

B. Selective Neutrality

At the other extreme, some polymorphisms may exist almost indefinitely precisely because the alleles that are involved have little or no differential effect on fitness. Such selectively neutral alleles are subject only to genetic drift. Daniel Dykhuizen and Daniel Hartl sought to determine whether some polymorphic loci in natural populations of *E. coli* might exist because of selective neutrality, or whether other explanations are needed. To that end, naturally occurring alleles at particular loci were transferred into a common genetic background, and the fitness effects associated with the various alleles were determined. Even when the bacteria were grown under conditions where growth was directly dependent on the particular enzymatic steps encoded by these loci, there were often no discernible effects on fitness due to the different alleles. These studies support the hypothesis that random genetic drift is responsible for some of the genetic variation that is present within natural populations.

C. Frequency-Dependent Selection

In the course of growth and competition in a particular environment, microorganisms modify their environment through the depletion of resources, the secretion of metabolites, and so on. When this happens, the relative fitnesses of genotypes may depend on the frequency with which they are represented in a population, and selection is said to be frequency-dependent. Frequency-dependent selection can give rise to several different patterns of genetic variation.

1. Stable Equilibria

Two (or more) genotypes can coexist indefinitely when each one has some competitive advantage that disappears as that genotype becomes more common. In that case, each genotype can invade a population consisting largely of the other genotype, but cannot exclude that other genotype, so that a stable equilibrium results (Fig. 4A).

Several distinct ecological interactions can promote these stable equilibria. For example, an environment may contain two carbon sources. If one genotype is better at exploiting one resource and another genotype is superior when in competition for the second resource, then whichever genotype is rarer will tend to have more of the resource available to it, thereby promoting the genotypes' stable coexistence. In some cases, a resource that is essential for one genotype may be produced as a metabolic byproduct of growth by another genotype; such interactions are termed cross-feeding. Stable coexistence of genotypes in one population can also occur when the environment contains a population of predators (or parasites); predator-mediated coexistence requires that one of the prey genotypes be better at exploiting the limiting resource while the other prey genotype is more resistant to being exploited by the predator. The evolution of two or more stably coexisting bacterial genotypes from a single ancestral type has been demonstrated in several experiments involving both cross-feeding and predator–prey interactions.

A striking example of the rapid evolution and stable coexistence of several genotypes comes from a recent experiment with *Pseudomonas fluorescens* by Paul Rainey and Michael Travisano. Starting from a single clone, placed in a static flask that contained a rich liquid medium, three morphologically distinct genotypes emerged within a matter of days, and they all coexisted with one another thereafter. By reconstituting the various combinations of these three genotypes, it was shown that each one had a selective advantage when it was rare relative to another genotype. As a consequence of its advantage when rare, each genotype could invade the others, so that a

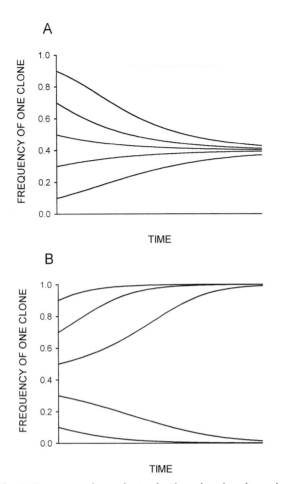

Fig. 4. Frequency-dependent selection, showing dynamics of competition between two clones at different initial frequencies. (A) Stable equilibrium. Each clone has a selective advantage when it is rare, but a disadvantage when it is common, relative to the other clone. Thus, each clone can increase in frequency when it is initially rare, which leads to a stable equilibrium in which the two clones coexist indefinitely. (B) Unstable equilibrium. Each clone has a selective advantage when it is common, but not when it is rare. Neither clone can become established in a population that is dominated by the other type, and they cannot coexist locally.

stable community was formed. However, if the flask was physically agitated, the coexistence was disrupted and only one of the genotypes prevailed. The three genotypes coexisted in the static flask because they had different abilities to exploit gradients (e.g., oxygen concentration) that were generated by the organisms' metabolic activities in concert with the physical environment. But when these gradients were

eliminated by continually shaking the flask, stable coexistence of the three genotypes was impossible.

2. *Unstable Equilibria*

Those ecological interactions that promote the stable coexistence of two or more genotypes contribute to the maintenance of genetic variation in populations. However, certain ecological interactions give rise to unstable equilibria. An unstable equilibrium exists when each of two genotypes prevents the other from increasing in number (Fig. 4B). Such interactions do not promote polymorphisms within a particular population. However, they may contribute to the maintenance of genetic differences between populations because neither genotype can invade a resident population of the other.

One form of ecological interaction that can give rise to an unstable equilibrium is interference competition. This occurs when one genotype produces an allelopathic (toxic) substance that inhibits the growth of competing genotypes; this is distinguished from exploitative competition, which occurs by simply depleting resources. Many microorganisms secrete allelopathic compounds, including fungi, which produce antibiotics. Some strains of *E. coli* produce colicins, which kill competing strains of *E. coli* but to which the producing types are immune. Producing genotypes, when common, make so much toxin that they can wipe out a sensitive strain that is more efficient in exploitative competition. When the producing type is rare, however, the cost of colicin production is greater than the benefit of the resource that becomes available by the killing of sensitive cells, and so the producing type loses out to the more efficient colicin-sensitive competitor. The outcome of competition between colicin-producing and sensitive strains also depends on the physical structure of the environment, as shown by Lin Chao and Bruce Levin. In particular, the advantage shifts to the colicin-producing types on agar surfaces, even when they are rare, because the resources made available by the killing action of colicins accrue locally to the producers, rather than being dispersed evenly as in a liquid medium.

3. *Nontransitive Interactions*

In some cases, frequency-dependent interactions among three or more genotypes are so complex as to be nontransitive. For example, genotype *A* outcompetes genotype *B,* and genotype *B* outcompetes genotype *C,* but genotype *C* outcompetes genotype *A.* Charlotte Paquin and Julian Adams found nontransitive interactions in populations of the yeast, *Saccharomyces cerevisiae,* that were evolving in chemostats fed with glucose as a sole carbon source. (A chemostat is a vessel used for the continuous culture of microorganisms. A steady state is achieved by the constant flow of fresh medium into the vessel and the simultaneous removal of an equal volume of spent medium.) Mathematical models of competition for a single limiting resource in a homogeneous environment (like a chemostat) predict transitive interactions among genotypes. Hence, this demonstration of nontransitivity suggests that other factors, such as the accumulation of allelopathic metabolites in the medium, must be involved.

Nontransitive interactions can lead to situations in which the average fitness of an evolving population relative to some distant ancestor declines, even though each genotype has increased fitness relative to its immediate predecessor. Indeed, Paquin and Adams observed precisely this phenomenon. Nontransitive interactions may also maintain genetic diversity within populations over time by recycling genotypes that would otherwise be lost.

V. COEVOLUTION OF INTERACTING GENOMES

Microorganisms in nature rarely, if ever, exist as single species, as they are usually studied in the laboratory. Rather, they exist in complex communities of many interacting populations. Some interactions are exploitative: One population makes its living by parasitizing or preying on another population. Other interactions are mutualistic: Each population obtains some benefit from its association with the other. In many cases, these interactions are plastic, both genetically and ecologically. For example, a single mutation in a bacterium may render it resistant to lethal infection by a bacteriophage. And a plasmid that confers antibiotic-resistance may be beneficial to its bacterial host in an antibiotic-containing environment but detrimental in an antibiotic-free environment.

As a consequence of this variability, microorganisms have proven useful for investigating questions about the coevolution of interacting populations. Are there evolutionary "arms races" between host defenses and parasite counterdefenses? Why are some parasites so virulent to their hosts, whereas others are relatively benign? How can mutualistic interactions evolve, if natural selection favors "selfish" genes that replicate themselves even at the expense of others?

A. Exploitative Interactions

A number of studies have demonstrated the stable coexistence of virulent bacteriophage (lytic viruses) and bacteria in continuous culture. In these studies, the virus population may hold the bacterial population in check at a density that is several orders of magnitude below the density that would be permitted by the available resource if viruses were not present. Typically, however, bacterial mutants eventually appear that are resistant to the virus, and these mutants have a pronounced selective advantage over their virus-sensitive progenitors. The proliferation of bacteria that are resistant to infection by the original virus provides a selective advantage to host-range viral mutants, which are capable of infecting the resistant bacteria. Thus, one can imagine, in principle, an endless "arms race" between resistant bacteria and extended host-range viruses.

In fact, there seem to be constraints that preclude this outcome. Bacterial mutants may appear, sooner or later, against which it is difficult or impossible to isolate corresponding host-range viral mutants. This asymmetry may arise because bacterial resistance can occur via mutations that cause either the structural alteration or the complete loss of certain receptors on the bacterial surface, whereas viral host-range mutations can counter only the former. Despite this asymmetry, the virus population can persist if the virus-resistant bacterial mutants are less efficient than their sensitive progenitors in competing for limiting resources. In such cases, a dynamic equilibrium is obtained in which the growth-rate advantage of the sensitive bacteria relative to the resistant type is offset by death due to viral infection. Such trade-offs between competitiveness and resistance commonly occur because the receptors that are used by viruses to adsorb to the cell envelope often serve also to transport nutrients into the cell or to maintain its structural integrity.

A widely held belief is that a predator or parasite that is too efficient or virulent will drive its prey or host population extinct, thereby causing its own demise. However, virulent phage often coexist with bacteria, even though the successful reproduction of the virus is lethal to the infected bacterium. Moreover, the process of natural selection neither requires nor permits foresight, so the mere prospect of extinction cannot deter the evolution of more efficient predators or more virulent parasites. Nevertheless, many viruses exist (lysogenic and filamentous bacteriophages) that are replicated alongside the host genome and whose infections, although deleterious, are not necessarily lethal. These viruses, as well as conjugative plasmids, have life cycles that include both horizontal (infectious) and vertical (intergenerational) transmission.

The evolutionary factors that favor these alternative modes of transmission are not fully understood. One factor that is thought to be important is the density of hosts. If susceptible hosts are abundant, then there are many opportunities for horizontal transmission. Therefore, selection favors those parasites that replicate and infectiously transmit themselves most rapidly, regardless of the consequences of these activities for the host's fitness. But if susceptible hosts are scarce, then horizontal transmission becomes infrequent. Vertical transmission, by contrast, does not depend on the parasite or its progeny finding another host. Instead, the success of a vertically transmitted parasite is determined by the success of its infected host. The greater the burden that such a parasite imposes on its host, the slower the host is to reproduce its own genome and that of the parasite. Hence, when the density of susceptible hosts is low, selection may favor those parasites that minimize their replicative and infectious activities, and thereby minimize their deleterious effects on the host.

Two recent studies sought to test this hypothesis by manipulating the supply of susceptible hosts. One experiment, which used a filamentous bacteriophage, supported the hypothesis, whereas the other study, which used a conjugative plasmid, did not (for reasons that are unclear). Perhaps more importantly, both studies demonstrated genetically encoded trade-offs between the parasites' rates of horizontal and vertical transmission. That is, parasites that were transmitted between individual hosts at higher rates reduced the host's growth rate—and hence their own vertical transmission—more severely than did parasites that were infectiously transmitted at lower rates.

B. Mutualistic Interactions

It has often been suggested that many mutualisms evolved from formerly exploitative interactions. Indeed, mathematical models predict that, at low host densities, genetic elements such as plasmids and phage can persist only if they are beneficial to the host. Many plasmids encode functions useful to their bacterial hosts, including resistance to antibiotics, restriction immunity to phages, production of bacteriocins, and so on. And some of these plasmids are unable to promote conjugation, relying exclusively on vertical transmission. Moreover, several studies have found unexpected competitive advantages for bacteria that are infected by plasmids, transposons, and even temperate phage, compared to cells that are not infected but are otherwise genetically identical.

Two studies have even demonstrated the evolution of mutualistic interactions from formerly antagonistic associations. In one study, the growth rate of the protist *Amoeba proteus* was initially greatly reduced by a virulent bacterial infection. The harmful effects of the bacteria were diminished by propagating the infected amebae for several years; in fact, the amebae eventually became dependent on the bacterial infection for their viability. In the other study, a plasmid initially reduced the fitness of its bacterial host in antibiotic-free medium; after 500 generations of coevolution, however, the plasmid enhanced the fitness of its host in this same medium. Interestingly, the genetic change responsible for the newly evolved mutualistic interaction was in the host chromosome, not in the plasmid genome. Both of these studies demonstrate that hosts can become dependent on, or otherwise benefit from, formerly parasitic genomes, thus giving rise to mutualistic interactions.

VI. EVOLUTION OF NEW METABOLIC FUNCTIONS

Microorganisms exhibit a tremendous diversity of metabolic activities, some of which function in biodegradative pathways (catabolism), whereas others work in biosynthetic pathways (anabolism). How has this diversity evolved? One area of research in the field of experimental evolution seeks to elucidate the various processes by which microorganisms can acquire new metabolic functions. This research is timely as humans seek to harness microorganisms that degrade toxic pollutants in the environment.

A. Acquisition by Gene Transfer

Perhaps the simplest way in which a microorganism can acquire some new metabolic function is by gene transfer from another microorganism that already possesses that function. For example, antibiotic resistance functions are often encoded by plasmids, which are transmitted from donors to recipients by conjugation. Acquisition by gene transfer is not always so simple a solution, however. Effective biodegradation of certain recalcitrant compounds may require the complex coordination of several steps in a biochemical pathway, which are encoded by complementary genes from two (or more) different microorganisms. The acquisition of activities that depend on such pathways may require not only genetic exchange, but also subsequent refinement of the new function by mutation and natural selection.

B. Changes in Regulatory and Structural Genes

In several cases, microorganisms have been shown to acquire new metabolic activities without any genetic exchange. Instead, the acquisition of a new metabolic function may occur by selection for one or more mutations in existing regulatory or structural genes, which normally have some other function. For example, the bacterium *Klebsiella aerogenes* cannot normally grow on the sugar D-arabinose, although it does possess an enzyme, isomerase, that is able to catalyze the conversion of D-arabinose into an intermediate, D-ribulose, which can be further degraded to provide energy to the cell. This isomerase is normally expressed at a very low level that does not permit growth on D-arabinose. Mutations that increase the level of expression of this isomerase are sufficient to enable growth by *K. aerogenes* on D-arabinose. This ability to grow on D-arabinose is further improved by certain mutations in the structural gene, which change the amino acid sequence of the isomerase so as to improve the efficiency of the conversion of D-arabinose to D-ribulose.

In essence, the evolution of new metabolic activities may involve "borrowing" gene products that were previously used for other functions. It is not surprising that this may sometimes also encroach on, and disrupt, these previous functions. Such encroachment could, in turn, favor gene duplication, a type of mutation in which a single copy of an ancestral gene gives rise to two homologous copies, each of which may then evolve toward different metabolic capabilities.

C. Reactivation of Cryptic Genes

Selection for novel metabolic activities has occasionally revealed the existence of "cryptic" genes, which are apparently nonfunctional but can be made functional by one or a few mutations. Cryptic genes are presumably derived from once-active genes that have been silenced by mutations that destroyed their functions. In the course of experiments, such cryptic genes are discovered by selection for new mutations that reverse or suppress these earlier mutations, thus restoring the lost metabolic activities.

VII. EVOLUTION OF GENETIC SYSTEMS

The process of adaptation by natural selection requires genetic variation in those characteristics that influence the survival and reproduction of organisms. The two sources of genetic variation are mutation and mixis. The rates of mutation and mixis depend not only on environmental factors (e.g., ultraviolet irradiation), but also on properties of the genetic system intrinsic to the organism in question. Here, genetic system refers to all those aspects of the physiology, biochemistry, and reproductive biology of an organism that influence the rates of mutation and mixis. For example, organisms have mechanisms of varying efficacy that promote the accurate replication and repair of their DNA. And while sex is an integral part of reproduction for some organisms, many others reproduce asexually, so that progeny are usually genetically identical to their parent and to one another.

Some of the most interesting questions in evolutionary biology concern the significance and evolutionary consequences of alternative genetic systems. Why do some organisms reproduce sexually, whereas others reproduce asexually? If mutation generates variation that is necessary for adaptation by the population, but most mutations have deleterious effects on the individual, then what mutation rate is optimal? Might organisms somehow be able to choose only those mutations that are beneficial to them, given their present circumstances?

A. Sexuality and Mixis

Numerous hypotheses have proposed evolutionary advantages for sexuality. All of these depend, in one way or another, on the genetic variation that is produced by mixis. Most efforts to test these hypotheses have relied on comparing patterns of the distribution of sexual versus asexual organisms. However, several experimental studies have examined the evolutionary consequences of mixis using microbes in which one can manipulate the extent of intergenomic recombination.

For example, mixis in viruses can be manipulated by varying the multiplicity of infection (MOI) of host cells because the recombination of viral genotypes can occur only if two or more viruses infect the same host cell. One study compared the rate of adaptive evolution of a bacterial virus at high and low MOI; the total size of the virus population was standardized for both treatments. The average fitness increased more rapidly under the high-MOI (high-recombination) treatment than under the low-MOI (low-recombination) treatment. This result is consistent with the hypothesis that sexual populations can adapt more rapidly than asexual populations because two or more advantageous mutations can be incorporated

simultaneously in the former, but only sequentially in the latter.

Some experiments have suggested that another advantage of mixis may occur when the rate of deleterious mutation is high and the effective population size is very small. Such conditions may apply especially to microorganisms with high error rates during replication (e.g., RNA viruses) or those with relatively large genomes (e.g., protozoa), if their populations are subject to periodic "bottlenecks." In these cases, deleterious mutations tend to accumulate indefinitely in asexual lineages, a process called "Muller's ratchet" (after the geneticist, H. J. Muller, who first described this phenomenon). However, even occasional mixis can purge lineages of their accumulated load of deleterious mutations, as was demonstrated by Lin Chao and his colleagues using a segmented RNA virus that infects bacterial cells. This effect occurs because two recombining genomes may each complement the deleterious mutations that are present in the other, thereby producing some progeny with a reduced load of deleterious mutations (as well as other progeny with an increased load, which will be removed by natural selection).

In still other cases, mixis appears to be less an adaptation to recombine genes per se than a coincidental consequence of the movement between cells of parasitic entities. In many bacteria, for example, recombination of chromosomal genes occurs only when cells are infected by viruses (transduction) or plasmids (conjugation). The new combinations of chromosomal genes that may result from such infections will occasionally be advantageous. But one should not regard phages and plasmids as benevolent agents of bacterial carnal pleasure because their infections are more often deleterious to the host.

B. Evolutionary Effects of Mutator Genes

Mutator genes increase the mutation rate throughout an organism's genome by disrupting DNA repair functions. (Mobile genetic elements may also behave as mutator genes because their physical transposition can alter the expression of other genes.) Several studies have investigated the effect of mutator genes on bacterial evolution. These studies have revealed a pattern that seems, at first glance, rather curious. When a mutator gene is introduced into a population above a certain initial frequency (e.g., 0.1%), it tends to increase in frequency over the long term. But if that mutator gene is introduced at a frequency below that threshold, then it typically becomes extinct.

What causes this curious effect? In a sense, there is an evolutionary race between two clones, with and without the mutator gene, to see which one gets the next beneficial mutation. The rate of appearance of new mutations for each clone depends on the product of its population size, N, and its mutation rate, u. When the ratio of mutation rates for the mutator and nonmutator clones, u'/u, is greater than the inverse ratio of their population sizes, N/N', then the mutator clone is more likely to have the next favorable mutation and thus prevail over the long term. When u'/u is less than N/N', the nonmutator clone, by virtue of its greater numbers, is likely to produce the next beneficial mutation and thereby exclude the mutator clone.

But this explanation presents a problem for understanding the evolution of mutators in nature, where they are moderately prevalent. If mutator genes are useful only when they are common, then how do they become common in the first place? Two recent studies, one theoretical and the other experimental, suggest that a process called "hitchhiking" may provide the explanation. In hitchhiking, a deleterious mutation (such as one that disrupts DNA repair) gets carried along to high frequency if it is genetically linked to a beneficial mutation. In bacteria, which mostly have only a single chromosome and are usually asexual, the entire genome is effectively linked. Moreover, a mutator gene is more likely than any other deleterious mutation to be associated with a beneficial mutation because of the high mutation rate it causes. It is unlikely that any particular mutation that produces a mutator will yield a beneficial mutation that allows the mutator to hitchhike; but given enough time one "lucky" mutator may do so, hence increasing the mutation rate for the entire population.

It has also been proposed that mutator genes may be more common in pathogenic bacteria than in their nonpathogenic counterparts. The idea is that patho-

genic bacteria face especially rapidly changing selective conditions due to immunological and other host defenses. By having a higher mutation rate, pathogens would have a greater likelihood of evolving a counterdefense. This explanation also fits well with the hitchhiking hypothesis because every change in the host environment that selects a new mutation in the pathogen population creates an opportunity for a mutator allele to hitchhike to high frequency. In contrast, for an organism living in a constant environment to which it is already adapted, beneficial mutations would be much rarer and hence there would be fewer opportunities for a mutator gene to hitchhike.

Thus, aspects of genetic systems that increase variation—whether by mutation or mixis—may accelerate adaptive evolution. But mutation and mixis can also break down genotypes that are already well adapted to particular environments. The evolution of genetic systems may often reflect a balance between these opposing pressures.

C. Directed Mutations?

The experiments of Luria and Delbrück and Lederberg and Lederberg showed that mutations arose before bacteria were exposed to a selective agent; hence, the mutations were not a response by the bacteria to that agent. But in 1988, John Cairns and his colleagues called into question the generality of random mutation in bacteria. Their paper and several subsequent studies purported to show that certain mutations occurred only (or more often) when the mutants were favored. Such mutations were called "directed" or "Cairnsian." Other studies showed that some of the evidence for directed mutation was flawed or misinterpreted, and this subject became very controversial. However, there is now widespread agreement that the most radical hypotheses put forward to explain this phenomenon—a reverse flow of information from the environment through proteins and RNA back to the DNA—are incorrect.

This controversy has generated new interest in the mechanisms of bacterial mutation, especially in the ways in which a cell might exercise some control over the mutational process. Attention has now focused on understanding the effects of stress, such as starvation, on DNA repair and mutation, as well as the extent of variation among local DNA sequences in their mutability. For example, numerous studies have documented unusually high mutation rates in short repeated sequences (e.g., TTTT); the mutations are typically frameshift events involving the loss or gain of a repeated element, and they apparently occur via a slippage of DNA strands during replication. Hypermutable sequences are not distributed randomly throughout bacterial genomes. Rather, they are found more often in genes encoding products (such as fimbriae and lipopolysaccharides on the cell surface) that are involved in pathogenicity and the evasion of host immune surveillance. This distribution suggests that bacteria may have evolved a simple but effective strategy to increase the mutation rate in genes that deal with unpredictable aspects of the environment, without inflating the load of deleterious mutations in "housekeeping" genes whose products interact predictably with the environment. These mutations are apparently random insofar as a particular mutation does not occur as a direct response to an immediate and specific need, but they are nonrandom in their genomic distribution and may thereby promote a more favorable balance between evolutionary flexibility and conservatism.

See Also the Following Articles

DNA SEQUENCING AND GENOMICS • MUTAGENESIS • TRANSFORMATION, GENETIC

Bibliography

Baumberg, S., Young, J. P. W., Wellington, E. M. H., Saunders, J. R. (eds.) (1995). "Population Genetics of Bacteria." Cambridge University Press, Cambridge.

Bell, G. (1997). "Selection: The Mechanism of Evolution." Chapman & Hall, New York.

Chadwick, D. J., and Goode, J. (eds.) (1997). "Antibiotic Resistance: Origins, Evolution, Selection and Spread." John Wiley, Chichester, UK.

Chao, L. (1992). Evolution of sex in RNA viruses. *Trends Ecol. Evol.* 7, 147–151.

Dykhuizen, D. E. (1990). Experimental studies of natural selection in bacteria. *Annu. Rev. Ecol. Syst.* 21, 373–398.

Dykhuizen, D. E., and Dean, A. M. (1990). Enzyme activity and fitness: evolution in solution. *Trends Ecol. Evol.* 5, 257–262.

Futuyma, D. J. (1998). "Evolutionary Biology," 3rd ed. Sinauer Associates, Sunderland, MA.

Lederberg, J., and Lederberg, E. (1952). Replica plating and indirect selection of bacteria mutants. *J. Bacteriol.* 63, 399–408.

Lenski, R. E. (1988). Dynamics of interactions between bacteria and virulent phage. *Adv. Microb. Ecol.* 10, 1–44.

Michod, R. E., and Levin, B. R. (eds.) (1988). "The Evolution of Sex." Sinauer Associates, Sunderland, MA.

Mortlock, R. P. (ed.) (1984). "Microorganisms as Model Systems for Studying Evolution." Plenum Press, New York.

Moxon, E. R., Rainey, P. B., Nowak, M. A., and Lenski, R. E. (1994). Adaptive evolution of highly mutable loci in pathogenic bacteria. *Curr. Biol.* 4, 24–33.

Sniegowski, P. D., and Lenski, R. E. (1995). Mutation and adaptation: the directed mutation controversy in evolutionary perspective. *Annu. Rev. Ecol. Syst.* 26, 553–578.

Exobiology

Gerald Soffen

National Aeronautics and Space Administration

GLOSSARY

astrobiology A term used to describe a new field which will expand on the limited studies embraced within the confines of exobiology. The definition is the study of life in the universe; the chemistry, physics, and adaptations that influence its origins, evolution, and destiny. It addresses the question, "Is life a cosmic imperative?"

exobiology A term originally used to refer to life that is not indigenous to the earth, or extraterrestrial biology. Use of the term expanded to refer to studies dealing with the origin of terrestrial life, investigations to determine the possibility of life, or its precursor chemistry of non-terrestrial origin and related technical issues.

life detection An experiment or observation that demonstrates the presence of any self-replicating biological organism.

organic Describing or referring to any carbon-bearing molecule other than free carbon, carbon monoxide, or carbon dioxide.

planetary protection The provisions taken to prevent the contamination of any planet by biological organisms from another planet due to spacecraft missions. This also includes the possible contamination of Earth from non-terrestrial samples from another planet.

THE TERM "EXOBIOLOGY" was used initially to refer to the study of putative life whose origin is non-terrestrial—hence "extraterrestrial biology." The term (coined by Joshua Lederberg) was used and developed in the late 1950s by scientists who were interested in the possibility of life on other planets.

As the field emerged in the 1960s, the term exobiology was used by scientists mostly associated with the National Aeronautics and Space Administration (NASA) to cover both the search for life on other planets and understanding the origin of terrestrial life. Exobiology broadened to include the following: the study of organic compounds in meteorites, methods of life detection by remote sensing and *in situ* tests for life, planetary protection and quarantine against microbial contamination, laboratory studies of possible chemical prebiotic paths that might have led to the origin of terrestrial life, and simulations of planetary environments to study the survival of microorganisms under varying physical and chemical conditions. The term continues to be used in this manner and is being incorporated as a segment of astrobiology.

Fundamental to exobiology is the definition of what is meant by the word "life." Biologists have discussed this considerably. Textbooks have used a variety of concepts depending on the experience and beliefs of the authors. Contemporary dogma of most biologists is based on the view that terrestrial life was initiated only once, and that all known terrestrial living organisms on the earth have been the results of the evolution of that ancestor or set of precursors. There is considerable biochemical and genetic evidence to support this theory. Most biologists agree that there are two essential characteristics of terrestrial life: the ability to self-reproduce and the ability to evolve by mutation and natural selection. This concept can be traced back to H. J. Muller (1926).

I. HISTORY OF EXOBIOLOGY

Prior to NASA's entrance into the field of exobiology, there were many biologists, theories, and experiments that contributed to the concepts that make up exobiology. Prominent scientists include Pasteur, Darwin, Huxley, Haldane, Oparin, and J. D. Bernal. Louis Pasteur's famous paper delivered to the Sorbonne in 1864 challenging the concept of spontaneous generation and establishing the doctrine of microbial reproduction shown by his experiments was a critical milestone. This led to thinking by other naturalists and to observations of biologists such as Charles Darwin, who believed that life began "in some warm little pond," that were years ahead of their time. In 1868, Thomas Huxley in a lecture entitled "The Physical Basis of Life" suggested that protoplasm was substantially the same for all living organisms. He believed that the existence of life depends on certain molecules, such as carbonic acid, water, and nitrogenous compounds, as the building blocks of life which, when brought together, gives rise to life. Considerations about the origin of terrestrial life, during the last part of the nineteenth century, were very conservative. A process called "panspermia," transport of germ material from one place to another, gained great popularity. Late in the century, organic chemistry began to flourish. Following the discovery of the laboratory synthesis of urea and the synthesis of indigo and alizarin, there was a blossoming of synthetic organic compounds which prepared the way for future laboratory experiments to address the chemical question of the origin of terrestrial life on the earth.

II. LABORATORY EXPERIMENTS

In the 1920s, A. I. Oparin (from Russia) and J. B. S. Haldane (from Britain) independently formulated theories to explain the origin of life. Oparin established the concept of chemical evolution preceding life. He viewed three phases of chemical evolution,

from inorganic chemistry to organic chemistry to biochemistry. In Oparin's theory, colloidal gels, "coazervates," prior to biological enzymes played a major role in the origin of life. In 1928, Haldane theorized that ultraviolet light (UV) from the sun was a critical energy source for biogenesis. He believed that UV acting on the earth's early atmosphere resulted in synthesis of a vast array of organic compounds, including amino acids and sugars. In addition, he believed that these compounds accumulated in a primitive ocean as a "hot dilute soup," the primordial broth of life.

In 1953, Stanley Miller, working with Nobelist Harold Urey, performed a seminal experiment in which he synthesized amino acids from methane, ammonia, hydrogen, and water (the presumed gases of the primitive Earth's atmosphere), and an electric discharge was used as the energy source to simulate lightning. Glycine, alanine, and aspartic and glutamic acids were formed along with numerous other compounds, including fatty acids and acetic and formic acids. This pioneering work led to numerous other laboratories setting up different feasible chemistries and a wide variety of energy sources, including UV light, electrons from a linear accelerator, the use of tesla coils to produce a spark, and the use of shock tubes and heated metal balls to mimic meteorites. The chemical path believed plausible is the "Strecker synthesis," which is the synthesis of an aldehyde and a nitrile that, when in the presence of ammonia, forms an intermediate which combines with water to form an amino acid.

In 1959, Juan Oro synthesized adenine and the purine precursors from hydrogen cyanide and ammonia solutions in water. He showed that aminoimidazole–carboxamidine and formamidine could form spontaneously. Later, James Ferris suggested that diaminomalonitrile formation is a likely pathway of reactions that leads to purines. The importance of the purines and pyrimidines for the formation of nucleic acids is obvious.

During this same period, Sidney Fox heated dry amino acids in carbon dioxide to form polypeptides which reacted with water to form "proteinoids," which form spheres the size of microbes that increase in size by combining and appear to divide. He was able to incorporate the inclusion of a variety of substances to mimic the behavior microorganisms.

Since the 1960s, numerous laboratories (mostly supported by NASA) have been developing possible biogenic pathways from prebiotic chemistry to the precursors of the first living ancestor. Many laboratories have synthesized *in vitro* organic chemical reactions that mimic the catalytic and genetic properties of living systems, but life in a test tube has not been initiated. The synthesis of viral DNA is important knowledge in filling in the gaps between chemistry and biology.

A central problem is that we do not know the exact conditions of the primitive earth. William Schopf estimates the earliest terrestrial microbial fossils to be approximately 3.5 billion years old, and recently Minik Rosing moved the date back to 3.7 bya. This was after both the earth's geological differentiation and the period of heavy bolide bombardment. Among the many other unknowns, we still do not know the state of oxidation at the time of life's first ancestor. Since the earth is believed to be 4.5 billion years old, it appears that life arose on the earth within a period less than the first billion years. For the past several decades, many laboratories have been exploring the most likely chemical paths.

Notable laboratory milestones, besides the first nucleic acids and proteins, include photochemical formation of formaldehyde, the formation of sugar molecules, the first high-energy phosphates, the importance of inorganic ions, the possible involvement of clay and other minerals, phospholipids and membrane formation, the formation of bioinformational molecules, and the genetic concepts that support the evolutionary "tree of life."

In contemporary work, scientists are examining a possible RNA world, the enzymatic properties of nucleic acids, electron transfer and redox via inorganic ions, chemical and thermal energy, the origin of chirality, alternative bases and backbone in informational molecules, polymerization on mineral/clay surfaces, organic synthesis in hydrothermal vents, and the evolutionary path of microorganisms.

III. THE SEARCH FOR EXTRATERRESTRIAL LIFE

NASA was initiated in 1958 in response to the Soviet Union's launching of Sputnik. On May 21, 1961, President Kennedy announced a decision to "put a man on the moon before the end of the decade," despite the urgings of many U.S. scientists who argued for instrumented planetary missions. Homer Newell headed the first Space Science Office and realized the needs of the new agency to invoke both a practical and pure biology into the program. In his book on the history of space science, titled *Life Sciences: No Place in the Sun,* Newell states.

… NASA[is] concerned with life sciences in a variety of ways … medical support for manned spaceflight … life support systems … aviation medicine … exobiology [the search for and the study of extraterrestrial life] … . Only space biology and exobiology could be regarded as pure science.

The 1960s was used by the "exobiologists" mostly to develop techniques of life detection.

The exobiologists agreed that Mars was the most likely planet, other than the earth, to support living organisms, and that if there were any Martian life it would most likely be microbial in nature. At that time, there was serious attention to what was called "the wave of darkening" on Mars, which some planetologists interpreted as possible biological microorganisms that were quiescent during the dry part of the season and became active when water became available. Another spectroscopic observation was the "Sinton Bands," absorption at 3.58 and 3.69 μ, which suggested possible organic material; the bands were subsequently discovered to be caused by deuterium in the earth's atmosphere.

Wolf Vishniac is credited as the first to invent a device for monitoring microbial growth on another planet. Under a NASA grant, in 1961 Vishniac built a laboratory model of a machine that would self-inoculate a small soil sample into liquid growth media. This was illuminated, and the subsequent increase in the number of organisms was measured by the changing opacity of the growing inoculation. This technique of measuring cloudiness by forward light scattering (nephelometry) to measure growth rate had been previously developed. Changes in pH as a function of time were also measured. The device called the "Wolf Trap" was developed and field tested on Earth, but it was not included on board the Viking missions that landed the first "life-detection" experiments on Mars in 1976 (Fig. 1).

Fig. 1. This is a photo of the Viking lander that was sent to Mars in 1975 and landed in 1976. This robot had two cameras for taking pictures of the landing site, a meteorology boom projecting from the side, an internal life-detection instrument containing three experiments to measure metabolism or growth, an internal organic chemical analyzer, and an instrument for measuring the elemental composition of the surface material. The data from this robot were sent to an overhead Mars orbiter that relayed the information to Earth. Two identical instruments were landed on Mars in the northern hemisphere and gave similar results.

Another life-detection instrument funded by NASA utilized a redioactive tracer, C^{14}, to measure the respiratory products of a growing culture of microorganisms. Dr. Gilbert Levin, having used this technique for the rapid detection of coliform organisms in sewerage, modified the growth media and the inoculation technique and built the first C^{14} detector/instrument for measuring respiration and growth of microbes on a foreign planet. He named this "Gulliver," and the concept was used as one of the three life-detection experiments onboard the 1976 Viking missions. This was called the "labeled release" (LR)

part of the experiment. The results are still subject to interpretation and strongly debated.

Many other concepts were developed during the 1960s, including the abbreviated vidicon microscope, gas chromatography, mass spectroscopy, J-band (detection of a reaction between an organic dye and organic macromolecules such as proteins or nucleic acids), UV spectroscopy (absorption at 1800 Å due to peptide linkages), optical activity, multivator (photomultiplier to detect biochemistry of biological activity using substrates), ATP detection using luciferase to measure light, detection of redox potential, O^{18} and N^{15} as tracers in metabolic reactions, gas exchange (to measure metabolism), and $C^{14}O_2$ uptake (to measure organic formation or photosynthesis). There was no shortage of ideas, but the actual development of these ideas into practical experiments to be carried out on Mars required a more directed project than is normally done under research and development fiscal constraints.

In 1968, NASA decided that Mars was the most likely planet to have developed indigenous life. The agency commissioned the two Viking missions to perform instrumented exploratory missions on Mars to take pictures of the surface, determine the nature of its indigenous organic compounds, perform life-detection experiments, analyze the atmosphere, and perform measurements of the meteorology and seismometry of Mars.

Venus was considered too hot (temperatures $>600°$ C) for the existence of living organisms. The moon, with its intense ionizing irradiation, was not considered a likely site. Also, there were plans for lunar surface samples to be returned to Earth by the astronauts in 1969. In the same year, the Viking missions to Mars (two identical spacecraft) were planned. Each of the spacecraft consisted of an orbiter to serve as both the bus and the communication relay and a soft lander. The lander contained cameras, life-detection experiments, and other instruments to measure the physical nature of the surface and atmosphere, and it also had the ability to perform a chemical analysis of the atmosphere and soil. In anticipation of finding organic material on the Martian surface, great emphasis was placed on the organic analysis of collected surface samples because this might bear on the question of Martian biogenesis.

In addition to the camera, which might have detected macroscopic life, there were three experiments performed on the Mars surface and sub-surface samples to detect viable microbial activity. Since Mars is closer to the asteroid belt (where organic-bearing carbonaceous chrondrites come from), it was logical that the Mars surface must contain organic material from the asteroid belt. The challenge then was to distinguish the indigenous organic matter on Mars biologically or chemically through *de novo* synthesis from matter deposited by meteoric infall.

The three life-detection experiments on the Viking missions were performed to detect metabolism, growth, or organic synthesis. They used C^{14} tracers and gas exchange techniques. In the LR experiment, soil samples were inoculated with a dilute solution of nutrients (formic, gylcolic, and lactic acids and glycine and alanine and both of their optical isomers). In a second experiment, pryolytic release (PR), Mars soil samples were inoculated with gaseous $C^{14}O$ and $C^{14}O_2$ while the samples were exposed to simulated Martian sunlight. If these gases were incorporated into the putative life, then by measuring the pyrolisate following exposure and subsequent combustion of the soil sample for C^{14} this would be a test for metabolic processes (perhaps photosynthesis). The third test, gas exchange (GEX), was performed to examine the gases that had evolved from a rich mixture of nutrients which was inoculated with Mars soil samples.

Each of these experiments was performed repeatedly on both of the Mars landing sites in the northern hemisphere (7000 km apart). The organic analysis (related to the biological search) of the Mars soil sample was performed with a sensitive pyrolytic–gas chromatograph/mass spectrometer (mass 12–250 mass units and sensitivity in the range of parts per billion) on the Martian atmosphere and on soil samples similar to those used for the biological tests. Despite the success of the operation of this instrument, no organic material was found on Mars at either of the landing sites.

In the case of the three biology experiments, all gave results, but interpretation of the results was inconclusive. The PR experiment gave one anomalous unrepeatable result and several negative ones. The LR experiment yielded some evolution of CO_2 but not enough to be explained by reproducing organisms. The GEX evolved O_2 from the soil sample when exposed to water vapor, but no metabolic gases when the nutrient was added. This result led to an interpretation that the Mars surface includes a superoxide material over most of its surface deposited by global dust storms. This oxidizing material (highly reactive) presumably is formed by the action of the unfiltered solar UV radiation splitting the H_2O and freeing the oxygen atom to form a metal peroxide. This would account for the absence of organic material (from meteorite deposition) due to its destruction by the Martian peroxide.

The conclusion reached by the Viking biology team was that there is not conclusive evidence for life on Mars. They left open the door to other experiments in other places and the possibility of fossils, should there be no contemporary life on Mars. In the succeeding decade, the interest in life on Mars waned, and those interested in exobiology began to focus on a better understanding of how life got started on the Earth.

IV. POST-VIKING INTERESTS

In the decades following the Viking missions of 1976, attention in exobiology was paid mostly to prebiotic chemistry. This included a great variety of laboratory experiments dealing with possible routes for the formation of precursor compounds. In addition to the interest in purely organic substances synthesized under various conditions, there was interest in how phosphates might have entered the biological processes and the role of inorganic ions such as iron and zinc. The possible role of clay in the synthesis of macromolecules and lipid formation and its importance in membranes were recognized as important.

Of particular interest were the concepts of self-assembly employing hydrogen bonds, nonpolar forces to stabilize supramolecular assembly, amphiphilic molecules to form membranes, bilayer lipid membranes that can readily encapsulate large molecules, permeability constraints on primitive cell functions, and possible protocellular processes such as the active transport of nutrients, energy transduction, and catalytic polymers which could store information and support growth and reproduction.

A classical paper in 1986 by T. Cech and coworkers describe the catalytic activity as a polymerase of a shortened form of a self-splicing ribosomal RNA molecule. In this paper, he addresses the question of which was the antecedent—the protein or the nucleic acid. A molecule that has both catalytic and information-bearing properties could be imagined as a critical step in the origin of life. This was followed by much speculation (as well as laboratory experiments) in what has become known as the "RNA world." Many authors have contributed to a wealth of literature on this subject.

Fossil evidence of living cells has been found dating back to about 3.5 bya. The RNA world is said to have begun about 4.2–4.0 bya and ended 3.6–3.8 bya. During this period there is evidence of carbonate enrichment of $C^{12/13}$ in ancient rocks. Large microfossil colonies in the form of stromatolites have been dated back to 3.5 bya. There is strong evidence for nucleic acid proteinaceous life by about 3.4 bya.

Organic compounds in both meteorites and comets have been reported. Comets are known to contain HCN. Among the organic compounds found in meteorites, it is of especial interest whether the amino acids are present. The composition of these substances resembles those resulting from a Miller–Urey synthesis. Calculation of the amount of infall to the early Earth is surprisingly large—10^{20} g of organic carbon over a period of only several hundred million years. Organic synthesis in the cosmos must be extensive inasmuch as carbon and hydrogen are among the most abundant elements which produced the stars and planets. The question of how much organic infall was taking place on the early earth was calculated by Chyba *et al.*, whose model deals with organic-bearing cometary ice, ablation, aerobraking, shock heating, and pyrolysis. They determined values of 10^6–10^7 kg/year on average for organic infall for a 10-bar atmosphere.

Chirality of the stereoisomers produced by living terrestrial biota is of significance in considering extraterrestrial life. Is this aspect of biogenesis universal (or unique for terrestrial life)? Many physical explanations for the chiral selectivity have been offered, including cosmic asymmetry and light polarization, but this issue is unresolved.

V. PLANETARY PROTECTION

An issue of particular importance in the field of exobiology is that of planetary contamination and protection. Studies have been performed to determine what steps must be taken to prevent contamination of other planets by terrestrial organisms and that of the earth by putative organisms from another body. Steps were taken for the Viking missions to

Mars to ensure that no terrestrial organisms would be carried to the planet. The Mars lander was heat sterilized (118° C for 48 hr). The Mars orbiters were retained at a high altitude to prevent their entry. Plans for a Mars sample return in the early twenty-first century include special precautions for quarantine of the samples to ensure containment. The major concern is the environment of the earth. The likelihood of a human infection by a Martian organism is considered extremely low, but the consequences of extraterrestrial biological contamination are of global concern. This has gained international recognition.

VI. RECENT CONCEPTS AND ACTIVITY

Since the Viking missions of 1976, much has been learned about organisms that live in exotic regions of the Earth. In addition to the extensive microbial populations that abound at the bottom of the oceans, there is an extraordinarily rich variety of macroorganisms, including tube worms and mollusks, that live in the vicinity of the volcanic vents and smokers at the bottom of the deep oceans. We have also learned about cryptoendolithic microorganisms that live several millimeters beneath the surface of porous rocks. The extent of biota living below the surface of the Earth has been recognized in the past two decades. These new views of microbial ecology have opened up the possibility of the existence of extraterrestrial life persisting in some niche that has no terrestrial parallel. In addition to Mars, planetologists have begun to consider Europa, and possibly Encedadus, two of the moons of Jupiter, as possible harbors for living organisms. This idea is based on the likely existence of liquid water beneath the surface of these bodies.

In 1996, McKay and co-workers reported that they might have discovered microfossils in a recovered meteorite known to have come from Mars (AH84001) (Fig. 2) This extraordinary report, including chemical evidence and photomicrographs of the putative organisms in the nanometer range, has met with both skepticism and enthusiasm and continues to be investigated.

NASA has recently developed a new program which shall include exobiology; this is called astrobi-

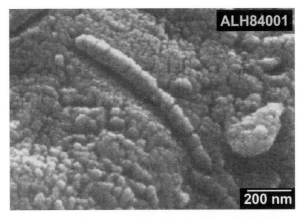

Fig. 2. Fossil microorganism. This photomicrograph was taken of a sample of a meteorite known to have come to Earth from Mars. The meteorite, identified as Alan Hills 84001, was recovered from the Antarctic and analyzed under very carefully controlled conditions by Dr. David McKay and co-workers. The date of the formation of the rock is about 4.0 bya. The image was taken by a scanning electron microscope. Other evidence supports the theory that this inclusion may be a fossil of a primitive Martian organism.

ology. The program is intended to cover a broad interdisciplinary set of research efforts to link the biology to astronomy and astrochemistry on one hand and evolutionary biology and biogeochemical cycles on the other hand. The question of life in the cosmos has become a major scientific issue. Many scientists believe that the answer to the question of the universality and extent of life in the cosmos requires disciplines besides biology. Astrobiology is aimed at developing the path of scientific inquiry that could lead to major milestones. This may include new instruments or missions that have not been formulated.

VII. SEARCH FOR EXTRATERRESTRIAL INTELLIGENCE

One high-risk/high-payoff effort in exobiology is the search for extraterrestrial intelligence. This employs the use of large antennas to listen for coherent signals of extraterrestrial origin. The argument for this is that if other living systems have developed a technological capability of communicating using the electromagnetic spectrum, then we should be able to detect this by listening in at the wavelengths being used. Given the dimensions of space, a search of the whole cosmos would be extremely ambitious. The detection of an unquestioned directed coherent signal would be considered by many as one of the greatest discoveries of mankind. There have been many efforts during the past several decades to detect coherent signals that are extraterrestrial. To date, none have been detected. Since they require unusual types of technology, most of these efforts have been supported through non-government funds.

VIII. SUMMARY

Exobiology is the study of unknown organisms, whose existence has yet to be proven. It is closely linked to the study of the origin of life on the Earth. Many scientists have speculated about the subject. Numerous contemporary laboratory experiments have led to plausible explanations for terrestrial biogenesis. In 1976, the first *in situ* tests for extant life were made on Mars by automated spacecraft. No conclusive evidence resulted from these experiments. NASA, the main agency supporting this work, has recently expanded the effort under its astrobiology program.

See Also the Following Articles

History of Microbiology • Microbes and the Atmosphere • Origin of Life

Bibliography

Bonner, W. A., Blairand, N. E., and Darbis, F. M. (1981). Experiments on the abiotic amplification of optical activity. *Origins of life* **11**, 119–134.

Chela-Flores, J., and Raulin, F. (1998). "Proceedings of the Fifth Trieste Conference on Chemical Evolution, Exobiology: Matter, Energy, and Information in the Origin and Evolution of Life in the Universe, Kluwer, Dordrecht.

Chyba, C., Thomas, P. J., Brookshaw, L., and Sagan, C. (1990). Cometary delivery of organic molecules to the early earth. *Science* **249**, 366–373.

Cooper, H. S. F. (1976). "The Search for Life on Mars." Holt, Rinehart & Winston, New York.

Deamer, D., and Fleischaker, G. R. (1994). "Origins of Life." Jones & Bartlett Boston.

Dick, S. J. (1998). "Life on Other Worlds: The 20th Century Extraterrestrial Life Debate." Cambridge Univ. Press, Cambridge, UK.

Ezell, E. C., and Edzell, L. N. (1984). On Mars, exploration of the red planet 1958–1978, NASA SP 4212. NASA, Scientific and Technical Information Branch, Washington, DC.

Fajszi, C., and Czege, J. (1981). Critical evaluation of mathematical models for the amplification of chirality.

Goldsmith, D., and Owen, T. (1980). "The Search for Life in the Universe." Benjamin/Cummings, Menlo Park, CA.

Haldane, J. B. S. (1967). *In* "The Origin of Life." (J. D. Bernal, Ed.), Reprint, pp. 242–249. Weidenfeld & Nicolson, London. (Original work published 1929)

Horowitz, N. H. (1986). "To Utopia and Back." Freeman, New York.

Joyce, G. F. (1991). The rise and fall of the RNA World. *New Biol.* **3**, 399–407.

Klein, H. P. (1977). The Viking Biological Investigation: General aspects. *J. Geophys. Res.* **82**, 4677–4680.

Newell, H. E. (1980). Beyond the atmosphere, NASA SP-4211. NASA, Scientific and Technical branch, Washington, DC.

Oparin, A. I. (1938). "The Origin of Life." Dover, New York.

Schopf, W. J. (1999). "The Cradle of Life." Princeton Univ. Press, Princeton, NJ.

Soffen, G. A., *et al.* (1976). Scientific results of the Viking mission. *Science* **194**, 1274–1353.

Soffen, G. A., *et al.* (1977). Scientific results of the Viking project. *J. Geophys. Res.* **82**, 3959–4681.

Space Studies Board of the National Research Council (1990). "The Search for Life's Origins," (H. P. Klein, Ed.). National Academy Press, Washington, DC.

Exotoxins

Joseph T. Barbieri

Medical College of Wisconsin

GLOSSARY

AB structure–function The structure–function organization of most bacterial exotoxins; the **A** domain comprises catalytic activity and the **B** domain comprises the binding and translocation domains.

exotoxin A soluble protein produced by a microorganism that can enter a host cell and catalyze the covalent modification of a cellular component to alter host-cell physiology.

heat-stable enterotoxins Soluble peptides that are secreted by bacteria, which bind to host cells and stimulate a signal transduction pathway within the host cell.

mechanism of action The specific reaction by which each exotoxin catalyzes to covalently modify a host-cell component, such as the ADP-ribosylation of elongation factor-2 by diphtheria toxin.

pore-forming toxins Soluble proteins that are secreted by bacteria, which bind to the surface of the host cell and oligomerize to form a pore to release soluble components from the host cell.

posttranslational modification A covalent modification to a cellular component that occurs subsequent to its synthesis.

proenzyme The form in which exotoxins are secreted by bacteria; these are processed to exhibit catalytic activity.

super antigens Soluble proteins that are secreted by bacteria and bind to the major histocompatibility complex of T lymphocytes. This stimulates antigen-independent proliferation of T lymphocytes.

toxoid The detoxified form of an exotoxin that is used for immunization. Conventional toxoiding is achieved with chemicals such as formalin; genetic engineering approaches have also been used to produce toxoids.

type III secreted cytotoxins Soluble proteins that are translocated into the host-cell cytoplasm via a type III secretion apparatus by host-cell surface-bound bacteria.

vaccination The administration of an immunogen (toxoid) to stimulate an immune response that protects the host from infection by the microorganism that produces the immunogen.

EXOTOXINS are a group of soluble proteins that are secreted by the bacterium, enter host cells, and catalyze the covalent modification of a host-cell component(s) to alter the host-cell physiology. Both gram-negative and gram-positive bacteria produce exotoxins. A specific bacterial pathogen may produce a single exotoxin or multiple exotoxins. Each exotoxin possesses a unique mechanism of action, which is responsible for the elicitation of a unique pathology. Thus, the role of exotoxins in bacterial pathogenesis is unique to each exotoxin. *Corynebacterium diphtheriae* produces diphtheria toxin, which is responsible for the systemic pathology associated with diphtheria, whereas *Vibrio cholerae* produces cholera toxin, which is responsible for the diarrheal pathology associated with cholera. Exotoxins vary in their cytotoxic potency, with the clostridial neurotoxins being the most potent exotoxins of humans. Exotoxins also vary with respect to the host that can be intoxicated. Exotoxin A of *Pseudomonas aeruginosa* can intoxicate cells from numerous species, whereas other toxins, such as diphtheria toxin are more restricted in the species that can be intoxicated. Some bacterial toxins, such as pertussis toxin, can intoxicate numerous cell types, whereas other toxins, such as the

Clostridial neurotoxins, show a specific tropism and intoxicate only cells of neuronal origin. Bacterial exotoxins catalyze specific chemical modifications of host-cell components, such as the ADP-ribosylation reaction catalyzed by diphtheria toxins or the deamidation reaction catalyzed by the cytotoxic necrotizing factor produced by *Escherichia coli*. These chemical modifications may either inhibit or stimulate the normal action of the target molecule to yield a clinical pathology. Bacterial exotoxins possess an **AB** structure–function organization, in which the **A** domain represents the catalytic domain and the **B** domain comprises the receptor-binding domain and the translocation domain. The translocation domain is responsible for the delivery of the catalytic **A** domain into an intracellular compartment of the host cell.

Many bacterial exotoxins can be chemically modified to toxoids that no longer expresses cytotoxicity, but may retain immunogenicity. Studies have shown that bacterial toxins can also be genetically engineered to toxoids, which may lead to a wider range of vaccine products. Exotoxins have also been used as therapeutic agents to correct various disorders, including the treatment of muscle spasms by botulinum toxin. Nontoxic forms of exotoxins have been used as carriers for the delivery of heterologous molecules to elicit an immune response and as agents in the development of cell-specific chemotherapy. In addition, bacterial toxins have been used as research tools to assist in defining various eukaryotic metabolic pathways, such as G-protein-mediated signal transduction.

I. CLASSIFICATION OF EXOTOXINS

Exotoxins are soluble proteins produced by microorganisms that can enter a host cell and catalyze the covalent modification of a cellular component(s) to alter the host-cell physiology. The term "host cell" refers to either vertebrate cells or cells of lower eukaryotes, such as protozoa because some bacterial exotoxins intoxicate a broad range of host cells. The recognition that some pathogenic bacteria produced soluble components that were capable of producing the pathology associated with a particular disease was determined in the late nineteenth century. Roux

and Yersin observed that culture filtrates of *Corynebacterium diphtheriae* were lethal in animal models and that the pathology elicited by the culture filtrate was similar to that observed during the infection by the bacterium. Subsequent studies isolated a protein, diphtheria toxin, from the toxic culture filtrates and observed that the administration of purified diphtheria toxin into animals was sufficient to elicit the pathology ascribed to diphtheria. Diphtheria toxin is a prototype exotoxin and has been used to identify many of the biochemical and molecular properties of bacterial exotoxins.

The ability of a bacterial pathogen to cause disease frequently requires the production of exotoxins, but the mere ability to produce a toxin is not sufficient to cause disease. Cholera toxin is the principal virulence factor of *Vibrio cholerae*. Administration of micrograms of purified cholera toxin to human volunteers elicits a diarrheal disease that mimics the magnitude of the natural infection. Nonetheless, nonvirulent toxin-producing strains of *V. cholerae* have been isolated and shown to lack specific biological properties, such as motility or chemotaxis. Similarly, although anthrax toxin is the principal toxic component of *Bacillus anthracis,* nonvirulent toxin-producing strains of *B. anthrasis* have been isolated and shown to lack the ability to produce a polyglutamic acid capsule. An exception to this generalization is the intoxication elicited by the botulinum neurotoxins, in which ingestion of the preformed toxin is responsible for the elicitation of disease; food poisoning by botulinum neurotoxins is an intoxication, rather than an infection by a toxin-producing strain of *Clostridium botulinum.*

Bacterial exotoxins are classified according to their mechanisms of action. The covalent modifications of host-cell components, which are catalyzed by bacterial exotoxins, include ADP-ribosylation, deamidation, depurination, endoproteolysis, and glucosylation (Table I). Most cellular targets of bacterial exotoxins are proteins, although there are exceptions such as shiga toxin, which catalyzes the deadenylation of ribosomal RNA. In addition to exotoxins, there are several other classes of toxins that are produced by bacterial pathogens, including the pore-forming toxins, type III-secreted cytotoxins, heat-stable enterotoxins, and superantigens. Each of these toxins fails to perform one of the properties associ-

TABLE I
Properties of Bacterial Exotoxins

Modification	Exotoxin	Bacterium	AB	Target	Contribution to pathogenesis
ADP-ribosylation	Diphtheria toxin	*C. diphtheriae*	AB	Elongation factor-2	Inhibition of protein synthesis
	Exotoxin A	*P. aeruginosa*	AB	Elongation factor-2	Inhibition of protein synthesis
	Cholera toxin	*V. cholerae*	AB5	Gsα	Inhibition of GTPase activity
	Heat-labile enterotoxin	*E. coli*	AB5	Gsα	Inhibition of GTPase activity
	Pertussis toxin	*B. pertussis*	AB5	Giα	Uncoupled signal transduction
	C2	*C. botulinum*	A-B	actin	Actin depolymerization
Glucosylation	Lethal toxin	*C. sordelli*	AB	Ras	Inhibition of effector interactions
	Toxin A and B	*C. difficile*	AB	RhoA	Inhibition of Rho signaling
Endoprotease	Anthrax toxin	*B. anthrasis*	A-B	A, edema factor	Adenylate cyclase
				A, lethal factor	Endoprotease?
	Botulinum toxin (A–F)	*C. botulinum*	AB	Vesicle proteins	Inhibition of vesicle fusion
	Tetanus toxin	*C. tetani*	AB	Vesicle protein	Inhibition of vesicle fusion
Deamidation	Cytotoxic necrotizing factor	*E. coli*	AB	RhoA	Stimulation of RhoA
Deadenylation	Shiga toxin	*Shigella* spp.	AB5	28S RNA	Inhibition of protein synthesis
	Verotoxin	*E. coli*	AB5	28S RNA	Inhibition of protein synthesis

ated with exotoxins. The pore-forming toxins are not catalytic in their action, but instead disrupt cell physiology through the formation of pores in the host-cell plasma membrane. The type III-secreted cytotoxins can not enter host cells as soluble proteins, but instead are translocated directly into the host cell by the type III secretion apparatus of the cell-bound bacterium. The heat-stable enterotoxin and superantigens do not enter the intracellular compartment of the host cell, and elicit host-cell responses by triggering signal-transduction pathways upon binding to the host-cell membrane. In this article, initial emphasis will be placed on the molecular properties of bacterial exotoxins, with a subsequent description of the general properties of pore-forming toxins, type III-secreted cytotoxins, heat-stable enterotoxins, and superantigens.

The pathology elicited by a specific exotoxin results from the catalytic covalent modification of a specific host-cell component. Although diphtheria toxin and cholera toxin are both bacterial ADP-ribosylating exotoxins, the pathogenesis elicited by each exotoxin is unique. This is due to the fact that diphtheria toxin ADP-ribosylates elongation factor-2, resulting in the inhibition of protein synthesis and subsequent cell death, whereas cholera toxin ADP-

ribosylates the Gsα component of the heterotrimeric protein, which stimulates the activity of adenylate cyclase. The stimulation of adenylate cyclase elevates intracellular cAMP and the subsequent secretion of electrolytes and H_2O from the cell, resulting in the clinical manifestations of cholera.

II. GENERAL PROPERTIES OF EXOTOXINS

A. Genetic Organization of Exotoxins

The genes encoding bacterial exotoxins may be located on the chromosome or located on an extrachromosomal element, such as a plasmid or a bacteriophage. Elegant experiments characterizing diphtheria toxin showed that the gene encoding this exotoxin was located within the genome of the lysogenic β-phage. Although both nonlysogenic and lysogenic strains of *C. diphtheriae* could establish local upper-respiratory-tract infection, only strains of *C. diphtheriae* lysogenized with a β-phage that encoded diphtheria toxin were capable of eliciting systemic disease. This established a basic property for the pathology elicited by bacteria that produce exotoxins; bacteria establish a localized infection and subsequently produce an exotoxin, which is responsible for pathology distal to the site of infection.

Most exotoxins are produced only during specific stages of growth with the molecular basis for the regulation of toxin expression varying with each bacterium. This differential expression often reflects a complex regulation of transcription, including responses to environmental conditions, such as iron. Multisubunit toxins are often organized in operons to allow the coordinate expression of their subunit components.

B. Secretion of Exotoxins from the Bacterium

Most bacteria secrete exotoxins across the cell membrane by the type II secretion pathway. The secretion of exotoxins by the type II secretion pathway was predicted by the determination that the amino terminus of mature exotoxins had undergone proteolysis relative to the predicted amino acid sequence. Type II secretion is also called the general secretion pathway. Type II secretion involves the coordinate translation and secretion of a nascent polypeptide across the cell membrane. During the translation of the mRNA that encodes a type II-secreted protein, the nascent polypeptide contains an amino-terminal leader sequence that is targeted to and secreted across the cell membrane. After secretion across the cell membrane, the nascent polypeptide folds into its native conformation and the leader sequence is cleaved by a periplasmic leader peptidase to yield a mature exotoxin.

Some gram-negative bacteria export the assembled exotoxin from the periplasm into the external environment via a complex export apparatus. While the heat-labile enterotoxin of *Escherichia coli* remains localized within the periplasmic space, *V. cholerae* and *Bordetella pertussis* assemble their respective exotoxins, cholera toxin and pertussis toxin, within the periplasm and then transport the mature exotoxin into the external environment. Although the multiple protein components of the export apparatus have been identified, the mechanism for export across the outer membrane remains to be resolved.

C. Bacteria Produce and Secrete Exotoxins as Proenzymes

Although one property of a bacterial exotoxin is the ability to intoxicate sensitive cells, early biochemical studies observed that *in vitro* many bacterial exotoxins possessed little intrinsic catalytic activity. These perplexing observations were resolved with the determination that bacteria produce and secrete exotoxins as proenzymes, which must be activated (processed) to express catalytic activity *in vitro*. Because exotoxins intoxicate sensitive cells, the requirements for *in vitro* activation reflect the activation steps *in vivo*. Each exotoxin requires specific conditions for activation, including proteolysis, disulfide-bond reduction, or association with a nucleotide or a eukaryotic accessory protein. Some activation processes result in the release of the catalytic **A** domain from the **B** domain, whereas other activation processes appear to result in a conformational change in the catalytic **A** domain, rendering it catalytically active. Some exotoxins require sequential activation steps. Diphtheria toxin is activated by limited proteolysis, followed by disulfide-bond reduction (see Fig. 1).

The determination of the activation mechanism of exotoxins has also provided insight into several physiological pathways of host cells. The eukaryotic

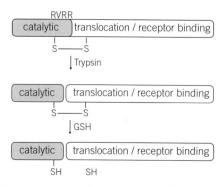

Fig. 1. Bacterial exotoxins are produced as proenzymes. Most bacterial exotoxins are produced as proenzymes that undergo processing to express catalytic activity. The sequential processing of diphtheria toxin involves protein cleavage at a trypsin-sensitive site in the carboxyl terminus of the **A** domain (Arg-Val-Arg-Arg, RVRR). The **A** and **B** domains are connected by a disulfide bond, which is reduced by agents such as reduced glutathione (GSH).

protein, ARF (*ADP-ribosylation factor*), which activates cholera toxin *in vitro*, was subsequently shown to play a central role in vesicle fusion within the eukaryotic cell. The ability of a host-cell extract to activate cholera toxin is often used as a sign of the presence of ARF. Similarly, the characterization of the mechanisms that pertussis toxin and cholera toxin use to intoxicate eukaryotic cells has provided insight into the pathways for eukaryotic G protein-mediated signal transduction. The ability of pertusis toxin to inhibit the action of a ligand in the stimulation of a signal-transduction pathway is often used to implicate a role for G proteins in that signaling pathway.

D. AB Structure–Function Properties of Exotoxins

Most bacterial exotoxins possess AB structure–function properties. (see Fig. 2). The **A** domain is the catalytic domain, whereas the **B** domain includes the translocation and binding domains of the exo-

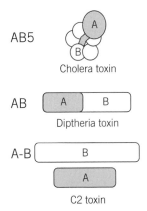

Fig. 2. Bacterial exotoxins posses **AB** structure–function organization. There are three general **AB** organizations of bacterial exotoxins. The **A** domain (shaded) represents the catalytic domain whereas the **B** domain comprises the translocation and receptor-binding domains. **AB5** is represented by cholera toxin of *V. cholerae,* which is composed of six noncovalently associated proteins. **AB** is represented by diphtheria toxin of *C. diphtheriae,* in which the **A** and **B** domain are included in a single protein. **A-B** is represented by C2 toxin of *Clostridium botulinum,* which is composed of two nonassociated proteins; the two proteins associate after the binding and processing of the **B** component on the host-cell membrane.

toxin. Exotoxins are organized into one of several general types of **AB** organization. The simplest **AB** organization is represented by the diphtheria toxin, in which the **A** domain and **B** domain are contained in a single protein. Diphtheria toxin is the prototype for this class of **AB** exotoxin. Diphtheria toxin is a 535-amino-acid protein in which the amino terminus constitutes the ADP-ribosyltransferase domain and the carboxyl terminus comprises the translocation domain and receptor-binding domain. The **AB5** exotoxins are composed of six proteins that are noncovalently associated as an oligomer. Cholera toxin is the prototype for the **AB5** exotoxin. The **A** domain of cholera toxin constitutes the ADP-ribosyltransferase domain, whereas the **B5** domain is composed of five identical proteins, forming a pentamer. This is organized into a ring structure, on which the **A** domain is positioned. The five proteins that make up the **B** domain may be identical, as is the case for cholera toxin and the heat-labile enterotoxin of *E. coli,* or may be different proteins that form a nonsymmetrical ring structure, as observed with the **B** oligomer of pertussis toxin.

The third class of **AB** exotoxin is composed of proteins that are not associated in solution, but that do associate following the binding and processing of the **B** domain to the host cell. C2 toxin is an example of this class of **A-B** exotoxin. C2 toxin is a bipartile exotoxin composed of a protein that encodes the catalytic **A** domain and a separate protein that encodes the **B** domain. The **A** domain protein of C2 toxin ADP-ribosylates actin. The **B** domain protein of C2 binds to sensitive cells and is nicked by a eukaryotic protease. The processed **B** components oligomerize and are then capable of binding either of the **A** domain proteins.

A new class of toxin organization has recently been recognized in which the **A** domain is a protein and the bacterium is directly responsible for its delivery into the cell. The bacterium binds to the eukaryotic cell and uses a type III secretion apparatus to deliver cytotoxins, also called effector proteins, into the intracellular compartment of the cell. The YopE cytotoxin of *Yersinia* is the prototype of this group of toxins. These cytotoxins do not conform to the strict definition of exotoxins because the purified cytotoxins cannot directly enter to modify host-cell physiology and are not included with the family of bacterial exotoxins. They are termed "type III secreted cytotoxins", and are described later.

Although the **A** domain possesses the catalytic activity of the exotoxin, the **B** domain possesses two specific functions, receptor binding and translocation capacity. Each exotoxin uses a unique host-cell surface component as a receptor. The cell surface receptor for each exotoxin may be specific. The cell surface receptor for cholera toxin is the ganglioside, GM1, whereas diphtheria toxin binds directly to the epidermal growth factor precursor. In contrast, the binding of pertussis toxin appears to be less specific, as pertussis toxin is able to bind numerous cell surface proteins. The ability to bind its cell surface receptor is an absolute requirement for an exotoxin to intoxicate a host cell because the deletion of the receptor-binding domain renders the exotoxin essentially noncytotoxic. After binding to the cell surface, some exotoxins are proteolytically processed or are processed during endocytic vesicle transport.

The second function of the **B** domain includes translocation capacity, which is responsible for the delivery of the **A** domain across the cell membrane. The presence of a translocation domain was predicted from early structure–function studies of diphtheria toxin, which showed that in addition to the catalytic domain and receptor-binding domain, a third function was required for the efficient expression of cytotoxicity. This third function was subsequently shown to correspond to a region of diphtheria toxin that had the propensity to interact with membranes. The crystal structure of diphtheria toxin revealed the presence of three distinct domains, representing the catalytic, translocation, and receptor-binding functions.

E. Exotoxins Enter Host Cells via Distinct Pathways

Although **A** domain translocation is one of the least understood aspects of the intoxication process of exotoxins, there are several general themes that are involved in translocation of the **A** domain across the cell membrane. One translocation mechanism uses a pH gradient within the endosome to stimulate protein conformational changes in the **B** domain,

making it competent to interact with the endocytic vesicle. After insertion into the endocytic membrane, the **B** domain generates a pore that is believed to be involved in the translocation of the **A** domain across the vesicle membrane in an unfolded form. After the translocation across the endocytic membrane, the **A** domain refolds to its native conformation. Subsequent to translocation of the **A** domain across the vesicle membrane, reduced glutathione may reduce the disulfide that connects the **A** domain with the **B** domain, and release the **A** domain into the cytoplasm. The potency and catalytic potential of exotoxins was demonstrated by the observation that the introduction of one molecule of the catalytic domain of diphtheria toxin into the intracellular cytoplasm was sufficient to inhibit host-cell physiology, resulting in cell death.

Other toxins, such as cholera toxin and exotoxin A of *Pseudomonas aeruginosa,* appear to use retrograde transport to enter the interior regions of the cell. Movement appears to occur through retrograde transport from the endosome to the Golgi apparatus and ultimately to the endoplasmic reticulum. Many exotoxins that are ultimately delivered to the endoplasmic reticulum possess a KDEL (Lys-Asp-Glu-Leu)-like retention signal sequence on their carboxyl terminus. Although the details for the actual transport pathway remain to be determined, studies with chimeric proteins have shown that the introduction of a KDEL retention sequence is sufficient to retrograde transport a protein, which is normally delivered only to the early endosome, into the endoplasmic reticulum. Thus, there is physiological precedence for the use of the KDEL sequence to retrograde transport exotoxins toward the endoplasmic reticulum. One of the basic questions concerning the intoxication process of these exotoxins is the actual mechanism of translocation and whether or not eukaryotic proteins assist in the translocation process.

F. Covalent Modification of Host-Cell Components by Exotoxins

Exotoxins use several unique mechanisms to covalently modify host-cell components. The major classes of reactions are the covalent addition of a chemical group to the target protein, the cleavage of a chemical group from a target protein, or the endoproteolytic cleavage of a peptide bond of the target protein.

The ADP-ribosylation of host proteins is the prototype mechanism of action of bacterial exotoxins. Numerous bacterial exotoxins catalyze the ADP-ribosylation of specific host proteins and elicit physiological changes. In the ADP-ribosylation reaction, exotoxins use the oxidized form of nicotinamide adenine dinucleotide (NAD) as the substrate, and transfer the ADP-ribose portion of NAD to a specific amino acid via an *N*-glycosidic linkage of ADP-ribose onto the host target protein. The specific type of amino acid that is ADP-ribosylated within the target protein varies with the specific exotoxin. ADP-ribosylation may either inactivate or stimulate the activity of the target protein. Diphtheria toxin ADP-ribosylates elongation factor-2 on a post-translationally modified histidine residue called diphthamide. ADP-ribosylated elongation factor-2 is unable to perform its translocation of nascent polypeptides in the ribosome, which results in the inhibition of protein synthesis and subsequent cell death. In contrast, cholera toxin ADP-ribosylates the Gsα component of a heterotrimeric G protein. ADP-ribosylated Gsα is locked in an active conformation, which results in the stimulation of adenylate cyclase and the subsequent elevation of intracellular cAMP. Likewise, deamidation of Gln63 of RhoA by *E. coli* cytotoxic necrotizing factor (CNF) results in a constitutively active RhoA protein. Note that although most host targets for exotoxins are proteins, Shiga toxin catalyzes the deadenylation of a specific adenine on 28*S* RNA.

Recall that each exotoxin modifies a specific host-cell component, which is responsible for the specific pathology elicited by that exotoxin. Although there are no absolute rules for the types of proteins targeted for covalent modification, the most frequent targets are nucleotide-binding proteins that are involved in signal-transduction pathways, including both the heterotrimeric G-proteins and the small-molecular-weight GTP-binding proteins of the Ras superfamily. It is not clear whether this class of host protein is targeted for modification due to the presence of a common structural motif, or to its critical role in host-cell metabolism.

G. Molecular and Structural Properties of Bacterial Exotoxins

Early biochemical studies provided significant advances in defining the structure–function properties of exotoxins, resolving many of the exotoxin mechanisms of action and developing the concept that exotoxins have **AB** organization. Molecular genetics and structural biology have extended the earlier studies and provided a more detailed understanding of the biochemical and molecular relationships among the exotoxins. The biochemical characterization of diphtheria toxin and exotoxin A (ETA) of *Pseudomonas aeruginosa* showed that these two exotoxins catalyzed kinetically identical reactions during the ADP-ribosylation of elongation factor-2. In addition, both diphtheria toxin and ETA were shown to possess an active site glutamic acid, which was subsequently shown to be a signature property of exotoxins that catalyze the ADP-ribosyltransferase reaction. These observations predicted that ADP-ribosylating exotoxins would possess considerable primary amino acid homology. Thus, the determination that the genes encoding diphtheria toxin and ETA shared little primary amino acid homology was unexpected. This paradox was resolved after the analysis of the three-dimensional structures of ETA and the heat-labile enterotoxin (LT) of *E. coli*, and subsequently confirmed with diphtheria toxin. The three-dimensional structures of ETA and LT showed little similarity in their respective receptor-binding domains and translocation domains; however, the catalytic domains of ETA and LT, which are composed of seven discontinuous regions of each protein, could be superimposed on each other despite possessing homology at only three of the 43 amino acids. One of the homologous amino acids in ETA and LT was the signature active site glutamic acid. This was a remarkable finding because ETA and LT ADP-ribosylate different host target proteins and possess different **AB** organization. A common theme has evolved for describing the structure–function properties of this family of bacterial exotoxins in which the ADP-ribosylating exotoxins possess a conserved three-dimensional structure in their active sites, despite the lack of primary amino acid homology.

These findings have provided a framework for the study of other classes of exotoxins produced by divergent groups of bacteria.

III. CONVERSION OF EXOTOXINS INTO TOXOIDS

A. Chemical Detoxification of Bacterial Exotoxins

Shortly after the determination that toxic components were associated with bacterial pathogens, several studies showed that cell extracts or cell cultures of a pathogen could be treated with chemical denaturants, such as formalin, to produce nontoxic immunogenic material that could prevent the disease associated with that pathogen. In the case of diphtheria toxin and tetanus toxin, chemical modification with formalin produced toxoids that were used as acellular vaccines in large-scale immunizations. This resulted in a remarkable decrease in the incidence of both diphtheria and tetanus within the populations that were immunized. In areas where these toxoids are not administered, diphtheria and tetanus remain clinically important diseases. In addition to formalin, other chemicals have been used to detoxify bacterial exotoxins, including glutaraldehyde and hydrogen peroxide. In contrast, the chemical toxoiding of other exotoxins, such as cholera toxin and pertussis toxin, has been more difficult because the treatment of these toxins with denaturants often results in a reduction of immunogenicity. Thus, there is a need to develop alternative strategies for eliminating the cytotoxicity of certain exotoxins without compromising their immunogenicity.

B. Genetic Detoxification of Bacterial Exotoxins

Developments in genetic engineering have provided an opportunity to produce recombinant forms of bacterial exotoxins that possess greatly reduced toxicity, but retain immunogenicity. The use of genetic engineering to develop a toxoid of pertussis toxin has been successful. The whole-cell pertussis vaccine is composed of a chemically treated prepara-

tion of *Bordetella pertussis,* which is effective in the elicitation of a protective immune response after mass immunization. However, the whole-cell pertussis vaccine is acutely reactive when administered to children. Pertussis toxin, a primary virulence determinant of *B. pertussis,* is an exotoxin that ADP-ribosylates the $Gi\alpha$ component of heterotrimeric G proteins and effectively uncouples signal transduction between the G protein-coupled receptor and the G protein. Genetically engineered forms of pertussis toxin have been produced that possess essentially no catalytic activity or cytotoxicity, but that maintain native conformation and elicit a protective immune response when used as an immunogen. These recombinant noncytotoxic forms of pertussis toxin have been engineered with multiple mutations in their active site, virtually eliminating the risk of reversion to a cytotoxic form. Similar strategies are being applied to other bacterial exotoxins with the goal of engineering acellular vaccine candidates.

IV. THERAPEUTIC APPLICATIONS OF EXOTOXINS

One of the most exciting areas of bacterial exotoxin research has been the development of strategies to use exotoxins in therapeutic disciplines. Some therapies use the native cytotoxic form of the exotoxin. Other therapies use either the A or B domain, which is conjugated to a heterologous binding component or to effector elements, respectively, to produce a chimeric molecule with directed properties.

Botulinum toxin and tetanus toxin (BT/TT) are each a single protein that is organized as an **AB** exotoxin. The amino terminus of BT/TT expresses endopeptidase activity and constitutes the **A** domain, whereas the **B** domain possesses neuronal cell-specific receptor-binding activity. The specific association of the **B** domain with neuronal cells is responsible for the clinical manifestation of these neurotoxins. BT/TT appear to enter neuronal cells by receptor-mediated endocytosis and to deliver the **A** domain to the cytosol, where it catalyzes the endoproteolytic cleavage of host proteins that are involved in vesicle fusion. Studies have shown that botulinum toxin can be introduced into the muscles surrounding the eye to temporarily reduce muscle spasms associated with several clinical disorders.

Diphtheria toxin has been used as a carrier to stimulate an immune response against several epitopes. One epitope is polyribitolphosphate, a component of the polysaccharide capsule of *Haemophilus influenzae* type b (Hib). Early attempts to elicit an effective immune response to purified Hib antigen resulted in the production of a T-cell-independent immune response that did not yield an effective memory. A noncatalytic mutant of diphtheria toxin, CRM197, has been used as a carrier for the Hib epitope. Immunization with the CRM197-Hib conjugate yielded a strong T-dependent immune response. Mass immunization with Hib conjugates has resulted in a dramatic reduction in the number of cases of Hib in the immunized population.

Due to their potency, the catalytic **A** domain of exotoxins have been used in the construction of chimeric immunotoxins that are designed to target cancer cells. Early studies used conjugates that were composed of the **A** domain of the diphtheria toxin coupled to an antibody that recognized a cell surface-specific antigen. The **A** chain of the diphtheria toxin was used in the first generation of immunotoxins because it had been shown to possess impressive cytotoxic potential when introduced into the cytosol of eukaryotic cells. It was estimated that the introduction of a single molecule of the **A** chain of diphtheria toxin into the cytosol was sufficient to kill that cell. In cell culture, these chimera have proven to be both potent and antigen specific. Ongoing research involves the determination of clinical situations for the use of these chimeras in a therapeutic arena.

The **B** component of anthrax toxin, protective antigen (PA) and a truncated, noncytotoxic form of one of its **A** components (LF) has recently been used to deliver epitopes into antigen presenting cells to elicit a cytotoxic lymphocyte (CTL) response. Anthrax toxin is a tripartite toxin composed of three nonassociated proteins. After binding to cells, PA is proteolytically processed and undergoes oligomerization to form a heptameric structure on the cell surface. Processed PA is able to bind either LF or edema factor (EF) and the **AB** complex undergoes receptor-mediated endocytosis in which acidification in the early endosome stimulates the translocation of the **A** do-

main into the cell cytosol. In the nontoxic anthrax delivery system, PA is added to antigen-presenting cells with a nontoxic LF–CTL epitope chimera used to deliver the epitope into the host cell for antigen presentation. One of the more attractive aspects of this CTL-epitope delivery system is that small amounts of PA are required to present antigen.

V. OTHERS CLASSES OF BACTERIAL TOXINS

A. Pore-Forming Toxins

The lack of a catalytic **A** domain differentiates the pore-forming toxins from exotoxins. Thus, the pathology associated with pore-forming toxins is due solely to the generation of a pore within the membrane of the host cell. Several bacterial pathogens produce pore-forming toxins, some of which are secreted by a type I secretion pathway. Unlike type II-secreted proteins, the amino terminus of type I-secreted proteins is not processed. Type I-secreted proteins possess a polyglycine signal sequence in the carboxyl terminus of the mature toxin. There are several classes of pore-forming toxins, including members of the hemolysin family of pore-forming toxins, the aerolysin family of pore-forming toxins, and the α-toxin of *Staphylococcus aureus*. Host-cell specificity differs among pore-forming toxins. The crystal structures of several of the pore-forming toxins have been determined. The molecular events generating a pore in the membrane of a host cell have been proposed for the aerolysin family of pore-forming toxins. Aerolysin is exported by *Aeromonas hydrophilia* as a monomeric molecule, which binds to the host cell. The monomer is proteolytically processed and subsequently undergoes oligomerization. The oligomerized complex is inserted into the membrane and generates a pore in the center of the complex, causing the release of the cytoplasmic components of the host cell.

B. Type III-Secreted Cytotoxins

The lack of a **B** domain differentiates the type III-secreted cytotoxins from exotoxins. Thus, the organization of the type III-secreted cytotoxins may be represented as **A** domains that are specific effector proteins. Type III-secreted cytotoxins are transported directly into the host cells by cell surface-bound bacteria. Type III secretion of bacterial proteins is a recently defined pathway for the delivery of proteins into the cytoplasm of host cells. Type III-secreted proteins were initially recognized by the fact that the secreted mature cytotoxins were unique to proteins secreted by either the type I or II secretion pathways, whereas the amino terminus of type III-secreted proteins is not processed nor is there a polyglycine motif in their carboxyl terminus. Although it is clear that a complete type III secretion apparatus is required for the transport of type III-secreted proteins into the host cytoplasm, the mechanism for the delivery of type III-secreted proteins across the host-cell membrane remains to be resolved. Numerous bacteria have been shown to possess type III secretion pathways, including members of the genera *Escherichia, Pseudomonas, Shigella, Salmonella,* and *Yersinia.* Cytotoxicity elicited by type III-secreted cytotoxins has an absolute requirement for the type III secretion apparatus of the bacterium, as purified forms of the cytotoxins are not toxic to host cells.

The **A** domains of type III-secreted cytotoxins catalyze several unique mechanisms of action, including the depolymerization of the actin cytoskeleton, phosphatase activity, ADP-ribosyltransferase activity, and the stimulation of apoptosis. Each type III secretion apparatus appears capable of delivering numerous type III-secreted proteins into the host cell.

C. Heat-Stable Enterotoxins

The inability of the heat-stable enterotoxins to enter the host cell or possess catalytic activity differentiates the heat-stable enterotoxins from exotoxins. Several genera of bacteria produce heat-stable enterotoxins, including *Escherichia* and *Yersinia.* The heat-stable enterotoxin a (STa) of *E. coli* is the prototype toxin of this group. *E. coli* secrets STa into the periplasm as a 72-amino-acid precursor in which three intramolecular disulfide bonds are formed and processed into a 53-amino-acid form. The 53-amino-acid form of STa is exported into the environment, where a second proteolytic cleavage results in the production of an 18- or 19-amino-acid mature STa molecule. The mature STa binds to a protein receptor on the surface of epithelial cells, which results in an increase in the intracellular concentrations of cGMP. The intracellular increase in cGMP results in a stimulation of chloride secretion and net fluid secretion, resulting in diarrhea.

D. Superantigens

The inability of superantigens to enter the host cell or possess catalytic activity differentiates the superantigens from exotoxins. Superantigens are soluble proteins of approximately 30 kDa that are secreted by bacteria that possess mitogenic properties. Superantigens are produced by both *Streptococcus* and *Staphylococcus.* The superantigens bind to a component of the major histocompatibility complex of T lymphocytes through an antigen-independent mechanism, which stimulates proliferation of a large subset of T lymphocytes.

See Also the Following Articles

Clostridia • Outer Membrane, Gram-Negative Bacteria • Transcriptional Regulation in Prokaryotes • Transduction

Bibliography

Collier, R. J. (1975). Diphtheria toxin: Mode of action and structure. *Bacteriol. Rev.* **39**(1), 54–85.

Michie, C. A., and Cohen, J. (1998). The clinical significance of T-cell superantigens. *Trends Microbiol.* **6**(2), 61–65.

Moss, J., and Vaughan, M. (1990). "ADP-Ribosylating Toxins and G proteins: Insights into Signal Transduction." *Am. Soc. Microbiol,* Washington, D.C.

Pastan, I., Chaudhary, V., and FitzGerald, D. J. (1992). Recombinant toxins as novel therapeutic agents. *Annu. Rev. Biochem.* **61**, 331–354.

Saelinger, C. B., and Morris, R. E. (1994). Uptake and processing of toxins by mammalian cells. *Meth. Enzymol.* **235**, 705–717.

Sandvig, K., Garred, O., Holm, P. K., and van Deurs, B. (1993). Endocytosis and intracellular transport of protein toxins. *Biochem. Soc. Trans.* **21**(Pt 3), 707–711.

Sixma, T. K., Pronk, S. E., Kalk, K. H., Wartna, E. S., van Zanten, B. A., Witholt, B., and Hol, W. G. (1991). Crystal structure of a cholera toxin-related heat-labile enterotoxin from *E. coli. Nature* **35**(6326), 371–377.

Extremophiles

Ricardo Cavicchioli and Torsten Thomas

University of New South Wales

GLOSSARY

alkaliphile An organism with optimal growth at pH values above 10.

Archaea One of the three domains (highest taxon) of life, evolutionarily distinct from the Bacteria and Eucarya (eukaryotes).

autotroph An organism able to use CO_2 as a sole source of carbon.

barophile An organism that lives optimally at high hydrostatic pressure.

chemolithotroph An organism obtaining its energy from the oxidation of inorganic compounds.

chemoorganotroph An organism obtaining its energy from the oxidation of organic compounds. Also referred to as a heterotroph.

extreme acidophile An organism with a pH optimum for growth at or below pH 3.

extremophile An organism that is isolated from an extreme environment and often requires the extreme condition for growth ("extreme" is anthropocentrically derived).

growth optimum The conditions (e.g., temperature and salinity) at which an organism grows fastest.

halophile An organism requiring at least 0.2 M salt for growth.

hyperthermophile An organism having a growth temperature optimum of 80°C or higher.

methanogen An archaeal microorganism that produces methane.

mixotroph An organism able to assimilate organic compounds as carbon sources and use inorganic compounds as electron donors.

oligotroph An organism with optimal growth in nutrient-limited conditions.

phototroph An organism that obtains energy from light.

phylogeny The ordering of species into higher taxa and the construction of evolutionary trees based on evolutionary relationships rather than on general resemblances.

prokaryote A cell or organism lacking a nucleus and other membrane-enclosed organelles, usually having its DNA in a single circular molecule. Loosely used to describe archaeal and bacterial microorganisms in contrast to eukaryotic microorganisms.

psychrophile An organism having a growth temperature optimum of 15°C or lower, and a maximum temperature of 20°C or lower.

EXTREMOPHILES are organisms that require extreme environments for growth. Although this is perhaps self-evident, what constitutes extreme? Extreme is a relative term, with the point of comparison being what is normal for humans. Extremophiles are therefore organisms that are "fond of" or "love" (-phile) environments such as high temperature, pH, pressure, or salt concentration; or low temperature, pH, nutrient concentration, or water availability. Extremophiles are also organisms that can tolerate other extreme conditions, including high levels of radiation or toxic compounds, or living conditions that humans consider unusual, such as living in rocks 1.5 km below the surface of Earth. In addition, extremophiles may be found in environments with a combination of extreme conditions, such as high temperature and high acidity or high pressure and low temperature.

I. INTRODUCTION

Most extremophiles are microorganisms. For example, the upper temperature limits for archaeal, bacterial, and eukaryotic microorganisms are 113°C, 95°C, and 62°C, respectively, in contrast to most metazoans (multicellular eukaryotes, e.g., animals and plants), which are unable to grow above temperatures of 50°C. This example of thermal adaptation highlights an important distinction among the different classes of microorganisms (i.e., Archaea can grow at extremely high temperatures in comparison to their eukaryotic counterparts), and underscores the fundamentally different evolutionary origins of the members of the three domains of life, Archaea, Bacteria, and Eucarya. Although Archaea (formerly Archaebacteria) and Bacteria are both loosely defined as prokaryotes, they are by no means more similar to each other than Archaea are to eukaryotes. For example, archaea have a number of archaeal-specific traits (glycerol-1-phosphate–lipid backbones and methanogenesis), in addition to sharing many bacterial (metabolism, biosynthesis, energy generation, transport, and nitrogen fixation) and eukaryotic (transcription, translation, and replication) features. Due to the fact that Archaea are often found in extreme environments, the term "extremophile" is often used synonymously with "archaea", and many of the extremophiles described here are members of the Archaea. It should be noted, however, that Archaea are also found in a broad range of nonextreme marine and soil environments.

The first use of the term "extremophile" appeared in 1974 in a paper by MacElroy (*Biosystems* 6: 74–75) entitled, "Some Comments on the Evolution of Extremophiles." In the 1990s, studies on extremophiles have progressed to the extent that the First International Congress on Extremophiles was convened in Portugal, June 2–6, 1996, and the scientific journal *Extremophiles* was established in February 1997. These developments in the field have arisen due to the isolation of extremophiles from environments previously considered impossible for sustaining biological life. As a result, the appreciation of microbial biodiversity has been reinvigorated and challenging new ideas about the origin and evolution of life on Earth have been generated. In addition, the novel cellular components and pathways identified in extremophiles have provided a burgeoning new biotechnology industry.

In recognition of the importance of extremophiles to microbiology, the following sections describe the habitats, biochemistry, and physiology of a diverse range of extremophiles, followed by a concluding section on the biotechnological applications of extremophiles and their products.

II. HYPERTHERMOPHILES

A. Defining Temperature Classes of Microorganisms

Microorganisms can generally be separated into four groups with regard to their temperature optima for growth. In order of increasing temperature, they are psychrophiles (optimum of 15°C, maximum < 20°C), mesophiles (optimum between 20 and 45°C), thermophiles (optimum between 45 and 80°C), and hyperthermophiles (optimum of 80°C or higher). On the whole, these definitions are generally accepted terms; the reader should be aware, however, that some terms may have alternative meanings in specific fields. For example, the upper temperature for the growth of yeast is about 48°C, and the majority of yeasts can grow below 20°C. As a result, a thermophilic yeast is defined by its inability to grow below 20°C, but with no restrictions placed on the maximum temperature for growth.

B. Habitats and Microorganisms

Hyperthermophiles have been isolated from a range of natural geothermally heated environments (Table I). The terrestrial environments tend to be acidic (as low as pH 0.5) with low salinity (0.1–0.5% salt), whereas the marine systems are saline (3% salt) and generally less acidic (pH 5–8.5). Artificial

Geography	Type of habitat	Location
Terrestrial	Hot springs	U.S., New Zealand, Iceland, Japan, Mediterranean, Indonesia, Central America, Central Africa
Subterranean	Oil reservoir (3500 m)	North Sea, North Alaska
Submarine	Shallow depth (beaches)	Italy, Sao Michel (the Azores), Djibouti (Africa)
	Sea mounts	Tahiti
	Moderate depth (120 m)	mid-Atlantic Ridge (north of Iceland)
	Deep sea (>1000 m)	Guaymas Basin (2000 m) and East Pacific Rise (2500–2700 m) (both near Mexico), mid-Atlantic Ridge (3600–3700 m), mid-Okinawa Trough (1400 m), Galapagos Rift

sources of isolates include coal refuse piles and geothermal power plants.

The terrestrially based, volcanic systems include hot springs such as solfataras, which are sulfur-rich and generally acidic, or others that are boiling at a neutral pH, or that are iron-rich. Most of the heated soils and water contain elemental sulfur and sulfides, and most isolates metabolize sulfur. A by-product of oxidation tends to be sulfuric acid and as a result many of the hot springs are extremely acidic. Hyperthermophilic archaea have also been isolated from oil from geothermally heated oil reserves present in Jurassic sandstone and limestone 3500 m below the bed of the North Sea and below the Alaskan North Slope permafrost soil. As a by-product of their metabolism during oil extraction, they produce hydrogen sulfide; this condition is referred to as reservoir souring.

Numerous hyperthermophiles have been isolated from submarine hydrothermal systems, including hot springs, sediments, sea mounts (submarine volcanoes), fumaroles (steam vents with temperatures up to 150–500°C), and deep-sea vents. The hot deep-sea vents are often referred to as "black smokers" due to the thick plume of black material that forms from mineral-rich fluids at temperatures up to 400°C, precipitating on mixing with the cold (1–5°C) seawater. The submarine systems are a rich source of microbial and higher eukaryotic life that spans the temperature ranges suitable for hyperthermophiles through psychrophiles. Due to the extreme depth of some of the vents (3500 m), organisms are also exposed to extreme pressure, and as a result they tend to be barophilic or barotolerant (see Section V).

The tube worms, giant clams, and mussels that inhabit the vents are dependent on the activity of the chemolithotrophic microorganisms to fix CO_2 and use a broad range of inorganic energy sources emitted from the vents. The first eukaryotic organism living at very high temperatures was identified at the Axial Summit Caldera, west of Mexico. The Pompeii worm (*Alvinella pompejana*) was observed living in a hydrothermal system in which the posterior of the worm was in 80°C water. Interestingly, it is speculated that the ability of the worm to survive the heat may be due to the presence of microorganisms that cover its exterior and secrete heat-stable enzymes.

Most isolated hyperthermophiles are members of Archaea, and it is this characteristic that is often associated with this class of microorganism. However, it was the isolation of *Thermus aquaticus,* a bacterium, from a hot spring in Yellowstone National Park by Thomas Brock in 1969 that led to the discoveries of hyperthermophilic archaea. In 1972, Brock described *Sulfolobus,* an obligately aerobic archaeon that is able to oxidize H_2S or S^0 to H_2SO_4 and fix CO_2 as a carbon source. The hot springs that *Sulfolobus* spp. thrive in are typically as hot as 90°C, with an acidity of pH 1–5. Some hyperthermophiles have been found in unique locations, for example, *Methanothermus* spp., which have only been isolated from one solfataric area of Iceland. In contrast, *Pyrococcus* and *Thermococcus* spp. have been found in both submarine systems and subterranean oil reserves. Of all the isolates to date, *Pyrolobus fumarii* has the highest maximum temperature (113°C). Furthermore, it is restricted to temperatures above 90°C. Using a microscope slide suspended in the outflow from a black smoker at 125–140°C, microbial films have been detected on the glass surfaces. Although organisms capable of growth at such high temperatures have not been isolated, this is good evidence that microorganisms are capable of growth in this extreme temperature range.

C. Biochemistry and Physiology of Adaptation

A broad diversity of hyperthermophiles with varied morphologies, metabolisms, and pH requirements have been isolated, and the characteristics of a number of these are shown in Table II. Although organisms such as *Sulfolobus* spp. are aerobic chemoorganotrophs, most hyperthermophiles are anaerobic and many are chemolithotrophs. The extent of metabolic diversity in the hyperthermophiles is exemplified by the strict requirement of *Methanococcus jan-* *naschii* for CO_2 and H_2 only, in comparison to the broad spectrum of substrates used as electron donors by the sulfate-reducing archaeon, *Archaeoglobus fulgidus,* including H_2, formate, lactate, carbohydrates, starch, proteins, cell homogenates, and components of crude oil. This indicates that there are no particular carbon-use or energy-generation pathways that are exclusively linked to growth at high temperatures.

For any microorganism, lipids, nucleic acids, and proteins are all generally susceptible to heat, and there is, in fact, no single factor that enables all hyperthermophiles to grow at high temperatures.

1. Membrane Lipids and Cell Walls

Archaeal lipids all contain ether-linkages (as opposed to the ester-linkages in most bacteria), which provide resistance to hydrolysis at high temperature. Some hyperthermophilic archaea contain membrane-spanning, tetra-ether lipids that provide a monolayer type of organization that gives the membranes a high degree of rigidity and may confer thermal stability. The bacterium *Thermotoga maritima* also contains a novel ester lipid, which may increase stability at

TABLE II
Characteristics of Selected Hyperthermophiles[a]

Genus	Minimum temperature	Optimum temperature	Maximum temperature	DNA mol % G + C	pH Optimum	Aerobe (+) Anaerobe (−)	Energy yielding reactions	Morphology
Submarine isolates								
Aquifex (bacteria)	67	85	95	40	6	+	H_2 reduction	Rods
Archaeoglobus	64	83	95	46	7	−	SO_4 reduction to H_2S	Cocci
Methanococcus	45	88	91	31	6	−	$H_2 + CO_2$ reduction to CH_4	Irregular cocci
Methanopyrus	85	100	110	60	6.5	−	$H_2 + CO_2$ reduction to CH_4	Rods
Pyrobaculum	60	88	96	56	6	+/−	Organic/inorganic compounds + NO_3^- reduction	Rod
Pyrolobus	90	105	113	53	5.5	+/−	NO_3^- reduction	Lobed cocci
Thermodiscus	75	90	98	49	5.5	−	Organic compound + S^0 reduction	Discs
Thermotoga (bacteria)	55	80	90	46	7	−	Organic compounds	Sheathed rods
Terrestrial isolates								
Acidianus	65	85–90	95	31	2	+/−	Aerobic oxidation & anaerobic reduction of S^0	Sphere
Desulfurococcus	70	85	95	51	6	−	Organic compound + S^0 reduction	Sphere
Sulfolobus	55	75–85	87	37	2–3	+	Organic compound & H_2S or S^0 oxidation	Lobed sphere
Thermoproteus	60	88	96	56	6	−	Reduction of S^0	Rod

[a] All are archaeal species unless otherwise indicated. Data from Madigan, M. T., Martinko, J. M., and Parker, J. (1997). *Brock: Biology of Microorganisms,* 8th Ed. Prentice Hall; Stetter, K. O. (1998). Hyperthermophiles: Isolation, classification and properties. *In* "Extremophiles: Microbial Life in Extreme Environments" (Horikoshi, K., and Grant, W. D., Eds.), Wiley series in Ecological and Applied Microbiology, Wiley-Liss.

high growth temperatures. However, the archaeon *Methanopyrus kandleri,* which can grow at up to 110°C, contains unsaturated diether lipids that resemble terpenoids, and it is unclear how this may affect the thermal resistance of the cell. Lipid composition also varies with the growth temperature of individual organisms. For example, in *Methanococcus jannaschii* at 45°C, 80% of the lipid content is one lipid (archaeol), whereas at 75°C, two lipids (caldarchaeol and macrocyclic archaeol) account for 80% of the total core.

Most archaeal species possess a paracrystalline surface layer (S-layer) consisting of protein or glycoprotein. It is likely that the S-layer functions as an external protective barrier. In *Pyrodictium* spp., the highly irregularly shaped, flagellated cells are interconnected by extracellular glycoprotein tubules that remain stable at 140°C.

2. Nucleic Acids

The thermal resistance of DNA could conceivably be improved by maintaining a high mol % G + C content; however, many hyperthermophiles have between 30 and 40% (Table II), in comparison to, for example, the mesophile *Escherichia coli,* which has 50%. *Acidianus infernus* has a G + C content of 31%, which would rapidly lead to the melting of double-stranded DNA at its optimum growth temperature of 90°C. Histone-like proteins that bind DNA have been identified in archaeal hyperthermophiles. It is likely that the DNA is protected by the histones and that this enables processes such as open-complex formation during transcription to occur without subsequent DNA melting. In addition, hyperthermophiles contain reverse gyrase, a type 1 DNA topoisomerase that causes positive supercoiling and therefore may stabilize the DNA.

3. Proteins and Solutes

Heat-shock proteins, including chaperones, are likely to be important for stabilizing and refolding proteins as they begin to denature. When *Pyrodictium occultum* is heat-stressed at 108°C, 80% of total protein accumulated is a single chaperonin, termed the "thermosome," the cells' protein-folding machine. In addition to the differential expression of certain genes throughout the growth temperature range of

the organism, proteins from hyperthermophiles are inherently more stable than those from thermophiles, mesophiles, or psychrophiles. Higher stability is a result of increased rigidity (decreased flexibility) of the protein. Certain structural properties favor a more rigid protein (see psychrophilic proteins in Section IV.B.2), including a higher degree of structure in hydrophobic cores, an increased number of hydrogen bonds and salt bridges, and a higher proportion of thermophilic amino acids (e.g., proline residues that have fewer degrees of freedom). As a result, the proteins from hyperthermophiles are extremely heat stable. For example, proteases from *Pyrococcus furiosus* have half-lives of > 60 hr at 95°C and an amylase from *Pyrococcus woesii* is active at 130°C.

Protein stability may also be assisted by the accumulation of intracellular potassium and solutes, such as 2,3-diphosophoglycerate (cDPG). In *Methanopyrus kandleri* and *Methanothermus fervidus,* there is evidence that potassium is required for enzyme activity at high temperature and the potassium salt of cDPG acts as a thermal stabilizer.

D. Evolution

Earth is about 4.6 billion years old and life is believed to have evolved on Earth around 3.6–4 billion years ago. The atmosphere of early Earth was devoid of oxygen and contained gaseous H_2O, CH_4, CO_2, N_2, NH_3, HCN, trace amounts of CO and H_2, and large quantities of H_2S and FeS. For the first 0.5 billion years, the surface of the planet was probably devoid of water because the temperature was higher than 100°C. Subsequently, the planet cooled and water liquefied. The high temperatures imply that life evolving in these conditions must have possessed thermophilic properties. Due to the high temperature and the available carbon and energy substrates, a microorganism such as H_2-oxidizing sulfur-reducing *Methanopyrus kandleri* would conceivably thrive.

Studying the phylogenetic relationship of extant (living) microorganisms by the analysis of 16S-ribosomal ribonucleic acid (16S-rRNA) sequences, reveals that the hyperthermophilic Archaea and Bacteria have short evolutionary branches that occur near the base of the tree of life. Short branches indicate a low rate of evolution and deep branches reflect a close relationship with primordial life forms. This suggests that hyperthermophiles living today may resemble some of the earliest forms of life on Earth.

There is some evidence that microbial life may have existed (or still exists) on Mars and other terrestrial bodies (e.g., on Europa, one of Jupiter's moons). In association with the characteristics of hyperthermophiles mentioned, this has led to the consideration that life on Earth may have originated from the introduction of extraterrestrial life. A possible scenario involves a meteor carrying microbial life similar to hyperthermophilic methanogens plunging into the ocean billions of years ago and initiating the use of inorganic matter to generate biological matter and the subsequent evolution of extant species. These possibilities are driving new research endeavors to discover extraterrestrial life (see Section IV.A on psychrophiles and lakes on Europa).

III. EXTREME ACIDOPHILES

A. Habitats and Microorganisms

An extreme acidophile has a pH optimum for growth at or below pH 3.0. This definition excludes microorganisms that are tolerant to pH below 3, but that have pH optima closer to neutrality, including many fungi, yeast, and bacteria (e.g., the ulcer- and gastric-cancer causing gut bacterium *Helicobacter pylori*).

1. Natural Environments

Extremely acidic environments occur naturally and artificially. The hyperthermophilic extreme acidophiles *Sulfolobus, Sulfurococcus, Desulfurolobus*, and *Acidianus* produce sulfuric acid in solfataras in Yellowstone National Park from the oxidation of elemental sulfur or sulfidic ores. Other members of the Archaea found in these hot environments include species of *Metallosphaera*, which oxidize sulfidic ores, and *Stygiolobus* spp., which reduce elemental sulfur. The novel, cell-wall-less archaeon, *Thermoplasma volcanium*, which grows optimally at pH 2 and 55°C, has also been isolated from sulfotaric fields around the world. The bacterium *Thiobacillus*

caldus has been isolated in hot acidic soils. *Bacillus acidocaldarius, Acidimicrobium ferrooxidans,* and *Sulfobacillus* spp. have also been isolated from warm springs and hot spring runoff.

The most extreme acidophiles known are species of Archaea, *Picrophilus oshimae* and *Picrophilus torridus*, which were isolated from two solfataric locations in northern Japan. One of the locations, which contained both organisms, is a dry soil, heated by solfataric gases to 55°C and with a pH of less than 0.5. These remarkable species have aerobic heterotrophic growth with a temperature optimum of 60°C and a pH optimum of 0.7 (i.e., growth in 1.2 M sulfuric acid).

In addition to Archaea, the phototrophic red alga *Galdieria sulphuraria* (*Cynadium caldarium*), isolated in cooler streams and springs in Yellowstone National Park, has optimum growth at pH of 2–3 and 45°C, and is able to grow at pH values around 0. The green algae *Dunaliella acidophila* is also adapted to a narrow pH range from 0 to 3.

2. Artificial Environments

The majority of extremely acid environments are associated with the mining of metals and coal. The microbial processes that produce the environments are a result of dissimilatory oxidation of sulfide minerals, including iron, copper, lead, and zinc sulfides. This process can be written as $Me^{2+}S^{2-}$ (insoluble metal complex) $\rightarrow Me^{2+} + SO_4^{2-}$; where Me represents a cationic metal. As a result of the extremely low pH in these environments, and due to the geochemistry of the mining sites, cationic metals (e.g., $Fe^{2+}, Zn^{2+}, Cu^{2+}$, and Al^{2+}) and metalloid elements (e.g., arsenic) are solubilized; this process is referred to as microbial ore leaching.

Most mining sites tend to have low levels of organic compounds, and as a result chemolithoautotrophs, such as the bacteria *Thiobacillus ferrooxidans, Thiobacillus thiooxidans, Leptospirillum ferrooxidans,* and *Leptospirillum thermoferrooxidans*, are prolific. In addition, mixotrophic *Thiobacillus cuprinus* and heterotrophic *Acidiphilium* spp. have been isolated from acidic coal refuse and mine drainage. Thermophilic acidophilic bacterial species include *Thiobacillus caldus* from coal refuse, *Acidimicrobium ferrooxidans* from copper-leaching dumps, and *Sulfobacillus* spp.

from coal refuse and mine water. The archaeal microorganism *Thermoplasma acidophilum* is frequently isolated from coal refuse piles. Coal refuse contains coal, pyrite (an iron sulfide), and organic material extracted from coal. As a result of spontaneous combustion, the refuse piles are self-heating and provide the thermophilic environment necessary for sustaining *Thermoplasma* and other thermophilic microorganisms.

In illuminated regions (e.g., mining outflows and tailings dams), phototrophic algae, including *Euglena, Chlorella, Chlamydomonas, Ulothrix,* and *Klebsormidium* species, have been isolated. Other eukaryotes include species of yeast (*Rhodotorula, Candida,* and *Cryptococcus*), filamentous fungi (*Acontium, Trichosporon,* and *Caphalosporium*) and protozoa (*Eutreptial, Bodo, Cinetochilium,* and *Vahlkampfia*).

B. Biochemistry and Physiology of Adaptation

Acidophiles (and alkaliphiles; see Section VII) keep their internal pH close to neutral. Most extreme acidophiles maintain an intracellular pH above 6, and even *Picrophilus* maintains an internal pH of 4.6 when the outside pH is 0.5–4. As a result, extreme acidophiles have a large chemical proton gradient across the membrane. Proton movement into the cell is minimized by an intracellular net positive charge and as a result cells have a positive inside-membrane potential. This is caused by amino acid side chains of proteins and phosphorylated groups of nucleic acids and metabolic intermediates, acting as titratable groups. In effect, the low intracellular pH leads to protonation of titratable groups and produces a net intracellular positive charge. In addition to this passive effect, some acidophiles (e.g., *Bacillus coagulans*) produce an active proton-diffusion potential that is sensitive to agents that disrupt the membrane potential, such as ionophores.

The ability of lipids from the archaeon *Picrophilus oshimae* to form vesicles is lost when the pH is neutral, thus indicating that the membrane lipids are adapted for activity at low pH to minimize proton permeability. In *Dunaliella acidophila*, the surface charge and inside membrane potential are both positive, which is expected to reduce influx of protons into the cell. In addition, it overexpresses a potent cytoplasmic membrane H^+-ATPase to facilitate proton efflux from the cell.

The pH to which a protein is exposed affects the dissociation of functional groups in the protein and may be affected by salt and solute concentrations. Few periplasmic surface-exposed or -secreted proteins from extreme acidophiles have been studied to identify the structural features important for activity and stability. In *Thiobacillus ferrooxidans,* the acid stability of rusticyanin (acid-stable electron carrier) has been attributed to a high degree of inherent secondary structure and the hydrophobic environment in which it is located in the cell. A relatively low number of positive charges have been linked to the acid stability of secreted proteins (thermopsin, a protease from *Sulfolobus acidocaldarius,* and an α-amylase from *Alicyclobacillus acidocaldarius*), by minimizing electrostatic repulsion and protein unfolding.

IV. PSYCHROPHILES

A. Habitats and Microorganisms

Over 80% of the total biosphere of Earth is at a temperature permanently below 5°C, and it is therefore not surprising that a large number and variety of organisms have adapted to cold environments. These natural environments include cold soils; water (fresh and saline, still and flowing); in and on ice in polar or alpine regions; polar and alpine lakes, and sediments; caves; plants, and cold-blooded animals (e.g., Antarctic fish). Artificial sources include many refrigerated appliances and equipment.

Organisms thriving in low-temperature environments (0°C or close to 0°C) are commonly classified as psychrophilic or psychrotolerant. Psychrophiles (cold-loving) grow fastest at a temperature of 15°C or lower, and are unable to grow over 20°C. Psychrotolerant (also termed psychrotrophic) organisms grow well at temperatures close to the freezing point of water; however, their fastest rates of growth are above 20°C. Psychrophilic and psychrotolerant microorganisms include bacteria, archaea, yeast, fungi, protozoa, and microalgae.

It is generally found that psychrophiles predominate in permanently cold, stable environments that have good sources of nutrition (e.g., old, consolidated forms of sea ice exposed to algal blooms). It is likely that the permanency of the cold in stable environments obviates the need for cells to be able to grow at higher temperatures. In addition, at low temperature the affinity of uptake and transport systems decrease, and as a result, psychrophiles tend to be found in environments that are rich in organic substrates, thus providing compensation for less effective uptake and transport systems.

1. Natural Environments

Psychrophiles have been isolated from permanently cold, deep-ocean waters, as well as from ocean sediments as deep as 500 m below the ocean floor. As a consequence of the pressure in these environments, many of these psychrophiles are also barotolerant or barophiles (see Section V).

A unique source of cold-adapted microorganisms are lakes in the Vestfold Hills region in Antarctica. The Vestfold Hills lakes are only about 10,000 years old and differ in their salinity and ionic strength, oxygen content, depth, and surface ice coverage. As a result, they have proven to be a rich resource of diverse and unusual microorganisms, including a cell-wall-less Spirochaete, a coiled or "C-shape" bacterium, and the only known free-living, psychrophilic archaeal species. The archaeal species include the methanogens *Methanococcoides burtonii* and *Methangenium frigidum*, and the extreme halophile *Halobacterium lacusprofundi*. The only other low-temperature-adapted archaeal isolate that has been studied is the symbiont *Cenarchaeum symbiosum*, which was isolated from a marine sponge off the Californian coast. It is perhaps surprising that more low-temperature-adapted Archaea have not been isolated throughout the world, as 16S-rRNA analysis of numerous aquatic and soil samples have indicated the prevalence of Archaea in these cold habitats.

Gram-negative bacteria, including members of the genera *Pseudomonas, Achromobacter, Flavobacterium, Alcaligenes, Cytophaga, Aeromonas, Vibrio, Serratia, Escherichia, Proteus,* and *Psychrobacter* are more frequently found in cold environments than are gram-positive bacteria (e.g., *Arthrobacter, Bacillus,*

and *Micrococcus*). Psychrophilic yeast are of the genera *Candida* and *Torulopsis*, and psychrotolerant members are mostly of the genera *Candida, Cryptococcus, Rhodotorula, Torulopsis, Hanseniaspora,* and *Saccharomyces*. Low-temperature-adapted fungi and molds include isolates of the genera *Penicillium, Cladosporium, Phoma,* and *Aspergillus*. The most common snow alga is *Chlamydomonas nivalis,* which produces bright red spores, marking its location clearly against the white snow background.

A remarkable, and as yet unstudied, low-temperature environment has been discovered in Antarctica. Lake Vostok is a subglacial lake found about one kilometer below Vostok Station, East Antarctica. More than sixty smaller subglacial lakes also exist in central regions of the Antarctic Ice Sheet. Due to the isolation of the lakes, they are likely to contain many novel microorganisms, some of which could be expected to have developed along a separate evolutionary path from that of currently known life. The exploration of Lake Vostoc is being considered using specialized robots based on thermal-probe technology for ice penetration, and submersible technology for lake exploration. Interestingly, the ocean on Europa (a moon of Jupiter) is also located below a kilometer-thick covering of ice. The technology developed for Lake Vostoc will be a model for future exploration (in 2003–2018) of the ocean on Europa and provides the unprecedented potential to identify extraterrestrial life in an aquatic environment.

2. Artificial Environments

The abundance of microorganisms in cold environments was realized as early as 1887, when the first low-temperature-adapted bacterium was probably isolated from a preserved fish stored at 0°C. Since then, numerous psychrotolerant organisms have been found in artificial habitats and are frequently responsible for the spoilage of food. Members of the genera *Pseudomonas, Acinetobacter, Alcaligenes, Chromobacterium,* and *Flavobacterium* are often associated with spoilage of dairy products, whereas *Lactobacillus viridescens* and *Brochothrix thermospacta* contaminate meat products. Psychrotolerant bacterial pathogens include *Yersinia enterocolitica, Clostridium botulinum,* and certain *Aeromonas* strains.

B. Biochemistry and Physiology of Adaptation

Unicellular organisms are unable to insulate themselves against low temperatures and, as a consequence, psychrophilic and psychrotolerant microorganisms need to adapt the structures of their cellular components.

1. Membranes

It is well documented that microorganisms adjust the fatty acid composition of their membrane phospholipids in response to changes in the growth temperature. Normal cell function requires membrane lipids that are largely fluid. As temperature is decreased, the fatty acids chains in membrane bilayers undergo a change of state from a fluid disordered state to a more ordered crystalline array of fatty acid chains. In order to adapt to low temperature, microorganisms decrease the transition temperature of the disordered-to-ordered state by altering the fatty acyl composition. This alteration consists of one or a combination of the following changes to the membrane lipids: an increase in unsaturation, a decrease in average chain length, an increase in methyl branching, an increase in the ratio of *anteiso*-branching compared to *iso*-branching, and an isomeric alteration of acyl chains in *sn-1* and *sn-2* positions. The most common alterations occur in fatty acid saturation and chain length. Changes in the amount and type of methyl-branching occur mostly in gram-positive bacteria. Most bacteria only have monosaturated fatty acids; however a notable exception was found in some marine psychrophilic bacteria that increase membrane fluidity at low temperatures by incorporating polyunsaturated fatty acids into their membranes.

Alterations in the membrane lipid composition can be mediated in a rapid fashion through the increase of unsaturation catalyzed by desaturases. Changes in fatty acyl chain length and the amount and type of methyl branching, however, require *de novo* fatty acid synthesis and are therefore much slower processes.

2. Enzymes

The thermodynamic problems associated with enzymes in an environment of low kinetic energy (i.e., low temperature) relate to the lack of sufficient energy to achieve the activation state for catalysis. Low-temperature adaptation of proteins therefore lowers the free energy of the activated state by decreasing the enthalpy-driven interactions necessary for activation. The necessity for weaker interactions leads to a less rigid, more flexible protein structure, or parts thereof. Consequently, factors that confer a more "loose" or flexible structure are expected to be important for cold-adapted proteins. Based on the comparison of proteins from low-temperature-adapted Bacteria, Archaea, and Eukarya with those from mesophiles to hyperthermophiles, including comparisons of three-dimensional structures, a number of structural differences have been identified; including the reduction of the number of salt bridges, the reduction of aromatic interactions, the reduction of hydrophobic clustering, the reduction of proline content, the addition of loop structures, and an increase in solvent interaction. It should be noted that no cold-adapted protein has been studied that exhibits all of these features. This highlights the importance of the molecular context of the changes to stability, activity, or both. As a general rule, however, enzymes from psychrophilic organisms have reduced thermostability and a lower apparent temperature optima for their activity when compared to their mesophilic or thermophilic counterparts.

3. Cold Shock and Cold Acclimation

The response to a rapid decrease in temperature (cold shock) has been well studied in mesophilic laboratory microorganisms, including *Escherichia coli*, *Bacillus subtilis*, and *Saccharomyces cerevisiae*. In response to cold shock, the pattern of gene expression is altered (cold-shock response). In bacteria, a class of small (7–8 kDa) acidic proteins are transiently induced. These cold-shock proteins (CSPs) have been well characterized and shown, in some cases, to function as transcriptional enhancers and RNA-binding proteins. In *S. cerevisiae*, TIP1 (temperature-shock-inducible protein 1) is a major cold-shock protein; it is targeted to the outside of the plasma membrane and appears to be heavily glycosylated with *O*-mannose, therefore invoking a role for TIP1 in membrane protection during low-temperature adaptation. Following a cold shock, cells resume growth, albeit at a lower growth rate, indicating that the cold-shock response is an adaptive response aimed at maintaining growth, rather than a stress response aimed only at cell survival.

In general, the process of protein synthesis is temperature sensitive. As a result, psychrophilic microorganisms must have specially adapted ribosomes and accessory factors (e.g., initiation and elongation factors). In addition, psychrophilic and psychrotolerant microorganisnms (particularly those that are food-borne) may be exposed to sudden changes in their environmental temperature. A cold-shock response has been demonstrated in a number of psychrophilic and psychrotolerant organisms, including *Trichosporon pullulans, Bacillus psychrophilus, Aquaspirrillum arcticum,* and *Arthrobacter globiformis*; and homologs of a major *E. coli* cold-shock protein (CspA) have been identified in a number of these. The extent of the response (i.e., the number of proteins induced) in these microorganisms is dependent on the magnitude of the temperature shift. Some of the CSPs are also cold-acclimation proteins (proteins expressed continuously during growth at low temperature), suggesting that this class of CSP is important for cold-shock adaptation and growth maintenance at low temperatures.

V. BAROPHILES

A. Habitats and Microorganisms

Barophiles ("weight lovers") are organisms that thrive in high-pressure habitats. An etymologically more accurate term is "piezophiles" ("pressure lovers"); however, it is less frequently used in the literature.

There is apparent confusion and contradiction in the literature concerning the defining traits of pressure-adapted organisms. This mainly stems from the complicating effects of temperature on growth rate. The effect of both temperature (T) and pressure (P) on the growth rate (k) of different organisms has been thoroughly investigated using "PTk-diagrams." With all other conditions held constant, there is a unique pressure, P_{kmax}, and temperature, T_{kmax}, at which the growth rate of an individual microorganism is maximal (k_{max}). These values have been used to define barophiles with $P_{kmax} > 0.1$ MPa and extreme

barophiles with $P_{kmax} \gg 0.1$ MPa. In addition, the values for T_{kmax} have been used to delineate psychro-, meso-, or thermo- (extreme) barophiles. Other terms used are obligate (extreme) barophiles, for organisms that cannot grow at atmospheric pressure (0.1 MPa), irrespective of temperature, and barotolerants, for those that grow best under atmospheric pressure but that can also grow at up to 40 MPa.

Barophiles were first isolated from the deep sea (deeper than 1000 m), and this environment still represents the most thoroughly studied habitat. Other high-pressure habitats include the deep-Earth and the deep-sediment layers below the ocean floor. The deep sea is a cold ($< 5°C$), dark, oligotrophic (low in nutrients) environment with pressure as high as 110 MPa, as is found in the Mariana Trench (almost 11,000 m). In contrast, fumarole and black-smoker hydrothermal vents, produced by the extrusion of hot subsurface water, represent a hyperthermal environment with high metabolic activity (see hyperthermophiles in Section II). The microbiota of Atlantic and Pacific vents are remarkably similar, indicating that the (hyper-) thermobarophilic microorganisms are able to survive for long periods in cold waters, thus facilitating their effective dispersal throughout the oceans. Supporting this, these microorganisms have been isolated from ocean waters far removed from hydrothermal vents.

Psychrophilic deep-sea (extreme) barophilic isolates predominantly belong to five genera of the γ-Proteobacteria–*Photobacterium, Shewanella, Colwellia, Moritella,* and a new group containing the strain CNPT3. One *Bacillus* species (strain DSK25) has also been isolated. Hyperthermobarophilic archaeal isolates include *Pyrococcus* spp. It is noteworthy that the difficulty in isolating microorganisms from the deep sea, and the specific enrichment and culturing techniques that are used, are likely to produce a bias in the types of barophiles that are isolated; a complete analysis of the phylogenetic distribution of barophilic microorganisms is yet to be attempted.

B. Biochemistry and Physiology of Adaptation

1. General Physiological Adaptations

Membrane lipids from barophilic bacteria have been well studied. In response to high pressure, the relative amount of monounsaturation and polyunsaturation in the membrane is increased. The increase in unsaturation produces a more fluid membrane and counteracts the effects of the increase in viscosity caused by high pressure. This response is analogous to adaptations caused by temperature reduction (see Section IV). Note, however, that an increase in unsaturation was not observed in two extreme barophiles that have been studied, thus indicating that alternative mechanisms of adaptation exist.

An interesting observation for many barophiles is that they are extremely sensitive to UV light and need to be grown in dark or light-reduced environments. This adaptation is not unexpected, considering the darkness that prevails in the deep sea.

There are presently no studies on barophilic microorganisms investigating the adaptation of molecular processes such as chromosomal replication, cell division, transcription, or translation. Studies of *E. coli*, however, have revealed that when cells are grown at their upper pressure limit, DNA synthesis is completely inhibited, protein synthesis is slowed down, and mRNA synthesis and decay appears to be unaffected. This demonstrates that specific cellular processes are affected by pressure, and therefore barophiles are likely to have adapted mechanisms to compensate. Pressure may stabilize proteins and retard thermal denaturation. In support of this, thermal inactivation of DNA polymerases from the hyperthermophilic microorganisms *Pyrococcus furiosus*, *Pyrococcus* strain ES4, and *Thermus aquaticus* is reduced by hydrostatic pressure.

2. Gene and Protein Expression

Pressure-regulated gene and protein expression has been observed and investigated in deep-sea bacteria. *Photobacterium* sp. SS9 expresses two outer-membrane proteins (porins) under different pressures, the OmpH protein at 28 MPa and OmpL at 0.1 MPa. The genes are regulated by a homolog of the *toxRS* system from *Vibrio cholerae*. ToxR and ToxS are cytoplasmic membrane proteins that are thought to be pressure sensors controlled by membrane fluidity.

A pressure-regulated operon has been identified in the barophilic bacterium DB6705. A complex promoter region, which is controlled by a variety of regulatory proteins expressed under different pressure conditions, was identified upstream of three open reading frames (ORFs). The function of the first two ORFs is unknown, while the third ORF encodes the CydD protein. CydD is required for the assembly of the cytochrome-bd complex in the aerobic respiratory chain. The membrane location of this protein highlights the apparent importance of membrane components in high-pressure adaptation.

VI. HALOPHILES

A. Habitats and Microorganisms

The first recorded observation of organisms adapted to high salt concentrations (halophiles and halotolerants) probably dates to 2500 BC, when the Chinese noted a red coloration of saturated salterns. What they detected was most likely a bloom of extreme halophilic Archaea that possess red or orange C_{50} carotenoides.

The definition of a hypersaline environment is one that possesses a salt concentration greater than that of seawater (3.5% w/v). For water-containing environments, the salt composition depends greatly on the historical development of the habitat, and the environments are normally described as thalassohaline or athalassohaline. Thalassohaline waters are marine derived and therefore contain, at least initially, a seawater composition; however, with increasing evaporation the concentration of various salts alters depending on the thresholds for the crystallization and precipitation of different minerals (see Table III). Athalassohaline water may also be influenced by the influx of seawater; however, the chemical composition is mainly determined by geological, geographical, and topographical parameters. Examples of athalassohaline environments include the Great Salt Lake in Utah and the Dead Sea.

In addition to lakes formed by evaporation in moderate climate conditions, hypersaline Antarctic lakes (e.g., Vestfold Hills; see Section IV) have been formed from the effects of frost and dryness in this environment. Antarctic and moderate temperature soils also contain salinities between 10 and 20% (w/v) and efforts have been directed at characterizing these comparatively poorly studied ecosystems. Less obvi-

TABLE III
Concentration of Ions in Thalassohaline and Athalassohaline Brines[a]

Ion	Seawater	Seawater at onset of NaCl saturation	Seawater at onset of K⁺-salt saturation	Great Salt Lake	Dead Sea	Typical soda lake
			Concentration (g/liter)			
Na^+	10.8	98.4	61.4	105.0	39.7	142
Mg^{2+}	1.3	14.5	39.3	11.1	42.4	<0.1
Ca^{2+}	0.4	0.4	0.2	0.3	17.2	<0.1
K^+	0.4	4.9	12.8	6.7	7.6	2.3
Cl^-	19.4	187.0	189	181.0	219.0	155
SO_4^{2-}	2.7	19.3	51.2	27.0	0.4	23
CO_3^{2-}/HCO_3^-	0.34	0.14	0.14	0.72	0.2	67
pH	8.2	7.3	6.8	7.7	6.3	11

[a] Data from Grant, W. D., Gemmell, R. T., and McGenity, T. J. (1998). Halophiles. In "Extremophiles: Microbial Life in Extreme Environments" (Horikoshi, K., and Grant, W. D., Eds.), Wiley series in Ecological and Applied Microbiology, Wiley-Liss.

ous saline habitats known to be colonized by microorganisms are animals skins, plant surfaces, and building surfaces. In addition, interest is also being focused on subterranean salt deposits as a habitat for extreme halophilic Archaea, and as a possible source of ancient prokaryotic lineages preserved in fluid inclusion bodies of salt crystals.

Organic compounds in hypersaline lakes are mostly produced by cyanobacteria, anoxygenic phototrophic bacteria, and by species of the green algae *Dunaliella* spp. They are found in most natural and artificial hypersaline lakes around the world. With seasonal and transient contributions from animals or plants, these environments can contain dissolved organic carbon levels up to 1 g/liter. Due to the low solubility of oxygen in saline solutions, however, many habitats become anaerobic and as a result aerobic growth is often restricted to the upper layers.

Organisms living in saline habitats exhibit different levels of adaptation to salt. To account for the variety of tolerances, an extensive set of definitions exist and are summarized in Table IV.

A salt concentration of about 1.5 M is the upper limit for vertebrates, although some eukaryotes such as the brine shrimp (*Artemia salina*) and the brine fly (*Ephydra*) can be found in habitats with higher salinity. For salinities above 1.5 M, prokaryotes be-

come the predominant group with moderate halophilic and haloversatile bacteria at salt concentrations between 1.5 and 3.0 M, and extreme halophilic Archaea (halobacteria) at salinities around the point of sodium chloride precipitation.

Aerobic gram-negative chemoorganotrophic bacteria are abundant in brines of medium salinity, and many strains have been isolated, including members of the genera *Vibrio, Alteromonas, Acinetobacter, Deleya, Marinomonas, Pseudomonas, Flavobacterium, Halomonas,* and *Halovibrio.* Gram-positive aerobic bacteria of the *Marinococcus,*

TABLE IV
Defining Terms for Microorganisms Ranging in Tolerances to Salt

Category	Salt concentration (M) Range	Optimum
Nonhalophilic	0–1.0	<0.2
Slightly halophilic	0.2–2.0	0.2–0.5
Moderate halophilic	0.4–3.5	0.5–2.0
Borderline extreme halophilic	1.4–4.0	2.0–3.0
Extreme halophilic	2.0–5.2	>3.0
Halotolerant	0–1.0	<0.2
Haloversatile	0–3.0	0.2–0.5

Sporosarcina, *Salinococcus* and *Bacillus* species have been found in saline soils, salterns, and occasionally in solar salterns. Members of the genera *Halomonas* and *Flavobacterium* have been isolated from Antarctic lakes. Two of the most remarkable isolates from Antarctica are the extreme halophilic archaeal species *Halobacterium lacusprofundi* and a *Dunaliella* spp., which are found in association in Deep Lake. These are the only two microorganisms growing in this lake, which has 4.8 M salt and whose temperature for 8 months of the year is less than 0°C. With the extremes of temperature, salt, and ionic strength, and with primary production rates of 10 g C m^{-2} year^{-1}, Deep Lake has been described as one of the most inhospitable environments on Earth.

Wherever light reaches the anoxic layer of hypersaline brines, anoxygenic phototrophs such as *Chromatium salexigens*, *Thiocapsa halophila*, *Rhodospirillum salinarum*, and *Ectothiorhodospira* are commonly found and represent the primary producers in these environments. Sulfate reducers *Desulfovibrio halophilus* and *Desulfohalobium retbaense* have been isolated from anaerobic sediments and are thought to perform dissimilatory sulfate reduction to H$_2$S. H$_2$S is subsequently used for growth by most anoxygenic phototrophs (except for members of the *Rhodospirillaceae* family). Anaerobic fermentative halophiles from the bacterial lineage of *Haloanaerobiaceae* have also been described.

Most of the bacterial species mentioned here are predominant in environments up to 2 M salt. At higher concentrations, extremely halophilic Archaea (including the confusingly named genus, *Halobacteria*) are more abundant. Most of these Archaea require at least 1.5 M salt for growth and for retaining their structural integrity. Members of the genera *Haloarcula*, *Halobacterium*, and *Halorubrum* are frequently found in hypersaline waters reaching the sodium chloride saturation point, and the proteolytic species *Halobacterium salinarium* is often associated with salted food. Other Archaea are also found in saline lakes, marine stromatolites, and solar ponds, and these tend to be methanogens (see Section X). In addition, the eukaryote *Dunaliella* can adapt to a wide range of salt concentrations, from less than 100 mM to saturation (5.5 M).

B. Biochemistry and Physiology of Adaptation

Cells exposed to an environment with a higher salinity than the one inside the cytoplasm inevitably experience a loss of water and undergo plasmolysis (Fig. 1). Microorganisms living in high-salt environments generally adopt one of two strategies (either salt-in-cytoplasm or compatible-solute adaptation) to prevent the loss of cytoplasmic water and to establish osmotic equilibrium across their cell membranes.

1. Salt-in-Cytoplasm Adaptation

Extremely halophilic archaea and anaerobic halophilic bacteria use a salt-in-cytoplasm strategy. This involves cations flowing through the membrane into the cytoplasm (Fig. 1A). Archaea accumulate intracellular potassium and exclude sodium, whereas Bacteria accumulate sodium rather than potassium. As a consequence of the high salt, intracellular components (e.g., proteins, nucleic acid and cofactors) require protection from the denaturing effects of salt. The most common protective mechanism is the presence of excess negative charges on their exterior surfaces. Malate dehydrogenase from *Haloarcula marismortui* has a 20 mol% excess of acidic residues over basic amino acid residues, compared to only 6 mol% excess for a nonhalophilic counterpart. Structural analysis shows that the acidic residues are mainly located on the surface of the protein and are either involved in the formation of stabilizing salt bridges or in attracting water and salt to form a strong hydration shell. This malate dehydrogenase is able to bind extraordinary amounts of water and salt (0.8–

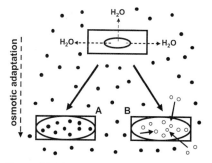

Fig. 1. Effect of high salinity on intracellular water, and two strategies used for adaptation. Solid circles represent salt molecules and open circles compatible solutes.

1.0 g water and 0.3 g salt/g protein) compared to a nonhalophilic malate dehydrogenase (0.2–0.3 g water and 0.01 g salt/g protein). Due to the adaptation mechanism in these cells, they have an obligate requirement for high concentrations of salt. In low-salt environments, the lack of salt removes the shielding effect of cations from the proteins and leads to the rapid denaturation of the three-dimensional structure.

2. *Compatible-Solute Adaptation*

Most halotolerant and moderately halophilic organisms, including the anoxygenic phototrophic bacteria, aerobic heterotrophic bacteria, cyanobacteria, and methanogens, maintain a salt-minimized cytoplasm. They accumulate small organic and osmotically active molecules, referred to as compatible solutes. These compounds can be synthesized *de novo* or imported from the surrounding medium (Fig. 1B). The latter mechanism is also used by nonhalophilic organisms; they adapt to increased salt concentrations by importing extracellular compatible solutes using transporters involved in amino acid or sugar uptake.

A large range of compatible solutes has been identified in a broad range of halophiles (Table V). All these molecules are polar, highly soluble, and uncharged or zwitterionic at physiological pH values. They are strong water-structure formers and as such are probably excluded from the hydration shell of proteins (preferential exclusion), and therefore exert a stabilizing effect without interfering directly with the structure of the protein. In addition to being stabilizers against salt stress, they have also been shown to prevent the denaturation of proteins caused by heating, freezing, and drying.

3. *Membranes*

Even though the cytoplasmic interior is protected from the effects of external salt by cytoplasmic-compatible solutes, the outer surface of the cytoplasmic membrane is permanently exposed to high salt concentrations. To protect the membrane in halophilic bacteria, the proportion of anionic phospholipids (often phosphatidylglycerol and glycolipids) increases with increasing salinity at the expense of neutral zwitterionic phospholipids. These alterations produce additional surface charges to the membrane and

TABLE V
Compatible Solutes Used by Halophilic Microorganisms

Class of compatible solute	Specific compatible solute	Microorganism
Polyoles	Glycerol and arabite	Algae, yeast, fungi
	Glucosyl-glycerol	Cyanobacteria
Betaines	Glycine-betaine	Cyanobacteria, anoxygenic phototrophic bacteria, methanogens and *Actinopolyspora halophila*
	Dimethylglycine	*Methanohalophilus* sp.
Amino acids	Proline	*Bacillus* spp. and other *Firmicutes*
	α-glutamine	Corynebacteria
	β-glutamine	*Methanohalophilus* sp
Dimethylsulfoniopropionate		Marine cyanobacteria
Glutamine amides	N-α-carbomoyl-glutamine amide	*Ectothiorhodospira marismortui*
	N-α-acetyl-glutaminyl-glutamine amide	Anoxygenic phototrophic proteobacteria, *Rhizobium meliloti*, and *Pseudomonas* spp.
Acetylated diamino acids	N-δ-acetyl-ornithine and N-ε-acetyl-α-lysine	*Bacillus* spp. and *Sporosarcina halophila*
	N-ε-acetyl-α-lysine	Methanogens
Ectoines	Ectoine and hydroxyectoine	Proteobacteria, *Brevibacteria*, gram-positive cocci, and *Bacillus* spp.

help to maintain the hydration state of the membrane.

Most halophilic Archaea posses an S-layer consisting of sulfated glycoproteins, which surrounds the cytoplasmic membrane. The sulfate groups confer a negative charge to the S-layer and possibly provides structural integrity at high ionic concentrations. In addition, archaeal ether lipids have been shown to be more stable at high salt concentrations (up to 5 M) compared to the ester lipids found in the membranes of bacteria.

VII. ALKALIPHILES

A. Habitats and Microorganisms

Alkaliphiles are defined by optimal growth at pH values above 10, whereas alkalitolerant microorganisms may grow well up to pH 10, but exhibit more rapid growth below pH 9.5. Alkaliphiles are further subdivided into facultative alkaliphiles (that grow well at neutral pH) and obligate alkaliphiles (that grow only above pH 9). In addition, due to their natural habitats, many alkaliphilic organisms are adapted to high salt concentrations and are referred to as haloalkaliphiles.

Alkali environments include soils where the pH has been increased by microbial ammonification and sulfate reduction, and water derived from leached silicate minerals. These environments tend to have only a limited buffering capacity and the pH of the environments fluctuates. As a result, alkalitolerant microorganisms are more abundant in these habitats than are alkaliphiles. Artificial environments include locations of cement manufacture, mining, and paper and pulp production. Probably the best studied and most stable alkaline environments are soda lakes and soda deserts (e.g., in the East African Rift Valley or central Asia). The formation of soda lakes and deserts is similar to the formation of athalassohaline salt lakes (see Section V), with the exception that carbonate is the major anion in solution, due to the lack of divalent cations (Mg^{2+}, Ca^{2+}) in the surrounding environment. A typical soda lake composition is shown in Table III.

Bacteria, Archaea, yeast, and fungi have been isolated from alkali environments. Archaeal alkaliphiles include members of the *Halobacteriaceae* (*Halorubrum*, *Natronbacterium*, and *Natronococcus* spp.) and *Methanosarcinaceae* (*Methanohalophilus* spp.). Cyanobacterial genera *Spirulina* and *Synechococcus* represent the dominant primary producers in the aerobic layers of soda lakes. Other alkaliphilic bacteria are members of the *Actinomyces*, *Bacillaceae*, *Clostridiaceae*, *Haloanaerobiales*, and γ-Proteobacteria (*Ectothiorodospira*, *Halomonadaceae*, and *Pseudomonas*). Anaerobic thermophilic alkaliphiles (alkalithermophiles) include *Clostridium* and *Thermoanaerobacter* spp., and the only representative of a new taxon, *Thermopallium natronophilum*.

B. Biochemistry and Physiology of Adaptation

1. pH Homeostasis

Studies of alkaliphiles (particularly *Bacillus* spp.) have demonstrated that they maintain a neutral or slightly alkaline cytoplasm. This is reflected by a neutral pH optimum for intracellular enzymes compared to a high pH optimum for extracellular enzymes. The intracellular pH regulation has been shown to be dependent on the presence of sodium. The Na^+ ions are exchanged from the cytoplasm into the medium by H^+/Na^+ antiporters (Fig. 2). Electrogenic proton extrusion is mediated in aerobic cells by respiratory chain activity and protons are transported back into the cell via antiporters that are

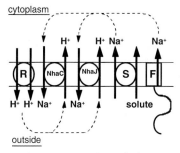

Fig. 2. Schematic representation of elements involved in pH homeostasis in alkaliphilic microorganisms. R represents respiratory chain-mediated proton translocation; NhaC and NhaJ represent two separate H^+/Na^+ antiporters found in *Bacillus* spp.; S represents a Na^+-dependent solute transporter; and F represents Na^+-dependent flagella motility.

efficient at transporting H$^+$ into the cell at the expense of Na$^+$ export from the cell. The resulting net production and influx of protons creates a more acidic cytoplasm. In addition to controlling protons, Na$^+$-dependent pH homeostasis requires the reentry of Na$^+$ into the cell. Na$^+$-coupled solute symporter and sodium-driven flagella rotation ensure a net sodium balance. The combined action of the antiporters coupled with respiration provides the cell with a means to control its internal pH while maintaining sufficient Na$^+$ levels through symport and flagella rotation.

The exterior surfaces of the cell are also important for maintaining a pH differential. This is supported by evidence that the protoplasts of alkaliphilic *Bacillus* spp. are unstable in alkaline conditions. The peptidoglycan in these strains has a higher cross-linking rate at higher pH values, which may provide a shielding effect by "tightening" the cell wall. Large amounts of acidic compounds, including teichuronic acid, teichoic acid, uronic acids, and acidic amino acids, are evident in alkaliphilic cell walls compared to the cell walls of nonalkaliphilic microorganisms. The negative charge of these acidic substances may create a more neutral layer close to the outer surface of the cell.

2. Bioenergetics

Alkaliphiles have unique bioenergetic properties. Nonalkaliphilic respiring bacteria energize their cytoplasmic membrane with a chemiosmotic driving force (Δp) by generating an electrochemical gradient of ions that has two parameters—acidic conditions outside (caused by the extrusion of protons and described by the term ΔpH) and a positive charge outside (described by the transmembrane electrical potential, $\Delta\Psi$). The Δp is used for proton-coupled symport of solutes, proton-driven mobility (flagella) and ATP synthesis. In alkaliphilic environments, the contribution of ΔpH to Δp becomes smaller with increasing extracellular pH. However, with increasing extracellular pH, sodium ion export is increased (see Section VII.B.1) and may contribute to an increased $\Delta\Psi$. This partially compensates for the decrease of Δp by the reduction of ΔpH. Interestingly, the sodium gradient is used to energize solute transport and flagella movement, but not for ATP synthesis. No sodium-dependent ATP-synthases have been

identified in alkaliphiles. In addition, ATP-synthases in alkaliphilic *Bacillus* spp. have been shown to be exclusively proton translocating.

VIII. OLIGOTROPHS

A. Habitats and Microorganisms

A most important environmental factor in microbial ecology is the availability of energy, and virtually all microbial cells in nature are limited in their growth by the availability of one or more essential growth nutrients. For example, in the intestinal tract, the number of *E. coli* doubles about twice per day, whereas in ideal laboratory conditions it doubles in 20 min. Similarly, soil bacteria are estimated to grow in soil at about 1% of the maximal rate of growth observed in the laboratory. This highlights that in natural ecosystems nutrient limitation is the rule, rather than the exception.

In addition to the overall nutrient status, in aquatic and soil environments, nutrient levels are often transient; for example, a fallen leaf provides nutrients to soil microorganisms, or a dead fish provides nutrients in an aquatic environment. As a result of this apparent "feast or famine," microorganisms have adopted two main strategies for surviving in nutrient-depleted (oligotrophic) environments. Eutrophic microorganisms (also referred to as copiotrophs, saphrophytes, and heterotrophs) grow in bursts when nutrients are available and produce resting-stage cells when nutrients are in short supply (referred to as r-strategy). In contrast, oligotrophic microorganisms (also referred to as oligocarbophiles; low-nutrient, LN, bacteria; low-K_s bacteria; and dilute-nutrient-broth, DNB, organisms) grow slowly, using low concentrations of nutrients (referred to as K-strategy).

It is important to note that whereas oligotrophs grow slowly in oligotrophic environments, eutrophs are in a resting stage. In contrast, when members of these two classes are subcultured from oligotrophic environments into rich media, the eutrophs resume rapid growth, whereas the oligotrophs do not grow at all. The cellular responses that prevent growth of the oligotrophs in rich media is not understood; however, the response highlights the physiological differences between these two classes of microorganisms. In addition, the difficulty in growing oligotrophs in the laboratory has proven problematic for examining their physiology. Fortunately, some isolates have adapted (facultative oligotrophs) to growth in rich media and are amenable to laboratory studies.

Although the definition of an oligotroph and an oligotrophic environment remain the subject of debate, it is generally accepted that an oligotroph is able to grow in a medium containing 0.2–16.8 mg dissolved organic carbon/liter. The terms "obligate oligotroph" (implying the inability to grow in high concentrations of nutrients) and "facultative oligotroph" (indicating the ability to grow in low and high concentrations of nutrients) are also used to further clarify the nutritional requirements for growth.

In natural ecosystems, oligotrophs and eutrophs coexist, and the proportion of each varies depending on their individual abilities to dominate in the particular environment. For example, when the marine oligotrophic ultramicrobacterium *Sphingomonas* sp. strain RB2256 was isolated from Resurrection Bay, Alaska, it was a numerically dominant species in a place where the total bacterioplankton population was 0.2 to 1.07×10^6 cells/ml. In contrast, significantly lower numbers (<1%) of larger, faster-growing cells (typical of eutrophs) were able to be immediately cultured in rich media and on plates.

B. Biochemistry and Physiology of Adaptation

A characteristic that is often associated with oligotrophic bacteria is their ultramicro size (<0.1 μm^3). Ultramicrobacteria (or dwarf cells) are commonly found in aquatic and soil environments. Oligotrophic ultramicrobacteria, such as *Sphingomonas* RB2256, retain their ultramicro size irrespective of growth phase, carbon source, or carbon concentration. In contrast, eutrophic microorganisms, such as *Vibrio angustum* S14 and *Vibrio* ANT-300, undergo reductive cell division during starvation, but resume their normal size (>1 μm^3) when grown in rich media.

The characteristics that are thought to be important for oligotrophic microorganisms include a substrate uptake system that is able to acquire nutrients from its surroundings, and the capacity to use the nutrients in order to maintain its integrity and

growth. As a result, oligotrophs would ideally have large surface-area-to-volume ratios, and high-affinity uptake systems with broad substrate specificities. Consistent with this, a number of microorganisms that are adapted to low-nutrient environments produce appendages to enhance their surface area. These include members of the bacterial genera, *Caulobacter*, *Hyphomicrobium*, *Prosthecomicrobium*, *Ancalomicrobium*, *Labrys*, and *Stella*. Other bacteria have tiny cell volumes (ultramicrobacteria: < 0.1 μm^3) to maximize their surface-area-to-volume ratio.

The most comprehensive physiological studies of oligotrophic marine isolates have been performed on *Sphingomonas* sp. strain RB2256. The characteristics that distinguish it from typical marine eutrophs include constant ultramicro size irrespective of growth or starvation conditions, a mechanism for avoiding predation (ultramicro size), a relatively slow maximum specific growth rate (< 0.2 hr^{-1}), a single copy of the rRNA operon compared to 8–11 copies for *Vibrio* spp., a relatively small genome size compared to faster-growing heterotrophs, an ability to use low concentrations of nutrients, high-affinity broad-specificity uptake systems, an ability to simultaneously take up mixed substrates, an ability to immediately respond to nutrient addition without a lag in growth, and an inherent resistance to environmental stresses (e.g., heat, hydrogen peroxide, and ethanol).

The small (0.2 μm^3) oligotrophic bacterium *Cycloclasticus oligotrophicus* RB1, which was also isolated from Resurrection Bay, shares some properties similar to *Sphingomonas* RB2256 (e.g., single copy of the rRNA operon, and relatively small cell size and genome size). Interestingly, while this chemoorganotroph is unable to grow using glucose and amino acids, it can use acetate and a few aromatic hydrocarbons such as toluene.

IX. RADIATION-RESISTANT MICROORGANISMS

A. Habitats and Microorganisms

Radiation from the sun drives photosynthetic reactions, thus ensuring primary production throughout the global ecosystem. Although the visible spectrum leads to biomass production, visible light and other portions of the electromagnetic spectrum (particularly short-wavelength) also cause cellular damage. Damage to cells primarily occurs directly to nucleic acids (e.g., UV-induced thymine–dimer formation and strand breakage) or indirectly through the production of reactive oxygen species (e.g., H_2O_2, O_2^-, ^0OH, and $^1O_2$), which cause damage to lipids, proteins, and nucleic acids. Due to the prevalence of natural radiation, most cells have developed a range of DNA-repair and other protective mechanisms to facilitate their survival.

In contrast to natural forms of relatively low-level radiation, microorganisms may be exposed to intense sources of radiation in the form of γ-irradiation (^{60}Co and ^{137}Cs) as a means of sterilization or by being in close proximity to nuclear reactors. It is mainly these forms of radiation that have been the sources of highly radiation-resistant extremophiles. The most well-studied of these is the bacterium *Deinococcus radiodurans*, which was first isolated in 1956 from tins of meat that had been irradiated with γ-rays. Although some microorganisms escape radiation damage by forming spores (e.g., *Clostridium botulinum*), *D. radiodurans* is resistant while in the exponential growth phase. The degree of resistance of these cells is illustrated by their ability to survive 3,000,000 rad, a dose that is sufficient to kill most spores (a lethal dose for humans is about 500 rad). As a consequence of the extreme radiation resistance of *D. radiodurans* and other microorganisms, biocides are routinely added to the cooling waters of nuclear reactors to prevent the proliferation of microorganisms. By virtue of their resistant properties, this class of extremophile can be readily isolated by exposing samples to intense UV or γ-irradiation and then plating them out on a rich medium.

Although *Deinococcus* spp. have been found in dust, processed meats, medical instruments, textiles, dried food, animal feces, and sewage, their natural habitats have not been clearly defined. Thermophilic species have, however, been isolated from hot springs in Italy, and deinococci have been identified in many soil environments. It therefore appears that this class of radiation resistant extremophiles exists in a broad range of environmental niches.

In addition to *Deinococcus* spp., some hyperthermophilic Archaea (e.g., *Thermococcus stetteri* and *Pyrococcus furiosus*) are able to survive high levels of γ-irradiation.

B. Biochemistry and Physiology of Adaptation

Resistance to radiation could conceivably occur by two main mechanisms, prevention of damage or efficient repair of damage. In *D. radiodurans*, it has been clearly shown that the DNA is severely damaged during γ-irradiation (e.g., ~110 double-strand breaks per cell when exposed to 300,000 rad); however, within 3 hr of recovery, the fragmented DNA is replaced by essentially intact, chromosomal DNA. Furthermore, even though the DNA within the cells has been extensively degraded, viability is unaffected. *P. furiosus* is also able to repair fragmented DNA after exposure to 250,000 rad when the cells are grown at 95°C. As a result, it has been suggested that active DNA-repair mechanisms may be important determinants of survival of hyperthermophiles in their natural environments.

In addition to γ-radiation resistance, *D. radiodurans* is resistant to highly mutagenic chemicals, with the exception of those that cause DNA deletions (e.g., nitrosoguanadine, NTG). *D. radiodurans* is also extremely resistant to UV irradiation, surviving doses as high as 1000 J/m^2. The dose that is required to inactivate a single colony-forming unit of an irradiated population is for *D. radiodurans* 550–600 J/m^2 compared to just 30 J/m^2 for *E. coli*.

The DNA-repair systems in *D. radiodurans* are so effective that they have proven to be a hindrance for genetic studies, that is, it is difficult to isolate stable mutants. However, through the combined use of chemical mutagenesis and screens for mitomycin C-, UV radiation-, and ionizing radiation-sensitive strains, genes involved in nucleotide-excision repair, base-excision repair, and recombinational repair have been identified. In addition, *D. radiodurans* is multigenomic (e.g., there are 2.5–10 copies of the chromosome depending on growth rate) and a novel mechanism of interchromosomal recombination has been proposed. This process would help to circumvent problems of reassembling a complete and contiguous chromosome from the chromosome fragments that have been generated as a result of irradiation. It appears that *D. radiodurans*'s ability to use its genome multiplicity to repair DNA damage is the most fundamental reason for this species's extraordinary radioresistance.

Almost all *Deinococcus* spp. that have been isolated (even those isolated without selection for radiation resistance) are radiation resistant, thus demonstrating that the extreme resistance is not a result of selection by irradiation, but a normal characteristic of the genus. Desiccation leads to DNA damage in all cells and prevents DNA repair from occurring. For a cell to be viable when it is rehydrated, it needs to be able to repair the damaged DNA. It is therefore likely that the evolutionary process that has led to the inherent radiation resistance in *Deinococcus* is natural selection for resistance to desiccation. This is supported by evidence that *D. radiodurans* is also exceptionally resistant to desiccation.

X. OTHER EXTREMOPHILES

As a group, methanogenic Archaea (methanogens) are often considered extremophiles. They are the most thermally diverse organisms known, inhabiting environments from close to freezing in Antarctic lakes to 110°C in hydrothermal vents. Those isolated from Antarctica include *Methanococcoides burtonii* and *Methanogenium frigidum,* and those from hydrothermal vents include *Methanopyrus kandleri, Methanothermus fervidus, Methanothermus sociabilis, Methanococcus jannaschii,* and *Methanococcus igneus.* Methanogens can also be isolated from a diverse range of salinities, from fresh water to saturated brines. *Methanohalobium evestigatum* was isolated from a microbial mat in Sivash Lake and grows in pH neutral, hypersaline conditions from 2.6–5.1 M. Alkaliphilic (*Methanosalsus zhilinaeae,* pH 8.2–10.3) and acidophilic (*Methanosarcina sp.,* pH 4–5) methanogens have also been isolated.

Other microorganisms that may be considered to be extremophiles include toxitolerants (those tolerant to organic solvents, hydrocarbons, and heavy metals, such as *Rhodococcus* sp., which can grow on benzene as a sole carbon source), and xerophiles and xerotolerant microbes that survive very low-water activity (e.g., extremely halophilic Archaea, fungi such as *Xeromyces bisporus,* and endolithic microorganisms that live in rocks).

XI. BIOTECHNOLOGY OF EXTREMOPHILES

A major impetus driving research on extremophiles is the biotechnological potential associated with the microorganisms and their cellular products. In 1992, of the patents related to Archaea, about 60% were for methanogens, 20% for halophiles, and 20% thermophiles. Examples of "extremozymes" that are presently used commercially include alkaline proteases for detergents. This is a huge market, with 30% of the total worldwide enzyme production for detergents. In 1994, the total market for alkaline proteases in Japan alone was ~15,000 million yen. DNA polymerases have been isolated from the hyperthermophiles *Thermus aquaticus, Thermotoga maritima, Thermococcus litoralis, Pyrococcus woesii,* and *Pyrococcus furiosus* for use in the polymerase chain reaction (PCR). A eukaryotic homolog of the *myc* oncogene product from halophilic Archaea has been used to screen the sera of cancer patients. Its utility is demonstrated by the fact that the archaeal homolog produced a higher number of positive reactions than the recombinant protein expressed in *E. coli.* β-carotene is commercially produced from the green algae *Dunaliella bardawil.* The applications in industry are still limited; however, the potential applications are extensive. Some examples of their uses and potential applications are listed in Table VI.

The biotechnology potential is increasing exponentially with the isolation of new organisms, the identification of novel compounds and pathways, and the molecular and biochemical characterization of cellular components. Major advances are likely in the area of protein engineering. For example, the identification of the structural properties important for thermal activity and stability will enable the construction of proteins with required catalytic and thermal properties. Recently, a metalloprotease from the moderately thermophilic bacterium *Bacillus stearothermophilus* was mutated using a rational design process in an effort to increase its thermostability.

TABLE VI
Extremophiles and Their Uses in Biotechnology

Source	Use
Hyperthermophiles	
DNA polymerases	DNA amplification by PCR
Alkaline phosphatase	Diagnostics
Proteases and lipases	Dairy products
Lipases, pullulanases, and proteases	Detergents
Proteases	Baking and brewing and amino acid production from keratin
Amylases, α-glucosidase, pullulanase, and xylose/glucose isomerases	Starch processing, and glucose and fructose for sweeteners
Alcohol dehydrogenase	Chemical synthesis
Xylanases	Paper bleaching
Lenthionin	Pharmaceutical
S-layer proteins and lipids	Molecular sieves
Oil degrading microorganisms	Surfactants for oil recovery
Sulfur oxidizing microorganisms	Bioleaching, coal, and waste gas desulfurization
Hyperthermophilic consortia	Waste treatment and methane production
Psychrophiles	
Alkaline phosphatase	Molecular biology
Proteases, lipases, cellulases, and amylases	Detergents
Lipases and proteases	Cheese manufacture and dairy production
Proteases	Contact-lens cleaning solutions, meat tenderizing
Polyunsaturated fatty acids	Food additives, dietary supplements
Various enzymes	Modifying flavors
β-galactosidase	Lactose hydrolysis in milk products
Ice nucleating proteins	Artificial snow, ice cream, other freezing applications in the food industry
Ice minus microorganisms	Frost protectants for sensitive plants
Various enzymes (e.g., dehydrogenases)	Biotransformations
Various enzymes (e.g., oxidases)	Bioremediation, environmental biosensors
Methanogens	Methane production
Halophiles	
Bacteriorhodopsin	Optical switches and photocurrent generators in bioelectronics
Polyhydroxyalkanoates	Medical plastics
Rheological polymers	Oil recovery
Eukaryotic homologues (e.g., *myc* oncogene product)	Cancer detection, screening antitumor drugs
Lipids	Liposomes for drug delivery and cosmetic packaging
Lipids	Heating oil
Compatible solutes	Protein and cell protectants in a variety of industrial uses (e.g., freezing, heating)
Various enzymes (e.g., nucleases, amylases, proteases)	Various industrial uses (e.g., flavoring agents)
γ-linoleic acid, β-carotene and cell extracts (e.g., *Spirulina* and *Dunaliella*)	Health foods, dietary supplements, food coloring, and feedstock
Microorganisms	Fermenting fish sauces and modifying food textures and flavors
Microorganisms	Waste transformation and degradation (e.g., hypersaline waste brines contaminated with a wide range of organics)
Membranes	Surfactants for pharmaceuticals
Alkaliphiles	
Proteases, cellulases, xylanases, lipases and pullulanases	Detergents
Proteases	Gelatin removal on X-ray film
Elastases, keritinases	Hide dehairing
Cyclodextrins	Foodstuffs, chemicals, and pharmaceuticals
Xylanases and proteases	Pulp bleaching
Pectinases	Fine papers, waste treatment, and degumming
Alkaliphilic halophiles	Oil recovery
Various microorganisms	Antibiotics
Acidophiles	
Sulfur-oxidizing microorganisms	Recovery of metals and desulfurication of coal
Microorganisms	Organic acids and solvents

The mutant protein was 340 times more stable than the wild-type protein and was able to function at 100°C in the presence of denaturing agents, while retaining wild-type activity at 37°C.

Advances are also likely to arise from the construction of recombinant microorganisms for specific purposes. A recombinant strain of *Deinococcus radiodurans* has been engineered to degrade organopollutants in radioactive mixed-waste environments. The recombinant *Deinococcus* expresses toluene dioxygenase, enabling it to oxidize toluene, chlorobenzene, 2,3-dichloro-1-butene, and indole in a highly irradiating environment (6000 rad/hr), while remaining tolerant to the solvent effects of toluene and trichloroethylene at levels exceeding those of many radioactive waste sites. In recognition of the number of waste sites contaminated with organopollutants plus radionuclides and heavy metals around the world, and the safety hazards and cost involved in clean up using physicochemical means, the potential use of genetically engineered extremophilic microorganisms is an important and exciting prospect.

See Also the Following Articles

ALKALINE ENVIRONMENTS • CRYSTALLINE BACTERIAL CELL SURFACE LAYERS • EVOLUTION, THEORY AND EXPERIMENTS • EXOBIOLOGY • HIGH-PRESSURE HABITATS • LOW-TEMPERATURE ENVIRONMENTS

Bibliography

Atlas, R. M., and Bartha, R. (1998). "Microbial Ecology: Fundamental and Applications," 4th ed. Benjamin/Cummings.

Battista, J. R. (1997). Against all odds: The survival strategies of *Deinococcus radiodurans*. *Annu. Rev. Microbiol.* **51**,

DeLong, E. F., Franks, D. G., and Yayanos, A. A. (1997). Evolutionary relationship of cultivated psychrophilic and barophilic deep-sea bacteria. *Appl. Environ. Microbiol.* **63**, 2105–2108.

First International Congress on Extremophiles. (1996). *FEMS Microbiol. Rev.* **18(2–3)**.

Galinski, E. A. (1995). Osmoadaptation in bacteria. *Adv. Microb. Physiol.* **37**, 273–327.

Horikoshi, K., and Grant, W D. (1998). "Extremophiles: Microbial Life in Extreme Environments". Wiley-Liss,.

Kato, C., and Bartlett, D. H. (1997). The molecular biology of barophilic bacteria. *Extremophiles* **1**, 111–116.

Krulwich, T. A. (1995). Alkaliphiles: 'Basic' molecular problems of pH tolerance and bioenergetics. *Mol. Microbiol.* **15**, 403–410.

Madigan, M. T., Martinko, J. M., and Parker, J. (1997). "Brock: Biology of Microorganisms," 8th ed. Prentice Hall, Upper Saddle River, NJ.

Matin, A. (1990) Keeping a neutral cytoplasm: The bioenergetics of obligate acidophiles. *FEMS Microbiol. Rev.* **75**, 307–318.

Morita, R. Y. (1997). "Bacteria in Oligotrophic Environments: Starvation-Survival Lifestyle." Chapman and Hall, New York.

Pick, U. (1998). *Dunaliella:* a model extremophilic alga. *Israel J. Plant Sci.* **46**, 131–139.

Russel, N. J. (1998). Molecular adaptations in psychrophilic bacteria: Potential for biotechnological applications. *Adv. Biochem. Eng. Biotechnol.* **61**, 1–21.

Schut, F., Prints, R. A., and Gottschal, J. C. (1997). Oligotrophy and pelagic marine bacteria: facts and fiction. *Aquatic Microb. Ecol.* **12**, 177–202.

Second International Congress on Extremophiles. (1998). *Extremophiles* **2(2)**.

Eyespot

Paul S. Dyer
University of Nottingham

GLOSSARY

anamorph A reproductive state involving the production of asexual spores identical to the parent.

apothecia Cup- or saucer-shaped sexual reproductive structures formed by certain fungi.

epidemiology The study of factors causing and influencing the outbreak of disease.

fungicide A chemical used to kill fungi or inhibit fungal growth.

hypha A tubular filament constituting the basic fungal growth unit.

mycelium The collective term for a mass of hyphae.

pathogen A disease-causing organism.

tillers Side shoots that develop from the base of a stem.

EYESPOT is a damaging stem-base disease of cereal crops and other grasses caused by fungi of the genus *Tapesia*. It occurs in temperate regions worldwide including Europe, the former USSR, Japan, South Africa, North America, and Australasia. In many countries, eyespot can be found on the majority of autumn-sown barley and wheat crops, and may cause an average 5–10% loss in yield, although low rates of infection do not generally have a significant effect. However, eyespot is potentially of great economic importance because yield losses of up to 50% can occur when epidemics become severe.

I. DISEASE SYMPTOMS

Eyespot is so named because the lesions formed on the leaf sheaves and stems of mature plants often have a central black "pupil," consisting of a mass of compacted hyphae, surrounded by a diffuse lighter-brown oval margin. Such lesions may be 15–30 mm in length and occur on the stem base, generally below the first node (Fig. 1). Grey mycelium is often visible in the stems of colonized plants. This contrasts with less distinct superficial brown lesions seen in younger plants. Slight infection causes little damage, but well-developed lesions may sufficiently weaken the stem to cause it to bend or break. This can result in "straggling" (a few tillers collapsing) or "lodging" (whole plants and areas of the crop falling over), particularly in strong wind or rainfall. Weakened plants are also liable to subsequent attack by other fungal ear and leaf pathogens. Infection can, in addition, cause the premature death of plants with the formation of bleached, shrivelled ears known as "whiteheads." As a result of these symptoms, eyespot is also known by other names, including footrot and strawbreaker (United States); piétin-verse (France); and Halmbruchkrankheit (Germany).

II. CAUSAL ORGANISMS

Eyespot results from infection by two closely related fungi, *Tapesia yallundae* and *Tapesia acuformis*, both of which can cause symptoms of the disease, either alone or in combination. The two species were known jointly by the anamorph name *Pseudocercosporella herpotrichoides* until the late 1980s, and were formerly distinguished by the designations W

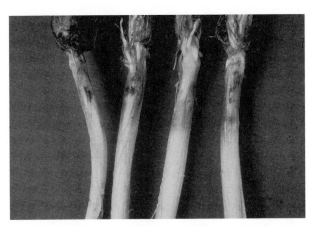

Fig. 1. Characteristic eyespot lesions formed on wheat stem base by *Tapesia yallundae.* Courtesy of J. A. Lucas.

Fig. 2. Apothecia of *Tapesia yallundae* formed on straw stubble.

and R type or var. *herpotrichoides* and var. *acuformis*, respectively. However, discoveries in Australia and England in 1986 and 1994 by Wallwork, Dyer and their co-workers revealed that the pathogens are able to reproduce sexually (producing apothecia, provided that sexually compatible MAT1 and MAT2 isolates are present; Fig. 2), and that the two species were genetically distinct with no interbreeding possible. This led to the adoption of the current names *T. yallundae* and *T. acuformis*. The species may be separated on the basis of a variety of growth characteristics (Table I; Fig. 3) and by the use of DNA-based molecular markers. They also have slightly different host plant ranges, with both being able to infect wheat and barley but only *T. acuformis* being able to infect certain cultivars of rye. It is unknown whether similar differences in host range exist in wild grasses. However, specialized pathotypes of *T. yallundae* are thought to infect couch grosses *Agropy-*

uon pens and *Aegilops squarrosa* (the C and S types, respectively). Field surveys have indicated that the relative proportion of each species on cereal crops varies from year to year. There may also be significant geographic variation, with *T. yallundae* being the only species found in South Africa whereas *T. acuformis* predominates in cooler regions of northern Europe. The use of azole fungicides may also lead to population shifts with an increased proportion of *T. acuformis* (compared to *T. yallundae*) observed in the UK during the 1990s. Apothecia of *T. yallundae* have been discovered at a variety of field locations in New Zealand, South Africa, and many European countries, indicating that sexual reproduction is an intrinsic part of the life cycle. In contrast, there have been only two confirmed reports of apothecia of *T. acuformis*, suggesting that this species reproduces almost exclusively by asexual reproduction.

TABLE I
Differences between *Tapesia yallundae* and *Tapesia acuformis*

Growth characteristic	Tapesia yallundae	Tapesia acuformis
Colony morphology	Even-edged colonies	Feathery-edged colonies
Growth rates at 15°C	1.6–2.6 mm/day	0.7–1.5 mm/day
Spore morphology	Curved or straight spores	Straight spores

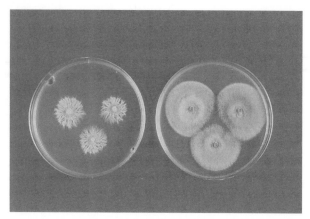

Fig. 3. Colonies of *T. yallundae* and *T. acuformis* with smooth and feathery margins, respectively. See color insert.

III. EPIDEMIOLOGY

The severity of disease varies from year to year, with epidemics favored by early autumn sowing of dense crops, wet and mild winters, and cool springs. Both *T. yallundae* and *T. acuformis* overwinter on infected plant debris and secondary weed hosts, on which they may survive up to 3 years. Asexual spores are produced from late autumn onward and are splash-dispersed over relatively short distances (up to 1 m) by rain droplets to healthy stem bases (Fig. 4). The coleoptile is the main site of infection, with runner hyphae forming infection plaques on the surface. Both spore production and infection are favored by high humidity and temperatures between 5 and 10°C. Meanwhile, sexual spores are ejected from apothecia (which develop on straw stubble) from early spring onward. Such wind-dispersed ascospores may be transported over long distances (over 1 km) and have been shown to be capable of infecting plants, although there is little evidence as yet of their dispersal in the field. The role of the sexual cycle in generating variation and producing airborne inoculum may be increasing in importance in Europe as a result of the EC "setaside" policy, which has led to the creation of large areas of standing straw stubble in many European countries during the 1990s. Lesion development then involves the fungus penetrating successive leaf sheaths during the growing sea-

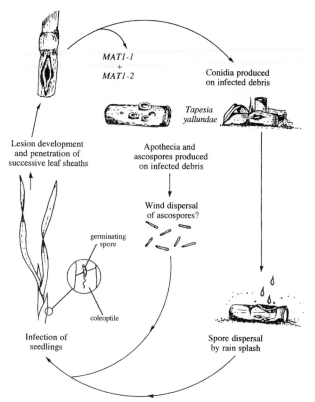

Fig. 4. The disease cycle of eyespot fungi. MAT1-1 and MAT1-2 represent complimentary mating types of the fungi that must both be present to allow sexual reproduction to occur. Courtesy of J. E. Hardy.

son. Spores may be produced on infected plants to initiate secondary infections, but this process is unlikely to occur above 16°C. Infected material left after harvest continues the disease cycle in the following growing season, explaining why eyespot can cause severe problems in continuous cereals in places where inoculum builds up from year to year.

IV. CONTROL MEASURES

A. Crop Management

The risk of development of eyespot disease can be reduced by attention to certain aspects of crop management. A noncereal break crop, or a 3-year break from wheat or barley will minimize the possibility of a serious outbreak of disease in subsequent crops. The sowing date is critical because eyespot has a tendency to be less severe in late-sown crops. Both the excessive use of nitrogen fertilizers and very high seed rates produce lush crops that favor eyespot. Finally, eyespot tends to be less severe where straw debris has been chopped and incorporated rather than burned.

B. Use of Resistant Varieties

The careful choice of cereal varieties, especially where cereals are grown intensively, provides an important way of controlling eyespot disease, and this has been the main method used in the Pacific Northwest of the United States. Although no wheat varieties exhibit total resistance, some modern lines (e.g., those derived from the variety Cappelle Desprez) are available that have increased resistance to lodging, having short, stiff stems and thickened walls that may hinder pathogen attack; such resistance appears to be durable. Breeding programs are being used to derive novel resistance from wild grasses, and this technique has already been used to incorporate resistance from the wild oat grass *Aegilops ventricosa*. Unfortunately, the resulting wheat variety 'Rendevous' has not been commercially successful due to poor-quality returns, but it provides hope for future resistance breeding. Winter barley varieties are, in general, less severely attacked than winter wheat.

C. Chemical Control by Fungicides

Fungicides are widely used to control eyespot disease; crops are often sprayed as a precautionary measure even if there is no visible evidence of disease. Total eradication of eyespot is rarely possible and fungicides are instead used to attempt to suppress the development of severe disease symptoms. This involves spraying once fairly early in the growing season to stop lesions developing on the coleoptiles or outer leaf sheaths. Further applications of fungicide may then be required, depending on the outcomes of field surveys and disease forecasting. The first fungicides to be used for eyespot control were the benzimidazoles (MBCs), which gave excellent

control when they were introduced in the 1970s. However, isolates resistant to the MBCs appeared in many countries in the early 1980s, and an alternative class of fungicides that inhibit sterol biosynthesis (the DMIs) has now generally replaced (or is used in conjunction with) MBC use. DMIs, such as prochloraz, have proved to be effective in controlling eyespot but there are recent reports of resistance to these fungicides in areas of northern France and Belgium. Fortunately, a new fungicide, cyprodinil (an anilinopyrimidine), was launched in 1994 for the control of eyespot, and this is able to suppress the growth of DMI-resistant strains.

V. FUTURE PROSPECTS

The eyespot fungi *T. yallundae* and *T. acuformis* are capable of causing significant damage to cereal crops worldwide. Eyespot disease has increased in importance in many countries as a result of changes in crop management, such as the earlier sowing of winter crops (which favors infection) and the creation of large areas of standing straw stubble (which allow the sexual cycle to occur). Although eyespot can be controlled by measures such as the use of resistant cultivars and application of novel fungicides, the *Tapesia* species have proven their ability to adapt, as shown by the evolution of resistance to MBC (and now possibly to DMI) fungicides. The discovery that they are able to reproduce sexually has added an extra threat, with the possibility of wind-dispersed spores exhibiting genetic variation being spread over long distances. The best hope for continued control of eyespot disease lies with the use of integrated control measures, varietal breeding programs and novel research into the molecular biology of the causal organisms aimed at interfering with the infection process and sporulation.

See Also the Following Articles

Plant Disease Resistance • Plant Pathogens • Sporulation

Bibliography

Daniels, A., Papaikonomou, M., Dyer, P. S., and Lucas, J. A. (1995). *Phytopathology* **85**, 918–927.

Dyer, P. S., and Lucas, J. A. (1995). *Plant Pathol.* 44, 796–804.

Dyer, P. S., Nicholson, P., Lucas, J. A., and Peberdy, J. F. (1996). *Mycol. Res.* 100, 1219–1226.

Fitt, B. D. L. (1992). Eyespot of cereals. *In* "Plant Diseases of International Importance" (U. S. Singh et al., eds.), Vol. 1, pp. 336–354. Prentice Hall, Englewood Cliffs, NJ.

Hollins, T. W., Scott, P. R., and Paine, J. R. (1985). *Plant Pathol.* 34, 369–379.

Parry, D. (1990). "Plant Pathology in Agriculture." Cambridge University Press, Cambridge.

Wallwork, H., and Spooner, B. (1988). *Trans. Br. Mycol. Soc.* 91, 703–705.

Fermentation

August Böck
University of Munich

GLOSSARY

electron transport-coupled phosphorylation ATP synthesis by the membrane-bound ATP synthase with the electrochemical gradient across the membrane as the driving force.

fermentation An anaerobic type of metabolism in which organic compounds are degraded in the absence of external electron acceptors. The organism produces a mixture of oxidized and reduced compounds.

fermentation balance The sum of the oxidized and the reduced compounds arising during fermentation in which the oxidation state is calculated in arbitrary units.

substrate-level phosphorylation ATP synthesis from a phosphorylated organic compound synthesized as an intermediate during substrate degradation.

MANY ENVIRONMENTS on the surface of Earth are devoid of oxygen, due to its chemical consumption or to the respiratory activity of aerobic organisms. This creates microenvironments (e.g., in soil particles and lake sediments) in which anaerobic microorganisms can thrive. Instead of using O_2 as the terminal electron acceptor, these anaerobes use alternate acceptors, such as nitrate, nitrite, sulfur, sulfate, or fumarate, or they carry out fermentions when such acceptors are not available. The variety of substrates that can be degraded via fermentation is broad and includes carbohydrates, amino acids, short-chain organic acids, purines, and a few others. Fermentation can be an obligatory process, as in the diverse group of clostridia, which lack the ability to switch to aerobic or anaerobic respiration when the relevant electron acceptors are provided, or it may be facultative, as in staphylococci and enterobacteria, which have complex regulatory mechanisms that ensure that fermentative enzymes are only induced when they are deprived of oxygen.

I. FERMENTATION BALANCES

The average energy gain in fermentation is in the range of 2–2.5 mol ATP/mol glucose degraded. Because of the low energy gain, most of the carbon of the substrate is not converted into the cell mass but rather appears in fermentation products. Because the reducing equivalents withdrawn from the substrate must be transferred to an internally generated acceptor, the sum of the oxidation state of the reduced products and of the oxidized products must equal the oxidation state of the substrate. Fermentations, therefore, are disproportionate conversion reactions when one neglects that a minor part of the carbon is incorporated into the cell mass. The calculation of a fermentation balance, therefore, can give important clues about the metabolic route by which the products are formed and may shed light on the question of whether an unknown oxidant or reductant participate. An example is shown Fig. 1.

An important role in the fermentation balances of many organisms is played by the hydrogenases. These enzymes either reduce protons with electrons

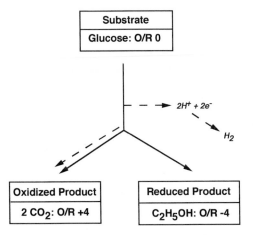

Fig. 1. Fermentation balance of the ethanol fermentation. In the calculation, oxygen is arbitrarily assigned a value of +1, hydrogen of −0.5, and nitrogen of +1.5. The production of hydrogen shifts the ratio of the other fermentation end products to the side of the oxidized ones.

or they oxidize molecular hydrogen. In fermentation, hydrogen production helps to dissipate the reducing equivalents and thus enables a shift of the ratio of reduced carbon products to the oxidized side. Because hydrogen has a low solubility in aqueous solutions, its formation also "pulls" the kinetics of unfavorable reaction equilibria, especially when hydrogen-producing organisms live in syntrophy with hydrogen-consuming ones.

II. FERMENTATION OF CARBOHYDRATES

Carbohydrates in the form of polysaccharides are the most abundant electron donors and thus are sources of energy for heterotrophs in nature. Their products of hydrolysis are important fermentation substrates for many anaerobes. The types of fermentation are classified in most cases according to the products that are formed. Figure 2 schematically presents the fermentation routes followed by selected biological systems.

Pyruvate is the predominant central metabolite, which is reduced to lactate in the homo- and hetero-lactic fermentations and in the mixed-acid fermentation of enterobacteria, or is transformed into intermediates that serve as acceptors for the reducing

equivalents generated during glucose degradation via the glycolytic, the hexosemonophosphate, or Entner–Doudoroff routes. These intermediates are compounds of the reductive branch in the citric acid cycle, in succinate formation during mixed-acid fermentation, and in propionic acid fermentation by *Propionibacterium*. In the butyric acid fermentation by saccharolytic clostridia, pyruvate is converted to acetyl-CoA and subsequently to acetoacetyl-CoA, which delivers butyrate and butanol by investing two and four reducing equivalents, respectively. Acetyl-CoA can also be reduced to ethanol in mixed-acid fermentations or converted to acetate with the generation of one mole of ATP by substrate-level phosphorylation. This reaction is particularly relevant because it contributes to energy conservation. Because acetate formation is not accompanied by the consumption of reducing equivalents, some other reaction has to compensate for this deficiency by regenerating two oxidized coenzyme equivalents. In mixed-acid fermentations, these are the routes to ethanol or succinate formation. In other systems, acetate formation is balanced by the production of molecular hydrogen via gas-evolving hydrogenases or by the reduction of 2 mol CO_2 to 1 mol acetate (as in homoacetogenic clostridia). In summary, the fermentation of carbohydrates frequently follows a route in which pyruvate is the central metabolite; pyruvate either acts as an electron acceptor or it is converted into compounds serving as the electron acceptors for the reoxidation of the reduced coenzymes.

III. FERMENTATION OF ORGANIC ACIDS, AMINO ACIDS, AND PURINES

A. Organic Acids

In addition to carbohydrates, organic acids and amino acids are the major substrates for fermentations. Organic acids are generated from carbohydrates via the reactions described in Section II or by the oxidative deamination of the amino acids arising from proteolysis. Among the short-chain organic acids, acetate is the most abundant substrate. Its fermentation by methanogenic bacteria (e.g., *Metha-*

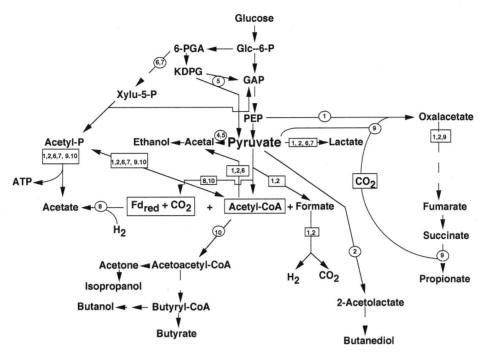

Fig. 2. Scheme of carbon flux in selected fermentation types with carbohydrate as the substrate. Only characteristic intermediates are shown; the reducing equivalents that are formed and consumed are omitted. Abbreviations: Fd_{red}, reduced ferredoxin; GAP, glyceraldehyde-3-phosphate: Glc-6-P, glucose-6-phosphate; KDPG, 2-keto-3-desoxy-6-phosphogluconate; PEP, phosphoenolpyruvate. 6-PGA, 6-phosphogluconate; Xylu-5-P, xylulose-5-phosphate;. The specific fermentation pathways are indicated by numbers at the arrows: (1) mixed-acid fermentation (*E. coli* type); (2) mixed-acid fermentation (*Enterobacter* type); (3) homolactic fermentation; (4) alcoholic fermentation (yeast); (5) alcoholic fermentation (*Zymomonas*); (6) heterolactic fermentation (*Leuconostoc* type); (7) heterolactic fermentation (*Lactobacillus* type); (8) homoacetic (acetic) fermentation; (9) propionic acid fermentation (*Propionibacteria*); (10) butanol–acetone fermentation (saccharolytic clostridia).

nosarcina, Methanospirillum, or *Methanothrix*) is widespread in nature. About 70% of the methane produced biogenically derives from the reduction of the CH_3 group of acetate by the following reaction:

$$CH_3-COOH \rightarrow CH_4 + CO_2$$

The major amount of methane production from acetate occurs in the rumen of ruminants and in lake sediments. Lactic acid is fermented to propionate, acetate, and CO_2 via two convergently evolved pathways. *Propionibacterium* uses the pathway depicted in Fig. 2 (reaction 9), whereas *Clostridium propionicum* forms propionate by the dehydration of lactyl-CoA to acrylyl-CoA and then reduction to propionate.

Citrate fermentation has been studied extensively because citrate is the major substrate for the generation of flavor compounds in fermented milk, especially diacetyl. Citrate degradation is initiated by its cleavage into oxalacetate and acetate by citrate lyase. Oxalacetate decarboxylase, a membrane-bound enzyme complex, catalyzes biotin-dependent pyruvate formation coupled to the extrusion of a Na^+ ion across the membrane, causing a build up of a sodium gradient that can be used for ATP synthesis. Further metabolism of pyruvate takes place along the routes depicted in Fig. 2 (e.g., in enterobacteria via reactions 1 and 2). Similar Na^+-dependent decarboxylases are involved in succinate or methylmalonate fermentation by *Propionigenium modestum* and in glutamate fermentation by *Acidaminococcus* and *Peptococcus* species.

B. Fermentation of Amino Acids

Amino acids are readily fermentated by many bacteria, especially by members of gram-positive anaerobes belonging to the peptolytic clostridia and nonsporogenic *Peptococcus,* as well as by *Eubacterium* species. Table I lists some examples in which a single amino acid can serve as a substrate of fermentation. Because of the diversity of chemical structures, the fermentation of amino acids, unlike that of most carbohydrates, does not follow a route convergently leading to some central metabolite; instead it follows individual pathways that involve a plethora of chemically unusual (and novel) reactions. Frequently, different organisms follow different pathways. Examples of this are listed in Table I. Glutamate fermentation of *C. tetanomorphum* proceeds through mesaconate as an intermediate, whereas that of *Acidaminococcus* and *Peptostreptococcus* involves oxidative deamination to 2-oxyglutarate, reduction and dehydratation to glutaconate, and decarboxylation to crotonyl-CoA. The last is then dismutated to butyrate and acetate, yielding the same end products as glutamate fermentation by *C. tetanomorphum*. The pathways of threonine fermentation by *C. propionicum* and *E. coli* differ in that the 2-oxobutyrate generated from threonine by the threonine dehydratase is oxidatively cleaved by a ferradoxin-dependent oxidoreductase in *C. propionicum,* whereas an isozyme of pyruvate formate lyase is involved in the reaction by *E. coli.*

Most of the peptolytic *Clostridia* also are able to ferment pairs of amino acids in the Stickland reaction. In this type of fermentation, one of the amino acids serves as the electron donor and the other as the electron acceptor. Preferred electron donors are alanine, the branched-chain amino acids, and histidine. Preferred acceptors are glycine, proline, arginine, and tryptophan. The oxidation of the donor involves oxidative deamination to the 2-oxo-acid, followed by oxidative decarboxylation to the acyl-CoA derivative, which can be used in substrate-level phosphorylation for ATP synthesis via the phosphotransacetylase–acetokinase reaction sequence with acetyl-phosphate as intermediate. The reduction of the acceptor amino acid involves at least three proteins, two of which are selenoproteins. Formally, the amino acid is reduced to the corresponding organic acid plus NH_4^+ with the generation of another molecule of ATP, again by substrate-level phosphorylation.

C. Fermentation of Purines and Pyrimidines by Specialists

Purines and pyrimidines are degradation products of nucleic acids; they can be degraded by a few specialists. Some of them are so restricted in their substrate spectrum that they accept a single compound only. Thus *C. acidi-urici* and *C. cylindrosporum* only ferment guanines, the end products being formate, CO_2, NH_4^+, acetate, and, in the case of the latter organism, glycine. Other organisms ferment pyrimidines; *C. oroticum,* for example, degrades orotic acid to acetate, CO_2, and NH_4^+.

TABLE I
Examples of Selected Amino Acid Fermentations

Amino acid	Overall reaction products	Organisms
Alanine	Acetate, propionate, NH_4^+	*Clostridium propionicum*
Glycine	H_2, NH_4^+, CO_2, acetic acid	*Peptococcus anaerobius*
Glutamate	NH_4^+, H_2, CO_2, acetate, butyrate	*Clostridium tetanomorphum*
Glutamate	NH_4^+, CO_2, acetate, butyrate	*Acidaminococcus fermentans, Peptostreptococcus* sp.
Threonine	NH_4^+, H_2, CO_2, propionate	*Clostridium propionicum*
Threonine	NH_4^+, formate ($H_2 + CO_2$), propionate	*E. coli*
Arginine	CO_2, NH_4^+, (ornithine)	Halophiles, *Pseudomonas, Mycoplasma*

IV. ENERGY CONSERVATION REACTIONS

The classic energy conservation mode in fermentations is substrate-level phosphorylation. Immediate phosphoryl donors for the phosphorylation of ADP are acetyl-phosphate, propionyl-phosphate, butyryl-phosphate, carbamyl-phosphate, 1,3-diphosphoglycerate, and phosphoenolpyruvate. Other energy-rich intermediates originating in fermentative pathways are acetyl-CoA, propionyl-CoA, butyryl-CoA, and succinyl-CoA. ATP is generated by transfer from CoA to acetate via the activities of phosphotransacetylase and acetokinase. Similar reaction sequences exist for propionyl-CoA and butyryl-CoA.

It is now generally accepted that Y_{ATP} (the cell yield obtained from 1 mol ATP after growth on monomeric substrates in minimal salts medium) is about 10.5 g. The determination of Y_{ATP} allowed the conclusion that in most fermentations with glucose as substrate, no more than 3 mol ATP/ mol glucose degraded are formed. The ATP that is formed is then used to build up a proton-motive force across the cytoplasmic membrane via ATP cleavage coupled with the extrusion of protons or sodium ions.

Substrate-level phosphorlation is frequently accompanied by direct membrane energy reactions. In the case of the sodium-dependent decarboxylation of oxalacetate in enterobacteria or methylmalonate by *Propioniogenium,* sodium ions are translocated to the outside of the cell, resulting in the build up of a sodium gradient used for ATP synthesis. Propionibacteria and enterobacteria conserve energy also by electron transport-coupled phosphorylation during the reduction of fumarate. Thus, in addition to the substrate-level ATP formation, a respiratory chain is used during fermentation. Finally, in several fermentation routes the export of certain acidic fermentative end products produced in symport with protons can contribute to the gain of energy.

V. REGULATORY ASPECTS

The regulation of fermentation has attracted considerable attention because it sheds light on the adaptation of microorganisms to environmental conditions, such as the amount of oxygen or the pH. Most of the work has dealt with the metabolic consequences of a shift from aerobiosis to anaerobiosis in facultative organisms, mainly *E. coli* and other enterobacteria.

A. Regulation of Pyruvate Cleavage

The key reaction involved in the shift of enterobacteria from respiration to fermentation is the cleavage of pyruvate. Aerobically, as well as under conditions of anaerobic respiration, this reaction is carried out by pyruvate dehydrogenase. Under fermentative conditions, the synthesis of this enzyme is repressed and its task taken over by pyruvate formate lyase, which cleaves pyruvate nonoxidatively into acetyl-CoA and formate. Pyruvate formate lyase is present in aerobically grown cells, but at a low basal level only and in an inactive state. The enzyme is activated by introducing a radical located at a C-terminal glycine residue by means of an activase enzyme, with S-adenosyl-methionine as cofactor and flavodoxin as electron donor (Fig. 3), according to the following reaction:

$$E - H + \text{S-adenosyl-methionine}$$
$$\rightarrow E^\circ + 5'\text{-desoxyadenosine} + \text{methionine}$$

The radical form of the enzyme is highly sensitive to oxygen; it is cleaved at glycine 734 of the protein backbone during exposure to oxygen, with a glycyl group becoming converted into an oxalyl residue.

Concomitant with the activation of the oxygen-stable precursor, about a 15-fold induction takes place. The *pfl* structural gene lies in a transcriptional unit with *focA*, probably coding for a formate transporter. The operon is transcribed from seven promoters. The *trans* elements involved are the fumarate-nitrate-reductase regulatory (FNR) protein, the ArcA/B two-component regulatory system, the catabolite activation protein (CAP), and the integration host factor (IHF) (Fig. 3). After full induction, pyruvate formate lyase is the major protein in anaerobically grown *E. coli* cells. The structural gene for the activase is located downstream of the *focA-pfl* transcriptional unit and is expressed constitutively.

When fermenting cells of *E. coli* encounter oxygen, the oxygen-sensitive radical form of pyruvate formate lyase is rapidly converted into the nonradical oxygen-

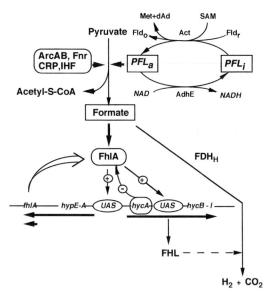

Fig. 3. The mixed-acid fermentation of *E. coli* and the principles of its regulation. Abbreviations: dAd, 5'-desoxyadenosine. Fld$_r$ and Fld$_o$, reduced and oxidized forms of flavodoxin; *hyc*, genes for hydrogenase 3, *hyp*, genes for hydrogenase maturation; Met, methionine; PFL$_a$ and PFL$_i$ active radical form and inactive form of pyruvate formate-lyase, respectively; SAM, S-adenosyl-methionine. FhlA binds to the upstream regulatory sequences (UAS) of the formate regulon genes and activates transcription. HycA counteracts this activation. Bold arrows indicate transcripts. The formate hydrogen–lyase (FHL) system disproportionates formate into H$_2$ and CO$_2$.

stable form. Deactivation is accomplished by the AdhE protein, which also catalyzes the reduction of acetyl-CoA to ethanol with acetal as intermediate. AdhE forms stringlike multimers, which are called spirosomes. Their physiological relevance and the mechanism of deactivation of the radical form of pyruvate formate lyase are unresolved.

B. Regulation of the Downstream Branches of Mixed-Acid Fermentation

As Fig. 3 indicates, acetyl-CoA formed in the anaerobic pyruvate cleavage reaction by enterobacteria can follow two alternative routes. One of them, to acetate via the phosphotransacetylase–acetokinase route, is coupled to ATP synthesis but does not contribute to reduced nicotinamide adenine dinucleotide (NADH) reoxidation. The other one, reduction

to ethanol, can compensate for this deficiency because two reducing equivalents are consumed. Formate, on the other hand, is converted to CO$_2$ and H$_2$ via the formate hydrogen–lyase complex.

The route to acetate appears to be formed constitutively. On the other hand, the route to ethanol (catalyzed by AdhE) and the formate hydrogen–lyase complex are regulated at the level of gene expression. The level of AdhE is expanded considerably under fermentation conditions. Several mechanisms, such as the involvement of FruR or dependence on the NAD/NADH ratio have been put forward, but there is no definite resolution of the question.

More information is available on the regulation of the formate hydrogen–lyase system. The expression of the structural genes and of the genes involved in hydrogenase maturation (*hyp* genes) is under the control of the activator protein FhlA, which requires formate as a ligand (Fig. 3). FhlA binds to an upstream regulatory sequence and induces transcription activation at the $-12/-24$ type of promoters. FhlA thus works like a regulator of two-component systems but, unlike regulators in such systems, does not require phosphorylation for activity.

The structural gene for FhlA is under the control of its own gene product. Under anaerobic conditions, therefore, a buildup on the cellular level of the activator takes places in an autocatalytic manner. Two counteracting mechanisms exist, however, that balance the expression level. The first one is the concomitant induction of the expression of the structural genes for the formate hydrogen–lyase system, which lowers the cellular concentration of formate (which is the low-molecular-weight inducer). The second one relies on the fact that the product of the first gene of the hydrogenase 3 operon (HycA) is an antagonist of FhlA and inactivates it, possibly by direct protein–protein interaction.

In summary, the anaerobic expression of *adhE* and of formate hydrogen–lyase genes is subject to different control mechanisms. AdhE formation appears to be controlled by the redox demands of metabolism, whereas the formate hydrogen–lyase genes are controlled by the pool level of formate and the acidity of the medium. Acidification is counterbalanced by the disproportionate conversion of formate into neutral end products.

C. Regulation of the Butanediol Operon in *Klebsiella*

Klebsiella and *Enterobacter* species can convert 2 mol pyruvate into butanediol via the intermediates 2-acetolactate and acetoin. The three-gene operon involved is under the control of the LysR-type regulator BudR. BudR requires acetate as a ligand for activity. Expression is increased at acidic pH values so that the physiological role of butanediol formation may again reside in the pH homoeostatis. Pyruvate is preferentially converted into the neutral end products butanediol and acetoin at the expense of the acidic fermentation products lactate, acetate, formate, and succinate.

D. Regulation of Butyrate–Butanol Fermentation in *Clostridium acetobutylicum*

Because of the potential industrial interest, a considerable amount of work has been invested in increasing the yield of solvents in the butyrate–butanol–acetone fermentation of *C. acetobutylicum*. When the pH is kept at neutral or slightly alkaline values, this organism predominantly produces a mixture of acetate and butyrate, in addition to CO_2 and H_2. When the pH is allowed to drop to values between 4 and 5, acetone and butanol are formed, and the amount of butyrate is concomitantly reduced. Solventogenesis appears to be another mechanism maintaining pH homoeostasis and thereby preventing lethal acidification. It is unresolved how the organisms monitor the pH and the mechanism, by which pH controls gene expression.

See Also the Following Articles

ANAEROBIC RESPIRATION • CARBOHYDRATE SYNTHESIS AND METABOLISM • ENERGY TRANSDUCTION PROCESSES

Bibliography

Böck, A., and Sawers, G. (1997). Fermentation. *In* "*Escherichia coli* and *Salmonella*" (F. C. Neidhardt, ed.), 2nd ed. pp. 262–282. ASM Press, Washington, DC.

Gottschalk, G. (1986). "Bacterial Metabolism," 2nd ed. Springer Verlag Berlin.

Kessler, D., and Knappe, J. (1997). Anaerobic dissimilation of pyruvate. *In* "*Escherichia coli* and *Salmonella*" (F. C. Neidhardt, ed.), 2nd ed., pp. 199–205. ASM Press, Washington DC.

Thauer, R. K., Jungermann, K., and Decker, K. (1977). Energy conservation in chemotrophic, anaerobic bacteria. *Bacteriol. Rev.* **41**, 100–180.

Fermented Foods

Keith H. Steinkraus
Cornell University

I. Classification
II. Safety of Fermented Foods
III. Fermentation Processes
IV. Nutritional Aspects of Fermented Foods

GLOSSARY

acid fermentations Fermentations in which lactic acid, acetic acid, or other organic acids are produced, lowering the pH sufficiently to make the substrate resistant to the growth of microorganisms that might otherwise spoil the food.

alcoholic fermentations Fermentations in which carbon dioxide flushes out residual oxygen, rendering the substrate anaerobic and producing ethanol at levels that make the substrate resistant to the development of food poisoning and spoilage microorganisms.

alkaline fermentations Fermentations in which the pH rises to 7.5 or higher, allowing the growth of the essential microorganisms and inhibiting the growth of organisms that might otherwise spoil the food.

bioenrichment The enrichment of the protein, essential amino acid, or vitamin content of food through the growth and synthetic activities of microorganisms overgrowing the substrate.

cereal–legume steamed breads and pancakes Breads and pancakes made from mixtures of cereal grains (other than wheat and rye) and legumes and without the need for gluten to retain the carbon dioxide needed for leavening.

high-salt savory-flavored amino acid–peptide sauces and pastes Savory meat-flavored protein hydrolysates, such as Chinese soy sauce, Japanese shoyu, and miso, fermented in NaCl concentrations above 13%.

nutritious Supplying calories, energy, protein, essential amino acids and peptides, essential fatty acids, vitamins, and minerals required to satisfy the metabolic needs of the consumer.

FERMENTED FOODS are food substrates that are invaded or overgrown by edible microorganisms whose enzymes, particularly amylases, proteases, and lipases, hydrolyze the polysaccharides, proteins, and lipids to nontoxic products with flavors, aromas, and textures that are pleasant and attractive to the human consumer. If the products of enzyme activities have unpleasant odors or undesirable unattractive flavors, or the products are toxic or disease producing, the foods are described as spoiled.

I. CLASSIFICATION

Fermented foods can be classified in a number of ways. They can be classified by category of microorganism (alcoholic beverages fermented by yeasts, vinegars fermented with *Acetobacter,* milks fermented with lactobacilli, pickles fermented with lactobacilli, fish or meat fermented with lactobacilli, and plant proteins fermented with molds with or without lactobacilli and yeasts), by class (beverages, cereal products, dairy products, fish products, fruit and vegetable products, legumes, and meat products) or by commodity (fermented starchy roots, fermented cereals, alcoholic beverages, fermented vegetable proteins, and fermented animal protein). Alternatively, they can be classified, as the Sudanese traditionally do, not on the basis of microorganism or commodity but on the basis of function as Kissar (staples), porridges and breads, such as aceda and kissra; Milhat, sauces and relishes for the staples; marayiss, 30 types of opaque beer, clear beer, date wines and meads, and other alcoholic drinks; and Akil-munasabat (foods for special occasions).

Steinkraus (1996) classified fermentations according to the following categories, which will serve as the basis for this article.

1. Fermentations producing Textured vegetable protein meat substitutes in legume–cereal mixtures. Examples are Indonesian tempe and ontjom.

2. High-salt meat-flavored amino acid–peptide sauce and paste fermentations. Examples are sauces (Chinese soy sauce, Japanese shoyu and miso, Indonesian Kecap and tauco Malaysian kicap, Taiwanese inyu, Philippine taosi, and Korean kanjang, doenjang and Kochujang), fish sauces (Vietnamese nuocmam, Philippine patis, and Malaysian budu) and fish pastes (Philippine bagoong, Malaysian belachan, Vietnamese mam, Cambodian prahoc, Indonesian trassi, and Korean jeotkal). These are predominately Asian fermentations, but these products are being used more and more in the United States.

3. Lactic acid fermentations. Examples of lactic acid-fermented vegetables are sauerkraut, cucumber pickles, olives, Egyptian pickled vegetables, Indian pickled vegetables, Korean kim-chi, Thai pak-sian and pak-dong, Chinese hum-choy, and Malaysian pickled vegetables and tempoyak. Examples of lactic acid-fermented milks are yogurt, Russian kefir, Iraqi liban, Indian dahi, Egyptian laban rayab and laban zeer, and Malaysian tairu (soybean milk). Examples of Lactic acid-fermented cheeses are cheese, and Chinese sufu and tofu-ru. Examples of lactic acid-fermented yogurt–wheat mixtures are Egyptian kishk, Greek trahanas, and Turkish tarhanas. Examples of Lactic acid-fermented cereals and tubers (cassava) are Mexican pozol, Ghanian kenkey, Nigerian gari. Examples of boiled rice–raw shrimp–raw fish mixtures are Philippine balao balao, and burong dalag. Examples of lactic fermented-leavened breads are sour-dough breads, Indian idli, dhokla, and khaman; Sri Lankan hoppers; Ethiopian enjera; Sudanese kisra; and Philippine puto. Examples of lactic acid-fermented meats are Western fermented sausages and Thai nham (fermented fresh pork).

4. Alcoholic fermentations. Examples are grape wines; Mexican puique; honey wines; South American chica; beers; Egyptian bouza; palm wines; Indian jackfruit wines, rice beer, madhu, and ruhi; Ethiopian tej and talla; Kenyan muratina; palm wines, urwage and busaa; Kaffir–Bantu beers; Nigerian ugba; Zambian maize beer; sugar cane wines; Japanese sake; Indonesian tape and brem; Malaysian tapuy; Chinese lao-chao; Thai rice wine; and Philippine tapuy.

5. Acetic acid fermentations. Examples are apple cider vinegars; wine vinegars; palm wine vinegars; Philippine coconut water vinegar, nata de pina, and nata de coco; and tea fungus and Kombucha.

6. Alkaline fermentations. Examples are Nigerian dawadawa; Ivory Coast soumbara; African iru and ogiri; Indian kenima; Japanese natto; and Thai thua-nao.

7. Leavened breads. Examples are yeast and sour-dough breads; and Middle East breads.

8. Flat unleavened breads.

These classes of fermented foods are found around the world. The lines between the classifications are not always distinct. Tempe involves a lactic acid fermentation during the soaking of the soybeans. Yeast (alcoholic)–lactobacilli (lactic acid) interactions are rather frequent, for example, in sour-dough breads; in beers and wines; and in Chinese soy sauce; Japanese shoyu, and Japanese miso fermentations. Nevertheless, the Steinkraus classification is useful and provides a way to predict what microorganisms may be involved and what chemical, physical, and nutritive changes may occur in new unfamiliar fermented foods. The classification also relates well to safety factors for fermented foods.

Fermented foods were originally produced by each household, and expanded to a cottage industry as consumer demand required. Some food fermentations, such as Japanese shoyu, miso, and sake; South African maize and sorghum beers, mageu, and mahewu; and Nigerian ogi and gari have been industrialized.

II. SAFETY OF FERMENTED FOODS

Fermented foods generally have a very good safety record, even in the developing world where the foods are manufactured by people without training in microbiology or chemistry in unhygienic, contaminated environments. They are consumed by hundreds of millions of people every day in both the developed and the developing world.

Although fermented foods are generally safe, it should be noted that fermented foods by themselves do not solve the problems of contaminated drinking water, environments heavily contaminated with human waste, improper personal hygiene in food handlers, flies carrying disease organisms, unfermented foods carrying food poisoning or human pathogens, and unfermented foods (even when cooked) that have been handled or stored improperly. Also, improperly fermented foods can be unsafe. However, the application of principles that lead to safe fermented foods could lead to an improvement in the overall quality and the nutritional value of the food supply, a reduction of nutritional diseases, and a greater resistance to intestinal and other diseases in infants.

A. Safety Principles

There are several principles the lead to safe food fermentations. The first principle is that food substrates overgrown with desirable, edible microorganisms become resistant to invasion by spoilage or toxic microorganisms. Other, less desirable, possibly disease-producing organisms find it difficult to compete. An example is Indonesian tempe kedelee. Indonesian tempe, illustrating a first safety principle of food fermentation, is an example of a major classification of fermented food, vegetable-protein meat substitutes. The substrate is soaked, dehulled, partially cooked soybeans. During the initial soaking, the soybeans undergo an acid fermentation that lowers the pH to 5.0 or below, a pH that is inhibitory to many organisms but highly acceptable to the mold. Following cooking, the soy cotyledons are surface dried, which inhibits the growth of bacteria that might spoil the product. The essential microorganism is *Rhizopus oligosporus* or related *Rhizopus* species, which knit the cotyledons into a compact cake that can be sliced or cut into cubes and used in recipes as a protein-rich meat substitute. These molds have the ability to grow very rapidly at relatively high temperatures (i.e., 40–42°C) that are too high for many bacteria and molds. The mold uses the available oxygen and produces CO_2, which inhibits some of the potential spoilage organisms. The mold also produces spore dust that permeates the environment

and contributes to the inoculation of the desired microorganisms. The combination of a relatively low pH, no free water, and a high temperature in the fermenting bean mass enables *Rhizopus oligosporus* to overgrow the soybeans in 18 hours. Organisms that might otherwise spoil the product are unable to compete with the mold. In addition, the mold also produces some antibiotics that inhibits other organisms that might invade and spoil the product. As soon as the substrate is overgrown by the desired organism(s), it is resistant to invasion by other, possibly dangerous microorganisms. An additional safety factor is that the raw tempe is sliced or cut into cubes and cooked before consumption, which destroys vegetative microorganisms that are present.

Soybean tempe (tempe kedelee) has an excellent record of safety in Indonesia and Malaysia, where the product has been used for centuries. In temperate countries including the United States, soybeans do not undergo a natural lactic acid fermentation during soaking. The pH remains closer to 6.0, a level that permits the growth of contaminating bacteria. It is known that unacidified soybeans can be invaded by a variety of food-poisoning organisms, including *Staphylococcus aureus, Bacillus cereus,* and *Clostridium botulinum.* This can be prevented by artificially acidifying the soybeans or inoculating the soak water with *Lactobacillus plantarum.* However, American producers have often been careless and have used unacidified or insufficiently acidified soybeans for tempe production. In some cases, this has led to spoilage and it could lead to food poisoning.

Although soybean tempe has an excellent safety record, there is a type of tempe that has a reputation for causing sickness and even death in the consumer. This is tempe bongkrek, made from the coconut residue from coconut milk production. Again, if the substrate is properly acidified, it is generally safe and the mold overgrows the substrate, knitting it into a compact cake. But if the substrate is not sufficiently acidified or if the fungal inoculum is insufficient, the bacterium *Pseudomonas cocovenenans* can grow in the coconut residue producing two toxic compounds—bongkrek acid and toxoflavin. These toxic compounds are inhibitory to the mold and it cannot overgrow the substrate properly. Bongkrek acid is colorless and it can be lethal to the consumer. Toxoflavin is yellow, and so if tempe bongkrek has a yellow color or if the mold has not overgrown the substrate properly, it should not be consumed. Quite a number of Indonesians in central Java die every year from the consumption of improperly fermented tempe bongkrek. This is an example of a fermented food that could become toxic, but it is amply documented and there is no good excuse for people consuming it and becoming ill or dying.

The mycotoxins that are present in many cereal grain and legume substrates before fermentation present a valid safety concern. They are a problem; however, it is not the fermentation that produces the mycotoxins. They are produced when the cereal grains or legumes are improperly harvested or stored. Furthermore, it has been found that during the tempe fermentation, mycotoxin levels are reduced. The ontjom mold *Neurospora* and the tempe old *Rhizopus oligosporus* decrease the aflatoxin content of peanut press-cake 50% and 70%, respectively, during fermentation. And during the soaking and cooking of raw substrates before fermentation, many potential toxins such as trypsin inhibitor, phytate, and hemagglutinin are destroyed. So, in general, fermentation tends to detoxify the substrates.

An infant formula developed in Indonesia containing tempe as 39.6% by weight along with wheat flour (31.6%), sugar (23.3%), and vegetable oil (2.9%) allowed infants suffering from acute diarrhea who consumed the formula to recover more rapidly than infants who did not receive the formula; also the infants gained weight rapidly. It would appear that the inclusion of tempe in infant formula on a broader scale would not only decrease the overall incidence of diarrhea but also improve infant and child growth rates and nutrition.

III. FERMENTATION PROCESSES

A. Lactic Acid Fermentations

Fermentations involving the production of lactic acid are generally safe. Lactic acid fermentations include those in which the fermentable sugars are converted to lactic acid by organisms such as *Leuconostoc mesenteroides, Lactobacillus brevis, Lactobacillus plan-* *tarum, Pediocccus cerevisiae, Streptococcus thermophilus, Streptococcus lactis, Lactobacillus bulgaricus, Lactobacillus acidophilus, Lactobacillus citrovorum,* and *Bifidobacterium bifidus.*

Lactic acid-fermented foods are a major classification of fermented foods, containing a vast quantify of safe human food. We are all aware of the excellent safety record of sour milks and yogurts, cheeses, and pickles. In addition, the lactic acid fermentations provide the consumer with a wide variety of flavors, aromas, and textures to enrich the human diet.

Probably the most ancient lactic fermentation is fermented or sour milk. Raw, unpasteurized milk rapidly sours because the lactic acid bacteria present in the milk ferment the milk sugar, lactose, to lactic acid. In the presence of acid and a low pH, other microorganisms that might invade the milk and transmit disease-producing organisms are less able to do so. If the whey is allowed to escape or evaporate, the residual curd becomes a primitive cheese.

Vegetable foods and vegetable–fish–shrimp mixtures are preserved around the world using lactic acid fermentation. The classic Western lactic acid vegetable fermentation is sauerkraut. Fresh cabbage is shredded and 2.25% salt is added. A sequence of lactic acid bacteria develops. First, *Leuconostoc mesenteroides* grows, producing lactic acid, acetic acid, and CO_2, which flushes out any residual oxygen and makes the fermentation anaerobic. Then, *Lactobacillus brevis* grows, producing more acid. Finally, *Lactobacillus plantarum* grows, producing still more lactic acid and lowering the pH to below 4.0. At this pH and under anaerobic conditions, the cabbage or other vegetables will be preserved for long periods of time.

Korean kimchi is a fermentation similar to sauerkraut but it includes not only Chinese cabbage but also radishes, red pepper, ginger, and garlic. It is less acid than sauerkraut and is consumed while it is still carbonated. It is a staple of the Korean diet and the consumption of 100 g or more of it per capita per day is not uncommon. Kimchi is still a household-produced fermentation in Korea, although it is also produced commercially.

Nearly all vegetables, such as cucumbers, radishes, and carrots; and even some green fruits, such as olives, papaya, and mango, are acid-fermented in the presence of salt around the world. Indian farmers can safely preserve their surplus vegetables by lactic acid fermentation on the farm. This improves the supply and availability of vegetable foods throughout the year and improves the nutrition of the Indian population.

1. Pit Fermentations

Lactic acid fermentations include the pit fermentations in the south Pacific islands. They have been used for centuries by the Polynesians to store and preserve breadfruit, taro, banana, and cassava tubers. The fermented pastes or whole fruits, sometimes punctured, are placed in leaf-lined pits. The pits are covered with leaves and the pits are sealed. The low pH and anaerobic conditions account for the stability of the foods. An abandoned pit estimated to be about 300 years old contained breadfruit still in edible condition.

In Ethiopia, the pulp of the false banana, *Ensete ventricosum*, is also fermented in pits. It undergoes lactic acid fermentation and is preserved until the pit is opened. Then the mash is used to prepare the flat bread Kocho—a staple in the diet of millions of Ethiopians.

2. Lactic Acid-Fermented Rice–Shrimp–Fish Mixtures

Philippine balao balao is a lactic acid-fermented rice–shrimp mixture prepared by mixing boiled rice, raw shrimp, and solar salt (about 3% w/w), packing this mixture in an anaerobic container, and allowing it to ferment for several days or weeks. The chitinous shell of the shrimp becomes soft, and when the product is cooked the whole shrimp can be eaten. The process provides a method of preserving raw shrimp or pieces of raw fish. The products are well preserved by the low pH and anaerobiosis until the containers are opened. Then they must be cooked and consumed.

3. Yogurt–Cereal Mixtures

Another household-produced lactic acid fermentation of considerable nutritional importance includes Egyptian kishk, Greek trahanas, and Turkish tarhanas. These products are parboiled wheat–yogurt mixtures that combine the high nutritional value of

wheat and milks while attaining excellent preservative qualities. The processes are rather simple. Milk is fermented to yogurt and the yogurt and wheat are mixed and boiled together until the mixture is highly viscous. The mixture is than allowed to cool, formed into biscuits by hand and sun-dried. Trahanas can be stored on the kitchen shelf for years and used as a base for highly nutritional soups. In the Egyptian kishk process, tomatoes, onions, and other vegetables are sometimes combined with the yogurt and wheat in the biscuits.

4. Cereal–Legume Sour Gruels, Porridges, and Beverages

Cereal–legume sour gruels, porridges, and beverages include Nigerian ogi; Kenyan uji; South African mahewu and magou; and Malaysian soybean milk yogurt (tairu). In Nigerian ogi, maize, millet, or sorghum grains are washed and steeped for 24–72 hr, during which time they undrgo lactic acid fermentation. They are drained, wet-milled, and finally wet-sieved to yield a fine, smooth slurry with about an 8% solids content. The boiled slurry, called pap, is a porridge. Pap is a very important traditional food for weaning infants and a major breakfast food for adults. Infants at 9 months old are introduced to ogi by being fed it once a day as a supplement to breast milk. Unfortunately, the nutritional value of ogi is poor. It is vastly improved by the addition of soybean to produce a soy-ogi. Kenyan uji is a related product, except that the grains are ground to a flour before being mixed with water and fermented. The initial slurry is 30% solids. This is fermented for 2–5 days, yielding 0.3–0.5% lactic acid. The slurry is then diluted to 10% solids and boiled. It is then diluted to 4–5% solids and 6% sucrose is added before consumption.

South African mageu, mahewu, or magou is a traditional sour nonalcoholic maize beverage popular among the Bantu people. Corn flour is slurried with water (8–10% solids), boiled, cooled, and inoculated with 5% w/w wheat flour (household-produced fermentation), which serves as a source of microorganisms, and incubated at ambient temperature; or the boiled slurry is inoculated with *Lactobacillus delbrueckii* (industrial process) and incubated at 45°C.

The slurries are fermented to a pH of 3.5–3.9 and then are ready for consumption.

Lactic acid-fermented gruels inhibit the proliferation of gram-negative pathogenic bacteria, including toxicogenic *Escherichia coli, Campylobacter jejuni, Shigella flexneri,* and *Salmonella typhimurium.* Thus, lactic acid fermentation has very profound effects on the nutritive value of cereal gruels for feeding infants and children in the developing world. The ingestion of foods containing live lactic acid bacteria improves the resistance of the gastrointestinal tracts of infants and children to invasion by organisms that cause diarrhea.

5. Cereal–Legume Steamed Breads and Pancakes

Indian idli, a sour steamed bread, and dosa, a pancake, are examples of a household-produced fermentation that could be useful around the world. Polished rice and black gram dahl in various proportions (i.e., 3 : 1 to 1 : 3) are soaked separately during the day. In the evening, the rice and black gram are ground in a mortar with a pestle with enough water added to yield a batter with the desired consistency; the batter is thick enough to require the use of the hand and forearm to mix it properly. A small quantity of salt is added. The batter ferments overnight, during which time *Leuconostoc mesenteroides* and *Streptococcus faecalis,* naturally present on the grains, legumes, and utensils, grow rapidly to outnumber the initial contaminants and dominate the fermentation. The organisms produce lactic acid (total acidity, as lactic acid can reach above 1.0%) and carbon dioxide, which makes the batter anaerobic and leavens the product. In the morning, the batter is steamed to produce small white muffins or fried as a pancake. Soybean cotyledons, green gram, or Bengal gram can be substituted for the black gram. Wheat, maize, or kodri can be substituted for the rice to yield Indian dhokla. A closely related fermentation is Ethiopian enjera that yields the large pancake that serves as the center of the meal.

B. Alcoholic Fermentations

Fermentations involving the production of ethanol are generally safe foods and beverages. These include wines, beers, Indonesian tape ketan and tape ketella, Chinese lao-chao, South African Kaffir and sorghum beer; and Mexican pulque. These are generally yeast fermentations, but they also involve yeast-like molds such as *Amylomyces rouxii,* mold-like yeasts such as *Endomycopsis,* and sometimes bacteria such *Zymomonas mobilis.* The substrates include diluted honey, sugarcane juice, palm sap, fruit juices, germinated cereal grains or hydrolyzed starch, all of which contain fermentable sugars that are rapidly converted to ethanol in natural fermentations by yeasts in the environment. Nearly equal weights of ethanol and carbon dioxide are produced, and the CO_2 flushes out the residual oxygen and maintains the fermentation as anaerobic. The yeasts multiply and ferment rapidly, and other microorganisms, most of which are aerobic, cannot compete. The ethanol is germicidal and, as long as the fermented product remains anaerobic, the product is reasonably stable and preserved.

With starchy substrates such as cereal grains, it is necessary to convert some of the starch to fermentable sugar. This is done in a variety of ways, for example, chewing the grains to introduce ptyalin (in the Andes region of South America) where maize is a staple, or germinating (malting) barley or the grains themselves in most of the world where beers are produced. In parts of Africa, a young lady cannot get married until she is capable of making Bantu beer for her husband.

In Asia, there are at least two additional ways of fermenting starchy rice to alcoholic foods. The first is the use of a mold such as *Amylomyces rouxii,* which produces amylases that convert starch to sugars, and a yeast such as *Endomycopsis fibuliger,* which converts the glucose and maltose to ethanol. The sweet-sour product of alcoholic rice fermentation is called tape ketan in Indonesia. It is consumed as a dessert. If cassava is used as substrate, the product is called tape ketella. This process can also be used to produce rice wines, Chinese lao-chao, and Malaysian tapuy and tapai.

Another method of alcoholic fermentation is the Japanese koji process that is used to ferment rice for rice wine (sake). In this process, boiled rice is overgrown with an amylolytic mold *Aspergillus oryzae* for about 3 days at 30°C. The mold-covered rice, called koji, is then inoculated with a culture of the yeast *Saccharomyces cerevisiae* and water is added. Saccharification by the mold amylases and alcoholic fermentation by the yeast proceed simultaneously. The results of this slow fermentation are high yeast populations and an ethanol content as high as 23% v/v.

C. Acetic Acid or Vinegar Fermentations

If the products of alcoholic fermentation are not kept anaerobic, bacteria belonging to genus *Acetobacter,* which are present in the environment, oxidize portions of the ethanol to acetic acid or vinegar.

Fermentations involving the production of acetic acid yield foods or condiments that are generally safe. Acetic acid also is bacteriostatic to bactericidal, depending on the concentration. In most cases, palm wines and kaffir beers contain not only ethanol but acetic acid. Acetic acid is even more preservative than ethanol. Vinegar is a highly acceptable condiment used in pickling and preserving cucumbers and other vegetables. The alcoholic and acetic acid fermentations can be used to insure safety in other foods.

D. Bread Fermentations

Yeast breads are closely related to alcoholic fermentation and have an excellent reputation for safety. Ethanol is a minor product in bread because of the short fermentation time; but carbon dioxide produced by the yeasts leavens the the bread, producing anaerobic conditions, and baking produces a dry surface resistant to invasion by organisms in the environment. Baking also destroys many of the microorganisms in the bread itself.

Yeast breads are made by the fermentation of wheat and rye flour doughs with yeasts, generally *Saccharomyces cerevisiae.* Sour-dough breads are fermented with lactic acid bacteria and yeast. Indian idli and dosa are rice–legume mixtures fermented with *Leuconostoc mesenteroides* and *Streptococcus faecalis* and sometimes added yeasts (see Section III.A.5).

E. Alkaline Fermentations

Fermentations involving highly alkaline fermentations are generally safe. Africa has a number of very important foods and condiments that are not only used to flavor soups and stews, but also serve as low-cost sources of protein in the diet. Among these are Nigerian dawadawa, Ivory Coast soumbara, and West African iru, made by fermentation of soaked, cooked locust bean *Parkia biglobosa* seeds with bacteria belonging to genus *Bacillus,* typically *Bacillus subtilis;* Nigerian ogiri, made by the fermentation of melon seed (*Citrullus vulgaris*); Nigerian ugba, made by the fermentation of the oil bean (*Pentacletha macrophylla*); Sierra Leonean ogiri-saro, made by the fermentation of sesame seed (*Sesamum indicum*);Nigerian ogiri-igbo, made by the fermentation of castor bean (*Ricinus communis*) seeds; and Nigerian ogirnwan, made by the fermentation of the fluted pumpkin bean (*Telfaria occidentale*) seeds. Soybeans can be substituted for locust beans. This group also includes Japanese natto, Thai thua-nao, and Indian kenima, all produced from soybean. They are all very nutritious and a source of low-cost nitrogen for the diet.

The essential microorganisms are *Bacillus subtilis* and related bacilli. The organisms are very proteolytic and the proteins are hydrolyzed to peptides and amino acids. Ammonia is released and the pH rapidly reaches as high as 8.0 or higher. The combination of high pH and free ammonia, along with very rapid growth of the essential microorganisms at relatively high temperatures (above 40°C), makes it very difficult for other microorganisms that might spoil the product to grow. Thus, the products are quite stable and well preserved, especially when dried. They are safe foods, even though they may be manufactured in an unhygienic environment.

These are all household-produced fermentations that depend on *Bacillus subtilis* spores present in the environment. No deliberate inoculum is needed. The seeds are soaked and the seed coats are removed. The seeds are then cooked and drained. Shallow pots or pans are used for the fermentation. The pans are sources of the required spores, which germinate and overgrow the seeds forming a sticky mucilaginous gum on the surfaces of the beans or seeds.

Interestingly, the Malaysians ferment locust beans

using a similar alkaline process to yield garlic-flavored products. The Japanese ferment soybeans with *Bacillus subtilis* after soaking and cooking to yield a protein-rich food called natto. The Thais ferment soybeans using a similar process to produce a product called thua-nao, which is consumed in northern Thailand as a substitute for fish sauce. The Indians ferment soybeans using similar processes to produce Kenima. Thus, these alkaline fermentations involving bacilli fermenting protein-rich beans and seeds are of considerable importance in widely separated parts of the world. They are all household-produced fermentations; however, Japanese natto has been commercialized.

F. High-Salt Savory-Flavored Amino Acid–Peptide Sauces and Pastes

The addition of salt in ranges from 13% w/v or higher results in a controlled protein hydrolysis that prevents putrefaction, prevents the development of food poisoning, such as botulism, and yields meaty savory amino acid–peptide sauces and pastes that provide very important condiments, particularly for those unable to afford much meat in their diets.

Meaty savory-flavored amino acid–peptides sauces and pastes include Chinese soy sauce, Japanese shoyu, Japanese miso, Indonesian kecap, Taiwan inyu, Korean kan-jang, and Philippine taosi, made by fermentation of soybeans or soybean–rice–barley mixtures with *Aspergillus oryzae*; and fish sauces and pastes made by the fermentation of small fish and shrimp principally using the proteolytic gut enzymes of the fish and shrimp.

The ancient discovery of how to transform bland vegetable protein into meat-flavored amino acid–peptide sauces and pastes was an outstanding human accomplishment. As are many discoveries, it was probably initially accidental. The earliest substrates were probably meat or fish mixed with a millet koji. In the most primitive process known today, soybeans are soaked and thoroughly cooked. They are mashed, formed into a ball, covered with straw, and hung to ferment under the rafters of the house. In 30 days, they are completely overgrown with the mold *Aspergillus oryzae*. Packed into a strong salt brine (about 18% w/v) the beans are hydrolyzed by the proteases, lipases, and amylases in the mold, yielding an amino acid–peptide liquid sauce with a meaty flavor and a similar meat-flavored paste or residue—a primitive miso. One can only guess what a dramatic effect this meat-flavored sauce and paste had on consumers of the typical bland rice diet in Asia. The process not only enriches the flavor but it also retains the essential amino acids in the soybean. Because of the relatively high lysine content of the soybean, soy sauce or miso is an excellent adjunct to the nutritional quality of rice. Soy sauce started as an Asian food; but, in recent years it has been adopted by the West.

In the presence of a high salt concentration (above 13% w/v, but generally in the range of 20–23%) fish-gut enzymes will hydrolyze fish proteins to yield amino acid–peptide, meaty-flavored fish sauces and pastes similar to soy sauces and pastes. Halophilic pediococci contribute to the typical desirable fish sauce flavor and aroma. Both soy sauce and fish sauce fermentations are household-produced as well as being produced by large commercial manufacturers.

IV. NUTRITIONAL ASPECTS OF FERMENTED FOODS

The nutritional impact of fermented foods on nutritional diseases can be direct or indirect. Food fermentations that raise the protein content or improve the balance of essential amino acids or their availability will have a direct curative effect. Similarly fermentations that increase the content or availability of vitamins, such as thiamine, riboflavin, niacin, or folic acid, can have profound direct effects on the health of the consumers of such foods. This is particularly true for people subsisting largely on maize, in which niacin or nicotinic acid is limited and for whom pellagra is incipient, and for people subsisting principally on polished rice, which contains limited amounts of thiamine and for whom beri-beri is incipient. The biological enrichment of foods via fermentation can prevent this.

On the other hand, fermentation does not generally increase the number of calories unless it converts substrates unsatisfactory for humans to human-quality foods, as when mushrooms are grown on straw or waste paper or peanut and coconut presscakes are converted to edible foods such as Indonesian ontjom (oncom).

Fermentation enriches the human dietary through the development of a wide diversity of flavors, aromas, and textures in food; preserves substantial amounts of food through lactic acid, alcoholic, acetic acid, alkaline, and high-salt fermentations; enriches food substrates biologically with vitamins, protein, essential amino acids, and essential fatty acids; decreases toxins, and decreases cooking times and fuel requirements.

A. Enrichment with Protein

In Indonesian tape ketan fermentation, rice starch is hydrolyzed to maltose and glucose and than fermented to ethyl alcohol. The loss of starch solids results in a doubling of the protein content (from about 8–16% in rice) on a dry solids basis. Thus, this process provides a means by which the protein content of high-starch substrates can be increased for the benefit of consumers needing higher protein intakes. This is particularly important for people who principally consume cassava, which has a protein content of about 1% (wet basis). If the tape ketella fermentation is applied to cassava, as it is in Indonesia, the protein content can be increased to at least 3% (wet basis)—a very significant improvement in nutrition to the consumer. In addition, the flavor of the cassava becomes sweet and sour and alcoholic, flavors consumers may prefer to the bland starting substrate.

B. Enrichment with Essential Amino Acids

The Indonesian tape Ketan and tape ketella fermentation not only enriches the substrate with protein, but the microorganisms also selectively enrich the rice substrate with lysine, the first essential limiting amino acid in rice. This improves the protein quality.

Several researchers have reported a 10.6–60.0% increases in methionine during the Indian idli fermentation. An increase in methionine, a limiting essential amino acid in legumes, greatly improves the protein value.

C. Enrichment with Vitamins

In the wealthy Western world, nutrients, particularly vitamins, are added to selected formulated manufactured foods as a public health measure. Example are the addition of vitamins A and D to milk and butter, and riboflavin to bread. Fruit juices may be fortified with ascorbic acid (vitamin C). Although enrichment and fortification are within the means of the Western world, they are far beyond the means of the developing world. Thus, much of the world must depend on biological enrichment via fermentation to enrich their foods.

Highly polished white rice is deficient in vitamin B-1 (thiamine). Consumers subsisting principally on polished rice are in danger of developing beri-beri, a disease characterized by muscular weakness and polyneuritis, leading eventually to paralysis and heart failure. Infants nursed by mothers suffering from thiamine deficiency may develop infantile beri-beri in which sudden death occurs at about 3 months of age due to cardiac failure. The microorganisms involved in the tape ketan fermentation synthesize thiamine and restore the thiamine content to the level found in unpolished rice. This can be a very significant contribution to the nutrition of rice-eating people.

Tempe is a protein-rich meat substitute in Indonesia made by overgrowing soaked dehulled partially cooked soybeans with *Rhizopus oligosporus* or related molds. The mold knits the cotyledons into a firm cake that can be sliced and cooked or used in place of meat in the diet. During the fermentation, proteins are partially hydrolyzed, the lipids are hydrolyzed to their constituent fatty acids, stachyose (a tetrasaccharide undigestible in humans) is decreased (decreasing flatulence in the consumer), riboflavin nearly doubles, niacin increases sevenfold and vitamin B-12 (usually lacking in vegetarian foods) is synthesized by a bacterium that grows along with the essential mold. The tempe process can be used to introduce texture into many legume–cereal mixtures, yielding products with decreased cooking times and improved digestibility. The bacterium responsible for vitamin B-12 production is a nonpathogenic strain of *Klebsiella pneumoniae*. It can also be inoculated into the Indian idli fermentation, produce vitamin B-12.

Mexican pulque is the oldest alcoholic beverage on the American continent. It is produced by the fermentation of juices of the cactus plant *Agave*. Pulque is rich in thiamine, riboflavin, niacin, panthothenic acid, p-amino benzoic acid, pyridoxine, and biotin. Pulque is of particular importance in the diets of low-income children in Mexico.

Kaffir beer is an alcoholic beverge with a pleasantly sour taste and the consistency of a thin gruel. It is a traditional beverage of the Bantu people of South Africa. The alcohol content ranges from 1–8% v/v. Kaffir beer is generally made from kaffir corn (*Sorghum caffrorum*) malt and unmalted kaffir corn meal. Maize or millet may be substituted for kaffir corn. During the fermentation, the thiamine content remains about the same, but the riboflavin more than doubles and the niacin and nicotinic acid nearly doubles, which is very significant in people who consume principally maize. Consumers of the usual amounts of kaffir beer are not in danger of developing pellagra.

Palm sap is a sweet clear colorless liquid that contain about 10–12% fermentable sugar and is neutral in reaction. Palm wine is a heavy, milk-white opalescent suspension of live yeasts and bacteria with a sweet taste and vigorous effervescence. Palm wines are consumed throughout the tropics. Palm wine contains as much as 83 mg ascorbic acid/liter. Thiamine increased from 25 to 150 μg/liter, riboflavin increased from 35 to 50 μg/liter, and pyridoxine increased from 4 to 18 μg/liter during fermentaiton. Palm wine also contains considerable amounts of vitamin B-12, 190 to 280 pg/ml. Palm toddies play an important role in nutrition among the economically disadvantaged in the tropics. They are the cheapest sources of B vitamins.

D. Reduction of Toxins

During the soaking and hydration that raw substrates undergo in various fermentation process and the usual cooking, many potential toxins such as the trypsin inhibitor, phytate, and hemagglutinin and cyanogens in cassava are reduced or destroyed. Even aflatoxin, frequently found in peanut and cereal grains substrates, is reduced in the Indonesian ontjom fermentation. The ontjom mold *Neurospora* and the tempe mold *Rhizopus oligosporus* decrease the aflatoxin content of peanut presscake 50 and 70%, respectively, during fermentation.

E. Reduction of Cooking Times and Fuel Requirements

Economizing fuel requirements is very important in the developing world, where women may spend hours every day collecting enough leaves, twigs, wood, and dried dung to cook the day's food. Lactic acid-fermented foods generally require little, if any, heat in their fermentation and can be consumed without cooking. Examples are pickled vegetables, sauerkraut, and kimchi. Indonesian tempe fermentation converts soybeans that would require as much as 5–6 hr cooking time to a product that can be boiled in soup for 5–10 min.

See Also the Following Articles

Food Spoilage and Preservation • Lactic Acid Bacteria • Vitamins and Related Biofactors, Microbial Production

Bibliography

Campbell-Platt, G. (1987). "Fermented Foods of the World: A Dictionary and Guide." Butterworths, London.

Dirar, M. (1993). "The Indigenous Fermented Foods of the Sudan." CAB International University Press, Cambridge.

Steinkraus, K. H. (1979). Nutritionally significant indigenous fermented foods involving and alcoholic fermentation. *In* "Fermented Food Beverages in Nutrition" (C. F. Gastineau et al., eds.), pp. 36–50. Academic Press, New York.

Steinkraus, K. H. (1989). "Industrialization of Indigenous Fermented Foods." Marcel Dekker, New York.

Steinkraus, K. H. (1991). African alkaline fermented foods and their relation to similar foods in other parts of the world. *In* "Traditional African Foods—Quality and Nutrition" (A. Westby and P. J. A. Reilly, eds.), pp. 87–92. International Foundation for Science, Stockholm.

Steinkraus, K. H. (1993). Comparison of fermented foods of

East and West. *In* "Fish Fermentation Technology" (C. H. Lee et al., eds.), pp. 1–12. United Nations University Press, Tokyo.

Steinkraus, K. H. (1994). Nutritional significance of fermented foods. *Food Res. Int.* **27**, 259–267.

Steinkraus, K. H. (1996). "Handbook of Indigenous Fermented Foods," 2nd ed. Marcel Dekker, New York.

Steinkraus, K. H. (1997). Classification of fermented foods: Worldwide review of household fermentation techniques. *Food Control* **8**, 311–317.

Steinkraus, K. H. (1998). Bio-enrichment: Production of vitamins in fermented foods. *In* "Microbiology of Fermented Foods" (B. J. B. Wood, ed.), Vol. 2, pp. 603–622. Blackie Academic, London.

Wood, B. J. B. (1998). "Microbiology of Fermented Foods," Vols. 1–2. 2nd ed. Blackie Academic, London.

Fimbriae, Pili

Matthew A. Mulvey, Karen W. Dodson, Gabriel E. Soto, Scott J. Hultgren

Washington University School of Medicine

GLOSSARY

curli A class of thin, irregular, and highly aggregated adhesive surface fibers expressed by *Escherichia coli* and *Salmonella enteritidis*.

periplasmic chaperones A class of proteins localized in the periplasm of gram-negative bacteria that facilitates the folding and assembly of pilus subunits, but which are not components of the final pilus structure.

phase variation The reversible on-and-off switching of a bacterial phenotype, such as pilus expression.

pilin The individual protein subunit of a pilus organelle (also known as a fimbrin). Immature pilins, containing leader signal sequences that direct the transport of pilins across the inner membrane of gram-negative bacteria, are called propilins or prepilins.

usher Oligomeric outer-membrane proteins that serve as assembly platforms for some types of pili. Usher proteins can also form channels through which nascent pili are extruded from bacteria.

PILI, also known as fimbriae, are proteinaceous, filamentous polymeric organelles expressed on the surface of bacteria. They range from a few fractions of a micrometer to greater than 20 μm in length and vary from less than 2 to 11 nm in diameter. Pili are composed of single or multiple types of protein subunits, called pilins or fimbrins, which are typically arranged in a helical fashion.

Pilus architecture varies from thin, twisting thread-like fibers to thick, rigid rods with small axial holes. Thin pili with diameters of 2–3 nm, such as K88 and K99 pili, are often referred to as "fibrillae." Even thinner fibers ($<$ 2 nm), which tend to coil up into a fuzzy adhesive mass on the bacterial surface, are referred to as thin aggregative pili or curli. High-resolution electron microscopy of P, type 1, and S pili of *Escherichia coli* and *Haemophilus influenzae* pili has revealed that these structures are composite fibers, consisting of a thick pilus rod attached to a thin, short distally located tip fibrillum. Pili are often expressed peritrichously around individual bacteria, but some, such as type 4 pili, can be localized to one pole of the bacterium.

Pili expressed by gram-negative bacteria have been extensively characterized, and the expression of pili by gram-positive bacteria has also been reported. The numerous types of pili assembled by both gram-negative and gram-positive organisms have been ascribed diverse functions in the adaptation, survival, and spread of both pathogenic and commensal bacteria. Pili can act as receptors for bacteriophage, facilitate DNA uptake and transfer (conjugation), and, in at least type 4 pilus, function in cellular motility. The primary function of most pili, however, is to act as scaffolding for the presentation of specific adhesive moieties. Adhesive pilus subunits (adhesins) are often incorporated as minor components into the tips of pili, but major structural subunits can also function as adhesins. Adhesins can mediate the interaction of bacteria with each other, with inanimate surfaces, and with tissues and cells in susceptible host organisms. The colonization of host tissues by bacterial pathogens typically depends on a stereochemical fit between an adhesin and complementary receptor architecture. Interactions mediated by adhesive pili can facilitate the formation of bacterial communities such as biofilms and are often critical to the successful colonization of host organisms by both commensal and pathogenic bacteria.

I. HISTORICAL PERSPECTIVE AND CLASSIFICATION OF PILI

Pili were first noted in early electron microscopic investigations as nonflagellar, filamentous appendages of bacteria. In 1955 Duguid designated these appendages "fimbriae" (plural, from Latin for thread or fiber) and correlated their presence with the ability of *E. coli* to agglutinate red blood cells. Ten years later Brinton introduced the term "pilus" (singular, from Latin for hair) to describe the fibrous structures (the F pilus) associated with the conjugative transfer of genetic material between bacteria. Since then "pilus" has become a generic term used to describe all types of nonflagellar filamentous appendages, and it is used interchangeably with the term "fimbria."

Historically, pili have been named and grouped based on phenotypic traits such as adhesive and antigenic properties, distribution among bacterial strains, and microscopic characterizations. In the pioneering work of Duguid and co-workers, pili expressed by different *E. coli* strains were distinguished on the basis of their ability to bind to and agglutinate red blood cells (hemagglutination) in a mannose sensitive (MS) as opposed to a mannose resistant (MR) fashion. Pili mediating MS hemagglutination by *E. coli* were designated type 1 pili and these pili have since been shown to recognize mannose-containing glycoprotein receptors on host eukaryotic cells. Morphologically and functionally homologous type 1 pili are expressed by many different species of Enterobacteriaceae. Despite their similarities, however, type 1 pili expressed by the various members of the Enterobacteriaceae family are often antigenically and genetically divergent within their major structural subunits. In contrast to type 1 pili, most other pili so far

identified are either nonhemagglutinating or mediate MR hemagglutination. These pili are very diverse and possess a myriad of architectures and different receptor binding specificities and functions.

Since the discovery and initial characterization of pili in the 1950s, substantial advances have been made in our understanding of the genetics, biochemistry, and structural and functional aspects of these organelles. A vast number of distinct pilus structures have been described and new types of pili continue to be identified. Pili are now known to be encoded by virtually all gram-negative organisms and are some of the best-characterized colonization and virulence factors in bacteria. Here we classify pili that are expressed by gram-negative bacteria into six groups according to the mechanisms by which they are assembled. This classification scheme is not all inclusive, but provides a convenient means for discussing the diverse types of pili, their functions, structures and assembly. Representatives of various pilus types assembled by the various pathways discussed in the following sections are listed in Table I and electron micrographs of the various pilus types are shown in Fig. 1.

II. CHAPERONE–USHER PATHWAY

All pilins destined for assembly on the surface of gram-negative bacteria must be translocated across the inner membrane, through the periplasm, and across the outer membrane. To accomplish this, various adhesive organelles in many different bacteria require two specialized assembly proteins, a periplasmic chaperone and an outer membrane usher. Chaperone–usher assembly pathways are involved in the biogenesis of over 30 distinct structures, including composite pili, thin fibrillae, and nonfimbrial adhesins. Here, we focus on the structure and assembly mechanisms of the prototypical P and type 1 pilus chaperone–usher systems.

A. Molecular Architecture

P and type 1 pili are both composite structures consisting of a thin fibrillar tip joined end to end to a right-handed helical rod (Fig. 2A–B). Chromosomally located gene clusters that are organizationally

TABLE I
Pilus Assembly Pathways

Assembly pathway	Structure	Assembly gene products[a]	Organism	Disease(s) associated with pilus expression
Chaperone–Usher pathway				
Thick rigid pili	P pili	PapD/PapC	*E. coli*	Pyelonephritis/cystitis
	Prs pili	PrsD/PrsC	*E. coli*	Cystitis
	Type 1 pili	FimC/FimD	*E. coli*	Cystitis
			Salmonella sp.	
			K. pneumoniae	
	S pili	SfaE/SfaF	*E. coli*	UTI
				Newborn meningitis
	F1C pili	FocC/FocD	*E. coli*	Cystitis
	H. influenzae fimbriae	HifB/HifC	*H. influenzae*	Otitis media
				Meningitis
	H. influenzae biogroup aegyptius fimbriae	HafB/HafE	*H. influenzae*	Brazilian purpuric fever
	Type 2 and 3 pili	FimB/FimC	*B. pertussis*	Whooping cough
	MR/P pili	MrpD/MrpC	*P. mirabilis*	Nosocomial UTI
	PMF pili	PmfC/PmfD	*P. mirabilis*	Nosocomial UTI
	Long polar fimbriae	LpfB/LpfC	*S. typhimurium*	Gastroenteritis
	Pef pili	PefD/PefC	*S. typhimurium*	Gastroenteritis
	Ambient-temperature fimbriae	AftB/AftC	*P. mirabilis*	UTI
	987P fimbriae	FasB/FasD	*E. coli*	Diarrhea in piglets
	REPEC fimbriae	RalE/RalD	*E. coli*	Diarrhea in rabbits
Thin flexible pili	K99 pili	FaeE/FaeD	*E. coli*	Neonatal diarrhea in calves, lambs, piglets
	K88 pili	FanE/FanD	*E. coli*	Neonatal diarrhea in piglets
	F17 pili	F17D/F17papC	*E. coli*	Diarrhea
	MR/K pili	MrkB/MrkC	*K. pneumoniae*	Pneumonia
Atypical structures	CS31A capsule-like protein	ClpE/ClpD	*E. coli*	Diarrhea
	Antigen CS6	CssC/CssD	*E. coli*	Diarrhea
	Myf fimbriae	MyfB/MyfC	*Y. enterolitica*	Enterocolitis
	pH 6 antigen	PsaB/PsaC	*Y. pestis*	Plague
	CS3 pili	CS3-1/CS3-2	*E. coli*	Diarrhea
	Envelope antigen F1	Caf1M/Caf1A	*Y. pestis*	Plague
	Non-fimbrial adhesins I	NfaE/NfaC	*E. coli*	UTI
				Newborn meningitis
	SEF14 fimbriae	SefB/SefC	*S. enteritidis*	Gastroenteritis
	Agregative adherence fimbriae I	AggD/AggC	*E. coli*	Diarrhea
	AFA-III	AfaB/AfaC	*E. coli*	Pyelonephritis
Alternate chaperone pathway	CS1 pili	CooB/CooC	*E. coli*	Diarrhea
	CS2 pili	CotB/CotC	*E. coli*	Diarrhea
	CS4 pili		*E. coli*	Diarrhea
	CS14 pili		*E. coli*	Diarrhea
	CS17 pili		*E. coli*	Diarrhea
	CS19 pili		*E. coli*	Diarrhea
	CFA/I pili	CfaA/CfaC	*E. coli*	Diarrhea
	Cable type II pili		*B. cepacia*	Opportunistic in cystic fibrosis patients

continues

Continued

Assembly pathway	Structure	Assembly gene products[a]	Organism	Disease(s) associated with pilus expression
Type II secretion pathway	Type-4A pili	General secretion spparatus (Main terminal branch)	*Neisseria* sp. *P. aeruginosa* *Moraxella* sp. *D. nodosus*	Gonorrhea, meningtitis Opportunistic pathogen Conjuntivitis, respiratory infections Ovine footrot
		14 to >20 proteins	*E. corrodens* *Azoarcus* sp.	
	Type-4B pili: bundle forming pili longus CFA/III R64 pili toxin co-regulated pili	General secretion apparatus (Main terminal branch) 14 to >20 proteins	*E. coli* *E. coli* *E. coli* *E. coli* *V. cholera*	Diarrhea Diarrhea Diarrhea Cholera
Conjugative pilus assembly pathway (Type IV secretion pathway)	F pili (IncF1) IncN, IncP, IncW-encoded pili T pili	Type IV export apparatus, 12–16 proteins	*E. coli* *E. coli* *A. tumefaciensi*	Antibiotic resistance Crown gall disease
Extracellular nucleation/pre-cipitation pathway	Curli	CsgG/CsgE/CsgF	*E. coli* *S. enteritidis*	Sepsis
Type III secretion pathway	Hrp pili EspA pilus-like structures	Type III secretion apparatus, ~20 proteins	*P. syringae* *E. coli*	Hypersensitive response (in resistant plants) Diarrhea

[a] Chaperone/usher for chaperone–usher and alternate chaperone pathway.

and functionally homologous encode P and type 1 pili (Fig. 2D–E). The P pilus tip is a 2-nm-wide structure composed of a distally located adhesin (PapG), a tip pilin (PapE), and adaptor pilins (PapF and PapK). The PapG adhesin binds to Galα(1-4)Gal moieties present in the globoseries of glycolipids found on the surface of erythrocytes and kidney cells. Consistent with this binding specificity, P pili are major virulence factors associated with pyelonephritis caused by uropathogenic *E. coli*. The minor pilin PapF is thought to join the PapG adhesin to the tip fibrillum, the bulk of which is made up of a polymer of PapE subunits. PapK is thought to terminate the growth of the PapE polymer and to join the tip structure to the rod. The pilus rod is composed of multiple PapA subunits joined end to end and then coiled into a right-handed 6.8-nm-thick helical rod having a pitch distance of 24.9 Å and 3.28 subunits per turn.

The rod is terminated by a minor subunit, PapH, which may serve to anchor the pilus in the membrane.

Similar to the P pilus structure, the type 1 pilus has a short, 3-nm-wide fibrillar tip made up of the mannose-binding adhesin, FimH, and two additional pilins, FimG and FimF. The FimH adhesin mediates attachment to mannosylated receptors expressed on a wide variety of cell types and has been shown to be a significant virulence determinant for the development of cystitis. The type 1 tip fibrillum is joined to a rod composed predominantly of FimA subunits arranged in a 6- to 7-nm-diameter helix with a pitch distance of 23.1 Å and 3.125 subunits per turn. Both type 1 and P pilus rods have central axial holes with diameters of 2–2.5 Å and 1.5 Å, respectively. Despite the architectural similarities, type 1 pili appear to be more rigid and prone to breaking than P pili. Some

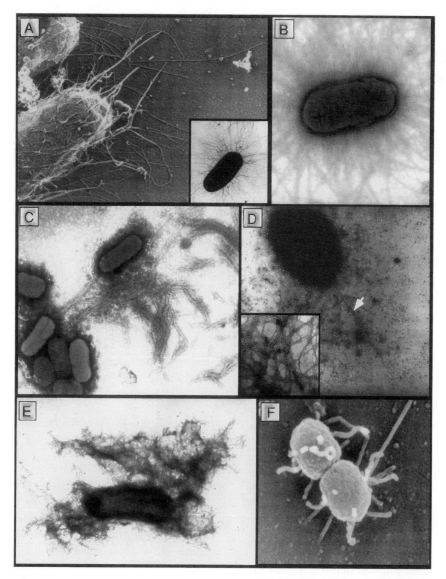

Fig. 1. Montage of various pilus structures. (A) High-resolution transmission electron micrograph of *E. coli* expressing P pili (~2 nm-thick pili), the prototypical structures assembled by the chaperone or usher pathway. Inset shows a typical lower-resolution micrograph of a negatively stained bacterium expressing P pili. (B) CS2 pili (~2 nm-thick pili) assembled by the alternate chaperone pathway in *E. coli.* Photo courtesy of Harry Sakellaris and June R. Scott. (C) The polar type 4B bundle forming pili (~6 nm-wide pili) of enteropathogenic *E. coli* (EPEC). Photo courtesy of Dave Bieber and Gary Schoolnik. (D) Arrow indicates a single, polar T pilus (10 nm-thick structure), the promiscuous conjugative pilus assembled by *A. tumefaciens.* Inset shows a field of purified T-pili. Photos provided by Clarence I. Kado. (E) *E. coli* expressing curli, < 2 nm-wide structures assembled by the extracellular nucleation–precipitation pathway. Photo courtesy of Stafan Normark. (F) Scanning electron micrograph of EPEC elaborating ~50 nm-thick bundles of pili containing EspA, a protein exported by a type III secretion pathway. The individual 6- to 8-nm-thick pili making up the bundles are not resolved in this micrograph. Photo provided by S. Knutton (Knutton, S., Rosenshine, I., Pallen, M. J., Nisan, I., Neves, B. C., Bain, C., Wolff, C., Dougan, G., and Frankel, G. (1998). *EMBO J.* **17,** 2166–2176) and reprinted with permission from Oxford University Press.

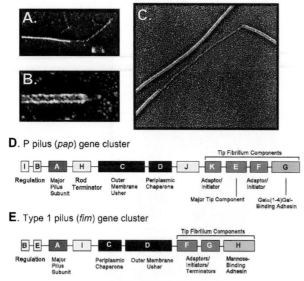

D. P pilus (*pap*) gene cluster

| I | B | A | H | C | D | J | K | E | F | G |

Regulation | Major Pilus Subunit | Rod Terminator | Outer Membrane Usher | Periplasmic Chaperone | | Adaptor/ Initiator | | Adaptor/ Initiator |

Tip Fibrillum Components

Major Tip Component | Galα(1-4)Gal-Binding Adhesin

E. Type 1 pilus (*fim*) gene cluster

| B | E | A | I | C | D | F | G | H |

Regulation | Major Pilus Subunit | Periplasmic Chaperone | Outer Membrane Usher | Adaptors/ Initiators/ Terminators | Mannose-Binding Adhesin

Tip Fibrillum Components

Fig. 2. High-resolution electron micrographs showing the pilus rod and tip fibrillum structures of (A) P and (B) type 1 pili. Unraveling of a portion of a P pilus rod into a linear fiber is shown in (C). The images shown in (A), (B), and (C) are at different magnifications. The P (*pap*) and type 1 (*fim*) gene clusters are depicted in (D) and (E), respectively. These gene clusters share organizational as well as functional homologies.

reports have argued that, unlike the P pilus system in which the tip subunits are thought to be located only within the tip, some of the type 1 tip subunits may also be occasionally intercalated within the rod structure.

In both P and type 1 pili, the major pilin subunits making up the rods are organized in a head-to-tail manner. Additional quaternary interactions between subunits in adjacent turns of the helical rod appear to stabilize the structure and may help drive the outward growth of the organelle during pilus assembly (see later). The disruption of these latter interactions by mechanical stress or by incubation in 50% glycerol can cause the pilus rod to reversibly unwind into a 2-nm-thick linear fiber similar in appearance to the tip fibrillum (Fig. 2C). Bullitt and Makowski (1995) have proposed that the ability of the pilus rods to unwind allows them to support tension over a broader range of lengths. This may help P and type 1 pili better withstand stress, such as shearing forces from the bulk flow of fluid through the urinary tract, without breaking.

In addition to composite structures exemplified by P and type 1 pili, chaperone–usher pathways also mediate the assembly of thin fibrillae such as K88 and K99 pili and nonfimbrial adhesins. K88 and K99 pili are 2- to 4-nm-thick fibers that mediate adherence to receptors on intestinal cells. They are significant virulence factors expressed by enterotoxigenic *E. coli* (ETEC) strains that cause diarrheal diseases in livestock. These pili were given the "K" designation after being mistakenly identified as K antigens in *E. coli*. In contrast to P and type 1 pili, the adhesive properties of K88 and K99 pili are associated with the major pilus subunits. The receptor-binding epitopes on the individual major pilus subunits are exposed on the pilus surface and available for multiple interactions with host tissue. In general, pili with adhesive major subunits, such as K88 and K99 pili, are thin flexible fibrillar structures. In comparison, pili with specialized adhesive tip structures, such as P and type 1 pili, are relatively rigid and rod-like.

B. Assembly Model

The assembly of P pili by the chaperone–usher pathway is the best understood of any pilus assembly pathway. PapD is the periplasmic chaperone and PapC is the outer-membrane usher for the P pilus system. These proteins are prototypical representatives of the periplasmic chaperone and outer-membrane usher protein families. Figure 3 presents the current model for pilus assembly by the chaperone–usher pathway, as depicted for P pili.

1. Periplasmic Chaperones

The PapD chaperone, the PapC usher, and all of the P pilus structural subunits have typical signal sequences recognized by the *sec* (general secretion) system. The signal sequences are short, mostly hydrophobic amino-terminal motifs that tag proteins for transport across the inner membrane by the *sec* system. This system includes several inner-membrane proteins (SecD to SecF, SecY), a cytoplasmic chaperone (SecB) that binds to presecretory target proteins, a cytoplasmic membrane-associated ATPase (SecA) that provides energy for transport, and a periplasmic signal peptidase. As the P pilus structural subunits emerge from the *sec* translocation machin-

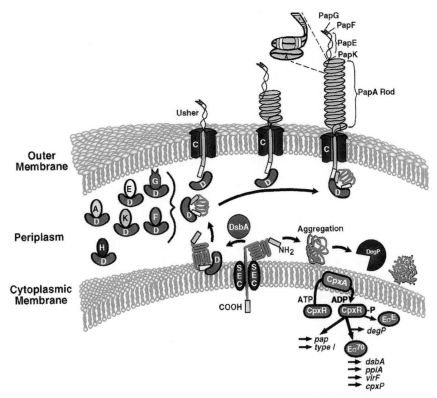

Fig. 3. Model of P pilus assembly by the chaperone–usher pathway. Structural subunits for the pilus tip (PapG, PapF, PapE, and PapK) and for the pilus rod (PapA) are translocated across the inner membrane of *E. coli* via the *sec* system. On the periplasmic side of the inner membrane they interact with the chaperone PapD, which facilitates the folding and release of the subunits from the inner membrane. DsbA is required for the proper folding of both the subunits and the PapD chaperone. In the absence of the chaperone, the subunits aggregate and are degraded by the periplasmic protease DegP. The buildup of unchaperoned subunits results in the activation of the Cpx system, increasing the production of periplasmic protein-folding and -degradation proteins. Subunit–chaperone complexes are targeted to the outer membrane usher, PapC, where the chaperone is released and subunit–subunit interactions occur. The PapC usher forms a 2-nm-wide pore through which the assembled pilus structure is extruded as a linear fiber across the outer membrane. Once on the exterior of the cell, the linear PapA polymer forms a thick right-handed helical rod that is unable to slip back through the usher pore. The formation of the coiled PapA rod is thought to help drive the outward growth of the pilus.

ery into the periplasm, PapD binds to each subunit, facilitating its release from the inner membrane. Each subunit forms an assembly competent, one-to-one complex with PapD. Proper folding of the subunits requires PapD and involves the action of the periplasmic disulfide bond isomerase DsbA. In the absence of PapD, subunits misfold, aggregate, and are subsequently degraded by the periplasmic protease DegP.

The misfolding of P pilus subunits is sensed by the CpxA–CpxR two-component system in which CpxA is an inner membrane-bound sensor or kinase and CpxR is a DNA-binding response regulator (Jones, 1997). Activation of the Cpx system alters the expression of a variety of genes and may help regulate pilus biogenesis.

The three-dimensional crystal structures of both

the PapD and FimC chaperones have been solved. Both chaperones consist of two Ig (immunoglobulin)-like domains oriented into a boomerang shape such that a subunit-binding cleft is created between the two domains. A conserved internal salt bridge is thought to maintain the two domains of the chaperone in the appropriate orientation. Using genetics, biochemistry, and crystallography, PapD was found to interact with pilus subunits, in part, by binding to a highly conserved motif present at the carboxyl terminus of all subunits assembled by PapD-like chaperones. The finer details of how PapD-like chaperones interact with pilus subunits were unveiled by the recent determination of the crystal structures of the PapD–PapK and the FimC–FimH chaperone–subunit complexes (Sauer, 1999, and Choudhury, 1999) (Fig. 4). This work demonstrated that the

PapD and FimC chaperones both make similar interactions with their respective subunits. Only the PapD–PapK structure is considered here. PapK has a single domain comprised of an Ig fold that lacks the seventh (carboxy-terminal) β-strand that is present in canonical Ig folds. The absence of this strand produces a deep groove along the surface of the pilin subunit and exposes its hydrophobic core. The carboxy-terminal F strand and the A2 strand of PapK form the groove. The PapD chaperone contributes its G_1 β-strand to complete the Ig fold of the PapK subunit by occupying the groove, an interaction termed donor strand complementation. This interaction shields the hydrophobic core of PapK and stabilizes immature pilus subunits within the periplasm. Similar interactions are thought to have a central role in the maturation of subunits assembled by a

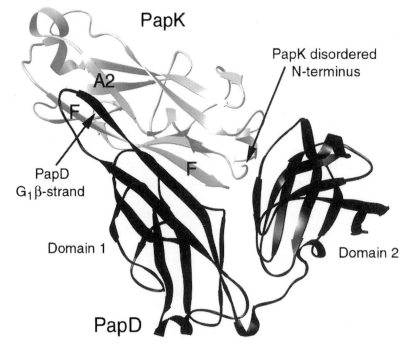

Fig. 4. Ribbon model based on the crystal structure of the PapD periplasmic chaperone complexed with the PapK pilus subunit. PapD is black and the single domain of PapK is gray. The G_1 β-strand in domain 1 of PapD completes the Ig fold of the pilin, occupying the groove formed between the A2 and F strands of PapK. This interaction has been termed donor strand complementation. The eight amino-terminal amino acids of PapK are disordered in the structure. These residues have been implicated in mediating subunit–subunit interactions within the mature pilus organelle. During pilus biogenesis, it is proposed that the amino-terminal strand of a pilin can displace the G_1 β-strand of the chaperone and insert into the carboxy-terminal groove of the neighboring subunit. The mature pilus would thus consist of a linear array of canonical Ig domains, each of whose fold is completed by a strand from the neighboring subunit.

variety of different chaperone–usher pathways. The residues that make up the carboxy-terminal groove formed by subunits and bound by PapD-like chaperones have been shown by mutagenesis studies to be involved in subunit–subunit interactions within the final pilus structure. Thus, in addition to stabilizing immature pilus subunits, the donor strand complementation interaction also caps one of the interactive surfaces of the subunit and prevents premature oligomerization and aggregation of pilus subunits within the periplasm.

2. Outer-Membrane Ushers

After being formed in the periplasm, chaperone–subunit complexes are targeted to the outer-membrane usher where the chaperone is released, exposing interactive surfaces on the subunits that facilitate their assembly into the pilus. Studies in the P and type 1 pilus systems have demonstrated that the adhesin–chaperone complexes, PapDG or FimCH, bind tightest and fastest to the usher and that the adhesins are the first subunits assembled into the pilus. The binding of the chaperone–adhesin complex induces a conformational change in the usher, possibly priming it for pilus assembly. Additional subunits are incorporated into the pilus depending, in part, on the kinetics with which they are partitioned to the usher in complex with the chaperone. Conserved amino-terminal regions, in addition to the conserved carboxy-terminal motif of the pilus subunits, mediate subunit–subunit interactions within the mature pilus. Differences in the complementary surfaces in these conserved regions from one subunit to another may help dictate which of the subunits can be joined to one another during pilus assembly. Thus, the order of the subunits within the final pilus structure is determined by the specific contacts made between the various pilus subunits and also by the differential affinities of the various chaperone–subunit complexes for the usher.

In addition to acting as an assembly platform for the growing pilus, the usher protein appears to have additional roles in pilus biogenesis. High-resolution electron microscopy revealed that the PapC usher is assembled into a 15-nm-diameter ring-shaped complex with a 2-nm-wide central pore (Thanassi, 1998). PapC and other usher family members are thought to have a predominantly β-sheet secondary structure, typical of outer-membrane pore-forming proteins, and they are predicted to present large regions to the periplasm for interaction with chaperone–subunit complexes. After dissociating from the chaperone at the usher, subunits are incorporated into a growing pilus structure that is predicted to be extruded as a 1-subunit-thick linear fiber through the central pore of the usher complex. The packaging of the linear pilus fiber into a thicker helical rod on the outside surface of the bacterium may provide a driving force for the translocation of the pilus across the outer membrane, possibly acting as a sort of ratcheting mechanism to force the pilus to grow outward. Combined with the targeting affinities of the chaperone–subunit complexes for the usher and the binding specificities of the subunits for each other, this may provide all the energy and specificity needed for the ordered assembly and translocation of pili across the outer membrane.

III. ALTERNATE CHAPERONE PATHWAY

A variation of the chaperone–usher pilus assembly pathway has been identified in strains of ETEC. These bacteria are major pathogens associated with diarrheal diseases of travelers, infants, and young children. ETEC strains produce several types of uniquely assembled adhesive pili that are considered to be important mediators of bacterial colonization of the intestine. The best studied of these pili is CS1, which appears to be composed predominantly of a major subunit, CooA, with a distally located minor component, CooD. Several CS1-like pili have been identified and include CS2, CS4, CS14, CS17, CS19, and CFA/I pili expressed by various ETEC strains and the cable type II pili of *Burkholderia cepacia*, an opportunistic pathogen of cystic fibrosis patients. Four linked genes, *CooA*, *CooB*, *CooC*, and *CooD*, are the only specific genes required for the synthesis of functional CS1 pili. Homologous genes required for the production of CS2 and CFA/I pili have also been cloned and sequenced. Electron-microscopic examination reveals that the CS1-like pili are architecturally similar to P and type 1 pili assembled by the chaperone–usher pathway (Fig. 1B), although none of the proteins involved in the biogenesis of CS1-like pili have any significant sequence homologies to those of any other pilus system.

The assembly of CS1-like pili depends on a specialized set of periplasmic chaperones that are distinct from those of the chaperone–usher pathway described previously. Therefore, we refer to this mode of pilus assembly as the alternate chaperone pathway. In the case of CS1 pili, the chaperone CooB binds to and stabilizes the major and minor pilin subunits, CooA and CooD, which enter into the periplasm in a *sec*-dependent fashion (Fig. 5). Both CooA and CooD share a conserved sequence motif near their carboxy-termini that may function as a chaperone-recognition motif. One of the functions of CooB appears to be the delivery of the pilin subunits to an outer-membrane protein, CooC, which may function as a channel, or usher, for the assembly of pilus fibers. In addition to the pilin subunits, CooB also binds to and stabilizes CooC in the outer membrane. In the absence of the CooB chaperone, CooC and the pilin subunits are degraded. Although less well defined, the assembly of CS1 and related structures appears similar in many respects to the assembly of pili by the classic chaperone–usher pathway. Because CS1-like pili do not appear to be related to those assembled by the chaperone–usher pathway, it has been suggested that these two pilus assembly systems arose independently through convergent evolution.

IV. TYPE II SECRETION PATHWAY FOR TYPE 4 PILUS ASSEMBLY

Type 4 pili are multifunctional structures expressed by a wide diversity of bacterial pathogens. These include *Pseudomonas aeruginosa*, *Neisseria gonorrhoeae* and *N. meningitidis*, *Moraxella* species, *Azoarcus* species, *Dichelobacter nodus*, and many other species classified in these and other genera. Type 4 pili are significant colonization factors and have been shown to mediate bacterial interactions with animal, plant, and fungal cells. In addition, these pili can modulate target-cell specificity, function in DNA uptake and biofilm formation, and act as receptors for bacteriophage. Type 4 pili are also associated with a flagella-independent form of bacterial locomotion, called twitching motility, which allows for the lateral spread of bacteria across a surface.

Type 4 pili are 6-nm-wide structures typically assembled at one pole of the bacterium. They can extend up to several micrometers in length and are made up of, primarily if not completely, a small subunit usually in the range of 145–160 amino acids. These subunits have distinctive features, including a short (6–7 amino acids), positively charged leader sequence that is cleaved during assembly, *N*-methyl-phenylalanine as the first residue of the mature subunit, and a highly conserved, hydrophobic amino-terminal domain. The adhesive properties of type 4 pili are, in general, determined by the major pilus subunit. Additional minor components, however, may associate with these pili and alter their binding specificities. In the case of *Neisseria*, a tip-localized adhesin, PilC1, appears to mediate bacterial adherence to epithelial cells.

Recently, a second class of type 4 pili, referred to as class B or type 4B, has been defined. Type 4B pili were initially characterized in enteric pathogens and

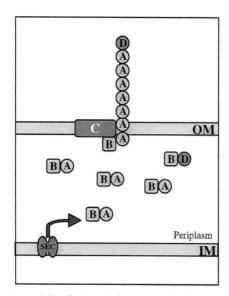

Fig. 5. Assembly of CS1 pili from *E. coli* via the alternate chaperone pathway. The CooB chaperone forms periplasmic complexes with the main components of the pilus, CooA and CooD. It also appears to bind and stabilize the outer-membrane protein CooC in the absence of subunits. CooC appears to function as an outer-membrane (OM) channel for passage of the pilin fiber IM indicates inner-membrane.

include the toxin-coregulated pilus (TCP) of *Vibrio cholera,* the bundle-forming pilus (BFP) of entero-pathogenic *E. coli* (EPEC), and the longus and CFA/III pili of ETEC. Compared to the typical type 4 pilins (referred to as class A or type 4A), the known type 4B pilins are somewhat larger and have a longer (13–30 amino acids) leader sequence. Also, in place of *N*-methyl-phenylalanine as the first amino acid in the mature pilus subunit, type 4B subunits have other methylated residues such as *N*-methyl-methionine for TCP and *N*-methyl-leucine for BFP. TCP, BFP, and longus pili form large polar bundles over 15 μm in length (Fig. 1C). In contrast, CFA/III pili are 1–10 μm long and are peritrichously expressed. The number of pili classified as type 4B is increasing and now includes the R64 thin pilus, an organelle involved in bacterial conjugation.

Parge and coworkers solved the crystal structure of the type 4A pilin subunit (PilE) from *N. gonorrhoeae* in 1995. This work greatly advanced understanding of the structure, function, and biogenesis of type 4 pili. PilE contains 158 amino acids and was determined to have an overall ladle shape, being made up of an α–β roll with a long hydrophobic amino-terminal α-helical spine (residues 2–54) (Fig. 6). All type 4 pilins (type 4A and 4B) are predicted to have a fairly similar structure. The carboxy-terminal domain of type 4 pilins possesses hypervariable regions that affect the binding specificities and antigenicity of type 4 pili. In PilE, these hypervariable regions include a sugar loop (residues 55–77) with an O-linked disaccharide at position Ser-63 and a disulfide-containing region (residues 121–158), which, despite having a hypervariable nature, adopts a regular β-hairpin structure (β_5–β_6) followed by an extended carboxy-terminal tail. A disulfide-containing carboxy-terminal hypervariable region is common among the type 4 pilins. The remainder of PilE was shown to consist of two β-hairpins forming a four-stranded antiparallel β-sheet (residues 78–93 and 103–122) with a connecting β_2–β_3 loop region (residues 94–102).

Through systematic modeling, Parge et al. showed that PilE was probably assembled into pili as monomers arranged in a helix with about five PilE subunits per turn and a pitch distance of approximately 41 Å. PilE subunits are predicted to be packed into pili

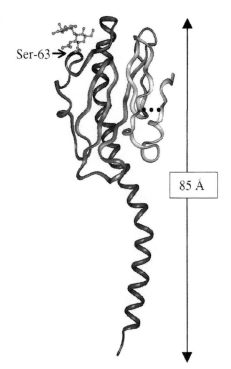

Fig. 6. Ribbon model of the type 4A pilin PilE from *N. gonorrhoeae*. Secondary structural elements include a hydrophobic, conserved amino-terminal α-helical spine connected to a variable domain containing an extended sugar loop with an O-linked disaccharide at position Ser-63 and a disulfide-containing carboxy-terminal region. The disulfide bridge is indicated by a dotted line.

as a three-layer assemblage consisting of an inner core of coiled conserved hydrophobic α-helices surrounded by β-sheets and an outermost layer composed of the disaccharide and hypervariable regions. Hydrophobic packing of the inner core of α-helices along with the flexibility of these helices may permit type 4 pili to bend and adopt twisted, bundled conformations, as seen in Fig. 1C. Hydrogen bonds throughout the middle layer of β-sheets may provide much of the mechanical stability for the pilus. The hypervariable outermost layer is not an integral part of the pilus structure and associates with the middle layer of β-sheets through only a few conserved interactions. Thus, the outermost region can be structurally pliant and accommodate extreme amino acid changes that can lead to antigenic variation and altered binding specificities without disrupting the assembly of the pilus. The antigenic characteristics of type 4 pili synthesized by *N. gonorrhoeae* can be modified extensively by a remarkable mechanism. This pathogen encodes more than 15 distinct silent pilin genes, termed PilS, that lack the invariant amino-terminal domain present in PilE. By recombination of silent PilS genes with the PilE locus, a single neisserial strain can theoretically express greater than 10 million PilE variants.

The biogenesis of type 4 pili is substantially more complicated than pilus assembly by the chaperone–usher or alternate chaperone pathways. Type 4 pilus assembly requires the expression of a myriad of genes that are usually located in various unlinked regions on the chromosome. Exceptions include TCP, BFP, and the R64 thin pilus, which are currently the only type 4 pili for which the majority of the genes required for pilus biogenesis are located within a single genetic locus. Although chromosomally located genes encode most type 4A pili, all known type 4B pili, with the exception of TCP, are encoded by plasmids. The number of genes essential for type 4 pilus biogenesis and function ranges from 14 (for pili such as BFP) to over 20 (for structures such as the type 4A pili of *N. gonorrhoeae*). In *P. aeruginosa*, it is estimated that about 0.5% of the bacterium's genome is involved in the synthesis and function of type 4 pili. Among the various bacterial species expressing type 4 pili, the genes encoding the type 4 pilus structural components are similar, whereas the regulatory components surrounding them are typically less conserved.

Several gene products are currently known to be central to the assembly of type 4 pili. These include a prepilin peptidase that cleaves off the leader peptide from nascent pilin subunits; a polytopic inner membrane protein that may act as a platform for pilus assembly; a hydrophilic nucleotide-binding protein located in the cytoplasm or associated with the cytoplasmic face of the inner membrane, which may provide energy for pilus assembly; and an outer-membrane protein complex that forms a pore for passage of the pilus to the exterior of the bacterium. Many of the components involved in type 4 pilus assembly share homology with proteins that are part of DNA uptake and protein secretion systems, collectively known as the main terminal branch of the general secretory (*sec*-dependent) pathway, or type II secretion. These secretion pathways encode proteins with type 4 pilin-like characteristics and other proteins with homology to type 4 prepilin peptidases and outer-membrane pore-forming proteins. Whether type II secretion systems can assemble pili or pilus-like structures is not known.

Meyer and colleagues have described a model for type 4 pilus assembly in *N. gonorrhoea* (Fig. 7). The PilE propilin subunits are transported into the periplasm by the *sec* translocation machinery. Following translocation, the propilin subunits remain anchored in the inner membrane by their hydrophobic amino-terminal α-helical domains, with their hydrophilic carboxy-terminal heads oriented toward the periplasm. The removal of the positively charged PilE propilin leader sequence by the PilD signal peptidase drives the hydrophobic stems of the PilE subunits to associate to form a pilus. An inner-membrane

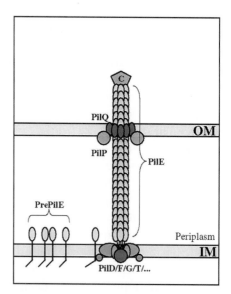

Fig. 7. Model of type 4 pilus assembly by *N. gonorrhoeae*. The PilE prepilin is translocated into the periplasm aided by the *sec* machinery. PilE is processed by the PilD signal peptidase, which cleaves the positively charged leader sequence from the amino-terminus of the pilin subunit. An inner-membrane assembly complex then assembles the mature PilE subunit into a pilus fiber. PilQ mediates translocation of the pilus through the outer membrane, possibly with the assistance of other factors, such as PilP. The PilC adhesin, which appears to be incorporated at the tip of the growing pilus fiber, also seems to be required for translocation of the pilus through the outer membrane.

assembly complex made up of several proteins including PilD, PilF, PilG, and PilT aids this process. The assembled pilus penetrates the outer membrane through a gated pore formed by the multimeric complex Omc. The PilC adhesin associated with the tips of type 4 pili produced by *N. gonorrhoea* may facilitate passage of the nascent pili through the Omc pore. One implication of this assembly model is that the amino-terminal region of PilE resides in a continuous hydrophobic environment during both inner-membrane transport and pilus assembly. This may allow polymerization and, interestingly, depolymerization of the pilus to proceed with only minimal energy requirements. Continued extension and retraction of type 4 pili by rounds of polymerization and depolymerization reactions are proposed as the basis of twitching motility, one of the functions of type 4 pili. This process could be controlled by nucleotide-binding proteins such as PilT, PilB, and PilU associated with the inner-membrane assembly complex. The capacity of type 4 pili to depolymerize may also provide a means for transforming DNA, which could potentially interact with the type 4 pilus fiber, to enter into the bacterial cell.

V. CONJUGATIVE PILUS ASSEMBLY PATHWAY

In gram-negative bacteria, certain pili, collectively known as conjugative pili, facilitate the interbacterial transfer of DNA. These pili allow donor and recipient bacteria to make specific and stable intercellular contacts before DNA transfer is initiated. In some cases, conjugative pili may also form the conduits for intercellular DNA transfer. Horizontal gene transfer, or conjugation, mediated by conjugative pili is inextricably associated with the spread of antibiotic resistance among bacterial pathogens. Conjugative pili are generally encoded by self-transmissible plasmids that are capable of passing a copy of their genes to a recipient bacterium. Closely related plasmids, with similar replication control systems, are unable to coexist in the same cell. This property has been termed incompatibility, and provides the primary basis for cataloging conjugal plasmids and the pili that they encode. Thus far, in *E. coli* alone, over 25

incompatibility groups made up of well over 100 distinct plasmids have been defined. Plasmids of a particular incompatibility group usually encode conjugative pili with similar antigenic properties, sensitivities to pilus-specific phages, and morphologies.

Among the multitude of known incompatibility groups, three morphologically and functionally distinct types of conjugative pili have been defined: rigid, thick flexible, and thin flexible pili. Rigid conjugative pili are 8- to 11-nm-wide structures that are usually specified by conjugal DNA transfer systems that function well only on solid surfaces. Thick flexible pili, on the other hand, are 8- to 11-nm-wide structures that typically, but not always, promote conjugation on solid surfaces and in liquid media equally well. Conjugal DNA transfer promoted by rigid or thick flexible pili can be enhanced, in some cases, by the presence of thin flexible pili. These pili are similar in appearance to type 4 pili and at least one member of the thin flexible pilus group (the R64 thin pilus) has been identified at the molecular level as a type 4B pilus (described previously). Thin flexible pili appear to function primarily in the stabilization of bacterial mating pairs, increasing the rate of DNA transfer. Conjugation does not occur in the presence of thin flexible pili alone, or in the absence of rigid or thick flexible pili.

The most thoroughly studied conjugative pilus is the F pilus encoded by the self-transmissible, broad host range F (fertility) plasmid, a member of the F1 incompatibility (IncF1) group of plasmids borne by *E. coli*. The F pilus system is prototypical for numerous other conjugation systems and F-pilus biogenesis is distinct from type 4 and other pilus assembly pathways. F pili are 8-nm-thick, flexible helical filaments composed primarily, if not completely, of repeating 7.2 kDa (70-amino-acid) TraA pilin subunits. Donor (F$^+$) cells typically express one to three F pili that are usually 1–2 μm long. Each F pilus possesses a 2-nm-wide central channel that is lined by basic hydrophilic residues, which could potentially interact with negatively charged DNA or RNA molecules during conjugation. TraA is organized into pentameric, doughnut-like discs that are stacked in the pilus such that successive discs are translated 1.28 nm along the pilus axis and rotated 28.8° with respect to the lower disc. The TraA pilin has two hydrophobic domains located toward the center and at the carboxy-terminus of the pilin. The hydrophobic domains are thought to extend as antiparallel α-helices from the central axis to the periphery of the pilus shaft. These domains are separated by a short basic region that appears to form the hydrophilic wall of the central channel of the pilus. The amino-terminal domain of TraA is predicted to face the exterior of the pilus. However, this domain is antigenically masked when the amino-terminal residue of TraA is acetylated during maturation of the pilin (see later). This modification is common among all known F-like pilins and appears to cause the amino-terminal domain to be tucked back into or along the pilus shaft. Acetylation is not essential for F-pilus assembly or function, but does help prevent aggregation of F-like pili and affects the phage-binding characteristics of these organelles. Phage are also known to recognize the carboxy-terminal hydrophobic domain of TraA. Although masked within the pilus shaft, the acetylated amino-terminal domain of TraA appears to be exposed in unassembled pilin subunits and at the distal tips of pili. The F pilus tip is believed to initiate contact between donor and recipient cells during conjugation. Alterations in the amino-terminal sequence of TraA provide the primary basis for the antigenic diversity observed among various F-like pili.

At least 16 gene products encoded by the F plasmid are involved in F pilus assembly and an additional 20 or more are needed for conjugation. Two gene products, TraQ and TraX, mediate the processing of the TraA pilin to its mature form. TraA is synthesized as a 12.8-kDa (121-amino-acid) cytoplasmic propilin that is translocated across the inner membrane where it is proteolytically processed by host signal peptidase I to yield the 7.2-kDa pilin form. TraQ, an inner-membrane protein, facilitates the translocation process and may help position the TraA propilin for processing into mature pilin. In the absence of TraQ, the translocation of TraA is disrupted and most of the pilin is degraded. After processing, the amino-terminal residue (alanine) of TraA is acetylated by TraX, a polytopic inner-membrane protein. Whereas TraQ and TraX are involved in the maturation of the TraA pilin, 13 additional gene products (TraL, TraB, TraE, TraK, TraV, TraC, TraW, TraU, TraF, TraH,

TraG, TrbC, and TrbI) affect the assembly of TraA into the pilus filament. Most of these proteins appear to associate with either the inner or outer bacterial membrane and may constitute a pilus assembly complex that spans the periplasmic space.

The exact mechanism by which TraA is assembled into pili is not defined. The mature TraA pilin accumulates in the inner membrane with its amino-terminus facing the periplasm. Both hydrophobic domains of TraA span the inner membrane, with the hydrophilic region of TraA connecting them on the cytoplasmic side. Small clusters of TraA also accumulate in the outer membrane and these may function as intermediates in F pilus assembly and disassembly. Large portions of the TraA sequence have the propensity to assume both β-sheet and α-helical structures, although the α-helical conformation is known to predominate in assembled pili. Frost and co-workers (1984, 1993) have suggested that a shift between β-sheet and α-helical conformations may drive pilus assembly and disassembly. F pilus assembly is energy dependent and the depletion of ATP levels by respiratory poisons such as cyanide results in F pilus depolymerization and retraction. It has been postulated that TraA is normally cycled between pili and periplasmic and inner-membrane pools by rounds of pilus outgrowth and subsequent retraction. During conjugation, F pilus retraction is thought to serve a stabilizing function by shortening the distance between bacterial mating pairs and allowing for more intimate contact.

Several components of the F pilus assembly machinery share significant homology with proteins encoded by other conjugative systems. These include proteins specified by broad host range plasmids in other incompatibility groups (such as IncN, IncP, and IncW) and many of the proteins encoded by the Ti plasmid-specific *vir* genes of the plant pathogen *Agrobacterium tumefaciens*. These bacteria elaborate 10-nm-wide promiscuous conjugative pili, called T pili (Fig. 1D), which direct the interkingdom transfer of a specific genetic element, known as T-DNA, into plant and yeast cells. The introduction of T-DNA into plant cells induces plant tumor formation. T pilus assembly by *A. tumefaciens* requires the expression of at least 12 *vir* gene products encoded by the Ti plasmid. VirB2 is the major, and possibly only, component of the T pilus and it is predicted to be structurally homologous to the F pilus subunit TraA.

Other than possibly stabilizing donor-recipient interactions, it is not clear how F and T pili or any pilus structures function in conjugative DNA-transfer processes. However, substantial evidence exists, at least in the case of F pili, suggesting that pilus components or the pilus itself can serve as a specialized channel for the transmission of DNA and any accompanying pilot proteins across the donor and possibly the recipient cell membranes. In light of this possibility, it is interesting to note that many components of the conjugative pilus systems encoded by the IncF, IncN, IncP, and IncW plasmids and by the *vir* genes of *A. tumefaciens* are similar to the Ptl proteins responsible for the export of the multiple subunit toxin of *Bordetella pertussis*. Furthermore, these secretion systems seem to be distantly related to transport systems used by *Legionella pneumophila* and *Helicobacter pylori* to inject virulence factors into host eukaryotic cells. Conjugative pilus systems such as those encoding F and T pili thus appear to be representative of a larger family of macromolecular transport systems. These type IV secretion systems (not to be confused with the secretion of autotransporters, such as IgA proteases, which is also known as type IV secretion) represent a major pathway for the transfer of both nucleic acid and proteins between cells (Zupan *et al.*, 1998). The understanding of how conjugative pili help mediate the intercellular transfer of macromolecules remains a significant challenge.

VI. EXTRACELLULAR NUCLEATION–PRECIPITATION PATHWAY

Many strains of *E. coli* and *Salmonella enteritidis* produce a class of thin (< 2 nm), irregular, and highly aggregated surface fibers known as curli (Fig. 1E). These distinct organelles mediate binding to a variety of host proteins, including plasminogen, fibronectin, and human contact-phase proteins. They are also involved in bacterial colonization of inert surfaces and have been implicated in biofilm forma-

tion. Curli are highly stable structures and extreme chemical treatment is required to depolymerize them. The major component of *E. coli* curli is a 15.3-kDa protein known as CsgA, which shares over 86% primary sequence similarity to its counterpart in *S. enteritidis,* AgfA.

The formation of curli represents a departure from the other modes of pilus assembly. Whereas structures exemplified by P, CS1, type 4, and F pili are assembled from the base, curli formation occurs on the outer surface of the bacterium by the precipitation of secreted soluble pilin subunits into thin fibers (Fig. 8). In *E. coli*, the products of two divergently transcribed operons are required for curli assembly. The *csgBA* operon encodes the primary fiber-forming subunit, CsgA, which is secreted as a soluble protein directly into the extracellular environment. The second protein encoded by the *csgBA* operon, CsgB, is proposed as inducing polymerization of CsgA at the cell surface. In support of this model, it has been demonstrated that a CsgA$^+$CsgB$^-$ donor strain can secrete CsgA subunits that can be assembled into curli on the surface of a CsgA CsgB$^+$ recipient strain.

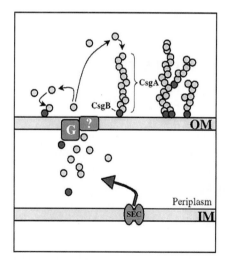

Fig. 8. Model of curli assembly by the extracellular nucleation–precipitaion pathway. CsgA, the main component of curli from *E. coli,* is secreted across the outer membrane. Surface-localized CsgB serves to nucleate CsgA assembly. CsgB is also found distributed along the curli fiber, where it may serve to initiate branching of the fiber. CsgG is an outer membrane-localized lipoprotein that is required for the secretion of CsgA and CsgB, although its function is not clear.

Furthermore, CsgB appears to be interspersed along the length of the curli fiber, where it may initiate branching of the curli structures. In the absence of CsgA, overexpressed CsgB is able to form short polymers on the bacterial cell surface.

The *csgDEFG* operon encodes a gene for a transcriptional activator of curli synthesis (CsgD), and three genes encoding putative assembly factors. One of these factors, CsgG, has been shown to be a lipoprotein that is localized to the outer membrane. In the absence of CsgG, curli assembly does not take place and CsgA and CsgB are rapidly degraded. The precise role of CsgG is not known. Normark and colleagues have suggested that CsgG might act as a chaperone that facilitates the secretion of the CsgA and CsgB and protects them from degradation within the periplasm. It is also possible that CsgG assembles into multimers that could function as a Csg-specific channel within the outer membrane. The roles of the *CsgE* and *CsgF* gene products are not known. It has been reported, however, that a strain lacking these two factors can export assembly-competent CsgA, suggesting that the production of CsgG alone is sufficient for functional maturation of the CsgA subunit of curli.

VII. TYPE III SECRETION PATHWAY

The various pilus assembly pathways described in the previous sections all rely on components of the *sec* machinery for the translocation of their respective pilus subunits across the inner membrane. Two new types of pili that are assembled by a *sec*-independent pathway known as type III secretion have been identified. The type III secretion system is encoded by numerous gram-negative pathogens and enables these bacteria to secrete and inject pathogenic effector molecules into the cytosol of host eukaryotic cells. About 20 gene products, most of which are inner-membrane proteins, make up the type III secretion system. The components mediating type III secretion are conserved in pathogens as diverse as *Yersinia* and *Erwinia,* but the secreted effector proteins vary significantly among species. The type III secretion apparatus, which appears to span the periplasmic space, resembles the basal body of a flagellum connected to a straight rod that extends across the outer membrane. Interestingly, all type III secretion systems encode some components with homologies to proteins involved in flagellar assembly. The secretion of proteins by the type III system is an ATP-dependent process that involves no distinct periplasmic intermediates. Type III-secreted proteins of EPEC and the plant pathogen *Pseudomonas syringae* have recently been shown to assemble into piluslike structures.

EPEC encodes four proteins, EspA, EspB, EspD, and Tir, that are secreted by a type III pathway. These proteins facilitate intimate contact between the pathogen and host intestinal cells and are required for the formation of specific (attaching and effacing) lesions. Knutton and colleagues (1998) showed that one of these proteins, EspA, can assemble into 7- to 8-nm-thick peritrichously expressed pilus-like fibers that are organized into ~50-nm-wide bundles that extend up to 2 μm from the bacterial surface (Fig. 1F). These fibers appear to be made up of only EspA molecules. Interestingly, EspA shares substantial sequence identity with a flagellin from *Y. enterolitica*. During the infection process, the EspA fibers appear to mediate contact between EPEC and the host-cell surface prior to the establishment of more intimate bacterial attachment. The EspA fibers seem to assist the translocation of EspB effector molecules into host cells, where they can subvert host signal-transduction pathways.

In *P. syringae* and other plant pathogens, hypersensitive response and pathogenicity (*hrp*) genes control the ability of these bacteria to cause disease in susceptible plants and to elicit the hypersensitive response in resistant plants. The hypersensitive response is a phenomenon characterized by rapid localized host-cell death at the site of infection that appears to limit the spread of a pathogen in an infected plant. A subset of the *hrp* genes, recently renamed *hrc* genes, encode components of a type III secretion system. In 1997, Roine and co-workers showed that one of the proteins, HrpA, secreted by the Hrp type III secretion system is assembled into 6- to 8-nm-wide, peritrichously expressed pili. It was proposed that these pili, known as Hrp pili, are involved in mediating bacteria–plant interactions in the intercellular spaces of the host plant. In addition, Hrp pili may assist the delivery of effector proteins into host-plant cells. The exact nature and functions of the Hrp pili of *P. syringae* and the EspA-containing pili of EPEC remain to be elucidated.

VIII. REGULATION OF PILUS BIOGENESIS

Pilus biogenesis, in general, is a tightly regulated process. Ideally, the costs in energy and other resources required for pilus assembly must be balanced with any potential benefits that pilus expression might provide a bacterium. For example, by producing pili in a nutritionally poor environment a bacterium will tax its available resources, but with pili the same bacterium may be able to gain access to a more favorable location. Pathogenic and other bacteria must also control pilus expression, in some cases, to avoid attachment to unfavorable sites (tissues) within their hosts. Furthermore, pathogenic bacteria may need to modulate pilus expression to escape detection by the host immune system. Whether a bacterium expresses pili is greatly affected by environmental factors. Changes in temperature, osmolarity, pH, oxygen tension, carbon source, and nutrient availability may either increase or decrease pilus expression. The presence of iron, aliphatic amino acids, and electron acceptors other than oxygen may also influence the expression of pili. A combination of these environmental cues can stimulate (or repress) pilus synthesis and alter the expression of a variety of other factors, all of which can influence the tropism of bacteria for specific niches in the environment or in host organisms.

Environmental signals affect pilus biogenesis through global regulator proteins that can modify the transcription of pilus genes. Various global regulators have been identified and include H-NS, a DNA-binding histone-like protein that often mediates temperature regulation of pilus synthesis. H-NS appears to alter DNA topology and typically functions as a negative regulator. Regulation by carbon source can occur through the catabolite activator protein (CAP), whereas the leucine-responsive regulatory protein (Lrp) can modulate pilus expression in response to aliphatic amino acids. The CAP and Lrp regulators can control sets of pilus operons, enabling the expression of different types of pili to be coordinated and integrated with the metabolic state of the bacterial cells. In addition to these and other global regulators, specific regulator proteins encoded by genes within some pilus operons may also modulate pilus biogenesis. Multiple regulatory factors can act upon the same promoter region, switching pilus gene expression from on to off and vice versa. This on-and-off switching, known as phase variation, can also be modulated by the methylation status of a promoter region and by the inversion of sequence elements within a promoter. Two-component systems, such as the Cpx system described for P pilus assembly, also appear to be involved, at least tangentially, in the regulation of the assembly and function of a large number of pilus types.

IX. ROLE OF PILI IN DISEASE PROCESSES

The expression of pili can have substantial impact on the establishment and persistence of pathogenic bacteria in their hosts. For many bacterial pathogens, adhesive pili play a key role in the colonization of host tissues. Uropathogenic *E. coli*, for example, require type 1 pili to effectively colonize the bladder epithelium. These pili attach to conserved, mannose-containing host receptors expressed by the bladder epithelium and help prevent the bacteria from being washed from the body with the flow of urine. P pili may serve a similar function in the kidneys, inhibiting the clearance of pyelonephritic *E. coli* from the upper urinary tract. Enteric pathogens produce a wide variety of adhesive pili that facilitate bacterial colonization of the intestinal tract. These include the K88, K99, and 987P pili made by ETEC strains, the long-polar fimbriae (LPF) and plasmid-encoded fimbriae (PEF) of *S. enterica,* and the aggregative adherence fimbriae (AAF) of enteroaggregative *E. coli.* In the small intestine, TCP are essential for the attachment of *V. cholera* to gut epithelial cells. These pili also act as receptors for the cholera toxin phage (CTXΦ), a lysogenic phage that encodes the two subunits of the cholera toxin. This phage, with its encoded toxin, is transferred between *V. cholera* strains via interactions with TCP in the small intestine. Other pili also function in the acquisition of virulence factors. The uptake of DNA facilitated by type 4 pili and DNA transfer directed by conjugative pili can provide pathogens with accessory genes enabling them to synthesize a wider repertoire of virulence factors and giving them resistance to a greater number of antibiotics. Biofilm formation, which in some cases appears to require pili such as type 1, type 4, or curli, can also increase the resistance of bacteria to antibiotic treatments and may aid bacterial colonization of tissues and medical implants.

Pili are not necessarily static organelles and dynamic alterations of pilus structures during the infection process may influence the pathogenicity of piliated bacteria. For example, electron-microscopic studies (Mulvey, 1998) of mouse bladders infected with type 1-piliated uropathogenic *E. coli* showed that the pili mediating bacterial adherence to the bladder epithelial cells were 10–20 times shorter than typical type 1 pili. It is possible that the shorter type 1 pili observed are the result of pilus retraction, or breakage, during the infection process. The shortening of type 1 pili may provide a means for reeling bacteria in toward their target host cells, allowing the bacteria to make intimate contact with the bladder epithelium after the initial attachment at a distance. Within the gut, type 4B BFP promote the autoaggregation of EPEC strains, a phenomenon that probably facilitates the initial adherence of EPEC to the intestinal epithelium. Work by Bieber and co-workers (1998) suggests that, after initial attachment, an energy-dependent conformational change in the quaternary structure of BFP is needed for the further dispersal of EPEC over human intestinal cells and for the full virulence of this pathogen.

During the infection process, adhesive pili are often situated at the interface between host and pathogen where they can potentially mediate cross-talk between the two organisms. A few examples of pilus attachment inducing signal-transduction pathways in host eukaryotic cells have been reported. The binding of the type 4A pili of *Neisseria* to host receptors (probably CD46) on target epithelial cells has been shown to stimulate the release of intracellular Ca^{2+} stores, a signal known to control a multitude of eukaryotic cellular responses. Similarly, the attachment

of P pili to Galα(1-4)Gal-containing host receptors on target uroepithelial cells can trigger the intracellular release of ceramides, important second-messenger molecules that are capable of activating a variety of protein kinases and phosphatases involved in signal transduction processes. The signals induced in uroepithelial cells upon the binding of P-piliated bacteria eventually result in the secretion of several immunoregulatory cytokines. The binding of type 1-piliated bacteria to mannosylated receptors on uroepithelial cells can similarly induce the release of cytokines, although apparently through different signaling pathways than those stimulated by P pilus binding. Pili can also transduce signals into bacterial cells. This was demonstrated by Zhang and Normark who, in 1996, showed that the binding of P pili to host receptors stimulated the activation of iron-acquisition machinery in uropathogenic *E. coli*. This probably increases the ability of uropathogens to obtain iron and survive in the iron-poor environment of the urinary tract. An understanding of how pili can transmit signals into bacterial cells, and the consequences of such signaling, awaits future studies. Continued research into the biogenesis, structure, and function of pili promises not only to advance our basic understanding of the role of these organelles in pathogenic processes, but may also aid the development of a new generation of antimicrobial therapeutics and vaccines.

See Also the Following Articles

Adhesion, Bacterial • *Escherichia coli*, General Biology • Horizontal Transfer of Genes between Microorganisms • Protein Secretion

Bibliography

Abraham, S. N., Jonsson, A.-B., and Normark, S. (1998). Fimbria-mediated host-pathogen cross-talk. *Curr. Opin. Microbiol.* **1**, 75–81.

Dodson, K. W., Jacob-Dubuisson, F., Striker, R. T., and Hultgren, S. J. (1997). Assembly of adhesive virulence-associated pili in gram-negative bacteria. *In* "*Escherichia coli*: Mechanisms of Virulence," (M. Sussman, ed.), pp. 213–236. Cambridge University Press, Cambridge.

Edwards, R. A., and Puente, J. L. (1998). Fimbrial expression in enteric bacteria: a critical step in intestinal pathogenesis. *Trends Microbiol.* **6**, 282–287.

Forest, K. T., and Tainer, J. A. (1997). Type-4 pilus-structure:

outside to inside and top to bottom—A minireview. *Gene* **192**, 165–169.

Fusseneggar, M., Rudel, T., Barten, R., Ryll, R., and Meyer, T. F. (1997). Transformation competence and type-4 pilus biogenesis in Neisseria gonorrhoeae—A review. *Gene* **192**, 125–134.

Hultgren, S. J., Jones, C. H., and Normark, S. (1996). Bacterial adhesins and their assembly. *In "Escherichia coli and Salmonella"* (F. C. Neidhardt, ed.), Vol. 2, pp. 2730–2756. ASM Press, Washington, D.C.

Klemm, P. (ed.) (1994). "Fimbria: Adhesion, Genetics, Biogenesis, and Vaccines." CRC Press, Boca Raton, FL.

Low, D., Braaten, B., and van der Woude, M. (1996). Fimbriae. *In "Escherichia coli* and *Salmonella"* (F. C. Neidhardt, ed.), Vol. 1, pp. 146–157. ASM Press, Washington, D.C.

Sakellaris, H., and Scott, J. R. (1998). New tools in an old trade: CS1 pilus morphogenesis. *Mol. Microbiol.* **30**, 681–687.

Silverman, P. M. (1997). Towards a structural biology of bacterial conjugation. *Mol. Microbiol.* **23**, 423–429.

Zupan, J. R., Ward, D., and Zambryski, P. (1998). Assembly of the VirB transport complex for DNA transfer from *Agrobacterium* tumefaciens to plant cells. *Curr. Opin. Microbiol.* **1**, 649–655.

Flagella

Shin-Ichi Aizawa

Teikyo University

GLOSSARY

flagellar basal body The major structure of the flagellar motor, consisting of ring structures and a rod.

flagellar motor A molecular machine that converts the energy of proton flow into rotational force.

master genes Flagellar genes that regulate the expression of all the other flagellar genes, sitting at the top of the hierarchy of flagellar regulons.

polymorphic transition The interconversion of helical forms on a flagellar filament in a discrete or stepwise manner.

type III export system One of the protein export systems that does not use the general secretory pathway. It consists of many protein components.

THE FLAGELLUM is an organelle of bacterial motility. It consists of several substructures: the filament, the hook, the basal body, the C ring, and the C rod. The flagellar motor, an actively functional part of the flagellum, can generate torque from proton-motive force. The structural aspects of the flagellum are described here, revealed in pursuit of the identity of the flagellar motor.

I. STRUCTURE

A. Filament

The flagellum is a complex structure composed of many different kinds of proteins. However, the term flagellum, especially in earlier studies, often indicates the flagellar filament only, because the filament is the major portion of the entire flagellum. In this section, I describe the filament and occasionally call the filament flagellum.

1. Number of Flagella per Cell

The number and location of flagella on a cell are readily discernible traits for the classification of bacterial species.

The number ranges from one to several hundred, depending on the species; hence the nomenclature, monotrichous (one) or multitrichous (two or more). Occasionally the term "amphitrichous" is used for two flagella.

There are three possible locations on a cell body for flagella to grow—polar (at the axial ends of the cell body), lateral (at the middle of the cell body), or peritrichous (anywhere around the cell body). In some cases, "lateral" is used as the counterpart of "polar," as in the two-flagellar systems of *Vibrio alginolyticus*, polar sheathed flagellum and lateral plain flagella (although the latter are actually peritrichous). A tuft of flagella growing from a pole is called lophotrichous. In most cases, flagella can be named by a combination of number and location, for example, polar lophotrichous flagella of *Spirillum volutans*.

Although ordinary flagella are exposed to the medium, some flagella are wrapped with a sheath derived from the outer membrane (e.g., in *Vibrio chol-*

erae). In an extreme case such as spirochetes, flagella are confined in a narrow space between the outer membrane and the cell cylinder. The flagella still can rotate; the helical cell body works as a screw, and the flagella counterbalance the torque of the cell body.

2. Filament Shape

The filament shape is helical. In theory, there are two types of helices, right-handed and left-handed; in reality, *Salmonella* spp. have left-handed filaments and *Caulobacter crescentus* has a right-handed filament. However, it should be noted that the shapes of these two helices are not mirror images of each other.

There are several detailed filament shapes, and it will be convenient to use the names of the typical shapes found in *Salmonella* spp.—normal (left-handed), curly (right-handed), coiled (left-handed), semi-coiled (right-handed), and straight. The helical parameters of these helices are discrete and distinguishable from one another (Fig. 1).

Flagella can switch among a set of helical shapes under appropriate conditions; both helical pitch and helical handedness are interchangeable. The transformation of shapes can be induced by physical perturbation (torque, temperature, pH, and salt concentration of medium). Genetic changes such as point mutations in the flagellin (the component protein of the flagellar filament) gene also result in transforma-

tion of helices, but some mutant flagella, such as straight flagella, are too stiff to transform into another helix. This phenomenon, called polymorphism of flagella, is a visible example of conformational changes in proteins and therefore has evoked an idea of a functional role of flagella in motility. Could polymorphism of the flagellum by itself cause the motion? The answer is No. Flagella are passive in terms of force generation. Polymorphism of flagella is observed to occur naturally on actively motile cells with peritrichous flagella. The helical transformation is necessary for untangling a jammed bundle of tangled flagella. When normal flagella in a jammed bundle are transformed into curly flagella, knots of tangled flagella run toward the free end of each flagellum to untangle the jammed bundle.

A theoretical model that explains the polymorphism successfully was presented by Dr. Chris Calladine (University of Cambridge). Twisting and bending a cylindrical rod gives rise to a helix. This model predicts 12 shapes, and 8 of them have been found in existing filaments: straight with a left-handed twist, f1, normal, coiled, semi-coiled, curly I, curly II, and straight with a right-handed twist. Only a small energy barrier seems to lie between any two neighboring shapes. Polymorphic transition occurs from one shape to its neighbors; for example, in a transition from normal to curly, the filament briefly takes on coiled and semi-coiled forms.

3. Flagellin

The component protein of the filament is called flagellin. Although the flagellum of many bacteria is composed of one kind of flagellin, some flagella consist of more than two kinds of closely related flagellins. The molecular size of flagellin ranges from 20 kDa to 60 kDa. Enterobacteria tend to have larger molecules, whereas species living in freshwater have smaller molecules.

One of the most characteristic features of flagellin is evident even in the primary structure of the molecule; the amino acid sequences of both terminal regions are well conserved, whereas that of the central region is highly variable even among species or subspecies of the same genera. As a matter of fact, this hypervariability of the central region gives rise to hundreds of serotypes of *Salmonella* spp.

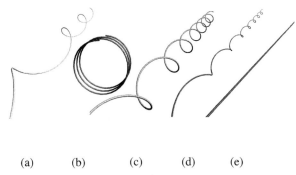

(a)   (b)   (c)   (d)   (e)

Fig. 1. Helical forms of flagellar filaments. Helices are seen from a position slightly off their axial direction so that the handedness can be easily visualized. The figure shows five typical forms with their helical parameters (p, pitch; d, diameter): (a) normal (p = 2.55 μm, d = 0.6 μm), (b) coiled (p = 0, d = 1.0 μm), (c) semi-coiled (p = 1.26 μm, d = 0.5 μm), (d) curly (p = 1.20 μm, d = 0.2 μm), and (e) straight filament (p = ∞, d = 0).

The terminal regions are essential for binding of each molecule to another to polymerize them into a filament. Complete folding of flagellin occurs during assembly; although the terminal regions do not take on any specific secondary structure in solution, they are converted into α-helix after polymerization.

In the filament, the terminal regions are located at the innermost radius of a cylindrical structures, whereas the central region is exposed to the outside. Note that the filament is extremely stable; it does not depolymerize in water, in contrast to actin filaments or tubulin filaments, which depolymerize in the absence of salts.

This description of the flagellin molecule is not applicable to that found in archaeobacteria. Archaic flagella seem to have a system totally different from those of eubacteria; archaic flagellins have signal sequences and the flagellum grows from the proximal end in the outer membrane, whereas the flagellum of eubacteria grows distally (see Section IV).

4. Cap Protein

Flagella have been regarded as a self-assembly system. Indeed, flagellin can polymerize into flagella under conditions that commonly promote protein crystalization *in vitro*. However, *in vivo* flagellin assembly requires another protein, without which the flagellin is secreted into the medium as monomers. The protein that helps filament formation is located at the tip and is thus called the cap protein, or HAP2 or FliD.

The cap proteins assemble in a pentamer, forming a star-shaped structure. The star hands fit in the grooves of flagellin subunits at the tip of flagellum, leaving a small gap for a nascent flagellin to insert.

B. Hook

1. Shape

Hook, as the name suggests, is more sharply curved (almost in a right angle) than the filament and is much shorter. The curvature indicates the flexibility of the hook, although it has to be stiff enough to transmit the torque generated at the basal structure to the filament. From these physical properties, the hook has been regarded as a universal flexible joint.

The length of the hook is 55 nm with a standard deviation of 6 nm, which is rather well controlled when compared with the length of filament. The hook length is not controlled by any molecular rulers, as might have been expected. The mechanism is not known.

A polyhook is a hook of indefinite length, obtained in certain kinds of mutants. Its shape is a right-handed superhelix. The wild-type hook has the same superhelix, but consists of about one-fourth of the helical pitch.

2. Hook Protein

The hook is a tubular polymer made of a single kind of protein, hook protein or FlgE. The molecular size of hook protein varies from 264 amino acids (a.a)(*Bacillus subtilis*) to 718 a.a. (*Helicobacter pylori*), but it is around 400 a.a. for most species.

The architecture of hook protein resembles that of flagellin—the amino acid sequence in both terminal regions is well conserved, but in the central region it is variable. Hook protein folding also completes on assembly.

3. Scaffolding Protein FlgD

The hook does not self-assemble; it requires a helper protein, FlgD, which functions in a similar way that FliD does for filament formation. FlgD sits at the tip of the nascent hook to polymerize the hook protein coming out from the central channel. When the hook length reaches 55 nm, FlgD is replaced by HAP1 (FlgK), which remains in the mature flagellum. Because of its temporary existence, FlgD is regarded as a scaffolding protein.

4. HAPs

There are two minor proteins between the hook and filament. They are called hook-associated proteins (HAPs) because they were found at the tip of the hook in several filamentless mutants. There were originally three HAPs, called HAP1, 2, and 3, in the order of their molecular size. HAP2 turned out to be located at the tip of filament as described above, leaving HAP1 and 3 between the hook and filament. They are, therefore, better termed hook–filament junction proteins.

The number of subunits of HAP1 and 3 in a filament is estimated to be 5 or 6, indicating that they form one-layer rings sitting one on another.

The roles of these two HAPs have been ambiguous. The idea of a connector to smooth the junction between the two polymers is blurred by the question, "Why are not one but two kinds necessary?" In a mutant of HAP3, filaments underwent polymorphic transitions so easily that cells cannot swim smoothly, suggesting a specific role of HAP3 as a stabilizer of filament structure.

C. Basal Structure

Flagella have to be anchored in the cell wall. The structural entity for the anchoring was called the basal structure or basal granule, hinted at by vague images by electron microscopy. Since DePamphilis and Adler in 1974 defined the details of the basal structure, it has been called the basal body. The basal body typically consists of four rings and one rod.

The basal body does not contain everything necessary for motor function. Fragile components were detached from the basal body during purification. In 1985, one such fragile structure was found attached to basal bodies purified by a modified method; it was named the C (cytoplasmic) ring. In 1990, another rod-like structure was found in the center of the C ring. Therefore, the basal structure (as of 1999) consists of the basal body, the C ring, and the C rod, but there may be more.

1. Basal Body

The basal body contains rings and a rod penetrating them. The number of rings varies depending on the membrane systems; there are four rings in most gram-negatives, and two rings in gram-positives, exemplified by *B. subtilis*. Some variation in the number (such as five rings in *C. crecentus*) has been occasionally seen, but the purpose and function of a fifth ring is unclear because the cell's membranes are supposed to be the same as those of *S. typhimurium* or *E. coli*.

The structure of the basal body of *S. typhimurium* has been extensively analyzed. The physical and biochemical properties of the substructures of the basal body described later are from *S. typhimurium,* unless otherwise stated.

2. LP-Ring Complex

The outermost ring, the L ring, interacts with the lipopolysaccharide layer of the outer membrane, and the P ring just beneath the L ring may bind to the peptidoglycan layer. The LP-ring complex works as a bushing, fixed firmly enough to hold the entire flagellar structure stably in the cell surface.

The component proteins, FlgH for the L ring and FlgI for the P ring, have signal peptides, indicating that they are secreted through the general secretory pathway (GSP), which is the exception for flagellar proteins (see Section IV). FlgH undergoes lipolysis modification.

After the L and P rings have bound together to form the LP-ring complex, this complex is resistant to extremes of pH or temperature. Subjecting it to pH 11, pH 2, or boiling for 1 min does not destroy the complex, confirming that the complex serves as a rigid bushing in the outer membrane.

The essential roles of the complex are still ambiguous because mutants lacking the complex still can swim, though poorly, and because no corresponding structure has been found in gram-positive bacteria.

3. MS-Ring Complex

Earlier studies of flagellar motor function assumed that torque would be generated between the M and S rings, which face each other on the inner membrane. However, in 1990, it was shown that a single type of protein, FliF, self-assembles into a complex consisting of the M and S rings and part of the rod.

FliF is 65 kDa, the largest of the flagellar proteins. It contains no cysteine residues. It consists of several regions—terminal regions and two distinguishable central regions. Overproduction of FliF in *E. coli* gives rise to numerous MS-ring complexes packed in the inner membrane.

The S ring has been seen in the basal bodies from all the species studied so far (at least seven examples). It stays just above the inner membrane (supramembrane) and has no apparent interaction with any other structures. Besides, it is very thin (~1 nm), and the role of the S ring remains mysterious.

Although the MS-ring complex is no longer regarded as the functional center of the flagellar motor, it is still the structural center of the basal structure and plays an important role in flagellar assembly (see Section IV).

4. Rod

The rod is not as simple as its name suggests; it consists of at least four distinct proteins. It breaks at the midpoint when external physical force is applied to the filament, which is not expected in a structure that transmits torque to the filament.

Rod formation seems complicated because of the four component proteins. No intermediate rod structure has been observed; either there is a whole rod or no rod at all.

Because the P ring is formed on the rod, some of the rod proteins could be the target of interaction with FlgI, the subunit of the P ring.

5. C Ring

The C ring is a fragile component of the basal structure. It is resistant to the nonionic detergent Triton X-100, but it is destroyed by the alkaline pH and high salts concentration employed by conventional purification methods. The dome shape of the C ring is easily flattened on a grid during preparation for electron microscopy.

The C ring consists of the switch proteins (FliG, FliM, and FliN) and so is sometimes called the switch complex. It is not known whether other proteins are present or not. FliG directly binds to the cytoplasmic surface of the M ring. FliM binds to FliG, and FliN to FliM. The stoichiometry of these molecules in the C ring is still controversial; 20–40 copies of FliG, 20–40 copies of FliM, and several 100 copies of FliN.

Genetic studies revealed that the switch complex plays important roles in flagellar formation, torque generation, and the switching of rotational direction. The C ring directly binds signal molecules, CheY, produced in the sensory transduction system, but the mechanism of the switching is ambiguous.

6. Export Apparatus

Flagella have been regarded as having a self-assembly system, similar to that of bacteriophages. However, flagellar assembly is quite different from phage assembly in many ways. First, the flagellum, being an extracellular structure, assembles not in the cytoplasm but outside the cell. Second, the component proteins therefore, have to be transported from the cytoplasm to the outside. Third, assembly consequently proceeds in a one-by-one manner, at the distal end of the nascent structure.

For this kind of assembly, a protein excretion system must play an important role. As a matter of fact, among the 14 genes required in the very first step of flagellar assembly, at least seven gene products are necessary to form a protein complex, called an export apparatus. One of them, FliI, has an ATPase activity, suggesting that one step in the export process requires ATP hydrolysis as an energy source. The physical body of the export apparatus has not been identified; the C rod is a strong candidate, judging from its location in the C ring.

II. FUNCTION

The function of flagella is described hare briefly so that the meaning of the structure can be understood.

Bacterial flagella rotate. There is no correlation between bacterial flagella and eukaryotic flagella, either in function or in structure; the type of movement, the energy source, and the number of component proteins differ greatly between the two. No evolutionary correlation between these two types of flagella has been shown.

Among motile bacterial species, swimming by flagellar rotation is the most common. However, some cyanobacteria can move on a solid surface in a gliding motion; the motile organ of gliding bacteria is not known.

A. Torque

The rotational force (torque) of the flagellar motor is difficult to measure directly, but can be estimated from the rotational speed of flagella. The method most widely employed is the tethered-cell method, in which the rotation of a cell body caused by a tethered filament can be observed with an ordinary optical microscope. A more sophisticated method, which employs a laser as the light source of a dark-field microscope, allows one to measure the rotation of a flagellum on a cell stuck on the glass surface at a time resolution of millisecond.

1. *Rotational Direction*

The flagella of many species (e.g., enterobacteria) can rotate both clockwise (CW) and counter-clockwise (CCW). Under ordinary circumstances, around 70% of the time is occupied by CCW rotation, which causes smooth swimming. A brief period of CW rotation causes a tumbling motion of the cell. There is no perceptible pause in switching between the two modes.

In some bacterial species such as *Rhodobacter sphaeroides,* a lateral flagellum on a cell rotates in the CW direction only, with occasional pauses. During the pauses, the filament takes on a coiled form and curls up near the cell surface. Upon application of torque to this filament, the coiled form extends to a semi-stable right-handed form that closely resembles a curly form. The CW rotation of this right-handed helix causes a forward propulsive force on the cell.

2. *Rotational Speed*

The rotational speeds of flagella correlate directly with the torque and inversely with the viscosity of the solution. The correlation appears as a straight line in a speed–torque diagram; the higher the viscosity, the slower the speed. This indicates that the torque of the flagellar motor is constant over a wide range of speeds.

The highest speed measured is 1700 Hz, for *Vibrio alginolyticus.* However, only 10% of the torque derived from the speed is used as a propulsive force, the rest being lost in slippage of flagella in the medium.

B. Energy Source

The energy source of torque generation in the flagellar motor is not ATP but proton-motive force (PMF). PMF is the electrochemical potential of the proton, and results in the flow of protons from outside to inside the cell. PMF consists of two forms of energy, membrane potential, and entropy caused by a difference in pH between outside and inside the cell. Because these two parameters are independent and separable from each other, either one can, in principle, be abolished without affecting the other. Note that the polar flagellum of *V. alginolyticus* substitutes sodium ion (Na^+) for proton. One of the

goals of flagellar research is the elucidation of the mechanism by which PMF or NaMF is converted into torque in the motor.

C. Switching of Rotational Direction

Switching the rotational direction of flagella is the primary basis of chemotaxis, one of the most important behaviors shown by bacteria. Damage in the switching mechanism results in a rotation biased to either CCW only or CW only.

In a strict sense, the switching mechanism will not be solved until the mechanism of rotation is solved. However, the factors involved in the mechanism are known; an effector binds to the switch complex in the flagellar motor. The effector is the phosphorylated form of CheY, a signalling protein in the sensory transduction system.

III. GENETICS

Flagellar genetics has been most extensively studied in *S. typhimurium*, especially using the enormous number of strains that Dr. Shigeru Yamaguchi (Meiji University) has collected for more than 30 years. The discussion in this section is, unless otherwise indicated, based on results obtained from these strains.

A. Flagellar Genes

There are more than 50 flagellar genes, which are divided into three types, according to the null mutant phenotype.

1. *The fla Genes*

Defects in the majority of the flagellar genes result in flagellar deficient (Fla⁻) mutants. These genes were originally called *fla* genes. In 1985, when the number of genes exceeded the number of letters in the alphabet, a unified name system for *E. coli* and *S. typhimurium* was introduced: *flg, flh, fli,* and *flj*; one for each of the clusters of genes scattered in several regions around the chromosome (see Section III.B; Fig. 2).

2. *The mot Genes*

Mutants that produce paralyzed flagella are called motility deficient (Mot⁻) mutants. There are only two *mot* genes (*motA* and *motB*) in *S. typhimurium*, but four *mot* genes (*motA*, *motB*, *motX*, and *motY*) in *R. sphaeroides* and in some other species such as *V. alginolyticus*.

3. The che Genes

Mutants that can produce functional flagella but that cannot show a normal chemotactic behavior are called chemotaxis deficient (Che⁻) mutants. These are divided into two types, general chemotaxis mutants and specific chemotaxis mutants. The former involve the proteins working in the sensory transduction (CheA, CheW, CheY, CheZ, CheB, and CheR), and the latter involve the receptor proteins (e.g., Tsr, Tar, Trg, and Tap).

B. Gene Clusters in Four Regions

Most flagellar genes are found in gene clusters on the chromosome. They are in four regions: the *flg* genes in region I (at 26 min), the *flh* genes and *mot* and *che* genes are in region II (41.7 min), and the *fli* genes are in regions IIIa (42.4 min) and IIIb (42.7 min) (Fig. 2).

The *flj* operon (including *fljA* and *fljB*) at 60 min involves an alternative flagellin gene to *fliC* and is only found in *Salmonella*. Either FliC flagellin or FljB

flagellin is produced at any time. The *hin* gene inverts the transcriptional direction at a certain statistical frequency; if the *flj* operon is being expressed, FljA represses *fliC*, allowing FljB flagellin alone to be produced. This alternate expression of two flagellin genes is called phase variation.

C. Transcriptional Regulation

Flagellar construction requires a well-ordered expression of flagellar genes not only because there are so many genes, but also because flagellar assembly requires only one kind of component protein at a time, as described previously. There is a strict hierarchy of expression among the flagellar genes. The hierarchy is controlled or maintained by a few prominent regulatory proteins.

1. Hierarchy: Three Classes

The hierarchy of flagellar gene expression is divided into three classes; class 1 regulates class 2 gene expression, and class 2 regulates class 3. Class 1 contains only two genes in one operon, *flhD* and *flhC*.

Class 2 consists of 35 genes in eight operons. There are two regulatory genes, *fliA* and *flgM*; the rest are component proteins of the flagellum or the export apparatus.

Class 3 genes code flagellin, *motA* and *motB*, and all the proteins involved in sensory transduction. Flagellin is one of the most abundant proteins in a cell, suggesting that the tight regulation in the hierarchy guarantees the economy of the cell.

2. Master Genes, flhDC

Master gene products form a tetrameric complex of FlhD/FlhC, which works as a transcriptional activator of the class 2 operons.

The master operon (*flhDC*) is probably transcribed with the help of the "housekeeping" sigma factor, σ^{70}. The master operon has also been shown to be activated by a complex of cyclic AMP and catabolite activator protein (cAMP–CAP), which binds to a site upstream of the promoter.

3. Sigma Factor F (σ^F, FliA) and Antisigma Factor (FlgM)

The FliA and FlgM proteins expressed from the operon competitively regulate the class 3 operons.

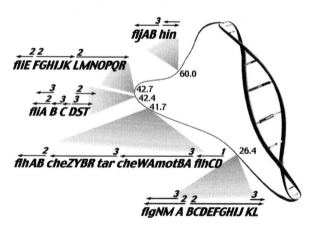

Fig. 2. Genetic map of *S. typhimurium*. Flagellar genes are distributed in several clusters on the chromosome. Arrows over the genes indicate the size of operons and their transcriptional directions. The numbers on the arrows indicate classes of transcription. The regulation of class 2 and 3 is not simple; some operons are expressed twice, in class 2 and class 3.

FliA is the sigma factor that enhances the expression of the class 3 operons, whereas FlgM is an antisigma factor against FliA.

If the hook and basal body have been constructed normally, FlgM is secreted into the medium through the basal body and the complete hook, allowing free FliA proteins to work on the class 3 operons. However, if the hook and basal body construction is somehow halted in the middle of process, FlgM stays in the cytoplasm in a complex with FliA, maintaining shut off of the expression of the class 3 operons.

4. Global Regulation versus Internal Regulation

There are several external genes or factors that affect the flagellar gene expression through the master operon *flhDC*. Some of the factors show pleiotropic effects on many cellular events such as cell division, suggesting that flagellation is finely tuned with the cell division cycle due to well-organized tasks of global regulation systems.

As described, the master operon (*flhDC*) is probably transcribed using the "housekeeping" sigma factor, σ^{70}. In the 1990s, other factors regulating or modulating *flhDC* expression have been identified mainly in *E. coli*. The motility of *E. coli* cells is lost at temperatures higher than 40°C as a result of reduced *flhDC* expression. It has been shown that some of the heat-shock proteins are involved in both class 1 and class 2 gene expression. This strongly suggests that flagellar genes are under global regulation in which the heat-shock proteins play a major role;

probably the proper protein folding (or assembly) mediated by these chaperones is essential for flagellar construction.

Other adverse conditions such as high concentrations of salts, carbohydrates, or low-molecular-weight alcohols, also suppress *flhDC* expression, resulting in lack of flagella. The regulation by all these factors is independent of the cAMP–CAP pathway.

It is unknown which factors directly turn on the master *flhDC* operon in accordance with the cell cycle, and how. After the roles of global regulators on flagellar gene expression are specified, the complex regulatory network connecting flagellation and cell division will be uncovered.

IV. MORPHOLOGICAL PATHWAY

A. Steps in the Morphological Pathway

The order of the steps of the construction of a flagellum (the morphological pathway) has been analyzed in the same way as for bacteriophages, analyzing the intermediate structures in various flagellar mutants and aligning them in size from small to large. Flagellar construction starts from the cytoplasm, progresses through the periplasmic space, and finally extends to the outside of the cell (Fig. 3).

1. Cytoplasm

The smallest flagellar structure recognizable by electron microscopy is the MS-ring complex; there-

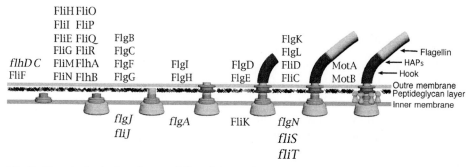

Fig. 3. Morphological pathway of flagellation. Flagellar construction proceeds from left to right. Each step requires several gene products, shown above the membranes. Roles of the genes shown under the membranes are still ambiguous or controversial; some of them are regarded as chaperones for flagellar proteins belonging to class 3.

fore, the MS-ring complex is regarded as the construction base onto which two other flagellar substructures attach, the C ring and the C rod.

2. *Periplasmic Space*

The second-smallest structure consists of a rod on the MS-ring complex. When the rod has grown large enough to reach the outer membrane, the hook starts to grow. However, the outer membrane physically hampers the hook growth until the outer ring complex makes a hole in it. Among flagellar proteins, FlgH and FlgI, the component proteins of the outer ring complex, are exceptional in terms of the manner of export; these two proteins have cleaved signal peptides and are exported through the GSP.

3. *Outside the Cell*

Once the physical block by the outer membrane has been removed, the hook resumes growth with the aid of FlgD until the length reaches 55 nm. Then, FlgD is replaced by HAPs, which is followed by the filament growth. The filament growth proceeds only in the presence of FliD (HAP2 or filament cap protein); without this cap, exported flagellin molecules are lost to the medium.

The number of genes necessary to proceed through each step of the construction varies; in early stages there are many genes whose roles are unidentified. Between the MS-ring complex and the rod, more than 10 genes are required. Some of those gene products form the C ring, and others are involved in the formation of the flagellar protein-specific export system.

B. Flagellar Protein Export as a Type III Export System

There are several ways to transport proteins outside bacterial cells. The best known pathway is the GSP. However, many flagellar proteins cannot pass through this system, because they do not have the signal sequences that are necessary for recognition by GSP.

As the number of examples of other types of protein transport has increased, the name became inappropriate. GSP is now categorized as the type II export system. There are now more than five export

systems, but here I will briefly explain the type III system.

The flagellar protein export system is now regarded as a type III system. The flagellar export apparatus consists of at least eight components. The amino acid sequences of these proteins share homology with those used for export of virulence factors from many pathogenic bacteria.

Even the structures of these two distinguishable systems resemble each other. The needle complex found in *S. typhimurium*, albeit the only example thus far identified as the export apparatus for the virulence factors, looks like the basal body, consisting of several ring structures and a rod or needle.

C. The Kinetics of Morphogenesis

The morphological pathway of the flagellum described indicates the order of the construction steps but ignores the time consumed at each step. In order to achieve coherent cell activities, flagellar construction has to be synchronized with cell division. The most time-consuming step of flagellation seems to be the filament elongation because filaments can grow over generations.

The growth process of the filament and the hook have been carefully analyzed. By taking a closer look at elongation modes of these two polymers, we will get a glimpse of the whole kinetic process of flagellar construction.

1. *Filament Growth*

In bacteria with peritrichous flagella, the number and the length of flagella are fairly well defined; there are 7–10 flagella per cell, and the average length of filament is 5–8 μm.

A defined number of flagella have to be supplied at each cell division. A large deviation from this number will cause disastrous results to the cell—either no flagella at all or too many flagella to swim. The number of flagella must be genetically controlled.

On the other hand, filament growth seems free from genetic control, because it continues over generations. From statistical analysis of the length distribution, the elongation rate of filaments is estimated to vary inversely to the length; thus, a filament grows rapidly in the beginning and gradually slows down to a negligible rate.

2. Hook Growth

In contrast to the wide distribution of filament lengths, the hook length is rather well controlled at 55 nm with a deviation of 6 nm. Mutations in the *fliK* gene result in hooks with unlimited length, called polyhooks. FliK is not a molecular ruler because a truncation in FliK also gives rise to polyhooks but not short hooks, as would be expected.

Statistical analysis of the length distribution of polyhooks reveals that the hook grows in a manner similar to the filament; it starts out growing at 40 nm/min and exponentially slows down to reach a length of 55 nm. After the length is 55 nm, the hook grows at a constant rate of 8 nm/min. It takes many generations for polyhooks to grow as long as several micrometers.

Studies of the correlation between flagellation and cell division are underway, but no definite schemes have been found.

V. CONCLUSION

The analysis of the flagellar structure is almost complete; most components of the flagellum have been identified, and the pathway of flagellar construction has been revealed. That is, the roles of ~40 flagellar genes in the flagellar construction are now known.

We will continue searching for the detailed mechanism of flagellation, including the relationship to cell division. One of the immediate goals is to answer a simple but important question, What is rotating against what? This question stems from the controversy that started at the beginning of the flagellar research. Without knowing the rotor and the stator, the mechanism of motor function will never be understood.

And then, we want to answer a more intriguing and difficult question, What is the ancestor of the flagellum? The question arose from the recent discovery of a similarity between the flagellum and the pathogenicity—not only are the gene sequences between the two systems homologous, but also their supramolecular structures resemble each other. This also leads us to the most basic question, What is the flagellum?

See Also the Following Articles

Escherichia coli and Salmonella, Genetics • Growth Kinetics, Bacterial • Microscopy, Electron

Bibliography

Aizawa, S.-I. (1996). Flagellar assembly in *Salmonella typhimurium. Mol. Microbiol.* **19**, 1–5.

Aizawa, S.-I., and Kubori, T. (1998). Bacterial flagellation and cell division. *Genes Cells* **3**, 625–634.

Kutsukake, K., Ohya, Y., and Iino, T. (1990). Transcriptional analysis of the flagellar regulon of *Salmonella typhimurium. J. Bacteriol.* **172**, 741–747.

Macnab, R. M. (1996). Flagella. *In "Escherichia coli and Salmonella typhimurium:* Cellular and Molecular Biology," (F. C. Neidhardt et al., eds.), pp. 123–145. American Society for Microbiology, Washington, D.C.

Namba, K., and Vonderviszt, F. (1997). Molecular architecture of bacterial flagellum. *Quart. Rev. Biophys.* **30**, 1–65.

Food-borne Illnesses

David W. K. Acheson

Tufts University and New England Medical Center

GLOSSARY

food-borne disease An illness resulting from the ingestion of harmful products in food.

food poisoning A condition that occurs following the ingestion of a preformed toxin.

gastroenteritis An inflammation of the intestinal wall, often as the result of exposure to a food-borne microbe or toxin.

Guillain–Barré syndrome An ascending paralysis associated with *Campylobacter jejuni*.

hemolytic uremic syndrome A triad of renal failure, thrombocytopenia, and hemolytic anemia, frequently associated with Shiga toxin-producing *E. coli*.

incubation period The time between exposure to a food-borne pathogen and the time at which symptoms begin.

infectious dose The number of microbes needed to cause symptoms.

reactive arthropathy Painful joint inflammation that can follow infection with a number of food-borne pathogens.

toxin A molecule that causes pathogenic changes to the host cells that may result in disease.

A FOOD-BORNE ILLNESS, at its broadest definition, can encompass any illness that is contracted from the consumption of, or exposure to, food. This could involve exposure to infective, toxic, or even carcinogenic agents that may or may not have a microbiological origin. For the purposes of this article, the definition will largely be restricted to microbiologically derived food-borne illness. However, there will be some discussion of other factors that cause food-borne illness and generally fall under the heading of natural toxins.

Typically, when one thinks of a food-borne illness, it is in a patient with acute gastrointestinal signs and symptoms, typified by nausea, vomiting, diarrhea, and abdominal pain. However, it is important to remember that some of the consequences of food-borne infections are systemic and may occur several days or weeks after the initial exposure to the food-borne pathogen. Examples include the association of Shiga toxin-producing *E. coli* with hemolytic uremic syndrome, *Campylobacter* and the development of Guillain-Barré syndrome, and the association of a number of enteric bacterial pathogens with the development of reactive arthritis. A variety of food-borne illnesses will be discussed, including those due to bacterial, viral, and protozoal infections, as well as some natural toxins. Some of the basic diagnostic microbiology will be addressed, as well as the clinical presentation and potential complications from the various infections.

I. FOOD-BORNE ILLNESS EPIDEMIOLOGY

The total burden of food-borne illnesses in the United States and other parts of the world is essentially unknown. Yet, death from food- or water-related exposure to pathogens worldwide is considered to be enormous. Several million children are thought to die each year from diarrheal disease, the majority of which is most likely due to contaminated food or water. In the United States, the number of deaths from food-borne disease is estimated to be between 1000 and 9000 per year. Until recently, these numbers were little more than extrapolations from small epidemiological studies in various parts of the country. The development of FoodNet by the Centers for Disease Control and Prevention has, for the first time, offered an opportunity for us to determine what the epidemiology of food-borne disease is in large numbers of people in the United States. The Food-borne Diseases Active Surveillance Network (FoodNet) is the main food-borne disease component of the CDC's Emerging Infections Program (EIP) and is a collaborative venture with EIP program sites, the United States Department of Agriculture, and the United States Food and Drug Administration. FoodNet has undertaken population-based active surveillance for confirmed cases of *Campylobacter, Escherichia coli* O157:H7, *Listeria, Salmonella, Shigella, Vibrio,* and *Yersinia* infections. In 1997, they had active surveillance in Minnesota, Oregon, parts of California, Connecticut, and Georgia. The total population under study was 15.9 million. More sites came on-line in 1998 and the plan is to expand even further in the future.

During 1997, which is the most recent year in which the data has been tabulated, FoodNet found a total of 8557 confirmed cases of infections caused by the pathogens listed above, as well as a number of *Cryptosporidium* and *Cyclospora* infections, as shown in Table I. There were seasonal variations, with the summer months (June, July, and August) being the peak isolation times for *Vibrios* (66%), *E. coli* O157:H7 (52%), *Campylobacter* (35%) and *Salmonella* (32%), thus illustrating the need for heightened awareness for food-borne pathogens during the warmer times of the year. Interestingly, there were large variations in the incidence rates of specific pathogens in the different sites. For example, the incidence rates for *Campylobacter* varied from 14/100,000 in Georgia to 50/100,000 in California. The incidence of *S. enteritidis* was highest in Connecticut. It is unclear what these geographic differences mean but they suggest that regional variations among food-

TABLE I
Cases of Infections Caused by Specific Pathogens Reported to FoodNet Sites in 1997

Campylobacter	3974
Salmonella	2207
S. *typhimurium*	641
S. *enteritidis*	359
S. *Heidelberg*	140
S. *Newport*	77
S. *Montevideo*	68
Shigella	1263
Cryptosporidium	463
E. coli O157:H7	340
Yersinia	139
Listeria	77
Vibrio	51
Cyclospora	43

Note. Cryptosporidium and *Cyclosopora* were looked for in only four of the five sites (CDC, 1997).

borne pathogens do occur, and yet there do not appear to be any areas in which the incidence of these major pathogens is so low that they can be dismissed as unimportant. In terms of age and sex distribution of food-borne illnesses, there is a clear preponderance for the young and the elderly. The incidence rates were much higher in the FoodNet study in children under the age of one year, especially in relation to *Salmonella* infection. Not all food-borne pathogens are equal in terms of the clinical severity of the disease and this is well illustrated from the FoodNet data. *Listeria* infection was associated with the highest hospitalization rate (88%) and the highest fatality rate (20%). This was followed by *E. coli* O157:H7 (29%), *Salmonella* (21%), *Yersinia* (15%), *Shigella* (13%), *Campylobacter* (10%), and *Vibrio* (10%).

FoodNet has offered an opportunity to take a glimpse at the causes of food-borne disease in those patients from whom a sample is obtained for microbiological analysis. It does not, however, address many other food-borne pathogens that were not part of the study. Nor does the FoodNet surveillance address what is going on in those patients who do not have a sample sent for testing. As part of the adjunctive studies reported by CDC, it was determined by phone consultations that, of 10,000 residents within the FoodNet sites, 11% reported a diarrheal illness during the previous month. This translates to 1.4 episodes of diarrhea per year, which, if multiplied roughly by the population of the United States, represents approximately 350 million cases per year. Of greater interest, only 8% of those that were ill sought medical care, and of those 8%, only 20% reported submitting a stool sample for culture. Thus, our best data on the causes of acute gastrointestinal disturbance are based on cultures of <2% of diarrheal episodes. There are still no clear data on the numbers of deaths from food-borne disease in the United States, but original estimates of around 9000 per year are probably too high and the real number may be in the 1000–2000 per year range. However, globally there are still over 3 million children dying each year from diarrheal diseases, many of which are probably related to contaminated food or water. Thus, on a global basis, food-borne illness remains a huge problem.

II. BACTERIAL FOOD-BORNE PATHOGENS

There are many different bacterial pathogens that have been associated with food-borne disease, as shown in Table II. Despite this large number of bacterial pathogens, they generally cause disease by a limited number of mechanisms, such as via a preformed toxin, the production of a toxin in the intestine, invasion of the intestinal epithelial cells, or some other process. The various organisms are broken down into these mechanistic groups and will be described in more detail.

A. Bacteria Causing Disease Primarily Mediated by a Preformed Toxin

This group of organisms includes bacteria that usually make a toxin in food prior to ingestion of the food. This is in contrast to other toxin-producing bacteria (discussed in a later section), which produce a toxin after they have been ingested.

TABLE II

Bacterial Enteric Pathogens Associated with Food-borne Illnesses Categorized by Principal Pathogenic Mechanism

Bacteria causing disease primarily mediated by a preformed toxin
- *Clostridium botulinum*
- *Staphylococcus aureus*
- *Bacillus cereus*

Bacteria causing disease by production of a toxin within the intestine
- *Vibrio cholerae*
- *Vibrio parahemolyticus*
- *Clostridium perfringens*
- Shiga toxin-producing *E. coli*
- Entero toxigenic *E. coli*

Bacteria causing disease primarily by invading the intestinal epithelial cells
- *Salmonella* spp.
- *Campylobacter* spp.
- *Yersina* spp.
- *Listeria monocytogenes*
- *Shigella* spp.
- Enteroinvasive *E. coli*

Others
- *Aeromonas hydrophila*
- *Plesiomonas shigelloides*
- Enteropathogenic *E. coli*
- Enteroaggregative *E. coli*

1. *Clostridum botulinum*

C. botulinum is a gram-positive, spore-forming, toxin-producing, obligate anaerobe. Its natural habitat is soil and, therefore, its spores are frequently present on fresh fruits and vegetables. The disease associated with ingestion of the toxins from *C. botulinum* is known as botulism. There are seven antigenically distinct types of botulinus toxin, each of which is designated by a letter (A–G). Types A, B, E, F, and G are the ones that are associated with human disease, with type A accounting for about 25% of outbreaks and type B 8%, although some regionality is seen with the various types within the United States. Botulinus toxin is heat-labile (5 minutes of boiling will destroy the toxin) and is produced at any temperature in which bacterial growth occurs. The majority of outbreaks have been traced to home-produced foods, especially vegetables, fish, fruits, and condiments. Commercial food sources are occasionally incriminated. The key to preventing this disease is to destroy the *C. botulinum* spores. Food needs to be subjected to a temperature of 120° for 30 minutes, usually with the aid of a pressure cooker in the home environment, to destroy spores. Any surviving spores may germinate in the anaerobic environment of the food and produce their toxins.

Botulinus toxin is absorbed from the proximal intestine and spreads via the bloodstream to the peripheral cholinergic nerve synapses. The toxin then irreversibly blocks acetylcholine release, resulting in flaccid paralysis. The toxin does not cross the blood–brain barrier. Typically, symptoms occur 18–36 hr after toxin ingestion, but incubation may be as short as a few hours or as long as 8 days. Cranial nerves are usually affected first but, subsequently, respiratory muscles may be affected, resulting in respiratory paralysis and death if left untreated. This food-borne infection does not typically cause any major gastrointestinal symptoms. Diagnostic confirmation is dependent on detection of the toxin or the organism in the patient or the implicated food. Samples, such as the food, vomitus, serum, gastrointestinal washings, and feces, are all reasonable specimens to test for the toxin. The toxin can be detected in specialist centers, using the mouse neutralization test. The organisms themselves or the spores can also be cultured, but this takes longer. Molecular methods, based on polymerase chain reaction (PCR) for the neurotoxin genes, are also available on a research basis. In terms of therapy, there are several approaches one can take. If the diagnosis is made quickly, emetics or gastric lavage may be helpful to remove unabsorbed toxin. A trivalent (ABE) horse serum antitoxin is available and should be used. However, supportive therapy is the mainstay of treatment. Botulism carries a significant mortality, with up to 25% of patients affected by the type A toxin dying. The longer the incubation period, the better the prognosis, since the shorter incubation periods are generally associated with a greater initial toxin load. Eventually, if patients survive the acute symptoms, the prognosis is good and most patients recover completely.

2. *Staphylococcus aureus*

S. aureus has the capacity to produce a variety of toxins (A, B, C1–C3, D, E) that are all small proteins with similar biological activities. Other *Staphylococcus* species, such as *S. hyicus* and *S. intermedius*, as well as coagulase negative species, have, on rare occasion, been found to produce the enterotoxins also. Strains that produce type A toxin, either alone or in combination with type D, are the ones most frequently associated with outbreaks. It is these toxins that are responsible for the disease manifestations and ingestion of as little as 100–200 ng is considered to be enough to cause disease in humans. Typically, patients become symptomatic within 6 hr of ingesting these toxins, with nausea (73–90%), vomiting (82%), and abdominal cramps (64–74%). Diarrhea occurs in a significant proportion of patients (41–88%), but fever is rare. This disease is often associated with outbreaks that may be very large, with over 1000 individuals involved. A number of different foods have been associated with staphylococcal food poisoning, especially dairy-based products, such as cream-filled desserts and cakes. Other foods that are often incriminated include egg products, meat products, poultry, mayonnaise, and tuna. Typically, the course of events is that a food-handler contaminates the food product with *S. aureus* from their fingers or nose. The organisms then grow in the food if it is held at room temperature for any length of time. Thus, disease is not often associated with commercially prepared food but is more frequently associated with food handled in a food-service establishment or in the home. The bacteria produce the toxin as they grow and bacterial levels in the region of 10^5 per gram will produce enough toxin in food to cause disease. The toxin is heat-stable and is not inactivated by heating or even boiling the food product. The toxin is also considered to be stable against pH extremes, proteases, and radiation. So, once formed in food, these toxins are almost impossible to remove. The pathogenetic mechanisms whereby these toxins cause their effects are not fully understood, but they are thought to act via stimulation of the autonomic nervous system. The inflammatory response is also thought to play a role and elevation of certain cytokines may be a critical factor. *S. aureus* enterotoxins are also classified as superantigens and are potent T-cell activators. The toxins are not absorbed systemically; therefore, protective immunity is not induced following exposure to them. Culturing the implicated food, stools, or vomitus can make a definitive diagnosis. Phage typing is considered to be a useful epidemiological tool, with type III being the one most often associated with outbreaks. Latex agglutination and enzyme immunoassays are available for the direct detection of *S. aureus* enterotoxins in food in the ng/g food range. Treatment is supportive and dehydration may be a significant problem. Symptoms usually abate within 8 hr and there is no need to treat patients with antibiotics directed toward the *S. aureus* strain. The outcome is usually good.

3. *Bacillus cereus*

B. cereus is a gram-positive, spore-forming aerobe that causes two distinct clinical syndromes. One is a short incubation period emetic syndrome and the other a longer incubation period diarrheal syndrome. The emetic syndrome has an incubation period and clinical presentation much like *S. aureus*-related food-borne disease, with nausea and vomiting being the principal features, occurring in a 1–6 hr incubation period following exposure. Occasionally, very short incubation periods (15–30 min) or longer ones (6–12 hr) have been reported. Fever is not associated with this disease and recovery is usual, although there has been at least one report of acute hepatic necrosis associated with exposure to *B. cereus* emetic toxin. The emetic toxin is a preformed toxin (5–10 kDa) that is typically produced when the organism is present in starchy foods–especially rice. The toxin is heat-stable and resists proteases and variations in pH. So reheating food in which the toxin has formed will not ensure its destruction and may lead to a false sense of security. *B. cereus* is present in soil and water and in most raw foods; even 10–40% of humans are colonized with this bacteria. Diagnosis is dependent on finding the organism in the food or vomitus of the patient. The frequency of asymptomatic carriage in the stool makes stool cultures alone unhelpful. Heating food may destroy the organisms, thus making culture confirmation difficult, but, as already mentioned, heat will not destroy the toxin. The toxin associated with the diarrheal syndrome is heat-labile and trypsin-sensitive and is thought to

be made up of three components (designated B, L_1, and L_2, for proposed binding and cell lysis properties). There have also been hemolytic activities described in proteins from *B. cereus* and this may be part of the activity of the three-component toxin. Like the emetic toxin, the diagnosis requires culture confirmation, preferably from the incriminated food source. Recovery is usual and therapy is supportive.

B. Bacteria Causing Disease by Production of Toxins within the Intestine

This group of organisms generally causes disease through the production of a toxin following intestinal colonization. Some of the bacteria in this group have other virulence attributes as well, and the disease process may be a combination of the toxin and some of these other factors.

1. *Vibrio Species*

Currently, there are over 30 *Vibrio* species, of which about one-third are known to cause disease in humans. Vibrios are gram-negative facultative aerobes and, in this section, a couple of the more important ones will be discussed.

a. *Vibrio cholerae* O1

V. cholerae O1 is a gram-negative organism that is divided into two biotypes: classical and E1 Tor. There are various ways to differentiate between the two but they were initially differentiated following the discovery that the E1 Tor strains make hemolysins. From a food-borne illness perspective, the main factor that is involved in the spread of *V. cholerae* is contaminated water. There have been six previous pandemics of cholera and we are currently in the seventh, which began in 1961. Although previous pandemics have been due to classical biotypes, the current pandemic is due to E1 Tor. *V. cholerae* is a facultative anaerobe and thiosulfate–citrate bile salts–sucrose (TCBS) agar is frequently used for isolation because it inhibits the growth of most other enteric organisms. *V. cholerae* is spread via contaminated water and food and cleaning up water supplies can have a major impact on the prevalence of this disease. Its survival in foods is affected by factors such as temperature, pH, and inoculum size, but it may survive for a couple of days on contaminated food. Foods such as seafood, pickled or dried fish, coconut milk, lettuce, and rice have all been associated with carriage of the organism. Clinically, the incubation period is usually 1–3 days but may be as short as a few hours or as long as 5 days. The infectious dose is around 10^6 organisms, but decreased gastric acidity appears to predispose individuals to infection with *V. cholerae*. The organisms principal virulence factor is the production of cholera toxin that is an activator of cyclic AMP in intestinal epithelial cells. This results in profound intestinal secretion of fluid, profuse watery diarrhea, and, if untreated, death secondary to dehydration. This is known clinically as cholera. The organisms can be cultured from stools of infected patients, but treatment in the form of rehydration, which can be life-saving, should not wait for culture confirmation. Antibiotics are not required but may shorten the duration of illness.

b. Non-O1 V. cholerae

In recent years, there has been clinical disease, predominantly in the Indian subcontinent, associated with *V. cholerae* O139. This is the first time a non-O1 strain has been associated with epidemic disease. It appears to be spread and to cause disease via the production of cholera toxin in the same way as do the O1 strains. Other non-O1 *V. cholerae* strains have also been associated with food-borne disease—especially in relation to consumption of raw oysters—but such disease is usually sporadic in nature.

c. Vibrio parahemolyticus

This organism requires NaCl for growth, and, like the other members of the Vibrio species, lives in water. It is frequently found in shellfish and consumption of shellfish is usually associated with clinical disease. In the past, this has been a major food-borne pathogen in Japan, but is less common in the United States. Clinically, symptoms may appear in as little as 4 hr, but are typically present 12–24 hr after exposure. Diarrhea, abdominal cramps, nausea, and vomiting are the principal features. Fever and chills may occur in about 25% of cases and there have been reports of a dysentery-like diarrhea associated with this organism. The major factor produced by *V. parahemolyticus* that is considered to be important in pathogenesis is a 23 kDa protein called thermostable direct hemolysin (TDH). Isolation from stool can be undertaken with TCBS agar and *V. parahemolyticus* is differentiated from *V. cholerae* by the fact that it will not grow on 0% NaCl. The disease is usually self-limiting and attention to fluid replacement is important. Antibiotics may be required if the intestinal symptoms become persistent.

d. Vibrio vulnificus

Like *V. parahemolyticus*, *V. vulnificus* is a free-living estuarine organism that is frequently isolated from shellfish. The majority of illnesses related to this organism are thought to be acquired by eating raw oysters. Individuals who are alcoholic or who have underlying illnesses, such as liver disease or hemochromatosis, are especially susceptible. The consumption of even small amounts of alcohol may increase the risk of becoming infected with this bacterium. *V. vulnificus* is the most common cause of serious Vibrio-related disease in the United States and, typically, presents as a bacteremia, bullous skin lesions, and hypotension. The mortality rate in those with bacteremia is around 50% and rises to 90% in those with hypotension. Gastrointestinal symptoms may occur in about 25% of patients. Fully virulent *V. vulnificus* is encapsulated, and these encapsulated organisms are resistant to the bactericidal activity of normal human serum. They are also very sensitive to the amount of transferrin-bound iron in the host, which is almost certainly linked to the increased susceptibility in patients with hemochromatosis. A definitive diagnosis requires culture of the organism from blood or, occasionally, from bullous fluid. Blood agar or other nonselective media can be used for isolation from blood. The organism may also be isolated from stool in an infected patient, using TCBS agar. In view of the severity and the high mortality, it is important to initiate antibiotic therapy as early as possible; thus, a rapid clinical diagnosis may be life-saving. *V. vulnificus* is susceptible to a variety of antimicrobials, including tetracycline, ciprofloxacin, trimethoprim–sulfamethoxazle, chloramphenicol, and ampicillin.

e. Other Vibrios

A number of other members of the Vibrio species have been associated with food-borne disease, including *V. hollisae, V. furnissii, V. fluvialis,* and *V. mimicus.* These share similar epidemiology with the Vibrios already discussed, and, typically, cause gastroenteritis. Some of these other Vibrios have occasionally been associated with bloody diarrhea. Biochemical differentiation is the best way to tell them apart.

2. Clostridum perfringens

C. perfringens is an anaerobic, spore-forming, gram-positive rod that is associated with at least two types of food-borne disease. Of the various types of *C. perfringens,* type A is the one predominantly associated with food-borne disease. Type A results in a noninflammatory diarrhea that is usually linked to the consumption of meat or poultry (typically, foods that are high in protein) that has been prepared and allowed to remain between 15° and 60°C for more than 2 hrs. During this period, clostridial spores germinate and begin vegetative growth. When around 10^5 vegetative cells are ingested, they are present in large enough numbers to transiently colonize portions of the intestine and produce enough toxin to cause disease. Ingestion of preformed toxin or nongerminated spores will not usually result in disease. *C. perfringens* enterotoxin (CPE) is a heat-labile 35-kDa protein encoded by the *cpe* gene. In fact, *C. perfringens* types A, C, and D all have this gene but, for some reason that is not fully understood, only type A is frequently associated with food-borne diseases. CPE has a complex mechanism of action and appears to insert itself into the host cell membrane to form a protein complex. This results in membrane permeability changes, loss of intracellular potassium, and other intracellular disruptions. Clinically, typical symptoms include diarrhea and severe abdominal cramps developing 6–14 hrs after exposure; vomiting and fever are less common. Making a definitive diagnosis is complicated, in view of the fact that *C. perfringens* is found in the normal bowel flora of many individuals. However, a number of tests have been developed to detect the toxins in stool, using either enzyme immunoassays or latex agglutination. Ideally, one should attempt to recover large numbers ($>10^5$/g) of organisms from the suspected food. *C. perfringens* type C is also a food-borne cause of disease but the clinical picture is very different from type A disease discussed above. *C. perfringens* type C causes a necrotizing enterocolitis that is seen in the context of poor nutrition and so is almost exclusively a disease of developing countries. The type C strains produce three toxins (an enterotoxin and an α and β toxin). It is the β toxin that appears to be primarily responsible for the necrosis seen following infection with type C strains. β toxin is usually inactivated by proteolytic enzymes in the intestine and so the disease is associated with conditions in which the enzymes are inadequate (e.g., in malnutrition) or in the presence of trypsin inhibitors, such as undercooked pork or sweet potatoes.

3. Shiga Toxin-Producing E. coli

Shiga toxin-producing *E. coli* (STEC) are relative newcomers to the scene of food-borne pathogens. The first STEC to be associated with disease in humans was *E. coli* O157:H7, following two outbreaks of hemorrhagic colitis in 1982. Since then, it has been learned that there are, in fact, many different serotypes of STEC and at least 60 different types have been associated with clinical disease. Recent studies have suggested that around 1% of samples submitted to clinical microbiology laboratories in the United States contain STEC, of which around two-thirds are O157:H7, the remainder being non-O157. STEC are present in the gastrointestinal tracts of many mammalian species but appear to be especially common in ruminants (cattle, sheep, and goats). Therefore, the main source of STEC in our food supply is from bovine products. Recently, there have been an increasing number of reports associating STEC infection with fresh produce (lettuce, alfalfa sprouts, apple cider) and water. This is thought to be mainly due to contamination with fecal material from cattle pasture. Clinically, STEC cause a variety of diseases, ranging from diarrhea, which may or may not be bloody, hemorrhagic colitis, and the development of hemolytic uremic syndrome (HUS). HUS is a triad of renal failure, thrombocytopenia, and hemolytic anemia. Acutely, HUS has a mortality rate around 5% and up to 50% of HUS patients may have some degree of permanent renal insufficiency.

The main virulence factor from STEC is the production of one or more bacteriophage-encoded Shiga toxins (Stx). Shiga toxins are of two main types, Stx1 and Stx2, and there are at least 4 subtypes of Stx2 (Stx2, 2c, 2d, and 2e). All members of the Stx family have been associated with human disease, although Stx2e mainly causes disease in pigs. Following ingestion of the bacteria, they colonize portions of the lower intestinal tract and produce the toxins. Stx is then thought to cross the intestinal epithelial cell barrier and damage distant target sites, especially the kidney and brain, by a direct effect on endothelial cells in the microvasculature. The infectious dose of STEC may be very low, in the region of 10–100 organisms, in some instances. Symptoms typically develop 2–4 days following ingestion but may occur in as little as 1 day or as long as 8 days. The diarrhea can be of variable type (bloody or nonbloody) and may contain leukocytes. The type of diarrhea is not a reliable indication of who will go to develop HUS. *E. coli* O157:H7 has the unique biochemical characteristic of being a slow fermenter of sorbitol. This has been utilized as a means to diagnose O157:H7, using sorbitol MacConkey agar (SMAC). The disadvantage of this is that the other non-O157 STEC usually ferment sorbitol and so will be missed. An alternative strategy is to detect STEC in stool by looking either for Shiga toxins themselves directly or for the presence of Shiga toxin in broth cultures of stool samples. A positive result indicates the presence of STEC in the original sample. Enzyme immunoassays are now available to detect Shiga toxins under both conditions. The mainstay of treatment for STEC and its major complications is supportive. There is a degree of controversy over the use of antibiotics and a number of studies have suggested that certain antimicrobials (e.g., trimethoprim–sulfamethoxazole) actually increase the likelihood that a patient will go on to develop serious complications. In view of this, routine antimicrobial testing of STEC isolates is not recommended.

4. Enterotoxigenic E. coli

Enterotoxigenic *E. coli* (known as ETEC) are a common cause of disease in developing countries and are frequently associated with travelers' diarrhea. Like many other *E. coli* strains, they are transmitted by the consumption of contaminated water and food. It is not clear how frequently ETEC are a cause of sporadic disease in the United States because they are not routinely looked for. However, they have caused a number of large outbreaks due to the consumption of contaminated food. Incubation periods range from 12 hr to 2 days and, typically, symptoms are diarrheal (usually watery and without blood) with abdominal discomfort; fever is unusual. The diarrhea is not inflammatory and the illness is usually self-limiting within a few days. However, persistent diarrhea (defined as greater than 14 days) has been associated with ETEC infection. ETEC have two significant virulence characteristics: the ability to colonize the intestine and the capacity to produce toxins. A variety of colonization factor antigens (CFA) have been found in ETEC and at least four major groups CFAI–IV are described with several subgroups. ETEC produce two different types of toxin known as heat-stable toxins (ST) and heat-labile toxins (LT). The ST group is further subdivided into two major types and they have the characteristics of being small peptides that mediate their effects by increasing the intracellular concentration of cyclic GMP. The LT toxins are also part of a family of toxins but are structurally and functionally much like cholera toxin from *V. Cholerae*. The diagnosis of ETEC is usually clinical because ETEC do not have unique microbiological or biochemical properties to differentiate them from other *E. coli* strains. While a number of *E. coli* O serogroups are more frequently associated with STEC, this is not an absolute association. Detection of the toxin genes, using PCR or DNA probes, is effective but impractical for routine diagnostics. A variety of immunological assays, using latex bead agglutination or reversed passive latex agglutination, have been described. Oral rehydration is the mainstay of treatment of ETEC and may be life-saving. Antibiotic therapy is not routinely required.

C. Bacteria Causing Disease Primarily by Invading the Intestinal Epithelial Cells

This group of food-borne bacteria causes disease by a variety of mechanisms, with the common theme that invasion of the intestinal epithelial cell barrier is part of their pathogenic process. Some of them also produce toxins that may contribute to the disease process, but toxin production is not necessarily the primary mechanism by which this group of bacteria makes people sick.

1. Salmonella spp.

Salmonella are one of the most common causes of food-borne illness in humans. There are many types of *Salmonella* but they can be divided into two broad categories: those that cause typhoid and those that do not. The typhoidal *Salmonella*, such as *S. typhi* and *S. paratyphi*, only colonize humans and are usually acquired by the consumption of food or water contaminated with human fecal material. The much broader group of nontyphoidal *Salmonella* are found in the intestines of other mammals and, therefore, are acquired from the consumption of food or water that has been contaminated with fecal material from a wide variety of animals and poultry. With a few exceptions (*S. typhi* and *paratyphi* C, which express a polysaccharide capsule called Vi), *Salmonella* are nonencapsulated gram-negative motile rods. They ferment glucose, not lactose, and generally produce H_2S, that is seen as a black pigment in triple sugar–iron agar. Their somatic (O) antigens and flagella (H) antigens differentiate the more then 2300 *Salmonella*. Many of these different isolates have been named after the towns or by the individuals who first discovered them. Using newer DNA-based classification systems, the number of species has been reduced to two: *S. bongori*, containing nonhuman organisms, and *S. enterica*, consisting of six groups, including human pathogenic strains (Table III). *S. typhi* and *S. paratyphi* are the causes of typhoid fever that continues to be a global health problem but is less of a problem in North America. Having said that, the most likely explanation for cases of typhoid in the United States are either from acquisition overseas or from food that has been contaminated by a chronic carrier. In contrast, the number of cases of nontyphoidal *Salmonella* has increased steadily over the last four decades. *S. entertidis*, particularly, has become a growing problem, especially in hen eggs. Currently, the estimate is that around 1 in 10,000 eggs is contaminated with *Salmonella*. It is now known that *Salmonella* can penetrate intact eggs lying

TABLE III
Taxonomy and Serogroups of Some Common Human Pathogenic Salmonella

Species	Subspecies	Serogroup	Common serovars
S. enterica	enterica	A	Paratyphi A
		B	Typhimurium, Agona, Derby, Heidelberg
		C	Cholerasuis, Infantis, Virchow
		D	Dublin, Enteritidis, Typhi
	salamae		
	arizone		
	diarizonae		
	houtenae		
	indica		
S. bongori			

in fecally contaminated material, and can also infect eggs transovarially during egg development, before the shell is formed. Other than eggs, common sources of nontyphoidal salmonellosis are milk, foods containing raw eggs, meat and poultry, and fresh produce. Essentially, as with many of the other food-borne bacterial infections, *Salmonellae* are frequently transmitted through fecal contamination of food because of the large numbers of animals that carry the organism. The infectious dose of *S. typhi* is thought to be around 10^5 organisms. The infective dose of nontyphoidal *Salmonella* may vary from <100 to 10^6, depending on the host and on the actual type of *Salmonella*. Irrespective of the type of *Salmonella*, the most critical virulence determinant of these bacteria is their ability to invade the intestinal epithelium, following which they interact with underlying lymphoid tissue. A variety of bacterial genes are involved in the invasion process, and much has now been determined about the ways in which various *Salmonella* proteins then subvert the intestinal epithelial cell biology, discussion of which is beyond the scope of this article.

Clinically, *Salmonella* can cause a multitude of symptoms. Typhoid fever typically causes prolonged high fever, associated with bacteremia and, if untreated, a mortality rate as high as 30%. The incubation period is usually around 5–21 days, shorter if the inoculum is high and the patient is high risk. The disease may begin with gastroenteritis, which often will disappear by the time the fever begins— patients may even be constipated. If left untreated, intestinal perforation and hemorrhage are a concern. Health providers in developed countries are much more familiar with the clinical presentation of nontyphoidal *Salmonella*. This typically presents with gastroenteritis 24–48 hr after exposure to the organisms. There is usually nausea, vomiting, abdominal cramps, and diarrhea, which may be watery or, occasionally, bloody. It is not unusual for there to be associated symptoms of fever, chills, headache, and myalgia. Occasionally, there can be long-term consequences following *Salmonella* infection, such as reactive arthritis (especially in individuals who are HLA B27 positive), endocarditis, and localized infections, such as osteomyelitis, septic arthritis, and soft tissue infections. Diagnosing infection with *Salmonella* is dependent on culturing the organism, usually from either stool or blood cultures. In the case of nontyphoidal *Salmonella*, it is also worth trying to culture organisms from the incriminated food. Serological tests, such as the Widal test that measures agglutinating antibody titers to the O-antigen, are also available. They are simple and may be helpful in the diagnosis of typhoid fever, but are usually not specific or rapid enough to be of major value. Microbiologically, most laboratories use MacConkey media (the vast majority of *Salmonella* do not ferment lactose but about 1% do), deoxycholate agar, salmonella shigella (SS), or Hektoen agar. Following primary isolation, either commercial identification systems are used or a screening media, such as triple sugar iron and lysine iron agar.

In terms of therapy, gastroenteritis secondary to nontyphoidal *Salmonella* is usually self-limiting and rehydration is the most critical aspect of treatment. Antibiotic therapy is not routinely required for this aspect of *Salmonella* infection and has, in some instances, been thought to promote chronic carriage. When there is systemic invasion with the bacteria and in cases of enteric fever, antibiotic therapy is important. Third-generation cephalosporins and quinolones are most frequently used, although chloramphenicol has been the mainstay of treatment for typhoid fever for many years and is still used in many developing countries, but chloramphenicol does carry a risk of complications. Antibiotic resistance has become a major problem with many *Salmonella* serovars. Of particular concern is the recent emergence and spread of *Salmonella* DT104 that carries resistance to multiple antibiotics, including, in some instances, to the fluoroquinolones, such as ciprofloxacin.

2. *Campyllobacter spp.*

Food-borne disease due to *Campylobacter,* which was not recognized until the mid-1970s, is now one of the most frequently diagnosed causes of food-related gastroenteritis. In fact, *Campylobacter* and *Salmonella* probably account for 70% of diagnosed cases of bacterial food-borne disease in the United States. Campylobacters are gram-negative, spiral organisms. They are strictly microaerophilic and need oxygen for growth but the level of oxygen in air is too high for them, in view of their sensitivity to oxygen free radicals. An atmosphere of 5–10% O_2 and 1–10% CO_2 is ideal for their growth, which is better at 42–43°C and gives additional selectivity for *Campylobacter*. There are two main species of *Campylobacter* responsible for most of the illness seen in humans. *C. jejuni* accounts for the vast majority (~90%) and *C. coli* accounts for the majority of the remainder. Other *Campylobacter* species that have been associated with gastroenteritis in humans include *C. fetus, C. upsaliensis, C. hyointestinalis,* and *C. lari. Campylobacter* are fragile organisms and tend to die in transport media; therefore, rapid plating of sample is an advantage. Enriched media is usually used that contains a variety of antibiotics to reduce competitive flora. *Campylobacter* isolates can be typed using biotyping methods. Serotyping is probably the most widely used method currently and is based on the somatic lipopolysaccharide antigens or the heat-labile surface protein antigens. The infectious dose of *Campylobacter* can be very wide. Low numbers of *Campylobacter* may be all that is needed to cause disease and, in some instances in human volunteers, less than 100 organisms can cause disease. These organisms are more frequently associated with sporadic disease rather than outbreaks and person-to-person spread does not appear to be common, although it can occur in a family environment, for example. This sporadic epidemiology suggests that the infectious dose may actually be higher in many situations. *C. jejuni* and *C. coli* are commensals in the intestines of many animals and birds, including domestic pets. The main vehicles for infection are raw meats, especially poultry, milk, and water. Surface water is frequently contaminated with campylobacters and water-borne outbreaks have occurred. The pathogenesis of *Campylobacter* is dependent on its motility. *In vitro,* nonmotile strains are not capable of invading intestinal epithelial cells. It is assumed that invasion in the intestine is one of the principal pathogenic mechanisms of these bacteria, although the genes involved are not known. However, following infection with campylobacters there is typically an inflammatory response, with marked inflammatory infiltration of the lamina propria, resulting in leukocytes being present in the stool. Clinically, symptoms usually occur within 2–3 days following exposure; this may be as long as 7 days after exposure. Nongastrointestinal symptoms, such as fever (which may be high), headache, and myalgia may precede the onset of nausea, vomiting, and diarrhea. Diarrhea is usually the predominant symptom and it may be watery or bloody and is usually associated with severe cramping abdominal pain. Interestingly, the disease is occasionally biphasic, with an apparent settling of symptoms after 4–5 days, only to be followed by a recrudescence. A number of complications are associated with *Campylobacter,* including cholecystitis, hepatitis, acute appendicitis, pancreatitis, and focal extraintestinal infections. Longer-term complications include reactive arthritis, Reiter's syndrome, uveitis, and Guillain-Barré syndrome; molecular mimicry may be involved in some of these complications. A definitive diagnosis requires culture using selective media as previously outlined; however, newer methodologies, such as enzyme immunoassays, are now available and are designed to detect specific campylobacter antigens. Such enzyme immunoassays have the advantage of being able to detect dead organisms, if the transport to the laboratory has been delayed. However, they do not allow typing or assessment of antibiotic sensitivity. The majority of infections are self-limiting and require supportive

therapy only, especially in the form of rehydration. Antibiotic therapy is not routinely required; however, patients with prolonged or severe symptoms or who have other significant risk factors (AIDS, cirrhosis, diabetes, etc.) should be treated. Erythromycin is a reliable therapy, although thee is a general move toward using quinolone antibiotics. Despite the use of erythromycin for many years, the resistance levels have remained low. This is not the case with the quinolones and there are increasing reports of ciprofloxacin resistance, which may become more of a problem, in view of the expanded use of these antibiotics in agriculture.

3. *Yersinia spp.*

Of the three members of the genus *Yersinia, Y. enterocolitica* and *Y. pseudotubeculosis* are considered to be food-borne; *Y. pestis* is not. *Y. enterocolitica* and *Y. pseudotuberculosis* are nonlactose fermenting gram-negative organisms that are distinguished by biochemical and serological criteria. *Y. enterocolitica* is divided into biogroups according to biochemical properties. Serotyping based on the more than 50 described O-antigens is used to designate strains. Most human isolates are of serotypes O3, O8, or O9, although there are geographic variations in the serotype prevalence. Six serotypes and 4 subtypes of *Y. pseudotuberculosis* have been described. Serotype O1 is associated with about 80% of human cases. *Yersinia* are not very commonly found as causes of food-borne illness, compared with *Salmonella* or *Campylobacter*. However, they are clearly transmitted in food and can cause significant gastrointestinal illness. The food most frequently associated with Yersiniosis is pork. Swine are a major reservoir of these organisms, and, although they have been found in many other animals (e.g., sheep, dogs, cats, cattle) consumption of undercooked pork is a common association. Milk is another frequently reported source and since *Y. enterocolitica* can survive and, indeed, multiply in milk at 4°C small numbers of organisms can become a significant health threat, even if the milk is refrigerated. Infection with *Y. pseudotuberculosis* has been associated with consumption of contaminated water or unpasterurized milk. Clinically, *Y. enterocolitica* may cause a variety of symptoms. A gastrointestinal presentation is the most common with diarrhea, low-grade fever, and abdominal pain—usually in children. In about 25% of patients, the stools are bloody and leukocytes are present in about half the patients. *Y. enterocolitica* is a cause of "pseudo-appendicitis," in which patients have fever, abdominal pain, and a leukocytosis. Such patients frequently end up having a laparotomy, in which the appendix is typically normal but there is inflammation in the terminal ileum. *Y. pseudotuberculosis* typically causes an acute mesenteric lymphadenitis, that also presents much like appendicitis. Symptoms in *Y. enterocolitica* infection can be prolonged, lasting several weeks or even longer. Most infections are, however, self-limiting, although complications may occur, such as ulceration and intestinal perforation. A classic long-term complication following yersiniosis is the development of reactive arthritis. As with other enteric pathogens, this is more likely in patients who are HLA B27 positive. *Y. enterocolitica* are invasive organisms and at least three bacterial proteins (Invasin, Ail (attachment invasion locus), and YadA) are involved in this process. Not all *Y. enterocolitica* strains are equally pathogenic and this may be related to levels of expression of some of the various virulence factors. Following invasion across the intestinal epithelial cell barrier (which is likely occurring via M cells), the bacteria localize in the Peyer's patches and mesenteric lymph nodes. *Y. enterocolitica* may be isolated from stool or other body tissue on commonly used enteric media, such as MacConkey agar, on which it appears as lactose-negative colonies after 48 hrs of growth at 25–28°C. Cold enrichment and selective agars have also been used but they may result in isolation of nonpathogenic strains. A variety of tests can be used to determine if a strain is pathogenic, including congo red absorption, salicin fermentation, and esculin hydrolis. Uncomplicated cases will settle spontaneously and antibiotic therapy is not routinely required. Many antimicrobials are effective, but ceftriaxone or fluoroquinolones are recommended for serious infection.

4. Listeria monocytogenes

L. monocytogenes is a gram-positive motile rod that is one of the most frightening food-borne pathogens because of the high mortality rate associated with infection. Of the seven *Listeria* species, only *L. monocytogenes* is pathogenic for humans. *L. monocytogenes* is a common environmental organism and is frequently present in soil and water, on plants, and in the intestinal tracts of many animals. It is has been found in 37 different types of mammals and at least 17 species of birds. Between 1 and 10% of people are carriers of *L. monocytogenes*. This organism is associated with both sporadic disease and outbreaks. Incriminated foods include milk, cheese, raw vegetables, undercooked meat, and foods prepared for instant use, such as hot dogs. The infectious dose is not really known; some studies have suggested that it may be very high (up to 10^9 organisms) and others that it may be as low as several hundred. In practice, the most critical aspect is probably individual susceptibility, rather than the infectious dose per se. Clinically, it usually begins with nonspecific symptoms, such as fever, myalgia, and gastrointestinal upset in the form of diarrhea and nausea. What makes *L. monocytogenes* exceptional as a food-borne pathogen is its very high mortality rate. Of the around 1800 cases per year that are estimated to occur in the United States, there are over 400 deaths. This gives a case fatality rate of over 20%. There are certain groups that are especially at risk of developing listeriosis and these include pregnant women, the elderly, and the immunocompromised. Transplacental transmission in pregnant women is a major concern, although it does not inevitably lead to major consequences. Spontaneous abortion, prematurity, neonatal sepsis, and meningitis are all complications of transplacental transmission. Although *L. monocytogenes* is readily killed by heat and cooking, the fact that it is so ubiquitous makes recontamination a real risk. This then poses a major health problem, because the organism will grow and multiply at standard refrigerator temperatures. Thus, even minor contamination of a product may, after storage, result in high levels of bacteria, even if the product has been adequately refrigerated. *L. monocytogenes* grows well on blood or other nutrient agar; it can be separated from other bacteria by growing at cold temperatures (cold enrichment). Selective media may also be used and is better than cold enrichment. Following diagnosis, *L. monocytogenes* is readily treated by penicillins or aminoglycosides.

5. *Shigella spp.*

There are four species of *Shigella* (*S. dysenteriae*, *S. felxneri*, *S. boydii,* and *S. sonne*). Group-specific polysaccharide antigens of LPS, biochemical properties, and phage or colicin susceptibility differentiate them. All closely resemble *E. coli* at the genetic level and will grow readily on many routine laboratory media, but selective media are usually used. MacConkey agar, deoxycholate or eosin–methylene blue, and salmonella–shigella media are all adequate, but xylose–lysine–deoxycholate (XLD) or Hektoen agars are better for the recovery of *Shigella*. *Shigella* is typically spread via the fecal–oral route, through contamination of food or water, or by person-to-person spread. They are highly host-adapted and infect only humans and some nonhuman primates. Therefore, when food becomes contaminated with *Shigella,* it usually originates from some other infected human. The low infectious dose (of the order of 10–100 organisms, in some cases) makes person-to-person spread a real problem and has resulted in large outbreaks in institutions. Shigellosis is a major problem in developing countries and results in significant morbidity and mortality. Clinically, the disease usually begins with fever, fatigue, anorexia, and malaise. Soon thereafter, watery diarrhea develops, which may or may not become bloody with the development of dysentery in hours or days. The classic dysenteric stool consists of small amounts of blood and mucus, sometimes grossly purulent, and is typically associated with severe abdominal cramps and tenesmus. The severity of the disease varies with the type of *Shigella, S. dysenteriae* being the worst, followed by *S. flexneri* and *S. sonne* or *S. boydii*—the latter is unusual in most settings. *S. dysenteriae* type 1 carries the added clinical complication of hemolytic uremic syndrome (a triad of renal failure, thrombocytopenia, and hemolytic anemia), because it is able to express Shiga toxin. Shiga toxin from *S. dysenteriae* type 1 strains is almost identical to Shiga toxin 1 from *E. coli,* and the clinical diseases that the two toxins produce may be very similar. However, *S. dysenteriae* is rare in developed countries, where *S. sonne* or *S. flexneri* is more commonly seen. Other complications of *Shigella* infection include encephalopathy, reactive arthritis, Reiter's syndrome, toxic megacolon, and perforation. The pathogenesis of *Shigella* relates largely to their capacity to invade intestinal epithelial cells and induce an inflammatory response. Many of the genes responsible for the invasion are on a virulence plasmid. Only *S. dysenteriae* type 1 strains are able to produce Shiga toxins that are chromosomally encoded and, unlike *E. coli*, are not transmitted on bacteriophages. The diagnosis can be confirmed microbiologically; however, *Shigella* of all types are fastidious organisms, and transporting fecal samples to the laboratory rapidly is important. As with all enteric infections, rehydration is the most critical therapeutic modality. In patients with *S. flexneri* or *S. dysenteriae,* antibiotic therapy is usually recommended. However, resistance is becoming a significant problem in many developing countries, but most strains are sensitive to quinolones and naladixic acid is frequently used in developing countries. Alternatives include cefixime or ceftriaxone. *S. sonne* infection is usually mild and self-limiting and requires supportive therapy only.

6. *Enteroinvasive E. coli*

Enteroinvasive *E. coli* (EIEC) are not frequently recognized as food-borne pathogens. However, they are not routinely diagnosed in patients and so little data are available. EIEC have been associated with several outbreaks linked to both water and other foods, such as cheese. A number of prominent serogroups found to be EIEC have been described, including O28, O112, O124, O136, O143, O144, O147, and O164. Clinically, they produce a disease much like shigellosis, described previously, with a watery diarrhea that may lead to dysentery. As with *Shigella,* EIEC are invasive; however, they do not produce Shiga toxins. Microbiologically, they are lactose-fermenting and indistinguishable from other *E. coli*. This makes the diagnosis difficult to confirm. Therefore, in the context of a patient with a dysenteric-like clinical picture, in whom other invasive organisms have been ruled out, one should consider EIEC, especially if there are large numbers of leukocytes in the stool. Isolating and typing the strains may help, but the only way to confirm the diagnosis is to determine the presence of specific virulence genes using molecular techniques (PCR or DNA probes) or to show that the organisms are invasive in the Sereny test. The latter is a biological assay in which invasion of a guinea pig conjunctiva is demonstrated.

D. Other Bacterial Causes of Food-borne Illness

1. *Aeromonas spp.*

Aeromonads are gram-negative, facultatively anaerobic, motile, oxidase-positive bacilli. They are frequently present in soil and freshwater, and tend to peak during the summer months in water, which may lead to contamination of fresh produce, meat, and dairy products via contaminated water. Of the various Aeromonas spp., *A. hydrophila, A. caviea, A. veonii,* and *A. jandaei* are the ones that are most frequently associated with gastroenteritis and foodborne infections. They grow on most media used for fecal testing, including MacConkey, XLD, desoxycholate, and blood agar, and can be differentiated biochemically. Clinically, this group of organisms causes watery diarrhea that may become persistent. Fecal leukocytes and red cells are usually absent (although dysenteric-like symptoms have been associated with *Aeromonas* infection) and the patients usually have abdominal pain. Nausea, vomiting, and fever may occur in up to 50% of patients. Simply finding Aeromonas in the stool may not be enough to confirm the diagnosis since asymptomatic carriage has been reported. Therefore, immunological tests to confirm recent infection may be needed. In practice, however, infection with Aeromonas is usually self-limiting and previously healthy people are likely to recover fully, do not require antimicrobial therapy, and the diagnosis is often of academic interest only. The exception to this may be in a patient with persistent diarrhea in whom no other cause has been identified.

2. *Plesiomonas shigelloides*

Plesiomonas shigelloides was placed in its own genus in 1962. It is primarily a freshwater organism and, like Aeromonas spp., the isolation rates increase in warmer months. Plesiomonads are gram-negative, motile, facultative anaerobes and will grow on a variety of media. About 30% ferment lactose and, if one is examining specifically for *Plesiomonas*, inositol–brilliant green–bile salts may be used. Once isolated, *Plesiomonas* can be differentiated from other Vibrionaceae and Enterobacteriaceae biochemically. As with the *Aeromonas* spp., contamination of a variety of food products with water that contain *Plesiomonas* may cause disease. Data that *Plesiomonas* are truly associated with enteric disease is variable. Some volunteer studies have not been very convincing that it is a true pathogen in humans. However, treatment of infected patients with antibiotics has been associated with an improvement, suggesting that it is contributing to a disease process in some way. This is supported by outbreaks in Japan associated with contaminated oysters and water. Symptoms usually occur within 24–48 hrs of exposure and are thought to be linked to the production of an enterotoxin. Typically, infection with *Plesiomonas* is associated with abdominal cramping; nausea, vomiting, and fever are less common, but bloody stools have been reported. Therapeutically, rehydration is key and, if antibiotic therapy is required, most strains are sensitive to quinolones and trimethoprim–sulfamethoxazole.

3. *Enteropathogenic E. coli*

Enteropathogenic E. coli (EPEC), like *Shigella* species, are transmitted mainly by the fecal–oral route from one infected individual to another. Therefore, they may be transmitted via food and water but are usually introduced into the food chain from another infected person. EPEC are not associated with an animal reservoir; therefore, meat, poultry, and fresh produce—which are the typical sources of foodborne pathogens—are not usually contaminated at the source, but may become contaminated during handling. EPEC are an important cause of enteric disease in developing countries, but their role in developed nations is questionable. They have caused major outbreaks in the past in various developed countries, but the lack of routine diagnostics for EPEC makes it difficult to know what their role in sporadic disease may be. Clinically, EPEC infection usually presents with a watery diarrhea in the absence of blood. Low-grade fever and vomiting are common, and in developing countries, the mortality rates may be very high, especially in infants, although the factors that are actually causing death in these children are not clear. Microbiologically, EPEC are like any other *E. coli*. However, there are certain serotypes that may be considered classical EPEC serotypes, including O55, O86, O111, O114, O119, O127, and O142, that are closely associated with EPEC virulence factors. Finding one of these serotypes does not mean that it is an EPEC; indeed, many of the same serotypes have been shown to carry Shiga toxins and to be associated with HUS. Confirmation, using molecular tools, is the only way to be sure an isolate is a true EPEC. Much is known about the pathogenesis of EPEC, and localized adherence, intimate attachment to intestinal epithelial cells followed by activation of a number of epithelial cell signal pathways, all seem to be important in EPEC virulence. Treatment of EPEC requires appropriate rehydration and supportive therapy. Antibiotics may be required, but EPEC are frequently resistant to many antimicrobials and, therefore, susceptibility testing is suggested.

4. Enteroaggregative E. coli

Enteroaggregative *E. coli* (EAEC) get their name from the way in which they adhere to epithelial cells in an aggregative, or "stacked brick," pattern. EAEC have been associated with persistent diarrhea in many developing countries and in immunocompromised patients in developed nations. There are no known animal reservoirs for EAEC, which is more a reflection of a lack of detailed studies than accurate epidemiological information. Fecal–oral spread from one person to another is considered to be the usual route of transmission. As with EPEC, contamination of food and water from infected individuals is probably important. Clinically, EAEC cause watery diarrhea without major associated symptoms, Microbiologically, they are like any other *E. coli* and can only be reliably differentiated by their characteristic aggregative phenotype. A number of virulence factors have been found in EAEC, including the production of fimbriae important in aggregation and the expression of a plasmid-encoded toxin. However, not all EAEC have these and it is not clear what the critical virulence determinants really are. At least one study has suggested that treating HIV-positive patients who have persistent diarrhea and are colonized with EAEC results in clearing of the organisms and im-

provement in symptoms, suggesting that they are true pathogens, but may be more opportunistic than some other food-borne bacteria.

III. VIRAL FOOD-BORNE PATHOGENS

Viral agents are considered to be an increasingly important cause of food-borne illness. A number of different viral agents have been associated with food-borne disease and cause a variety of illnesses, varying from a simple gastroenteritis to major systemic upset, such as hepatitis. Food and water are vehicles for viruses, but viruses do not reproduce in food and nor do they produce toxins in food. Some viruses, such as Norwalk, cause large outbreaks; others seem to be more frequently associated with sporadic disease. Overall, the difficulty in diagnosing viral illness has precluded the development of large amounts of epidemiological data. However, the advent of rapid tests (e.g., enzyme immunoassays for the detection of rotavirus) is beginning to change this, and will eventually lead to a better understanding of the epidemiology and disease burden caused by the various food-borne viral pathogens.

A. Hepatitis A Virus

Hepatitis A is an RNA virus belonging to the family *Picornaviridae* and is present throughout the world. Hepatitis A is usually spread by direct contact from one person to another via the fecal–oral route. There have been many examples of outbreaks of hepatitis A due to contaminated food or water. Some of the incriminated foods include raw shellfish, milk, potato salad, and orange juice. The incubation period following exposure is around 30 days, with a range of 15–50 days. This long incubation period can make tracing the source of infection very difficult. Clinically, hepatitis A may be mild or even asymptomatic, especially in children. In adults, it usually begins with flulike symptoms of headache, myalgia, anorexia, nausea, vomiting, and headache. This is then followed by the development of jaundice. In hepatitis A, the rule is that the disease is self-limiting and long-term consequences are unusual, but occasionally the hepatitis can become fulminant. While during the acute illness, it is usually possible to detect hepatitis A virus in the stool. The virus rapidly disappears after that. Patients usually excrete the virus for 3 weeks prior to the onset of illness and for about 1 week after. Therefore, there is a prolonged period when an individual is infectious and does not realize it. Diagnosis is usually based on the detection of antihepatitis A IgM serum antibodies. IgG antibodies usually appear after resolution of the clinical disease.

B. Hepatitis E Virus

Hepatitis E virus was first described in 1977 and is a small RNA virus from the *Caliciviridae* family. It is usually transmitted in contaminated drinking water, and probably in food as well, but this has not been documented. Person-to-person spread also occurs. Hepatitis E has an incubation period of 2–9 weeks, although most people develop symptoms around 40 days postexposure. Clinically, the disease is very much like hepatitis A, with constitutional symptoms followed by jaundice. Patients usually recover but the mortality rate has been reported to be as high as 3% in some instances, especially in pregnant women. The diagnosis is made serologically.

C. Rotavirus

Rotaviruses have become a very common cause of gastrointestinal disease in both developed and developing nations. In the United States, an estimated 1 million cases per year occur and the numbers are much higher in other countries, with over half a million deaths associated with rotavirus infection. Rotavirus is primarily a disease of children and, like the other enteric food-borne viruses, it is transmitted by the fecal–oral route. In some animal models, as few as one viral particle is enough to cause disease. Rotaviruses have a double-stranded RNA genome and are classified by group, subgroup, and serotype. There are three main groups, designated A, B, and C, and it is group A that is the most common cause of human disease. Groups B and C have also been linked to food-borne transmission, again by the fecal–oral route. Rotaviruses are lytic and cause diarrhea by invading and then destroying villous intestinal epithelial cells—the cells primarily involved in

absorption of fluid. The incubation period is usually between 1 and 3 days and is followed by watery diarrhea, vomiting, and fever. The diarrhea may persist for a week or more, resulting in severe dehydration if untreated. Occasionally, the disease can be fulminant and, even with appropriate rehydration, lead to death of the infected infant. Currently, the most rapid way to diagnose the infection is by enzyme immunoassay, although genetic-based and electron microscopy-based tests are also available. Supportive care is required for patients infected with rotavirus, especially rehydration, and the disease will usually be self-limiting with no long-term problems.

D. Other Food-borne Viruses

Norwalk virus is a small round structured virus (SRSV) and was the first virus to be clearly associated with gastroenteritis. Norwalk is a calicivirus and this group of viruses causes disease worldwide and they have been associated with some large outbreaks, often in confined environments, such as cruise ships. Outbreaks have been associated with contaminated drinking water, swimming water, consumption of undercooked shellfish, ice, and salads. As with the other enteric viruses, fecal contamination of food or water is usually found to be the ultimate source. The incubation time following exposure is around 48 hrs and the clinical illness usually consists of vomiting and diarrhea. The diarrhea is watery without red cells, leukocytes or mucus. The disease is usually self-limiting, settles in 24 hr, and requires no specific therapy. Specific diagnosis is difficult. A number of assays are available, including electron microscopy, enzyme immunoassays, and RT–PCR.

A number of other viruses have also been associated with outbreaks of gastroenteritis and are, therefore, thought to be spread by the fecal–oral route. The list of potential food-borne viruses is long and includes the viruses previously discussed as well as enteric adenovirus, coronaviruses, toroviruses, reoviruses, and the smaller-sized viruses, such as caliciviruses, astroviruses, parvoviruses, and picobirnaviruses. All of these viruses cause a similar type of gastroenteritis, which typically consists of a noninflammatory watery diarrhea, which is self-limiting.

Diagnosis is difficult and usually requires direct electron microscopy of stool samples.

IV. PROTOZOAL FOOD-BORNE PATHOGENS

A. *Cryptosporidium parvum*

C. parvum is an apicomplexan protozoan parasite, capable of causing disease in both immunocompetent and immunocompromised individuals. It has gained a reputation in recent years as being a major problem in AIDS patients, but it is important to realize that it can infect normal hosts as well. It is a cause of diarrhea in cattle and so water that is close to cattle pasture may be contaminated with this organism. Other domestic and wild animals are also reservoirs for *C. parvum*. Both water and food are sources of *C. parvum* and one of the largest outbreaks, infecting over 400,000 people, was due to contamination of a municipal water supply. Clinically, *C. parvum* causes watery diarrhea, which may be profuse, abdominal cramping, nausea, and vomiting. Fever is not seen and neither is intestinal bleeding. The organisms are ingested as cysts, which then excyst to release four sporozoites. The sporozoites then attach to and invade intestinal epithelial cells. The invasion remains superficial and *C. parvum* does not penetrate beyond the intestinal epithelial cell barrier. *C. pravum* is diagnosed by microscopy of stool to look for cysts or sporozoites, by immunofluorescene microscopy, or by enzyme immunoassay. In immunocompetent patients, the natural history of *C. parvum* infection is for the disease to be self-limiting and recovery is the rule after a week or two. The pattern is very different in immunocompromised hosts in which the infection is not cleared, and malabsorption becomes a significant and life-threatening problem. There is currently no treatment that is effective for the eradication of *C. parvum*.

B. *Giardia lamblia*

G. lamblia is probably the most frequently found enteric protozoan worldwide. This organism does not cause a dramatic enteric disease or systemic complications, yet infection with it can lead to profound malabsorption and misery in the patient. Like other enteric protozoa, it is found in fecally contaminated water and food and is yet another example of the fecal–oral transmission route. There are different *Giardia* types but only *Lamblia* are known to infect humans. The disease is initiated by ingestion of the cysts, and as few as 10–100 may be an infectious dose. Following ingestion, the cysts excyst in the proximal small intestine and release trophozoites. The trophozoites divide by binary fission and attach intimately to the intestinal epithelium, via a ventral disc on the trophozoite. Clinically, giardiasis may be extremely variable. At one extreme, there may be asymptomatic infection, and at the other, severe chronic diarrhea, leading to intestinal malabsorption. Acutely, infection usually results in watery diarrhea and abdominal discomfort. *G. lamblia* can be diagnosed by fecal microscopy, looking for either the cysts or the trophozoites. Currently, many laboratories look for the parasites using commercially available kits, that utilize either fluorescence microscopy with specific antibodies or enzyme immunoassays. In terms of therapy, metronidazole is the drug of choice.

C. *Entamoeba histolytica*

E. histolytica is one the leading causes of parasitic death in the world. It is usually spread by the fecal–oral route, either directly or via contamination of food, e.g., lettuce or water. The cyst is the infective form and cysts may survive weeks in an appropriate environment. Following ingestion, the cysts, excyst in the small bowel and form trophozoites, which then colonize the large bowel, and multiply or encyst, depending on local conditions in the intestine. There are various types of *E. histolytica,* some of which are pathogenic to humans and some of which are not. They have been differentiated based on zymodeme analysis, which is a determination of the electrophoretic mobility of certain isoenzymes. *E. histolytica* causes amebiasis, which may have various clinical manifestations. The trophozoites have the capacity to invade the host from the intestinal lumen, which in the colon results in ulceration of the mucosa, causing amebic dysentery. When the trophozoites invade further, they gain access to the portal blood vessels and are transmitted to the liver. Once in the liver, they are then able to destroy the parenchyma, resulting in a hepatic amebic abscess (this occurs in about 1% of patients with intestinal amebiasis). Intestinal infection with *E. histolytica* is diagnosed by microscopic examination of the stool, either by wet mount or trichrome stain. Serology is useful in the diagnosis of patients with invasive amibiasis. Patients with *E. histolytica* need to be treated, and metronidazole is the drug of choice. Luminal drugs that are poorly absorbed are an alternative therapy for carriers of cysts. Drugs such as paromomycin or iodoquinol are used for this purpose.

D. *Cyclospora cayetanensis*

C. cayetanensis is a recently described apicomplexan parasite that has been found in food. Most recently, it has been responsible for a number of outbreaks in North America associated with consumption of imported raspberries. It has also been associated with undercooked meat and poultry, and contaminated drinking water and swimming water. Clinically, it causes a self-limiting diarrhea, with nausea, vomiting, and abdominal pain in immunocompetent patients but may lead to a more persistent diarrhea in immunocompromised individuals. *C. cayetanesis* is diagnosed by direct stool microscopy and oocyst autofluorescence, which appears blue by Epi-illumination and a 365-nm dichroic filter and green by a 450–490-nm dichroic filter. In the laboratory, oocysts may be induced to sporulate, even in the presence of potassium dichromate used to preserve the specimens. After 1–2 weeks, approximately 40% of oocysts contain two sporocysts with two sporozoites in each. Excystation requires a number of steps and occurs when oocysts are subjected to bile salts and sodium taurcholate plus mechanical pressure. Susceptible humans are infected by ingesting sporulated oocysts and, in view of the complexity of the process and the time required, direct person-to-person spread is considered unlikely. The infection can be successfully treated with trimethoprim–sulfamethoxazole.

E. Other Protozoan Infections

A number of other protozoa have been associated with food- and water-borne infections in humans. These include Microsporidium, that causes watery diarrhea and malabsorption, and are becoming an increasingly recognized problem in the immunocompromised. Various microsporida, including *Enterocytozoon bieneusi* and *Septata intestinalis,* cause human disease. *Isospora belli,* another apicomplexan protozoon, is an opportunistic pathogen in immunocompromised patients. Sarcocystosis is a rare zoonotic infection in humans that can, on occasion, cause necrotizing enteritis. *Dientamoeba fragilis* was originally thought to be a commensal but now appears to be associated with a variety of gastrointestinal symptoms, including abdominal pain, nausea, diarrhea, and anorexia. *Balantidium coli* is the only ciliate known to parasitize humans. Although most infections are asymptomatic, the disease may present itself as dysentery. *Blastocystis hominis* is a strict anaerobic protozoon that infects both immunocompetent and immunocompromised hosts and results in a variety of gastrointestinal symptoms, including diarrhea, abdominal pain, nausea, vomiting, anorexia, and malaise.

V. CESTODES AND WORMS

A. *Taenia saginata*

Taenia saginata, the beef tapeworm, is highly endemic in certain areas of the world, such as parts of South America, Africa, south Asia, and Japan. Humans are the definitive host for the adult tapeworm, which is one of the largest human parasites. They may live as long as 20 years and grow up to 25 meters in length. Consumption of undercooked or raw beef containing living larval forms is how it is acquired. Cattle are the intermediate hosts, in which the hexacanth embryos emerge from the eggs and pass by blood or lymph to muscle, subcutaneous tissue, or viscera. Then, when humans eat the undercooked animal tissue, the life cycle is completed. Clinically, the worms are remarkably quiescent and nausea or a feeling of fullness may be the only symptoms. Vomiting, nausea, and diarrhea may occur. Diagnosis depends on detecting the proglottids in stool and treatment with either praziquantel or albendazole should be curative.

B. *Taenia solium*

T. solium is the pork tapeworm, and, unlike the *T. saginatum,* the larval stage can invade humans and cause infections of the central nervous system. *T. solium* is distributed worldwide and is acquired by ingesting pork meat that is infected with cysticerci. The adult worm in humans sheds proglottids, which are then eaten by pigs, after which the hexacanth embryos emerge and penetrate the pig's intestinal wall, where they migrate to muscle and other tissues. Humans then eat the larval forms, which completes the cycle. *T. solium* is usually smaller than *T. saginatum* so the clinical symptoms are even less remarkable. Infected humans may just notice the proglottids in stool. The diagnosis is dependent on identifying the proglottids and the treatment is with praziquantel or albendazole.

C. *Diphyllobothrium latum*

D. latum is a fish tapeworm and is most common in Northern Europe and Scandinavia. Eating raw fish is the greatest risk factor and the increased consumption of sushi in the United States has resulted in more cases than in the past. The life cycle is complex; after the eggs are passed in human feces, they must hatch and be eaten by copepods, which are freshwater crustaceans. They then develop into larval forms and subsequently are eaten by fish. The procecoids then invade the stomach wall of the fish and, finally, reside in the muscle of the fish. Humans then become infected by eating a fish that harbors a viable plerocercoid larva. Clinically, the infection is usually asymptomatic, but diarrhea, fatigue, and distal paresthesia are well described, as is pernicious anemia, since the tapeworm actively absorbs free vitamin B12. Diagnosis is made by identifying the ova or proglottids in stool. Treatment with praziquantel or niclosamide is effective.

D. *Hymenolepis nana*

H. nana, the dwarf tapeworm, is very common in humans and is transmitted via the fecal–oral route. This worm does not require an intermediate host. The majority of infections are asymptomatic but diarrhea, abdominal cramping, and anorexia may occur. The diagnosis is dependent on identifying the typical double lumen eggs in the stool and treatment with praziquantel should be curative.

E. Ascariasis

Ascaris is the most common intestinal helminth worldwide. *Ascaris lumbricoides* is specific for humans, who are infected by ingesting food containing the mature ova. The larvae are then released in the small intestine, enter the circulation, and then reach the pulmonary alveoli, where they develop. This pulmonary infestation may cause pneumonitis and allergic manifestations. Finally, larvae migrate up the bronchial tree and are swallowed. Humans are the definitive hosts, but soil is necessary for development of the eggs and also acts as a reservoir. Humans then ingest the developed eggs to complete the cycle. Thus, food or water that are contaminated are sources that infect humans. The diagnosis is made by finding adult worms, larvae, or eggs in the stool. Acarisis may be treated with mebendazole or pyrantel pamoate.

F. Trichuriasis

Trichuris trichiura is commonly known as whipworm and is frequently found in the same parts of the world as ascaris. Humans are the definitive host and eggs that are passed in stool mature in warm moist soil to become infective. They then contaminate food and are ingested by a new host. Clinically, the worms remain associated with the intestine and may be either asymptomatic or result in chronic diarrhea. In the context of a heavy worm burden, dysentery may develop along with malnutrition. The diagnosis is made by finding adult worms or eggs in stool. Treatment with mebendazole is usually curative.

G. Trichinella spiralis

This nematode begins its infection in humans following the ingestion of the first-stage larvae and its nurse cell in striated skeletal muscle tissue. The larvae are released from tissue in the stomach and pass to the small intestine, where they infect epithelial cells. The larvae then develop into adult worms and are shed in the stool. Larvae also penetrate into lymph or blood vessels and then on into muscle cells, where a nurse cell begins to form. The principal mode of transmission to humans is through the consumption of undercooked meat, usually pork. The major clinical features of this disease relate to cellular destruction secondary to the parasitic penetration of cardiac or nervous tissue. Gastrointestinal symptoms are also common and include diarrhea and vomiting. The diagnosis is dependent on the histologic identification of nurse cells containing larvae within infected muscle tissue. Serological tests are also of value. The infection may be treated with thiabendazole.

VI. NATURAL TOXINS

There are a number of naturally occurring toxins that may be present in various types of food. Many of these are associated with consumption of seafood but others are related to different and specific foods.

A. Ciguatera

Ciguatera poisoning is due to the ingestion of a neurotoxin from fish. The toxin is produced in dinoflagellates (e.g., *Gambierdiscus toxicus*). It then accumulates in the flesh of the fish. This occurs mainly in tropical and subtropical marine fin fish, including mackerel, groupers, barracudas, snappers, amberjack, and triggerfish, although not all of these types of fish are infected all of the time. In humans, the incubation period ranges from as little as 5 min up to 30 hr (with a mean of about 5 hr). There are usually gastrointestinal and neurological symptoms, including nausea, vomiting, watery diarrhea, parasthesias, ataxia, vertigo, and blurred vision. Some patients may go on to develop cranial nerye palsies or even respiratory paralysis. The symptoms may last up to a week, but then usually resolve. The initial diagnosis is usually clinical and confirming the presence of the toxin is difficult. Toxin detection can be undertaken using a mouse bioassay and enzyme immunoassays for toxin detection are being developed.

B. Scrombroid

Scrombroid poisoning typically occurs after the ingestion of spoiled fish, especially tuna and mackerel. Excess levels of histamine in the flesh of the fish are thought to be the cause of this poisoning. Histamine and other amines are formed in food by the action of decarboxylases, produced by bacteria that act on the histidine or other amino acids. Scrombroid has been associated with a number of other foods, such as Swiss cheese. However, it is more frequently associated with fish, especially if the fish has not been frozen rapidly after being caught. Clinically, symptoms may begin within 10 min to 3 hr following ingestion. Nausea, vomiting, diarrhea, flushing, and headache may all occur. Respiratory distress is a rare complication. The natural history of this disease is that it will resolve in a few hours. The diagnosis is usually clinical and can be confirmed by detecting elevated levels of histamine in the suspected food.

C. Shellfish Poisoning

Four main types of shellfish poisoning have been described, including paralytic shellfish poisoning, neurotoxic shellfish poisoning, diarrheic shellfish poisoning, and toxic–encephalopathic shellfish poisoning. Shellfish poising is due to toxins made by algae (usually dinoflagellates) that accumulate in the shellfish. Paralytic shellfish poisoning is due to saxitoxin, a sodium channel toxin. Clinically, symptoms occur usually within an hour and consist of nausea, vomiting, and paralysis that may be limited to the cranial nerves or, in more severe cases, involve the respiratory muscles. Neurotoxic shellfish poisoning is due to brevitoxin, which is a lipophilic, heat-stable toxin that stimulates postgaglionic cholinergic neurons. Symptoms usually occur within 3 hr of exposure and consist of nausea, vomiting, and parasthesisa. Paralysis does usually not occur. Diarrheic shellfish poisoning, as the name implies, causes a mainly gastrointestinal disturbance with nausea, vomiting, and diarrhea. Toxic–encephalopathic shellfish poisoning (also known as amnesic shellfish poisoning) has caused outbreaks of disease in association with consumption of mussels. Symptoms include nausea, vomiting, diarrhea, server headache, and, occasionally, memory loss. With all these types of poisoning, symptoms usually occur rapidly (within 2 hr) and will usually resolve spontaneously.

The exception to this is toxic–encephalopathic poisoning, when the symptoms may not occur for 24–48 hr following exposure. The diagnosis in humans is clinical. However, it may be possible to detect the presence of the toxins using either mouse bioassays or by high performance liquid chromatography (HPLC).

D. Tetrodotoxin

Tetrodotoxin is present in certain organs within puffer fish and, if ingested, can cause rapid paralysis and death. Symptoms may occur in as little as 20 minutes or after several hours. Symptoms progress from a gastrointestinal disturbance to almost total paralysis, cardiac arrhythmias, and, finally, death within 4–6 hr, after ingestion of the toxin. The diagnosis is clinical and based on history of exposure. Mouse bioassays and HPLC have been used to detect these toxins in food.

E. Mushroom Toxins and Aflatoxins

There are a large variety of toxins from different mushrooms that cause a wide variety of diseases in humans. These toxins can be divided into four general groups as follows: protoplasmic poisons (e.g., amatoxins, hydrazines, orellanine) that cause cellular damage and organ failure (e.g., hepato–renal syndrome); neurotoxins (e.g., muscarine, ilbotenic acid, muscimol, psilocybin) that cause coma, convulsions, and hallucinations, etc; gastrointestinal irritants that produce nausea, vomiting, and diarrhea; and disulfiramlike toxins that only cause a problem if the person ingesting the mushroom has had exposure to alcohol in the previous 48–72 hr. The diagnosis of mushroom poisoning is based largely on the clinical picture and the history of exposure. There is, however, a commerical radioimmunoassay available for the amanitins. Therapy is largely supportive but interventions to reduce toxin absorption from the intestine, such as lavage or administration of activated charcoal, may help. Plasmapheresis also helps to reduce the mortality rate.

Aflatoxins are produced by certain strains of fungi, e.g., *Aspergillus flavus* and *A. parasiticus,* that grow on various types of food and produce toxins. Nuts, especially tree nuts (Brazil nuts, pecans, pistachio nuts, and walnuts), peanuts, and other oilseeds, including corn and cottonseed, have been implicated most often. There are various types of aflatoxins (B1, B2, G1, and G2), of which B1 is the most common and the most toxic. Clinically, these toxins cause liver damage that may be in the form of cirrhosis or hepatic malignancy. Occasionally, the ingested dose of aflatoxin is so high that an acute condition develops, known as aflatoxicosis, in which there is fever, jaundice, abdominal pain, and vomiting. Diagnosis in humans is clinical, but there are assays available for the detection of the toxins in food.

F. Other Natural Toxins

A number of other naturally occurring toxins have been reported. These include grayanotoxin, which is from eating honey made from rhododendrons. This usually causes nausea, vomiting, and weakness soon after the honey is ingested and, typically, is self-limiting in 24 hr. Akee fruit from Jamaica contains hypoglycin A, that causes hypoglycemia and vomiting in 4–10 hr. Curcurbitacin E, from bitter cucumber, can cause cramps and diarrhea within 1–2 hr of ingestion. Hydrogen cyanide may be present in lima beans or cassava root and can lead to death within minutes. Castor beans can contain a hemagglutinin that may cause nausea and vomiting. Red kidney beans also produce a hemagglutinin, known as phytohemagglutinin. This is associated with eating raw or undercooked red kidney beans and usually causes symptoms in 1–3 hr following exposure. Patients may develop severe nausea and vomiting that can be followed by diarrhea. The toxin is heat-sensitive but needs to reach a high enough temperature to be inactivated. A number of cases have been linked to beans cooked in slow cookers, in which the temperature does not get high enough. The outcome is usually good and supportive therapy is all that is needed. Many other agents that may occur in food, such as the pyrrolizidine alkaloids causing liver damage, and a variety of chemicals and heavy metals can be included in the long list of agents that cause food-borne illness; however, it is beyond the scope of this article to discuss them.

VI. SUMMARY

As time moves on, we are finding more and more microbes of multiple types that are associated with food-borne illness. The various bacteria, protozoa, viruses, and natural chemicals discussed in this chapter include the major ones—but there may be others that we do not even know about yet. It is disconcerting that the majority of cases of gastroenteritis that may be related to food and water go undiagnosed. The majority of our epidemiological data is based on less than 5% of cases, and many laboratories do not routinely look for many of the enteric viruses that are probably causing much of the disease. In recent years, food safety has become a major issue, following a number of highly publicized outbreaks involving a variety of enteric pathogens, including *E. coli, Salmonella, Listeria, Cyclospora,* and Hepatitis A. The food industry has made major efforts to improve the safety of food-processing and we are beginning to see the benefits of this with lower levels of bacterial pathogens in poultry. One of the key unanswered questions in relation to food-borne illness is a determination of the outcome, in most cases. We presume that, in the vast majority of food-borne disease, the outcome is good, with no long-term problems. However, we do not know this for a fact. Occasionally however, the outcome is disastrous, with the development of conditions such as HUS, Guillain-Barré, or reactive arthropathy following infection with enteric pathogens. The story is very different in developing countries, where sanitation may be poor and where fecal contamination of food and water is frequent. In such places, food-borne illness is probably killing hundreds of thousands of children each year and resulting in nutritional deficiency in many more. In developed nations, our immediate goals are to further our understanding of how these organisms cause disease, and how to reduce the load in the animals and fresh produce that provide our food supply. In developing countries, simply instituting basic sanitation to reduce fecal–oral spread can have a huge impact on morbidity and mortality. Maintaining good personal hygiene and sanitation will reduce the chance of fecal–oral spread. Similarly, paying attention to food handling, proper cooking, and the avoidance of temperature abuse will all go a long way to reduce the burden of food-borne illness.

See Also the Following Articles

Escherichia Coli and *Salmonella*, Genetics • Intestinal Protozoan Infections in Humans • Mycotoxicoses • Surveillance of Infectious Diseases • Water, Drinking

Bibliography

Blaser, M. J., Smith, P. D., Ravdin, J. I., Greenberg, H. B., and Guerrant, R. L. (eds.) (1995). "Infections of the Gastrointestinal Tract." Raven Press, New York.

Doyle, M. P., Beuchat, L. P., and Montville, T. J. (1997). "Food Microbiology: Fundamentals and Frontiers." ASM Press, Washington.

LaMont, J. (ed.) (1997). "Gastrointestinal Infections." Marcel Dekker, Inc., New York.

Lorber, B. (1997). Listeriosis. *Clin. Infect. Dis.* **24,** 1–11.

Nataro, J. P., and Kaper, J. B. (1998). Diarrheagenic *Escherichia coli. Clin. Microbiol. Rev.* **11,** 142–201.

Paton, J. C., and Paton, A. W. (1998). Pathogenesis and diagnosis of Shiga toxin-producing *Escherichia coli* infections. *Clin. Microbiol. Rev.* **11,** 450–479.

Sears, C. L., and Kaper, J. B. (1996). Enteric bacterial toxins: Mechanisms of action and linkage to intestinal secretion. *Microbiol Rev.* **60,** 167–215.

Smith, J. L. (1998). Foodborne illness in the elderly. *J. Food Protect.* **61,** 1229–1239.

Food Spoilage and Preservation

Daniel Y. C. Fung
Kansas State University

I. Food Spoilage
II. Microbial Food Spoilage Activities
III. Food Preservation
IV. Food Preservation Methods

GLOSSARY

extrinsic parameters of foods Environmental storage conditions which may influence the stability of food: gas, temperature, relative humidity, physical stress, and time.

food-borne microbes Living organisms less than 0.01 mm in diameter (mainly bacteria, yeast and mold) which can grow in foods and beverages and cause undesirable changes (food spoilage and food-borne diseases) or desirable changes (food preservation and fermentation).

food fermentation Utilization of bacteria, yeast, and molds to convert solid or liquid food substrates into desirable liquid, semi-solid, and solid foods and beverages with long-lasting value for human consumption.

food preservation Physical, chemical, biochemical, and microbiological treatments of foods and beverages to prevent food spoilage and food-borne diseases.

food spoilage Undesirable physical, chemical, microbiological changes in foods and beverages with or without the growth of undesirable or pathogenic microorganisms.

intrinsic parameters of foods Naturally occurring attributes of foods which may influence food spoilage, preservation, and fermentation: pH, oxidation/reduction potential, moisture contents, nutrient contents, antimicrobial agents, and biological structures.

FOOD FOR HUMAN CONSUMPTION consists basically of proteins, carbohydrates, fats and lipids, minerals, vitamins, and water in various amounts and proportions which provide nutrition and energy to sustain health and growth. Bacteria, yeasts, molds, protozoa, nematodes, insects, rodents, and higher animal forms also need these nutrients for survival; thus, they compete with humans for the food supply. When food is no longer suitable for human consumption, the food is termed "spoiled." By trial and error through the ages, humans have found methods to prevent spoilage and preserve foods. Modern food scientists and technologists and microbiologists have been investigating methods to prevent food spoilage and develop ways to preserve foods scientifically.

I. FOOD SPOILAGE

Food spoilage can be caused by undesirable physical, chemical and microbiological changes in the food, such that the food becomes unacceptable to the person consuming it. It must be emphasized that due to cultural differences, some food that is totally acceptable to one culture is not acceptable to another. For example, *pidan* (hundred-year-old egg) is a delicacy in China but is very objectionable to many Americans. Thus, it is difficult to generalize what constitutes food spoilage. Basically there are three areas in determining spoilage: (A) aesthetic differences, (B) economic signifiance, and (C) actual health hazards.

(A) Aesthetic differences. The above example of *pidan* illustrates the matter of cultural tastes and preferences. There are endless examples in this category.

(B) Economic significance. Many foods with small physical, chemical, and even microbiological defects may be perfectly safe to eat but, due to regulations and consumer appeal, are considered "spoiled" and are unnecessarily discarded, which may cause large economic loss. In times of famine, people will eat almost anything. Thus, economics sometimes dictate whether a food is spoiled or not. Stale bread is a very good example. It is perfectly safe to eat but is usually discarded by retail stores, due to lack of consumer acceptance. Various estimates indicate that 25 to 50% of food is considered spoiled worldwide and is lost for human consumption.

(C) Actual health hazard. Many spoiled foods are actual health hazards, due to presence of physical objects (metal, stones, etc.), poisonous chemicals (arsenic, lead, mercury, etc.), and pathogenic bacteria, yeast, mold, and viruses. These foods should not reach consumers and definitely should be considered spoiled. The problem is that some of these foods actually show no sign of visible defects and are consumed by unsuspecting persons.

Another way of looking at food spoilage is to define food as "fresh" versus "not fresh." For example, fruits can be defined as fresh (nonmoldy and crispy) versus not fresh (moldy and soggy), vegetables as fresh (newly harvested and firm) versus not fresh (stored and wilted), or bread as fresh (day-fresh and moist) versus not fresh (stale, old, and dry).

II. MICROBIAL FOOD SPOILAGE ACTIVITIES

From a microbiological standpoint, food spoilage can be the result of one or more of the following microbial activities.

(A) Lipolysis. Lipolytic enzymes (lipases) cause rancid flavor and odor changes; increase free fatty acids.

(B) Proteolysis. Proteolytic enzymes (proteases) cause flavor and odor changes; in advanced cases, cause putrefication of food products.

(C) Saccharolysis. Breakdown of carbohydrates and sugar. This will cause souring of food products.

(D) Pigment production. At refrigerated temperatures, psychrotrophs tend to produce more pigments than at higher temperatures. *Pseudomonas fluorescens* produces a fluorescent compound in food and causes chicken to "glow" under ultraviolet light.

(E) Production of polysaccharides. Many organisms produce large amounts of polysaccharides, resulting in slime formation in foods, highly undesirable in most foods.

(F). Other activities. As a result of a combination of undesirable chemical and biochemical activities in the food by microorganisms, a variety of compounds may be generated, such as hydrogen sulfide, mercaptans, trimethylamines, indole, skatol, cadaverine, and putrescine, resulting in fruity, rancid, putrid, ropy, cheesy, soapy, bitter, "sweaty feet," skunklike, fishy taint, etc. odors. In order to keep food from these undesirable changes, modern food scientists use a variety of preservation methods to keep food in stable, nutritious, safe, palatable, and appealing conditions.

As a general rule, food containing 1–100 microorganisms/gram, ml, or cm² are considered as having low counts; 1000 to 10,000/g, ml, cm² are considered intermediate counts, 100,000 to 1,000,000/g, ml cm² are considered high counts, and 10 million/g, ml cm² are considered as very high counts (index of spoilage). Food with 50 million/g, ml cm² will start to have bad odor and, with 100 million/g, ml cm², will have slime. Most ground beef in supermarkets has about 1 million bacteria per gram.

III. FOOD PRESERVATION

Before discussing food preservation one must consider the intrinsic and extrinsic parameters of foods.

A. Intrinsic Parameters of Food

This involves the natural state of the food. All food has its unique intrinsic parameters. Disturbing these parameters will affect the growth and death of microorganisms in the food. To preserve food properly, one needs to know these parameters in detail to adequately design processing conditions and equipment.

1. pH

This includes acidity, alkalinity, total dissociated and nondissociated inorganic and organic acids, and buffer capacity of the food. pH 4.5 is considered the neutral pH for food. Food with a higher pH is considered basic food and that with a lower pH is considered acid food. Yeast and mold can grow and survive from pH 1 to 11, whereas bacteria generally can survive and grow between 3.2 to about pH 10. Spoilage of food with pH 4.6 and higher will be mainly by bacteria and spoilage of food with pH 4.4 and lower will be mainly by yeast and mold.

2. Oxidation/Reduction Potential

This is expressed as + or − millivolts (mV). The interior of foods such as a piece of thick beef steak is usually considered as anaerobic and sterile (i.e., center) and the exterior is considered aerobic and contaminated (i.e., surface). Hamburger, however, is far more aerobic and contaminated than a beef steak due to grinding and exposure to air throughout the mass. Thus, bacteria will not grow inside the steak, which is sterile and anaerobic, but will grow very well throughout the hamburger. Aerobes can grow between +300 and −50 mV; facultative anaerobes, +300 to −420 mV; aerotolerant anaerobes, +180 to −350 mV; and obligate anaerobes, −150 to −420 mV.

3. Moisture Content

This includes bound water, free water, and the water activities of the food. Water activities are expressed as a_w. Pure water has a_w 1.00. Most fresh food has a_w of 0.99. Dry foods will drop to 0.95, 0.90, 0.85, 0.70, or even 0.60. Minimum a_w of most spoilage bacteria is 0.90; yeast, 0.88; and mold, 0.8. Some exceptions include halophilic bacteria, with a low a_w of 0.75, and xerophilic yeast and mold, 0.61.

4. Nutrient Content

Foods containing a complement of all necessary nutrients will be spoiled by microorganisms far more than will nonnutritious foods.

5. Antimicrobial Agents

Natural foods contain a variety of compounds which can inhibit microbial growth, such as lactenin in milk and lysozyme in eggs, which prevent growth of Gram-positive bacteria. Benzoic acid in cranberries prevents mold growth and eugenol in cloves prevents bacterial growth, etc.

6. Biological Structure

Foods are biological entities and are protected one way or another by some form of external structures. When these structures are intact, contamination will be minimal, but when these are damaged microorganisms will invade the food. For example, the shell of eggs, skin of fruits, shell of nuts, and hide of animals protect the internal entity. When these are broken or damaged, contamination will occur.

B. Extrinsic Parameters

Foods are also exposed to the environment and are influenced by extrinsic parameters. There is an intricate interaction between intrinsic and extrinsic parameters of food in studying food spoilage and food preservation.

1. Gaseous Environment

The presence of various gases, either naturally or purposely introduced into the food environment (either in a large room or in packages), makes a big difference in preservation of food. The presence and level of oxygen, nitrogen, carbon dioxide, ethylene oxide, etc. will determine the storage lives of many foods, since different microorganisms have different gaseous requirements for growth.

2. Temperature

Microorganisms can be mesophilic (growth range, 20–45°C; optimum, 37°C), thermophilic (growth range 45–80°C; optimum 55–56°C), psychrophilic (range, 0–10°C; optimum, 8°C), or psychrotrophic (range, 0–35°C; optimum, 21°, but will grow at 10°C and below). Thus, food storage temperature make a big difference in spoilage and preservation potential.

3. Relative Humidity

Moist and dry foods should be stored at the appropriate relative humidity environment to prevent undesirable migration of moisture to and from the food.

4. Physical Stress

Any damage to the structure of the food will result in greater chances of spoilage. This is linked to the biological structure of foods.

5. Time

In general, storage time will affect the physical, chemical, and microbiological changes in foods, since these activities either increase or decrease with time and, thus, will affect the food under storage.

IV. FOOD PRESERVATION METHOD

Food preservation techniques can be grouped under drying, low-temperature and freeze-drying, high temperature, ionizing and non-ionizing radiations, gaseous environments, chemical treatments, and food fermentation.

A. Drying

Drying is considered the oldest form of food preservation. The purpose is to remove moisture in order to retard and inhibit enzyme activities and the growth of organisms in foods. Sun-drying is the oldest method of drying. It is a form of uncontrolled drying needing a large amount of space and subject to contamination from the environment, but is the least expensive and most practical way of drying, especially in developing countries. Dehydration is removing moisture using controlled temperature, relative humidity, air flow, and heat transfer systems. This requires less space but is more expensive to operate. The products are more uniform and suffer less contamination. Heat is applied to the food to remove moisture by using either heated air as a medium or hot solid contact surfaces. Cabinet, tunnel, spray, and kiln dryers are examples of using heated air as a medium. Drum dryer and hot plates with or without application of vacuum is considered hot solid contact surface drying. Application of vacuum reduces the temperature of drying and is less damaging to food structures. Bacteria play very little role in the spoilage of dried foods with water activity of 0.90 or less. Food with a_w between 0.80–0.85 are spoiled mainly by yeast and mold. Food can be held for several years at a_w of 0.65 to 0.75. Drying itself will not sterilize the food but will retard or prevent the growth of organisms. Rehydrated food will be spoiled if not kept at proper temperatures. Some problems of drying include shrinkage of food, browning (Maillard reaction or nonenzymatic), case hardening (due to more rapid evaporation at the surface than diffusion from interior), and loss of desirable volatile compounds (flavor chemicals).

B. Low Temperature Preservation

Microbial activities are highly related to optimal temperature of growth and metabolism. As a general rule, an increase of 10°C will result in doubling of enzyme reaction rates (temperature coefficient, $Q_{10} = 2$). Consequently, a reduction of 10°C will cut the rate in half. Thus, as temperature is reduced, microbial reaction rates reduce concomitantly and stop at some point below the freezing temperature. Some enzymes will still be active at the freezing temperature. This is why many foods are blanched (100°C for a few s) in water to inactivate food enzymes before freezing for long-term storage. These are basically three ranges of low temperature for food preservation: chilling temperature: 10–15°C, good for fruits and vegetables and chilling of meat products; refrigeration temperature: 1–2°C and 5–7°C, used for perishable and semiperishable foods; and freezing temperature: −20°C. Most household freezers are at this temperature, which will prevent growth of all common food-borne microorganisms. Food can be quick-frozen by having the food reach −20°C in 30 minutes. This is achieved by direct immersion, indirect immersion, or air blast systems. Slow freezing refers to reaching −20°C within 3 to 72 hours, such as in the case of a home freezer. During freezing, at the onset, there is a 90% reduction of microbial population but, thereafter, the population exhibits a slow decline during frozen storage. During fast freezing, small ice crystals are formed intracellularly and cause less damage to microbes. During slow freezing, large ice crystals are formed extracellularly and cause more damage due to disruption of cell membranes. There are a variety of physical and chemical changes during freezing which will cause injury and death of microbes.

In low-temperature preservation of food, the important group of organisms is the psychrotrophs. This group actually grow better at 21°C but they can grow well and cause food spoilage when foods are kept below 10°C for prolonged incubation times, such as a week. The most accurate way to enumerate psychrotrophic population is to incubate agar plates with bacteria at 7°C for 10 days before counting the colonies. Psychrotrophic bacteria include *Pseudomonas* (the most important genus), *Acinetobacter, Alcaligenes,* and *Flavobacterium.* Psychrotrophic molds and yeasts include *Penicillium, Mucor, Geotrichum, Cladosporium, Candida, Torulopsis, Rhodotorula,* and *Debaryomyces.* That is why prolonged storage of food such as tomatoes, oranges, lemons, cheeses, etc. in the refrigerator will result in visible growth of mold in the form of blue, green, or black patches. They may still be edible after trimming the mold patches but most people will discard these food items.

Thawing is the reversion of frozen foods to room temperature. In rapid thawing, ice crystal melt rapidly results in dripping of liquid from foods. In slow thawing, more liquid is absorbed back into the food with less dripping. It is advisable to thaw food in a refrigerator rather than at room temperature, because bacteria can grow at the surface of a slowly thawing food while the center may still be frozen. Freezer burn is the result of loss of moisture at the unprotected surfaces of food, resulting in browning and drying of food. Good packaging materials will reduce freezer burn.

C. Freeze-Drying or Lyophilization

Freeze-drying or lyophilization of food relies on the knowledge of the triple point of water, where water exists in liquid, solid, and gaseous forms simultaneously; this occurs when the pressure is at 4.5 mm Hg and the temperature is at 0.01°C. Food is first frozen (moving food from liquid phase to solid phase) and then a vacuum is applied. When the vacuum is low enough, water molecules will sublime out of the food mass, leaving the structure of the food intact but without water. This type of food will be very light, which is excellent for long-distance transportation, such as space travel. Foods will not spoil for a very long time even without refrigeration, since microorganisms will not grow without water. Freeze-dried food rehydrates very quickly and retains normal odor, texture, and color. So far, freeze-dried food is still expensive and finds its way mainly to specialized markets (as freeze-dried coffee, camping and outdoor food markets, and space travel food programs). It is worthwhile to note that the best way to preserve bacterial cultures is to freeze-dry them (lyophilization). These cultures can last almost indefinitely.

D. High Temperature Preservation

The main aim of high temperature preservation is to kill pathogenic and spoilage organisms without destroying nutritive and other desirable properties of foods. Heat kills microorganisms by denaturing protein and enzyme systems in the cell. Pasteurization and sterilization are two ways to preserve foods using high heating temperatures.

1. Pasteurization

Pasteurization is the process of applying just enough heat to kill food-borne pathogens without grossly affecting the delicate flavor and quality of foods, such as wine, beer, vinegar, cider, and milk. Low Temperature Long Time (LTLT) pasteurization refers to heating milk at 62.8°C/145°F for 30 minutes. This is used in batch pasteurization of milk for cheese making. High Temperature Short Time (HTST) pasteurization heats milk at 72°C/161°F for 15 s in a continuous unit. Most commercial dairy companies use HTST to process fluid milk for human consumption. Although Louis Pasteur invented the pasteurization process for wine, beer, and vinegar, it was Soxhlet who, in 1888, used the procedure to pasteurize milk. The original time and temperature of milk pasteurization was designed to kill *Mycobacterium tuberculosis* and *Coxiella burnetii.* In recent years, very high temperature (VHT) and ultra-high temperature (UHT) pasteurization were introduced. There is no official definition of VHT, while UHT is defined as treatment of milk at 130°C for 1 s or more. This method is used extensively in Europe for pasteurization of milk. The UHT milk can be kept at room temperature for several weeks without defect. U.S. consumers are less likely to accept UHT milk because of the "burnt" flavor associated with high temperature treatment of milk. Although pasteurization kills most food-borne pathogens, some organisms such as *Streptococcus thermophilus*, *Lactobacillus bulgaricus*, *Microbacterium*, and *Micrococcus* and the spores of *Bacillus* and *Clostridium*, will survive. Common psychrotrophs will not survive pasteurization, and if they were found in pasteurized milk, it indicates post-pasteurization of milk. The problem can be solved with proper food hygiene and sanitation of the processing plant.

2. Sterilization

In order to destroy all microorganisms in food, the procedure of sterilization is applied. Canning was the starting point of high temperature preservation of food. The inventor of canning, Nicholas Appert, had no knowledge of microbiology but he perfected the method by first sealing food in glass bottles and then the filled bottles were immersed in boiling water. The heat sterilized both the bottles and their contents, so the food remained edible until the bottles were opened. Napoleon Bonaparte offered a 12,000-franc prize for the development of ways to preserve food to supply a large and moving army in 1795. It took Appert more than 10 years to perfect the canning process and he won the 12,000-franc award in 1809.

Modern canning processes rely on the knowledge of food composition (intrinsic parameters of food), heat transfer in the food throughout the cans, time and temperature of the process, and the heat resistance of bacteria and bacterial spores in the food. The main target is to design conditions to kill the most heat resistant spores of *Clostridium botulium* which might be in the can. The D-value is the time in minutes to reduce a population of microorganisms by 1 log unit (90% reduction) by heat or other physical or chemical methods, such as radiation or acid treatment. Different microorganisms at different temperatures have different D-values. For example, the D-value of *Bacillus stearothermophilus* is 4–5 minutes (very resistant), *Clostridium sporogenes* is 0.1 to 1.5 minutes, and *Clostridium botulinum* is 0.21 minutes at 250°F. The 12 D concept is to heat food such that a reduction of 12 D cycles occurs. This is a highly safe process designed to kill *Clostridium botulinum*

spores. It should be mentioned that *C. botulinum* is an anaerobic organism and produces heat-resistant spores as well as botulin, which is the most toxic compound produced by a biological system. It has been estimated that one ounce of this toxin can kill 200 million people. Thus, if spores of *C. botulinum* are in the food and the food is not adequately heat treated in the canning process, the spores can survive while vegetative cells will be destroyed by the heat. Later, the surviving spores can germinate in the anaerobic environment of the can and, since there are no competitors in the canned food, *C. botulinum* can grow and produce toxins in the can. If this food is consumed by a susceptible person, a fatal case of food-borne disease may occur. Fortunately, this toxin is sensitive to heat. The standard advice to consumers is that, when in doubt concerning the safety of any canned food, the food should be boiled for 10 minutes and discarded safely.

Proper canning also depends on knowing the D value (decimal reduction time), F value (the equivalent time, in minutes at 250°F of all heat considered, with respect to its capacity to destroy spores or vegetative cells of a particular organism), and Z value (the degrees in Fahrenheit required for the thermal destruction curve to traverse one log cycle) related to the thermal death time curve. Modern canning processors use computers to control all these factors in the canning process to ensure the safety and palatability of the food being canned.

Since the concept of heat sterilization is based on log reduction of cells, it is not possible to claim absolute "zero" reduction of microbes. Therefore, the term "commercial sterilization" has been introduced. The term indicates that, after the proper heating time and temperature and using the most appropriate bacterial cultivation medium, incubation time, and temperature, no living microorganisms can be found in the processed food. This does not guarantee the absolute absence of microorganisms in the food but gives a margin of safety that will pose no health hazard to consumers.

Besides the traditional tin cans used in canning, in high temperature heating of food, various kinds of materials, such as glass, ceramic, metal, and pouches made from different materials, can be used as containers for sterilization of food by heat.

This is a highly complex subject with much research going on worldwide for consumer protection; suffice it to say that commercially canned foods have an excellent safety record. There have been about 100 times more people killed by botulism by improper home canning than by faulty commercial canning in the past century. Thus, it is essential to promote better consumer education concerning home canning to avoid food poisoning by botulism.

G. Ionizing Radiation

Ionizing radiations include alpha rays, beta rays, gamma rays and x-rays. Alpha rays are made up of helium nuclei, have poor penetration, and are not used on food. Beta rays are electrons and can pass through paper but cannot penetrate aluminum. Gamma rays and x-rays can penertrate aluminum but are blocked by lead. For practical purposes, cobalt 60 is the only isotope used for food preservation. It has good penetration power, a disintegration constant of 0.121/year, and a half-life of 5.7 years. Recently, electron beams and accelerated electrons have been researched as means of radiation preservation of foods.

The unit of radiation for food preservation is the rad (radiation absorbed dosage). One rad is 100 erg/g absorption of soft tissue. More recently, the term Gray has been used. One Gray (Gy) = 100 rad. Killing power of ionizing radiation is attributed to either target theory or indirect theory. The target theory suggests that the rays hit DNA or other important molecules in the cell, thereby killing or mutating the cell. The indirect theory suggests that, as ionizing radiation passes through food, the water molecules are ionized and form free radicals, such as $H^+ + OH^- + H_2O_2 + O_2H$. These free radicals break H–H bonds of important molecules in living systems, thereby killing the organisms. This form of sterilization of food does not generate heat; therefore, it is also called "cold sterilization." Important terms in radiation preservation of food include:

Radappertization: equivalent to commercial sterilization by a heating process, at dose levels of 30–40 kGy.

Radicidation: equivalent to pasteurization; destroys pathogens in food, at dose levels of 2.5 to 10 kGy.

Radurization: equivalent to pasteurization in killing spoilage organisms, at dose levels of 0.75 to 2.5 kGy.

Irradiation of a variety of food has been studied all over the world. Threshold doses at which detectable irradiated odor occurs in some foods are: turkey, 1.5 kGy; pork, 1.75 kGy; beef, 2.5 kGy; chicken, 2.5 kGy; shrimp, 2.5 kGy; frog, 4.0 kGy; lamb, 6.25 kGy; and horse, 6.56 kGy. These food were irradiated at 5° to 10°C (Farkas, 1997). Foods that have been irradiated for prevention of sprouting, microbial control, and disinfection include potatoes, onions, garlic, wheat, strawberries, shallots, spices, beef, rabbit, and poultry meats. In the case of pork, radiation is designed to control *Trichinella spiralis*. D-values of radiation of some food-borne pathogens are as follows: *Vibrio parahemolyticus*, 0.02–0.36 kGy; *Yersinia enterocolitica*, 0.04–0.21 kGy; *Campylobacter jejuni*, 0.08–0.20 kGy; *Escherichia coli* 0157:H7, 0.24–0.43 kGy; *Staphylococcus aureus*, 0.26–0.86 kGy; *Salmonella*, 0.18–0.92 kGy; *Listeria monocytogenes*, 0.27–1.0 kGy. These are irradiated in the nonfrozen state (Farkas, 1997). Countries which have received unconditional or conditional clearances for radiation decontamination of spices/herbs include Argentina, Bangladesh, Belgium, Brazil, Canada, Chile, Croatia, Cuba, Czech Republic, Denmark, Finland, France, Hungary, India, Indonesia, Iran, Israel, Italy, Korea, Mexico, Netherlands, Norway, Pakistan, Philippines, Poland, South Africa, Thailand, United Kingdom, United States, Vietnam, and Yugoslavia.

There is no question that irradiation can kill microorganisms and preserve food. Widespread acceptance is still not a reality because of costs of operation, limitation of facilities for large-scale treatment of foods nationwide and worldwide, potential of dangerous accidents, occupational health hazards for operators, and, most important of all, consumer education and acceptance. The future of this form of preservation appears bright.

H. Nonionizing Radiation

This form of radiation does not generate enough energy to ionize water molecules and, thus, the killing effects are different from those of ionizing radiation. The two main forms are ultraviolet radiation and microwave treatment for food preservation.

1. *Ultraviolet Radiation*

Ultraviolet radiation covers the range between 136 to 4000 Å in wave length. 2600 Å waves are absorbed by the DNA of living organisms. This causes breakage of DNA which kills the cell or causes mutation by altering the base composition and arrangements of DNA. Ultraviolet light has poor penetration and can be used only for surface application. Since the radiation causes "radiation odor" in foods, it has not been used for food treatment. The most appropriate applications are in sterilization of air and treatment of thin layers of water.

2. *Microwaves*

The use of microwaves for food preservation and processing has enjoyed tremendous popularity in the past two decades worldwide. More than 70% of U.S. households have microwave ovens and many of these have more than one unit. Microwaves are a form of nonionizing radiation. Generation of heat by microwaves occurs when the rapidly alternating waves pass through food containing "dielectric" molecules (molecules with positive and negative charges) such as water; the molecules align themselves with the fluctuating wave, thus generating friction, which is translated to heat. Microwaves are measured in MHz. One MHz is one million cycles per second. Commonly used frequencies are 2450 and 915 MHz for treatment of food and liquid. This rapid vibration of waves generates heat in food very quickly; therefore, microwave heating of food occurs in seconds and minutes, compared with conventional heating which occurs in minutes and hours. Also, microwaves usually start heating food from the center, whereas conventional heating starts from the surface and penetrates to the interior. Another important difference is that microwave heating is far more dependent on the composition of the food than is conventional heating, because food with large numbers of dielectric molecules, such as water, will be heated much faster than will food containing less water. Food or materials without water, such as dried flour or a coffee cup, will not be heated at all. Uneven heating and lack of browning are some disadvantages of microwave heating but manufacturers are constantly improving the ovens to make them safer and make them function better for cooking of food. Microwave ovens find many uses at home and in industrial settings. From a microbiological standpoint, microwaves kill microorganisms mainly by heat (95%). There is still debate on the possibility of a "microwave" effect on reduction of microbes in liquid and solid food (5%).

I. Preservation by Gases

Oxygen promotes the growth of most spoilage organisms in foods. Vacuum packaging by removing air reduces the growth of many spoilage organisms and, thus, preserves food longer, especially in conjunction with cold storage. When the carbon dioxide level is above 10% of the gases in the storage environment, it is called "modified atmosphere" or "controlled atmosphere" storage. Modified atmosphere packaging has been an important technological advancement in food preservation. The three components being modified are oxygen, nitrogen, and carbon dioxide. A great variety of combinations of these gases have been studied. The concentration of any one of these gases can range from 100 to 0%, with all sorts of ranges in between. Some reported mixtures include 15% CO_2 + 85% O_2, or 50% CO_2 + 50% N_2, or 15% CO_2 + 40% O_2 + 45% N_2. One of the major concerns of vacuum packaging and modified atmosphere packaging is the potential growth of *Clostridium botulinum*. Much work has been done to manipulate other food components, such as pH, antimicrobials, preservatives, and temperature control to prevent the growth of this food-borne pathogen in this type of anaerobic food package.

J. Chemical Preservation of Food

Many organic and inorganic compounds can kill microorganisms associated with foods. It is imperative to select those compounds that can kill microorganisms and yet will not hurt the consumer or cause gross changes of the food. Food additives are a subject of much debate since many people consider that food additives are not natural and may be harmful. In fact, many so-called food additives actually do occur naturally in foods. Any chemical that may cause cancer or is poisonous is not allowed to be added to food. There is a list of compounds called GRAS—Generally Recognized As Safe—which are allowed to be used in foods. Compounds such as salt, sugar, acetic acid, lactic acid, propionic acid, sorbic acid, and benzoic acid are among those on the list. The list is exceedingly comprehensive and, periodically, the compounds are tested for their safety to animals. Some familiar chemical food preservatives include propionic acid used at 0.32% to prevent molds in bread, cakes, and cheese; sorbic acid at 0.2% to prevent molds in cheese, salad dressing, and jellies; benzoic acid at 0.1% to prevent yeast and mold in pickle relishes, apple cider, soft drinks, and tomato catsup; parabens at 0.1% to prevent yeast and mold in bakery products and pickles; sulfites at 200–300 ppm to prevent insects and microorganisms, used in winemaking, dry food, and molasses; ethylene and propylene at 700 ppm to inhibit yeast, mold, and vermin, used in fumigation of spices and nuts; nisin at 1% to prevent lactic acid bacteria and clostridia in pasteurized cheese spreads; sodium nitrite at 120 ppm to inhibit clostridia in meat-curing preparations, and many other examples.

K. Food Fermentation

One of the most effective means of preservation of food is by fermentation. When the proper substrate is used (grape juice, milk, wort, flour, cabbage, etc.) and in the presence of desirable microorganisms (wine yeast, starter cultures, bread and beer yeast, lactic acid bacteria) and conditions (anaerobic), the substrate will be converted to highly stable fermented foods, such as wine, cheese, beer, bread, and sauerkraut. Food fermentation is indeed a natural way to preserve excess foods. For example, during fall harvest, there will be a great abundance of produce that cannot be consumed or sold immediately. It will either be spoiled or will need more costly preservation processes, such as refrigeration, canning, pasteurization, or irradiation, to keep it from deterioration. Food fermentation is the use of microbial activity, usually anaerobic, on suitable substrates under controlled or uncontrolled conditions, resulting in the production of desirable foods or beverages that are characteristically more stable, palatable, and

nutritious than the raw substrate. After fermentation, people either consume the end products, such as alcohol and acids; the microorganisms themselves, such as mushrooms and single-cell protein; or a combination of the substrate and the cells, such as cheese, yogurt, sour and acidophilus milk, and a great variety of fermented soy products all over the world.

In conclusion, this article discussed briefly the important points concerning food spoilage and most of the common ways to preserve food. Detailed knowledge in this field will help to promote physical and public health, prevent unnecessary waste of valuable food, and provide the world with an adequate and plentiful supply of nutritious food for the growing population of planet Earth.

See Also the Following Articles

FERMENTED FOODS • FREEZE-DRYING OF MICROORGANISMS • REFRIGERATED FOODS

Bibliography

Bourgeois, C. M., Leveau, J. Y., and Fung, D. Y. C. (1995). "Microbiological Control for Food and Agricultural Products." VCH Publishers, New York.

Doyle, M. P., Beuchat, L. R., and Montville, T. J. (1997). "Food Microbiology: Fundamentals and Frontiers." ASM Press, Washington, DC.

Erickson, L. E., and Fung, D. Y. C. (1988). "Anaerobic Fermentation." Marcel Dekker, New York.

Fung, D. Y. C., and Matthews, R. F. (1991). "Instrumental Methods for Quality Assurance in Foods." Marcel Dekker, New York.

Fung, D. Y. C., and Vichienroj, K. (1998). "Applied Food Fermentation." Kansas State Univ. Press, Manhattan, KS.

Jay, M. J. (1996). "Modern Food Microbiology" (6th ed.). Chapman & Hall, New York.

Mossel, D. A. A., Corry, J. E. L., Struijk, C. B., and Baird, R. M. (1995). "Essentials of the Microbiology of Foods." John Wiley & Sons, New York.

Ray, B. (1996). "Fundamental Food Microbiology." CRC Press, New York.

Vanderzant, C., and Splittstoesser, D. F. (1992). "Compendium of Methods for the Microbiological Examination of Foods." APHA, Washington, DC.

Foods, Quality Control

Richard B. Smittle

Silliker Laboratories

GLOSSARY

American Public Health Association A group of scientists, physicians, health professionals, and health administrators interested in improving the health of humankind.

AOAC International An association of scientists who strive to evaluate, standardize, and recommend methods of proven accuracy and reliability.

Food and Drug Administration The U.S. government agency responsible for regulating food, drugs, and cosmetics.

International Commission on Microbiological Specifications for Foods A chosen group of microbiologists under the aegis of the International Union of Microbiological Societies who attempt to improve the microbiological safety and quality of foods in international trade.

National Academy of Science A nongovernmental body of scientists and engineers who advise the executive branch of the government; the National Research Council is the working arm of the National Academy of Sciences.

sanitizer A chemical or physical agent used to eliminate or control microorganisms.

U.S. Department of Agriculture The government agency responsible for regulating the meat and poultry industry.

MICROBIOLOGICAL QUALITY CONTROL OF FOODS is essential for a safe, wholesome, consistent food supply. Food quality is defined by microbiological criteria that are developed by manufacturers and government agencies. To have a full appreciation of food quality, a thorough understanding of the chemical, physical, and biological aspects of microbial ecology is necessary. The principles of statistical quality control is essential for properly evaluating ingredients and products from a production system with appropriate sampling plans. Adequate evaluation of production systems, ingredients, finished products, and environments requires methods that are accurate, reliable, and convenient to use. The key to producing safe, wholesome food is using properly engineered food equipment and employing comprehensive personal hygiene, cleaning, sanitation, and pest control programs. To maintain the highest possible quality, it is essential to control hazardous microorganisms and detect indicator microorganisms of sanitation and spoilage agents. The Hazard Analysis and Critical Control Point (HACCP) system is a recently developed systematic approach to food safety. It is the foundation for producing safe and stable food.

I. INTRODUCTION

With the advent of microbiology, it became clear that microorganisms, especially bacteria, play an important role in food quality. Some of the earliest studies by Louis Pasteur, one of the fathers of microbiology, dealt with the spoilage of wine and milk. Since these first studies, microbiologists have made great advances in food safety and wholesomeness. With increased knowledge of disease-producing microorganisms, microbial ecology, and physiology, it became possible to produce large quantities of safe, wholesome food. Various modern schemes for ensuring a safe food supply have been proposed. The traditional quality control relied on finished product testing. However, one of the earliest proponents for controlling the quality of food during production was Sir Graham Wilson, who proposed "intervention for prevention," which involved control of a process at designated critical steps in the manufacture and distribution. This has been recently refined in Europe, where risk assessment and preventative control are emphasized. In the United States, a scheme based on the intervention for prevention was developed by the Pillsbury Co. and the National Aeronautics and Space Administration (NASA) to prevent health hazards in food for space flights—HACCP, which was first introduced in 1971. An extensive description of HACCP was presented by the International Commission on Microbiological Specifications for Foods (ICMSF) of the International Union of Microbiological Societies in 1988. More recently, this has been altered by the National Advisory Committee on Microbiological Criteria for Foods (NAC) of the National Academy of Science for a rational approach to regulatory agencies in the United States.

II. MICROBIOLOGICAL CRITERIA

Microbiological criteria are defined by the microorganisms of concern, the method, the sampling plan, and the decision criteria. They are determined by municipal, state, and federal government regulatory agencies, companies, trade associations, and scientific societies. Owing to the importance of microorganisms to food stability and safety, establishing food microbiological criteria for quality and hazards is necessary. The three types of criteria are as follows:

1. "Standard" is a criterion that is a mandatory government requirement.
2. "Specification" is a mandatory criterion where the acceptance of a food is dictated by the buyer.
3. "Guideline" is a criterion used by a manufacturer or regulatory agency to evaluate a process.

Infectious agents and toxins are not permitted in processed foods and are standards. Generally, infectious agents have a zero tolerance, as defined by their absence in some specific quality of food. As a rule, they are not permitted in any quantity in food that will not undergo a final heating step for their destruction. Specifications are useful in determining the microbiological state of ingredients sold by one company to another. In some cases, they may be eventually agreed upon in contracts or by reference to trade, scientific, or government publications. Guidelines are used mainly by manufacturers to steer production; however, some are published by government agencies and used to evaluate food in trade. Table I contains a list of references for published microbiological criteria.

TABLE I
Reference Organization and Specific Foods with Microbiological Criteria (United States)

Product	Reference organization
All types	International Commission on Microbiological Specifications for Foods
Starch and sugar	National Food Processors
Granulate sugar	American Bottlers of Carbonated Beverages
Liquid sugar	American Bottlers of Carbonated Beverages
Dairy products	U.S. Public Health Service
Certified milk	American Association of Medical Milk Comm., Inc.
Milk for manufacturing and processing	USDA
Dry milk	U.S. Public Health Service
Dry milk	USDA
Dry milk	American Dry Milk Institute, Inc.
Frozen desserts	U.S. Public Health Ordinance Code
Tomato juice and products	FDA
All processed food pathogens	FDA and USDA
All foods in international trade	Codex Alimentarius Commission FAO/WHO

III. PRINCIPLES OF FOOD MICROBIAL ECOLOGY

To establish a quality control program for foods, a firm knowledge of the ecological aspects must be understood. The chemical, physical, and biological properties of a food, the processing it requires, and the conditions under which it is stored, distributed, and handled before consumption dictates the type and number of microorganisms and their response to the food.

The important intrinsic characters of the food to consider are (1) pH, (2) type acid, (3) available water (a_w), (4) moisture content, (5) nutrient content, (6) oxidation–reduction potential (E_h), (7) biological structures, (8) natural inhibitors, and (9) added inhibitors. The processing of the food can alter many of these intrinsic characteristics but, most importantly, the microflora that may be added or altered to include pathogens, indicators, or spoilage agents. The extrinsic parameters of paramount importance are (1) temperature of storage and processing, (2) relative humidity, and (3) gaseous environment.

IV. SAMPLING AND SAMPLING PLANS

A. Statistical Sampling

Probability is the long-term proportion of positive samples to the total number of samples, assuming uniform distribution. For example, 100 positive samples out of 1000 total samples is 0.100, or 10%. Samples taken from a population can only approximate the true probability of their frequency. Because it is impractical to sample the whole food lot, a microbiologist must depend on an estimate of the true probability. Samples must be taken to estimate the character of the population. The larger the number of sample units tested, the larger number frequency distribution units should be obtained. To eliminate bias in determining the true population of a food, random choices must be made. Where possible, random number tables must be used to assign randomness to the food to be sampled. Random choice will reduce the risk of accepting or rejecting a good or bad lot.

Operating characteristic (OC) curves are used to determine the risk of acceptance or rejection of a food or bad unit of food. These curves are generated by calculations of binomial distribution or data taken from Poisson distribution tables. OC curves are used to determine the frequencies of numbers with various levels of defective units. Figure 1 is an operating characteristic curve for 10 sampled units (n) with 2 defective units (c). The probability of accepting a lot of food is 0.68 under these conditions. A lot of food with a 20% defective level will be accepted 68% of the time or, conversely, rejected 32% of the time.

Before sampling, what is to be sampled must be determined. Generally, the food produced is defined in terms of a "lot" in simplest terms: a lot is "the total amount of food produced and handled under uniform conditions." However, this definition does not always fit. First, the microbial population in non-liquid foods is notoriously heterogeneous. Second, continuous systems are difficult to define, given the conditions. Continuous systems would be better defined as per length of time of production. Alternatively, a lot could be defined as "any uniform batch

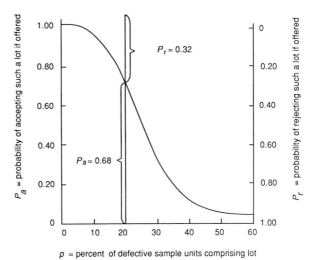

Fig. 1. The operating characteristic curve for n = 10, c = 2, i.e., the probability of accepting lots, in relation to the proportion defective among the sample units comprising the lots. (From International Commission on Microbiological Specifications for Foods (1986). "Microorganisms in Foods. 2. Sampling for Microbiological Analyses: Principles and Specific Applications." 2nd. ed. University of Toronto Press, Toronto.)

TABLE II
ICMSF Definitions for Sampling and Testing

n = number of samples taken per lot
c = maximum number of allowable defects per lot
m = lower specification
M = maximum allowable specifications

of food produced over a short period of time with identifiable assigned code numbers." A lot should be uniform and contain as little variation as possible. Any food produced over a prolonged period where there is a loss of quality—microbiological, chemical, sensory, color, etc.—is an unacceptable practice. Therefore, a lot can be defined in several ways, depending on the food itself, process, storage, who is to consume it, sensitivity to production, etc.

What is a representative sample? Samples must be drawn to represent the product being investigated. Sampling at random, where random numbers are used to take samples, is universally recognized as the means for eliminating bias. In some circumstances, random sampling is undesirable. Is the food uniformly mixed? Has stratification occurred? Are there records for consistency? What is the value of the product versus the testing cost? What are the consequences of a wrong decision?

B. Sampling Plan

Traditional food manufacturing has resulted in Quality Control acting as police. This adversarial conflict between production and quality was common. Repeated sampling of out-of-specification lots continued until a desirable result was obtained with little design to sampling. Consequently, better sampling plans were needed to address these issues. The recommended sampling plan for resolving these conflicts is that proposed by the ICMSF. It is a two-class Attribute Plan and a three-class Attribute Plan. The two-class plan is for decision-making on good and bad lots ($c = o$). The three-class plan is for decision-making on acceptable, marginal, and unacceptable products ($n \geq c \geq o$). Table II contains the definition and symbols for this plan. Two-class plans are generally used for infectious pathogens where an accept/reject decision is required.

To choose a sampling plan, a knowledge of how a food is to be handled and the hazards involved is essential. Table III presents these concerns for

TABLE III
Suggested Sampling Plans for Combinations of Degrees and Conditions of Use (i.e., the 15 "Cases")

Conditions in which food is expected to be handled and consumed after sampling, in the usual course of events[a]

Degree of concern relative to utility and health hazard	Conditions reduce degree of concern	Conditions cause no change in concern	Condition may increase concern
No direct health hazard	Increase shelf-life	No change	Reduce shelf life
Utility (e.g., shelf-life and spoilage)	Case 1	Case 2	Case 3
	3-class $n = 5$, $c = 3$	3-class $n = 5$, $c = 2$	3-class $n = 5$, $c = 1$
Health hazard	Reduce hazard	No change	Increase hazard
Low, indirect (indicator)	Case 4	Case 5	Case 6
	3-class $n = 5$, $c = 3$	3-class $n = 5$, $c = 2$	3-class $n = 5$, $c = 1$
Moderate, direct, limited spread[b]	Case 7	Case 8	Case 9
	3-class $n = 5$, $c = 2$	3-class $n = 5$, $c = 1$	3-class $n = 10$, $c = 1$
Moderate, direct, potentially extensive spread[b]	Case 10	Case 11	Case 12
	2-class $n = 5$, $c = 0$	2-class $n = 10$, $c = 0$	2-class $n = 20$, $c = ?$
Severe, direct	Case 13	Case 14	Case 15
	2-class $n = 15$, $c = 0$	2-class $n = 30$, $c = 0$	2-class $n = 60$, $c = 0$

[a] More stringent sampling plans would generally be used for sensitive foods destined for susceptible populations.

[b] See Conclusions, p. 46, for explanation of extensive and limited spread. [From International Commission on Microbiological Specifications for Foods (1974). "Microorganisms in Foods 2. Sampling for Microbiological Analyses: Principles and Specific Applications." Univ. of Toronto Press, Toronto, Canada.]

TABLE IV
Sampling Plans and Recommended Microbiological Limits

		Plan			Limit/g		
Product	Test	Case	Class	n	c	m	M
Roast beef	*Salmonella*	12	2	20	0	0	–
	Listeria monocytogenes	15	2	60	0	0	–
Dried pasta	*Salmonella*	10	2	5	0	0	–
	Staphylococcus aureus	8	3	5	1	10^2	10^4

choosing a plan, and they are listed as cases 1–15, where case 15 is the most stringent. After having chosen the case, the table contains suggested sampling plans for each. These sampling plans are for normal foods. Where problems have been encountered before, suspicions of abuse, new and/or unknown sources, or food to be consumed by susceptible populations, more stringent sampling is dictated. For class 2 illustration purposes, *Salmonella* concerns in roast beef would fall into case 12 and *Listeria (L.) monocytogenes* into case 15. *Salmonella* is considered a direct, moderate health hazard with potentially extensive spread. In a food that, when handled improperly in refrigerated storage, it can grow and increase the hazard (Table IV).

Pasta is a good example for a class 3 and class 2 plan (Table IV). The hazards are from *Salmonella* (case 10, class plan 2) and *Staphylococcus (Staph.) aureus* (case 8, class plan 3). *Staphylococcus aureus* in excess of 100/g but less than 10,000/g in one sample would be a marginal lot but acceptable. When any one sample is >10,000/g or two or more samples >100/g are encountered, the product is rejected.

In the United States, *Salmonella* and *L. monocytogenes* sampling programs are outlined by the Food and Drug Administration (FDA) and the U.S. Department of Agriculture (USDA). Briefly, the FDA *Salmonella* plan categorizes foods as Category I foods, which would normally be in Category II, except that they are intended for consumption by the aged, the infirm, and infants; Category II foods, which would not normally be subjected to a process lethal to *Salmonella* between the time of sampling and consumption; and Category III foods, which would normally be subjected to a process lethal to *Salmonella* between the time of sampling and consumption. The sampling plan is as follows:

Category I:	60–25 g units
Category II:	30–25 g units
Category III:	15–25 g units

The maximum composite is 15 sample units × 25 g or 375 g/test.

The USDA *L. monocytogenes* and *Salmonella* sampling plan for cooked, ready-to-eat meat or poultry products requires five samples from each lot and 5 g from each to be composited for one analytical test.

V. METHODS

In determining microbiological criteria, choosing a standard method is essential. Standard and/or officially recognized methods provide (1) better reproducibility between laboratories for comparison purposes, (2) proven sensitivity, (3) security for legal disputes, and (4) standard reference for historical comparisons. The method must be validated to ensure its adequacy and must be sensitive enough to detect the levels of microorganisms required. Where possible, the method should be rapid and easily performed. Table V contains a list of references for the detection and enumeration of microorganisms in foods.

VI. SANITATION

Sanitation is the total effort made to control the contamination and growth of undesirable microorganisms and the prevention of adulteration with filth from insects, rodents, birds, pets, other animals, and humans. It is the scientific application of microbiology, chemistry, engineering, physics, biology, and

TABLE V

Some References for the Detection and Enumeration of Microorganisms in Foods

Association of Official Analytical Chemists (AOAC) (1995). "Official Methods of Analysis." 16th ed. AOAC, Washington, DC.

International Dairy Federation (IDF). (1981–1982). IDF, Brussels, Belgium.

International Organization for Standardization (ISO) (1983). "International Standards Organization Catalogue. 1983." ISO Geneva, Switzerland.

International Commission on Microbiological Specifications for Food (1978). "Microorganisms in Foods, 1. Their Significance and Methods of Enumeration," 2nd ed. Univ. of Toronto Press, Toronto.

Marth E. H. (Ed.), (1992). "Standard Methods for the Examination of Dairy Products," 16th ed. American Public Health Association, Washington, DC.

National Food Processors (1968). "Laboratory Manual for Food Canners and Processors." Vol. 1, "Microbiology and Processing." National Canners Association, Washington, DC.

Vanderzant, C., and Splittstoesser, D. F. (Eds.). (1992). "Compendium of Methods for the Microbiological Examination of Food," 3rd ed. American Public Health Association, Washington, DC.

U.S. Department of Agriculture (1998). "Microbiology Laboratory Guidebook," 3rd ed. Food Safety Inspection Service, Washington, DC.

U.S. Food and Drug Administration. (1995). "Bacteriological Analytical Manual," 8th ed. AOAC, Arlington, Virginia.

management to the hygienic control of adulteration. Commitment to an effective sanitation program is a responsibility of management that must be communicated to all employees. The keys to its success are proper training and education along with adequate supervision to ensure effectiveness.

The code by which foods are produced in the United States is outlined in the Current Good Manufacturing Practice in Manufacturing, Packing, or Holding Human Food. These are commonly called Good Manufacturing Practices (GMPs). The definition and the enforcement are described in the Food, Drug, and Cosmetic Act. Basically, they describe when a food is considered adulterated during manufacture or held under unsanitary conditions that may lead to a contaminated product. Although in recent years the general GMPs have not had the full force of the law, they are still useful as good outlines for sanitation. However, the more specific GMPs for food such as low-acid canned foods are enforceable and very effective in preventing food-borne illness. The GMPs are broken down into (1) definitions, personnel, and personal hygiene, (2) sanitary considerations of building and facilities, (3) sanitary design of equipment and utensils, and (4) production and process controls.

Personal hygiene and foodhandling are pivotal in sanitation. Humans who are involved in food handling and manufacture must be clean and healthy. Human contamination of foods with infectious agents and toxin-producing microorganisms is common. The microorganisms may be transmitted from the skin, hands, hair, eyes, nose, mouth, respiratory tract, and excretal organs. *Staph. aureus* can be transmitted from the skin and nasal passages; *Streptococcus pyogenes* from the throat and respiratory tract; and *Salmonella, Shigella, Vibrio cholera,* enteropathogenic *Escherichia coli,* protozoans, hepatitis, and Norwalk virus from excretal organs. Unhealthy employees must be segregated from direct handling and manufacture of food. Proper systems and facilities must be in place and available to prevent disease transmission, especially for asymptomatic and convalescing carriers.

The food production area and equipment must be chosen and designed to prevent disease and adulteration. It must be easily cleaned, sanitized, and made from materials that are not toxic. Equipment for storing and conveying must be designed to exclude or minimize the growth of microorganisms.

Food held in the temperature danger zone between 45° and 140°F for 2 hours or longer should be avoided. This also includes food buildup on contact surfaces.

Proper cleaning, followed by sanitizing, is essential to control microbial hazards, spoilage organisms, and indicator bacteria. Chemical sanitation has traditionally been applied to food contact surfaces, especially in small operations and food service establishments. However, modern food plants employ Clean-in-Place systems where applicable. They are particularly effec-

TABLE VI
Common Sanitizers and Recommended Concentrations for Some Specific Food Processing Areas

Processing area	Sanitizer	Concentration
Hand dips	Quaternary compounds	25 ppm
	Iodopor	25 ppm
Rubber belts	Quaternary compounds	25–50 ppm
	Iodopor	25–50 ppm
Stainless steel equipment	Quaternary compounds	200 ppm
	Active chlorine	200 ppm
	Iodophor	200 ppm
Clean-in-place system	Acid sanitizer	100–150 ppm
	Active chlorine	200 ppm
Floors (concrete)	Active chlorine	1000 ppm
	Quaternary compounds	

tive in liquid foods, such as milk and beverages. Some common chemical sanitizers and their recommended concentrations are found in Table VI.

Adequately designed sanitary facilities are crucial to a food manufacturing facility. The elements of well-designed sanitary facilities are (1) adequate and potable water supply, (2) proper plumbing for waste removal and cleaning, (3) convenient toilet facilities, (4) convenient hand washing and sanitization stations, and (5) a segregated efficient rubbish and offal disposal system. The facilities must be designed to eliminate or minimize the presence of insects, birds, rodents, and pets that might lead to food adulteration from filth and/or microbial contamination.

VII. MICROBIOLOGICAL QUALITY

A. Indicators

Indicators of lack of sanitation in foods have been used in assessing the quality of water and foods since the turn of the century. Table VII contains a list of tests commonly used for this purpose. Coliforms and *E. coli* are the most frequently used to detect problems with sanitation. Coliforms are gram-negative, facultative anaerobic nonsporeforming rods that ferment lactose to acid and gas in 24–48 hours on liquid or solid media at 32–35°C. They are in the genera *Escherichia, Enterobacter, Klebsiella,* and *Citrobacter.*

Coliforms are commonly found in feces of warm-blooded animals, vegetable material, and soil. However, *E. coli* is more specifically used as a direct indicator of fecal pollution in foods because of its closer association with feces. Coliforms capable of growing at 44.5°C in the Official Association of Official Analytical Chemists procedures EC broth are fecal coliforms. Perhaps, a more accurate term would be thermotolerant coliforms. Many are *E. coli* but not all. Being somewhat more specific than the standard coliform test, they are frequently used. The fecal streptococci or enterococci are Lancefield's serologic Group D streptococci, which are *Streptococcus (S.) faecalis, Streptococcus faecium, Streptococcus bovis,* and *Streptococcus equinus.* These are intimately associated with the intestinal tracts of animals. They are more fastidious than the coliform group, with *S. faecalis* being the most common to the intestinal

TABLE VII
Commonly Used Indicator Microorganisms

Coliforms
Escherichia coli
Mesophilic aerobes
Fecal streptococci enterococci
Fecal coliforms
Enterobacteriaceae count

tract of humans but less specific than *E. coli*. The Enterobacteriaceae family may be used as an indicator group, although it is infrequently used. On the other hand, total counts (aerobic plate count, standard plate count) are frequently used to indicate the handling of certain foods, especially those that have an extensive history. They are particularly useful in evaluating frozen, refrigerated, and dried products.

B. Spoilage

Table VIII contains a list of the more commonly used tests for spoilage organisms. These would be employed in products where the chemical, physical, and biological factors, both intrinsic and extrinsic, determine the spoilage flora. For example, salad dressing, with a low pH due to acetic acid, would be tested for yeast and mold, lactobacilli, lactic acid bacteria, and/or acetophiles.

C. Health Hazards

A variety of microorganisms associated with food cause disease (Table IX). They are broken down into groups, with the bacteria being the most prevalent, particularly in developed countries. The bacteria are further divided into infection producers, toxin producers, and enterotoxigenic types. All are classified

TABLE VIII
Commonly Used Tests for Spoilage Microorganisms

Aerobic plate count
Standard plate count
Psychrophilic plate count
Anaerobic mesophilic plate count
Lactic acid bacteria count
Yeasts and molds
Lactobacillus
Osmophilic yeasts and molds
Lipolytic counts
Pectinolytic counts
Acetophilic counts
Thermophilic counts
Canned food enrichment procedures
Spore counts

TABLE IX
Disease Microorganisms Associated with Foods

Microorganisms	Hazard
Infections, Bacterial	
Salmonella typhi, paratyphi A and B	*
Salmonella	**
Shigella dysenteria	*
Shigella sp.	**
Listeria monocytogenes	*
Brucella abortus	*
Brucella melitensis	*
Enteropathogenic Escherchia coli	**
Enteroinvasive Escherichia coli	**
Enterotoxigenic Escherichia coli	**
Enterohemorrhagic Escherichia coli	*
Aeromonas hydrophila	***
Streptococcus pyogenes	***
Vibrio parahaemolyticus	***
Coxiella burnettii	***
Yersinia enterocolitica	***
Yersinia pseudotuberculosis	***
Campylobacter jejuni and coli	**
Vibrio vulnificus	***
Plesiomonas shigelloides	***
Toxins, Bacterial	
Clostridium botulinum	*
Staphylococcus aureus	***
Enterotoxigenic bacterial	
Clostridium perfringens	***
Bacillus cereus	***
Infections protozoan and parasitic	
Trichinella spiralis	***
Helminthic parasites	***
Taenia saginata	***
Taenia solium	***
Isospora (Toxoplasma) gondii	***
Entamoeba histolytica	***
Giardia lamblia	***
Cryptosporidium parvum	***
Echinoccoccus sp.	***
Capillaria phillippinensis	***
Anisakiasis (nemotodes)	***
Viral	
Rotavirus	**
Norwalk virus	**
Echovirus	**
Hepatitis virus	*
Polio	*
Food–Microorganism interaction	
Fish and shellfish toxins	*
Vasoactive amines	***
Mold	
Aflatoxins	*
Mycotoxins (predominately produced by the genera Aspergillus, Penicillium, Fusarium, and Mucor)	***

Note. *, severe hazard; **, moderate hazard and potentially extensive spread; ***, moderate hazard with limited spread.

as to their relative hazard, which is determined according to the organism's pathogenicity, potential spread, and the population at risk.

VIII. HAZARD ANALYSIS AND CRITICAL CONTROL POINT

HACCP is a scientific and systematic approach to food safety, where hazards are identified with risks assigned and control measures implemented. It is a preventative system for controlling microbiological safety. Although designed primarily for safety, it can also be applied to spoilage control. Traditional microbiological quality control programs rely heavily on inspection and finished product testing for health and quality hazards. This approach results in information of little significance to safety, and test results are available long after the product is produced, consequently providing no flexibility in adjusting production for prevention. Obviously, a system such as HACCP, where control is placed on hazardous situations during production, is superior. Intervention at critical control points (CCPs) during production makes it a proactive program with great flexibility.

Since the introduction of HACCP in 1971, it has steadily gained in application. It has been gradually adopted by food manufacturers as a rational approach to food safety. Recently, it was outlined and adopted by the NAC on Microbiological Criteria for Foods, part of the NAC/National Research Council, as an effective quality assurance program. The most successful application of HACCP has been with the low-acid canned foods, where the critical control approach, includes at least a microbiologist, an engineer, a production supervisor, and a quality control specialist, as well as a sanitation supervisor. Depending on the food product, other areas of a business may be involved, such as sales and distribution. As proposed recently by the NAC, the HACCP system consists of seven principles.

Principle 1. Conduct a hazard analysis. All significant hazards are identified, including microbiological hazards, that could reasonably result in illness or injury if not effectively controlled. To fully understand the scope of these hazards and their potential risks, probing questions must be answered concerning the nature of ingredients, intrinsic factors of food during and after processing, procedures used for processing affecting the microbial content of the food, facility design, equipment design and use, packaging sanitation, employee health, hygiene and education, conditions of storage between packaging and the end user, intended use, and intended consumer.

Principle 2. Determine critical control points (CCPs). A critical control point is any step in food production that would control a reasonable hazard. Conversely, this represents any step that, if not in control, would result in the increased potential risk of injury or illness. Some typical control points that would eliminate or reduce the risk of a microbiological hazard would be heating, cooling, freezing, salting, drying and/or acidification. An integral part of a HAACP plan is the detailed development of a schematic diagram of the process flow identifying the CCPs.

Principle 3. Establish critical limits. Maximum and/or minimum limits are set on physical, chemical, sensory, or biological attributes of a food in process. They can be temperature, pH, moisture, viscosity, visual appearance, available water, salt concentration, tritralable acidity, aroma, or any other parameter used to determine whether or not a CCP is under control. When these limits or values are not met, the loss of control of a CCP is considered to result in an unacceptable health or spoilage risk.

Principle 4. Establish monitoring procedures. Monitoring procedures must measure and quantify the control necessary in a CCP to ensure that it does not exceed the limits established for adequate control. This measurement should be continuous to be effective; however, spot checks can be useful where continuous monitoring is not practical. Monitoring can be chemical, physical, visual, sensory, and/or microbiological testing. The most rapid and useful are the chemical or physical tests such as pH, time, temperature, total acidity, salt, a_w, humidity, moisture, and viscosity. Microbiological tests may also be applicable when rapid enough to provide control, especially with regard to pathogens in critical environmental areas of continuous processes. Monitoring is a planned management activity of measuring and recording CCPs to facilitate the tracking of critical steps in processing, to determine control or lack of control of a CCP, and to provide written documentation for the use in verification of a HAACP plan.

Principle 5. Establish corrective actions. When a product that is out of the established limits is produced, it must be placed on hold. The HAACP plan must contain provisions for out-of-limit food products. It must contain the action plans to be taken to include reprocessing, reexamination, or disposal. Action must be taken to correct the out-of-control CCP and recorded.

Principle 6. Establish verification procedures. A procedure must be in place to determine if the HAACP plan is performing as intended. When the system is first introduced, extensive testing may be involved to verify the plan's efficacy. Auditing of the finished product will be a part of the HAACP. Changes in processing technology, suspicion of foodborne illness, and organisms or new food safety issues will require revalidation. The HAACP plan must be constantly verified through records review, plant inspections, and random testing, all of which must be written to confirm compliance to the HAACP plan.

Principle 7. Establish record-keeping and documentation procedures. According to the National Advisory Committee on Microbiological Criteria for Foods (August 14, 1997), the HAACP system should include the following:

(1) A summary of the hazard analysis, including the rationale for determining hazards and control measures.
(2) The HAACP Plan to include
   - Listing of the HAACP team and assigned responsibilities.
   - Description of the food, its distribution, intended use, and consumer.
   - Verified flow diagram.
   - HAACP Plan Summary Table that includes information for
     Steps in the process that are CCPs
     The hazard(s) of concern
     Critical limits
     Monitoring
     Corrective actions
     Verification procedures and schedule
     Record-keeping procedures
(3) Support documentation such as validation records.
(4) Records that are generated during the operation of the plan.

See Also the Following Articles

Escherichia Coli and Salmonella, Genetics • Food-borne Illnesses • Meat and Meat Products • Temperature Control

Bibliography

Doyle, Michael P. (1989). "Foodborne Bacterial Pathogens." Marcel Dekker, Inc., New York.

Food and Drug Administration (1976). "Current Good Manufacturing Practice in Manufacturing. Packaging, or Holding Human Food; Revised Current Good Manufacturing Practices." Federal Register 21 CFR Parts 20 and 110.

ICMSF. (1980). "Microbial Ecology of Foods I. Factors Affecting Life and Death of Microorganisms." Academic Press, New York.

ICMSF. (1980). "Microbial Ecology of Foods II. Food Commodities." Academic Press, New York.

ICMSF. (1986). "Microorganisms in Foods 2. Sampling for Microbiological Analysis: Principles and Specific Applications" (2nd ed.). Univ. of Toronto Press, Toronto, Canada.

ICMSF. (1988). "Microorganisms in Foods 4. Application or the Hazard Analysis Critical Control Point (HACCP) System to Ensure Microbiological Safety and Quality." Blackwell Scientific Publications, Oxford, UK.

Jay, J. M. (1986). "Modern Food Microbiology" (3rd ed.). Van Nostrand Reinhold, New York.

Marriott, N. G. (1989). "Principles of Food Sanitation" (2nd ed.). Van Nostrand Reinhold, New York.

Mossell, D. A. A., van der Zee, H., Corry, J. E. L., and van Nenen, P. (1984). Microbiological quality control. *In* "Quality Control in the Food Industry", Vol. I (S. M. Herschdoerfer, ed.), pp. 79–168. Academic Press, New York.

National Advisory Committee on Microbiological Criteria for Foods. (1997), "Hazard Analysis and Critical Control Point Principles and Application"—Adopted August 14, 1997. National Academy of Science, Washington, DC.

Freeze-Drying of Microorganisms

Hiroshi Souzu

Hokkaido University

GLOSSARY

bound water Small amounts of cellular water that remain in an unfrozen state under normal freezing temperatures.

protective substance Smaller or larger molecular weight compounds that are dispersed into the suspension to protect biological materials from injury arising from freezing or drying.

secondary drying Desorption of bound water which binds firmly to the cellular materials and exerts a force which keeps the tissue structure.

THE FUNCTIONS OF LIVING ORGANISMS are basically dependent upon water molecules, either involved in their metabolic reactions or contributing to the structural stability of a cell's constituents or organelles. Freeze-drying is a multistage process of scrupulous drying of frozen biological materials. When freeze-drying is applied to microorganisms, cellular function is suspended temporarily, due to the reduced water activity on the metabolic pathways which are mediated by the action of water molecules. As a result, the organisms can be stored at room temperature for a prolonged period if protected from oxygen, moisture, and light. In addition, due to the highly porous structure of the products, they can easily reabsorb water and be restored to their original state. However, the freeze-drying procedure involves many critical stages or steps and, if a single step is not appropriately carried out, the organisms cannot survive.

Freeze-drying of microorganisms is highly empirical. However, it is absolutely evident that living organisms require a specific amount of water and a narrow range of temperature. Therefore, water state and temperature are the most important considerations in the process of freeze-drying. The freeze-drying procedure consists of separate processes: solidification, drying of the frozen materials under reduced pressure, storage, and reconstitution of the dried product. The drying stage can be further divided into two successive steps: (1) the stage of ice crystal sublimation and (2) the stage of isothermal desorption of the remaining liquid phase in the cellular material. In the following sections, detailed procedures to achieve efficient preservation of microbial cells will be described.

I. OPERATIONAL METHODOLOGY OF FREEZE-DRYING

A. Preliminary Preparation

Freeze-drying of microorganisms is carried out in an aqueous suspension. An appropriate suspending solution is required, because the properties of the medium sometimes have a significant effect on cellular activity throughout the freeze-drying process. The suspending medium should undergo sterilization, filtration, and degasification before the freeze-drying process begins.

The determination of the cell concentration in the suspension will also have a serious effect on the subsequent processes. A high concentration of cells will prevent an effective sublimation of ice crystals by the formation of a thick interstitial network. In contrast, when the cell concentration is too low, the cells, because they lack effective networks joining each of the cells, will disperse during the ice crystal sublimation. In order to correct for these situations, the use of a bulking or connecting substance is recommended. Incorporation of appropriate protective agents is also recommended, especially where very sensitive organisms are handled.

B. Cooling Procedure

The various cooling procedures employed include contact with a cold surface, immersion in a cold bath, direct spraying into liquid nitrogen, and utilization of liquefied gas. Cooling rates that are slower than 1°C/min are usually called slow freezing, and rates faster than 100°C/ min are referred to as rapid freezing; however, no precise definitions exist for "slow" and "rapid" freezing.

During cooling of the specimen, freezing of the suspending solution commences at a temperature slightly lower than 0°C; however, at such temperature, the cellular water still remains in a supercooled state. In slow freezing, the higher vapor pressure of the supercooled water, compared to that of ice at the same temperature, causes the cellular water to move outside of the cells, and the water will freeze externally. This mode of freezing is called extracellular freezing. In extracellular freezing, specimens, precooled in a bath adjusted to a temperature approximately 1 or 2°C lower than the freezing point of the suspending solution, are inoculated with ice. After complete ice crystallization at this temperature, the cooling bath temperature is lowered at the rate designated for the particular specimen.

In rapid freezing, specimens are generally frozen by direct immersion of the vessel into a cooling bath, which consists of either liquid nitrogen, dry ice–acetone, dry ice–alcohol, or any other medium previously cooled to the appropriate temperature range. Because the rate of cooling is significantly higher than the rate at which intracellular water molecules can move from the cells, the supercooled water in the cells freezes *in situ*. The result is called intracellular freezing. A simple and frequently utilized technique in freeze-drying is as follows. An appropriate amount of the cell suspension is dispensed into a spherical container. The container, touched with a surface of freezing medium at a tilt angle, is manually rotated until the specimen freezes completely, spreading the cell suspension evenly over the inner surface of the vessel.

In any freezing procedure, a substantial amount of cellular water remains unfrozen in the cells. This water is called unfreezable or bound water, because it is firmly bound to the cellular materials and helps to stabilize their structure.

C. Drying

1. Sublimation and Isothermal Desorption

Freeze-dryers are usually classified into two types: a chamber type, which is generally used for industrial work and/or mass production, and a manifold type, which is suitable for experimental use or production of small amounts of materials. An apparatus consists mainly of three parts: a drying chamber (drying manifolds), a vacuum system, and a cold trap which contains a coolant or refrigerant. The drying operation is carried out under reduced pressure to increase the velocity of vapor flow throughout the system. The vapors released from the products can be eliminated directly by a pumping system; however, its capacity is usually not sufficient. Thus, a cold trap is utilized to capture and condense the vapor on its surface. The vaccum pump is used for the initial evacuation of the dryer system and to handle small leaks in the apparatus.

Drying consists of two different processes which progress in tandem in the same specimen: sublimation of crystallized ice and successive isothermal desorption of unfrozen water molecules that remain in the cells. The force of the drying process arises from the difference of vapor pressures between the surface of the specimen and the cold trap; thus, utilization of a cold trap equipped with a lower temperature coolant is more effective.

The process of sublimation is the most important stage in the freeze-drying process and a strict temperature balance between the heating site and the drying surface of the specimen must be maintained. As

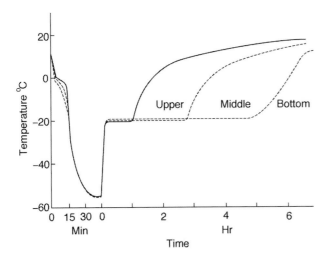

Fig. 1. Progress of the specimen temperature during the drying of *Saccharomyces cerevisiae* cells at room temperature (20°C).

shown in Fig. 1, throughout the sublimation stage, a nearly constant subzero temperature is maintained in the portion of the specimen in which most of the ice crystals remain, but in the part of the specimen where the drying front has passed, the temperature rises quickly in accordance with ambient temperatures. For efficient sublimation, a considerable amount of energy is required. However, excessive amounts of energy may bring about melting in the parts of the sample where ice crystals still remain and heat denaturation of the products in the areas where sublimation has been completed. To promote the sublimation rate while avoiding heat denaturation, the use of a water bath of appropriate temperature is recommended. In chamber drying, specimens are usually heated from the bottom by placing the vessels on shelves that are equipped with electric heating systems. Changing the ambient temperature by using a water bath or some other device affects the specimen temperature as follows. Warming the specimen slightly increases the specimen temperature, indicating that the sublimation rate of ice is promoted by heating. In contrast, cooling the specimen to below −40°C decreases the specimen temperture due to the lowering of the ambient temperature. The results indicate that the sublimation rate is repressed in these lower temperature regions

(Fig. 2). To achieve effective warming of the specimen without overheating, monitoring of the specimen during the sublimation period is critical. Many different systems to monitor and regulate temperature have been proposed. For instance, a special monitoring procedure, in which heat and vapor pressures in the system are automatically controlled by the electric characteristics of the specimen itself, has been introduced. When heating creates interstitial softening, the electric resistance of the specimen drops sharply, which triggers reduction of heating and lead to a better vacuum. When the electric resistance climbs again, heating is increased and stimulates the sublimation rate again.

In the initial stage of the so-called secondary drying process, the energy requirement drops sharply and the specimen temperature rises quickly in the part of the specimen where the sublimation of ice is complete. In this situation, the residual moisture in this part of the specimen might represent a substantial share of the initial moisture content (approximately 15%). During the successive drying, the residual moisture level decreases gradually in the areas farthest from the remaining ice crystal surface. However, still higher moisture levels remain in the specimen just after the disappearance of the ice crystals (Fig. 3). The presence of such high moisture levels does not allow long-term preservation of material,

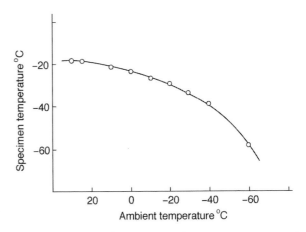

Fig. 2. Relationship between specimen temperature and ambient temperature during freeze-drying of aqueous suspension of *Escherichia coli* cells.

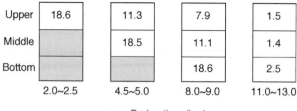

Upper	18.6	11.3	7.9	1.5
Middle		18.5	11.1	1.4
Bottom			18.6	2.5
	2.0~2.5	4.5~5.0	8.0~9.0	11.0~13.0

Drying time (hrs)

Fig. 3. Variation of the residual moisture content at the different parts of the specimen during the drying of aqueous suspension of *Saccharomyces cerevisiae* cells. The residual moisture content is obtained in the equation $A - B/B \times 100$, where A = the weight of the specimen immediately after drying, and B = the weight of the specimen, which is obtained after drying of an additional 3 hr at 60°C. Hatched part shows the bulk of ice.

and the excess moisture should be removed by successive drying.

The secondary drying stage is also endothermic. It is well known that unfreezable water extracted during this stage is more tightly bound to the cell constituents. In addition, the porous interstitial network of dried materials plays a role in preventing the migration of water vapor from the specimen. In order to promote drying, a specifically devised heating procedure must be employed. Supplying energy to the required portions of the specimen through a porous system under vacuum is also very difficult to manage, because both have excellent insulating properties. Microwaves or infrared radiation with a wave length adequate to penetrate the products might be appropriate for reaching a specific target, namely, the adsorbed water molecules.

2. Termination of Drying

It is important to know the correct termination point of the freeze-drying process. Higher moisture content in the specimen will produce damage during storage, but over drying living cells will result in a serious disruption of the structure of the cellular materials and an accompanying loss of cell viability. In order to determine an end-point of the process, many methods have been suggested, including pressure rise measurement, the Karl Fischer method, weight curves, vapor tension, moisture equilibration, nuclear magnetic resonance, and dielectric measurement.

Termination of the freeze-drying process presents another difficulty. The freeze-dried products are highly porous and hygroscopic structures, and if the vacuum is broken in ambient air, the dried material immediately absorbs atmospheric moisture and oxygen. To avoid undesirable contamination, a direct sealing of the sample by fusing the ampule at the neck under vacuum is recommended. In a situation where the direct sealing cannot be achieved, the vacuum should be broken with a dry inert gas introduced directly into the drying chember. Stopping, sealing, and even packing should then be performed in an airtight and closed area filled with the same inert gas. Most commonly, nitrogen, carbon dioxide, and, in some cases, argon have been used for that purpose.

3. Storage and Reconstitution

Freeze-dried products can theoretically withstand storage over a wide temperature range, avoiding denaturation. However, oxidative reactions and chemical degradation, including fatty acid oxidation, are very temperature dependent. Thus, storage within a moderate temperature range will improve the cell viability over a longer storage period. Light is also highly detrimental. Therefore, storage in the dark or in an opaque container is recommended.

In most cases, rehydration is performed by the addition of an exact amount of aqueous phase that was previously extracted from the system. This rapid rehydration has an advantage in that the specimens can avoid long-term exposure to concentrated solution. However, in some cases where such rapid rehydration produces deleterious effects, the specimens are first rehydrated by a hypertonic solution, then the solution is diluted by dialysis, changing the tonicity of the outer solutions gradually.

The rehydration temperature must also be considered. Generally, rehydration is carried out at room temperature. In some cases, it may be carried out at low temperatures to reduce the deleterious chemical and/or enzymatic reactions that will inevitably be introduced to the specimens following rehydration. It is recommended that rehydrated specimens be held at ice-cold temperatures until the next treatment can be started.

II. PREVENTION OF CELLULAR DAMAGE DURING FREEZE-DRYING

A. Operational Processes That Affect Viability

Although there is no decisive evidence indicating that ice crystallization in the cells causes lethal damage, it cannot also be denied that freezing conditions which bring about ice crystallization in the cells are related to reduced cell viability. For example, during freeze-thawing of aerobically or anaerobically cultivated *E. coli* cells, reduction of viability in both types of cells increases when the cooling rate exceeds 10°C/min. The reduction of viability in nonaerated cells increases even more rapidly (Fig. 4). These results indicate that anaerobically cultivated cells are more sensitive to higher freezing rates due to lower permeability of their membrane to the water molecules. The composition of the suspension medium is also a factor in the freezing damage to many kinds of microorganisms. For instance, when freezing *E. coli* cells, alkaline metal salts are known to have an extremely lethal effect on the cells during the freezing and storing periods. It has also been demonstrated that Na-glutamate provides a remarkable protective action against damage, even in the presence of alkaline metal salts.

The viability reduction of cells during the drying

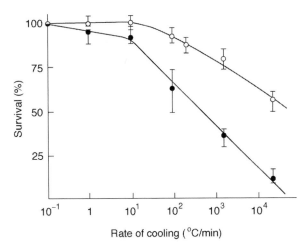

Fig. 4. Survival rate of aerated (open circle) and nonaerated (solid circle) cultures of *Escherichia coli* cells frozen at various rates.

process is thought to be concentrated mostly in the secondary drying stage. Secondary drying commences immediately after in the area of the specimen where a drying front has passed. Indeed, the reduction of the cell viability in these areas is known to increase considerably while the sublimation of ice in the whole specimen is still progressing actively. The experimental results have suggested that certain portions of the cells have already become overdried in this stage (Fig. 3). Overdrying will not only result in damage to the cellular structure but will also trigger the oxidation reactions of lipids or proteins, which are theoretically protected by the presence of hydrated water layers. In order to overcome these obstacles, care should be taken to prepare a specimen that can be dried homogeneously. Thin layers of specimens with large surface areas may yield better results.

The hydration temperature is another important factor. Experimental results suggest that the temperature range (in which the maximum recovery of several freeze-dried microorganisms occurs) is between 30° and 40°C. The recovery rates were lower when hydration was carried out at both higher and lower temperatures. In a similar experiment, it was demonstrated that the maximum recovery of freeze-dried *Lactobacillus bulgaricus* cells was facilitated by rapid rehydration at 20–25°C, and maximum levels of recovery varied, depending on the presence of protective additives.

B. Utilization of Protective Additives

The protective additives commonly used in freeze-drying of microorganisms are classified roughly into (1) low molecular weight compounds, such as amino acids, organic acids, sugars, and sugar alcohol and (2) high molecular weight substances, such as proteins, polysugars, polyvinylpyrrolidone, and synthetic polymers. It has been suggested that the low molecular weight protective additives for freeze-drying have a common characteristic in that each molecule contains a total of more than three hydrogen-bonding functions ($-OH$, NH_2, $=NH$, $=O$) or ionic function groups (either acidic or basic) in specifically organized structures. The action of these substances is thought to specifically protect the cell surface structure from the damage. The use of high molecular weight protective substances alone can protect the cells from the damage in the freeze-drying process; however, this mode of protection is not sufficient for protection from the damage that occurs during storage. The use of solutions of a combination of both low and high molecular weight additives will have more benefit in the maintainance of higher stability of the cells throughout drying and storage.

It is also known that oxidation reactions, which are associated with the secondary drying and subsequent storage, lead to cellular death. In order to prevent oxidation reactions in freeze-dried specimens, incorporation of an appropriate free-radical scavenger or antioxidant, such as monosodium glutamate, is recommended.

C. The Relationship between Growth Environments and Tolerance of Cells to Freeze-Drying

In many species of microorganisms, cell growth phases and growth environments exhibit a strong correlation to their tolerance to freeze-drying. Generally, logarithmically growing cells are very susceptible to damage, and the tolerance of the cells increases as cell growth approaches its stationary phase. For instance, *E. coli* cells grown in two different temperature ranges showed similar increases in resistance to freeze-drying as cell growth phases progressed from logarithmic to early stationary and to the stationary phase, although the cultivation length necessary to approach each growth phase differed considerably between these cultures (Table I). In a similar experiment, the growth temperature for *E. coli* logarithmic phase cells was shifted from 30° to 42°C and subsequently incubated for 40 min. The tolerance of the cells to freeze-drying increased significantly in the specimens incubated at 42°C when compared to the results obtained with the cultures incubated at 30°C for the same length of time.

The growth circumstances of the cell can also affect the tolerance of the cells to different rates of rehydration. The cells grown under aeration show a good recovery after both rapid and slow rehydration. In contrast, the cells grown anaerobically show a similar

TABLE I

Viability of Freeze-Thawed or Freeze-Dried *Escherichia coli* Cells Cultivated at Different Temperatures to Differing Growth Phases

Cultivations for:		Survival of the cells (%)		
		Freeze-thawed		
°C	Hr[a]	−5°C	−30°C	Freeze-dried
37	2*	4.0	3.8	4.0
	3**	52.3	50.4	24.9
	5***	93.9	92.5	38.6
17	8*	6.2	5.7	0.2
	20**	52.2	35.5	8.2
	48***	82.8	51.4	47.8

[a] Cells growing in logarithmic (*), early stationary (**), and stationary phase (***).

[From Souzu, H. (1985). The structural stability of *Escherichia coli* cell membranes related to the resistance of the cells to freeze-thawing and freeze-drying. *In* "Fundamentals and Applications of Freeze-Drying to Biological Materials, Drugs and Foodstuffs," pp. 247–253. International Institute of Refrigeration, Commission C1, Tokyo, Japan.]

recovery in slow rehydration but a significantly reduced recovery after rapid rehydration.

III. CONCLUSIONS

In the field of microbiology, freeze-drying is utilized for the long preservation and subsequent reproduction of microorganisms. To attain this objective, the specimen should be restored to its original state. In many instances, cell recovery decreases significantly after freeze-drying treatments. In order to obtain higher recovery and stability of freeze-dried products, the selection and utilization of appropriate protective substances are necessary. In general, the selection of the freeze-drying equipment itself is difficult. However, there are many effective protective substances identified. The efforts to find the most effective protective additives for each individual microorganism and to establish a generalized definition of protective substances are of great importance. At the same time, an understanding of the mechanism underlying the physical and chemical changes that occur during freeze-drying of microbial cells provides a basis upon which one could devise a successful application of freeze-drying for many species of microorganisms.

Bibliography

De Luca, P. P. (1985). Fundamentals of freeze-drying pharmaceuticals. *In* "Fundamentals and Applications of Freeze-Drying to Biological Materials, Drugs and Foodstuffs," pp. 79–85. International Institute of Refrigeration, Commission C1, Tokyo, Japan.

Rey, L. R. (1963). "Princips generaux de la liophilisation et l'humidite residulle des produits lyophilises," Publications de la Sociedad Espanola de Farmacotecnica, Barcelona, Spain.

Souzu, H. (1992). Freeze-Drying of Microorganisms. *In* "Encyclopedia of Microbiology," Vol. 2. (J. Lederberg, ed.), pp. 231–243. Academic Press, San Diego.

Takano, M., Takemura, H., and Tsuchido, T. (1985). Freeze-drying tolerance of *Escherichia coli* rec A mutants caused by growth temperature shift. *In* "Fundamentals and Applications of Freeze-Drying to Biological Materials, Drugs and Foodstuffs," pp. 279–284. International Institute of Refrigeration, Commission C1, Tokyo, Japan.

Willemer, H. (1985). Freeze-drying and advanced technology. *In* "Fundamentals and Applications of Freeze-Drying to Biological Materials, Drugs and Foodstuffs," pp. 201–207. International Institute of Refrigeration, Commission C1, Tokyo, Japan.

Freshwater Microbiology

Louis A. Kaplan

Stroud Water Research Center

Allan E. Konopka

Purdue University

riparian zone   The terrestrial area along the stream banks.

thermal stratification   The separation of lake water maintained by density gradients based on water temperature.

turnover   The mixing of a thermally stratified lake accomplished by wind-driven turbulence when the density gradients diminish.

GLOSSARY

allochthonous   Arising from outside the system, from another biotope.

authochthonous   Arising from within the system, from the same biotope.

Aufwuchs   All organisms firmly attached to, but not penetrating a streambed substratum.

benthos   Bottom-dwelling organisms.

biotope   A place or environment containing a biological community.

ecotone   A zone of transition between two habitats.

epilimnion   The surface layer above the metalimnion in a lake that thermally stratifies.

hypolimnion   The bottom layer below the metalimnion in a lake that thermally stratifies.

hyporheic zone   The ecotone between the streambed and the groundwater characterized by a mixture of surface water and groundwater.

lentic   Still or standing waters, as in a lake or pond.

lotic   Flowing waters, as in a stream or river.

metalimnion   The water layer in a thermally stratified lake between the epilimnion and the hypolimnion, characterized by steep vertical gradients of water temperature and dissolved oxygen.

plankton   Organisms of the free water that float, swim, or resist sinking.

riffle   A stream zone of high water velocity, characterized by shallow depth and flow over rocks.

FRESHWATERS ON THE SURFACE OF THE EARTH   are broadly separated into two major categories, lentic or standing waters, which include lakes, ponds, temporary pools, bogs, marshes, and swamps, and lotic or flowing waters, which include springs, streams, and rivers. All of these ecosystems are considered open systems, in that an exchange of species, nutrients, and energy occurs across air/water and land/water interfaces. Lakes and streams are the most commonly recognized freshwaters and embody the general principle that physical and chemical gradients within the ecosystem establish conditions that influence the types of microorganisms present, their abundances, and their activities. The microbial communities of both lentic and lotic ecosystems are complex assemblages that include representatives of the viruses, bacteria, algae, fungi, and protozoa.

Lakes and streams experience spatially and temporally dynamic physical and chemical gradients that create niches (that is, habitats or functions) for a diverse assemblage of microbial species. Physical processes, such as solar insolation and wind-driven turbulent mixing in lakes, contribute to their dynamic nature and drive thermal stratification and turnover. Streams experience currents, hydrograph fluctuations generated by storms or snowmelt, and interactions with the landscape through hydrologic exchanges with groundwater and the riparian zone. Some consequences of these physical processes and microbial activities in freshwaters are spatial and temporal heterogeneities in the chemical environment. These are important environmental factors to which microorganisms must respond and in which different microbial species may have a selective advantage.

The range of microbial species in lakes and streams includes organisms that use different resources (energy sources, carbon and other nutrient sources, terminal electron acceptors and habitats). Because many of these resources are produced by some species and consumed by others, there are important interactions between different metabolic groups. In addition, the transfer of energy through the food chain from photosynthetic microalgae to fish carnivores may be channeled through heterotrophic bacteria and protozoa (the microbial loop) or returned to the water following viral infection and lysis.

I. IMPORTANCE OF MICROBES IN LAKES AND STREAMS

Less than 1% of the water on earth is contained in freshwater systems. However, the impact of freshwater microbial processes is significant, due to the magnitude and diversity of nutrient inputs from the surrounding watershed. The bacteria in lakes and streams function as decomposers that recycle nutrients from organic back to inorganic forms necessary for photosynthetic production. In addition, they also channel dissolved organic substrates and inorganic nutrients into higher trophic levels through predation by protozoa and metazoans. Therefore, the rates of material flow and nutrient cycling in aquatic ecosystems are highly dependent upon microbial activities.

Lakes and streams do not operate as homogeneously mixed reaction vessels, but rather experience spatial and temporal heterogeneities in chemistry caused by physical and biological processes. In lakes, spatial and temporal heterogeneities in the water column are caused by physical processes, such as heating and turbulent mixing. In streams, the water column is always well mixed, but spatial and temporal heterogeneities in chemistry exist within the streambed sediments, the riparian zone soils, and within benthic habitats that are distinguished by the presence of algal mats and leaf packs. Additionally, the physical process of current flow creates erosional zones with high shear stress, such as rock surfaces in riffles, and depositional zones with aggrading fine-grain sediments and organic particles in pools and back eddies. Streams also possess longitudinal profiles and, with increasing stream size, there are concomitant changes in discharge, width, depth, and current velocity. Temporal heterogeneities in freshwaters range across broad scales from minutes to decades. These heterogeneities affect the distribution and consequences of microbially mediated and purely chemical reactions that affect the productivity of the aquatic ecosystems.

Lakes and streams, taken on average, are considered moderately productive ecosystems with net primary productivity of approximately 200 g $C/m^2/y$ (Whitaker and Likens, 1973). These values lie between the highly productive ecosystems, such as swamps, marshes, and tropical rain forests (1000 g $C/m^2/y$) and the very low productivity desert scrub ecosystem (30 g $C/m^2/y$). For both lakes and streams, there is considerable variation around these averages, including oligotrophic lakes (20 g $C/m^2/y$), eutrophic lakes (400 g $C/m^2/y$), small forested streams (15 g $C/m^2/y$), and midsized open stream channels (350 g $C/m^2/y$), representing different points of the spectrum. Streams differ from lakes in their close association with the terrestrial environment, and that includes allochthonous inputs of organic matter. These carbon subsidies often exceed authochthonous primary productivity in streams and can be as high as 700 g $C/m^2/y$. The density of algae in lakes and streams is difficult to compare because of the planktonic nature of the former and benthic nature of the latter; however, phytoplankton densities typically range from less than 10^3 cells/l to over 10^6 cells/l (Hutchinson, 1967) and periphyton densities range from 10^6 to 10^{12} cells/m^2 (Lock, 1981). The algal assemblages in lakes and streams can be very diverse, with greater than 100 species represented in the community at any given time.

Far fewer studies of chemoheterotrophic bacteria have been reported for lakes and streams. Estimates of bacterial generation times in lakes and streams range from a few hours to more than one month and cell densities in lakes and streams range from approximately 10^7 to 10^{10}/l. In streams, most of the bacterial biomass is attached to the streambed, and the densities of bacterial cells attached to sediments range from 10^7 to 10^9 cells/g dry weight, while the densities attached to rocks range from 10^{10} to 10^{12} cells/m^2. There is little information about the species of bacteria in lakes and streams, especially information that is not biased by culture methods. However, bacterial communities are probably as diverse, or more so, than the algal community, and this remains an important area for future investigation.

II. MICROBIAL PROCESSES IN FRESHWATERS

In order to grow, a microbe must be present where all of its nutrient resources are available. Organisms require an energy source, carbon source, terminal electron acceptor (nonphototrophs) or electron donor (phototrophs), as well as sources of N, P, S, and other chemical elements required in small supplies. At least one of each of these types of resources is essential for growth. However, aquatic ecosystems contain complex mixtures of molecules, and some may be substituted for others. For example, a heterotrophic bacterium might be capable of using either glucose or acetate as its carbon and energy source and oxygen or nitrate as its terminal electron acceptor.

The set of microbial species dominant in a habitat will often be determined by the outcome of competition among the entire suite of microbial species present. These species compete for the spectrum of available resources; the outcome of the competition can change with time and with depth in lake waters or streambed sediments as environmental conditions change. In addition to competition, the metabolic products of some microbes may be resources used by other microbes, so that synergistic interactions may occur. For example, an alga may excrete photosynthate that stimulates growth of a bacterium, and the bacterium may enhance algal growth through the production of CO_2 or an organic molecule that promotes algal growth. Within bacterial assemblages present in biofilms, there may be the formation of a consortium of species that contribute to metabolism

TABLE I
Metabolic Types of Microbes in Freshwaters

Nutritional type	Energy source	Carbon source	Electron donor	Terminal electron acceptor	Metabolic products
Photoautotrophs					
Eukaryotic microalgae and cyanobacteria	Light	CO_2	H_2O	–	O_2
Photosynthetic sulfur bacteria	Light	CO_2	H_2S	–	S^0, $SO_4^=$
Chemoheterotrophs					
Aerobic bacteria	DOC[a]	DOC	DOC	O_2	CO_2
Fermentative bacteria	DOC	DOC	DOC	DOC	DOC, CO_2; H_2
Denitrifying bacteria	DOC	DOC	DOC	NO_3^-	$CO_2 + N_2$
Sulfate-reducing bacteria	DOC	DOC	DOC	$SO_4^=$	DOC, CO_2, H_2S
Chemoautotrophs					
Sulfur-oxidizing bacteria	H_2S, S^0, $S_2O_3^=$	CO_2	H_2S, S^0, $S_2O_3^=$	O_2	$SO_4^=$
Nitrifying bacteria	NH_4^+, NO_2^-	CO_2	NH_4^+, NO_2^-	O_2	NO_3^-
Hydrogen oxidizing bacteria	H_2, DOC	CO_2	H_2, DOC	O_2	H_2O
Methane oxidizing bacteria	CH_4	CH_4	CH_4	O_2	CO_2
Methane-producing bacteria	H_2, Acetate	CO_2	H_2, Acetate	CO_2	CH_4, CO_2

[a] DOC, dissolved organic carbon.

of complex organic molecules. Table I lists some of the resource requirements and metabolic products of microbes found in aquatic ecosystems.

In the epilimnion of lakes, the most important microbial processes are photosynthetic primary production (mediated by oxygenic microalgae and cyanobacteria) and heterotrophic mineralization of this primary production by bacteria. Contributions to the carbon cycle can be located in the near-shore, littoral zone, or the deeper waters of the profundal or pelagic zone, as well as input from the watershed via tributary streams or groundwater. In streams, organic carbon production occurs by oxygenic microalgae within the Aufwuchs, those organisms firmly attached to rocks, sediments, or other organisms. Phototrophic production is most often limited in lakes by the supply rate of inorganic phosphate. However, N-limited conditions may occur if P inputs to a lake are excessive, or lake habitats may become light-limited when N and P inputs are high. Light limitation in streams can occur under a dense forest canopy, nutrient limitation by N or P depends greatly on the geology of the watershed, and nutrient levels can influence whether gross primary production exceeds respiration (Allan, 1995).

Chemoheterotrophic bacteria and fungi require organic compounds for carbon and energy. Fungi play an important role in streams that receive significant inputs of allochthonous detritus but, as a group, they are not very important in lakes. Organic substrates exist as particulate or dissolved organic C (POC or DOC). Bacteria with high affinity enzyme systems outcompete fungi for dissolved molecules present in low concentrations. POC occurs either as living or dead matter, and fungi, with the ability to penetrate different types of tissues, and thus gain access to higher concentrations of organic molecules, have a competitive advantage over bacteria on particles, especially those particles with a protective cuticle layer, such as deciduous leaves. When an organism dies, heterotrophic bacteria rapidly colonize it, and hydrolytic enzymes on the surface of heterotrophic bacteria decompose its constituent macromolecules. These enzymes convert the macromolecules to low molecular weight molecules (monomeric amino acids and carbohydrates) which can be transported into the heterotrophic cell and metabolized. Colonization of detritus in streams by bacteria is often fol-

lowed by fungi, which, in time, can become the dominant microbial decomposer.

DOC is the major carbon and energy source available for chemoheterotrophic bacteria. The concentration of DOC in lakes and streams may range from less than 1 mg/l to over 35 mg/l. However, this may be 6- to 10-fold higher than the amount of organic C tied up in biomass in a lake or stream. DOC arises through excretion by living organisms, autolysis of cells, or inputs of allochthnous DOC from the watershed. DOC is a complex mixture of molecules. Although the total concentration of DOC is relatively high, this is somewhat misleading. Most DOC is in the form of fulvic and humic acids. These complex, heterogeneous macromolecules are thought to be poorly metabolized by bacteria, and though some humic substances persist for years in aquatic habitats, others contribute to bacterial production. The heterotrophic microbes make their living rapidly metabolizing a small pool of easily degraded molecules, such as amino acids and proteins, carbohydrates, and polysaccharides, and slowly metabolizing a large pool of less easily degraded humic substances. The concentrations of individual amino acids or carbohydrates is estimated to be usually < 10 $\mu g/l$. This is due to the tremendous ability of bacteria to transport these substances even at very low concentrations.

III. SPATIAL HETEROGENEITIES IN LAKES

Two important physical processes in lakes are solar heating and turbulent mixing of the water column due to wind action. In some lakes in temperate latitudes, these processes lead to thermal stratification of the water column during the summer (Fig. 1). The upper layer (the epilimnion) comprises water that is uniformly warm and mixed freely by wind action. The metalimnion is a transition zone in which there is a steep thermal gradient. The bottom zone of cold and relatively undisturbed water is called the hypolimnion.

The significance of these thermal discontinuities is that the density of water is a function of its temperature. Water heated at the surface is less dense than deeper, colder water, and, therefore, work must be done to mix these waters. If the density gradient is

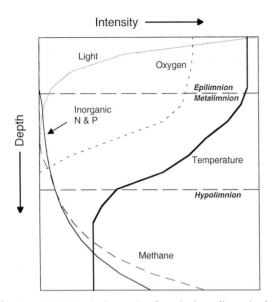

Fig. 1. An idealized schematic of vertical gradients in thermally stratified lakes. Stable strata are the result of thermal stratification, which generates a density gradient in the water column. Differences in biological processes (see Fig. 2) then result in the creation of chemical gradients, which provide specialized niches for other microbes.

large enough to reduce vertical turbulence, turbulent mixing will be confined to the epilimnion. During thermal stratification, gradients of chemical and microbial species can become established in the stable strata of the water column.

The density gradient produced by thermal stratification leads to chemical gradients in the metalimnion and hypolimnion. These chemical gradients are often a product of microbial metabolism. The vertical distribution of oxygen provides an example. After thermal stratification, phytoplankton may settle out of the epilimnion, and their organic matter is then mineralized by oxygen-consuming aerobic chemoheterotrophic bacteria in the hypolimnion or, more importantly, at the sediment–water interface (Fig. 2). The consumed oxygen cannot be replenished, because the hypolimnion is physically isolated from the atmosphere and, therefore, transport occurs primarily by molecular diffusion. If the mass of organic matter delivered to the bottom strata is large enough, anaerobiosis will develop.

There are a number of other vertical gradients that

are relevant to microbial processes in lakes (Fig. 1). These include:

A. Light

Solar energy is the driving force for the ecosystem. Phototrophs capture light via photosynthesis and use it to produce organic matter from CO_2 (primary production). Light intensity decreases exponentially with depth. As a rule of thumb, the depth of the euphotic zone (z_{eu}), the region where net photosynthesis can occur, is the depth at which the irradiance is 1% of that at the lake surface (I_0). In some lakes, light penetrates deeper than the epilimnion, and stable, stratified layers of phototrophic microorganisms (cyanobacteria, motile eucaryotic algae, or photosynthetic sulfur bacteria) may develop in these layers, occasionally at depths where the irradiance is < 1% I_0. Although the phototrophs in these layers are exposed to rather low irradiances, they may have access to higher concentrations of inorganic macronutrients or essential electron donors (hydrogen sulfide in the

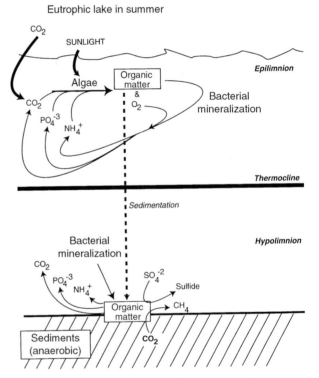

Fig. 2. Some important microbial processes which occur in different regions of a thermally stratified lake.

case of photosynthetic sulfur bacteria) in the metalimnion or hypolimnion than do phototrophs in the epilimnion.

B. Inorganic N and P

In the epilimnion, these nutrients are depleted by phototrophs. Therefore, the rate of primary production in the epilimnion is often limited by one of these elements (most often P). Although the major fraction of these nutrients may be recycled in the epilimnion, the remainder sinks to the stable regions of the water column and is remineralized. Therefore, a vertical gradient of macronutrient availability develops during summer stratification.

C. Reduced Substances

Once the sediment–water interface has become anaerobic, the reduced products of anaerobic microbial metabolism in the sediments (hydrogen sulfide and methane) will diffuse from the sediments and form gradients in the water column.

The vertical gradients in the metalimnion and hypolimnion provide a much broader opportunity for specialized niches. These niches develop at depths where these vertical gradients intersect, and can be exploited by specific physiological groups of microbes. When the overlying water is oxic, these intersections occur at the sediment–water interface. In lakes with anaerobic hypolimnia, the interface between the metalimnion and hypolimnion is of most importance. The intersecting gradients provide a narrow zone in which certain microbial processes can occur; as a consequence, discrete layers of physiologically specialized microbes may arise in stratified lakes. Gorlenko *et al.* (1983) have compiled a number of studies of the vertical distribution of microbial activities in lakes.

D. The Light–Macronutrient Gradient

For oxygenic phototrophs (cyanobacteria and eukaryotic algae), the major determinants of production rate and biomass are light and macronutrients (especially ammonia and phosphate). If adequate light for photosynthesis penetrates deeper than the epilim-

nion, phototrophs can stratify and grow in the metalimnion, where N and P availability are much greater than in the epilimnion. However, organisms that are larger than a few μm in diameter will sink out of the optimal depth for growth and, therefore, must have a means of maintaining their vertical position in the water column. The mechanisms are flagellar motility (eukaryotic algae such as dinoflagellates, chrysophytes, and cryptomonads) and buoyancy regulation employing intracellular gas vesicles (filamentous cyanobacteria).

E. The Light–Hydrogen Sulfide Gradient

Anoxygenic phototrophs (the purple and green sulfur bacteria) require light, the absence of oxygen, and a suitable electron donor (hydrogen sulfide) for photosynthesis and growth. Net growth of anoxygenic phototrophs can occur at irradiances much less than 1% of the surface irradiance. Stratified layers of phototrophic bacteria are most often found in small deep lakes protected from wind mixing. Layers may contain enough biomass to be visibly colored purple or green and may contain more than 100 g bacteriochlorophyll/l.

These strata often contain several different groups of anoxygenic phototrophs that stratify at different depths. The primary environmental factors that influence what species occur appear to be light intensity, the spectral quality of the light, and sulfide concentration. If the light intensity reaching the sulfide-containing layer is relatively high, a layer of purple sulfur bacteria occurs nearest the surface; purple bacteria have been shown to require higher irradiances to grow at the same rate as green bacteria. The spectral quality of the light which passess through the purple *Chromatium* layer will only support the growth of green *Chlorobium* species.

F. Gradients of Oxygen and Reduced Substances

Just as overlapping gradients of light and hydrogen sulfide create a niche for anoxygenic phototrophs, an overlap of gradients of oxygen (diffusing from the epilimnion) and reduced inorganic compounds (diffusing from the sediment–water interface) results in layers of aerobic chemoautotrophic bacteria within the stable region of the water column. The reduced substances (ammonia, hydrogen sulfide, or methane) are the products of microbial metabolism in the sediments and arise by processes such as ammonification, sulfate reduction, and methanogenesis. The factors which impact the spatial and temporal changes in methane oxidation have been investigated in large, eutrophic Lake Mendota and in very small eutrophic Lake 227. These two studies illustrate the importance of lake morphometry and physical and chemical processes upon microbial metabolism. In Lake 227, methane oxidation during summer stratification was restricted to a zone 1–2 m in depth, whereas in Lake Mendota, methane oxidation rates were higher and extended over a 5 m zone in the thermocline. Lake Mendota has a much larger surface area than Lake 227, and, therefore, is more susceptible to wind turbulence. The wind turbulence more effectively mixes oxygen from the epilimnion with methane diffusing through the hypolimnion. The narrow zone of methane oxidation in Lake 227 was a consequence of inorganic nitrogen limitation in the thermocline. Therefore, the methane-oxidizing bacteria were dependent upon nitrogen fixation, and this oxygen-sensitive process restricted their distribution to strata where the oxygen concentration was <1 mg/l.

The vertical stratification of bacterial species in the metalimnion and hypolimnion is sometimes much greater than indicated by the above considerations. This is especially true in meromictic lakes (those which remain thermally stratified throughout the year). Specific bacteria are found in discrete layers 10–20 cm in breadth. Many of these bacteria contain gas vesicles, which could provide neutral buoyancy and permit microstratification in the hypolimnion. Many of the morphological forms observed in these samples have never been grown in culture, and their specific contributions to the chemical cycles in lakes are unknown.

IV. TEMPORAL HETEROGENEITIES IN LAKES

Temporal changes in environmental factors occur on a broad range of time scales. Some occur over decades (long-term changes in climate), some are seasonal (annual changes in thermal stratification), and others occur on a daily basis (solar irradiance, local meteorological events). There are even events which occur on a time scale of minutes (consequences of turbulent vertical mixing in the water column). The biological processes affected by these various scales differ. For surface waters, Harris (1986) has interrelated the scales of temporal and spatial variability in surface waters to the time over which specific biological processes operate. Whereas physiological processes, such as nutrient uptake, may respond to environmental changes that occur on time scales of minutes to days, changes in community composition are expected to reflect environmental changes which occur over periods of weeks to months. Some examples of temporal heterogeneities that occur on different time scales are given below.

In lake ecosystems, the seasonal changes in phytoplankton species composition have been of great interest, and an idealized mechanistic model has been developed (Sommer *et al.*, 1989). This model identifies seasonal changes in light and nutrient availability, as well as the population dynamics of herbivores and carnivores, as important factors. Thermal stratification also plays an important role in these processes, because it alters light availability and creates a sink for inorganic macronutrients, such as ammonia or phosphate, which may become limiting in the epilimnion.

As an example, consider a eutrophic lake found at temperate latitudes. During early spring, concentrations of dissolved nutrients are maximal. The water column is isothermal, so it can be completely mixed by turbulent wind action. The intensity of solar radiation is low and the length of the day is short. As a consequence, primary production is limited by the average irradiance to which phytoplankton are exposed. The average irradiance is related to the ratio of the depth of the photic zone to the depth of the lake and will, therefore, be lower in deeper lakes. Day length and insolation increase during the spring. This stimulates primary production. Diatoms often dominate because of their capacities to grow efficiently at low average irradiances and temperatures and because sinking of these nonmotile forms is not a selective disadvantage in the turbulently mixed water column.

The spring diatom bloom may end due to depletion of silicon, an essential element for diatom growth (but not for other phototrophs which do not have a silica-containing cell wall). If there is a period of low wind turbulence after the onset of thermal stratification, diatoms can be lost from the productive zone of the lake due to sedimentation. The sinking of the spring diatom bloom can represent the major loss of phosphorus to the sediments in a lake during the year.

Lake stratification stimulates the growth of small algae which have rapid growth rates. Members of the Chlorophyceae or Cryptophyceae often dominate the phytoplankton during these periods. Dissolved macronutrient concentrations are still relatively high, and the average irradiance to which cells are exposed increases substantially, because the depth of the mixed zone decreases by a factor of two or more. However, these species are readily grazed by zooplankton and a large increase in the zooplankton population leads to the clear water phase. This creates a selection pressure for phytoplankton species which are resistant to zooplankton grazing. Large size prevents grazing, but also has the negative consequence of increasing the organisms' sinking rate. There are adaptations which decrease sinking; these include mucilaginous sheaths (which decrease sinking rate by reducing overall particle density), flagellar motility, or buoyancy conferred by gas vesicles in cyanobacteria.

Cool fall weather causes thermal stratification to deteriorate in the fall, as convectional cooling deepens the epilimnion. Inorganic macronutrients (ammonia and phosphate) present in the hypolimnion are reintroduced into the nutrient-depleted surface waters by this process. This stimulates primary production of the macronutrient-limited phytoplankton. However, photosynthesis subsequently becomes limited by the decrease in average irradiance that occurs because daily insolation and day length decrease during the fall. Eventually, wind power is sufficient to overcome the weakened density gradient, and the lake is completely mixed (fall overturn). In winter, under ice cover, the lake can again stratify, but it is the colder water ($<4°C$), which is nearer the surface.

Primary production is low, especially if light penetration is attenuated by snow on the lake ice.

The annual changes in thermal structure affect microbial groups other than the phytoplankton. As discussed above, chemoautotrophic methane oxidizers form a metalimnetic layer during summer stratification in the zone where oxygen and methane overlap. However, after fall turnover the methane that had accumulated in the anaerobic hypolimnion is distributed throughout the aerobic water column. More than 90% of the total annual oxidation of methane has been measured to occur during and just after fall overturn. The chemoautotrophic oxidation of ammonia to nitrate may follow a pattern similar to methane oxidation, as ammonia is converted to nitrate after fall overturn and during the winter. The subsequent fate of this nitrate depends upon thermal structure the following summer. Nitrate in the epilimnion can serve as an inorganic nitrogen source for phytoplankton growth, whereas nitrate in the anaerobic hypolimnion is used as a terminal electron acceptor by heterotrophic bacteria (denitrification).

V. SPATIAL HETEROGENEITY IN STREAMS

Two linked aspects of lotic environments distinguish them from other aquatic environments, current and unidirectional flow. These physical factors contribute to the spatial heterogeneity in streams, which influences the distribution and activity of microorganisms. Spatial heterogeneity in streams or rivers can be identified at three different scales, decreasing in size from the watershed, the reach, and the habitat. This vast spatial dimension ranges over at least 9 orders of magnitude from the microbial colony in micrometers to the river system in tens or hundreds of kilometers. These scales are layered or nested, providing a high level of complexity.

A. Watershed Scale

At the watershed scale, streams generally increase in size in a downstream direction as tributaries and groundwater inputs augment the discharge. Discharge is a function of width, depth, and velocity, and all three parameters increase with downstream direction. With increasing stream size, the direct influence of the terrestrial environment recedes, including such factors as shading from the riparian zone, which impacts temperature and light levels, and leaf litter inputs, which influence the kinds and amounts of organic carbon present. Also, as streams get larger, the relative importance of the stream bottom as a habitat, the benthic influence, diminishes and water column or planktonic processes become increasingly important (Vannote *et al.*, 1980).

Changes in the quantity and quality of organic matter inputs are closely associated with light levels. In small forested streams, in-stream (autochthonous) primary production can contribute less than 1% of the total organic matter inputs. Nearly all of the microbial biomass and activity is associated with the streambed or the benthic environment. Attachment to a solid substratum helps microorganisms prevent displacement downstream, and some algae, such as diatoms, use stalks of mucilage (Fig. 3A) or glue their entire length (Fig. 3B) to rocks or sediments in the streambed. Bacteria attach to the streambed with mucopolysaccharide. In midsized streams, stream width increases, the forest does not shade the entire stream, and significant levels of algal production, as well as rooted vascular plants called macrophytes, can occur. In large streams or rivers, a truly planktonic existence for microorganisms dominates as the reduction of light due to depth and sediment load limits benthic primary production. One result of that change is that diatom communities in lotic waters shift from pennate forms attached to surfaces in small streams to centric forms floating in rivers; the bacterial community experiences these changes as well, from attached cells within the Aufwuchs of small streams to planktonic cells in large rivers, many washed in from the flood plain.

The base of the dissolved organic matter pool also changes from the dominance of terrestrial inputs in small streams to a mix of algal, macrophyte, and terrestrial inputs in midsized streams, to algal inputs or flood plain inputs in larger streams. When floodwaters engulf a broad floodplain they can deliver large amounts of terrestrially derived organic matter to the river as the flood recedes. Given the longitudi-

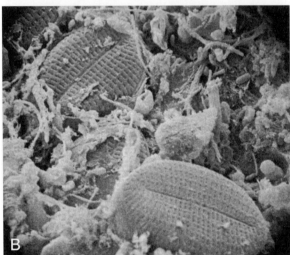

Fig. 3. Scanning electron micrograph of stream Aufwuchs communities showing the attachment of (A) stalked diatoms of the genus *Gomphonema* and (B) prostrate diatoms of the genus *Cocconeis* and spatial association with unidentified bacteria of coccus-, rod-, and filament-shapes.

nal connections in stream systems, food resources from upstream can be transported downstream if they are not completely metabolized. DOC concentrations increase with downstream direction in the headwaters of some streams where groundwaters that have low DOC concentrations rise to the land surface and accumulate DOC from surface-dwelling organisms and detritus. Beyond the headwaters, DOC concentrations remain relatively constant with downstream direction, suggesting that processes that consume or dilute DOC balance the inputs.

Streams in temperate climates experience annual temperature fluctuations of more than 25°C and diel temperature fluctuations that can exceed 10°C. While most temperature change is seasonally influenced, there are longitudinal influences on temperature as well. Headwater streams near groundwater inflows have a low diel temperature range because of the constancy of groundwater temperatures. Large rivers have such mass that temperature changes occur slowly. The streams with the greatest diel temperature ranges are midsized streams that are beyond the control of groundwaters, yet small enough to respond to increased solar inputs as the forest canopy influence diminishes.

B. Stream Reach Scale

Water velocity has a profound impact on the microbiology in streams. Changes in slope, channel roughness, and channel shape all influence water velocity and are seen in the riffle/pool sequence of a stream. Riffles are areas of fast velocities, typically over 0.5 m/s with shallow water flowing over a substratum of small cobble-sized rock or larger. Pools are characterized by slower flow, deeper waters, and smaller substrata. The substrata in riffles tend to be more stable than the sediments often found in pools, and this provides a degree of stability to the riffle community, compared to the pool community. Algal mats or dense biofilms can develop and lead to layers of microorganisms and even anaerobic zones beneath a layer of active aerobic organisms or within particles. Scour during high flows associated with storms will remove some of the microbial community attached to rocks in riffles, but there are surviving organisms. Sediments in pools typically undergo scour and transport during storms, and new sediments from the upstream channel or the watershed with different microorganisms settle in the pools as the storm waters recede.

Water velocity also has an impact on the exchange of surface waters with the hyporheic zone of the streambed, a zone characterized by the mixture of surface waters and groundwaters. Streambed areas where there is downwelling of surface waters can be zones conducive to microbially mediated processes involving alternative electron acceptors. These include denitrification, iron reduction, sulfate reduction, and methanogenesis. The size and influence of the hyporheic zone varies both within and between streams, so general descriptions of their influence on microbial processes are limited. Some streams that developed on glacial till have very extensive hyporheic zones, with stream-dwelling insects and microorganisms found in subterranean zones hundreds of meters from the main channel. Other streams, including those that were not glaciated during the most recent glacial maximum, can have shallow confined hyporheic zones with limited exchange of water or organisms. While there is scant knowledge of microbial populations and processes within hyporheic zones, some depth profiles have been described. For example, in streams where there is a downwelling of surface water enriched with DOC, a decline of O_2, DOC, microbial biomass, and activity occurs with depth into the sediments. Other streams exhibit a more complex profile, where particulate organic matter is deposited in pools or a back eddy and buried during storms. These accumulations of organic matter and the resulting microbial metabolism generate an anaerobic layer with an associated liberation of highly labile organic products of fermentation or anaerobic respiration when an organic molecule is the electron acceptor.

C. Habitat Scale

Habitats in streams include the hyporheic zone, the streambed surface, and the water column. As described above, the hyporheic zone can be an important interface or ecotone between the groundwater and surface water environments. The resulting gradients of organic molecules and oxygen provide for a range of heterotrophic and autotrophic metabolic types within the microbial community. These same gradients can occur over very small distances at the streambed surface where Aufwuchs or biofilms form on rock surfaces and sediments. The surface habitat is generally considered to be well oxygenated and, thus, inhabited by aerobic microorganisms. However, the presence of facultative anaerobes has been demonstrated within these habitats, suggesting that at least transient oxygen depletion occurs within microbial assemblages surrounded by highly oxygenated waters.

For photoautotrophic microorganisms in streams, access to light is an obvious constraint that limits the colonization of the undersides of rocks or growth of planktonic forms in the deeper regions of turbid rivers. However, there are other more subtle differences in the distribution of organisms within the community of photoautotrophs in streams. Considerable variation in the shape and vertical layering of algal growth forms attached to rocks on the streambed, often referred to as the periphyton community, occurs as a consequence of fine-scale variations in current velocities. Growth forms of algae have been described, without consideration of species composition, as crustose, prostrate, stalked, filamentous, and gelatinous.

When considering the taxonomy of algae attached to rocks in streams, descriptions of growth zones are numerous. In one example from a northern temperate stream, a green alga *Ulothrix* grew in a narrow band on the upper surface of stones that break the water surface, and growth was limited to the upstream-facing surfaces directly impacted by the current. Below the *Ulothrix* band was a band of diatoms (*Gomphonema*) that was also limited to the upstream-facing surfaces, and then a band of another diatom, *Diatoma*, that extended to the downstream-facing areas of the rock grew on the lower regions of the rock in deeper water, but still above the sediment surface. Such observations of distributions remain mostly descriptive, with little mechanistic understanding about the niche requirements of individual algal species within the periphyton.

Algal mats like those just described can grow wherever there is a stable substratum, nutrients, heat, and light. So accumulations of algae are found on rocks, sediments, woody vegetation, leaves, and on other living plants and animals. When the substratum is organic, these habitats not only provide a substratum for microbial attachment, but a nutritional substrate as well for heterotrophic microorganisms. This means that heterotrophic bacteria do not necessarily rely on DOC in the water column, but can obtain carbon and energy from a living or dead plant to which they attach or grow near. Microbial colonization of living and detrital plant material changes its nutritional quality. Dead leaves falling into streams provide an important food resource for stream-dwelling insect larvae. The palatability and nutrition of C-rich, but N-poor, leaves increase when they become colonized by aquatic hyphomycetes and other microorganisms. These microorganisms also colonize woody debris in streams.

VI. TEMPORAL HETEROGENEITY IN STREAMS

A. Decades

Variation on the temporal scale in streams, like lakes, can range from decades to minutes. From the perspective of the microbial community, long-term variations, such as meandering of the streambed within its floodplain or wetter or dryer than normal years, result in incremental alterations of the physical environment. These variations are on time scales that allow adequate time for responses at the community and population levels and, typically, do not alter or eliminate a microbial niche from the system.

B. Seasonal

In contrast, seasonal variations, such as changes in light levels, temperature, and annual snowmelt in temperate environments, or monsoon seasons and dry seasons in tropical climates, can profoundly alter the microbiology of streams. Seasonal changes in algal abundance and activity are relatively predictable in temperate streams and are associated with incident levels of solar radiation. There are numerous cases where an algal maximum occurs in the spring of the year just prior to leaf out. Once the canopy closes and blocks sunlight, production and abundance decline, even though water temperatures increase. A secondary annual peak is observed following leaf fall. In arid environments, streams can support extensive algal production, but the algae are susceptible to a nearly complete wash-out during seasonally predictable, but irregularly occurring, torrential rains.

Concentrations of dissolved organic carbon under baseflow or nonstorm conditions change seasonally in small temperate streams. Peak concentrations coincide with peak litter inputs, but this is also the period when a secondary algal peak is observed. Seasonally low concentrations are observed under conditions of hard freezes that eliminate some of the shallow, carbon-rich flow paths water takes in getting to a stream. The combination of low food resources and low temperatures suppress the activities of bacterial heterotrophs and the microzooflagellate and ciliate protozoa that feed on the bacteria. Seasonally predictable snowmelt in mountainous regions also suppresses bacterial activity and abundance due to scour and transport, even though DOC concentrations can peak as organic matter that has accumulated in the soils during the growing season is flushed from the watershed.

C. Diel or Shorter

Temporal heterogeneity on short time scales typically relates to the activities of organisms living in streams. For example, diel cycles in algal activity associated with the solar cycle can result in pulses of algal exudates, as excretion is closely tied to photosynthetic activity. Whether or not the released DOC is metabolized by a bacterium within the biofilm or diffuses into the overlying water and is carried downstream depends upon the nature of the exudate, the proximity of heterotrophic bacteria, and the temperature. The diel behavior of higher organisms can also influence microbial abundance and activity. For example, many stream-dwelling insects are nocturnal. Feeding on biofilms, releasing DOC through "sloppy feeding," and excreting DOC as part of digestion generate resources for microbial heterotrophs in a narrow window of time and space.

An extremely pervasive form of short-term temporal heterogeneity, and one which has a major impact on microbial communities in streams, is the influence of hydrology. In the absence of storms, stream water originates entirely from groundwater inputs. Groundwater inputs are typically highly processed, having passed through the unsaturated soil horizons en route to the water table. Storms alter the pathways that water takes to the stream, including flow through shallow soil regions and overland flow. As a result, higher concentrations of DOC, including DOC that has not been extensively degraded, enter the stream during a storm. At the same time that new resources for microbial metabolism enter the stream with storm waters, the discharge of the stream increases, and microorganisms can be dislodged from their substrata and transported downstream. Only organisms firmly attached to stable substrata can withstand scour by storm flows. In small streams, the amount of DOC transported during storms can be very large. Consider the combined effects of a 5-fold increase in DOC concentration from 2 to 10 mg C/l and an increase in discharge by two orders of magnitude, and the result is a 500-fold increase in DOC flux. Headwater streams with extensive drainage networks can respond very rapidly to storm surges and also return to baseflow shortly after the storm passes. In larger rivers, the storm flows are buffered, so the river flow increases more slowly and stays elevated for a longer period of time. In temperate environments where storms occur once every week or two, the microbial community in streams is nearly always in a state of recovery from the previous storm disturbance.

VII. THE MICROBIAL LOOP

In classic ecological food chains, organic C produced by photosynthetic organisms is consumed by herbivores, which are subsequently consumed by carnivores. In lentic aquatic ecosystems, how does matter move up the food chain to fish in systems where the primary producers (phytoplankton) are resistant to herbivory by zooplankton? The microbial loop mediates the transfer process (Fig. 4). Phytoplankton production in lakes enters the DOC pool due to excretion or lysis of algal cells. Most of the molecules are readily metabolized by heterotrophic bacteria and fuel the growth of these microbes. Heterotrophic protozoa (often flagellates) graze these bacteria. These protozoans are large enough to be grazed by metazoan zooplankton. In this way, photosynthetic primary production enters the classic food chain, mediated by intermediate steps involving heterotrophic bacteria and protozoa.

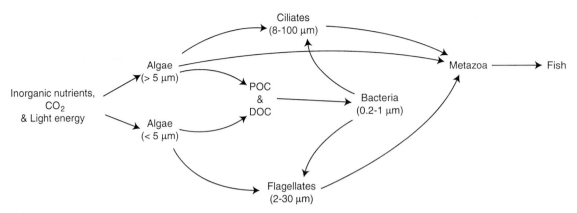

Fig. 4. The transfer of photosynthetic primary production to higher trophic levels in the food chain. Whereas production by large algae (>5 μm) can be consumed directly by metazoan zooplankton, which are, in turn, eaten by fish, dissolved organic carbon (DOC) excreted by large algae and cells of small algae (<5 μm) must be processed through a microbial loop consisting of bacteria and protozoa (ciliates and flagellates) before they can be consumed by metazoans. POC (particulate organic carbon) consists of deal algal cells or other particulate organic matter.

In streams, the role of the microbial loop is less obvious. Bacteria do consume DOC derived from algae and are, in turn, consumed by protozoa. However, the linkage beyond protozoan bacteriovores is not clear. Early instar insects and other stream meiofauna, such as nematodes, rotifers, and copepods can consume protozoa, but the quantitative importance of that link as a mechanism for transferring energy from microbes to insects and then fishes is not well understood. The microbial loop could function as a means of transferring energy to large consumers, or it could dissipate the energy through catabolic pathways and recycle organically bound minerals.

See Also the Following Articles

Biofilms and Biofouling • Heterotrophic Microorganisms • Sulfide-Containing Environments

Bibliography

Allan, J. D. (1995). "Stream Ecology. Structure and Function of Running Waters." Alden Press, Oxford, UK.

Ford, T. E. (1993). "Aquatic Microbiology: An Ecological Approach." Blackwell, Boston.

Gorlenko, V. M., Dubinina, G. A., and Kuznetsov, S. I. (1983). "The Ecology of Aquatic Micro-organisms." E. Schweizerbart'sche Verlagbuchhanadlung, Stuttgart, Germany.

Harris, G. P. (1986). "Phytoplankton Ecology." Chapman and Hall, London, UK.

Hutchinson, G. E. (1967). "A Treatise on Limnology. Vol. 2. Introduction to Lake Biology and the Limnoplankton." J. Wiley and Sons, Inc., New York.

Hynes, H. B. N. (1970). "The Ecology of Running Waters." University of Toronto Press, Toronto.

Lock, M. A. (1981). River Epilithon—a light and organic energy transducer. In "Perspectives in Running Water Ecology" (M. A. Lock and D. D. Williams, ed.), pp. 3–40. Plenum Press, New York.

Sommer, U. (1989). "Plankton Ecology: Succession in Plankton." Springer-Verlag, Berlin, Germany.

Vannote, R. L., Minshall, G. W., Cummins, K. W., Sedell, J. R., and Cushing, C. E. (1980). "The River Continuum Concept." *Canadian J. of Fisheries and Aquatic Sciences* 37, 130–137.

Whittaker, R. H., and Likens, G. E. (1973). Carbon in the biota. In "Carbon and the Biosphere" (G. M. Woodwell and E. V. Pecan, ed.), pp. 182–302. [Proceedings of the 24th Brookhaven Symposium in Biology, Upton, New York, May 16–18, 1972].

Fungal Infections, Cutaneous

Peter G. Sohnle and David K. Wagner

Medical College of Wisconsin and Milwaukee VA Medical Center

I. Cutaneous Host Defenses
II. Description of the Diseases

GLOSSARY

candidiasis An infection caused by fungal organisms of the genus Candida.

dermatophytosis A condition resulting from infection of keratinized structures, including hair, nails, and stratum corneum of the skin, by three genera of fungi termed the dermatophytes.

dermis A thick layer of skin tissue below the epidermis consisting of loose connective tissue and containing blood and lymph vessels, nerves, sweat and sebaceous glands, and hair follicles.

epidermis The outer layer of the skin, consisting of dividing and maturing epidermal cells with an outermost layer of dead, keratinized cells called the stratum corneum.

keratin An epidermal cell protein that makes up the hair, nails, and stratum corneum of the skin.

mycetoma Chronic cutaneous and subcutaneous infection resulting from direct implantation of actinomycetes or true fungi.

sporotrichosis A chronic infection of the skin, subcutaneous tissue, and sometimes deep tissues with the fungus *Sporothrix schenckii.*

tinea (pityriasis) versicolor An infection of the stratum corneum of the skin with the yeastlike fungal organism *Malassezia furfur.*

CUTANEOUS FUNGAL INFECTIONS encompasses a discussion of the major cutaneous host defense mechanisms and a description of the various superficial and deeper infections of the skin. The skin has a number of physical and chemical factors which make it difficult for microorganisms to survive and grow in this location. The immune system, particularly cell-mediated immunity and the activity of phagocytic cells, appears to be important in the defense against cutaneous fungal infections. Mechanisms involved in generating cutaneous immunologic reactions are particularly complex, with epidermal Langerhans cells, other dendritic cells, lymphocytes, microvascular endothelial cells, and the keratinocytes themselves all playing important roles. Infections involving the skin include both superficial and deep mycoses. The superficial mycoses are infections of the epidermis and cutaneous appendages with a number of yeasts and filamentous fungi that are well adapted for growth at this location. The resulting diseases include cutaneous candidiasis, dermatophytosis, tinea (pityriasis) versicolor, and some related mycoses. Deeper cutaneous mycoses, such as sporotrichosis, chromoblastomycosis, and mycetoma, may begin with direct implantation of the organisms into the skin through accidental punctures involving contaminated objects. The skin may also be involved by disseminated fungal infections such as North and South American blastomycosis, cryptococcosis, and several other types of fungal infections.

I. CUTANEOUS HOST DEFENSES

A. Structure of the Skin

The physical and chemical structure of the skin represents a form of defense against fungal pathogens. The skin surface is relatively inhospitable to fungal growth because of exposure to ultraviolet light, low moisture conditions, and competition from the normal bacterial flora of this site. Therefore, this surface acts as a barrier to the entry of fungi. The stratum corneum is made up of keratin, which most microorganisms cannot use for nutrition. However, *Candida albicans* and the dermatophytic fungi produce keratinases, which hydrolyze this substance and facilitate the growth of these organisms in the stratum corneum itself. This very superficial site of infection may protect the infecting organisms from direct contact with at least some of the effector cells of the immune system. Although neutrophils and small numbers of lymphocytes may enter the epidermis, the major infiltrates of cell-mediated immune responses are generally confined to the dermis.

B. Keratinization and Epidermal Proliferation

The process by which the stratum corneum is continually renewed through keratinization of the epidermal cells may also present a form of defense against organisms infecting this site. The basal epidermal cells produce continued growth of the epidermis as they undergo repeated cell divisions that move the resulting daughter cells (keratinocytes) outward, toward the surface. As they mature, these cells lose their nuclei and become flattened to form the keratinized cells. This process results in continuous shedding of the stratum corneum, which also may remove infecting fungal microorganisms residing there. Inflammation, including that produced by cell-mediated immune reactions, appears to enhance epidermal proliferation so that rates of transit of epidermal cells towards the stratum corneum are increased. A number of studies have demonstrated that epidermal proliferation is important in the defense against superficial mycoses.

C. Antifungal Substances

Lipids of adult hair contain saturated fatty acids that are fungistatic against *Microsporum audouini*, formerly a common cause of hair and scalp infections. In particular, various types of sphingosines have recently been found to be active against certain dermatophytes and *C. albicans*. Whereas the sebum of adults may not be significantly more fungistatic than that from children on a weight per weight basis, older individuals appear to produce quantitatively more of this material than do children. In addition, fungicidal proteins have been isolated from normal epidermis and could play some role in the defense against cutaneous fungal infections.

D. Unsaturated Transferrin

In contrast to other fungal pathogens such as *C. albicans*, *Malassezia furfur*, and *Trichosporon beigelii*, the dermatophytic fungi appear to be relatively incapable of causing disseminated disease, except for occasional local abscesses or granulomas in severely immunosuppressed patients. Thus, infections with dermatophytes are generally confined to the keratinized stratum corneum and the cutaneous appendages, such as the hair and nails. This phenomenon has been related to presence in the dermis of unsaturated transferrin, which may prevent growth of the organisms in the deeper layers of the skin by competition for iron.

E. The Inflammatory Response

With various kinds of superficial fungal infections, there appears to be an inverse relationship between the degree of inflammation produced by a particular fungal pathogen and the chronicity of that infection. *M. furfur* and the anthropophilic dermatophytes, *Trichophyton rubrum* and *Epidermophyton floccosum*, generally produce little inflammation in their cutaneous lesions and frequently cause infections that persist for long periods. On the other hand, many of the geophilic or zoophilic dermatophytes, e.g., *T. verrucosum*, produce highly inflammatory infections that are usually self-limited. Thus, the local inflammatory processes may indeed be involved in the defense against this group of pathogens.

A variety of mechanisms have been described by which inflammatory cells are attracted into the sites of cutaneous fungal infections. Fungal organisms are generally capable of activating complement by the alternative pathway to produce chemotactic activity for neutrophils and can also produce low molecular weight chemotactic factors analogous to the ones made by growing bacteria. Keratinocytes themselves can generate chemotactic cytokines that could also be responsible for some of the inflammation in the lesions of cutaneous fungal infections.

Neutrophils and monocytes/macrophages appear to be important in the defense against fungi, including those involved in the cutaneous mycoses. Neutrophils can directly attack pathogens using a variety of microbicidal processes. These processes depend upon either microbicidal oxidants or nonoxidative granule microbicidal substances. Most of these compounds have been studied primarily for their ability to kill the organisms, although lactoferrin may have both microbistatic and microbicidal effects. Macrophages have an additional antimicrobial mechanism whereby they can use production of nitric oxide to inhibit growth of ingested fungal pathogens. Neutrophils also appear to have significant growth inhibitory activity in addition to their microbicidal processes. These cells contain large amounts of a calcium- and zinc-binding protein, called calprotectin, that has potent microbistatic activity against *C. albicans* and other fungi.

F. The Cutaneous Immune System

Since cutaneous fungal infections are more frequent and more severe in patients with immunologic defects, immune responses to fungal antigens would seem to play an important role in the host defense against these infections. Immunologic host defense mechanisms in normal hosts seem to be effective even when the infections are limited to superficial locations, such as the stratum corneum. A number of studies suggest that the epidermis not only represents a passive barrier against entry of infecting organisms, but also acts as an immunologic organ with some unique elements. An hypothesis regarding the skin-associated lymphoid tissue (SALT) has been advanced wherein the skin acts as an immune surveillance unit. A variety of cell types are believed to have involvement in this cutaneous immune system, including epidermal Langerhans cells, dermal dendritic cells, epidermal T-lymphocytes, keratinocytes, and microvascular endothelial cells. The mechanisms employed are complex, involving a network of fixed or mobile cells interacting either by the trafficking of the cells themselves from one site to another or by the production of cytokines that influence the function of other cells. Such skin-initiated immune responses act against a broad spectrum of foreign antigens including contact allergens, tumors, and transplants, and it is likely that they are also active against the fungal pathogens of interest here. Therefore, this system is probably responsible for initiating immune responses that work to eliminate the infecting organisms in the immune host. In addition, such responses may also produce some of the inflammation that results in much of the symptomatology of these infections.

II. DESCRIPTION OF THE DISEASES

A. Superficial Fungal Infections

These infections are limited to the most superficial layers of the epidermis and/or its keratinized appendages, such as the hair and nails. The major cutaneous structures are shown in Fig. 1 and the most common pathogens causing superficial mycoses are listed in Table 1.

1. Cutaneous Candidiasis

Cutaneous candidiasis is an infection of the skin that is generally caused by the yeast-like fungus *C. albicans* and which can be either acute or chronic in nature. *C. albicans* is part of the normal flora of the gastrointestinal tract, rather than that of the skin, although it can be found on the skin on occasion. This organism can grow as either yeast cells or filamentous forms, with mixtures of the two phases generally seen in tissue infections.

Acute cutaneous candidiasis may present as intertrigo, producing intense erythema, edema, creamy exudate, and satellite pustules within folds of the skin. Other infections may be more chronic, as in the feet where there can be a thick white layer of infected stratum corneum overlaying the epidermis of the interdigital spaces. Candida paronychia is marked by infection of the periungual skin and the nail itself, resulting in the typical swelling and redness of this type of candida infection.

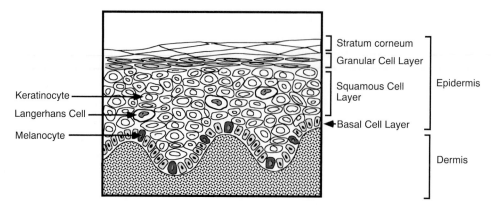

Fig. 1. Anatomy of the normal skin (reprinted with permission from Wagner and Sohnle, 1995, *Clin. Microbiol. Rev.* **8**, 317–335, American Society for Microbiology).

In some cases, superficial *C. albicans* infections may be particularly severe and recalcitrant to treatment, producing the uncommon disorder known as chronic mucocutaneous candidiasis. This condition consists of persistent and recurrent infections of the mucous membranes, skin, and nails, along with a variety of other manifestations. The superficial infections last for years in affected patients unless they are properly treated, although deep candida infections are very rare in this situation. Oral thrush and candida vaginitis are fairly common in patients with chronic mucocutaneous candidiasis. There is often infection of the esophagus, although further extension into the viscera is unusual. Epidermal neutrophilic microabscesses, which are common in acute cutaneous candidiasis, are rare in the lesions of chronic mucocutaneous candidiasis. The oral lesions are generally tender and painful. A number of other disorders are associated with the syndrome of chronic mucocutaneous candidiasis, including endocrine dysfunction, vitiligo, dysplasia of the dental enamel, congenital thymic dysplasia, thymomas, and certain other infections. Chronic mucocutaneous candidiasis, no doubt, represents a group of syn-

TABLE I
Superficial Cutaneous Mycoses and the Common Responsible Pathogens[a]

Type of infection	Pathogens
A. Cutaneous candidiasis	*Candida albicans*
B. Dermatophytosis	*Trichophyton, Microsporum, Epidermophyton*
1. Tinea pedia	*T. rubrum, T. mentagrophytes, E. floccosum*
2. Tinea cruris	*E. floccosum, T. rubrum, T. mentagrophytes*
3. Tinea barbae	*T. rubrum, T. verrucosum, T. mentagrophytes*
4. Tinea unguium (onychomycosis)	*T. rubrum, T. mentagrophytes, E. floccosum*
5. Tinea capitis	*T. tonsurans, T. schoenleini* (favus), *T. violaceum, M. canis*
6. Tinea corporis	*T. rubrum, T. mentagrophytes, T. concentricum, T. verrucosum, T. tonsurans, M. canis, M. gypseum, E. floccosum*
C. Tinea versicolor	*Malassezia furfur (Pityrosporum orbiculare)*
D. Malassezia folliculitis	*M. furfur*
E. Tinea nigra	*Phaeoannellomyces werneckii (Exophialia werneckii)*
F. White piedra	*Trichosporon beigelii*
G. Black piedra	*Piedraia hortae*

[a] Reprinted with permission from Wagner and Sohnle, 1995, *Clin. Microbiol. Rev.* 8, 317–335, American Society for Microbiology.

dromes with a variety of predisposing or secondary abnormalities in host defense function, most commonly deficient cell-mediated immune responses against candida antigens.

The diagnosis of superficial candidiasis is usually suspected on clinical grounds and can be confirmed in skin scrapings by demonstrating the organism using potassium hydroxide preparations and/or culture on appropriate antifugal media. Long term (3–9 months) treatment with azole antifungal drugs can produce good results in chronic mucocutaneous candidiasis, although occasional failures have occurred due to the development of resistant strains of *C. albicans*. Patients who present with chronic mucocutaneous candidiasis should be evaluated for the presence of infection with the human immunodeficiency virus and, if presenting as adults, for the possibility of thymoma.

2. Dermatophytosis

Dermatophytoses are infections of keratinized structures, such as the nails, hair shafts, and stratum corneum of the skin, by organisms of three genera of fungi termed the dermatophytes. Although they are not part of the normal human skin flora, these organisms are particularly well adapted to infecting this location because they can use keratin as a source of nutrients—unlike most other fungal pathogens. The different types of dermatophytosis are classified according to body site, using the word "tinea," followed by a term for the particular body site. The major types of dermatophytosis and the most frequent organisms associated with them are listed in Table 1. The degree of inflammation produced in the lesions appears to depend primarily on the particular organism and perhaps also to some extent on the immunological competence of the patient.

Tinea pedis (athlete's foot) is probably the most common form of dermatophytosis. This condition is a chronic toe web infection that can be scaly, vesicular, or ulcerative in form and which can sometimes produce hyperkeratosis of the sole of the foot. Tinea cruris is an expanding dermatophyte infection in the flexural areas of the groin and occurs much more frequently in males than in females. Dermatophytosis of the major surface areas of the body is termed tinea corporis. These infections frequently take the

classical annular, or "ringworm," shape. Involvement of the beard area in men, a condition known as tinea barbae, is often caused by zoophilic organisms such as *T. verrucosum*. Tinea unguium is a form of onychomycosis, or fungal infection of the nails, and is most frequently caused by *T. rubrum*. Nail infections, particularly of the toenails, are among the most difficult type of dermatophytosis to treat. Infection of the hair and skin on the scalp is called tinea capitis and is more common in children than adults. Dermatophytes rarely invade the deep tissues or produce systemic infections, even in severely immunocompromised patients.

Diagnosis of dermatophytosis is made in a similar manner to that of cutaneous candidiasis, with examination of skin scrapings by potassium hydroxide preparations and culture on appropriate fungal media. The treatment of this condition has improved markedly in recent years with the development of new antifungal agents for topical application or oral administration.

However, certain kinds of dermatophytosis, including widespread infections and those of hair and nails, will often respond poorly to topical therapy and will require prolonged courses of an oral antifungal agent, such as griseofulvin, ketoconazole, itraconazole, fluconazole, or terbinafine. There is an opportunistic fungal organism, *Scytalidium dimidiatum* (*Hendersonula toruloidea*), that can produce conditions clinically mimicking those caused by the usual dermatophyte species, but which does not respond well to conventional antifungal therapy.

3. Tinea (Pityriasis) Versicolor and Malassezia Folliculitis

Tinea versicolor is a chronic superficial fungal infection of the skin generally affecting the trunk or proximal parts of the extremities and caused by the yeast *M. furfur* (*Pityrosporum orbiculare*). The organism is lipid-requiring and will not grow on most laboratory media. The lesions resulting from infection with *M. furfur* are macules that may coalesce into large, irregular patches characterized by fine (pityriasiform) scaling, along with hypopigmentation or hyperpigmentation. These infections can persist for years unless treated appropriately. *M. furfur* has also been postulated to play a role in certain other diseases, including atopic dermatitis, seborrheic dermatitis, psoriasis, and reticulate papillomatosis. Malassezia folliculitis is a condition that resembles several other cutaneous infections, including acne vulgaris, the macronodular lesions of disseminated candidiasis in immunosuppressed patients, the candidal papular folliculitis of heroin addicts and graft versus host disease in bone marrow transplant recipients. The papules of this condition begin as an inflammation of the hair follicles, instead of the macules typical of tinea versicolor, and may progress to frank pustules.

In tinea versicolor, potassium hydroxide preparations of skin scrapings reveal the typical grape-like clusters of yeast and tangled webs of hyphae of the causative fungus, yielding the diagnosis of this condition. The organism is not usually cultured because of the requirement for specialized media. Tinea versicolor can be treated topically with lotions or creams containing selenium or sodium thiosulfate, specific antifungal agents, or sulfur–salicylic acid shampoo. Oral azole antifungal drugs can also be used for more difficult cases. Malassezia folliculitis can be treated using topical antifungal agents or an oral azole antifungal drug.

4. Miscellaneous Superficial Fungal Infections

Tinea nigra is a superficial mycosis of the palms that is most often caused by *Phaeoannellomyces werneckii* (*Exophiala werneckii*). The lesions are generally dark colored, non-scaling macules that are asymptomatic, but can be confused with melanomas and perhaps result in unnecessary surgery. Tinea nigra is most often seen in tropical or semitropical areas of Central and South America, Africa, and Asia, although some cases do occur in North America. This condition can be treated effectively with either keratinolytic agents or topical azoles. White piedra is an asymptomatic fungal infection of the hair shafts that is caused by *Trichosporon beigelii*. This infection produces light-colored, soft nodules on the hair shafts and may cause the involved hairs to break. Otherwise, this condition appears to be asymptomatic, although the causative fungus can produce serious infections in immunocompromised patients. Black piedra is similar to white piedra in that it is a

nodular, generally asymptomatic, fungal infection of the hair shafts. It is caused by *Piedraia hortae* and most commonly affects the scalp hair. Black and white piedra are generally treated by clipping off the affected hairs.

B. Deeper Cutaneous and Subcutaneous Mycoses

The dermis and subcutaneous tissues can be infected by a variety of fungal agents that are directly implanted into the skin by punctures with sharp objects contaminated by the organisms. The most common organisms causing the deep cutaneous and subcutaneous mycoses are listed in Table II.

1. Sporotrichosis

This condition is generally caused by accidental implantation of the causative fungus *Sporothrix schenckii* into the skin. The lesions most often consist of cutaneous and subcutaneous nodules extending up the limb from the site of inoculation. However, spread may occur through the lymphatics or blood vessels to the bones, joints, or other organs. It is also possible to develop lesions in the lungs by inhalation of the fungal elements. The causative organism is a dimorphic fungus that exists as either hyphae or elongated yeast cells.

The most common reservoir of the fungus in nature is on vegetation such as rose bushes, sphagnum moss, or in soil. The site of implantation may develop into a papule or pustule and cutaneous nodules may then develop proximally in a linear fashion. If the fungus is inhaled, it may cause a granulomatous pneumonitis that can cavitate and produce a clinical picture similar to tuberculosis. Immunosuppressed patients are more likely to develop disseminated disease.

Demonstration of the characteristic small, cigar-shaped yeast cells is diagnostic but often difficult. Multiple sections may have to be examined. Stellate, periodic acid-Schiff (PAS) positive eosinophilic material surrounding the organisms are known as asteroid bodies. The diagnosis of sporotrichosis is best made by culture of material from the lesions on appropriate fungal media. Isolation of this organism is usually indicative of sporotrichosis in that the fungus

TABLE II
Deep Cutaneous and Subcutaneous Mycoses and the Common Responsible Pathogens

Type of infection	Pathogens
A. Sporotrichosis	*Sporothrix schenckii*
B. Chromoblastomycosis	*Fonsecaea pedrosoi, F. compacta, Phialophora verrucosa, Cladophialophora carrionii, Botryomyces caespitosus, Rhinocladiella aquaspersa*
C. Eumycotic mycetoma	*Pseudallescheria boydii, Madurella mycetomatis, M. grisea, Acremonium spp., Leptosphaeria senegalensis*
D. North American blastomycosis	*Blastomyces dermatitidis*
E. South American blastomycosis	*Paracoccidioides braziliensis*
F. Histoplasmosis	*Histoplasma capsulatum*
G. Cryptococcosis	*Cryptococcus neoformans*
H. Infections with immunosuppression	*Trichosporon beigelii, Blastoschizomyces capitatus, Fusarium spp.*

is not part of the normal flora of humans. Iodides may be given for cutaneous sporotrichosis, with oral itraconazole or fluconazole being used if these measures fail. For disseminated disease, either amphotericin B or itraconazole are generally effective, although relapse is common.

2. Chromoblastomycosis

Certain species of the dematiaceous (darkly pigmented) fungi can cause chronic cutaneous and subcutaneous infections. A number of genera can be involved, but *Fonsecaea, Phialophora, Cladophialophora*, and *Rhinocladiella* are most common. The dark pigment of these organisms is dihydroxynaphthalene melanin, which is different from the dihydroxyphenylanine melanin associated with *Cryptococcus neoformans*. Dematiaceous fungi can also cause mycetoma. Chromoblastomycosis is characterized by the presence of sclerotic (muriform) bodies in the tissues. When yeastlike cells, pseudohyphae, or hyphae of the dematiaceous fungi are present in the tissues, the term "phaeohyphomycosis" is used. Chromoblastomycosis usually results from implantation of the organisms from local trauma, usually to the feet or legs. Usually, the first lesion is an erythematous papule, followed by scaling and crusting, with eventual development into a warty structure. The pathology is characteristic of a suppurative granuloma, often with overlying pseudoepitheliomatous hyperplasia. The distribution of cases is worldwide, although most come from Central and South America.

The finding of the characteristic cross-walled, pigmented sclerotic bodies is pathognomonic of chromoblastomycosis. However, since those formed by all the relevant dematiaceous fungal species are similar, culture of the infecting organism on fungal media containing cycloheximide and antibiotics is necessary to identify it. Treatment may be difficult in that the organisms may not be sensitive to antifungal agents. Surgery or local heat may be other options in the early stages of the disease.

3. Mycotic Mycetoma

Mycetomas are swellings with draining sinuses and grains. They usually affect the feet, legs, or hands and begin with direct implantation of the causative organisms. The latter are either actinomycetes (actinomycotic mycetoma) or the true fungi (eumycotic mycetoma). About half of the cases of mycetoma are caused by true fungi, including the genera of *Madurella, Leptosphaeria, Pseudoallescheria, Acremonium*, and several others. Initially, pain and discomfort develop at the implantation site, followed weeks or months later by induration, abscess development, granulomas, and draining sinuses. The lesions may extend to bone and cause severe bony destruction. Eumycotic mycetoma are rare in the United States, although those caused by *Pseudallescheria boydii* are more common in the latter location.

Specimens of exudate or biopsy material should be examined by the naked eye for the presence of grains. The latter can be gram-stained and examined microscopically to differentiate actinomycotic from eumycotic mycetoma. The grains should be washed with saline containing antibacterial compounds and then cultured on Sabouraud dextrose agar containing chloramphenicol and cycloheximide, as well as on media for bacterial and actinomycotic organisms. Identification of the organisms is based on gross colonial morphology, pigmentation, and mechanism of conidiogenesis. Treatment of eumycotic mycetoma is often unsatisfactory because the causative organisms generally show poor sensitivity to available antifungal agents. Amputation of an infected limb or surgical debridement of infected tissue may be necessary. Amphotericin B or azole antifungal drugs can be used if the particular fungal strain is sensitive. If not treated effectively, mycetomas may progress for years and produce marked tissue damage, deformity, and even death.

C. Systemic Mycoses with Cutaneous Manifestations

A number of deep fungal infections may produce cutaneous lesions as part of a disseminated disease process. In these infections, the portal of entry is usually the lung, with development of a pneumonia and spread to other organs. Dissemination is most likely to occur in patients with compromised host defenses, although blastomycosis has a very high incidence of skin lesions in noncompromised individuals. The group of infections under discussion here is different from those in the last section, where direct implantation into the skin or subcutaneous tissues is the usual mode of entry.

North American blastomycosis a systemic fungal infection well known to have cutaneous dissemination. Skin involvement occurs in approximately 60% of cases at some time during the illness, even though the pulmonary lesions may have healed at the time the skin manifestations develop. The latter generally consist of either verrucous lesions that look like squamous cell carcinomas or ulcerative lesions that begin as pustules or ulcerating nodules. The patient may either be quite ill with symptoms and signs from other manifestations of the infection or may have only skin lesions and be relatively asymptomatic otherwise. Coccidioidomycosis is a fairly similar disease

which can produce papules, pustules, plaques, nodules, ulcers, abscesses, or large proliferative lesions. Dissemination to the skin is less common in coccidioidomycosis than in blastomycosis, although chronic meningitis is a more prominent feature of the former. The respective organisms of the two diseases are *Blastomyces dermatitidis* and *Coccidioides immitis;* they are endemic fungi found in different parts of the United States. A similar endemic fungal infection is histoplasmosis, caused by *Histoplasma capsulatum;* however, this disease is less likely to have cutaneous manifestations.

South American blastomycosis (paracoccidioidomycosis) is an important systemic mycosis in Latin America; this infection begins in the lungs and then frequently disseminates to the skin, mucous membranes, reticuloendothelial system, and elsewhere. It is caused by *Paracoccidioides braziliensis*. The cutaneous lesions tend to appear around the natural orifices and may be verrucous, ulcerating, crusted, or indurated and granulomatous. Cryptococcosis is an infection beginning in the lungs and caused by *Cryptococcus neoformans*. Like coccidioidomycosis, this disease tends to disseminate to the central nervous system to cause a chronic meningitis, but it can also cause skin lesions. The latter are generally papules or nodules with surrounding erythema; they are often found on the face. Severely immunosuppressed patients are at risk for disseminated disease with a number of other fungi such as *T. beigelii* (the cause of white piedra, discussed previously), *Blastoschizomyces capitatus*, and *Fusarium spp*. Each of these agents causes erythematous papules or nodules, and, in some cases, these lesions may break down to form ulcers.

Diagnosis of these various infections is usually obtained by demonstrating the organisms in specimens of sputum, blood, or biopsy material from the skin or other organs. Often, the individual fungi can be identified on appropriately stained smears or on histologic sections from biopsy material; otherwise, they may be cultured on appropriate fungal media. The latex agglutination test for cryptococcal antigen in serum or cerebrospinal fluid is very helpful for diagnosing and following cryptococcosis. Skin tests and other serological tests are generally less useful in diagnosis of the other fungal infections discussed above than is demonstration of the organisms. Treatment usually can be accomplished through the use of amphotericin B or the azole antifungal drugs. The various organisms vary somewhat in susceptibility, with *Coccidioides* and *Fusarium* being relatively resistant and *Cryptococcus* and *Blastomyces* being more sensitive to the various agents. The presence of immunosuppression makes treatment of the various diseases much more difficult.

See Also the Following Articles

ANTIFUNGAL AGENTS • SKIN MICROBIOLOGY

Bibliography

Ajello, L. (1974). Natural history of the dermatophytes and related fungi. *Mycopathol. Mycol. Appl.* **53**, 93–110.

Aly, R., and Berger, T. (1996). Common superficial fungal infections in patients with AIDS. *Clin. Infect. Dis.* **22** (Suppl 2), S128–132.

Elgart, M. L., and Warren, N. G. (1992). The superficial and subcutaneous mycoses. *In* "Dermatology," 3rd ed. (S. L. Moschella and H. J. Hurley, eds.), pp. 869–912. Saunders, New York.

Filler, S. G., and Edwards, J. E., Jr. (1993). Chronic mucocutaneous candidiasis. *In* "Fungal Infections and Immune Responses" (J. W. Murphy, H. Friedman and M. Bendinelli, ed.), pp. 117–133. Plenum Press, New York.

McGinnis, M. R., and Rinaldi, M. G. (1997). Some medically important fungi and their common synonyms and obsolete names. *Clin. Infect. Dis.* **25**, 15–17.

Stingl, G., Hauser, C., and Wolff, K. (1993). The epidermis: An immunologic microenvironment. *In* "Dermatology in General Medicine," 4th ed. (T. B. Fitzpatrick, A. Z. Eisen, K. Wolff, I. M. Freedberg, and K. F. Austen, eds.), pp. 172–197. McGraw-Hill, New York.

Wagner, D. K., and Sohnle, P. G. (1995). Cutaneous defenses against dermatophytes and yeasts. *Clin. Microbiol. Rev.* **8**, 317–335.

Fungal Infections, Systemic

Arturo Casadevall

Albert Einstein College of Medicine

GLOSSARY

antifungal drugs   Drugs used to treat fungal infections. Antifungal agents can be fungicidal or fungistatic, depending on their ability to kill or inhibit the growth of the fungal pathogen, respectively. Examples of currently used antifungal drugs include Amphotericin B, fluconazole, ketoconazole, itraconazole, and 5-fluorocytosine.

impaired immunity   A defect in host defense from any cause. Most systemic fungal infections occur in patients with disorders of cellular immunity. However, host defense against fungi is complex and requires both specific and nonspecific humoral and cellular immunity. Individuals with impaired immunity are said to be immunocompromised or immunosuppressed.

opportunistic   Causing disease in patients with impaired immunity. Many fungal infections discussed in this article have been termed "opportunistic" because they are more likely to cause disease in patients with impaired immunity. However, several of the major fungal pathogens can also cause life-threatening systemic infection in patients with no obvious immune disorder, although this happens rarely.

portal of entry   The site at which the pathogen enters the host and establishes the primary infection. For example, for *Cryptococcus neoformans, Histoplasma capsulatum, Blastomyces dermatitides,* and *Coccidioides immitis,* the portal of entry is generally considered to be the lung. Most fungal infections are contained at the portal of entry, but dissemination can occur, especially in patients with impaired immunity.

systemic fungal infection   An infection that extends to involve internal organs or tissues. For example, candidal vaginitis or oral thrush would not be considered systemic fungal infections because they are limited to the vaginal and oral mucosa, respectively. However, under certain conditions, *Candida* can invade the bloodstream and disseminate to various organs to cause disseminated candidiasis, a systemic fungal infection. Many fungal pathogens can cause both localized and disseminated (or systemic) infections. The probability that the infection becomes systemic is usually dependent on the immunological status of the host. Systemic fungal infections are more common in patients with impaired immunity.

SYSTEMIC FUNGAL INFECTIONS are a major cause of mortality and morbidity in patients with impaired host defenses. A systemic fungal infection is one where the fungus invades past the skin and/or mucosal membranes to involve internal organs. The fungi responsible for systemic fungal infections are a biologically diverse group of species. Of the more than 100,000 fungal species known, only about 150 have been associated with human disease and, of these species, only 10 to 15 are commonly encountered in clinical practice. Table 1 lists some of the major fungal systemic fungal infections.

Systemic fungal infections were relatively rare until the mid-20th century when advancements in medicine led to the discovery of immune suppressive therapies and broad spectrum antibiotics. After 1950, the use of corticosteroids for the treatment of several medical conditions was associated with an increase in serious fungal infections as a consequence of corticosteroid-induced immune suppression. The use of

TABLE I
Some Systemic Fungal Infections and Their Causative Agents

Infection	Pathogen	Risk factors	Geographic prevalence
Aspergillosis	*Aspergillus* sp.	Neutropenia, late stage HIV infection, immunosuppressive therapy	Worldwide
Blastomycosis	*Blastomyces dermatitides*	Immunosuppressive therapy, late stage HIV infection. Can occur in normal individuals	North America and Africa
Candidiasis	*Candida* sp.	Neutropenia, immunosuppressive therapy, antibiotic use, surgical procedures, indwelling catheters	Worldwide
Coccidioidomycosis	*Coccidioides immitis*	Late stage HIV infection, pregnancy, certain ethnic groups. Can occur in normal individuals	Certain areas of North and South America
Cryptococcosis	*Cryptococcus neoformans*	Late stage HIV infection, corticosteroid use, lymphoproliferative malignancies. Can occur in normal individuals	Worldwide
Histoplasmosis	*Histoplasma capsulatum*	Late stage HIV infection, corticosteroid therapy. Can occur in normal individuals	Prevalent in certain areas of North and South America and Africa.
Mucormycosis	*Rhizopus* sp.	Diabetic ketoacidosis, neutropenia	Worldwide
Paracoccidioidomycosis	*Paracoccidioides brasiliensis*	Late stage HIV infection. Can occur in normal individuals	South America and Mexico
Penicilliosis	*Penicillium marneffei*	Late stage HIV infection, immunosuppressive therapy	Southeast Asia
Sporotrichosis	*Sporothrix schenckii*	Late stage HIV infection. Can occur in normal hosts	Worldwide

broad spectrum antibiotic drugs contributed to an increased frequency of Candidiasis because these drugs can alter the balance of the host microbial flora and promote the growth of fungi. Similarly, medical progress allowed some individuals with life-threatening illnesses to survive at the price of impaired host defenses. For example, patients with major surgery, malignancies, organ transplants, and autoimmune diseases were often at high risk for fungal infection as a consequence of therapeutic interventions. Another major contributor to the increased prevalence of fungal infections in recent years has been the epidemic of HIV infection. Patients with advanced HIV infection have profound immunological derangements that place them at high risk for acquiring any of a variety of fungal pathogens (Table I). Since systemic fungal infections are rare in patients with normal immune function, the prevalence of systemic fungal infections in a human population may be an index of its overall immunological health.

The label "opportunistic" is often applied to systemic fungal infections because these tend to occur more frequently in patients with impaired immunity. However, this label is inexact because many fungal pathogens cause systemic infections in patients with normal immune function. For example, *Cryptococcus neoformans* and *Histoplasma capsulatum* cause infections in apparently normal individuals but are much more likely to cause systemic infections in populations with immune disorders. In contrast, systemic candidiasis is almost always associated with a breakdown in host defenses and *Candida* is a true opportunistic pathogen. The use of the term opportunistic is more applicable when used to refer to a fungal infection in the setting of impaired immunity than to specific types of fungal pathogens.

The problem of invasive fungal infections is compounded by a relative paucity of antifungal drugs and, for many fungal pathogens, inadequate diagnostic tests. Amphotericin B remains the most effective antifungal drug against many fungal infections four decades after its introduction. Because systemic fungal infections tend to occur in individuals with immunological deficits, many become chronic and respond slowly to antifungal therapy. In fact, some infections like cryptococcosis, histoplasmosis, and coccidioidomycosis are essentially incurable when they occur in the setting of advanced HIV infection.

I. ACQUISITION AND INFECTION

Unlike other infectious diseases human-to-human transmission of systemic fungal infections is exceedingly rare. Except for *C. albicans* infections, which are usually acquired from the endogenous microbial flora, the overwhelming majority of fungal infections are acquired from the environment. Epidemics and outbreaks are often associated with unusual exposures to the pathogen or disruptions in the ecosystem. For example, there have been well documented descriptions of local epidemics of histoplasmosis associated with cutting trees or construction projects. The probability of acquiring a systemic fungal infection is usually a function of exposure and host immune function. Because systemic fungal infections are rarely communicable and because they tend to occur in groups with defined risk factors, most fungal pathogens have attracted relatively little attention from public health authorities. As a result, most fungal infections are not reportable to or the subject of active surveillance by public health authorities. Therefore, we have incomplete information on the incidence and prevalence of most types of systemic fungal infections.

Among the common systemic fungal infections only candidiasis, cryptococcosis, and aspergillosis have worldwide distribution. Since candidiasis is usually an endogenous infection, it can be found wherever humans live. For cryptococcosis and asper-gillosis, the worldwide occurrence of these infections reflects the fact that these fungi are prevalent throughout the world. In contrast, histoplasmosis, penicilliosis, paracoccidioidomycosis, blastomycoses, and coccidioidomycosis are found only in geographic areas of the world where those pathogens are found in the environment. Many fungal pathogens can induce a latent asymptomatic infection after primary infection. Latent infections can reactivate years later, especially if the individual develops a condition associated with impaired immunity. Latent infections combined with routine air travel means that individuals can present with systemic fungal infections in nonendemic areas. For example, an individual could develop coccidioidomycosis or histoplasmosis in New England (a nonendemic area) after living or visiting in the Southwest or Midwest, respectively.

II. HOST DEFENSES AND SUSCEPTIBILITY TO INFECTION

With some notable exceptions, which include candidiasis and sporotrichosis, most systemic fungal infections are initially acquired by inhaling infectious particles. Inhalation of infectious particles initially results in a localized pulmonary infection, which is usually contained in the lung by a granulomatous inflammatory response. However, dissemination can occur, especially in the setting of impaired immune function. The pathogenesis of some fungal, infections, such as histoplasmosis, cryptococcosis, blastomycosis, and coccidioidomycosis, has similarities to tuberculosis, including a pulmonary portal of entry and the potential for latent infection. Host defenses include humoral and cellular specific and nonspecific immune mechanisms. Animal experimentation has established that multiple components of the immune system are important for protection against fungi, including complement, antibody, neutrophils, macrophages, and T lymphocytes. Invasive fungal infections are usually associated with derangements of cellular immunity. For example, patients with HIV infection are at high risk for cryptococcosis, histoplasmosis, and coccidioidomycosis when their blood CD4+ T cell lymphocyte counts are below 200/mm^3.

Invasive aspergilloses and candidiasis are associated with neutrophil deficiencies, such as those caused by cancer therapy. There is general consensus in the field that cellular immune mechanisms play a critical role in containment and eradication of most fungal infections. Invasive fungal infections can occasionally occur in hosts with no apparent immunological deficits but it is not clear whether these cases are the result of unrecognized immunological disorders, unusual exposures, or more virulent fungal strains.

III. INDIVIDUAL SYSTEMIC FUNGAL INFECTIONS

A. Aspergillosis

Systemic aspergillosis is caused by *Aspergillus* species (sp.) of which the most common are *A. fumigatus* and *A. flavus*. Most cases of human aspergillosis are believed to occur after inhalation of airborne conidia. However, infection following surgery can also follow the deposition of airborne spores in exposed tissues. *Aspergillus* sp. are very common in the environment and can be found in soil, plants, and decaying vegetation. Conidia germinate in the lung and can invade host tissue if not controlled by local defenses. The invasion can also originate from colonization of paranasal sinuses, the gastrointestinal tract, or the skin. Invasive aspergillosis is a devastating disease of severely immunocompromised patients. Individuals at high risk include those with leukemia, burns, and late-stage HIV infection, as well as those using intravenous drugs and receiving immunosuppressive therapies. Since most *Aspergillus* infections are acquired from the ambient air, precautions to limit the number of airborne spores, such as laminar flow rooms and filtered air, may reduce the risk of infection. In patients with neutropenia, *Aspergillus* hyphae often invade and extend along blood vessels, causing tissue infarction and necrosis. A diagnosis of aspergillosis is usually made by culture and pathological examination of tissue samples in the appropriate clinical setting. A major problem with aspergillosis is the absence of good diagnostic tests for the detection of early infection. Systemic aspergillosis has high mortality despite antifungal therapy.

B. Blastomycosis

Systemic blastomycosis is caused by the dimorphic fungus *Blastomyces dermatitides*. In North America, the infection is endemic in the Mississippi and Ohio river valleys and areas adjoining the Great Lakes and the St. Lawrence River. Blastomycosis also occurs in Africa, Central America, and the Middle East. *B. dermatitides* is believed to reside in soils and decaying vegetation but its exact environmental niche remains to be defined. Initial infection is believed to occur from inhalation of conidia that germinate to yeast cells in the lung. Several outbreaks of blastomycosis have been associated with recreational activities in sites near rivers. In immunologically normal individuals, the infection is contained in the lung and humans appear to have a high level of resistance to disseminated infection. Systemic blastomycosis results when pulmonary blastomycosis disseminates to other organs. Dissemination can occur to practically any organ but the skin is the site most commonly involved. The diagnosis of blastomycosis infection is made by culture or pathological examination of involved tissue in the appropriate clinical setting. Blastomycosis is also a major fungal infection in dogs.

C. Candidiasis

Systemic candidiasis can be caused by any of several *Candida* sp. and is the most common systemic fungal infection. Systemic candidiasis is distinguished from the more common type of candidal infections, such as thrush and vaginal candidiasis, by involvement of the blood stream and internal organs. The name *Candida* is used to refer to more than one hundred fungal species, of which about a dozen are important human pathogens. *C. albicans* is the most common *Candida* sp. that causes human disease. However, other species, such as *C. glabrata, C. parapsilosis* and *C. krusei,* commonly cause serious infections in certain patient groups. *Candida* sp. are components of the human microbial flora and systemic candidiasis differs from the other systemic fungal infections in that this infection almost always originates from the endogenous microbial flora. Candidal infections are very common in hospitalized patients and *Candida* species are frequent causes of bacteremia and deep-seated organ infections. Most cases of systemic *C. albicans* infection are iatrogenic and are related to the use of antibiotics, the presence of indwelling catheters, and surgical implantations of prosthetic devices. Antibiotics predispose to infection by suppressing the normal bacterial flora and this, presumably, allows *Candida* to proliferate and invade the gut mucosa. Indwelling catheters provide a break in the skin that allows *Candida* access to the bloodstream and the internal skin layers. Prosthetic devices provide surfaces where the fungus can attach and escape clearance by host defense mechanisms. Systemic candidal infections can affect virtually any organ and the eye, liver, spleen, and kidneys are frequently involved. The diagnosis of systemic candidiasis is usually made by culture of *Candida* sp. from a normally sterile body site. However, in many patients, making a diagnosis of systemic candidiasis is difficult because the blood cultures are negative and there are few diagnostic tests to detect deep-seated organ infections.

D. Coccidioidomycosis

The causative agent of coccidioidomycosis is *Coccidioides immitis,* a dimorphic soil fungus. This infection is found in certain areas of North and South America. In the United States, coccidioidomycosis is highly prevalent in the Southwest. This infection is believed to be acquired by the inhalation of arthroconidia that swell in tissue to become spherule-containing internal spores. Primary infection is usually either asymptomatic or reassembles a common upper respiratory infection. Individuals with occupations that involve exposure to soils in endemic areas are at higher risk. Outbreaks of coccidioidomycosis have followed archeological investigations. Systemic coccidioidomycosis occurs in less than 1% of primary infections and can involve virtually any organ. Extrapulmonary coccidioidomycosis is more common in individuals with impaired immunity. However, pregnant women and individuals of Filipino, African, and Mexican ancestry might be at increased risk for disseminated infection. Coccidioidomycosis can present as acute infection, chronic pulmonary disease, or systemic infection. The diagnosis is made

by culture, histological examination of tissue, and/or serological testing.

E. Cryptococcosis

The causative agent of cryptococcosis is *Cryptococcus neoformans.* The prevalence of *C. neoformans* infection has risen dramatically in recent years in association with HIV infection. In New York City, there were more than 1200 cases of cryptococcosis in 1991, of which most occurred in patients with AIDS. Cryptococcal meningitis occurs in 5–10% of all patients with advanced HIV infection. Cryptococcal strains have been divided into two varieties known as *neoformans* and *gattii.* Variety *neoformans* is found worldwide, is associated with bird (usually pigeon) excreta, and is the predominant agent of cryptococcosis in patients with AIDS. Variety *gattii* is found in the tropics, is associated with eucalyptus trees, and can cause infections in apparently normal hosts. *C. neoformans* is unusual among fungal pathogens, in that it has a polysaccharide capsule that is important for virulence (Fig. 1). *C. neoformans* is acquired by inhalation, where it usually causes an asymptomatic pulmonary infection. In patients with impaired immunity, extrapulmonary dissemination can results in cryptococcal meningitis, the most common clinical presentation of cryptococcosis. Pathological examination of tissues infected with *C. neoformans* often reveals little or no inflammatory response and this phenomenon is believed to be caused, in part, by the immunosuppressive effects of the capsular polysaccharide. The diagnosis of cryptococcosis is usually made by culture from cerebrospinal fluid or blood. The capsular polysaccharide is shed into body fluid where it can be detected by serological assays. Detection of cryptococcal polysaccharide antigen is useful in diagnosis and in following the response to therapy.

F. Histoplasmosis

The causative agent of histoplasmosis is *Histoplasma capsulatum,* a soil organism that is common in the Ohio and Mississippi river valleys and in various parts of South America. *H. capsulatum* is often

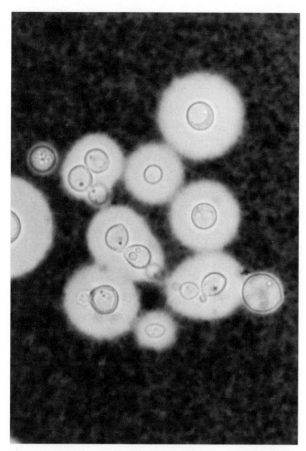

Fig. 1. India-ink preparation showing *C. neoformans* cells. The cell bodies are surrounded by a polysaccharide capsule which is important for virulence. Note differences in cell size and capsule size among the various cells shown. Photograph obtained at a magnification of ×200.

found in soils contaminated by bird excrement. Small epidemics have occurred because of large exposures created by disturbing contaminated sites during constructions, excavations, tree cuttings, etc. In 1978–1980, two major outbreaks of histoplasmosis occurred in the city of Indianapolis that may have been related to construction projects. *H. capsulatum* is a dimorphic fungus that grows as a mycelial form at environmental temperatures and as a yeast at mammalian body temperature. Two varieties of *H. capsulatum* are known: variety *capsulatum* is found in the Americas and variety *duboisii* is found in Africa. Infection presumably occurs by the inhalation of conidia and small mycelial fragments that convert to

yeast forms in lung tissue. The clinical presentation of pulmonary histoplasmosis is similar to that of pulmonary tuberculosis, such that many cases of histoplasmosis were confused with tuberculosis until specific diagnostic methods became available. The overwhelming majority of primary pulmonary infections are asymptomatic. However, many infections become chronic and some disseminate. The probability of acquiring disseminated disease ranges from about 1 in 2000 for normal individuals to up to 27% for patients with advanced HIV infection. Disseminated disease can affect virtually any organ, with adrenal, skin, gastrointestinal, and central nervous system involvement being particularly common. The diagnosis of systemic histoplasmosis is made by culture of the fungus from blood, bone marrow, sputum, cerebrospinal fluid, or the affected body site. However, culture methods can yield false-negative results. A presumptive diagnosis can be made by visualizing yeast cells in infected tissues. In this regard, *H. capsulatum* can sometimes be detected in peripheral blood leukocytes of patients with disseminated histoplasmosis. Serological tests for histoplasma antibody and antigen can provide important clues to the presence of *H. capsulatum* infection. Antigen detection in urine samples is particularly useful in cases where the diagnosis is suspected but cultures are negative.

G. Other Systemic Fungal Infections

Many other fungal species can cause invasive infection besides the more common systemic fungal infections listed above. In Southeast Asia, *Penicillium marneffei* is a major cause of invasive fungal infection in patients with advanced HIV infection. *Paracoccidioides brasiliensis* is a major fungal pathogen in some areas of South America. *P. brasiliensis* causes asymptomatic infection in normal individuals that can remain latent and disseminate if the immune system subsequently becomes impaired. Recently, there have been several reports of systemic fungal infection with the *Saccharomyces cerevisiae,* which is commonly known as brewer's or baker's yeast. Although rare, these cases illustrate how a usually nonpathogenic organism can cause serious infection if it colonizes a susceptible host. *Pseudallescheria boydii* is a mold that causes severe infections in patients with prolonged neutropenia or who are receiving high-dose corticosteroid therapy. *P. boydii* infections are similar to those caused by *Aspergillus* sp. *Sporothrix schenckii* is a dimorphic fungus, found in soils and plants, that can cause sporotrichosis after inoculation in the skin. Sporotrichosis has been reported throughout the world but is associated primarily with activities that result in exposure to plants, such as gardening and farming. Mucormycosis is caused by a variety of fungal species with a complex taxonomy, including *Rhizopus, Absidia,* and *Mucor.* In diabetic patients with poorly controlled hyperglycemia, mucormycosis is a devastating and often incurable infection. Phaeohyphomycosis is caused by a variety of fungal species that have dark cell walls. Cerebral phaeohyphomycosis is a rare but rapidly lethal brain infection that has been commonly associated with the fungus *Clasdosporium trichoides.*

IV. TREATMENT AND PREVENTION OF SYSTEMIC FUNGAL INFECTIONS

Most systemic fungal infections are fatal without antifungal therapy. Unlike the situation for bacterial pathogens, the antibiotic arsenal against the fungi is small and consists of no more than half a dozen drugs. Since the late 1950s, Amphotericin B has been the mainstay of therapy for many invasive mycoses. Amphotericin B is a powerful fungicidal agent that binds to fungal sterols and kills the fungal cell by disrupting cellular membranes. Amphotericin B may also have important immunomodulatory effects that could contribute to its therapeutic efficacy. Amphotericin B has significant toxicity that can be lessened by incorporating it into liposomal preparations but these are significantly more expensive. In recent years, other agents that target the sterol metabolic pathways have been introduced, including fluconazole and itraconazole. These agents are usually fungistatic but are much less toxic than amphotericin B and have the added advantages of being available in oral formulations. 5-Fluorocytosine is another antifungal drug that is effective when used in combination with other antifungal drugs. Several antifungal agents are in preclinical and clinical development

and newer agents with enhanced efficacy and reduced toxicity may be available in the future.

A major problem in the therapy of systemic fungal infections is that antifungal chemotherapy is less effective in the setting of defective immunity. For example, antifungal therapy cannot usually eradicate *C. neoformans, H. capsulatum,* and *C. immitis* infections in patients with advanced HIV infection. As a result, affected individuals must be given lifelong suppressive therapy to reduce the likelihood of clinical recurrence of infection. The difficulties associated with the therapy of systemic fungal infections have stimulated interest in immunotherapy but this therapeutic strategy is still experimental.

There are ongoing efforts to develop vaccines against coccidioidomycosis, histoplasmosis, and cryptococcosis but none is currently available. At this time, the two main strategies for the prevention of systemic infection in patients at risk for infection include avoidance of infection and the use of prophylactic antifungal drugs. For many systemic fungal infections, prevention is difficult because the fungal pathogen is highly prevalent in the environment. For example, *C. neoformans* is found in high concentration in pigeon excreta in urban areas such as New York City, where many patients with advanced HIV infection live. Similarly, *H. capsulatum* and *C. immitis* are prevalent in soils of specific geographic areas of the world and avoiding exposure may be difficult for residents in those regions. Nevertheless, it is prudent for individuals with immunological disorders to avoid sites likely to contain high concentrations of aerosolized fungal pathogens such as construction sites, aviaries, chicken farms, and compost sites. Prophylactic administration of antifungal drugs has been shown to reduce the incidence of certain fungal infections, such as cryptococcosis, in patients at high risk for infection. However, there are concerns that prophylactic drug use will encourage the selection of drug-resistant fungi, and drug prophylaxis is not used routinely for the prevention of fungal infection. Another preventive strategy to reduce invasive fungal infections in patients with neutropenia is to administer colony-stimulating factors that reduce the neutropenic interval by stimulating leukocyte production.

See Also the Following Articles

Antifungal Agents • Fungal Infections, Cutaneous • Surveillance of Infectious Diseases

Bibliography

Clemons, J. V., McCusker, J. H., Davis, R. W., and Stevens, D. A. (1994). Comparative pathogenesis of clinical and nonclinical isolates of *Saccharomyces cerevisiae. J. Infect. Dis.* **169**, 859–867.

Dixon, D. M., McNeil, M. M., Cohen, M. L., Gellin, B. G., and LaMontagne, J. R. (1996). Fungal infections. A growing threat. *Public Health Rep.* **111**, 226–235.

Fridkin, S. K., and W. R. Jarvis. (1996) Epidemiology of nosocomial fungal infections. *Clin. Microbiol. Rev.* **9**, 499–511.

Hazen, K. C. (1995). New and emerging yeast pathogens. *Clin. Microbiol. Rev.* **8**, 462–478.

Kwon-Chung, K. J., and Bennett, J. E. (1992)," Medical Mycology." Lea & Fabiger, Philadelphia.

Romani, L., and Howard, D. H. (1995). Mechanisms of resistance to fungal infections. *Curr. Opin. Immunol.* **7**, 517–523.

Wheat, J. (1992). Histoplasmosis in Indianapolis. *Clin. Infect. Dis.* **14**(Suppl 1), S91–99.

Vartivarian, S. E. (1992). Virulence properties and nonimmune pathogenetic mechanisms of fungi. *Clin. Infect. Dis.* **14**(Suppl 1), S30–36.

Fungi, Filamentous

J. W. Bennett

Tulane University

GLOSSARY

ascus (pl. asci)   A saclike cell containing ascospores (usually eight in number), formed after nuclear fusion (karyogamy) and meiosis.

basidium (pl. basidia)   An enlarged clublike cell that bears basidiospores (usually 4 in number), formed after nuclear fusion (karyogamy) and meiosis.

dimorphism   The ability of a fungus to grow as a yeast or a filament; sometimes also used to describe fungi that produce two kinds of motile spores.

heterokaryon   Genetically dissimilar nuclei found in a common cytoplasm.

hydrophobin   A small cysteine-rich protein found associated with the aerial hyphae of filamentous fungi.

hypha (pl. hyphae)   The threadlike tubular cell that constitutes the growing form of filamentous fungi.

mycelium (pl. mycelia)   A mass of hyphae together.

mycology   The study of fungi.

pleomorphic   The ability of fungi to produce more than one type of spore and to grow in more than one form.

sclerotium (pl. sclerotia)   a mass of hyphae with a protective, hard covering; a resistant propagule that may remain dormant for long periods.

septum (pl. septa)   A hyphal cross wall.

spitzenkorper   A refractive region near the hyphal tip in certain fungi, filled with many small vesicles.

yeast   A single-celled fungus that reproduces by budding or fission.

THE FUNGI are eukaryotic (possess true nuclei), heterotrophic (do not fix carbon), and osmotrophic (exhibit absorptive nutrition). Most fungi possess cell walls, lack motility, and reproduce by spores. Further, most fungi are aerobic, although some grow under microaerophilic conditions and a few species are obligate anaerobes (e.g., rumen microflora). Fungi possess strikingly different morphologies, ranging from inconspicuous microscopic forms to those with large, macroscopic, fruiting bodies such as the mushrooms, truffles, and shelf fungi. Although yeasts and chytrids are single-celled, most fungi are multicellular, growing in branching, threadlike cells called hyphae. This filamentous cellular architecture is characteristic of the vast majority of fungal species, so, in the broadest sense, the term "filamentous fungi" encompasses almost the entire kingdom. However, common usage is narrower and limits filamentous fungi to the microscopic molds, mildews, rusts, and smuts.

I. INTRODUCTION

Fewer than 100,000 fungal species have been described, but it has been estimated that there may be as many as 1.5 million species worldwide. Although widely cited, the latter figure is a supposition based on a number of simple assumptions and extrapolations and should be viewed merely as an educated guess. It is more accurate to say that the total number of fungi is unknown and that it is likely that very many species remain to be discovered.

In popular culture, filamentous fungi have a largely negative reputation, reminding people of the green molds growing on fruits stored too long in the refrigerator, the black discoloration on shower curtains, and the unpleasant musty odor of damp cellars and enclosed attics. Mushroom poisoning, athlete's foot, corn smut, and the Irish potato famine add to their notoriety. In reality, though, fungi are largely benign creatures. Without filamentous fungi, there would be no penicillin, no blue cheese, no soy sauce. Many commercial alcohols, organic acids, enzymes, and pharmaceuticals rely on the metabolism of filamentous fungi for their manufacture. Most importantly, fungi drive essential carbon and nitrogen cycles, especially in acid environments, and play a major role in recycling nutrients in the biosphere.

II. THE HYPHAL CELL

The distinctive feature of filamentous fungi is the hyphal cell. "Hypha" comes from the Greek word for thread; and hyphae are variously described as threadlike, hairlike, filamentous, and fibrous. Each hypha consists of a series of thin, tubular cells in linear succession. As hyphae grow, they branch, differentiate, and sometimes fuse with other hyphae. Hyphal fusion is called anastomosis. A group of hyphae together is called a mycelium. Sometimes, in the older literature, the word "thallus" is used to refer to the mycelium, particularly in forms that can be seen with the naked eye.

Hyphae are particularly well adapted to penetrating solid substrates. Individual hyphae are rarely noticed in nature because they are microscopic and are more likely to be found growing inside than outside of their food sources. Nevertheless, aggregations of hyphae can sometimes be observed as a white, fluffy mass. Sporulating mycelia can regularly be seen on spoiling food or deteriorating fabrics and wood. The word "mold" originally described such a colony consisting of a mycelium and its spores; now "mold" is often used loosely to refer to filamentous species in

general, especially common genera such as *Aspergillus* and *Penicillium*.

Sometimes, otherwise hardly noticeable fungal colonies produce characteristic developmental structures that are easily seen with the naked eye. The sclerotium, for example, is a pigmented multihyphal resistant structure that protects cells from desiccation. Species of *Claviceps* infect rye, barley, and other grasses and form sclerotia that are associated with human and veterinary disease. This sclerotium is a spur-shaped structure, often called an "ergot" (from the French word for spur), that contains a cocktail of bioactive alkaloids. The sclerotia lodge in the seed heads of infected plants and can be carried along during harvest and milling. When people eat grains contaminated with ergot alkaloids, they may be afflicted with ergotism, also called St. Anthony's fire, a disease that caused widespread suffering and death in medieval times.

The hyphal cell wall confers cell shape and is largely composed of chitin and β-glucans. Some hyphae are regularly divided by cross walls and are called septate, while others possess cross walls only at the base of reproductive structures and are called nonseptate (= aseptate or coenocytic). When septa occur, they possess a small central pore through which cellular contents can flow.

Each hyphal cell contains one or more nuclei. It is believed that most fungal nuclei are haploid. When the nuclei are genetically identical, the mycelium is said to be homokaryotic; when the nuclei are different, the mycelium is heterokaryotic. Many basidiomycetous fungi possess two nuclei in each cell, i.e., they are dikaryotic. Dikaryotic hyphae frequently have a specialized hyphal projection called a clamp connection involved in nuclear transfer.

Hyphae grow almost exclusively at their tips, a consequence of polarized assembly of cellular components, requiring cytoskeletal proteins such as actin and tubulin. This apical growth allows fungi to explore their environment by putting out new cells, a strategy particularly useful in penetrating solid substrates. Growing hyphal tips are packed with vesicles clustered into a dynamic microscopic structure called the *Spitzenkorper* (literally, the "apical body"). The mechanisms by which hyphal tips expand, lay down the matrix of the fungal cell wall, sample their environment, and make navigational decisions are not well understood; however, it is known that almost all exchange of nutrients takes place at or near these growing points.

Hyphal tips not only sense their environment, they also influence it. By secreting organic acids and powerful enzymes, by mechanically penetrating their substrate, and by absorbing nutrients that may be translocated to distant parts of the mycelium, filamentous fungi are a dynamic constituent of many substances. Humus, for example, is composed of partially decomposed organic matter and living fungal hyphae, and helps bind soils into a crumb state. Similarly, blue and Camembert cheeses have mold hyphae as integral components. The Japanese word *koji* also captures the sense of a partly digested substrate pervaded with living hyphae, pleasantly aromatic with fungal metabolic products.

Specialized hyphal types include haustoria, which are branched or lobed hyphal extensions formed by plant parasitic fungi. Haustoria penetrate host cells and absorb nutrients. Some nematode-trapping species form looped branches. Rhizoids are rootlike anchoring hyphae found in chytrids. Appressoria are specific infection structures formed at the tips of germ tubes. Rhizomorphs (mycelial cords) are thick strands of hyphae joined together, which resemble roots, and which may extend for long distances.

Aerial growth of hyphae is associated with the production of hydrophobins, a family of small, cysteine-rich proteins. Once exported out of the cell they polymerize to form a highly insoluble (hydrophobic) rodlet layer on the surface of hyphae. Hydrophobins are developmentally regulated and are associated with both the aerial growth of mycelia and the formation of certain reproductive structures. Hydrophobins have been detected in all filamentous fungi examined. They have not been found in fungal species that grow only a yeasts or in organisms from other kingdoms.

Fungi are among the most plastic of living things. There are many examples in which a single fungus may grow, at different times in its life cycle, as either a yeast or a filament and produce either sexual or asexual spores. This phenomenon is called pleomorphy. The different phases of pleomorphic fungi may be separated in time and space and may look ex-

tremely unlike one another. Therefore, it is not uncommon for different phases of a single species to be isolated separately, described separately, and named separately. Sometimes, it takes years of experimental research to "connect" the different phases of a pleomorphic species (see the following for problems associated with dual nomenclature.)

III. REPRODUCTION

Fungi produce a wide variety of spores for survival and dispersal. Sexual reproduction is the only known form of sporulation is some species, while asexual reproduction is the only known form in others. However, it is common for a single fungal species to produce both a meiotic (sexual) spore and one or more mitotic (asexual) spore types in morphologically separate phases. Mycologists have developed a large and specialized vocabulary to describe these spores and the structures on which they are borne. This jargon can be intimidating for nonmycologists. Some of the more commonly used spore names are listed in Table I. Unfortunately, these terms are not always applied as carefully as they should be, and imprecise and overlapping conventions are encountered. For example, a single propagule may be variously described as a conidiospore, a conidium, or a chlamydospore. Another problematic issue is the profusion of vocabulary concerning asexual and sexual spores. Asexual spores, i.e., those formed in the absence of meiosis, are variously termed anamorphic, imperfect, mitosporic, and vegetative. Sexual spores, those formed in association with meiosis, are variously called teleomorphic, perfect, and meiosporic. In this jargon, the fungus producing asexual spores may be called an anamorph, the sexual phase the teleomorph, and "the whole fungus" the holomorph.

The nomenclature for developmental structures is often derived from the spore name. A sporocarp is an organ that contains or bears spores. An ascoma or ascocarp produces ascospores, while a basidioma or basidiocarp produces basidiospores. Similarly, a modified hyphae that bear spores uses the suffix "-phore." A conidiophore bears conidiospores, a sporangiophore bears sporangiospores, and so forth. Other confusing constructs are encountered. Special-

TABLE I
Major Fungal Spore Types

aeciospore: binucleate infectious spore of a rust

arthrospore: asexual spore formed through hyphal fragmentation

ascospore: sexual spore borne within a saclike structure called an ascus, defining character of ascomycetes

basidiospore: sexual spore borne on the outside of a club-like structure called a basidium; defining character of basidiomycetes

chlamydospore: nonsexual, thick-walled spore formed through the transformation of portions of hyphae

conidiospore (= conidium): an asexual thin-walled spore formed on the ends of hyphae or from a preexisting hyphal cell, not in a sporangium (pl. = conidia)

meiospore: a spore formed after meiosis; a sexual spore

mitospore: a spore formed after mitosis; an asexual spore

oidiospore (= oidium): asexual thin-walled spore found in some basidiomycetes (pl. = oidia)

sporangiospores (= sporangia): asexual spores produced by cleavage divisions within a sporangium

teliospore (= teleutospore): thick-walled resting (over-wintering) spore that produces the basidium in the rusts and smuts

uredospore (= urediniospore): a binucleate spore produced by rusts

zoospore: asexual motile spore with one or two flagella

zygospore: sexual spore formed by fusion of two gametangia, characteristic of zygomycetes

---

ized spore-containing cells might be described as mitosporangia (the anamorph) or meiosporangia (the teleomorph).

IV. FUNGAL TAXONOMY

Most fungal taxonomy is based on the morphology of sexual spores and their associated developmental structures. Experts concoct differing schemes and frequently disagree about classification systems. Most modern taxonomists utilize phylogenetic classifications, i.e., ones based on the genealogy of a group and its hypothetical ancestors. A monophyletic group consists of an ancestor (usually conjectural) and all its descendants. Computer-generated phylogenetic schemes are constructed using DNA sequences of ribosomal subunits and/or other conserved genetic regions. One important outcome of the use of such data is that many contemporary systematists now view fungi and animals as monophyletic sister lineages. This phylogenetic viewpoint has created somewhat of an identity crisis for mycology (the scientific study of fungi), a field that has traditionally been a subdiscipline of botany. Despite their lack of photosynthesis, early taxonomists thought that fungi were more plant-like than animal-like. Fungi do not move, they do have rigid cell walls, and they reproduce by spores. Nevertheless, even before comparative DNA analyses were available, considerable ultrastructural, physiological, and biochemical data had convinced most twentieth century taxonomists of the unique status of fungi. The entire group was moved out of the plant kingdom into its own kingdom and, concurrently, a number of organisms once classified as fungi were placed in separate groups.

Although taxonomists no longer view fungi as part of the plant kingdom, they do adhere to nomenclatural conventions that are governed by the International Code of Botanical Nomenclature. The Code is updated and revised at regular intervals so, in reading mycological literature, it is important to note the year of publication. Moreover, depending on the authority and the philosophy of classification, different systems are encountered. For mycological hierarchies, Divisions/Phyla end in "-mycota," Subdivisions in "-mycotina," Classes in "-mycetes," Orders in "-ales," and Families in "-aceae." Different authorities may endow the same group with disparate ranks. Thus, the fungi with flagellated spores are sometimes viewed as a Class, Chytridiomycetes, or can be treated as a Phylum, Chytridiomycota.

The traditional concept of fungi recognized two groups: the slime molds (those lacking a cell wall and possessing an amoeboid plasmodium or pseudoplasmodium) and the so-called "true fungi" or Eumycota (those with cell walls and yeastlike or filamentous assimilative phases). The true fungi were, in turn, divided into the "lower fungi" and the "higher fungi". The lower fungi included those with nonseparate hyphae, formation of asexual spores by cleavage of cytoplasm in sporangia, and contained two groups; the Mastigomycotina (those with flagellated spores) and the Zygomycotina (those with nonmotile spores). Based on a theory that the lower fungi were degenerate algae, this group was sometimes called the "phycomycetes" (algal fungi).

The higher fungi in this scheme have septate hyphae and are divided into three subdivisions: the Ascomycotina (those with reproductive spores borne in a saclike ascus), the Basidiomycotina (those with reproductive spores borne on a clublike basidium), and the Deuteromycotina (an artificial assemblage, distinguished by the absence of any known sexual form).

The traditional circumscription of fungi unites organisms that share morphological, ecological, and nutritional features. These organisms continue to be studied together by modern mycologists. However, modern mycologists recognize that the traditional view of fungi embraces organisms with different evolutionary histories, i.e., the traditional concept of fungi is polyphyletic.

A contemporary phylogenetic scheme is given in Table II. In this scheme, the idea of "lower," "higher," and "true" fungi is dropped, and several lineages are no longer grouped with the fungi. While the

TABLE II
Classification of Fungi[a]

Taxonomic group	Representative forms
Kingdom: Fungi	
Chytridiomycota	*Allomyces, Blastocladiella*
Zygomycota	*Mucor, Rhizopus, Pilobolus*
Ascomycota	*Neurospora, Podospora, Saccharomyces,* truffles
Basidiomycota	Chanterelles, mushrooms, polypores, rusts, smuts, stinkhorns
Kingdom: Stramenopila	
Oomycota	*Achlya, Peronospora, Pythium, Saprolegnia*
Hyphochytriomycota	*Hyphochytrium*
Labyrinthulomycota	*Thraustochytrium*

[a] The "slime molds" do not form a monophyletic group and are currently classified in separate protist phyla: Plasmodiophoromycota (= "endoparasitic slime molds," e.g. *Plasmodiophora*); Dictyosteliomycota (= "dictyostelid cellular slime molds," e.g. *Dictyostelium*); Acrasiomycota (= "acrasid cellular slime molds," e.g. *Acrasis*); and Myxomycota (= "true slime molds," e.g. *Physarum*).

After Alexopoulos, Mims, and Blackwell (1996).

plasmodial cell types encompassed by the term "slime molds" have never been considered filamentous fungi, the stramenopiles (e.g., Oomycota) include many well-known filamentous organisms. The Oomycota have a number of distinguishing characteristics: they produce asexual biflagellate zoospores, possess a diploid thallus, and have β-glucan-cellulosic cell walls that contain little or no chitin. Many well-known "water molds" (e.g., *Achlya, Saprolegnia*) are oomycetes. Several devastating plant pathogens are also oomycetes, e.g., *Pythium* species causes various "damping off" diseases that afflict young seedlings, *Phytophthora infestans* causes the late blight of potatoes associated with the Irish potato famine; and *Plasmopora viticola* causes the downy mildew disease of grape. Other fungal-like organisms in the Stramenopila include the Hyphochytriomycota, a small group with superficial resemblance to the chytrids, and the Labyrinthulomycota, commonly called "net slime molds."

Zygomycetes are characterized by the formation of modified hyphae called gametangia that fuse to form a thick-walled zygosporangium containing zygospores. Well-known zygomycete genera include *Mucor* and *Rhizopus*, sometimes called "pin molds", dung inhabiting genera such as *Pilobolus*; and several important insect pathogens such as *Entomophthora*.

The Ascomycetes or "sac fungi" are the largest class of fungi. They produce sexual spores in a saclike structure called an ascus; these spores are called ascospores. The structure enclosing the ascospores may be enclosed (a cleistothecium), possess a single opening (perithecium), or be an open, disklike form (apothecium). Included in the ascomycetes are several well-known model species, such as *Aspergillus nidulans* and *Neurospora crassa*, as well as macroscopic forms that make subterranean fruiting structures such as the truffles. Other familiar ascomycetes are the powdery mildews (e. g., *Erysiphe*) and numerous other destructive plant pathogens, such as the causative agents of both Dutch Elm Disease (*Ophiostoma ulmi*) and the American Chestnut Blight (*Cryphonectria parasitica*).

Basidiomycetes or "club fungi" produce their sexual spores on a club-shaped basidium. The majority of species that produce fleshy, macroscopic fruiting bodies are basidiomycetes, e.g., agarics (gilled mushrooms), bird's-nest fungi, chanterelles, earth stars, jelly fungi, puffballs, shelf fungi, stinkhorns, coral fungi, and so forth. Many agarics establish mutualistic relationships called mycorrhizae with the roots of trees and other plants. Also included in the Basidiomycotina are the rusts (Uridinales), a plant pathogenic group that often has complex life cycles involving two host species and up to five spore types, and the smuts (Ustilaginales), a group named for its black, sooty teliospores. *Ustilago maydis* is the causative agent for the common and economically destructive disease called corn smut.

Two prominent fungal groups are not included in the scheme presented in Table II because their use has fallen out of favor with modern taxonomists. The first of these is the Deuteromycotina or Fungi Imperfecti. These are organisms that do not possess any known sexual phase and reproduce entirely by asexual spores. Many economically important genera are deuteromycetes including *Acremonium, Aspergillus, Candida, Fusarium, Histoplasma, Penicillium,* and *Trichoderma.* Based on DNA sequence data, we know that most deuteromycetes are phylogenetically related to ascomycetes. However, in most cases, no sexually reproducing stage is known. To nonmycologists, one of the most perplexing aspects of both fungal terminology and fungal taxonomy concerns the fact that the asexual and sexual forms of a single organism can have two names. This dual nomenclature is allowed by the formal rules of the Botanical Code because so many species are known largely by one phase or the other. The sexual name takes precedence in formal taxonomy, but many common species are best known by their asexual names.

Dual nomenclature has generated a lot of jargon. A single organism may be known by its asexual (= mitosporic, anamorphic, imperfect) name as *Aspergillus nidulans,* as well as by its sexual (= meiosporic, teleomorphic, perfect) name as either *Emericella nidulans,* or *Emericella nidulellus.* Similarly, *Fusarium graminearum* is also named *Gibberella zeae.* Many taxonomists, in a rigorous attempt to create "natural taxonomies," do not acknowledge asexual names (or alert unsuspecting readers to the synonymous nomenclature), which can make it difficult to link the taxonomic literature with many other published sources. A useful Web site, called the Anamorph/Holomorph Connection, is available for matching asexual (anamorphic) and sexual (teleomorphic/holomorphic) names.

The second group not presented in Table II is the lichens, occasionally accorded status as a Phylum called the Mycophycophyta, but now usually classified based on the fungal component. Lichens are dual organisms, consisting of a filamentous fungus and its enslaved algal partner. The hyphae and the algae grow together to form distinct morphological types with enough phenotypic stability to be designated as discrete species. Usually, the bulk of the lichen thallus consists of the fungal component. Slow-growing and highly sensitive to air pollution, lichens are not common in urban areas, but they do form a major part of the living things found in rocky, harsh environments inhospitable to other organisms. Reindeer "moss" (*Cladonia rangiferina*) is actually a lichen, as is "oakmoss" (*Evernia prunastri*), a species widely used in the perfume industry. Well over 10,000 species of filamentous fungi, mostly ascomycetes, are known only as lichens.

V. DIMORPHISM AND MEDICALLY IMPORTANT FUNGI

In addition to producing different spore types, some fungi can grow as either a yeast or a hyphal colony. Species that exhibit this kind of pleomorphism are called dimorphic; the phenomenon is called dimorphism. Many disease-causing animal parasites are dimorphic. The ability to interconvert between yeast and mycelial growth forms is an important virulence trait, but the molecular mechanisms are poorly understood.

Healthy people are resistant to attack by fungi, and fungal diseases are rarer than those caused by bacteria and viruses. Collectively, the diseases caused by fungi growing on or in human beings are called mycoses. The most common mycoses are infections of the skin (cutaneous mycoses), such as ringworm and athlete's foot, associated with species that grow on the outer layers of the skin, commonly called dermatophytic ("skin plant"). A number of effective antifungal drugs, such as nystatin, are available for treating cutaneous mycoses.

Systemic mycoses are more serious and may involve frankly pathogenic, dimorphic species. Examples of virulent species that grow as molds at 25–30 °C and as yeasts at 37°C include *Blastomyces dermatitidis, Coccidioides immitis, Histoplasma capsulatum,* and *Paracoccidioides brasiliensis.* They usually enter through the lungs and later spread throughout the body. Both *Histoplasma* and *Blastomyces* have sexual stages (teleomorphs) in the genus *Ajellomyces.*

Another group of normally benign filamentous fungi is involved in so-called opportunistic mold infections (e.g., species of *Aspergillus, Candida,* and *Cryptococcus*). As with the more frankly pathogenic fungi that cause systemic mycoses, these molds usually take advantage of an immunocompromised host. Individuals with AIDS or diabetes, undergoing cancer chemotherapy, taking immunosuppressive drugs after organ transplantation, the very old, and all people with poorly functioning immune systems are susceptible to fungal infections. Because human and fungal metabolisms are so similar, there is a shortage of good drugs. Currently, systemic mycoses are treated with polyenes (which disrupt membrane structure) and azoles (which inhibit ergosterol synthesis) but both have toxic side effects. Deaths due to fungal-related diseases are increasing at an alarming rate, and effective antifungal agents are badly needed.

Finally, the spores of filamentous fungi can be allergenic. Under appropriate growing conditions, these spores are produced in astronomical numbers. Enclosed spaces with high spore density (e.g., barns filled with moldy hay) can expose susceptible individuals to problematic levels of inhaled spores. "Sick building syndrome" is often associated with massive quantities of fungi growing in insulation or air conditioning vents.

VI. PHYSIOLOGY AND ECOLOGY

The fungi cannot fix carbon, as plants do, nor do they ingest it like animals. Rather, they absorb their food. They must grow in or on a food supply and then excrete digestive enzymes across their cell walls. Their extracellular enzymes break down macromolecular foodstuffs into smaller molecules,

which are then reabsorbed back into the mycelium. The process requires water. Although fungal spores and other resistant structures can tolerate severe desiccation, fungal growth is dependent on adequate hydration.

As a group, fungi are able to degrade almost any carbon source, from feathers to kerosene, lignin to munitions, by excreting a variety of extracellular enzymes including amylases, cellulases, chitinases, laccases, pectinases, peroxidases, and proteinases. They are particularly adapted to break down plant polymers. The most common molds, obviously, are those with the broadest substrate range and include species of *Aspergillus, Penicillium* and *Fusarium.* Despite the wide range of degradative enzymes they produce and their ability to degrade many recalcitrant substrates, in their own metabolism most filamentous fungi can only utilize simple sugars, alcohols, and acids as carbon sources. In general, glucose is the preferred form of carbon. The breakdown of lignin, as well as a variety of pesticides and pollutants, is usually dependent on the presence of glucose or other simple sugar, a phenomenon referred to as co-metabolism.

Filamentous fungi are able to use a great variety of nitrogen sources, with ammonium often the preferred form. Nitrate is unavailable to many chytrids, oomycetes, and basidiomycetes but can be accessed by numerous other fungi. A few species require organic nitrogen sources. Often, in laboratory and industrial fermentations, fungi are fed with materials such as yeast extract and casein hydrolysate to maximize growth. However, it is dangerous to generalize about the physiological requirements of individual species based on generalities about the entire fungal kingdom. No experimental data exist for most fungi.

Sometimes, fungal nutritional modes are classified as saprophytic (nutrition from dead organic matter), parasitic (from living organic matter), and symbiotic (in mutualistic relationships, e.g., lichens and mycorrhizae). Species that have never been cultured apart from a living host are termed obligate parasites (=biotrophs), while those that are able to grow parasitically or saprophytically are called facultative. When viewed from the human perspective, those obligate and facultative parasites that cause plant and animal disease to agricultural species and to ourselves are viewed negatively and called pathogens. Other parasites that kill weeds and animal pests are categorized positively as biocontrol agents. Similarly, when fungal saprophytic action breaks down toxic pollutants, the phenomenon is labeled bioremediation. When fungal enzymes consume materials and structures of human value, it is called biodeterioration. Fungal destruction of trees and lumber causes huge economic losses (e.g., blue rot, brown rot, dry rot, heart rot), as does fungal breakdown of fabrics, paper, and other valuable substances. Agricultural losses due to fungi, both before and after harvest, are vast and impossible to calculate.

In nature, most fungi lead a feast or famine existence. Spore germination and filamentous growth are associated with times when nutrients and water are plentiful. When nutrient deprivation, desiccation, or other stress occurs, the organism usually responds by sporulating and/or producing a resistant structure such as a sclerotium. Fungal spores are variously adapted to dispersal by air, water, or animal vectors and tend to be produced in astronomical numbers. When suitable habitats become avilable, these spores germinate and form new colonies. With the bacteria, fungi are nature's most significant decomposers and recyclers.

VII. THE ROLE OF FILAMENTOUS FUNGI IN BIOTECHNOLOGY

Long before the invention of the microscope, people knew indirectly about microscopic fungi through their metabolic activities. Major fermentation processes harnessed early in human history involved both yeasts (bread, wine, and beer) and filamentous species (cheese, soy sauce, sake). Although western cultures utilized filamentous fungi largely for the flavors they imparted to cheese (e.g., Camembert and Stilton) and cured meats (e.g., certain salamis and hams), it was Asian cuisine that developed a larger repertoire of fungal food fermentations. Filamentous fungi were used to process foods, making them more palatable and more nutritious. *Ang-kak,* otherwise known as Chinese red rice, is made by fermenting cooked rice with *Monascus purpureus* and is used as

both a food and a medicine. *Ontjam* (also spelled *oncom*) is an Indonesian specialty prepared from pressed peanut cakes and *Neurospora intermedia* or *Rhizopus* sp. Tempe (also spelled tempeh) is another Indonesian dish, prepared with cooked soybeans fermented with *Rhizopus* and a lactic acid bacterium. The best studied of the Asian fungal food fermentations are Japanese. Soy sauce, miso, and sake involve the growth of *Aspergillus oryzae* or *A. sojae* on steamed rice, wheat, or other starchy substrates in a warm, humid environment. As the hyphae grow, they secrete hydrolytic extracellular enzymes that partially degrade the grain. This mixture of hyphae and partially degraded substrate is called a *koji*. For soy sauce, the *koji* is mixed with about equal proportions of soybeans and wheat, salted, and subjected to an anaerobic fermentation with yeasts and lactobacilli.

The *koji* molds were among the first microorganisms to be exploited in industrial enzymology. In the late nineteenth century, the early Japanese biotechnologist, Jokichi Takamine, patented a diastase (amylase) that could be used in alcohol manufacture as a substitute for the amylase made by germinating barley. Since then, thousands of enzymes from all life forms have been characterized by biochemists, but only about three dozen are in large-scale commercial use. Of these commercial enzymes, about half are of fungal origin, mostly amylases, pectinases, and proteinases. With the advent of molecular biotechnology, several fungal species have been engineered for the production of heterologous proteins, such as calf chymosin, used in making cheese.

Citric acid is another fungal product with an important place in the development of early biotechnology. Originally isolated from citrus fruits, citric acid is valued for its flavor as well as its properties in detergents, cosmetics, and other commodities. Since the early part of the twentieth century, the major process for citric acid production has utilized molds such as *Aspergillus niger*. Moreover, many of the early biological engineering advances for growing large volumes of mycelia, later perfected in antibiotic fermentations, got their start in the context of citric acid production.

One of the most important developments in modern medicine was the discovery and development of penicillin. Penicillin is active against gram positive bacteria and can cure infectious diseases such as diphtheria, gonorrhea, and rheumatic fever. Produced by *Penicillium notatum* and *P. chrysogenum*, penicillin and related β-lactams transformed infectious disease therapeutics and led to the Golden Age of Antibiotics. Following World War II, industrial microbiology became a robust field as numerous pharmaceutical companies funded research to find new and improved antibiotics, not only from fungi, but also from actinomycetes and other microorganisms. Moreover, the field of chemical engineering also had a renaissance. Batch and continuous culture technologies were perfected for sterile industrial-scale fermentations of filamentous microorganisms.

In addition to antibiotics, filamentous fungi produce secondary metabolites with other biological activities. The mevalonins, such as lovastatin, produced by *Aspergillus terreus,* lower blood pressure; and the cyclosporins produced by *Tolypocladium inflatum* are immunosuppressive and facilitate organ and bone marrow transplants. Griseofulvin, produced by *Penicillin griseofulvum*, has use as an antifungal antibiotic, and gibberellic acid, produced by *Gibberella fujikuroi* is a plant growth regulator. Only a few of the thousands of known fungal secondary metabolites have been developed as drugs or growth factors. It is believed that many other useful therapeutic agents remain to be discovered. Fungal biodiversity is a promising target for "drug prospecting."

Not all bioactive fungal metabolites are beneficial. Mycotoxins are mold poisons capable of causing toxic effects in humans and other animals. Aflatoxins, produced by *Aspergillus flavus* and *A. parasiticus*, are among the most carcinogenic compounds known. They are common contaminants of stored grains, legumes, and nuts, and have been associated with primary liver cancer in Africa and Asia, as well as massive veterinary deaths in poultry and trout. Fumonisins, produced by *Fusarium moniliforme* on feed corn, can cause terrible brain degeneration in horses. In fact, dozens of mycotoxins are known, mostly associated with veterinary disease, but also contributing to human suffering in times of famine when people are reduced to eating mold-contaminated food.

Two important filamentous fungi have figured in the development of modern genetics: *Neurospora crassa* and *Aspergillus nidulans* (= *Emericella nidulans*.) *N. crassa* was used for the elucidation of the "one gene–one enzyme" theory by George Beadle and Edwin Tatum, who were awarded a Nobel Prize for their research. *A. nidulans* was used to elucidate the parasexual cycle, an eukaryotic "alternative to sex." Both of these species have been important in the definition of intragenic complementation, the molecular biology of tubulin and other motor proteins in the cell cycle, in elucidating aspects of cytoplasmic inheritance and mitochondrial genetics, in the study of circadian rhythms, and as models for developmental biology. Projects to sequence the genomes of both species are under way, with expectations that this work will bring new insights to the mechanisms of fungal pathogenicity and accelerate antifungal drug discovery by providing new targets for screening. It is also hoped that genome research will bring new understanding to the fundamental biological nature of the filamentous life style.

VIII. CULTURE COLLECTIONS AND OTHER RESOURCES

Culture collections are essential for the preservation and distribution of living cultures and aid workers in the identification of species. In the United States, the three most important resources for filamentous fungi are the American Type Culture Collection in Manassas, Virginia; the Northern Regional Research Laboratory in Peoria, Illinois; and the Fungal Genetics Stock Center in Kansas City, Kansas. Other major collections for filamentous fungi are the Centraalbureau voor Schimmelculture in Baarn, The Netherlands; the Commonwealth Mycological Institute in Kew, England; and the University of Alberta Microfungus Collection, Alberta, Canada.

Information about fungal culture collections, as well as a wealth of other facts about fungal biology, is available on the World Wide Web. This resource is in constant flux and is expanding in volume and complexity. By using an appropriate search engine and entering the word "mycology" or "fungus," you will find numerous resources. In many cases, information about individual groups and species can also be accessed by entering key words.

IX. SUMMARY

The colloquial and traditional scientific use of the term "filamentous fungi" encompasses a paraphyletic assemblage of nonphotosynthetic organisms equipped with cell walls, absorptive nutrition, apical growth, and the capacity to form spores. They range from microscopic molds and mildews to macroscopic mushrooms and truffles. Many species have negative consequences to human welfare as causes of plant and animal disease or as agents of biodeterioration. Still more fungal species are beneficial as nature's primary recyclers and as producers of drugs, enzymes, and commodity chemicals. All filamentous fungi live by the secretion of enzymes into the substrate and subsequent absorption of nutrients. Because of their life style, filamentous fungi literally become part of their substrates, modifying the composition of the materials in which they grow, dynamically participating in the great ecological cycles of nature.

Acknowledgments

The author thanks John Taylor University of California, Berkeley, and Ronald Bentley, University of Pittsburgh, for their helpful suggestions.

See Also the Following Articles

Bibliography

Alexopoulos, C. J., Mims, C. W., and Blackwell, M. (1996). "Introductory Mycology," 4th ed. John Wiley & Sons, Inc., New York.

Bennett, J. W., and Lasure, L. L., eds. (1991). "More Gene Manipulations in Fungi." Academic Press, San Diego.

Bos, C. J., ed. (1996). "Fungal Genetics. Principles and Practice." Marcel Dekker, New York.

Esser, K., and Lemke, P. A., eds. (1994–1996). "The Mycota. A Comprehensive Treatise on Fungi as Experimental Systems for Basic and Applied Research," Vols. I–VI. Springer-Verlag, Berlin, Germany.

Godfrey, T., West, S., eds. (1996). "Industrial Enzymology," 2nd ed. Macmillan Press, London, UK.

Griffin, D. H. (1994). "Fungal Physiology," 2nd ed. Wiley-Liss, New York.

Hawksworth, D. L., Kirk, P. M., Sutton, B. C., and Pegler, D. N., eds. (1995). "Ainsworth & Bisby's Dictionary of the Fungi." 8th ed. CAB Publishing, Wallingford, UK.

Hudler, G. W. (1998). "Magical Mushrooms, Mischievous Molds." Princeton Univ. Press, Princeton, NJ.

Isaac, S. (1992). "Fungal–Plant Interactions." Chapman & Hall, London, UK.

Lancini, G., Parenti, F., and Gallo, G. G. 1995. "Antibiotics. A Multidisciplinary Approach." Plenum, New York.

Moore-Landecker, E. (1996). "Fundamentals of the Fungi." 4th ed. Prentice Hall, Upper Saddle River, NJ.

Peruski, L. F., and Peruski, A. H. (1997). "The Internet and the New Biology." ASM Press, Washington, DC.

Reynold, D. R., and Taylor, J. W., eds. 1993. "The Fungal Holomorph: Mitotic, Meiotic, and Pleomorphic Speciation in Fungal Systematics." C. A. B. International, Wallingford, UK.

Wainwright, M. (1992). "An Introduction to Fungal Biotechnology." John Wiley & Sons, Chicester, UK.

Wessels, J. G. H. (1997). Hydrophobins, proteins that change the nature of the fungal surface. *Adv. Microb. Physiol.* **38**, 1–45.

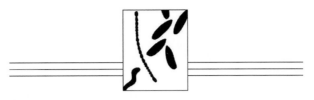

Gaeumannomyces graminis

Joan M. Henson

Montana State University

Henry T. Wilkinson

University of Illinois

I. Taxonomy
II. Parasite or Pathogen?
III. Epidemiology
IV. Disease Management

GLOSSARY

ectotrophic On the surface of the root.

endotrophic Inside the root.

hyphopodium (pl. hyphopodia) An attachment and infection structure that develops from short hyphal branches.

perithecium (pl. perithecia) Flasklike fruiting body, usually dark in color and thick-walled, containing ascospores within asci.

phialide A cell that develops one or more open ends from which conidia (phialospores) are formed.

GAEUMANNOMYCES GRAMINIS is a filamentous fungus that colonizes the root and crown tissue of many members of the grass family (Poaceae). The interaction of the fungus with its host is dynamic and can change from a parasitic to a pathogenic relationship. *Gaeumannomyces graminis* is the etiologic agent of take-all, a devastating root disease of cereals (Fig. 1). The disease, first observed over 150 years ago in wheat, is named take-all because affected plants fail to fill their grain, and the crop yield is taken. *G. graminis* also causes black sheath rot of rice, Bermuda grass decline, take-all patch of bentgrass (Fig. 2), and Bermuda grass spring dead spot (Fig. 3).

Gaeumannomyces graminis is a member of the *Gaeumannomyces–Phialophora* complex of fungi that includes many related species. *Gaeumannomyces graminis* is currently the most widely distributed species in this complex and displays the greatest variation among *Gaeumannomyces* species. Other species of the complex include *G. incrustans*, *G. cylindrosporus*, *G. caricis*, *G. leptosporus*, *G. oryzinus*, and several asexual states of *Gaeumannomyces* species assigned to the genus *Phialophora*. Although this list is still tentative, there is conclusive evidence that *G. graminis* and related fungi have a significant impact on both the ecology and the economy of global agriculture.

I. TAXONOMY

The take-all fungus was named *Ophiobolus graminis* Sacc. until 1952, when descriptions of both its hyphopodium and unitunicate (single-sheathed) ascus provided the rationale for reclassification as a separate species, *Gaeumannomyces graminis* (Sacc.) von Arx & Olivier. *G. graminis* is classified among the Ascomycotina in the order Pyrenomycetes, which produce perithecia during their sexual cycle. *G. graminis* is differentiated taxonomically from other *Gaeumannomyces* and related species by its morphological characteristics, including the production, color, and shape of fungal structures, such as hyphae, hyphopodia, conidia, and perithecia. *G. graminis* perithecia are darkly pigmented, or melanized, and usually develop on diseased root and crown tissues in the field or on infected roots in laboratory culture. Each perithecium contains hundreds of asci and each ascus has eight randomly ordered, filiform ascospores (Fig. 4). Ascospores are hyaline (colorless), with several transverse septa that divide genetically

Fig. 1. A wheat field with take-all disease caused by *G. graminis* var. *tritici*. Affected plants are yellowing and stunted due to root rot.

Fig. 3. Bentgrass turf with take-all patch caused by *G. graminis* var. *avenae*.

identical cells. *G. graminis* may also produce asexual spores, or phialospores, on short, modified hyphae called phialides that branch singly or in clusters from vegetative hyphae (Fig. 5). In culture, hyphae are septate and hyaline when young and, at colony margins, they characteristically curl backward toward the colony center. As they age, hyphae turn grey to dark brown, black, or green. On the surface of host roots, crowns, and lower stems, *G. graminis* myce-

Fig. 2. A golf course with Bermuda grass spring dead spot caused by *G. graminis* var. *graminis*. Root rot causes plant necrosis with consequent yellowing. See color insert.

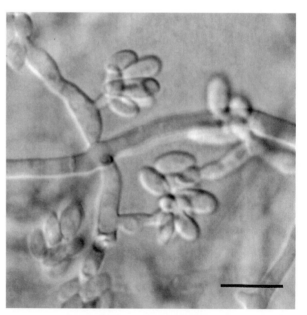

Fig. 4. Phialospores and phialides of *P. graminicola* in culture. Bar = 10 μm. See color insert.

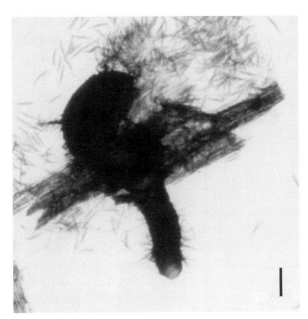

Fig. 5. A ruptured perithecium with individual ascospores released from their asci (not visible). Bar = 0.2 mm.

lium develops as a network of thick (4–7 μm diameter), melanized "runner" or "macro" hyphae that often fuse together (Fig. 6). Hyphopodia and infection plates or cushions, composed of masses of hyphopodia, also form on host root and crown tissue surfaces (Fig. 7). Infection hyphae that penetrate the plant epidermis may originate from runner hyphae, hyphopodia, or infection cushions.

Varieties of *G. graminis* are also distinguished by their morphological and molecular DNA differences in laboratory culture. The variety *tritici* is usually isolated from diseased wheat affected by take-all and has similar morphology to the *avenae* variety, which is usually isolated from oats or bentgrass affected by take-all. Both varieties produce simple, melanized hyphopodia on host surfaces and are difficult to distinguish from each other without examination of the perfect (sexual) state, where the longer, 110–130 μm ascospores of variety *avenae* distinguish it from the shorter, 70–100 μm ascospores of variety *tritici*. Variety *graminis* is usually isolated from diseased turfgrass and is distinguished from *avenae* and *tritici* by its melanized, lobed hyphopodia that develop on host tissue or hydrophobic surfaces in the laboratory (Fig. 8). A proposed fourth variety, *maydis*, was recently isolated from maize in China, but is not well characterized.

The *Gaeumannomyces-Phialophora* complex includes asexual, parasitic, but nonpathogenic *Phialophora* species, *Phialophora* are usually more pigmented (melanized) than *Gaeumannomyces* and,

Fig. 6. Runner hyphae of *G. graminis* var. *avenae* on a bentgrass root. Bar = 0.3 mm.

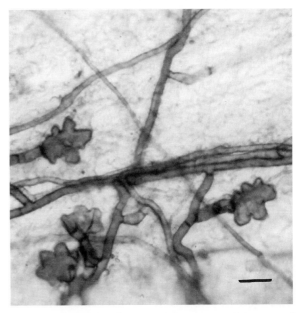

Fig. 7. Hyphopodia of *G. graminis* var. *graminis* on centipede grass. Bar = 10 μm. See color insert.

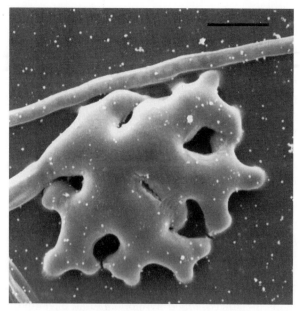

Fig. 8. A scanning electron micrograph of a lobed hyphopodium of *G. graminis* var. *graminis* on a mylar membrane. Bar = 10 μm.

unlike most *G. graminis*, they produce phialospores. Some *Phialophora* species may actually be the imperfect (asexual) state of a *Gaeumannomyces* species. For example, *Phialophora graminicola* inoculated on laboratory-grown wheat produces perithecia of *G. cylindrosporus*, and their relatedness was further supported by their similar rRNA gene sequences. Protein comparisons between *G. graminis* var. *tritici* and *Phialophora* spp. suggest that these species are closely related. Conversion between the sexual and asexual states of *Gaeumannomyces* and *Phialophora* also is believed to occur under field conditions. However, the genetic basis for this change, or the environmental conditions that favor one state over the other, remain unknown.

II. PARASITE OR PATHOGEN?

A. Survival

G. graminis can survive as mycelia, perithecia, or ascospores in plant debris or soil. There is no evidence that the fungus can survive as a seed contaminant, and it is unlikely that it does so, since the fungus is soil-borne and does not progress beyond the crown of the plant during infection. Take-all disease may worsen each year in a mono-cultured wheat field infested with *G. graminis* var. *tritici*, a fact that demonstrates that the fungus can survive harsh winter environmental conditions.

B. Inoculation and Ectotrophic Colonization

Take-all diseases initiate at the roots in early seedling development. *G. graminis* is a primary colonist and, thus, requires newly formed roots and root branches for infection. Inoculation occurs either passively, when roots encounter the fungus in soil, or actively, by saprophytic hyphal growth through the soil toward the host root, which may be more than a centimeter away from the inoculum. After reaching host roots, hyphae grow ectotrophically on root surfaces. These hyphae thicken, melanize, and often fuse to become runner hyphae. Hyphopodia and infection cushions may form on short branches of these hyphae.

C. Penetration

Penetration of the host is likely achieved by a combination of both physical forces and chemical reactions. The fungus traverses several layers of the host, including the cuticle, plant cell wall and plasmalemma, each of which is chemically different and complex. Penetration is accomplished by infection hyphae originating from runner hyphae. Hyphal tips secrete degradative enzymes that hydrolyze host tissue, including pectinases, cellulases, and proteases, softening it and providing nutrients for the invasive hyphae. Additionally, infection pegs (penetration hyphae) may extend from the adhesive surface of a hyphopodium or infection cushion. Because of their strong, melanized cell walls, hyphopodia build osmotic (turgor) pressure that promotes mechanical penetration of the host surface by the infection peg. Infection pegs also secrete hydrolytic enzymes.

D. Endotrophic Colonization

Infection of epidermal and cortical host tissue may be asymptomatic. The host responds to this parasitism by depositing lignin, or lignitubers, around invasive hyphae. The fungus may be contained in the outer cortical layers of the root by host lignification and other plant defense responses. Alternatively, if the host is additionally stressed, or if the fungus is an aggressive isolate, hyphae will either degrade the lignitubers or grow more rapidly than lignin can be deposited and, thus, advance into the vascular tissue (stele) of the host. Stelar infection leads to necrosis and/or occlusion of the vascular tissue, with subsequent root death or dysfunction.

How and why *G. graminis* either remains parasitic or becomes pathogenic is incompletely understood, but it depends on the interplay among host defenses, fungal virulence, and environmental parameters. For example, bentgrass asymptomatically infected with *G. graminis* rapidly displays wilt or necrosis during drought, and irrigated wheat is more aggressively colonized than is nonirrigated wheat because the added moisture favors fungal growth.

III. EPIDEMIOLOGY

A. Geographical Distribution

The geographical distribution of *G. graminis* coincides with the cultivation and/or habitation of grasses and cereals, whereas the distribution of each of the *G. graminis* varieties depends on their host ranges and soil and climatic conditions. For example, varieties *avenae* and *tritici* are more likely to be isolated from soils in temperate climatic zones, while variety *graminis* is found in both warm and temperate climates. In general, for inoculation and infection to occur, the soil must be damp or moist, but not saturated. *G. graminis* is an obligate aerobe and will asphyxiate during prolonged submersion, especially at soil temperatures >15°C. Cereal or turfgrass populations that are irrigated or receive frequent, intermittent moisture are more likely to develop take-all disease. The range of soils in which take-all disease will develop is as varied as the soils where grasses grow, but the severity of take-all is greater in irrigated soils that drain well. *G. graminis* requires free moisture to infect and colonize, and the fungus is most effective in colonizing roots of grass plants that are growing rapidly, which occurs in soils that drain water rapidly and exchange oxygen readily. As with most soil-borne pathogens, the severity of the disease *G. graminis* causes also depends on soil conditions, such as pH, organic matter content, and general microbial activity.

G. graminis is also sensitive to soil temperature. Variety *avenae* infects roots at soil temperatures of ~12–15°C, but is less active at >20°C. Variety *tritici* initially infects wheat roots at ~15°C and continues to colonize at >25°C. Variety *graminis* can infect and colonize over a wide range of soil temperatures from 12–30°C.

Parasitism of grass roots is a finely balanced process in nature and generally does not result in noticeable necrosis of the host. However, in cultivated populations of hosts, i.e., cereals or turfgrasses, the plants are often stressed due to unnatural growth conditions, and it is believed that these stresses predispose plants to severe take-all disease.

B. Host Range

To plant pathologists, host range is synonymous with the range of hosts on which the pathogen can cause a disease, but, in fact, grasses can host *G. graminis* without developing disease. Nevertheless, the varieties of *G. graminis* can also be characterized by the severity of the disease they can cause on various hosts. For example, *G. graminis* var. *avenae* is pathogenic on oats (*Avena sativa* L.) and bentgrass (*Agrostis palustris* Huds.), but only parasitic on wheat (*Triticum aestivum* L.). *G. graminis* var. *tritici* is pathogenic on wheat, but not on oats or bentgrass. However, it is parasitic on bromegrass (*Bromus* spp. L.) and quackgrass (*Agropyron* spp. L.). What limits the host range of *G. graminis* varieties is not known, with the exception of variety *avenae*. Unlike the other varieties, *G. graminis* var. *avenae* produces an enzyme, avenacinase, that allows it to inactivate avenacin, a preformed fungal inhibitor synthesized by oats but not other cereals. Thus, variety *avenae* can cause take-all on oats, whereas variety *tritici* usually does not.

IV. DISEASE MANAGEMENT

Take-all disease is most effectively managed by preventative cultivation methods. These include crop rotation with nonsusceptible plants; application of acidifying fertilizers; infrequent irrigation; and reduction of plant stress from physical damage, drought, or soil compaction. Crops with significant levels of genetic resistance to *G. graminis* are not available; however, fungicidal compounds will effectively reduce disease severity. Finally, there are naturally occurring bacteria that suppress *G. graminis*, and these are currently being formulated for commercial use as biocontrol agents.

See Also the Following Articles

Fungi, Filamentous • Plant Disease Resistance • Plant Pathogens

Bibliography

Asher, M. J. C., and Shipton, P. J. (eds.). (1981). "Biology and Control of Take-all." Academic Press, London, UK.

Bowyer, P., Clarke, B. R., Lunness, P., Daniels, M. J., and Osbourn, A. (1995). A saponin detoxifying enzyme. *Science* **267**, 371–374.

Cook, R. J., and Veseth, R. J. (1991). "Wheat Health Management." American Phytopathological Society, St. Paul, MN.

Smiley, R. W., Dernoeden, P. H., and Clarke, B. B. (1992). "Compendium of Turfgrass Diseases" (2nd ed.). American Phytopathological Society, St. Paul, MN.

Webster, R. K., and Gunnell, P. S. (1992). "Compendium of Rice Diseases." American Phytopathological Society, St. Paul, MN.

Wilkinson, H. T. (1998). "Interactive Turf." Univ. of Illinois Press, Urbana-Champaign, IL.

Gastrointestinal Microbiology

T. G. Nagaraja

Kansas State University

GLOSSARY

anaerobes Microbes that are capable of generating ATP without the use of oxygen and exhibit various degrees of oxygen sensitivity.

allochthonous Nonindigenous, dormant, in transit, and not characteristic of the habitat.

autochthonous Indigenous, present during evolution of the host, and characteristic of the habitat.

cecum The proximal blind portion of the hindgut.

colon The mid-portion of the hindgut.

competitive exclusion The protective function of the normal flora of the gut to prevent entry and colonization of pathogens.

epimural bacteria Bacteria attached to the epithelial cells lining the gut.

foregut fermentation Microbial fermentation prior to the gastric or peptic digestion.

hindgut fermentation Microbial fermentation after the gastric or peptic digestion.

hydrogenosomes Cytoplasmic organelles present in anaerobic protozoa and fungi and containing enzymes that produce hydrogen from oxidation of reduced cofactors.

normal flora The population of microbes that normally reside in the host and for the most part live in harmony.

peristalsis A wave of contraction followed by relaxation that propels the digesta down the gastrointestinal tract.

rumen The largest of the four compartments of the ruminant stomach, inhabited by a myriad of microbes.

volatile fatty acids Short-chain fatty acids that are major products of microbial fermentation in the gut.

zoospore A free living and flagellated reproductive structure of fungi.

THE GASTROINTESTINAL TRACT OR DIGESTIVE TRACT is essentially a tubular organ of varying diameter extending from the mouth to the anus. The gut is an open ecosystem, because the lumen is essentially external to the body. The gastrointestinal tract has five major regions: mouth, esophagus, stomach, small intestine (duodenum, jejunum, and ileum), and large intestine (cecum, colon, and rectum). Some of the regions, depending on the species, may be enlarged with or without sphincters or further compartmentalized. Such enlarged or compartmentalized regions slow down the transit of contents, allowing for microbial growth and fermentation.

The gastrointestinal tracts of animals, including humans, and of birds are complex microbial ecosystems. The complexity is attributable to differences in anatomical features, diet, and the health of the animal. The gastrointestinal tract contains distinct microbial populations with diverse compositions. Many of these organisms colonize and grow and are considered indigenous and, hence, are termed "normal flora," also called autochthonous microbiota. These microbes, for the most part, live in harmony with the host. Additionally, the gut flora include allochthonous microbiota that do not get established (colonization and growth) and are dormant and in passage. These are derived largely from ingested food and water and, to a small extent, from swallowed air or from another habitat of the host (e.g., skin, respiratory tract, or reproductive tract). The nonindigenous microbes also include a variety of gastrointestinal pathogens that may colonize and grow to establish infections. Also, some members of the normal flora could assume pathogenic roles (opportunistic pathogens) when the ecosystem is perturbed in some way or when a breach occurs in the integrity of the gut wall.

I. THE GASTROINTESTINAL ECOSYSTEM

Most of the tract offers conditions that are conducive for microbial growth. The temperature remains relatively constant (36–40°). Water and exocrine secretions (saliva and other digestive secretions) provide a moist environment. Ingested food provides the energy and other nutrients needed for microbial growth. Normal gut motility (peristalsis) helps mix the digesta, which brings microbes into contact with fresh substrate. End products of fermentation (mainly acids) are removed by absorption into the blood. Absorption, coupled with the buffering effect provided by digestive secretions (mainly saliva), helps regulate gut pH. Only in the gastric stomach and the duodenum is the low pH inhibitory to microbial growth. The gut ecosystem often is referred to as a continuous or semicontinuous culture system with more or less continuous availability of substrate, removal of end products (by absorption or passage), and passage of undigested and waste products.

The composition of the gut contents of animals is extremely complex. Physical and chemical conditions within the gut differ among species of animals but are relatively constant within species. However, within species, the composition is influenced by the diet. Among the gut ecosystems, the reticulo-rumen of cattle and the colon of humans are the most extensively investigated (Table I).

The main sites of microbial fermentation differ from species to species and are broadly classified into pregastric, or foregut, and postgastric, or hindgut, fermentations (Table II). Ruminants, such as cattle,

sheep, goats, and buffaloes, have a capacious chamber where food is retained for a long time to allow microbial fermentation. Also, there are nonruminant foregut fermenters that have large stomachs but do not ruminate their food for secondary mastication. Hindgut fermenters rely primarily on the cecum and colon for microbial fermentation and vary considerably in their dependence on that fermentation. Herbivorous monogastrics, such as horses and elephants, have capacious hindguts, and their primary sources of energy are fermentation products of microbial digestion.

II. TYPES OF GUT MICROBIAL AND HOST INTERACTIONS

The interactions between the gut microbial flora and the host are complex and, basically, include competition, cooperation, and a combination of both (Table III). In the competitive interaction, host and the microbes are competing for the same food. This is found primarily in carnivorous animals whose food is usually of animal origin, consisting largely of protein, fat, and carbohydrate. These animals have a highly acid stomach and duodenum, in which microbes are inhibited, allowing the host enzyme to digest the foods first and absorb the sugars, amino acids, and fatty acids.

A cooperative relationship predominates in animals whose food is primarily of plant origin, containing relatively small amounts of protein, fat, and digestible carbohydrates (starch and sugars) but a much larger amount of plant fiber, with cellulose and hemicellulose as the chief components. The plant fiber is indigestible by mammalian enzymes and, hence, the host is absolutely dependent on the gut microbes to convert the fiber into fermentation products and microbial cells to be used by the host. The cooperative model is exemplified best by ruminants equipped with a capacious foregut populated by a myriad of microbes. One negative aspect of this cooperative relationship is the inefficiency associated with the fermentation of protein, resulting in the loss of ammonia (it is absorbed and excreted as urea in the urine), as well as potential reduction in the quality of dietary protein. The model also fits marsupials and

TABLE I
Physical, Chemical, and Microbiological Characteristics of the Rumen of Cattle and the Colon of Humans

Characteristics	Rumen	Colon
Physical		
Capacity	30–70 liters	0.5–1.2 liters
pH	5.5–7.5	6.5–7.5
Temperature	39°C	37°C
Redox potential	−250–350 mV	−200–300 mV
Osmolality	250–350 mOsm/kg	300–400 mOsm/kg
Dry matter	10–18%	10–25%
Chemical		
Acids—Volatile fatty acids	Acetate, propionate, butyrate, valerate, and branched-chain fatty acids	Acetate, propionate, butyrate, and others
—Nonvolatile acids (mM)	Lactate, succinate	Lactate, succinate
Gases	CO_2, CH_4, H_2, H_2S, N_2, O_2	CO_2, CH_4, H_2, H_2S, N_2, O_2
Ammonia	Present	Present
Amino acids and peptides	Small amounts detectable 1 to 3 h after feeding	Trace amounts
Soluble carbohydrates	Small amounts detectable 1 to 3 h after feeding	Trace amounts
Complex carbohydrates		
Dietary (cellulose and hemicellulose)	Always present	Always present
Endogenous (mucopolysaccharides)	Present but minimal	Present
Minerals	Present, high sodium	Present, high sodium and chloride
Vitamins	Present and serve as major supply of B vitamins	Present but does not serve as the source of B vitamins
Microbiological		
Bacteria	10^{10}–10^{11}/g	10^9–10^{10}/g
Protozoa	10^4–10^6/g	Absent
Fungi	Present (not quantifiable)	Possibly present
Bacteriophages	Up to 10^{12} particles/ml	Possibly present

TABLE II
Sites of Microbial Fermentation in the Gastrointestinal Tract

Site	Animals	Microbes
Foregut (pregastric or prepeptic)		
Ruminants	Cattle, sheep, goats, buffaloes, deer, antelopes, camel, llama	Bacteria, protozoa, fungi
Nonruminants	Colobid and langur monkeys, marsupials, hamsters, voles	Bacteria, protozoa, fungi
Hindgut (postgastric or postpeptic)		
Cecal and colonic fermenters	Horses, zebras, rabbits, rats	Bacteria, protozoa, fungi
Colonic fermenters		
Sacculated	Humans, pigs	Bacteria
Unsacculated	Dogs, cats	Bacteria

TABLE III
Types of Animal–Microbe Nutritional Interrelationships

Interaction	Examples	Characteristics of food	Fermentation products used as source of energy	Microbial cells digested and used as source of protein
Competition	Carnivores	Chiefly animal protein, carbohydrate and fats	No	No
	Omnivores	Animal and plant foods	No	No
Cooperation				
Offset fermentation chamber	Ruminants and tylopods Hippopotamus	Plant food including fiber	Yes	Yes
Inline fermentation chamber	Marsupials and leaf-eating monkeys	Plant food including fiber	Yes	Yes
Cooperation and competition	Herbivores	Plant food including fiber	Yes	No
	Rodents	Plant food including fiber	Yes	Yes (Coprophagy)

leaf-eating monkeys, in which the fermentation chamber is not a separate chamber but is an enlarged and partially separated proximal region of the stomach, with the distal region being the site for glandular digestion.

Animals with a well-developed cecum, such as herbivorous monogastrics (e.g., horses, elephants, and rodents), have combined competitive and cooperative interactions. Host enzymes break down the food first, and the components indigestible by the host are fermented by the microbes in the hindgut. The microbial cells cannot be digested, because they are formed distal to the stomach and small intestine. Some animals (rodents, lagomorphs) overcome this limitation by consuming their feces (coprophagy), thus salvaging the microbial cell protein.

III. TYPES OF GUT MICROORGANISMS

The gastrointestinal tracts of all animals are inhabited by a mixed population of microorganisms that generally include bacteria, protozoa, fungi, and bacteriophages. The establishment of a microbial population commences soon after birth. The digestive tracts of fetuses of all animals are sterile. But within hours after birth, the gut gets colonized by bacteria derived from the immediate vicinity, including the birth canal of the mother. Subsequent development of the microbial population is the result of interactions of diet, environment, and host physiology. The processes involved are complex, involving a succession of microorganisms and eventually resulting in a dense, more or less stable population.

IV. GUT BACTERIA

The gastrointestinal tract contains a dense population of bacteria with numbers ranging from 10^4 to 10^{11}/g, depending on the site. The counts are highest in the enlarged regions of the gut, such as the rumen of ruminants, the cecum of herbivorous nonruminants, and the colon of other animals and man. The predominant flora is anaerobic, with various degree of aerotolerance.

Based on their distribution, the bacteria in the gut are grouped into three population types: the fluid population, the food particle-associated population, and the epithelial cell-associated population called epimural bacteria. The first group includes bacteria that float freely in the fluid milieu of the gut and are usually not attached to any surface and a large number of usually adherent cells that are detached temporarily from digesta particles or epithelial cells. The second group is the largest component of the gut population that adheres rapidly and avidly to digesta particles. The fluid population digests soluble nutrients in the food, whereas those associated with food particles digest insoluble polymers. The inside of the gut, extending from the keratinized, stratified, squamous epithelial cells in the upper region to the convoluted mucus membrane of the lower region, provides an enormous surface area for bacterial attachment. The ability to attach onto the gut epithelium permits many bacteria to survive the peristaltic flow of gut contents and inoculate fresh ingesta. Some host specificity may be involved in the adherent flora. For example, certain lactobacilli strains that adhere to pig cells will not adhere to chicken cells and vice versa. The mechanisms of adherence to epithelial cells include surface glycocalyx or proteins (adhesins) or pill. Some of these organisms penetrate into or through the epithelial cells. Attachment of microbes to the gut epithelium may or may not induce structural changes in the epithelial cells. Generally, most of the adherent bacteria do not induce structural changes. The maintenance of the epithelial-associated bacteria is not dependent on the diet. The close association of bacteria with the epithelial cells permits access to potential nutrients that diffuse from blood (e.g., urea). The pH of the epithelial surface is close to neutrality and, therefore, more suitable for bacterial growth in some regions of the gut. Oxygen diffusing from blood also may permit the growth of facultative or microaerophilic bacteria.

A. The Esophagus and the Stomach

The esophageal and gastric ecosystems generally are not conducive to microbial growth. In the case of the esophagus, the flow of solids, fluids, and oral secretions washes away the microbes. The gastric stomach is inhospitable for most microbes because of the acidic environment. Yet the esophagus and stomach do host many different microbes, more in some animals than others. Lactic acid bacteria (*Bifidobacterium, Lactobacillus,* and *Streptococcus*) and yeasts are the most common isolates from the esophagus and stomachs of most animals. The stomach in rodents and in other animals is compartmentalized incompletely into two regions. One region is lined with stratified squamous epithelial cells, and the other is lined with columnar epithelial cells. In rodents, the squamous epithelial cells are colonized

by lactobacilli, and the columnar cells usually are colonized by yeasts. The human stomach is lined entirely by columnar cells, but, nevertheless, Lactobacilli are isolated consistently from gastric contents. Lactobacilli also are known to colonize the squamous epithelium of the crops of chickens. Lactobacilli and yeasts are sufficiently acid-tolerant to be able to grow at the low pH of the stomach.

A spiral-shaped, gram-negative bacterium, named *Helicobacter pylorii,* is associated with gastric mucosa of humans and primates. The organism is implicated as the cause of chronic gastritis, peptic ulcers, and gastric adenocarcinoma. The organism is able to colonize the acidic environment because of its high ureolytic activity, which creates a microenvironment rich in ammonia. Species of *Helicobacter* also have been isolated from the stomachs of cats and dogs (*H. felis*), ferrets (*H. mustelae*), and other wild animals.

B. Ruminal Bacteria

The bacterial flora of the rumen of cattle is the most extensively investigated and clearly described among the gut bacterial ecosystems. A large variety of bacterial species exist in the rumen. The numbers range from 10^8 to 10^{11}/g of ruminal contents and, with the type of diet the animal consumes, profoundly impact the numbers and species composition. Generally, the numbers are reflective of the digestibility of foodstuffs. Typically, bacterial numbers are up to tenfold higher in grain-fed cattle than in forage-fed animals. The majority of the bacteria are obligate anaerobes. Facultative bacteria comprise a very small fraction of the flora. Ruminal bacteria are predominantly gram negative and rod shaped but vary greatly in their substrate specificity and nutritional requirements (Table IV). Some ruminal bacteria (*Butyrivibrio*) are considered "generalists" because of their ability to ferment a range of substrates, including cellulose, hemicellulose, pectin, starch, and protein. Others are "specialists" that utilize a narrow range of substrates. Bacteria capable of utilizing cell wall (cellulose and hemicellulose) and storage polysaccharides (starch) comprise a large fraction of the flora.

Ruminal bacteria are classified into functional groups based on the substrate fermentation (e.g.,

TABLE IV
Some of the Predominant Bacterial Species of the Rumen of Cattle and Sheep

Species	Major substrates	Major fermentation product
Anaerovibrio lipolytica	Lipid, lactate, glycerol	Acetate, propionate
Butyrivibrio fibrisolvens	Cellulose, hemicellulose, pectin, protein	Acetate, butyrate, H_2
Eubacterium ruminantium	Sugars	Acetate, butyrate, lactate
Fibrobacter succinogenes	Cellulose	Acetate, succinate
Lachnospira multiparus	Pectin, protein	Acetate, lactate, formic, H_2
Lactobacillus vitulinus	Sugars, starch	Lactate
Megasphaera elsdenii	Lactate, sugars, starch	Acetate, propionate, butyrate, H_2
Methanobrevibacter ruminantium	Hydrogen, formate	Methane
Prevotella rumincola	Hemicellulose, pectin, protein	Acetate, succinate
Ruminobacter amylophilus	Starch, protein	Acetate, succinate
Ruminococcus albus	Cellulose, hemicellulose	Acetate, formate, ethanol, H_2
Ruminococcus flavefaciens	Cellulose, hemicellulose	Acetate, succinate, H_2

cellulolytics, amylolytics, or proteolytic) and on the types of fermentation products produced (e.g., lactate producers or methanogens). In pure cultures, ruminal bacteria produce a number of fermentation products, such as lactate, succinate, formate, and hydrogen, which are normally present in low concentrations in the rumen. In addition to interspecies hydrogen transfer, which reduces the need to produce the electron-sink products, cross-feeding among ruminal bacteria allows utilization of the product of one bacterial species by another. Interspecies hydrogen transfer in the ruminal habitat involves almost exclusively methanogens as the hydrogen utilizers. Although acetogens have been shown to be present, they are unable to compete effectively with methanogens for hydrogen. Rapid utilization of hydrogen by methanogens keeps the partial pressure of hydrogen extremely low, which permits direct oxidation of reduced cofactors in nonmethanogenic bacteria. Therefore, production of hydrogen-sink products is not required. Interspecies hydrogen transfer decreases lactate, succinate, propionate, and ethanol and increases acetate.

Ruminal bacteria are nutritionally fastidious. Besides requirements for ammonia, amino acids or peptides, and B vitamins, certain ruminal bacteria, particularly fiber digesters, require one or more branched chain fatty acids (isobutyrate, isovalerate, and 2-methylbutyrate). The branched-chain fatty acids are required for the synthesis of branched amino acids or long-chain fatty acids. These acids cannot be replaced by amino acids. Therefore, culture medium employed to cultivate ruminal bacteria traditionally contains clarified ruminal fluid.

The ruminal habitat also is inhabited by certain unusual bacteria. Many of the morphological descriptions are based on electron microscopic examination of ruminal contents. The morphological types include large oval forms; crescentic cells (selenomonads); large cigar-shaped, septate, and spore-forming cells (*Oscillospira*); and large cocci in tetrads or sheets (*Sarcina* and *Lampropedia*). Relatively little is known of their physiology and ecological roles.

C. Human Intestinal Bacteria

In the human gastrointestinal tract, the esophagus, stomach and duodenum usually are not colonized by bacteria. The small intestine becomes increasingly colonized along its length, so that the bacterial flora of the ileal region of the small intestine resembles that of the colon. Obligate anaerobic counts from ileal contents have ranged from 10^6 to 10^{10}/g, whereas in the colon the numbers of anaerobes could range from 10^9 to 10^{11}/g.

Knowledge of human intestinal bacteria has been gained mainly from compositional analysis of feces because of the difficulty in obtaining *in situ* luminal samples. *In situ* samples generally are taken during surgery or from people dying from accidental injuries. Samples collected during surgery are likely to come from persons with some disease condition, and patients also are likely to have been starved and often have received antibiotics. The most representative samples are likely from persons dying from trauma, provided collection occurs soon after death. However, bacterial composition of feces generally is regarded as being fairly representative of the colonic flora.

The colonic bacterial flora in adult humans is extremely complex, and obligate anaerobes comprise 98 to 99% of the total flora. As many as 300 species have been isolated, but many of them have not been described fully. The dominant organisms belong to the genera *Bacteroides*, *Bifidobacterium*, *Eubacterium*, *Fusobacterium*, *Lactobacillus*, *Prevotella*, and *Ruminococcus* (Table V). Species of *Clostridium* are only minor components. The major factor affecting the composition of the flora is the diet. Also, antibiotic administration, particularly if prolonged, can have a major effect on the bacterial flora, often resulting in adverse changes. In such instances, the protective effect of the normal flora is compromised, leading to proliferation of pathogens such as *C. difficile*.

The colonic microflora in infants is less complex. At birth, the gastrointestinal tract is sterile but rapidly becomes colonized with bacteria from the surround-ings, particularly from the mother. In breast-fed infants, the colonic flora is composed largely of *Bifidobacterium* and *Lactobacillus* species, which may have the protective function of preventing colonization by enteropathogens. When infants are weaned and start to consume dry food, the flora gradually changes to resemble that of the adult human.

Despite differences in flora between the human colon and the rumen of ruminants (absence of protozoa in the colon, for example), the fermentative metabolism is very similar. Acetic, propionic, and butyric acids are the major products, with interspecies hydrogen transfer preventing the accumulation of electron-sink products like lactate or ethanol. Methanogens are present in the colon of some humans, and in their absence, sulfate-reducing bacteria are present.

The human colonic flora are not essential to the nutrition of the host, although volatile fatty acids (VFA) and vitamins may be absorbed into the blood and contribute to nutrition. The importance of gut bacteria lies in their involvement in diseases, either directly or indirectly. A significant function of the intestinal flora is to exclude enteropathogens. Use of broad spectrum antibiotics can cause overgrowth of pathogens, often resistant to the antibiotics, such as *Staphylococcus aureus*, *Salmonella typhimurium*, and

TABLE V
Some Bacteria Frequently Isolated from the Intestinal Contents of Humans, Pigs, and Chickens

Human colon	Pig colon	Poultry cecum
Actinomyces naeslundii	*Bacteroides uniformis*	*Bacteroides fragilis*
Bacteroides distanosis, B. fragilis, B. theta-	*Butyrivibrio fibrisolvens*	*Bifidobacterium bifidum*
iotaomicron	*Clostridium perfringens*	*Clostridium perfringens, C. beijerinckii*
Bifidobacterium adolscentis, B. infantis,	*Eubacterium aerofaciens*	*Eubacterium* spp.
B. longum	*Fibrobacter succinogenes*	*Fusobacterium* spp.
Clostridium bifementans, C. perfringens,	*Lactobacillus acidophilus, L. brevis,*	*Gemmiger formicillis*
C. ramosum	*L. cellobiosus, L. fermentans,*	*Lactobacillus acidophilus, L. fermentans,*
Eubacterium aerofaciens, lentum	*L. salivarius*	*L. salivarius*
Fusobacterium necrophorum, F. mortiferum	*Peptostreptococcus productus*	*Ruminococcus obeum*
Lactobacillus spp.	*Prevotella ruminicola*	*Streptococcus faecium, S. faecalis*
Peptostreptococcus productus, P. prevotii	*Ruminococcus flavefaciens*	
Propionibacterium acnes	*Selenomonas ruminantium*	
Ruminococcus albus, R. bromii	*Streptococcus bovis, S. equinus,*	
Streptococcus intermedius	*S. faecalis, S. salivarius*	
Veillonella spp.	*Veillonella* spp.	

Candida albicans. Another pathological condition called "small bowel syndrome" or "contaminated small bowel syndrome" results from overgrowth of bacterial population in the upper small intestine. The counts of bacteria increase 1 to 2 logs or more higher than in healthy humans. Anatomical defects (diverticulitis) or physiological abnormalities (e.g., lack of acid secretion in the stomach or stasis) will encourage overgrowth. Pathophysiological consequences include malabsorption and nutritional deficiencies resulting directly from bacterial activity (mucosal damage by bacterial products) and indirectly by competing for nutrients (vitamin B_{12}). Pseudomembranous colitis, caused by *C. difficile,* which almost always occurs in association with antibiotic use, has been well investigated. The proliferation of *C. difficile* is attributed mainly to the removal of competition from the normal flora. Approximately 3% of healthy adults harbor *C. difficile* in the intestinal tract. Surprisingly, healthy infants harbor both the organism and the toxin with no pathological consequences, suggesting age-related susceptibility.

Another aspect of gut flora that has been the focus of intensive research is their potential etiological role in colon cancer. The high incidence of colon cancer in North America and Western Europe is related to altered bacterial flora and production of carcinogens because of consumption of diets rich in fats and proteins of animal origin. Nitrosamines, derivatives of steroids and bile acids, and other compounds produced by the anaerobic colonic bacteria are suspected to be involved in colon cancer.

D. Intestinal Bacteria of Pigs

Studies concerning intestinal bacteria of pigs are of interest because of their potential role in protection against diseases. Gram positive bacteria outnumber gram negative organisms in colonic contents and feces of pigs. Species of gram positive bacteria belong to the genera *Lactobacillus, Streptococcus, Peptococcus, Peptostreptococcus,* and *Eubacterium* (Table V). In contrast, the flora of the cecum is predominantly gram negative, with *Bacteroides* and *Selenomonas* being the major constituents. The significance of hindgut fermentation in pigs is difficult to assess and, undoubtedly, is related to the diet. Absorption of

VFA from the pig colon has been well established, and evidence indicates that the rate of absorption for pig cecal and colonic mucosa is higher than that for equine hindgut mucosa and bovine ruminal epithelium. In pigs fed a grain diet, hindgut fermentation is of little value. However, microbial fermentation may have a significant impact on the energy metabolism of pigs eating a more natural diet that is higher in roughage. Significant methanogenic activity also occurs in the hindgut of pigs.

E. Intestinal Bacteria in Chickens

The gastrointestinal tracts of birds, like those of humans and animals, are inhabited by a variety of microorganisms. Because it is an important source of human food, the domestic chicken has been studied in some detail. Although crop, proventriculus, and gizzard harbor bacteria, the cecum is the region that provides a stable habitat for microorganisms. Ceca, like fermentation chambers in herbivores, have a dense population of obligate anaerobes, ranging in number from 10^9 to 10^{11}/g of cecal contents. The flora includes gram positive and gram negative cocci and rods, and some of the predominant genera include *Bacteroides, Eubacterium, Clostridium Peptostreptococcus,* and *Propionibacterium* (Table V). Two major factors known to affect cecal flora are diet and age of the chicken. Many studies have been conducted to evaluate effects of the major dietary components, carbohydrates and proteins, and also growth-promoting supplements, including antibiotics, on the cecal flora of chickens. Prior to hatching, the intestinal tract is sterile but within a few hours of hatching, streptococci, coli-aerogenes bacteria, and *Clostridium* colonize the region. However, within a few days, these facultative species are replaced by an anaerobic flora, which continues to change for several weeks and becomes increasingly complex. One consistent feature throughout the microbial development of the cecum is the presence of a large number of organisms capable of utilizing uric acid, a dominant nitrogenous compound in the urine of birds. Uric acid-degrading bacteria range from 10^8 to 10^9/g of contents. These bacteria may be important in the recycling of nitrogen in birds. The colon is quite short in chickens, and its floral composition is similar to that of the ceca.

V. PROTOZOA

The protozoa are highly specialized eukaryotic cells that are able to compete and coexist with bacteria in the gastrointestinal tract of many animals. Because of their relatively large size and active motility, a protozoan was the first microorganism to be discovered in the gastrointestinal tract. Unlike bacteria, which exist somewhere in the gastrointestinal tract of all animals, protozoa are not part of the normal flora in all animals. The occurrence of protozoa in the digestive tract is dependent on an environmentally compatible region of the tract (close to neutral pH) and a retention time for contents that exceeds the protozoan is generation time (6 to 24 hr). Protozoa are found in the rumens of all wild and domestic ruminants and camelids and also in the cecum and colon of some nonruminant herbivores, such as horses, elephants, and hippopotamus.

Protozoa that occur in the digestive tract of animals are grouped broadly into flagellates and ciliates. In most animals, flagellates occur in low numbers ($<10^3$/g); therefore, their contribution to overall gut fermentation is considered to be nonsignificant. However, intestinal flagellates do play a significant role in the hindgut digestion in termites and other insects, such as wood roaches, that thrive largely on cellulosic materials.

In ruminants and herbivorous nonruminants, ciliated protozoa constitute an important component of the gut flora and play a significant role in the nutrition of the host. Ciliated protozoa are classified further into holotrichs and entodiniomorphs based on certain morphological features, e.g., ciliary arrangement and shape and location of nucleus. Holotrichs have cilia covering the entire or almost the entire surface of the cell, whereas the entodiniomorphid ciliates have restricted zones of cilia. Holotrichs are generally a smaller fraction of the total ciliates and primarily ferment soluble carbohydrates. Entodiniomorphid ciliates, the dominant fraction in most animals, digest starch and structural polysaccharides (cell wall polysaccharides).

The total and generic and species compositions of ciliated protozoa are dependent on the type of host, its geographical location, nature of the diet, and frequency of feeding. Most of the studies on gut ciliated protozoa have been with the ruminal habitat. The majority of the ciliates are entodiniomorphids and many of the species are unique to ruminants and camelids. The concentration varies markedly between animals, ranging from 10^4 to 10^6/g of contents. Generally, information on hindgut protozoa is limited. Quantitative and species compositions are available from horses and elephants and show numbers ranging from 10^3 to 10^5/g of contents. As many as 70 species of ciliated protozoa have been identified in the hindgut of the horse (Table VI).

Protozoa are anaerobic and can ferment carbohydrates, proteins, and lipids. Fermentation products include VFA, lactic acid, CO_2, and hydrogen. Many of the anaerobic protozoa contain cytoplasmic organelles called hydrogenosomes, which are somewhat analogous to mitochondria of higher cells. These membrane-bound structures provide compartmentation to protect O_2-sensitive enzymes and produce hydrogen from the reoxidation of reduced cofactors. The hydrogenosomes also confer ciliates a certain degree of tolerance to oxygen.

Information on enzymes and metabolic pathways

TABLE VI
Ciliated Protozoa in Gastrointestinal Tract of Herbivores

Location	Number per ml or g of contents			
	Cattle	Sheep	Horse	Elephant
Rumen	10^4–10^6	10^4–10^6	–	–
Cecum	0	0	2–223 $\times$ 10^4	401 $\times$ 10^4
Colon	0	0	2–706 $\times$ 10^4	4.6 $\times$ 10^3
No. of ciliate genera	15–17	15–17	10–15	10–15
No. of ciliate species	50–55	50–55	40	17

in protozoa is very limited because of the difficulty of culturing them *in vitro*. Ciliates also are colonized by bacteria on their surface, so obtaining an axenic culture of ciliates is difficult. Therefore, much of the information has been obtained from either washed cell suspensions treated with antibacterial agents or cell-free extracts for studies on enzymatic activities. Another approach frequently used to elucidate the role of protozoa in gut ecosystems is to achieve ciliate-free status. The technique is popularly called defaunation. A number of physical or chemical treatments and dietary manipulations have been described. The chemicals used are selectively toxic to ciliates. Another method to achieve defaunation is to isolate the newborn from the mother immediately after birth and raise it in complete isolation from other animals. However, in reality, achieving and maintaining ciliate-free status for a long period of time are not easy. Elimination of ciliated protozoa from the gastrointestinal ecosystem impacts profoundly on bacterial and fungal populations because of predatory activity of ciliates. Therefore, fermentative changes observed in defaunated animals are the results of absence of protozoa and increases, and possibly compositional changes, in bacterial and fungal populations.

One question that has been of interest to gut microbiologists concerned with ciliated protozoa is how essential they are to the host animal. Many of the protozoal activities do benefit the host. However, bacteria can provide equally well all the fermentative activities that ciliates can. Therefore, ciliates have no unique contribution to make to the host. Growth and digestion trials with defaunated and normal, faunated animals indicate that protozoa are not essential to the host. However, ciliated protozoa may play a significant role, either negative or positive for the host, depending on the diet.

VI. FUNGI

The existence of fungi as part of the gut microbial flora was recognized relatively recently. Early researchers had documented the existence of flagellated protozoa, but further study showed that some of the flagellates were actually fungal spores released from sporangia of the rhizoids. The presence of chitin in the cell walls of these organisms confirmed that they were true fungi. The first report of isolation of an anaerobic fungus, a species of *Neocallimaltix*, from the rumen of sheep was published in 1975. Until

TABLE VII
Anaerobic Fungi Isolated from Herbivores

Genus	Thallus type	Flagellation	Species
Anaeromyces (*Ruminomyces*)	Polycentric	Single	*A. elegans*
			A. mucronatus
Caecomyces	Monocentric or polycentric	Single	*C. communis*
			C. equi
Neocallimastix	Monocentric	Multi	*N. frontalis*
			N. patriciarum
			N. hurleyensis
			N. variabilis
Orpinomyces	Polycentric	Multi	*O. bovis*
			O. joyonii
Piromyces	Monocentric	Single	*P. communis*
			P. mae
			P. dumbonica
			P. rhizinflata
			P. minutus
			P. spiralis

then, all known fungi were aerobes or facultative anaerobes, and the detection of fungus that was a strict anaerobe was a novel discovery.

Three groups of fungi occur in the gastrointestinal tract. The first group includes yeasts (*Saccharomyces*, *Candida*) and aerobic molds, which are transient, do not grow anaerobically, and are nonfunctional. However, some of the aerobic fungi could become opportunistic pathogens when conditions become favorable (immune compromised) and invade the gut wall to set up mycotic infection. The second group consists of species that parasitize certain ophryoscolecid protozoa. Two species described as chytrid fungi have been identified. Little is known about these species, and their significance remains to be elucidated. However, heavy infection of the protozoal cell with these species may result in death of the host cell. The third group includes fungi that colonize on plant material, are considered to be indigenous to the gut, and are believed to make a significant contribution to fermentation in the gastrointestinal tract. The life cycle of anaerobic fungi consists of alternating stages of motile, flagellated zoospores and vegetative, mycelial cells. The zoospores are free-living in the liquid phase of the digesta, and the vegetative stage colonizes the digesta fragments. Generally, the alternation between the stages takes about 24 to 32 hr.

All known anaerobic fungi are zoosporic, with the zoospores being either monoflagellated or multiflagellated. The vegetative growth, the thallus, may be either monocentric or polycentric. The genera of anaerobic fungi are defined on the basis of thallus morphology, rhizoid type (filamentous or bulbous), and number of flagella per zoospore, and the species are differentiated mainly on zoospore ultrastructure. In a monocentric fungus, either the encysted zoospore retains the nucleus and enlarges into a sporangium (called endogenous zoosporangial development) or the nucleus migrates out of the zoospore, and the zoosporangium is formed in the germ tube or rhizomycelium (called exogenous zoosporangial development). In both types of monocentric development, only one zoosporangium is formed per thallus, and only the zoosporangium contains nuclei. In a polycentric fungus, the nucleus migrates out of the encysted zoospore (exogenous sporangial develop-

ment) and undergoes mitosis in the rhizomycelium, which subsequently forms multiple sporangia. Thus, in polycentric fungi, both the zoosporangia and the rhizomycelium contain nuclei. Like many eukaryotes adapted to anaerobic growth conditions, anaerobic fungi lack mitochondria. They obtain energy by the fermentation of carbohydrates, which act as both electron acceptors and electron donors. They have mixed acid fermentation profiles. They ferment sugars to formate, acetate, lactate, succinate, ethanol, CO_2, and H_2. Anaerobic fungi contain hydrogenosomes, organelles containing enzymes capable of transfering reducing power from glycolytic products to hydrogen, which is then excreted to permit more ATP production via glycolysis.

Until recently, anaerobic fungi were isolated only from the gut contents or the feces of animals. However, organisms similar to *Neocallimastix* and *Orpinomyces* species were isolated from anoxic regions of a pond in a cow pasture. The significance of anaerobic fungi outside the gastrointestinal tract is not known.

Fungal biomass in the gut is difficult to quantify. Based on the chitin content, it has been estimated to be about 8 to 10% in the rumen of cattle. Zoospore counts can be obtained easily but, because of diurnal variation, counts do not provide adequate estimates of the fungal biomass. Also, colony counts are not reflective of the number, because fragments of mycelium of polycentric fungi can develop into colonies. Therefore, the term "thallus-forming units" is used for enumeration. An endpoint dilution procedure, based on most-probable number, has been developed to enumerate thallus-forming units as an estimate of the activity of the fungi in gut contents. Counts usually range from 10^3 to 10^7/g of ruminal contents.

The fungi produce a wide array of enzymes that can hydrolyze a range of glycosidic linkages, digest the major structural polysaccharides of plant cell walls, and allow the fungi to grow on a number of polysaccharides. Many of the polysaccharide-hydrolyzing enzymes are localized on rhizoids and rhizomycelia (vegetative stage). Anaerobic zoosporic fungi produce some of the most active polysaccharidases and phenolic acid esterases yet reported. Therefore, there is considerable interest in the biology of anaerobic fungi because of their potential application in biomass conversions.

VII. BACTERIOPHAGES

Bacteriophages have been shown to be present in the ruminal contents of cattle, sheep, and reindeer and in the cecum and colon of horses and may occur in the gut contents of all animals. Bacteriophages are the least studied microbes of the gut. Initial reports of phagelike particles were based on electron microscopic examination of gut contents. Therefore, in many instances a specific phage host was not identified. Both temperate and lytic phages have been demonstrated. More than 125 morphologically distinct phagelike particles have been documented in the ruminal contents of cattle. Some of the varieties may be degenerate forms of mature phages. It is estimated that 20–25% of ruminal bacteria may harbor temperate phages. Only a few studies have identified the specific bacterial host and have led to molecular analysis.

In a gut ecosystem, the presence of prophages in bacteria may confer a competitive advantage, outweighing the burden of additional DNA. Possible advantages could be protection against superinfection by the same phage or against infection by related phages. The possession of prophages also may enhance the genetic potential of bacteria to adapt to change in their environment.

VIII. ROLE OF GUT MICROBES

The association of microbes with the gastrointestinal tract has resulted in the development of a balanced relationship between the host and the microbes. The interaction, for the most part, is beneficial; however, the association also could result in some negative or harmful interactions. The beneficial roles include nutritional interrelationships between the gut microbes and the host and the influence of microbes in preventing the establishment of pathogens in the gastrointestinal tract.

A. Gut Microbes and Host Nutrition

The importance of gut microbes to the host's nutrition is well documented in ruminants with foregut fermentation and in herbivorous nonruminants with hind gut fermentation. Because of microbial activity, ruminants and herbivorous nonruminants are able to utilize cell wall polysaccharides of plants as sources of energy. Another feature that makes ruminants unique is their ability to convert nonprotein nitrogen into protein. In monogastrics, the products of microbial energy-yielding metabolism, such as short-chain fatty acids, lactate, and ethanol are absorbed and utilized as carbon and energy sources by the animal tissues. Furthermore, products from lysis or digestion of microbial cells, such as proteins and vitamins, are absorbed and utilized by the animal tissues. This is supported by the observation that germ-free animals have higher requirements of B vitamins than do conventional animals.

Besides nutritional interrelationships, the normal gut flora also influences various physiological functions of the gastrointestinal tract. Evidence in support of such influences has come primarily from studies involving comparison of animals with normal gut flora and germ-free animals. Some of the intestinal functions influenced by the gut include:

1. **Transit time in the tract.** Digesta passes more rapidly in conventional animals compared to germ-free animals. The mechanism is not fully understood but possibly involves microbial products influencing smooth muscle activity.

2. **Cell turnover and enzymatic activities in the small intestine.** The cell turnover is higher and enzymatic activity is lower in conventional animals than in germ-free animals.

3. **Water, electrolyte, and nutrient absorption.** The water content of the digesta is lower and electrolyte contents (e.g., Cl^- and CHO^{-3}) are higher in the lumen of conventional animals than in that of the germ-free animals.

B. Normal Flora and Prevention of Infections

The presence of normal flora reduces the chances for pathogens to get established in the gastrointestinal tract. The dense population of normal flora occupies the niches and space, thus making it difficult for small numbers of pathogens to compete and get established. This phenomenon is referred to as the

Nurmi concept (named after the Finnish scientist who first described it); colonization resistance; bacterial antagonism; microbial interference; or, most commonly, competitive exclusion. Interest in competitive exclusion has existed for many years. However, the topic is receiving greater attention because of *C. difficile* infections in humans and potential application of the concept to reduce foodborne pathogens in chickens, pigs, and other animals. For example, *Campylobacter* or *Salmonella* infections could be prevented or minimized by feeding chicks anaerobic cecal flora from adult birds. Poultry products contaminated with *Salmonella* or *Campylobacter* are the major sources of foodborne infections in humans. The mechanisms involved in competitive exclusion are multifactorial and include regulatory forces exerted by the host, the diet, and the microbes. Some of the major microbial factors include competition for attachment sites, low pH, low redox potential, and elaboration of antimicrobial substances (e.g., VFA, lactic acid, and bacterocin).

C. Normal Flora as the Cause of Infections

Members of the intestinal flora, which, for the most part, are beneficial and nonpathogenic as long as they reside in the intestinal tract, can cause infections in other sites in the body. Such infections often are associated with a predisposing condition that allowed the bacteria to cross the gut wall barrier (e.g., perforation of the intestines, inflamed gut wall, and surgical procedures). The infections generally are caused by mixtures of anaerobic and facultatively anaerobic bacteria and can occur in any site of the body, with abdominal cavity and liver being the common sites. Liver abscesses in cattle are classic examples of a disease condition caused by a bacterium that is a normal inhabitant of the gut but becomes a pathogen when it reaches the liver. Liver abscesses are observed frequently in beef cattle fed high-grain diets, and the primary etiologic agent is *Fusobacterium necrophorum*. The organism is a normal component of the flora of the rumen of cattle and is involved in fermentations of proteins and lactate. In cattle fed high-grain diets, the ruminal wall integrity is compromised because of high acidic conditions, and *F.*

necrophorum invades and colonizes the ruminal wall, then enters the portal circulation and reaches the liver. The organism gets trapped in the capillary system of the liver, where it grows and causes abscesses in the hepatic parenchyma.

Although hundreds of species of anaerobic bacteria reside in the gastrointestinal tract, only a small fraction of these are capable of causing infections. Generally, the species isolated from the infections belong to the genera *Bacteroides*, *Fusobacterium*, *Clostridium*, and *Peptostreptococcus*. The facultative species include *Escherichia coli*, *Enterobacter aerogenes*, *Klebsiella* spp., and *Streptococcus* spp. The pathogenic mechanisms and virulence factors involved in intestinal anaerobic infections are not understood fully. All gram negative anaerobes have endotoxic lipopolysaccharide, and their biological effects are similar to those of aerobic gram negative bacteria. Also, certain fementative products of anaerobes, such as succinic acid and amines, could contribute to the virulence. In the case of *B. fragilis*, a frequent isolate in abdominal sepsis of humans, the virulence factors include a polysaccharide capsule and an array of extracellular enzymes.

IX. CONCLUSIONS

The gastrointestinal tracts of all animals harbor a myriad of microbes performing a variety of metabolic activities that play a vital role in the health of the host. The gastrointestinal tract is a complex anaerobic microbial ecosystem. The microbial species include bacteria, protozoa, fungi, and bacteriophages. The functionally important and dominant microbes are anaerobes with fermentative metabolism. Among the gut ecosystems the rumen of domestic ruminants and the colon of the human have been investigated extensively. The interaction between the microbes and the host is complex and includes various degrees of cooperation and competition, depending on the species of animal. The cooperative interactions include nutritional interdependence and the role of normal flora in preventing the establishment of enteric pathogens. Members of the normal flora that are nonpathogenic as long as they reside in the intestinal tract can cause infections in other sites, particularly the abdominal cavity and liver, if situations allows the microbes to cross the gut wall barrier.

See Also the Following Articles

HELICOBACTER PYLORI • INTESTINAL PROTOZOAN INFECTIONS IN HUMANS • LACTIC ACID BACTERIA • RUMEN FERMENTATION

Bibliography

Bonhomme, A. (1990). Rumen ciliates: Their metabolism and relationships with bacteria and their hosts. *Anim. Feed Sci. Technol.* **30**, 203–266.

Cheng, K. J., and Costerton, J. W. (1986). Microbial adhesion and colonization within the digestive tract. *In* "Anaerobic Bacteria in Habitats Other than Man" (E. M. Barnes and G. C. Mead, eds.), pp. 239–261. Blackwell Scientific Publ., Boston, MA.

Clarke, R. T. J., and Bauchop, T. (1977). "Microbial Ecology of the Gut." Academic Press, New York.

Dehority, B. A. (1986). Protozoa of the digestive tract of herbivorous mammals. *Insect Sci. Applic.* **7**, 279–296.

Hentges, D. J. (1983). "Human Intestinal Microflora in Health and Disease." Academic Press, New York.

Hill, M. J. (1986). "Microbial Metabolism in the Digestive Tract." CRC Press, Boca Raton, FL.

Hobson, P. N., and Stewart, C. S. (1997). "The Rumen Microbial Ecosystem." Blackie Academic Publication, New York.

Hobson, P. N., and Wallace, R. J. (1982). Microbial ecology and activities in the rumen: Parts I and II. *Crit. Rev. Microbiol* **9**, 165–295.

Gibson, G. R., and Macfarlane, G. T. (1995). "Human Colonic Bacteria." CRC Press, Boca Raton, FL.

Mackie, R. I., and White, B. A. (1997). "Gastrointestinal Microbiology. Vol. 1. Gastrointestinal Ecosystems and Fementations." Chapman and Hall, New York.

Mountfort, D. O., and Orpin, C. G. (1994). "Anaerobic Fungi: Biology, Ecology, and Function." Marcel Dekker, Inc., New York.

Mackie, R. I., White, B. A., and Isaacson, R. E. (1997). "Gastrointestinal Microbiology. Vol. 2. Gastrointestinal Microbes and Interactions." Chapman and Hall, New York.

Savage, D. C. (1986). Gastrointestinal microflora in mammalian nutrition. *Ann. Rev. Nutr.* **6**, 155–78.

Trinci, A. P. J., Davies, D. R., Gull, K., Lawrence, M. I., Nielsen, B. B., Rickers, A., and Theodorou, M. K. (1994). Anaerobic fungi in herbivorous animals. *Mycol. Res.* **98**, 129–152.

Williams, A. G., and Coleman, G. S. (1991). "The Rumen Protozoa." Springer-Verlag, New York.

Wolin, M. J. (1981). Fermentation in the rumen and human large intestine. *Science* **213**, 1463–1468.

Wubah, D. A., Akin, D. A., and Borneman, W. S. (1993). Biology, fiber degradation, and enzymology of anaerobic zoosporic fungi. *Crit. Rev. Microbiol.* **19**, 99–115.

Genetically Modified Organisms: Guidelines and Regulations for Research

Sue Tolin

Virginia Polytechnic Institute and State University

Anne Vidaver

University of Nebraska

I. Concern over Genetically Modified Organisms
II. History of Guidelines and Regulations
III. Oversight Mechanisms
IV. Compliance and Approvals for Research in the United States
V. Appropriateness of Guidelines and Regulations for Research
VI. Conclusions

GLOSSARY

confinement Procedures to keep genetically modified organisms within bounds or limits; usually, in the environment, with the result of preventing widespread dissemination.

containment Conditions or procedures that limit dissemination and exposure of humans and the environment to genetically modified organisms in laboratories, greenhouses, and some animal-holding facilities.

genetically modified organism (GMO) Any organism that acquires heritable traits not found in the parent organism; while traditional scientific techniques such as mutation can result in a GMO, the term is most frequently used to refer to modified plants, animals, and microorganisms that result from deliberate insertion, deletion, or other manipulation of deoxyribonucleic acid (DNA); also referred to as genetically engineered organisms or as organisms with modified hereditary traits.

oversight Application of appropriate laws, regulations, guidelines, or accepted standards of practice to control the use of an organism, based on the degree of risk or uncertainty associated with that organism.

recombinant DNA Broad range of techniques in which DNA, usually from different sources, is combined *in vitro* and then transferred to a living organism to assess its properties.

THE TERM GENETICALLY MODIFIED ORGANISM (GMO) is most frequently used to refer to an organism that has been changed genetically by recombinant DNA techniques. Historically, research with GMOs has been subject to special oversight that, to this day, differs, depending on the location of the research with the organism, whether inside (contained) or outside (so-called field research), type of organism (e.g., plant, animal, microorganism) or use (e.g., medical, agricultural, environmental), and country in which one works. The oversight mechanism for contained research is principally through guidelines developed by scientists and endorsed by the private and public sector. Outside research is currently overseen by a number of federal agencies. In some countries, such as the United States, there can be overlapping jurisdictions, differing interpretations of legal statutes, and different requirements or standards for compliance by scientists who do research with GMOs in the outside environment. Scientific issues deal with differences in perspective on the risks of introductions of GMOs into the environment, the types of data required prior to the introduction to conclude the experiment is of low risk, and the types of monitoring and mitigation practices, if necessary. Nonscientific issues are also considered and include those dealing with legal and social concerns. These differences in interpretation have resulted in few introductions into the environment of microorganisms.

I. CONCERN OVER GENETICALLY MODIFIED ORGANISMS

A. The Concern over Safety

The new biology, dating to the 1970s and usually encompassing recombinant DNA techniques, enabled scientists to perform modifications of organisms with great precision and to combine DNA of organisms that can not, in current time, combine; yet these combinations are derived from components of naturally occurring organisms. The scientific community raised hypothetical questions about the safety of genetically engineered organisms, and the public questioned the potential adverse effects of organisms with the new combinations of genetic information on humans and the environment. It was argued that, as such, the organisms have not been subjected to evolutionary pressures, including dissemination and selection, and may pose a risk to humans or the environment. However, it was recognized that genetic modifications can arise by classical or molecular methods, ranging from selection of desirable combinations by farmers or bakers since antiquity to nucleotide insertion or deletion by molecular biologists.

The new biology, often called biotechnology, has generated fear that the new organisms may be unpredictable in survival and dissemination and, particularly, that transfer of the gene for the modified trait to nontarget organisms might occur. However, gene transfer occurs whether or not humans intervene. Such gene transfers are expected to have minimal consequences unless selection is imposed. Increasing evidence supports the conclusion that microorganisms, particularly bacteria, usually maintain their fundamental characteristics and their essential identities and moderate the amount of change that can be absorbed by known and unknown mechanisms. Deleterious changes can occur, and will, whether or not microorganisms are manipulated. The preliminary testing under contained conditions that is requisite and standard practice in science should, however, identify most of such gross changes. Principles for assessing risk of GMOs as developed in scientific and public forums, are based on the premise that, if one begins with a beneficial organism and imparts a neutral or beneficial trait, the probability of harm from transfer of genetic information is minimal. Some scientists are more concerned about the widespread adoption of a beneficial organism in commerce, rather than about small-scale field research, because greater exposure is likely to increase the probability of risk.

The concern over GMOs resides partly on a perceived increase in ability to survive or persist. Thus, some scientists and consumers argue that oversight of GMOs should be as stringent as that for toxic chemicals, physical disruptions such as water control projects, or exotic organism introductions. The strongest arguments for these concerns are voiced by persons who compare the risks of GMOs with that of introducing exotic organisms. The appropriateness of this analogy can be questioned because there is generally a familiarity with the organism being modified and the trait being introduced, and the fraction of the new genetic information is quite small (Table 1). In contrast, exotic organisms are unfamiliar and are entire genomes.

There is also a perception by some that, should there be a problem with survival or dissemination of a microorganism, nothing can be done. Essentially, the assumption is that once the gene(s) is out, it cannot be recalled. However, orderly and inadvertent movement and dissemination of microorganisms occur repeatedly because of their presence on humans, plants, and animals that are moving throughout the world at increasing rates. There are also long-standing and environmental practices that are in use to decontaminate or mitigate unwanted effects of microorganisms. Such practices are known for microorganisms associated with plants and animals, as well as for free-living microorganisms. Immediate decontamination methods include, among others, burning, chemical control, and sanitation by various means. Short-term and long-term methods are abundant for plant- and animal-associated microorganisms, since a great deal of research on developing mitigation methods is conducted by scientists in the disciplines of plant pathology, veterinary medicine, and human medicine. Many of these deal with management practices and the use of genetic resistance and application of biological control organisms. Immunization of humans and animals is another type of long-term management practice to mitigate the effects of microorganisms.

TABLE I
Comparison of Exotic Species and Genetically Engineered Organisms

	Exotic organism[a]	Engineered organism[b]
No. of genes introduced	4000 to >20,000	1 to 10
Evolutionary tuning	All genes have evolved to work together in a single package	Organism has several genes it may never have had before. These genes will often impose a cost or burden that will make the organism less able to compete with those not carrying the new genes.
Relationship of organism to receiving environment	Foreign	Familiar, with possible exception of new genes

[a] "Exotic organism" is used here to mean one not previously found in the habitat.

[b] "Engineered organism" is used here to mean a slightly modified (usually, but not always, by recombinant DNA techniques) form of an organism already present in the habitat.

[From U.S. Congress, Office of Technology Assessment (1988). "New Developments in Biotechnology: Field Testing Engineered Organisms: Genetic and Ecological Issues." Washington, DC.]

B. Concerns over Genetically Modified Domesticated Organisms in Agriculture and the Environment

Virtually all domesticated organisms used in the production of food and fiber have been genetically modified over long periods of time, including certain live domesticated microorganisms used in making bread, beer, wine, various types of cheese, yogurt, and other foods. Selected microorganisms that have been shown to be beneficial are also widely used in the environment. These uses include, among others, microorganisms that fix nitrogen and provide nutrients for trees (mycorrhizae), as well as those used in sewage treatment plants and oil drilling. Also, naturally occurring pathogenic microorganisms are used in the testing of domesticated plants to ascertain their disease resistance. In such critical tests with known deleterious organisms, there has been no documented case of untoward effects, such as a plant disease epidemic, arising from such standard field trials.

It is widely accepted that the first step in risk assessment, whether in containment or in confined field trials, is identifying the risk by determining how much is known about the parental organism. It is also recognized that the risk can be minimized by the preferential selection of parental organisms that are generally recognized as safe because of their long history of use. In the oversight of food, such foods are categorized as GRAS, or generally recognized as safe. A similar category can be considered for microorganisms that would be introduced into the environment: GRACE, or those microorganisms that are generally recognized as compatible with the environment.

Examination of the food safety issues associated with genetic modifications has led to the conclusion that potential health risks are not expected to be any different in kind than with traditional genetic modifications. All such evaluations rest on knowledge of the food, the genetic modification, the composition, and relevant toxicological data. A recent international body concluded that rarely, if ever, would it be necessary to pursue all such evaluations exhaustively. There is reasonable agreement that a threshold should exist for regulation, or even of concern below which further evaluations on a genetically modified food product or its individual components need not be conducted. The International Food Biotechnology Council recommends flexible, voluntary procedures between food producers and processors and a regulatory agency.

II. HISTORY OF GUIDELINES AND REGULATIONS

The concern over the potential risks of GMOs led to the initiation of various mechanisms for the oversight of research conducted throughout the world. This oversight was in the form of guidelines, a set of principles and practices for scientists to follow, and regulation by laws applicable to certain processes or organisms used for certain purposes. The legal profession claims that this is the first case in which hypothetical or speculative risks have become the basis for regulation.

Codified guidelines date back to the 1970s, when the previously described concerns were raised. This led the United States National Institutes of Health (NIH), under the Department of Health and Human Services, to develop guidelines for containment of research involving recombinant DNA molecules. The first guidelines were first published in 1976 and have been updated and republished numerous times, most in recently 1998 (63 FR 26018). A public-meeting body of peers and nonscientists was assembled as the NIH's Recombinant DNA Advisory Committee (RAC) by the Office of Recombinant DNA Activities (ORDA) and was given the task to review all recombinant DNA experiments within the United States. Other countries soon followed suit.

The RAC was to assess the risk of the experiment and recommend containment conditions under which they thought the risk would be minimized for the laboratory worker and the environment. The resulting guidelines, which are now available from the NIH ORDA Website (http://www.nih.gov/od/orda), spelled out the recommended facilities and procedures for safety to individuals and to the environment. They included such specifics as the type of pipetting one should undertake, sterilization procedures, air filtration procedures, and decontamination and mitigation procedures.

Within a short time, most of the microorganisms and experiments had been assigned a containment level, and the responsibility for overseeing such experiments was decentralized and delegated to local institutional biosafety committees (IBCs) and to other institutions in other countries. Many experiments to modify common laboratory strains of bacteria and yeast were judged to pose no risk and were exempted from the guidelines or any containment requirements. Research with other microorganisms, including viruses, required containment no greater than one would use for research with the microorganisms that did not involve genetic modification experiments. These assignments were generally consistent with the recommendations of the biomedical

authorities, such as the Centers for Disease Control and Prevention in the United States.

Experiments involving introduction of GMOs into the environment were first reviewed and approved by the RAC in the early 1980s. However, they were not conducted until 1986, after investigators received approval for their research from regulatory agencies. This action signaled the beginning of oversight for such experiments by centralized regulatory authorities, rather than through guidelines that describe principles and practices for confinement of the GMO to minimize risk. In the United States, regulatory oversight currently includes research conducted by any party and is under the jurisdiction of either the United States Department of Agriculture's Animal and Plant Health Inspection Service (APHIS) or the Environmental Protection Agency (EPA). The former generally oversees plants and microorganisms that are or might be considered as plant pests, while the latter oversees research with so-called pesticidal organisms and microorganisms for other uses (industrial, manufacturing). Where there is research with microorganisms and plants, both agencies may be involved, as well as the Food and Drug Administration. More detailed descriptions of legal and jurisdictional issues can be found in publications included in the bibliography.

At the present time, there is no decentralized body in any nation for oversight of field research that is comparable to the IBC for contained research (Table II). The United States Department of Agriculture (USDA) published (Federal Register, Feb. 1, 1991) a draft of guidelines for conducting research under confinement in the open environment, prepared by an advisory committee of scientists. However, these guidelines were never implemented even though they provided generalized principles for assessing and managing the risks of microorganisms, as well as plant and animals that have been genetically modified, particularly by recombinant DNA. The principles laid out in these guidelines were sufficiently generic that they were to be applicable throughout the world. Later, in 1995, this USDA committee developed performance standards specifically for research with genetically modified fish and shellfish, since no statutory authority existed for oversight by a regulatory agency. These standards can be accessed

TABLE II
**Types of Oversight of Genetically
Modified Organisms[a]**

Stage of development	Oversight
Laboratory, greenhouse, animal pen[b]	Decentralized oversight from federal research or regulatory agency in the form of guidelines
Small-scale field research[c]	Guidelines and regulations: Combination of decentralized and federal oversight
Scale-up or large-scale testing	Federal regulations
Commercial products	Federal, state, international regulations

[a] Reflects current practices: The degree of oversight differs in each country and among different funding and regulatory agencies.

[b] For unmodified organisms (naturally occurring, chemically altered, spontaneous or selected mutants), standards of practice apply in research, whether conducted by the public or private sector.

[c] Includes tests on land and in enclosed waters.

through the Information Systems for Biotechnology website (http://www.isb.vt.edu) and give key points relevant for safety, are useful for researchers in designing containment systems and for IBCs in evaluating these designs.

In many countries, the oversight of GMOs is essentially the same as for unmodified organisms, except for the contentious issue of planned introduction into the environment. In countries such as the United States and Canada, a sizable bureaucracy has built up to oversee both the research and product development. Even though the risks remain speculative, the fear of litigation and unknown hazards has served to minimize the actual number of introductions, particularly of microorganisms. Of the approximately 800 tests of plants, animals, and microorganisms introduced into the environment for research, approximately 10% have been microorganisms. Most of these microorganisms were modified to have marker genes that enabled them to be monitored in the environment. The first functional genes added to microorganisms and field-tested in the early 1990s were those encoding an insecticidal toxin from *Bacillus thuringiensis* added both to a pseudomonad and a coryneform bacterium. In the former case, the modified

organism was killed before it was released. In the latter case, in field trials to test for insecticidal activity, plants inoculated with the GMO were successfully protected from the corn borer. But commercialization of this gene succeeded only when it was introduced into the genomes of corn (maize) and cotton.

III. OVERSIGHT MECHANISMS

A. Standards of Practice

All trained professionals learn and adhere to certain accepted procedures and practices with respect to safety and to the appropriate scientific methods for the profession. Most persons trained in microbiology and related disciplines using microorganisms are exposed to the same standards of practice that are used in training medically oriented professionals. These include safe preparation, use, and disposal of inocula and inoculated organisms, whether plants or animals, and appropriate decontamination/mitigation procedures, such as sterilization in contained facilities or incineration of animals or burial of plant material.

Standards of practice for all scientists include procedures appropriate for conduct of experiments; the use of data, including statistical analysis techniques; publication ethics and standards; and sharing of biological materials after publication of results. Standards of practice in the open environment are particularly evident in agricultural and forestry research; the sites are evaluated, plots are designed to enable statistical analysis of results, criteria for evaluation of results are widely distributed and agreed upon through peer review, and results are disseminated through various publications. Practices to preclude significant risk to the environment and to mitigate possible untoward effects are routinely considered and used by researchers.

B. Guidelines and Directives

Guidelines may be considered a statement of policy by a group having authority over that policy. Sometimes, such guidelines are also published as "points to consider." These guidelines offer assistance to investigators and do not have legal authority, except if required by a funding or regulatory agency. Guidelines are generally considered far more flexible than are directives or regulations. Guidelines that have been adopted worldwide for contained research with GMOs are those originating from the U.S. NIH. The NIH RAC guidelines have been considered *de facto* regulations, as commercial concerns have also adopted them as standards of good manufacturing practice. Since NIH is not a regulatory agency, it can require compliance only for its grantees.

Directives, particularly those issued by governments, are orders or instructions. Directives are currently an overarching method of oversight implemented by multinational groups, such as the European Union (EU). Countries are being urged to implement the directives within the next few years that would ideally harmonize oversight within the EU. However, individual countries would still have the prerogative of overseeing the details of such directives or even making them more stringent. Groups such as the United Nations, the World Health Organization (WHO), the Organization for Economic Cooperation and Development (OECD), and others also make informed statements for governments to consider as they develop their own oversight policies and mechanisms.

C. Regulations

Regulations are laws or rules to control or govern procedures or acts. In the case of GMOs, some countries have implemented new laws for oversight of field tests, especially in Europe. In other countries, no laws are currently applicable. In the United States, legal interpretations of current statutes have resulted in extensive oversight of research involving field trials. The laws have legal penalties for noncompliance, whether in the public or private sector. This type of oversight in the United States has resulted in an elaborate permitting, evaluation, interagency coordination and reporting system that may be viewed as time-consuming, cumbersome, and costly to some and inadequate to others. In addition to federal and national laws, other governmental entities, such as states or cities, have enacted their own regulations, compounding the difficulty of conducting field re-

search. Confidential business information is protected in all cases, including oversight by guidelines or by competent authorities, unless such information relates to safety issues.

IV. COMPLIANCE AND APPROVALS FOR RESEARCH IN THE UNITED STATES

A. Contained Research under the NIH Guidelines

The current version of the "Guidelines for Research Involving Recombinant DNA Molecules" is available from NIH Office of Recombinant DNA Activities (http://www.nih.gov/od/orda). Since 1978, compliance with these guidelines has been mandatory for all research utilizing recombinant DNA techniques at an institution receiving federal research funding. The guidelines are promulgated only for NIH-funded research, but all other Federal agencies funding research also require compliance with these guidelines. An investigator working with or constructing a GMO must first determine whether or not the research is exempt from or subject to the guidelines. If the research is exempt, no further action is needed but it is recommended that the containment conditions in Biosafety Level 1 (see following) be followed during the course of the research. If the particular research is not exempt from the guidelines, the Biosafety Level recommended for the research will be specified therein, and certification of compliance is the responsibility of the local IBC before or at the time the research is begun (notification). Only experiments with human gene therapy, and the occasional new organism or special request for change in recommended containment level, now require review by the RAC and approval of the NIH.

Appendix A of the guidelines lists organisms exempt from oversight because they naturally exchange genetic information with each other. Only bacteria are currently listed. Sublist A is the largest and includes all species in the genera *Escherichia, Shigella, Salmonella, Enterobacter, Citrobacter, Erwinia,* and certain species in the genera *Pseudomonas, Serratia,* and *Yersinia.* Sublist B includes eight species of *Bacillus,* Sublists C and D each list three species of *Strepto-*

myces. Sublists E and F include certain species of *Streptococcus.* If the research is not exempt, the investigator determines the recommended containment, depending on the type of experimentation and the risk level of the organism and vector system being used. The IBC certifies compliance for this level, in accordance with Section III of the guidelines.

Risk levels for many microorganisms can be found in Appendix B, "Classification of Human Etiologic Agents on the Basis of Hazard" and are summarized in Table III. Included are those biological agents known to infect humans, as well as selected animal agents that may pose theoretical risks if inoculated into humans. The listing in Appendix B reflects the current state of knowledge and serves as a resource document for commonly encountered agents, but is not intended to be all inclusive. The classification is based on agent risk assessment information compiled by the Centers for Disease Control and Prevention; further guidance on agents not listed in Appendix B may be obtained from the CDC. The list is to be reviewed annually by a special committee of the American Society for Microbiology, and its recommendations for changes are to be presented to the Recombinant DNA Advisory Committee for consideration as amendments to the NIH Guidelines.

Organisms that are not agents of human disease are not included in the listing in the guidelines. Those microbial agents associated with plant and animal diseases are subject to the regulations of APHIS. Permits from USDA are currently required for transporting any plant pathogenic agent between laboratories and specific containment conditions are specified on a case-by-case basis, often with on-site inspection by USDA personnel.

In Appendix C, exemptions are explained and listed. Exempt research includes using, as the host organisms for cloning DNA, *E. coli* K-12, *Saccharomyces cerevisiae* and *S. uvarum,* asporogenic *Bacillus subtillis* and *B. licheniformis,* and also certain extrachromosomal elements of gram positive organisms propagated and maintained in them. Using DNA from organisms that are Risk Group 3, Risk Group 4, or otherwise restricted is not exempt. Cloning potent vertebrate toxin genes is also not exempt, except under conditions explained in Appendix F. Appendix E lists certified cloning vectors, including plasmid, bacteriophage, and others, for each of the above host systems and for *Neurospora crassa, Streptomyces,* and *Pseudomonas putida.* The rationale for the safety of these host–vector systems, and the process for certification of new systems for a particular degree of biological containment based on the survival of the host and the potential for transfer of the

TABLE III
Classification of Biohazardous Agents by Risk Group

Risk Group	Definition	Examples
Risk Group 1 (RG1)	Agents that are not associated with disease in healthy adult humans	*Bacillus subtilis, E. coli* K–12, Adeno Associated Virus-types 1–4
Risk Group 2 (RG2)	Agents that are associated with human disease which is rarely serious and for which preventive or therapeutic interventions are often available	Forty bacterial genera, including *E. coli* O157:H7; 12 fungal genera; 31 parasites; common viruses in 18 families
Risk Group 3 (RG3)	Agents that are associated with serious or lethal human disease for which preventive or therapeutic interventions may be available (high individual risk but low community risk)	Nine types of bacterial agents, including *Ricksettsia;* 2 fungi; several viruses in 8 families including Hanta, HIV; CJD and other prions
Risk Group 4 (RG4)	Agents that are likely to cause serious or lethal human disease for which preventive or therapeutic interventions are not usually available (high individual risk and high community risk)	Extremely severe and emerging viruses in 6 families, including Lassa, Ebola, undefined hemorrhagic fever agents

Adapted from NIH Guidelines for Research Involving Recombinant DNA Molecules, Appendix B (1998).

vector to another organism, is explained in Appendix I—Biological Containment.

Detailed descriptions of physical facilities and practices to be followed by researchers to assure containment of organisms and safety to workers and the environment in standard laboratory research are given in Appendix G. Four Biosafety Levels, BL-1 through BL-4, represent increasing stringency and control of the environment. Their use generally corresponds to research with organisms in the Risk Group 1–4 classification of etiologic agents (Table III), whether or not recombinant DNA technology is being used. Appendix K specifies physical containment facilities and practices for large-scale uses of recombinant DNA molecules, and Appendices P and Q specify physical and biological containment facilities and practices for experiments involving plants and animals in greenhouses and animal pens, respectively. In practice, APHIS assists the investigators and the IBCs in certifying containment in these facilities. Other appendices deal with shipment, committees, and human gene therapy.

B. Research Outside of Containment

Any research with GMOs that is to be conducted in facilities that are not specified in the NIH guidelines is considered a planned introduction into the environment. Under current policies, such research, regardless of the funding source of the researcher, is subject to regulations of either the USDA or EPA. Neither agency has a codifications similar to that of NIH for contained research. Instead, each introduction is treated as a distinct case and application must be made by the investigator to the appropriate federal agency for permission to conduct the research. This includes all research with genetically modified microorganisms, plants, animals, and vaccines. The current status of each agency and of approval processes can be obtained most readily by viewing their home pages. Each of the agencies has recently published policy statements or has issued Final Rules describing how regulations they enforce are applicable to research with GMOs.

For example, EPA's Final Rule under the Toxic Substances Control Act, published in the April 11, 1997, Federal Register (62 FR 17910-17958), is ac-

cessible at the Office of Pollution Prevention and Toxics site at http://www.epa.gov/opptintr/biotech/. An 80-page document entitled "Points to Consider in the Preparation of TSCA Biotechnology Submission for Microorganisms" is available from EPA. In this regulation, research and development activities are subject to TSCA if funded, in whole or in part, by a commercial entity; or if the researcher intends to obtain an immediate or eventual commercial advantage from the research. Microorganisms that are covered are those "new micoorganisms" that are intergeneric and have been formed by the deliberate combination of genetic material originally isolated from organisms of different taxonomic genera. Intergeneric microorganisms also include those containing a mobile genetic element (including plasmids, transposons, viruses) first identified in a microorganism in a genus different from the recipient microorganism. Reporting to or obtaining approval from EPA is required prior to manufacturing such organisms in containment or releasing them experimentally. However, an exemption is provided for researchers conducting small-scale field tests with the nitrogen-fixing bacteria *Bradyrhizobium japonicum* and *Rhizobium meliloti,* providing certain conditions are met. Researchers complying with the NIH Guidelines may also be granted an exemption from EPA for contained research.

V. APPROPRIATENESS OF GUIDELINES AND REGULATIONS FOR RESEARCH

A. Contained Research

The conditions described in the NIH guidelines have served as a codification of practices for conducting research with both unmodified and modified organisms within a traditional laboratory setting and for large-scale fermentations with many microorganisms to produce many products of biotechnology. There is no indication that exempting certain organisms from containment requirements, or conducting most other experiments at the lowest containment conditions, has caused any problem to individual workers or the environment. The guidelines have served both national and world interests well for over 20 years. Although mandatory only for federally funded research, private companies have embraced them and established their own IBCs.

Conditions for conducting research with other organisms that require other conditions for optimum growth, such as plants, animals, and miooorganisms associated with them, have also been described in the guidelines. The practices are based on those developed as standards of practice in research, and there is no evidence of adverse effects upon the environment when these practices have been followed.

Rapid advances in plant and animal biotechnology, however, have practically made the guidelines for contained research with certain genetically modified plants and animals absolete. For example, engineered soybean and corn grow extensively in farmers' fields, yet when grown in a greenhouse for research purposes, such plants become subject to the NIH guidelines and treated as a potential threat to the environment. The Guidelines need to be updated and exemptions made for those plants and animals, as well as microorganisms, judged by regulatory agencies to be of no risk for commercial uses in the environment.

B. Research Involving Planned Introduction into the Environment

The United States raised questions about the safety of planned introductions and examined such introductions in multiple ways. In 1987, the National Academy of Sciences made major conclusions regarding risk, one of which stated that there is no evidence that unique hazards exist, either from the use of recombinant DNA techniques or from the movement of genes between unrelated organisms. Further, the risks of the introduction of GMOs carrying recombinant DNA are the same in kind as those associated with the introduction of unmodified organisms and organisms modified by other methods. The final conclusion was that assessment of risks of introducing GMOs carrying recombinant DNA into the environment should be based on the phenotype of the organism and the environment into which it is introduced, not on the method by which it was produced—product, not process.

In going further with the assessment of risk, an-other study by the National Academy of Sciences in 1989 posed three fundamental questions to assess risk. (1) Are we familiar with the properties of the organism and the environment into which it may be introduced? (2) Can we confine or control the organism effectively? and (3) What are the probable effects on the environment should the introduced organisms or genetic traits persist longer than intended or spread to nontarget environments? Similar questions for focusing on risk were elaborated by the Ecological Society of America in greater detail. Questions still remain on how to scale regulatory oversight on the basis of risk.

The principles espoused in the preceding types of publications have played a major role in the risk assessment policies developed by international bodies such as the European Union, the OECD, and others. Various bodies have provided for the responsible oversight of recombinant DNA research and field trials in countries in which formal oversight mechanisms are lacking.

C. In Principle: Needs and Options

The objectives of a sound oversight policy are to develop a sensible, scientifically based policy that is consistent with a reasonable and accepted degree of safety (rarely absolute or "ensured" safety). Regulatory agencies also recognize procedures that do not pose an unreasonable risk to humans or the environment, i.e., one cannot be absolutely certain that no deleterious effects can occur. Oversight ideally should balance risks with expected benefits. Many issues that deal with GMOs, particularly introduction into the environment, are subject to change and remain unresolved. These include the following:

1. The scope of oversight: what organisms or parts of organisms (genetic elements and sequences) should be subject to oversight?
2. By whom and at what level (e.g., professional organization, standards of practice, institutional biosafety committees, local regulatory bodies, federal agencies, or some combination) should oversight be conducted? Should all experiments be examined at a federal or national level or can some be decided upon at a local level? Decentralization of authority to

make decisions on field releases has not yet occurred, although APHIS has expedited approvals for many plants.

3. Can consistent definitions be developed? Definitions differ among agencies and countries and can lead to problems in legal interpretations. The meaning of the phrase "release or planned introduction into the environment" and even "pathogen" or "pesticide" are yet to be agreed upon. For example, the EPA has proposed a new class of plant-pesticides based on use of microbial components engineered into plants with the intent of making disease- or insect-resistant plants. This also affects whether different considerations apply to plant pathogens that are applied as beneficial biological control agents.

4. To what degree should the manner of open and peer review be part of policy making?

5. What appeal procedures, if any, should there be for scientists and others disagreeing with the oversight bodies?

6. Decision-making: Who should decide whether approval of field-testing is needed? Is it the public, the scientists, or the courts? What role is there for common sense? Should the decision be based on risk alone or include other factors, such as socioeconomic considerations?

7. Should there be a "sunset" clause on termination of oversight of research with some organisms or certain types of experiments?

There has been general agreement that a centralized database for field trials would be desirable in order to compare information, including negative results that are not always published. The USDA through its Information Systems for Biotechnology program (http://www.isb.vt.edu) and the OECD (http://www.oecd.org) through its Biotrack monitoring database have compiled a great deal of this information. Whether or not these activities serve the purpose of the scientific and commercial community and allay the concerns of the public remains to be seen. Thousands of tests worldwide have been conducted up to the present time and no unpredictable effects have been detected. However, it can be argued that such effects may occur in later years, and, hence, monitoring will be necessary to assess any problems that might arise. Another question that remains

unanswered is how long monitoring should occur, compared with naturally occurring organisms.

Given the different views of different countries and applicable laws, global agreements on oversight likely will not be forthcoming. However, there is reasonable general agreement on standards of practice through the scientific and professional societies of the world for conducting research. There are also areas of reasonable agreement in principle, acknowledging that the process by which a genetic modification is made is not as significant as the effects of that modification, i.e., the phenotype. The same degree of oversight is not applied to unmodified organisms and organisms modified by traditional approaches. Hence, the process of modification is still the "trigger" for oversight. A second area of agreement is that familiarity or knowledge of the organism and its modification are likely to be good predictors of the characteristics of the modified organism. A third is that knowledge of the ability to confine an organism or mitigate its effect, if need be, offers a reasonable indicator of expected risk and of the potential for its management.

VI. CONCLUSIONS

Differing perspectives remain on the safety to humans and the environment of genetically modified organisms, particularly those that have been modified by recombinant DNA techniques. The concerns are particularly high with respect to the use of microorganisms in the environment. Discussions are likely to continue for several more years. It is too early to predict whether or not such tests will go forward with reasonable ease, given the stringency of the requirements to conduct the tests. The scientific concerns may not warrant the expenditure of time, effort, and money to conduct field research, since risks unique to GMOs have not yet been identified. The same question could be asked about the oversight of contained research.

Several potential oversight mechanisms would be commensurate with risk assessment and risk management and have been demonstrated as effective for laboratory research. These could include (1) categorical exclusions, (2) only notification requirements,

(3) review and approval by a local organization (e.g., institutional biosafety committees), (4) review and approval by a federal agency with an advisory group consisting of members familiar with the relevant research area, or (5) review and approval by an international agency, in cooperation with a member country.

A reasonable policy of oversight will encourage research with GMOs, especially those modified by recombinant DNA techniques. Competing perspectives may occur in different countries, and within a country, and may not be reconciled. There is no perfect oversight mechanism for any human activity, including environmental releases for research, development, or commercial purposes. There is also the recognition that various viewpoints or perspectives cannot always be accommodated or reconciled. Thus, persons of reason and broad vision will be needed to resolve some of the contentious issues dealing with planned introduction of GMOs into the environment.

See Also the Following Articles

INTERNATIONAL LAW AND INFECTIOUS DISEASE • PATENTING OF LIVING ORGANISMS AND NATURAL PRODUCTS • RECOMBINANT DNA, BASIC PROCEDURES

Bibliography

Baumgardt, B. R., and Martens, M. A. (eds.) (1991). "Agricultural Biotechnology: Issues and Choices." Purdue Univ. Agricultural Experiment Station USA.

Cordle, M. K., Payne, J. H., and Young, A. L. (1991). Regulation and oversight of biotechnological applications for agriculture and forestry. *In* "Assessing Ecological Risks of Biotechnology" (L. R. Ginzberg, ed.), pp. 289–311. Butterworth-Heinemann, Boston.

James, C., and Krattiger, A. F. (1996). Global review of the field testing and commercialization of transgenic plants, 1986 to 1995: The first decade of crop biotechnology.

"ISAAA Briefs No. 1." The International Service for the Acquisition of Agri-Biotech Applications, Ithaca, NY.

Levin, M., and Strauss, H. (eds.) (1991). "Risk Assessment in Genetic Engineering." McGraw-Hill, Inc., NY.

MacKenzie, D. R., and Henry, S. C. (eds.) (1991). "Biological Monitoring of Genetically Engineered Plants and Microbes." Agricultural Research Institute, MD.

Miller, H. I. (1997). "Policy Controversy in Biotechnology: An Insider's View." R. G. Landes Co. and Academic Press, Austin TX.

Miller, H. I., Burris, R. H., Vidaver, A. K., and Wivel, H. A. (1990). *Science* **250**, 490–491.

National Academy of Sciences (1987). "Introduction of Recombinant DNA-Engineered Organisms into the Environment." Committee on the Introduction of Genetically Engineered Organisms into the Environment, National Academy Press, Washington, DC.

National Research Council (1989). "Field Testing Genetically Modified Organisms: Framework for Decisions." Committee on Scientific Evaluation of the Introduction of Genetically Modified Microorganisms of Plants into the Environment. National Academy Press, Washington, DC.

Organization for Economic Cooperation and Development (1986). "Recombinant DNA Safety Considerations. Safety Considerations for Industrial, Agricultural and Environmental Applications of Organisms Derived by Recombinant DNA Techniques." OECD, Paris.

Organization for Economic Cooperation and Development (1990). "Good Development Practices for Small-Scale Field Research with Genetically Modified Plants and Micro-Organisms. A Discussion Document." OECD, Paris.

Organization for Economic Cooperation and Development (1993). "Safety Considerations for Biotechnology: Scale-up of Crop Plants." OECD, Paris.

Tiedje, J. M., Colwell, R. K., Grossman, Y. L., Hodson, R. E., Lenski, R. E., Mack, R. N., and Regal, P. J. (1989). *Ecology* **70**, 298–315.

Tolin, S. A., and Vidaver, A. K. (1989). *Annual Review of Phytopathology* **27**, 551–581.

U.S. Congress, Office of Technology Assessment (1988). "New Developments in Biotechnology—Field-Testing Engineered Organisms: Genetic and Ecological Issues." U.S. Government Printing Office, Washington, DC.

Genomic Engineering of Bacterial Metabolism

J. S. Edwards, C. H. Schilling, M. W. Covert, S. J. Smith, and B. O. Palsson

University of California, San Diego

GLOSSARY

feasible set The metabolic flux distributions that satisfy the mass balance constraints, in addition to the inequality constraints from the physical–chemical properties of the metabolic processes.

metabolic engineering The manipulation of metabolic processes through the use of recombinant DNA technology, with the goal of improving the properties of a cell.

in silico biology The study of biological processes using computer simulations.

linear programming An optimization technique that can be used to determine the maximal value of an objective function, subject to a set of constraints.

flux balance analysis A methodology, used to study metabolic systems, that defines capabilities of a metabolic network based on the mass balance and physical–chemical constraints.

metabolic flux A central concept in metabolic engineering, defined as the rate at which material is converted via metabolic reactions and pathways.

metabolic genotype The complete set of metabolic genes identified in a genome.

metabolic phenotype The utilization of the metabolic genotype under a given condition characterizes the metabolic phenotype.

CONSIDERABLE INTEREST in the redirection of metabolic fluxes for medical and industrial purposes has developed in recent years. As a result, the field of *metabolic engineering* has been developed with the primary goal of implementing desirable metabolic behavior in living cells. Advances and applications of several scientific disciplines, including computer technology, genetics, and systems science, lie at the heart of metabolic engineering.

I. INTRODUCTION

The engineering approach to analysis and design is to have a mathematical or computer model, e.g., a dynamic simulator, of metabolism that is based on fundamental physical–chemical laws and principles. The metabolic engineer then hopes that such models can be used to systematically "design" a new strain. The methods of recombinant DNA technology should then be applied to achieve the desired changes in the genotype of the cell of interest. However, a recent review in the field has concluded that "despite the recent surge of interest in metabolic engineering, a great disparity still exists between the power of available molecular biological techniques and the ability to rationally analyze biochemical networks" (Stephanopoulos, 1994). How have we arrived at this juncture?

The mathematical modeling of metabolic reaction networks dates back to the mid-1960s. With the availability of analog computers and the knowledge of metabolic regulation, dynamic simulations of simple metabolic and genetic control loops appeared. The systemic nature of metabolic function was apparent and so was its complexity. Attention turned to developing methods that could shed light on the relative importance of various metabolic events. Methods for sensitivity analysis of metabolic regulation began in the 1960s (Savagean, 1969) and continued into the 1970s (Kacser and Burns, 1973; Heinrich *et al.,* 1977), and resulted in the biochemical systems theory (BST) and the popular metabolic control analysis (MCA).

In the early 1970s, recombinant DNA technology was developed and the first report of bacterial gene splicing appeared in 1972 (Jackson *et al.*). This development was of major historical significance. An important ramification was that the cellular components in a bioprocess were now subject to design. No longer did biochemical engineers have to be content with designing equipment and operating strategies, but they could actually change underlying cellular determinants and alter the characteristics of the industrial strain used. Consequently, the field of metabolic engineering was born, with its first major review appearing in 1991 (Bailey, 1991).

Establishing complete kinetic models of cellular metabolism became a scientific goal early on in the history of metabolic engineering with the intended use of elucidating the systemic behavior of metabolic networks. This desire to have such comprehensive dynamic models of metabolism for use in strain design still remains unfulfilled due to the fragmented nature of enzyme kinetic information available on *in vivo* metabolic processes. However, with genomics developing rapidly, we now face the prospect of having "parts catalogues" of the metabolic components in particular strains (Edwards and Palsson, 1998, 1999). From this information, it is possible to reconstruct cellular metabolic networks and ascertain information regarding the structure and stoichiometry of the metabolic network and its reactions. With such information at hand, we are now in a position to build genome-scale metabolic models. This chapter focuses on how steady-state models can be built based on annotated genomic data and analyzed using flux balance analysis.

II. METABOLIC ENGINEERING AND APPLICATIONS

Decades of investigation of the metabolic machinery that drives the living process has revealed the intricate network of metabolic reactions present in a number of organisms or their metabolic map. Further analysis has revealed the respective genes that code for the enzymes performing these reactions, the regulatory mechanisms that control the expression of many of the genes that code for products with metabolic activity, and the mechanisms that control the activity of the individual reactions. It would appear that the process is completely understood. Although the reductionist approach has revealed a tremendous amount of information regarding metabolic processes, it has not revealed a key aspect of cellular systems, namely, that the physiology of complex networks (such as metabolism or signal transduction) is a function of multiple interacting components. Thus, to understand the function of metabolism as it relates to overall cellular physiology, a holistic approach to the study of metabolism must be pursued. The understanding of the systemic function of a metabolic network is imperative to the success of the field of metabolic engineering.

In the early years of metabolic engineering, successive rounds of mutagenesis and selection of strains with desirable qualities was the approach by which cellular processes were improved. Although this approach has been successful in a number of cases, the theoretical yield of a product is not always attained through random mutagenesis and selection procedures. With the advent of recombinant DNA technology, we now have the potential to manipulate the cellular genetic content almost at will. Therefore, specific changes (identified through an "engineering" design process) can be made in the genotype, thus producing a strain with superior qualities.

The idea of utilizing a living cell to perform a biochemical task is not new. Living cells have been utilized to carry out industrial processes since well before the structure of DNA was solved. For example, people have been utilizing yeast for thousands of years (before 2000 BC) for winemaking, and the industrial use of microorganisms grew out of the fermentation of alcoholic beverages. As another example, one of the most influential advances in medical science was brought about through the use of a *Penicillium* mold for the production of penicillin in the 1940s. At that time, the penicillin yield was low; however, since then, the yield has been increased

500-fold, using the basic technique of random mutagenesis and strain selection. The yield has been further increased using recombinant DNA techniques but only slightly compared to the gain by random mutagenesis and selection. However, as the structure of DNA was determined, restriction enzymes were discovered, PCR was invented, and as the biological processes, in general, became known, utilizing living cells to carry out industrial processes became more widespread and productive.

Controlling metabolic flux is the primary goal of metabolic engineering. Metabolic flux is a central concept in metabolic engineering and is defined as the rate at which material in converted via metabolic reactions and pathways. However, in order to control metabolic flux, the factors influencing the flux must be understood; thus, a central theme of metabolic engineering is to determine and understand the control of metabolic fluxes. The critical importance of metabolic flux is justified because the fluxes are a determinant of the physiological state of the cell. Thus, once one understands the metabolic fluxes within a cell, a major step is taken toward understanding cellular metabolism, placing the metabolic engineer in a position to alter the genotype of a cell with the hope of imparting a favorable phenotype.

Next, we will discuss a few specific applications of metabolic engineering principles. For presentation clarity, the goal of metabolic engineering can be classified into several distinct objectives (Cameron and Tong, 1993): a production increase of the host's natural products or the addition of novel production capacity to the cell, the addition of catabolic processes that the cell normally doesn't possess, and the general modification of cellular properties to improve the cell's potential utility.

Utilizing living cells, such as microorganisms, to produce a chemical is an ultimate goal of metabolic engineering. Metabolic engineering is typically used to design organisms which produce chemicals that are too expensive to produce by chemical synthesis, e.g., vitamins such as riboflavin. The cell can naturally produce the product, and metabolic engineering strategies are employed with the aim of increasing the native production of the compound. The product yield of many compounds has been improved by this technique. Such products include acetic acid,

ethanol, butanol, acetone, various antibiotics, and amino acids. Furthermore, some protein products are overproduced by microorganisms for use in the pharmaceutical and food industries. In addition to the utilization of metabolic engineering techniques to improve the productivity of metabolites that are already produced by the cell, metabolic engineering has been used to impart the ability to produce products that are not intrinsically produced. The introduction of new production capabilities has been applied for the production of biopolymers, antibiotics, pigments, and other compounds (Cameron and Tong, 1993). Metabolic engineering is proving to be a useful application to economically produce many different metabolites.

The ability to design organisms that possess added catabolic processes is of particular interest, especially for environmental applications. For example, a strain of *E. coli* has been genetically engineered, such that the enzymes for the degradation of trichloroethylene are expressed and active. This strain could potentially be used to degrade this common pollutant. Furthermore, additional catabolic processes can be added, such that different (or less expensive) substrates can be used in an industrial process for the production of different metabolites, amino acids, vitamins, antibiotics, etc.

Metabolic engineering can be used to alter the genotype, such that the phenotype exhibits cellular properties that are beneficial for the organisms utilized in industrial processes. This area is a very broad classification that describes any alterations in the cellular operation that leads to an improvement in an industrial strain. The redirection of metabolic flux away from a toxic product (e.g., acetate) to a less toxic product (e.g., ethanol or acetolactate) is an example of an improvement in cellular properties that can be used to increase the density of cultured cells and increase the productivity of the cells. Additionally, *E. coli* cells have been engineered to produce oxygen-carrying compounds to improve the capacity for product formation and growth in oxygen-limited conditions. It has been shown that improvements in cellular processes that are not directly linked to the metabolic pathway involved in the formation of a product can have a significant influence on the industrial value of a strain.

Metabolic engineering has been successful in improving many industrial processes; however, the theoretical yield has not been approached. Additionally, unexpected results are common when metabolic enzymes are overexpressed or repressed, and our ability to manipulate the cellular genotype to produce specific changes in the metabolic network outweighs our ability to rationally design the manipulations so that the desirable qualities are imparted to the organism. The unpredictability in the metabolic response to the genetic alterations is due to the multigeneic nature of metabolic function. Thus, understanding the function of the individual enzymes, or even the metabolic pathway, is not sufficient to describe the holistic response of the metabolic system due to a genetic perturbation. Hence, for metabolic engineering to benefit from the applications of recombinant DNA technology, our understanding of multigenic functions must be increased. Today, through recent advances in genomics, we are in a position to study metabolic characteristics as a function of the entire genome.

III. GENOMICS

During the first half of the twentieth century, genetics was based on the premise that an invisible information-containing unit called a gene existed. Throughout this time period, genetics focused on the functional aspects of a gene, as little was known about its molecular structure. Since then, science has unraveled the intricacies and mechanistic details of genetic information transfer and determined the structure of DNA and the nature of the genetic code, establishing DNA as the source of heredity containing the blueprints from which organisms are built. With this in mind, it is of no surprise that the scientific community has devoted considerable effort toward determining the genome of many microorganisms and even more complex eucaryotes, such as the nematode worm, the mouse, and even humans. These endeavors have rapidly spawned the development of new technologies and emerging fields of research based around genome-sequencing efforts (i.e., functional genomics, structural genomics, proteomics, and bioinformatics).

The goal of sequencing an entire genome finds its roots in the initiation of the Human Genome Project (HGP). Setting the stage for modern genomics, the HGP sparked sequencing initiatives for a number of organisms from all the major kingdoms of life. Small genome sequencing is becoming routine and in the future, studies of nearly every organism will be aided by the knowledge and availability of the complete DNA sequence of their genomes.

Upon completion of a genome-sequencing project, the first step is to identify the location of the genes or open reading frames (ORFs). There are basically two fundamentally different techniques used to search for genes within a DNA sequence: search by content and search by signal approaches as well as integrated methods. An identified ORF can then be compared, via databases, with known protein and DNA sequences to determine its function. In fully sequenced genomes, approximately 45–80% of their ORFs have been given functional assignments and these percentages are increasing.

Therefore, we are at the brink of having a complete "parts catalogue" for many organisms, including ourselves. DNA sequence data now needs to be translated into functional information, both in terms of the biochemical function of individual genes, as well as their systemic role in the operation of multigeneic functions. The study of function is encompassed by the field of functional genomics, which can be viewed as a two-tiered approach to the study of genetic information transfer. The first level is the prediction and determination of the function of individual gene products from their sequence through comparison with other known genes and genomes. The second level involves the use of this functional information of individual genes to analyze, interpret, and predict cellular functions resulting from their coordinated activity.

It is becoming clear that there is not a linear relation between individual genes and the overall cellular functions. This is demonstrated by the large number of knockout mice that exhibit null phenotypes following the removal of particular genes; the development of the animal is more or less normal without the deleted gene. This is a demonstration of the complex, nonlinear genotype–phenotype relation, which cannot be predicted by simply cataloging and assigning functions to genes identified in a genome.

Since cellular functions rely on the coordinated activity of multiple genes and proteins, the interrelatedness and connectivity of these elements becomes critical. The coordinated action of multiple genes and proteins can be viewed as a network, or a "genetic circuit," which is the collection of different gene products that together are required to execute a particular function. To understand the systemic nature of cellular functions, every gene must be studied relative to its contribution to the overall cellular function. Strategic approaches need to be formulated and methods developed which utilize the wealth of data provided by genomics and other facets of functional genomics.

Currently, microorganisms offer the best test-bed for establishing *in silico* methods that can be used to analyze, interpret, and predict the genotype–phenotype relation starting from complete genomic DNA sequences. The best physiological function from which to launch the development of this line of study is metabolism, where the systemic properties have already been studied in detail. In addition, major portions of the genes in microbial cells encode products with metabolic functions. The ORF assignments are most advanced for metabolic genes, due to the long history of metabolic study. Thus, once the ORF assignments have been completed for the metabolic genes, the entire metabolic map representing the stoichiometry of all the metabolic reactions taking place in the cell can be constructed. In fact, there are now a number of extensive on-line databases detailing the metabolic content and pathway diagrams for most of the fully sequenced microbes.

The annotated genomic sequence provides a list of genes found in a genome. A subset of the genes that encode metabolic enzymes can be selected from this "parts list," and the metabolic reaction network can be constructed based on reactions known to be catalyzed by the related gene products. A flux balance can be written on each metabolite (as discussed in the next section), and over time, in a steady state, the fluxes must balance to avoid a significant accumulation of a metabolite in the network. The stoichiometry of all the reactions in a metabolic network can be represented in a stoichiometric matrix (Eq. 2). Unlike the kinetic properties of a system, the stoichiometry is an invariant property of the system. This matrix contains all the information about how substances are linked through reactions within the network. DNA microarray technology can also be applied to study the metabolic and genetic control of gene expression on a genomic scale. Thus, *in silico* metabolic modeling, based on the primary invariants of a system and recent experimental advances in genomics, will provide the capability to analyze cellular systems from a systemic perspective, such that one can analyze, interpret, and predict an organism's metabolic phenotype from its genotype.

IV. FLUX BALANCE ANALYSIS

In the previous section, a parts catalogue of the metabolic network has been generated. The next step is to convert the genomic data into functional information, with the objective of understanding the metabolic flux in the metabolic network. We will discuss flux balance analysis (FBA), which is an approach that is well suited to utilize the genomic data. FBA has been developed to capitalize on the well-known metabolic stoichiometry and is based on the principle of mass conservation.

The rate of change in the number of molecules of a metabolite = Sum of the reactions creating the metabolite − Sum of the reactions consuming the metabolite + The net transport of the metabolite into the cell

In terms of a mathematical equation, a flux balance can be written for each metabolite (X_i) within a metabolic system to yield the dynamic mass balance equations that interconnect the various metabolites. The transport processes are also written as metabolic reactions. Then the mass balances can be written in mathematical form

$$\frac{dX_i}{dt} = \sum_{j=1}^{n} S_{ij} v_j \text{ for } 1 \le i \le m, \tag{1}$$

where X_i is a metabolite in the metabolic network (i.e., X_{GLC}) and m is the total number of metabolites in the system. The element v_j, is the jth metabolic flux (i.e., $v_{Glucokinase}$) and there are n different metabolic fluxes in the system. S_{ij} is the stoichiometric coefficient that indicates the amount of the ith compound produced per unit flux of the jth reaction. For instance, in the reaction catalyzed by glucokinase, one molecule of glucose is consumed, and thus $S_{glucose, glucokinase} = -1$.

The time constants characterizing metabolic transients are typically vary rapid compared to the time constants of cell growth and process dynamics; therefore, the mass balances can be simplified to only consider the steady-state behavior. Eliminating the time derivative and representing the equation in matrix form, we arrive at the equation that is central to flux balance analysis:

$$\mathbf{S} \cdot \mathbf{v} = \mathbf{0}, \qquad (2)$$

where $\mathbf{v}$ is the vector of n metabolic fluxes and $\mathbf{S}$ is the $m \times n$ stoichiometric matrix. This equation simply states that, over long times, the formation fluxes of a metabolite must be balanced by the degradation fluxes. Otherwise, significant amounts of the metabolite will accumulate inside the metabolic network. Note that this balance is formally analogous to Kirchhoff's current law, used in electrical circuit analysis.

Mathematically, the metabolic fluxes are restricted to be positive ($0 \leq v_i \leq \infty$). Thus, if a reaction is known to be reversible under physiological conditions, the reverse reaction and forward reactions are written as two separate positive reactions. Furthermore, specific information regarding the metabolic reaction fluxes is incorporated ($\alpha_j \leq v_j \leq \beta_j$, for known v_j). The flux levels are typically known for metabolic uptake reactions. This process limits the metabolites that are capable of being transported across the cell membrane.

The linear inequalities discussed above form a set of physical–chemical constraints. The physical–chemical constraints and the mass balance constraints [Eq. (2)] define the *feasible set* of the metabolic network, and thus define what the cell can and cannot do.

To determine the metabolic state of an organism, we need to solve Eq. (2) for the metabolic fluxes ($\mathbf{v}$), which will then indicate the metabolic state of the cell under the defined conditions. The matrix $\mathbf{S}$ is the key component to flux balance analysis, and importantly, this matrix can be derived from the genomic information data that is generated by the genome-sequencing projects (as discussed in the previous section).

The system of linear homogeneous equations defined by Eq. (2) is typically underdetermined; thus, the number of metabolic fluxes ($\mathbf{v}$) is greater than the number of mass balances (i.e., $n > m$), resulting in a plurality of feasible flux distributions to Eq. (2). There are several techniques for determining the metabolic flux distributions. A number of investigators have experimentally measured the required number of metabolic fluxes such that there is a single solution to Eq. (2). Also, additional flux measurements can be made and statistics (least-squares analysis) can be used to determine the best fit to the data. Measuring metabolic fluxes can be a simple procedure (as for a transport reaction) or a very difficult procedure (as for internal fluxes), but the measurement of only the external fluxes is rarely sufficient to uniquely define the metabolic flux distributions without other assumptions or measurements of enzyme expression levels. There are many examples in the literature of using flux measurements and a flux balance model to determine the metabolic flux distributions.

The undetermined system results in a plurality of feasible flux distributions, and this range of solutions is indicative of the flexibility in the flux distributions that can be achieved with a given set of metabolic reactions. Although infinite in number, the solutions to Eq. (2) lie in a restricted region, the feasible set. The feasible set can be defined as the capabilities of the metabolic genotype of a given organism, since it defines all the metabolic flux distributions that can be achieved with a particular set of metabolic genes. The particular use of the metabolic genotype under a given condition can be defined as the *metabolic phenotype* that is displayed under those particular conditions. To explore the capabilities of a metabolic system, with the hope to understand and, ultimately, modify the metabolic network, one must explore the feasible set.

In addition to using metabolic flux measurements

to calculate the remaining metabolic fluxes, other approaches have been used to examine the capabilities of the metabolic network based solely on the stoichiometric, reversibility, and physical–chemical constraints placed on the reactions comprising the network. The capabilities of the metabolic network are defined by the feasible set. The feasible set can be studied using a branch of linear algebra known as convex analysis. Furthermore, the metabolic phenotype can be examined by using linear programming (LP), which is an applied branch of linear algebra and convex analysis.

A particular solution to Eq. (2) that optimally satisfies the linear inequality constraints can be determined by using LP, in which the flux distribution that minimizes or maximizes a particular objective is identified. Objectives for metabolic function can be chosen to explore the "best" use of the metabolic network within a given metabolic genotype. Mathematically, this optimization is stated as

$$\text{Minimize } Z \tag{3}$$

$$\text{where } Z = c_i v_i = \langle \mathbf{c} \cdot \mathbf{v} \rangle, \tag{4}$$

where Z is the objective which is represented as a linear combination of metabolic fluxes, v_i. The optimization can also be stated as the equivalent maximization problem, i.e., by changing the sign on Z.

This general representation of Z enables the formulation of a number of diverse objectives. These objectives can be design objectives for a strain, exploitation of the metabolic capabilities of a genotype, or physiologically meaningful objective functions (such as maximum cellular growth). For instance, growth can be defined in terms of biosynthetic requirements. Thus, biomass generation is defined as a reaction flux draining intermediate metabolites in the appropriate ratios and represented as an objective function Z.

A number of different objective functions have been used for metabolic analysis, including minimize ATP production, minimize nutrient uptake, minimize redox potential production, minimize the Euclidean norm, maximize metabolite production, maximize biomass and metabolite production, and maximize growth rate. There are a number of diverse metabolic objectives that can be represented by Eq. (4). Thus, this formulation allows us to obtain answers to a number of important questions.

V. EXAMPLE APPLICATIONS OF FLUX BALANCE ANALYSIS

As we have seen, genomics provides us with information regarding the structure and content of an organism's metabolic network. This is captured in the stoichiometric matrix, which can be generated for any given organism and is defined by the ORF assignments. FBA can be then be used to explore the theoretical production capabilities and general fitness of the metabolic genotype. Following is a discussion of some general areas of application for flux balance analysis.

A. Gene Deletions

FBA can be used to interpret and predict the phenotypic effects of alterations to the genotype. Any loss in function of a gene product due to events, such as genetic deletions or enzymatic inhibition, can be readily studied by constraining the flux through the associated reactions to zero. Single or multiple deletions can be simulated. The ability of the metabolic network to redistribute metabolic resources to meet cellular growth demands in the face of any set of alterations to the genotype can be compared to the wild type. The maximum value of the objective function relative to that for the wild type will indicate if the deletion(s) should be expected to negatively influence the growth rate or show no influence at all. The inability of the network to absorb an alteration in the genotype will immediately lead to the identification of auxotrophic growth requirements. This allows the cellular engineer to quickly determine the auxotrophic requirements that are necessary for a genetically modified strain, perhaps, designed to route metabolic resources through particular reactions occurring within the cell.

FBA has been used to explore the relation between the *E. coli* metabolic genotype and phenotype in an *in silico* knockout study. In this study, all possible combinations of single, double, and triple deletions of the genes involved in central intermediary metabolism were analyzed. Based on the results, these gene products were classified as essential, important/nonessential, and redundant. It was found that a surprising number of the metabolic gene products were determined to be redundant for growth on glucose minimal media. Even when multiple knockouts were performed, a surprising number of the double and triple deletions resulted in no change in the ability of the cell to meet its growth demands. The predicted metabolic phenotypes were also consistent with experimentally characterized mutants. This deletion study therefore has immediate implications to the interpretation of the cellular metabolic physiology. Thus, FBA allows the ability to ask very direct questions regarding the structure and capabilities of the metabolic network of an organism.

As an extension to the previous application and example, FBA can also be used to examine expected shifts in metabolic routing that occur when subjected to the loss of the function of a gene product through genetic deletions or inhibition of activity. The cellular metabolic network has the stoichiometric flexibility to redistribute its metabolic fluxes with remarkably little change in its ability to support biomass synthesis, even if faced with the loss of key enzymes. Additionally, FBA can be used to predict metabolic phenotypes under different growth conditions, such as substrate and oxygen availability. In this case, rather than changing the actual structure of the metabolic network, the inputs to the network are changed to reflect substrate limitations and changing growth environments. In these situations, it is necessary to perform a detailed analysis on the resulting changes that are suggested to occur as the metabolic phenotype shifts. From studying aspects of the solution, such as the metabolite shadow prices and the reduced costs of the various fluxes in the network, one is able to rationalize the redistribution of metabolic resources to meet biomass requirements presented in the objective function.

In studying shifts in metabolic behavior due to alterations in the genotype and/or changing growth conditions, FBA can be combined with recent experimental techniques developed to examine gene expression on a genomic-wide scale. The relation between the flux value and the gene expression levels is nonlinear. However, FBA can give qualitative (on/off) information, as well as the relative importance of gene products under a given condition. Based on the magnitude of the metabolic fluxes, qualitative assessments of gene expression can be inferred. This can then be compared to corresponding experimental data sets, such as those generated from using high-density oligonucleotide arrays and cDNA microarrays monitoring changes in gene expression for the entire metabolic genotype. Recently, it has been shown how the metabolic and genetic control of gene expression can be studied on a genomic scale using DNA microarray technology (DeRisi *et al.*, 1997). The temporal changes in genetic expression profiles that occur during the diauxic shift in *S. cerevisiae* were observed for every known expressed sequence tag (EST) in this genome. FBA can be used to qualitatively simulate shifts in metabolic genotype expression patterns due to alterations in growth environments.

Thus, FBA can serve to complement current studies in metabolic gene expression by providing a fundamental approach to analyze, interpret, and predict the data from such experiments. Additionally, through the use of more detailed theoretical approaches, a pathway-based perspective can be used to interpret genome-scale expression data and may provide insight into the regulatory logic imposed by the cell to control the function of the entire metabolic genotype. The structure of the stoichiometric matrix itself can reveal the answers to many important questions regarding the connectivity of the metabolic network and the capabilities of the system and general fitness of an organism under changing conditions.

B. *Penicillium chrysogenum*

The metabolism of *Penicillium chrysogenum* has been studied with FBA (Jorgensen *et al.*, 1995). The stoichiometric model in this study considered 61 internal fluxes and the uptake of glucose, lactate, γ-aminobutyrate and 21 amino acids. Since 49 intracellular metabolites were considered and 82 fluxes including uptake rates, the number of degrees of freedom was 33. Exactly 33 fluxes were measured, the same as the number of degrees of freedom. These included the uptake rates of the aforementioned substrates and the formation rates of penicillin V and 5 other key intermediates. Rates of formation of cellular RNA and DNA, protein, lipid, carbohydrate, and amino carbohydrate were also measured.

The maximum theoretical yield of penicillin production was calculated for two different pathways of cysteine biosynthesis, one by direct sulfhydrylation and the other by transsulfuration. The study proposed that an increase of 20% of the maximum theoretical yield of penicillin V on glucose was possible if cysteine is synthesized by direct sulfhydrylation rather than by transsulfuration. The introduction of the pathway for direct sulfhydrylation was suggested as a possible metabolic engineering approach to increase penicillin yield.

C. Flux Distributions in *C. glutamicum*

FBA has been used to determine the flux distributions in *C. glutamicum* during growth and lysine biosynthesis (Vallino and Stephanopoulos, 1993). They measured metabolite uptake and secretion rates to reduce the number of unknowns; however, the number of unknowns was still greater than the number of independent equations. Measurements of enzymatic activity were used to define fluxes that were not active in the flux distribution. Using the intracellular assays, the glyoxylate bypass was determined not to operate during growth on glucose or glucose and acetate. The pyruvate carboxylase and the malic enzyme were also determined not to operate in the metabolic fluxes. Thus, the corresponding reactions were removed from the stoichiometric matrix. The enzyme oxaloacetate decarboxylase was also removed from the stoichiometric matrix to eliminate a singular group, even though its activity was determined to be high. The removal of this reaction was based on the argument that growth on glucose could not occur if oxaloacetate decarboxylase carried a significant flux. The researchers also removed one pathway leading from meso-DAP to lysine and effectively lumped parallel pathways into one.

Based on the experimental results, lysine fermentation was grouped into four phases and the metabolic flux distributions were calculated for each phase. Phase I exhibited balanced growth and produced little by-product formation. Phase II was initiated by the depletion of the threonine from the medium. This depletion removed the inhibition of the metabolic enzymes that prevented the overproduction of lysine.

During phase II, a high growth rate and by-product formation were observed. Phase III continues the high lysine formation, but no growth was observed. In phase IV, a decrease in biomass and lysine production occurred, as well as additional by-product formation of pyruvate, acetate, alanine, and valine.

The flux distributions computed by the FBA were checked for thermodynamic violations. Second, the percentage of carbon metabolized by the PPP was calculated, and they found the results quantitatively described the metabolic flux distributions during growth and lysine biosynthesis. These theoretical findings are fundamental in identifying the key branch points and can be used to engineer metabolism for the overproduction of biochemicals.

D. Metabolic Flux in *Bacillus subtilis* Using ^{13}C NMR

FBA and ^{13}C NMR techniques have been used to determine the metabolic flux distributions in riboflavin producing *Bacillus subtilis* under different dilution rates (Sauer *et al.*, 1997). The underdetermined system of equations was converted into an overdetermined system by placing additional constraints on the metabolic network based on the ^{13}C NMR results. This technique provided key information regarding the intracellular metabolic fluxes.

The ^{13}C labeled amino acids were used as probes to determine the metabolite of origin for the precursor metabolites. The procedure is to uniformly label glucose in a minimal media, then randomly label the glucose, and the probability that any two successive carbon atoms are labeled is 1%; however, if they originated from the same precursor metabolite, the probability is increased to 10%. Based on these probabilities, bounds can be placed on the intracellular metabolic fluxes.

The results for the ^{13}C NMR study showed that approximately 50% of the oxaloacetate was produced in the anaplerotic reaction known as the pyruvate shunt. A significant amount of phosphoenolpyruvate was formed from the decarboxylation of oxaloacetate, which is a reaction assumed to be inactive during the growth on glucose. Finally, an inequality constraint was placed on the pyruvate formation from malate.

The bounds determined from the ^{13}C NMR results were used as constraints in the least-squares calculation. The metabolic flux distributions in the metabolic network were determined. It was found that the biomass yield was between 36 and 52%, depending on the dilution rate, and the riboflavin yield was up to 3.7%. The oxidative branch of the PPP was found to carry between 65 and 72% of the carbon flux, and the pyruvate shunt was shown to support high fluxes, 33 to 54%. At low dilution rates, there was found to be a futile cycle involving phosphoenolpyruvate, pyruvate, and oxaloacetate.

Based on these results, the classical assumption regarding the purpose of the PPP was challenged. The PPP seemed to be a major pathway for the catabolism of glucose. The transhydrogenase is probably used to reoxidize the NADPH. In summary, their conclusions show how FBA can be used to elucidate the metabolic physiology of microbial cells. Specifically, they determined that the biosynthesis is not limited in the central catabolic pathways, thus, defining a starting point for developing, metabolically engineered *Bacillus subtilis* strains.

VI. CONCLUDING REMARKS

As previously mentioned, genomic data is compositional in nature and contains limited information about the dynamic behavior of integrated cellular processes. However, there is a clear shift under way from "structural genomics" to "functional genomics," which is expanding the scope of biological investigation from a one-gene approach to a more systems-based "holistic" approach. With recent advances, such as oligonucleotide chip technology and DNA microarrays, it is possible to take genomic sequences and begin analyzing the dynamic events which occur in gene expression as a consequences of operational shifts in genetic circuits and cellular systems. In addition to mountainous genomic sequence databases, we will soon be confronted with burgeoning gene expression databases. Just as experimental technologies will be critical to growth in the postgenomic era, so too will the development of systems science-

based approaches and computational strategies making sense of all the information provided thus far from genomics.

Here we have illustrated the use of FBA as a tool for the analysis of the massive amounts of genomic data currently being generated. This method can be used for a broad range of scientific interests, all related to the deepest goal of understanding the genotype–phenotype relationship. FBA offers an example of the type of techniques and applications that lie ahead if we are to truly profit from genomics. Future methods will need to be developed and tailored for specific uses that move beyond the management of sequence data.

See Also the Following Articles

DNA Sequencing and Genomics • Mapping Bacterial Genomes • Strain Improvement

Bibliography

Stephanopoulos, G. (1994). Metabolic engineering. *Curr. Opin. Biotechnol.* **5**, 196–200.

Savageau, M. A. (1969). Biochemical systems analysis. II. The steady state solutions for an n-pool system using a power-law approximation. *J. Theoretical Biol.* **25**, 370–379.

Kacser, H., and Burns, J. (1973). The control of flux. *In* "Rate Control of Biological Processes." (D. Davies, ed.). Cambridge Press, Cambridge.

Heinrich, R., Rapaport, S. M., and Rapaport, T. A. (1977). Metabolic regulation and mathematical models. *Prog. Biophysics Mol. Biol.* **32**, 1–82.

Jackson, D. A., Symons, R. H., and Berg, P. (1972). Biochemical methods for inserting new genetic information into DNA of Simian Virus 40: Circular SV40 DNA molecules containing lambda phage genes and the galactose operon of *Escherichia coli. Proc. Nat. Acad. Sci. U.S.A.* **69**, 2904–2909.

Bailey, J. E. (1991). Toward a science of metabolic engineering. *Science* **252**, 1668–1675.

Edwards, J., and Palsson, B. (1999). Properties of the Haemophilus influenzae Rd metabolic genotype. *J. Biol. Chem.*

Edwards, J. S., and Palsson, B. O. (1998). How will bioinformatics influence metabolic engineering? *Biotechnol. Bioeng.* **58**, 162–169.

Cameron, D. C., and Tong, I. T. (1993). Cellular and metabolic engineering. *Appl. Biochem. Biotechnol.* **38**, 105–140.

DeRisi, J. L., Iyer, V. R., and Brown, P. O. (1997). Exploring the metabolic and genetic control of gene expression on a genomic scale. *Science* **278**, 680–686.

Jorgensen, H., Nielsen, J., and Villadsen, J. (1995). Metabolic flux distributions in *Penicillium chrysogenum* during fed-batch cultivations. *Biotechnol. Bioengin.* **46**, 117–131.

Vallino, J., and Stephanopoulos, G. (1993). Metabolic flux distributions in *Corynebacterium glutamicum* during growth and lysine overproduction. *Biotechnol. Bioengin.* **41**, 633–646.

Sauer, U., Hatzimanikatis, V., Bailey, J., Hochuli, M., Szyperski, T., and Wuthrich, K. (1997). Metabolic fluxes in riboflavin-producing *Bacillis subtilis. Nature Biotechnol.* **15**, 448–452.

Germfree Animal Techniques

Bernard S. Wostmann

University of Notre Dame

GLOSSARY

chemically defined (CD) diet A diet consisting of only chemically defined small molecules that can be sterilized by ultrafiltration with a minimal loss of nutrients.

conventional Describing or referring to an animal raised under normal animal house conditions.

germfree (GF) Indicates freedom from all known microorganisms, though not always indicating freedom from virus for all species.

gnotobiotic (GN) Indicating the presence of well-defined microbial associates.

hexaflora A stable, six-member microflora often introduced to normalize specific germfree anomalies, such as the enlarged cecum. They most often consist of *Lactobacillus brevis*, *Streptococcus fecalis*, *Staphylococcus epidermidis*, *Bacteroides fragilis*. var. *vulgatus*, *Enterobacter aerogenes*, and a *Fusibacterium* sp.

ALTHOUGH PASTEUR in 1885 had expressed the opinion that germfree life would be impossible, 10 years later Nutthal and Thierfelder produced the first germfree guinea pig. Their ingenuously designed but complicated glass-based equipment was soon replaced by the much more simple metal isolator systems. Basically these consisted of a main compartment equipped with gloves to handle animals, presterilized food, and other necessities, with an attached two door entry port. Germfree animals were obtained by caesar-

ean section and entry into the sterile compartment either directly or via a germicidal bath. During the fifties plastic started to replace metal in the isolator systems, leading to the well-known systems used not only for germfree experimentation, but nowadays for protection of prematures and patients in need of special protection against environmental contamination.

I. INTRODUCTION

As early as 1885, Louis Pasteur had mentioned to his students the desirability of being able to study metabolism without the interference of an actively metabolizing intestinal microflora, mentioning at the same time that, in his opinion, life under those conditions would be impossible (Pasteur, 1885). However, before the end of that century, Nutthal and Thierfelder (1895) had built equipment that eventually enabled a newborn caesarean-derived guinea pig to survive (Fig. 1). Thus, the study of germfree (GF) life had begun.

II. GERMFREE ISOLATORS

Soon, the above contraption was replaced by simpler two-compartment steel autoclave-type equipment, in which one compartment served as a two-way entry port for sterilizing food and bedding, the other being the sterile compartment fitted with (nowadays) neoprene gloves, long enough for handling animals, cages, and equipment (Fig. 2). Gloves were often fitted with thin latex hand gloves to make necessary manipulations easier. Equipment was sterilized beforehand either by steam or, later, by chemi-

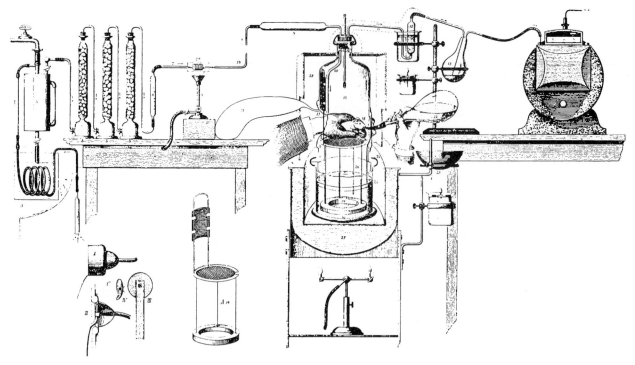

Fig. 1. Equipment used by Nutthal and Thierfelder (1895) to raise Cesarean-derived germfree guinea pigs. At left: sterilized air intake; in middle: guinea pig being fed a sterilized milk formula; at right: air exhaust system with air flow meter.

Fig. 2. Steel isolator with entry port capped with plastic in preparation for chemical sterilization of materials.

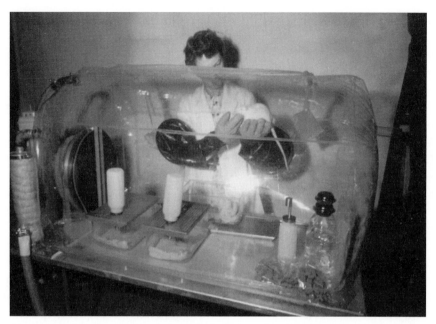

Fig. 3. Plastic isolator with air filter in foreground and outlet trap partially visible behind the filter.

cal means. In another approach, lightweight steel units with all equipment in place were sterilized in a large autoclave.

In the 1950s, plastic gradually replaced steel, although steel units are still in use in many institutions. While keeping the same two-compartment layout, the main compartment was now plastic, easily sterilized by chemical means, while food and other necessities were sterilized separately in round-bottom steel containers capped on one side with plastic. These were then connected to the main compartment via a plastic sleeve which, after chemical sterilization of the sleeve and removal of the plastic separators, allowed for transfer of the sterile material. Thus, the plastic isolator was born, now in use all over the world.

Plastic isolators used for experimental purposes usually have an effective work area of 4″ × 6″ or 4″ × 8″. They resemble the well-known incubators used in hospitals to protect premature infants (Fig. 3). For colony breeding, much larger units are used, constructed to fit exigent requirements (Latuada, 1981). Caging requirements for GF animals generally resemble those of their CV counterparts.

The versatility of the plastic material made it possible to adjust the isolators to requirements of experimentation and for industrial animal production. Pro-

duction units are often larger, although the efficiency of larger units must be weighed against the potential loss due to contamination. Some very large units have been built for specific purposes, even one with a manhole in the supporting table with an attached "halfsuit," to make it possible for the caretaker to be within the unit and be able to reach any point necessary. The ultimate use has been the housing of a child with severe combined immunodeficiency (see Section V).

In a different approach, GF mice were transferred in the isolator to a standard mouse shoebox cage, which was then capped with a tight-fitting bonnet, the top consisting of HEPA-like filter material. The cage could then be removed and placed in a laminar-flow bench outfitted with HEPA filter. Properly protected personnel could now service the mice (Sedlacek *et al.*, 1981).

III. PRODUCTION OF GERMFREE ANIMALS

Germfree animals are obtained via Cesarean section. In one approach, a specially constructed steel autoclave-type unit (as has been described) is used, which is divided into two compartments by a horizontal separator with a circular opening that is closed with a sheet of plastic. After autoclaving the units with the surgical equipment in the upper part, the lower part is opened to the environment. This (now nonsterile) part contains a small platform on which the anesthetized pregnant female can be raised so that the belly of the animal is tight against the plastic. The babies are then removed by hysterectomy. With a cauterizing knife, a cut is made through the plastic and abdominal wall, heat-sealing the abdominal cavity to the GF compartment. The babies can now be removed and placed in a separate isolator for cleaning and feeding. In this way, GF mice, rats, guinea pigs, and rabbits have been produced.

After birth, the newborn had to be fed a specifically adjusted formula, using a nipple attached to a suitable syringe-type container. Germfree guinea pigs had to be fed for only a few days because of their advanced development at birth. Other species, however, had to be fed in this way for several weeks. This turned out to present no special problems, except for the feeding of newborn GF mice. After many trials, a pair of baby C3H mice were obtained that grew to maturity, bred, and produced offspring. Thereafter, newborn mice of any strain could be foster-nursed by the GF mice already available. All of the species mentioned above reproduced well under GF conditions, rabbits only after an anomaly in Fe absorption was recognized, which needed some dietary adjustment. Germfree gerbils have been obtained by foster nursing on GF mice that were 2 to 3 days postpartum. However, these animals did not reproduce because of the size of the cecum (Bartizal *et al.,* 1982).

All GF rodents and leporidae have significantly enlarged ceca. In GF rats, the distension starts about two weeks after birth. Depending on diet, the GF cecum could be from 4 to 8% of body weight, versus around 1% in the CV rat, and more in GF rabbit and guinea pigs. The enlarged cecum is caused by the absence of a microflora that normally will digest the intestinal mucus which would otherwise accumulate in the cecum. This accumulation of acidic mucus changes the local mineral balance, leaving insufficient sodium ions necessary for water removal (Gordon and Wostmann, 1973). As a result, the water content of the cecal material resembles that of the ilium. Although comparable absorption of water oc-

curs in the colon, water content of rat GF feces is more than 20%, against about 4% in its CV counterpart. While the GF rat excretes the larger part of its water intake with the feces, the CV rats excretes most in the urine. The rather fluid feces of the GF animal may create a substantial housekeeping problem.

As mentioned earlier, GF gerbils had to be associated with a microflora consisting of six well-defined microbial species before reproduction would occur (the hexaflora). Association with this hexaflora has been used for cecal reduction, the major abnormality of the above-mentioned GF species. The hexaflora (or other multiflora) components were introduced by bringing sterile monocultures into the isolator after sterilization the outside of the usual test tubes.

Germfree animals have also been obtained via Cesarean section in a clean, minimally contaminated environment, subsequently passing the newborn via a germicidal trap into the GF unit. This method had been originally used for rats and has later been used to produce larger GF animals like pigs, lambs, and calves (Miniats, 1984).

Depending on the requirements of the study, either noninbred or inbred strains have been used to produce the GF animal. When noninbred strains were used, regular introduction of GF animals derived from the CV stock was used to assure genetic comparability between GF and CV animals. In case inbred strains were used, it would seem that this might be unnecessary but, presumably, still advisable after prolonged periods to prevent genetic drift from occurring.

Germfree chickens have been produced from fertilized eggs that had been incubated for at least 18 days. The eggs are then placed in the entry port of a GF isolator, chemically sterilized, and taken inside. They are spread out and temperature is kept at 37–38°C, humidity between 70 and 80%. After hatching, the birds are kept at 37°C for a few days. They rapidly learn to eat and drink. Thereafter, the temperature can be gradually lowered to normal room temperature (Coates, 1984).

IV. DIETS FOR GERMFREE ANIMALS

Diets for GF animals are nowadays well established. They fall into three categories: natural ingre-

TABLE I
Composition of Natural Ingredient Diet L-485

Ingredients	g/kg	Nutrient composition, %	
Ground corn	590	Protein	20.0
Soy bean meal, 50% CP[a]	300	Fat	5.3
Alfalfa meal, 17% CP	35	Fiber	3.0
Corn oil	30	Ash	5.5
Iodized NaCl	10	Moisture	11.2
Dicalcium phosphate	10	Nitrogen-free material	55.0
Calcium carbonate	5		
Lysine (feed grade)	5	Gross energy 3.9 Kcal/g	
Methionine (feed grade)	5		
Vitamin and mineral mixes	0.25[b]		
BTH	0.125		

[a] Crude protein.

[b] Kellogg, T. F., and Wostmann, B. S. (1969). Stock diet for colony production of germfree rats and mice. *Lab. Anim. Care* **17**, 589.

dient diets fortified for the losses that occur during sterilization (Table I); diets with minimal antigenicity, used mainly for immunological studies; and the chemically defined antigen-free diets, used for immunological and other studies where absolute definition of environment and dietary intake is required, again used mainly in immunology. For more defined work in the first category, diets based on rice starch, a well-defined source of protein like casein or soy protein, some fat, generally corn oil, and well-defined vitamin and mineral mixes have been developed (Table II) and are commercially available. Diets of mini-

TABLE II
Composition of Casein–Starch Diet L-488F, Fortified with Cholesterol

	g/100 g
Casein	24.0
DL-Methionine	0.3
Starch	60.4
Cellophane spangles	5.0
Corn oil[a]	5.0
Cholesterol	0.05
Vitamin and mineral mixes[b]	5.3

[a] Contains fat-soluble vitamins.

[b] For details of the vitamin and mineral mixes, see Reddy *et al.* (1972). Studies on the mechanisms of calcium and magnesium absorption of germfree rats. *Ann. Biochem. Biophys.* **149**, 15.

mal antigenicity have been developed for the study of the very early phases of immune response in young colostrum-deprived gnotobiotic (GN) piglets (Kim and Watson, 1969).

Extreme definition was obtained with the development of the GF mouse reared on a totally chemically defined, water-soluble antigen-free diet (CD diet), consisting of dextrose, amino acids, water-soluble vitamins, minerals, and a fat supplement, sterilized by ultrafiltration (Table III) (Pleasants *et al.*, 1986). This successful colony was started with inbred GF BALB/cAnN mice obtained from the GF colony maintained at the University of Wisconsin. Pregnant GF mice of this colony originally were fed natural ingredient diet L-485 (Table I), then transferred to another isolator and fed CD diet. Their offspring, never having been in contact with other than the CD diet, were housed in plastic isolators in shoebox-type mouse cages, with lids modified to hold the inverted diet and water bottles. The plastic bottoms of the cages had been replaced with stainless steel wire mesh above removable drip pans. Fortified soy-derived triglycerides provided essential fatty acids plus readily available calories. Purified vitamins A, D, E, and K were added. Again, this mixture was ultrafiltered. It was then fed in stainless steel planchets that were welded to the stainless steel dividers in the cages (Fig. 4). Whatman Ashless filter paper pro-

TABLE III

Composition of Chemically Defined Diet L-489E14SE, per 100 g Water-Soluble Solids in 300 ml Ultrapure H₂O

L-Leucine	1.90 g	Ca glycerophosphate	5.22 g
L-Phenylalanine	0.74 g	CaCl₂·2H₂O	0.185 g
L-Isoleucine	1.08 g	Mg glycerophosphate	1.43 g
L-Methionine	1.06 g	K acetate	1.85 g
L-Tryptophan	0.37 g	NaCl	86.00 mg
L-Valine	1.23 g	Mn(acetate)₂·4H₂O	55.40 mg
Glycine	0.30 g	Ferrous gluconate	55.00 mg
L-Proline	1.48 g	ZnSO₄·H₂O	40.60 mg
L-Serine	1.33 g	Cu(acetate)₂·2H₂O	3.70 mg
L-Asparagine	1.03 g	Cr(acetate)₃·H₂O	2.50 mg
L-Arginine·HCl	0.81 g	NaF	2.10 mg
L-Threonine	0.74 g	KI	0.68 mg
L-Lysine.HCl	1.77 g	NiCl₂·3H₂O	0.37 mg
L-Histidine·HCl·H₂O	0.74 g	SnCl₂·2H₂O	0.31 mg
L-Alanine	0.59 g	(NH₄)₆Mo₇O₂₄·4H₂O	0.37 mg
Na L-Glutamate	3.40 g	Na₃VO₄	0.22 mg
Ethyl L-Tyrosinate·HCl	0.62 g	Co(acetate)₂·4H₂O	0.11 mg
α-D-Dextrose	71.40 g	Na₂SeO₃	0.096 mg

Note. B vitamins (in mg): thiamine·HCl, 1.23; pyridoxine·HCl, 1.54; biotin, 0.25; folic acid, 0.37; vitamin B₁₂ (pure), 1.44; riboflavin, 1.85; niacinamide, 9.2; i-inositol, 61.6; Ca pantothenate, 12.3; choline·HCl, 310.

Amounts of lipid nutrients in one measured daily adult dose of 0.25 mL: purified soy triglycerides, 0.22 g; retinyl palmitate, 4.3 μg (7.8 IU); cholecalciferol, 0.0192 μg (0.77 IU); 2-ambo-α-tocopherol, 2.2 mg; 2-ambo-α-tocopheryl acetate, 4.4 mg; phylloquinone, 48.0 μg. The fatty acid content is 12% palmitate, 2% stearate, 24% oleate, 55% linoleate, 8% linolenate.

vided indigestible fiber while also serving as bedding and nesting material. This was autoclaved for 25 min. at 121°C or irradiated at 4.5 Mrad before being taken into the isolator.

Here, the original purpose had been the establishment of nutritional requirements in the absence of a metabolizing intestinal microflora. But again, without any original exposure to antigen or microflora, this proved an ideal tool for the study of the early and later development of the immune system. For these studies, the inbred GF BALB/c mouse proved to be the animal of choice, because of excellent breeding results through at least 6 generations. This was in contrast to earlier studies with inbred C3H mice, which did not reproduce beyond the second generation.

While, obviously, the GF animal maintained on a chemically defined diet is the animal of choice to study the various aspects of immunology, the GF

Fig. 4. Mouse cage used in isolator containing mice fed chemically defined diet. See text.

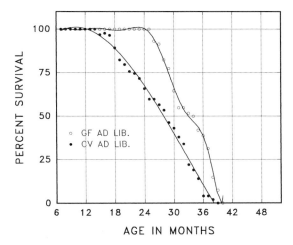

Fig. 5. Life span. Percent survival of male germfree and conventional Lobund–Wistar rats fed diet natural ingredient diet L-485.

animal maintained under less rigorous conditions will be of great value for the study of the effects of the complete absence of a microflora on function and metabolism. Comparing GF and CV rats, it was found that the stress of a metabolizing microflora reduces life span to an extent that otherwise can only be achieved by an involuntary reduction of dietary intake by 30% (Fig. 5).

V. APPLICATION OF GERMFREE TECHNIQUES: GNOTOBIOTES

A. The Boy in the Bubble

In general, most controlled studies have been carried out with GF and GN rats and mice. The GF animal technique has the potential for selective association of the originally GF animal with a specific microflora element, including parasites. Mentioned earlier is the potential of bringing its distended cecum back within acceptable limits by association with a well-defined cocktail of microbial species. These are introduced by bringing monocultures grown in test tube into the main unit after chemical exterior sterilization in the entry compartment. In this way, studies have become possible of what is called the gnotobiote, the animal that harbors a desired but defined combination of microflora ele-

ments. However, the term gnotobiote per se also includes the GF animal.

Once the GF animal had been produced, the logical extension was the study of function and metabolism of this gnotobiote. Apart from anomalies mentioned earlier, the absence of a metabolizing microflora resulted in the absence of certain nutrients normally produced by the flora (e.g., vitamin K) and certain flora-produced stimuli (e.g., LPS). Because of this, diet had to be adjusted, apart from adjustment necessitated by heat or radiation sterilization. The effects of the GF state on immune function are substantial, although all GF species proved to be eventually immunologically competent. Upon antigenic stimulation, the GF animal generally shows a delayed, but eventually adequate, immune response. However, antibacterial response may be inadequate, due to slow response of macrophages, which originally are defective in chemotaxis and destructive and lytic capacity, leading to a delayed presentation of antigenic material to other elements of the immune system (Wostmann, 1996).

The possibility of the production of gnotobiotes with a stable and defined microflora opened many possibilities and can be regarded as the originally GF animal's major potential. It soon became obvious that GF rats did not develop that scourge of our society, dental caries, thereby indicating its microbial origin. GF rats were then associated with a number of different bacteria. This resulted in the recognition of *Streptococcus mutans* as the major originator of caries and put the acid-producing lactobaccili in second place (Orland *et al.*, 1955). Similarly, it was shown that GF guinea pigs inoculated with the parasite *Entomoebia histolytica,* the cause of a potentially lethal intestinal infection, do not develop any symptoms. The animals retained the amoeba for only a few days. A microflora is obviously needed to change intestinal conditions to the point where the infection could take hold (Phillips, 1964). On the other hand, it was found that a CV microflora, by its stimulation of the immune system, may affect a certain amount of protection against Schistosomiasis (Bezerra *et al.*, 1985). Studies of the various factors involved in the establishment of the nematode *Trichinella spiralis* were carried out by Przyjalkowski (Przyjalkowski *et al.*, 1983) and Despommier (Despommier, 1984). In the pork industry, GF pigs have been used to solve problems of bacterial and viral infection.

Gnotobiotes are important to the study of colonization resistance, which seeks to determine which microflora elements may be important for the flora's stability and which for its potential to resist pathogens. In a similar way, gnotobiotes enable the study of microbial translation to determine what microflora composition will enhance or inhibit certain of its members to pass the intestinal barrier and possibly cause disease.

Recently, it has become possible to establish a "normal" human microflora in originally GF mice. After the actual composition of this flora has been established, and its stability ascertained, this could open the door to a multitude of studies pertaining to human health and disease.

The plastic isolator, already in general use to protect premature infants, found its culmination in its use to protect David, the "Boy in the Bubble." Before birth diagnosed as having SCID (severe combined immune deficiency), he was born via Cesarean section and placed in an isolator, which eventually grew to a four-room apartment. Over the years, he accumulated a number of non-life-threatening organisms, but death came at the age of about $12\frac{1}{2}$ years, apparently caused by the sequela of an unsuccessful bone marrow transplant (Bealmear *et al.*, 1985).

Bibliography

Bartizal, K. F., *et al.* (1982). Cholesterol metabolism in gnotobiotic gerbils. *Lipids* **17**, 791.

Bealmear, P. M., *et al.* (1985). David's story: A gift of 12 years, 5 months and 1 day. *In* "Progress in Clinical and Biological Research" Vol. 181 (B. S. Wostmann, ed.), p. 485. Alan R. Liss, Inc., New York.

Bezerra, J. A., *et al.* (1985). The life cycle of *Schistosoma mansoni* under germfree conditions. *J. Parasitol.* **71**, 519.

Coates, M. E. (1984). Production of germfree animals. Part 3. Birds. *In* "The Germfree Animal in Biomedical Research" (M. E. Coates and B. E. Gustafsson, eds.). Laboratory Animal Handbooks 9, p. 79. Laboratory Animals Ltd., London, UK.

Coates, M. E. (ed.). (1987). "ICLAS Guidelines on the Selection and Formulation of diets for Animals in Biomedical Research" ICLAS, Institute of Biology, 20 Queensberry Place, London, UK.

Despommier, D. D. (1984). Antigens from *Trigenella spiralis* that induce a protective response in the mouse. *J. Immunol.* **132**, 898.

Fitzgerald, R. F., *et al.* (1989). Cariogenicity of a lactate dehydrogenase-deficient mutant of *Streptococcus mutans* serotype c in gnotobiotic rats. *Infect. Immun.* **57**, 823.

Gordon, H. A., and Wostmann, B. S. (1973). Chronic mild diarrhea in germfree rodents: A model portraying host–flora synergism. *In* "Germfree Research, Proc. IV Internat. Symp. Germfree Research" (J. B. Heneghan, ed.), p. 593, Academic Press, New York.

Hashimoto *et al.* (ed.). (1996). Germfree life and its ramifications. "Proceedings of the XII International Symposium on Gnotobiology" Sense Printing, Japan.

Kim, Y. B., and Watson, D. W. (1969). *In* "Germfree Biology, Experimental and Clinical Aspects" (E. A. Mirand and N. Back, eds.), p. 259. Plenum Press, New York.

Latuada, C. P. (1981). Large isolators for rearing rodents. *In* "Recent Advances in Germfree Research, Proc. VII Internat. Symp. On Gnotobiology" (Sasaki *et al.*, eds.), p. 53, Tokay University, Tokyo.

Miniats, O. P. (1984). Production of germfree animals. Part 2. Farm animals. *In* "The Germfree Animal in Biomedical Research" (M. E. Coates and B. E. Gustafsson, eds.), Laboratory Animal Handbooks 9, p. 49. Laboratory Animals Ltd., London, UK.

Nutthal, G. H. F., and Thierfelder, H. (1895). Thierisches Leben ohne Bakterien im Verdauungskanal, *Z. Physiol. Chem.* **21**, 109.

Orland, F. J., *et al.* (1955). Experimental caries in rats inoculated with *Enterococci. Am. Dental Assoc.* **50**, 259.

Pasteur, L. (1885). Observations relative à la note précédente de M. Duclaux. *CR Acad. Sci. Paris* **100**, 68.

Phillips, B. F. (1964). Studies of the amoeba–bacteria relationships in amebiasis III. Induced amoebic lesions in the germfree guinea pig. *Am. J. Trop. Med. Hyg.* **13**, 301.

Pleasants *et al.* (1986). Adequacy of chemically defined, water-soluble diet for germfree BALB/c mice through successive generations and litters. *J. Nutr.* **116**, 1949.

Przyjalkowski *et al.* (1983). Intestinal *Triginella spiralis* and *Triginella pseudospiralis. Prog. Fd. Nutr. Sci.* **7**, 117.

Sedlacek, R. S., *et al.* (1981). A flexible barrier at cage level for existing colonies. *In* "Recent Advances in Germfree Research, Proc. VII Internat. Symp. on Gnotobiology" (S. Sasaki *et al.*, eds.), p. 65, Tokay University, Tokyo.

Wostmann, B. S. (1981). The germfree animal in nutritional studies. *Ann. Rev. Nutr.* **1**, 257.

Wostmann, B. S. (1996). "Germfree and Gnotobiotic Animal Models 101." CRC Press, New York.

Global Burden of Infectious Diseases

Catherine M. Michaud

Harvard School of Public Health

GLOSSARY

***disability-adjusted life year* (*DALY*)** The measurement unit that was developed for the Global Burden of Disease Study. The DALY is a time-based measure, and thus a form of QALY, in which the value choices have been standardized. The DALY is one lost year of healthy life. It captures in a single indicator the impact of both premature death and nonfatal health outcomes of diseases and injuries. DALYs from a condition are the sum of years lost to premature death and years lived with a disability adjusted for the severity of the disability. The DALY incorporates explicit and transparent value choices for the duration of life lost, the value of life lived at different ages, comparison of time lived with a disability with time lost due to mortality, and time preference. It is built on the principle that only two characteristics of individuals that are not directly related to their health—their age and sex—should be taken into consideration when calculating the burden of a given health outcome in that individual, whereas socioeconomic status, race, and education are not considered. Time lived with a disability and time lost to premature death are age weighted to reflect the greater social role played by adults in caring and providing for the young and the old. Time has been discounted at 3% so that a year lost in the future is less valuable than a year lost today.

***quality-adjusted life year* (*QALY*)** A time-based measure that incorporates judgments about the value of time spent in different health states. Since the late 1940s, researchers have generally agreed that time is an appropriate currency: time (in years) lost through premature death and time (in years) lived with a disability. The term QALY does not imply any specific set of value choices or methods used to elicit preferences for health states. A range of such time-based measures has been developed in different countries.

***years lived with disability* (*YLD*)** Years lived with a disability of known severity and duration. Although death is clearly defined, disability is not. It is difficult to define because nonfatal health outcomes differ from each other in their causes, nature, impact on the individual, and the way in which the surrounding community responds. To compare different disabilities, time lived with various short-, medium-, and long-term disabilities is weighted by a severity weight that is based on the measurement of social preferences for time lived in various health states. Severity weights range from 0 (perfect health) to 1 (equivalent to death). Two methods were used to formalize social preferences for different states of health. Both ask people to make trade-offs between quantity and quality of life. Results of time trade-off exercises showed a surprisingly wide agreement between cultures on what constitutes a severe or mild disability. For example, a year lived with blindness appears to most people as more severe than a year lived with watery diarrhea but less severe than a year lived with quadriplegia.

***years of life lost* (*YLL*)** A year of life lost due to premature death, defined as a death occurring before the age to which the dying person could have expected to survive if he or she was a member of a standardized model population with a life expectancy at birth equal to that of one of the world's longest surviving populations (Japan), or 82.5 years for females and 80 years for males.

SYSTEMATIC EFFORTS to quantify and monitor the burden of specific health conditions in populations, at the national level, started in the mid-1950s for malaria, poliomyelitis, and influenza in the United States. Comprehensive surveillance of morbidity and mortality for dozens of conditions has since been well established in the United States and in other industrialized countries. However, despite the clear need for epidemiological data to inform health policies, reliable and comprehensive health statistics are not available in many developing countries. International efforts to assess and monitor the burden of certain diseases have been limited in the past to a small number of infectious diseases in the context of global eradication programs—smallpox, poliomyelitis, guinea worm, and, recently, HIV/AIDS.

The Global Burden of Disease study (GBD), published in 1996, filled an important gap in our knowledge of population health status. It was conducted by researchers at the Harvard School of Public Health and the World Health Organization, with the help of more than 100 collaborators throughout the world. The GBD provided the first comprehensive, internally consistent and comparable set of estimates of patterns of mortality and disability from disease and injury for all regions of the world in 1990, with projections to the Year 2020. The disability-adjusted life year (DALY) is the unit of measurement that was developed to estimate the global burden of disease and injury. It combines the number of years lost to premature death (YLL) and the number of years lived with a disability (YLD) by age and sex for more than 100 causes of deaths.

Causes of deaths were categorized into three main groups: group I (infectious diseases and maternal, perinatal, and nutritional conditions), group II (noncommunicable diseases), and group III (injuries). Accordingly, estimates of the global burden of infectious diseases are provided in the context of the overall burden from other conditions, diseases, and injuries. The relative importance of the burden of infectious diseases was forecasted to change by the Year 2020. As the epidemiological transition progresses worldwide, a decline in the burden of infectious diseases is expected as the burden of non-communicable diseases and injuries gradually increases. The pace of the epidemiological transition, however, varies greatly among regions so that the projected decreases in the burden of infectious diseases are expected to vary between regions. Trends in the global burden due to specific infectious diseases pro-

jected to the Year 2020 also vary among specific conditions. The global burden of HIV/AIDS, for instance, is expected to greatly increase, whereas the global burden due to respiratory infections, diarrheal diseases, and malaria is expected to decrease.

I. THE GLOBAL BURDEN OF DISEASE STUDY

A. Objectives

The GBD had three main objectives. The first objective was to add information about nonfatal health outcomes to debates of national and international health policy. International data sets which are based on similar diagnostic and reporting procedures are almost exclusively focused on mortality and fail to incorporate comparable information on nonfatal health outcomes. As a result, nonfatal health outcomes have been for the most part neglected in the international health policy debate.

The second objective was to produce objective, independent, and demographically plausible epidemiological assessments of health status in order to decouple epidemiology from advocacy. Estimates of the numbers killed or affected by particular conditions or diseases may be exaggerated beyond demographically plausible limits by well-intentioned epidemiologists who also act as advocates for the affected populations in competition for scarce resources.

The third objective was to provide an outcome measure for cost-effectiveness analyses of interventions that could reduce the burden of either proximal biological causes or the more distal risk factors and socioeconomic determinants, in terms of cost per unit of burden averted.

B. GBD Regions

The GBD provides internally consistent estimates of deaths, YLLs, YLDs, and DALYs for 107 causes of deaths disagreggated by age and sex for the world and eight regions in 1990 and projected to the Year 2020. The eight regions were the established market economies (EME), the formerly socialist economies of Europe (FSE), India (IND), China (CHN), other Asia and islands (OAI), sub-Saharan Africa (SSA), Latin America and the Caribbean (LAC), and the Middle Eastern crescent (which includes North Africa, the Middle East, Pakistan, and the central Asian republics of the former Soviet Union) (MEC). The criteria used to define these regions included the level of socioeconomic development, epidemiological homogeneity, and geographic contiguity. EME and FSE were grouped as developed regions and the other six regions as developing regions.

C. GBD Classification System for Diseases and Injuries

The selection of the classification scheme to represent mortality by cause for the GBD was driven by the intent to provide information that would be useful for the health policiy debate. The challenge was to strike the proper balance between too little and too much detail in the selection of the final list of causes. At the simplest end of the spectrum, cause of death can be aggregated in just two categories: infectious and parasitic diseases and chronic diseases. This level of aggregation, however, would not be useful to inform the choice of specific health priorities or to assess the potential to improve survival through specific intervention strategies. At the other end of the spectrum, an overly detailed list of causes would make cross-national and intertemporal comparisons difficult to interpret. Accordingly, the classification adopted follows a tree structure of causes of death. At the first level of disaggregation, overall mortality was divided into three broad groups: group I (infectious diseases and maternal, perinatal, and nutritional conditions), group II (noncommunicable diseases), and group III (injuries). Each group was further subdivided into several major subcategories. All infectious diseases were included under the first two major subcategories of group I: infectious and parasitic diseases (IA) and respiratory infections (IB). Other group I subcategories are maternal conditions (IC), perinatal conditions (ID), and nutritional conditions (IE). Group II was subdivided into 14 subcategories—i.e., cancers, cardiovascular diseases, and neuropsychiatric conditions. Group III was subdivided into two categories—intentional injuries and

nonintentional injuries. Each second-level category was further disaggregated into two levels that include a total of 107 individual causes, such as HIV, lung cancer, or motor-vehicle accidents. Table I provides the detailed classification used in the GBD for infectious diseases.

II. THE GLOBAL BURDEN OF INFECTIOUS DISEASES: MAIN FINDINGS

A. Mortality from Infectious Disease

1. *Deaths Due to Infectious Diseases for the World and by Region, 1990*

Worldwide an estimated 50.5 million people died from all causes in 1990. Almost one-third of these deaths (27%) were due to infectious causes, and virtually all were in developing regions (13.2 million out of 13.7 million). Mortality from infectious diseases was highest in sub-Saharan Africa, with 4.5 million deaths, and India, with 3.9 million deaths, or 61% of all infectious disease deaths (Table II).

2. *Leading Causes of Deaths*

Five infectious causes ranked among the top 10 killers in developing regions: lower respiratory infections (4.3 million deaths), diarrheal diseases (2.9 million deaths), tuberculosis (1.9 million deaths), measles (1.1 million deaths), and malaria (0.9 million deaths) (Table III). Lower respiratory infections ranked fourth in developed regions.

Lower respiratory infections, diarrheal diseases, measles, and malaria deaths affected predominantly children under 5 years of age: 59% of all deaths from infectious causes occurred in this age group. Tuberculosis affected mostly young adults: 57% of all deaths occurred between ages 15 and 59 years.

B. Years of Life Lost to Infectious Causes

In contrast to crude number of deaths, a time-based measure such as YLLs takes into account the age at which a death occurs and thus quantifies the loss of life resulting from premature deaths. YLLs give a greater weight to deaths occurring at younger

Global Burden of Disease Study Classification System for Diseases and Injuries: Group I—Communicable, Maternal, Perinatal, and Nutritional Conditions[a]

Title of GBD cause

I   Communicable, maternal, perinatal, and nutritional conditions

    A. Infectious and parasitic diseases
        1. Tuberculosis
        2. Sexually transmitted diseases, excluding HIV
            a. Syphilis
            b. Chlamydia
            c. Gonorrhea
        3. HIV
        4. Diarrheal diseases
        5. Childhood cluster diseases
            a. Pertussis
            b. Polyomielitis
            c. Diphteria
            d. Measles
            e. Tetanus
        6. Bacterial meningitis and meningococcemia
        7. Hepatitis B and hepatitis C
        8. Malaria
        9. Tropical-cluster causes
            a. Trypanosomiasis
            b. Chagas' disease
            c. Schistosomiasis
            d. Leishmaniasis
            e. Lymphatic filariasis
            f. Onchocerciasis
      10. Leprosy
      11. Dengue
      12. Japanese encephalitis
      13. Trachoma
      14. Intestinal nematode infections
            a. Ascariasis
            b. Trichuriasis
            c. Ancylostomiasis and necatoriasis

    B. Respiratory infections
        1. Lower respiratory infections
        2. Upper respiratory infections
        3. Otitis media

    C. Maternal conditions

    D. Conditions arising during the neonatal period

    E. Nutritional deficiencies

[a] Data from Global Health Statistics.

Regional Distribution of Deaths Due to Infectious Causes[a]

	No. of deaths (in thousands)			
Region	Infectious and parasitic diseases (IA)	Respiratory infections (IB)	All infectious causes	All causes
EME	111	275	386	7,121
FSE	52	114	166	3,791
IND	2647	1229	3,876	9,371
CHN	544	474	1,018	8,885
OAI	1176	552	1,728	5,534
SSA	3456	1023	4,479	8,202
LAC	473	180	653	3,009
MEC	871	534	1,405	4,553
World	9330	4381	13,711	50,466
Developing regions	9167	3992	13,159	39,554
Developed regions	163	389	552	10,912

[a] Data from The Global Burden of Disease.

ages. The younger the age at which a death occurs, the greater the number of YLLs. The proportion of all YLLs that were lost as a result of premature deaths from lower respiratory infections and diarrheal diseases exceeds the proportion of all YLLs lost as a result of deaths due to ischemic heart disease and cerebrovascular disease which occur mostly in older age groups (Fig. 1).

C. Disability Due to Infectious Diseases

The necessity of quantifying and analyzing non-fatal health outcomes, in addition to mortality, is essential because the leading causes of disability differ substantially from the leading causes of deaths. The GBD's findings demonstrate that the burden of disability is dominated by a short list of causes. First, neuropsychiatric conditions were estimated to account for 28% of all YLDs. Five of the top 10 causes of disability were neuropsychiatric conditions: unipolar major depression, alcohol use, bipolar disorder, schizophrenia, and obsessive-compulsive disorders.

Second, maternal conditions and sexually transmitted diseases and HIV, which are largely related

TABLE III
The 10 Leading Causes of Death, 1990[a]

Developed regions	*Deaths (in thousands)*	*Developing regions*	*Deaths (in thousands)*
All causes	10,912	All causes	39,554
1. Ischemic heart disease	2,695	1. Lower respiratory infections	3,915
2. Cerebrovascular disease	1,427	2. Ischemic heart disease	3,565
3. Trachea, bronchus, and lung cancer	523	3. Cerebrovascular disease	2,954
4. Lower respiratory infections	385	4. Diarrheal disease	2,940
5. Chronic obstructive pulmonary disease	324	5. Conditions arising during the perinatal period	2,361
6. Colon and rectum cancers	277	6. Tuberculosis	1,922
7. Stomach cancers	241	7. Chronic obstructive pulmonary disease	1,887
8. Road traffic accidents	222	8. Measles	1,058
9. Self-inflicted injuries	193	9. Malaria	856
10. Diabetes mellitus	176	10. Road traffic accidents	777

[a] Data from The Global Burden of Disease.

to sexual activity, account for 6.6% of global YLDs: 2.5% in developed regions and 7.3% in developing regions. Other conditions included in the top 10 leading causes of YLDs were anemia, falls, road traffic accidents, chronic obstructive pulmonary disease (COPD), and osteoarthritis.

D. The Global Burden of Infectious Diseases

1. Distribution of Total DALYs

The global burden of disease is expressed as the number of DALYs, which is the sum of YLLs and YLDs for each of the 107 conditions that were included in the GBD classification.

Worldwide, a total of 1.4 billion DALYs were lost from all causes as a result of premature death and disability—88% in developing regions (which comprise 78% of the world's population) and 12% in developed regions (which comprise 22% of the world's population). Group I causes (communicable, maternal, perinatal, and nutritional conditions) still accounted for almost half of total DALYs in developing regions. Group II causes (noncommunicable diseases) were the leading cause of burden in developed regions, whereas group II DALYs represented 77% of the total burden. In developing regions, group II accounted for about one-third of total DALYs. There was much less variation in the proportion of DALYs lost to group III (injuries) causes, about 15%

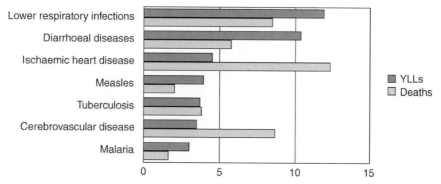

Fig. 1. Leading causes of deaths and premature deaths throughout the world, 1990.

TABLE IV
Percentage Distribution of DALYs among Specific Causes (Level-Two Categories), 1990[a]

Group/cause	Developed regions (%)	Developing regions (%)	World (%)
All causes	100.0	100.0	100.0
I. Communicable, maternal, perinatal, and nutritional conditions	7.8	48.7	43.9
A. Infectious and parasitic diseases	2.7	25.6	22.9
B. Respiratory infections	1.6	9.4	8.5
C. Maternal conditions	0.6	2.4	2.2
D. Conditions arising during the perinatal period	1.9	7.3	6.7
E. Nutritional deficiencies	0.9	4.1	3.7
II. Noncommunicable diseases	77.7	36.1	40.9
III. Injuries	14.5	15.2	15.1

[a] Data from The Global Burden of Disease.

of the total in both developed and developing countries (Table IV).

There were great inequalities in the distribution of the burden of disease within developing regions as well. Peoples of sub-Saharan Africa and India together bore more than four-tenths of the global burden of disease in 1990, although they made up only 26% of the world's population. China emerged as the developing region which had the least burden for the size of its population (Fig. 2).

2. DALYs Lost to Infectious Diseases

In 1990, 432.7 million DALYs were lost as a consequence of infectious diseases worldwide, or 31.4% of total DALYs worldwide (groups IA and IB). The distribution of the burden of infection follows the distribution of group I: Burden of infections affect mostly populations in developing regions, in which

98.4% of the total burden of infections occurred, with only 1.6% occurring in the EME and FSE combined. Sub-Saharan Africa and India, with respectively 36.2 and 27.2% of the burden of infections, were the two regions with the largest contributions. China, with 21.5% of the world's population, accounted for only 6.5% of the total burden of infections (Fig. 3). The ranking and number of DALYs for specific infectious diseases for the world and developed and developing regions are presented in Tables V–VII.

3. DALYs Lost from Infectious Diseases That Are Powerful Risk Factors for Other Diseases

Traditional methods of assessing deaths by cause fail to consider the fact that some infectious diseases can be powerful risk factors for other diseases.

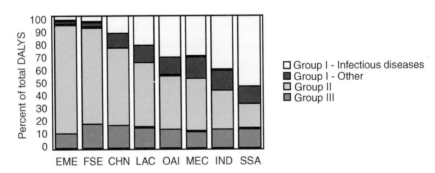

Fig. 2. The burden of disease by broad cause group, 1990 (data from The Global Burden of Disease).

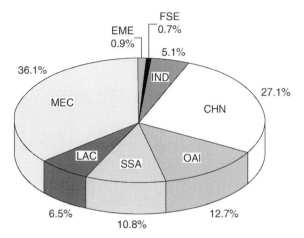

Fig. 3. Burden of infectious diseases, 1990.

Several diseases occur from prior or current exposure to an infectious agent. The approach adopted in the global burden of disease study to overcome this limitation and provide a more complete picture of the actual burden related to a small number of infectious diseases was to treat each condition as a risk factor. The study provides estimates of how much of the total burden would be averted in each region if the conditions were eliminated. The largest difference between directly coded and total burden was for hepatitis B and hepatitis C. Infection with hepatitis B virus increases the risk of developing liver cancer and cirrhosis of the liver (Table VIII).

4. The Burden of Risk Factors for Infectious Diseases

The GBD assessed the burden of disease associated with the following major risk factors: malnutrition; poor water supply, sanitation, and hygiene; unsafe sex; tobacco use; alcohol use; occupation; hypertension; physical inactivity; illicit drug use; and air pollution. Exposures that underlie the major infectious diseases in young children—malnutrition and poor water supply, sanitation, and hygiene—cause an estimated 20–25% of the total burden of disease and injury. Unsafe sex is a major risk factor for HIV and sexually transmitted diseases, as well as for other maternal conditions (Table IX).

5. Projecting Mortality and Morbidity from Infectious Diseases, 1990–2020

Projections of mortality and disability are essential to guide investments in the health sector worldwide. The GBD provided three alternative scenarios of future burden from 1990 to 2020.

The projection model used for the study was based on the observed relation between cause-specific mortality and three socioeconomic variables: income per capita, average years of schooling, and tobacco use and time. Three sets of projections of these independent variables formed the basis of the baseline, optimistic, and pessimistic scenarios presented in the GBD.

Dramatic changes in rank order of deaths for the 15 leading causes worldwide are expected to occur between 1990 and 2020. From 1990 to 2020, the baseline projection scenario suggests that ischemic heart disease, unipolar major depression, cerebrovascular disease, chronic obstructive pulmonary disease, HIV, war, violence, suicide, and lung cancer will increase in the relative ranking of causes. The most striking change is projected to occur for deaths due to infection with HIV. Although it ranked 28th as a cause of deaths worldwide in 1990, infection with HIV is projected to rank 10th in 2020. Major declines in relative rankings are expected for lower respiratory infections, diarrheal diseases, measles, malaria, anemia, and protein-energy malnutrition. Malaria, which ranked 11th as a cause of death worldwide in 1990, will rank 24th in 2020.

Even in the pessimistic scenario, lower respiratory infections and diarrheal diseases would decrease in the relative rankings but would remain much larger causes in absolute terms. Lower respiratory infections will remain the leading cause of infectious disease deaths No change is expected to occur in the ranking for tuberculosis under the baseline scenario (Table X).

Projections of future health trends were developed using a parsimonious model which included only three variables: (i) income per capita, (ii) the average number of years of schooling in adults, and (iii) time as a proxy measure for secular improvements in health this century that resulted in part from accumulating knowledge and technological development.

It follows that forecasts in any of the scenarios

TABLE V
**Causes of DALYs (Percentage of Total) in Descending Order, 1990—World Top 10
and Ranking of Infectious Diseases[a]**

Rank	Disease or injury	DALYs (in thousands)	% of total
	All causes	1,379,238	
1	Lower respiratory infections	112,898	8.2
2	Diarrheal diseases	99,633	7.2
3	Conditions arising during the perinatal period	92,313	6.7
4	Unipolar major depression	50,810	3.7
5	Ischemic heart disease	46,699	3.4
6	Cerebrovascular disease	38,523	2.8
7	Tuberculosis	38,426	2.8
8	Measles	36,520	2.7
9	Road traffic accidents	34,317	2.5
10	Congenital anomalies	32,921	2.4
	Other infectious diseases		
11	Malaria	31,706	2.3
18	Tetanus	17,517	1.3
23	Pertussis	13,403	1.0
28	HIV	11,172	0.8
38	Chlamydia	7,169	0.5
39	Syphilis	6,596	0.5
43	Bacterial meningitis and meningoccemia	6,242	0.5
49	Gonorrhea	4,909	0.4
55	Lymphatic filariasis	3,997	0.3
60	Poliomyelitis	3,371	0.2
67	Otitis media	2,163	0.2
68	Hepatitis B and hepatitis C	2,136	0.2
69	Leishmaniasis	2,092	0.2
72	Trichiuriasis	1,788	0.1
74	Ascariasis	1,750	0.1
78	Schistosomiasis	1,519	0.1
79	Ancylostomiasis and necatoriasis	1,484	0.1
80	Trypanosomiasis	1,467	0.1
84	Upper respiratory infections	1,311	0.1
87	Trachoma	1,024	0.1
88	Onchocerciasis	884	0.1
89	Dengue	750	0.1
90	Japanese encephalitis	744	0.1
92	Chagas' disease	641	0.1
94	Leprosy	384	0.0
95	Diphteria	361	0.0

[a] Data from The Global Burden of Disease.

TABLE VI
Causes of DALYs (Percentage of Total) in Descending Order, 1990—Developing Regions Top 10 and Ranking of Infectious Diseases[a]

Rank	Disease or injury	DALYs (in thousands)	% of total
	All causes	1,218,244	
1	Lower respiratory infections	110,506	9.1
2	Diarrheal diseases	99,168	8.1
3	Conditions arising during the perinatal period	89,193	7.3
4	Unipolar major depression	41,031	3.4
5	Tuberculosis	37,930	3.1
6	Measles	36,498	3.0
7	Malaria	31,705	2.6
8	Ischemic heart disease	30,749	2.5
9	Congenital anomalies	29,441	2.4
10	Cerebrovascular disease	29,099	2.4
	Other infectious diseases		
17	Tetanus	17,513	2.3
21	Pertussis	13,353	1.1
26	HIV	9,864	0.8
35	Syphilis	6,586	0.5
36	Chlamydia	6,524	0.5
39	Bacterial meningitis and meningococcemia	5,833	0.5
45	Gonorrhea	4,788	0.4
49	Lymphatic filariasis	3,997	0.3
56	Poliomyelitis	3,368	0.3
66	Leishmaniasis	2,091	0.2
67	Hepatitis B and hepatitis C	2,045	0.2
68	Otitis media	2,031	0.2
70	Trichiuriasis	1,788	0.2
71	Ascariasis	1,750	0.1
76	Schistosomiasis	1,519	0.1
78	Ancylostomiasis and necatoriasis	1,484	0.1
79	Trypanosomiasis	1,467	0.1
80	Upper respiratory infections	1,233	0.1
82	Trachoma	1,024	0.1
83	Onchocerciasis	884	0.1
85	Dengue	750	0.1
86	Japanese encephalitis	744	0.1
90	Chagas' disease	641	0.1
92	Leprosy	382	0.0
93	Diphteria	360	0.0

[a] Data from The Global Burden of Disease.

TABLE VII
Causes of DALYs (Percentage of Total) in Descending Order, 1990—Developed Regions Top 10 and Ranking of Infectious Diseases[a]

Rank	Disease or injury	DALYs (in thousands)	% of total
	All causes	160,994	
1	Ischemic heart disease	15,590	9.9
2	Unipolar major depression	9,780	6.1
3	Cerebrovascular disease	9,425	5.9
4	Road traffic accidents	7,064	4.4
5	Alcohol use	6,446	4.0
6	Osteoarthritis	4,681	2.9
7	Trachea, bronchus, and lung cancers	4,587	2.9
8	Dementia and other degenerative and hereditary CNS disorders	3,816	2.4
9	Self-inflicted injuries	3,768	2.3
10	Congenital anomalies	3,480	2.2
	Other infectious diseases		
17	Lower respiratory infections	2,392	1.5
27	HIV	1,303	0.8
41	Chlamydia	645	0.4
49	Tuberculosis	496	0.3
50	Diarrheal diseases	495	0.3
54	Bacterial meningitis and meningoccemia	409	0.3
65	Otitis media	132	0.1
66	Gonorrhea	121	0.1
68	Hepatitis B and hepatitis C	91	0.1
69	Upper respiratory infections	79	0.1
73	Pertussis	50	0.0
77	Measles	22	0.0
78	Syphilis	10	0.0
79	Poliomyelitis	4	0.0
80	Tetanus	3	0.0
81	Malaria	2	0.0
82	Leprosy	1	0.0
83	Leishmaniasis	1	0.0
84	Diphteria	1	0.0
85	Japanese encephalitis	<1	0.0
86	Schistosomiasis	<1	0.0
87	Trypanosomiasis	<1	0.0
88	Trachoma	<1	0.0
90	Ancylostomiasis and necatoriasis	<1	0.0
91	Dengue	<1	0.0
92	Trichuriasis	<1	0.0
93	Ascariasis	<1	0.0
94	Onchocerciasis	<1	0.0
95	Chagas' disease	<1	0.0
96	Lymphatic filariasis	<1	0.0

[a] Data from The Global Burden of Disease.

TABLE VIII
Total Burden for Major Infectious Diseases That Are Risk Factors for Other Diseases—DALYs (World), 1990

Disease	Both sexes (millions)	Male (millions)	Female (millions)	Direct/total (millions)
Onchocerciasis	1.2	0.7	0.5	0.76
Chagas' disease	1.6	0.9	0.7	0.40
Trachoma	2.0	0.6	1.4	0.51
Hepatitis B and hepatitis C	13.3	9.2	4.1	0.16
Sexually transmitted diseases	25.3	9.1	16.2	0.74

TABLE IX
Burden Attributable to Selected Risk Factors for Infectious Diseases

Risk factor	Deaths (in thousands)	% of total deaths	DALYs (in millions)	% of total DALYs
Malnutrition	5881	11.7	219.6	15.9
Poor water supply, sanitation, and hygiene	2668	5.3	93.4	6.8
Unsafe sex	1095	2.2	48.7	3.5

TABLE X
Change in Rank Order of Disease Burden for the 15 Leading Causes—World, 1990–2020[a]

1990			2020 (Baseline scenario)
Disease or injury			Disease or injury
Lower respiratory infections	1	1	Ischaemic heart disease
Diarrheal diseases	2	2	Unipolar major depression
Conditions arising during the perinatal period	3	3	Road traffic accidents
Unipolar major depression	4	4	Cerebrovascular disease
Ischaemic heart disease	5	5	Chronic obstructive pulmonary disease
Cerebrovascular disease	6	6	Lower respiratory infections
Tuberculosis	7	7	Tuberculosis
Measles	8	8	War
Road traffic accidents	9	9	Diarrheal diseases
Congenital anomalies	10	10	HIV
Malaria	11	11	Conditions arising during the perinatal period
Chronic obstructive pulmonary disease	12	12	Violence
Falls	13	13	Congenital anomalies
Iron deficiency anemia	14	14	Self-inflicted injuries
Protein energy malnutrition	15	15	Trachea, bronchus and lung cancers
		24	Malaria
		25	Measles
HIV	28		

[a] Disease burden measured in DALYs.

outlined previously rely on two major implicit assumptions: (i) Improvements in hygiene observed during the past decades in developing regions will continue or at least be sustained in the future, and (ii) no major emerging infection or novel pandemics will occur.

III. POLICY IMPLICATIONS

The main goal of the GBD was to provide the objective evidence base on the magnitude of the burden of disease due to premature deaths and to nonfatal health outcomes needed to inform international public health policy. As such, the choice of the DALY indicator to quantify the GBD was not a neutral exercise. The vast policy implications of the findings of the study set forth an extensive debate, which challenged the explicit social value choices embedded in the DALY: the sex difference in the standard length of life, the choice of severity weights for different disabilities, the introduction of age weights, and the discounting of future health outcomes. The following major recommendations for public health policy were derived from the GBD:

1. The magnitude of the remaining burden due to infectious diseases in developing countries, despite the sustained efforts to reduce child mortality during the past few decades, underscore the need for lower respiratory infections, diarrheal diseases, tuberculosis, malaria, and hepatitis B and C to remain a key priority for global public health action.

2. Although the high mortality due to infectious diseases in developing countries was already well-known, the finding that deaths from two major cardiovascular causes (ischemic heart disease and cerebrovascular disease) were the second and third cause of death in developing countries was a significant surprise. Also noteworthy was the finding that deaths from road traffic accidents were among the top 10 causes in all regions. The major implication of these findings is that they need to be included in the international public health policy debate.

3. It is essential to include the burden of nonfatal health outcomes in the global assessment of health problems because it notoriously shifts the ranking of priorities. The burden of neuropsychiatric conditions and sexually transmitted diseases had been greatly underestimated.

4. Although some diseases and injuries occur without prior exposure to health hazards, review of the contribution of one or more major risk factor for respiratory infections and diarrheal diseases illustrates the importance of exposures that underlie major infectious diseases mostly in developing countries. Their control must remain a priority.

In the absence of sustained efforts to improve hygiene and sustained vigilance about emerging infections and novel pandemics, forecasts about the pace of the epidemiological transition from a predominance of infectious diseases to a predominance of chronic conditions may well be self-negated. It has been well documented that average improvement in GNP per capita does not always translate into good health for all.

5. Evidence-based health policy formulation would require regular updates of global and regional information to better monitor trends for planning purposes.

See Also the Following Articles

Economic Consequences of Infectious Diseases • International Law and Infectious Disease • Malaria • Smallpox

Bibliography

Anand, S., and Hanson, K. (1997). Disability-adjusted life years: A critical review. *J. Health Econ.* **16**(6).

Baker, C., and Green, A. (1996). Opening the debate on DALYs. *Health Policy Planning* **11**(2), 179–183.

Murray, C. J. L., and Acharya, A. K. (1997). Understanding DALYs. *J. Health Econ.* **16**, 703–730.

Murray, C. J. L., and Lopez, A. D. (1996a). "The Global Burden of Disease: A Comprehensive Assessment of Mortality and Disability from Diseases, Injuries and Risk Factors in 1990 and Projected to 2020." Harvard Univ. Press, Cambridge, MA.

Murray, C. J. L., and Lopez, A. D. (1996b). "Global Health Statistics. A Compendium of Incidence and Prevalence Estimates for over 200 Conditions." Harvard Univ. Press, Cambridge, MA.

Murray, C. J. L., and Lopez, A. D. (1996c). Evidence-based health policy—Lessons from the Global Burden of Disease study. *Science* **274**, 740–743.

Glycogen Biosynthesis

Jack Preiss

Michigan State University

GLOSSARY

csr A Carbohydrate storage regulator gene.

glg C' - 'lac Z Translational fusion of glg C promoter region to β-galactosidase gene.

pstI A restriction nuclease.

rel A ATP:GTP 3'pyrophospho-transferase or stringent factor.

spo T Guanosine 3',5'-bis (diphosphate) 3'pyrophosphatase gene.

MANY BACTERIA ACCUMULATE GLYCOGEN, a polysaccharide that is considered to function as an energy reserve. Glycogen may accumulate during growth or at the end of the growth phase and provide an energy source no longer available from the environment. Glycogen usually accumulates when there is a carbon excess in the medium, i.e., growth is limited by a nutrient other than the carbon source. The concentration of accumulated glycogen is dependent on the nutrient content of the medium as well as the growth phase of the organism and can, at times, be very high, at least 50% of the cell dry weight. An advantage in using glycogen as a reserve storage compound is that, because of its high molecular weight and physical properties, it has little effect on the internal osmotic pressure in the cell.

Various aspects of the enzymology and regulation of bacterial glycogen synthesis, including genetic regulation, have been extensively reviewed (Preiss, 1978, 1984, 1996; Krebs and Preiss, 1975; Preiss and Romeo, 1989).

I. CRITERIA FOR DESIGNATING AN ENERGY-STORAGE FUNCTION TO A COMPOUND

Wilkinson (1959) proposed three criteria for classification of compounds having energy-storage function. First, the compound accumulates intracellularly under conditions in which the energy supply for growth of the organism is in excess. Second, the reserve polymer should be used when the components of energy or carbon in the media are no longer available for sustaining growth or other processes required for maintenance of viability. Wilkinson (1959) pointed out various cell functions that would require energy or carbon, e.g., maintenance of a functioning semipermeable cytoplasmic membrane, energy for replacement of proteins and nucleic acids during turnover, maintenance of intracellular pH, or other processes induced for bacterial survival, such as sporulation and encystment.

Third, the storage compound should be used by the cell for energy production that would enable it to survive in a nonsupportive environment. This energy requirement for survival is known as energy of maintenance. This last criterion is perhaps the most important, for, as Wilkinson notes, a number of other substances may be produced for other functions such as, "To detoxicate end products of metabolism which otherwise accumulate too rapidly and prove toxic."

II. OCCURRENCE OF GLYCOGEN IN BACTERIA

Many bacterial species accumulate glycogen, either in stationary phase or under limited growth conditions with excess carbon in the medium. For *Escherichia coli*, the rate of growth and the quantity of accumulated glycogen are inversely related when cells were grown in media containing glucose as the carbon source and limited in nitrogen. Some bacteria, however, can optimally synthesize glycogen during exponential growth. Glycogen does not seem to be required for growth, since bacterial mutants, having defective structural genes for the glycogen biosynthetic enzymes and thereby unable to synthesize glycogen, can grow as well as their normal parent strains.

Glycogen or other similar α-1,4 glucans have been reported in more than 40 different bacterial species. The polysaccharide is not restricted to any class of bacteria as many gram-negative and gram-positive bacteria, as well as archaebacteria, have been reported to accumulate glycogen.

In many studies, it has been shown that glycogen accumulates in stationary phase when growth ceases because of depletion of a nutrient, either nitrogen in the form of ammonia or amino acid, or because of diminished growth due to unfavorable pH. When studied in detail, it is usually found that the largest accumulation of glycogen occurs under nitrogen-limited conditions.

III. BIOLOGICAL FUNCTIONS OF BACTERIAL GLYCOGEN

Various strains of *E. coli*, *Enterobacter aerogenes*, and *Streptococcus mitis* not containing glycogen have been compared with respect to their viability in media having no exogenous carbon source with the strains containing glycogen. A more prolonged survival rate was observed with the strains containing glycogen. Moreover, it was shown that under starvation conditions, the glycogenless enteric cells degraded their RNA and protein constituents to NH_3. This either did not occur or occurred at a lower rate in cells containing glycogen. It was postulated that, when glycogen was available, its use as an energy source minimized RNA and protein degradation for production of energy.

In various clostridia and *Bacillus* and in *Streptomyces viridochromogenes*, a glycogenlike molecule is accumulated to up to 60% of the cell's dry weight prior to the onset of sporulation. The polysaccharide is then degraded during formation and maturation of the spore. Thus, it is believed that glycogen serves as an endogenous source of carbon and energy for spore formation.

Although many experiments suggest that glycogen plays a role in the survival of bacteria, its precise function remains unclear. Further clarification of the role of bacterial glycogen is needed.

The synthesis and later degradation of glycogen by oral bacteria may be an important factor in the development of dental caries. Consistent with this postulate is the finding that these organisms are capable of synthesizing glycogen, produce more acid when exogenous carbohydrate is present, and can produce acid in the absence of exogenous carbohydrate when compared with oral bacteria unable to synthesize glycogen. The acid formed from polysaccharide catabolism may be of significance in production of dental caries. It is formed over a considerable period of time and may consequently be responsible for the lower resting pH observed in plaque from individuals with active caries.

IV. STRUCTURAL STUDIES OF GLYCOGEN

Microbial glycogen is very similar to mammalian glycogen, in that it is composed mainly of (α-1,4-glucosyl linkages and is a branched polysaccharide with about 8–12% of the glucosyl linkages being α-1,6).

Most bacterial glycogen studied in this manner have chain lengths of ~10–13 glucose units and I_2 spectra with maximum absorption of ~410–480 nm. However, the bacterial α-glucans can differ to some extent in their structural properties, and the variation is attributable to a number of factors, including the relative amounts and types of branching and debranching activities.

V. ENZYMATIC REACTIONS INVOLVED IN GLYCOGEN SYNTHESIS

A. Sugar Nucleotide Pathway

In 1964 (Greenberg and Preiss, 1964), it was also shown that extracts of several bacteria contained both an ADP glucose synthetase, Eq.(1), as well as an ADP-glucose-specific glycogen synthase, Eq. (2).

$$ATP + \alpha\text{-glucose-I-P} \Leftrightarrow ADPglucose + PPi \quad (1)$$

$$ADPglucose + \alpha\text{-glucan}$$
$$\Leftarrow \alpha\text{-1,4-glucosyl-glucan} + ADP \quad (2)$$

Subsequently, it was shown that branching enzyme activity, Eq. (3), was also present in many bacterial extracts and various photosynthetic organisms.

$$\text{Elongated } \alpha\text{-1,4-glucan}$$
$$\Rightarrow \text{Branched } \alpha\text{-1,4, } \alpha\text{-1,6-glucosyl-glucan} \quad (3)$$

B. Synthesis of Glycogen Directly from Disaccharides

α-Glucans similar to glycogen have been shown to be formed from either sucrose or maltose. When *Neisseria* strains are grown on sucrose, they accumulate large amounts of a glycogen-type polysaccharide. The direct conversion of sucrose to an α-1,4 glucan, Eq. (4), is catalyzed by the enzyme amylosucrase. The glucosyl moiety of sucrose is transferred to a primer to form a new α-1,4-glucosidic linkage. The gene for the *N. polysaccharea* amylosucrase has been cloned (Büttcher *et al.,* 1997) and can be expressed in *E. coli. In vitro,* the product formed is unbranched. Thus, in *Neisseria* strains, the α-1,6 linkages formed must be due to a branching enzyme activity.

$$\text{Sucrose} + \alpha\text{-glucan} \Rightarrow \text{D-fructose}$$
$$+ \alpha\text{-1,4-glucosyl-glucan} \quad (4)$$

Amylosucrase is limited to a few bacterial strains and is only induced in the presence of sucrose. Moreover, there are no reports that indicate that either *Neisseria* or any other bacteria are able to synthesize sucrose from other metabolites. Thus, accumulation of glycogen in *Neisseria* and in other bacteria grown on other carbon sources besides sucrose cannot be mediated by amylosucrase action.

A number of enteric and other organisms, when grown on maltose or maltodextrins as a carbon source, can synthesize low-molecular weight α-1,4-glucan via transfer of the glucosyl moiety of maltose to a growing α-1,4-glucan chain [Eq. (5)]. The enzyme catalyzing this reaction is amylomaltase.

$$\text{Maltose} + \alpha\text{-glucan} \Leftrightarrow \alpha\text{-D-glucose}$$
$$+ \alpha\text{-1,4-glucosyl-glucan} \quad (5)$$

The polymer is of very low molecular weight and is quickly degraded by another enzyme that is also induced by the presence of maltose in the medium maltodextrin phosphorylase. The synthesis of these two enzymes, however, is repressed by glucose and, thus, neither amylomaltase nor maltodextrin phosphorylase activity would be responsible for synthesis of glycogen when the organisms are grown on carbon sources other than maltose.

Glycogen phosphorylase, as well as maltodextrin phosphorylase, is found in many bacteria and can catalyze synthesis or phosphorolysis of α-1,4-glucosyl linkages either in maltodextrins or in glycogen.

$$\alpha\text{-1,4-(glucosyl)}_n + Pi \Leftrightarrow \alpha\text{-glucose 1-P}$$
$$+ \alpha\text{-1,4-(glucosyl)}_{n-1} \quad (6)$$

However, maltodextrin phosphorylase is usually induced only in the presence of maltose or maltodextrins. *E. coli* mutants deficient in this enzyme accumulate maltodextrins, indicating that the maltodextrin phosphorylase is involved in degradation of the α-glucans rather than in synthesis. The glycogen phosphorylase activity found in organisms is usually insufficient to account for the synthetic rate observed.

In summary, it appears that glycogen synthesis in bacteria occurs mainly by the sugar nucleotide pathway. Both *E. coli* and *Salmonella typhimurium* mutants, either glycogen deficient or containing glycogen in excess of that observed in the parent wild-type strain, have been isolated. They are affected either in their glycogen synthase or in ADPglucose synthetase activity, implying that, in these organisms, the ADPglucose pathway is the major, if not exclusive, route for glycogen formation.

VI. PROPERTIES OF THE GLYCOGEN BIOSYNTHETIC ENZYMES

A. ADPglucose Pyrophosphorylase

ADPglucose pyrophosphorylase (ADPGlc PPase) has been purified from a number of microorganisms and, with one exception, the enzyme has been found to be homotetrameric in structure, with a subunit molecular size of about 50 kDa. The ADPGlc PPase from *Bacillus stearothermophilus* is also a tetramer (Takata *et al.*, 1997) but is composed of two subunits, one of 387 amino acids, GlgC, 43.3 kDa, the other, GlgD, 343 amino acids, 39 kDa. The *Bacillus* enzyme is, thus, similar to the higher plant ADPGlc PPases, which have been shown to be also heterotetrameric, $\alpha_2\beta_2$). The plant small subunit is also known as the catalytic subunit and the large subunit, the regulatory subunit. The potato tuber ADPGlc PPase catalytic subunit, expressed alone in *E. coli*, is highly active, while the large subunit by itself is inactive. Expression of both potato tuber ADPGlc PPase subunits results in a heterotetramer having higher affinity for the allosteric activator and lower affinity for the inhibitor. Therefore, the large subunit is regarded as the regulatory subunit. The *B. stearothermophilus* GlgC subunit shows a 42–70% similarity with other ADPGlc PPases and, when expressed in *E. coli* alone, has catalytic activity. The *Bacillus* GlgD subunit has no activity and its function is unknown. However, it seems to increase the Vmax of the GlgC activity and slightly increases the apparent affinity of GlgC for the substrates.

An important facet of both bacterial and plant ADPGlc PPases is that they are allosteric enzymes and the allosteric function is important for the regulation of bacterial glycogen and plant starch syntheses. Over 50 ADPGlc PPases, mainly bacterial but also plant, have been studied with respect to their regulatory properties. In almost all cases, glycolytic intermediates activate ADPGlc synthesis while AMP, ADP, or Pi are inhibitors. Glycolytic intermediates in the cell can be considered as indicators of carbon excess and, therefore, under conditions of limited growth with excess carbon in the media, accumulation of glycolytic intermediates would be signals for the activation of ADPGlc synthesis. For most of the ADPGlc PPases studied, the activator glycolytic intermediate increases the enzyme's apparent affinity for the substrates, ATP, and glucose-1-P, and increasing concentrations of activator reverse the inhibition caused by the inhibitors, AMP, ADP, or Pi.

The activator specificity of the bacterial and plant ADPGlc PPases can be categorized into seven groups (Table I) on the basis of their specificity of activation by the various glycolytic intermediates. The variation of activator specificity observed correlates with the nature of carbon assimilation dominant in the bacterium or plant tissue. *E. coli* or *S. typhimurium* obtain their energy mainly through glycolysis. The primary activator for their ADPGlc PPases is fructose 1,6-bisP and 5′-adenylate is the major inhibitor, and their ADPGlc synthetic activity is regulated by the [fructose 1,6-bisP] / [AMP] ratio. In plants and in the cyanobacteria, the activator is 3-P-glycerate (3PGA) and inhibitor is Pi. 3PGA is the primary product in the carbon assimilation pathways of these photosynthetic organisms/tissues.

Evidence has accumulated to indicate, that the *in vitro* observed kinetic allosteric activation and the inhibitor effects also occur *in vivo* in bacterial cells. There is a class of mutants of *E. coli* and of *S. typhimurium* LT-2 affected in their ability to accumulate glycogen. This mutant class has ADPGlc PPases with altered regulatory properties. Generally, those mutants with ADPGlc PPases with higher affinity for the activator, fructose 1,6-bisphosphate, and/or a lower affinity for the allosteric inhibitor, AMP, accumulate glycogen at a faster rate than the parent wild-type strain. Mutants with enzymes with a lower affinity for the activator accumulate glycogen at a slower rate than the parent strain.

Table II summarizes the allosteric properties of the mutant ADPGlcPPases that have been studied

TABLE I

ADP Glucose Pyrophosphorylases Classified with Respect to Activator Specificity

Activators(s)	Microorganisms	Predominant carbon assimilation pathway
Pyruvate	*Rhodospirillum sp.* *Rhodocyclus purpureus*	Reductive pyruvate cycle
3-P-Glycerate	Cyanobacteria (higher plants)	Photosynthetic Calvin cycle
Pyruvate, Fructose-6-P	Some anaerobic photosynthetic bacteria, *Agrobacterium*, *Arthrobacter*	Reductive pyruvate cycle Entner-Doudoroff pathway
Pyruvate, Fructose-6-P	Some *Rhodopseudomonas* species	Reductive pyruvate cycle Entner-Doudoroff pathway Glycolysis
Fructose-6-P, Fructose-1,6-bis-P	*Rhodopseudomonas viridis*, Aeromonads, *Mycobacterium smegmatis*	Glycolysis
Fructose-1,6-bis-P	Enterics	Glycolysis
None	*Serratia* sp., *Enterobacter hafniae*, *Clostridium pasteuranium*	Glycolysis

and their ability to accumulate glycogen in the stationary phase. With respect to *E. coli*, there is a direct relationship between the affinity of the enzyme for the activator and the ability of the mutant to accumulate glycogen. If the apparent affinity for the activator, fructose-1,6-bis-P, is higher, glycogen accumulation by the mutant is higher. If the apparent affinity for activator is lower, as seen for mutant SG14 enzyme,

glycogen accumulation is lower in the mutant than the parent strain. The two *S. typhimurium* mutant strains have ADPGlcPPases that are more affected in the affinity of the inhibitor. Both JP23 and JP51 enzymes have lesser affinity for the inhibitor, and these mutants accumulate higher amounts of glycogen than does the parent strain. Of interest is that JP23 ADPGlc Ppase is fully active in the absence of

TABLE II

Allosteric Kinetic Constants of Wild-Type *E. coli* and *S. typhimurium* LT2 and Allosteric Mutant ADPGlc PPases, Their Mutation and Their Glycogen Accumulation Rates

Strain	Maximal glycogen accumulation[a] mg/gram-cell	Fructose 1,6-bisP[a] $A_{0.5}$, μM	AMP[b] $I_{0.5}$, μM	Mutation
E. coli B	20	68	75	—
Mutant SG14	8.4	820	500	A44T
Mutant SG5	35	22	170	P295S
Mutant 618	70	15	860	G336D
Mutant CL1136	74	5	680	R67C
S. typhimurium				
LT2	12	95	110	
Mutant JP23	15	No activation	250	
Mutant JP51	20	84	490	

[a] $A_{0.5}$ is the fructose 1,6-bis-P giving 50% of maximal activation.

[b] $I_{0.5}$ is the AMP concentration required for 50% inhibition.

[c] The bacterial strains were grown in minimal media with 0.75% glucose and the data is expressed as maximal mg of anhydroglucose units per gram (wet wt) of cells in stationary phase.

activator. Addition of activator Fru-1,6-bis-P causes no further increase in activity.

The above studies, plus one showing a direct relationship between the fructose 1,6-bisP concentration in the *E. coli* cell and the rate of glycogen accumulation (Dietzler *et al.*, 1974), clearly point out that fructose 1,6-bisP is an allosteric activator of ADPglucose pyrophosphorylase and a physiological activator of glycogen synthesis in *E. coli* and *S. typhimurium*.

Chemical modification and site-directed mutagenesis studies of the ADPGlc PPases have provided evidence for the location of the activator binding site, the inhibitor binding site, and the substrate binding sites. These experiments have used pyridoxal-P as an analog for either the activator, fructose 1,6-bisP, or as subsequently shown for the substrate, glucose-1-P. For an ATP analog, the photoaffinity reagent 8-azido-ATP(8N$_3$ATP) proved to be a substrate for the *E. coli* enzyme, whereas 8-azido AMP (8N$_3$AMP) was an effective inhibitor analog. Since the *E. coli* ADPGlc PPase gene, *glgC*, had been cloned and its sequence determined, the identification of the amino-acid sequence about the modified residue enabled one to determine the location of the modified residue in the primary structure of the enzyme. The amino-acid residue involved in binding the activator was Lys-39 and the amino-acid involved in binding the adenine portion of the substrates (ADPGlc and ATP) was Tyr-114. Tyr-114 was also the major binding site for the adenine ring of the inhibitor, AMP. Lys-195 is protected from reductive phosphopyridoxylation by the substrate, ADPGlc. It was proposed that it is also a part of the substrate binding site.

Similar experiments were also done to elucidate the activator, 3PGA, inhibitor, Pi, and substrate, Glc-1-P sites in *Anabaena* PC7120 and in the potato tuber and spinach leaf ADPGlc PPases. Table III shows the amino-acid sequences of the substrate and allosteric sites identified via chemical modification in the bacteria, *E. coli* and *Anabaena*, as well as in the higher plant enzymes. The sequence for the Glc-1-P substrates site is highly conserved for both bacterial and higher plant ADPGlc PPases.

The ADPGlc PPase of *Agrobacterium tumefaciens* has been cloned and expressed in *E. coli* (Uttaro *et al.*, 1998). This enzyme has as activators, fructose-6-P and pyruvate, an activator specificity different

TABLE III
Amino Acid Residues and Sequences Involved in Binding of Substrates and Effectors of ADPGlc PPase

Organism/tissue	Activator sites	Sequences
E. coli	Fru-1,6-bis-P	[38]N**K**RAKPAV
Anabaena PC7120	3-P-glycerate	
	Site 1	[412]SGIVVVL**K**NAV
	Site 2	[375]QRRAIID**K**NAR
Potato tuber	3-P-glycerate	
	Site 1	[434]SGIVTVI**K**DAL
	Site 2	[397]IKRAIID**K**NAR
	Inhibitor sites	
E. coli	5′-AMP	[113]W**Y**RGTADAV
Anabaena PC7120	Phosphate	[291]TRA**R**YLPPTK
	Substrate sites	
E. coli	Glucose-1-P	[190]IEFVE**K**PAN
Potato tuber	Glucose-1-P	[193]IEFAE**K**PQG
E. coli	ATP	[113]W**Y**RGTADAV

Note. The amino acids shown by chemical modification and site-directed mutagenesis to be involved in binding are in bold and underlined.

from the *E. coli* ADPGlc PPase. It would be of interest to determine the activator site(s) of this enzyme and compare it with the *E. coli* enzyme.

B. Cloning of ADPGlc PPases with Altered Allosteric Properties from *E. coli* Mutants Affected in Glycogen Synthesis

As indicated previously (Table II), a class of mutants of *E. coli* and of *S. typhimurium* with altered rates of glycogen accumulation was found to have ADPGlc PPases that were affected in their allosteric properties. To gain insight with respect to amino-acid residues or domains involved in maintaining allosteric function, the allosteric mutant ADPGlcPPases were cloned.

Table II shows the various amino-acid substitutions in the allosteric mutants that have been cloned and analyzed. Of interest is that the mutations causing large changes in the allosteric properties of the enzyme occur throughout the sequence of the ADPGlc PPase. These changes of amino acids in the enzyme affect not only the affinity of the allosteric effectors (Table II) but also the apparent affinities for the substrate ATP and Mg^{+2}. Thus, there are many domains affected by the mutations.

Site-directed mutagenesis of the ADPGlc PPase gene and analysis of various allosteric mutant genes have provided much information in the structure–function relationships of the substrate and catalytic sites. What is needed for greater clarification is knowledge of the three-dimensional structure of the enzyme.

C. Glycogen Synthase

The bacterial glycogen synthases are specific for the sugar nucleotide, ADPglucose. In cases where it has been studied, the enzyme subunit size is about 50 kDa and the enzyme in its native form is either a dimer or homotetramer.

An affinity analog of ADPGlc, adenosine diphosphopyridoxal (ADP-pyridoxal), was used to identify the ADPGlc binding site. Incubation of the enzyme with the analog plus sodium borohydride led to an inactivated enzyme. The degree of inactivation correlated with the incorporation of about one mol of analog per mol of enzyme subunit for 100% inactivation. After tryptic hydrolysis, one labeled peptide was isolated and the modified Lys residue was identified as Lys15. The sequence Lys-X-Gly-Gly, where Lysine is the amino acid modified by ADP-pyridoxal, has been found to be conserved in the mammalian glycogen synthase and the plant starch synthases.

The structural gene for glycogen synthase, *glgA*, has been cloned from *E. coli. S. typhimurium, A. tumefaciens,* and *B. stearothermophilus* and their nucleotide sequences have been determined. The *E. coli* glycogen synthase sequence consists of 1431 bp specifying a protein of 477 amino acids with a MW of 52,412. Its availability has enabled researchers to perform site-directed mutagenesis experiments to determine structure–function relationships for a number of amino acids in the *E. coli* glycogen synthase. Substitution of other amino acids for Lys at residue 15 suggested that the Lys residue is mainly involved in binding the phosphate residue adjacent to the glycosidic linkage of the ADPGlc and not in catalysis. The major effect on the kinetics of the mutants at residue 15 was the elevation of the Km of ADPGlc, of about 30- to 50-fold, when either Gln or Glu were

the substituted amino acids. Substitution of Ala for Gly at residue 17 decreased the catalytic rate constant, k_{cat}, about 3 orders of magnitude compared to the wild-type enzyme. Substitution of Ala for Gly 18 only decreased the rate constant 3.2-fold. The Km effect on the substrates, glycogen, and ADPGlc, were minimal. It was postulated by the researchers that the 2glycyl residues in the conserved Lys-X-Gly-Gly sequence participated in the catalysis by assisting in maintaining the correct conformational change of the active site or by stabilizing the transition state.

Since there is still binding of the ADPGlc and appreciable catalytic activity of the Lys15Gln mutant, the ADP-pyridoxal modification was repeated and, in this instance, about 30-times higher concentration was needed for inactivation of the enzyme. The enzyme was maximally inhibited about 80% and tryptic analysis of the modified enzyme yielded one peptide containing the affinity analog and with the sequence, Ala-Glu-Asn-modified Lys-Arg. The modified Lys was identified as Lys277. Site-directed mutagenesis of Lys277 to form a Gln mutant was done and the Km for ADPGlc was essentially unchanged but K_{cat} was decreased 140-fold. It was concluded that Lys residue 277 was more involved in the catalytic reaction than in substrate binding.

D. Branching Enzyme

The structural genes of various branching enzymes (BE) have been cloned from many bacteria. The nucleotide and deduced amino-acid sequences of the *E. coli glg* B gene consisted of 2181 bp specifying a protein of 727 amino acids and with a MW of 84,231.

The relationship in amino-acid sequences between that of branching enzyme (BE) and amylolytic enzymes, such as α-amylase, pullulanase, glucosyltransferase, and cyclodextrin glucanotransferase, have been compared and there is a marked conservation in the amino-acid sequence of the four catalytic regions of amylolytic enzymes in the branching enzymes, whether they are of bacterial, plant, or mammalian source (Svensson, 1994). Four regions that putatively constitute the catalytic regions of the amylolytic enzymes are conserved in the plant branching isoenzymes and the glycogen branching enzymes of *E. coli*. Analysis of this high conservation in the α-

amylase family has been pointed out and greatly expanded with respect to sequence homology but also in the prediction the $(\beta/\alpha)_8$-barrel structural domains with a highly symmetrical fold of eight inner, parallel β-strands, surrounded by eight helices, in the various groups of enzymes in the family. The $(\beta/\alpha)_8$-barrel structural domain was determined from the crystal structure of some α-amylases and cyclodextrin glucanotransferases.

The conservation of the putative catalytic sites of the α-amylase family in the glycogen and starch branching enzymes would be expected, as the BE catalyzes two consecutive reactions in synthesizing α-1,6-glucosidic linkages by cleavage of an α-1,4-glucosidic linkage in an 1,4-α-D-glucan to form a non-reducing end oligosaccharide chain that is transferred to a C-6 hydroxyl group of the same or other 1,4-α-D-glucan. The eight highly conserved amino-acid residues of the α-amylase family are also functional in branching enzyme catalysis. Further experiments, such as chemical modification and analysis of the three-dimensional structure of the BE, would be needed to determine the precise functions and nature of its catalytic residues and mechanism. The regions of the C-terminus and N-terminus are dissimilar in sequence and in size in the various branching isoenzymes. Recent studies (Kuriki *et al.*, 1997) suggest that these amino-acid sequence regions are important with respect to BE specificity, with respect to substrate specificity (amylose or amylopectin), as well as in size of chain transferred and extent of branching.

VII. GENETIC REGULATION OF GLYCOGEN SYNTHESIS IN *ESCHERICHIA COLI*

As indicated before, the glycogen biosynthetic enzymes are induced in the stationary phase. The rate of glycogen synthesis is inversely related to growth rate when growth is limited for certain nutrients, e.g., nitrogen. Consistent with this are the findings that the levels of glycogen biosynthetic enzymes in *E. coli* increase as cultures enter the stationary phase.

When *E. coli* or *S. typhimurium* cells are grown in an enriched medium and 1% and glucose, the specific

activities of ADPGlc PPase and glycogen synthase increase 11- to 12-fold, while glycogen branching enzyme increases 5-fold, as cultures enter stationary phase. However, when the organisms are grown in a defined medium, the ADPGlc PPase and glycogen synthase activities are relatively high in the exponential phase. Branching enzyme in defined media is fully induced in the exponential phase, with only about 2-fold increase in specific activity of the ADPGlc PPase and glycogen synthase occurring when the cells reach the stationary phase. These experiments suggest that the gene encoding the branching enzyme is regulated differently from the genes for ADPGlc PPase and glycogen synthase. The addition of inhibitors of RNA or protein synthesis to prestationary phase cultures prevents the enhancement of glycogen synthesis in the stationary phase and this is expected for a pathway that is under transcriptional control.

A. Location of the Structural Genes for Glycogen Biosynthesis

The structural genes for glycogen biosynthesis are clustered in two adjacent operons, which also contain genes for glycogen catabolism. The structural genes for glycogen synthesis were shown to be located at approximately 75 min on the *E. coli* K-12 chromosome, and the gene order at this location was shown to be *glgA-glgC-glgB-asd*. These genes encode the enzymes glycogen synthase, ADPGlc PPase, glycogen branching enzyme, and are close to *asd*, the structural gene for the enzyme, aspartate semialdehyde dehydrogenase (EC 1.2.1.11).

The *E. coli glg* structural genes were cloned into pBR322 via selection with the closely linked essential gene *asd*. Among several *asd*+ plasmid clones isolated, pOP12 was found to contain a 10.5-kb PstI fragment, which encoded the structural genes *glgC*, *glgA*, and *glgB*. The arrangement of genes encoded by pOP12 has also been determined by deletion-mapping experiments and the nucleotide sequence of the entire *glg* gene cluster has been determined. The genetic and physical map of the *E. coli* K-12 *glg* gene cluster is shown in Fig. 1. The continuous nucleotide sequence of over 15 kb of this genome region includes the sequences of the flanking genes

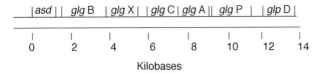

Fig. 1. Structure of the glycogen cluster in *E. coli*. The restriction map is constructed from known contiguous sequences. All of the genes are transcribed from left to right counterclockwise on the genome except for *glp* D.

asd and *glpD* (glycerol phosphate dehydrogenase; EC 1.1.99.5; 1.1.1.8).

Nucleotide sequence analysis indicated that, in addition to the *glgC*, *glgA*, and *glgB* genes, pOP12 contains an open reading frame (ORF), *glgX*, located between *glgB* and *glgC*, and a second ORF, *glgP*, located downstream from *glgA*. *glgX* has recently been shown to express isoamylase activity (Yang *et al.*, 1996).

The *glgP* gene is identified by homology with rabbit muscle glycogen phosphorylase. This gene encodes glycogen phosphorylase, as shown via the expression and characterization of its gene product. Neither *glgX* or *glgP* is needed for glycogen synthesis, suggesting that both may be more involved in glycogen catabolism.

The organization of the gene cluster suggests that the *glg* genes may be transcribed as two randomly arranged operons, *glgBX* and *glgCAP* (Fig. 1). The coding regions of *glgB* and *glgX* ORF overlap by one base pair, *glgC* and *glgA* are separated by two base pairs, and genes *glgA* and *glgP* are separated by 18 base pairs. The close proximity of these genes suggests translational coupling within the two proposed operons. Studies of the regulation of the *glg* structural genes, with '*lacZ* translational fusions and other approaches, are consistent with a two-operon arrangement for the *glg* gene cluster, in which the *glgCAP* and *glgBX* operons may be preceded by growth phase-regulated promoters. Transcriptions initiating upstream of *glgC* have been analyzed by S1 nuclease mapping.

B. Genetic Loci Affecting Glycogen Biosynthetic Enzyme Levels in *Escherichia coli*

Studies of glycogen-excess *E. coliB* mutants SG3 and AC70R1, which exhibit enhanced levels of the enzymes in the glycogen synthesis pathway, suggested that glycogen synthesis is under negative genetic regulation. The mutations in these strains, *glgR* and *glgQ*, respectively, affect *glg* A and C transcription, although these mutations have not been isolated and sequenced.

Four 5′ termini *in vivo* transcripts were identified within 0.5 kilobases of the upstream region of the *glgC* coding region by S1-nuclease protection analyses (Figs. 2, 3). Three of these transcripts were mapped to high resolution and their sequences are as shown in Fig. 2.

The *glgR* mutation is closely linked to the glycogen structural genes by P1 transduction analysis and the mutation results in 8- to 10-fold higher levels of ADPGlc PPase, 3- to 4-fold higher levels of glycogen synthase in exponential phase, but does not alter the level of branching enzyme in minimal media. Analysis of RNA transcripts for *glgC* in strain SG3 having the *glgR* mutation leads to an increase in transcript B only (Figs. 2, 3). Therefore, it appears

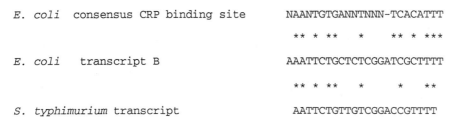

Fig. 2. The cAMP-CRP binding sites and the upstream sequences of the *glg* C gene of *E. coli* and *S. typhimurium*. The asterisks indicate the nucleotide identities between the proposed CRP-binding sites.

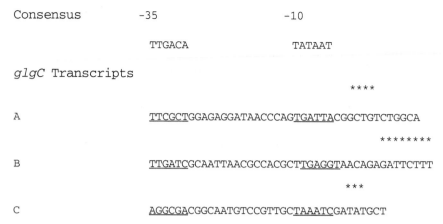

Fig. 3. Comparison of the 5′-flanking regions of the *E. coli glg* C transcripts with the consensus sequence of the *E. coli · σ*70 promotor. The *glg* C transcripts, A, B, and C. They are located, respectively, at −60, −130, and −245 from the *glg* C initiation codon. The asterisks indicate the 5′ termini of the transcripts. The best −10 and −35 regions are underlined.

that the *glgR* mutation may alter a *cis*-acting site involved in the regulation of transcript B. This effect might be mediated via a negative regulatory site, but the current experimental evidence is also consistent with an overexpressed phenotype or a higher affinity CRP-binding site.

The *glgQ* mutation is not linked to the glycogen gene cluster in P1 transduction and results in 11-fold, 5.5-fold, and 2-fold increases in ADPGlc PPase, glycogen synthase, and glycogen branching enzyme, respectively. Therefore, *glgQ* appears to affect one or more *trans*-acting factors for the expression of the genes in the two glycogen operons. Levels of the four transcripts for the *glgC* gene are elevated in the *glgQ* mutant, AC70R1 (Figs. 2, 3). Transcript A was affected the most dramatically, with approximately 25-fold higher levels being present in AC70R1 versus the wild-type *E. coli* B or the SG3 strain. Since the levels of branching enzyme are also elevated in AC70R1, it was not considered likely that *glgQ* was a mutation in the cAMP-CRP or ppGpp regulatory systems (see following), which do not affect *glgB* expression. The expression of the chromosomal *lac* Z gene in AC70R1 (*glgQ* mutant) and in the wild-type *E. coli* B strain was also similar, providing further evidence for the idea that *glgQ* affects a different regulatory system for the *glg* genes. In summary,

glgQ affects mainly Transcript A and *glgR* affects mainly Transcript B.

C. Sigma Factor E · *σ*70 Transcribes *glgCAP*

The DNA sequences immediately preceding the 5′-ends of the four transcripts in Fig. 2 are weakly related to consensus sequences for *E. coli* promoters (Fig. 2). Although positively regulated promoters typically show weak similarity to the consensus sequence, it is also possible that one or more of the *glg* promoters is recognized via an alternative sigma factor. Therefore, the dependence of *glgC* expression on three sigma factors was tested in coupled transcription-translation, using monoclonal antibodies that selectively inhibit transcription by specific and selective recognition of the sigma factors. A monoclonal antibody directed against *E. coli σ*70 inhibited up to 85% of the *glgC* expression, while antibody directed against *σ*54 or *σ*32 did not inhibit expression of *glgC*, relative to *glnAP* 2 (*σ*54-dependent) and *dnaK* (*σ*32-dependent) controls. Therefore, even though nitrogen limitation enhances glycogen synthesis *in vivo*, the expression of *glgC* is not regulated by the nitrogen starvation, *σ*54-dependent transcriptional controls. This result agrees with the experiment that *NtrC* and *NtrA* (σ54) did not enhance expression of *glg* genes in S-30 experiments. The heat shock regulatory system (σ32-dependent) also appears to have no involvement in control of *glgC* expression. Therefore, the major active form of RNA polymerase (E · σ70) is apparently utilized for *glg* expression in the S-30 transcription-translation system. However, the S-30 extracts used in the experiments were prepared from exponential cells to obtain optimal translational activity, and would not have contained endogenous activity from a sigma factor or an accessory factor synthesized in stationary phase. Therefore, expression via one or more other possible *glgC* transcripts may not have been detected by the S-30 experiments.

D. Possible Regulation by *kat* F

Interruption of the gene for a stationary phase sigma factor, *kat* F, also leads to decreased levels of glycogen. Several studies provide evidence for a class of stationary phase-induced genes depending on an alternative sigma factor for expression. The gene for this sigma factor is referred to as *katF* or *rpoS*. Iodine staining of *kat*F$^+$ and *kat*F$^-$ strains suggests that a functional *kat*F allele is required for optimum accumulation of glycogen in *E. coli* strain MC4100. The effect of *katF* on glycogen synthesis does not function via regulation of the structural genes for the enzymes of the glycogen pathway.

Hengge-Aronis and Fisher (1992) have isolated, cloned, and sequenced a gene in *E. coli* K-12 that stimulates glycogen synthesis, *glg S*, and that requires *kat* F for optimum expression. The *glg* S gene appears to be transcribed via a cAMP-dependent promoter and a *kat* F-requiring promoter. Based upon iodine staining of colonies, a null mutant in *glgS* accumulates more glycogen than a *kat* F mutant, indicating that *kat*F may have additional effects on glycogen synthesis besides *glgS* induction. The *glgS* mutation did not affect the expression of *glgC* or *glgA* gene fusions.

The function of *glgS* and the role of *katF* in glycogen synthesis remain to be elucidated and these factors may regulate a processor metabolic step prior

to the glycogen biosynthesis pathway that can affect the ultimate level of glycogen.

E. Genetic Regulation of Glycogen Synthesis by Cyclic AMP and Guanosine 5′-diphosphate 3′-diphosphate and by the *Csr* A Gene

Glycogen biosynthesis is under the direct control of at least three global regulatory systems, catabolite repression (stimulation of *glg* ABC expression by cAMP); stringent response (stimulation of *glgABC* expression by pGpp), and repression of the *glgABC* expression by the *csrA* gene product, a 61 amino acid polypeptide.

F. Regulation by cAMP

The genes *cya*, encoding adenylate cyclase (EC 4.6.1.1) and *crp*, encoding cAMP-receptor protein (CRP), are required for optimal synthesis of glycogen. Exogenous cAMP can restore glycogen synthesis in a *cya* strain but not in a *crp* mutant. cAMP and CRP are strong positive regulators of the expression of the *glgC* and *glgA* genes, but do not affect *glgB* expression. The addition of cAMP and CRP to S-30 extracts having *in vitro* coupled transcription-translation reactions and containing pOP12 as the genetic template resulted in up to 25 and 10-fold increase in the expression of *glgC* and *glgA*, respectively. cAMP and CRP also enhanced the expression of *glgC* and *glgA* encoded by either plasmids or restriction fragments in *in vitro* reactions of protein synthesis. A restriction fragment that contained *glgC* and 0.5 kb of DNA from the upstream noncoding region of *glgC* was sufficient to permit cAMP–CRP regulated expression, suggesting that the *glgC* gene contains its own cAMP-regulated promoter(s).

Evidence for a CRP-binding site on a 243 bp restriction fragment from the upstream region of *glgC* was obtained using gel retardation analysis. There are also potential consensus CRP-binding sequences within the *glg* C upstream region preceding both the *E. coli* and the *S. typhimurium glg* C genes.

G. Regulation by ppGpp

Glycogen biosynthesis in *E. coli* is positively regulated by ppGpp as *rel* A strains are glycogen-deficient. Expression of the *glg* C and *glg* A genes is stimulated by ppGpp. Expression of *glg* C in transcription-translation reactions was increased three- to four-fold in the presence of ppGpp; *glg* A expression exhibited approximately a two-fold enhancement. The expression of *glg*B was not affected by ppGpp.

cAMP and ppGpp are also have independent effects on *glg* CA expression in *in vitro* transcription-translation experiments. Actually, their combined effects on *glg*C expression in transcription-translation experiments can be synergistic. The addition of cAMP-CRP or ppGpp results in an increase of 6.3- or 1.6-fold, respectively, in the expression of *glg* C over the basal or unactivated level of expression, while their addition, together, leads to an 18.8-fold stimulation.

Evidence for positive regulation of *glg* C expression *in vivo* by ppGpp was obtained using the *glg* C′- ′*lac* Z translational fusion in pCZ3-3. This gene fusion was introduced into strains comprising an isogenic series that varied in basal levels of ppGpp due to increasingly severe mutations in *spo* T. The *spo* T gene affects the levels of ppGpp in the cell.

H. The csrA Gene Product Affects Glycogen Synthesis by Regulating the *glg*s Operons

A gene from *E. coli* K-12, *csrA*, encoding a negative factor for *glg* transcription, has recently been identified. A mutant TR1-5 accumulated about 24-fold more glycogen than an isogenic wild-type strain. The gene affected by the TR1-5 mutation, *csrA* (for "carbon storage regulator"), has been cloned, sequenced, and mapped on the *E. coli* genome and some of its regulatory effects have been studied.

The TR1-5 mutation was also shown to affect glycogen levels by causing elevated expression of genes representative of both glycogen operons, *glg* C and *glg* B. Levels of ADPGlc PPase expressed from the chromosome were approximately 10-fold higher in the TR1-5 mutant than in an isogenic *csrA*⁺ strain in the stationary phase. The TR1-5 mutation affects glycogen levels and expression of the *glg* B and *glg*

C genes in the exponential as well as the stationary phase.

The *csrA* gene also appears to regulate the expression of the gluconeogenic enzyme, PEP carboxykinase (EC 4.1.1.38). Expression of a PEP carboxykinase operon fusion (*pckA′*-′*lacZ*) was increased about 2-fold in exponential and stationary phases in the TR1-5 mutant, suggesting that gluconeogenesis may also be under negative control by *csrA*. When several isogenic strains were grown on synthetic media, it was found that *csr* A⁺ and *csr* A::*kanR* strains (transductants with the TR1-5 mutation) were capable of growth on a wide variety of carbon sources. However, a strain that contained the functional *csr* A gene encoded on a multicopy (pUC19-based) plasmid, pCSR10, could grow on glucose and fructose but not on any of the gluconeogenic substrates, succinate, glycerol, pyruvate, and L-lactate. When strains were plated on a richer defined medium, support of growth of a pCSR10-containing strain using some gluconeogenic substrates, including acetate, as a major carbon source did occur. However, the pCSR10-containing strain formed only pinpoint colonies on succinate, whereas each of the other strains grew well. The *csrA* gene may affect succinate utilization or transport independently of its effect on gluconeogenesis.

The *csrA* gene is located at 58 min on the physical map of the *E. coli* K-12 genome. The *csrA* gene is between the gene *alaS*, which encodes alanyl tRNA synthetase (EC 6.1.1.7), and the *serV* operon of tRNA genes and is transcribed counterclockwise on the chromosome. The *csrA* ORF encodes a 61-amino-acid polypeptide, which was strongly expressed from the plasmid pCSR10 in S-30 transcription-translation experiments. Deletion mapping experiments of the plasmid-encoded *csrA* gene demonstrated that the ORF is required to mediate the inhibitory effects on the glycogen synthesis phenotype *in vivo*.

The *csrA* gene product negatively modulates post-transcriptionally by facilitating decay of *glgCA* mRNAs (Romeo *et al.*, 1993). The *csrA* gene product is a specific mRNA-binding protein binding to *csrB* RNA (Liu *et al.*, 1995, 1997). Binding of the *csrA* gene product to *csrB* RNA inhibits the repression of the *glgCA* operon and the *glgB* genes. Since *csrB* RNA increases in stationary phase, it is believed that repression of the glycogen biosynthetic genes is relieved by the binding of *csrA* gene product by the *csrB* RNA.

Analysis of *glgC* transcripts by S1-nuclease protection mapping showed that the steady-state levels of four *glgC* transcripts (Fig. 3) are elevated in the TR1-5 mutant and are severely depressed in apCsr10-containing strain, indicating that *csrA* affects the transcriptional regulation of *glgC*.

VIII. A PROPOSED INTEGRATED MODEL FOR THE GENETIC REGULATION OF THE GLYCOGEN BIOSYNTHETICS PATHWAY IN *E. COLI* AND POSSIBLY IN OTHER BACTERIA

Regulation of glycogen metabolism involves a number of factors coordinating the glycogen synthetic rate with the physiology of the cell. Genetic regulation of the glycogen biosynthesis pathway by cAMP and ppGpp allows *E. coli* to adjust its metabolic capacity for converting available carbon substrate into glycogen in response to the availability of carbon or energy or amino acids, respectively. When cells are rapidly mutiplying, the levels of the enzymes are repressed and, although the energy and carbon are available for glycogen synthesis, the glycogen synthetic rate is low. Upon nutrient deficiency, synthesis of ADP glucose pyrophosphorylase and glycogen synthase are induced and the capacity for glycogen synthesis is greater. The level of glycogen that is ultimately accumulated will be dependent upon carbon availability and is subject to allosteric regulation of the ADP glucose pyrophosphorylase activity.

Regulation of glycogen metabolism involves a multitude of factors coordinating the glycogen synthetic rate with the physiology of the cell. The genetic regulation of the glycogen biosynthesis pathway by cAMP and ppGpp allows *E. coli* to adjust its metabolic capacity for converting available carbon substrate into glycogen in response to the availability of carbon or energy or amino acids, respectively. When cells are rapidly mutiplying, the levels of the enzymes are repressed and, although the energy and carbon are available for glycogen synthesis, the rate of glycogen synthesis is low. Upon nutrient deficiency, synthesis of ADP glucose pyrophosphorylase and glycogen synthase are induced and the capacity for glycogen synthesis is greater. The level of glycogen that is ultimately accumulated will be dependent upon substrate availability and is subject to allosteric regulation of the ADP glucose pyrophosphorylase activity.

Genetic regulation, thus, determines the capacity for glycogen synthesis and this should be distinguished from regulation of the absolute glycogen levels. As an example, the glycogen biosynthetic enzymes are induced in stationary phase when cells are grown on enriched medium lacking glucose. Yet, glycogen synthesis does not occur because of insufficient carbon source. However, in media with excess glucose, where nitrogen is limiting, the expression of the genes for the biosynthetic enzymes is somewhat weaker, as cAMP is at a low level. However, the glycogen synthetic rate is relatively greater because of the carbon availability and the conditions are such that allosteric activation occurs.

Mutants that are affected in either negative (*glgR*, *glgQ*, *csrA*) or positive control systems (*cya*, *crp*, *relA*, *spoT*) for *glgCA* gene expression clearly demonstrate that the genetic regulation of the levels of glycogen biosynthetic enzymes is most important in determining the ultimate level of glycogen synthesized and accumulated under any given physiological condition.

The structural and regulatory genes involved in glycogen metabolism in *E. coli* are listed in Table IV and the effects of both positive and negative regula-

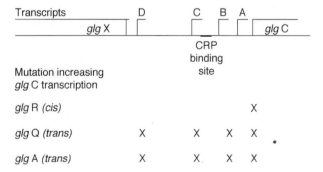

Fig. 4. Regulatory sites for transcription of *glg* C. The location of the Crp-binding region and the *glg* upstream transcription sites is reviewed in Preiss (1996).

TABLE IV
Genes Affecting Glycogen Metabolism in *E. coli*

Regulatory gene	Gene product	Map. site (min)	Comments
cya	Adenylate cyclase	85	Regulates catabolite repression
crp	Cyclic AMP receptor protein	74	Regulates catabolite repression
relA	(p)ppGpp synthase I	60	Mediates stringent response & response to carbon/energy
spoT	(p)ppGpp3′-pyrophosphophohydrolase	60	
csrA	6.8 KDa polypeptide (CsrA)	58	Regulation of gluconeogenesis & *glgCA* & *glgB* transcription
kat F	(rpoS) σs	59	Required for expression of *glgS*, not *glgCA*, pleiotropic
glgQ	Transacting factor (unidentified)	(?)	Transcriptional regulation of *glgCA* & *B*
glgR	Cis-acting site	75	Transcriptional regulation of *glgCA*

Structural gene	Gene product	Map site (min)	Comments
I. Biosynthetic			
glgC	ADP-glucose pyrophosphorylase	75	Synthesis of glucosyl donor
glgA	Glycogen synthase	75	α-1,4-glucosyl transferase
glgB	Glycogen branching enzyme	75	α-1,6 branch synthesis
glgS	7.9 kDa polypeptide	67	Function?
II. Degradative			
glgP (glg Y)	Glycogen phosphorylase	75	Phosphorolysis
amy Aᵃ	α-amylase	43	Hydrolysis
glgX	Glucan hydrolase/transferase	75	Isoamylase

[a] Raha *et al.*, 1992.

tory factors which control the expression of the *glg* genes of the glycogen biosynthetic pathway are listed. Many important questions remain to be solved, particularly the regulatory role of factor, *CsrA*. The physiological states which the *CsrA* regulatory system responds to remain to be determined. Moreover, actual functions of the *glgS*, *katF*, and many carbon starvation induced genes in glycogen synthesis at the biochemical and molecular levels remain to be established.

See Also the Following Article

CARBOHYDRATE SYNTHESIS AND METABOLISM

Bibliography

Büttcher, V., Welsh, T., Willmitzer, L., and Kossmann, J. (1997). Cloning and characterization of the gene for amylosucrase from *Neisseria polysaccharea*: Production of a linear α-D-glucan. *J. Bacteriol.* **179**, 3324–3330.

Dietzler, D. N., Leckie, M. P., Lais, C. J., and Magnani, J. L. (1974). Evidence for the regulation of bacterial glycogen biosynthesis *in vivo*. *Arch. Biochem. Biophys.* **162**, 602–606.

Greenberg, E., and Preiss, J. (1964). The occurrence of adenosine diphosphateglucose: Glycogen transglucosylase in bacteria. *J. Biol. Chem.* **239**, 4314–4315.

Hengge-Aronis, R., and Fischer, D. (1992). Identification and molecular analysis of *glgS*, a novel growth-phase-regulated and *rpo*S-dependent gene involved in glycogen synthesis in *Escherichia coli*. *Mol. Microbiol.* **6**, 1877–1886.

Krebs, E. G., and Preiss, J. (1975). Regulatory mechanisms in glycogen metabolism. *In* "Biochemistry of Carbohydrates" (W. J. Whelan, ed.), *MTP International Review of Science*, Vol. 5. pp. 337–389. Univ. Park Press, Baltimore, MD.

Kuriki, T., Stewart, D. C., and Preiss, J. (1997). Construction of chimeric enzymes out of maize endosperm branching enzymes I and II: Activity and properties. *J. Biol. Chem.* **272**, 28999–29004.

Liu, M. Y., Gui, G., Wei, B., Preston, III, J. F., Oakford, L., Yüksel, Ü., Giedroc, D. P., and Romeo, T. (1997). The RNA molecule Csr B binds to the global regulatory protein CsrA and antagonizes its activity in *Escherichia coli*. *J. Biol. Chem.* **272**, 17502–17510.

Liu, M. Y., Yang, H., and Romeo, T. (1995). The product of the pleitropic *Escherichia coli* gene *csrA* modulates glycogen biosynthesis via effects on mRNA stability. *J. Bacteriol.* **177**, 2663–2672.

Preiss, J. (1978). Regulation of ADP-glucose pyrophosphorylase. *In* "Advances in Enzymology and Related Areas of Molecular Biology" (A. Meister, ed.), Vol. 46, pp. 317–381. John Wiley and Sons, Inc.

Preiss, J. (1984). Bacterial glycogen synthesis and its regulation. *Annu. Rev. Microbiol.* **38**, 419–458.

Preiss, J. (1996). Regulation of glycogen synthesis. *In* "Escherichia coli and Salmonella typhimurium: Cellular and Molecular Biology" (2nd ed.), (F. C. Neidhardt, ed.), pp. 1015–1024. Amer. Soc. Microbiol., Washington, DC.

Preiss, J., and Romeo, T. (1989). Physiology, biochemistry and genetics of bacterial glycogen synthesis. *In* "Advances in Bacterial Physiology," Vol. 30, pp. 184–238.

Preiss, J., and Sivak, M. N. (1998). Biochemistry, molecular biology and regulation of starch synthesis. *In* "Genetic Engineering, Principles and Methods," Vol. 20, (J. K. Setlow, ed.), pp. 177–223, Plenum Press, Inc.

Romeo, T., Gong, M., Liu, M. Y., and Brun-Zinkernagel, A.-M. (1993). Identification and molecular characterization of *csrA*, a pleiotropic gene from *Escherichia coli* that affects glycogen biosynthesis, gluconeogenesis, cell size, and surface properties. *J. Bacteriol.* **175**, 4744–4755.

Svensson, B. (1994). Protein engineering in the α-amylase family: Catalytic mechanism, substrate specificity and stability. *Plant Molec. Biol.* **25**, 141–157.

Takata, H., Takaha, T., Shigetaka, O., Takagi, M., and Imanaka, T. (1997). Characterization of a gene cluster for glycogen biosynthesis and aheterotetrameric ADP-glucose pyrophosphorylase from *Bacillus stearothermophilus*. *J. Bacteriol.* **179**, 4689–4698.

Uttaro, A. D., Ugalde, R. A., Preiss, J., and Iglesias, A. A. (1998). Cloning and expression of the *glg* C gene from *Agrobacterium tumefaciens*. Purification and characterization of the ADP glucose synthetase. *Arch. Biochem. Biophys.* **357**, 13–21.

Wilkinson, J. F. (1959). The problem of energy-storage compounds in bacteria. *Exptl. Cell Res. Suppl.* **7**, 111–130.

Yang, H., Liu, M. Y., and Romeo, T. (1996). Coordinate genetic regulation of glycogen catabolism and biosynthesis in *Escherichia coli* via the CsrA gene product. *J. Bacteriol.* **178**, 1012–1017.

Glyoxylate Bypass in Escherichia coli

David C. LaPorte, Stephen P. Miller, and Satinder K. Singh

University of Minnesota, Minneapolis

GLOSSARY

IclR A repressor protein which inhibits the glyoxylate bypass operon (*aceBAK*) and its own structural gene (*iclR*).

IDH kinase/phosphatase A bifunctional regulatory protein which catalyzes both the phosphorylation (kinase) and dephosphorylation (phosphatase) of isocitrate dehydrogenase.

integration host factor (IHF) A histonelike protein which activates *aceBAK* expression by opposing repression by IclR. IHF serves many other functions in *E. coli*.

isocitrate dehydrogenase (IDH) The Krebs cycle enzyme, which competes with isocitrate lyase, the first enzyme of the glyoxylate bypass. IDH is regulated by reversible phosphorylation.

GLYOXYLATE BYPASS is a pathway, composed of isocitrate lyase and malate synthase, which bypasses the CO_2-producing steps of the Krebs cycle. The glyoxylate bypass of the bacterium *Escherichia coli* has been used as a model system for the study of metabolic regulation. This pathway is regulated by two distinct mechanisms: repression of transcription and phosphorylation of isocitrate dehydrogenase, the Krebs cycle enzyme which competes with the glyoxylate bypass. The long-term goal of these projects is to determine the mechanisms for each of these regulatory processes individually and how they interact to control the overall process.

I. THE GLYOXYLATE BYPASS

The Krebs cycle of *E. coli* serves the dual role of energy production and biosynthesis during growth on many carbon sources (Fig. 1). Most of these growth substrates can follow a variety of routes to enter the central metabolic pathways. This metabolic flexibility allows the cell to partition these substrates efficiently between the need to produce energy and the need to synthesize cellular components.

Compounds such as acetate and fatty acids present an unusual problem for cells because they can only enter metabolism as acetyl-CoA. Acetate carbons entering the Krebs cycle as acetyl-CoA at citrate synthase could not support growth because the CO_2 loss at isocitrate dehydrogenase and α-ketoglutarate dehydrogenase would prevent the net accumulation of most carbon containing compounds. With no mechanism for accumulating these metabolites, biosynthesis would be impossible and the cells would not grow.

Nature has addressed this problem for many plants and microorganisms with the glyoxylate bypass, which short-circuits the Krebs cycle before any CO_2 is lost. Isocitrate lyase and malate synthase combine with a subset of the Krebs' Cycle enzymes to assimilate two acetyl-CoAs. This pathway also produces significant metabolic energy, although it does so less efficiently than does the Krebs cycle.

Two fundamentally different mechanisms control the glyoxylate bypass. Coarse control is achieved by regulating the transcription of *aceBAK*, the operon which encodes malate synthase, isocitrate lyase, and isocitrate dehydrogenase (IDH) kinase/phosphatase, respectively. Once induced, the flow of isocitrate through the bypass is regulated by the phosphoryla-

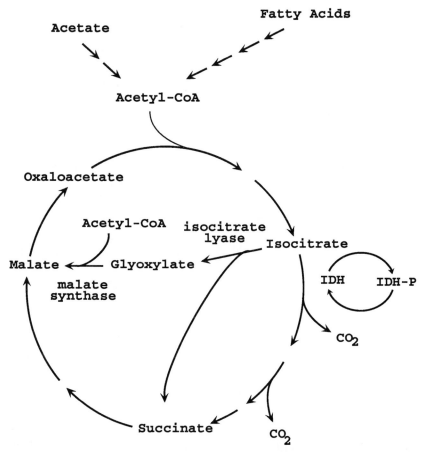

Fig. 1. The Krebs cycle and the glyoxylate bypass. The glyoxylate bypass is catalyzed by isocitrate lyase and malate synthase. The velocity of this pathway is regulated by the phosphorylation of isocitrate dehydrogenase (IDH).

tion of IDH, the Krebs cycle enzyme which competes with isocitrate lyase.

II. REGULATION OF TRANSCRIPTION

A. Control of *aceBAK*

Transcription of the *aceBAK* operon appears to respond to metabolic stress. Full operon induction is observed only when the glyoxylate bypass is essential, such as during growth on acetate or fatty acids. However, the presence of these substrates is not sufficient for induction, since the operon is repressed by the simultaneous presence of preferred carbon sources, such as glucose or pyruvate. Poor carbon

sources other than acetate and fatty acids produce a partial induction of *aceBAK*, even though the glyoxylate bypass is not required for growth on these compounds. There is a roughly linear relationship between the culture doubling times and the extents of induction, suggesting that this partial induction is a response to general metabolic stress. Finally, partial induction of the operon occurs upon growth into stationary phase on rich media, a condition which also represents a metabolic stress.

The *aceBAK* operon is transcribed from a single, σ70-type promoter (Fig. 2). This promoter is repressed by IclR, a protein which binds to a site which overlaps the "−35" element of the promoter. Preliminary evidence indicates that IclR represses transcription by competing with RNA polymerase (unpub-

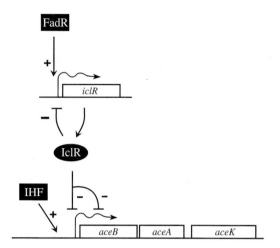

Fig. 2. Control circuits for the glyoxylate bypass operon. The glyoxylate bypass operon, *aceBAK*, encodes malate synthase, isocitrate lyase, and IDH kinase/phosphatase, respectively. This operon is repressed by IclR and activated by IHF. The gene encoding IclR is repressed by its own product and activated by FadR.

lished data). The mechanism which regulates IclR repression has not been determined.

B. A Remote Operator for IclR

In addition to the promoter-proximal IclR operator for *aceBAK*, a distal operator lies within the coding region of *aceB*. Binding of IclR to this site does not directly repress transcription, but this does increase repression mediated through the promoter-proximal operator (unpublished data). Our working model proposes that this cooperativity results from simultaneous binding of IclR at both sites with looping out of the intervening DNA. This type of cooperativity between remote sites has been most extensively characterized for the *lac* operon, but appears to occur in other operons as well.

C. Integration Host Factor "Activates" *aceBAK*

IHF is a histonelike protein which has been implicated in a wide variety of processes in *E. coli*. Examples include plasmid replication, lambda phage recombination, and transcriptional activation. IHF binds to a site upstream of the *aceBAK* promoter,

activating expression under inducing conditions by preventing IclR repression. Since IHF does not affect IclR activity under repressing conditions, the carbon source specificity of this effect amplifies regulation of *aceBAK* (Resnik *et al.*, 1996).

IHF plays a novel role in the regulation of *aceBAK* expression, since it "activates" transcription by blocking repression and does so in a carbon-source specific manner. Our working model proposes that IHF and IclR bind independently near the *aceBAK* promoter under repressing conditions. Under inducing conditions, we suspect that IclR undergoes a structural change, which causes the binding of IHF and IclR to become competitive. IclR repression is relieved under this condition because the IHF site is occupied.

D. Regulation of *iclR* Expression

IclR inhibits its own transcription by binding to an operator which overlaps the promoter of *iclR*. This provides a feedback control mechanism which helps to maintain a constant level of IclR.

FadR, a transcriptional regulator involved in control of fatty acid metabolism, also regulates *iclR*. FadR represses the genes encoding the enzymes of fatty acid degradation and activates the transcription of *fabA*, a gene whose product participates in unsaturated fatty acid biosynthesis. FadR activates *iclR* expression and may provide a link between *aceBAK* and fatty acids, a carbon source which requires the glyoxylate bypass. Activation of IclR expression allows FadR to negatively regulate *aceBAK* without binding to this operon's control regions.

E. Differential Regulation by IclR

Autorepression of IclR expression provides the cell with an efficient means for maintaining a constant level of IclR, but it also creates a potential problem. Since *aceBAK* and *iclR* are both repressed by IclR, conditions which require the induction of the operon could also increase the levels of the repressor protein which opposes induction. This problem has been avoided because, while the expression of *aceBAK* is exquisitely sensitive to the culture conditions, expression of *iclR* is virtually constant on all carbon sources tested thus far.

IHF, which is required for carbon source sensitivity of IclR, appears to be responsible for differential control by this repressor. *aceBAK* responds to the carbon source because it has an associated site for IHF (see preceding). In contrast, *iclR* does not respond to the carbon source because it does not have a site for IHF.

F. The IclR Family

IclR is the prototype for an expanding family of repressor proteins. Members of this family share about 25% identity throughout their sequences. The sequence homology is particularly high near the N terminus, a region in which we identified sequences which match the consensus for helix-turn-helix DNA binding motifs. An "IclR signature" in the C-terminal regions of these proteins has also been defined in the PROSITE database.

The IclR family is distinct from other families of transcriptional regulators in bacteria. The IclR family currently includes 19 members from 12 different gram-positive and gram-negative bacterial species. Family members regulate a wide variety of catabolic genes. IclR also fits this pattern, since the glyoxylate bypass is essential for the catabolism of acetate (See The NCBI database at www.ncbi.nlm.nih.gov).

III. REGULATION OF PROTEIN PHOSPHORYLATION

A. Regulation by Protein Phosphorylation

Fine control of the glyoxylate bypass is achieved by the phosphorylation of IDH, the Krebs cycle enzyme which competes with isocitrate lyase. This phosphorylation cycle is catalyzed by a bifunctional protein, IDH kinase/phosphatase (reviewed in LaPorte, 1993).

B. Structure and Function in the Regulation of IDH

Phosphorylation of Ser113 completely inactivates IDH by blocking the binding of the substrate isocitrate. Formation of a hydrogen bond between Ser113

and isocitrate contributes to binding in the active, dephosphorylated form of IDH. Phosphorylation blocks binding by electrostatic and steric effects (reviewed in LaPorte, 1993).

C. IDH Kinase/Phosphatase: Structure and Mechanism

IDH kinase/phosphatase appears to catalyze both the kinase and phosphatase reactions at the same active site. This active site is capable of efficiently hydrolyzing ATP, both in the absence of IDH and when this protein substrate is bound. Our working model proposes that the IDH phosphatase reaction results from the kinase back reaction tightly coupled to ATP hydrolysis (reviewed in (LaPorte, 1993) (Fig. 3).

D. Control of IDH Kinase/Phosphatase

A wide variety of metabolites control the IDH phosphorylation cycle. Examples include isocitrate, NADP, 3-phosphoglycerate, pyruvate, and AMP. There appear to be two distinct mechanisms by which these ligands regulate IDH kinase/phosphatase (Fig. 4) (reviewed in LaPorte, 1993).

One class of effectors acts through the active site of IDH. Occupation of this site by ligands, such as

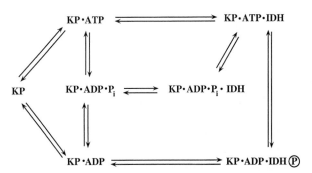

Fig. 3. Model for the catalytic mechanism of IDH kinase/phosphatase. This model proposes that the kinase and phosphatase reactions occur at the same active site. The intrinsic ATPase activity of IDH kinase/phosphatase, which occurs even in the absence of IDH, is also catalyzed in this site. The phosphatase reaction results from the back reaction of IDH kinase coupled to the hydrolysis of ATP. KP, IDH kinase/phosphatase; IDH, isocitrate dehydrogenase; IDH-P, the phosphorylated form of IDH.

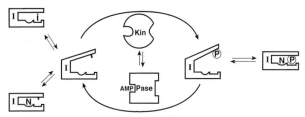

Fig. 4. Model for the regulation of the IDH phosphorylation cycle. This model proposes that IDH kinase/phosphatase can exist in two conformations, one favoring the kinase reaction (Kin) and the other favoring the phosphatase (Pase). This conformational equilibrium is controlled by metabolites, such as AMP, which bind preferentially to the phosphatase form. IDH (I) can exist in either the open or closed conformation, depending on whether the substrates isocitrate (i) or NADP⁺/NADPH (N) are bound. The open conformation is a substrate for IDH kinase/phosphatase, while the closed conformation is not. Ligands such as isocitrate may also inhibit IDH kinase/phosphatase by direct steric effects when they bind to the IDH active site. The site of phosphorylation, Ser-113, in indicated with a tick mark for dephospho-IDH. Phosphoserine-113 is indicated by "P."

isocitrate and NADP(H), blocks access to the phosphorylation site. Inhibition of IDH kinase/phosphatase may result from direct steric effects, since the site of phosphorylation is also in this active site. Indeed, Ser113 forms a hydrogen bond with isocitrate. Alternatively, inhibition may result, because these substrates may convert IDH from an open to a closed conformation. While Ser113 appears to be accessible in the open conformation, it is buried in the closed form.

Other ligands appear to regulate a conformational equilibrium in IDH kinase/phosphatase. Metabolites like AMP may control an equilibrium between kinase and phosphatase conformations by binding preferentially to the phosphatase form. The resulting shift in the conformational equilibrium would activate IDH phosphatase and inhibit IDH kinase.

IV. CONTROL OF METABOLISM

A. Growth on Glucose

The glyoxylate bypass is essential for growth on acetate but serves no purpose during growth on car-

bon sources, such as glucose or pyruvate. To avoid wasteful protein biosynthesis, *aceBAK* is repressed by IclR under these conditions. These conditions also favor the dephosphorylation of IDH by the basal level of IDH kinase/phosphatase which is present. Pyruvate, which activates IDH phosphatase and inhibits IDH kinase, appears to be the primary regulator of this phosphorylation cycle under these conditions.

B. Growth on Acetate

The glyoxylate bypass operon is induced during adaptation to growth on acetate (see preceding). Induction is essential for growth on these conditions, a conclusion based on the observation that *aceA* mutations prevent growth.

Adaptation to acetate also requires phosphorylation of IDH. Under these conditions, isocitrate must be partitioned between the Krebs cycle for the production of energy and the glyoxylate bypass for the production of metabolic intermediates. Cells growing on acetate maintain about 75% of the IDH in the inactive, phosphorylated form. Phosphorylation inhibits the Krebs cycle and so forces isocitrate through the glyoxylate bypass. The primary regulators of IDH kinase/phosphatase during growth on acetate appear to be isocitrate, 3-phosphoglycerate and AMP. Isocitrate and 3-phosphoglycerate act as indicators of the levels of metabolites, while AMP reflects the energy needs of the cell (reviewed in LaPorte, 1993).

C. Metabolic Transitions

The flux through the glyoxylate bypass responds dramatically to the culture conditions. Addition of glucose to a culture growing on acetate produces a rapid dephosphorylation of IDH, increasing this enzyme's activity and virtually eliminating the flux through the bypass. Long-term adaptation to the carbon source results from control of *aceBAK* transcription.

Two mechanisms amplify regulation of the IDH phosphorylation cycle. Zero-Order Ultrasensitivity results when one of the modification enzymes (e.g., the kinase or phosphatase) is saturated by its protein substrate. Multi-Step Effects, which can either enhance or reduce sensitivity, result when an effector controls more than one step in a protein modification cascade. These phenomena had been predicted from theoretical studies, but the glyoxylate bypass provided the first experimental demonstrations. Both phenomena have since been found in other systems.

An additional mechanism amplifies the regulation of the glyoxylate bypass. The "Branch Point Effect," which was discovered during studies on the glyoxyate bypass, results from the competition between enzymes for limiting amount of a common substrate. During growth on acetate, isocitrate levels are nearly saturating for IDH but are within the pseudo first-order range of isocitrate lyase. A consequence of this difference in saturation is that the flux carried by isocitrate lyase is ultrasensitive to the phosphorylation state of IDH. Thus, a modest change in IDH phosphorylation can yield a dramatic change in the bypass flux.

Bibliography

Bairoch, A., Bucher, P., and Hofmann, K. (1997). The PROSITE database, its status in 1997. *Nucl. Acids Res.* **25**, 217–221.

Brice, C. B., and Kornberg, H. L. (1968). Genetic control of isocitrate lyase activity in *Escherichia coli. J. Bacteriol.* **96**, 2185–2186.

Chung, T., Klumpp, D. J., and LaPorte, D. C. (1988). Glyoxylate bypass operon of *Escherichia coli*: Cloning and determination of the functional map. *J. Bacteriol.* **170**, 386–392.

Finer-Moore, J., Tsutakawa, S. E., Cherbavaz, D. B., LaPorte, D. C., Koshland, D. E., Jr., and Stroud, R. M. (1997). Access to phosphorylation in isocitrate dehydrogenase may occur by domain shifting. *Biochemistry* **36**, 13890–13896.

Freundlich, M., Ramani, N., Mathew, E., Sirko, A., and Tsui, P. (1992). The role of integration host factor in gene expression in *Escherichia coli. Mol. Microbiol.* **6**, 2557–2563.

Friedman, A. M., Fischmann, T. O., and Steitz, T. A. (1995). Crystal structure of lac repressor core tetramer and its implications for DNA looping. *Science* **268**, 1721–1727.

Goosen, N., and van de Putte, P. (1995). The regulation of transcription initiation by integration host factor. *Mol. Microbiol.* **16**, 1–7.

Gui, L., Sunnarborg, A., and LaPorte, D. C. (1996a). Regulated expression of a repressor protein: FadR activates transcription of *iclR. J. Bacteriol.* **178**, 4704–4709.

Gui, L., Sunnarborg, A. R., Pan, B., and Laporte, D. C. (1996b). Autoregulation of iclR, the repressor of the glyoxylate bypass operon. *J. Bacteriol.* **178**, 321–324.

Henry, M. F., and Cronan, J. E. Jr. (1991). *Escherichia coli* transcription factor that both activates fatty acid synthesis and represses fatty acid degradation. *J. Mol. Biol.* **222**, 843–849.

Kornberg, H. L. (1966). The role and control of the glyoxylate cycle in *Escherichia coli. Biochem. J.* **99**, 1–11.

LaPorte, D. C. (1993). Isocitrate dehydrogenase kinase/phosphatase: regulation and enzymology. *J. Cell. Biochem.* **51**, 14–18.

LaPorte, D. C., and Koshland, D. E., Jr. (1983). Phosphorylation of isocitrate dehydrogenase as a demonstration of enhanced sensitivity in covalent regulation. *Nature* **305**, 286–290.

LaPorte, D. C., Thorsness, P. E., and Koshland, D. E., Jr. (1985). Compensatory phosphorylation of isocitrate dehydrogenase: a mechanism for adaptation to the intracellular environment. *J. Biol. Chem.* **260**, 10563–10568.

LaPorte, D. C., Walsh, K., and Koshland, D. E., Jr. (1984). The branch point effect: ultrasensitivity and subsensitivity to metabolic control. *J. Biol. Chem.* **259**, 14068–14075.

Maloy, S. R., and Nunn, W. D. (1982). Genetic regulation of the glyoxylate shunt in *Escherichia coli* K12. *J. Bacteriol.* **149**, 173–180.

Miller, S. P., Chen, R., Karschinia, E. J., Romfo, C., Dean, A., and LaPorte, D.C. (2000) Locations of the regulatory sites for isocitrate dehydrogenase kinase/phosphatase. *J. Biol. Chem.*, in press.

Negre, D., Cortay, J.-C., Galinier, A., Sauve, P., and Cozzone, A. J. (1992). Specific interactions between the IclR repressor of the acetate operon of *Escherichia coli* and its operator. *J. Mol. Biol.* **228**, 23–29.

Nunn, W. D. (1986). A molecular view of fatty acid catabolism in *Escherichia coli. Microbiol. Reviews* **50**, 179–192.

Pan, B., Unnikrishnan, I., and LaPorte, D. C. (1996). The binding site of the IclR repressor protein overlaps the promoter of *aceBAK. J. Bacteriol.* **178**, 4704.

Resnik, E., Pan, B., Ramani, N., Freundlich, M., and LaPorte, D. C. (1996). Integration host factor amplifies the induction of *aceBAK* operon by relieving IclR repression. *J. Bacteriol.* **178**, 2715.

Sunnarborg, A., Klumpp, D., Chung, T., and LaPorte, D. C. (1990). Regulation of the glyoxylate bypass operon: cloning and characterization of iclR. *J. Bacteriol.* **172**, 2642–2649.

Gram-Negative Anaerobic Pathogens

Arthur O. Tzianabos, Laurie E. Comstock, Dennis L. Kasper

Harvard Medical School

I. Normal Flora and Epidemiology
II. Clinical Syndromes Caused by Anaerobes
III. Pathogenesis
IV. Immunity
V. Genetics
VI. Genetics of Antibiotic Resistance by *Bacteroides*

GLOSSARY

abscess A classic host response to bacterial infection in which many anaerobic species predominate.

anaerobe A bacterium that can survive and proliferate in the absence of oxygen.

antibiotic resistance The ability of bacteria to withstand the killing effect of antimicrobial agents.

cellular immunity A T cell-dependent host response to bacteria.

intraabdominal sepsis A disease process in humans that occurs following spillage of the colonic contents within the peritoneal cavity.

GRAM-NEGATIVE ANAEROBIC ORGANISMS constitute an important group of pathogens that predominate in many infectious processes. These organisms generally cause disease subsequent to the breakdown of mucosal barriers and the leakage of indigenous flora into normally sterile sites of the body. The predominance of anaerobes in numerous clinical syndromes can be attributed to the elaboration of a variety of virulence factors, the ability to resist oxygenated microenvironments, synergy with other bacteria, and resistance to certain antibiotics.

I. NORMAL FLORA AND EPIDEMIOLOGY

Hundreds of species of anaerobic organisms make up the human microflora. Mucosal surfaces such as the oral cavity, gastrointestinal tract, and female genital tract are the major reservoirs for this group of organisms. Infections involving anaerobes are polymicrobial and usually result from the disruption of mucosal surfaces and the subsequent infiltration of resident flora. However, gram-negative anaerobic bacilli are the most commonly isolated anaerobes from clinical infections. The clinically important gram-negative anaerobic bacilli belong to the genera *Bacteroides*, *Fusobacterium*, *Porphyromonas*, *Prevotella*, and *Bilophila* (Table I).

Anaerobes normally reside in abundance as part of the oral flora, with concentrations ranging from 10^9/ml in saliva to 10^{12}/ml in gingival scrapings. The indigenous oral anaerobic flora primarily comprise *Prevotella* and *Porphyromonas* species, with *Fusobacterium* and *Bacteroides* (non-*Bacteroides fragilis* group) present in fewer numbers. Aerobic or microaerophilic and anaerobic bacteria reside in approximately equal numbers in the oral cavity.

Anaerobic bacteria are present only in low numbers under the normally acidic conditions in the stomach and upper intestine. In people with decreased gastric acidity, the microflora of the stomach resemble that of the oral cavity. The upper intestine contains relatively few organisms until the distal ileum, where the flora begins to resemble that of the colon. In the colon, there are up to 10^{12} organisms/stool, with anaerobes outnumbering aerobes by approximately 1000:1. *Bacteroides* and *Fusobacterium* are the predominant gram-negative species in the

TABLE I
Disease Processes Caused by Gram-Negative Anaerobic Bacteria

Anaerobic species	*Colonization site*	*Typical disease*
Bacteroides fragilis	Colon	Intra-abdominal abscesses; bacteremia, mixed soft tissue infections
Other *Bacteroides* spp.	Colon; vagina	Abscesses and necrotizing fasciitis
Fusobacterium spp.	Oral cavity; colon; vagina	Abscesses; internal jugular venous thrombophlebitis
Prevotella spp.	Oral cavity; vagina	Lung abscess; pelvic infections
Porphyromonas spp.	Oral cavity	Gingival infection
Bilophila wadsworthia	Oral cavity; vagina; colon	Abscesses; bacteremia; necrotizing fasciitis

colonic flora. The *Bacteroides* species present in the colon are often referred to as the *B. fragilis* group, based on their earlier classification as subspecies of *B. fragilis*. This group encompasses several species, including *B. fragilis, B. thetaiotaomicron, B. ovatus, B. vulgatus, B. uniformis,* and *B. distasonis* (although the last organism may be reclassified as a *Porphyromonas* species). These *Bacteroides* species are all considered to be penicillin resistant, whereas the *Fusobacterium* species remain penicillin sensitive.

Prevotella, Bacteroides, and *Fusobacterium* are part of normal vaginal flora (10^6 organisms/g secretions). The most common isolates from clinical specimens are *P. bivia* and *P. disiens,* although *B. fragilis* is also frequently isolated from this site. *Bacteroides* species are found in approximately 50% of women, whereas *B. fragilis* is found in < 15% of this population.

Bilophila species are found in normal stool specimens and are occasionally part of the normal oral or vaginal flora. *B. wadsworthia* exhibits a fairly broad resistance to β-lactam antibiotics and is of interest because of its reported occurrence in human infections.

Anaerobic gram-negative bacilli reside on mucosal surfaces and cause infection following the contamination of normally sterile sites. For example, intraabdominal infection develops after fecal spillage into the peritoneum as a result of trauma or necrosis of the intact bowel wall, and severe infections of the head and neck may arise from an abscessed tooth. It is remarkable that, despite the identification of scores of anaerobic species in normal flora, relatively few species seem to play a role in infection. After the contamination of normally sterile sites by mucosal microflora, the relatively few anaerobic bacteria that

survive are those few that have resisted changes in oxidation reduction potential and host defense mechanisms. The hallmark of infection due to these gram-negative anaerobic bacteria is abscess formation, although some sepsis syndromes have been described. Typically, abscesses form at sites of direct bacterial contamination, although distant abscesses resulting from hematogenous spread are not uncommon with the more virulent anaerobes.

II. CLINICAL SYNDROMES CAUSED BY ANAEROBES

A. Anaerobic Infections of the Mouth, Head, and Neck

Anaerobes contribute to infection associated with periodontal disease and to disseminated infection arising from the oral cavity. Locally, infection of the periodontal area may extend into the mandible, causing osteomyelitis of the maxillary sinuses or infection of the submental or submandibular spaces. Other infections associated with oral anaerobes include gingivitis, acute necrotizing ulcerative mucositis, acute necrotizing infections of the pharynx, Ludwig's angina, fascial infections of the head and neck, sinusitis, and otitis.

B. Pleuropulmonary Infections

Anaerobes are associated with aspiration pneumonia, lung abscesses, and empyema. In these cases, the organisms isolated generally reflect the oral flora, and more than one bacterial species is routinely involved. For example, in anaerobic lung abscesses, it is not uncommon to find up to 10 species of organisms.

C. Intraabdominal Infections

Infections originating from colonic sites, such as intraabdominal abscesses, most frequently involve *Bacteroides* species, among which *B. fragilis* is the most common isolate. *B. fragilis* has also been associated with watery diarrhea in a few studies. Enterotoxin-producing strains are more prevalent in patients with diarrhea than in control groups. Secondary bacterial peritonitis arises when organisms from the intestine contaminate the peritoneum. The terminal ileum and colon are the most common sites of origin because of the high numbers of organisms they contain. Patients typically develop acute peritonitis following contamination, and intraperitoneal abscesses may result.

D. Pelvic Infections

Anaerobes are encountered in pelvic abscess, septic abortion, endometritis, tubovarian abscess, pelvic inflammatory disease, and postoperative infections. In addition to the major anaerobic gram-negative isolates already mentioned, *P. melaninogenica,* clostridia, and peptostreptococci are commonly found in infected pelvic sites. These organisms are most often isolated when infection is not due to sexually transmitted agents.

E. Central Nervous System Infections

Either a single species or a mixture of anaerobic or aerobic bacteria may be found in brain abscesses; prominent among the anaerobes are *Fusobacterium, Bacteroides,* and anaerobic gram-positive cocci. Anaerobic brain abscesses may arise by hematogenous dissemination from an infected distant site or by direct extension from otitis, sinusitis, or tooth infection.

F. Skin and Soft-Tissue Infections

Bacteroides species are found in necrotizing fasciitis, usually as part of a mixed anaerobic–aerobic infection. Approximately five species are typically isolated, with a 3:2 ratio of anaerobes to aerobes. These infections usually occur at sites that can be contaminated from oral secretions or feces; they may spread rapidly and be very destructive. Gas may be found in the infected tissues.

G. Bone and Joint Infections

These infections typically arise from infected adjacent soft-tissue sites. Both osteomyelitis of bone and septic arthritis are seen. *Fusobacterium* species are the most common gram-negative anaerobes isolated from infected joints, whereas infected bone may yield a wider variety of isolates.

H. Bacteremia

Anaerobic organisms, particularly *B. fragilis*, have been detected in up to 5% of blood cultures. When *B. fragilis* is isolated, patients are frequently ill with rigors and fever.

III. PATHOGENESIS

Infections involving gram-negative anaerobes are generally due to the breakdown of a mucosal barrier and the subsequent leakage of indigenous flora into closed spaces or tissue. The introduction of bacteria into otherwise sterile sites leads to a polymicrobial infection in which certain organisms predominate. These include *B. fragilis, Prevotella* species, *Fusobacterium* species, and *Porphyromonas* species. Although some of these organisms are numerically dominant in the normal flora, others (such as *B. fragilis*) make up a much smaller proportion. The greater ability of these organisms to cause disease more often than numerically dominant anaerobes usually indicates the possession of one or more virulence factors. These factors include the ability to evade host defenses, adhere to cell surfaces, produce toxins or enzymes, or display surface structures that contribute to pathogenic potential.

A. Synergy

The ability of different anaerobic bacteria to act synergistically during polymicrobial infection has been described but remains poorly characterized. It has been postulated that facultative organisms function in part to lower the oxidation–reduction potential in the microenvironment and that this change allows the propagation of obligate anaerobes. Conversely, studies have shown that anaerobes, including *B. fragilis,* can produce compounds such as succinic acid and short-chain fatty acids that inhibit the ability of phagocytes to clear facultative organisms. Further, it is clear that facultative and obligate anaerobes synergistically potentiate abscess formation in experimental models.

B. Role of *B. fragilis* Capsular Polysaccharide in Abscess Induction

The anaerobe most commonly isolated from clinical infections is *B. fragilis*. The most frequent sites of isolation of *B. fragilis* are the bloodstream and abscesses associated with intraabdominal sepsis. The characteristic host response to *B. fragilis* infections is the development of intraabdominal abscesses. Although the development of an abscess limits the initial spread of the organism, the host usually cannot resolve the abscess, therefore requiring surgical drainage. The high frequency of abscess formation associated with *B. fragilis* led to studies investigating this organism's pathogenic potential in relevant animal models of disease. This work identified the capsular polysaccharide as the major virulence factor of *B. fragilis* and defined the capsule's role in the induction of abscesses in an animal model of intraabdominal sepsis.

Further studies have delineated the structural attributes of the *B. fragilis* capsular polysaccharide that promotes abscess formation in animals. The capsule of strain NCTC 9343 comprises two distinct ionically linked polymers, termed PS A and PS B. Each of these purified polymers induces abscess formation when implanted with sterilized cecal contents (SCC) and barium sulfate as an adjuvant into the peritonea of rats. Historically, this adjuvant is included to simulate the spillage of colonic contents that occurs in intraabdominal sepsis. The implantation of PS A without SCC does not induce abscess formation in animals. PS A is the more potent of the two polysaccharides; less than 1 μg is required for abscess induction in 50% of challenged animals. The structures of both saccharides have been elucidated. Each polymer consists of repeating units whose possession of both positively and negatively charged groups is rare among bacterial polysaccharides. The ability of PS A and PS B to induce abscesses in animals depends on the presence of these charged groups. Numerous fecal and clinical isolates of *B. fragilis* have been examined, and this dual polysaccharide motif has been found in every strain.

The capsule of *B. fragilis* probably acts to regulate the host response within the peritoneal cavity to initiate the steps leading to abscess formation. Studies in mice have shown that the capsule mediates bacterial adherence to primary mesothelial cell cultures *in vitro*. In addition, the capsule promotes the release of the proinflammatory cytokines TNF-α and IL-1β, as well as the chemokine IL-8, from macrophages and neutrophils. The release of TNF-α from peritoneal macrophages potentiates the increase of cell-adhesion molecules such as ICAM-1 on mesothelial cell surfaces, and this potentiated response in turn leads to an increase in the binding of neutrophils to these cells. These events are the first steps leading to the accumulation of neutrophils at inflamed sites within the peritoneal cavity and are likely to lead to the formation of abscesses at these sites.

C. Abscess Formation and T Cells

T cells are critical in the development of intraabdominal abscesses, but little is known about the mechanisms of cell-mediated immunity underlying this host response. Attempts to define the immunologic events leading to abscess formation have been made in athymic or T cell-depleted animals. The results from these studies show that T cells are required for the formation of abscesses following bacterial challenge of animals. Studies by Sawyer *et al.* (1995) have documented a role for CD4+ T cells in the regulation of abscess formation.

B. fragilis produces a host of virulence factors that allow this organism to predominate in disease. Although the lipopoly saccharide (LPS) of *B. fragilis* possesses little biologic activity, this organism synthesizes pili, fimbriae, and hemagglutinins that aid in attachment to host-cell surfaces. In addition, *Bacteroides* species produce many enzymes and toxins that contribute to pathogenicity. Enzymes such as neuraminidase, protease, glycoside hydrolases, and superoxide dismutases are all produced by *B. fragilis.* Recent work has shown that this organism produces an enterotoxin with specific effects on host cells *in vitro.* This toxin, termed BFT, is a metalloprotease that is cytopathic to intestinal epithelial cells and induces fluid secretion and tissue damage in ligated intestinal loops of experimental animals. Strains of *B. fragilis* associated with diarrhea in children (termed enterotoxigenic *B. fragilis,* or ETBF) produce a heat-labile 20-kDa protein toxin. BFT specifically cleaves the extracellular domain of E-cadherin, a glycoprotein found on the surface of eukaryotic cells.

The pathogenesis of *P. gingivalis* relies on a broad range of virulence factors. This organism is a prominent etiologic agent in adult periodontitis. The progression of this disease is hypothesized to be related to the production of a variety of enzymes (particularly proteolytic enzymes), fimbriae, capsular polysaccharide, LPS, hemagglutinin, and hemolytic activity. *P. gingivalis* has been shown to invade and replicate host cells, a mechanism that may facilitate its spread. A class of trypsin-like cysteine proteases, termed gingipains, have recently been implicated as a major virulence factor contributing to the tissue destruction that is the hallmark of periodontal disease.

The capsular polysaccharide of *P. gingivalis* acts as a potent virulence factor facilitating a spreading infection in mice greater than that seen with unencapsulated strains. The LPS of *P. gingivalis* has been implicated in the initiation and development of periodontal disease. It has been shown that the LPS activates human gingival fibroblasts to release IL-6 via CD14 receptors on host cells. In addition, neutrophils stimulated with *P. gingivalis* LPS release IL-8.

F. necrophorum causes numerous necrotic conditions (necrobacillosis) and human oral infections. Several toxins, such as leukotoxin, endotoxin, and hemolysin, have been implicated as virulence factors. Among these, leukotoxin and endotoxin are believed to be the most important. *F. nucleatum* is a major contributor to gingival inflammation and is isolated from sites of periodontitis. This organism can co-aggregate with other oral bacteria to promote attachment to plaque; in addition, it produces several adhesins that facilitate attachment. Both *F.*

nucleatum and *F. necrophorum* produce a potent LPS that is responsible for the release of numerous proinflammatory cytokines and other inflammatory mediators.

Virulence factors associated with *Prevotella* species are poorly defined. The organism's ability to interact with other anaerobes has been reported. Among their prominent virulence traits is the production of proteases and metabolic products, such as volatile fatty acids and amines. This group of organisms is particularly noted for secretion of IgA proteases. The degradation of IgA produced by mucosal surfaces allows *Prevotella* to evade this first line of host defense. A study has demonstrated that *P. intermedia* can invade oral epithelial cells and that antibody specific for fimbriae from this organism inhibits invasion.

IV. IMMUNITY

Although relatively little is known about immunity to anaerobic organisms in general, the immune response to *B. fragilis* has been studied in detail. Prior treatment of animals with PS A from this organism prevents the formation of intraabdominal abscesses after challenge with *B. fragilis* or other abscess-inducing bacteria. This protection depends on the presence of positively charged amino and negatively charged carboxyl groups associated with the saccharide's repeating unit structure. Attempts to define the immunologic events regulating abscess formation have suggested an important role for cell-mediated immunity.

Studies on rats have shown that administration of PS A shortly before or even after bacterial challenge protects against abscess formation induced by a heterologous array of organisms. This protective activity is dependent on T cells. In other words, these studies suggested that PS A elicits a rapid, broadly protective immunomodulatory response that is dependent on cell-mediated immunity.

The capsular polysaccharide of *B. fragilis* has also been shown to inhibit opsonophagocytosis, and an antibody specific for its capsule activates both the classical and alternative pathways of the complement system. Significant increases in antibody titers in patients with *B. fragilis* bacteremia have been reported, but hyperimmune globulin generated in animals specific for *B. fragilis* does not protect against abscesses. Studies have shown that specific capsular antibody does reduce the incidence of bacteremia in experimentally infected animals.

Immunity to *P. gingivalis* infections has been described and can be generated to various degrees in animals models by the capsular polysaccharide, LPS, hemagglutinin, or gingipains. The involvement of T cells in regulating the immune response to this organism has also been demonstrated. Immune T cells derived from mucosal and systemic tissues of rats given live *P. gingivalis* yielded an increase in serum and salivary responses compared to control animals. These results indicate a role for serum IgG and salivary IgA in protection against periodontal disease in which a balance between Th1 and Th2-like cells occurs in humoral immune responses to *P. gingivalis*. In a recent clinical study, patients with periodontal disease develop a significant antibody response to *P. gingivalis,* but this response does not eliminate infection.

Relatively little is known about the host immune response to *Fusobacterium* species and *Prevotella* species. *F. nucleatum* produces a protein that inhibits T cell activation *in vitro* by arresting cells in the mid-G1 phase of the cell. It is hypothesized that the suppressive effects of this protein enhance the virulence of *F. nucleatum*. In addition, *F. nucleatum* or its purified outer membrane can induce a potent humoral response in mice. Several investigators have attempted to investigate mechanisms of protective immunity against *F. necrophorum*, but potential immunogens isolated from the organism have not afforded satisfactory protection.

In studies of the host response to *P. intermedia*, a polysaccharide surface component exerted a strong mitogenic effect on splenocytes and a cytokine-inducing effect on peritoneal macrophages from both C3H/HeJ and C3H/HeN mice; this polysaccharide also stimulated human gingival fibroblasts to produce cytokines. The immunization of nonhuman primates with *P. intermedia* resulted in the production of significantly elevated levels of specific IgG. The level of serum IgA antibody also increased. Finally, coculture of *P. intermedia* with T cells significantly upregulated the expression of specific T-cell receptor-variable regions, a result suggesting that this organism has significant impact on T cells.

V. GENETICS

Knowledge of the genetic makeup of the gram-negative anaerobic pathogens lags far behind their aerobic counterparts. Sequence for only a few hundred nonredundant genes have been deposited in the databases for all these genera combined. An accurate understanding of the genetic makeup of these organisms awaits complete genome sequencing.

The G + C content of these bacteria shows some dissimilarity among genera, with variation for *Bacteroides* species reported at 41–46%, *Porphyromonas* species 41–45%, *Prevotella* species 39–51%, and *Fusobacterium* species 26–34%. Little genetic information is available for *Bilophila*. The only reported sequence for this genus is that of the 16S rDNA, which demonstrates the relatedness of this organism to other sulfur-reducing organisms such as the *Desulfovibrio* species. Similarly, few genes aside from rDNA have been sequenced from *Fusobacterium* and few factors implicated in virulence of the human *Prevotella* species have been characterized genetically.

A large proportion of the genes sequenced from *Porphyromonas* encode the various and diverse types of proteases or hemagglutinins produced by these organisms that are involved in virulence. In addition, mutational analysis of *fimA* demonstrated that fimbriae are essential for the interaction of the organism with human gingival tissue cells.

The most extensive area of genetic research in *Bacteroides* is the study of antibiotic resistance and the elements involved in the transfer of resistance genes. Aside from this area of research, the genetic analysis of other virulence factors are being performed. The gene encoding the metalloprotease toxin (*bft*) of *B. fragilis* is contained on a pathogenicity island that is present only in enterotoxigenic strains. Other products involved in aerotolerance of *B. fragilis,* such as catalase and superoxide dismutase, have also been studied at the molecular level.

VI. GENETICS OF ANTIBIOTIC RESISTANCE BY *BACTEROIDES*

Antibiotic resistance is an increasing problem in the treatment of *Bacteroides* infections: The bacteria continue to acquire genes that make them resistant to multiple antibiotics. A drastic increase in resistance to antibiotics such as tetracycline, cephalosporins, and clindamycin over the last 2 decades has necessitated the use of carbapenems, metronidazole, and β-lactamase inhibitors. Resistance to these latter agents, however, is also increasing. The genetic elements responsible for the evolution of antibiotic resistance in *Bacteroides* are the topic of this section.

A. β-Lactamases

The majority of *Bacteroides* species display some level of resistance to β-lactam antibiotics. The genes encoding the enzymes responsible for resistance vary among the *Bacteroides* species. The best studied are the β-lactamases of *B. fragilis*. Two distinct classes of β-lactamases have been described in *B. fragilis*, the active-site serine enzymes encoded by *cepA* and the metallo-β-lactamases encoded by *cfiA*. Any given *B. fragilis* strain contains only one of these β-lactamase-encoding genes, and taxonomic investigations have revealed that *cfiA+* strains and *cepA+* strains form two genotypically distinct groups. Although these groups cannot be differentiated phenotypically, *cfiA+* strains exhibit a distinctive and homogeneous ribotype, can be distinguished from *cepA+* strains by arbitrarily primed PCR, and preferentially contain most of the insertion-sequence (IS) elements described for *B. fragilis* (i.e., IS 4351, IS 942, and IS 1186).

1. *cepA*

The most prevalent β-lactamase of *B. fragilis* is the active-site serine enzyme encoded by the chromosomal *cepA*. This cephalosporinase does not confer protection against the carbapenems (such as imipenem) and is sensitive to the action of β-lactamase inhibitors. Not all *cepA+* strains produce a high level of the cephalosporinase. The analysis of the sequence of *cepA* from seven *B. fragilis* strains that produce high or low levels of cephalosporinase demonstrated that the *cepA* coding regions for all strains were identical. Therefore, structural differences in the enzyme do not account for the differing levels of resistance conferred by *cepA+* strains. Rather, the production of the enzyme is regulated at the transcriptional level by sequences upstream of *cepA*.

2. *cfiA*

Only approximately 3% of *B. fragilis* strains contain *cfiA*, which encodes a metallo-β-lactamase that varies in properties, substrate specificity, and activity from the *cepA* gene product. The emergence of *cfiA* has been closely monitored, as the enzyme it encodes is not affected by β-lactamase inhibitors and is active against a variety of β-lactams, including carbapenems. Only one-third of *cfiA+* strains have been reported to produce the enzyme; the inability of other *cfiA+* strains to do so is due to the lack of a promoter driving the transcription of *cfiA*. These silent *cfiA* genes become active when an IS element (IS 1168 or IS 1186) is inserted into the chromosome just upstream of *cfiA*. These IS elements contain outward-oriented promoters that drive the transcription of *cfiA*, leading to a 100-fold increase in the amount of β-lactamase. Given the small fraction of *B. fragilis* strains that produce the metallo-β-lactamase, carbapenems are still effective for the treatment of *Bacteroides* infections.

B. Conjugative Elements Involved in the Transfer of Antibiotic Resistance

Unlike the β-lactamase genes of *B. fragilis*, the resistance of *Bacteroides* to clindamycin, tetracycline, and 5-nitroimidazole is conferred by genes carried on elements that are self-transmissible. Both conjugative transposons and conjugative plasmids are involved in the transfer of these antibiotic-resistance genes in *Bacteroides*.

1. *Conjugative Transposons*

The conjugative transposons of *Bacteroides* are 70- to 80-kb elements that are normally integrated into the chromosome or on a plasmid. In addition to carrying genes for resistance to tetracycline and clindamycin, conjugative transposons contain all the necessary genes for their excision and transfer. Once the transposon has been excised from the chromosome, the element forms a covalently closed circle. Transfer of a single strand of the transposon then begins at *oriT* through a mating pore to the recipient cell. The single strand is replicated in the recipient cell and integrated into the chromosome in an orientation- and site-specific manner.

Several factors account for the ability of the conjugative transposons to propagate antibiotic-resistance genes so successfully. Their broad host range allows for their transfer between species that are only distantly related. The conjugative transposons can also mediate the mobilization of coresident plasmids and the excision and mobilization of unlinked chromosomal segments of 10–12 kb, termed nonreplicating *Bacteroides* units (NBUs). These elements contain an origin of transfer that permits their transfer by the mating pore of the conjugative transposon. Therefore, conjugative transposons allow for the spread of antibiotic-resistance genes contained on unlinked elements. Lastly, the transfer capabilities of the conjugative transposons are inducible by subinhibitory concentrations of tetracycline as described later.

2. Conjugative Plasmids

Of the several conjugative plasmids described in *Bacteroides*, some contain genes conferring resistance to clindamycin or the 5-nitroimidazoles. The regions involved in transfer have been studied for many *Bacteroides* conjugative plasmids and usually involve the products of one to three *mob* genes. As are the conjugative transposons, the conjugative plasmids have been transferred across genera; however, their range is probably more restricted than that of the conjugative transposons, as recognition of the origin of replication is necessary for maintenance. Because conjugative plasmids can be mobilized by coresident conjugative transposons, the spread of 5-nitroimidazole resistance conferred by genes on conjugative plasmids can be induced by tetracycline pretreatment of cells containing conjugative transposons. Therefore, the use of tetracycline may lead to the spread of resistance to various antibiotics.

C. Antibiotic Resistance Genes Contained on Conjugative Elements

1. Tetracycline Resistance

Tetracycline, once effective against *Bacteroides* infections, now encounters resistance in the majority of clinical isolates. Most tetracycline-resistant *Bacteroides* contain the gene *tetQ*, which is believed to confer widespread resistance among *Bacteroides*

strains. Although the *tetQ* product is most similar to the TetM and TetO classes of resistance mediated by ribosomal protection, the degree of similarity is low enough (40%) to merit a separate class of ribosomal resistance genes.

The gene *tetQ* is carried by the conjugative transposons of *Bacteroides*. What is unique about these conjugative transposons is that their transfer is increased by 100- to 1000-fold by pretreatment of the donor with subinhibitory concentrations of tetracycline. This finding led to the discovery of three regulatory genes downstream of *tetQ* on the conjugative transposon. The corresponding gene products are probably involved in the tetracycline-dependent transcriptional activation of genes involved in transfer.

Two other tetracycline-resistance determinants have been identified in *Bacteroides*. *tetX*, first cloned in 1991, encodes a product that inactivates tetracycline. An additional tetracycline-resistance determinant was found on a *Bacteroides* transposon that leads to tetracycline efflux. Neither of these products actually confers resistance to tetracycline in *Bacteroides*, and both are probably remnants of DNA transfer from other organisms.

2. Clindamycin Resistance

Two reports have shown the frequency of clindamycin resistance among *Bacteroides* species at various institutions to be as high as 21.7% and 42.7%. The first clindamycin-resistance gene from *Bacteroides* (*ermF*) was sequenced in 1986. Since then, two additional genes have been sequenced and found to encode products that are 98% identical to the *ermF* product. These genes are contained on transposons, conjugative transposons, and conjugative plasmids, which accounts for their widespread distribution among *Bacteroides* strains.

The mechanism of resistance conferred by the *ermF* product involves neither inactivation nor efflux of the drug. Instead, resistance to clindamycin occurs at the level of the ribosome. The strong identity of the *ermF* product with *erm* genes from gram-positive bacteria suggests that the resistance is mediated by methylation of 23S rRNA, which prevents binding of the antibiotic.

The clindamycin resistance gene *ermG*, which was

recently sequenced from a *Bacteroides* conjugative transposon, encodes a product that is only 46% similar to the *ermF* product. The *ermG* product is extremely similar to the *erm* products of gram-positive organisms, whose functions as 23*S* rRNA methylases have been established. Therefore, *Bacteroides* species contain two distinct genes, probably of different origins, that both confer resistance to clindamycin by the same mechanism.

3. 5-Nitroimidazole Resistance

The vast majority of *Bacteroides* strains are sensitive to the 5-nitroimidazole antibiotics, and metronidazole remains the drug of choice for the treatment of *Bacteroides* infections. However, genes conferring resistance to 5-nitroimidazoles have been described in various *Bacteroides* species. Four distinct resistance genes have been sequenced (*nimA–nimD*), the products of which exhibit 67–91% similarity. *nimA*, *nimC*, and *nimD* are present on conjugative plasmids, and *nimB* is present in the chromosome of a *B. fragilis* clinical isolate. As are the other three *nim* genes, *nimB* is transferable by conjugation to other *B. fragilis* strains. IS elements are present just upstream of each of the *nim* genes. Some of these IS elements are highly homologous to IS 1168 and IS 1186, which control the transcription of *cfiA* of *B. fragilis*. It is likely that transcription of the *nim* genes, as with *cfiA*, is controlled by outward-oriented promoters the IS elements.

The mechanism of the resistance conferred by the *nimA* product has been studied. The gene probably encodes a 5-nitroimidazole reductase that prevents the formation of the toxic form of the drug.

See Also the Following Articles

Anaerobic Respiration • Cellular Immunity • Oral Microbiology • T Lymphocytes

Bibliography

Dorn, B. R., Leung, K. L., and Progulske-Fox, A. (1998). Invasion of human oral epithelial cells by Prevotella intermedia. *Infect. Immun.* **66**, 6054–6057.

Gorbach, S. L., and Bartlett, J. G. (1974). Anaerobic infections. *N. Eng. J. Med.* **290**, 1237–1245.

Lamster, I. B., Kaluszhner-Shapira, I., Herrera-Abreu, M., Sinha, R., and Grbic, J. T. (1998). Serum IgG antibody response to Actinobacillus actinomycetemcomitans and Porphyromonas gingivalis: Implications for periodontal diagnosis. *J. Clin. Periodontol.* **25**, 510–516.

Polk, B. J., and Kasper, D. L. (1977). *Bacteriodes fragilis* subspecies in clinical isolates. *Ann. Int. Med.* **86**, 567–571.

Rasmussen, J. L., Odelson, D. A., and Macrina, F. L. (1986). Complete nucleotide sequence and transcription of ermF, a macrolide- lincosamide-streptogramin B resistance determinant from Bacteroides fragilis. *J. Bacteriol.* **168**, 523–533.

Sawyer, R. G., Adams, R. B., May, A. K., Rosenlof, L. K., and Pruett, T. L. (1995). CD4+ T cells mediate preexposure-induced increases in murine intraabdominal abscess formation. *Clin. Immunol. Immunopathol.* **77**, 82–88.

Sugita, N., Kimura, A., Matsuki, Y., Yamamoto, T., Yoshie, H., and Hara, K. (1998). Activation of transcription factors and IL-8 expression in neutrophils stimulated with lipopolysaccharide from Porphyromonas gingivalis. *Inflammation* **22**, 253–267.

Thadepalli, H., Gorbach, S., Broido, P., Norsen, J., and Nyhus, L. (1973). Abdominal trauma, anaerobes, and antibiotics. *Surg. Gynecol. Obstetrics* **137**, 270–276.

Tzianabos, A. O., Kasper, D. L., Cisneros, R. L., Smith, R. S., and Onderdonk, A. B. (1995). Polysaccharide-mediated protection against abscess formation in experimental intra-abdominal sepsis. *J. Clin. Invest.* **96**, 2727–2731.

Tzianabos, A. O., Onderdonk, A. B., Rosner, B., Cisneros, R. L., and Kasper, D. L. (1993). Structural features of polysaccharides that induce intra-abdominal abscesses. *Science* **262**, 416–419.

Wu, S., Lim, K. C., Huang, J., Saidi, R. F., and Sears, C. L. (1998). Bacteroides fragilis enterotoxin cleaves the zonula adherens protein, E- cadherin. *Proc. Natl. Acad. Sci. U.S.A.* **95**, 14979–14984.

Gram-Negative Cocci, Pathogenic

Emil C. Gotschlich

The Rockefeller University

I. Infection by *Neisseria meningitidis* and *Meisseria gonorrhoeae*
II. Molecular Mechanisms of Infection
III. Antigens
IV. Natural Immunity
V. Prevention
VI. Summary

GLOSSARY

capsule   An external layer, usually consisting of a complex polysaccharide, coating the surface of many species of pathogenic bacteria.

meningitis   Inflammation of the meninges, which are membranes covering the brain and the spinal cord, resulting from a bacterial or viral infection.

meningococcemia   An infection of the blood stream by meningococci in the absence of meningitis.

petechial skin lesions   Small purplish spots on the skin caused by a minute hemorrhage.

porins   Protein molecules found in outer membranes of gram-negative bacteria that serve as channels for the diffusion of water and small-molecular-weight solutes.

GRAM-NEGATIVE COCCI   are almost invariably isolated from nasopharyngeal cultures of human beings. The majority of these are commensal species, which are classified principally by sugar-fermentation reactions performed on the organisms following their isolation from the primary throat culture. Thus, *Neisseria lactamica* derives its name from its ability to ferment lactose. On rare occasions bacteria belonging to a few of these species are able to invade the human host and cause disease. Table I lists the pathogenic gram-negative cocci.

I. INFECTION BY *NEISSERIA MENINGITIDIS* AND *NEISSERIA GONORRHOEAE*

A. Local Infection

Among the *Neisseria* and related organisms listed, *Neisseria meningitidis* and *Neisseria gonorrhoeae* are the primary pathogens, that is, organisms that are able to cause disease in an otherwise healthy host. Hence, this discussion focuses on these species. DNA hybridization has indicated that *Neisseria meningitidis* and *Neisseria gonorrhoeae* are extremely closely related organisms. It is therefore not surprising that the pathogenetic strategies of the organisms are for the most part very similar. Both are able to cause infection of the mucous membranes. *Neisseria meningitidis*, transmitted by droplets from the respiratory tract of an infected individual, usually infects the mucous membranes of the nasopharynx, but has been isolated on occasion from genitourinary sites. The nasopharyngeal infection with *Neisseria meningitidis* is most often asymptomatic and is self-limited, with the infection lasting for a few weeks to months. This colonization is referred to as the carrier state and is quite common. During the winter months, the frequency of carriers is usually 10% or greater. In populations that live in close contact, such as in boarding schools and military recruit camps, the carrier rate not infrequently will exceed 50%.

Gonococci are transmitted most often by sexual contact and infect the genitourinary tract, but are also

TABLE I
Commensal and Pathogenic Gram-Negative Cocci

Species	Distribution	Pathogenicity
Commensal species		
Neisseria lactamica	Very commonly colonizes nasopharynx of young children	Very rare instances of sepsis and meningitis
Neisseria subflava family: flava, sicca, perflava	Commonly found in nasopharyngeal cultures	
Neisseria cinerea	Nasopharynx	
Moraxella catarrhalis	50% of children are carriers during winter months	Third most common cause of acute otitis media and sinusitis in young children; sepsis very rare
Moraxella lacunata	Nasopharynx	Formerly common cause of conjunctivitis and keratitis; now rare
Moraxella bovis	Human and bovine nasopharynx	Causes outbreaks of bovine keratoconjunctivitis
Moraxella nonliqefaciens	Nasopharynx	
Moraxella osloensis	Nasopharynx	Occasionally causes invasive disease
Moraxella canis	Canine and feline oral cavity	Infections following dog and cat bites
Kingella kingi	Commonly in nasopharyngeal cultures	Occasionally causes arthritis, osteomyelitis and sepsis in children below age 2
Primary pathogens		
Neisseria meningitidis	Nasopharynx, rarely genitourinary tract	Sepsis and meningitis
Neisseria gonorrhoeae	Genitourinary tract, less commonly rectum and pharynx	Gonorrhea, sepsis, septic arthritis, and dermatitis; rarely meningitis

quite frequently isolated from rectal and pharyngeal cultures. The infections of the genitourinary tract are most often symptomatic, whereas pharyngeal infections generally cause no symptoms. The conjunctivae are susceptible to gonococcal infection, particularly in neonates who acquire it by passing through an infected birth canal. Historically this infection, ophthalmia neonatorum, was extremely common and was the major cause of acquired blindness until the general acceptance of the preventive Credè procedure consisting of the instillation of drops of a 1% silver nitrate solution in the eyes of all newborns. This has been replaced by the use of less irritating antibiotic ointments.

B. Extension of Local Infection

Genitourinary gonococcal infection in women usually involves the endocervix and with lesser frequency the urethra, the rectum, and the pharynx. As many as 50% of infections may be asymptomatic or with insufficient symptoms to motivate the person to seek medical attention. In at least 10% of infections, there is early extension to the uterine cavity (endometritis), and subsequently ascending infection of the fallopian tubes (pelvic inflammatory disease, PID). PID is the major complication of gonorrhea. Most often it requires hospitalization for differential diagnosis and treatment, and causes tubal scarring resulting in infertility and tubal pregnancies.

Infections in men, although sometimes asymptomatic, most often cause sufficient symptoms to drive the affected individual to seek medical attention. In the days prior to availability of antibiotic therapy, epidydimitis occurred in about 10% of cases, but this complication as well as prostatitis are rarely seen today.

C. Bacteremic Infection

Both *N. meningitidis* and *N. gonorrhoeae* first colonize the epithelial surface and are able to traverse epithelial cells by mechanisms to be described here. Once the organisms are in the subepithelial space, both the meningococcus and the gonococcus may invade the bloodstream. With *N. meningitidis* invasion of the bloodstream and the subsequent invasion of the meninges of the brain with resulting meningitis are very dangerous bacterial infections that can very rapidly lead to death, even with optimal antibiotic and supportive therapy. In the United States there are about 3000 reported cases each year of meningococcal meningitis, and now that meningitis due to *Haemophilus influenzae* type b is a vanishing disease as a result of the widespread acceptance of vaccination, the meningococcus is the leading cause of meningitis. In contrast with other organisms that cause meningitis, *N. meningitidis* has the ability to cause epidemic outbreaks with incidences approaching 200 or more per 100,000 per year. Epidemics of meningitis have occurred in all parts of the world, but have been a particular problem in sub-Saharan Africa; for instance in the winter of 1995–1996, there were 250,000 cases in west Africa.

In the case of the gonococcus, it is estimated that 1–2% of patients with gonorrhea do have invasion of the bloodstream, which can in rare instances be as fulminant as meningococcal infection, but usually has a much more benign course. The disease most often presents with fever, arthritis, and petechial, hemorrhagic, pustular, or necrotic skin lesions. If untreated, this can progress to septic arthritis that may cause severe damage to the affected joint.

D. Treatment

Meningococcal infection is a medical emergency and the earlier effective antibiotic treatment is initiated the better the prognosis. The mortality rate for meningococcal meningitis is generally about 10%, but it is much higher in meningococcemia and shock. The drug of choice for treatment remains intravenously administered penicillin G in very high doses. However, the pneumococcus and *Haemophilus influenzae* can at times present a very similar clinical picture, and because these organisms frequently are resistant to penicillin, third-generation cephalosporins are recommended as initial therapy until the meningococcal etiology has been established. Patients need to be hospitalized because the course of the disease is unpredictable and supportive therapy is frequently needed.

The treatment of gonococcal infection has changed over the past decades due to the development of partial or complete resistance of this organism to many antibiotics. The list of recommendations proposed by the CDC (Centers for Disease Control) in 1993 take into account the need to treat this disease at the time of the clinic visit with a single dose because many patients are not compliant with regimens that require repeated administration of the medication. In addition, coinfection with *Chlamydia trachomatis* is a very common occurrence, and the treatment should also eliminate this organism.

II. MOLECULAR MECHANISMS OF INFECTION

One of the problems in the study of neisserial disease is that these organisms are restricted to human beings, and animal models have provided very little information on the pathological events occurring during the various stages of the infection. Because of the excellent response of uncomplicated gonorrhea in men to treatment with modern antibiotics, challenge studies of volunteers have been accepted as ethically justifiable. Such scientific experiments, because of their cost, can be performed only infrequently and therefore correlations have to be made with *in vitro* models that mimic the *in vivo* or natural conditions as closely as possible. The human disease has been most closely simulated by an organ culture system employing fallopian tubes that are obtained from women undergoing medically indicated hysterectomies. The epithelium lining the fallopian tubes consists principally of two kinds of cells, mucus-secreting cells, and cells that bear cilia and beat in unison to move the mucus layer. When gonococci are added to the explants *in vitro*, the first discernible interaction is that the gonococci attach by means of long hair-like projections known as pili to the surface of the mucus-secreting cells, but not to the ciliated cells. This is a distant attachment between the bacteria and the cell surface and occurs about 6 hr following inoculation. Then over the next 12–18 hr this distant attachment converts to a very close attachment in which the membranes of the host and the parasite come into extensive and intimate contact. The close attachment is believed to be mediated by a set of outer-membrane proteins, named opacity proteins and discussed later. These interactions initiate a signaling cascade that causes the epithelial cells to engulf the gonococci, transport the bacteria through the body of the cell in vacuoles, and egest them in an orderly fashion on the basal part of the cells onto the basement membrane. Later in the infection (24–72 hr) toxic phenomena occur that result in expulsion of the ciliated cells from the epithelium. If meningococci are placed on human fallopian tubes, the same events occur, but over a shorter period of time. However, these events are not seen with commensal *Neisseria* species or with fallopian tubes that are not of human origin. The appearance of biopsies of cervical tissue taken for cancer-diagnostic purposes that were inadvertently obtained from infected patients show a picture that is quite similar.

The events transpiring in the course of the model infection have been the focus of research in order to understand meningococcal and gonococcal disease in molecular terms. Many bacterial species are able to invade epithelial cells, and in the case of Yersinia, *Salmonella, Shigella,* and *Listeria* quite a lot is known about this process because of the genetic tractability of these species. In the case of the *Neisseria,* this exploration is not as far advanced. Obviously the establishment of the mucosal infection depends on cross-talk between the bacterium and the host cells. It is noteworthy that compared to the other pathogens mentioned, for which invasion occurs quite promptly, there is a long lag in the invasion by the *Neisseria* as if some slow inductive events need to occur in the host cell or the organism or both.

III. ANTIGENS

The surface antigens of the *Neisseria* have been extensively studied to gain an understanding of the molecular steps underlying the pathogenesis of these diseases as well as to identify candidate molecules for inclusion in vaccines.

A. Pili

Most peripheral on the surface are pili, which are hair-like appendages with a diameter of about 8 nm that emanate from the outer membrane of gonococci and are several bacterial diameters long. Pili consist of the helical aggregation of a single kind of protein subunit of about 18,000 MW, known as pilin. The study of pili on gonococci is enormously simplified by the fact that their presence imparts a distinctive appearance to the gonococcal colony as it grows on agar. Piliated colonies are smaller and have sharp edges when viewed with a colony microscope. On laboratory media, isolates with a different colonial appearance, namely larger colonies that are flatter and have an indistinct edge, appear; if the organisms are subcultured nonselectively, these become the predominant colonial form. This colonial form of gonococci is no longer piliated, but in some instances these strains can revert to the piliated state. This ability to turn on and off pilus expression occurs at very high frequency, on the order of 1 in a 1000 cells per generation.

Gonococci that are freshly isolated from patients invariably are piliated. It is known from challenge studies of volunteers that only piliated strains are capable of causing infection. Stable nonreverting pilus-negative strains cannot cause infections. Thus, pili seemed at first to be an ideal vaccine candidate, but it soon became evident that pili are antigenically very variable. It was noted that no two strains of gonococci appeared to have the same pili, and later it was found that the pili expressed by a single strain of gonococcus maintained in the laboratory would over time repeatedly change their antigenicity. Following the cloning of the pilin structural gene, this problem could be approached on a molecular level and the mechanism of antigenic variation is summarized in a very simplified form in Fig. 1. Generally, the gonococcus possesses a single genetic locus expressing the pilin, which is called *pilE*. This locus contains a complete pilin structural gene with its promoter. In addition to the expression locus, gonococci also have eight or more other loci referred to

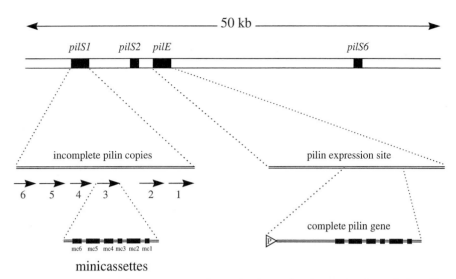

Fig. 1. Genetics of pilin variation. *pilE* is the locus for the expression of the pilin protein and contains a complete pilin gene including the promoter region, indicated as a triangle. There are a number of *pilS* regions, some as indicated within 50 kb of the expression site, which contain incomplete pilin genes lacking promoter sequences and the N-terminal portion of the coding frame. As indicated, *pilS1* contains several incomplete pilin genes. As indicated at the bottom of the figure, the variable portions of pilin are distributed as minicassettes with intervening conserved sequences that serve as targets for homologous recombination.

as silent loci. One of these, *pilS1*, is shown enlarged to indicate that the silent loci contain a large number of distinct pilin genes. This particular locus contains five silent pilin genes. These pilin genes are characterized by the fact that all of them are incomplete. They lack the promoter sequences and portions corresponding to the beginning of the protein coding frame. These incomplete genes are efficiently shuttled by homologous recombination into the *pilE* expression site, causing the production of a large number of serologically variant pili by a single strain of gonococcus over a period of time. The recombination events can occur in the conserved N-terminal portion of the coding frame and in the conserved region immediately following the termination codon. But it can also occur in several islands of conserved sequences scattered in the variable portion of the pilin genes between the minicassettes. Such antigenic changes occur at a rate of about 1 in a 1000 cells per cell division. If the new antigenic variant pilin can be assembled into intact pili, then an antigenic variation step has occurred. If the new pilin cannot be assembled into an intact pilus, the organism is pilus negative, but is able to revert to pilus positive

as soon as a gene copy that can be assembled is recombined into the expression locus. Thus, the recombinational mechanism accounts for both on–off variation and for antigenic variation. This is obviously a remarkably complex genetic mechanism for varying this protein and the only conceivable evolutionary pressure to force the development of this system is, of course, the human immune system. The volunteer studies have demonstrated that the antigenic variation does occur in *vivo* and that in fact almost all of the reisolates from the infected volunteers had a different pilus type than that expressed by the infecting strain. Recent crystallographic studies have shown that pili consist of a helical aggregate of five pilin subunits per turn and that, remarkably, the exposed surface of the pilus cylinder consists of the variable domains, whereas the constant regions of the pilin molecule are buried in the cylinder.

Pili are associated with a protein of about 110,000 MW, named PilC, which is involved in the assembly of pili and also appears to be present at the tip of the pili and imparts the ability to adhere to epithelial cells. Evidence has been provided that the host antigen recognized by PilC is CD46, a widely distributed cell-surface protein that acts a complement regulatory factor. This antigen also serves as the receptor for measles virus. Meningococci also bear pili that are very similar to gonococcal pili with a similar ability for antigenic variation.

B. Capsules

Meningococci are classified into serogroups on the basis of the chemical nature of the capsular polysaccharide they express (see Table II). Epidemiologically, groups A, B, and C are the most important because they are the cause of over 90% of cases of meningitis and meningococcemia. Meningococcal disease occurs worldwide as an endemic disease principally in infants 3 months or older and in young children with a rate of two to five cases per 100,000 population. The incidence is seasonal, with winter and spring having most cases. However, meningococcal disease can also occur in epidemic form where the rate of disease can rise as high as 200–500 cases per 100,000. During epidemics, the peak incidence shifts to an older age group, children 5–7 years of age. During the first half of the twentieth century, epidemics caused by group A meningococci occurred in the United States about every 12 years. Since World War II, there has not been a major epidemic in the United States, but epidemics have occurred in many other parts of the world, notably Brazil, China, Finland, Russia, and Mongolia. However, the area of the world most severely affected is Africa in the meningitis belt that extends through all of the sub-Saharan countries from the Sahel to the rain-forest region. In this region, major epidemics affecting tens of thousands of inhabitants have occurred every 3–4 years, and the problem has worsened recently, so that in the winter season of 1995–1996 there were 250,000 cases in west Africa.

The presence of the capsular polysaccharides protects the organism from the natural defense of the host, such as phagocytosis by white cells and killing mediated by complement via the alternative pathway. However, the presence of antibodies to the capsular polysaccharides is protective and these antigens are the basis of the meningococcal vaccines to be discussed later. The locus for the biosynthesis of capsular polysaccharide has been characterized and is encompassed by about 25 kb of the genome. The right and left sides of the locus are conserved among serogroups and are concerned with common biosynthetic steps, such as the addition of lipid carriers and the export of the product from the cytoplasm to the exterior. The middle portion of the locus differs between serogroups and contains the enzymes responsible for the biosynthesis of the activated sugar intermediates for the particular serogroup, as well as the specific polymerase assembling the polysaccharide.

C. Outer-Membrane Proteins

The outer membrane of the pathogenic *Neisseria*, as is that of other gram-negative bacteria, consists of a lipid bilayer with the outer leaflet consisting principally of lipopolysaccharide (LPS). The outer membrane contains a number of integral membrane proteins, of which the porins are quantitatively predominant. The nomenclature of the neisserial outer-

---

TABLE II
Chemical Structure of Meningococcal Capsular Polysaccharides

Serogroup	Repeating unit in capsular polysaccharide
A	→ 6)-α-N-acetyl mannosamine-1-phosphate; O-acetyl C3
B	→ 8)-α-N-acetyl neuraminic acid-(2 →
C	→ 9)-α-N-acetyl neuraminic acid-(2 →; O-acetyl C7 or C8
X	→ 4)-α-N-acetyl glucosamine-1-phosphate
Y	→ 6)-α-glucose-(1 → 4)-N-acetyl neuraminic acid-(2 →; contains O-acetyl groups
Z	→ 3)-α-N-acetyl galactosamine (1 → 1)-glycerol-3-phosphate
29E	→ 3)-α-N-acetyl galactosamine (1 → 7)-β-KDO (2 →; O-acetyl C4 or C5 of KDO
W-135	→ 6)-α-galactose-(1 → 4)-α-N-acetyl neuraminic acid-(2 →

membrane proteins has evolved with the increasing knowledge of these proteins and is summarized in Table III.

1. Porins

The porins of the pathogenic *Neisseria*, as are those of *Escherichia coli*, are postulated to consist principally of β-pleated sheets arranged perpendicularly to the membrane, with loops exposed to the cytoplasm and eight loops exposed on the surface of the organism. Each functional porin consists of a trimer of the porin subunit. The meningococcus contains two genetic loci that code for the production of outer membrane porins, originally referred to as the class 1 and class 2/3 proteins and now called *porA* and *porB*. The gonococcus, although lacking a homolog of the first porin, possesses the *porB* locus that gives rise to porins that are very similar in amino acid sequence to the class 2/3 proteins. The gonococcal porins vary antigenically to a limited extent primarily in the surface exposed loops and fall into two main classes, referred to as PIA and PIB. PIA strains predominate among gonococci isolated from the bloodstream and apparently have an increased capacity to invade the bloodstream and cause disseminated gonococcal disease. PIA strains also tend to be resistant to the bactericidal action of normal human serum. Strains that cause ascending infection of fallopian tubes are invariably of the PIB type.

There are a number of indications that gonococcal porins not only serve as channels through which water and solutes of less than 1000 MW can diffuse through the outer membrane, but also play an active role in pathogenesis. Biophysical studies in artificial lipid membranes indicate that the gonococcal porins are unusual among gram-negative porins in that they are somewhat anion selective and very voltage sensitive. Voltage sensitivity means that when the protein is in a membrane, the channel will be modulated by the potential across that membrane, such that at low membrane potentials the porin molecules will be open and as the voltage is raised, the probability that the porin molecule is closed increases. In addition, it has been shown that the porins are able to bind GTP and certain other phosphate compounds and that this binding favors the closing of the porin channel.

The neisserial porins readily transfer from the outer membrane of living gonococci to foreign membranes, including human cells. It obviously becomes very interesting to ask what the functional consequences of a newly inserted voltage-dependent ion channel in the host-cell membrane may be. This has been studied in detail with human polymorphonuclear leukocytes (PMN) using purified gonococcal porin. Within seconds after the addition of porin to the leukocytes, the membrane potential of these cells becomes hyperpolarized due to chloride ion movement. Shortly thereafter, the membrane potential returns to baseline, presumably because the porin channels adjust to this hyperpolarization by closing, and the active ion pumps of these cells reestablish their baseline potential. Even though the initial effects on the cell's membrane potential by porin addition are short-lived, the consequences of a foreign voltage-regulated channel in the membrane of these cells are much longer lasting. This is seen when these cells are subsequently exposed to a stimulus such as fMLP (formyl methionyl-leucyl-proline). Normally, fMLP causes an immediate depolarization of the membrane. However, with porin channels present in the membrane, this depolarization is replaced with a prolonged hyperpolarization. Porin also markedly inhibits the aggregation of PMN. Degranulation in response to fMLP, LTB4, or the complement component C5a is also blocked, but is normal when induced with PMA. However, there is no inhibition of superoxide generation in response to these signals.

2. Rmp

All strains of gonococci produce an outer-membrane protein originally designated PIII. This protein migrates on a SDS-PAGE with an apparent 31,000 MW unreduced, and with an apparent 32,000 MW when exposed to reducing agents such as β-mercaptoethanol. Hence, the protein has been named *reduction modifiable protein* (Rmp). In contrast to other outer-membrane proteins, Rmp is a highly preserved antigen showing little if any variation among strains. The sequence of Rmp has substantial homology with OmpA, a protein that is universally present in all enterobacterial species. Rmp is also present in meningococci, where it was originally named class 4 protein and it is almost identical to the Rmp of gonococcus.

It has been found that complement-fixing IgG antibodies to Rmp are present in the sera of at least 15% of normal human beings with no history of prior gonococcal infection. These antibodies arise in response to the meningococcal carrier state and also by contact with the enterobacterial flora. Surprisingly, these antibodies do not mediate serum killing or opsonization of gonococci, but instead block the ability of normally bactericidal antibodies directed to other surface antigens to exert their function. These results have been substantiated with monoclonal antibodies. It was found that an anti-Rmp antibody was a powerful blocking antibody, inhibiting the activity of other bactericidal monoclonal antibodies directed to a number of different surface proteins or LPS (lipopolysaccharides). In an epidemiologic study of a population at very high risk for acquiring sexually transmitted diseases, it has been demonstrated that the presence of anti-Rmp antibodies significantly increases the risk of gonococcal infection, demonstrating the inhibitory role of blocking antibodies in the local mucosal infection. The blocking activity of anti-Rmp antibodies is not seen with meningococci, perhaps because this organism expresses quantitatively less Rmp. The molecular mechanism by which anti-Rmp antibodies act as blocking antibodies is not understood.

3. Opa Proteins

Pathogenic *Neisseria* express another surface-exposed outer-membrane protein that in the meningococcus was called class 5 protein and in the gonococcus protein II. The presence of this class of protein leads to distinctive changes in the morphology of the colonies on agar. Gonococci that do not express protein II give rise to colonies that are transparent and resemble a beaded water droplet, whereas gonococci expressing this antigen give rise to colonies that are opaque and have a ground-glass appearance. Hence, they have been named opacity or Opa proteins. The expression of Opa can turn off and on at high frequency, and a clone of gonococcus can express at least five or six recognizably different opacity proteins over a period of time. Gonococci freshly isolated from the blood of patients with disseminated gonococcal infection do not express Opas. The same is true of isolates from pelvic inflammatory

TABLE III
Outer-Membrane Proteins of Pathogenic *Neisseria*

Meningococcal proteins	Gonococcal proteins	Genetic designation
Class 1	No homolog	*porA*
Class 2	Protein I, PI, PIB	*porB*
Class 3	Protein I, PI, PIA	
Class 4	PIII, rmp	*rmp*
Class 5	PII, opa	*opaA–opaJ*, or *opa$_{xx}$–opa$_{yy}$*
Opc	No homolog	*opc*

disease. Strains from males with genitourinary disease usually express Opa protein. Most remarkably, in young women not on birth control pills, the gonococci that can be isolated from the cervix vary so that at the time of ovulation the isolates express Opas, although at the time of menses they do not. So, as with pili, there is phase and antigenic variation both *in vitro* and *in vivo*. However, the mechanism is entirely different than that seen with pili. Gonococci possess 11 copies of complete *opa* genes in the gonococcal genome, and all of them have a variable number of pentameric repeats of the sequence CTCTT between the ATG initiation codon and the remainder of the protein. This repeat codes for leucine, serine, and phenylalanine, amino acids that are normally contained in the hydrophobic portion of signal sequences. The number of repeats is subject to rapid change due to slipped-strand mispairing during replication. The consequence is that with some number of repeats the beginning of the gene will be in frame with the remainder of the gene, and with a different number of repeats it will be out of frame. Thus, the expression of this class of proteins is controlled at the level of protein translation. The same pertains to meningococci, although they possess only three *opa* genes. This mechanism of genetic variation has also more recently been described in a number of other mucosal pathogens, notably *Haemophilus influenzae* and *Helicobacter pylori*.

The different *opa* genes of a few strains have been sequenced and have been distinguished either by naming them *opaA–opaJ* or by adding a numerical subscript (see Table III). The loci code for mature proteins about 250 amino acids long. The genes are highly homologous, except for two regions that are very variable and a smaller region that has lesser variation. The differences among the proteins in the content of basic amino acids is noteworthy and the pI of the proteins range from about 7.0–10.0.

Since the discovery of this class of proteins, it has been noted that they increased the adhesiveness of gonococci to epithelial cells in tissue culture or to human PMN, and it was inferred that they mediated the close attachment phase in the fallopian tube model. The ligand specificity of the Opa proteins has been defined on a molecular level. It has been shown that a particular Opa protein (OpaA protein of strain MS11) recognizes heparan sulfate on the surface of epithelial cells and that heparin is able to inhibit the binding. The heparan sulfate occurs mainly on the syndecan class of molecules, of which four have been described and two of these (1 and 4) are expressed on epithelial cells. The syndecans are believed to act as receptors or coreceptors for interaction between cells and the extracellular matrix.

The other Opa proteins react with several proteins that are members of the carcinoembryonic antigen CEA family. CEA was originally described as a colon cancer associated antigen and tests for blood levels of CEA antigen are used to clinically monitor the progression of colon cancer. There are now about 20 related proteins known; their genes are clustered on human chromosome 19, they belong to the immunoglobulin (Ig) superfamily, and they have a N-terminal domain homologous to immunoglobulin variable (IgV) domain and a variable number of domains with homology to Ig constant regions. Some are transmembrane proteins with cytoplasmic tails, whereas CEA is GPI-linked. The proteins are heavily glycosylated. Several of the genes are subject to alternative splicing and various family members are expressed on a wide variety of cells, including epithelial cells. The Opa proteins of both the gonococcus and the meningococcus react with the IgV domain of the molecules irrespective of its state of glycosylation.

D. LPS

As do other gram-negative bacteria, the pathogenic *Neisseria* carry lipopolysaccharide (LPS) in the exter-

nal leaflet of their outer membranes. In contrast to the high-molecular-weight LPS molecules with repeating O-chains seen in many enteric bacteria, the LPS of *Neisseria* is of modest size and therefore is often referred to as lipooligosaccharide or LOS. Although the molecular size of the LPS is similar to that seen in rough LPS mutants of *Salmonella* ssp., this substance has considerable antigenic diversity. In the case of the meningococcus, a serological-typing scheme has been developed that separates strains into 12 immunotypes and the detailed structure of the majority of these has been determined. The LPS of the pathogenic *Neisseria* is heterogeneous and LPS preparations frequently contain several closely spaced bands by SDS-PAGE. Using monoclonal antibodies, it is evident that gonococci are able to change the serological characteristics of the LPS they express and that this antigenic variation occurs at a frequency of 10^{-2} to 10^{-3}, indicating that some genetic mechanism must exist to achieve these high frequency variations.

The structure of the largest fully characterized gonococcal LPS molecule is shown in Fig. 2. To the lipid A are linked two units of keto-deoxyoctulosonic acid (KDO) and two heptoses (HEP). This inner core region as shown in Fig. 2 can carry three oligosaccharide extensions that have been named the α-, β- and γ-chains. The γ-chain consisting of N-acetyl glucosamine (GlcNAc) appears to be always present. The β-chain, when present, consists of a lactosyl group; when it is absent, the position is substituted with ethanolamine phosphate. The α-chain in its full form consists of the pentasaccharide shown in Fig. 2. An alternative α-chain structure consisting of a trisaccharide is also shown. However, as indicated in Table IV the sugar composition of the α-chain can vary and in every instance it is identical to human cell-surface oligosaccharides, most often part of the glycosphingolipids that in some instances are the determinants of blood-group antigens.

Gonococci possess a very unusual sialyl transferase activity, which *in vitro* is able to use exogenously supplied cytosine monophosphate-NANA (CMP-NANA) and add N-acetyl neuraminic acid to the LPS if the organism is expressing the lacto-N-neotetraose α-chain (see Table IV). In the human infection *in vivo*, the concentration of CMP-NANA found in vari-

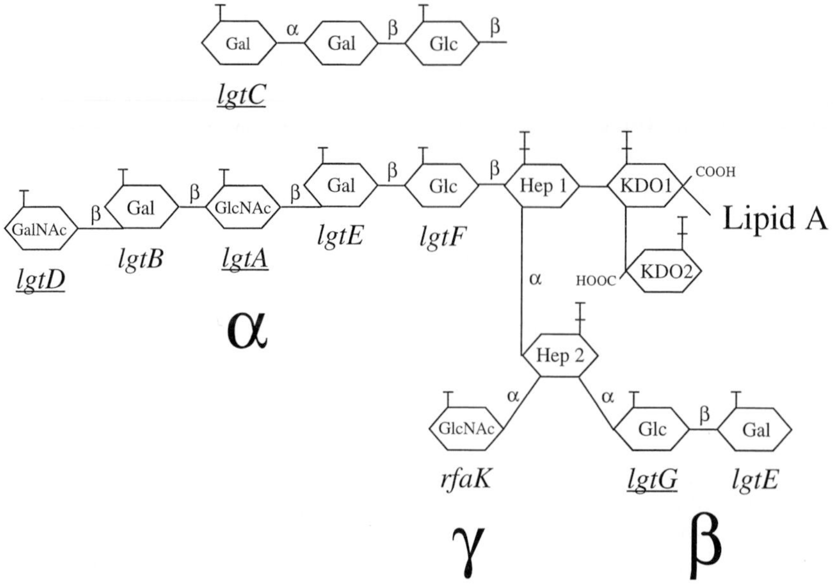

Fig. 2. Genetics of gonococcal LPS synthesis. The LPS contains a lipid A portion with two residues of keto-deoxyoctulosonic acid (KDO) and linked to these are two residues of heptose (HEP) to form the inner core. This structure can bear three additional extensions, indicated as the α-, β-, and γ-chains. The largest structurally characterized α-chain is indicated in the figure and consists of glucose (Glc), galactose (Gal), N-acetyl glucosamine (GlcNAc), Gal, and N-acetyl galactosamine (GalNAc). An alternative α-chain has been characterized and is the trisaccharide shown at the top of the figure. The glycosyl-transferases responsible for the addition of each of the sugars are indicated by their genetic designation. The genes that are underlined are subject to high frequency variation. If the organism grows *in vivo* or *in vitro* in medium supplemented with cytosine monophosphate-N-acetyl neuraminic acid (CMP-NANA), part or all of the terminal GalNAc is replaced by NANA.

ous host environments is sufficient to support this reaction. The sialylation of the LPS causes gonococci to become resistant to the antibody complement-dependent bactericidal effect of serum. The resistance is to the bactericidal effect mediated by not only antibodies to LPS, but also to other surface antigens as well. Group B and C meningococci have the capacity to synthesize CMP-NANA as the precursor of their capsule biosynthesis and frequently sialylate their LPS without requiring exogenous CMP-NANA.

In the late 1990s, most of the glycosyl transferases responsible for the biosynthesis of gonococcal and meningococcal LPS have been identified, and they are shown in Fig. 2. This has provided an understanding of the genetic mechanism that underlies the high frequency variation in the LPS structures expressed by these organisms. Note that four of these genes (*lgtA*, *lgtC*, *lgtD*, and *lgtG*) are underlined to indicate that they contain in their coding frames homopolymeric tracts of nucleotides. In the case of *lgtA*, *lgtC*, and *lgtD*, these are stretches of 8–17 deoxyguanosines (poly-G) that can vary in size due to errors resulting from slipped-strand mispairing during replication. In *lgtG*, there is homopolymeric poly-C tract. When the number of bases in the tracts is such that the coding frames are not disrupted, the respective glycosyl transferases are produced, but, if the number changes, premature termination occurs and no functional enzyme is produced. Thus, the presence of the β-chain depends on whether functional LgtG glucosyl transferase is produced. Simi-

TABLE IV
Molecular Mimicry by Gonococcal LPS

Human antigen mimicked	α-chain oligosaccharide
Lactosyl ceramide	Galβ1 → 4Glcβ1 → 4-R
Globoside, p^k blood group antigen	Galα1 → 4Galβ1 → 4Glcβ1 → 4-R
Lacto-*N*-neotetraose, paragloboside	Galβ1 → 4GlcNAcβ1 → 3Galβ1 → 4Glcβ1 → 4-R
Gangliosides, X$_2$ blood group antigen	GalNAcβ1 → 3Galβ1 → 4GlcNAcβ1 → 3Galβ1 → 4Glcβ1 → 4-R
Sialyl-gangliosides	NANAα2 → 3Galβ1 → 4GlcNAcβ1 → 3Galβ1 → 4Glcβ1 → 4-R

larly in the instance of the α-chain synthesis, if *lgtA* is on, then the lacto-*N*-neotetraose chain will be formed and whether the terminal GalNAc is added depends on whether *lgtD* is on or off. If *lgtA* is off, then the globoside structure is synthesized if *lgtC* is on, and only the lactosyl structure is synthesized if *lgtC* is off. The gonococcus and the meningococcus have evolved a very elegant system to shift readily between a large number of different LPS structures, all of them mimics of human glycolipids. This ability to shift the expression among a number of different LPS structures is not peculiar to the pathogenic *Neisseria*, but also occurs in *Haemophilus influenzae* in which at least four genes are subject to phase variation. In this organism, the mechanism is also by slipped-strand mispairing, but occurs in repeated tetrameric sequences that can be either CAAT or GCAA. Thus, it is likely that LPS antigenic variation is important because it is an attribute of a number of mucosal pathogens.

How does this molecular mimicry, listed in Table IV, benefit the organism? It has been proposed that the human host may find it difficult to produce antibodies to any of these structures and that the ability to change to a different one may compound this problem. Although immune evasion is attractive as an idea, it is clear that the LOS does serve as a target for bactericidal antibodies, and, at least *in vitro*, perhaps the majority of bactericidal antibodies are directed to this antigen, rather than to other surface structures. *In vivo* this is, of course, very different because the sialylation of the LOS very effectively inhibits the bactericidal reaction and interferes with phagocytosis as well. However, only the lacto-*N*-neotetraose structure is effectively sialylated to produce the serum-resistant phenotype. Why does the organism then have a genetic mechanism to alter away from this structure? Perhaps the answer lies in the observation that sialylation of the LOS interferes with invasion of epithelial cells *in vitro*. There is also evidence that sialylated gonococci are significantly less infectious when used to challenge volunteers. It is clear that the gonococcus can circumvent LOS sialylation, either by the addition of the terminal *N*-acetyl galactosamine or by the truncation of the chain. It is also possible that the mimicry may benefit the organism by allowing it to be recognized by human carbohydrate-binding molecules such as the C-lectins, the S-lectins, and the sialoadhesins.

IV. NATURAL IMMUNITY

A. Bactericidal Antibody

In the case of the meningococcus, there is clear evidence that the major predisposing factor for bloodstream invasion is the lack of biologically active antibodies to surface components and resultant failure to mediate an antibody–complement bacteriolytic reaction. This was first demonstrated in 1969 by Goldschneider and his colleagues by using two lines of evidence. The first is based on a study done in an adult population. Nearly 15,000 sera were collected from military recruits within the first week of training and stored in anticipation that a number of these would develop meningococcal meningitis during the 8-week basic training. In fact, 60 cases occurred in this cohort and in 54 of these the *Neisseria meningitidis* causing the infection could be isolated. Each of these sera, as well as 10 sera obtained from unaffected recruits serving in the same training platoons, were tested for bactericidal activity against the strain of *Neisseria meningitidis* isolated from the patient. Only 5.6% of the patients' sera were able to kill the disease causing *Neisseria meningitidis,* whereas 82% of sera obtained from unaffected recruits demonstrated bactericidal activity. The second line of evidence is the demonstration that there is an inverse relationship between the incidence of meningococcal disease and the prevalence of bactericidal antibody, and age. The disease is very rare during the first 3 months of life when maternally derived antibodies are still present. Incidence rises to a peak during between 6–12 months of age, when the nadir of bactericidal activity is seen. Thereafter, the incidence progressively diminishes as the prevalence of antibodies rises with age. This is the same relationship that was reported for *Haemophilus influenzae* meningitis by Fothergill and Wright in 1933. Finally, it is evident that the antibody–complement-dependent bactericidal reaction is clearly important in protection against neisserial systemic infections because deficiencies of late complement components (C6 or C8) impart a specific susceptibility to blood-borne neisserial infections, but not to other bacterial infections.

Is there natural immunity to gonococcal infection? It is established that individuals with no known immunological defective can acquire gonorrhea multiple times. In some instances, it has been documented that a single untreated consort may represent the source of the repeated infections. Thus, it has been suggested that there is no such thing as natural immunity to this disease. However, there is another side to the coin and that is the clear evidence that gonococcal infection before the days of antibiotic therapy was as a rule a self-limited disease lasting for a few weeks. This spontaneous elimination of the infection applied not only to the genitourinary disease, but also to disseminated gonococcal infection, to gonococcal arthritis (albeit with bad sequelae), and even in some instances to gonococcal endocarditis. Hence, there is ample evidence that after a period of time gonococci are killed effectively *in vivo.* In the face of this ability to self-cure, how can we explain the apparent lack of natural immunity? The most likely explanation is that gonococci are inherently so antigenically variable that the im-

mune system requires considerable time to catch up with the repertoire of the gonococcus and eliminate the infection.

V. PREVENTION

Since the beginning of the twentieth century, attempts have been made to prepare vaccines for the prevention of meningococcal disease. Vaccines based on whole-cell preparations proved to be ineffective. In the late 1960s and 1970s, methods were developed to purify the capsular polysaccharides of the meningococcus in a form that maintained their high molecular weight. It was shown that injection of school-age children and adults with 25–50 μg of group A, C, Y, or W-135 polysaccharide resulted in a strong and long-lasting antibody response and that *in vitro* these antibodies were opsonic and bactericidal. Large-scale field trials both in the United States and overseas demonstrated that both group A and group C polysaccharide vaccines were highly effective in preventing the disease and that the protection lasted for at least 2 years. These vaccines were introduced in the U.S. military over 20 years ago and have essentially eliminated the problem of meningococcal disease among recruits. Vaccination is employed in the military of many other countries and is required for Muslim pilgrims participating in the Hadj.

As a general rule, the immune response to purified polysaccharides is age-related, but the response varies with the antigen. Thus, responses to the group C antigen are very low at ages younger than 18 months. Children between 2–4 years do respond to the group C antigen, but the response is short-lived, lasting only a few months. After age 6, the responses are similar to adults. As the experience with the *Haemophilus influenzae* vaccine has demonstrated, the immune responses in this age group can be markedly improved by covalently linking the polysaccharide to a protein carrier to enhance T-cell help in the immune response. Conjugate group C vaccines are being tested.

The response to group A antigen among young children is unusually favorable. Infants who are vaccinated twice, at 3 months and again at 6 months of age will show a brisk booster immune response to the second injection that is sufficient to provide protection. This booster response has not been seen with any other polysaccharide antigen. It has been demonstrated that a protective level of group A antibodies can be maintained by immunization twice in the first year of life, then again at age 2 and upon entry to school. Unfortunately this property of the group antigen has not been taken advantage of in prevention of epidemic disease in Africa.

The group B capsular polysaccharide is a homopolymer of $\alpha(2-8)$-linked N-acetyl neuraminic acid (see Table II). This structure is present in mammalian tissues, notably on the neural cell-adhesion molecule (N-CAM), and the degree of sialylation is particularly elevated during embryonic life. Although the majority of adults have some level of antibodies to this antigen, the injection of the purified antigen generally does not raise additional antibodies. There has also been concern that engendering a strong immune response to this antigen may have deleterious effects on infants during fetal life. Therefore, group B meningococcal vaccines based on partially purified outer membranes with their LPS content reduced by detergent extraction have been prepared and have proved to be able to prevent disease under epidemic conditions. However, as noted before, there is considerable antigenic heterogeneity in meningococcal outer-membrane proteins, and a broadly effective vaccine group B is not available.

No vaccine exists for the prevention of gonorrhea, and the problem is formidable because of the extraordinary antigenic variability of this organism. Nevertheless, the experience in several European countries has demonstrated that prevention of this disease can be very effective if public education is combined with rapid treatment of infected individuals and their contacts.

VI. SUMMARY

The discrete steps that occur in the mucosal infection by the pathogenic *Neisseria* are increasingly be-

ing explained on molecular and cell biological level. It is evident that the gonococcus has developed very elaborate mechanisms to evade the immune response of human beings. With pili it has chosen the path of antigenic variation. This is an evasion mechanism that is highly developed in eukaryotic parasites such as trypanosomes, and is also seen in prokaryotes such as *Borrelia*. In the case of Rmp, the gonococcus has chosen the path of antigenic constancy as a target for blocking antibodies. With Opa proteins, the variation may be more a way to succeed in various environments in the host, rather than being an immune evasion. The biological significance of LOS variation is not yet clear, but it must be very useful because *Neisseria* and *Haemophilus influenzae* have developed, in principle, the same variation mechanism, although the specific details are quite different. In the era before ready treatment with antibiotics, self-cure of gonorrhea over a period of weeks was commonly seen, and this slow acquisition of natural immunity was probably a reflection of the time needed for the immune response to finally catch up with the variability of the particular strain infecting the human host.

See Also the Following Articles

ANTIGENIC VARIATION • FIMBRIAE, PILI • LIPOPOLYSACCHARIDES • OUTER MEMBRANE, GRAM-NEGATIVE BACTERIA • SEXUALLY TRANSMITTED DISEASES

Bibliography

Blake, M. S., and Wetzler, L. M. (1995). Vaccines for gonorrhea: Where are we on the curve? *Trends Microbiol.* **3**, 469–474.

Jennings, H. J. (1990). Capsular polysaccharides as vaccine candidates. *Curr. Top. Microbiol. Immunol.* **150**, 97–128.

Preston, A., Mandrell, R. E., Gibson, B. W., and Apicella, M. A. (1996). The lipooligosaccharides of pathogenic gram-negative bacteria. *Crit. Rev. Microbiol.* **22**, 139–180.

Swanson, J., Belland, R. J., and Hill, S. A. (1992). Neisserial surface variation: How and why? *Curr. Opin. Genet. Dev.* **2**, 805–811.

van Putten, J. P. M., and Duensing, T. D. (1997). Infection of mucosal epithelial cells by Neisseria gonorrhoeae. *Rev. Med. Microbiol.* **8**, 51–59.

Growth Kinetics, Bacterial

Allen G. Marr

University of California, Davis

I. Growth and Reproduction of Individuals
II. Populations
III. Unrestricted Growth
IV. Balanced Growth
V. The Growth Curve
VI. Crop Yield
VII. Effect of Nutrient Concentration on Growth Rate
VIII. Continuous Culture and the Chemostat
IX. Effect of Temperature on Growth Rate

GLOSSARY

binary fission   Division of a mother cell into two daughters or in some cases into a mother and daughter.

chemostat   A type of continuous culture. Fresh medium, with one nutrient limiting, is continuously added to and mixed with a bacterial culture; an equivalent volume of the culture is withdrawn. In time, the culture will reach a constant population that is thereafter maintained.

exponential growth   If a bacterial population is in an environment without restriction by nutrients or metabolic products, the number of bacteria increases as an exponential function of time. If so, the logarithm of the number increases as a linear function of time.

total and viable counts   Estimates of the number of bacterial cells (individuals) in a population. After suitable dilution, total counts are made by microscopic observation or electronically. Viable counts are made by determining the number of colonies formed on a suitable medium. In general, total counts exceed viable counts.

KINETICS OF GROWTH   means the rate at which the number of individuals in a population changes. The kinetics (or rate) of growth of bacteria has fascinated scientists for more than a century for two reasons. First, the rate of growth of bacteria can be much more rapid than that of other organisms; bacterial populations can double in a few minutes. Second, the kinetics of growth of bacterial populations is often exponential; that is, the rate of increase is that of true compound interest.

I. GROWTH AND REPRODUCTION OF INDIVIDUALS

The most common means of reproduction in bacteria is binary fission. A (mother) cell grows by increasing all of its cytoplasmic constituents, enlarging its wall and membrane, and doubling its DNA. Most bacteria enlarge by lengthening of the long axis. After replication the circular molecules of DNA are decatenated and by an unknown means move toward opposite poles. The cell then forms a septum and divides in half, producing identical sister cells. Gram-positive bacteria form a septem that splits in division; gram-negatives such as *Escherichia coli* divide by constrictive growth of the wall.

A few bacteria reproduce asymmetrically. Rather than two identical daughters, one can distinguish a mother from a daughter. An example is *Caulobacter crecentus*. The mother cell has a stalk extending from one pole and may be attached by that stalk to a surface. The cell lengthens, producing a daughter cell with a polar flagellum. After division the daughter swims chemotactically, loses its flagellum, and grows its own stalk. Mother and daughter cells differ not only in morphology but also in physiology.

II. POPULATIONS

Populations of microbes, either natural or in culture, consist of enormous numbers of individuals. A modest-sized culture can easily contain 10,000 million individuals—more than all *Homo sapiens* over all of time. Two approaches to the measurement of microbial populations are counting and estimation of biomass.

Counting may be used to measure populations either in pure culture or in nature. Prior to approximately 1940 almost all studies of the kinetics of growth were based on measurement by counting.

Total counts are determined microscopically using a ruled chamber of known depth or electronically. Electronic counting depends on the fact that cells have higher impedance than surrounding electrolyte. A measured volume of a sample is pumped through a small pore though which an electric current also passes. As a cell passes through the pore, the increase in impedance causes a pulse in voltage which is tallied. Alternatively, cells may be counted electronically based on detection of pulses of scattered or fluorescent light.

Viable counts are counts of the number of colonies that develop after a sample has been diluted and spread over the surface of a nutrient medium solidified with agar and contained in a petri dish.

Biomass in culture is usually estimated from turbidity. Liquid cultures are turbid because microbes have a higher refractive index than water and, thus, scatter light. One measures turbidity by the attenuation of light by bulk scattering of a sample of a liquid culture. Measuring turbidity is quick, easy, and accurate.

Biomass in natural environments is often estimated by determining the amount of a chemical constituent. Since living cells always have ATP, and since ATP can be measured with exquisite sensitivity by counting photons emitted by luciferase, ATP is often used as an estimate of biomass. Other methods include the uptake of radioactive nutrients or the use of microelectrodes.

III. UNRESTRICTED GROWTH

Imagine a small population, n, of bacteria in a medium which can support a much larger population. One could expect the rate of growth (the increase in n with time, t) to be proportional to n. We can express this mathematically as

$$dn/dt = kn \qquad (1)$$

where the parameter k is the specific growth rate. We can solve Eq. (1) by integration to give

$$n(t) = n(0)e^{kt} \qquad (2)$$

where the notation $n(t)$ is the value of n at any elapsed time t. Fig. 1 is a plot of Eq. (2) with $n(0)$ taken as 1000 and k as 0.5 hr^{-1}. Taking natural logarithms of Eq. (2) gives

$$\log_e n(t) = \log_e n(0) + kt \qquad (3)$$

According to Eq. (3), a plot of the logarithm of the number as a function of time (Fig. 2) should be a straight line with a slope of k. Experimental results are insignificantly different from Eq. (3). Linearity of such plots means that the population is growing exponentially (not logarithmically as some microbiologists state incorrectly). That microbial populations can grow exponentially was expected from theory and demonstrated experimentally more than a century ago.

Plots of the logarithm of number (or mass) against

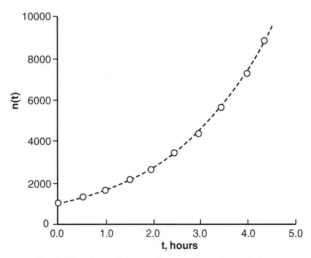

Fig. 1. Number of bacteria as a function of time.

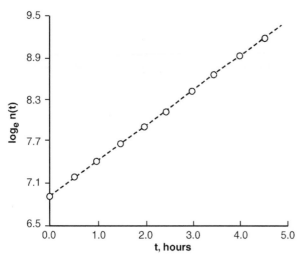

Fig. 2. Logarithm of number of bacteria as a function of time.

time are used to estimate values of k from experimental data. If natural logarithms are plotted as in Fig. 2, the slope is an estimate of k. Most microbiologists plot $\log_{10} n(t)$, in which case the specific growth rate is given by the slope divided by $\log_e 10$ or 2.303. Many microbiologists use a parameter equivalent to specific growth rate—the time required for the population to double. The relationship between specific growth rate and the doubling time, g, is obtained by substituting g for t in Eq. (2) to give

$$2n(0) = n(0)e^{kg} \qquad (4)$$

Cancelling the term $n(0)$ and taking logarithms gives

$$g = \log_e 2/k \qquad (5)$$

The value of $\log_e 2$ is 0.693. Thus, if the specific growth rate is 0.9 hr^{-1}, and the doubling time is 0.77 hr or 46.2 min.

The value of the specific growth rate at a given temperature in a given medium is characteristic of a particular microbe. For *E. coli* growing in basal salts medium at 37°C the value of k depends on the carbon source: acetate, 0.3 hr^{-1}; succinate, 0.4 hr^{-1}; and glucose, 0.9 hr^{-1}. In rich complex medium the value of k is 2.0 hr^{-1}. Cells of enteric bacteria such as *E. coli* and *Salmonella typhimurium* change in size and composition systemically with the specific growth rate of the population varied nutritionally. At low growth rates the cells are small and relatively poor in ribosomes; at high growth rates, as in complex medium, the cells are large and rich in ribosomes. This variation has been used to estimate the growth rate of populations in natural environments.

IV. BALANCED GROWTH

If the growth of a bacterial population is balanced, all extrinsic variables (such as protein and RNA per milliliter) increase in the same proportion and all intrinsic variables (such as mean cell volume) are constant. Unrestricted growth is ordinarily balanced. In principle, the values of intrinsic variables could oscillate, and, over short intervals of time, the increases in extrinsic variables could be disproportionate. Regulatory controls, particularly feed-back inhibition of enzymes, provide stability that results in balanced growth. Bacterial populations at low density cannot alter rapidly the concentrations of nutrients in the medium; this, too, contributes to stability and, thus, to balanced growth.

V. THE GROWTH CURVE

Unresricted growth of populations cannot proceed without limit. Ultimately, the population will either exhaust a nutrient or accumulate toxic metabolic products. The specific growth rate declines and finally growth ceases, a status called stationary phase. In stationary phase the cells of the population change, becoming smaller and more resistant to chemicals such as oxidizing agents and to physical conditions such as high temperature or ultraviolet light. The viable count declines, often quite slowly but in some cases rapidly. If, after some time in stationary phase, cells are transferred to fresh medium, growth does not begin immediately; the new culture is said to be in lag phase. After adjustment (mechanism usually not understood), the specific growth rate increases, ultimately attaining the value typical of unrestricted growth.

VI. CROP YIELD

As growth of a population proceeds, nutrients are used and metabolic products accumulate. The rate

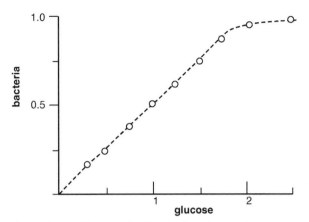

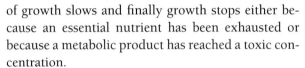

Fig. 3. Crop of bacterial cells as a function of the concentration of glucose as the crop-limiting nutrient.

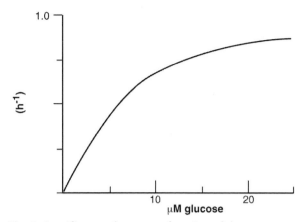

Fig. 4. Specific growth rate as a function of the concentration of glucose as the rate-limiting nutrient.

of growth slows and finally growth stops either because an essential nutrient has been exhausted or because a metabolic product has reached a toxic concentration.

Imagine an experiment in which we prepare a set of media and vary the concentration of one nutrient. Perhaps we make basal salts medium and vary the concentration of glucose. We inoculate each medium with a small number of bacterial cells, incubate until growth of the population has consumed the nutrient, and measure the final crop of bacteria by turbidity. Figure 3 shows the result expressed as milligrams dry weight of bacteria and milligrams glucose per liter. If we denote the final concentration of bacteria as x and the initial concentration of glucose as c, the results in the linear region are in accordance with the equation

$$x = Yc \qquad (6)$$

where Y is the yield coefficient for the organism, the nutrient, and the growth conditions. For aerobic growth of *E. coli* the value of Y is about 0.5 mg bacteria per milligram glucose; for anaerobic growth it is about 0.15 milligram.

VII. EFFECT OF NUTRIENT CONCENTRATION ON GROWTH RATE

Instead of measuring crop as a function of nutrient concentration, we measure the specific growth rate. We would find that over most of the range of concentration of, for example, glucose, the specific growth rate is almost invariant. Only at very low concentrations would we find an effect. This results from the fact that the phospho-transport system that pumps glucose into the cell works at near maximum efficiency if the concentration of glucose is higher than approximately 10 mM.

The result shown in Fig. 4 is in accordance with the following equation:

$$k(c) = k_{max}c/(K_S + c) \qquad (7)$$

where c is the concentration of a limiting nutrient (glucose in this case); $k(c)$ is the specific growth rate at that concentration; k_{max} is the specific growth rate approached at high concentration; and K_S is the concentration at which $k(c) = k_{max}/2$. Equation (7) is a classic description of the rate of an enzymatic reaction as a function of the concentration of the substrate of that enzyme. It is not precisely the relationship between specific growth rate and nutrient concentration, but it is often a close approximation.

VIII. CONTINUOUS CULTURE AND THE CHEMOSTAT

Imagine a microbial culture which is continuously diluted with fresh medium, well mixed, and maintained at a constant volume by overflow (Fig. 5). In time the population ordinarily will reach a constant concentration because the rate of growth will just

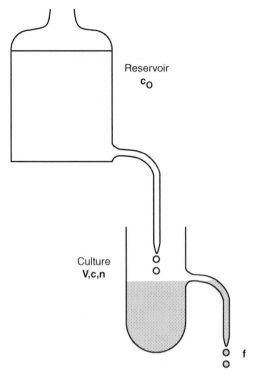

Fig. 5. Schematic representation of a continuous culture.

equal the rate of dilution. This condition is called a steady state. In a steady state the concentration not only of microbes but also of all nutrients and metabolic products will be constant with time. The volume of the culture is denoted as V; the flow rate as f; and the concentration of bacteria as n.

For a continuous culture in the steady state or not we may write

$$dn/dt = kn - (f/V)n \qquad (8)$$

Since in the steady state $dn/dt = 0$,

$$k = f/V \qquad (9)$$

If we increase or decrease f, k will increase or decrease accordingly. We can change growth rate literally by turning a tap. What is intrinsically controlling growth rate? If a nutrient is at a crop-limiting concentration c_0, the concentration of that nutrient in the culture will be reduced to the concentration c, low enough to control specific growth rate according to Eq. (7).

Although continuous cultures are often stable, they may be unstable. Continuous cultures are usually operated at high population density for industrial production of a useful products. Such cultures rapidly consume nutrients and may overwhelm the controls that otherwise provide stability. Continuous cultures with even dilute populations can be unstable, particularly if variations alter the yield coefficient.

IX. EFFECT OF TEMPERATURE ON GROWTH RATE

The specific growth rate of bacteria varies with temperature. Over much of the range of temperature compatible with growth, specific growth rate can be fit by an equation of the same form as the Arrhenius equation that describes the rate of simple chemical reactions as a function of temperature:

$$k = e^{-\mu/RT} \qquad (10)$$

where μ is the temperature characteristic, T is the absolute temperature, and R is the caloric gas constant. The form of the plot is similar for almost all bacteria. Over a certain interval of temperature, commonly called the normal range, the logarithm of the specific growth rate is linear with the reciprocal of absolute temperature in accordance with Eq. (10). At temperatures below or above the normal range, growth rate decreases approaching a vertical asymptote at both the minimum and the maximum temperatures for growth.

For most bacteria the normal range spans approximately 40°C. For *E. coli* the normal range extends from 21 to 37°C, and μ; is about 14 kcal/mol (ca. 59 kJ/mol). The maximum temperature at which balanced growth can be sustained is approximately 49°C; the minimum is between 7.5 and 7.8°C. The normal range varies among the species of bacteria: *E. coli* is representative of species called mesophiles; species with a lower range are called psychrophiles; and those with a higher range are called thermophiles.

Unlike nutritional variation of specific growth rate, thermal variation within the normal range results in no differences in the size of cells or their cellular composition.

See Also the Following Articles

Bibliography

Duclaux, E. (1898). "Traite de Microbiologie," Vol. 1. Mason, Paris.

Herbert, D., Elsworth, R., and Telling, R. C. (1956). The continuous couture of bacteria; A theoretical and experimental study. *J. Gen. Microb.* 14: 601.

Monod, J. (1942). "Reserches sur la Croissance des Cultures Bacteriennes." Herman et Cie, Paris.

Neidhardt, F. C., Ingraham, J. L., and Schaechter, M. (1990). "Physiology of the Bacterial Cell: A Molecular Approach: Sinauer, Sunderland, MA. (See Chapters 7 and 8).

Rahn, O. (1932). Physiology of Bacteria." Blakeston's, Philadelphia.

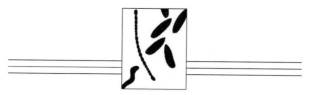

Haemophilus influenzae, Genetics

Rosemary J. Redfield

University of British Columbia

GLOSSARY

capsule A polysaccharide coating external to the outer membrane.

lipooligosaccharide A lipopolysaccharide that lacks the O-specific side chain present in enteric bacteria.

nontypable Lacking in a capsule, and therefore unreactive with standard antisera.

natural competence The genetically specified ability of bacterial cells to take up DNA fragments.

paralogs Genes which began diverging from each other after an ancestral gene duplication (as distinguished from orthologs, which began diverging after an ancestral speciation).

phase variation High-frequency genetic variation resulting in different modifications of the cell surface.

transformation A change in genotype caused by recombination with DNA brought in by a competent cell.

uptake signal sequence (USS) The DNA sequence AAGTGCGGT or its complement, which causes fragments containing it to be preferentially taken up by competent cells.

HAEMOPHILUS INFLUENZAE is a small gram-negative rod that occurs primarily as a commensal in the mucosa of the human upper respiratory tract. It is also an opportunistic pathogen, causing meningitis in infants and small children, otitis media in children, and pneumonia in the elderly and immunocompromised. Its genetic systems have long been studied because of its natural competence, which facilitates chromosomal manipulations. In 1995, its genome was the first to be completely sequenced, and the resulting databases of genes and pathways are invaluable tools for genetic research.

This article considers both the genetic relationships and the naturally occurring genetic processes of bacteria assigned to *H. influenzae*, and the genetic tools and techniques used to investigate these processes, concentrating on the ways in which *H. influenzae* differs from better-known model bacteria, especially *E. coli*.

I. PHYLOGENETIC RELATIONSHIPS

The family Pasteurellaceae is well established as a sister branch to the Enterobacteriaceae, in the gamma subdivision of the proteobacteria. Bacteria in the Pasteurellaceae are assigned to three genera, *Haemophilus*, *Pasteurella*, and *Actinobacillus*; however, this classification has not proved to be very informative about actual evolutionary relationships. Dewhirst, using ribosomal RNA sequences, found that different *Pasteurella* species occurring in a single type of host were more closely related than their genus assignments and metabolic capabilities would suggest. For example, *Haemophilus paragallinarum* was much closer to *Pasteurella gallinarum* than to *H. influenzae*.

Isolates are assigned to *H. influenzae* if they require both heme and NAD. If such cells have a capsule (polysaccharide coating), its antigenic properties are used to assign the strain to one of the six *H. influenzae* serotypes, a–f. However, serotype may not be a good guide to overall genetic relatedness, as the gene cluster specifying the capsular type is mobile (see following). Many isolates lack a capsule altogether; these "nontypable" strains are a distinct evolutionary branch, not sporadic derivatives of capsulated strains. The term "rough" (R) is usually applied to capsule-free strains derived in the laboratory from capsulate strains, because their colonies lack the glossy sheen of "smooth" capsulate colonies (e.g., strain Rd is a nonencapsulated derivative of a serotype d strain). The nontypable isolates formerly called *H. aegyptius* have been recognized as falling within the cluster of strains assigned to *H. influenzae* and are now assigned to *H. influenzae* biogroup aegyptius.

Most molecular and genetic investigations have used strain Rd, a rough derivative of a type d strain isolated by Alexander and Leidy in the mid-1940s. Until recently, most clinically oriented research was done on serotype b strains, which are responsible for most cases of infant meningitis. With the successful development of a type b-specific vaccine, the focus of clinical research has moved to other serotypes and to nontypable strains, which cause a broad range of respiratory infections in healthy and immunocompromised hosts.

II. GENETIC ELEMENTS NATIVE TO *H. INFLUENZAE*

A. Chromosome

Because the *H. influenzae* Rd genome was completely sequenced in 1995 by TIGR (The Institute for Genomic Research), it is the major source of our knowledge of *H. influenzae*'s genetic structure and capabilities. The Rd genome is a single 1,830,137 bp circle, with a base composition of 38% G + C. Apart from its small size (approximately 40% that of *E. coli*), the basic features of the *H. influenzae* genome are similar to those of the enteric bacteria. There is a single origin of replication and six rRNA operons transcribed divergently from the origin.

H. influenzae strains show substantial differences in their genome organization. Although the DNA sequences of different alleles are very similar (96–98% identity), their locations on the chromosome may vary, as do restriction patterns on pulsed-field electrophoresis.

B. Genes and Gene Families

The genetic complement of *H. influenzae* Rd has been extensively analyzed since the publication of the complete genome sequence. Most of the resulting databases are available on the Internet (a detailed URL-ography is provided below). Genes were first identified in the completed sequence as open reading frames with appropriate patterns of codon usage. These have been checked for similarity to other sequences in the databases, giving about 85% of the 1743 Rd genes which are clearly related to genes known from other organisms. Most of these similarities are to genes of known function, allowing unambiguous functional prediction for the *H. influenzae* homolog; the rest are to "hypothetical proteins," genes whose function is unknown (usually, genes also identified by genome-sequencing projects). Many of the approximately 275 *H. influenzae* genes with no clear database match have recognizable motifs that allow classification into broad functional categories such as "permease" or "nucleotide binding."

The genome contains relatively few gene families and functional superclusters, only about one third as many as *E. coli*. For example, there are only 82 permeases (201 in *E. coli*) and only 49 DNA-binding proteins (145 in *E. coli*). However, this provides only a partial explanation for its small genome size and gene complement, as there are multiple paralogs of genes involved in pathogenicity, including seven transferrin-binding protein genes.

C. Repeated and Mobile Sequences

A single 9 bp repeated sequence is grossly overrepresented in the *H. influenzae* Rd genome; the sequenced strand contains 1465 copies of this sequence, AAGTGCGGT or its complement, whereas only 8 copies would be expected for a random sequence of the same base composition. This sequence has been named the uptake signal sequence (USS) because DNA fragments containing it are preferentially taken up by the natural competence system described below. The genome also contains 764 copies of sequences differing from the USS at one position. Taken together, perfect and imperfect USS comprise more than 1% of the genome; adding to these the conserved bases flanking most USSs (corrected for the degree of conservation) brings the total to about 2%, a substantial fraction of the genome. The abundance of USS sequences in the genome is thought to be related to their uptake by competent cells, either because their preferential uptake causes them to accumulate in the genome or because they help cells take up homologous DNA that may contain beneficial new alleles.

The only other substantial repeats in the *H. influenzae* Rd genome are the 8 bp recombination-promoting Chi$_{Hi}$ sequences, which occur about once per 5 kb, similar to the frequency of Chi$_{Ec}$ in *E. coli*. Unlike Chi$_{Ec}$, Chi$_{Hi}$ is partially degenerate and shows less evidence of an orientation bias. Apart from the Mu-like phage and the transposable capsulation complex mentioned below, no transposons or other mobile elements have been found in *H. influenzae*.

D. Native Plasmids and Phages

Between 10 and 40% of clinical isolates carry plasmids conferring resistance to beta-lactam antibiotics. Resistance to other antibiotics is less common and usually mediated by chromosomal genes. Many strains harbor cryptic plasmids.

The only well-studied *H. influenzae* phage is HP1, which is related to the P2 family of temperate phages and which has been thoroughly investigated and completely sequenced by Scocca's group. This temperate phage integrates into the chromosome—the high frequency of USS in its genome suggests that it has either been subject to the same forces that cause USS abundance in the chromosome (i.e., has spent much of its evolutionary history as a prophage) or that it exploits the USS to gain access to host cells by the competence pathway of DNA uptake.

Analysis of the Rd genome sequence revealed a possibly defective copy of a 30kb prophage similar in sequence and organization to the *E. coli* transposable phage Mu. This phage is thought to be one previously identified as induced by DNA damage and by transformation with heterologous DNAs, and responsible for extensive cell death and release of phage particles after UV irradiation. However, because no sensitive host has been identified, the biology of the phage has not been studied. Several other phages have been identified over the years, but little is known about them.

III. PROCESSES THAT GENERATE GENETIC VARIATION

Spontaneous mutation rates in *H. influenzae* are typical of gram-negative bacteria. The standard pathways of mutation prevention are also present: mismatch repair, recombinational repair, and the *mutT* pathway (prevention of mutations caused by oxidized guanosine). However, *H. influenzae* Rd lacks photoreactivation repair of photodimers and the UV-inducible mutagenesis genes *umuDC*, making it both quite sensitive to and not mutable by UV irradiation.

Tandem oligonucleotide repeat sequences undergo frequent slippage errors in DNA replication, resulting in insertion or deletion of single repeats. Such repeats are often found at the 5′ ends of genes specifying cell-surface components, where it is thought that their rapid mutation allows the cell to evade the host immune response. In *H. influenzae*, tetrameric repeats are found in several genes required for lipo-oligosaccharide production (*lic1*, *lic2*, *lic3*, *lgtC*) and in four homologs of the *Neisseria* hemoglobin receptor. Similar repeats are also found in a defective adhesin gene (*yadA*), in a restriction methyltransferase, and in two genes of unknown function. A triplet repeat occurs in the promoter of the heme-utilization locus *hxuC*, where changes are likely to affect the efficiency of transcription initiation. A TA repeat occurs in the promoter region of a fimbriation gene cluster. This cluster is absent from Rd but present in many other *H. influenzae* strains, and flanking direct repeats suggest that it may be horizontally transferred between strains and between species.

H. influenzae has an efficient natural competence pathway which takes up free DNA from the cell's environment. If this DNA has homology to chromosomal sequences, it may replace its chromosomal homolog by recombination. However, there is little evidence that the genetic exchange permitted by transformation actually randomizes gene combinations between different *H. influenzae* lineages. Rather, *H. influenzae* isolates show a clonal population structure, possibly because most of the available *H. influenzae* DNA comes from sibling cells in singly infected hosts. Even when homologous DNA is taken up, most fails to recombine with the chromosome, being instead degraded intracellularly and providing nucleotides that are rapidly reused for DNA synthesis. The regulation of competence and the mechanism of DNA uptake are discussed in Section IV.

The competence system might be expected to have transferred genes from a vertebrate host into an ancestral *H. influenzae* genome, especially because respiratory tract mucus contains abundant host DNA. Two candidate genes were found in the genome sequence, both homologs of vertebrate neurotransmitter-type amino-acid transporters. However, many bacterial homologs of these genes have now been revealed in the genome sequences of other bacteria, and no other candidates have been identified.

Little is known about gene transfer by plasmid conjugation or phage transduction in *H. influenzae*. Both are likely to occur. However, because *H. influenzae* lacks the resident insertion sequences that facilitate transfer of *E. coli* chromosomal genes to conjugative plasmids, the frequency may be lower than in *E. coli*.

IV. GENETIC REGULATORY MECHANISMS

Although the *H. influenzae* Rd genome encodes most of the regulatory factors that have been well characterized in enteric bacteria, few of their mechanisms have been directly investigated. Where operon structures resemble those in *E. coli* and other enterics, it may be reasonable to begin by assuming that regulation is also similar. The cAMP receptor protein CRP (also known as CAP) is present, recognizing a similar sequence to its *E. coli* counterpart and regulating both genes involved in sugar uptake and catabolism and the competence pathway. Adenylate cyclase is regulated by a simpler version of the phospho-

transferase (PTS) sugar-uptake system of enteric bacteria, apparently in response to changes in the availability of fructose, the only sugar taken up by the *H. influenzae* PTS. The RNA polymerase factors sigma-70, sigma-32, and sigma-24 are present, but there is no sigma-54 factor. Homologs of five *E. coli* two-component systems are present; these are known to not regulate competence, but other functions have not been examined. The *sxy* (*tfoX*) gene product is required for transcription of genes involved incompetence but has not been shown to directly regulate transcription.

V. GENETIC TOOLS FOR RESEARCHERS

In 1991, Barcak and coworkers described the genetic tools and methodology available for *H. influenzae*; here these tools are summarized and information added which has subsequently become available.

A. Cloning

Plasmids containing inserts of *H. influenzae* DNA are usually well tolerated by *E. coli*. Although the standard *E. coli* plasmids (e.g., the pUC and pGEM series) do not replicate in *H. influenzae*, plasmids of the pSU series provide convenient restriction sites, a readily selectable chloramphenicol-resistance marker, and moderately high copy numbers in both *H. influenzae* and *E. coli*. Useful *H. influenzae*-derived shuttle vectors are the pRSF0885 derivatives such as pHK (kanamycin resistant) and pGJB103 (ampicillin resistant).

B. Mutagenesis

Despite the availability of the genome sequence, random mutagenesis remains an essential tool for dissection of genetic pathways. Point mutations are easily induced using EMS. Antibiotic resistance cassettes and mini-transposons are both very useful for creating gene knockouts, because fragments containing them can be efficiently introduced into the chromosome by transformation (see following). A number of cassettes are available: Kan (Tn*903*-derived), Spc, Cm (Tn9-derived), Tet. Transposon-insertion mutations can be created by transposition into *H. influenzae* fragments cloned in *E. coli* (the mini Tn*10* series developed by Kleckner's laboratory work very well) and by *in vitro* transposition with Tn7. The conjugative transposon Tn*916* can be introduced into *H. influenzae*, either by conjugation from an *E. coli* plasmid or by introduction of a Tn*916*-carrying plasmid into *H. influenzae*. It usually produces a single insertion, which remains stably inserted at its original location, but its strong site specificity makes it unsuitable when random insertions are required.

C. Selectable Markers

Nutritional markers (e.g., amino-acid auxotrophy, or inability to use a specific sugar) are not very useful in *H. influenzae* because of its relatively complex nutritional requirements (defined media are troublesome and often unreliable). However, many antibiotic resistance markers are available, and some can be conveniently introduced at any desired location by cassette mutagenesis. For measuring transformation frequencies, the most useful selectable markers are chromosomal point mutations conferring antibiotic resistance to novobiocin, kanomycin, streptomycin, spectinomycin (MAP7, a strain containing these mutations, is a convenient source of transforming DNA). Resistance to antibiotics other than novobiocin and kanamycin requires 60 minutes or more preincubation; to achieve this, cells may either be plated in agar and later overlaid with antibiotic agar or incubated in broth before plating. Counterselectable markers, such as *sacB*, have not been tested in *H. influenzae*.

D. Transformation

The major advantage *H. influenzae* offers for genetic analysis is its natural competence, which greatly facilitates introduction and recombination of chromosomal markers (see Fig. 1). Depending on the level of competence needed, cells may be simply grown to late-exponential stage in rich medium (brain–heart infusion broth supplemented with hemin and NAD) or transferred to the "MIV" competence-maximizing starvation medium, which lacks sugars, nucleotide precursors, and some amino acids.

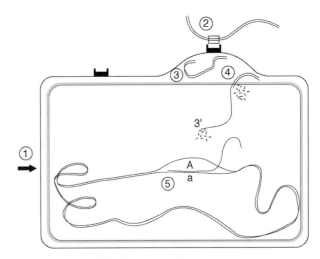

1. Induction of competence:
   cya = adenylate cyclase
   crp = catabolite regulatory protein
   sxy = competence activator
   *top*A = topoisomerase 1
   Other genes unknown

2. Sequence-specific DNA binding:
   Genes unknown
   Recognizes AAGTGCGGT

3. Initial uptake:
   Genes unknown

4. DNA translocation:
   rec–2
   *com*101A
   dprA
   Other genes unknown

5. Recombination by strand displacement:
   rec–1 = *rec* Ahomolog
   Other genes unknown

Fig. 1 Competence and transformation in *H. influenzae*. The genes known to be involved in each numbered step are indicated.

Transformation frequencies (i.e., number of transformed cells/total number of cells) with saturating chromosomal DNA are about 10^{-4} and 10^{-2}, respectively.

DNA uptake proceeds by two stages. DNA fragments containing the above-mentioned USS repeats are preferentially bound and taken up. DNA is first taken up into a poorly defined "compartment," where it is inaccessible to both external nucleases and intracellular restriction enzymes, and then translocated into the cytoplasm. One DNA strand undergoes concomitant or subsequent degradation, and the other may recombine with the chromosome, if homology permits. Large insertions and deletions do not prevent this recombination, provided there is about 1 kb of flanking homology on each side.

Transformation frequencies using short cloned DNA fragments may be higher than those obtained with chromosomal DNA, because all of the DNA fragments carry the marker, or may be lower if the fragments are short or carry large insertions. Transformation with high concentrations of cloned fragments also can give rise to artefactual gene duplications, apparently because DNA fragments often undergo ligation if they are taken up simultaneously. This can be minimized by using the lowest DNA concentration that will give the desired transformants. However, the genotypes of critical transformants should always be checked by Southern blotting or PCR.

Because the cell's uptake machinery requires DNA with a free end, plasmids can normally only be taken up by naturally competent cells if they contain nicks or breaks, which must be repaired within the cell to give an intact plasmid. If the plasmid contains an insert of *H. influenzae* DNA, repair usually is by recombination with the chromosomal homolog, which may alter the plasmid or chromosomal genotype. However, addition of 32% glycerol bypasses the need for a free end and reduces, but does not eliminate, recombination with the chromosome. Plasmids may be readily introduced into cells that have been made artificially competent with CaCl₂ (using the standard *E. coli* protocols) or by electroporation. The efficiencies of these procedures are roughly comparable.

E. Restriction Systems

The efficient *Hind*III and *Hind*II (*Hinc*II) restriction systems present in the Rd strain limit some genetic manipulations. These enzymes do not interfere with transformation using the normal pathway, because the incoming DNA is single-stranded and, therefore, immune to cleavage. However, restriction can reduce the efficiency of transformation by *E. coli*-grown plasmids, and *H. influenzae* Rd genomic DNA cannot be cleaved with these enzymes. Genes specifying restriction-modification systems appear to be somewhat mobile, and enzymes with different specificities have been isolated from other *H. influenzae*

strains (e.g., *Hae*I, *Hae*II, and *Hae*III from *H. influenzae* biogroup aegyptius and *Hinf*1 from *H. influenzae* f).

See Also the Following Articles

Antigenic Variation • Chromosome, Bacterial • DNA Restriction and Modification • Natural Selection, Bacterial

Bibliography

Barcak, G. J., Chandler, M. S., Redfield, R. J., and Tomb, J.-F. (1991). Genetic systems in *Haemophilus influenzae*. *Meth. Enzymol.* **204**, 321–342.

Dewhirst, F. E., Paster, B. J., Olsen, I., and Fraser, G. J. (1992). Phylogeny of 54 representative strains of species in the family *Pasteurellaceae* as determined by comparison of 16S rRNA sequences. *J. Bacteriol.* **174**, 2002–2013.

Esposito, D., Fitzmaurice, W. P., Benjamin, R. C., Goodman, S. D., Waldman, A. S., and Scocca, J. J. (1996). The complete nucleotide sequence of bacteriophage HP1 DNA. *Nucleic Acids Res.* **24**, 2360–2368.

Fleischmann, R. D., Adams, M. D., White, O., Clayton, R. A., Kirkness, E. F., Kerlavage, A. R., Bult, C. J., Tomb, J.-F., Dougherty, B. A., Merrick, J. M., McKenney, K., Sutton, G., FitzHugh, W., Fields, C., Gocayne, J. D., Scott, J., Shirley, R., Liu, L.-I., Glodek, A., Kelley, J. M., Weidman, J. F., Phillips, C. A., Spriggs, T., Hedblom, E., Cotton, M. D., Utterback, T. R., Hanna, M. C., Nguyen, D. T., Saudek, D. M., Brandon, R. C., Fine, L. D., Fritchman, J. L., Fuhrmann, J. L., Geoghagen, N. S. M., Gnehm, C., McDonald, L. A., Small, K. V., Fraser, C. M., Smith, H. O., and Venter, J. C. (1995). Whole-genome random sequencing and assembly of *Haemophilus influenzae* Rd. *Science* **269**, 496–512.

Hood, D. W., Deadman, M. E., Jennings, M. P., Bisercic, M., Fleischmann, R. D., Venter, J. C., and Moxon, E. R. (1996). DNA repeats identify novel virulence genes in *Haemophilus influenzae*. *Proc. Natl. Acad. Sci. USA* **93**, 11121–11125.

Macfadyen, L. P., and Redfield, R. J. (1996). Life in mucus: Sugar metabolism in *Haemophilus influenzae*. *Res. Microbiol.* **147**, 541–551.

Smith, H. O., Tomb, J.-F., Dougherty, B. A., Fleischmann, R. D., and Venter, J. C. (1995). Frequency and distribution of DNA uptake signal sequences in the *Haemophilus influenzae* Rd genome. *Science* **269**, 538–540.

Tatusov, R. L., Mushegian, A. R., Bork, P., Brown, N. P., Hayes, W. S., Borodovsky, M., Rudd, K. E., and Koonin, E. V. (1996). Metabolism and evolution of *Haemophilus influenzae* deduced from a whole genome comparison with *Escherichia coli*. *Curr. Biol.* **6**, 279–291.

Internet Resources:

http://www.tigr.org/tdb/mdb/hidb/hidb.html (Complete genome sequence at TIGR)

http://www.ncbi.nlm.nih.gov/cgi-bin/Entrez/framik?db= Genome&gi=25 (NCBI's Entrez Genomes interface)

Koonin and co-workers have developed an *H. influenzae* genome database with many attractive features at *http:// www.ncbi.nlm.nih.gov/Complete_Genomes/Hin.* (This otherwise very useful database uses an idiosyncratic numbering system for its nucleotide positions and gene numbers, which creates many traps for the unwary.)

http://www.genome.ad.jp/kegg (Kyoto Encyclopedia of Genes and Genomes [KEGG] metabolic pathway database)

http://pedant.mips.biochem.mpg.de/hi/ (PEDANT allows the *H. influenzae* genome to be searched for protein domains and motifs)

Heat Stress

Christophe Herman and Carol A. Gross

University of California, San Francisco

GLOSSARY

chaperone A protein that helps other proteins fold correctly by interacting preferentially with nonnative states of the protein.

protease An enzyme that hydrolyzes peptide bonds between amino acids of a protein, thus leading to the degradation of that protein.

sigma factor The subunit of transcriptase (RNA polymerase) that directs RNA polymerase to the promoter region of DNA, so that transcription originates from the appropriate point on DNA.

THE BACTERIAL HEAT STRESS RESPONSE refers to the mechanism by which bacteria adapt to a sudden increase in the ambient temperature of growth. The precise components of the signal–transduction system that senses and responds to heat stress vary among bacteria. Even within a single bacterial species, several different mechanisms are used to respond to heat stress. However, the logic of the response is universal. During the induction phase, the cell senses increased temperature and produces a signal that activates a transcription factor to increase the transcription of a group of genes. Accumulation of unfolded proteins contributes to signal generation, and a majority of the proteins produced during this stress response help to restore the normal folding state of the cell. During the adaptation phase, as the folding state of the cell returns to normal, the signal is damped down and transcription of the heat shock genes declines. Thus, in general, the heat stress response is self-limiting. Only when organisms are switched to lethal temperatures does the response continue unabated for as long as the cells are able to synthesize protein. In the present article, we first consider the general inputs and outputs to emphasize the universal logic of the bacterial heat shock response. We then describe several specific responses in detail, emphasizing how different components are used to execute this logic.

I. INPUTS TO THE HEAT STRESS RESPONSE

What is the thermometer that allows the cell to sense even small changes in temperature? Unfolded or partially folded proteins are, by far, the best characterized inducers of the heat shock response, and they must comprise part of the cellular thermometer. Normally, the low levels of unfolded or partially folded proteins result either from newly synthesized proteins or those maintained in a partially folded state prior to transport across membranes. Upon heat shock, some fully folded proteins partially or completely denature, increasing the pool of unfolded proteins and the need for the cellular proteins that maintain folding state. This is a matter of crucial concern for the cell. Partially folded proteins may not simply lose their activity; they may, in fact, cause toxic outcomes for the cell through a variety of mechanisms. It is, therefore, a top cellular priority to decrease the extent of protein unfolding whenever it occurs. Additional stresses, such as alcohol, also increase protein unfolding, thereby activating the same response. Other consequences of heat stress, besides the general increase in level of protein unfolding, are sensed by the cell. However, most of these inputs are currently either unknown or uncharacterized, with one exception. As we shall see later, in *E. coli*, the cell directly senses the folding state of a critical RNA molecule, and this provides a secondary thermometer to sense temperature.

How is the cellular thermometer constructed? Surprisingly, the nature of the primary thermometer sensing heat stress is not known in detail for any system. However, a variety of circumstantial evidence, described later, suggests that titration of the outputs of the response, the heat shock proteins themselves, may provide a thermometer of sufficient calibration to explain the response (Fig. 1). Some (or many) of the heat shock proteins have dual roles in the cell: they interact with unfolded protein substrates to promote folding and they interact with the heat stress transcription factor to regulate its activity. During the induction phase of the heat stress response, an increase in the cellular concentration of unfolded proteins increases the ratio of unfolded substrates to heat shock proteins. This could titrate them away from their "homeostatic" regulatory role vis-à-vis the heat stress transcription factor. As a consequence, the amount or activity of transcription factor will increase, resulting in an increased concentration of heat shock proteins. During the adaptation phase of the heat stress response, the ratio of unfolded protein substrates to heat shock proteins normalizes, the heat shock proteins resume negative regulation of the transcription factor, and the response is damped down (Fig. 1).

II. OUTPUTS OF THE HEAT STRESS RESPONSE

A. Chaperones

Chaperones, or proteins that help other proteins fold, are a major class of the proteins produced after

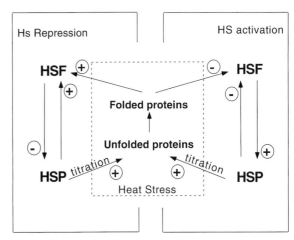

Fig. 1. Mechanism of heat shock regulation in bacteria. The heat shock response can be regulated either by a repressor (left) or an activator (right). In both cases, the activity of the heat shock transcription factor is controlled by a heat shock protein; this feedback control insures homeostasis between the amount of heat shock proteins and the activity of the heat shock factor during steady-state conditions. During heat stress, the level of free heat shock proteins decreases as they are titrated by unfolded proteins. This perturbs homeostasis, resulting in increased activity of the heat shock transcription factor and more heat shock proteins. Once the unfolded proteins are repaired or degraded, homeostasis will be restored.

heat stress. In the test-tube, under very dilute conditions, small proteins fold by themselves, demonstrating that the polypeptide chain itself encodes the information necessary for proper folding. However, in the cell, proteins are present at very high concentrations, and the nascent, unfolded protein has a very high potential to aggregate with other nascent chains via hydrophobic interactions, rather than proceeding on its folding pathway. By interacting with nascent and partially folded proteins, the molecular chaperones successfully thwart the tendency toward aggregation.

There are several different families of chaperones, most of which are highly conserved throughout evolution. One of the most prominent molecular chaperones is the Hsp70 family, called DnaK in bacteria, which work together with a co-chaperone Hsp40, called DnaJ in bacteria. The DnaKJ chaperone family is conserved in almost all organisms and members of each family are found in every compartment of the eukaryotic cell. Interestingly, in bacteria or in eukaryotic organelles, a third protein, GrpE, is necessary for this chaperone machine to function. GroEL, which cooperates with GroES, is a second major chaperone in bacterial cells and eukaryotic organelles, but is not found in the eukaryotic cytosol. The other major chaperones, Hsp90 and the small heat shock proteins, are conserved among most organisms; however, their role in the bacterial cell has not been completely elucidated.

Some chaperones are very large protein machines, consisting of multiple subunits of one or more proteins. For example, the GroELS chaperone machine has a central cavity consisting of two seven-membered rings of GroEL subunits. One or both ends is "capped" by a single seven-membered ring of the smaller GroES subunits. Proteins first bind to an exposed hydrophobic cavity in GroEL. Then, a combination of GroES binding and ATP hydrolysis drives conformational changes, which first expose hydrophilic residues and then drive release of the protein from the cavity. This allows at least the first critical steps of protein folding to take place within the GroEL cavity, in total isolation from other unfolded molecules in the bacterial cytoplasm.

By contrast, the DnaK chaperone machine consists of a single molecule of DnaK, which transiently interacts with its co-chaperone, a dimer of DnaJ molecules. For DnaK, its binding site for unfolded proteins is a small cleft located in the C-terminus of the molecule, which interacts preferentially with hydrophobic stretches in protein chains. The peptide-binding domain of DnaK is connected via a linker region to the N-terminal ATPase domain of the molecule. Interestingly, as was the case for the GroES co-chaperone, DnaJ both alters the rate of ATP hydrolysis and substrate release. GrpE works as a nucleotide release factor, promoting dissociation of whatever nucleotide is bound to DnaK. Despite its different construction, for both the DnaK and GroEL chaperone machines, an ATP-driven cycle of binding to and release from the chaperone underlies the action of the chaperone.

B. Proteases

Proteases, or proteins that degrade other proteins, are the second major class of proteins produced fol-

lowing heat stress. When proteins cannot be refolded by chaperone machines, the quality control system, composed of a variety of proteases, degrades the damaged proteins. This process serves two purposes. It releases the amino acids for reuse in making new proteins and eliminates proteins that cannot be repaired. Proteases use a variety of recognition systems, including exposure of hydrophobic C-terminal tails, internal hydrophobic stretches, and N-terminal signals.

Interestingly, many proteases are also large machines, composed of one of more types of subunit. The active sites of these proteases are contained within the central cavity, thus confining protein degradation to this cavity, where it is insulated from the cytoplasm. Such proteases also have either separate subunits or separate domains that carry out protein recognition. These portions of proteases can serve as chaperones when removed from the cleavage machinery of the protease. Many, but not all, proteases require ATP, which, most likely, drives the unfolding process.

C. Other Heat Shock Proteins

The two slowest steps in protein folding are isomerization around the cis-trans bond of proline and the making and breaking of disulfide bonds. Protein folding catalysts are proteins that speed up these slow steps in protein folding. Some of the peptidyl prolyl isomerases (catalyzing proline isomerization) and the disulfide bond proteins are heat shock proteins.

More systematic study of the various proteins induced by heat has indicated that a number of heat shock proteins do not fall into these simple categories. The function of many such proteins is still being elucidated.

III. THE *ESCHERICHIA COLI* CYTOPLASMIC HEAT SHOCK RESPONSE: REGULATION BY σ^{32}

A. Description of the *E. coli* Heat Shock Response

E. coli cells can rapidly sense both temperature upshifts and temperature downshifts. The induction phase begins within one minute of shift from 30°C to the higher growth temperature of 42°C, and the peak rate of synthesis of the 30 or more heat shock proteins is attained by 5 to 10 min after temperature upshift. At this point, the adaptation phase begins, and synthesis of the heat shock proteins declines until the new steady-state rate of synthesis is attained. At higher growth temperatures, both the maximal and steady-state rates of synthesis of the heat shock proteins are higher and the induction phase is prolonged. At temperatures so high that they are lethal to cells, the heat-shock response is maintained at its maximal point for as long as the cells can synthesize protein. Indeed, at these temperatures, synthesis of heat shock proteins constitutes the major protein synthetic activity of the cell.

In the converse temperature shift from the high growth temperature of 42°C to a lower growth temperature of 30°C, the synthesis of heat shock proteins begins to decline within 1 to 2 min, reaching a point of minimal synthesis at about 10 min after temperature downshift. Synthesis of heat shock proteins resumes slowly, reaching the rate normal for that temperature within 50 to 100 min after downshift. Note that this response on shift to lower growth temperature is distinct from the cold shock response. The cold shock response ensues when cells are shifted to very low temperatures (20°C or below) and involves the induction of a distinct set of proteins. The cold shock response will not be discussed here.

B. A Transcriptional Factor Regulates the *E. coli* Heat Shock Response

In bacterial cells, the sigma subunit of RNA polymerase directs the transcriptase to the promoter region of the gene. Most bacteria contain multiple sigmas. In addition to one or more housekeeping sigmas, responsible for the bulk of the transcription, cells have several alternative sigma factors, which allow them to respond to environmental or developmental signals. Each sigma factor recognizes a distinctive set of promoters in the cell. RNA polymerase is directed to the promoters of the heat shock genes by one such alternative sigma, called σ^{32}. During the heat shock response, the rate of transcription of the heat shock genes and, consequently, the rate of

synthesis of the heat shock proteins, changes in response to alterations in the amount and/or activity of σ^{32}.

Upon temperature upshift, the amount of σ^{32} increases rapidly, reaching its peak just prior to the peak rate of synthesis of the heat shock proteins, accounting for the induction phase of the response (Fig. 2). The amount of σ^{32} increases both because this normally very unstable molecule (usually degraded with a half-time of only 1 min) is transiently stabilized and because the rate of translation of σ^{32} transiently increases. During the adaptation phase, σ^{32} becomes unstable again and its rate of translation decreases, resulting in a decline in the amount of σ^{32} in the cell. Eventually, σ^{32} reaches a new steady-state level characteristic of the particular growth temperature. The adaptation phase may be sharpened by some control over the activity of σ^{32}. Thus, the response of the cell to temperature upshift is primarily governed by changes in the amount of σ^{32} (Fig. 2). In contrast, there is little change in the amount of σ^{32} upon temperature downshift. In this case, a dramatic decrease in the activity of σ^{32} accounts for the dramatic shutoff of transcription of the heat shock genes and the consequent decrease in heat shock protein synthesis.

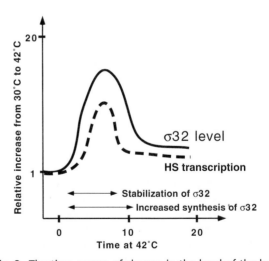

Fig. 2. The time course of change in the level of the heat shock transcription factor, σ^{32}, and the transcription of heat shock genes after a shift of temperature from 30°C to 42°C. The amount of σ^{32} increases both because translation of the protein increases and the protein itself becomes more stable.

C. Two (or More) Thermometers Control the *E. coli* Heat Shock Response

At least two different thermometers control expression of the heat shock proteins. One thermometer controls the translation of σ^{32} mRNA in a positive way: increased physical (environmental) stress leads to increased translation (Fig. 3). This pathway is induced by exposure to heat but not by accumulation of unfolded proteins. A second thermometer regulates the stability of σ^{32} itself (Fig. 3). This pathway is induced by accumulation of unfolded proteins, as well as by exposure to heat. Elements of this thermometer may also regulate the activity of σ^{32}.

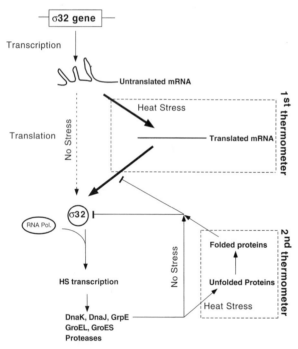

Fig. 3. Model for regulation of the σ^{32} heat shock factor in *Escherichia coli*. During steady-state conditions, the mRNA of σ^{32} is poorly translated. The few σ^{32} molecules translated are subject to rapid degradation and may also be negatively regulated by the DnaK chaperone machine. During heat shock, the structure of the mRNA changes, allowing increased σ^{32} synthesis. Unfolded proteins created during the shock titrate away the DnaK machine from σ^{32}, resulting in its stabilization and increased activity. When unfolded proteins are repaired or degraded, the free DnaK machine will again promote σ^{32} degradation and inactivation.

The nature of the thermometer that controls translation of σ^{32} during the induction phase has now been defined (Morita, 1999). The only player is the σ^{32} message itself. At low temperatures, the translation start site is occluded by base-pairing with other regions of the σ^{32} mRNA. Raising the temperature destabilizes this base-pairing, allowing increased translation of σ^{32}. Both mutational studies (which changed the base-pairing in the critical region) and chemical probing (which assessed base-pairing as a function of temperature) are consistent with this idea. Most importantly, by using an *in vitro* assay that determines accessibility of mRNA to translation by the ribosome, it has been possible to show that no components other than σ^{32} mRNA are necessary for this regulatory system (Fig. 3). During the adaptation phase, translation of σ^{32} mRNA declines. The DnaK chaperone machine has been involved in this translational repression, but the mechanism by which this chaperone operates is still unclear.

The nature of the thermometer that regulates the stability and possibly the activity of σ^{32} is more complex and less defined (Fig. 3). Since increased production of unfolded proteins in the cell stabilizes σ^{32}, unfolded proteins are at least part of the signal that calibrates this thermometer. We also know that the DnaK, DnaJ, and GrpE heat shock proteins are required for the instability of σ^{32} and its inactivation. In other words, the DnaK chaperone machine is negatively regulating both the stability and activity of σ^{32}. A homeostatic mechanism coupling the occupancy of the DnaK chaperone machine with unfolded proteins to the amount and activity of σ^{32} has been proposed. The key concept is that the thermometer is set by competition between σ^{32} and all other unfolded or misfolded proteins that bind to the DnaK chaperone machine. During the induction phase, the increased amounts of unfolded or misfolded proteins titrate the DnaK chaperone machine away from σ^{32}, relieving their negative regulatory effects on σ^{32} stability. As a consequence, σ^{32} is stabilized and its amount will rise. This response is self-limiting. During adaptation phase, overproduction of DnaK, DnaJ, and GrpE will restore the free pool of these chaperones to an appropriate level. This notion can be extended to account for inactivation of σ^{32} during temperature downshift. Here, a decreased amount of unfolded or misfolded proteins will allow σ^{32} to compete better for the DnaK, DnaJ, GrpE chaperone machine, with a consequence of inactivation of σ^{32}. In this model, the amount of free DnaK, DnaJ, and GrpE is a "cellular thermometer" that measures the "folding state" of the cell (Fig. 3). A key prediction of this model, that the amount of the DnaK chaperone machinery in the cell directly correlates with the activity of σ^{32}, has recently been verified. However, at present, this model is far from established, as many of the critical experiments to test the model have not been carried out.

D. Additional Mechanisms for the Cytoplasmic Heat Shock Response in *E. coli*

Although σ^{32} controls expression of many of the genes that respond to heat stress, induction of the σ^{32} regulon is not synonymous with the cellular response to temperature upshift. Another alternative sigma factor, σ^{s}, primarily responsible for transition into stationary phase, is also induced by heat. Induction of the σ^{s} regulon after exposure to high temperature is slower than induction of the σ^{32} regulon. The signal transduction system has not been studied in detail. However, like σ^{32}, the activity and amount of σ^{s} is controlled on multiple levels, including stability, translation, and, possibly, activity. In addition, one operon (*psp*) is controlled by a dedicated activator protein that promotes transcription by yet another sigma factor, σ^{54}, after shift to very high temperature. Finally, yet another sigma factor, σ^{E}, is responsive to heat stress generated in the periplasm. That response will be described in detail in Section V.

E. Conservation and Divergence of the σ^{32} Paradigm for the Cytoplasmic Heat Shock Response

Homologues of σ^{32} have been identified in diverse gram-negative bacterial species. Including *E. coli*, ten gamma proteobacterial species contain σ^{32}. In addition, σ^{32} has been found in seven alpha and one beta proteobacteria. With one exception, these bacterial species have only one gene encoding σ^{32}. The exceptional species, *Bradyrhizobium japonicum*, has three

genes encoding σ^{32} with distinct functional and regulatory features. Where it has been examined, the amount of σ^{32} increases with temperature, although the mechanism for accomplishing this is variable. The *E. coli* paradigm for regulating σ^{32} is most closely conserved among gamma bacteria. Here, when the temperature increases, generally, the stability and translation of σ^{32} increase as well. For some alpha bacteria, transcription of σ^{32} appears to increase with temperature. Regardless of the mechanisms for accomplishing this, the increased cellular concentration of σ^{32} gives the increased transcription of heat shock genes characteristic of the heat shock response. No homologues of σ^{32} have been identified in gram-positive bacteria.

IV. THE *BACILLUS SUBTILIS* HEAT SHOCK RESPONSE: THE CIRCE/HrcA CONTROLLED REGULON

A. The Heat Shock Response

The *dnaK* and *groE* operons of *B. subtilis* exhibit a heat shock response similar to the one exhibited by these operons in *E. coli*. Upon shift to high temperature, transcription of the *dnaK* and *groE* operons rapidly increases, leading to a rapid increase in the rate of synthesis of the seven heat shock proteins encoded by the *B. subtilis dnaK* operon (DnaK, DnaJ, GrpE, the HrcA repressor, and three other proteins) and the two heat shock proteins encoded by the *B. subtilis groE* operon (GroEL and GroES). This response reaches its peak about 5 to 10 min after shift to high temperature. Transcription of these operons then declines, reaching a new steady-state rate by 30 to 60 min after the high temperature shift. However, the components generating this response are completely different from those responsible for the heat shock in *E. coli*.

The *dnaK* and *groE* operons of *B. subtilis* are transcribed by its housekeeping sigma (SigA). The heat shock response of these operons is controlled by a negative regulatory system, which is composed of an operator site, called CIRCE, and its cognate repressor, called HrcA (Fig. 4). CIRCE is a nine base inverted repeat sequence separated by nine bases

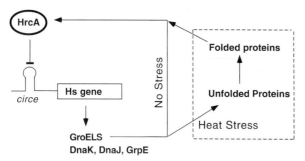

Fig. 4. Model of regulation of HrcA heat shock repressor in *Bacillus subtilis.* During steady-state conditions, the GroELS chaperone machine promotes folding of the heat shock repressor, HrcA, which then binds to the CIRCE element, located upstream of the heat shock genes, and inhibits their transcription. During a heat shock, unfolded proteins titrate away the GroELS chaperone machine, preventing it from helping HrcA to fold properly and function as a repressor. Therefore, the level of heat shock proteins increases until free GroELS can again promote proper folding of the HrcA repressor.

(TTAGCACTC-N9-GAGTGCTAA), which is located close to or overlapping the *dnaK* and *groE* promoters. The HrcA repressor (encoded in the *dnaK* operon) binds to the CIRCE element to shut off transcription of these operons. Upon temperature upshift, repression is temporarily relieved and synthesis of the CIRCE-controlled heat shock proteins increases. The major elements of this regulatory system have been verified by genetic analysis and some *in vitro* studies. Mutating conserved bases in CIRCE results in constitutive expression of the operon downstream of the mutated element, as expected for an operator. Likewise, mutating *hrcA* results in constitutive expression of both the *groE* and *dnaK* operons. Direct binding studies, demonstrating that HrcA binds to CIRCE, have been performed with variable success. HrcA is very prone to aggregation *in vitro*, limiting the ability to characterize this reaction extensively. As will be seen in the next section, the propensity of HrcA to aggregate is likely to be for its function.

B. The CIRCE/HrcA Thermometer

Heat shock regulation in this system requires that the HrcA repressor be transiently removed from its CIRCE operator upon shift to high temperature. The mechanism for accomplishing this constitutes the cellular thermometer. A priori, two solutions seem possible. First, HrcA itself may undergo a temperature-mediated transition. Second, achieving the native state of HrcA may require the intervention of some other molecule in the cell, which is itself responsive to temperature. A variety of evidence suggests that the latter solution has been chosen.

The current idea is that HrcA requires constant interaction with a chaperone to maintain its native state. The GroEL/S chaperone is considered to play this role and, as for the *E. coli* heat shock response, a homeostatic mechanism has been proposed (Fig. 4). The occupancy of the *B. subtilis* GroEL/S chaperone machine with unfolded proteins is suggested to be coupled to the fraction of HrcA in the native state. During the induction phase, the increased amounts of unfolded or misfolded proteins titrate the GroEL/S chaperone machine away from HrcA, allowing it to aggregate or otherwise misfold in the cell. As a consequence, it will be unable to bind to CIRCE and repress transcription. This response is self-limiting. During the adaptation phase, overproduction of GroEL/S will restore the free pool of these chaperones to an appropriate level, HrcA will be maintained in its native state and repression from the CIRCE operator will be resumed.

Two key predictions of this model have been verified qualitatively. First, this model implies that repression by HrcA will be responsive to the amount of GroEL/S in the cell and that is true. When GroEL/S is depleted from cells, the heat shock response is induced at low temperature. Conversely, overexpressing GroEL/S represses basal synthesis at low temperature and decreases the absolute amount of induction at high temperature. Second, this model implies that overexpression of known GroEL/S substrates, but not other proteins, will induce the heat shock response. This is also true. Finally, accumulating evidence suggests that GroEL/S and HrcA interact as required by this model. GroEL/S decreases aggregation of HrcA *in vitro* and promotes binding of HrcA to the CIRCE operator, which suggests that the two proteins interact directly, rather than through intermediate molecules. However, a great deal more work remains to be done to understand precisely how this thermometer works.

Two observations make it unlikely that HrcA itself senses temperature. First, in GroEL/S depleted cells, HrcA is inactive even at low temperature, indicating that the protein is not intrinsically native at low temperature. Second, when HrcA from a thermophilic organism (*B. stereothermophilis*) is used to reconstitute the heat shock response in *B. subtilis*, heat shock occurs at the low temperature characteristic of *B. subtilis*, rather than the high temperature characteristic of *B. stereothermophilis*. Thus, it is the cellular milieu, rather than the HrcA protein per se, that senses temperature.

C. Distribution of CIRCE/HrcA Regulation

The CIRCE/HrcA regulatory system is very widespread. It occurs in more than 40 different eubacteria, including gram-positive organisms, gram-negative organisms, cyanobacteria, and more distantly related eubacteria, such as chlamydia and spirochaeta. The particular genes regulated by the system vary, depending upon the organism. In gram-positive organisms with low G+C content examined thus far, CIRCE/HrcA regulates the *dnaK* and *groE* operons, but in those with high G+C content, and in α-proteobacteria, only *groE* is regulated. In addition, in some bacteria, several other genes have been reported to be regulated by this system. To add to the regulatory diversity, preliminary experiments suggest that in at least one gram-positive organism (*Lactococcus lactis*), DnaK regulates HrcA. Moreover, several cases where regulation by HrcA and σ^{32} coexist have been documented.

D. Additional Mechanisms for Response to Heat Stress in *B. subtilis*

Although understanding of the CIRCE/HrcA regulatory system is most advanced mechanistically, this system controls only nine of the many genes induced after shift to high temperature. A great many others are controlled by σ^B, an alternative sigma factor that carries out the general stress response and the starvation response in *B. subtilis*. Heat is one of the many stimuli that induce this system. σ^B is controlled by a phosphorylation cascade that affects the ability of an anti-sigma to bind to σ^B. This system is currently under intense study but, at present, there are few details about how heat stress alters this regulatory cascade. Additional heat shock proteins in *B. subtilis* are controlled by unknown mechanisms. However, recent studies in other organisms have identified several repressors in addition to HrcA that are responsive to heat stress (OrfY, HspR) and such repressors may be involved in regulating additional heat stress proteins in *B. subtilis* as well.

V. THE *ESCHERICHIA COLI* EXTRACYTOPLASMIC HEAT SHOCK RESPONSE: REGULATION BY σ^E

A. Description of the *E. coli* Extracytoplasmic Heat Shock Response

The hallmark of the gram-negative bacterial cell is the existence of two membrane layers, the inner or cytoplasmic membrane and the outer membrane, which, in turn, form the boundaries of two aqueous subcellular compartments, the cytoplasm and the periplasm. The conditions within each of these compartments differ markedly. The cytoplasm is an energy-rich, highly regulated reducing environment, in which basic cellular processes, such as transcription, DNA replication, and translation, are carried out. In contrast, the extracytoplasmic compartment is a relatively energy poor, oxidizing environment in which conditions vary with those of the external environment, due to the existence of pores in the outer membrane which allow the free exchange of small molecules and some specific substrates. Given the disparity in conditions between the two compartments, it is not surprising that the heat stress response in *E. coli* is compartmentalized. Exposure to high temperature activates both the cytoplasmic and extracytoplasmic responses, whereas conditions that specifically perturb protein folding in the periplasmic compartment activate only the extracytoplasmic response.

B. The Thermometer Controlling the Extracytoplasmic Heat Shock Response

The extracytoplasmic heat shock response is controlled by the alternative sigma factor σ^E, which is responsible for directing transcription of the >10 genes that comprise its regulon. Only some of these genes have been identified. These include the periplasmic protease DegP and the periplasmic peptidyl prolyl cis/trans isomerase, FkpA, which are involved, respectively, in protein degradation and folding. Interestingly, transcription of the heat shock factor σ^{32} is also under the control of σ^E, suggesting an interconnection between the two stress regulons. The induction phase of the σ^E heat stress response is slow, with maximum synthesis occurring about 20 min after shift to high temperature. The adaptation phase of this response has not been studied.

Studies on this response have advanced to the point where many of the players are known; however, the signaling mechanisms coupling the activity of σ^E to temperature are currently unknown. The central problem facing this regulatory system is to transduce a signal generated in the periplasm to the cytoplasm, where activity of σ^E is regulated. That problem has been solved by utilizing a protein chain consisting of two negative regulators. RseA, the major negative regulator of σ^E, is a membrane spanning protein whose cytoplasmic face acts as an anti-sigma factor. During steady-state growth conditions, RseA binds to σ^E, thereby preventing this sigma factor from binding to RNA polymerase. The periplasmic face of RseA binds the second negative regulator, RseB. RseB is located completely in the periplasm and is a weak negative regulator of σ^E.

The activity of σ^E is controlled by regulated proteolysis. Upon shift to high temperature, RseA is destabilized, relieving negative regulation of σ^E. At least one function of RseB is to enhance the stability of RseA. Various agents that specifically enhance unfolded proteins in the periplasmic compartment (lack of folding agents, overexpression of outer membrane proteins, interruption of lipopolysaccharide biosynthesis) all induce σ^E. How generation of unfolded proteins is coupled to destabilization of RseA is currently unknown. Moreover, additional yet to be discovered mechanisms regulate σ^E.

VI. SUMMARY AND PROSPECTS

Study of the heat stress response has led us to realize the extraordinary importance of controlling the state of protein folding in the cell. Organisms have evolved multiple, graded responses to cope with this problem. The thermometers that calibrate the response to the level of stress are composed of different materials. However, almost all of them are induced, at least in part, by unfolded proteins and respond by overexpressing a set of universally conserved heat shock proteins. Currently, many, but not all, of the types of responses to heat stress have been identified. However, on the most basic level, the way in which the thermometer controls the gradual response to temperature is not really understood for any system. Moreover, the interaction of the various heat responsive systems with each other has not yet been studied. Clearly, considerably more will be learned about this response in the next few years.

See Also the Following Articles

Protein Secretion • Temperature Control

Bibliography

Bukau, B., and Horwich, A. L. (1998). The Hsp70 and Hsp60 chaperone machines. *Cell* **92**(3), 351–366.

Gross, C. A. (1996). "Function and Regulation of the Heat Shock Proteins. Vol. 1, *Escherichia coli* and Salmonella: Cellular and Molecular Biology" (F. C. Neidhardt, ed.), pp. 1382–1399. American Society for Microbiology, Washington, DC.

Morita, M., *et al.* (1999). Translational induction of the heat shock transcription factor sigma 32: Evidence for a built-in RNA thermosensor. *Genes Devel.* **13**, 655–665.

Narberhaus, F. (1999). Negative regulation of bacterial heat shock genes. *Molec. Microbiol.* **31**, 1–8.

Yura, T., and Nakahigashi, K. (1999). Regulation of the heat-shock response. *Curr. Opin. Microbiol.* **2**(2), 153–158.

Heavy Metal Pollutants: Environmental and Biotechnological Aspects

Geoffrey M. Gadd

University of Dundee, Scotland

I. Environmental Aspects of Heavy Metal Pollution
II. Biotechnological Aspects of Heavy Metal Pollution

GLOSSARY

biosorption Removal of metal or metalloid species, compounds and particulates, radionuclides, and organometal (loid) compounds from solution by physicochemical interactions with microbial or other biological material.

desorption Nondestructive elution and recovery of, for example, metal or metalloid species, radionuclides, and organometal(loid) compounds from loaded biological material by physicochemical treatment(s).

heavy metals Ill-defined group of biologically essential and nonessential metallic elements, generally of density >5, exhibiting diverse physical, chemical, and biological properties with the potential to exert toxic effects on microorganisms and other life forms.

metal resistance Ability of a microorganism to survive toxic effects of heavy metal exposure by means of a detoxification mechanism, usually produced in response to the metal species concerned.

metal tolerance Ability of a microorganism to survive toxic effects of heavy metal exposure because of intrinsic properties and/or environmental modification of toxicity.

metallothioneins Low-molecular-weight cysteine-rich proteins capable of binding essential metals (e.g., Cu and Zn), as well as nonessential metals (e.g., Cd).

organometallic compound Compound containing at least one metal–carbon bond, often exhibiting enhanced microbial toxicity. When such compounds contain "metalloid" elements (e.g., Ge, As, Se, and Te), the term "organometalloid" may be used.

phytochelatins Metal-binding γ-Glu-Cys peptides of general formula $(\gamma\text{Glu-Cys})_n\text{Gly}$ (n generally 2–7). Now designated as class III metallothioneins which are atypical, nontranslationally synthesized metal thiolate polypeptides. Cd-binding γ-Glu-Cys peptides from some yeasts are also called cadystins.

siderophores Low-molecular-weight Fe^{3+} coordination compounds excreted by microorganisms which enable accumulation of iron from the external environment.

HEAVY METALS comprise an ill-defined group of over 60 metallic elements, of density greater than 5, with diverse physical, chemical, and biological properties but generally having the ability to exert toxic effects toward microorganisms. Many metals are essential for microbial growth and metabolism at low concentrations (e.g., Cu, Fe, Zn, Co, Mn), yet are toxic in excess, and both essential and nonessential metal ions may be accumulated by microbial cells by physicochemical and biological mechanisms. Thus "toxic metal" and "potentially toxic metal" are also useful general terms. In this article, the term "heavy metal" will be used in a broad context and discussion will include actinides and organometal(loid) compounds. All these substances have a common potential for microbial toxicity and bioaccumulation and are of environmental significance as pollutants or because of introduction as biocides and other substances.

I. ENVIRONMENTAL ASPECTS OF HEAVY METAL POLLUTION

A. Heavy Metals in the Environment

Although elevated levels of toxic heavy metals can occur in natural locations (e.g., volcanic soils, hot springs), average environmental abundances are generally low, with most of that immobilized in sediments and ores being biologically unavailable. However, anthropogenic activities have disrupted natural biogeochemical cycles, and there is increased atmospheric release, as well as deposition into aquatic and terrestrial environments. Major sources of pollution include fossil fuel combustion, mineral mining and processing, nuclear and other industrial effluents and sludges, brewery and distillery wastes, biocides, and preservatives including organometallic compounds. In fact, almost every industrial activity can lead to altered mobilization and distribution of heavy metals in the environment. Because of the fundamental microbial involvement in biogeochemical processes, as well as in plant and animal productivity and symbioses, toxic metal pollution can have significant short- and long-term effects and, ultimately, affect higher organisms, including humans, e.g., by accumulation and transfer through food chains.

B. Effects of Heavy Metals on Microbial Populations

For toxicity to occur, heavy metals must directly interact with microbial cells and/or indirectly affect growth and metabolism by interfering with, e.g., nutrient uptake, or by altering the physicochemical environment of the cell. A variety of nonspecific and specific mechanisms (e.g., biosorption and transport, respectively) determine entry of mobile metal species into cells and, if toxic thresholds are exceeded, cell death will result unless mechanisms for detoxification are possessed. The plethora of intracellular metal-binding ligands ensures that many toxic interactions are possible. Thus, practically every index of microbial activity can be adversely affected by toxic metal concentrations, including primary productivity, methanogenesis, nitrogen fixation, respiration, motility, biogeochemical cycling of C, N, S, P, and other elements, organic matter decomposition, and enzyme synthesis and activity in soils, sediments, and waters.

Despite this, many microorganisms from all major groups can survive in the presence of toxic metals and can be isolated from polluted habitats. However, general conclusions about heavy metal effects on natural populations are difficult to make because of the diversity in speciation, chemical behavior, and toxicity of a given metal, the interactions that occur between metal species and environmental components, and the morphological and physiological biodiversity encountered in microorganisms. Furthermore, environmental perturbations associated with industrial metal pollution (e.g., extremes of pH, salinity, and nutrient limitations) may also contribute to adverse effects on microbial communities. Nevertheless, it is commonly stated that heavy metals affect natural microbial populations by reducing numbers and species diversity and selecting for a "resistant" or "tolerant" population. Resistance and tolerance are arbitrarily defined, frequently interchangeable, and often based on whether particular strains and isolates can grow in the presence of selected heavy metal concentrations in laboratory media. It is probably more appropriate to use "resistance" to describe a direct mechanism resulting from heavy metal exposure, e.g., bacterial reduction of Hg^{2+} to Hg^0, and metallothionein synthesis by yeasts. "Tolerance" may rely on intrinsic biochemical and structural properties of the host, such as possession of impermeable cell walls, extracellular slime layers or polysaccharide, and metabolite excretion, as well as environmental modification of toxicity. However, distinctions are difficult, in many cases, because several direct and indirect mechanisms, both physicochemical and biological, can contribute to microbial survival. Thus, although heavy metal pollution can qualitatively and quantitatively affect microbial populations in the environment, it may be difficult to distinguish metal effects from those of environmental components, environmental influence on metal toxicity, and the nature of the microbial resistance/tolerance mechanisms involved.

C. Environmental Modification of Heavy Metal Toxicity

The physicochemical characteristics of a given environment determine metal speciation and, therefore, chemical and biological properties of heavy metals. Because major mechanisms of metal toxicity are a consequence of strong coordinating properties, a reduction in bioavailability may reduce toxicity and enhance microbial survival. Such parameters as pH, temperature, aeration, soluble and particulate organic matter, clay minerals, and salinity can all influence heavy metal speciation, mobility, and toxicity. Acidic conditions may increase metal availability, although H^+ may successfully compete with and reduce or prevent binding and transport. Environmental pH also affects metal complexation with organic components and inorganic anions (e.g., Cl^-). With increasing pH, there may be enhanced entry of metal cations into cells, as well as the formation of hydroxides, oxides, and carbonates of varying solubility and toxicity. Some hydroxylated species may associate more efficiently with microbial cells than corresponding metal cations. In other cases, a reduction in availability leads to a reduction in toxicity. A reduction in toxicity in the presence of elevated concentrations of anions such as Cl^-, CO_3^{2-}, S^{2-}, and PO_4^{3-} is frequently observed. Mono-, di-, and multivalent cations may affect heavy metal toxicity by competing with binding and transport sites, while clay minerals can bind metal cations and reduce bioavailability. Other synthetic and naturally produced soluble and particulate organic substances, including microbial metabolites, may influence toxicity by binding and complexation. The removal of heavy metal species by intact living and dead microbial biomass, by physicochemical and/or biochemical interactions, may also be significant in some locations. In more general terms, microbial growth and activity is influenced by environmental parameters, including the availability of organic and inorganic nutrients, and this can clearly affect responses to potentially toxic metals.

D. Mechanisms of Microbial Heavy Metal Detoxification

Extracellular metal complexation, precipitation, and crystallization can result in detoxification. Polysaccharides, organic acids, pigments, proteins, and other metabolites can all remove metal ions from solution and/or convert them to less toxic species. Iron-chelating siderophores may chelate other metals and radionuclides and possibly reduce toxic effects. The production of H_2S by microorganisms, e.g., by *Desulfovibrio* spp., results in the formation of insolu-

ble metal sulfides and also disproportionation of organometallics to volatile products as well as insoluble sulfides, e.g.,

$$2CH_3Hg^+ + H_2S \rightarrow (CH_3)_2Hg + HgS$$
$$2(CH_3)_3Pb^+ + H_2S \rightarrow (CH_3)_4Pb + (CH_3)_2PbS$$

Many other examples of metal crystallization and precipitation are known and mediated by processes dependent and independent of metabolism. Some of these are of great importance in biogeochemical cycles and involved in, for example, microfossil formation, iron and manganese deposition, silver and uranium mineralization, and the formation of stable calcareous minerals.

Decreased accumulation, sometimes the result of efflux, and impermeability may be important survival mechanisms. Impermeability may be a consequence of cell wall and/or membrane composition, lack of a transport mechanism, or increased turgor pressure. Some bacterial metal resistance mechanisms, e.g., resistance to Cd, Cu, and As, depend on efflux. Reduced heavy metal uptake has been observed in many tolerant microbes, including bacteria, algae, and fungi, although this is also dependent on environmental factors including pH and ion competition. However, some resistant strains may accumulate more metal than sensitive parental strains because of more efficient internal detoxification. Inside cells, metal ions may be detoxified by chemical components, which include metal-binding proteins, or compartmentalized into specific organelles. Metal-sequestering granular material, including cyanophycin granules (in cyanobacteria) and polyphosphate have been implicated in bacteria, while metal-binding proteins, including metallothioneins and γ-Glu-Cys peptides (phytochelatins, cadystins), have been detected in all microbial groups examined. Metallothioneins are small, cysteine-rich polypeptides that can bind essential metals (e.g., Cu and Zn), in addition to nonessential metals (e.g., Cd). Metal binding γ-glutamyl-cysteinyl peptides are short peptides of general formula $(\gamma Glu\text{-}Cys)_n\text{-}Gly$. Peptides where $n = 2$ to 7 are most common, and these are important detoxification mechanisms in algae, as well as in several fungi and yeasts. In eukaryotic microorganisms, metal ions (e.g., Co, Zn, Mn) may preferentially accumulate in vacuoles, the vacuolar membrane (tonoplast) possessing transport systems for their accumulation from the cytosol.

Chemical transformations of heavy metal species by microorganisms may also constitute detoxification mechanisms, e.g., bacterial Hg^{2+} reduction to Hg^0. In addition to this, other examples of reduction are carried out by bacteria, algae, and fungi (e.g., Au^{3+} to Au^0, Ag^+ to Ag^0, and Cr^{6+} to Cr^{3+}). Methylated metal and metalloid species may be volatile and lost from a given environment. Methylation of Hg^{2+}, by direct and indirect microbial action, involves methylcobalamin (CH_3CoB12; vitamin B12) with the formation of two products, CH_3Hg^+ and $(CH_3)_2Hg$. S-adenosylmethionine (SAM) is the methylating agent, by transfer of carbonium ions (CH_3^+), for arsenic and selenium.

Organometallic compounds may be detoxified by sequential removal of alkyl or aryl groups. Organomercurials can be degraded by organomercurial lyase, whereas organotin detoxification involves sequential removal of organic groups from the tin atom:

$$R_4SN \rightarrow R_3SnX \rightarrow R_2SnX_2 \rightarrow RSnX_3 \rightarrow SnX_4$$
$$(X = \text{counter-ion, e.g., } Cl^-)$$

It should be stressed that abiotic mechanisms of metal methylation and organometal(loid) degradation also contribute to their transformation and redistribution in aquatic, terrestrial, and aerial environments. The relative importance of biotic and abiotic mechanisms is often difficult to discern.

II. BIOTECHNOLOGICAL ASPECTS OF HEAVY METAL POLLUTION

A. Microbial Processes for Metal Removal and Recovery

Certain microbial processes can solubilize metals, thereby increasing their bioavailability and potential toxicity, whereas others immobilize them and thus reduce their bioavailability (Fig. 1). The relative balance between mobilization and immobilization varies, depending on the organisms and their environment. As well as being an integral component of biogeochemical cycles for metals, these processes may be exploited for the treatment of contaminated

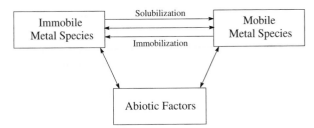

Fig. 1. Diagram depicting the major roles of microorganisms in effecting transformations between soluble and insoluble metal phases. The relative balance between such processes will depend on the environment and associated physicochemical conditions and the microorganism(s) involved, as well as relationships with plants, animals, and anthropogenic activities. The chemical equilibria between soluble and insoluble phases is influenced by abiotic components, including dead biota and their decomposition products, as well as other physicochemical components of the environmental matrix, e.g., pH, water, inorganic and organic ions, molecules, compounds, colloids, and particulates. Solubilization can occur by autotrophic leaching; heterotrophic leaching; metabolite excretion including H^+ and organic acids; enzymic activity; siderophores and other complexing agents; oxidative–reductive reactions; alkylation; biodegradation of organo-radionuclide complexes. Immobilization can occur by biosorption to cell walls, exopolymers, other structural components and derived/excreted products; precipitation as a result of sulfide formation or reduction; other oxidative–reductive transformations; transport, accumulation, intracellular deposition, localization and sequestration; adsorption and entrapment of colloids and particulates. Overall scheme also affected by reciprocal interactions between biotic and abiotic components of the ecosystem, such as abiotic influence on microbial diversity, numbers, and metabolic activity; ingestion of particulates and colloids (including bacteria) by phagotrophs; biotic modification of physicochemical parameters including E_h, pH, O_2, CO_2, other gases and metabolites, temperature, and nutrient depletion.

solid and liquid wastes. Metal mobilization can be achieved by autotrophic and heterotrophic leaching, chelation by microbial metabolites and siderophores, and methylation which can result in volatilization. Similarly, immobilization can result from sorption to cell components or exopolymers, transport and intracellular sequestration or precipitation as insoluble organic and inorganic compounds, e.g., oxalates, sulfides, or phosphates. In addition, microbiologically mediated reduction of higher-valency species may effect either mobilization, e.g., Mn(IV) to Mn(II), or immobilization, e.g., Cr(VI) to Cr(III) and U(VI) to U(IV). In the context of bioremediation, solubilization of metal contaminants provides a route for removal of the metals from solid matrices such as soils, sediments, and industrial wastes. Alternatively, immobilization processes enable metals to be transformed *in situ* and in bioreactors into insoluble, chemically inert forms. Biotechnological development of microbial systems may provide an alternative or adjunct to conventional physicochemical treatment methods for contaminated effluents and wastewaters. Growing evidence suggests that some biomass-related processes are economically competitive with existing treatments in mining and metallurgy.

B. Metal Solubilization

1. Autotrophic Leaching

Most autotrophic metal leaching is carried out by chemolithotrophic, acidophilic bacteria which obtain energy from oxidation of Fe(II) or reduced sulfur compounds and solubilize metals by several possible mechanisms, including Fe(III)/Fe(II) cycling via thiosulfate or polysulfide/sulfur intermediates which lead to the production of H_2SO_4. Autotrophic leaching of metal sulfides by *Thiobacillus* species and other acidophiles is well established for industrial scale biomining processes but, for bioremediation, autotrophic production of sulfuric acid can also be used to remove metals from contaminated solid materials, including sewage sludge and soil. One two-stage process used a mixture of sulfur-oxidizing bacteria to acidify metal-contaminated soil before treatment of the metal-loaded leachate with sulfate-reducing bacteria.

2. Heterotrophic Leaching

Many heterotrophic fungi (and bacteria) can leach metals from industrial wastes, low grade ores, and metal-bearing minerals. This occurs as a result of proton efflux, siderophores (for Fe(III)), and organic acids, e.g., citric and oxalic. Organic acids provide a source of protons and a metal-complexing anion, e.g., citrate, oxalate, with complexation being dependent on the metal/anion concentrations, pH, and

metal complex stability constants. Organisms such as *Aspergillus niger* and *Penicillium simplicissimum* have been used to leach Zn, Cu, Ni, and Co from a variety of solid materials, including industrial filter dust, copper-converter slag, lateritic ores, red mud (the waste product of the extraction of Al from bauxite), manganiferous minerals, and municipal waste fly ash.

C. Metal Immobilization

1. Biosorption

Biosorption (the microbial uptake of organic and inorganic metal species, both soluble and insoluble, by physicochemical mechanisms) may be influenced by metabolic activity (in living cells) and may also provide nucleation sites for the formation of stable minerals including phosphates, sulfides, and oxides. Crystallization of elemental gold and silver may occur as a result of reduction, whereas the formation of hydrolysis products can enhance precipitation of U and Th. All biological macromolecules have some affinity for metal species with cell walls and associated materials being of the greatest significance in biosorption (Table I). In addition, mobile cationic species can be accumulated by cells via transport systems of varying affinity and specificity and internally bound, transformed, precipitated, localized within organelles, or translocated to specific structures, depending on the metal concerned and the organism.

a. Biosorption by Cell Walls and Associated Components

Extracellular polymeric substances (EPS), a mixture of polysaccharides, mucopolysaccharides, and proteins, can bind significant amounts of potentially toxic metals and entrap precipitated metal sulfides and oxides. One process uses floating mats of cyanobacteria, the metal-binding process being due to large polysaccharides (>200,000 Da). In bacteria, peptidoglycan carboxyl groups are main cationic binding sites in gram-positive species with phosphate groups contributing significantly in gram-negatives. Chitin, phenolic polymers, and melanins are important structural components of fungal walls and these are also effective biosorbents for metals and radionu-

TABLE I
Selected Examples of Microbial Metal and Actinide
Accumulation to Industrially Significant Levels

Microorganism	Metal	Accumulation (% dry weight)
Bacteria		
Streptomyces sp.	Uranium	2–14
S. viridochromogenes	Uranium	30
Thiobacillus ferrooxidans	Silver	25
Bacillus cereus	Cadmium	4–9
Zoogloea sp.	Cobalt	25
	Copper	34
	Nickel	13
Citrobacter sp.[a]	Lead	34–40
	Cadmium	170
	Uranium	900
Pseudomonas aeruginosa	Uranium	15
Mixed culture	Silver	32
Cyanobacteria		
Anabaena cylindrica	Cadmium	0.25
Anacystis nidulans	Nickel	1
Spirulina platensis	Gold	0.52
Plectonema boryanum	Zirconium	0.16
Nostoc sp.	Cadmium	1
Algae		
Chlorella vulgaris	Gold	10
	Lead	8.5
Chlorella regularis	Uranium	15
	Zinc	2.8
	Manganese	0.8
Scenedesmus sp.	Molybdenum	2.3
Euglena sp.	Aluminum	1.5
Sargassum natans[b]	Gold	25
	Lead	8
	Silver	7
	Uranium	4.5
	Copper	2.5
	Zinc	2
	Cobalt	6
	Cadmium	8.3
Ascophyllum nodosum[b]	Gold	4
	Cobalt	15
	Cadmium	10

continues

Continued

Microorganism	Metal	Accumulation (% dry weight)
Fungi		
Phoma sp.	Silver	2
Penicillium sp.	Uranium	8–17
	Lead	0.6
Rhizopus arrhizus	Cadmium	3
	Lead	10
	Uranium	20
	Thorium	19
	Silver	5
	Mercury	6
Aspergillus niger	Thorium	19
	Uranium	22
	Gold	6–18
	Zinc	1–10
	Silver	10
Saccharomyces cerevisiae	Thorium	12
	Uranium	10–15
	Cadmium	7
	Copper	1–3
Ganoderma lucidum	Copper	1
Mucor miehei	Zinc	3.4
	Uranium	18

[a] Phosphatase-mediated metal removal.

[b] Macroalgae.

Data derived from a number of sources and presented without reference to important experimental conditions, e.g., metal and biomass concentration; pH; and whether freely suspended, living, dead, or immobilized; or the mechanism of accumulation. In most cases, highest uptake levels are due to general biosorptive mechanisms.

clides. Variations in the chemical behavior of metal species, as well as the composition of microbial cell walls and extracellular materials, can result in wide differences in biosorptive capacities (Table I).

b. Biosorption by Free and Immobilized Biomass

Both freely suspended and immobilized biomass from bacterial, cyanobacterial, algal, and fungal species has received attention, with immobilized systems appearing to possess several advantages including higher mechanical strength and easier biomass/liquid separation. Living or dead biomass of all groups has been immobilized by encapsulation or cross-linking using supports which include agar, cellulose, alginates, cross-linked ethyl acrylate–ethylene glycol dimethylacrylate, polyacrylamide and silica gels, and cross-linking reagents such as toluene diisocyanate and glutaraldehyde. Immobilized living biomass has mainly taken the form of bacterial biofilms on inert supports and has been used in a variety of configurations, including rotating biological contactors, fixed-bed reactors, trickle filters, fluidized beds, and airlift bioreactors.

c. Metal Desorption

Biotechnological exploitation of biosorption may depend on the ease of biosorbent regeneration for metal recovery. Metabolism-independent processes are frequently reversible by nondestructive methods and can be considered analogous to conventional ion exchange. Most work has concentrated on nondestructive desorption, which, for maximum benefit, should be efficient, cheap, and result in minimal damage to the biosorbent. Dilute mineral acids ($\cong 0.1M$) can be effective for metal removal, although more concentrated acids or lengthy exposure times may result in biomass damage. It may be possible to apply selective desorption of metalloid species from a loaded biosorbent using an appropriate elution scheme. For example, metal cations (e.g., Cu^{2+}, Cr^{3+}, Ni^{2+}, Pb^{2+}, Zn^{2+}, Cd^{2+}, and Co^{2+}) were released from algal biomass using eluant at pH 2, whereas at higher pH values, anionic metal species (e.g., SeO_4^{2-}, CrO_4^{2-}, and MoO_4^{2-}) were removed. Au^{3+}, Ag^+, and Hg^{2+}, however, remained strongly bound at pH 2, and these were removed by addition of ligands that formed stable complexes with these metal ions. Carbonates and/or bicarbonates are efficient desorption agents with the potential for cheap, nondestructive metal recovery. Operating pH values for bicarbonates cause little damage to the biomass, which may retain at least 90% of the original uptake capacity.

2. Metal-Binding Proteins, Polysaccharides, and Other Biomolecules

A diverse range of specific and nonspecific metal-binding compounds are produced by microorganisms. Nonspecific metal-binding compounds are metabolites or by-products of microbial metabolism and range in size from simple organic acids and alcohols to macromolecules, such as polysaccharides and humic and fulvic acids. Specific metal-binding compounds may be produced in response to external levels of metals. Siderophores are low-molecular-weight Fe(III)-coordination compounds (500–1000 Da) excreted under iron-limiting conditions by iron-dependent microorganisms to enable accumulation of iron. Although specific for Fe(III), siderophores can also complex Pu(IV), Ga(III), Cr(III), scandium (Sc), indium (In), nickel, uranium, and thorium. Specific, low-molecular-weight (6000–10,000 Da) metal-binding metallothioneins are produced by animals, plants, and microorganisms in response to the presence of toxic metals. Metal binding γ-Glu-Cys peptides (phytochelatins, cadystins) contain glutamic acid and cysteine at the amino-terminal position and have been identified in plants, algae, and

several microorganisms. The metal-binding abilities of siderophores, metallothioneins, phytochelatins, and other similar molecules may have potential for bioremediation of waters containing low metal concentrations, though few examples have been rigorously tested.

3. Transport and Accumulation

Microbial metal transport systems are of varying specificity and essential and nonessential metal(loid) species may be accumulated. Rates of uptake can depend on the physiological state of cells, as well as the nature of the environment or growth medium. Integral to the transport of metal ions into cells are transmembrane electrochemical gradients, e.g., of H^+, resulting from the operation of enzymatic pumps (ATPases) that transform the chemical energy of ATP into this form of biological energy. ATPases are also involved in ion efflux in a variety of organisms and organellar ion compartmentation in eukaryotes via operation across vacuolar membranes. Metals may also enter (and leave) cells via pores or channels. With toxic heavy metals, permeabilization of cell membranes can result in exposure of intracellular metal-binding sites and increase passive accumulation. Intracellular uptake may result in death of sensitive organisms unless a means of detoxification is possessed or induced. Other mechanisms of microbial metal accumulation include iron-binding siderophores and the co-transport of metals with organic substrates.

4. Metal Precipitation
a. Precipitation by Metal-Reducing Bacteria

A diverse range of microorganisms can use oxidized metallic species, e.g., Fe(III), Cr(VI), or Mn(IV), as terminal electron acceptors. Many can utilize more than one metal or anions such as nitrate or sulfate. Fe(III) and Mn(IV) appear to be the most commonly utilized metals as terminal electron acceptors in the biosphere. However, since the solubility of both Fe and Mn is increased by reduction, other metals have been targeted in waste-treatment, e.g., molybdenum(VI) and Cr(VI) where reduction leads to a decrease in solubility. Reduction of, e.g., Cr(VI) to Cr(III), by organisms including *Enterobacter cloacae* and *Escherichia coli* may facilitate subsequent removal by biosorption or (bio)precipitation. One promising potential application of dissimilatory biological metal reduction is uranium precipitation by reduction of soluble U(VI) compounds to U(IV) compounds, such as the hydroxide or carbonate, which have low solubility at neutral pH. Strains of *Shewanella* (*Alteromonas*) *putrefaciens* and *Desulfovibrio* sp. can produce a very pure precipitate of U(IV) carbonate. Such bacterial uranium reduction can also be combined with chemical extraction methods. The solubility of some other radionuclides, e.g., Ra or Pu, may be increased by reduction, which may favor removal from, for example, soil.

b. Sulfate-Reducing Bacteria

Sulfate-reducing bacteria (SRB) are strictly anaerobic heterotrophic bacteria found in environments where carbon substrates and sulfate are available. SRB utilize an energy metabolism in which the oxidation of organic compounds or hydrogen is coupled to the reduction of sulfate as the terminal electron acceptor, producing sulfide, which may reach significant concentrations in sediments or chemostat cultures. The main mechanism whereby sulfate-reducing bacteria remove toxic metals from solution is by the formation of metal sulfide precipitates:

$$M^{2+} + SO_4^{2-} + 2\ CH_3CH_2OH \rightarrow$$
$$2\ CH_3COOH + 2H_2O + MS\downarrow$$
$$M^{2+} + SO_4^{2-} + 2\ CH_3CHOHCOOH \rightarrow$$
$$2CH_3COOH + 2CO_2 + 2H_2O + MS\downarrow$$
$$M^{2+} + SO_4^{2-} + CH_3COOH \rightarrow$$
$$2CO_2 + 2H_2O + MS\downarrow$$

(M = divalent metal)

The solubility products of most heavy metal sulfides are very low, in the range of 4.65×10^{-14} (Mn) to 6.44×10^{-53} (Hg), so that even a moderate output of sulfide can remove metals to levels permitted in the environment. Sulfate-reducing bacteria can also create extremely reducing conditions which can chemically reduce metals such as uranium(VI). In addition, sulfate reduction partially eliminates acidity from the system, which can result in the further precipitation of metals, such as Cu or Al, as hydroxides, as well as increasing the efficiency of sulfide precipitation.

c. Processes Utilizing Metal Sulfide Precipitation

Sulfate-reduction can provide both *in situ* and *ex situ* metal removal from acid mine drainage and can contribute to the removal of metals and acidity in artificial and natural wetlands, together with other mechanisms such as biosorption. Large-scale bioreactors have also been developed using bacterial sulfate reduction for treating metal-contaminated waters, e.g., at Budel-Dorplein in the Netherlands. This plant removes toxic metals (primarily Zn) and sulfate from contaminated groundwater at a long-standing smelter site by precipitation as metal sulfides. A process integrating bacterial sulfate reduction with bioleaching by sulfur-oxidizing bacteria has been developed to remove contaminating toxic metals from soils. In this process sulfur- and iron-oxidizing bacteria are employed to liberate metals from soils by the breakdown of sulfide minerals and production of sulfuric acid, which liberates acid-labile forms, such as hydroxides, carbonates, or sorbed metals. Metals are liberated in the form of an acid sulfate solution, which enables both a large proportion of the acidity and almost the entirety of the metals to be removed by bacterial sulfate reduction.

d. Phosphatase-Mediated Metal Precipitation

In this process, metal or radionuclide accumulation by bacterial (*Citrobacter*) biomass is mediated by a phosphatase enzyme, induced during metal-free growth, which liberates inorganic phosphate from a supplied organic phosphate donor molecule, such as glycerol 2-phosphate. Metal/radionuclide cations are then precipitated as phosphates on the biomass, often to high levels.

e. High Gradient Magnetic Separation (HGMS)

Metal ion removal from solution is achieved using bacteria rendered susceptible to magnetic fields. "Nonmagnetic" bacteria can be made magnetic by the precipitation of metal phosphates (aerobic) or sulfides (anaerobic) on their surfaces as described previously. For those organisms producing iron sulfide, it has been found that this compound is not only magnetic but also an effective adsorbent for metallic elements. Solutions treated by HGMS can have very low residual levels of metal ions remaining in solution.

D. Metal, Metalloid, and Organometal Transformations

Microorganisms can transform certain metal, metalloid, and organometallic species by oxidation, reduction, methylation, and dealkylation. Biomethylated derivatives are often volatile and may be eliminated from a system by evaporation. The two major metalloid transformation processes described are reduction of metalloid oxyanions to elemental forms and methylation.

1. Microbial Reduction of Metalloid Oxyanions

Reduction of selenate (Se(VI)) and selenite (Se(IV)) to elemental selenium can be catalyzed by numerous microbes, which often results in a red precipitate deposited around cells and colonies. In addition, some organisms can also use SeO_4^{2-} as an electron acceptor to support growth. Reduction of TeO_3^{2-} to Te^0 is also an apparent means of detoxification found in bacteria and fungi, the Te^0 being deposited in or around cells, and resulting in black colonies.

2. Methylation of Metalloids

Microbial methylation of metalloids to yield volatile derivatives, e.g., dimethylselenide, dimethyltelluride, or trimethylarsine, can be effected by a variety of bacteria, algae, and fungi. Selenium methylation appears to involve transfer of methyl groups as carbonium (CH_3^+) ions via the S-adenosyl methionine system. Arsenic compounds such as arsenate (As(V), AsO_4^{3-}), arsenite (As(III), AsO_2^-), and methylarsonic acid ($CH_3H_2AsO_3$) can be methylated to volatile dimethyl-($(CH_3)_2HAs$) or trimethylarsine ($(CH_3)_3As$).

3. Microbial Metalloid Transformations and Bioremediation

In situ immobilization of SeO_4^{2-}, by reduction to Se^0, has been achieved in Se-contaminated sediments. Microbial methylation of selenium, resulting in volatilization, has also been used for *in situ* bioremediation of selenium-containing land and water at Kesterson Reservoir in the United States. Selenium volatilization from soil was enhanced by optimizing soil moisture, particle size, and mixing, while in waters it was affected by the growth phase, salinity,

pH, and selenium concentration. The selenium-contaminated agricultural drainage water was evaporated to dryness until the sediment concentration in the sediment approached 100 mg Se kg^{-1} dry weight. Conditions such as carbon source, moisture, temperature, and aeration were then optimized for selenium volatilization and the process continued until selenium levels in sediments declined to acceptable levels. Some potential for *ex situ* treatment of selenium-contaminated waters has also been demonstrated.

4. Mercury and Organometals

Key microbial transformations of inorganic Hg^{2+} include reduction and methylation. The mechanism of bacterial Hg^{2+} resistance is enzymic reduction of Hg^{2+} to nontoxic volatile Hg^0 by mercuric reductase. Hg^{2+} may also arise from the action of organomercurial lyase on organomercurials. Since Hg^0 is volatile, this could provide one means of mercury removal. Methylation of inorganic Hg^{2+} leads to the formation of more toxic volatile derivatives: the bioremediation potential of this process (as for other metals and metalloids (besides selenium) capable of being methylated, e.g., As, Sn, Pb) has not been explored in detail, although trapping of volatile derivatives is a possibility. As well as organomercurials, other organometals may be degraded by microorganisms. Organoarsenicals can be demethylated by bacteria, while organotin degradation involves sequential removal of organic groups from the tin atom. In theory, such mechanisms and interaction with the bioremediation possibilities described previously may provide a means of detoxification.

E. Concluding Remarks

Microorganisms play important roles in the environmental fate of toxic metals, metalloids, and radionuclides with physicochemical and biological mechanisms effecting transformations between soluble and insoluble phases. Such mechanisms are important components of natural biogeochemical cycles for metals and associated elements, e.g., sulfur and phosphorus, with some processes being of potential application to the treatment of contaminated materials. Removal of such pollutants from contaminated solutions by living or dead microbial biomass and derived or excreted products may provide an economically feasible and technically efficient means for element recovery and environmental protection. Although the biotechnological potential of most of these processes has only been explored in the laboratory, some mechanisms, notably bioleaching, biosorption, and precipitation, have been employed at a commercial scale. Of these, autotrophic leaching is an established major process in mineral extraction but has also been applied to the treatment of contaminated land. There have been several attempts to commercialize biosorption using microbial biomass but success has been short-lived, primarily due to competition with commercially produced ion exchange media. Bioprecipitation of metals as sulfides has achieved large-scale application, and this holds out promise of further commercial development. Exploitation of other microbiological processes will undoubtedly depend on a number of scientific, economic, and political factors.

See Also the Following Article

HEAVY METALS, BACTERIAL RESISTANCES

Bibliography

Bosecker, K. (1997). Bioleaching: metal solubilization by microorganisms. *FEMS Microbiol. Rev.* **20**, 591–604.

Burgstaller, W., and Schinner, F. (1993). Leaching of metals with fungi. *J. Biotechnol.* **27**, 91–116.

Cooney, J. J., and Gadd, G. M. (eds.) (1995). Metals and microorganisms. Special Issues: *J. Indust. Microbiol.* **14**, 59–199 and **14**, 201–353.

Ehrlich, H. L. (1996). "Geomicrobiology." Marcel Dekker, New York.

Ehrlich, H. L. (1997). Microbes and metals. *Appl. Microbiol. Biotechnol.* **48**, 687–692.

Gadd, G. M. (1993). Interactions of fungi with toxic metals. *New Phytol.* **124**, 25–60.

Gadd, G. M. (1993). Microbial formation and transformation of organometallic and organometalloid compounds. *FEMS Microbiol. Rev.* **11**, 297–316.

Gadd, G. M. (1996). Influence of microorganisms on the environmental fate of radionuclides. *Endeavour* **20**, 150–156.

Gadd, G. M., and White, C. (1993). Microbial treatment of metal pollution: a working biotechnology? *Trends Biotechnol.* **11**, 353–359.

Hughes, M. N., and Poole, R. K. (1989). "Metals and Microorganisms." Chapman and Hall, London, UK.

Karlson, U., and Frankenberger, W. T. (1993). Biological alkylation of selenium and tellurium. *In* "Metal Ions in Biological Systems" (H. Sigel and A. Sigel, eds.), pp. 185–227. Marcel Dekker, New York.

Lovley, D. R., and Coates, J. D. (1997). Bioremediation of metal contamination. *Curr. Opinion Biotechnol.* **8**, 285–289.

Macaskie, L. E. (1991). The application of biotechnology to the treatment of wastes produced by the nuclear fuel cycle—Biodegradation and bioaccumulation as a means of treating radionuclide-containing streams. *Crit. Rev. Biotechnol.* **11**, 41–112.

Rauser, W. E. (1995). Phytochelatins and related peptides. *Plant Physiol.* **109**, 1141–1149.

Rawlings, D. E., and Silver, S. (1995). Mining with microbes. *Biotechnol.* **13**, 773–778.

Silver, S. (1998). Genes for all metals—A bacterial view of the Periodic Table. *J. Indust. Microbiol. Biotechnol.* **20**, 1–12.

White, C., and Gadd, G. M. (1998). Reduction of metal cations and oxyanions by anaerobic and metal-resistant organisms: chemistry, physiology and potential for the control and bioremediation of toxic metal pollution. *In* "Extremophiles: Physiology and Biotechnology" (W. D. Grant and T. Horikoshi, eds.), pp. 233–254. John Wiley and Sons, New York.

White, C., Sayer, J. A., and Gadd, G. M. (1997). Microbial solubilization and immobilization of toxic metals: key biogeochemical processes for treatment of contamination. *FEMS Microbiol. Rev.* **20**, 503–516.

Winkelmann, G., and Winge, D. R. (eds.) (1994). "Metal Ions in Fungi." Marcel Dekker, New York.

Heavy Metals, Bacterial Resistances

Tapan K. Misra
University of Illinois at Chicago

I. Antimony and Arsenic Resistances
II. Cadmium Resistance
III. Chromium Resistance
IV. Mercury and Organomercurial Resistance
V. Concluding Remarks

GLOSSARY

chemiosmosis An electron transport process that pumps protons across the inner membrane to the outer aqueous phase, generating an H^+ gradient across the inner membrane. The osmotic energy in this gradient supplies the energy for ATP synthesis.

constitutive gene The transcription of the gene is dependent on RNA polymerase only and no other regulatory factor.

efflux Pumping out.

operon Segment of DNA consisting of genes that are cotranscribed and the DNA element(s) that are recognized by the regulatory gene product.

promoter Specific segment of DNA recognized by RNA polymerase for initiation of transcription.

"−10" sequence RNA polymerase recognition sequence which is located close to 10 bp upstream of mRNA start site. The consensus sequence in bacterial DNA is TATAAT.

"−35" sequence Approximately 35 bp upstream of mRNA start site. The consensus sequence in bacteria is TTGACA.

transposon Mobile genetic element which can replicate and then transfer a copy at new locations in the genome.

trans-acting element An element that can function on a physically separate expression system.

transcribe To synthesize RNA using the coding strand of a DNA.

transcription Synthesis of RNA using DNA as a template.

uncouplers Chemicals that can disrupt linkage between phosphorylation of ADP and electron transport.

MOST HEAVY METALS are toxic to cells. Generally, it is the cations and anions formed from the metals and not the reduced metallic material (which is inert) that are toxic. Resistant bacteria are prevalent in environments (soil, water) enriched with toxic compounds.

Some of the heavy metals are components for many cellular enzymes and, therefore, are needed by cells in trace amounts. Examples of this group of metals are cobalt, manganese, nickel, and zinc. Another group of metals—antimony, arsenic, bismuth, cadmium, chromium, lead, mercury, and tellurium—are not needed by bacterial cells; they are very toxic. Some heavy metal compounds are used in industry as catalysts, some are used as preservatives (on seeds and wood products) to prevent microbial growth, some are used as insecticides and herbicides, and some had been used as disinfectants in clinics, hospitals, and homes.

Apparently, separate genetically determined mechanisms of resistances exist against essentially each toxic metal ion. Due to their structural similarities, some metal ions are transported into the cells through specific transport systems designed for normally required metal ions. A few examples can be cited: arsenate (AsO_4^{2-}) uses the phosphate transport system(s), cadmium (Cd^{2+}) enters bacterial cells via manganese (Mn^{2+}) or zinc (Zn^{2+}) transport systems, and chromate (CrO_4^{2-}) uses the sulfate (SO_4^{2-}) transport system. One known exception is the mercuric ion (Hg^{2+}), which is transported into bacterial cells by an active transport process mediated by the protein products of genes that are cotranscribed with the other genes involved in the detoxification of Hg^{2+}.

Encyclopedia of Microbiology, Volume 2
SECOND EDITION

618

Two basic mechanisms of resistances by cells against toxic ions can be envisaged: (1) specific alteration of ion transport (inward, preventing entry into the cell; or outward, pumping out of the cell) of the toxic ion, and (2) by chemical modification or by binding to cellular factor(s), resulting in a form that is no longer toxic to the cell. These mechanisms will be discussed with specific examples later.

Genes conferring resistances to heavy metals are often found on mobile genetic elements, such as plasmids and transposons. Coexistence of different metal ion resistance genes and antibiotic resistance genes on one plasmid is not rare. This is advantageous from an evolutionary point of view. Genetic exchange between different strains, species, or genera is facilitated by relatively small extrachromosomal elements, rather than by complex rearrangement within the chromosome. Recent genome sequence information revealed that some of the heavy metal resistance genes are also present on the bacterial chromosome. The level of resistance to heavy metal ions conferred by resistance genes on the chromosome is generally low as compared to that found on plasmids. The expression of many known metal ion resistance genes is inducible by the metal ion itself. When bacterial cells sense the metal ion, increased expression of genes conferring resistances to the metal ion takes place. Thus, these processes are efficiently controlled; enzymes and other proteins involved in chemical transformations and transport of the toxic ions are produced only when they are required by cells (in a contaminated environment). The toxic ion "induces" the expression of ion-specific resistant gene(s).

I. ANTIMONY AND ARSENIC RESISTANCES

Arsenicals have widely been used in cattle feed supplements as herbicides and insecticides, and also in printing fabrics. Bacteria have developed distinct mechanisms of resistances against the oxyanions of arsenic (arsenite and arsenate) and antimony (antimonite). Arsenate and arsenite resistances have been found in *Escherichia coli*, *Staphylococcus aureus*, *Pseudomonas pseudomalleli*, *Synechoccous*, and some strains of *Alcaligenes*. The biochemical and molecular basis of arsenite (AsO_2^-), arsenate (AsO_4^{3-}), and antimonite (SbO_2^-) resistances are most thoroughly known in *E. coli* and *S. aureus*. In both organisms, arsenite, arsenate, and antimonite resistances are determined by a set of genes present in a single operon; therefore, they are coinduced by any of those ions.

A. Biochemistry of Antimony and Arsenic Resistances

The biochemical mechanism of plasmid-determined arsenate, arsenite, and antimonite resistances are similar in *E. coli* and *S. aureus*. Resistances result from reduced net accumulation of these oxyanions in induced resistant cells. The chromosomally encoded arsenic resistance, recently found in *E. coli*, is also governed by the same basic mechanism. Arsenate is accumulated by virtue of its structural similarity to phosphate. Phosphate is essential for cellular growth. Many bacteria appear to have two separate chromosomally encoded phosphate transport systems. Arsenate enters the cell via both of these. Resistant cells quickly reduce arsenate to arsenite, which is exported out of the cell by special pumping mechanisms, aided by proteins encoded on the resistant operons.

Arsenite, arsenate, and antimonite resistances were reported in *E. coli* and in *S. aureus* cells harboring resistance plasmids. Cells preloaded with radioactive arsenate ($^{74}AsO_4^{3-}$) efficiently pumped out ^{74}As, when the cells were induced with arsenate (or arsenite or antimonite) prior to loading with $^{74}AsO4^{3-}$; the uninduced cells failed to extrude ^{74}As. A protein named ArsC reduces arsenate to arsenite. Arsenate reductase (ArsC) of *S. aureus* uses thioredoxin as the source of reducing potential, while that of *E. coli* uses glutaredoxin as the reducing potential. A protein named ArsB functions as a membrane-potential driven arsenite efflux transporter or, in combination with another protein, ArsA, the complex functions much more efficiently as an ATPase-driven arsenite pump. Genetic and molecular biological studies advanced the biochemical understanding of the mechanism of arsenite and antimonite efflux.

B. Molecular Genetics of Arsenic and Antimony Resistances

The best characterized system in gram-negative bacteria are those of the plasmids R773 and R46 of *E. coli* and the one present on the *E. coli* chromosome. Plasmid PI258-encoded and pSX267-encoded systems from the gram-positive bacteria *S. aureus* and *S. xylosus*, respectively, have been studied. The arsenite, arsenate, and antimonite resistance operons (*ars*) consist of three to five genes. The DNA sequences of these operons are known and the organization of the genes of the plasmid R773-encoded *ars* operon is shown in Fig. 1. All the *ars* genes are cotranscribed from a single DNA site "promoter." The product of the regulatory gene, ArsR, is a transacting regulatory protein, which represses its own synthesis as well as the expression of the other downstream *ars* genes (in the absence of arsenicals and antimonite). ArsR induces gene expression from the *ars* promoter in the presence of arsenite, arsenate, antimonite, and bismuth ion. The *ars* operon does not confer resistance to Bi^{3+}, but Bi^{3+} functions as an inducer, hence, it is known as a gratuitous inducer. Following the *arsR* gene, there are four other genes in R773- and R46-encoded operons (Fig. 1): *arsD, arsA, arsB,* and *arsC*. *arsA* and *arsB* are necessary and sufficient for arsenite and antimonite resistance, and arsC is also needed for arsenate resistance. The ArsC protein is a reductase enzyme that specifically reduces arsenate to arsenite. ArsD in R773 and R46 is a secondary negative regulator which binds to the same site as the ArsR on the *ars* operator, but with about 100-fold less efficiency as compared to that of ArsR. ArsR controls the basal level of expression, while ArsD functions in a concentration-dependent fashion, setting a limit on the maximal level of expression of *ars* genes.

The 5' and 3' halves of the *arsA* gene have extensive sequence similarity, suggesting evolution of the gene through a gene duplication step and fusion. Thus, the ArsA protein is basically a fusion dimer with two glycine-rich adenylate binding sites, one near the N terminus of the polypeptide and the other near the middle. The purified ArsA protein functions as an arsenite- or antimonite-stimulated soluble ATPase. The protein exists in an equilibrium between monomeric and dimeric forms in solution, with equilibrium favoring dimerization upon binding with the anionic inducers.

How does the soluble ArsA ATPase export oxyanions of antimony and arsenic? The product of the *arsB* gene is a 45.6 kDa hydrophobic protein localized in the inner membrane of the cell. When the ArsA and ArsB proteins form a membrane-bound complex, this functions as an antimonite/arsenite specific efflux pump, thus exporting out these otherwise toxic anions, resulting in reduced net accumulation of these ions inside the cell. The ArsC, an arsenate reductase protein, is needed for arsenate resistance.

II. CADMIUM RESISTANCE

Cadmium salts are commercially used as lubricants, as ice nucleating agents, in dry cell batteries, in photography, in dyeing and calico printing, in electroplating and engraving, and in the manufacture of special mirrors. Cadmium resistance has been shown in certain strains of *Staphylococcus, Bacillus, Alcaligenes, Pseudomonas, Synechococcus,* and *Listeria*. The best studied cadmium (Cd^{2+}) resistance determinants are those in *S. aureus* and in *Alcaligenes eutrophus*. There are at least two distinct Cd^{2+} resistance determinants in *S. aureus,* and they are both found on plasmids. One of them, known as the cadA system, confers about 100-fold greater tolerance to resistant cells over that for sensitive cells (without the cadmium resistance determinant). The mechanism of Cd^{2+} resistance conferred by cadB system in *S. aureus* is not clearly understood. *A. eutrophus* plasmid-borne Cd^{2+} resistance system *czc* confers Cd^{2+}, Zn^{2+}, and Co^{2+} by effluxing these ions using special mechanisms. In the resistant cynobacterium *Synechococcus,* a Cd^{2+}-binding metallothionein-like protein, SmtA, adsorbs Cd^{2+} and works as sink, masking the harmful effect of Cd^{2+} on biological processes.

P	*ars* **R**	*ars* **D**	*ars* **A**	*ars* **B**	*ars* **C**
	117 aa	120 aa	583 aa	429 aa	141 aa

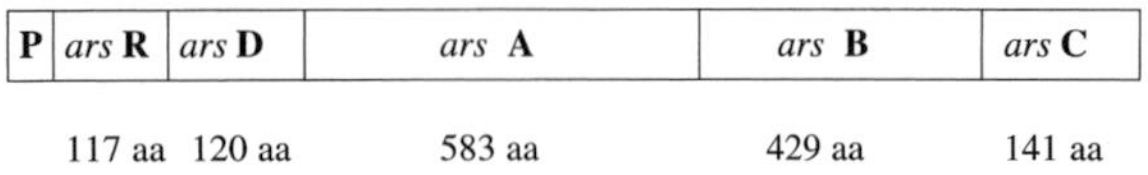

Fig. 1. Schematic representation of arsenic resistance operon of the plasmid R773 from *E. Coli*. P, promoter; aa, amino acid.

A. Biochemistry of *cadA* Cadmium Resistance

Cadmium (Cd^{2+}) is transported into the cell by a chromosomally encoded manganese (Mn^{2+}) transport system (in gram-positive bacteria). The uptake of these ions is membrane potential dependent and accumulation of cadmium inside the cells results in respiratory arrest. The resistant cells rapidly pump Cd^{2+} out. The efflux was proposed as a chemiosmotic electroneutral process with the exchange of one Cd^{2+} for two protons. The evidence supporting this conclusion was that Cd^{2+} efflux is sensitive to agents that block or accelerate proton movement but not to agents that disrupt membrane potential. However, recent DNA sequence analysis and subsequent biochemical studies demonstrated that an ATP is involved in pumping Cd^{2+} out from resistant cells.

B. Molecular Genetics of *cadA* Cadmium Resistance

Two structural genes, *cadC* and *cadA*, that constitute the *cadA* determinant are cotranscribed from a single promoter. The products of these two genes are necessary and sufficient for the maximum level of resistance to Cd^{2+} and to Zn^{2+}. *S. aureus* cells containing the *cadA* resistance determinant also confers resistance to zinc. The expression of the Cd^{2+}/Zn^{2+} resistance genes are inducible by both Cd^{2+} and Zn^{2+} as well as by Bi^{3+}, Co^{2+}, or Pb^{2+} (gratuitous inducers). CadC, the product of the *cadC* gene, negatively regulates (repress) transcription from the cadA promoter by binding with the promoter region (operator). *cadC* gene and the corresponding gene product CadC have recently been renamed to *cadR* and CadR, respectively, to reflect the known regulatory role of the gene and the gene product. CadR is a small, highly charged soluble polypeptide of 122 amino acids. The binding of CadR to the operator DNA is reversed by Cd^{2+}, Zn^{2+}, and other divalent cations, which function as gratuitous inducers. CadR is significantly homologous in amino-acid sequence and similar in function to ArsR and the ArsR family of proteins, including SmtB. The *cadA* gene product, CadA, is a 727 amino acid membrane polypeptide with recognizable sequence similarity with the E1E2

class of cation-translocating ATPases of bacteria and all eukaryotes, including the well-known K^+ ATPases of *E. coli* and *E. faecalis*, H^+ ATPases of yeast and plant cells, Na^+/K^+ ATPases of humans and electric eel, and Ca^{2+} ATPase of rabbit muscle. Everated membrane vesicles (inside-out), prepared from *Bacillus subtilis*, containing cloned *cadA* could accumulate radioactive Cd^{2+} and Zn^{2+} in an ATP-dependent fashion. The same membrane vesicles were shown to incorporate ^{32}P-ATP into a protein corresponding to the size of CadA.

III. CHROMIUM RESISTANCE

Chromium is used in industry, primarily in the manufacture of alloys and in tanning. Chromium (VI) is the most toxic state of chromium salts and Cr (III) is less toxic. Chromate (CrO_4^{2-}) resistance has been observed among certain strains of *Pseudomonas*, *Alcaligenes*, *Salmonella*, *Streptococcus*, *Aeromonas*, and *Enterobacter*.

A. Biochemistry of Chromate Resistance

Chromate enters bacterial cells by the sulfate uptake system(s). There are at least two potential mechanisms of chromate resistance: (1) by reduction of Cr(VI) to Cr(III), which subsequently precipitates as $Cr(OH)_3$ (extracellular), and (2) by reduced uptake of chromate governed by a plasmid system in resistant cells. Is the reduced net uptake of CrO_4^{2-} due to inhibition of CrO_4^{2-} uptake directly or due to rapid efflux of CrO_4^{2-} by a mechanism parallel to those described above for arsenic and cadmium? At the present, there is no evidence to distinguish between the two models.

Chromate reductase activity has been studied in *P. Fluorescens*, *P. aeruginosa*, and in *Enterobacter cloacae*. In *Pseudomonas*, chromate reductase is a soluble enzyme. On the other hand, *E. cloacae* chromate reductase is located in the membrane. Interestingly, *E. cloacae* is resistant to chromate under both aerobic and anaerobic growth conditions, but chromate reduction occurs only under anaerobic conditions.

Chromate reductase activity from *E. cloacae* is inhibited by cyanide and membrane uncouplers.

B. Molecular Genetics of Chromate Resistance

The genes conferring resistance to chromate in *P. aeruginosa* and *A. eutrophus* have been cloned and sequenced. These genes were originally found on plasmids. Plasmid-determined chromate resistance is inducible in *A. eutrophus*, whereas chromate resistance is expressed constitutively in *P. aeruginosa* after cloning. In both cases, a single gene, *chrA*, is responsible for chromate resistance. The gene product is a 401 or 416 amino acid hydrophobic polypeptide. The ChrA proteins from the two sources are structurally related (primary amino acid sequence). It is hypothesized that the ChrA protein either limits the transport of CrO_4^{2-} in cells or is responsible for chromate efflux.

IV. MERCURY AND ORGANOMERCURIAL RESISTANCE

Mercury compounds are used in industry as catalysts in the oxidation of organic compounds. Other uses are in the extraction of gold and silver (currently not used in developed countries), in dry cell batteries, in electric rectifiers, etc. Mercury forms amalgams with other metals and silver–mercury amalgams are still heavily used today in dental restorations. Phenylmercury, merthiolate, and mercurochrome had been used as household and hospital disinfectants. Phenylmercury and methylmercury were used in agriculture as fungicides on seeds and to keep golf-course grass green. The largest source of environmental release of mercury compounds from human activities is that from burning of coal and petroleum products, which are naturally enriched with mercury compounds. Mercury compunds are also leached from (weathering) natural deposits on rocks and soil. Mercury resistance is known in a very wide number of bacterial genera. A few examples of gram-negative mercury resistant bacteria are *Escherichia*, *Pseudomonas*, *Shigella*, *Serratia*, *Thiobacillus*, *Yersinia*, *Acinetobacter*, and *Alcaligenes*. Some examples of resistant gram-positive bacteria are *Staphylococcus*, *Bacillus*, *Mycobacterium*, *Streptococcus*, and *Streptomyces*. Of all the heavy metal resistances studied to date, mercury is certainly the most thoroughly studied system at the biochemical and molecular level.

A. Narrow- and Broad-Spectrum Mercury Resistance

Those bacteria which are resistant primarily to inorganic mercury salts are called narrow-spectrum mercury resistant, and those that are resistant to organomercurial compounds in addition to inorganic mercury salts are called broad-spectrum mercury resistant.

B. Biochemistry of Mercurial Detoxification by Resistant Bacteria

All, or essentially all, of the mercury resistant bacteria isolated from mercury-polluted soil or water have the same basic biochemical mechanism of detoxification of mercury compounds involving soluble enzymes. All produce mercuric reductase that reduces Hg^{2+} to Hg^0. Hg^0 is lipophilic; it is also relatively biologically inert, volatile, and less toxic. Mercuric reductase is a flavin adenine dinucleotide (FAD)-containing enzyme and reduced nicotinamide adenine dinucleotide phosphate ($NADPH^+_2$)-dependent. $NADPH^+_2$ functions as an electron donor, initially reducing the FAD. Broad-spectrum mercury resistant bacteria produce mercuric reductase and a second enzyme, organomercurial lyase, which cleaves the carbon–mercury bond in such compounds as CH_3Hg^+ and $C_6H_5Hg^+$; Hg^{2+} is released and subsequently reduced by the mercuric reductase. These reactions are illustrated in Fig. 2. Recently, the crystal structure of the mercuric reductase from a *Bacillus* species has been solved by x-ray diffraction studies.

Organomercurial lyase isolated from broad-spectrum mercury resistant bacteria is a smaller monomeric polypeptide protein of approximately 23 kDa. Its enzymatic activity is relatively poor, both *in vivo* and *in vitro*, as compared to that of the mercuric ion reductase. Thus, mercuric ions (Hg^{2+}) do not

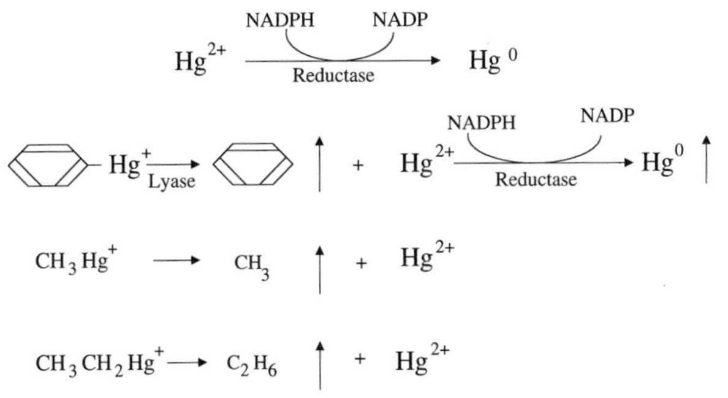

Fig. 2. Enzymatic detoxification of mercury compounds.

accumulate in the cell when both enzymes are present.

An alternative mechanism of mercury resistance by precipitation of insoluble HgS ($Hg^{2+} + H_2S = HgS + 2H^+$) has occasionally been proposed but never studied. If this occurs, then it is most likely a fortuitous by-product of a metabolic process.

1. Mercuric Reductase: Inferences from the Crystal Structure

Mercuric reductase is a dimer consisting of two identical monomers of approximately 65 ± 5 kDa polypeptide monomers, which has three pairs of conserved cysteine residues. However, the NH_2-terminal 160 amino acids region is excluded in the crystal structure and this region may lack a fixed position in the crystals. There are two putative Hg^{2+}-binding domains in this region of the reductase, isolated from most bacteria with some exceptions. Sequence analysis suggested that the mercuric reductase and the human glutathione reductase structures might be similar. Glutathione reductase is a flavin-containing disulfide oxide-reductase. The structure of the glutathione reductase was solved from x-ray diffraction studies even before the DNA sequence of the mercuric ion reductase gene was studied. The NH_2-terminal 80 or so amino acids of mercuric reductase are the only part of the enzyme that bears no relationship to glutathione reductase, and it appears as if these represent an additional Hg^{2+}-related domain. NADPH binds to the two cysteines Cys207 and Cys212 and reduces the dithiolate. The charge is transferred to FAD, which, in turn, reduces Hg^{2+}. Hg^{2+} binds cooperatively between the two subunits of the reductase enzyme involving Cys628 and

Cys629 on one subunit and Tyr264 and Tyr605 on the other subunit.

C. Molecular Genetics of Mercury Resistance

The genes conferring resistance to mercury compounds are physically clustered in an operon. Mercury resistance operons from different bacteria have been cloned and their DNA sequences have been determined. The organization of the genes in gram-positive and gram-negative bacteria is somewhat different. Let us discuss these systems separately.

1. Organization of the Mercury Resistance Genes in Gram-negative Bacteria

The DNA sequence of the mercury resistance genes is 80–90% identical when compared in several different systems (with one known exception in *Thiobacillus ferrooxidans*). The physical organization of the broad-spectrum mercury resistance genes from the *Serratia* plasmid pDU1358 is shown in Fig. 3. The regulatory gene, *merR*, is separated from the other *mer* genes by an operator/promoter region. The operator sequence is the one with which the product of the *merR* regulatory gene interacts, regulating expression of the remaining genes in the operon. Two messenger RNA (mRNA) are synthesized from two divergently oriented promoters. One mRNA is translated into the regulatory protein, MerR, and the other mRNA is translated into the remaining *mer* proteins. The first two structural genes are *merT* and *merP*, whose products transport mercury ion across the cell membrane into the cytoplasm. Then follows the *merA*, *merB*, and *merD* genes. *merA* encodes mercuric reductase; *merB* encodes organomercurial lyase; and *merD* encodes a coregulatory protein, whose influence on the rate of operon expression is much smaller than the effects of MerR. The mercury resistance determinant of the plasmid R100 (from *Shigella flexneri*) contains an additional transport gene, *merC*, located between *merP* and *merA*. In a chromosomally encoded mercury resistance determinant from *Thiobacillus ferroxidans*, only the *merC* gene encodes the mercury transport protein. The *merC* gene, and the *merT* and *merP* genes, encode proteins for independent transport pathways.

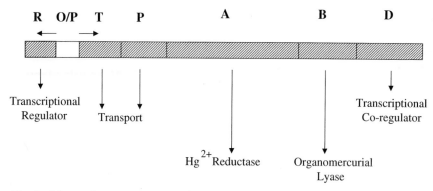

Fig. 3. Schematic representation of the broad-spectrum mercury resistance genes of the *Serratia* plasmid pDU1358. O/P, operator/promoter. The horizontal arrows show the direction of transcription.

2. Organization of Mercury Resistance Genes in Gram-Positive Bacteria

Chromosomally encoded mercury resistance genes from a *Bacillus* species or those from *S. aureus* plasmids are organized in a single operon and they are expressed from a single promoter. The regulatory gene is the first gene after the mRNA start site and it is followed by 3 or 4 genes which are thought to encode transport proteins. Then there are genes for mercuric reductase and organomercurial lyase. No gene equivalent of *merD* of gram-negative bacteria has been identified in the mercury resistance operons of gram-positive bacteria. In one of the *Bacillus mer* operons, a second *merB* gene is located on a separate operon, which is positioned approximately 1.1 kb apart from the major *mer* operon on the chromosome. DNA sequence analyses, and mutational and biochemical studies, lead to a better understanding of the transport system, detoxifying enzymes and the regulation of the operon expression. We have discussed the detoxifying enzymes. Other two aspects are discussed in the following section.

3. Transport of Mercury Ion into the Cell

There is no biochemical data to support transport of Hg^{2+} into the cell. The transport function was first defined in plasmid R100 by cloning and mutation studies. Insertion of a transposon disrupted the expression of the *merA* gene where it was inserted and all other genes downstream from the site of insertion of the transposon in the operon. Such transposon-generated mutation in the mercuric reductase gene,

merA, resulted in hypersensitivity to mercury salts. In other words, cells containing the *mer* operon with a mutation in *merA* are sensitive to lower concentrations of mercury salt, compared to cells containing no *mer* operon. When incubated with $^{203}Hg^{2+}$, the hypersensitive cells accumulated higher concentrations of $^{203}Hg^{2+}$ as compared to cells containing no *mer* operon or those containing the intact *mer* operon. In these mutant cells, Hg^{2+} was transported inside but was not reduced to volatile Hg^0 because of the absence of the reductase enzyme. This accounts for the increased intracellular accumulation of Hg^{2+} and the hypersensitivity to mercury salts.

Sequence prediction suggests that MerT is a hydrophobic, integral membrane protein. MerP is a periplasmic protein and has a leader sequence of nineteen amino acids. Signal peptidase cuts the leader peptide, leaving the 72 amino acid periplasmic Hg^{2+}-binding protein. Mature MerP then delivers the bound Hg^{2+} to the MerT protein and, subsequently, MerT delivers the Hg^{2+} to the mercuric reductase enzyme. Genetic studies revealed that a defective MerP, in which cysteine at position 36 is replaced by Ser (Cys36Ser), confers higher sensitivity to Hg^{2+} as compared to a *merP* deletion mutant in whole-cell experiments. MerC is a hydrophobic protein and is associated with the inner membrane of the cell. It is not known how organomercurial compounds are transported inside the cell.

In gram-negative bacterial cells, the mercury transport protein(s) are essential for resistance and, in their absence, the cells are Hg^{2+} sensitive. In gram-positive bacterial cells, the role of the transport proteins (there appear to be three membrane proteins both in *S. aureus* and *Bacillus*) is less distinct: with the transport proteins present but mercuric reductase missing, the cells are somewhat hypersensitive to Hg^{2+}, indicating that a transport system is functioning. However, with the mercuric reductase present but the transport proteins absent, the cells are still relatively resistant to Hg^{2+}, indicating that Hg^{2+} can cross the membrane into the cell by some alternative pathway.

4. Regulation of Expression of the Mercury Resistance Operon

Mercury resistance gene cluster has become one of the best understood metalloregulatory operons. The expression of the operon, as measured by mercuric reductase activity (^{203}Hg volatilization activity), is increased about 200-fold when cells are grown (induced) with subinhibitory concentrations of mercury salts. Regulation of the gram-positive and gram-negative *mer* operons is somewhat different. Let us focus first on regulation of the gram-negative mercury resistance operon.

Mutation in the *merR* gene leads to constitutive expression of the operon, but at a very reduced level of approximately 2% of the fully induced rate. This suggests that MerR is needed as a repressor to suppress activity from the 2% level down to the 1/200 repressed level. Mutations in *merR* are complemented *in trans* (when intact *merR* gene is cloned in a different plasmid vector and introduced in cells containing the *mer* operon with a mutant *merR* gene). Full (100%) expression of the intact operon is induced in the presence of inorganic mercury salts. Thus, MerR protein also functions as a transcriptional activator in the presence of Hg^{2+}. The MerR protein has been purified and its interaction with the operator/promoter region of DNA has been determined. Note that *merR* gene and the other genes are expressed from two divergently oriented promoters (Fig. 3). MerR protein binds to the promoter region, which contains a 7 base pair inverted repeat sequence. The site of binding to the DNA is not influenced by the presence or absence of Hg^{2+}. MerR occupies the *mer* promoter (P*mer*, promoter for the transcription of the structural genes and *merD*) as a homodimer and represses the synthesis of the *merR* mRNA as well as *mer* operon mRNA. The "−10" and "−35" canonical sequences (Fig. 4) of P*mer* are separated by 19 bases, which is longer than the 17 bases ideal for proper binding of RNA polymerase and efficient transcription. Binding of Hg^{2+} to the MerR protein results in a conformational change of the MerR protein itself, which, in turn, bends the P*mer* DNA. The "−10" and "−35" sequences are brought closer together, thereby facilitating interaction of the promoter sequence with the RNA polymerase. Binding of MerR in the presence of Hg^{2+} results in unwinding of approximately 16 bp downstream from the −10 sequence, facilitating formation of an open DNA complex with the RNA polymerase and allowing efficient transcription.

The narrow-spectrum *mer* operon is inducible by subinhibitory concentrations of inorganic mercury

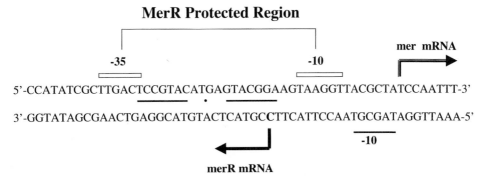

Fig. 4. Nucleotide (DNA) sequence of the promoter region of gram-negative mercury resistance operon. Directions of mRNA synthesis are shown by arrows. The inverted repeat sequences are underlined.

salts but not by organomercurial compounds. On the other hand, broad-spectrum *mer* operon of the *Serratia* plasmid pDU1358 is inducible by both inorganic and organic mercury salts. The amino acid sequences of the narrow- and broad-spectrum MerR are very similar except for the C-terminal nine amino acid residues (Fig. 5). Deletion of seventeen amino acids from the C-terminal end of the broad-spectrum MerR results in complete loss of inducibility with organomercurial compounds, but not with inorganic mercury salts. This suggests that the amino acid sequence at the C terminus of the broad-spectrum MerR is involved in the recognition of organomercurial inducers. This is a good example of how nature cleverly modifies gene structure to respond to changing environmental stimuli. Note in Fig. 5 that there are three conserved cysteine residues (underlined), which are essential for Hg^{2+} binding and transcriptional activation of the *mer* operon. These are conserved in all currently available sequences for MerR proteins.

The *merD* gene (Fig. 3) is present in at least four gram-negative mercury resistance operons. Its product acts in *trans* and down-regulates transcription from the *mer* promoter (P*mer*). The MerD protein binds to the same site as the MerR in experiments with purified DNA and purified protein. The level of MerD in the cell is low (about 15-fold less than the mercuric reductase). The *in vivo* role of MerD is unclear. It appears to fine tune *mer* operon expression as a coregulatory protein in a concentration-dependent manner.

In gram-positive bacteria, the *merR* gene and the other *mer* genes are cotranscribed, unlike the gram-negative *mer* operon, where *merR* gene and all the other *mer* structural genes are transcribed as two separate units. The gram-positive MerR and gram-negative MerR amino acid sequences are 35% identical, and there are three conserved cysteine residues. The operator sequence, where MerR binds the DNA (Fig. 4) is very highly conserved in both gram-positive and gram-negative *mer* operons. gram-positive MerR represses *mer* operon transcription in the absence of Hg^{2+} and activates transcription in the presence of Hg^{2+}.

V. CONCLUDING REMARKS

The foregoing examples of heavy metal resistance mechanisms are the ones that are best known. Genes governing resistances against many additional heavy metal ions, including those of cobalt, copper, lead, nickel, silver, tellurium, and zinc, also have been identified and have been sequenced in many systems. The mechanisms of resistance of these ions are not understood, or much less understood, at this time. The processes by which bacteria convert more toxic forms of heavy metal ions to less toxic forms have the potential for application in bioremediation. Among the different heavy metals discussed, the reduction of mercury and chromate by specific reductases is applicable for bioremediation. There have been trials to detoxify mercury salts and organomercurials in bioreactors, where resistance organisms reduce Hg^{2+} to Hg^0 and Hg^0 is adsorbed on glass or alginate beads. Organisms that synthesize macromolecules, capable of adsorbing high concentrations of heavy metal ions (e.g., metallothioneins are able to adsorb many divalent metal ions), could be used as biomasses to clean up contaminated metals from water and soil.

Fig. 5. Comparison of amino acid sequences of the broad-spectrum (pDU1358) and narrow-spectrum (R100) MerR protein. Asterisks indicate identical amino acids. One letter code for amino acids used. Cysteine residues are underlined.

See Also the Following Articles

ARSENIC • HEAVY METAL POLLUTANTS • MERCURY CYCLE

Bibliography

Ansari, A. Z., Bradner, J. E., and O'Halloran, T. V. (1995). *Nature* **374**, 371–375.

Cervantes, C., Ohtake, H., Chu, L., Misra, T. K., and Silver, S. (1990). *J. Bacteriol.* **172**, 287–291.

Cervantes, C., Ji, G., Ramirez, J. L., and Silver, S. (1994). *FEMS Microbiol. Rev.* **15**, 355–367.

Hobman, J. L., and Brown, N. L. (1997). *Metal Ions Biol. Syst.* **34**, 527–568.

Ji, G., and Silver, S. (1995). *J. Indust. Microbiol.* **14**, 61–75.

Misra, T. K. (1992). *Plasmid* **27**, 4–16.

Nies, D. H., and Silver, S. (1995). *J. Indust. Microbiol.* **14**, 186–199.

Rosen, B. P. (1996). *J. Biol. Inorg. Chem.* **1**, 273–277.

Schiering, N., Kabsch, W., Moore, M. J., Distefano, M. D., Walsh, C. T., and Pai, E. F. (1991). *Nature* **352**, 168–171.

Silver, S. (1996). *Annu. Rev. Microbiol.* **50**, 753–789.

Xu, C., Zhou, T., Kuroda, M., and Rosen, B. P. (1998). *J. Biochem* (Tokyo) **123**, 16–23.

Summers, A. O. (1992). *J. Bacteriol.* **174**, 3097–3101.

Wang, P., Mori, T., Toda, K., and Ohtake, H. (1990). **172**, 1670–1672.

Helicobacter pylori

Sebastian Suerbaum

University of Würzburg

Martin J. Blaser

*Vanderbilt University School of Medicine and
VA Medical Center*

GLOSSARY

chronic active gastritis The infiltration of the gastric mucosa with lymphocytes, plasma cells, and granulocytes. Essentially, all people colonized with *H. pylori* develop chronic active gastritis, but the intensity varies.

mucosa-associated lymphoid tissue (MALT) Tissue that is frequently acquired in the course of *H. pylori* gastritis. The "germ-free" stomach does not have MALT. The presence of gastric MALT can give rise to malignant B-cell lymphomas.

panmictic population structure A population structure that arises when recombination is so frequent in a given bacterial species that no remnants of clonal descent are discernible.

pathogenicity island (PAI) A large fragment of DNA in the genome of a pathogenic microorganism that contains virulence-related genes and has been acquired by horizontal gene transfer. Hallmarks of PAI are GC content that differs from that of the rest of the genome, insertion in proximity to tRNA genes, and mobility genes (e.g., insertion sequences).

peptic gastric/duodenal ulcer A breach in the epithelium of stomach or duodenum caused by an imbalance between aggressive factors (acid and pepsin) and mucosal protection mechanisms. Ulcers have a strong tendency to relapse and can progress to the potentially fatal complications of bleeding and perforation; removal of *H. pylori* colonization ameliorates ulcer disease.

urease An enzyme abundantly produced by *H. pylori* that hydrolyzes urea. It is essential for gastric colonization. Detection of urease activity is used to diagnose *H. pylori* by the rapid biopsy urease test and the ^{13}C or ^{14}C urea breath tests.

vacuolating cytotoxin (VacA) A novel protein that affects epithelial cell function. Multiple alleles exist that vary in VacA production *in vitro*. Colonization with particular genotypes affects the risk of disease development.

HELICOBACTER PYLORI is one of the most common bacteria affecting humans, colonizing more than half of the world's population. *H. pylori* colonizes the gastric mucosa, frequently persisting for the entire life of the host. This colonization invariably causes a chronic tissue response of the gastric mucosa, termed chronic or chronic active gastritis. Although this process is frequently asymptomatic, up to 10% of those colonized develop clinical sequelae, including gastric and duodenal ulcer, gastric adenocarcinoma, and malignant lymphoma of the mucosa-associated lymphoid tissue. *H. pylori* was first isolated in culture in 1982 and *H. pylori* research has since become one of the most rapidly moving fields in medical microbiology. Although the focus of this article is on data that are generally accepted, we have integrated a number of areas that are still controversial.

I. HISTORICAL INTRODUCTION

Pathologists noted the presence of spiral bacteria in the human stomach as early as 1906 (Krienitz).

Although similar observations were repeatedly reported during the subsequent decades, they did not receive much attention because the bacteria could not be cultured. The introduction of flexible fiberoptic endoscopes, and the establishment of microaerobic and selective culture techniques for *Campylobacter* species were important prerequisites for the first isolation in culture of *H. pylori* from human gastric biopsies by two Australian researchers, Barry Marshall and Robin Warren in 1982. The most important argument for a pathogenic role of *H. pylori* came from clinical trials showing that the elimination of *H. pylori* substantially changes the clinical course of ulcer disease. The elimination of *H. pylori* with antibiotic-containing regimens significantly reduced the high relapse rate of gastroduodenal ulcer disease (Hentschel, 1993; Sung, 1995). Thus, peptic ulcer disease, which previously could only be controlled by long-term treatment with inhibitors of gastric acid secretion or by surgery, became a condition that can be substantially improved by a short course of antibiotic treatment. However, the long-term consequences of *H. pylori* eradication are not known.

In the early 1990s, three large prospective seroepidemiological studies each indicated that *H. pylori* is a major risk factor for the development of gastric noncardiac adenocarcinomas and *H. pylori* was classified as a definitive carcinogen by the World Health Organization in 1994. Subsequently, *H. pylori* has been strongly associated with malignant non-Hodgkin's lymphomas of the stomach.

II. MICROBIOLOGY

A. General Microbiology

H. pylori was first designated *Campylobacter pyloridis*, then *Campylobacter pylori*; the new genus *Helicobacter* was established in 1989. It is a motile, urease-, catalase-, and oxidase-producing gram-negative eubacterium that is classified in the epsilon subdivision of the proteobacteria on the basis of 16S rRNA-sequence analysis. *H. pylori* requires rich media supplemented with blood, serum, or cyclodextrin and grows best at 37°C in a microaerobic atmosphere (5% oxygen is optimal). Colonies become visible after 2–5 days of incubation. *In vivo H. pylori* cells are spiral-shaped (see Fig. 1), whereas after *in vitro* culture cells with curved-rod shape predominate. After prolonged incubation or after treatment of cultures with subinhibitory concentrations of certain antibiotics, *H. pylori* cells assume a coccoidal form. Whether these coccoid cells are in viable-but-non-culturable (VBNC) state and of epidemiological relevance, or are the morphological correlate of cell degeneration and death is still the subject of controversy. *H. pylori* cells have 6–8 unipolar sheathed flagella.

H. pylori is the type species of the genus *Helicobacter*, which comprises more than 20 species, with more being discovered every year. *Helicobacter* sp. colonize the gastric mucosa of particular mammals; some have been widely used in animal models of gastric *Helicobacter* colonization. The best studied among these are *H. mustelae*, which naturally colonizes ferrets, and *H. felis*, which in nature colonizes cats and dogs, but experimentally also colonizes mice. *H. heilmanni* is a related organism uncommonly found in the human stomach. The genus *Helicobacter* also contains nongastric species, such as the human enteric pathogens *H. cinaedi* and *H. fennelliae*, intestinal colonizers (*H. muridarum* and *H. pametensis*), and bile-resistant organisms that first were isolated from the biliary tract and the liver of rodents (*H. hepaticus, H. bilis*), but which may in fact represent normal colonic biota.

B. Genome Sequence

H. pylori (strain 26695) has a circular chromosome of 1.668 Mb and an average G + C content of 39%; the genome contains 1590 predicted coding sequences (Tomb *et al.*, 1997). A second complete genome sequence (strain J99) was published in 1999 (Alm). Each of the two genomes contains a significant number of genes (89 -and 117, respectively) that are not present in the other strain, most of them located in a highly variable chromosome region (the "plasticity region"). Most strains isolated from patients with symptomatic disease harbor a 37-kb (presumed) pathogenicity island, the *cag* PAI. The genome is rich in homopolymeric tracts and dinucleotide repeats that permit phase variation of gene

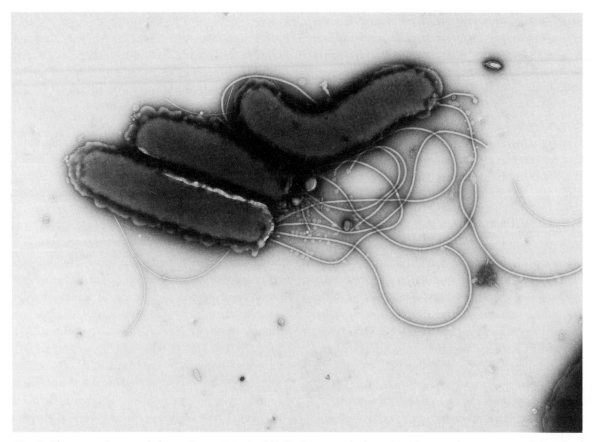

Fig. 1. Electron micrograph (negative staining) of *Helicobacter pylori,* strain N6. Note the presence of sheathed flagella attached to one cell pole. At the tip of the flagella, the sheath forms a characteristic terminal bulb. The length of the bar corresponds to 0.5 μm. EM courtesy of Dr. Christine Josenhans (Ruhr-University Bochum, Germany).

expression by frameshifts due to slipped-strand mispairing. Two fucosyl transferases involved in the synthesis of unusual oligosaccharides (Lewis[x], Lewis[y]) in the lipopolysaccharide (LPS) O-specific side chains may undergo such phase variation. The Lewis antigens are identical to human gastric epithelial cell antigens and may play a role in immune evasion by *H. pylori* (Appelmelk, 1998). Many *H. pylori* strains contain plasmids, whose biological functions are unknown.

C. Population Genetics and Evolution

The age of the species *H. pylori* and the point at which the *cag* PAI was acquired have not been determined, although most investigators believe that *H. pylori* has coevolved with humans for a very long

time. *H. pylori* is highly diverse, as indicated by unusually high sequence variation of both housekeeping and virulence-associated genes, as well as in variability of gene order. Much of this diversity is due to very frequent intraspecific recombination, leading to the shuffling of alleles and individual mutations among different strains, and thus to the rapid disruption of clonal groupings, generating linkage equilibrium (Suerbaum *et al.,* 1998). This high rate of recombination is, probably at least in part, due to the natural transformability exhibited by many strains of *H. pylori*. The population structure of *H. pylori* is largely panmictic, but weakly clonal groupings have been identified in *H. pylori* strains from different geographic regions (Achtman, 1999). For reasons unknown, significantly different allelic types have been conserved at some gene loci and some of these are statistically associated with an elevated risk of developing clinical disease (e.g., a fragment of *vacA* encoding the signal sequence) (Atherton, 1995).

D. Colonization and Virulence Factors

H. pylori colonizes the gastric mucus layer and epithelium, a niche of the human body that normally is not colonized by other bacteria (except for *H. heilmanni*). Humans and nonhuman primate species are the only known natural hosts of *H. pylori*; no reservoir in the inanimate environment has been identified. *H. pylori* has a strong tropism for the gastric epithelium. In the duodenum, it only colonizes areas where the normal duodenal mucosa has been replaced by gastric-type epithelium (gastric metaplasia). The colonized stomach contains two *H. pylori* subpopulations, one that moves freely in the viscous mucus layer that covers the gastric epithelium and others that are attached to gastric epithelial cells. The relative contributions of these subpopulations to persistence and tissue damage are unknown. Intracellular bacteria are rarely observed, and *H. pylori* is considered an extracellular organism. The pathophysiology underlying *H. pylori* colonization is incompletely understood. Progress was hampered by the lack of tools for the genetic analysis of *H. pylori*. The techniques for constructing isogenic mutants of *H. pylori* were first established by Agnès Labigne in 1990, and were pivotal to the development of virulence factor analysis in *H. pylori*. Although a number of bacterial factors have been implied in the pathogenesis of *H. pylori*, there is little experimental evidence to support a specific pathogenetic role for many of these. Here, we review only those factors for which there is strong experimental or clinical evidence that supports their role in colonization and tissue injury: urease, motility, adherence, the vacuolating cytotoxin, and the *cag* pathogenicity island.

1. Urease

H. pylori produces large amounts of urease, a nickel-containing metalloenzyme that catalyzes the hydrolysis of urea. *H. pylori* urease is a hexapolymer composed of two subunits, UreA and UreB, present in a 1:1 stoichiometry, to which are bound two Ni^{2+} ions (Mobley, 1988; Dunn, 1990; Labigne 1991).

Urease has multiple functions. In the presence of urea, urease permits *H. pylori* to maintain a constant internal and periplasmic pH, even when the external pH is strongly acidic, thereby preventing a collapse of the transmembrane potential difference. Isogenic urease-negative mutants of *H. pylori* are incapable of colonizing the gastric mucosa in several experimental challenge models in animals (Eaton, 1995). This inability to colonize could not be overcome by blocking the host animal's acid secretion, indicating that protection of the bacteria from acid is not the sole function of urease. Urease may permit the use of urea as a nitrogen source and contribute to tissue injury by the generation of ammonia as well as by the recruitment and activation of inflammatory cells.

2. Motility

H. pylori is a highly motile bacterium, due to a bundle of 6–8 unipolar flagella. Each flagellar filament is covered by a membraneous flagellar sheath, shielding the inner filament from low pH. The flagellar filaments of *H. pylori* are copolymers of two subunits, the flagellins FlaA and FlaB, whose expression is differentially regulated. Concomitant expression of both flagellin proteins is essential for full motility (Josenhans, 1995). In all, more than 60 genes are involved in the biogenesis of flagella, the assembly of the flagellar motor, and the chemotaxis system. Motility, as is urease, is essential for *H. pylori* to colonize of its host. Mutants that are defective in the synthesis of either one of the two flagellins are severely impaired in colonization efficiency, whereas nonmotile double mutants are completely avirulent.

3. Adherence

H. pylori can adhere tightly to human gastric epithelial cells, causing effacement of microvilli. More than 10 distinct *H. pylori* binding specificities for host-cell glycoproteins, carbohydrates, and phospholipids have been reported. Two adhesins have been characterized by molecular cloning and construction of isogenic mutants. The BabA2 adhesin (Ilver, 1998) is an outer-membrane protein that mediates *H. pylori* binding to the Lewis[b] tissue antigen. *babA2* is part of a large family of closely related genes encoding similar proteins (*Helicobacter* outer-membrane proteins, Hop). The Hop protein family comprises at least two further adhesins, the lipoproteins AlpA and AlpB (Odenbreit, 1999), whose receptor(s) on the host cell are not known.

4. Vacuolating Cytotoxin (VacA)

Approximately 50% of *H. pylori* strains produce a protein toxin that induces vacuole formation in eukaryotic cells (Leunk, 1988). In Western countries, toxin-producing *H. pylori* strains are more frequently isolated from patients with ulcers than from patients with uncomplicated *H. pylori* colonization, although toxin production is not essential for ulcerogenesis. In Asia, nearly all strains produce the toxin. The protein responsible for this vacuolization, VacA, first purified in 1992 by Cover and Blaser, is produced as a 140-kDa protoxin and actively secreted by means of a C-terminal autotransporter domain similar to that used for export of *Neisseria gonorrhoeae* IgA protease. The mature 87-kDa subunits present extracellularly form multimeric complexes that resemble flowers with six or seven petals. Exposure to acid induces conformational changes in VacA that increase activity. VacA perturbs intracellular-membrane trafficking by an unknown mechanism, inducing the formation of intracellular vacuoles derived from the late endosomal compartment. Although the extraordinary resistance of VacA to degradation by acid and pepsin, and its activation by acid indicate that it is highly adapted to an acidic environment, the *in vivo* role of VacA is not known.

5. cagA and the cag Pathogenicity Island

CagA is a large (120–140 kDa) highly immunogenic protein of unknown function (Tummuru, 1993; Covacci, 1993). About 70% of all *H. pylori* strains possess *cagA*, whereas > 90% of strains from patients with duodenal ulcer or noncardial gastric adenocarcinoma are *cagA+*. In Western countries, *cagA* is positivity statistically associated with more severe disease (more severe gastritis, more induction of pro-inflammatory cytokines, such as IL-8, a higher risk of ulcers, mucosal atrophy, and gastric carcinoma). Because the function of CagA itself is not known, these differences between *cagA+* and *cagA−* strains were not understood until it was discovered that *cagA* is a part of (and a marker for) a pathogenicity island (PAI). A number of the 29 proteins encoded

by the *cag* PAI are similar to the Vir proteins of the plant pathogen *Agrobacterium tumefaciens,* and Ptl proteins of *Bordetella pertussis*. These proteins form a type IV secretion system that transfers DNA into the cells of an infected plant or enables pertussis toxin secretion, respectively. It is conceivable that proteins encoded by the *cag* PAI also are involved in transfer of proteins or DNA to gastric epithelial cells, but this is hypothetical. In total, *cag+* strains have a much greater interaction with the host than do *cag−* strains, which is consistent with their greater association with disease states.

III. EPIDEMIOLOGY

H. pylori occurs in all parts of the world. The major determinants of prevalence are socioeconomic conditions and age. In developed countries, the overall prevalence is 40–50%, whereas it is much higher in developing countries. In most industrialized countries, the incidence of *H. pylori* acquisition has been decreasing, probably due to improved hygienic conditions. *H. pylori* is mostly acquired in childhood, and in developed countries the risk of an adult acquiring *H. pylori* is low (0.5%/year). The route of acquisition and infectious dose are unknown; there is evidence for fecal–oral and oral–oral transmission. Contaminated water supplies also may contribute to transmission. Direct transmission from person to person has been documented within families, between spouses, and in communities where people live together in close contact, such as orphanages. Young children may be the most important amplifiers for transmission of the organisms. *H. pylori* is equally frequent in women and men. Twin studies have shown that susceptibility to *H. pylori* contains a hereditary component.

IV. DISEASES ASSOCIATED WITH *H. PYLORI*

A. Acute and Chronic Active Gastritis

The acute acquisition of *H. pylori* is rarely diagnosed. It is characterized by no or nondiagnostic abdominal symptoms, including nausea and vomiting. Essentially all colonized persons develop a tissue response that is termed chronic active gastritis, characterized by the infiltration of the gastric mucosa with lymphocytes and plasma cells (chronic component) as well as neutrophils (active component). This response varies substantially in its intensity and anatomic distribution. However, most investigators believe that it is the specific nature of the response that affects clinical consequences, such as the risk of ulcer disease or cancer. *H. pylori* colonization has complex effects on gastric physiology (gastrin and somatostatin secretion, and acid secretion) that probably are dependent both bacterial and host factors.

B. Ulcer Disease

Peptic ulcers of stomach or duodenum are a common and potentially fatal condition. Duodenal ulcers are usually associated with *H. pylori,* although medication (nonsteroid anti-infammatory drugs, NSAID)-associated ulcers are becoming more common. Elimination of *H. pylori* substantially reduces ulcer relapses. The same holds true for gastric ulcers, although NSAID-induction is proportionally more common. It is in the treatment of ulcer disease that the discovery of *H. pylori* has had the most significant clinical impact. The recommendation is for *H. pylori* to be eliminated in all patients with ulcer disease, but there is preliminary evidence that this may increase the risk of esophageal disease.

C. Gastric Carcinoma

Large seroepidemiological studies have shown that the presence of *H. pylori* increases between three- and ninefold the risk for subsequent development of non-cardial gastric adenocarcinoma (Nomura, 1991; Parsonnet, 1991; Forman, 1991). Between 53 and 60% of all gastric cancers, or 500,000 new cases per year, can be attributed to the presence of *H. pylori.*

D. Gastric Lymphoma

Gastric B-cell non-Hodgkin's lymphomas are relatively rare gastric malignancies (about 1 per 1 million in the population annually) that in most cases arise from acquired mucosa-associated lymphoid tissue (MALT). *H. pylori* almost always induces the development of lymphoid follicles in the submucosa; in the absence of *H. pylori,* the stomach is usually devoid of MALT. Gastric MALT lymphomas are therefore very rare in patients without *H. pylori.* In low-grade MALT lymphoma, the proliferation of the malignant B-cell clones appears to be dependent on stimulation by bacterial antigens, probably explaining why *H. pylori* eradication may induce tumor remissions.

E. Esophageal Diseases

In the twentieth century, although *H. pylori* prevalence has dropped in developed countries, the incidence of esophageal diseases such as gastroesophageal reflux disease (GERD) and its sequelae, including Barrett's esophagus and adenocarcinoma of the esophagus, have increased dramatically, and the trend continues. Is there any relationship between the decline in *H. pylori* and the rise of these diseases? Preliminary evidence indicates that people carrying *cag+* strains have a substantially lower risk of GERD sequelae, including adenocarcinoma, than persons who are *H. pylori*-free. If these findings are confirmed, they will substantially change our evaluation of the clinical significance of *H. pylori.*

V. DIAGNOSIS

Diagnosis of *H. pylori* can be established by a gastric biopsy obtained during an endoscopy of the upper gastrointestinal tract, as well as by noninvasive techniques. The simplest biopsy-based test for *H. pylori* is the rapid urease test, which detects the abundant urease activity. The bacteria also can be visualized in tissue sections when specific histological-staining techniques (e.g., the Warthin–Starry stain) are used. Culturing *H. pylori* is rarely done in clinical settings, but it may be advisable to do so if the determination of antibiotic-susceptibility is important in particular patients.

Essentially all colonized people develop humoral (especially serum IgG) immune responses to *H. pylori* antigens, a property that can be used to diagnose its presence without endoscopy. An alternative way to diagnose *H. pylori* presence is by the ^{13}C or ^{14}C urea breath test.

VI. TREATMENT

The recommendation is to eliminate *H. pylori* in all patients with gastric or duodenal ulcer disease with low-grade gastric MALT lymphoma, and who have undergone surgery for resection of early-stage gastric carcinoma. Although *H. pylori* is susceptible to many antibiotics *in vitro,* no single agent achieves eradication rates above 55%, and thus combination therapies must be used. The standard treatment to eliminate *H. pylori* in 1999 consists of a combination of two antibiotics (most commonly the macrolide clarithromycin, in combination with either a nitroimidazole or amoxicillin) with an inhibitor of gastric-acid secretion (such as a proton-pump inhibitor), which are given for 7 days. These short-term regimens have low rates of side effects, and in clinical practice achieve eradication rates exceeding 80%. Increasing *H. pylori* resistance to nitroimidazoles and macrolides has reduced their efficacy; this resistance is due to the acquisition of point mutations in the nitroreductase gene *rdxA* and 23S ribosomal RNA genes.

VII. VACCINE DEVELOPMENT

When *H. pylori* was considered to be exclusively a pathogen, vaccine development appeared to be desirable. Accumulating evidence suggesting a protective role against esophageal diseases has tempered the recommendations concerning worldwide elimination of these organisms. Vaccine strategies in the future may be aimed at preserving colonization, but reducing the specific disease risks.

VIII. CONCLUSION

The isolation of *H. pylori* and analysis of its relation to disease has had an important impact on the clinical practice of gastroenterology, a phenomenon that continues to evolve. Because peptic ulcer disease had been considered for so long to be a medical disease without the consideration of a microbial role, the elucidation of the contributions of *H. pylori* also has reinvigorated the search for microbial participation in other diseases whose pathogenesis is at present poorly understood (e.g., inflammatory bowel disease, atherosclerosis, and biliary tract diseases). As our understanding of the interaction of *H. pylori* with humans matures, it provides answers to ecological questions about relationships with our commensal organisms.

See Also the Following Articles

ADHESION, BACTERIAL • ENTEROPATHOGENIC BACTERIA • GASTRO-INTESTINAL MICROBIOLOGY

Bibliography

Blaser, M. J. (1997). Ecology of *Helicobacter pylori* in the human stomach. *J. Clin. Invest.* **100,** 759–762.

Blaser, M. J. (1998). *Helicobacter pylori* and gastric diseases. *Br. Med. J.* **316,** 1507–1510.

Dunn, B. E., Cohen, H., and Blaser, M. J. (1997). *Helicobacter pylori. Clin. Microbiol. Rev.* **10,** 720–741.

Farthing, M. J. G., and Patchett, S. E. (eds.) (1998). *Helicobacter* infection. *Br. Med. Bull.* **54,** (1).

Hentschel, E. *et al.* (1993). Effect of ranitidine and amoxicillin plus metronidazole on the eradication of *Helicobacter pylori* and the recurrence of duodenal ulcer. *N. Engl. J. Med.* **328,** 308–312.

Marshall, B. J., and Warren, J. R. (1983). Unidentified curved bacilli on gastric epithelium in active chronic gastritis. *Lancet* **1,** 1273–1275.

Suerbaum, S. *et al.* (1998). Free recombination within *Helicobacter pylori. Proc. Natl. Acad. Sci. U.S.A.* **95,** 12619–12624.

Tomb, J.-F. *et al.* (1997). The complete genome sequence of the gastric pathogen *Helicobacter pylori. Nature* **388,** 539–547.

Hepatitis Viruses

William S. Mason

Fox Chase Cancer Center, Philadelphia

Allison R. Jilbert

Institute of Medical and Veterinary Science, Adelaide, Australia

I. Introduction
II. Viral Hepatitis
III. Enterically Transmitted Hepatitis Viruses
IV. Viruses Transmitted by Blood and Body Fluids

GLOSSARY

ALT (*alanine amino transferase*) Enzyme liberated into the bloodstream when hepatocytes are destroyed or damaged; serum levels are determined for evidence of hepatitis.

cytocidal Causing cell death.

cytopathic virus Virus whose replication causes damage to infected cells.

endoplasmic reticulum (*ER*) System of cytoplasmic membrane compartments involved in transport of proteins and lipids.

enteric viruses Viruses transmitted by fecal/oral route and that generally replicate in the intestine; although HAV and HEV are transmitted by the fecal/oral route, they are not strictly enteric viruses, since the intestine is not their major site of replication.

hepatocytes Major cell type of the liver, comprising ~70% of liver cells; a specific target of infection by viruses which cause hepatitis.

hepatitis Inflammatory reaction in the liver, with infiltration of leukocytes generally leading to liver cell damage or destruction.

hepatocellular carcinoma Cancer originating in the liver, probably from neoplastic transformation of hepatocytes.

internal ribosome entry site Site for ribosome entry and cap-independent initiation of translation of proteins.

nucleocapsid (*core*) protein Virus protein(s) that associates with viral nucleic acids to form the internal structure of an enveloped virus.

quasispecies Viral strains within the viral population in a single host. Derived by mutation during viral replication.

RNA-directed DNA polymerase; reverse transcriptase Virally encoded reverse transcriptase that synthesizes complimentary strands of DNA using an RNA template. Also active on DNA templates.

RNA-directed RNA polymerase Virally encoded enzyme that synthesizes complementary strands of RNA using an RNA template.

FIVE UNRELATED VIRUSES from diverse families are now recognized as major causes of human hepatitis. These are designated hepatitis A, B, C, D, and E viruses (HAV, HBV HDV, HDV, and HEV, respectively). All but HEV are endemic throughout the world, and two (HBV and HCV) have been proven to cause not only liver disease, but also liver cancer. Worldwide, at least 2,000,000 deaths per year can be attributed to these infectious agents. Vaccines have been developed against HAV and HBV, but are not yet in universal use. Three viruses (HBV, HCV, and HDV) cause chronic infections in a proportion of infected hosts; current therapies have been successful in eliminating virus from a relatively small fraction of chronically infected patients.

I. INTRODUCTION

The liver plays essential roles in many physiological processes, including energy storage and conversion, blood homeostasis, chemical detoxification, and innate immunity against microbial infections. The organ itself is composed of many cell types, including hepatocytes, which constitute almost 70% of the liver, bile duct cells, endothelial lining cells, and macrophages (Kupffer cells) (Gartner and Hiatt, 1997). Among them, two differentiated cell types that are abundant and unique to the liver are hepatocytes and bile ductule epithelial cells, which originate during embryonic life from a common progenitor. Many different viruses may infect cells of the liver (Zuckerman and Howard, 1979). It would be expected that viral infections that are unique to this organ might be restricted to hepatocytes and bile duct cells, which is generally the case. The ability of these viruses to infect other cell types in the liver and throughout the body is generally limited and, in some cases, the evidence for extrahepatic infections is contradictory.

At present, five viruses from diverse virus families are known to target human hepatocytes of healthy individuals and result in serious inflammatory disease (hepatitis). Of these, hepatitis A virus (HAV), a member of the *Picornaviridae*, and hepatitis E virus (HEV), a member of the *Caliciviridae*, are enterically transmitted viruses, which, though transiently present in blood, are normally transmitted via the fecal/oral route. This route of transmission results from secretion of virus into the bile, in which it is transported to the intestine. Viral transmission is facilitated by inadequate hygiene and poor sanitation, so that virus is eventually ingested and transported to the liver, probably by uptake from the intestine. The remaining three viruses, hepatitis B virus (HBV), a member of the *Hepadnaviridae*, hepatitis C virus (HCV), a member of the *Flaviviridae*, and hepatitis delta virus (HDV), of the genus *Deltavirus*, are transmitted from blood and mucosal fluids of infected individuals. This occurs via transfusion, needle sticks, bites, and high-risk sexual activities. These routes of transmission are the consequence of the secretion of viruses from infected hepatocytes into the bloodstream, where they may reach very high titers. New liver infections require the introduction of virus into the bloodstream of the recipient.

All five viruses cause major public health problems throughout the world. Transient, acute infections can lead to serious illness and, in some cases, death. Hepatitis B, C, and D viruses also have the ability to cause chronic, lifelong infections that may lead to death from progressive liver injury, cirrhosis, and/or liver cancer (hepatocellular carcinoma). These viruses cause at least two million deaths per year. The total numbers of deaths from acute HAV and HEV infections are much lower.

These five viruses do not account for all cases of hepatitis reported and other still uncharacterized viruses must exist. Perhaps 3% of the current cases of acute hepatitis, including 10% occurring subsequent to transfusion, are of unknown etiology. Searches for other transmissible agents recently led to the isolation of a new serum-borne virus, hepatitis G virus (HGV), or GBV-C. This virus is related to HCV in overall genetic organization, but lacks an obvious nucleocapsid–protein encoding region. However, subsequent work has called into question the claim that HGV causes hepatitis and whether it even infects the liver. What is clear is that this virus is found throughout the world and that it is especially prevalent in individuals who are also at risk for HBV and HCV infection. Even less is known about another recently described virus, TTV (transfusion-transmitted virus). These two viruses will not be considered any further because there is no convicing evidence that they can cause acute or chronic hepatitis.

II. VIRAL HEPATITIS

Infections by the five well-characterized hepatitis viruses can not be distinguished based on initial clinical observations. For all, the average time between infection and the appearance of overt liver disease is quite long, ranging from 2 weeks to several months. In general, the latent period is longer for HBV, HDV, and HCV, but this is not necessarily helpful for diagnostic purposes because of the wide range of incubation periods for patients infected with any of the serum-borne viruses.

In rare patients, there is a rapid loss of nearly the entire hepatocyte population (fulminant hepatitis), a situation which results in about 50% mortality. Fortunately, most infections are milder and most are probably not recognized as viral hepatitis. Rather, the patients experience mild flulike symptoms that resolve uneventfully within a few weeks. A time course for a "typical" bout of acute viral hepatitis is shown in Fig. 1. HAV and HEV infections resolve with lifelong immunity to reinfection. HBV, HCV, and HDV infections may become chronic, especially if they are not cleared

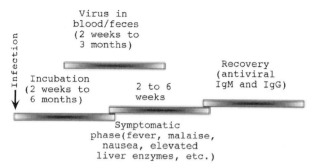

Fig. 1. Transient viral infections. Exposure is followed by an incubation period, in which virus migrates to the liver and spreads through the hepatocyte population. Incubation periods for HAV and HEV range from 2 to 6 weeks, HBV and HDV/HBV infections from 8 to 10 weeks, HDV superinfection of an HBV carrier, 4 to 6 weeks. Clinical symptoms then appear, preceded a few days earlier by a rise in liver enzymes in the blood. Symptoms for HAV and HEV infections last up to a few weeks, while the symptomatic phase of HBV and HDV infections may last 2–6 weeks. For all of the viruses, the symptomatic phase may be shorter, or not occur at all. The symptomatic phase signals the beginning of virus clearance and the appearance of IgM and IgG antibodies to the virus. The nature of transient HCV infections has not been determined. Fulminant hepatitis can occur following infection with any of the hepatitis viruses and is a clinical syndrome of sudden and severe liver dysfunction, resulting from submassive liver necrosis with impaired regeneration of hepatocytes.

within 6–12 months. Patients with such infections do not produce detectable neutralizing antibodies, have lower cell mediated responses than patients with acute hepatitis caused by transient infections, and may remain viremic for their lifetime.

It is curious that hepatitis viruses are not eliminated during the incubation period preceding the development of overt liver disease. There is no evidence that any of these viruses undergo an eclipse-phase in which virus proteins are not expressed; therefore, it might be assumed that the immune system would effectively recognize and kill infected cells before virus could spread throughout the liver. The fact that these viruses can cause serious liver disease, infecting a large fraction of the hepatocytes, suggests that this assumption is incorrect. Rather, it appears that the immune response often becomes effective only after a large fraction of hepatocytes are infected and are shedding virus.

A factor that could help these viruses establish

infection and spread throughout the hepatocyte population is that they are not cytocidal. Rather, they are capable of replicating and shedding progeny without killing the host hepatocyte. This lack of cell death mitigates the rapid induction of any nonspecific inflammatory response that would occur if there was a large amount of tissue destruction and which would also help to trigger the adaptive immune response. The consequence is that individuals can experience an extended period of well-being prior to the onset of liver disease, during which large amounts of virus are shed from infected hepatocytes and end up in blood or feces.

The events that trigger the symptomatic phase of acute viral hepatitis are unknown. However, it is generally agreed that the acute disease results from the immune response to the infected cells, with production of virus specific cytotoxic T lymphocytes (CTL) and antibodies and significant levels of hepatocyte destruction. Killing may be caused by CTLs, through their recognition of viral antigens on the cell surface, or may be nonantigen specific, due to the accompanying inflammatory response initiated by the CTLs (Chisari and Ferrari, 1997). Destruction of hepatocytes and the accompanying production of cytokines by the immune system are accompanied by a number of clinical conditions, including fever, loss of appetite, malaise, nausea, diarrhea, and abdominal discomfort or pain. During this phase of the infection, there is often a release of cytoplasmic proteins into the blood stream as a result of hepatocyte death. The detection of these proteins is a good indicator of acute liver disease in patients who otherwise have nonspecific symptoms. Standardized assays are available for a number of these proteins, including alanine amino transferase (ALT), aspartate aminotransferase (AST), and glutathionine S transferase (GST). A 10- to 100-fold rise in the level of ALT in the blood is not uncommon during acute viral hepatitis. On the other hand, these assays can not be interpreted as quantitative indicators of cell death.

Other serologic signs of liver dysfunction may be present. For instance, there may be an increase in bilirubin levels, perhaps progressing to jaundice, and a reduction in blood-clotting factors, which are produced by the liver.

The role of cell killing in clearance of liver infections is unknown. The fact that these can resolve rapidly, with minimal sequelae, after infection of the entire hepatocyte population, suggests that cell killing is not the sole basis for virus clearance. Otherwise, the host would die. It has been proposed that the cytokines produced during the inflammatory response may induce noncytocidal clearance of virus from infected cells. However, the relative roles of cell death and noncytocidal clearance are still unresolved.

HAV and HEV cause transient infection only. In contrast, the risk of developing chronic infection with HCV is at least 80%, irrespective of age, while the corresponding risk of developing chronic HBV

and HDV infection declines with age. All chronic infections probably induce some degree of liver damage throughout their course, mediated by the response of the immune system (Fig. 2). Carriers may be defined as "healthy," based on normal levels of liver enzymes in the bloodstream, or may have elevated levels of these enzymes, generally associated with more progressive liver disease. However, even "healthy" carriers can have ongoing liver damage. Irrespective of whether symptoms are present, cirrhosis and/or liver cancer may eventually develop. Also, in adult-acquired infections, a chronic active hepatitis may result in severe liver damage and death

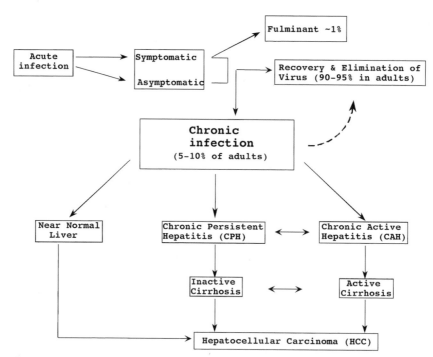

Fig. 2. Chronic hepatitis virus infection. Exposure of humans to each of the hepatitis viruses HAV, HBV, HDV, and HEV can result in transient infection, with or without symptoms. While infections with HAV and HEV are always transient and result in immunity to infection, infection with HBV, HDV, and HCV may either be transient or may progress to chronic (persistent) infection. In adult humans, HBV infection and HBV and HDV coinfections are transient in 90–95% of cases, while superinfection of HBV carriers with HDV and infection with HCV usually result in chronic infection. Those patients who develop persistent infection show varying degrees of liver damage that range from near normal liver to chronic persistent and chronic active hepatitis, identified by the extent of infiltration of inflammatory cells into the liver. Continued liver damage results in the deposition of fibrous connective tissue in the liver, resulting in cirrhosis. Persistent infection with HBV and HCV also predisposes to the development of liver cancer (hepatocellular carcinoma).

within a few years. Long-term carriers have a lifetime risk of ~25% of dying as a direct consequence of their infection.

III. ENTERICALLY TRANSMITTED HEPATITIS VIRUSES

Identification of the viruses that cause hepatitis depended initially upon the recognition of their different routes of transmission and only later on the development of more specific diagnostic assays. Epidemic hepatitis, caused by the enterically transmitted viruses, has been known for thousands of years, though the recognition of a viral etiology was only established in the early part of the 20th century. The hepatitis viruses transmitted by blood are not normally responsible for large epidemics of acute hepatitis, and their existence was only recognized during World War II, with the introduction of massive yellow fever vaccination programs, in which contaminated human serum had inadvertently been used as a stabilizing agent. The introduction of human plasma and blood transfusion at about the same time led to the recognition of posttransfusion hepatitis, which was assumed to be caused by the same agent. During the 1940s, the enterically transmitted virus was designated type A and the blood-borne, type B. A serological assay for HBV infection was not widely available until the late 1960s. HAV was finally visualized in feces extracts in the 1970s. The introduction of reliable screening assays for these two viruses confirmed the suspicion that there were additional enterically transmitted and blood-borne viruses prevalent in the human population, though their isolation did not come until much later.

A. Hepatitis A Virus

The prevalence of HAV infections in the United States is monitored by the Centers for Disease Control (CDC) in Atlanta (Table 1). Among the estimated annual caseload, about 100 are fatal, or about 0.1%. The seriousness of HAV infection generally increases with age, with the highest risk of liver failure and death occurring in older patients (Hollinger and Ticehurst, 1996). Generally, the rate of infection is

TABLE I

CDC Summary of Hepatitis Virus Infections in the USA*

Virus	Infections per year	Deaths per year
HAV	125,000–200,000	100
HBV	140,000–320,000	140–320 (acute infection) 5000–6000 (chronic infection)
HCV	28,000–180,000	8000–10,000 (chronic infection)
HGV	900–2000	0

* Data from the Hepatitis Branch, Centers for Disease Control, Atlanta, GA, Web site.

highest in the 5- to 30-year-old age group, and this early exposure confers lifelong immunity. Thirty percent of adults in the United States have detectable levels of antibodies to HAV, indicative of past infections. HAV is transmitted by the fecal/oral route and the main risk factors for infection are household contact with an infected person, sexual activity, and close contact with young children attending day-care centers.

The percent of previously infected adults is probably much higher in countries with lower levels of sanitation, and consequently, epidemics of hepatitis A among susceptible adolescents and adults are most common in countries with the highest levels of sanitation. Such epidemics of hepatitis A are often initiated by infected food handlers. Infected individuals may secrete very large amounts of virus (10^9 virus particles per gm of feces) during the incubation period, before symptoms of hepatitis develop.

1. HAV Replication Strategy

HAV is a member of the picornavirus family, which includes such well-studied agents as poliovirus and rhinovirus. Like other picornaviruses, HAV is a nonenveloped spherical virus with a diameter of ~30 nm. The virus has an icosahedral shape that is formed by 60 repeated units (protomers) that each contains all 4 viral capsid proteins, VP1, VP2, VP3, and VP4, in a tightly interlocking structure. Contained within the virus is the single-stranded, positive sense RNA genome of 7478 bases (Fig. 3). The viral genome has a virus encoded protein VPg attached to its 5′ end

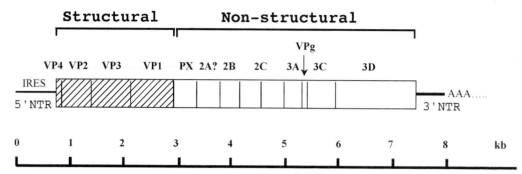

Fig. 3. Genome of the hepatitis A virus. The HAV genome is single-stranded positive sense RNA molecule of 7478 bases. The genome contains a 5′ nontranslated region (5′ NTR; solid line), followed by a large polyprotein open reading frame. The 5′ NTR contains an internal ribosome entry site (IRES). The HAV polyprotein contains 3 domains, P1 (structural), P2, and P3 (nonstructural). The shaded genes (VP4, VP2, VP3, VP1) encode the virus capsid or structural proteins. The genes that encode the nonstructural proteins include PX, 2A, 2B, 2C (putative helicase), 3A, VPg, 3C (cysteine protease), and 3D (polymerase). The genome also contains a 3′ nontranslated region (3′ NTR) and is polyadenylated.

and has a long 5′ nontranslated region (5′ NTR) that contains an internal site (IRES) for ribosome entry during cap-independent initiation of translation of viral proteins. Downstream of the 5′ NTR the viral genome contains a long open reading frame (ORF) that encodes a single polyprotein. Upon infection of cells the positive sense RNA genome is able to function directly as mRNA to form this viral polyprotein that undergoes posttranslational cleavages to yield the structural and nonstructural viral proteins. The HAV polyprotein contains 3 domains, P1 (structural), and P2 and P3 (nonstructural) (Fig. 3). P1 includes the genes VP4, VP2, VP3, VP1 that encode the viral capsid proteins. The nonstructural genes include PX (a virion associated protein of unknown function), 2A, 2B, 2C (a helicase), 3A, VPg, 3C (a cysteine protease), and 3D (the RNA dependent RNA polymerase). The HAV genome also contains a 3′ nontranslated region (3′ NTR) and is polyadenylated.

Although the polyprotein processing patterns and cleavage sites have been characterized in detail for some members of the picornavirus family, the sequential steps and cleavage sites of the HAV polyprotein have not yet been determined. Cleavage is thought to be mediated by the product of the 3C region, a cysteine protease that can function both in *cis* and *trans*. The first cleavage site within the polyprotein is thought to be at the carboxy terminus of PX, with a second cleavage leading to the production of VP1 (Lemon *et al.*, 1994).

Replication of HAV includes production of new strands of genomic RNA that are packaged in capsids and are exported from the cell. Genomic RNA strands are formed by copying of the incoming positive-sense genomic RNA by the viral RNA dependent RNA polymerase to form complimentary negative-sense RNA. This newly formed RNA is then used as a template for the synthesis of new positive-sense genomic RNA strands. The helicase encoded by the 2B–2C region is presumably involved in separation of RNA strands following transcription.

2. Unique Clinical Features

HAV is transmitted by the fecal/oral route, with rapid passage of virus by an unknown mechanism from the gastrointestinal tract to the liver. The specific receptor used by HAV has not been identified, but viral antigen expression has been detected in the small intestine in occasional crypt cells and in the liver in both Kupffer cells and hepatocytes. Replication of HAV occurs entirely within the cytoplasm of the hepatocyte, and unlike many picornaviruses HAV replication is noncytopathic. This is evidenced by an absence of cytopathic changes in HAV infected cells *in vitro* and the fact that peak shedding of HAV by infected tamarins *in vivo* precedes significant rises in ALT levels due to the immune response. After assembly, viral particles are released from the liver into the serum and the bile and are transferred to the feces.

Studies of the time course and tissue specificity of HAV infection have been recently performed by oral inoculation of owl monkeys (Asher *et al.*, 1995). In these studies, residual HAV from the inoculum was present in the feces for 1–3 days postinfection, no virus was detected for the next 2–3 days, then virus reappeared with highest concentrations from days 17–26 postinfection and declined to undetectable levels by day 39. The decline in virus in stools coincided with the development of anti-HAV antibodies on days 21–28 postinfection and elevated levels of ALT from day 35–38 postinfection.

The severity of disease due to HAV infection in humans is directly related to the age at which infection occurs. Acute HAV infection in children is generally asymptomatic, while the majority of adults have symptomatic infection and jaundice. Fulminant hepatitis and death due to HAV infection occur almost exclusively in the over-50 age group, with a reported death rate due to HAV infection of ~30 deaths per 1000 reported cases. Nearly all adult patients with clinically apparent disease experience complete recovery within 6 months. However, during recovery from acute HAV infection, single mild relapses have been reported to occur in 3 to 20% of patients. In these patients, relapse is milder and usually occurs within 3 weeks of the first clinical phase. Relapse is usually accompanied by continuing viremia and shedding of virus in the feces. The course of relapsing HAV is almost always benign and uneventful recovery is, with few exceptions, the rule.

Wild-type HAV strains have a slow and nonlytic growth cycle in cultured cells. However, strains of HAV have been adapted to growth in cell culture and have been used to develop formaldehyde-inactivated HAV vaccines. Intramuscular injections of children and adults with inactivated HAV vaccines have been found to produce minimal side effects and to be highly immunogenic, resulting in seroconversion in 100% of recipients.

B. Hepatitis E Virus

The development of diagnostic procedures for HAV and HBV led to the demonstration, by 1980, that large epidemics of acute hepatitis in developing countries, resulting from a contaminated water supply, were caused by a different virus or viruses (Bradley, 1995; Purcell, 1996; Krawczynski, *et al.*, 1997). The causative agent, Hepatitis E virus (HEV), was identified in 1990 and was found to be endemic and a frequent cause of sporadic hepatitis in these countries. Like HAV, HEV transmission occurs mostly by the fecal/oral route. However, the pattern of infection in endemic areas is different from that for HAV, with outbreaks appearing to target primarily, but not exclusively, teenagers and young adults, up to about 40 years of age. Thus, the prevalence of antibodies to HEV infection in endemic areas rises more slowly than the prevalence of anti-HAV immunoglobulins. Anti-HEV antibodies have also been detected in pigs and rodents in endemic areas, which raises the possibility of an epizootic component to these outbreaks.

Recently, a virus sharing about 80% sequence similarity to human HEV was partially cloned from domestic pigs in the midwestern United States, where it appears to be endemic. This virus does not have an obvious role in human HEV infections, as clinically recognized HEV is almost unknown in the U.S. However, it is not clear whether this virus, or some other virus, explains why 2% of the U.S. population have antibodies that react with HEV (as compared to, perhaps, 20–30% in HEV endemic areas).

1. HEV Replication Strategy

HEV is currently classified as a member of the Calicivirus family, with which it shares similarities in genome organization and virus structure. Like other members of this family, and also like HAV, HEV is a nonenveloped icosahedral virus, with a diameter of 27–34 nm. Also like HAV, HEV is stable at low pH and is presumed to be stable enough to pass through the acid environment of the stomach

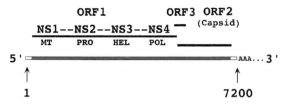

Fig. 4. Genome of the hepatitis E virus. The viral genome of a Chinese isolate of HEV (accession number L25547) is shown. This isolate includes a 26 base untranslated region at the 5′ end and a 63 base untranslated region at the 3′ end. This is followed by a polyA stretch of variable length (not shown). ORF 1 can encode a 1668 amino acid protein, ORF 2 a 660 amino acid protein, and ORF 3 a 123 amino acid protein. Studies in which the ORF 2 region that encodes the capsid protein has been expressed by itself suggest it contains a signal sequence, for transport into the endoplasmic reticulum, and is post-translationally modified by one or more cleavages, including removal of the N-terminal 111 amino acids, and glycosylation. A 50 kDa cleavage product was able to self-assemble into viruslike particles. The uncleaved and cleaved products of ORF 1 have not been characterized. The overall organization of functional domains includes a methyltransferase (MT), a cysteine protease (PRO), a helicase (HEL), and an RNA polymerase (POL). There is suggestive evidence for an additional domain between the methyltransferase and protease domain. The ORF 1 protein is presumed to undergo post-translational cleavages, but the cleavage sites have not been mapped.

and to be taken up in the intestine. HEV has a ~7.5 kb single-stranded, positive sense RNA genome that contains 3 ORFs (Fig. 4). Based on sequence comparisons, tentative functions have been assigned to the proteins that are encoded by the viral ORFs. ORF 1 encodes the nonstructural proteins that carry out genome replication, while ORF 2 encodes the viral capsid protein. The function of ORF 3 is not yet clear.

The ORF 1 polyprotein contains domains for a methyltransferase activity, presumably for capping of the viral RNA, a protease, for protein processing, a helicase, and an RNA polymerase (Fig. 4). Studies with Togaviruses (Schlesinger and Schlesinger, 1996), which are similar in genetic organization to HEV, have shown that uncleaved products of ORF 1 initiate negative strand synthesis. Positive strand synthesis is carried out by the more abundant, cleaved products of ORF 1; thus, posttranslational

processing plays a key role in regulating which strand of viral RNA is synthesized. However, it is still not known if this model is true for HEV.

The ORF 2 protein product is predicted to be 72 kDa with an amino terminal consensus signal sequence followed by a capsidlike region with a high content of basic amino acids. Cells transfected with ORF 2 contain viral polypeptides of 72, 63, 56, and 53 kDa and secrete a protein of 53 kDa. ORF 3 overlaps the 3′ end of ORF 1 and the 5′ end of ORF 2. Recent studies indicate that the ORF 3 protein may associate with the cytoskeleton of the cell, suggesting a role in the organization of replication complexes and/or polarized transport of viral particles to the apical cell surface.

The cell receptor used by HEV and the precise mechanism of replication, like that of many other viruses that are difficult to propagate in the laboratory, is unknown. Again by analogy to the more thoroughly studied Togaviruses, it is possible that only ORF 1 is translated from full-length viral RNA, while ORF 2 and ORF 3 may be translated from one, or perhaps two, subgenomic RNAs. Some evidence for the synthesis of such RNAs was obtained by sequencing of HEV-specific cDNAs derived from infected liver (Bradley, 1995; Krawczynski *et al.*, 1997). However, direct evidence of subgenomic transcripts and the existence of internal promoters of transcription have not yet been reported.

Finally, like HAV, HEV infection does not appear to be directly cytopathic. The mechanism for noncytolytic release of virus, presumably through the apical surface of hepatocytes into the bile canaliculi, is unknown.

2. Unique Clinical Features

The course of infection is similar to that of HAV, with some exceptions. HEV is generally believed to produce a higher mortality than HAV, at about 1–2% of infected patients. The infections are particularly serious to pregnant women, especially in the third trimester, with about 20% mortality. An increase in fetal loss may occur among surviving women. The reason for the high death rate in pregnant women is unknown; it has not been observed with the other hepatitis viruses. In addition to humans, a number of other primate species have been shown to be susceptible to HEV. However, despite shedding of virus-like particles in the feces, only mild hepatitis-like disease was observed and fulminant hepatitis was not seen. At present, there is no specific therapy for human HEV infections and an effective vaccine has not yet been described.

IV. VIRUSES TRANSMITTED BY BLOOD AND BODY FLUIDS

The major push to identify and isolate the blood-borne viruses came with the realization that exposure to blood and blood products could cause acute hepatitis. HBV was a major cause of acute viral hepatitis following blood transfusion, prior to the introduction of appropriate assays in 1970. During the 1960s, Blumberg and his colleagues discovered a novel blood-borne antigen in an Australian aborigine, which they named Australia antigen. This antigen was found to be common throughout southeast Asia. The detection of this antigen in the blood, coincident with an attack of viral hepatitis, led to the discovery that the Australia antigen was the envelope protein of the virus that causes hepatitis B.

With the introduction by blood banks of wide-spread testing for the Australia antigen, there was a major reduction in the incidence of posttransfusion hepatitis. However, even with careful screening and the development of increasingly more sensitive assays, posttransfusion hepatitis continued to affect about 5% of transfused patients. Hepatitis C virus, discovered in 1989, was found to be the cause of a vast majority of the remaining cases. A small fraction of transfused patients continue to develop hepatitis of unknown cause.

A. Hepatitis B Virus

Among the hepatitis viruses, HBV is the most prevalent, with estimated cases of chronic infection probably exceeding 200–300 million worldwide. The major causes of this high prevalence are perinatal transmission from chronically infected mothers to their newborns, and horizontal transmission to, and among, young children. Perinatal transmission predominates in Southeast Asia, where 5–10% of the

population is chronically infected. A similar prevalence is found in parts of Africa, although there, horizontal transmission to children beyond one year of age appears to play a more significant role than perinatal transmission.

An effective vaccine for HBV has been available for almost 20 years, though universal vaccination is only now being attempted in most countries. Originally, the vaccine consisted of Australia antigen that was purified from the blood of HBV carriers and extensively treated to eliminate infectious viruses. The current vaccine is similar to the original, but is produced in yeast by recombinant DNA technology.

The titers of virus in the serum of carriers range from 10^9 to 10^{10} per ml to undetectable (Mason, Evans, and London, 1998). The lower limit of detection may be about 10^6 per ml by nucleic acid hybridization techniques (to detect the viral DNA genome) to only a few per ml by the polymerase chain reaction. Interestingly, noninfectious virus envelope protein containing particles may persist in serum after virus becomes undetectable. Even when virus is detectable, envelope proteins are produced in excess of what is needed for virus assembly. The excess envelope protein assembles into "surface antigen" particles (HBsAg), which are also secreted from the cell. In serum, these particles are always present in at least 100-fold excess over the virions (Fig. 5, arrows). It is not known for certain why high titers of surface antigen particles, up to 100 μg per ml, persist after virus nearly or entirely disappears from the blood. The current belief is that this surface antigen is expressed from mRNA transcribed from viral DNA that has randomly integrated into host DNA while the virus was actively replicating. This is consistent with the observation that the liver may be essentially free of replicating viral DNA, despite the continued production of HBsAg. Some studies have suggested that nearly all hepatocytes may contain integrated HBV DNA at this stage of an infection.

Some studies suggest that virus production ceases, or at least drops below the detection limit of the available assays, in 5–10% of viremic carriers per year. This change is generally associated with the coincident appearance of antibodies to the viral e-antigen (HBeAg). HBeAg is a secretory protein encoded in part by the same ORF as the viral nucleocap-

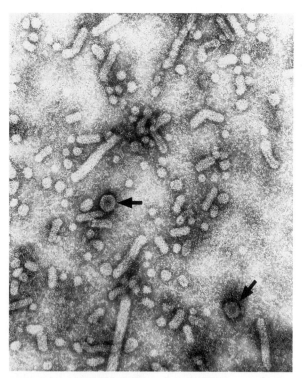

Fig. 5. Electron micrograph of a hepatitis B-like virus and surface antigen particles. Virus and surface antigen particles of woodchuck hepatitis virus, a close relative of HBV. Shown in the micrograph are the abundant spherical surface antigen particles, ~22 nm in diameter, the less abundant rodlike surface antigen particles, and the rare virus particles (arrows), 40 nm in diameter. (Micrograph courtesy of Tracy Gales, Fox Chase Cancer Center).

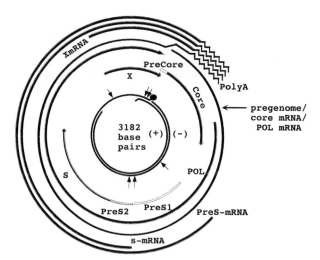

Fig. 6. Genome of the hepatitis B virus. The genome structure of HBV is compact, with overlapping open reading frames (out of register in the overlap). The double stranded DNA genome (inner circle) has the viral polymerase (pol) covalently attached to the 5′ end of the negative strand, reflecting its dual role as a primer of reverse transcription. A short RNA is attached to the 5′ end of the plus strand. In virus, the minus strand is complete, while the plus strand is generally less than full length. (Plus-strand completion occurs during initiation of new rounds of infection.) The large open reading frame, spanning nearly the entire viral genome, encodes the polymerase. The three envelope proteins L, M, and S all contain the S domain. The S protein spans S alone, while the M protein includes PreS2 and the L protein also includes PreS1. Host range specificity is encoded in the PreS1 domain. X encodes a protein which is required for viral infection, but is of uncertain function. X has displayed pleotropic activities, including transcriptional transactivation, in cell culture. The product of the Core domain is the viral nucleocapsid subunit. The product of preCore + core is a secretory protein (HBeAg) that is thought to somehow blunt the immune response to infection. Arrows indicate the start sites of viral mRNAs, all of which terminate at a polyadenylation site in the Core domain. One of the largest mRNAs, the core and pol mRNA, also called the pregenome, is the template for reverse transcription. Production of full-length negative strand DNA via reverse transcription is facilitated because the pregenome is terminally redundant.

sid protein (Fig. 6) and is secreted into the blood from hepatocytes that are producing virus. However, irrespective of whether or not anti-HBe immunoglobulins are present, and even if virus is not detected, all individuals with serological evidence of past infection (for example, antibodies to the viral nucleocapsid or core protein) should be considered as potential sources of new infection and should not donate blood. Indeed, about 30% of anti-HBeAg positive mothers may still transmit HBV to their unvaccinated newborns.

1. HBV Replication Strategy

HBV is a member of the hepadnavirus family that also includes viruses that specifically infect woodchucks (WHV), ground squirrels (GSHV), arctic squirrels (ASHV), ducks (DHBV), and herons (HHBV). HBV is a enveloped, spherical virus with a diameter of 42 nm. The virus nucleocapsid has an icosahedral shape and is formed by the viral core protein and contains the viral genome and polymerase. The genome of HBV is a 3.2 kb relaxed circular

DNA molecule that is converted in the nucleus of the cell to covalently closed circular DNA (cccDNA). This circular episome serves as the template for transcription of the viral mRNAs in the nucleus of infected cells (Seeger and Mason, 1996). Subsequent steps of virus replication take place in the cytoplasm. The virus encodes a reverse transcriptase, which binds to a terminally redundant RNA copy of the complete viral genome, known as the pregenome. The complex is then packaged into viral nucleocapsids assembled using the viral core protein. Viral DNA synthesis takes place within the immature nucleocapsids. These are then formed into virions, containing the three viral envelope proteins, by budding into the endoplasmic reticulum (ER), from which virus is then exported from the cell. It is thought that envelope protein is overproduced to prevent delivery of newly synthesized viral DNAs directly to the nucleus. The demonstration with DHBV that a block in envelope protein synthesis causes a rapid accumulation of viral DNA in the nucleus, resulting within a few days in cell death, supports this view.

2. Unique Clinical Features

One of the most striking attributes of chronic HBV infection is the high lifetime risk of liver cancer. It is estimated that 10–25% of individuals infected with HBV as children will develop primary liver cancer, which is almost always fatal. A classic prospective study of male civil servants in Taiwan, demonstrated that the risk of primary liver cancer was increased 100-fold in carriers vs noncarriers (Beasley, 1982). This association between chronic infection and liver cancer has been even more strikingly demonstrated in a related animal disease (Buendia, 1992). Woodchucks of the eastern United States are naturally infected with a virus, woodchuck hepatitis virus (WHV), which is closely related to HBV. Woodchucks infected as neonates have an almost 100% lifetime risk of liver cancer, which generally appears by the age of 2 years. In contrast, for humans, the interval between HBV infection and liver cancer is estimated to be up to 30–50 years.

Carcinogenesis by HBV is generally ascribed to chronic liver cell death and injury caused by the immune response to the infection. This idea is fostered by the failure to demonstrate that the virus is cytopathic for hepatocytes, and the observations that carriers develop a chronic liver disease that can most easily be attributed to a persistent, sometimes aggressive immune response to infected hepatocytes. Thus, in addition to mutagenic free-radicals that may be generated during a chronic inflammatory response, there is an elevated proliferation of the normally quiescent hepatocytes, to replace those destroyed by the immune system. This could provide the stimulus for mutagenized hepatocytes to proliferate into tumors. Most of these tumors also contain clonally integrated HBV DNA, which led to the hypothesis that the integration events are involved in cell transformation, possibly by a promoter insertion mechanism or other form of insertional mutagenesis. However, with only a few exceptions, the integration sites have not provided insights into the mechanism of carcinogenesis and, indeed, are scattered over many different chromosomes. Moreover, they are seldom expressed to produce any viral gene products except, in some tumors, the viral envelope proteins. This is in contrast to the woodchuck, where nearly all of the tumors contain viral DNA that is integrated near certain proto-oncogenes, N-myc1 and N-myc2, in association with a high expression of the respective gene in the tumor. Integrations have also been detected near the C-myc gene and in a locus (win) about 170 kbp from N-myc2. Integration at win induces expression of N-myc2.

The possibility that HBV encodes a weak oncogene has also been considered. The favorite candidate has been the X protein, which can activate transcription from many different promoters in cell culture, and which induced liver cancer in one strain of transgenic mice. However, the long duration of infection that generally precedes cancer and the lack of evidence that X is expressed in the majority of tumors, which usually produce little or no virus, makes it difficult to assess the role of X in carcinogenesis.

In addition to the risk of liver cancer, chronic HBV infection also carries a high risk of death from liver failure as a result of chronic injury, often leading to cirrhosis. Overall, the lifetime risk of death from liver disease is about 25% of the risk from primary liver cancer. The risk is even greater for patients who develop persistent infection as adults. Though the risk of chronic infection declines with age, from

>90% at birth to ~5% in adults, such infections tend to have more severe and rapidly progressing consequences in adults, primarily because of the aggressive immune response to the infection. While hepatocytes in normal adults have lifetimes that may exceed 6–12 months, hepatocytes in patients with a very active hepatitis may have life spans as short as 1–2 weeks. While this rapid turnover serves as an indicator of the severity of the attack by the immune system, it is the loss of tissue organization and scarring, with fibrosis and cirrhosis and the resulting disruption of the blood flow through the liver, that ultimately harms the patient the most.

Most patients show an early response to therapy with interferon alpha. However, at the end of treatment, after correcting for the spontaneous rate of clearance, only about 10–20% of patients sustain that response as a direct result of the therapy, with permanent loss of virus from the serum and normalization of ALTs. The therapy appears to work best in patients with active liver disease and it is presumed that the major effect of therapy is to stimulate the host's immune response to the virus. Work with mice transgenic for HBV expression suggests that the therapy could also induce noncytolytic loss of virus from hepatocytes (Chisari and Ferrari, 1997).

More recently, clinical use of lamivudine has been introduced. Lamivudine is a nucleoside analog that specifically blocks replication of HBV DNA and facilitates normalization of liver function in about 50% of patients. Selection of drug-resistant virus is a problem in these patients, but continued administration of lamivudine may still be beneficial. It is thought that the drug-resistant virus replicates less efficiently than the wild type and is, therefore, less pathogenic. Withdrawing therapy permits re-emergence of wild type virus which is still present in the liver.

B. Hepatitis D Virus

HDV was discovered in the 1970s after unexpected numbers of chronic HBV patients in Italy began presenting with acute hepatitis. Seeking a new variant of HBV as a causative agent, investigators looked for antibodies in their patients that would react with liver tissue collected during the exacerbation of liver disease. Antibodies were discovered which reacted strongly with the nuclei of a large percent of hepatocytes. A similar reaction was not observed with liver tissue collected from normal individuals or from typical HBV carriers. This antigen, termed delta, was later discovered to be a determinant of the only protein encoded by a new agent of serum hepatitis, hepatitis delta virus (HDV). HDV is a defective virus and is always found as a coinfection or superinfection with HBV, from which it obtains its envelope.

1. HDV Replication Strategy

HDV is unique among animal viruses (Taylor, 1996; Purcell and Gerin, 1996). It has a 1.7 kb, negative-sense, single-stranded circular RNA genome. This RNA replicates in much the same way as the viroids, the small infectious RNAs of plants. In particular, HDV does not produce its own RNA polymerase, but instead is thought to utilize the host RNA polymerase II. Following infection, the HDV RNA genome is transcribed in the nucleus into a 1.7 kb positive stranded, antigenomic RNA and a smaller, 0.8 kb polyadenylated mRNA. The mRNA is translated into the small 195 amino acid delta antigen (s∂Ag) (Fig. 7). The s∂Ag migrates to the nucleus where it associates with antigenomic RNA to form replication complexes for production of genomic RNA. Additional replication complexes also form with the newly synthesized genomic RNA. Some of the mRNA that is produced is modified as the result of an RNA editing event that changes a nucleotide in the stop codon of the s∂Ag ORF. This modification allows the synthesis of the 214 amino acid large delta antigen (L∂Ag). L∂Ag has two functions. First, it suppresses viral RNA synthesis. Second, it is essential for assembly, as it, along with s∂Ag, associates with viral RNA to form the ribonucleoprotein (RNP) core of the virus. Virus assembly is thought to occur by migration of viral RNP from the nucleus to the cytoplasm and budding into the ER. Budding requires the presence of the HBV envelope proteins, which form the outer coat of HDV. Thus, HDV can not be formed in hepatocytes which are not infected with HBV and/or producing the HBV envelope proteins.

The production of full-length antigenomic and genomic RNAs occurs by a rolling-circle mechanism, in which the RNA polymerase continuously traverses

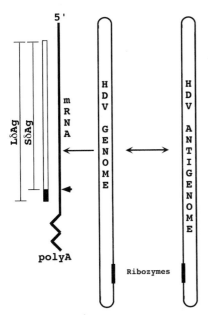

Fig. 7. Genome of the hepatitis delta virus. HDV has a 1.7 kb, negative-sense, single-stranded, circular RNA genome that can form a rodlike structure with 70% base pairing. A promoter element is located near the top of this rod and directs transcription of a 0.7 kb mRNA. This RNA is translated into the 195 amino acid s∂Ag, which is needed for viral RNA replication. As the result of an RNA editing event (arrowhead), the stop codon is changed from UAG to UGG, which encodes tryptophan and allows synthesis of the 214 amino acid LδAg, which shuts down RNA synthesis and facilitates packaging of newly synthesized genomes into virus particles. The same promoter element involved in mRNA synthesis may also serve to initiate synthesis of full-length antigenomic HDV RNA, which functions as template for new genome synthesis. The promoter for synthesis of genomic RNA has not been determined. Both the genome and the antigenome contain ribozymes (which cause self-cleavage), near the bottom of the rod, which allow conversion of greater than full-length precursors into the circular unit length RNAs. Folding of the cleaved RNAs into the rod structure inactivates the ribozymes.

around the circular templates to produce greater than unit-length RNAs. These are self-cleaved by a ribozyme that is present on both the positive and negative strands of viral RNA. Ligation to form circular RNAs then occurs. The ligation can occur *in vitro* in the absence of any proteins. However, it remains possible that the *in vivo* ligation is facilitated by a host-cell RNA ligase.

Of all the five hepatitis viruses, HDV may be the most cytopathic. Though mice are not a normal host for HDV and HBV, inoculation of HDV/HBV stocks will result in HDV infection of a small fraction of hepatocytes. This infection disappears within 1–2 weeks, even in severe combined immunodeficiency (SCID) mice, suggesting a direct killing of infected cells. On the other hand, the virus does not inevitably destroy a host cell, as it has been possible to produce transfected cell lines and transgenic mice that replicate HDV.

While HDV infections may become chronic, the greatest level of virus replication occurs during the acute phase, suggesting that subsequent replication is suppressed by the immune response. During this acute phase, HDV titers in the blood may be as high as 10^{11} to 10^{12} per ml. The immune response to HDV associated with the acute phase of infection may be responsible for the coincident suppression of HBV replication which has been observed. As pointed out earlier, there is compelling evidence that cytokines produced during an acute inflammatory reaction strongly suppress HBV (Chisari and Ferrari, 1997).

2. Unique Clinical Features

Transient coinfections with HBV and HDV follow the general course shown in Fig. 1. Chronic infection by the combination of HBV and HDV is more serious than chronic infection with HBV alone. About $\frac{2}{3}$ of these patients will develop cirrhosis, compared to perhaps $\frac{1}{3}$ of patients with HBV infection alone. Chronic infection with the two is also more likely to produce chronic active hepatitis (Fig. 2) than chronic HBV infection alone. There is no specific treatment for chronic HDV infections. However, vaccination against HBV will also prevent infection by HDV, and treatments that inhibit HBV and/or block HBV envelope protein synthesis should also inhibit spread of HDV within the liver.

C. Hepatitis C Virus

The attention focused on the relationship between chronic HBV infection and liver cancer led to careful screening for HBV in the serum and liver of patients who developed primary liver cancer without any other risk factors. In the mid-1980s, it was realized that a vast majority of new patients with primary liver cancer in Japan were not HBV carriers. Thus, it was assumed that these patients were infected with a nonA nonB hepatitis virus. Studies on liver cell morphology in nonA nonB patients raised the possibility that a member of the Flavivirus family might be responsible. The virus responsible for most of the unexplained cases of chronic nonA nonB hepatitis progressing to cirrhosis and liver cancer was finally identified in 1989.

1. HCV Replication Strategy

HCV is a member of the Flaviviridae and has similar genomic organization and mode of replication to the pestiviruses, which belong to a different genus of the same virus family (Houghton, 1996). HCV is an enveloped virus with a diameter of 50 nm. The virus nucleocapsid is spherical and is composed of the product of the core gene and an outer envelope that contains the two viral glycoproteins E1 and E2. Like HAV and HEV, it is a positive-sense single stranded RNA virus. The genome size is 9.5 kb. Translation of the genome gives rise to a single polyprotein that is cleaved to form individual virus proteins (Fig. 8). Besides the three structural proteins,

there are several nonstructural proteins (NS2, NS3, NS4a, NS4b, NS5a, and NS5b).

Studies of HCV replication have focused on the function of the nonstructural proteins rather than on RNA synthesis, since efficient virus replication has not been obtained in a cell culture system. Sequence homology studies suggested that NS3 should have both helicase and protease activities, while NS5b should be an RNA polymerase. These predictions have been borne out by cell culture experiments and experiments with purified proteins. However, virtually nothing is known about the exact mechanisms for synthesis of positive and negative strands of viral RNA and the differential packaging of positive strands into viral particles. Virus production is generally believed to be inefficient. Virus titers in serum generally do not exceed 10^8 per ml and can be much lower in some carriers. These titers are generally determined using quantitative RT-PCR or other very sensitive techniques.

A number of studies have attempted to define the roles of the HCV proteins in pathogenesis. These studies are preliminary and have yielded interesting but conflicting results. Mutations in the NS5a region of the genome have been linked to a lack of response

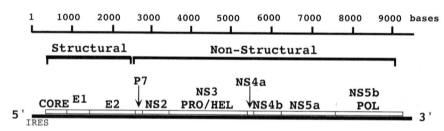

Fig. 8. Genome of the hepatitis C virus. The 9.3 kb genome of HCV has an ~340 base 5'-untranslated region that contains an internal ribosome entry site (IRES). Following this is a single ORF and then a 3' untranslated region of variable length, including a short homopolymeric region about 40 bases downstream of the ORF and, then, an additional, nonhomopolymeric, 98 bases. The structural gene region is cleaved to produce a 20 kDa core (nucleocapsid) protein and ~35 kDa (E1) and 70 kDa (E2) envelope glycoproteins. Cleavage of the nonstructural (NS) domain region gives rise to 7 and 21 kDa (NS2) proteins of unknown function, though NS2 includes part of a protease domain involved in self-cleavage. NS3 includes helicase and protease domains. While the upstream cleavages, except NS2, utilize host protease, the NS3 protease is required for cleavage at the remaining downstream sites. Proper function of NS3 requires association with NS4a. The functions of NS4b and NS5a are not clearly defined. NS5b has RNA polymerase activity. It is not known what role cleavages within the NS region play in regulating virus replication.

to interferon alpha therapy. The product of NS5a was found, in cell culture experiments, to inhibit PKR protein kinase, a host enzyme that is activated by interferon to shut down protein synthesis. Inhibition of PKR would presumably enhance virus survival. In contrast, other cell culture experiments have suggested that the core protein of HCV may be directly cytocidal, which would not enhance virus survival in an obvious way. Yet other cell culture studies have suggested that the core protein and the NS3 protease may function as oncogenes in hepatocarcinogenesis. It is not yet clear if any of these observations will be helpful in understanding how HCV function in humans who are chronically infected.

2. Unique Clinical Features

HCV causes primarily chronic infection, even in adults. HCV populations (quasi-species) show a great deal of genetic variation within patients over time, which probably contributes to their escape from immune surveillance. Chronic HCV infections also carry a high lifetime risk (at least 25%) of death from liver disease or primary liver cancer. The outbreak of primary liver cancer in Japan in the 1980s was traced to HCV dissemination by the widespread medical use of blood transfusion in the 1950s. Thus, as for patients with chronic HBV infections, primary liver cancer only becomes prevalent among carriers after many decades.

Antiviral therapies for HCV infection have been extensively sought because of the life-threatening outcomes of infection and the large numbers of infected patients world-wide. However, there is no highly effective therapy for chronic HCV infections. Interferon alpha has been extensively evaluated and induces a prolonged loss of virus from the serum and normalization of liver enzymes in about 10–20% of patients. Recent efforts have focused on combined therapy with interferon alpha and Ribavirin, a nucleoside analog. Treatment with Ribavirin (Virazole) alone does not markedly inhibit HCV replication, but the combination is reported to be more effective than interferon alone, especially in patients who have relapsed after a course of interferon alpha treatment. Ribavirin is thought to promote a cytotoxic T cell response to the infection.

See Also the Following Articles

GLOBAL BURDEN OF INFECTIOUS DISEASES • INTERFERONS • VIRUS INFECTION

Bibliography

Asher, L. V., Binn, L. N., Mensing, T. L., Marchwicki, R. H., Vassell, R. A., and Young, G. D. (1995). Pathogenesis of hepatitis A in orally inoculated owl monkeys (Aotus trivirgatus). *J. Med. Virol.* **47**, 260–268.

Beasley, R. P. (1982). Hepatitis B virus as the etiologic agent in hepatocellular carcinoma: Epidemiologic considerations. *Hepatology* **2**, 21S–26S.

Bradley, D. W. (1995). Hepatitis E virus: A brief review of the biology, molecular virology, and immunology of a novel virus. *J. Hepatology* **22**, 140–145.

Buendia, M. A. (1992). Hepatitis B viruses and hepatocellular carcinoma. *Adv. Cancer Res.* **59**, 167–226.

Chisari, F. V., and Ferrari, C. (1997). Viral Hepatitis. *In* "Viral Pathogenesis" (N. Nathanson, R. Ahmed, F. Gonzalez-Scarano, D. E. Griffin, K. V. Holmes, F. A. Murphy, and H. L. Robinson, eds.), pp. 745–778. Lippincott-Raven Publishers, Philadelphia, PA.

Gartner, L. P., and Hiatt, J. L. (1997). "Color Textbook of Histology." W. B. Saunders Company. Philadelphia, PA.

Hollinger, F. B., and Ticehurst, J. R. (1996). Hepatitis A virus. *In* "Fields Virology, 3rd ed." pp. 735–782. Lippincott-Raven Publishers, Philadelphia, PA.

Houghton, M. (1996). Hepatitis C viruses. *In* "Fields Virology, 3rd ed." pp.1035–1058. Lippincott-Raven Publishers, Philadelphia, PA.

Krawczynski, K., Mast, E. E., and Purdy, M. A. (1997). Hepatitis E: An overview. *In* "Viral Hepatitis and Liver Disease" (M. Rizzetto, R. H. Purcell, J. L. Gerin, and G. Verme, eds.), pp. 305–312. Edizioni Minerva Medica, Turin, Italy.

Lemon, S. M., Whetter, L. E., Chang, K. H., and Brown, E. A. (1994). Recent advances in understanding the molecular virology of Hepatoviruses: Contrasts and comparisons with hepatitis C virus. *In* "Viral Hepatitis and Liver Disease" (K. Nishioke, H. Suzuki, S. Mishiro, and T. Oda, eds.), pp. 22–27. Springer-Verlag, Tokyo.

Mason, W. S., Evans, A. A., and London, W. T. (1998). Hepatitis B virus replication, liver disease, and hepatocellular carcinoma. *In* "Human Tumor Viruses" (D. McCance, ed.), pp. 251–281. ASM Press, Washington, DC.

Purcell, R. H. (1996). Hepatitis E virus. *In* "Fields Virology, 3rd ed." pp. 2831–2843. Lippincott-Raven Publishers, Philadelphia, PA.

Purcell, R. H., and Gerin, J. L. (1996). Hepatitis delta virus. *In* "Fields Virology, 3rd ed." pp. 2819–2829. Lippincott-Raven Publishers, Philadelphia, PA.

Schlesinger, S., and Schlesinger, M. (1996). Togaviridae: The viruses and their replication. *In* "Fields Virology, 3rd ed." pp. 825–841. Lippincott-Raven Publishers, Philadelphia, PA.

Seeger, C., and Mason, W. S. (1996). Reverse transcription and amplification of the hepatitis B virus genome. *In* "DNA Replication in Eukaryotic Cells" (M. DePamphilis, ed.), pp. 815–831. CSHL Press, Cold Spring Harbor, NY.

Taylor, J. M. (1996). Hepatitis delta virus and its replication. In "Fields Virology, 3rd ed." pp. 2809–2818. Lippincott-Raven Publishers, Philadelphia, PA.

Zuckerman, A. J., and Howard, C. R. (1979). "Hepatitis Viruses of Man." Academic Press, Inc. New York, NY.

Heterotrophic Microorganisms

James T. Staley
University of Washington

GLOSSARY

autotroph An organism that uses inorganic carbon (carbon dioxide) as a carbon source for growth.

chemoheterotroph (chemoorganotroph) An organism that uses organic carbon both as a source of carbon and as a source of energy (most heterotrophs are chemoheterotrophs).

eutroph (copiotroph) An organism that grows well when the concentration of organic substances is high.

heterotroph An organism that uses organic material as a source of carbon for growth.

mineralization The complete decomposition of organic material to form inorganic compounds.

oligotroph An organism that grows well on low concentrations of organic carbon.

photoheterotroph An organism that uses organic material as a source of carbon for growth and light as an energy source.

HETEROTROPHIC MICROORGANISMS use organic carbon sources as their source of carbon for growth and synthesis of cell material. In contrast, *autotrophic* microorganisms derive their carbon from an inorganic source, carbon dioxide, from which they synthesize all their cellular material including proteins, fats, nucleic acids, and polysaccharides.

Heterotrophic organisms rely on autotrophic organisms, the primary producers, for the production of organic materials for their use. Therefore they occupy higher trophic levels in the carbon cycle. The Protozoa ingest particulate organic material in the form of other microorganisms or nonliving organic material. Many heterotrophic microorganisms are saprophytes, which derive their organic carbon nutrients from the decomposition of nonliving organic material. Some heterotrophic microorganisms live on or in close spatial association with other microorganisms, plants, or animals in various types of symbioses. An extreme of these associations occurs when the heterotrophic microbe is a parasite, predator, or pathogen and adversely affects the health or viability of its symbiotic partner or host.

The overall role of heterotrophic organisms in the carbon cycle is to recycle organic material into inorganic material, a process called mineralization. In this process, the organic carbon substrate is ultimately converted back into carbon dioxide, the substrate for autotrophic organisms, as well as other inorganic products such as ammonia, phosphate, and sulfate. This process of organic decomposition is accomplished by the concerted action of many different heterotrophic organisms including animals as well as microorganisms. But, because of their ubiquitous distribution, abundance, and diverse metabolic capabilities, it is the microbial heterotrophs that are primarily responsible for decomposition activities.

I. NUTRITIONAL GROUPS

A. Chemoheterotrophic versus Photoheterotrophic

Heterotrophic microorganisms are placed is subcategories depending upon their energy source

(Table I). Chemoheterotrophy is very common in the microbial world, so common, in fact, that the term heterotrophy normally refers to this group of organisms. *Most bacteria are chemoheterotrophic. All fungi and protists are chemoheterotrophic.* Both the bacteria and the fungi have rigid cell walls and, therefore, obtain their organic carbon in the soluble form from the environments in which they live. This type of heterotrophic nutrition is referred to as osmotrophic. In contrast, protists are microscopic organisms that lack rigid cell walls and feed by engulfing particulate materials. This type of feeding is referred to as phagotrophic. Therefore, the protists serve as microbial grazers that ingest bacteria, algae, and other microorganisms, as well as small particles of nonliving organic material (detritus) in the habitats in which they live.

Although bacteria and fungi have rigid cell walls that restrict them from ingesting particulate materials, some of them produce extracellular enzymes that allow them to degrade insoluble and refractory substances such as cellulose, lignin, and chitin (see following). The soluble by-products of degradation can then be transported into the cell as carbon and energy sources.

Photoheterotrophy is confined to representatives of three bacterial groups, the green gliding bacteria, the Gram-positive bacteria, and the photosynthetic Proteobacteria. These bacteria are able to use sunlight as their energy source but can use certain organic carbon sources as carbon for growth while growing anaerobically in the presence of light. *Chloroflexus* is an example of a green gliding bacterium; this is quite versatile nutritionally, being able to use inorganic as well as organic compounds as its carbon source. The genus, *Heliobacter,* is a Gram-positive bacterial group that is photoheterotrophic. Several genera of the so-called "non-sulfur" photosynthetic Proteobacteria use organic acids such as acetate or succinate or ethanol as a source of carbon and sunlight as their source of energy. Among these genera are *Rhodopseudomonas, Rhodosprillum, Rhodobacter, Rhodomicrobium,* and *Rhodocyclus.* In addition, some of the photosynthetic sulfur Proteobacteria can also use organic compounds during photoheterotrophic growth.

TABLE I
Nutritional Types of Microorganisms

Nutritional category	*Microbial group(s)*

I. Heterotrophic
  Derive carbon from organic sources
  A. Chemoheterotrophic
    Obtain energy from degradation of organic
    carbon sources
	Protozoa, fungi, most bacteria

B. Photoheterotrophic Obtain energy from sunlight	Some proteobacteria Some Gram positives

II. Autotrophic
  Derive carbon from inorganic sources
  A. Chemoautotrophic
    Obtain energy from oxidation of reduced inorganic compounds such as ammonia, nitrite, sulfide, sulfur, ferrous ion, or hydrogen gas
	Some bacteria

B. Photoautotrophic Obtain energy from sunlight	Algae, cyanobacteria, green filamentous bacteria, green sulfur bacteria, some proteobacteria

B. Oxygen Requirements for Growth

Many microorganisms (and all plants and animals) require oxygen for growth. Such organisms grow by aerobic respiration (see following) and are referred to as obligate aerobes. At the other extreme, the growth of some microbes is inhibited by oxygen. These are referred to as obligate anaerobes. Most of these organisms grow by fermentations, such as members of the bacterial genus *Clostridium.*

Facultative anaerobes are microorganisms that can grow either in the presence of oxygen or in its absence. Included in this group are such organisms as *Escherichia coli,* a common bacterium found in the intestinal tract of all humans, as well as many other warm-blooded animals. *Saccharomyces cerevisiae,* the common baker's and brewer's yeast, is another example of a facultative anaerobe. Facultative anaerobes may grow by fermentation when they grow anaerobically or they may use alternate terminal electron acceptors to oxygen in anaerobic respiration (thus, nitrate or sulfate can be used by some bacteria; see Metabolism section to follow).

Another group of microorganisms requires oxygen, but they cannot grow if the aquatic environment is saturated with oxygen at normal atmospheric pressures. These bacteria are called microaerophilic.

II. MICROBIAL GROUPS AND CLASSIFICATION

A. Prokaryotic versus Eukaryotic

Bacteria are **prokaryotic**, whereas fungi and protozoa are **eukaryotic**, microorganisms. They differ in several ways. First, and most distinctively, the cells of prokaryotic organisms are structurally more simple than eukaryotic organisms. The nucleus of prokaryotic cells contains only one chromosome (a double-stranded DNA molecule which carries the hereditary information of the organism in its genes) and this is not bound by a nuclear membrane (Fig. 1). In contrast, the nucleus of eukaryotic organisms contains more than one chromosome and is bound by a nuclear membrane (Fig. 2). Because eukaryotes have more chromosomes, their process of cell division, called mitosis, is more complex. Mitosis involves

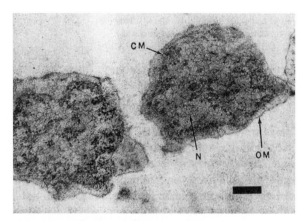

Fig. 1. An electron microscopic thin section showing the fine structure of a heterotrophic prokaryotic microorganism, the bacterium *Prosthecomicrobium enhydrum.* Note that the nuclear material (N) is not bound by a membrane. This is a Gram-negative bacterium and the outer cell-wall membrane (OM) appears as a double track structure. The peptidoglycan layer, not discernible here, is located between the OM and the cell membrane (CM). Bar = 0.2 μm.

replication and separation of the chromosomes from one another prior to cell division to ensure that each daughter cell receives a complete set. Mitosis does not occur in prokaryotic organisms.

A second major feature that distinguishes between these two groups is that eukaryotes have membrane-bound organelles within their cells, whereas prokaryotes do not. One example of this is the mitochondrion (Fig. 2). This organelle is actually about the size of a bacterial cell and is bound by its own protein/lipid cell membrane and even contains its own DNA and ribosomes. It is of interest to note that the DNA of the mitochondrion resembles that of bacteria in that it has no nuclear membrane around it, and, furthermore, its ribosomes are small, like the ribosomes of prokaryotes, which are smaller than eukaryotic ribosomes. Evidence from sequencing ribosomal RNA (rRNA) from the small subunit of the ribosome indicates that the mitochondrion evolved from a Proteobacterial ancestor. Likewise, the chloroplast, a membrane-bound organelle of phototrophic eukaryotes that contains the chlorophyll *a* for photosynthesis, evolved from a cyanobacterium.

Additional differences exist between prokaryotic and eukaryotic microorganisms (Table II). For example, the cells of eukaryotes are generally much larger

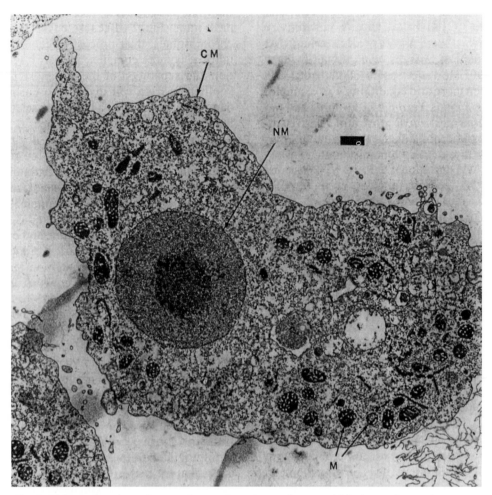

Fig. 2. Thin section through a protist, an *Acanthamoeba* sp., which shows the nuclear membrane (NM) around the nucleus as well as many mitochondria (M). Unlike fungi and most bacteria, there is no cell wall, only a bounding cell membrane (CM). Also, note that this eukaryotic microorganism is much larger than the bacterium shown in Fig. 1. Bar = 1.0 μm (courtesy of Thomas Fritsche).

compared to those of prokaryotes, and most eukaryotes are multicellular, a less common feature among prokaryotes.

Moreover, sexuality in bacteria is quite different from that of eukaryotic organisms. Eukaryotes have male and female mating types which produce special reproductive cells called sperm and eggs (or + and − cells). These cells fuse during mating (sexual reproduction) to form (either immediately or eventually) a fertilized diploid cell called a zygote, which receives one set of chromosomes from each parent. Prokaryotic organisms are never diploid but always

haploid, meaning that they do not have a paired chromosome. Some types of genetic exchange can occur between bacteria, but cell fusion, zygote formation, and diploidy are unknown.

B. Classification of Microorganisms

There are several groups of microorganisms. The bacteria and cyanobacteria are prokaryotic and the fungi, algae, and protozoa are eukaryotic microorganisms. Higher organisms, the plants and animals, are also eukaryotic.

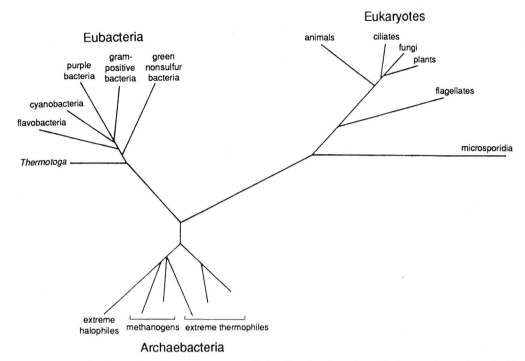

Fig. 3. Diagram of the three Domain system of classification based on phylogenetic studies of 16S rRNA sequences. The major groups include the Eubacteria (or Bacteria), the Archaebacteria (or Archaea), and the Eukaryotes (or Eucarya). Adapted from Woese, C. (1987) *Microbiol. Rev.* **51** 221–271.

It is now possible to place all organisms into a common classification. This breakthrough occurred through sequencing rRNA from the ribosome, a structure which all organisms have (Woese, 1987). The actual rRNA that has been sequenced from a variety of different organisms is that of the small subunit (16S rRNA of prokaryotes and 18S rRNA of eukaryotes). The resulting Universal Tree of Life (Woese *et al.*, 1990) indicates the vast diversity of microbial life. Three main branches of living organisms are evident: the Eubacteria and Archaea, which are prokaryotic, and the Eucarya, which are eukary-

TABLE II
Differences between Prokaryotic and Eukaryotic Cells

Characteristic	*Prokaryotic*	*Eukaryotic*
1. Nuclear features		
a. Nuclear membrane	–	+
b. Chromosome	1	>1
2. Mitosis	–	+
3. Cellular organelles		
a. Mitochondria	–	+
b. Chloroplasts	–	+ (if photosynthetic)
4. Ribosomes	Small	Large
5. Cell diameter (typical)	0.5–2 μm	>5 μm

otic. The Eubacteria contains typical bacteria and the Archaea contains some unusual bacteria that grow at high temperatures, at saturated salt concentrations, and produce methane gas (Table III). The Eukaryotic microorganisms include the Protists and Fungi.

Although they are biological entities, viruses are not considered organisms because they are not cellular, that is, they do not have a cytoplasm with enzymes and metabolic pathways and do not carry out reproduction in the same manner as organisms. Viruses will not be treated further here.

III. HETEROTROPHIC MICROORGANISMS

A. Prokaryotes

1. Characteristics of Eubacteria

As has been discussed, bacteria are prokaryotic organisms, and, based upon analyses of rRNA as well as other evidence, bacteriologists separate the bacteria into two different groups: Eubacteria (or Bacteria) and Archaea. The Eubacteria are more common bacteria, those with which most people are familiar. These are the bacteria that cause bacterial diseases (e.g., cholera, plague, bacterial pneumonia, whooping cough, tetanus, botulism, scarlet fever, bacterial dysentery, and diphtheria). They are also the most common organisms that grow in usual soil and aquatic habitats.

Almost all of the Eubacteria contain a cell wall constituent called peptidoglycan, which is not in Archaea or anywhere else in the biological world.

Peptidoglycan is an unusual polymer, consisting of a polymer of alternating amino sugars (glucosamine and muramic acid), that is interconnected with other amino sugar polymers by short segments of amino acids, called a peptide side-chain. Some of the amino acids in the peptidoglycan are D-amino acids, such as D-alanine and D-glutamic acid. These D-amino acids are found only in Eubacteria, and in Eubacteria they are never found in enzymes and other proteins, but only in the cell wall peptidoglycan or extracellular capsules (see Perry and Staley, 1997). It is of interest to note that the peptidoglycan is the site of action of the antibiotic penicillin (and its derivatives, e.g., ampicillin). Cells that contain peptidoglycan are susceptible to penicillin because they cannot synthesize their cell walls properly if this antibiotic is present in the environment. Since only Eubacteria have peptidoglycan, they are the only organisms that are inhibited by the antibiotic. This is an example of selective toxicity, a concept that explains how antibiotics can be used in human therapy to kill pathogenic bacteria without adversely affecting the patient (some humans may develop hypersensitivity reactions to antibiotic exposure).

There are many different groups of heterotrophic Eubacteria. Tables IV and V contain listings of them (for a more thorough treatment of them, please refer to *Bergey's Manual of Systematic Bacteriology,* which is a comprehensive compendium of the bacteria, or the more abbreviated treatment in *Bergey's Manual of Determinative Bacteriology*). *The Prokaryotes* (1992) should also be consulted for more information on the properties of specific groups. The Eubacteria are separated into two broad groups based upon their

TABLE III
Differences between Eubacteria and Archaea

Characteristic	Eubacteria	Archaea
1. Peptidoglycan	$+^a$	−
2. Membrane lipids	Fatty acids	Phytanyl groups
3. Unusual capabilities		
a. Methane production	−	Methanogens
b. Growth in saturated salts	−	Extreme halophiles
c. Growth at 70–115°C	−	Extreme thermophiles

[a] Some eubacteria do not produce peptidoglycan, e.g., the mycoplasmas lack cell walls entirely and, therefore, do not produce peptidoglycan.

TABLE IV
Phylogenetic Groups of Gram-Negative Heterotrophic Eubacteria

Phylogenetic group	*Representative genera*
Spirochetes	*Treponema, Leptospira, Borrelia*
Proteobacteria	
Aerobic spirilla	*Spirillum, Aquaspirillum, Campylobacter*
Aerobic rods and cocci	*Pseudomonas, Rhizobium, Agrobacterium, Methylomonas, Acetobacter, Legionella, Neisseria, Caulobacter, Myxococcus, Sphaerotilus, Leptothrix*
Facultative anaerobes	*Escherichia, Salmonella, Klebsiella, Erwinia, Yersinia, Proteus, Vibrio, Photobacterium, Haemophilus, Zymomonas*
Sulfate reducers	*Desulfovibrio, Desulfomonas*
Sulfur oxidizers	*Beggiatoa, Thiobacillus*
Cytophaga/Bacteroides	*Cytophaga, Flexibacter, Bacteroides*
Chlamydia	*Chlamydia*
Planctomycetales	*Planctomyces, Pirellula*
Verrucomicrobia	*Verrucomicrobium, Prosthecobacter*
Deinococcus group	*Deinococcus, Thermus*
Thermophilic groups	*Thermotoga, Thermomicrobium*

Gram stain reactions, a simple laboratory staining test which differentiates between two groups of bacteria on the basis of their cell wall structure. In gram-positive bacteria, the major cell component is peptidoglycan. Most gram-negative bacteria also have peptidoglycan, but they have another major cell wall layer in addition, called the outer membrane, which consists of protein, lipid, and some polysaccharide. Because of the simplicity and utility of the test in differentating bacteria, it is still commonly used by bacteriologists.

2. Characteristics of Archaea

Archaea do not produce peptidoglycan, a major difference between them and typical Eubacteria (Table III). They also differ from Eubacteria in the environments in which they live and activities in which they are involved. Most Archaea live in ex-

TABLE V
Phylogenetic Groups of Heterotrophic Gram-Positive Eubacteria

Group	*Representative genera*
Low mol% G + C Group	
Unicellular cocci	*Staphylococcus, Streptococcus, Sarcina, Leuconostoc*
Nonspore-forming rods	*Lactobacillus, Listeria, Caryophanon*
Spore-forming rods	*Bacillus, Clostridium, Desulfotomaculum*
High mol% G + C Group	
Irregular rods	*Arthrobacter, Corynebacterium, Propionibacterium, Cellulomonas, Actinomyces, Bifidobacterium*
Acid-fast bacteria	*Mycobacterium,* some *Nocardia*
Actinomycetes and streptomycetes	*Frankia, Actinoplanes, Streptomyces,*

treme environments and carry out unusual metabolic reactions.

There are three different physiological groups of Archaea (Table VI). One of the Archaeal groups, the methanogens, produce methane gas. These are obligate anaerobes that cannot grow in the presence of oxygen. Therefore, they are found in sediments or in special anaerobic environments, such as the rumen of cattle or the termite gut. There are two subgroups of the methanogens. One group produces methane gas from hydrogen gas and carbon dioxide. Since this group does not use organic material as carbon and energy sources, these bacteria are classified as autotrophs. The other group of methanogens are heterotrophic. They use methanol or acetate as carbon sources, from which they produce methane gas.

Another archaeal group is the extreme halophilic bacteria. These bacteria grow in saturated salt brines. They grow as ordinary heterotrophs and balance the external osmotic pressure by maintaining a high internal concentration of potassium ions. Some of these bacteria are able to carry out a primitive type of light-driven adenosine triphosphate (ATP) production by virtue of a special pigment they produce called bacteriorhodopsin, which is similar to the rhodopsin found in human eyes. In these bacteria, the pigment is associated with a membrane called the "purple membrane."

The final archaeal group contains extreme thermophiles as well as bacteria that grow at moderate and even cold temperatures. Some of these bacteria can grow at temperatures as high as 110–120°C. They grow in hot springs fed by volcanic fumaroles, such as those found in Yellowstone National Park or the deep sea hydrothermal vents. Some of them are autotrophic sulfur oxidizers, such as the genus *Sulfolobus*. *Sulfolobus* species oxidize sulfur to sulfuric acid and grow at temperatures approaching the boiling point of water. They are called thermoacidophiles because they grow at high temperatures and very low pH values (pH 1–6). This genus is facultatively autotrophic in that it can grow both autotrophically and heterotrophically. As an autotroph, it uses sulfur as an energy source and carbon dioxide as a carbon source; as a heterotroph, it uses complex organic carbon sources, such as yeast extract. Many of the sulfur oxidizing Eubacteria, such as in the genus *Thiobacillus,* are also facultative autotrophs.

Some of the extreme thermophiles use organic compounds that they oxidize while reducing sulfur compounds. For example, *Pyrococcus furiosus,* which grows at temperatures from 70–103°C, oxidizes organic components of yeast extract or peptone anaerobically and reduces elemental sulfur (S^0) to hydrogen sulfide in a type of anaerobic respiration.

Textbooks (e.g., Perry and Staley, 1997) provide more detailed information on the metabolism and physiology, as well as other properties of bacteria.

B. Eukaryotic Microorganisms

With the exception of the algae, which are eukaryotic photosynthetic microorganisms, all eukaryotic microorganisms are heterotrophic. These heterotrophic groups were originally classified as being either Protists or Fungi. However, recent work has resulted in their phylogenetic classification based on the sequence of small subunit rRNA. Table VII lists the various phylogenetic groups that are heterotrophic or have heterotrophic members.

Some of the most deeply branching protists lack mitochondria. Some of these are important parasites including *Giardia* (giardiasis) and *Trichomonas* (trichomoniasis). Most members of this group are beneficial symbionts that live in the digestive tracts of animals, where they feed on bacteria. The Euglenozoa contains some photosynthetic members such as *Euglena;* however, strains of *Euglena* that lack chloroplasts are heterotrophic and feed by phagotrophy.

TABLE VI
Phylogenetic Groups of Heterotrophic Archaea

Group	Representative genera
1. Extreme halophiles	*Halobacterium*
2. Extreme thermophiles[a] sulfur reducers	*Pyrococcus*
3. Methanogens[b]	*Methanosarcina*

[a] Many extreme thermophiles are autotrophic, using H_2S or hydrogen gas as energy sources

[b] Many methanogens are autotrophs that use CO_2 as their sole source of carbon and hydrogen gas as an energy source; however, some can use organic compounds, such as methanol or acetic acid.

TABLE VII
Major Groups of Heterotrophic Eukaryotic Microorganisms

Phylogenetic group	Representative genera or groups
Protists that lack mitochondria	*Giardia, Trichomonas, Euglena*[a], *Leishmania, Trypanosoma*
Euglenozoa	
Amoeboflagellates	*Naegleria*
Entamebids	*Entamoeba*
Other amoeboid forms	Foraminifera, Radiolaria, Heliozoa
Acellular slime molds	*Physarum*
Cellular slime molds	*Dictyostelium*
Ciliates	*Tetrahymena, Didinium, Paramecium*
Dinoflagellates	Most photosynthetic, but colorless phagotrophs exist
Apicomplexans	*Plasmodium*
Heterokonts	Most photosynthetic, but colorless types exist
Chytrids	*Synchytrium*
Fungi	
Zygomycetes	*Rhizopus*
Ascomycetes	*Saccharomyces,* truffles, morels,
Basidiomycetes	Puffballs, smuts, mushrooms, bracket fungi
Deuteromycetes	*Penicillium, Aspergillus*

[a] Colorless *Euglena*, which lack a chloroplast, feed entirely by phagotrophy.

This group contains tropical and subtropical parasites *Leishmania* (Leishmaniasis) and *Trypanosoma* (sleeping sickness).

The amoebae are quite diverse phylogenetically. Some free-living forms, such as the foraminiferans, are large enough to be seen by the naked eye. Some unicellular forms are pathogenic, such as *Entamoeba* (amoebic dysentery) and *Naegleria* (encephalitis). The cellular (e.g., *Dictyostelium*) and acellular (e.g., *Physarum*) slime molds graze on other microorganisms in natural environments.

The ciliate protozoa are quite diverse and live in a variety of aerobic and anaerobic environments. They are ideally suited for grazing on other microorganisms, particularly bacteria and other protists. Common genera include *Tetrahymena, Didinium,* and *Paramecium.*

The Dinoflagellates and Heterokonts are flagellated. Most are photosynthetic, however, colorless forms lacking chloroplasts are common and these are heterotrophic.

One of the most thoroughly studied protists is *Plasmodium*, which is the causative agent of malaria. Like many pathogenic protists, these organisms have

a complex life cycle and are difficult to treat by chemotherapy.

The chytrids and fungi are closely related phylogenetically. The chytrids are a group of aquatic heterotrophs. Like most fungi, they live as saprophytes on the organic remains of plants and animals. The fungi are more diverse. Some are plant or animal parasites or pathogens. All have rigid cell walls composed of cellulose or chitin or both. They derive their nutrition from soluble organic compounds.

The simplest type of fungi are the Zygomycetes, which produce zygospores in their sexual stage. One example of this group is *Rhizopus stolonifer,* also called the "black bread mold" because it is commonly found on stale bread. The Ascomycetes are very common. These include the unicellular fungi, the yeasts, as well as a group of macroscopic, mycelial types. The Ascomycetes produce their sexual spores in a distinctive structure called an ascus, or sac. The group of morels are ascomycetous mushrooms.

The Basidiomycetes are another very important group of the fungi. These comprise the most common mushroom and toadstool types. Thus, *Agaricus campestris,* the common edible store mushroom, is a basidiomycete. The basidiomycetes produce their sexual spores on a special fruiting structure called a basidium. These basidia are found in the aerial fructifications of these organisms. The basidiospores are dropped into the wind and are carried long distances in this manner.

The final group of the fungi are the Deuteromycetes (also called Fungi Imperfecti), a collection of organisms whose sexuality is not yet determined. This contains many of the common molds, such as the genera *Alternaria, Penicillium, Nigrospora,* and *Fusarium.*

IV. METABOLISM OF ORGANIC COMPOUNDS

A. Fermentation

The simplest metabolic process involved in the degradation of organic substances is fermentation. In this process, which occurs in the absence of oxygen, a variety of organic compounds can be broken down. For example, lactic acid bacteria can ferment sugars to produce lactic acid. This fermentation proceeds via glycolysis, a process referred to as the Emden–Meyerhof pathway. A sugar such as glucose is oxidized anaerobically to pyruvic acid and the chemical bond energy from the process is captured in the synthesis of adenosine triphosphate (ATP). The ultimate product of the pathway is lactic acid, not pyruvic acid. During the oxidation, a dehydrogenase enzyme containing the coenzyme, nicotinamide adenine dinucleotide (NAD^+), is reduced to $NADH + H^+$. As this accumulates in the cells, it, in turn, is used to reduce the pyruvic acid ($C_3H_4O_3$) to lactic acid ($C_3H_6O_3$). Therefore, the ultimate product of the pathway is lactic acid, not pyruvic acid. This fermentation is carried out by the gram-positive lactic acid bacteria, such as members in the genera *Streptococcus* and *Lactobacillus,* as well as in others. As in all fermentations, an organic compound (i.e., a sugar) is oxidized for energy and another one (i.e., pyruvic acid) is reduced.

Other sugar fermentations include the yeast fermentation by *Saccharomyces cerevisiae,* or brewers' yeast, in which ethanol and carbon dioxide are produced, or the mixed acid fermentation by the enteric bacterium, *Escherichia coli,* in which acetic, lactic, and succinic acids, as well as carbon dioxide and hydrogen, are formed. There are also bacteria that produce propionic acid (*Propionibacterium* from Swiss cheese), butyric acid, butanol as well as a variety of other products from carbohydrates. Amino acids can also be fermented by some bacteria, particularly the gram-positive spore-forming bacteria in the genus *Clostridium.* Even polymeric compounds such as cellulose and chitin can be broken down anaerobically by fermentative processes.

In fermentations, other organic products are usually formed, such as the organic acids already mentioned. In addition, gases such as hydrogen and carbon dioxide are frequent products. These substances can be used by other bacteria in the environments in which they are produced. For example, the hydrogen and carbon dioxide are substrates for methanogenic bacteria, which can obtain energy from the oxidation of hydrogen and form methane gas as a product of their metabolism.

B. Respiration

1. Aerobic

Aerobic respiration occurs in the presence of oxygen, O_2. This process is carried out by virtually all plants and animals. It is also a process used by many aerobic microorganisms that oxidize organic compounds. Many bacteria which respire have the enzymes of the glycolytic pathway and, therefore, produce pyruvic acid. If the organism is aerobic and has the tricarboxylic acid cycle (TCA cycle), then the pyruvic acid can be completely oxidized to carbon dioxide and water. The carbon dioxide is produced directly from decarboxylation of the intermediates, including pyruvic acid, citric acid, and apha-glutaric acid. Hydrogen ions and electrons are produced during the oxidation. The electrons are passed through an electron transport chain (cytochrome enzymes), which react with oxygen to produce the water. Hydrogen ions (protons) formed by this process are extruded from the cell to produce a proton gradient. ATP is produced in respiration from the ATPases that are located in the cell membrane and operate on the proton gradient established between the interior and exterior of the cell.

Therefore, it is possible for organisms that use

sugars as carbon and energy sources to oxidize these materials completely to inorganic products, i.e., CO_2 and H_2O, during the sequential processes of glycolysis and respiration.

Biological respiration is analogous to chemical combustion in that organic substrate is converted to carbon dioxide and water. However, in biological systems, the process occurs at much lower temperatures and part of the chemical bond energy is captured in the synthesis of ATP. This ATP can then be used in biosynthetic reactions to produce additional cellular material.

2. *Anaerobic Respiration*

A number of bacteria can grow in the absence of oxygen and carry out a process that resembles aerobic respiration. For example, nitrate is an oxidized inorganic substance (like O_2) that can accept electrons from the cytochrome respiration chain of some bacteria (like *Pseudomonas*). The result is that the nitrate is reduced to N_2 and N_2O. This process is referred to as a type of anaerobic respiration because it occurs in the absence of oxygen, yet organisms which carry it out have a TCA cycle and cytochrome enzymes. When nitrate is reduced to nitrogen gases, the process is called denitrification. Other bacteria can use oxides of iron and manganese as electron acceptors in respirations. These organisms can be found in the surface of sediments where oxygen is depleted.

Another type of anaerobic respiration occurs with sulfate-reducing bacteria, such as *Desulfovibrio* and *Desultomaculum*. These bacteria use organic compounds as energy sources and can oxidize them anaerobically in the presence of sulfate. The sulfate is reduced to hydrogen sulfide in this process, which is called sulfate reduction.

A final type of anaerobic respiration occurs with methanogens. One group of the methanogens uses carbon dioxide as an acceptor of electrons and reduces it to methane gas.

V. ECOLOGICAL CONSIDERATIONS

A. Microbial Loop

The roles of eukaryotic microorganisms in ecological processes are of considerable interest and impor-

tance. Because protists feed like animals, i.e., by phagotrophy, they are the simplest organisms that can ingest particulate food and degrade it internally. The microbial loop refers to the feeding of protists and a variety of larger micro-invertebrates that ingest microscopic particulate organic materials including bacteria and microbial phototrophs. Since most of the particulate organic material in aquatic habitats is either microbial or of microbial origin, the microbial loop is responsible for the degradation of most of the organic material.

B. Eutrophs and Oligotrophs

Bacteria and fungi are categorized further on the basis of their ability to use organic carbon sources, depending upon whether they grow in organic-rich environments or organic-poor environments. The most commonly studied bacteria, such as *Escherichia coli*, baker's yeast (*Saccharomyces cerevisiae*), and the animal and plant pathogens, grow in organic-rich environments, which can be readily simulated on artificial media, such as nutrient agar. These bacteria have been referred to as eutrophic or copiotrophic bacteria. Other bacteria grow in environments that have low concentrations of organic materials and these are referred to as oligotrophic bacteria.

In general, eutrophic bacteria have rapid growth rates (some can divide in less than 15 minutes) and low affinities for transport of nutrients into the cell (i.e., high K_s values). Oligotrophic bacteria, on the other hand, do not grow rapidly on any medium, rich or poor in nutrients, but have high affinities for their substrates (i.e., low K_s values). The eutrophic bacteria are referred to as r-strategists and the oligotrophs as K-strategists. The eutrophs can outcompete the oligotrophs in organic-rich habitats because their transport systems can provide nutrients to the organisms when they occur in high concentration and they can grow more quickly than the oligotrophs. However, in many natural habitats, such as the ocean and freshwater lakes and soils, organic carbon sources occur in low concentrations. In these habitats, the oligotrophs have the advantage because their nutrient uptake systems are effective at providing organic carbon sources to the organisms, and there is no advantage to rapid growth because the low concentrations of substrate will not support it. Not only do the oligotrophs grow well in such environments, but they are responsible for maintaining the low concentration of nutrients found in such habitats.

C. Substrates for Heterotrophic Microbes

1. Particulate Organic Material

Given the appropriate environment and microbial species, virtually all particulate organic materials can be degraded by microorganisms. Among the more refractory compounds found in nature are the polymers, such as cellulose and lignin, that are produced by plants. There are a number of microorganisms that can degrade cellulose. For example, some of the anaerobic ciliate protozoans that reside in the rumen, a stomach compartment of ruminant animals, can degrade cellulose. The basidiomycetes are a group of fungi that are also important in cellulose decomposition. Finally, many bacteria also produce cellulase. This includes both anaerobic bacteria and aerobic ones.

Until recently, it was not thought that microorganisms could break down lignin. However, it is now known that a certain group of fungi, called the "white rot fungi," are lignolytic. The ligninase enzyme is an unusual peroxidase that attacks this complex insoluble material and breaks several different bonds. Energy is not derived from the breakdown of lignin, but the enzyme is used by these fungi to allow the organism to remove the lignin from areas in the plant tissue to allow access to the cellulose, which is degraded by these organisms as an energy source.

Chitin is produced as the tough exoskeleton material of insects and crustaceans as well as certain other animals. It is similar to cellulose in that it is a hexose polymer whose subunits are linked by a β 1–4 bond. However, the subunit molecule is not glucose, but N-acetyl glucosamine. Chitin is degraded by some microorganisms, both bacteria and fungi, as well as by some higher animals.

2. Dissolved Organic Materials

As a group, microorganisms show great diversity in their metabolic capabilities. Thus, a large number of organic compounds can be degraded by them in-cluding polymeric materials such as polysaccharides, lipids, proteins, and nucleic acids.

The simplest organic compound is methane, which is a product of methane-producing archaebacteria. Methane can be broken down by bacteria called methanotrophs. This is a special group of bacteria that live in aerobic environments which receive methane, which is formed in an anerobic environment. Thus, these bacteria reside at the sediment (anaerobic)–water (aerobic) interface of many aquatic environments. In shallow ponds or bogs, bubbles of methane gas can be seen coming to the surface of the water and breaking. Therefore, not all of the methane is being oxidized in these habitats and some escapes into the atmosphere. In deeper habitats such as lakes, the bubbles are not seen because the methane oxidizers completely oxidize the methane. It is interesting to note that this group of methanotrophs are closely related to a group of autotrophic bacteria, the ammonia-oxidizing nitrifying bacteria. *Methylomonas* is an example of a genus of methanotrophic bacteria. They are members of the Proteobacteria, a large group of the gram-negative bacteria.

Many different types of bacteria can utilize organic acids, sugars, and amino acids as carbon sources for growth. Indeed, even organic compounds that are toxic to higher life forms, such as methanol, formaldehyde, and phenol, can be broken down by some bacteria, provided they are exposed to them at low concentrations. It is this principle that is used in the practice of bioremediation of toxic organic materials Thus, heterotrophic microorganisms are able to degrade organic compounds in contaminated sites, such as oil spills that have poorly soluble hydrocarbon compounds, and toxic waste dumps containing halogenated compounds.

See Also the Following Articles

AUTOTROPHIC CO$_2$ METABOLISM • BIOREMEDIATION • FERMENTATION • METHANOGENESIS

Bibliography

Balows, A., Trüper, H. G., Dworkin, M., Harder, W., and Schleifer, K. H. (eds.) "The Prokaryotes. A Handbook on the Biology of Bacteria. Ecophysiology, Isolation, Identification, Applications." (2nd ed.) (1992).

Holt, J. G., Krieg, N. R., Sneath, P. H. A., Staley, J. T., and Williams, S. T. (eds.), "Bergey's Manual of Determinative Bacteriology." (1994). Williams and Wilkins, 1994. Baltimore, MD.

Holt, J. G. (ed.), "Bergey's Manual of Systematic Bacteriology, Volumes I–IV." (1984–1989). Williams and Wilkins, Baltimore, MD.

Lee, J. L., Hutner, S. H., and Bovee, E. C. (eds.) "Illustrated Guide to the Protozoa." (1985). Allen Press, Inc. Lawrence, KS.

Moore-Landecker, E. (1990). "Fundamentals of the Fungi" (3rd ed.), Prentice Hall, Englewood Cliffs, NJ.

Perry, J. J., and Staley, J. T. (1997). "Microorganisms: Dynamics and Diversity." Saunders College Publishing, Harcourt Brace Jovanovich, Philadelphia, PA.

Woese, C. (1987). Bacterial evolution. *Microbiol. Revs.* **51**, 221–271.

Woese, C. R., Kandler, O., and Wheelis, M. C. (1990). Towards a natural system of organisms: Proposal for the domains Archaea, Bacteria, and Eucarya. *PNAS USA* **87**, 4576–4595.

High-Pressure Habitats

A. Aristides Yayanos

University of California, San Diego

GLOSSARY

barophile An organism growing or metabolizing faster at a pressure greater than atmospheric pressure.

hyperpiezophile An organism whose maximum rate of growth in all of its possible growth temperatures occurs at a pressure greater than 50 MPa. Only hyperpiezopsychrophiles have been isolated.

piezophile An organism whose maximum rate of growth in all of its possible growth temperatures occurs at a pressure greater than atmospheric pressure and less than 50 MPa.

pour tube A test tube containing nutrient medium that can be caused to gel by temperature change or the addition of another component. Bacteria added to the pour tube become immobilized by the gelling. Each immobilized bacterium grows and divides to form, eventually, a visible colony of cells. Pour tubes are well suited for high-pressure microbiology.

pressure The ratio of the force acting on a surface divided by the area of the surface on which the force acts. Atmospheric pressure at sea level is 0.101325 MPa. 1 Pa = 1 N/m^2 (Newton per square meter).

PTk diagram A graph of k, the exponential growth-rate constant of a microorganism, versus temperature T and pressure P.

THE INFLUENCE OF PRESSURE ON BIOLOGICAL SYSTEMS attracted the interest of the scientists at

Accademia del Cimento in Florence and of Robert Boyle in England for the first time during the seventeenth century. Science historian Stephen G. Brush credits Boyle for introducing "a new dimension—pressure—into physics." Euler provided the first mathematical definition of pressure in the eighteenth century. By the end of the nineteenth century, the concept of pressure had been developed into its present-day meaning. Pressure and temperature are today essential parameters for physical–chemical theory, for the description of environments, in industrial chemistry and biotechnology, and in both laboratory and ecological investigations of organisms. Fundamental to the analysis of organisms inhabiting high-pressure environments is that temperature and pressure are coordinate variables.

I. HISTORICAL BACKGROUND

Seventeenth-century biological research at the Accademia del Cimento and by Boyle quantified and described the effects of decompression from atmospheric pressure. Hot-air ballooning began in the late eighteenth century and provided another reason for studies in high-altitude physiology. Human problems similarly aroused interest in elevated pressures. Workers in diving bells, caissons, and in deep mine shafts often became ill and experienced physiological difficulties. Bert's (1943) classic treatise reviews many of these early studies. The interest in what we now term high-pressure habitats, epitomized by the deep sea, began in earnest in the nineteenth century. Certes, Regnard, and Roger in France did pioneering experimental work in biology. The question of how high pressure influences deep-sea life was, indeed, an impetus for their work. References to nineteenth-century and to early twentieth-century high-pressure research are in Johnson, Eyring, and Pollisar (1954).

Several oceanographic expeditions from the nineteenth century onward established the existence of life in the deep sea, the largest high-pressure habitat on Earth. Scientific research on H.M.S. *Challenger* during the first round-the-world oceanographic expedition from 1873–1876 showed the presence of animals throughout the seas to depths greater than 5000 m. Nearly 80 years later, participants in the Danish *Galathea* Expedition (1950–1952) demonstrated the presence of life in the greatest ocean depths. ZoBell published in 1952 the results of his work done on the Danish research vessel *Galathea*. He found that bacteria from 10,000-m depths in the Philippine Trench grow at pressures as great as 100 MPa, whereas bacteria from the upper ocean do not. This work, done with the most-probable-number method and microscopic examination of cultures incubated at high pressures, was with natural populations of microorganisms. Since 1979, pure cultures of deep-sea bacteria have been isolated by several investigators. Beginning in the late 1970s, expeditions using the deep submergence research vessel (DSRV) *Alvin* led to the discovery of hydrothermal vents at depths approaching 4000 m. At these vents, high temperatures and high flow rates of seawater containing copious nutrients result in remarkably localized and productive communities of microorganisms and animals.

High-pressure habitats in addition to those in the oceans are found in the seafloor and beneath the surface of the continents. These subterranean regions, also called subsurface environments, are inhabited by microorganisms. Geochemical analyses showed in the 1970s that microbes inhabit the seafloor to depth of hundreds of meters. The continental subsurface has also been studied for many years, as reviewed in Amy and Haldeman (1997). Microbiologists began only recently to sample a wide spectrum of subsurface environments. Although results unusual in terms of pressure have not been found, these will probably appear as deeper parts of the subsurface environment are explored. Petroleum reservoirs in the seafloor and in continental locales have yielded thermophilic bacteria. Continental sub-

surface habitats have been sampled by scientists in Sweden to understand and quantify microbial processes that could influence plans to bury radioactive waste. In summary, many regions of Earth beneath our feet are inhabited.

A. Definition of Pressure

A stress σ is the ratio of the magnitude of a force ΔF to the area ΔA, on which it is acting as $\Delta A \to 0$. Force acting on an area can be further resolved into components that are parallel and perpendicular to the area. The parallel components are the shear stresses and the perpendicular ones are the normal stresses. These are summarized by the stress tensor,

$$|\sigma_{ij}| = \begin{vmatrix} \sigma_{11} & \sigma_{12} & \sigma_{13} \\ \sigma_{21} & \sigma_{22} & \sigma_{23} \\ \sigma_{31} & \sigma_{32} & \sigma_{33} \end{vmatrix}$$

In fluids at rest and at a uniform temperature, shear stresses are absent or negligible ($\sigma_{ij} = 0$ for $i \neq j$), whereas the three normal stresses are equal to each other ($\sigma_{11} = \sigma_{22} = \sigma_{33}$). The normal stresses are called the hydrostatic pressure or simply the pressure, P, when dealing with fluids of uniform temperature and at rest. At any point in such a fluid, the pressure is the same in all directions. In solids, the pressure may or may not be hydrostatic. For example, some solids can have $\sigma_{11} \neq \sigma_{22} \neq \sigma_{33}$. The stress tensor is a comprehensive expression of stress distribution in a substance. Although the remainder of this article deals with hydrostatic pressure alone, the stress tensor is a reminder that the distribution of stress in organisms may not always be hydrostatic.

B. High-Pressure Environments on Earth and Other Planets

The pressure at the surface of Earth arises from the gravitational attraction between Earth and its atmosphere and is 0.101325×10^6 Pa = 0.101325 MPa (1 Pa = 1 Nm^{-2}). Because water is considerably denser than air, the pressure increases rapidly with depth in the oceans to reach approximately 110 MPa in its deepest trenches. The pressure in the ocean increases with depth according to the equation

$$dP = g\rho\, dz$$

where g is the gravitational constant, ρ is the density of the seawater, and z is the depth. On Earth at sea level, $g = 9.8$ ms^{-2}. Although the values of g and ρ vary with latitude, longitude, and depth, these variations in the oceans are small for microbiological considerations, so that the pressure in megapascals is approximately given by the depth in meters multiplied by 0.01013. At a depth of 10,000 m, the pressure in Earth's ocean is close to 101.3 MPa.

On Mars, $g = 3.71$ ms^{-2}. Therefore, if Mars once had an ocean with the same maximum depth as the one on Earth, then the maximum pressure in the Mars ocean would have been less than 38 MPa. Furthermore, if an ocean on Mars had been as cold or warmer than the one on Earth, then pressure would not have been a limiting factor for Earth-like life. However, if the Mars ocean had been substantially colder than the one on Earth, then the effects of pressure would have been pronounced even at 38 MPa. Obviously, if life on Mars had been substantially different from that on Earth, there is no way to surmise how pressure influenced it. New evidence suggests that there may be oceans on the Jovian satellites. Europa, a little smaller than our moon, has a $g \approx 1.23$. Callisto has a $g \approx 1.3$ and is about the size of the planet Mercury. The pressure in any oceans on these two moons of Jupiter will be also determined by both depth and the local gravitational field.

A list of high-pressure environments is given in Table I. There is no evidence that the Kola well at a depth of 8000 m or that hydrothermal vents at temperatures much greater than 115°C are inhabited. Figure 1 is a diagram of the *PT* plane, where *P* is the pressure and *T* is the temperature. Some of the environments listed in Table I are indicated in this figure as lines of constant pressure (isopiests) or constant temperature (isotherms) to show where life is found, as well as where it may be found. Only a small portion of all possible *PT* habitats have been studied. Chief among these are the atmospheric pressure isobar, the cold deep-sea isotherm, the Mediterranean Sea isotherm, hydrothermal vent isopiests up to 40 MPa, and a few subsurface habitats. Potentially

interesting habitats that have not been sampled enough are beneath the seafloor at water depths in excess of 5000 m. Figure 1 serves also to underscore the fact that the distribution of organisms is influenced and delimited by both temperature and pressure acting in concert. That is, the temperature range wherein life is found increases with pressure; and, quite possibly, the pressure range of life increases with temperature. There is a line of demarcation on the *PT* plane separating conditions compatible with life from abiotic ones. This *PT* envelope for life processes may be a continuous concave-inward curve. However, this is not known.

Life found is in most of the high-pressure environments listed in Table I. The world's oceans contain an abundant and diverse group of animals, bacteria, and archaea. The deep Earth is a home for mainly bacteria and archaea. Environments on Mars and Europa could contain microorganisms similar to those on Earth. However, incontrovertible evidence must await further exploration. Venus has a surface temperature of 227°C, which experiments and calculations suggest is far too high for life processes.

Table I and Fig. 1 show that Earth's oceans and seafloor have inhabited regions not easily reproduced on other planets and moons of the solar system. That is, almost all of the extreme and habitable pressure–temperature conditions are represented somewhere on Earth in a marine setting. It is interesting how oceans provide this variety of pressure–temperature habitats. The temperature of the sea surface is warmest at the tropical latitudes and coldest at the polar regions. The deep sea is a nearly uniformly cold habitat. Seawater at the poles is less than -1.5°C. The cold surface waters acquire a density less than that of the underlying ocean waters. Enough cold surface water forms and then sinks into the polar oceans to initiate and sustain deep-ocean circulation. Although sinking polar-water masses compress in a mostly adiabatic fashion, they become only slightly warmer as they sink. The net result is a deep sea that is preponderantly a cold environment with temperatures close to 2°C.

Only a few warm deep-sea environments exist and they arise in two distinct ways. One is through the presence of sills, elevated portions of the seafloor, that completely separate the deep parts of adjacent

TABLE I
High-Pressure Environments

	T (°C)	P (MPa)	Depth (m)
Earth's oceans			
Weddell Basin	−0.5	45.6	4500
Central South Pacific	1.2	50.7	5000
Central North Pacific	1.5	50.7	5000
Peru–Chile Trench	1.9	60.8	6000
Philippine Trench	2.48	101.3	10000
Tonga Trench	1.8	96.3	9500
Mariana Trench	2.46	110.4	10915
Celebes Sea	3.26	63.0	6300
Halmahera Basin	7.54	20.7	2043
Sulu Sea	9.84	56.5	5576
Mediterranean Sea	~13.5	50.7	5000
Red Sea	44.6	22.3	2200
Hydrothermal vent	~2–380	25.3	2500
Freshwater bodies			
Lake Baikal	3.5	16	>1600
Lake Vostok	>−5	*ca.* 35	3800 ice +500 water
Subsurface of Earth's continents			
Kola well in Russia	>155	88.3–205.9	8000
Subsurface of Earth's seafloor	0		
Nankai Trough, 30 m into sediments	~5	45.9	4530
6-km water depth, 1 km into seafloor	100	70.9	7000
Planetary environment			
Venus (at its surface)	227	9	0
Europa (conjectured values)	5	40	30000

basins. The sills block the entry of cold deep water derived from polar regions. Thus, the abyssal regions of these seas are warm, as shown in Table I. An example is the Sulu Sea, which receives its water from the South China Sea over a sill depth of 400 m, reckoned from the sea surface to the top of the sill. The temperature of the deep sea in the Sulu Sea is 9.8°C. Another example is the Mediterranean Sea, separated from the Atlantic Ocean by the Strait of Gibraltar sill at a depth of 320 m, which blocks the entrance of cold deep Atlantic water. Furthermore, the Mediterranean Sea has strong vertical mixing driven by evaporation, which forms a dense layer of warm water on the sea surface, which then sinks.

Thus, horizontal mixing as well as vertical mixing play a role in determining the temperature of about 13.5°C in the deep sea of the Mediterranean Sea. Figure 1 shows an isotherm at 13.5°C extending to a pressure of 50 MPa found at the greatest depth of the Mediterranean Sea.

The second cause for warm habitats in the otherwise cold deep ocean is hydrothermal circulation, especially at mid-ocean ridges. The temperature in geothermal water can be over 370°C. Mixing with cold ocean water occurs rapidly. A vent environment, called a vent field, is a highly localized region of hundreds of square meters to over a square kilometer. The water depth at most mid-ocean ridges is less

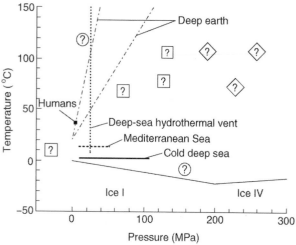

Fig. 1. Aspects of life in environments of different temperatures and pressures. The heavy black line is an isotherm at 2°C, which represents the cold deep sea inhabited to pressures greater than 110 MPa. The deep sea is populated with animals, archaea, and bacteria, and is the most prominent high-pressure habitat on Earth. The isotherm at 13.5°C, drawn as a dashed line, represents the habitats of the Mediterranean Sea and extends to 50 MPa. The vertical dashed line is an isopiest (line of constant pressure) at 25 MPa. This represents the pressure–temperature environment around a hydrothermal vent at 2500 m deep, where the maximum temperature observed, in excess of 350°C, undoubtedly exceeds the tolerance of organisms. The upper temperature limit for life has not been established. Many believe that it will be a temperature less than 150°C. Two lines show how temperature and pressure might increase along a depth profile into Earth on the continents. Similar lines could be drawn beginning, for example, at a pressure of 50 MPa and a temperature of 2°C to represent a temperature–pressure profile in the seafloor. The two question marks in circles are to indicate that we do not yet know the upper pressure–temperature limits of life. The question marks in squares represent extant temperature–pressure conditions on Earth where life could plausibly exist. The three upper squares are over conditions found deep in the seafloor. The square at negative pressures is at temperature–pressure conditions found in the xylem sap of tall trees. The question marks in diamond symbols are positioned at temperature–pressure conditions that possibly do not exist on Earth, but in which life could exist. The bottom line is approximately along the water–ice phase transition.

than 4000 m. Vents are inhabited to temperatures somewhat above 110°C.

Note that there is evidence that the deep sea was not always as cold as it is today. Estimates suggest that the temperature of the deep sea has alternated several times between 15 and 2°C over the past 800 million years. Thus, the comparative biology of organisms in the cold deep sea, the Sulu Sea, the Mediterranean Sea, and at hydrothermal vents has value for paleoecology.

II. INSTRUMENTS TO SURVEY AND SAMPLE HIGH-PRESSURE HABITATS

There are two distinct methods used to study life in high-pressure habitats. One is the *in situ* approach, whereby measurements, observations, and experiments are performed in the deep sea or other high-pressure habitats. The second is to recover an organism from the high-pressure habitat and conduct research on it in the laboratory. Organisms during the sampling process may suffer decompression, temperature change, contamination, exposure to light, and a change in habitat chemistry. Specially designed sampling instruments minimize or eliminate these changes. Instruments used in marine microbiology include Niskin bottles, pressure-retaining water samplers, pressure-retaining animal traps, pressure-retaining corers, and thermally insulated corers. Sediment microbiologists use cores collected mostly with decompression. Microbiologists sampling the terrestrial subsurface devise methods to minimize or detect contamination. Sometimes they flame the outer surface of a core to inactivate microbial contaminants. They also add to drilling fluids detectable particles whose presence in a core sample serves as a proxy for possible microbial contamination.

DSRV *Alvin* and DSRV *Shinkai 6500* are particularly effective for the collection of samples, both without the introduction of additional foreign microorganisms and from precisely identified locales. Scientists Kato, Li, Tamaoka, and Horikoshi of the Japanese Marine Science and Technology Center operated *Kaiko*, a ROV (remotely operated vehicle), with DSRV *Shinkai 6500* to successfully collect axenically sediments from the deepest part of the Marianas Trench in 1997. Axenic sampling of seafloor sediments and subterranean habitats is difficult because they are contaminated constantly with upper-ocean organisms by natural processes.

If every experiment required samples strictly maintained at habitat conditions, then research progress would be slow and expensive. Thus, the development of sampling strategies is important. Indeed, most deep-sea microbiology is on decompressed samples collected with Niskin bottles and a variety of coring devices deployed from ships. An inescapable sampling requirement for microbiology of the cold deep sea is the avoidance of sample warming. The justification for this is seen in Fig. 2, which shows the rapid thermal inactivation of a deep-sea bacterium at temperatures encountered in the upper tropical and temperate ocean through which sampling gear must pass. Also shown in Fig. 2 is the inactivation of a bacterium from the deepest part of the ocean while it is held at atmospheric pressure and 0°C. This shows that decompression per se, although lethal to this bacterium, does not instantly kill it. Death following decompression has so far been seen only in bacteria of the cold deep ocean. Most subsurface microbiology has been done with decompressed core samples from mesophilic and thermophilic habitats and with atmospheric pressure cultivation. Such samples have been adequate to establish enrichment cultures of mesophilic and thermophilic bacteria and archaea. Pressure-retention of the samples remains, however, as an essential aim in work addressing questions on community structure and function.

III. HIGH-PRESSURE APPARATUS FOR LABORATORY STUDIES OF MICROORGANISMS

The essential tools of microbiologists are enrichment culture technique, plating for the isolation of clones and assay of colony-forming ability (viability), replica plating for the isolation of mutants, and determination of growth rates in liquid culture. Each of these methods needs to be modified for the study of microorganisms having a strong preference or absolute necessity for high pressure.

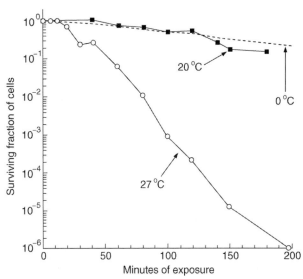

Fig. 2. Thermal inactivation kinetics for two bacterial isolates. The data are for bacterial isolate CNPT3, from a depth of 5782 m the Pacific Ocean. One suspension of cells was placed at 27°C and another at 20°C. At selected times, the cell suspensions were sampled and the pour-tube method (Fig. 3) was used to determine the fraction of cells that were still able to form colonies. This bacterium loses its viability (colony-forming ability) at atmospheric pressure when kept at temperatures greater than 10°C. Sensitivity to warming has not been determined for very many bacterial strains from the cold deep sea. Nevertheless, a high thermal sensitivity is probably one of their traits. One bacterial strain from the Marianas Trench, isolate MT41, loses its viability at 0°C at atmospheric pressure. The dashed line shows this slow rate of death for isolate MT41. This slow death rate may be due to decompression per se. Data from papers cited in Yayanos (1995).

The key components, commercially available, of a high-pressure apparatus are a pressure gauge, a pump (sometimes called an intensifier), a pressure vessel, and tubing connecting these components. Although pressure vessels can be purchased, the need to fabricate them with features otherwise unavailable often arises. The fluid used to compress a sample in a pressure vessel is usually water. Gases present problems as a hydraulic medium and are infrequently used. For example, they dissolve in biological phases and have effects other than those arising from compression. Also, the precautions needed to work safely with compressed gases make the work costly.

One of the technical difficulties in high-pressure

microbiology is that cultures must be isolated from the hydraulic fluid. Although liquid cultures have been incubated in syringes with success, the possibility of contaminants entering the culture around the syringe plunger must always be kept in mind, especially in long-term incubations. A great boon to high-pressure microbiological research has been the advent of heat-sealed plastic containers, such as bags and polyethylene transfer pipettes. These allow for pressure equilibration, isolation of the culture from microbes and chemicals in the hydraulic fluid, and a degree of control of gas composition in the culture. The latter is achieved through the selection of plastic-bag materials with appropriate gas permeability, allowing for the exchange of dissolved gases between the culture and the hydraulic fluid.

Clones of bacteria can be obtained at high pressure with pour tubes. Figure 3 shows colonies of a marine bacterium grown at high pressure. Parafilm, stretched tightly over the opening of a pour tube, sealed it from contamination and served to transmit the pressure to the medium. The pour-tube method can also be accomplished with heat-sealed polyethylene transfer pipettes rather than test tubes. The pour-tube technique works for high-pressure microbiology as petri plates do for atmospheric-pressure microbiology, with one exception; replica plating cannot be done with pour tubes.

Mesophiles and thermopiles from habitats having a pressure of less than 50 MPa grow in enrichment cultures incubated at atmospheric-pressure. For example, methanogens and sulfide oxidizers have been isolated through atmospheric-pressure enrichment cultures that were inoculated with samples from deep-sea hydrothermal-vent habitats. The enrichment-culture method is also known as the ecological method. Strict application of the ecological method would seem to require the incubation of the enrichment cultures at the pressure of the hydrothermal-vent habitat. If additional evidence, however, provides little doubt that an organism isolated at atmospheric pressure is an inhabitant of the sampled high-pressure habitat, then the need for conducting enrichments at high pressure seems diminished. All cultures of mesophilic and thermophilic organisms from deep-sea hydrothermal vents and the terrestrial subsurface have so far been found to be amenable

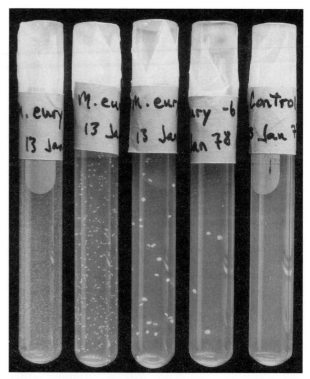

Fig. 3. Five pour tubes are shown. The one on the right was not inoculated. The other four, beginning with the tube on the left, are sere inoculated with a serial diultion of a culture of *Micrococcus euryhalis,* covered with Parafilm and incubated at high pressure. After several days, visible colonies developed around each immobilized bacterial cell. Note that the colonies are larger at the greater dilutions. This pour tube method is ideal for obtaining clones of bacteria in pressure vessels and for the assay of cell viability following the exposure of cells to chemical or physical stress.

to cultivation and plating at atmospheric pressure. Perhaps a requirement for a strict high-pressure technique will be necessary for the successful isolation of mesophiles and thermophiles from habitats having a pressure of 100 MPa. In contrast to mesophiles and thermophiles, the isolation and study of psychrophilic deep-sea bacteria is nearly impossible without the use of laboratory methods, especially the pour-tube technique, modified for high pressures.

An instrument called a pressurized temperature gradient (PTG) allows for the concurrent determination of growth parameters of bacteria incubated at different pressures and temperatures. It also allows for the isolation of clones of bacteria growing at different temperatures and pressures from a single inoculum. Sakiyama and Ohwada designed the latest model of a PTG and used it to study bacteria from the Japan Trench in 1997.

IV. DISTRIBUTION OF BACTERIA AND ARCHAEA IN HIGH-PRESSURE HABITATS

Bacteria and archaea are ubiquitous in the ocean, including its greatest depths with a pressure of approximately 110 MPa. Because animals are found at all ocean depths, as shown in Fig. 4, a highly plausible hypothesis is that animals, bacteria, and archaea

Fig. 4. Photograph of amphipods swarming to food (dead fish) placed on the seafloor at a depth greater than 10,500 m. The dead fish were tied to one end of an expendable pole and the 35 mm camera in a pressure-resistant housing was mounted at the other end of the pole by means of a time-release mechanism. The pole and other ballast were left on the seafloor when the time-release mechanism was activated to enable the floats to raise the camera to the sea surface. The amphipods are *Hirnodellea gigas*. Pictures such as this one suggest that the animals inhabit this depth. But this does not prove that the animals reproduce at this depth. These animals were not able to survive decompression to even 34 MPa in the few experiments conducted.

could exist at pressures even higher than 110 MPa. It is, further, likely that microorganisms could live at pressures beyond those limiting animal life. Perhaps an environment having a pressure of 200 MPa would be habitable by bacteria. If such an environment existed and were as cold or colder than the 2°C temperature of today's deep sea, then metabolic processes probably would be very slow and confined within narrow ranges of temperature and pressure.

Bacteria have been found beneath the surface of both the continents and seafloor. Among the principal factors bounding the distribution of life in subsurface environments are restricted space, as judged by the size of pores and the absence of cracks or fissures; and temperatures exceeding approximately 115°C.

V. PROPERTIES OF HIGH-PRESSURE INHABITANTS

Before any research had been done on deep-sea bacteria, it was known that physiological and biochemical processes in bacterial and eukaryotic cells were influenced by pressure. Single-cell organisms from atmospheric-pressure habitats typically show morphological and physiological aberrations when placed at 20–50 MPa. Given this background, the finding of pressure adaptation in deep-sea bacteria is expected. Exactly how adaptation is achieved, however, remains an active area of research. Current biochemical and molecular-biological research is primarily on pressure adaptation in gram-negative heterotrophic bacteria isolated from the cold deep ocean and on archaea from both the cold ocean and hydrothermal vents. The most obvious manifestation of adaptation to high pressure is seen in the organisms of the cold deep sea. Only there, for example, have microorganisms been found that grow exclusively at high pressure. Such bacteria have been isolated from samples of ocean depths between 6000 and 10,000 m. Nevertheless, pressure adaptation, although less conspicuous, is present in cold-deep-sea microorganisms inhabiting depths at least as shallow as 2000 m and in hyperthermophiles from submarine hydrothermal vents. The rest of this article describes some of the manifestations of pressure adaptation.

A. Rates of Growth

The kinetics of chemical transformations mediated by microbes and of microbial growth are of general interest because the associated rate constants are important parameters in models of food-web dynamics and biogeochemical cycles. The question of whether bacteria at a given depth in the sea grow more slowly than do their relatives at a shallower depth is difficult to answer. This is because the growth rate of bacteria is a multivalued function of not only temperature and pressure but also of other factors. These include pH of the medium, oxidation–reduction reactions, and nutrient types and levels. *Escherichia coli,* for example, grows 20 times more rapidly in complex nutrient medium than in a minimal medium with succinate as the sole carbon source. Experiments with a deep-sea bacterial isolate also show slower growth in a minimal medium. Remarkably, the growth rate of this bacterial isolate also exhibited a diminished response to pressure change when grown on glucose or glycerol minimal medium. To summarize, the pressure dependence of the growth rate is dependent on many factors.

Nevertheless, a comparison of the growth rates of various heterotrophic bacterial species grown in identical nutrient-rich medium at their respective habitat temperatures and pressures shows a trend for deeper-living species to have the slowest growth rates. Thus, the fastest growing deep-sea heterotrophic bacteria are usually from depths of less than 5 km. And, the slowest growing ones are generally from the greatest ocean depths of about 10 km. Growth rates (at habitat temperatures and pressures) of bacteria isolated from depths of 2–11 km are in the range of 3–35 hr, about as ZoBell (1952) reported.

B. *PTk* Diagrams

In pure culture, the growth rate of microorganisms can be exponential.

$$dN/dt = kN$$

where N is the number of cells at time t, and k is the exponential growth-rate constant. The pressure dependence of the growth rate of an exponentially growing bacterial species is widely used as an index of pressure adaptation. Thus, cells are called barophi-lic if they grow or metabolize more rapidly at high pressure than at atmospheric pressure. Bacterial growth rates in media having high levels of nutrients provide a view of the cellular adaptation to pressure in the cold deep ocean. In Fig. 5, the growth rates of five heterotrophic bacterial strains, isolated from six depths in the Pacific Ocean, are shown to be a function of both temperature and pressure. These plots, called *PTk* diagrams, where P is the pressure, T the temperature, and k is the specific growth-rate constant, summarize a large number of growth-rate determinations and reveal relationships otherwise difficult to see. Examples are as follows.

1. The pressure dependence of the growth rate is more pronounced in bacteria from the deepest habitats. Thus, strains MT199 and MT41 from deep habitats do not grow at atmospheric pressure, whereas strain PE36 from a relatively shallow habitat of 3.5 km grows at atmospheric pressure.

2. The *PTk* diagrams show that each bacterial species has a single pressure–temperature condition at which its growth rate is a maximum.

3. The pressure at which the growth rate has a maximum value is very nearly the pressure of the presumed habitat. This relationship between the pressure of maximum growth rate and the habitat pressure is not observed among heterotrophic bacteria from the warm deep-sea environments of the Sulu and Mediterranean Seas. This fact shows very clearly that growth rate as an index of pressure adaptation depends greatly on the temperature of the high-pressure habitat. It is conceivable that physiological parameters other than growth rate will provide useful indices of pressure adaptation for inhabitants of high-temperature, high-pressure environments.

4. A particularly curious characteristic of deep-sea psychrophilic bacteria is that the temperature at which the bacteria grow best is greater by 6–10 degrees than the constant habitat temperature of 2°C. This relationship, also evident in bacteria from the Sulu and Mediterranean Seas, is unexplained.

C. Physiological Classification of Bacteria and Archaea with Respect to Temperature and Pressure

Microorganisms are called psychrophiles, meso-philes, thermophiles, or hyperthermophiles based on

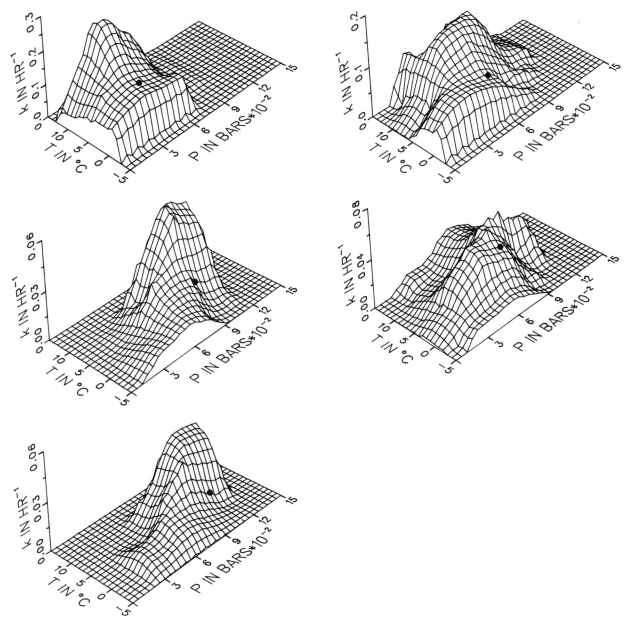

Fig. 5. PTk diagrams of five deep-sea bacteria. This shows that pressure adaptation becomes more pronounced as the habitat depth of an isolate increases. Each bacterial isolate has a maximum growth rate close to the pressure of its habitat (indicated by the black dot).

the temperature range in which they grow. Although there is probably a continuum of adaptations along the temperature scale, this classification is useful. There are two ways microorganisms are classified with respect to their growth as a function of pressure. The traditional classification is to call them barophiles if they grow best at a pressure greater than atmospheric pressure. This binary classification of

the pressure-adaptive trait is thus different from the one used for classifying organisms based on their response to temperature. Furthermore, growth studies of bacterial isolates from high-pressure habitats of different temperatures make it apparent that an organism can be barophilic at one temperature but not at another. Significantly, there are bacteria showing little or no barophilic character at the temperature of their habitat, but being distinctly barophilic at a greater temperature. This categorization of microorganisms with respect to pressure was created before there were any studies with axenic cultures of deep-sea bacteria and of bacteria and archaea from high-pressure habitats having different temperatures. A newly suggested categorization of a given bacterial isolate is based on its *PTk* diagram and is analogous to the scheme used for temperature adaptation. If the pressure P_{kmax} at which the maximum growth rate occurs, as determined with a *PTk* diagram, is

$$0.1 \text{ MPa} < P_{kmax} < 50 \text{ MPa}$$

then the isolate is a piezophile. If

$$P_{kmax} > 50 \text{ MPa}$$

then the isolate is hyperpiezophile. Table II shows the possible types of microorganisms based on where the pressure and temperature of the maximum growth appears on a *PTk* diagram. An inspection of Table II shows that this is a more useful classification than to simply state that an organism is a barophile. Also evident in Table II is that hyperpiezophiles have only been found in the cold deep ocean—that is, among the psychrophiles of a high-pressure habitat. As discussed in the context of Figure 1, the possibly inhabited environments having a pressure of approximately 100 MPa and a temperature around 100°C may exist only in the seafloor beneath the deepest parts of the ocean. It is there that hyperpiezomesophiles, hyperpiezothermophiles, and hyperpiezohyperthermophiles perhaps reside. To summarize, the scheme in Table II provides a succinct view of states of adaptation to both temperature and pressure.

D. Molecular Biology and Biochemistry of Adaptation to High Pressures

The structure of deep-sea bacteria viewed with light and electron microscopies is the same as that of bacteria in general. The adaptations making possible their existence at high pressure can be presumed, therefore, to be at the molecular level. Several macromolecular structures and interactions, as well as chemical reactions, are altered by pressure change. These include microtuble assembly, ribosome integrity, helix-coil transitions in nucleic acids and proteins, protein conformation, protein–protein interactions, protein–nucleic acid interactions, enzyme activity, transport across membranes, and membrane fluidity. The list of pressure-affected processes is long. Bacterial-membrane protein composition is also a function of pressure. The altered composition reflects pressure action on gene expression, on protein synthesis, and on membrane function. The changes observed are further dependent on the physiological state of the cell. For example, protein profiles not only change with pressure, but also in a different way when cells are grown on different carbon sources. Thus, the adaptation of marine microorganisms to nutritional states alters the manifestation of pressure adaptation. The final view of pressure and temperature adaptation will most likely be a combination of both an understanding of the pressure and temperature dependencies of many individual processes and a mathematical model of how the cell operates as a dynamic system. The latter is necessary because biological systems are quintessentially greater than the sum of their parts.

The fatty-acid composition of the membrane phos-

TABLE II
Categorization of Microorganisms[a]

	P of k_{max}		
T of k_{max}	$P \approx 0.1$ MPa	0.1 MPa $< P < 50$ MPa	$P > 50$ MPa
$T < 15°C$	Pychrophiles	Piezopsychrophiles	Hyperpiezopsychrophiles
$15 < T < 45°C$	Mesophiles	Piezomesophiles	*Hyperpiezomesophiles*
$45 < T < 80°C$	Thermophiles	Piezothermophiles	*Hyperpiezothermophiles*
$80°C < T$	Hyperthermophiles	Piezohyperthermophiles	*Hyperpiezohyperthermophiles*

[a] Representative bacteria or archaea in the categories shown in italics have not been found.

pholipids of deep-sea heterotrophic bacteria is a function of the growth temperature and pressure. In some bacterial isolates, the trend in fatty-acid composition change is consistent with the homeoviscous hypothesis. The crux of this hypothesis is that cells regulate their membrane composition to maintain the membrane in an appropriate physical state (fluidity, in particular). The regulation is done by altering membrane fatty-acid composition and phospholipid molecular species. The regulation is activated in response to any physical or chemical factor that causes a change in membrane fluidity. The bacterial-membrane lipid composition changes in response to temperature change have long been known. It is now well documented that the composition of the membranes of deep-sea bacteria varies with both the temperature and pressure of growth. Work remains to be done to show that the observed changes are along the lines of the homeoviscous hypothesis. Figure 6 shows the membrane fatty-acid composition of a bacterial isolate from the Philippine Trench and its dependence on growth pressure.

Deep-sea bacteria synthesize polyunsaturated fatty acids (PUFAs) for their membrane phospholipids. Figure 7 shows the chemical structure of the principal PUFAs found in many deep-sea bacteria. Al-

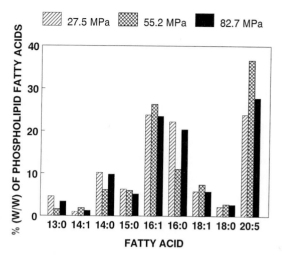

Fig. 7. Fatty-acid composition of membrane phospholipids of a deep-sea bacterium and how it changes with pressure. Plotted from data in DeLong and Yayanos (1985).

though not demonstrated for deep-sea animals, it is generally believed that animals do not synthesize PUFAs and fulfill the requirement for them through their diet. It is conceivable that deep-sea bacteria contribute to this dietary need in the deep sea.

The deep sea is also dark. The results in Fig. 8 show that one deep-sea bacterial isolate has an extraordinary sensitivity to UV light. Similar results were obtained with four other deep-sea bacterial

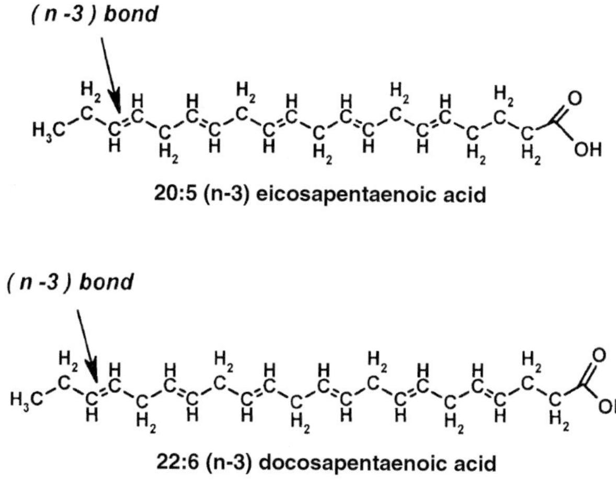

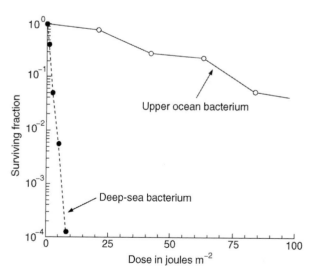

Fig. 8. The sensitivity of a deep-sea bacterium and of an upper-ocean bacterium to UV light (modified from Yayanos, 1989).

Fig. 6. The chemical structures of two polyunsaturated fatty acids found in deep-sea bacteria. Prior to the discovery of these fatty acids in the membrane phospholipids of deep-sea bacteria, it was believed that few if any bacteria could synthesize these lipids. They now appear to be common among heterotrophic deep-sea bacteria.

strains. However, it has not been determined whether these bacteria have lost the genes to repair UV-damaged DNA. The sensitivity of deep-sea bacteria to ionizing radiation, moreover, is no different than that observed with shallow-water bacteria. This is in accord with the background of natural radioactivity being similar in both shallow water and the deep sea. Because the radiation biology of deep-sea microorganisms is based on scanty information, more work is needed before general statements can be made confidently.

See Also the Following Articles

EXTREMOPHILES • GROWTH KINETICS, BACTERIAL • LOW-TEMPERATURE ENVIRONMENTS

Bibliography

Amy, P. S., and Haldeman, D. L. (eds.) (1997). "The Microbiology of the Terrestrial Deep Subsurface." CRC Lewis, Boca Raton, FL.

Bert, P. (1943). "Barometric Pressure. Researches in Experimental Physiology" (M. A. Hitchcock and F. A. Hitchcock, trans.) College Book Company, Columbus, OH.

DeLong, E. F. (1997). Marine microbial diversity: The tip of the iceberg. *Trends Biotechnol.* **15**, 203–207.

DeLong, E. F., Franks, D. G., and Yayanos, A. A. (1997). Evolutionary relationships of cultivated psychrophilic and barophilic deep-sea bacteria. *Appl. Environ. Microbiol.* **63**, 2105–2108.

DeLong, E. F., and Yayanos, A. A. (1986). Biochemical function and ecological significance of novel bacterial lipids in deep-sea prokaryotes. *Appl. Environ. Microbiol.* **51**, 730–737.

Johnson, F. H., Eyring, H., and Polissar, M. J. (1954). "The Kinetic Basis of Molecular Biology." John Wiley & Sons, New York.

Karl, D. M. (ed.) (1995). "The Microbiology of Deep-Sea Hydrothermal Vents." CRC Press, Boca Raton, FL.

Kato, C., and Bartlett, D. H. (1997). The molecular biology of barophilic bacteria. *Extremophiles* **1**, 111–116.

Kato, C., Li, L., Tamaoka, J., and Horikoshi, K. (1997). Molecular analyses of the sediment of the 11000-m deep Mariana Trench. *Extremophiles* **1**, 117–123.

Middleton, W. E. K. (1971). "The Experimenters: A Study of the Accademia del Cimento." Johns Hopkins, Baltimore, MD.

Sakiyama, T., and Ohwada, K. (1997). Isolation and growth characteristics of deep-sea barophilic bacteria from the Japan trench. *Fisheries Sci.* (Tokyo) **63**, 228–232.

Yayanos, A. A. (1980). Measurement and instrument needs identified in a case history of deep-sea amphipod research. *In* "Advanced Concepts in Ocean Measurements for Marine Biology" (F. D. Diemer *et al.*, eds.), pp. 307–318. University of South Carolina Press, Columbia, SC.

Yayanos, A. A. (1986). Evolutional and ecological implications of the properties of deep-sea barophilic bacteria. *Proc. Natl. Acad. Sci. U.S.A.* **83**, 9542–9546.

Yayanos, A. A. (1989). Physiological and biochemical adaptations to low temperatures, high pressures and radiation in the deep sea. *In* "Proceedings of the 5th International Symposium on Microbial Ecology" (T. Hattori *et al.*, eds.), pp. 38–42. Japan Scientific Societies Press, Tokyo.

Yayanos, A. A. (1995). Microbiology to 10,500 meters in the deep sea. *Ann. Rev. Microbiol.* **49**, 777–805.

Yayanos, A. A. (1998). Empirical and theoretical aspects of life at high pressures in the deep sea. *In* "Extremeophiles" (K. Horikoshi and W. D. Grant, eds.), pp. 47–92. John Wiley & Sons, New York.

Yayanos, A. A., and DeLong, E. F. (1987). Deep-sea bacterial fitness to environmental temperatures and pressures. *In* "Current Perspectives in High Pressure Biology" (H. W. Jannasch *et al.*, eds.), pp. 17–32. Academic Press, New York.

Yayanos, A. A., and Dietz, A. S. (1982). Thermal inactivation of a deep-sea barophilic bacterium, isolate CNPT-3. *Appl. Environ. Microbiol.* **43**, 1481–1489.

Yayanos, A. A., Dietz, A. S., and Van Boxtel, R. (1979). Isolation of a deep-sea barophilic bacterium and some of its growth characteristics. *Science* **205**, 808–810.

ZoBell, C. E. (1952). Bacterial life at the bottom of the Philippine Trench. *Science* **115**, 507–508.

History of Microbiology

William C. Summers

Yale University

GLOSSARY

adaptation Change in the ability of a microbe to utilize a specific nutrient after exposure to that nutrient.

animalcules Term given to microscopic organisms first described by Leeuwenhoek.

antisepsis Practice of using agents to kill microbes during surgical or other procedures.

asepsis Practice of surgical or other procedures carried out in the absence of microbes.

attenuation Reduction in the virulence of a microbe by certain growth conditions or by chemical or physical treatment.

chemotherapy Treatment for disease with specific drugs of known composition, usually a natural or synthetic organic compound, sometimes including the toxic, antimicrobial natural products of other organisms, termed antibiotics.

contagion The belief that an illness can pass from one individual to another by contact or by indirect transfer of germs.

cyclogeny A theory of life cycles of bacteria, usually in terms of an entire population of microbes.

dissociation The phenomenon by which bacteria with different properties, usually related to virulence, arise from a homogeneous population of bacteria. Now attributed to gene mutations.

episome An extrachromosomal element in cells which can sometimes become associated with the chromosome.

fermentation Classically, the process of conversion of sugar into alcohol.

infusion A suspension resulting from boiling material in water, for example, tea.

inoculation Deliberate transfer of material from one source, such as a sick animal or a laboratory culture, into another.

miasma Concept of contagion by some atmospheric influences, "bad airs" or emanations, not perceivable by the senses.

microbe Generic term for living organisms too small to be seen by the unaided eye.

monomorphism Nineteenth-century doctrine that held that bacteria are of a fixed form or morphology and that the different forms indicate distinct bacterial species.

mutation An abrupt change in a bacterial character which is stably heritable. Now known to be the result of a change in the genetic code of an organism.

operon A concept applied to a group of genes for functions which are regulated coordinately by means of control of the transcription, or messenger RNA synthesis, of those genes.

plasmid An extrachromosomal, self-replicating genetic element, which may or may not become associated with the chromosome. More general term than episome.

pleomorphism Nineteenth-century doctrine that held that bacteria are of variable form or morphology and that the different forms are not indicative of distinct bacterial species, in contrast to the concept of monomorphism.

Poisson distribution A stastical function which predicts the probability of occurrence of all-or-none rare events, such as mutations, under the assumption that their occurrence is random.

putrefaction The process of breaking down, decomposition, or rotting of organic matter, similar to, but often contrasted with, fermentation.

septicemia   A bacterial infection carried in the blood of the affected organism.

spontaneous generation   The process whereby living organisms arise from nonliving matter. The matter may be inorganic (abiogenesis) or organic matter from previously living organisms (heterogenesis).

transformation   A change from one form of bacteria to another, usually in a "single property," mediated by exposure to some material, now known to be DNA, from bacteria exhibiting the second property.

vaccine   A preparation (made in various ways) from an infectious organism which induces immunity, but not a full infection, when inoculated into a susceptible animal.

virulence   The capacity of a specific organism to cause disease when inoculated into a susceptible host.

vitalism   The concept that all living organisms possess a "life-force" or "vital principle" which accounts for the distinct properties of life, beyond the chemical and physical organization of the organism.

THE FIELD OF MICROBIOLOGY   is defined, more or less, by the physical scale of the living objects of its study. In this sense, it might be said to have begun with the invention of instrumentation to visualize objects below the limits of normal human visual acuity, that is, the microscope. In another sense, however, microbiology also includes the study of properties of these objects which can be observed macroscopically or indirectly, for example, the metabolic consequences, the diseases, and the products of microbial activity. In this latter sense, the history of microbiology includes fermentations and other processing of foods, preservation of materials from microbial decay, and concepts of disease and contagion, to mention just a few examples.

The term "microbe" is a broad, and somewhat general one, meant to embrace the biological organisms that are characterized principally by their small size. It was first suggested by Maximilien-Paul-Emile Littré (1801–1881), the great French lexicographer and linguist, proposed to the Academy of Medicine by the French surgeon Charles Sédillot (1804–1883), and quickly adopted by the French school. "Microbe"

allows an agnostic stance until one is sure of the precise biological nature of the organism under discussion, a particular virtue in the early days of microbiology. Bacteria, molds, viruses, protozoans, and sometimes even small multicellular parasites and mammalian cells in culture are subsumed under this designation. For historic reasons, no doubt, the single-cell algae seem to have remained the province of the botanists and rarely are considered as microbes.

This article will focus on the history of microbiology since the invention of the microscope for two reasons: first, the new mechanistic philosophy of the seventeenth century provided a context in which to understand microbes, and second, the ability to observe microbes, albeit indirectly with an optical instrument, was almost essential to the further development of the subject.

The history of microbiology is fortunate in that the sources are plentiful and accessible. We do not need to painstakingly reconstruct ancient texts in long-lost languages, nor is the culture of early microbiology so different from the present that it becomes nearly impossible to decipher the motivations, approaches, and methods of our predecessors.

Since any brief survey must be selective, this article will emphasize those events and discoveries which are considered common to all microbiological investigations. Some areas have been omitted or perhaps treated with less attention that they deserve (e.g., parasitology and mycology), and the history of some newer topics (such as genomics) have been left to the text articles on those subjects. The bibliography includes standard monographs of historical interest, key textbooks which give historical accounts of various topics, and a few references to the literature of the history of biology, science, and medicine. Articles on the history of microbiology are included in such searchable databases as Medline and History of Science and Technology (Research Libraries Group), both accessible on the Internet.

I. MICROSCOPY IN THE SEVENTEENTH CENTURY

The invention of the telescope at the beginning of the seventeenth century quickly led Galileo to his revolutionary astronomical discoveries and, shortly thereafter, experimenters in optics produced the first microscope, reportedly demonstrated for the first time in Paris in 1620. These first microscopes were "simple," in that they had a single lens of high curvature, often a glass bead, and were first exhibited as curiosities. By the mid-seventeenth century, several serious investigations were under way which employed the microscope to examine the invisible structures, if any, of matter. This interest in the structure of matter was stimulated by the broad seventeenth-century philosophical debates on the mechanistic concept of the universe, which included speculation that matter, animate and inanimate, is made of small particles, some of which function as tiny machines to impart function to the whole. Notable among the seventeenth-century microscopists were Robert Hooke, Nehemiah Grew, Marcello Malpighi, Jan Swammerdam, and Antony van Leeuwenhoek.

It was Leeuwenhoek who is most remembered today for his voluminous investigations and for the superior quality of his self-made instruments, and who was most concerned with observations on suspensions which contained objects we now consider as the subjects of microbiology. Leeuwenhoek (1632–1723) was a local cloth merchant in Delft, Holland, who had little formal education but who developed a passion for lens-making and microscopic observations and descriptions. Through the Delft anatomist Reinier de Graaf, Leeuwenhoek came to the attention of Henry Oldenburg, the secretary of the newly formed Royal Society of London, who began publishing Leeuwenhoek's descriptive studies in 1673. Leeuwenhoek's "Letters" to the Royal Society span over 50 years of microscopic observations on infusions, body fluids, and a wide range of other materials. His skill as a lensmaker became widely recognized, and his reports were appreciated for his careful descriptions, as well as for his technical virtuosity. His studies on human semen and description of spermatozoa (his own) contributed to the seventeenth-century debates on animal generation and the demise of various preformation theories of reproduction. In his own time, Leeuwenhoek was best known for such contributions, and his observations on microorganisms, now celebrated as the beginning of microbiology, were of relatively little interest and significance in his own time.

While Leeuwenhoek's observations on what he called "animalcules" (little animals) in rainwater and in various infusions indicate that he had indeed seen various common protozoa, yeasts, and some bacteria, he did not extend these observations very much beyond descriptions. Throughout the eighteenth century, dedicated microscopists pursued the study of these objects and refined and classified their descriptions, but all were limited by the optics of the simple microscope: chromatic and spherical aberration and relatively low magnification. Specimen preparation methods, such as sectioning and staining, were at a very early stage of development, as well. For example, prior to the introduction of the mechanical microtome in 1770, thin sections had to be cut freehand.

Ideas about life, matter, and disease at the time did not allow seventeenth- and eighteenth-century microscopic observations to be easily interpreted. Even in the late eighteenth century, Linnaeus was puzzled as to how to treat these little animals in his grand classification schemes. He assigned many of them to the genus and species: *Chaos infusoria*, perhaps indicative of the ambiguous status of microbes at that time.

II. SPONTANEOUS GENERATION AND MICROBES

In parallel to microscopic studies, complementary debates on the origin of living things were intense during the seventeenth and eighteenth centuries. Although these two topics started from quite different points, they would become intimately connected in the nineteenth century, so much so, that we now think of them as part and parcel of one tradition. Common observations of the apparently spontaneous appearance of insects, worms, and other small creatures (even mice) suggested to many observers that these living beings could arise from nonliving sources, such as mud or putrefying vegetable or animal matter. Even the famous chemist and philosopher J. B. van Helmont (1577–1644) believed that mice could arise from grain stored in granaries.

In 1668, Francesco Redi (1626–1697) published

his investigations to test the belief that maggots and flies arose spontaneously from meat. Redi was a courtier and natural philosopher under the patronage of the Medici Grand Duke in Tuscany and, as such, was expected to present "observations" and "demonstrations" for the enlightenment and amusement of the court. At one point, he conducted a series of observations on rotting animal flesh, some portions of which had been protected from the air or from other sources of infection. The interpretation of these "experiments" was that the maggots, flies, and ova seen in the exposed samples, but rarely in the protected samples, did not have their origin from the meat itself, but rather the rotting flesh only served as suitable nest for the growth and nourishment of the eggs of the animals which were deposited there. While Redi is often cited in textbooks as having used or even invented the "controlled experiment," when his work is viewed in the context of his own time, it is clear that his view of "experiment" was quite different from the modern interpretation. For Redi, experimentation meant visual demonstration of already "known" truths.

Despite work such as that of Redi and others, the idea that living beings were continuously being created from more basic material persisted. Georges Leclerc, Compte de Buffon (1707–1788), a famous French naturalist, in his monumental *Histoire naturelle* (1749 *et seq.*), described his beliefs that all living things contain an indestructible vital principle. Buffon incorporated newer ideas about the chemistry of "molecules" (small globules or particles) into his theories and regarded spontaneous generation as the result of the intrinsic properties of these organic molecules. The noted British naturalist, John Turberville Needham (1713–1781), employed microscopic studies of various infusions, both heated and unheated, both open and closed to the air, and consistently observed the development of many of Leeuwenhoek's little animals. He interpreted such results in terms of Buffon's hypotheses and was a strong supporter of the notion of spontaneous generation. When Needham's work was repeated and extended by the Italian naturalist Lazzaro Spallanzani (1729–1799), Spallanzani found that heating was quite effective in preventing the appearance of the little animals, and furthermore, that Buffon's ideas about the

vitality of the material present in the infusions, thought to be the precursor of the little animals, were untenable. Spallanzani did many careful experiments with heated sealed and unsealed flasks, from which he concluded that exposure to air was the source of the organisms which appeared in the infusions after heating. Neither Needham nor Spallanzani had the microscopical techniques which might have allowed resolution of their disputes. Furthermore, the biological status of the "little animals" remained unclear, with the result that well into the nineteenth century, the questions of spontaneous generation and "vital principles" remained unresolved.

As discussed below, motivated by practical concerns with fermentations and disease, Pasteur and others pursued these questions with newly developed techniques, concepts, and instruments and by the end of the nineteenth century, serious discussion of spontaneous generation waned. The belief in vitalism, however, has been more durable, and even the most current textbook writers feel a need to exorcise the ghosts of this eighteenth- and nineteenth-century doctrine by pro forma reference to Wöhler's famous synthesis of the "organic" compound urea, from an "inorganic" cyanate (a reductionistic view that claimed that only living organisms, because of the vital principle, could produce "organic" compounds).

III. CLASSIFICATION OF MICROBES

The biological nature of the animalcules began to attract the attention of naturalists in mid-eighteenth century with the classification of many of these forms by the Danish naturalist Otto Friderich Müller (1730–1784), who devised some of the terminology still in use today. He grouped the animalcules into two major groups: those without visible external organs and those with such structures. The first group was divided further into two groups, one of which included five genera, two of which (*Monas* and *Vibrio*) contained bacterial forms. One of Müller's species, *Monas termo*, would become the subject of much nineteenth-century attention, much as the colon bacillus, *Escherichia coli*, has been the paradigmatic species in the late twentieth century.

The physical theory and technology of optical science were sufficiently developed by the 1820s to allow construction of compound microscopes with achromatically corrected lenses, as well as with some correction for spherical aberrations. Better microscopy resulted in more detailed observations of the microbial world. In some sense, modern classification in microbiology started in 1838 with the work of Carl Gustav Ehrenberg (1795–1876), who published *Die Infusionsthierchen als vollkommene Organismen*, a massive folio of 574 pages with an atlas of 64 hand-colored plates. Many of his drawings reflect his belief that "Infusoria" (a category that included bacteria, protozoa, rotifers, and diatoms) were all animals and were organized on the same principles as larger animals; thus, he believed all these little organisms had tiny stomachs and were perfect and complete (*vollkommene*) animals. Three of Ehrenberg's 22 families of Infusioria include organisms which are recognizable as bacteria: Monadina, Cryptomonadina, and Vibrionia. Vibrionia was composed of five genera: *Bacterium, Vibrio, Spriochaeta, Sprillium,* and *Spirodiscus*. These groups were defined by gross morphology: e.g., *Bacterium* included filaments or threads of rodlike forms which showed transverse divisions. Although Ehrenberg gave detailed descriptions and drawings of many of his observations, it is difficult, if not impossible, to determine with certainty how these organisms would be classified today.

Subsequently, many classifications of microorganisms were proposed, based predominantly on morphologic characteristics with some attempts to incorporate growth characteristics, such as growth by transverse division or budding. After Ehrenberg, the next important classification scheme was that of Ferdinand Cohn (1828–1898). In 1854, Cohn published an important work in which he suggested that Ehrenberg's Vibrionia should be regarded as belonging to the vegetable kingdom rather than to the animal kingdom and suggested that bacteria were very similar to the well-known microscopic algae. He suggested that bacteria be classified as Mycophyceae or "Wasserpilze" (water fungus). Cohn's major work in bacteriology, *Untersuchungen über Bacterien* (1872, 1875, 1876), was highly influential and established the outlines of modern bacteriological thought. While Cohn was arguing for the establishment of a

taxonomy for bacteria which positioned them within the known biological realm, Carl von Nägeli (1817–1891) maintained that microscopic fungi may arise spontaneously. Nägeli was an important figure in nineteenth-century biology, and his views gave strong support to an alternative view of bacterial classification, namely, the idea that there would be little or no constancy of form in bacteria which were produced spontaneously from animal or plant precursors. This belief in the multiplicity of forms became known as di-, poly-, or pleomorphism and was supported by observations that many fungi seemed to change form, depending on conditions and stage in the growth cycle. The late nineteenth century saw intense investigation and debate about these changes or alteration of forms among the fungi and bacteria. In contrast to the pleomorphic hypothesis was the view of Cohn and his followers that each kind of organism had a definite and fixed form. This view became known as "monomorphism." Needless to say, the outcome of the monomorphism–pleomorphism controversy was crucial to the future of bacterial classification and taxonomy. An example of this confusion is given by the work of Wilhelm Zopf (1846–1909), who strongly sided with Nägeli in supporting the pleomorphic doctrine, but implicitly followed the monomorphists in his widely influential classification scheme published in 1885 (*Die Spaltpilze*).

Toward the end of the nineteenth century the classification of microbes was influenced by three major developments: improved morphologic methods (both better instruments and better staining methods); pure culture techniques; and better knowledge of the functional activities of bacteria (e.g., pathological actions, fermentations). By the turn of the century, classifications, such as those proposed by Migula and Orla-Jensen, included properties of bacteria beyond morphology, such as pathogenicity, growth requirements, and staining properties.

The filamentous fungi as well as the protozoa were also included in these early classification schemes, and their distinctive growth patterns, natural histories, and pathologies were recognized. The spore formation by fungi, the colonial growth forms, and the life cycles were characteristics that allowed classification of these organisms to proceed more surely than that of the smaller bacteria. Many of the animalcules of Leeuwenhoek were first called infusoria. The first genus for these infusoria (*Paramecium*) was introduced in 1752 and, in 1817, the generic term protozoa was employed by Goldfuss.

While classification methods, stemming from the work of Ehrenberg and culminating in the early work of David H. Bergey (1860–1937) and the Society of American Bacteriologists (*Bergey's Manual of Determinative Bacteriology*), have been of great pragmatic use in categorizing and recognizing bacterial isolates, another goal of classification and taxonomy is to delineate evolutionary relationships. At the light microscopic level, the observable structures of bacteria are relatively uninformative, however, and there has been little confidence that simple size and shape determinations carry much evolutionary weight. Only with the more detailed investigations and classification of bacteria using biochemical and genetic analyses (genomics) has it been possible to devise classifications that go significantly beyond those of the beginning of the twentieth century.

IV. CONTAGION

The periodic and widespread occurrence of disease in populations has been noted by historians for several millennia. Both popular and medical accounts abound with descriptions, explanations, and prophylactic advice about epidemic diseases. Many of these accounts suggest that disease is transmissible from an afflicted person to a healthy person, that is, the disease is contagious. Although "contagious," in modern thinking, almost always implies "infectious," this is not at all the way the notion of contagion should be read in these older accounts. Without the knowledge of microbes and their relation to disease, the concept of infection is especially problematic. Instead, contagion sometimes implied active maliciousness, such as the "evil-eye," hex-casting, or other such magical practice. Alternatively, the contagion could reside in local conditions, such as "foul" air, or on the clothing or other belongings of the afflicted person, or could emanate from decaying material or earth. Thus, the notion of contagion was broad and flexible.

While much has been written on the history of

epidemics, from the Plague of Athens in 430–425 BC (Thucydides) to AIDS, and many writers have tried their hands at retrospective diagnosis and search for precursors to modern ideas of infectious disease, the small book usually called *De Contagione* by Girolamo Fracastoro (1478–1553) is usually taken as the first more or less clear recognition that a contagious disease might be transmissible by some sort of infectious agent. Fracastoro treats three types of contagion: by contact alone, by fomites (things "which are not corrupted themselves, but are able to preserve the original germs and give rise to their transference to others"), and by contagion at a distance. In spite of Fracastoro's rather clear exposition, however, his ideas seem to have lain fallow for several centuries. Contagionist doctrines, while held by many scientists, did not develop in the direction of infectionist beliefs for a long time.

V. GERM THEORIES

While many of the early microscopists speculated that their little animals might be related to disease, fermentations, and putrefaction, such ideas did not take hold, partly because the existing concepts of these phenomena had no way to incorporate microbes into their explanatory schemes. While the diligent historian can find many examples of writings that appear to be precursors to modern germ theories, it is quite clear that they represent a rather silent or ignored tradition in medicine and biology in the seventeenth and eighteenth centuries.

However, as the biological understanding of bacteria developed in the nineteenth century, and as scientists turned more and more to physiology and "animal chemistry" for understanding of biology and medicine, the newer findings about bacteria found their applications. No longer content with explanations of diseases based on miasmas, supernatural retribution, and simple environmental conditions, late eighteenth- and early nineteenth-century thinkers looked for causes of disease based within the body itself, in its function, normal and deranged. These explanations often incorporated specific aspects to account for the contagiousness of certain diseases. The process of fermentation was so important and widely studied that it formed the model by which other biological processes were conceived. Thus, digestion was seen as analogous to fermentation. Likewise, the processes of putrefaction, and by extension, tissue degradations in abscesses, for example, were thought to be related to fermentation. The great German chemist Justus von Liebig (1803–1873) conceived of disease as a putrefaction of tissue produced by nonliving (yet "vital") substances (which we now recognize as enzymes; *ferments* in French) with the production of toxins. These toxins, when transferred to another individual, could induce further putrefaction "by contact." Such a mechanism could explain both contagion and physiological effects without the need for living, infectious agents. In Liebig's theory, fermentations, too, could be brought about by introduction of the nonliving *ferments* to suitable circumstances, e.g., grape juice in wine-making.

Debate on the theories of contagion thus evolved into investigation of the biological processes of both fermentation and digestion in the mid-nineteenth century. As might be inferred, too, these debates were intimately related to the controversies over spontaneous generation and vitalism mentioned previously. Did fermentation (and by extension, disease) require the presence of living organisms? Was fermentation a strictly chemical process or did it require some essentially "vital" component? If fermentations always required living organisms (yeast), where did the yeast come from in the usual, apparently spontaneous, fermentations seen in wine- and beer-making?

From his studies on the fermentation of beer and wine, Louis Pasteur (1822–1895) concluded that the agent which is responsible for alcoholic fermentation is a living organism, yeast, which must be introduced either accidentally or intentionally. Occasional accidents of fermentation ("diseases" of wine and beer) occur when the wrong organism is present. This pathway of experimentation led Pasteur to strongly attack the belief in spontaneous generation. His vigorous experimentation and even more vigorous rhetorical activities led to his fame as the slayer of the doctrine of spontaneous generation. The English physicist John Tyndall (1820–1893) provided strong support for Pasteur's ideas by his investigation of dust and small particles in the air as a source of the contaminations in experiments which claimed to show the existence of spontaneous generation.

Along with studies of "diseases" of fermentations, Pasteur undertook the study of a disease of silkworms that was causing severe economic difficulties for French silk producers, and in 1866, found the "cause" of this disease to be a specific microbe. By the mid-1870s, Pasteur had come to believe that all contagious diseases were caused by specific microbes, a grand generalization which had many versions, which now go under the rubric of "germ theories of disease."

Pasteur devised methods for preparation of pure liquid cultures of bacteria by serial transfers at high dilution and exploited such pure culture techniques to investigate the role of bacteria in disease and later to devise protective vaccines.

A related problem also engaged the attention of nineteenth-century physicians interested in the causes of disease: the nature and cause of diseases variously called putrid intoxication, septicemia, surgical fever, and pyemia. Experiments based on early belief in toxins showed that these diseases were often transmissible by inoculation of blood from one animal to another. Pus from a human wound could induce disease in animals, for example. The precise relationships between pus globules, white blood cells, and other components of blood were unclear. Casimir-Joseph Davaine (1812–1882) examined the blood of animals with anthrax and septicemia and reported that bacteria were only present in the blood of animals with these diseases. Much work of this sort from 1850 onward supported the association between the presence of bacteria and the occurrence of disease. Davaine extended this work with deliberate inoculation experiments with graded injections of material from sick animals and showed that the transmissible agent (called the "virus" in nineteenth-century terminology) could be titrated and that different animal species varied in their susceptibility to the virulent agent. Davaine's work attracted great interest in France and elsewhere, yet some investigators interpreted this work such that the bacteria were considered a consequence of the disease rather than a cause.

Theodor Albrecht Klebs (1834–1913), working in military hospitals in Karlsruhe, studied septic deaths from gunshot wounds and showed bacteria of different forms present in almost every case he examined. He added much to the knowledge of wound infections and devised several approaches to the study of wound infection, which were soon adopted by Koch.

Robert Koch (1843–1910) was a medical practitioner in the small town of Wollstein in Pozen, where he undertook the study of wound infections to determine if they were of parasitic (bacterial) origin. He conducted animal experiments and self-consciously considered just what kind of experiments he must do to prove, as conclusively as possible, that the wound infections that he produced in animals by injection of pus were, indeed, caused by the bacteria found in the pus. Koch was aided in his work by newly available aniline dyes for staining samples, by the newly available Abbé microscope illuminator, and by the fine oil-immersion lenses made in Jena by Carl Zeiss. In his publication *Aetiology of Traumatic Infectious Diseases,* he clearly raised the standards for the study and description of the relationship between bacteria and disease. In spite of the minor stir that this publication caused, many physicians remained unconvinced. Similar investigations were carried out by many others workers including Alexander Ogston (1844–1929) in Scotland and Daniel E. Salmon (1850–1914) in America.

Koch undertook the study of the life history of the anthrax bacillus, in which he noted the formation of spores which were especially heat-resistant. He immediately realized the relevance of the spore form both from an experimental point of view and the epidemiological standpoint. Koch communicated this work to Ferdinand Cohn, a leader in the German academic world, who recognized its importance and arranged for its publication. From this work on anthrax, Koch became widely recognized as a masterful experimentalist and creative thinker.

Early in his research, Koch recognized the need for cultivation and identification methods in bacteriology and he worked hard to develop the tools he saw as essential. Both his new staining methods and his improvements of the methods of solid surface cultures led to major new advances in his research work. He devised fixation methods which preserved the morphology of the bacteria and, thus, was able to study the organisms in a nonmotile form. He tested a wide variety of fixatives and stains (much of this work based on the early studies of Paul Erhlich) and was able to obtain preparations far superior to any previously available. Koch also pioneered the use of photomicrography, which had an important role in helping spread his theories and convince his critics.

Although there had been prior attempts to obtain pure cultures of bacteria by growth of individual "colonies" on solid medium, certainly as early as 1865, these were generally unsatisfactory. In 1881, Koch reported that he could grow individual colonies of bacteria on the sterilized cut surface of a potato slice and soon he followed up on prior work by Oscar Brefeld, who laid the theoretical and practical foundation for work with pure cultures. Koch realized that what was needed was a medium that was sterile, transparent, and solid. Initially, he found gelatin to be very useful and later employed the more stable carbohydrate agar-agar, as the gelling agent in the medium. At first, the medium was simply poured on a glass plate, but soon Koch's assistant R. J. Petri, made slight modifications and introduced the flat dish with an overhanging lid (Petri dish), still in use today.

The solid surface culture of bacteria was a revolutionary advance in technique that had major consequences. With this method for pure culture isolation, many of the ambiguities in the monomorphism–pleomorphism debate were resolved. Bacterial identification and classification became easier and more certain. And most importantly, Koch was finally able to affirm the identity of a specific bacterial type with the virulent agent in the blood, pus, or tissue extracts in his animal inoculation experiments.

Koch and several other thoughtful advocates of germ theory doctrines of disease were concerned about the knowledge claims they were making: what experimental evidence was needed to prove that a specific bacterium was the cause of a specific disease? Over a period of several years (1876, 1878, 1882), Koch evolved a set of criteria for such evidence. In his 1878 paper on wound infections, he gave a set of three rather weak criteria, but in 1882, in his landmark paper on the etiology of tuberculosis, he stated, "To prove that tuberculosis is a parasitic disease, that it is caused by the invasion of bacilli and that it is conditioned primarily by the growth and multiplication of the bacilli, it was necessary to isolate them [free] from any disease-product of the animal organism which might adhere to them; and, by administering the isolated bacilli to animals, to reproduce the same morbid condition which, as known, is obtained by inoculation with spontaneously developed tuberculous material." Later versions by Koch and textbook authors have combined this statement with some of Koch's other writing to synthesize three criteria which have come to be known as "Koch's postulates." (Apparently, however, he never referred to these criteria as "postulates.")

In America, Daniel Salmon was investigating the cause of hog cholera and had developed his own set of criteria for disease causation by bacteria. In addition to the criteria that Koch proposed, Salmon believed that the causal connection required the added demonstration that the killing of the bacteria was curative of the disease.

In the final decades of the nineteenth century, with the new methods available and the conceptual framework of Koch, Pasteur, Davaine, and others, advocates of germ theories of disease were hard at work "hunting microbes." With each success of associating another disease with a specific pathogenic organism, the rush to find bacterial causes for all disease increased. In addition to bacterial causes of cancer, there even were reports of bacterial causes for mental disorders.

VI. APPLICATIONS OF THE GERM THEORIES

With a clearer understanding of the causes of contagious diseases, there was a strong impetus to use this knowledge to treat or prevent the disease. These efforts took several forms: sanitation and public health; asepsis and antisepsis; and preventive vaccinations. Specific chemotherapy came a bit later.

Germ theories of disease fit well with the program of the nineteenth-century public health movement with its emphasis on "filth" as the cause of disease. From the midcentury work of the lawyer–sanitarian Edwin Chadwick (1800–1890) in England, it was believed that mortality rates, and perhaps health in general, could be affected by sanitary reform. Bacteri-

ologists noted that the very places and conditions considered filthy were generally the places and conditions where bacteria were likely to flourish. It is no surprise then, that many of the leading nineteenth-century bacteriologists were scientists interested in water supply sanitation, food quality control, and sewage and waste treatment and disposal. In America, William T. Sedgwick (1855–1921) was the most accomplished sanitary scientist and water bacteriologist of his day and, as director of the Massachusetts Institute of Technology, he educated a generation of public health-oriented bacteriologists.

In England, Joseph Lister (1827–1912) was the most active advocate for the application of Pasteur's germ theories to the practice of surgery. In 1868, he reported on his use of antisepsis (not asepsis) during surgery to prevent the occurrence of surgical wound infections. He employed phenol (carbolic acid) in an oil suspension. His results and his approaches to surgical cleanliness initiated a new era in surgical practice and led to a dramatic fall in postsurgical septic mortality. Lister's work led others to study antiseptics in detail, and Koch soon was able to make the important distinction between agents that simply arrest bacterial growth without killing (bacteriostatic agents) and those which are able to kill bacteria upon contact (bacteriocidal agents).

Germ theories did not find application only in medicine. Crop diseases, especially those involving fungi, were recognized as infectious in origin. As early as 1767, Targionni-Tozetti proposed that rusts of cereals might be caused by infection with microscopic fungi and, by 1807, Prevost had demonstrated experimental smut infections and showed that copper sulfate solutions could be used to disinfect plant seeds. The biological study of fungi led to better understanding of their life cycles and of the alternation of hosts required by some organisms. For example, by 1889, De Teste explained the role of barberry plants as an intermediate host in wheat rust, and recommended that rust epidemics could be controlled by elimination of the practice of using barberry hedges near wheat fields. Ergot of rye was another important pathogen which led to widespread human illness in certain years when the infection was prevalent.

Several fungi (mildews) were recognized as pathogenic to grape vines by the mid-nineteenth century and direct treatment of vineyards with agents known to kill the organisms were developed. A mixture of copper sulfate and lime suspended in water was widely employed and became known as Bordeaux Mixture.

Human diseases caused by fungi and protozoa were also recognized in the nineteenth century. Skin disease (dermatomycoses), such as favus (Schönlein, 1839) and thrush (Langenback, 1839), were shown to be related to specific fungi (*Achorion schonleinii* and *Candida albicans,* respectively), and Robin's monograph *Histoire Naturelle de Végétaux Parasites* (1853) was a landmark summary of early medical mycology. However, in 1910, when Sabouraud devised a medium on which pathogenic fungi could be easily cultivated, the development of medical mycology accelerated.

The role of protozoa in disease was suggested early by observations such as Alphonse Laveran's (1845–1922) discovery of a specific organism in the blood of malaria patients in 1880. Later, Battista Grassi (1855–1925) and Ronald Ross (1857–1932) were able to discover the role of the mosquito in the transmission of malaria (1890s). Ross and his collaboration with Patrick Manson exemplify the way the field of tropical medicine developed from the need to better understand newly encountered diseases during nineteenth-century European colonial expansion.

Other forms of microorganisms were found to be the cause of disease as well. For example, in the early twentieth century, small, bacterialike, obligate intracellular forms, known as *Rickettsia*, were found to be the infectious agent in disease such as louse-born epidemic typhus (Charles Nicolle, Howard Ricketts, and Stanislaus Prowazek) and Rocky Mountain spotted fever (Ricketts). Chlamydiae, another group of obligate intracellular pathogenic forms, thought to be distantly related to Gram-negative bacteria, were recognized in 1952 as distinct from large viruses.

Mycoplasmas, the smallest known free-living microbes, are a pleomorphic and widespread group of organisms. Since they often pass through conventional filters, they were originally classified as filterable viruses. The first known mycoplasma disease was pleuropneumonia of cattle, which was studied by such luminaries as Pasteur (1883), Nocard (1898), and Bordet (1910). From this example, mycoplasmas came to be called "pleuropneumonia-like-organisms" or PPLO, until recently. One isolate from primary atypical pneumonia in humans was propagated in chicken embryo tissue and was known eponymously as the Eaton agent (1944).

Perhaps the most dramatic advances that followed the new understanding of germ theories of disease were those involving preventive vaccination. The general phenomenon of resistance and immunity had been recognized for a long time, but beyond general beliefs, such as that a prior smallpox attack protected the individual in a subsequent epidemic and that some diseases seemed to be species-specific, there seemed to be no way to understand or manipulate these phenomena except in the unusual case of smallpox.

Smallpox had long been deliberately transmitted by contact from a sick person to a healthy person with the intent of inducing a (hopefully) mild case of the disease, which would then result in lifelong immunity. This practice was called variolation and was introduced to England and America in the early eighteenth century, apparently from the Middle East, although the practice was widespread in India and East Asia before that. In 1878, Edward Jenner (1749–1823) started his studies on the role of cowpox infection in humans as a protective experience against smallpox. He was following an apparently well-known local folk belief, but through his careful work and clear exposition in his report of 1798, he gave wide attention to the effectiveness of this procedure, called vaccination, in protection against smallpox.

Pasteur seemed to have been influenced by notions of such cross-immunity resulting from similar but slightly different diseases, and when he was called upon to study chicken cholera in France, he saw an opportunity to apply this notion to another disease. While studying the bacteria associated with chicken cholera, Pasteur transferred the cultures serially and noted that the virulence of the cultures for killing inoculated chickens decreased with the number of laboratory culture transfers. He termed this decrease "attenuation." Apparently, he sought to "challenge" some birds which had recovered from an inoculation of an attenuated culture with a highly virulent form of the bacterium and discovered that the attenuated inoculum had produced strong immunity to the lethal form of the disease. From these observations, Pasteur went on to formulate rather elaborate and complex theories of immunity and attenuation. These theories are no longer accepted, but they stimulated much important research on vaccines, immunity, and the nature of bacterial virulence. Out of this work came Pasteur's famous rabies vaccine, as well as his famous work on anthrax vaccine, which resulted in highly publicized trials of his vaccine on a herd of sheep, some cows, and a goat at Pouilly-le-Fort in 1881.

Vaccine development has proceeded from this time partly along the empirical approach, used by Pasteur and his colleagues, and partly along the more theory-based approaches, developed by Ehrlich, Emil von Behring (1854–1917), and later immunologist. In trying to explain the mechanisms by which vaccines induce immunity, early research focused on the interaction of the bacterium and the blood. Richard Pfeiffer (1858–1945) noted that the serum of immune animals was able to cause lysis of the specific bacterium against which the immunity was directed, and later, Jules Bordet (1870–1961) discovered that multiple serum components (complement) are involved in the immune cytolysis (cell lysis) phenomenon. The antibody molecules induced in response to vaccine treatment was studied intensely and Paul Ehrlich (1854–1915) explained the specificity of the antibody–antigen interaction in terms of the structural features of each of these components. The science of immunology took new directions with the landmark studies of Karl Landsteiner (1868–1943) on the specificity of serological reactions.

While germ theories of disease gradually gained adherents in the last two decades of the nineteenth century, and hunts were underway for microbes in every conceivable situation, doubts remained. For example, the discovery of the healthy carrier state in cholera by Koch and his colleagues provided a serious challenge to germ theories. When Max Von Pettenkofer (1818–1901), a major critic of the germ theories, drank a pure culture of cholera germs and remained healthy, Koch registered his worry that germ theory had suffered a serious setback. Vaccination campaigns were not always readily accepted by the public and the validity of vaccination for a variety of diseases was doubted not only by many lay people, but by many physicians as well. The new science of microbiology did not simply march triumphantly into the clinics and up to the bedsides to take over medicine by force of reason, superior science, and undeniable successes.

VII. ANTIMICROBIAL THERAPIES

Following on the work of Lister on antiseptics, many microbiologists undertook searches for agents which could be used to kill bacteria *in vivo*. Various attempts to use chloroform, iodine, thymol, and many other disinfectants were reported. Paul Ehrlich (1854–1915) reasoned that, just as there were dyes as well as antibodies that are specific for certain bacteria, there must be other kinds of chemicals that can bind to, and inactivate, specific microbes. His search for such compounds led to the discovery of several drugs useful in the treatment of protozoal and spirochetal diseases between 1905 and 1915. The arsenical compound, arsphenamine (salvarsan, compound 606), was used to treat syphilis until the advent of penicillin. In a continuation of Ehrlich's approach, another product of the dye industry, prontosil, was introduced as an antibacterial agent by Gerhard Domagk (1895–1964) in 1935. This compound was metabolized to sulfanilamide, a fact which when recognized, led to the development of many new sulfa drugs.

While investigating lysozyme, a bacteriolytic enzyme present in some body fluids and abundant in egg white, in 1929, Alexander Fleming (1881–1955) noted that cultures of the mold *Penicillium* produced a substance (penicillin) which inhibited the growth of *Staphylococcus* on culture plates. Although Fleming was unable to purify and more fully characterize penicillin, an accomplishment of Howard Flory and Ernst Chain in 1940, his finding suggested to others that saprophytic fungi might be useful source of antimicrobial agents. René J. Dubos (1901–1982) discovered one such product, tyrothricin, which has the peptide gramacidin as its active component. This success marked the beginning of a search for microbial products that can be used as antimicrobials. The discovery of streptomycin by Selman A. Waksman (1888–1973) and Albert I. Schatz (b. 1920) in 1944 was the beginning of a cornucopia of useful antibiotics. Potent microbial toxins, such as the highly carcinogenic aflatoxin from *Aspergillus flavus,* have also been discovered.

VIII. PUBLIC HEALTH

The origins of public health microbiology are rooted in the earlier concerns of public health with sanitation and control of epidemic diseases. The linkage of epidemiology and microbiology was a natural outgrowth of germ theories of disease and the general understanding of the extent of microbes in the environment. W. T. Sedgewick wrote a key text in 1902 (*Principles of Sanitary Science and Public Health*), which described the examination of drinking water for microscopic forms (bacteria, diatoms, algae, and infusoria), the use of coliform counts to assess the effectiveness of water filtration, and the processing of sewage involving microbial actions.

Major problems in public health included food microbiology, such as testing milk cows for tuberculosis (1889), epidemics, exemplified in "Sources and Modes of Infection" published in 1910 by Charles V. Chapin (1856–1941), and sanitation testing through the establishment of government microbiological laboratories.

Partly in response to the cholera epidemics in the United States in the nineteenth century, several major cities had established permanent, active boards of health by the end of the century. The U.S. Public Health Service established the Laboratory of Hygiene in a Marine Hospital on Staten Island in 1887, to serve as a cholera study unit. This laboratory evolved into the National Institutes of Health in 1930. Governmental laboratories, both municipal and state, served as diagnostic centers for physicians and public health officers to carry out microbiological isolations, diagnostic identification, and community surveillance. In New York, for example, the Public Health Laboratory under Herman Biggs, Haven Emerson, and William H. Park provided diagnostic services, bacteriological screening of food handlers and schoolchildren, and later, supervised immunization programs. These public health laboratories combined diagnostic microbiology, epidemiology, field work, quarantine, and research to define the current scope of public health microbiology.

IX. CLINICAL MICROBIOLOGY

Microbiology in medicine often was in the professional domain of pathologists, because in the period before effective antimicrobial therapies, many infections were studied in the autopsy room after the demise of the patient. Within the hospital, then, microbiology grew up along with pathology, and until the middle of the twentieth century were often practiced in the same department, frequently called "pathology and bacteriology," recognizing the seniority of pathology. Starting in the early twentieth century, however, bacteriology began to claim its independence from pathology. Educational programs in bacteriology were separated from pathology, laboratory practices differentiated bacteriology from anatomic pathology, and professional boundaries began to develop.

The birth of medical microbiology in association with pathology probably made it relatively easy for microbiology to take on service roles in examination of clinical specimens as had pathology. Many of the early microbiologists had been trained as physicians and they could relate well to their clinical colleagues and argue for the importance of increased bacteriological study of patients prior to death. Especially with the discovery of antimicrobial therapies, starting with serums and vaccines, and later with bacteriophages and then sulfas and antibiotics, the role of the bacteriological laboratory in clinical medicine became central to the practice of medicine. With the advent of specific antimicrobial therapy, it became crucial to know the identity of the infectious agent in the patient. This demand for accurate and rapid microbiological analysis of clinical samples continues to the present, with great emphasis put on technological innovation and the processing of massive numbers of samples.

X. VIROLOGY

One of the basic methods for "microbe hunting" from the very early period of microbiology was filtration. Very fine filters were designed to remove small particles, such as microbes, in order to sterilize fluids and to characterize the particulate nature of infectious materials. Two widely used filters were the Chamberland filter and the Berkfeld filter. The former, named for Charles Chamberland of the Pasteur Institute, is a tube of unglazed porcelain (a "candle" filter) through which liquids can be passed under pressure. The latter is a column of diatomaceous earth (Kieselguhr), named for the owner of the mine near Hannover which produced the material. If an infectious agent was not retained by such a filter, it was termed "filterable" (sometimes spelled filtrable) or "filter-passing." Of course, both particulate and soluble substances could be "filterable."

The term "virus" was initially used to designate the component of an inoculum which was the causative agent of the disease (Latin: *virus*: poison, venom). The term was generically applied to bacterial agents as well as other organisms. One of the main goals of the early germ theorists was to isolate and identify the virus which was present in blood, tissue extract, sputum, or stool sample, which could transmit disease when inoculated into susceptible hosts. Soon it was found that some of these viruses of disease could pass through the filters which were known to retain bacteria. These agents of disease came to be known as "filterable viruses" to distinguish them from all the other viruses. Since the nature of the filterable viruses was obscure in the early part of the twentieth century, the term "filterable virus" persisted in the literature until the 1930s; e.g., the classic 1928 text, edited by Thomas M. Rivers (1888–1962), was entitled *Filterable Viruses*. Shortly thereafter, common usage came to drop the qualifier "filterable" in favor of simply "virus," to designate the filter-passing forms of infectious agents.

The first disease that was recognized as being caused by a filter-passing agent was tobacco mosaic disease. This disease was economically devastating to Dutch tobacco growers and its cause was actively studied in Holland, starting with the work of Aldoph Meyer in 1879, who was able to transmit the infection but failed to find the causative virus, believed to be a bacterium. Martinus W. Beijerinck (1851–1931) worked on tobacco mosaic disease off and on for over a decade and, by 1898, he found that the virus was filterable, that it would diffuse through agar, and

that it was serially transmissible. For the virus of tobacco mosaic disease, Beijerinck proposed a new category of agent, living and nonparticulate, a *contagium vivum fluidum*. A few years earlier, Dimitri Ivanovski (1864–1920) had shown the filterability of the agent of tobacco mosaic, but because he was able to transfer the disease via bacterial colonies, Ivanovski believed that tobacco mosaic was a bacterial disease.

The nature of filterable viruses, as represented by the virus of tobacco mosaic (TMV), was controversial. Some thought of the agent as chemical, perhaps a cellular component, while others thought of it as an "ultramicrobe," an organized living being, too small to be seen in the microscope.

Subsequent attempts to study TMV were hampered by the inability to grow the agent outside of the infected plant and by difficulties in its detection and quantification. Some of these difficulties were overcome by the preparation of large amounts of infected plant material by Carl G. Vinson (1927) and an improved quantitative assay by Francis O. Holmes (1928). In 1935, Wendell Stanley (1904–1971) reported that he had crystallized TMV, a startling observation which forced reconsideration of the definition of "living" "infectious," and "microbe." Stanley and his coworkers, based on their chemical analyses at the time, believed that TMV was composed only of protein. Since crystallization was the standard criterion for purity of organic compounds and, by extension, of proteins, this work was interpreted to show that filterable viruses such as TMV were self-replicating, infectious proteins. Indirectly, of course, this conclusion strengthened the belief that genes, too, were protein molecules. The RNA component of TMV (about 7% by weight) was soon found by Frederick Bawden and N. W. Pirie in England. Still, the controversy over viruses remained. How did they replicate? Where did they originate? Were they autonomous or part of the cell? Should they be conceived as macromolecules or as microbes?

From the beginning of the twentieth century, filterable viruses were found as the causes of many contagious diseases that had resisted bacteriological etiologies. Herpes labialis (fever blisters), influenza, and poliomyelitis, to name a few human diseases; hog cholera, rabies, and cowpox among the mammalian diseases, and leukemia and sarcomas in birds. It was eventually realized that one of the defining, although "negative," characteristics of all these viruses is that they are obligate intracellular agents; they cannot be grown independently of their host, or at least host cells. This realization finally led to searches for better ways to grow and assay viruses apart from animal or plant inoculations. The tissues of the embryo in the chicken egg became a convenient and standardized growth and assay medium for many viruses and is still in use today for some purposes. Egg cultures of many viruses were studied by Goodpasture, who perfected this method in the 1930s. By the 1950s, the ability to grow explanted mammalian cells in various culture media ("tissue culture" or "cell culture") provided a new and improved way to grow, assay, and study many viruses. In particular, the growth of poliovirus in monkey kidney cells in culture in 1949 by John Enders (1897–1985) and his colleagues led to better understanding of this virus and its ultimate control through preventative immunizations, first with a killed preparation of poliovirus (the Salk vaccine) and then with a live, attenuated, orally effective vaccine (Sabin).

A special class of filterable agents was discovered in the first decade of the twentieth century which deserve special mention: the cancer-causing viruses. This group of viruses, not related by structure, pathogenesis, or genealogy, has provided important insights into the interplay between genes, viruses, and cells. In 1911, Peyton Rous (1879–1970) observed that a cellfree filtrate prepared from a chicken sarcoma could induce similar sarcomas when inoculated into other chickens. Likewise, in 1908, Ellerman and Bang suggested that leukemia in fowl could be caused by inoculation with a filterable agent. The notion that cancer might be an infectious disease was so potentially frightening to the public that Rous tried to avoid any unnecessary publicity of his work and eventually abandoned it. In searches for the causes of cancer, there were many reports of bacteria associated with various malignancies (including *Agrobacterium tumefaciens*, which causes tumorlike proliferation in plant tissues), but only the filterable viruses seemed to emerge as possible causal agents for animal cancers. By the mid-1950s, several classes of viruses had been identified which could induce cancers in experimental animals. However, it was not until much later that candidate human tumor viruses were isolated. Most of these tumor viruses have RNA genomes which can be copied into DNA and integrated into the cell genome to reside there in symbiosis, while causing a neoplastic transformation, in many cases. Interestingly, the organism associated with AIDS, human immunodeficiency virus (HIV), first identified in the 1980s, turned out to be in this same retrovirus group of viruses, although HIV does not cause tumors.

Bacteriophages are now recognized as viruses that infect bacteria; however, originally, they were not believed to be similar to the filterable viruses. In 1915, F. W. Twort in England reported "glassy transformation" of micrococci which contaminated his attempts to grow vaccinia virus on cellfree culture media. This phenomenon of glassy transformation was serially transmissible and killed the bacteria. Twort's interpretation of this phenomenon was unclear and ambiguous. In 1917, Félix d'Herelle, a French-Canadian working in Paris, independently observed lysis of dysentery cultures and noted that the lytic principle was filterable, and that something in the lysed culture could produce clear spots (plaques) on confluent bacterial cultures. This lytic principle was serially transmissible and, by d'Herelle's interpretation, particulate. He conceived of this agent as a microbe which parasitizes the bacteria, that is, a virus of bacteria. Not only did he devise the quantitative plaque-counting method, but he worked out the basic life cycle of this agent, which d'Herelle termed bacteriophage (although long known by the noncommittal term "Twort–d'Herelle Phenomenon"). The biological nature of bacteriophage was hotly debated for about 20 years, with the majority view in opposition to d'Herelle's ultravirus hypothesis and in favor of some sort of endogenous autocatalytic process. Only with the use of electron microscopy in 1940 did the particulate view of bacteriophage become widely accepted.

D'Herelle and Twort engaged in a 10-year polemic over the issue of priority of discovery of bacteriophage which tarnished the reputation of both scientists. Twort did not pursue phage research, and d'Herelle focused mostly on the application of phages as therapy and prophylaxis for infectious diseases in the era before antibiotics. While this application seemed to offer promise, and is now being reexamined, it was eclipsed by the marvels of the new antibiotics in the early 1940s.

The study of phage from the biological point of view has been central to the development of molecular biology. Viruses in general, and phages in particular, were recognized as very useful "probes" for cellular processes which they exploit during their life cycles. The problem of gene duplication, especially, seemed susceptible to study with phages. Both in the U.S. and in France, phage research since the 1940s was directed at understanding what happens during the half hour or so that it takes for one infecting phage to produce a hundred progeny in an infected bacterium. This was the basic research program that Max Delbrück set for the American Phage Group. The work of this loosely defined school of research has been instrumental in deciphering much about the nature of the gene, mutagenesis, recombination, and gene expression and regulation.

In virology, the electron microscope has had a major impact. Prior to about 1940, viruses were defined by their small size, by filtration and diffusion properties, and their invisibility in the light microscope. The electron microscope was invented in Germany in 1931 by Max Knoll and Ernst Ruska and, by the late 1930s, had sufficient resolving power to demonstrate the particulate nature of several viruses, including bacteriophage (Helmuth Ruska, 1940). The RCA company designed and constructed the first electron microscopes in the United States and Thomas Anderson and Salvador Luria (1942) used an RCA instrument to examine phages in detail. The ability to visualize viruses at last (albeit, indirectly) brought some long-awaited unity to virology, and bacteriophages were finally accepted as viruses of bacteria.

The history of virology is one of shifting paradigms: at one time, viruses are microbes, at another, they are macromolecules; at one time, agents of disease, at another time, vaccines to prevent disease.

XI. MICROBIAL PHYSIOLOGY

With the recognition of the role of microbes in ancient processes, such as fermentations and other types of food processing (cheese, bread, soy sauce, silage), a better understanding of these uses of microbes and their by-products was possible. Antibiotics represent just one aspect of this exploitation of microbial metabolism. Several groups of researchers, including Serge Winogradsky (1856–1953) and Martinus Beijerinck, emphasized the diversity of microbial forms and metabolism and pioneered the fundamental study of the physiology of bacteria. These studies led directly to the use of microbes to produce useful "secondary metabolites," compounds such as glycerol, lactic acid, pigments, and other intermediates which are products of the metabolism in specific organisms. Soil and dairy microbiology developed in colleges of agriculture and in the agriculture experiment stations established in the United States in the late nineteenth century. This work promoted the understanding of microbial processes used in cheesemaking, such as the use of specific molds to achieve the best products, in spoilage of foods, and in the use of manure, legumes, and composts to fertilize the soil. By the 1930s, this work on bacterial metabolism was central to the new science of biochemistry, representing, for example, a major focus in Gowland Hopkins' department in Cambridge under the leadership of Marjory Stephenson (1885–1948).

Interest in animal and plant nutrition, especially work on growth factors and vitamins, was extended to microbes in the 1920s and 1930s. In Paris, André Lwoff (1902–1994), for example, investigated the growth requirements of protozoa, while in England, Stephenson, B. C. J. G. Knight (1904–1981) and Paul Fildes (1882–1971) studied bacterial growth requirements, and in America, Edward Tatum (1909–1975) turned to fungal biochemistry. These kinds of investigations led to the concepts of essential nutrients and specific metabolic reactions in microbes.

XII. MICROBIAL GENETICS

When George Beadle (1903–1989) and Tatum were investigating the genes for eye color in the fruit fly (*Drosophila melanogaster*), they saw an opportunity to examine the biochemistry of gene action in more tractable system, the biosynthesis of vitamins in fungi. In 1941, they produced pyridoxine-requiring mutants of *Neurospora crassa* by x-ray mutagenesis, and showed that this mutant behaved in a Mendelian fashion in mating experiments. This result suggested to Beadle and Tatum that genes were more or less directly involved in the control of metabolic steps, i.e., of enzymes. While the exact chemical nature of the gene was unclear at the time (many scientists thought that the genes were the enzymes themselves), they formulated a theory of the gene that came to be known as the "one gene: one enzyme hypothesis," a basic principle of modern molecular genetics (notwithstanding the additional complexities introduced later by discovery of messenger RNA, splicing, operons, and oligomeric enzymes). This approach, employing both genetic and biochemical studies, was remarkably fruitful and, in a few years, the notion that genes controlled all the biochemical pathways in cells was widely (although not universally) accepted.

While some of the literature on cyclogeny (a theory based on bacterial growth cycles) suggested that mating, with the formation of zygotic pairs, might occur in bacteria, and some data were presented to support this concept, none of these studies led to new understanding and further progress in the genetics of bacteria. However, after the successful discovery of nutritional mutants in *Neurospora*, C. H. Gray and Tatum, in 1944, were able to produce nutritional mutants in bacteria and, in 1947, Joshua Lederberg (b. 1925) and Tatum carried out experiments with two nutritionally defective mutants of the K-12 strain of *E. coli* (a clinical isolate which had been used in the student laboratory at Stanford University, where Tatum had recently taught), and they found that they could obtain recombinant forms of *E. coli* K-12 in a clear demonstration of sexual mating in bacteria. By means of bacterial recombination, a genetic map of *E. coli* K-12 was constructed and, by 1952, William Hayes (1913–1994) clarified this process of conjugation by his explanation of unidirectional transfer of genetic material. Lederberg and Cavalli formulated bacterial mating as a rudimentary form of sexuality with the donor cell called F^+ (fertility plus, "male"), the recipient called F^- (fertility minus, "female"), and the recipient cell with the donated genes, called the zygote. The transfer of the genetic material from donor cell to recipient cell could be interrupted by mechanical agitation of the culture, and François Jacob (b. 1920) and Élie Wollman (b. 1917) were able to investigate the nature and sequence of the gene transfer by this "interrupted mating" (*coitus interruptus*) experiment. Since some of the genes they studied did not seem to be associated with the bulk of the cell genes, they hypothesized that some genes exist on extrachromosomal elements, not always present in all cells, which they termed episomes (1958).

While the understanding of microbial genetic processes was increasing during the 1940s and 1950s, the understanding of the nature of the gene itself, and of the processes involved in gene stability and change, that is, mutation, were also maturing, but from a somewhat different tradition. By the mid-1920s, two strands of investigation converged to direct attention to the problem of heredity in microbes. First, many studies on the virulence of bacterial isolates suggested, even from the time of Pasteur's work on attenuation of virulence, that supposedly pure strains of bacteria could give rise to variants with altered virulence in animal inoculation tests. Second, many bacteria with recognizable colony morphologies, e.g., smooth or rough colonies, pigmented or nonpigmented, were noted to throw off variants of the other type now and then. This phenomenon in which a "pure" culture "dissociated" into a mixture of two types was called "bacterial dissociation" and was widely studied.

Interestingly, explanations of dissociation first centered on the ideas that cultures had "life cycles" and that the different forms of bacteria represented stages in the life cycle of the culture. Contemporary work on sporulation, protozoal development, "phase variation" in some bacteria such as *Salmonella*, and fungal growth variations all supported this concept of bacterial "cyclogeny," as it was termed. The theory of cyclogeny led to a resurgence of the pleomorphism concept of the nineteenth century and was a rather widely held belief in the period between World War I and World War II. The more recent success in the alternative explanation, genetic variation, has all but obliterated memory of the heyday of cyclogeny.

While some bacteriologists, such as Paul Henry DeKruif (1890–1971), better known as the author of *Microbe Hunters,* advocated genetic mutation of individual cells as the explanation for bacterial dissociation, the continued focus on the entire culture rather than the individual cell obscured the genetic basis for this phenomenon. Indeed, the genetic status of bacteria and some other microorganisms was unclear. The science of genetics arose from breeding experiments, both in plants and animals, and its focus was on sexual transmission and reassortment of characters from parent to offspring. Without easily visible chromosomes and without a sexual phase of reproduction, bacteria did not fit easily into the existing genetic paradigms. As late as 1942, the eminent British biologist Julian Huxley wrote about bacterial heredity: "One guess may be hazarded: that the specificity of their composition is maintained by a purely chemical equilibrium, without any of the mechanical control supposed by the mitotic (and meiotic) arrangements of higher forms." This view was, no doubt, strongly influenced by Huxley's contact with Cyril Hinshelwood and his school, which saw bacterial physiology and genetics in terms of chemical kinetics and which minimized the centrality of the gene in the life of the bacterium.

Perhaps because of the well-known phenomena of adaptation and "training" of bacteria to different growth conditions, much of the discussion of mutation in bacteria during the 1930s and early 1940s has a strong neo-Lamarckian quality. The exposure of the organism to the agent somehow provoked the observed changes. Exposure of bacteria to bacteriophages led to the emergence of cultures resistant to the phage and often displaying new properties, such as altered virulence, changed colony morphology, and different antigenic types. The "adaptation" of cultures to exposure to antiseptics, drugs, extremes of pH, etc., was often seen as purposeful and directed response.

A minority view in the 1930s was, however, that the dissociation phenomenon was happening independently of the exposure to the agent used to detect the change, i.e., the phage, drug, or chemical. The major problem in this work was one of experimental design: how to observe a rare event that happened in a huge population prior to the selection for the outcome of that event. In a particularly clear and convincing work, Isaac M. Lewis (1873–1943) examined the mutational change from the inability to ferment lactose to the ability to use this sugar source in the *E. coli* strain *mutabile* (in 1907, Rudolf Massini (1880–1955) reported on a variant of *Bacterium coli* (*Escherichia coli*), which had lost its ability to utilize lactose but which frequently regained this ability, and called it *B. coli mutabile*). Lewis spread the parental (lactose-negative) bacteria on glucose-containing plates and also on lactose-containing plates. About 1 in 1,000,000 of the bacteria grew on the lactose-containing plates and, when retested, all the cells in these lactose-utilizing colonies bred true and could use lactose. By laboriously picking colonies from the glucose plates and testing them for their ability to utilize lactose, he estimated the frequency of lactose-using bacteria in the culture in the absence of exposure to lactose. This frequency was similar to that determined by plating of the mass culture on lactose medium, so he concluded that the mutations to lactose utilization occurred prior to the selection, not as a consequence of exposure to the selective conditions. This elegant and clear approach, however, did not change many minds, and it was not until the 1940s that two related experimental approaches gave results that comprise the canonical account of bacterial mutation.

In their work on the reproduction of bacteriophage, Max Delbrück (1906–1981) and Salvador Luria (1911–1991) were aware that bacteria often developed resistance to phage. If a bacterial culture was infected with phage and lysed, some time later, a "secondary" growth of bacteria would often appear, and these bacteria were resistant to infection with the original phage. While the occurrence of resistant bacteria was a drawback to the use of phage as an antibacterial therapeutic agent, it afforded Luria and Delbrück (1943) an opportunity to study the change in another hereditary property of bacteria. They were also aware of the problem in the interpretation of bacterial mutation experiments. No doubt because of their routine use of statistical models in their work on the inactivation of bacteria and phage by radiation, as well as their backgrounds in atomic physics, they devised a statistical approach to show that phage-resistant mutants exist in the bacterial population prior to exposure to the lethal effects of phage. Their method ("fluctuation test") was based on the properties of the Poisson distribution and the arrangement of the experiment to analyze the frequency of occurrence of mutants in numerous replicate cultures of bacteria. If the occurrence of the mutants took place prior to the analysis (i.e., prior to the application of selective conditions), the Poisson distribution predicted that the frequencies of mutants would fluctuate widely among the replicate cultures. Conversely, if the mutations all occurred during the analysis (i.e., induced by the selective conditions), the frequencies of mutants would be very similar among the replicate cultures. Not only did the frequencies fluctuate as predicted by the Poisson distribution, but Luria and Delbrück showed that this fluctuation could be used to estimate the mutation rate, that is, probability of a given type of mutation per cell division.

The method of Luria and Delbrück, and a related procedure devised in 1949 by Howard B. Newcome, were indirect, mathematical, and subject to suspicion. However, in 1952 Joshua Lederberg and Esther Lederberg (b. 1922) devised a simple and direct method to demonstrate that mutations were occurring in a random way, independent of the selection procedures. They reasoned that the approach of Lewis was logically correct, but that the methods for its execution had to be improved. They devised a method for transferring very large numbers of colonies from one plate to another by the use of velvet cloth as a transfer tool. The tiny fibers that stand out from the velvet acted as small inoculating needles and picked up some bacteria when pressed against the surface of a culture plate studded with colonies. When the velvet was pressed on a fresh, sterile plate, this plate was inoculated with bacteria in exactly the same pattern as the original plate. Thus, this "replica plate" could be used to test colonies in great numbers for mutant properties. Lederberg and Lederberg (1952) applied this technique to study of phage resistance, as well as streptomycin resistance, and in both cases, it was clear that the mutants had appeared before the application of the selective agent.

Although by midcentury it was generally recognized that bacteria have some sort of genetic apparatus, since they do not have cytologically visible nuclei and chromosomes, the exact organization of the genes was unclear. Perhaps influenced by the widespread interest in cytoplasmic inheritance in higher organisms, there was much discussion about the possibility that bacterial heredity was carried on plasmagenes or that bacteriophages were "raw genes" from bacteria. J. Lederberg (1952) introduced the term "plasmid" to explain and encompass the wide variety of genetic determinants which had been observed or hypothesized to exist in microbes. Thus, examples of cytoplasmic inheritance in protozoa, lysogeny in phage, and some time later, bacterial fertility factors, came under this rubric. This concept, in a variant form described by Jacob and Wollman (1958) under the name of "episome," nicely explained the genetic studies with fertility factors, specialized transfer of one or a few markers (e.g., the lactose-utilization genes) in some mating experiments, and later on, the "infectious" nature of antibiotic resistance (resistance transfer factors, RTF) and the variation in virulence and toxin production in many pathogens. Today, the term plasmid has come to signify a circular, extrachromosomal DNA molecule, which autonomously replicates in the cell and is not essential for that species (although culture conditions, such as the presence of certain nutrients or antibiotics, can make the plasmid essential under specific conditions).

The plasmid concept has facilitated the understanding, too, of the "gene flow" between the chromosome proper and exogenous agents, such as viruses. The ability of some plasmids to undergo genetic recombination with the chromosome explained the mechanisms by which genes can move between species in nature and also established some of the key steps in the process of lysogeny of bacteria by certain bacteriophages. As such, the plasmid concept has been important in understanding the genetics of cancer, as well as certain schemes for evolution of microbes. Of course, the role of plasmids as vectors for gene transfer in current biotechnology is too well known to need elaboration here.

XIII. BACTERIAL TRANSFORMATION AND DNA

With the beginning of the understanding of the process of bacterial mutation, as well as the discovery of ways of manipulation of bacterial genes by conjugation, the field of microbial genetics entered the modern period.

In spite of evolving understanding of the phenomenology of mutation and the process of bacterial mating, the nature of genes themselves and their expression remained to be clarified. Two disparate strands of research led to the current understanding of the nature of genes in bacteria. One strand comes from work on bacterial virulence and pathogenesis in pneumonia, and led to the identification of DNA and the chemical stuff of which genes are composed. The other strand comes from the study of bacteriophage replication.

In 1928, Frederick Griffith (1877–1941) was investigating the virulence of the organism that is responsible for pneumonia (*Streptococcus pneumoniae*, or "pneumococcus"), and he found that heat-killed bacteria of one form of pneumococcus could somehow convert live bacteria of another form to exhibit some of the antigenic and virulence properties of the heat-killed form. His work relied on passage of the mixed cultures through animals, so it was particularly difficult to get at detailed mechanisms of this conversion process. While these results were of no apparent interest or relevance to the few bacteriologists interested in microbial heredity, they were of real importance to the pathologists interested in pneumonia. Oswald T. Avery (1877–1955) was one of several scientists who confirmed and followed up Griffith's work. Almost immediately, Avery, working at the Rockefeller Institute, initiated a research program on what he called "transformation" of antigenic types. Contemporary research identified the changes in these experiments as variation in the capsular polysaccharides that surround the pneumococcus. Work in his laboratory, first by Martin Dawson, followed by James L. Alloway, and then by Colin MacLeod (1909–1972) and Maclyn McCarty (b. 1911), eventually led to characterization of the transforming material from the heat-killed bacteria as DNA itself. The final characterization relied on the use of the newly purified and characterized enzymes, DNase and RNase, as well as several well-known proteolytic digestive enzymes. When this work was published in 1944, it marked the beginning of new thinking and new possibilities. New research directions were possible and old ones were redirected. Still, appreciation of DNA was not universal: at the midcentury meeting of the Genetics Society of America, DNA was hardly mentioned. This reluctance to accept DNA as "the genetic material" may have had several origins. First, the chemical structure of DNA was poorly understood; since the 1930s, it was believed to be a repeating polymer of a tetranucleotide unit. At the same time, the complexities of proteins were beginning to be appreciated. Protein chemists had demonstrated the startling diversity and chemical individuality of proteins, and enzymologists had shown the exquisite specificity of enzymes. In the post-World War II enthusiasm for cybernetics and "information theory," geneticists had begun to develop a new concept of the gene as a unit of "information" rather than function or phenotype. Proteins exhibited the diversity expected of genes. However, without any plausible chemical basis for diversity, the "information content" of DNA was thought to be too low to have genetic potential. Two major works soon challenged that belief: using new analytical methods resulting from wartime research, Rollin D. Hotchkiss (b. 1911) showed that the base compositions of DNAs from different organism were not identical and Erwin Chargaff (b. 1905) noted certain regularities in the analyses of all DNAs: the number of purines always equal the number of pyrimidines, and the ratios of adenine to thymine and guanine to cytosine were always very near to one.

While the chemistry of the "transforming principle" in pneumococcus was being established, a second line of work was going on to understand the process of gene duplication. Max Delbrück saw bacterial viruses (bacteriophages) as a simple model for the process of gene replication and he recruited a group of like-minded associates to attack this problem. In the early stages of their work, the American Phage Group (as Gunther Stent has named Delbrück's school) treated the host bacterium more or less as a "black box" and studied phage replication as a simple input–output process. In the early 1950s, however, radioactive isotopes became available for general research use and were soon applied to study the biochemistry of phage growth and replication. In a widely cited experiment in 1952, Alfred Hershey (1908–1997) and Martha Chase labeled bacteriophage with two isotopes, ^{32}P in the DNA and ^{35}S in the protein, and in an attempt to follow the fate of the two major components of the phage through one life cycle, they noted that most of the ^{32}P label entered the cell and a significant fraction ended up in progeny phage, while very little of the ^{35}S label entered the cell and even less ended up in the progeny phage. While this result is often described in texts and reviews as if the results were "all or none," the experimental results given in the original paper, while certainly supportive of the "only DNA" hypothesis, are far from conclusive.

Another member of the American Phage Group was James D. Watson, a student of Luria in Indiana University. Watson's thesis was on the radiobiology of bacteriophage and he firmly believed that DNA was the chemical substance of the gene. Watson and Francis Crick, working at the Cavendish Laboratory in Cambridge, England, devised a plausible model for the three-dimensional structure of DNA, which finally provided the much-needed explanatory framework for the genetic role of DNA (1952). Accounts of this landmark research abound and do not need repetition here (e.g., Olby, 1994).

XIV. PHYSIOLOGICAL GENETICS

While the understanding of the role of the gene in the hereditary transmission of properties from parent to progeny organisms and in the explanation of the mutational process derived from research in such diverse areas of microbiology as pneumonia research, microbial growth factors, and bacteriophage reproduction, the role of the gene in the growth, development, and physiology of microorganisms was studied in different contexts.

Both basic and practical concerns focused attention on the biochemistry and physiology of microbial cultures and cells. The actions of microbes to produce products (chemicals, secondary metabolites), to carry out fermentations (alcohol, dairy products), and to cause trouble (decay, erosions) required study and control. Although as early as Pasteur microbiologists investigated microbial physiology, this field developed rapidly in the 1930s. In parallel with the general advances in biochemical understanding, mi-

crobial physiology was important in the discovery of growth factors, in understanding the complexities of nutrition in animals with symbiotic flora (e.g., ruminants), and in appreciating the role of bacteria in conversion of atmospheric nitrogen to organic forms of nitrogen in the process of "nitrogen fixation" that occurs in the bacteria-packed root nodules of leguminous plants.

A particularly interesting problem in bacterial physiology, one that represented a general phenomenon in microbes, was that of "enzymatic adaptation." It was widely observed that cultures growing on one substrate (e.g., glucose) and then shifted to another substrate (e.g., lactose) exhibited a short lag in growth and then "adapted" to the new substrate by the appearance of an enzyme (or several enzymes) which function to metabolize the new substrate (e.g., lactose-splitting enzymes). The mechanism by which the adaptation occurred was subject of intense discussion and debate up to the mid-1950s. By that time, the work of Jacques Monod (1910–1976) and his colleagues at the Pasteur Institute had focused on the specific case of the response of *E. coli* to the shifts between glucose and lactose as the sole carbon source for growth. Their early approach was classically physiological, that is, they studied the rates of growth, the kinetics of the adaptations, the levels of the relevant enzymes, and the concentration of substrates involved. A key question arose: when the enzyme appeared in the adaptation (termed "enzyme induction"), was it made *de novo*, perhaps upon the "instruction" of the inducer (lactose), or did the enzyme exist in a preformed state, which was somehow stabilized or activated by the inducer? These basically physiological questions were brilliantly reformulated by the group at the Pasteur Institute into genetic terms by the reconceptualization of the process as involving an intermediate system for synthesis of the enzyme which was distinct from the enzyme itself and by exploiting mutations affecting the induction phenomenon to identify the components in this hypothetical enzyme-forming system. This work resulted in the operon concept of gene function as first described in 1960 by Jacob, Perrin, Sanchez, and Monod and fully elaborated by Jacob and Monod in 1961. This synthesis of physiology and genetics provided a broad explanation of the biology of a bacterium, *E. coli,* yet it unified, as well, much of the rest of biological thought. It ended the long estrangement between transmission genetics (Morgan and the "nuclear monopoly") and the Continental embryologists, who viewed genes primarily as the agents directing growth and development, and rejuvenated interest in what is now known as cell biology.

Bibliography

Brock, T. (1988). "Robert Koch. A Life in Medicine and Bacteriology." Science Tech Publishers, Madison, WI.

Brock, T. (1990). "The Emergence of Bacterial Genetics." Cold Spring Harbor Laboratory Press, Cold Spring Harbor, NY.

Bulloch, W. (1938). "The History of Bacteriology." Oxford University Press.

Cairns, J., Stent, G. S., and Watson, J. D. (eds). (1966). "Phage and the Origins of Molecular Biology." Cold Spring Harbor Laboratory for Quantitative Biology, Cold Spring Harbor, NY.

Clark, P. F. (1961). "Pioneer Microbiologists of America." University of Wisconsin Press, Madison, WI.

Dobell, C. (1955). "Antony van Leeuwenhoek and His 'Little Animals'." Russell & Russell, New York.

Dubos, R. J., and Hirsch, J. G. (eds). (1965). "Bacterial and Mycotic Infections of Man (4th ed.)." Lippincott, Philadelphia.

Duclaux, E. (1920). "Pasteur: The History of a Mind." (E. F. Smith and F. Hedges, tr.). Saunders, Philadelphia.

Evans, A. S. (1993). "Causation and Disease." Plenum, New York.

Geison, G. L. (1995) "The Private Science of Louis Pasteur." Princeton Univ. Press, Princeton, NJ.

Helvoort, T. van. (1991). What is a virus? The case of tobacco mosaic disease. *Stud. Hist. Phil. Sci.* **22**, 557–588.

Jacob, F., and Wollman, E. L. (1958). Les episomes, éléments génétiques ajoutés. *Comptes rendus de l'Académie des sciences* **247**, 154–155.

Jacob, F., Perrin, D., Sanchez, C., and Monod, J. (1960). L'Operon: froup de gènes á expression coordonée par un opérateur. *Comptes rendus de l'Académie des sciences* **250**, 1727–1729.

Jacob, F., and Monod, J. (1961). Genetic regulatory mechanisms in the synthesis of proteins. *J. Mol. Biol.* **3**, 318–356.

Lechavalier, H. A., and Slotorovsky, M. (1965). "Three Centuries of Microbiology." McGraw-Hill, New York.

Lederberg, J. (1952). Cell genetics and hereditary symbiosis. *Physiol. Rev.* **32**, 403–430.

Maurois, A. (1959). "The Life of Sir Alexander Fleming." Jonathan Cape, London, UK.

McCarty, M. (1985). "The Transforming Principle." Norton, New York.

Monod, J., and Borek, E. (eds). (1971). "Of Microbes and Life/ Les microbes et la vie." Columbia Univ. Press, New York.

Olby, R. (1994). "The Path to the Double Helix. The Discovery of DNA." Dover, New York.

Topley, W. W. C., and Wilson, G. S. (1929). "The Principles of Bacteriology and Immunity." (2 vols). William Wood, New York.

Watson, J. D., and Crick, F. H. C. (1952). Molecular structure of nucleic acids. A structure for deoxyribose nucleic acid. *Nature* **171**, 737–738.

Wilson, C. (1995). "The Invisible World: Early Modern Philosophy and the Invention of the Microscope." Princeton Univ. Press, Princeton, NJ.

Horizontal Transfer of Genes between Microorganisms

Jack A. Heinemann

University of Canterbury, New Zealand

GLOSSARY

genome A collection of genetic material and genes normally dependent upon successful reproduction of the entire organism for its reproduction, e.g., chromosomes.

homology Related by descent from a common ancestor. Orthologous genes are homologs diverging since organismal speciation; paralogous genes are homologs diverging since duplication; analogous genes are structurally or functionally similar but not related by descent.

horizontal gene transfer (HGT) The movement of genetic material between organisms.

horizontal gene transmission The reappearance of genetic material received by HGT in the offspring of the original recipient.

horizontal reproduction An increase in genetic material (e.g., DNA), genes or sets of genes (e.g., viruses and plasmids) due to transfer between organisms rather than organismal reproduction (vertical reproduction).

horizontally mobile element (HME) A collection of genetic material or genes that reproduce by horizontal reproduction and that may also reproduce by vertical reproduction.

host range The particular group of species in which genes transferred from another particular organism can replicate.

transfer range The particular group of species to which genes can be transferred from another organism by a particular mechanism, e.g., conjugation.

vertical reproduction The concomitant reproduction of genetic material and host.

HORIZONTAL GENE TRANSFER (HGT) describes the lateral movement of genes between organisms. In contrast to the vertical transmission of genes during organismal reproduction, genes transferred horizontally do not always become genes that pass to organismal offspring. HGT is also known as infectious transfer, exemplified by viruses and plasmids. Three classical mechanisms for HGT in microbes are: transformation, the uptake of nucleic acids; transduction, virus-mediated gene transfer; and conjugation, plasmid-mediated gene transfer. Conjugation alone probably occurs between all bacteria, bacteria and plants, and bacteria and fungi. Possibly all organisms, microbial or not, are significantly affected by HGT.

I. THINKING ABOUT GENES, NOT GENOTYPES

It is difficult as biologists, and particularly as genetical thinkers, to consider genes separately from organisms and phenotypes. The conceptual independence of genotype and phenotype was only first introduced this century by W. Johannsen, who coined the term "gene." That revolution in thought is now most pertinent to those thinking about evolution. Although genes are still discovered by their effects on organismal phenotype, these effects are only indirectly related to the history and ancestry of the gene and organism.

Genes that are parts of chromosomes reproduce when the chromosome is replicated. Chromosomal replication is tightly associated with organismal reproduction. If the offspring die at any stage of the reproductive process, from incorporation of the first nucleotide at a repliction fork to their last encounter with a predator, then all the genes on the chromosomes of that organism are at an evolutionary end. Genes in chromosomes share the fate of the organism.

To the extent that gene reproduction is synchronous with organism reproduction, the evolution of particular genes and organisms is undoubtedly due to the effect of the gene on the organism. However, not all genes reproduce in this vertical fashion, that is, sychronously with organisms—at least, not all of the time. Some reproduce horizontally by transferring between organisms. When that transfer results in a gene that will be reproduced vertically, i.e., is inherited by the recipient organism, then the gene has been horizontally transmitted.

When genes can reproduce horizontally, then they can evolve somewhat independently of their effects on the host. If genes transfer horizontally at rates that exceed vertical reproduction, then it is possible for genes to evolve functions that cannot be determined by studying the effects of the gene on a host. This last point is especially relevant if the host–gene relationship is studied under conditions where the gene is mostly confined to reproducing at the rate of the host.

For HGT to contribute to the evolution of genes, it must occur (1) at least infrequently, but produce strongly selected phenotypes in organisms (and probably leave records of the event through maintenance of particular DNA sequences in organismal descendants), (2) so frequently that most often the effects on organisms are unimportant to the genes reproducing horizontally (and particular DNA sequences may not accumulate in offspring), (3) frequently, but leave short nucleotide sequence records in organisms, or (4) infrequently most of the time and extremely frequently for short periods of time (e.g., during the age when mitochondria and chloroplasts first entered the ancestors of most eukaryotes).

Of course, these various possibilities are not mutually exclusive.

The remainder of this article will be devoted to reviewing the mechanisms of HGT, barriers to gene transmission, estimates of transfer/transmission rates, and the difficulty of determining such rates. The article will conclude with considerations of the importance HGT has for studying evolution and assessing the risk of new biotechnologies, including the introduction of new antibiotics and genetically modified organisms.

II. THE WAY GENES REPRODUCE BETWEEN ORGANISMS

A. Mechanisms

Genes are transferred between organisms by three known routes: transformation, transduction, and conjugation.

Transduction and conjugation are conducted by viral and plasmid vectors, respectively. These vectors are themselves groups of genes that reproduce horizontally, possibly far more often than they do vertically. Transformation may have a vector of sorts, such as membrane-bound vesicles, escort proteins, or "uptake sequences." The vectors are sometimes also transmitted, but other times the vectors are only transferred with the genes. For example, transducing viruses can package chromosomal or plasmid DNA during infection or incorporate nonviral DNA into their own genome. Subsequent infections can result in a new host's receiving all to none of the virus, with subsequent incorporation and inheritance of the transferred nonviral DNA. Transmission by transformation and transduction is often limited to closely related organisms because these mechanisms usually require DNA–DNA recombination and, in the case of transduction, DNA delivery is mediated by viruses that may infect a small number of species. It would be premature, however, to exclude the contribution of the growing number of broad-host range viruses being described and to equate the transfer range of viruses with the more limited range of hosts that support their infectious cycle.

Of the three mechanisms, conjugation can move

the largest DNA fragments (as much as an entire bacterial genome may be moved in one conjugative encounter). Conjugation also has the broadest known transfer range, mediating exchanges between all eubacteria and between prokaryotes and eukaryotes. Conjugation, a process determined by plasmids or transposable elements, is not usually dependent on homologous recombination to achieve the formation of a recombinant.

B. Host Ranges

Comparing the sequence of particular genes in different organisms has become a taxonomic tool for inferring organismal homology. Comparisons are complicated by gene sequences that suggest a lineage different from other genes in the same organism. If that gene has been acquired by horizontal transmission, then, truly, it would be a rouge and its exceptional sequence signature could be explained. The origin of genes by HGT has often been discounted, however, when the host range of known vectors is thought to not overlap with the putative donor and recipient species involved and no other vector or ecological relationship is obvious. R. F. Doolittle (1998) calls this the "opportunity" factor. "[T]he possibility for gene transfer is often given wider berth whenever parasitism, symbiosis or endosymbiosis is involved."

Host range determinations are generally the result of studies that require a vector or gene to be transmitted to determine retrospectively if genes had transferred. Thus, when certain plasmids or viruses do not cause demonstrable infections in an organism, they are assumed not to have transferred to that organism. The history of the host range studies using the Ti plasmid of *Agrobacterium tumefaciens* and the conjugative plasmids, like IncF and IncP, of the Gram-negative bacteria illustrate the lesson well. The transfer of DNA from *A. tumefaciens* to dicotyledonous plants is determined by the Ti plasmid. Indeed, that process, which results in tumors in certain susceptible plant species, was thought to be limited to those species. However, when the relevant DNA was conferred with sequences that would maintain it in other species, the host range was extended to monocots and then to fungi. Clearly, Ti can mediate trans-

fer to more species than normally display the effects of that transfer because those effects are not frequently heritable or selectable.

The mechanism of Ti-mediated DNA transfer is biochemically and genetically equivalent to bacterial conjugation. Since the equivocation of the mechanisms, it has been demonstrated that even mundane bacterial conjugative plasmids transfer to eukaryotes. Once again, demonstration required engineering the plasmids with a selectable marker and a strategy for replication in the eukaryotic host (either replication autonomous from the chromosomes or by integration into the chromosomes). These changes had no obvious effect on transfer. Thus, the transfer range of genes can be remarkably different from host range. The "opportunity" factor in determining the likelihood of HGT is often less amenable to test than DNA sequence comparisons, making opportunity a very distant secondary consideration for evaluating the possibility of HGT.

III. GENE ARCHAEOLOGY

Determining which genes are most closely related, and how that reflects organismal relationships, is difficult. The underlying assumption in sequence comparisons is that sequence is the best indicator of homology. The difficulty in establishing relatedness independently of sequence information makes testing the proposition problematical. Even when sequence information is available, the identity and history of the gene are not always obvious (Fig. 1). Comparisons must be made between homologous genes and not just genes with homologous names— those with functional similarity (discussed by Doolittle in Horizontal Gene Transfer, 1998). For example, did two genes with similar structure and function diverge from a single sequence or converge from much different sequences? Defining genes with sufficiently similar sequences as homologous begs the question of the adequacy of the criteria for determining homology. Those who compare sequences have to beware several common problems, as will be discussed.

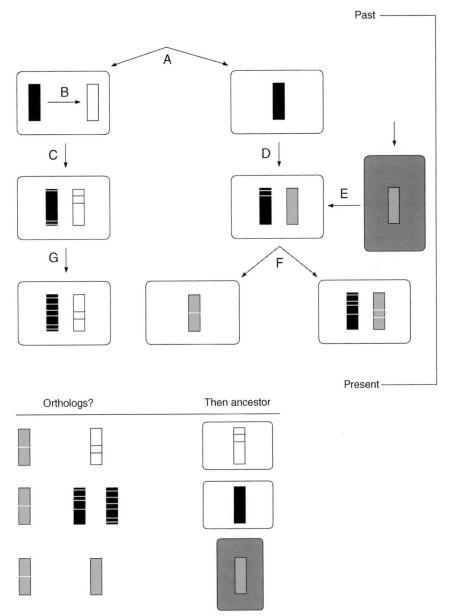

Fig. 1. The complexity of inferring organismal descent from homologous genes. (*Top*) Illustrated is the hypothetical history of two genes (small rectangles) with similar DNA sequences and/or functions. (A) A pair of closely related organisms (large open rectangles) begins to diverge. (B) The solid black gene duplicates in one of the two, producing a paralog (small open rectangle). (C) As each organismal lineage evolves, mutations (white lines in black and gray rectangles, black lines in white rectangles) accumulate independently in the orthologous copies of the gene. The paralogs also evolve relative to each other. (D) The other lineage acquires a structurally similar, possibly homologous, copy of the same gene (small gray rectangle) by (E) horizontal transmission from another organism (large gray rectangle). (F) Perhaps as a consequence of recombination with the similar sequence, some descendants lose the original copy of the gene while others maintain both copies. (G) All genes accumulate mutations independently until the present. (*Bottom*) Working back from the present using only DNA sequences, which genes are orthologs and which organisms are most closely related? Illustrated are the choices for a single gene, shown left (grey rectangle) and the possible orthologs shown center. Depending on the number of nucleotides introduced into genes by horizontal recombination events, the similarity of the introduced genes, and the number of mutations in the true orthologs, the ancestry of organisms can be difficult to determine.

A. Sequence Evidence of Ancestry

1. *Orthologous or Paralogous?*

Are the genes orthologous, diverged when the species diverged, rather than paralogous, diverged since duplication? The differences between orthologs can represent the divergence of the two organisms from a common ancestor. When gene duplication yields paralogs, each allele can change separately, reflecting the history of the genes within, instead of between, lineages.

Identifying which paralogs in different organisms reflect organismal histories is sometimes done by assuming that the most similar of the paralogs in the different species are the orthologs. However, accepting this a priori assumption can render the comparison redundant. Moreover, that test can never be independently verified because, no matter how close the sequence match, the possibility that one species lost the true ortholog since the speciation event can never be excluded. Indeed, the genes may be orthologous but, because the gene was transferred horizontally, the comparison misrepresents the ancestry of the organism. Mistakes have been made by the inadvertent comparisons of paralogs (discussed by Doolittle, 1998).

2. *Mutation or Recombination?*

Do the differences and similarities in sequences result from historical events other than time since divergence? Sequences can be maintained by selection or diverge rapidly through selection at rates that differ from other genes in the same lineage. Recombination can also maintain or disrupt sequences. Recombination, particularly with sequences obtained horizontally, can convert a portion of a gene sequence. The resulting mosaic could have an overall average sequence similarity that supported the phylogeny suggested by comparing other genes between the organisms but actually be the product of genes that evolved independently.

Reports of mosaic genes are becoming increasingly common. Perhaps one of the most instructive examples of the impact of HGT on the evolution of organismal phenotypes is the story of β-lactam resistance evolution in *Neisseria* and *Streptococcus* (Fig. 2). Alterations in their target penicillin-binding proteins

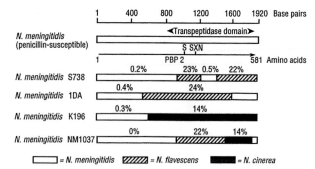

Fig. 2. Mosaic PBP 2 genes in penicillin-resistant meningococci. The open rectangle represents the PBP 2 gene of a penicillin-susceptible meningococcus. The line terminating in an arrow represents PBP 2; the active-site serine residue and the SXN conserved motif are shown. The percent sequence divergence between different regions of the genes and the corresponding regions in the susceptible strain are shown for four resistant meningococci. The origins of the diverged regions are illustrated. Reprinted with permission from Spratt, B. G. (1994). Resistance to antibiotics mediated by target alterations. *Science* **264**: 390. Copyright 1994 American Association for the Advancement of Science.

(PBP) are a common and growing means of resistance to the drugs. Not only should this type of resistance arise slowly, if at all, but it should be species-specific. The reasons for these expectations follow from the nature of the drugs and mechanisms of resistance. β-lactams bind to several different and essential PBPs; binding to any is usually sufficient for therapy. Thus, each PBP must change to confer phenotypic resistance. Moreover, a single point mutation that together conferred resistance and maintained protein function has never been observed. Thus, up to 4 genes (as in *Streptococcus pneumoniae*) must accumulate at least two changes *simultaneously* for the PBPs to retain function and for a cell to display phenotypic resistance. Based on a simple calculation using a reasonable mutation rate around 10^{-9}/base/generation, the absurd probability of a preexisting penicillin resistant strain would be 10^{-72}/cell/generation $[(10^{-9})^{4 \cdot 2}]$.

How did these pathogens become resistant? Individual PBPs of resistant pathogens are composed of segments of homologous PBPs from up to three different species! HGT has done what is a probabilistic impossibility for vertical evolution by mixing parts of different PBPs from various species of bacteria that have individual PBPs with low binding affinities to β-lactams.

3. Big Rare Events or Small Common Events?

How large does a horizontally transmitted sequence have to be to reveal itself as having been acquired horizontally? Herein lies the most difficult problem in both assessing the validity of gene comparisons for determinations of ancestry and determining the rate and extent of horizontal gene transmission. A horizontally transferred gene preserved in a vertically reproducing lineage may be identified by the anomalous phylogenetic tree its sequence creates, a significant deviation from the average G + C content of the host, an unusual organization of genes or deviation from the normal codon bias of the host. Most indicators are quantitative. Thus, they are useful if large tracts of sequences with one or more large deviations from accepted norms are being analyzed. The origin of sequences becomes increasingly difficult to determine the shorter they become or the closer to accepted ranges they appear to be. The proper conclusion in those cases is uncertainty as to how much of the sequences' reproductive history has been either horizontal **or** vertical. Moreover, some of the characteristic differences between transferred and endogenous genes, like codon bias and G + C content, can change to more closely resemble organismal norms with time. Sequence comparisons limit analysis both to large tracts of DNA and to recently transmitted genes.

How many bases are retained in the average HGT event? Horizontally transferred sequences are retained through vertical reproduction only by selection or chance. Sequences that reduce fitness will likely disappear with the organism. Protein coding sequences, the source of most DNA sequences from organisms, are the least likely repository of horizontally acquired sequences. Selectable changes could be preserved in gene-regulating sequences or in sequences important for chromosome structure. Those preserved by chance are most likely to accumulate in regions already known for their high rates of change: intergenic regions and junk. Overall, only sequences much shorter than necessary to encode a protein should normally be retained in organisms.

Transferred sequences that must ultimately recombine with those in chromosomes may also be excluded by the homologous recombination machinery. Although such enzymes are generally necessary for efficient recombination, they also discourage the retention of dissimilar sequences. For example, reducing the stringency of sequence comparisons made by such enzymes dramatically increases the frequency of recombination in interspecies crosses of *Escherichia coli* and *Salmonella typhimurium*. The E. coli RecA protein, for example, aborts recombination between sequences with less than an overall 90% identity or more than 3 mismatches in a row. Taking the biochemical stringency of RecA as roughly representative of most organisms, then the tracts of sequences incorporated will, in the main, be short, even if the donor and recipient are closely related. The tracts of heterologous sequences, as much as 24% diverged, in the mosaic PBPs of *Neisseria meningitidis* and *S. pneumoniae*, stretch hundreds of base pairs. Given the barriers to transmission, large mosaic structures should be created extremely rarely. Their existence is evidence of gene transfer frequencies on a scale large enough to produce such unlikely recombination events randomly.

4. Homology More Than Sequence

Is sequence structure *necessarily* conserved by evolution? The genomes of HMEs tend to be fluid. Over time, particular HMEs are replaced by relatives carrying different genes, such as plasmids with an expanding repertoire of antibiotic resistance genes. Tracing ancestry of HMEs by structure allows relationships to be determined over only very short periods of time. For example, the structure of infectious retroviruses can change 10^4–10^6 times the rate of other genes and defective retroviruses reproducing in synchrony with the host. The phylogeny of these viruses is confined to tracing the residues of defective viruses trapped in chromosomes of organisms or monitoring divergences on decade scales. This observation led Doolittle *et al.* (1989) to lament that "As is the case in the rest of the biological world, rapid evolutionary change appears to be associated with rapid extinction." A statement that cannot help but be true if the HME is only defined by its primary nucleic acid structure. Although contrary to their preferred conclusion, they did acknowledge that "the remarkable constellation of enzymes and structural proteins that constitute infectious retroviruses may have been assembled. . . in the quite recent past. . . [T]he erratic occurrence of retrovirus-like entities in the biological world could be the result of widespread distant horizontal transfers" (Doolittle *et al.*, 1989).

B. Genome Structure

How would ancestry of organisms be determined without, or with much less, reliance on nucleic acid or protein sequences? In the cases of ribonuclease H (Doolittle *et al., 1989*) and the A8 subunit of mitochondria (Jacobs, 1991), ancestry was inferred from three-dimensional structures or other biophysical characteristics because the primary nucleic or amino-acid sequences had lost such information.

Approaches that rely indirectly on sequences are emerging. The overall composition of genes carried by organisms, their relative positions, and their regulation establish grouping patterns. As discussed above, determining the impact and extent of HGT will require looking at the effects of the process rather than at conservation of particular sequences. The recent introduction of the "Competition Model" (discussed by Heinemann in "Horizontal Gene Transfer," 1998) and the "Selfish Operons Model" (Lawrence and Ochman, 1998), are producing robust tests of predictions of genome organization.

The Selfish Operon Model explains the organization of prokaryotic genes in operons as the result of HGT. The genes collected into operons are, and would need to be, nonessential for survival in at least one environment or only weakly selected, like traditional HME-borne genes, such as antibiotic resistance and novel virulence traits. Operon organization is not selected by clonal dissemination of hosts benefiting from this particular organization of its genes. Instead, the collection of genes in the operon reproduces faster by horizontal than by vertical reproduction and the genes are preserved in organisms when they transfer together.

The genes in operons are functionally codependent. Individually, they provide no selective benefit to a cell. So whereas they may be transferred horizontally as individual genes, they are lost in time from vertically reproducing lineages. As an example, imagine that the genes of the *lac* operon were once distributed around the chromosome of some ancient bacterium. That bacterium could survive occasional deletions of some intervening nonessential genes, rearrangements of genes by intrachromosomal recombination and transposition. Eventually, the genes of the *lac* operon may have come close enough together to be mobilized by a single HME or efficiently taken up by transformation. As the *lac* genes together could provide a selective benefit to a recipient cell that none of the genes could provide individually, then the recipient that maintained the cluster would be selected. Over time, the new lineage and other recipients of the transferred cluster could tolerate occasional deletions of material between the *lac* genes until the modern minimal structure of the operon emerged.

Thus, the genes were maintained vertically because of their contribution to the organismal phenotype but were organized in such a way as a result of HGT. For the model to work, HGT must occur frequently, with occasional retention of particular sequences in vertically reproducing lineages. The Competition Model, which seeks to explain HME organization, supports Selfish Operon expectations of high gene transfer frequencies. Normally, microbes are cultured under conditions that favor organismal reproduction. In tests of the Competition Model, conditions that favored HME reproduction were maintained. Sometimes gene transfer was allowed to occur as fast or faster than organismal reproduction. When multiple HMEs were mixed under such conditions, individuals with strategies to eliminate other HMEs emerged and dominated. The phenotypic expression of the genes (e.g., antibiotic resistance genes) that made HMEs successful during horizontal competition were sometimes detrimental and sometimes beneficial to the host. Studying those genes during clonal culture of the host, especially in the absence of competing HMEs, can lead to significantly different perceptions of their function and evolutionary history.

IV. ESTIMATING RATES

The emergence of microbial cells as legitimate entities in biology was delayed by centuries because of their size. The scale of the microbial community is still largely unknown. Similarly, the recalcitrance of horizontally transferred genes to study using existing technology has led to the untenable impression that they are rare. Judging from transmission rates, HGT could be quite common. For instance, it has been estimated that 18% of the *E. coli* genome was acquired horizontally since it diverged from *S. typhimurium* 100 million years ago (Lawrence and Ochman, 1998). As much as 5% of the mammalian genome is contained between copies of the two long terminal repeats characteristic of retroviruses. As much as 30% of the mammalian genome, and 10% of the human, was created by the action of reverse transcriptase. Plant genomes may be almost half the product of reverse transcriptase. Twenty-three percent of the human major histocompatibility complex class II region is of retroviral origin, strongly suggesting that transfer alters important genetic characteristics.

A. Organism Clonality

How frequently could genes be reproducing horizontally? The apparent clonality of many microgranisms amenable to the techniques used to estimate diversity would seem to support the perception of a vertical world. Evidence of clonal distribution is inappropriate evidence for conclusions about horizontal transfer, though, for several reasons. First, the number of clonal types is evidence, not of the *amount of recombination* that occurs between cells, but of the amount of recombination *and* subsequent selection of particular individuals. Second, the techniques for estimating diversity, which rely on alterations in genome sequences or protein conformations detectable by electrophoresis (e.g., restriction fragment length polymorphism (RFLP) and isozyme analysis), focus at a level of resolution that cannot detect recombination of short nucleotide sequences. Finally, the techniques are preoccupied with chromosomal characteristics. The mosaic structure of plasmids within bacteria argues for extensive interstrain recombina-

tion that is not preserved in the chromosomes. Whereas a lack of clonality, when it exists, can be concluded from such analyses, apparent clonality cannot at present preclude still enormous frequencies of HGT.

B. Viruses

Viruses in the environment provide some insight into the amount of horizontal gene flow. Since the viral life cycle can include a stable, extracellular period that other HME life cycles may not, the viruses are uniquely amenable to monitoring. The summertime free viral load of the world's oceans has been estimated at between 5×10^6 and 1.5×10^7/ml at up to 30 meters depth. These viral titers were from 5–100 times the estimated concentration of organisms. Up to 70% of marine prokaryotes are infected by viruses at any given time. Each of these infected hosts produce an estimated 10–100 new viruses. The absolute numbers vary somewhat between studies and season, but are consistent with counts in fresh water of 10^4 *Pseudomonas*- and chlorella-specific viruses/ml.

Finding viruses is difficult without the ability to culture each possible virus; counting the number of plasmid and transposon generations is, at present, impossible. Still, the ease of isolating plasmid-infected bacteria from natural habitats would suggest that plasmids and transposons are replicating no less than viruses. Summing the volume of water in the top 30m of oceans, adding a factor for viral turnover on land and the contribution of nonviral HMEs, produces an, albeit crude, estimated HGT frequency of $10^{30}\times$/day. The limited technologies available for surveying and then unambiguously culturing to purity the units that mostly reproduce horizontally leaves us with an exaggerated awareness of cellular life.

V. THE RISKS AND CONSEQUENCES

A. Effects of Transfer vs Transmission

The difference between transfer and transmission affects more than experimental design. The difference is the essence of the debate over the risk of gene escape from genetically modified organisms and the evolution of resistance to new generation antimicrobial agents. In both cases, we are interested in the formation of genotypes that produce organismal phenotypes better avoided. Experiments that have failed to detect the exchange of genes between organisms because no recombinants were detected have suffered from two important flaws. First, the scale of the experiments and the exposure to HME invasions were both too limited to represent the fate of genes in time and on global scales. The pace and extent of antibiotic resistance evolution is a clear indication of how extremely rare events (10^{-72}) can become certainties for populations as large as those of the HMEs and microbial cells. Second, preservation of horizontally transferred genes is the last step in the process of generating recombinant organisms. The rate limiting step in gene transmission is compromising the barriers to inheritance and expression of recombinant genes. Thus, recombinants are an inaccurate way of assaying gene transfer. Unless a risk assessment experiment can be conducted under constant selection with a flux of organisms and vectors, the effect of transfer on potentiating the creation of recombinant types cannot be estimated.

B. Potentiating Conditions

Other molecules are transferred concomitantly with nucleic acids. Viruses and conjugative plasmids carry proteins into recipients. In natural vesicle-mediated transformation, proteins and DNA enter recipient cells. Except for prions, these other types of transferred molecules are genetically inert because they do not direct their own replication. Nevertheless, escort molecules can instigate heritable states *de novo*. For example, both proteins and the methyl groups on DNA provoke heritable changes in gene expression patterns. The transfer of the *E. coli* RecA protein to *recA⁻* λ lysogens by conjugative plasmids can activate the latent virus, as can the transfer of damaged DNA. Once activated, the virus remains active until a new environmental signal causes its conversion to a prophage. In mammalian cells, the introduction of DNA fragments with preexisting methylation patterns different from the pattern on the homologous endogenous sequences can cause the methylation or demethylation of chromosomal sequences.

Thus, HGT processes are capable of more than the creation of recombinants. The impact, surely underestimated, includes the use of transferred nucleic acids for recombination with and repair of endogenous genes, the creation of recombinants with novel and potentially selectable phenotypes, and the potential to alter gene expression and heritable states.

Acknowledgments

I thank T. Cooper and N. Gemmell for critical reading of the manuscript. This effort was supported in part by Grant U6333 from the University of Canterbury and New Zealand Lotteries Health.

See Also the Following Articles

Conjugation, Bacterial • Transduction: Host DNA Transfer by Bacteriophages • Transformation, Genetic

Bibliography

Best, S., Le Tissier, P. R., and Stoye, J. P. (1997). Endogenous retroviruses and the evolution of resistance to retroviral infection. *Trends Microbiol.* **5**, 313–318.

Doolittle, R. F. (1998). The case for gene transfers between distantly related organisms. *In* "Horizontal Gene Transfer" (M. Syvanen and C. I. Kado, eds.), pp. 311–320. Chapman and Hall, London and New York.

Doolittle, R. F., Feng, D.-F., Johnson, M. S., and McClure, M. A. (1989). Origins and evolutionary relationships of retroviruses. *Quart. Rev. Biol.* **64**, 1–30.

Gray, M. W. (1998). Mass migration of a group I intron: Promiscuity on a grand scale. *Proc. Natl. Acad. Sci. USA* **95**, 14003–14005.

Heinemann, J. A. (1992). Conjugation, genetics. *In* "Encyclopedia of Microbiology" (J. Lederberg, ed.-in-chief), pp. 547–558. Academic Press, San Diego.

Heinemann, J. A. (1999). How antibiotics cause antibiotic resistance. *Drug Discovery Today* **4**, 72–79.

Kado, C. I., and Syvanen, M. (eds). (1998). "Horizontal Gene Transfer." Chapman and Hall, London and New York.

Huynen, M. A., and Bork, P. (1998). Measuring genome evolution. *Proc. Natl. Acad. Sci. USA* **95**, 5849–5856.

Jacobs, H. T. (1991). Structural similarities between a mitochondrially encoded polypeptide and a family of prokaryotic respiratory toxins involved in plasmid maintenance suggest a novel mechanism for the evolutionary maintenance of mitochondrial DNA. *J. Mol. Evol.* **32**, 333–339.

Kobayashi, I. (1998). Selfishness and death: Raison d'etre of restriction, recombination and mitochondria. *Trends Genet.* **14**, 368–374.

Lawrence, J. G., and Ochman, H. (1998). Molecular archaeology of the *Escherichia coli* genome. *Proc. Natl. Acad. Sci. USA* **95**, 9413–9417.

Matic, I., Taddei, F., and Radman, M. (1996). Genetic barriers among bacteria. *Trends Microbiol.* **4**, 69–73.

Whitman, W. B., Coleman, D. C., and Wiebe, W. J. (1998). Prokaryotes: the unseen majority. *Proc. Natl. Acad. Sci. USA* **95**, 6578–6583.

Identification of Bacteria, Computerized

Trevor N. Bryant

University of Southampton

GLOSSARY

classification Orderly arrangement of individuals into units (taxa) on the basis of similarity; each unit (taxon) should be homogenous and different from all others.

identification Matching of an unknown against knowns in a classification, using the minimum number of diagnostic characters.

identification matrix (also known as a **probability matrix**) Rectangular table containing the percentage positive character states for a range of taxa.

identification score Means of expressing the degree of relatedness of an unknown to a known taxon.

probabilistic identification Determination of the likelihood that the observed pattern of results of tests carried out on an unknown bacterium can be attributed to the results of a known taxon within an identification matrix.

taxon General term for any taxonomic group (e.g., strain, species, genus).

COMPUTERIZED IDENTIFICATION OF BACTERIA is still largely based on the determination of phenetic characters. These include morphological features, growth requirements, and physiological and biochemical activities. Other phenotypic methods include analysis of cell wall composition, cellular fatty acids, isoprenoid quinones, whole-cell protein analysis, polyamines, pyrolysis mass spectrometry, Fourier transformation infrared spectroscopy, and UV resonance Raman spectroscopy. These have not been incorporated into routine computer-based identification systems, although computers are used during the analytic process.

Recently, microbiologists have shown more interest in the use of genotypic methods to establish the taxonomic relationships between bacteria including DNA base ratio (moles percent $C+G$), DNA–DNA hybridization, rRNA homology studies, and DNA-based typing methods. Computers are employed in the collection and analysis of this data. Polyphasic identification is the integration of these various techniques for identification of unknown bacteria. Few, if any, computer-based polyphasic identification systems have been developed. Most computer-based identification systems use only a subset of taxonomic information available to the bacteriologist for a particular group of bacteria.

We concentrate on identification systems where the bacteriologist can enter the results of tests carried out on an unknown, obtain a suggested identification and, where identification has not been achieved, obtain a list of additional tests that should be carried out to enable identification.

I. PRINCIPLES OF BACTERIAL IDENTIFICATION

The taxonomy of any group of organisms is based on three sequential stages: classification, nomenclature, and identification. The first two stages are the prime concern of professional taxonomists, but the end product of their studies should be an identification system that is of practical value to others. Therefore, an identification system is clearly dependent on the accuracy and data content of classification schemes and the predictive value of the name assigned to the defined taxa.

The ideal identification system should contain the minimum number of features required for a correct diagnosis, which is predictive of the other characters of the taxon identified. However, the minimum number of characters required is dependent on both the practical objectives of the exercise and the clarity of the taxa defined in classification. Thus, many enterobacteria can be identified using relatively few physiological and biochemical tests, the numerous serotypes of *Salmonella* are recognized by their reactions to specific antisera, and the accurate identification of *Streptomyces* species requires determination of up to 50 diverse characters.

Workers at the Central Public Health Laboratory, United Kingdom, demonstrated the first practical computerized identification system in the 1970s, using a system for enterobacteria. During the same period, the application of computers for numerical classification was developed. This concept was subsequently applied to many bacterial groups, and these studies provided data that were ideal for the development of computerized identification schemes. Many bacterial taxonomists were slow to realize this potential, but these data now form the basis of many probabilistic identification schemes.

A wide and increasing range of computerized systems for the identification of bacteria is now reported in the scientific literature and allied to commercial kits. This reflects both the expansion of techniques used to determine characters for the classification of bacteria and the rapid developments in computer technology.

II. COMPUTER IDENTIFICATION SYSTEMS

The main approach to the identification of an unknown bacterium involves determination of its relevant characters and the matching of these with an appropriate database that defines known taxa. This database may be known as a probability matrix, or identification matrix. The ideal objective is to assign a name to the unknown that is not only correct but also predictive of some or all its natural characters. Computerized identification schemes provide a more flexible system than those of sequential systems (e.g., dichotomous keys) do.

Computerized identification can be achieved in several ways, numerical codes and probabilistic identification are the most popular approaches. Expert system and neural network have been investigated but their performance is no better than the probabilistic approach.

A. Numerical Codes

These are usually based on +/− character reactions. They are applied to a relatively small set of characters that have been selected for their good diagnostic value and are applied to clearly defined taxa. Numerical codes require determination of a series of character states and the conversion of the binary results into a code number that is then accessed against the identification database. Such identification systems are particularly appropriate for the analysis of test results obtained when commercial identification kits are used.

An example is the API 20E kit that generates a unique 7-digit number from a battery of 21 tests (Table I). The tests are divided into groups of three, and the results are coded 1, 2, 4 for a positive result for tests in each group. These values are then used to produce a score that reflects the test results, which can be accessed against the identification system. Organisms that generate profile numbers that are not in the identification system can be tested against appropriate computer assisted probabilistic identification systems. Numerical codes have proved to be convenient and effective, particularly for well-studied groups, such as the Enterobacteriaceae.

B. Probabilistic Identification

Probabilistic schemes are designed to assess the likelihood of an unknown strain's identifying to a

TABLE I
Steps in the Use of a System (API 20E) for the Rapid Computerized Identification of a Bacterium

			Kit test				Oxidase test
ABC	DEF	GHI	JKL	MNO	PQR	ST	
			Results				
+−+	+++	−−+	−−−	+−−	−+−	++	−
		Values allocated for a positive response					
124	124	124	124	124	124	12	4
		Cumulative scores for groups of three tests					
5	7	4	0	1	2	3	
		Input of scores to computer-based identification matrix					

[Modified from Austin, B., and Priest, F. (1986). "Modern Bacterial Taxonomy." Van Nostrand Reinhold, United Kingdom.]

known taxon. In theoretical terms, the taxa are treated as hyperspheres in an attribute space (*a-space*), in which the dimensions are the characters. The center of the hypersphere (taxon) is defined by the centroid (the most typical representative), and the critical radius encompasses all the members of each taxon. Ideally, each taxon will be distinct from any others if the identification matrix has been well constructed. To obtain an identification, the diagnostic characters for an unknown strain are determined and its position in the *a-space* calculated. If it falls within the hypersphere (taxon) of a known taxon, it is identified. Thus, in essence, probabilistic identification systems allow for an acceptable number of "deviant" characters in both the known taxa and the unknown strains.

Most computer assisted identification systems are based on Willcox's implementation of Bayes theorem,

$$P(t_i|R) = \frac{P(R|t_i)}{\sum P(R|t_i)}$$

where $P(t_i|R)$ is the probability that an unknown isolate, giving a pattern of test results R, is a member of taxon (group of bacteria) t_i and $P(R|t_i)$ is the probability that the unknown has a pattern R, given that it is a member of taxon t_i. Bayes theorem incorporates prior probabilities; these are the expected prevalence of strains included in the identification matrix. For bacterial identification, most authors give all taxa an equal chance of being isolated and, therefore, the prior probabilities for all taxa are set to 1.0 and omitted from the equation. The above equation therefore can be re-expressed as

$$L_i^* = \frac{L_i}{\sum L_i}$$

where the probabilities are now referred to as *Identification Scores,* or Willcox Scores. The identification scores for each taxon are normalized values and L_i^* for all taxa sums to one. Identification of an unknown isolate is achieved when L_i^* for one taxon exceeds a specified threshold value.

An example is shown with an identification matrix consisting of three taxa for which we have the probabilities for four tests (Table II). An unknown has been isolated and its results for the first three tests

TABLE II
Identification Matrix with Results of Unknown

		Tests			
		1	2	3	4
	a	0.01	0.20	0.99	0.90
Taxa	b	0.95	0.01	0.99	0.01
	c	0.99	0.10	0.85	0.99
Results of unknown		+	−	+	missing

TABLE III
Calculation of Likelihood of Unknown

		1		2		3		Likelihood
	a	0.01	*	(1–0.20)	*	0.99	=	0.00792
Taxa	b	0.95	*	(1–0.01)	*	0.99	=	0.93110
	c	0.99	*	(1–0.10)	*	0.85	=	0.75735
						Sum	=	1.69637

are positive, negative, and positive, respectively. The likelihoods that the taxa a, b, and c will give the pattern of results observed for the unknown is calculated by multiplying the probability of obtaining a positive result for test 1 by the probability of obtaining a negative result for test 2 by the probability of obtaining a positive result for test 3 for each taxon in turn (Table III). The original identification matrix (Table II) only gives the probabilities for positive results; in order to use the probability for a negative result, we must subtract the matrix entries for test 2 from 1. The Identification Scores are expressed as normalized likelihoods in Table IV. In this example, the unknown is not identified because a single taxon does not reach the identification threshold value. Taxa b and c are both still candidates for the identity of the unknown. Threshold values of 0.999 are typically used, for example, with the Enterobacteriaceae, but with other groups of bacteria, such as the streptomycetes, values as low as 0.95 have been used. In practical terms, a value of 0.999 means that the taxon which the unknown identifies with will have at least two test differences from all other taxa in the matrix.

Whatever type of identification system is used, there are four possible outcomes:

- The unknown is identified with the correct taxon.
- The unknown is misidentified, i.e., attributed to wrong taxon.
- The unknown is not identified at all, and correctly so, because the taxon to which it belongs is not present in the matrix.
- The unknown is not identified, but should have been identified with a taxon that is present in the matrix.

It is important that any system deal with these possibilities, although the last one is difficult to resolve. One problem with the identification score is that if an unknown is not represented in the matrix, but one strain within the matrix is closer to it (in *a-space*) than all others, the unknown may be identified as this strain. This is where additional criteria should be used to assist in the identification process. These include listing the differences in test results between the unknown and the strain it has been identified as, as well as the use of other numeric

TABLE IV
Willcox Probabilities (Normlized Likelihoods)

				Identification score
	a	0.00792/1.69637	=	0.004669
Taxa	b	0.93110/1.69637	=	0.548877
	c	0.75735/1.69637	=	0.446454
		Sum	=	1.000000

TABLE V
Maximum Possible Likelihoods

		1		2		3		Best likelihood
	a	(1−0.01)	*	(1−0.20)	*	0.99	=	0.78408
Taxa	b	0.95	*	(1−0.01)	*	0.99	=	0.93110
	c	0.99	*	(1−0.10)	*	0.85	=	0.75735

criteria, such as taxonomic distance, the standard error of taxonomic distance measures, or maximum likelihoods. Taxonomic distance is the distance of an unknown from the centroid of any taxon with which it is being compared; a low score, ideally less than 1.5, indicates relatedness. The standard error of taxonomic distance assumes that the taxa are in hyperspherical normal clusters. An acceptable score is less than 2.0 to 3.0, and about half the members of a taxon will have negative scores because they are closer to the centroid than average. The maximum, or best, likelihood is the maximum probability for a taxon calculated using those tests carried out on the unknown. The calculation uses the maximum of the probabilities of a negative and positive result of a test (Table V). This allows for taxa with several entries of 0.50 in a matrix. Some authors calculate the likelihood/maximum likelihood ratio, termed the model likelihood fraction (see Table VI), or its inverse and use it to decide whether to accept the identification offered by a Willcox score that has exceeded the identification threshold. In the IDENT module of the MICRO-IS program, for example, the Identification Score is not given if the best likelihood/likelihood is greater than 100.

III. GENERATION OF IDENTIFICATION MATRICES

One of three approaches can be used to generate identification matrices.

A. Cluster Analysis

A cluster analysis of fresh isolates and reference strains (taxa) is carried out. Phena (clusters of taxa) are selected from a dendrogram and their properties are summarized to create a starting or frequency matrix containing all characters used in the study. The identification matrix is developed by including the most useful characters and rejecting those that do not distinguish between phena. Some clusters produced by the cluster analysis are omitted from the starting matrix because they contain few members, one, two, or three isolates, or comprise a group that cannot be identified.

B. Grouping Known Strains

Known strains are characterized using a range of tests and the percentage of each exhibiting a charac-

TABLE VI
Modal Likelihood Fraction

				Modal likelihood
	a	0.00792/1.69637	=	0.010101
Taxa	b	0.93110/0.93110	=	1.000000
	c	0.75735/0.75735	=	1.000000

ter calculated. The identification of each strain is assumed to be accurate and no taxonomic analysis is carried out. The problem with this approach is that isolates that have traditionally been treated as strains of one species might be a grouping of two or more species. If characterization is carried out on strains that have been repeatedly subcultured, these strains may show less metabolic activity compared to fresh isolates. This could result in a matrix that is biased against fresh isolates.

C. Data Collected from the Literature

This is the least reliable approach. Data is collected from a variety of publications and merged to create a single matrix. The authors must resolve any conflicting test results and assign probabilities for positive, negative, and variable results. If the characterization methods are not adequately described in the literature, some characters may be misinterpreted and the results matrix may contain erroneous probabilities.

Whatever method is used to create the probability matrix, it is important that the characterization of unknowns is performed using the same techniques. For example, it would be inappropriate to create a matrix using miniaturized tests and use it for the identification of an unknown characterized using conventional tests.

D. Selection of Characters

Whatever the size and scope of the frequency matrix, by no means all the characters used will have sufficient diagnostic value for use in an identification matrix. Therefore, the major task is to determine a minimal battery of reliable tests that will distinguish between the taxa.

The ideal diagnostic character is one that is consistently positive or negative within one taxon, and this differentiates it from most of all selected taxa. This is seldom achieved with one character, but selected groups of characters may approach this ideal. How far this is achieved depends on the consistency of the taxa studied and the objectives of the identification, which, in turn, influence the principles and methods

used to select characters and to identify unknown strains.

Few characters can be regarded as entirely constant. Character variation may be real (e.g., strain variation) or occur from experimental error. Many tests and observations are difficult to standardize completely within or among laboratories, Therefore, any identification system should ideally take account of these sources of variation.

The first stage is to check the quality of the initial frequency matrix before proceeding further. Two criteria of particular relevance are (1) the homogeneity of the taxa and (2) the degree of separation or minimal overlap between them. These can be assessed using appropriate statistics calculated by the OVERMAT and OUTLIER programs.

Once it can be assumed that the frequency matrix is sound, the next step is to select the minimum number of characters from it that are required for the distinction between all the taxa included. There is some controversy about the minimum number of tests needed to separate a range of taxa effectively. One guideline is that the number of tests should at least equal the number of taxa. This may apply to relatively small and tightly defined taxa, particularly for genera rather than species. However, for many taxa (e.g., *Bacillus, Clostridium,* Enterobacteriaceae, and *Streptomyces,*) the large number of species would necessitate use of an excessive number of characters. Most of these problems can be solved if (1) the aims of the identification exercise are clear and (2) the selection of characters is approached objectively.

The ideal diagnostic character should be always positive for 50% of the taxa in the matrix and always negative for 50% of taxa in the matrix. Characters that are either always positive or always negative are clearly of no diagnostic value, nor are those that have a frequency of 50% within all or most taxa. Various separation indices have been devised that can rank characters in order of their diagnostic value. The CHARSEP program incorporates several of the indices and provides a useful means for selection characters. The use of one index, the variance separation potential (VSP), is illustrated in Table VII; this index is based on the variance within taxa multiplied by separation potential. Values greater than 25% indicate acceptable characters.

TABLE VII
Example of the Use of the CHARSEP Program to Determine the Most Diagnostic Characters of Streptomyces Species

Characters	No. of taxa in which character is predominantly +ve	−ve	VSP index (%)
Good diagnostic characters			
Resistance to phenol (0.1% w/v)	6	8	55.8
Spiral spore chains	7	6	54.9
Degradation of lecithin	13	4	48.6
Antibiosis to *Bacillus subtilis*	4	5	44.5
Poor diagnostic characters			
Production of blue pigment	22	0	0.07
Blue spore mass	22	0	0.20
Spore surface with hairy appendages	21	0	0.32
Use of l-arginine for growth	0	19	1.32
Proteolysis	0	14	3.02

[From Williams *et al.* (1983). *J. Gen. Microbiol.* **129**, 1815–1830.]

Another program (DIACHAR) ranks characters according to their diagnostic potential. The diagnostic scores of each character for each group in a frequency matrix are ordered and the sum of scores for all selected characters in each group (Table VIII) is also provided. The higher the score, the greater the diagnostic value of the selected characters. Sometimes, it is desirable to select a few characters that, although of low overall separation potential, are shown by DIACHAR to be diagnostic for a particular taxon. The program BEST uses similar methods to identify useful tests and can create an identification matrix from the frequency matrix.

IV. EVALUATION OF IDENTIFICATION MATRICES

Once an identification matrix has been constructed, it is important that its diagnostic value is

TABLE VIII
Examples of the Use of the DIACHAR Program to Evaluate Diagnostic Scores for Characters of Streptoverticillium Species

Species	Sum of scores
Streptoverticillium olivorteticuli	20.34
S. salmonis	19.01
S. ladakanum	18.81
S. hachijoense	18.10
S. abikoense	16.64
S. mobaraense	14.19

[From Williams *et al.* (1985). *J. Gen. Microbiol.* **131**, 1681–1689.]

assessed before it is recommended for use by microbiologists who are not necessarily expert taxonomists. Matrices can be evaluated by both theoretical and practical means.

A. Theoretical Evaluation

Evaluation of the matrix typically consists of determining the identification scores of the hypothetical median organism of each taxon in the matrix. This provides the best possible identification scores for each taxon included in the matrix. If any of these scores are unsatisfactory, practical identification of unknowns against such taxa will inevitably be unreliable. The MOSTTYP program does this using the Willcox probability, taxonomic distance, and the standard error of taxonomic distance as identification coefficients. The IDSC performs similar calculations and where test probabilities of 0.50 are encountered, three scores are calculated for positive, negative, and missing test results.

B. Practical Evaluation

Practical assessment involves entering the diagnostic character states of known taxa to the matrix. This should involve the redetermination of the diagnostic characters of a random selection of taxon representatives that have been included in the construction of both the frequency and identification matrices. It provides another assessment of experimental error

TABLE IX
Published Identification Matrices

Species/Groups covered	Taxa	Tests	Authors
Gram-negative, aerobic, nonfermenters	66	83	Holmes, B., Pinning, C. A., and Dawson, C. A. (1986). *Journal of General Microbiology* **132**, 1827–1842.
Gram-negative, aerobic rod-shaped fermenters	110	66	Holmes, B., Dawson, C. A., and Pinning, C. A. (1986). *Journal of General Microbiology* **132**, 3113–3135.
Enterobacteriaceae	41	16	Clayton, P., Feltham, R. K. A., Mitchell, C. J., and Sneath, P. H. A. (1986). *Journal of Clinical Pathology* **39**, 798–802.
Enterobacteriaceae	90	50	Holmes, B., and Costas, M. (1992). *In* "Identification Methods in Applied and Environmental Microbiology" (R. G. Board, D. Jones, and F. A. Skinner, eds.), pp. 127–149. Blackwell Scientific Publications, Oxford, UK.
Pseudomonas species	26	70	Costas, M., Holmes, B., On, S. L. W., and Stead, D. E. (1992). *In* "Identification Methods in Applied Environmental Microbiology" (R. G. Board, D. Jones, and F. A. Skinner, eds.), pp. 1–27. Blackwell Scientific Publications, Oxford, UK.
Vibrios	31	50	Dawson, C. A., and Sneath, P. H. A. (1985). *Journal of Applied Bacteriology* **58**, 407–423.
Vibrio and related species	38	81	Bryant, T. N., Lee, J. V., West, P. A., and Colwell, R. R. (1986). *Journal of Applied Bacteriology* **61**, 469–480.
Aeromonas species	14	30	Kämpfer, P., and Altwegg, M. (1992). *Journal of Applied Bacteriology* **72**, 341–351.
Aeromonas hybridization groups	15	32	Oakley, H. J., Ellis, J. E., and Gibson, L. F. (1996). *Zentralblatt fur bakteriologie-International Journal of Medical Microbiology Virology Parasitology and Infectious Diseases* **284**, 32–46.
Campylobacteria	23	42	Holmes, B., On, S. L. W., Ganner, M., and Costas, M. (1992). *In* "Proceedings of the Conference on Taxonomy and Automated Identification of Bacteria" (J. Schindler, ed.) pp. 6–9. Prague.
Campylobacter, Helicobacter, and others	37	67	On, S. L. W., Holmes, B., and Sackin, M. J. (1996). *Journal of Applied Bacteriology* **81**, 425–432.
Lactic acid bacteria	37	49	Cox, R. P., and Thomsen, J. K. (1990). *Letters in Applied Microbiology* **10**, 257–259.
Lactic acid bacteria	59	27	Döring, B., Ehrhardt, S., Lücke, F., and Schillinger, U. (1988). *Systematic and Applied Microbiology* **11**, 67–74.
Lactic acid bacteria	11	53	Maissin, R., Bernard, A., Duquenne, V., Baeten, S., Gerard, G., and Decallonne, J. (1987). *Belgian Journal of Food Chemistry and Biotechnology* **42**, 176–183.
Streptomyces species	23	41	Williams, S. T., Goodfellow, M., Wellington, E. M. H., Vickers, J. C., Alderson, G., Sneath, P. H. A., Sackin, M. J., and Mortimer, A. M. (1983). *Journal of General Microbiology* **129**, 1815–1830.
Streptomyces species (minor clusters)	26	50	Langham, C. D., Williams, S. T., Sneath, P. H. A., and Mortimer, A. M. (1989). *Journal of General Microbiology* **135**, 121–133.
	28	39	
Streptomyces species	52	50	Kämpfer, P., and Kroppenstedt, R. M. (1991). *Journal of General Microbiology* **137**, 1893–1902.
Streptoverticillum species	24	41	Williams, S. T., Locci, R., Vickers, J. C., Schofield, G. M., Sneath, P. H. A., and Mortimer, A. M. (1985). *Journal of General Microbiology* **131**, 1681–1689.
Actinoplanes species			Long, P. F. (1994). *Journal Industrial Microbiology* **13**, 300–310.

continues

Continued

Species/Groups covered	Taxa	Tests	Authors
Phytopathogenic corynebacteria	3	19	Firrao, G., and Locci, R. (1989). *Annuals of Microbiology* 39, 81–92.
(subclusters)	5	27	
Slowly growing mycobacteria	24	33	Wayne, L. G., Good, R. C., Krichevsky, M. I., *et al.*, (1991). *International Journal of Systematic Bacteriology* 41, 463–472.
Nontuberculous mycobacteria	27	23	Tortoli, E., Boddi, V., and Penati, V. (1992). *Binary-Computing in Microbiology* 4, 200–203.
Bacillus species	44	30	Priest, F. G., and Alexander, B. (1988). *Journal of General Microbiology* 134, 3011–3018.
Bacillus sphaericus	14	29	Alexander, B., and Priest, F. G. (1990). *Journal of General Microbiology* 136, 367–376.
Aerobic gram-positive cocci catalase positive	39	60	Feltham, R. K. A., and Sneath, P. H. A. (1982). *Journal of General Microbiology* 128, 713–120.
Aerobic gram-positive cocci catalase negative	33	60	Feltham, R. K. A., and Sneath, P. H. A. (1982). *Journal of General Microbiology* 128, 713–120.
Micrococcus	29	35	Alderson, G., Amadi, E. N., Pulverer, G., and Zai, S. (1991). *In* "The Staphylocci, Zentralblatt für Bakteriologie Supplement 21," (Jeljaszewicz/ Ciborowski, ed.) pp. 103–109. Gustav Fisher Verlag Stuttgart.
Staphylococcus species	12	15	Geary, C., Stevens, M., Sneath, P. H. A., and Mitchell, C. J. (1989). *Journal of Clinical Pathology* 42, 289–294.
Alaskan marine bacteria	86	61	Davis, A. W., Atlas, R. M., and Krichevsky, M. I. (1983). *International Journal of Systematic Bacteriology* 33, 803–810.
Medical bacteria to genus level	60	20	Feltham, R. K. A., Wood, P. A., and Sneath, P. H. A. (1984). *Journal of Applied Bacteriology* 57, 279–290.

in the determination of character states and its impact on the identification system. The selected representatives should then identify closely to their taxon when their identification coefficients are determined against the matrix. With a well-constructed matrix, bad identification scores are rare, but when they occur, they may reflect the random choice of an atypical representative of a poorly defined taxon rather than experimental error.

The final practical evaluation of a matrix clearly involves assessment of its success in identifying unknown strains. Therefore, the appropriate characters for unknowns are determined, and their identification scores are determined and assessed by the investigator.

To date most probabilistic identification systems have been tested against and applied to natural or "wild" isolates. However, there is an increasing use of genetic manipulation of such strains for scientific, medical, ecological, and industrial purposes. For a variety of reasons, not the least being patent laws, it is important to compare manipulated strains with each other and with their wild types. This is still a developing area in bacterial taxonomy, but probabilistic systems can be useful. For example, streptomycete strains that had been manipulated by various means, such as mutagens, plasmid transfer, and genetic recombination, were compared with their parent strains against an identification matrix. Most of the manipulated strains identified to the same species as their parents, indicating that most of the selected diagnostic characters were unchanged and that the identification matrix could accommodate some character state changes.

C. Computer Software

Most of the programs mentioned above were developed by P. H. A. Sneath (Department of Microbiology, Leicester University, United Kingdom). They are written in BASIC, and full details of the programs can be obtained from the following publications. Those by T. N. Bryant can be obtained from *http://staff.medschool.soton.ac.uk/tnb/pib.htm*.

Bryant, T. N. (1987). *Computer Applications in the Biosciences* **3**, 45–48.

Bryant, T. N. (1991). *Computer Applications in the Biosciences* **7**, 189–193.

Sneath, P. H. A. (1979). MATIDEN program. *Computers Geosci.* **5**, 195–213.

Sneath, P. H. A. (1979). CHARSEP program. *Computers Geosci.* **5**, 349–357.

Sneath, P. H. A. (1980). DIACHAR program. *Computers Geosci.* **6**, 21–26.

Sneath, P. H. A. (1980). MOSTTYP program. *Computers Geosci.* **6**, 27–34.

Sneath, P. H. A. (1980). OVERMAT program. *Computers Geosci.* **6**, 267–278.

Sneath, P. H. A., and Langham, C. D. (1989) OUTLIER program. *Computers Geosci.* **15**, 939–964.

Sneath, P. H. A., and Sackin, M. J. (1979). IDEFORM program. *Computers Geosci.* **5**, 359–367.

V. PUBLISHED IDENTIFICATION MATRICES

Many identification matices have been published (see Table IX). Most of these have been developed using the procedures described above. Success rates vary with the bacterial group under study. Probabilistic identification of gram-negative bacteria has been most effective; for example, 933 (98.2%) isolates of fermentative gram-negative bacteria and 621 (91.5%) isolates of nonfermenters were identified using a Willcox probability threshold of greater than 0.999. Of 243 vibrios isolated from freshwater, 71.6% were identified at a level greater than 0.999 and 79.4% at greater than 0.990.

When such stringent coefficient levels are applied to gram-positive bacteria, the results are often less impressive. For example, when a probability of greater than 0.999 was applied to coryneform bacteria, only 50% of the unknowns identified and, using a level of greater than 0.995, only 42% of streptomycete isolates were identified. If less stringent co-efficients are applied to take account of the heterogeneity of such groups, a higher rate of useful identifications can be achieved. Thus, 73% of 153 streptomycete isolates were identified using a Willcox probability of greater than 0.85.

VI. BACTERIAL IDENTIFICATION SOFTWARE

Several probabilistic identification programs have been published; these are presented below. In most instances, the programs can be used with various identification matrices. Examples from one program, Bacterial Identifier, are shown in Figs. 1, 2, 3, and 4.

Bacterial Identifier	*http://staff.medschool.soton.ac.uk/tnb/pib.htm*
BBACTID	Bryant, T. N., Capey, A. G., and Berkeley, R. C. W. (1985). *Computer Applications in the Biosciences* **1**, 23–27.
BACTID	Jilly, B. J. (1988). *International Journal of Biomedical Computing* **22**, 107–119.
CIBAC	Döring, B., Ehrhardt, S., Lücke, F. and Schillinger, U. (1988). *Systematic and Applied Microbiology* **11**, 67–74.
Gideon	*rmberger@ccsg.tau.ac.il.*
IDENTIFY	Jahnke, K. D. (1995). *Journal of Microbiological Methods* **21**, 133–142.
MICRO-IS	Portyrata, D. A. and Krichevsky, M. I. (1992). *Binary* **4**, 31–36.
MATIDEN	Sneath, P. H. A. (1979a). *Computers & Geosciences* **5**, 195–213.
Identmpm	Maradona, M. P. (1994). *Computer Applications in the Biosciences* **10**, 71–73.
no-name	Tortoli, E., Boddi, V., and Penati, V. (1992). *Binary* **4**, 200–203.
The Identifier	Gibson, L. F., Clarke, C. J. and Khoury, J. T. (1992). *Binary* **4**, 25–30.

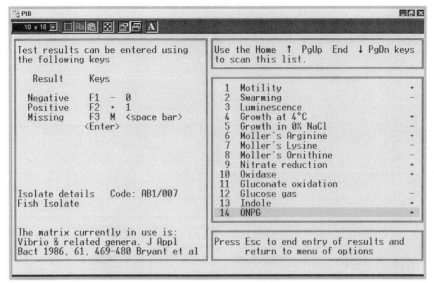

Fig. 1. The screen for entering results into the Bacterial Identifier program.

VII. OTHER APPLICATIONS OF COMPUTERS OF BACTERIAL IDENTIFICATION

We have concentrated on the use of computers to construct identification systems as well as to access them. However, computer programs are increasingly used solely to access taxonomic data for the identification of unknown strains. Developments in computer technology can also provide a more direct link between the determination of test results and their evaluation. An example is the use of so-called "breathprints," for identification of gram-negative, aerobic bacteria. This relies on a redox dye to detect the increased respiration when a carbon source is oxidized. A range of substrates in a microtiter plate are inoculated with a strain; if a substrate is used, a pigment is formed by the redox dye, indicating a

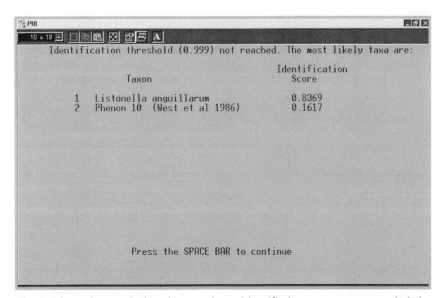

Fig. 2. The unknown isolate has not been identified, as no taxon exceeded the threshold of 0.999. The two possible taxa are listed (Bacterial Identifier).

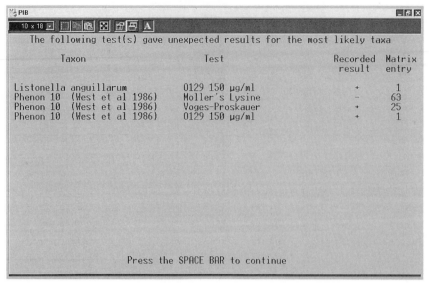

Fig. 3. Differences between test results for the unknown and possible taxa are listed (Bacterial Identifier). It is possible that the O129 150 μg/ml test has been carried out incorrectly if the unknown is an isolate of *Listonella anguiliarum.*

positive reaction. The pattern of these on the plate provides a breathprint, which can then be compared with those of known taxa using a system consisting of a microplate reader and a computer.

Databases for both the classification and identification of bacteria have been extended and improved by the inclusion of diagnostic characters provided by chemical analysis of cell components. These include cell wall amino acids, membrane lipids, and proteins, which have been particularly useful for the definition and identification of higher taxa, such as genera or families.

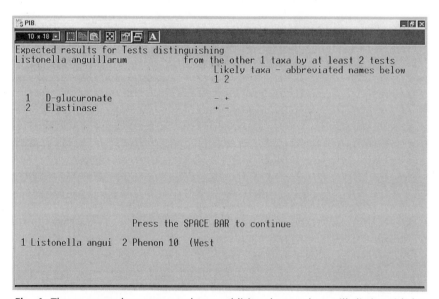

Fig. 4. The program has suggested two additional tests that will distinguish between the two possible identities for the unknown, *Listonella anguiliarum* and Phenon 10 (West *et al.,* 1986).

For example, analysis of bacterial lipids involves gas chromatography, which results in a printout of a set of peaks that are defined chemically but are difficult to evaluate and compare quantitatively by eye. Various programs have been used to transform the data for principal component analysis and to provide similarity and overlap coefficients for comparison of unknown strains.

The use of polyacrylamide gel electrophoresis (PAGE) to analyze protein patterns of bacteria is well established in bacterial taxonomy. The results are obtained in the form of stained bands on the gels. A variety of computer programs has been devised to facilitate their analysis. Typically, the stained gels are scanned by a computer-controlled densitometer. This digitizes the continuous output of the densitometer scan, removes or corrects for background effects, and permits comparison with other stored traces. Thus, an unknown can be compared with known taxa using various similarity coefficients.

Another method of bacterial identification is pyrolysis–mass spectrometry. This involves the thermal degradation of a small sample of cells in an inert atmosphere or vacuum, leading to production of volatile fragments. Under controlled conditions, these are characteristic of a taxon, and they are separated and analyzed in a mass spectrometer. Assessment of the traces obtained requires software for performing principal components, discriminant, and cluster analysis, to allow known taxa to be distinguished and unknowns identified.

The developments in nucleic acid techniques are having a marked and exciting impact on bacterial taxonomy, where they provide a genetic assessment of taxa, which can be used to supplement or revise the existing phenetic systems. Techniques such as DNA reassociation, DNA–rRNA hybridization, and DNA and RNA sequencing are increasingly used in bacterial classification, whereas nucleic acid probes and DNA fingerprints are of great potential in identification. Computation is used in the analysis and application of such data. Despite the relative novelty of these sources of taxonomic data, ultimately a con-venient and accurate means of comparing data for unknown strains with those of established taxa is still required. Thus, when determining DNA fingerprints, it is useful to have a permanent record of fragment size. Precise migration measurements on gels should not be spoiled by inaccurate assessments of fragment size. Fragments have a curvilinear relationship between the mobility of their bands on gels and their molecular sizes. A number of programs have been devised to transfer and assess such measurements.

Thus, computation has an established and developing role in all stages and aspects of bacterial identification.

Acknowledgment

The author and the publisher wish to thank Dr. Stanley T. Williams, contributor to the *Encyclopedia of Microbiology, First Edition*, whose article has been revised to produce this text.

See Also the Following Article

PATENTING OF LIVING ORGANISMS AND NATURAL PRODUCTS

Bibliography

Bryant, T. N. (1998). Probabilistic identification systems for bacteria. *In* "Information Technology, Plant Pathology and Biodiversity" (P. Bridge, P. Jefferies, D. R. Morse, and P. R. Scott), pp. 315–332. CAB International.

Canhos, V. P., Manfio, G. P., and Blaine, L. D. (1993). Software tools and databases for bacterial systematics and their dissemination via global networks. *Antonie van Leeuwenhoek, International Journal of General and Molecular Microbiology* 64, 205–229.

Langham, C. D., Williams, S. T., Sneath, P. H. A., and Mortimer, A. M. (1989). *J. Gen. Microbiol.* 135, 121–133.

Langham, C. D., Sneath, P. H. A., Williams, S. T., and Mortimer, A. M. (1989). *J. Appl. Bacterial.* 66, 339–352.

Priest, F. G., and Williams, S. T. (1993). Computer-assisted identication. *In* "Handbook of New Bacterial Systematics" (M. Goodfellow and A. G. O'Donnell, eds.), pp. 361–381. Academic Press, London, UK.

Willcox, W. R., Lapage, S. P., and Holmes, B. (1980). A review of numerical methods in bacterial identification. *Antonie van Leeuwenhoek* 46, 233–296.

Industrial Biotechnology, Overview

Erik P. Lillehoj and Glen M. Ford

Dexall Biomedical Labs, Gaithersburg, MD

I. Microbial Biotechnology Industry
II. Recombinant DNA Biotechnology Industry
III. Agricultural Biotechnology Industry
IV. Medical Biotechnology Industry
V. Social and Legal Issues

recombinant DNA DNA derived from different sources, generally used to isolate genes.

vaccine An antigen used to induce protective immunity against a pathogen.

GLOSSARY

antibiotic A chemical that inhibits the growth of or kills microorganisms.

antibody An immunoglobulin produced by B lymphocytes that binds to an antigen.

bioaugmentation Stimulation of microbial activity for production of a specific compound.

bioconversion Use of microorganisms for chemical modification.

bioremediation Use of microorganisms to remove environmental pollutants.

biosensor An instrument that combines a biological recognition unit with a physical transducer.

biotechnology Systematic use of organisms to manufacture products for human use.

genetic engineering Isolating and modifying genes using recombinant DNA techniques.

genome The entire genetic complement contained in the chromosomes of an organism.

herbicide A chemical compound used to kill plants, especially weeds.

immunoassay Use of antibodies in the detection or assay of compounds.

pesticide A chemical compound used to kill pests, especially insects.

proteome The entire protein content contained in an organism.

BIOTECHNOLOGY is the systematic manipulation of organisms to manufacture products for human use. Historically, humans have practiced biotechnology for over 8000 years, for example, using microorganisms to produce beer and wine. However, recently the term "biotechnology" has come to be more strictly concerned with genetic manipulation of organisms to achieve this goal, specifically with regards to recombinant DNA techniques.

Recombinant DNA is derived by procedures that combine DNA fragments from different sources. Because all organisms use DNA for their genetic material, individual genes from one organism can be removed, modified, and reinserted back into a different organism. The net effect is to place the transferred gene into a new host to accomplish a specific biotechnological purpose. Since the inception of recombinant DNA, commercial applications of biotechnology have grown enormously in both total volume and diversity. Presently, biotechnology is used to manufacture products for human and animal health care, food and agriculture, and environmental cleanup. In 1997, there were nearly 1300 biotechnology companies in the United States employing 120,000 workers. While most biotech companies record annual net financial losses, the yearly trend has been progressively improving and the industry as a whole is expected to reach the break-even point by the year 2000. The article highlights recent advance in the biotechnology industry and, where appropriate, provides examples of particular companies developing and commercializing state-of-the-art biotechnologies.

I. MICROBIAL BIOTECHNOLOGY INDUSTRY

A. Primary and Secondary Metabolites

Biotechnology began as industrial microbiology, using microorganisms for alcoholic fermentation. Subsequently, carefully selected microbial organisms were used to manufacture commodity and specialty chemicals. Commodity chemicals are produced in large quantity at low cost by microorganisms and are generally primary metabolites synthesized during the growth phase. These include ethanol, acetic acid, lactic acid, and glycerol. Specialty chemicals, such as amino acids, vitamins, and pharmaceuticals, are manufactured at substantially higher cost. The most common amino acids and vitamins produced by microorganisms are glutamic acid, lysine, tryptophan, riboflavin, vitamin B_{12}, and ascorbic acid (vitamin C).

Pharmaceuticals (antibiotics, steroids, and alkaloids) are generally produced as secondary metabolites, i.e., when the microbial culture enters the stationary phase. The salient features of secondary metabolism are (1) production by relatively few types of microorganisms, (2) nonessential for cell growth, (3) greatly influenced by growth conditions, (4) formed as a group of closely related, complex chemical structures, and (5) ability to be substantially overproduced *in vitro*. Microbial synthesis of antibiotics is the hallmark of secondary metabolism. More than 6000 antibiotics are known and more than 150 have been commercialized, with a worldwide market worth more than $10 billion annually. Most microbial-produced antibiotics are modified chemically to produce analogs with superior biological activity. By mutagenesis, the yields of antibiotics have been increased to a level where their production is commercially profitable. Penicillin, cephalosporin, streptomycin, erythromycin, tetracycline, rifampin, and lincomycin are produced in many industrial fermentations at 10–60 grams per liter.

B. Microbial Enzymes

The most economically important microbial enzymes include amylases, proteases, and lipases that have found applications in the food, paper, textile, and pharmaceutical industries. Microbial enzymes used in the food industry account for 50% of the $900 million annual value of industrial enzymes (Table I). The major suppliers of industrial enzymes include Alko, Amano Pharmaceutical, Bayer, Cultor, Gist-Brocades, Genencor International, Novo Nordisk, and Solvay and Cie. With the advent of genetic engineering, microbial enzymes have been successfully modified by mutagenesis to produce catalysts with more desirable reaction products. Examples of biotechnology companies that have commercialized modified enzymes by mutagenesis include Genencor (subtilisin) and Novo Nordisk (protease).

Extremozymes constitute a class of enzymes found in marine microorganisms isolated from extreme environments, such as superheated waters of geysers and midocean vents, brines, acidic environments, and the very cold ocean water of polar regions. These enzymes are active in their ambient conditions, either very hot, cold, acidic, or salty, and have recently begun to find commercial applications. For instance, hyperthermophilic enzymes can be used in commercial processes at high temperatures where normal enzymes are destroyed, such as refining corn syrup sweeteners and replicating DNA for biotechnology purposes.

Immobilized enzymes are enzymes bound to a solid support through which the substrate is passed and the product collected in a liquid state. Enzyme immobilization allows for not only increased enzyme stability but also enzymatic activity in organic solvents such as chloroform, dimethylformamide, dimethyl sulfoxide, ethyl acetate, and alcohols. Some reactions achieved with immobilized enzymes in organic solvents include stereospecific hydrolysis and esterfication of organic acids, peptide synthesis, hydroxylation of steroid hormones, and oxidation of cholesterol. Three types of enzyme immobilization are (1) cross-linkage of enzymes to each other with

TABLE I
Some Microbial Enzymes Used in the Food Industry

Enzyme	Product	Use
Alpha-amylase	Sweetners	Liquefaction of starch
	Beer	Removal of starch haze
	Bread	Flour supplementation
Amyloglucosidase	Sweetners	Saccharification
	Low carbohydrate beer	Saccharification
	Wine and fruit juice	Starch removal
	Bread	Improved crust color
Beta-galactosidase	Whey syrup	Greater sweetness
	Lactose-reduced milk	Lactose removal
	Ice cream	Prevention of "sandy" texture
Chymosin (rennin)	Cheese	Coagulation of milk proteins
Glucose isomerase	High fructose syrup	Conversion of glucose to fructose
Glucose oxidase	Fruit juice	Removal of oxygen
Invertase	Soft-centered sweets	Liquefaction of sucrose
Lipases	Cheese	Flavor development
Papain	Beer	Removal of protein
Pectinases	Wine and fruit juice	Clarification
	Coffee	Extraction of the bean
Proteases	Dairy products	Modification of milk proteins
	Caviar	Viscosity reduction
	Bread	Gluten weakening
	Meat	Tenderization

a bi-functional cross-linking agent such as glutaraldehyde, (2) bonding of the enzyme to a carrier through absorption, ionic bonding, or covalent attachment, and (3) enzyme inclusion into a semipermeable membrane consisting of microcapsules, gels, or fibrous polymers such as cellulose acetate.

C. Bioconversion, Bioaugmentation, and Bioremediation

Bioconversion is the use of microorganisms to achieve a specific chemical reaction. In an industrial setting, microbial cells are grown in a fermenter and, at an appropriate time, the chemical to be converted is added to the growth medium, the culture is subsequently incubated, the medium extracted, and the desired product purified. Production of vinegar by bioconversion of ethyl alcohol to acetic acid by acetic acid bacteria (*Acetobacter* and *Gluconobacter* spp.) is a classic example of this process. More recently, Genencor scientists engineered a microbial metabolic

pathway to create a novel *in vivo* route for the bioconversion of D-glucose to 2-keto-L-gulonic acid, a key intermediate in the synthesis of vitamin C. Briefly, a *Corynebacterium* gene that codes for an enzyme that stereospecifically reduces 2,5-diketo-D-gluconate to 2-keto-L-gluconate was expressed in *Erwinia herbicola*, a 2,5-diketo-D-gluconate producer. The recombinant cells produced 2-keto-L-gluconate when grown in the presence of D-gluconate. The final result was the conversion of D-glucose into vitamin C by a one microbial, one chemical step process that previously required four chemical steps.

Bioaugmentation is the stimulation of microbial metabolic activity for production of a specific compound. For example, production of citric acid by the fungus *Aspergillus niger* is performed in a growth medium with low iron content. Iron deficiency stimulates production of citric acid by the fungus as a chelator to scavenge the element. Bioremediation, or biodegradation, is the use of microorganisms to remove environmental pollutants such as halogenated aliphatics, aromatics, nitroaromatics, polychlorinated biphenyls, phthalate esters, polycyclic aromatic hydrocarbons, and nitrosamines. Bioenrichment refers to enhancing bioremediation by addition of oxygen or microbial growth nutrients. Many different types of microorganisms are capable of metabolizing chemical contaminants from the environment. Some of these organisms are found naturally in marine sediments and waters. Others can be selected *in vitro* and introduced into contaminated environments. The U.S. market for bioremediation of hazardous wastes is $2–3 billion per year.

II. RECOMBINANT DNA BIOTECHNOLOGY INDUSTRY

Bacteria are ideally suited to serve as hosts for recombinant DNA. They grow rapidly in nutrient broth and readily accept foreign DNA to be copied along with their own genes. Many bacteria possess extrachromosomal DNA (plasmids) that replicate separately from the bacterial genome. The first successful recombinant DNA techniques utilized plasmids isolated from the bacterial cell, modified *in vitro,* and reinserted back into the cell. When a foreign DNA is inserted into a selectable gene of a plasmid, that gene is inactivated, often allowing recombinant bacterial cells carrying the foreign DNA to be isolated from nonrecombinant cells. For example, recombinant cells with foreign DNA inserted into a plasmid gene conferring antibiotic resistance can be identified by replica plating the transformants from nonselective media to media containing the particular antibiotic. Colonies appearing on nonselective media, but absent from the selective media, will contain the recombinant plasmid. Plasmids also can be used to transfect eukaryotic cells and recombinants selected similarly. Once cloned into a homogeneous population of cells, the new gene can be expressed, purified, and utilized for novel purposes.

A. cDNA and Genomic Libraries

A recombinant DNA library is a collection of cells, usually bacterial, that have been transformed with recombinant plasmids carrying cDNA or genomic

DNA from a single source. cDNA is synthesized from an RNA template using reverse transcriptase (RT), while genomic DNA represents the entire genetic complement of the cell. Libraries made from both types of DNA have applications to particular needs. cDNA libraries contain only those genes transcribed by the cell and are useful for expression of recombinant proteins. As originally developed, cDNA is made using an oligo(dT) primer for RT to synthesize the first DNA strand followed by DNA polymerase to synthesize the second DNA strand. A number of biotechnology companies supply kits based upon this method (Amersham Life, Boehringer Mannhein, Clontech, Invitrogen, Life Technologies, Novagen, Amersham Pharmacia Biotech, Promega, Stratagene). Recently, a second approach was developed using RT in conjunction with the polymerase chain reaction (PCR), whereby second strand synthesis is followed by exponential amplification of cDNAs using thermostable DNA polymerase and short, sequence-specific oligonucleotides as primers. In addition to some of the companies mentioned, examples of companies supplying RT–PCR kits include Ambion, Bionexus, Tetra Link International, Epicenter, Hybaid, Perkin Elmer, and Roche Molecular Biochemicals. Technical difficulties in manipulating mRNA and RT have recently spawned commercialization of premade cDNA libraries, not only from different animal species, but also various cell types, tissues, and stages of development. The list of biotechnology companies supplying premade cDNA libraries includes Clontech, DNA Technologies, Genome Systems, Genpak, Invitrogen, Life Technologies, Novagen, Research Genetics, and Stratagene.

B. Prokaryotic and Eukaryotic Expression Vectors

Since the advent of modern biotechnology, many different *in vivo* systems have been developed for cloning and expression of recombinant proteins (Table II). *Escherichia coli* is the most common prokaryotic expression system. The newer high copy number plasmids replicating in *E. coli* possess highly inducible and repressible promoters, capable of expressing recombinant proteins constituting up to 30% of the total cellular protein. Efficient terminators

TABLE II
Recombinant Protein Expression Systems and Representative Suppliers

System	Major host cell(s)	Advantages	Disadvantages	Commercial vectors	Suppliers (Internet address)
Bacteria	*Escherichia coli*	Fast cell growth, well known promoters and terminators, high protein yields	No posttranslational modification, inclusion bodies	TriplEx	Clontech (*www.clontech.com*)
				pBAD, Thio-Fusion	Invitrogen (*www.invitrogen.com*)
				ProFEx HT	Life Technologies (*www.lifetech.com*)
				pET	Novagen (*www.novagen.com*)
				ABLE, pBK	Stratagene (*www.stratagene.com*)
Yeast	*Saccharomyces cerevisiae, Pichia pastoris*	High cell densities, genetically stable, soluble proteins	Difficult cell extraction	YEXpress, MATCHMAKER	Clontech
				P. pastoris, pYES2	Invitrogen
				ESP	Stratagene
Insect	*Spodoptera frugiperda, Drosophila,* silk worm	Large proteins, high expression levels, intron splicing, posttranslational modifications	Transient expression, high media cost, time consuming	BacPAK	Clontech
				MaxBac, DES	Invitrogen
				Bac-to-Bac	Life Technologies
				pBAC, pAcPIE1	Novagen
				BaculoGold	PharMingen (*www.pharmingen.com*)
				pPbac, pMbac	Stratagene
Mammal	CHO, BHK, HEK, Sp2/0 NIH3T3	Soluble proteins, intron splicing, posttranslational modifications	Slow growth, high media cost, complex expression mechanisms	Tet-Off, Tet-On	Clontech
				pShooter, pcDNA3.1	Invitrogen
				SFV	Life Technologies
				BacMan	Novagen
				LacSwitch	Stratagene

minimize background transcription and enhance mRNA stability. Recombinant proteins are often synthesized as fusion proteins or containing peptide tags to allow quick and efficient purification by antibody affinity chromatography. Prokaryotic expression systems, however, are incapable of posttranslational modifications, lead to poor protein refolding and disulfide bond formation, and often result in recombinant proteins synthesized as insoluble, denatured inclusion bodies. Eukaryotic expression systems circumvent some of these problems. These include yeast, insect, and mammalian cells. Additionally, several different reagents have been developed and commercialized for transfection of cloned genes into eukaryotic cells, including DEAE-dextran (5Prime-3Prime, Pharmacia Biotech, Promega, Stratagene),

calcium phosphate (5Prime-3Prime, Invitrogen, Life Technologies, Pharmacia Biotech, Promega, Stratagene), liposomes (Boehringer Mannheim, Invitrogen, Life Technologies, Mirus, Promega, Stratagene, Wako), activated dendrimers (Quigen), and recombinant human fibronectin (TaKaRa).

C. Fermenters and Bioreactors

Fermenters and bioreactors are used for large-scale growth of host cells producing recombinant proteins. Because the volume of large systems may exceed 100,000 gallons, instrument design features that have proven attractive to the biotechnology industry include dependability, flexibility, validatability, and engineering simplicity. Several types of technologies have been developed, including airlift bioreactors (Celltech), fluidized bed reactors (Verax), porous ceramic matrix reactors (Opticell), hollow fiber membranes (Cellex Biosciences, Unisyn Technologies, B. Braun Biotech), and stirred tanks (Associated Bioengineers and Consultants, B. Braun Biotech, New Brunswick Scientific, W.H.E. Bio-Systems). Hollow fiber bioreactors use capillary tubes with pores of about 10,000 molecular weight cutoff. Cells grow on the outside of the fibers in the extracapillary space, while nutrient media is constantly recirculated through the fibers in the intracapillary space. Nutrients diffuse through the fiber pores to the cells and metabolic wastes diffuse in the opposite direction. Expensive, high molecular weight growth factors need only be added to the extracapillary space, reducing their use 100-fold. Secreted recombinant proteins are retained in the extracapillary space, increasing their concentration 100-fold and substantially decreasing downstream processing costs. Stirred tanks use mechanical methods to constantly mix growth media, allowing maximum exposure of cells to nutrients. Bacterial cells grown in stirred tank fermenters may reach densities of 10^{10} cells/ml. Using the newly developed centrifugal bioreactor technology, Kinetic Biosystems has cultured *E. coli* to 10^{14} cells/ml. A rotary cell culture system designed by Synthecon allows mammalian cells to grow in complex three-dimensional arrays that mimic tissues.

III. AGRICULTURAL BIOTECHNOLOGY INDUSTRY

Since the development of agriculture over 10,000 years ago, world food production has supported the increasing human population. This has been due, in large part, to scientific advances in farming practices, particularly conventional plant and animal breeding techniques. However, the world's population is now increasing at the fastest rate in history and is forecast to reach 12 billion people by the year 2030. It is unclear whether traditional farming techniques will be able to supply the world's food requirements at that time. The agricultural biotechnology industry will play a major role in resolving this dilemma.

A. Plant Biotechnology

Plant biotechnology uses genetic engineering to improve traditional breeding techniques. Plants with reduced dependence on chemical pesticides and herbicides, increased disease resistance, improved survival during environmental stress (drought, high or low temperature), and improved quality and nutritional content have been produced and commercialized (Table III). More than 60 different plant species have been genetically modified to produce improved foods for human consumption. Additionally, totally new crops have been developed that produce novel

TABLE III
Commerical Examples of Plants Improved by Genetic Engineering

Plant	Improved genetic trait(s)	Company (Internet address)
Canola	Lower saturated fats, herbicide resistance	AgrEvo (*www.agrevo.com*), American Cyanamide, Monsanto (*www.monsanto.com*), Plant Genetic Systems
Corn	Insect resistance, herbicide resistance	AgrEvo, American Cyanamide, Ciba-Geigy (*www.ciba.com*), Monsanto, Mycogen (*www.mycogen.com*), Novartis (*www.novartis.com*)
Cotton	Insect resistance, herbicide resistance	Calgene (*www.calgene.com*), DuPont (*www.dupont.com*), Monsanto, Rhone-Poulenc (*www.rhone-poulenc.com*)
Potato	Insect resistance	Monsanto
Soybean	Lower saturated fats, herbicide resistance	Asgrow Seeds (*www.asgrow.com*), DuPont, Monsanto
Squash	Disease resistance	Asgrow Seeds
Tomato	Slower ripening, increased pectin	Agritope (*www.agritope.com*), Calgene, DNA Plant Technology (*www.dnap.com*), Monsanto, AstraZeneca (*www.astrazeneca.com*)

pharmaceuticals, oils, fuels, and other nonfood products.

In 1995, U.S. farmers used an estimated 265 million pounds of insecticides on agricultural crops, costing about $1 billion. Of this, cotton received 55% and corn 35% of the total. Due to unwanted side effects and development of insect resistance to chemical insecticides, novel genetically engineered insecticide-resistant plants have been developed. Monsanto developed Bollgard, a transgenic cotton plant containing an toxin gene from *Bacillus thuringiensis,* to control bullworn and budworm. As a first-generation product, Bollgard doesn't completely eliminate the need for insecticides, but does reduce the number of pesticide applications required per year. The benefits of using the *B. thuringiensis* toxin are that it rapidly degrades in the environment, does not present any human or animal health hazards, and is not harmful to beneficial insects. For protection against the European corn borer, one of the most devastating pests in the United States, Maximizer (Novartis), DeKalb Insect-Protected (DeKalb Genetics), Yield-Gard (Monsanto), and NatureGard (Mycogen) corns are available. G-Stac corn hybrids (Garst Seed) contain multiple transgenes in the same plant to simultaneously control the corn borer (*B. thuringiensis* toxin) and confer resistance to several herbicides.

Liberty Link corn and Liberty Link canola, both developed and produced by AgrEvo, allow farmers to apply Liberty herbicide during the growing season, resulting in weed control without effect on crop performance. IMI corn and IMI canola (American Cyanamid) allow growers to use environmentally safe imidazolinone herbicides. BXN cotton plants (Calgene) require fewer chemical herbicides. DeKalb GR hybrid corn (DeKalb Genetics) is tolerant to glufosinate herbicide. Roundup Ready cotton and Roundup Ready soybeans (Monsanto) tolerate Roundup herbicide, resulting in enhanced weed control without affecting crop yields.

The FlavrSavr tomato (Calgene) was one of the first agricultural biotechnology products approved by the U.S. Food and Drug Administration (FDA). The FlavrSavr was modified by antisense technology, allowing it to ripen on the vine, yet stay fresh for a prolonged period postharvest. In this case, the gene encoding the enzyme responsible for pectin degrada-

tion and fruit softening (polygalacturonase) was isolated and its DNA sequence elucidated. A reverse (antisense) synthetic DNA was then synthesized and inserted back into the tomato, where it inactivated the polygalacturonase gene. Increased pectin tomatoes have also been bioengineered by Zeneca Plant Sciences to delay softening and retain pectin during processing of tomato paste. Fresh World Farms Endless Summer tomato (DNAP Holding) was created through recombinant DNA technology to suppress production of ethylene, the hormone responsible for fruit ripening. Laurical (Calgene) is a less expensive source of high quality laurate from rapeseed oil for soaps and detergents. High oleic acid soybeans (DuPont) produce up to 85% oleic acid, compared to approximately 25% found naturally.

B. Animal Biotechnology

Traditional food-animal breeding techniques have done much to improve the productivity and health of livestock in the past 50 years. Milk production per cow has more than doubled. Feed efficiency for swine and poultry has increased almost 50%. Some major diseases (hog cholera) and pests (screwworm) have been eliminated. However, maximum productivity and complete disease eradication have yet to be realized. Biotechnology has begun to address these areas and is expected to assume a more prominent role in the future.

Gamete and embryo technologies represent venues of animal biotechnology already commercialized. *In vitro* maturation and fertilization of farm animal oocytes is a prerequisite for a system to mass produce cloned embryos. Animal embryos produced by *in vitro* fertilization can be cultured *in vitro* until they are transferred to the womb of a surrogate mother for completion of development. ABS Global and Advanced Cell Technology are two U.S. animal biotechnology companies that have successfully impregnated cows and pigs using cloned embryos. Furthermore, embryo gender selection can be achieved by separating X- and Y-chromosome-bearing spermatozoa prior to *in vitro* fertilization. Such separation has been accomplished in cattle by flow cytometric sorting of sperm cells, based upon DNA content.

Parallel to the human genome project, an organized international effort is under way to map and sequence the genomes of economically important animal species, particularly the bovine, swine, sheep, and poultry genomes. While these programs are not as advanced as the effort to sequence the human genome, it has been long known that most genetic traits in animals are multigenic and encoded by quantitative trait loci. These include traits for disease resistance, meat leanness and tenderness, and milk quality and quantity. In some animal species, quantitative trait loci have been separated into individual components, making them amenable to genetic manipulation. As a result of the bovine genome project, for example, genes conferring resistance to bovine trypanosomiasis could be identified and isolated from naturally resistant but low milk producing African breeds. Using these genes, disease resistant high milk producing Western breeds of transgenic cattle could be created.

Transgenic animals are produced by inserting a foreign DNA into the host cell nucleus, transmission to the germ line, and expression in the live animal. Transgenic biotechnology has been used to improve farm animal production traits, enhance food nutritional qualities, and produce pharmaceutical drugs for human medicine ("gene pharming"). Examples of improving production efficiency include better animal feed efficiency and growth rate, improved disease resistance, and better wool production in sheep. Transgenic dairy cattle producing milk with increased protein content, reduced fat, or bovine proteins replaced by human milk proteins ("humanized" milk more closely resembling human breast milk) have been envisioned. The list of biotech companies developing transgenic animals to produce human pharmaceutical drugs includes AltraBio, FinnGene, GenePharming Europe, Genzyme, NexTran, PPL Therapeutics, and Symbicon. Much interest has centered on using the mammary gland to express transgenes because of its ample protein synthesis capacity, the availability of milk fusion genes with easily controlled regulatory elements, and simplicity of harvesting milk. Human growth hormone, lysozyme, tissue plasminogen activator, urokinase, interleukin-2, follicle-stimulating hormone, alpha-1-antitrypsin, factor VIII, factor IX, gamma-interferon, serum albumin, and superoxide dismutase have been expressed as transgenes with mammary gland-specific regulatory elements in various animal species. In addition to specific proteins, transgenic animals (pigs) as a source of organs (heart, liver, pancreas) for human transplantation may become a reality in the future. Already realized is the production of transgenic animals by nuclear transfer technology, whereby cloning is achieved by transferring the nucleus from a somatic cell into an enucleated egg cell, allowing the embryo to begin development *in vitro*, and implanting the egg into a foster mother. In this manner, scientists at the Roslin Institute succeeded in cloning an adult sheep and adult mice and cattle have since been cloned by similar procedures. Somatic cell cloning will be used to produce large herds of genetically identical transgenic animals producing recombinant human therapeutic proteins in their milk. It also has the potential to save rare and endangered animal species from extinction.

C. Agricultural Biotechnology Industry in the 21st Century

The U.S. Department of Agriculture has identified five broad priority areas where additional research is needed to assure a constant supply of high quality agricultural products by the biotechnology industry during the next century. "First, continued mapping and sequencing of animal, plant, and microbial genomes to elucidate gene functions and regulation and to facilitate the discovery of new genes as a prelude to gene modification. Second, determination of the biochemical and genetic control mechanisms of metabolic pathways in animals, plants, and microbes that may lead to products with novel food, pharmaceutical, and industrial uses. Third, extended understanding of the biochemical and molecular basis of growth and development including structural biology of plants and animals. Fourth, elucidation of the molecular basis of interactions of plants and animals with their physical and biological environments as a basis for improving the organisms' health and well-being. Finally, enhanced food safety assurance methodologies such as rapid tests for identifying chemical and biological contaminants in food and water."

IV. MEDICAL BIOTECHNOLOGY INDUSTRY

Of the nearly 1300 biotechnology companies in the United States, approximately 80% have a primary interest in human health care. Much of the interest in medical biotechnology derives from the potential for new immunodiagnositc and therapeutic agents. The applicability of immunological techniques to the diverse disciplines of biology and chemistry has expanded dramatically over the last 20 years since the advent of monoclonal antibodies, development of elaborate immunosorbent techniques, identification of safe and effective adjuvants and vaccines, and sophisticated immunoassay techniques. Moreover, commercialization of the emerging technologies of biosensors, molecular diagnostics, genome and proteome analysis, and DNA and protein chips will continue to support the growth and development of the medical biotechnology industry.

A. Immunodiagnostics

Immunodiagnostics applies to the detection of biologically important substances in blood and other body fluids using antibody technology. An antibody is an immunoglobulin produced by B cells that binds to an antigen. The antigen can be a therapeutic drug, drug of abuse, infectious disease agent, hormone, autoimmune or cancer marker, or another antibody molecule. The method of detection can yield either a qualitative or quantitative result. During the last two decades, immunodiagnostics has experienced a continuum of technical revolutions, consisting of at least six milestone advances, all of which have been commercialized by the biotechnology industry. First is the introduction and success of immunoassay technology, which literally transformed the field of diagnostics. This transformation began in earnest in the 1960s and led to the development of accurate and sensitive radioimmunoassays (RIA) for hormones, viral, and bacterial antigens as well as specific antibodies. Second, nonisotopic methods (enzyme assays) slowly replaced the initial RIA technology as a safer and more environmentally friendly methodology. Third, even more sensitive technologies (e.g.,

fluorescence and chemiluminescence) replaced many of the conventional colorimetric substrates used in immunoassays. Fourth, development of monoclonal antibodies has had a profound effect on diagnostics due to the capability of producing exquisitely specific antibodies in large quantities. Fifth, the advent of manual and automated rapid testing delivered another technological infusion and created the ability to achieve assay results in minutes rather than hours. The sixth and most recent milestone is the use of biosensors to detect molecular interactions without the need to amplify the signal via secondary conjugates and substrates.

B. Polyclonal and Monoclonal Antibodies

A specific antibody reacts only with a small region (epitope) contained on the molecular structure of the antigen. Thus, the physical size and complexity of the antigen influences the number of different specific antibodies produced. Larger and more complex immunogens contain many different epitopes, each producing and reacting with its own specific antibody. A polyclonal antibody refers to the total set of antibodies reacting with multiple epitopes of a given antigen. In the analogy of injecting a microscopic airplane into a host animal, specific antibody would be divided among those antibodies reacting with antigenic regions on the wings, tail, nose, wheels, and fuselage. Since antibodies are able to distinguish optical isomers, they would differentiate the right and left wings of the airplane.

The technology used to develop polyclonal antibodies involves injecting animals (rabbits, sheep, goats, donkeys) with the antigen of interest and collecting the serum portion of the blood. Of the total serum immunoglobulins from an immunized animal only 0.1–10% contain specific antibodies reacting with the injected material. Various protein purification techniques are used to isolate specific antibody from nonspecific antibody and other serum proteins. In affinity purification, the serum is applied to a solid phase column containing a chromatographic resin with immobilized antigen to which the antibody was raised. This results in absorption of the antibody from the liquid to solid phase. By various chemical treatments, the specific antibody is eluted and separated from nonspecific serum proteins. Once specific antibody is obtained in sufficient quantity, immunodiagnostic tests can be constructed. All of these methods have been commercialized and examples of biotechnology companies providing custom polyclonal antibody services and immunoassay development include American Qualex, Biocon, Biodesign, Charles River Laboratories, Fitzgerald Industries, Harlan BioProducts, PerImmune, QED Biosciences, STI, and Zymed Laboratories.

A major advance in the medical biotechnology industry occurred in the late 1980s, with the widespread use of monoclonal antibodies. Briefly, monoclonal antibodies are extremely specific antibodies reacting with only one epitope on an antigen. They are derived by immunizing mice or rats with an antigen, isolating antibody-producing cells from the spleen or lymph nodes, screening the cells to determine antibody reactivity, and fusing them with an immortalized plasmacytoma cell to obtain a hybridoma cell that can be cultured *in vitro* or grown in mice. As the hybridoma cells replicate, continuous production of large quantities of monoclonal antibody occurs. The ability of monoclonal antibodies to react with a single epitope allows construction of immunoassays that have exquisite specificity. Monoclonal antibodies have also been used in cancer immunotherapy, where a radioactive or cytotoxic compound is attached to the antibody and injected into a patient. In this manner, the toxic material becomes concentrated primarily at the cancer site, leading to death of the cancerous cells. Humanized monoclonal antibodies are created by genetic engineering to contain only the antigen-binding region of the original mouse monoclonal antibody fused to a human antibody. In this way, many of the problems associated with using mouse antibodies for immunodiagnostics and immunotherapy are eliminated. In addition to the biotechnology companies indicated above, examples of commercialized monoclonal antibody services include abV Immune Response, Berkely Antibody, Cayman Chemical, HybriLogic, Lampire Biological Laboratories, Maine Biotechnology Services, Scantibodies Laboratory, and TSD BioServices.

C. Enzyme-Linked Immunoassays

An immunoassay is the use of "immune substances" (antibodies) in the detection or assay of a compound of interest. The compounds for clinical chemistry are typically hormones, serum and cancer proteins, drugs of abuse, therapeutic drugs, viruses, and bacteria. In addition, it is often desirable to detect and quantify human antibodies against infectious pathogens, allergens, and autoimmune markers as an indicator of current or past exposure. Immunoassays take advantage of the highly specific reaction between antibodies and antigens to which the antibody response was elicited in the host animal. The antibody–antigen reaction has been likened to a lock and key in terms of specificity. For any given lock (antigen), only a specific key (antibodies) will be able to fit. The market for enzyme immunoassays is very large and the list of biotechnology companies supplying commercial test kits is extensive. Some of the more well known include Abbott, Becton Dickinson, Boehringer Mannheim, Chiron, Dade International, DuPont, Organon Teknika, and Sigma Diagnostics.

The enzyme-linked immunosorbent assay (ELISA) is similar to RIA, except that the signal-generating reporter molecule is an antibody-linked enzyme (peroxidase, alkaline phosphatase, glucose oxidase). Two types of immunoassay are constructed, depending on whether the analyte is an antigen or antibody. For antigens, the test is a sandwich ELISA and for antibodies, it is an indirect ELISA (Fig. 1). In the sandwich ELISA, antibody is attached to a solid phase such as plastic (styrene beads, microspheres, microtiter wells), metal particles, cellulose or nylon fibers, or agar, through ionic electrostatic surface interactions, hydrophobic binding, or covalent bonding. After the antibody-bound surface is chemically stabilized, the analyte-containing sample is allowed to contact the solid phase, the solid phase washed to remove excess analyte, an enzyme-linked reporter molecule added and detected with a reactive substrate. The amount of analyte needed to generate a detectable signal is about 10^5–10^6 epitopes per milliliter for the more sensitive colorimetric substrates. Recently, fluorescent and chemiluminescent substrates have in-

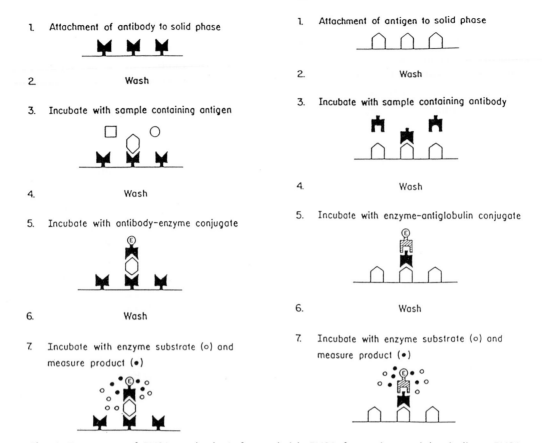

1. Attachment of antibody to solid phase

2. Wash

3. Incubate with sample containing antigen

4. Wash

5. Incubate with antibody-enzyme conjugate

6. Wash

7. Incubate with enzyme substrate (o) and measure product (•)

1. Attachment of antigen to solid phase

2. Wash

3. Incubate with sample containing antibody

4. Wash

5. Incubate with enzyme-antiglobulin conjugate

6. Wash

7. Incubate with enzyme substrate (o) and measure product (•)

Fig. 1. Two types of ELISA methods. Left, sandwich ELISA for antigens; right, indirect ELISA for antibodies.

creased sensitivity up to 100-fold. Examples of commercially available sandwich ELISAs include human immunodeficiency virus (Advanced Biotechnologies), hepatitis virus (Melotec), cytomegalovirus (Sigma Diagnostics), and antibodies (Dexall Biomedical Labs).

In the indirect ELISA method, antigen is coated on the solid phase. The antigen can be viral, bacterial, or parasitic proteins, or autoimmune markers (DNA, RNA, nuclear, and cytoplasmic proteins). A patient's sample is added to the coated solid phase and if specific antibody is present in the sample it will bind to the immobilized antigen. After washing, an enzyme-linked antibody directed against immunoglobulins binds to any antibody that is attached to the antigen. After washing, a colorimetric or fluorescent substrate is added and signal is generated proportionately to the amount of sample antibody bound. Commerical indirect ELISAs include autoantibodies

(R&D Systems), hepatitis virus (United Biotech), human immunodeficiency virus (Abbott), rubella virus (Hemagen Diagnostics), cytomegalovirus (Medix Biotech), and allergens (Dexall Biomedical Labs).

D. Rapid Manual and Rapid Automated Immunoassays

With the desire of health practitioners to obtain clinical results in minutes rather than hours, a major milestone occurred in the late 1980s with the development and optimization of rapid immunoassays. These can be of the manual or automated type. While the rest of the clinical diagnostic market remains sluggish, the use of rapid immunoassays is growing at an annual rate of 25%, with $1 billion in sales by biotechnology companies in 1997. Rapid manual immunoassays are used in point-of-care testing, where technician time is substantially reduced by use of coated polymer membranes as the solid phase. Two types have been developed, lateral flow and flow through assays. In lateral flow assays, the sample is added to an absorbent pad containing a colloidal colored conjugate. If analyte is present in the sample, it reacts with the conjugate. Both conjugate-bound analyte and ureactive material migrate by capillary action through the membrane in a chromatographic fashion. If analyte is present in sufficient concentration, the conjugate–analyte complex binds to the coated membrane, forming a visually detectable colored line or pattern on the membrane. Flow-through rapid manual assays are membrane tests where the sample flows transversely through the membrane and analyte is trapped on the surface with the colloidal conjugate. In both cases, total assay time is 3–5 minutes and the sensitivities approach the low nanogram range. Examples of biotechnology companies marketing rapid manual immunoassays include Dexall Biomedical Labs (C-reactive protein, human chorionic gonadotropin), Princeton BioMediTech (drugs of abuse), Boehringer Mannheim (cardiac markers), and Medix Biotech (infectious agents, human chorionic gonadotropin).

Many rapid automated tests are performed using membrane or microparticle solid surfaces. Because these have significantly greater surface areas than the wells, tubes, or macroparticles used in conventional ELISA, more capture antibody or antigen can be immobilized. Combined with the inherent property of membranes to channel analytes into close proximity with the coated solid phase, reaction rates occur significantly faster. Microparticles have the advantage of a mobile colloidal liquid phase that also brings the reactants into close proximity, thereby increasing the reaction rate. Since the reaction of analyte with the solid phase is usually complete after 30 minutes, high degrees of precision and reproducibility are realized. Microparticles used in rapid automated immunoassays include magnetic material or latex coated with antigen or antibody. A robotic arm removes a sample from a primary collection tube, dispenses a precise amount into a reaction well containing the microparticles, the reaction mixture is incubated, and the particles washed automatically. Magnetic particles are easier to wash since a magnet is used to pull the particles to the side of the reaction cup

during aspiration and washing. Latex microparticles need to be trapped on glass or cellulose fibers for washing. After washing, the particles are exposed to conjugate, washed again, substrate added, and the extent of reaction measured optically. The sensitivity of rapid automated tests for their corresponding analyte is in the sub-picogram range with linear dose-response over a 3 log range.

Several biotechnology companies manufacture and sell rapid automated immunoassay analyzers. The IMx system from Abbott uses microparticle capture enzyme immunoassay for large molecular weight, low concentration analytes, such as hormones, cancer markers, and infectious agents. The TDx system employs fluorescence polarization immunoassay for low molecular weight analytes, including therapeutic drugs and some hormones. Tosoh Medics markets the AIA-1200 system for analysis of hormones, cancer markers, and infectious diseases. Two formats are available, one using magnetic microparticles coated with specific antibody and enzyme-linked antibody (two-site immunoenzymatic assay) and the other antibody coated magnetic particles plus enzyme-linked antigen (competitive immunoassay). Recently, Boehringer Mannheim introduced the Elecsys 2010 system using electrochemiluminescent technology for cardiac markers, hormones, and infectious agents.

E. Biosensors, Molecular Imprinting, and Molecular Diagnostics

Although capable of performing many different immunoassays with minimal technician time, rapid automated immunoassay analyzers are expensive to purchase and maintain. Consequently, newer technologies using cheaper point-of-care testing instruments have become popular. Biosensors are "patient friendly" instruments that combine a selective binding unit with a physical transducer. Binding elements include antibodies (immunosensors), enzymes, and nucleic acids. Following binding of an analyte to the biological recognition unit, a biochemical event (e.g., enzymatic reaction) occurs that is subsequently converted into a measurable signal by the transducer. Because of the specificity of the binding event, biosensors can directly operate in heterogenous biological fluids, such as whole blood. Amperometric enzyme biosensors to quantify blood glucose levels are the most common biosensors on the market and are manufactured by many biotechnology companies including Biosen, Boehringer Mannheim, Ciba, Fuji Electric, i-Stat, Miles, Medisense, Nova Biomedical, Via Medical, and Yellow Springs Instruments. Other companies, such as Bioanalytical Systems and Thera-Sense, are in research and development stages of improved, third-generation glucose monitors. However, unlike the blood glucose monitoring market, commercialization of biosensors for other analytes has been slow. Future biosensors are expected to detect and measure infectious disease agents, pharmaceutical drugs, hazardous chemicals, and genetic defects.

Molecular imprinting is a chemical process in which a synthetic porous polymer network is formed around an analyte. Removal of the analyte leaves cavities with a three-dimensional structure complementary to the analyte. These cavities can subsequently rebind to the analyte with high specificity. Molecular imprinting was initially recognized as a novel method to detect and quantify molecules for which there are no known natural receptors. However, it soon became apparent that imprinting technology could be applied to any analyte and provide a chemically stable, cost-effective alternative to antibodies, enzymes, and other biologically based receptor molecules used in immunoassays, bioassays, and affinity chromatography resins. Molecular imprints of dozens of conventional analytes have been prepared, including drugs, hormones, steroids, amino acids, peptides, carbohydrates, coenzymes, pesticides, organic dyes, and bacteria.

Molecular diagnostic assays use nucleic acid technology for diagnostic purposes. Examples, include DNA hybridization and amplification procedures that have been applied to detection of genetic mutations and rearrangements, blood cell typing, and diagnosis of viral and bacterial infections. Oncor and Oncogene Science have developed Southern hybridization tests to detect rearrangements of B/T cell genes and the bcr gene, respectively. Roche has commercialized PCR tests for *Chlamydia trachomatis* and human immunodeficiency virus. PCR tests for blood cell typing are available from Gen-Trak and Biotest

Diagnostics. Other amplification-based tests are marketed by Abbott (*Chlamydia trachomatis, Neisseria gonorrhoeae*) and Gen-Probe (*Mycobacterium tuberculosis*). All of these tests have received FDA approval.

F. Genomics, Proteomics, and Bioinformatics

The human genome project is an international effort to completely map and sequence the approximately 80,000–100,000 genes in the 3 billion nucleotides of the 24 human chromosomes. This endeavor has three phases: (1) creation of a physical map, (2) complete sequence determination, and (3) determination of gene function. Presently, the project is in the initial stages of the second phase. One anticipated consequence that the information this project will provide is the design of new pharmaceutical drugs to control or cure some of the most common human diseases, including heart disease, cancer, Alzheimer's disease, and AIDS. Greater than 60% of new drug targets are predicted ultimately to be derived from genomics, compared to less than 10% in 1996. Not surprisingly, several biotechnology companies have recently joined this effort. In 1998, Celera Genomics and Incyte Pharmaceuticals separately disclosed plans to produce a complete human genetic map within 2–3 years at a cost of $200–250 million. It is currently unclear how the public and private endeavors will be coordinated.

Although genomics is expected to continue to provide the medical biotechnology industry with highly specific and sensitive DNA-based drugs and assays, expressed proteins are the ultimate regulator of cell function and activity. Recent developments in analyzing the total protein profile of a cell or tissue permit identification of specific alterations in protein expression associated with particular disease states. The proteome is defined as the entire protein content contained in a given cell or tissue and one of the challenges of genetic engineering has been to identify a single protein within the complex heterogeneous mixture of the cell. Two-dimensional polyacrylamide gel electrophoresis coupled with high resolution image analysis can be used to obtain cellular total protein profiles. By comparing fingerprints of normal and disease states, unique disease-associated proteins can be identified. Oxford GlycoSciences has applied this methodology to the identification of novel molecular markers in prostate cancer. Novartis Pharma and Large Scale Biology have applied proteomics to identify new patterns of protein expression following exposure to toxic agents, such as heavy metals. By improving two-dimensional gel electrophoresis technology, Proteome targets identification of low abundance proteins in microorganisms with fully characterized genomes, such as *Saccharomyces cerevisiae* (6100 proteins). Once identified, the protein can be isolated by robotic spot picking, digested with an appropriate protease, and the molecular weight of its peptides determined by mass spectrometry. Since every protein possess a unique peptide fingerprint, mass profiling allows unknown proteins to be matched to the theoretical mass profile generated from genomic databases and identified.

Bioinformatics is the rapidly emerging field of storage, retrieval, and analysis of data derived from genomics and proteomics. Gene and protein sequence computer databases are growing exponentially, some with a doubling time of less than 1 year. The number of total GenBank sequences is expected to exceed 1,000,000 by the year 2000. Several biotechnology companies have been founded whose sole mission is bioinformatics, including Pangea Systems, Base4 Informatics, Netgenics, Genomica, and Oxford Molecular Group. Additionally, Structural Bioinformatics provides three-dimensional protein structural models derived from genomic sequences and Acacia Biosciences furnishes *in vivo* activity profiles for accelerating drug discovery and development.

G. DNA Chips, Protein Chips, and Biochips

Affymetrix was one of the first biotechnology companies to commercialize DNA chips. Their patented GeneChip DNA probe array technology uses short synthetic oligonucleotides attached to microchips that serve as a substrate for hybridization with other DNA fragments in the sample to be analyzed. Since the oligonucleotides differ in sequence and are ar-

ranged in a two-dimensional array on the chip, the analyte DNA will only bind to a particular region on the chip's surface. Detecting where the DNA binds reveals its nucleotide sequence and thereby identifies it, for example, as a mutant allele of a genetic disease. A joint venture between Affymetrix and Roche Molecular Systems was formed recently to develop commercial applications of PCR on DNA chips. Ciphergen Biosystems has received the U.S. patent for its ProteinChip technology, where proteins are captured from crude sample extracts through affinity binding directly to the chip's surface. Surface-enhanced laser desorption/ionization is used to identify and quantify the bound proteins enabling receptor–ligand, epitope mapping, protein–protein, protein–DNA, and protein–drug interactions to be investigated directly on the chip. Biochips were developed by a joint venture between Motorola, Packard Instruments, and the Argonne National Laboratory. Essentially the size of a glass microscope size, each biochip contains 10,000 miniature (100 × 100 micron) polyacrylamide gel tabs containing separate target pieces of nucleic acid or protein. Liquid samples (200 picoliters) are dispensed onto individual gel tabs through a 75 μ-diameter capillary and binding activity monitored by fluorescence. Future commercial applications of the biochip technology are expected to revolutionize the speed, sensitivity, and cost of genetic analysis, disease detection, and drug discovery.

H. Therapeutics, Adjuvants, and Needle-Free Vaccination

Concurrent with the development of new medical diagnostic reagents, the biotechnology industry has been formulating new and improved therapeutic agents to treat human diseases. Many of these have been approved by the FDA. For example, antibody immunotherapy using RespiGam (MedImmune) holds the promise of averting hundreds of thousands of children from contacting respiratory illnesses, such as pneumonia and bronchiolitis. Rituxan (IDEC Pharmaceuticals) is a CD-20 chimeric antibody to treat non-Hodgkin's B-cell lymphoma. Zenapax (Protein Design Labs) is a humanized monoclonal antibody to control acute renal transplant rejection. In the arena of cancer therapy, recent discoveries in the mechanisms of carcinogenesis have lead to the development of new drugs presently in the FDA pipeline. FGN-1 (Cell Pathways) is a nonsteroidal anti-inflammatory compound that can selectively inhibit the growth of precancerous and cancerous cells. The tumor suppressor genes p53 (Introgen Therapeutics), mda-7 (GenQuest), E1A (Targeted Genetics), and Tg737 (Proctor and Gamble) arrest tumor growth and may one day be used in gene therapy treatments of local tumor masses. Alternatively, proteins that augment innate tumor suppressor gene activity, such as Neu differentiation factor/heregulin (Advanced Cellular Diagnostics), may be used therapeutically for the same purpose.

In addition to new drugs with direct activity on infectious agents and cancers, novel adjuvants to boost the immune response and improved vaccine delivery devices are being developed. For example, the adjuvant formulation SAF-1 (Syntex) use a threonyl analog of muramyl dipeptide in an emulsion of squalene. The Stimulon family of adjuvants from Aquila Pharmaceuticals are water soluble, nontoxic, and currently undergoing regulatory evaluation for both human (Quilimmune-P) and animal (Quilvax-M) applications. Methods of vaccination that stimulate the local mucosal immune response often provide protective immunity more closely resembling that produced during natural infections and, are such, are often superior to traditional vaccination by intravenous or intramuscular routes. Several different types of needle-free vaccination devices have been developed, depending on the organ systems to be targeted. Nasal administration stimulates the upper respiratory tract, oral administration for the gastrointestinal tract, and vaginal vaccination for sexually transmitted diseases. Scientists at Maxim Pharmaceuticals observed that nasal spray administration of cholera toxin subunit B induced a better mucosal immune response in the upper respiratory system than did oral vaccination. A nasal flu vaccine designed by Aviron protects children against both influenza and ear infections. Transcutaneous vaccinations using delivery devices such as skin patches or microneedle arrays are being developed at several companies including Iomai and Alza.

V. SOCIAL AND LEGAL ISSUES

Throughout recorded history, scientific achievements have created new moral issues debated in society and recent developments in biotechnology are no exception. Although governmental regulatory agencies exists in the United States and Europe to control the safety and efficacy of new products created through biotechnology, some areas are less well defined. In these areas, government regulations have lagged behind scientific advancements, often necessitating unilateral actions. For example, recent developments in the area of adult animal cloning have made the possibility of human cloning a likely event in the near future. Accordingly, U.S. President Clinton banned the use of public funding for human cloning experiments and the Congress has been recommended by the National Bioethics Advisory Commission to do the same for the private sector.

Since the 1970s, public knowledge about biotechnology has increased. Indeed, a recent survey found that the majority of the U.S. population had positive views concerning biotechnology and a willingness to use medicines or purchase food products developed through biotechnology. This poll reflects an evolving public perception, particularly with agricultural produce. Early consumer fears about genetically engineered foods caused some of the first products to be removed from the market. When these food were originally introduced, it was questioned whether or not the new plants could become pernicious weeds or transfer their new genes to wild-growing weeds. However, evidence from more than 1000 field trials of transgenic plants indicated that this possibility was exceedingly remote. In addition, it was questioned whether the newly altered plant genes could be transferred to other organisms, such as bacteria in the gastrointestinal tract of humans, with deleterious effects. Again, no scientific evidence for such a possibility was obtained. A more likely potentiality is transfer of unknown food allergens from one plant species to another and a major challenge for the food industry is predicting the allergenicity of transgenic plants. For example, naturally methionine-deficient soybeans, when made transgenic for a methionine-rich protein from Brazil nuts to increase their nutritional value, triggered an allergic response in people allergic to Brazil nuts. Comparison of the physico-chemical, biological, and immunological properties of known food allergens to the transferred protein is an important first step in predicting the safety of transgenic plants for human consumption.

Compared to food biotechnology, public opinion about animal biotechnology appears to be less accepted in the United States, Europe, and Japan. The major issues relating to this apprehension are (1) diminished biodiversity, (2) unexpected transfer of recombinant genes to other species, (3) creation of new, exotic animal species, (4) disruption of natural reproductive barriers between different species, (5) animal welfare rights, and (6) precursors to genetic manipulation of humans. Ultimately, consumer acceptance of the products of animal biotechnology will depend upon (1) the products satisfying consumer needs, (2) receiving governmental regulatory approval, (3) consumer education, and (4) acceptance of moral and religious objections.

Legal issues in the biotechnology industry are mostly centered around protection of intellectual property. As with any new industry, patent protection of pharmaceutical drugs is imperative to the survival of biotech companies. One indicator of the current status of the biotechnology industry is the number of patent applications submitted and double digit percentage increases have been recorded by the U.S. Patent and Trademark Office from 1994 to the present. Correspondingly, the average pendency time for a U.S. biotechnology patent has increased, for example, from 20.8 months in 1994 to 26.2 months in 1996. Also associated with increased biotechnology patent activity is increased patent litigation. While only 25 total patent litigation cases were filed over the 10-year period between 1977 and 1986, 44 were filed in 1996 alone. Part of the cause of the continual increase in litigation is due to a biotech company's need to recover the expense of bringing a new drug to market, now estimated to be $600 million per drug. As new biotechnology products are developed and commercialized, biotech patent law will maintain an important aspect of the industry.

See Also the Following Articles

Biosensors • Enzymes in Biotechnology • Recombinant DNA, Basic Procedures • Secondary Metabolites • Transgenic Animal Technology

Bibliography

Biotech '99: Bridging the Gap. (1999). Life Science's 13th Annual Biotechnology Report. Ernst and Young LLP, San Francisco, CA. This report can also be viewed by visiting website www.ey.com/publicate/life.

Brock, T. D., Madigan, M. T., Martinko, J. M., and Parker, J. (1995). "Biology of Microorganisms." Prentice Hall, Englewood Cliffs, NJ.

Bud, R., and Cautley, M. F. (1994). "The Uses of Life: A History of Biotechnology." Cambridge Univ. Press, Cambridge, UK.

Collins, R., and Galas, D. (1993). A new five-year plan for the U.S. human genome project. *Science* **262**, 43–46.

Coombs, J. (1986). "A Dictionary of Biotechnology." Elsevier, Amsterdam, The Netherlands.

Farkas, D. (1996). Today's molecular diagnostics: Applications and realities. *Clin. Lab. News* **22**, 10–11, December.

Houdebine, L. M. (1997). "Transgenic Animals. Generation and Use." Gordon and Breach Science Publishers, Newark, NJ.

Jameson, J. L. (1998). "Principles of Molecular Medicine." Humana Press, Totowa, NJ.

Kornberg, A. (1996). "The Golden Helix: Inside Biotech Ventures." University Science Books, Sausalito, CA.

Krimsky, S., and Wrubel, R. (1996). "Agricultural Biotechnology and the Environment: Science, Policy and Social Issues." Univ. of Illinois Press, Champaign, IL.

Malik, V. S., and Sridhar, P. (1992). "Industrial Biotechnology." Oxford and IBH Publishing Co., New Delhi, India.

Malik, V. S., and Lillehoj, E. P. (1994). "Antibody Techniques." Academic Press, San Diego.

Miller, R. H., Pursel, V. G., and Norman, H. D. (1996). "Biotechnology's Role in the Genetic Improvement of Farm Animals." American Society of Animal Science, Savoy, IL.

Moses, V., and Moses, S. (1995). "Exploiting Biotechnology." Harwood Academic Publishers, Newark, NJ.

Tijssen, P. (1993). "Practice and Theory of Enzyme Immunoassays." Elsevier, Amsterdam, The Netherlands.

Industrial Effluents: Sources, Properties, and Treatments

F. Vardar-Sukan

Ege University, Izmir, Turkey

N. Kosaric

University of Western Ontario, Canada

I. Nature and Origin of Industrial Wastewaters
II. Properties of Effluents from Selected Industries
III. Microbial Treatment of Industrial Effluents (Secondary Treatment)
IV. Sludge Treatment
V. New Trends in Wastewater Treatment

GLOSSARY

nutrients Organic and inorganic compounds that are necessary for microbial growth.

pollutants All compounds in industrial streams that exceed acceptable health levels.

recycling and reutilization Regarding waste materials as valuable raw materials for the production of high value-added products.

treatment Removal or elimination of the undesirable components and characteristics of wastewaters.

wastewater Industrial effluents.

INDUSTRIAL EFFLUENTS are liquid streams that accompany industrial production. These effluents usually contain compounds undesirable, and often detrimental, to the natural environment. Processing of industrial effluents is performed to decrease the concentration of contaminants to the level accepted by legislation. An important method for industrial effluent processing is the microbial wastewater treatment. Food industry wastewaters, due to their content of nutrients for microbial growth, are also most convenient for microbial treatment. Microbial wastewater treatment may be performed under aerobic or anaerobic conditions.

I. NATURE AND ORIGIN OF INDUSTRIAL WASTEWATERS

Clean drinking water is one of our most important resources. Total stock of water on earth is estimated to be 1.38×10^{18} m³ (Table I); however, 97.2% of this amount is in oceans and only 0.3% could be available as groundwater located down to 1 km.

Mass (industrial) production of goods demands a large quantity of water. Typical rates of water use for various industries are shown in Table II. During industrial processes, a certain amount of substrates, by-products, and final products pass to the water streams and contaminate them. Examples of different types of wastes generated by selected industries are shown in Table III.

The number and concentration of pollutants greatly depends on the technology used. Generally, new technologies generate less wastewater than the old (although sometimes more concentrated). Some wastewater streams may be treated or reused in the plant, discharged as effluent, or sent into other treatment processes. Others may be combined into a single large flow and treated biologically.

A more serious type of pollution is ground water contamination, due to the irreversible character of the damage caused and the large time scales involved, because of the slow movement of the ground waters. In addition, because of the complex interaction between the pollutants, soil and ground water, remediation of contaminated subsurface is a very difficult operation.

TABLE I
Total Stocks of Water on Earth

Location	Amount (10^{15} m^3)	Percentage of world supply
Oceans	1350	97.2
Icecaps and glaciers	29	2.09
Groundwater within 1 km	4.2	0.30
Groundwater below 1 km	4.2	0.30
Freshwater lakes	0.125	0.009
Saline lakes and inland seas	0.104	0.007
Soli water	0.067	0.005
Atmosphere	0.013	0.0009
Water in living biomass	0.003	0.0002
Average in stream channels	0.001	0.00007

[From Masters, G. M. (1991). "Introduction to Environmental Engineering." Prentice-Hall, Englewood Cliffs, NJ.]

TABLE II
Typical Rates of Water Use for Various Industries

Industry	Range of flow (m^3/100 kg of product)
Cannery	
Green beans	50–70
Peaches and pears	15–20
Other fruits and vegetables	4–35
Chemical	
Ammonia	100–300
Carbon dioxide	60–90
Lactose	600–800
Sulfur	8–10
Food and beverage	
Beer	10–16
Bread	2–4
Meat packing[a]	15–20
Milk products	10–20
Whisky	60–80
Pulp and paper	
Pulp	250–800
Paper	120–160
Textile	
Bleaching[b]	200–600
Dyeing[b]	30–60

[a] Live weight.
[b] Cotton.
[From Metcalf & Eddy, Inc. (1991). "Wastewater Engineering: Treatment, Disposal, Reuse." McGraw-Hill, New York.]

The object of any treatment is to remove certain waste materials, as specified by governmental regulations. Materials to be removed are as follows:

- Soluble organics (natural products of animal and plant life and certain synthetic chemicals)
- Suspended and colloidal solids
- Heavy metals (e.g., mercury, chromium, lead, cadmium)
- Acidity and alkalinity (must be neutralized)
- Oils, greases, surfactants, and other floating materials
- Excess inorganic nutrients, such as nitrogen and phosphorus
- Color, turbidity, and odor
- Priority pollutants (regulated by the U.S. Environmental Protection Agency and other government agencies)
- Toxic and radioactive chemicals
- Heated water
- Microorganisms

The minimum criteria for different applications vary from country to country. However, distinct divisions exist, such as stream quality (for public water supply and food processing industry), for industrial water supply, for aquatic life, for recreation, and for agricultural use and stock watering.

TABLE III
Examples of Wastes Generated by Selected Industries

Type of wastes	Type of plant
Oxygen consuming	Breweries, canneries, dairies, distilleries, packing houses, pulp and paper mills, tanneries, textile mills
With high-suspended solids	Breweries, canneries, coal washeries, coke and gas plants, iron and steel industries, distilleries, packing houses, pulp and paper mills, tanneries
With high-dissolved solids	Chemical plants, sauerkraut canneries, tanneries, water softening plants
Oily and greasy	Laundries, metal finishing, oil fields, packing houses, petroleum refineries, tanneries, wool-scouring mills, iron and steel industries
Colored	Electroplating, pulp and paper mills, tanneries, textile dye-houses
Toxic	Atomic energy plants, chemical plants, coke and oven by-products, petrochemical plants, electroplating, pulp and paper mills, tanneries
High acid	Chemical plants, coal mines, electroplating shops, iron and steel industries, sulphite pulp mills
High alkaline	Chemical plants, laundries, tanneries, textile-finishing mills
High temperature	Bottle-washing plants (dairies, beverage producers), electroplating, laundries, tanneries, power plants in all industries, textile-finishing mills

II. PROPERTIES OF EFFLUENTS FROM SELECTED INDUSTRIES

A. Definitions

The composition of industrial effluents varies with time and usually cannot be precisely defined. Several overall parameters are used to define quality of industrial effluents.

(a) Biochemical oxygen demand (BOD_5): The amount of oxygen (mg/liter) added to a wastewater to support microbial activity over a period of 5 days.

(b) Chemical oxygen demand (COD): The amount of oxygen (mg/liter) needed to chemically oxidize a given wastewater.

(c) Total organic carbon (TOC): The amount of carbon detected as CO_2. TOC measurement (using an infrared analyzer) lasts only a few minutes and, for simple chemical compounds, can be related to COD and BOD_5. For real industrial effluents, the TOC : COD and TOC : BOD ratios are estimated experimentally. Oxygen demand and TOC of industrial wastewaters for selected industries are presented in Table IV.

(d) Total suspended solids (TSS): All organic and inorganic suspended solids (mg/liter).

(e) Volatile suspended solids (VSS): Includes only those solids (mg/liter) that can be oxidized to gas at 550°C. At that temperature, most organics are oxidized to CO_2 and H_2O, while inorganics remain in ash. Estimate of the components of total (dissolved and suspended) solids in different types of wastewaters is shown in Table V.

(f) pH.

B. Composition and Properties of Effluents from Selected Industries

1. Petroleum Industry Waste

Petroleum refineries utilize from 0.5 to 40 volumes of water per unit volume of processed petroleum, depending on the types of processes used, modernization of the plant, integration with the petrochemical industry cooling system design, etc. A comparison of average waste flows and loadings from petroleum refineries for old, prevalent, and new technologies is shown in Table VI.

The major function of water in a refinery is for cooling purposes, while relatively small quantities are used for the boiler feed, direct processing, fire protection, sanitary, and other uses. Origin and nature of pollutants in refinery effluents are shown in Table VII. Because of its composition, the wastewater

TABLE IV
Oxygen Demand Organic Carbon of Industrial Wastewaters

Waste	BOD (mg/liter)	COD (mg/liter)	TOC (mg/liter)	BOD : TOC	COD : TOC
Chemical[a]	–	4,260	640	–	6.65
Chemical[a]	–	2,410	370	–	6.60
Chemical[a]	–	2,690	420	–	6.40
Chemical	–	576	122	–	4.72
Chemical	24,000	41,300	9,500	2.53	4.35
Chemical refinery	–	580	160	–	3.62
Petrochemical	–	3,340	900	–	3.32
Chemical	850	1,900	580	1.47	3.28
Chemical	700	1,400	450	1.55	3.12
Chemical	8,000	17,500	5,800	1.38	3.02
Chemical[b]	60,700	78,000	26,000	2.34	3.00
Chemical[b]	62,000	143,000	48,140	1.28	2.96
Chemical	–	165,000	58,000	–	2.84
Chemical	9,700	15,000	5,500	1.76	2.72
Nylon polymer	–	23,400	8,800	–	2.70
Petrochemical	–	–	–	–	2.70
Nylon polymer	–	112,600	44,000	–	2.50
Olefin processing	–	321	133	–	2.40
Butadiene processing	–	359	156	–	2.30
Chemical	–	350,000	160,000	–	2.19
Synthetic rubber	–	192	110	–	1.75

[a] High concentration of sulphides and thiosulphides.

[b] Data with a very high BOD (e.g., 60,000) refers to chemical industries containing a high proportion of biodegradable organics and also a high concentration of lignosulfates (COD around 100,000), such as the pulp and paper industry. Data with a high COD and no BOD refers to chemical industries that discharge effluents containing nonbiodegradable organics.

[From Eckenfelder, W. W., Jr. (1989). "Industrial Water Pollution Control." McGraw-Hill, New York.]

from the petroleum industry have a detrimental effect on soil and natural water reservoirs. Microbial treatment of wastewaters from the refinery industry is rather difficult, as compared to municipal wastewaters, as refinery wastewaters contain organics that are difficult to degrade microbiologically (e.g., chlorophenols).

Many processes in the petroleum refining and petrochemical industries use steam as a stripping medium in distillation and as a diluent to reduce the hydrocarbon partial pressure in catalytic or thermal cracking. The steam is eventually condensed as an aqueous effluent, commonly referred to as "sour or foul water." Condensation of the steam usually occurs simultaneously with the condensation of hydrocarbon liquids and in the presence of a hydrocarbon vapor phase containing H_2S. Thus, the condensed steam will usually contain H_2S, which imparts an unpleasant odor and, hence, the name "sour water." To differentiate these condensed steam effluent from other sources of sour water, they are referred to as "sour condensates." Typical sour condensate sources are shown in Table VIII.

Effluents produced by various process units differ, depending on the unit operation. In crude oil distillation, the overhead reflux drum produces an aqueous effluent. The main source of this effluent is stripping steam, which is used in the distillation system and which is condensed along with the overhead product, naphtha. This water, in sour condensate, contains H_2S and NH_3, most probably in the form of NH_4SH. It may contain a rather small amount of phenols.

TABLE V
Estimate of the Components of Total (Dissolved and Suspended) Solids in Wastewater

Component	Range, dry weight (g/capita · day)	Typical, dry weight (g/capita · day)
Water supply	10–18	14
Domestic wastes	–	–
Feces (solids, 23%)	30–70	40
Ground food wastes	30–80	45
Sinks, baths, laundries, and other sources of domestic wash waters		
Toilet (including paper)	15–30	20
Urine (solids, 3.7%)	40–70	50
Water softeners	*a*	*a*
Total for domestic wastewater, excluding water softeners	190–360	250
Industrial wastes	150–400	200[b]
Total domestic and industrial wastes	340–760	450
Nonpoint sources	9–40	18[c]
Storm water	18–40	27
Total for domestic, industrial, nonpoint, and storm water	360–480	860

[a] Variable.

[b] Varies with season.

[c] Varies with season.

[From Metcalf & Eddy, Inc. (1991). "Wastewater Engineering: Treatment, Disposal, Reuse." McGraw-Hill, New York.]

The pretreatment of crude oil to remove inorganic salts is known as "desalting" and produces another aqueous effluent, which contains a good amount of NaCl, a small amount of phenolics, and very little sulfide or free H_2S. However, this is usually contaminated with oil and has a significantly high oxygen demand.

The overhead product drum in a vacuum unit produces an aqueous effluent. The major part of this water is derived from condensation of the vacuum ejector motive steam. Condensed stripping steam and heater coil diluent steam are also contributing sources. This water may contain a small amount of H_2S if significant thermal cracking occurs in the vacuum unit feed heater. It may also contain some phenols for the same reason. It will very probably contain some oil.

The catalytic hydrogenation of naphthas to convert organic sulfur to free H_2S does not, per se, produce an aqueous effluent. However, subsequent fraction-

TABLE VI
Average Waste Flows and Loading from Petroleum Refineries for Old, Prevalent, and New Technology

Type of technology	Flow (liter/m³)[a] Average	Flow (liter/m³)[a] Range	BOD (g/liter) Average	BOD (g/liter) Range	Phenol (g/liter) Average	Phenol (g/liter) Range	Sulphide (g/liter) Average
Old	8000	5400–11,800	1.50	1.15–1.70	0.11	0.11–0.13	0.038
Prevalent	3000	2500–5,000	0.38	0.30–0.61	0.04	0.03–0.05	0.011
New	1500	650–1,900	0.19	0.08–0.23	0.02	0.004–0.023	0.011

[a] Liters per cubic meter of petroleum refined.

TABLE VII
Origin and Nature of Pollutants in Refinery Effluents

Process units	Pollutants
Handing of crude oil	Oil, sludge and oily emulsions, sulfur- and nitrogen-containing corrosion inhibitors, inorganic salts, suspended matter
Crude oil distillation	Hydrocarbons, coke, organic acids, inorganic salts, sodium chloride, phenols, sulfur, sour condensate (sulfides and ammonia)
Thermal cracking	Phenols, triphenols, nitrogen derivatives, cyanides, hydrogen sulfite, ammonia
Alkylation, polimerization, isomerization processes	Acid sludge, spent acid, phosphoric, sulfuric, hydrofluoric, and hydrochloric acids, oil, catalyst supports, aluminum or antimony chloride
Refining and reforming processes	Hydrogen sulfide, ammonium sulfide, gums, catalyst supports
Purification and extraction processes	Phenols, glycols, amines, acetonitrile, acids, spent caustic
Sweetening, stripping, filtration	Sulfur compounds (H_2S, mercaptans), nitrogen compounds, sulfonates, acids, inorganic salts, copper chloride, suspended matter

ation of the hydrotreated naphtha to remove the free H_2S may yield a sour aqueous stream. Free H_2S and any dissolved water in the naphtha have relative volatilities (in hydrocarbon systems) such that H_2S and water will appear wherever ethane and propane occur overhead in the fractionation scheme. Typically, the hydrotreated naphtha is debutanized and then depentanized or partially dehexanized. A usual procedure after hydrotreating (catalytic removal of sulfur from naphtha with hydrogen), the hydrocarbon liquid stream is further directed to fractionation columns, which separate butanes, pentanes, and hexanes. These may be further refined and represent a product or by-product in the refinery. Thus, in industry, they refer to debutanized, depentanized, and partially dehexanized naphtha. The overhead reflux drum of the debutanizer may, thus, be expected to produce a small stream of sour water, which has been in liquid–liquid contact with butane and lighter hyrocarbons.

The catalytic hydrotreating of diesel oil to reduce its sulfur content does not, per se, produce a sour water. However, the subsequent stream-stripping of the hydrotreated diesel to remove the free H_2S does yield a sour condensate overhead, along with a small amount of by-product sour naphtha. Since the hydrotreating process also produces some NH_3, the sour water most probably contains NH_4SH.

TABLE VIII
Typical Sour Condensate Source

Process	Effluent condensate source	Original steam source
Crude oil distillation	Distillation reflux drum	Stripping steam, water in crude oil
Petroleum coking (Thermal cracking)	Distillation reflux drum	Stripping steam, coke drum steam, heater injection steam
Petroleum visbreaking (Thermal cracking)	Distillation reflux drum	Stripping steam, heater injection steam
Vacuum distillation	Distillation overhead drum	Stripping steam, or diluent steam, ejector motive steam
Petroleum catalytic cracking	Distillation reflux drum	Stripping steam, catalyst stripping steam
Catalytic hydro-desulfurization	Product stripper reflux drum	Stripping steam
Petrochemical processes (Thermal cracking of hydrocarbons)	Quench systems and/or distillation reflux drum	Reaction diluent steam

Since the feed naphtha to a reformer should be hydrotreated and prefractioned, it is essentially free of H_2S and water. Catalytic reforming and subsequent fractionation of the reformed gasoline do not normally produce any aqueous effluents. In any event, if an aqueous effluent is produced it should not contain any H_2S, oil, or phenols.

Any catalytic or thermal cracking of heavy oils can be expected to produce phenols, thiophenols, H_2S, NH_3, and cyanides. The reaction effluent usually contains steam and the subsequent main fractionator uses stripping steam. Thus, the main fractionator overhead reflux drum produces a sour, foul condensate containing NH_4SH, phenols, and, perhaps, cyanides. The subsequent compression and fractionation of the residue gas from the main fractionator produces similar foul condensates. The foul condensates from heavy oil crackers are one of the most significant sources of aqueous pollutants in the typical refinery. Hydrocracking has only begun to be used extensively in the past 10 years. It combines catalytic hydrogenation and cracking functions. Any aqueous effluent can be expected to be rich in H_2S, and NH_3, probably in the form of NH_4SH. In fact, the amount of H_2S and NH_3 to be expected may make it economically attractive to recover them from the sour water as by-products.

In catalytic polymerization, high-quality gasoline components are produced by dimerization and trimerization of C_3 and C_4 mono-olefins over a phosphoric acid catalyst. The feed must be desulfurized to avoid formation of undesirable mercaptans and the water content must be controlled so that the phosphoric acid is neither dried nor diluted. Normally, a catalytic polymerization unit does not produce an aqueous effluent. In any event, the presence of H_2S, NH_3, or phenols is not to be expected.

Some catalytic polymerization processes employ liquid phosphoric acid, others employ the acid supported on inert solids, and still others use an acid-impregnated catalyst. In the liquid acid plants, caustic will probably be used to scrub the reaction effluent for removal of entrained acid. In the supported acid plants, caustic and some alkaline reagent may be required to scrub the light ends recovered in the product fractionation system.

Catalytic alkylation units produce branched-chain hydrocarbons ("alkylate") as a high-quality gasoline component by combining propylene, butylene, and amylenes with isobutane. The reaction is catalyzed by liquid sulfuric acid or by liquid hydrofluoric acid. Alkylation does not produce a sour process effluent and H_2S, NH_3, or phenols are not expected to be present. Sulfuric acid alkylation plants employ caustic and water washing of the reaction effluent, prior to fractionation of the effluent, as an intrinsic part of the process. The spent caustic should contain essentially sodium sulfate. The "spent" water will contain whatever caustic is entrained in the effluent from the caustic wash. Spent catalytic acid is continuously or intermittently withdrawn from the reaction system and must be disposed of. The spent acid may contain about 85–90% H_2SO_4 and is suitable for use in neutralizing phenolic spent caustic—or it can be processed for reconstitution to 98% H_2SO_4.

Catalytic isomerization processes convert normal paraffins into isoparaffins. The greatest application has been in the production of isobutane (for use in alkylation) from *n*-butane. Other installations and processes produce isopentane from *n*-pentane and isohexane from *n*-hexane.

The variety of isomerization processes available makes it difficult to generalize concerning their aqueous effluent problems, if any. However, the following can be said:

(a) In most cases, the feed must be carefully desulfurized. Therefore, H_2S and NH_3 very probably will not be present

(b) Many of the processes use a platinum metal catalyst and a hydrogen atmosphere in the reaction zone. These processes should produce no aqueous effluent problems.

(c) Some of the processes use aluminum chloride promoted with hydrochloric acid. Some involve the use of antimony trichloride. Other processes also use hydrochloric acid as a catalyst promoter. All of these processes require the subsequent stripping of HCl from the reaction effluent for return to the reaction zone. Thus, they may require some emergency neutralization facilities, which will produce aqueous effluents for disposal.

(d) In general, the isomerization processes should not be produce phenolics or any high oxygen demand effluents.

(e) Those processes involving aluminum chloride, antimony trichloride, HCl, etc. may create specific effluent problems.

Refineries and petrochemical plants employ a wide spectrum of solvent processes. This category includes azeotropic distillation, extractive distillation, liquid–liquid extraction, physical absorption, and chemical absorption.

Many of solvent processes may produce aqueous process effluents containing small amounts of the solvent. Obviously, there is an economic incentive to minimize such solvent "losses." It is also obvious that the local sewer system in a solvent process should segregate, and reuse, if practical, the unavoidable and inadvertent solvent losses from pump seals, flange leaks, etc.

Phenol is used as a solvent for the extraction of polycyclic aromatics from lubricating oil. Mixture of phenol–cresol propane are used to extract polycyclic aromatics and asphaltenes from lubricating oils (Duo-Sol process). In some older installations, phenol is used to extract C_6 to C_8 aromatics from gasoline boiling range hydrocarbons. Phenol has a high oxygen demand. Every effort should be made to limit the amount of phenol entering the plant effluent system.

Aqueous glycol solutions are used for the extraction of C_6 to C_8 aromatics from hydrocarbon mixtures (the Udex proces). Aqueous glycol and amine mixtures are used for removal of H_2S, CO_2, and water from sour gases. Glycols are also used as dehydrants in various low-temperature processes. Glycols can produce a high BOD.

Aqueous solution of mono- and di-ethanolamine are used, sometimes in conjunction with a glycol, to remove H_2S and CO_2 from gas or liquid streams.

In some cases, the water material balance in amine-treating units is such that there may be a continuous withdrawal of a process aqueous effluent. Since this water may be quite sour, it should be routed through a sour water stripper for removal of acid gases. The local sewer system in an amine unit is usually designed to maximize the recovery and reuse of leaks, drips, and drains. Amines can produce a high BOD.

Furfural is used to remove polycyclic aromatics from lubricating oils. It is also used to upgrade diesel oils and catalytic cracking recycle oils by removal of unstable, acidic, sulfur, organometallic, and nitrogen compounds. Aqueous solutions of furfural are also used in the extraction of butadiene from C_4 hydrocarbon mixtures.

Sulfuric acid is used in a wide variety of treating processes for the extraction of olefinics, asphaltenes, and organic sulfur compounds. Most of the processes result in an acid "sludge" for disposal. Since a subsequent caustic wash is usually required, a spent caustic disposal problem is also created.

Aqueous solution of acetonitrile are used to extract betadiene from C_4 hydrocarbons. Acetonitrile has the formula $CH_3 C{\equiv}N$ and any aqueous effluent containing acetonitrile may also contain amides. The acetonitrile may possibly hydrolyze to acetic acid under certain conditions.

Methyl ethyl ketone (MEK) and methyl isobutyl ketone (MIBK) are used in processes for the extraction and production of wax from the lubricating oil fractions of pertoleum.

Treating processes are a broad category of techniques for upgrading various intermediate and final product streams. In many cases, the treating process involves the use of a solvent. Listed below are some of the more common treating processes which are generally used for removing sulfur compounds or converting mercaptans to disulfides.

- **Caustic treating:** Aqueous NaOH solutions are used for the extraction and removal of acidic compounds, such as H_2S, mercaptans, phenols, and thiophenols. The resulting spent caustics fall into two broad classifications: phenolic spent caustics and sulfidic spent caustics.
- **Bender sweetening:** This is a process for converting objectionable mercaptans into less objectionable, odorless disulfides. In general, the conversion mercaptans to disulfides is termed "sweetening," to denote the improvement in odor. The process uses sulfur, caustic, and air in the presence of a lead sulfide catalyst. It produces sulfidic spent caustic for disposal.
- **Doctor sweetening:** This process uses sulfur, sodium plumbite (Na_2PbO_2), and caustic. The

spent doctor solution is regenerated, by air blowing, for reuse. However, there is either a continuous or intermittent discard stream of spent doctor solution, which contains emulsified caustic and oil, lead sulfide, and sodium plumbite.

2. Pulp and Paper Industry Waste

Enormous quantities of water are used in the pulp and paper industry. Depending on the process, a considerable proportion of the used water is released to the water streams carrying soluble and suspended solids made up of organic materials, such as lignin and carbohydrates, and inorganic salts derived from chemicals used in the cooking process. This material constitutes a tremendous pollution burden upon streams and is also an enormous waste of potentially valuable raw materials. The organic material is usually not recovered but is either burned (Kraft mills and very few sulfite mills) or released into the water streams.

The pulp and paper industry, representing the fifth largest industry in the United States and the largest single industry in Canada, is expanding at a steady rate of approximately 5% per year (statistics from Canada). The industry is one of the largest water consumers as well as polluters, and the problem of reducing pollution from old pulp and paper mills and design of processes and equipment to combat pollution is one of the most vexing problems confronting this industry. Effluent volume from the manufacture of pulp and paper products is presented in Table IX and untreated effluent loads in Table X. Waste liquors from the pulp and paper industry and, in particular, sulfite wastes liquors, could be considered as a potential raw material rather than waste. They contain about 50% of the processed wood. By-product recovery from sulfite liquors is practiced in many instances, although not to such large scale that would eliminate the water pollution problem.

Primarily due to contents of pentoses and hexoses (20–30% of the solids), the waste sulfite liquor is a good fermentation substrate. Both aerobic and anaerobic systems are practiced. In aerobic systems, production of yeast for food is most important and *Candida utilis* and *Monilia murmanica* are most cultivated. The protein content of *Candida* yeast is as high as

TABLE IX
Effluent Volumes from the Manufacture of Pulp and Paper Products

Process	m^3/1000 kg of product
Pulp manufacture	
Kraft and soda pulps	60–150
Sulfite pulp	170–250
Semichemical pulp	125–170
Groundwood pulp	15–40
Deinked pulp	85–145
Pulp bleaching	
Kraft and soda pulps	60–250
Sulfite pulp	125–210
Natural sulfite pulp	170–250
Paper manufacture	
White papers	85–170
Tissues	30–145
Kraft papers	8–40
Paperboard	8–60
Specialty papers	85–420

47–55%. Under efficient conditions, an overall 45% yield of dried yeast can be expected based on the sugar feed, yielding 27% by weight of the assimilated sugar as protein. Anaerobic fermentation comprises production of ethanol by *Saccharomyces cerevisiae*, that converts the hexoses to alcohol and CO_2. Up to 95% of the fermentable sugars can be converted to ethanol during a residence time of 15–20 hrs at 30–35°C. A pulp mill processing 500 tons/day may produce 10,000 U.S. gal/day (37.5m^3/day) of 95% alcohol.

3. Food Industry Waste

Waste from food industries is generally characterized by a very high organic content. When released to water streams, it can support an uncontrolled microbial proliferation (e.g., eutrophication, algae growth, sludge deposits). The waste varies considerably in strength and quantity (BOD from 100 to 100,000 mg/liter). A small seasonal plant can create a pollution load equivalent to 15,000–25,000 people, while larger plants in North America have a population equivalent to at least one-quarter of a million people. Suspended solids, almost completely absent from some waters, are found in others in concentrations as high as 120,00 mg/liter. The waste may be

TABLE X
Untreated Effluent Loads from Pulp and Paper Manufacture

Effluent	Suspended solids (kg/1000 kg of product) [range of design values*]	5-day BOD (kg/1000 kg of product) [range of design values][a]
Pulps		
Unbleached sulfite	10–20	200–350
Bleached sulfite	12–30	220–400
Unbleached kraft and soda	10–15	12–25
Unbleached groundwood	15–40	8–12
Bleached groundwood	22–42	12–30
Natural sulfite semichemical	40–90	125–250
Textile fiber	150–250	100–150
Straw	200–250	200–250
Deinked	200–400	30–80
Fine papers		
Bond-mimco	25–50	7–20
Glassine	5–8	7–12
Book or publication papers	25–50	10–25
Tissue papers	15–50	10–15
Coarse papers		
Boxboard	25–35	10–20
Corrugating brand	25–35	12–30
Kraft wrapping	8–12	2–8
Newspring	10–30	5–10
Insulating board	25–50	75–125
Specialty Papers		
Asbestos	150–200	10–20
Roofing felt	25–50	20–30
Cigarette papers	50–400	10–15

[a] Design values depend on field.

highly alkaline (pH = 11.0) or highly acidic (pH = 3.5). Mineral nutrients may be absent or present in excess of optimum ratio for microbial growth, but generally the wastes represent a good nutritional medium for microorganisms, so that biological treatment for this type of waste is recommended and practiced. Wastewater parameters from the canning industry are shown in Table XI, from slaughterhouses in Table XII, from poultry plants in Table XIII and from the dairy industry in Table XIV.

The canning industry in North America processes a tremendous volume of various food products. About 80% of tomatoes, 75% of beets, 65% of green peas, and more than 50% of sweet corn is being canned. As for fish, 99% of the tuna catch, 90% of the sardine catch, and 85% of the salmon catch is canned. Total production of the industry averages a billion cases per year—the average amount of clean water required being approximately 50 gal/case (50 billion gal water/year).

The process waste usually consists of wash water; solids from sorting, peeling, and coring operations; spillage from filling and sealing machines; and wash water from cleaning floors, tables, walls, etc.

Among fruits, the processing of peaches, tomatoes, cherries, apples, pears, and grapes presents the most common problem. The three main citrus fruits, oranges, lemons, and grapefruit, include waste from cooling water, pectin wastes, pulp-press liquors, processing plant wastes, and floor washings.

TABLE XI
BOD and Suspended Solids of Cannery Waste

Product	5-day BOD (mg/liter)	Suspended solids (mg/liter)
Apples	1680–5530	300–600
Apricots	200–1020	200–400
Cherries	700–2100	200–600
Cranberries	500–2250	100–250
Peaches	1200–2800	450–750
Pineapples	26	–
Asparagus	16–100	30–180
Beans, baked	925–1440	225
Beans, green wax	160–600	60–150
Beans, kidney	1030–2500	140
Beans, lima dried	1740–2880	160–600
Beans, lima fresh	190–450	420
Beets	1580–7600	740–2200
Carrots	520–3030	1830
Corn, cream-style	620–2900	300–675
Corn, whole kernel	1120–6300	300–400
Mushrooms	76–850	50–240
Peas	380–4700	270–400
Potatoes, sweet	1500–5600	400–2500
Potatoes, white	200–2900	990–1180
Pumpkin	1500–6880	785–1960
Sauerkraut	1400–6300	60–630
Spinach	280–730	90–580
Squash	4000–11000	3000
Tomatoes	180–4000	140–2000

The meat industry has three main sources of waste: stockyards, slaughterhouses, and packing houses. Animals are kept in the stockyards until they are killed. The killing, dressing, and some by-product processing are carried out in the slaughterhouse, or abattoir. To obtain the finished product—namely, the fresh carcass, plus a few fresh meat by-products, such as hearts, livers, and tongues—the following operations are performed in the slaughterhouse. The animals are stuck and bled on the killing floor (cattle being stunned prior to sticking). Carcasses are trimmed, washed, and hung in cooling rooms. Livers, hearts, kidneys, tongues, brains, etc. are sent to the cooling rooms to be chilled before being marketed. Hides, skins, and pelts are removed from the cattle, calves, sheep, and pigs, and are salted and placed in piles, until they are shipped to tanners or wool-processing plants. Viscera are removed and, together with head and feet bones, are sent to a rendering plant; other bones are shipped to glue factories. Many slaughterhouses are equipped to render their own inedible offal into tallow, grease, and tankage; other independent rendering plants convert inedible poultry, fish, and animal offal and waste products into animal feed and grease. This is accomplished by cooking the inedibles at a high temperature for several hours. The cooked material is pressed to remove the grease and the pressings are ground for feed.

TABLE XII
Approximate Range of Flows and Analyses for Slaughterhouses, Packing Houses, and Processing Plants

Operation	Waste flow (m³/1000 kg live weight slaughtered)	Typical analysis (mg/liter)		
		BOD	Suspended solids	Grease
Slaughterhouse	4–17	2200–650	3000–930	1000–200
Packing house	6–30	3000–400	2000–230	1000–200
Processing plant	8–33	800–200	800–200	300–100

	Approximate waste loadings (kg/1000 kg live weight slaughtered)		
	BOD	Suspended solids	Grease
Slaughterhouse	9.2–10.8	12.5–15.4	4.2–3.3
Packing house	18.7–11.7	12.5–0.7	6.3–5.8
Processing plant	6.7	6.7	2.5–3.3

TABLE XIII
Composition of Poultry Plant Waste

	Range
5-day BOD (mg/liter)	150–2400
COD (mg/liter)	200–3200
Suspended solids (mg/liter)	100–1500
Dissolved solids (mg/liter)	200–2000
Volatile solids (mg/liter)	250–2700
Total solids (mg/liter)	350–3200
Suspended solids (% of total solids)	20–50
Volatile solids (% of total solids)	65–85
Settleable solids (mg/liter)	1–20
Total alkalinity (mg/liter)	40–350
Total nitrogen (mg/liter)	15–300
pH	6.5–9.0

The milk industry is one of the most widely spread of all food industries. There is a considerable variation in the size and type of dairy products (whole homogenized, defatted, or partially defatted milk; cheeses, sour and sweet cream, butter and other similar products, condensed and dry milk products, etc.). Dairy waste for the most part is made of various dilutions of milk and milk products. A typical composition of the waste and various milk products is given in Table XV.

Dairy wastes are largely neutral or slightly alkaline but have a tendency to become acid quite rapidly because of the fermentation to lactic acid. Cheese plant waste is acid. Because of the presence of whey, suspended solids are generally low. The high BOD values (about 100 lb. of whole milk results in approx-

TABLE XIV
Dairy Wastewaters

Product	BOD (kg/100 kg)	Volume (liter/100 kg)
Creamery butter	0.34–1.68	3410–11,300
Cheese	0.45–3.0	10,780–19,300
Condensed and evaporated milk	0.37–0.62	2590–3500
Ice cream[a]	0.15–0.73	5180–1000
Milk	0.05–0.26	1670–4180

[a] Per 375 liters of product.

imately 10 lb. of BOD and a population equivalent of 60) are due to highly fermentable organics, causing a rapid oxygen depletion and, sometimes, complete exhaustion in receiving waters. As milk wastes also contain phosphorus and nitrogen, they are excellent substrates for microbial growth. Generally, milk waste pollution is characterized by heavy black sludge and strong butyric acid odors, caused by decomposition of casein.

A particular problem in dairy industries in the cheese whey, which is a greenish-yellow waste fluid, being either "acid" (pH 4–5, from cottage cheese, usually produced from skimmed milk) or "sweet" (pH 5–7 from processed cheese, such as cheddar, made from whole milk). Both wheys have a high BOD (30,000 to 45,000 ppm), 100 lb. being equivalent to waste produced by 21 people/24 hrs a cheese plant producing 100,000 lbs. of cheese has a waste equivalent to a population of 22,000 people. As it is difficult to handle whey in one unit, many municipalities refuse to accept it into their disposal systems.

Its quantity varies, but approximately 5–10 lb. of whey is left from each pound of cheese processed (22 billion lbs. produced yearly in the United States). About 1/3 of this is being used in human food or animal supplements. Its composition is approximately as follows:

- Water 93 to 94%

- Solids 6 to 7% from which 74% is lactose
  13% is protein
  8% is ash
  3% is lactic acid
  1% is fat,

which makes it a highly nutritious food and methods for whey recovery and utilization are encouraged.

III. MICROBIAL TREATMENT OF INDUSTRIAL EFFLUENTS (SECONDARY TREATMENT)

Primary treatment of industrial effluents implies removal of suspended solids or conditioning of wastewaters for discharge into either a receiving body

TABLE XV
Average Composition of Milk, Milk By-products, and Cheese Waste

Characteristics	Whole milk, ppm	Skim milk, ppm	Butter milk, ppm	Whey, ppm	Process wastes, ppm	Separated whey, ppm
Total solids	125,000	82,300	77,500	72,000	4516	54,772
Organic solids	117,000	74,500	68,800	64,000	2698	49,612
Ash solids	8000	7800	8700	8000	1818	5160
Fat	36,000	1000	5000	4000		
Soluble solids					3956	54,656
Suspended solids					560	116
Milk sugar	45,000	46,000	43,000	44,000		
Protein (casein)	38,000	39,000	36,000	8000		
Total organic nitrogen					73.2	1300
Free ammonia					6.0	31
Na					807	648
Ca					112.5	350
Hg					25	78
K					116	1000
P					59	450
BOD_5	102,500	73,000	64,000	32,000	1890	30,100
Oxygen consumed	36,750	32,200	28,600	25,900		

of water or a secondary treatment facility. Primary treatment usually includes equalization, neutralization, sedimentation, oil separation, and flotation. Thus, a minimum of secondary treatment should be required for all domestic, commercial, and industrial wastes discharged into fresh or salty waters, in order to enhance, or at least maintain, the quality of receiving waters.

A. Microorganisms Employed for Secondary Treatment

All microbial treatment processes are conversion processes in that they convert, in a matter of minutes to hours, the readily biodegradable organic contaminants (soluble/colloidal and, thus, nonsettleable in a gravity clarifier) into two fractions: (a) a gas which escapes from the liquid and (b) The excess biomass which will not readily degrade. (See Table XVI.)

Consequently, the fundamental principle of biological processes is to convert/degrade soluble/colloidal organic pollutants. In microbial treatment, the organic contaminants serve as the energy source (electron donor) in the reaction and oxygen, nitrite/

nitrate, sulfate, or carbon dioxide serve as the electron acceptor. Aerobic processes use oxygen as the electron acceptor, and anaerobic processes use sulfate, nitrate, ferric iron, or carbon dioxide as electron acceptor.

The microorganisms metabolize the organic contaminants and utilize the energy obtained for cell growth and maintenance. The microorganisms covert some of the carbon in the organic contaminants to CO_2 and CH_4 under anaerobic conditions. These gases are subsequently stripped from the liquid phase. The relative biodegradability of a number of synthetic and naturally occurring compounds is given in Table XVI.

B. Aerobic Treatment of Industrial Effluents

1. Theory of Aeration

The solubility of oxygen in liquids follows Henry's law. Oxygen saturation for distilled water at standard pressure is 14.6 mg/liter at 0°C and 7.6 mg/liter at 30°C. Presence of contamination decreases the saturation level. Oxygen demand for industrial ef-

TABLE XVI
Relative Biodegradability of Certain Organic Compounds

Biodegradable organic compounds[a]	Compounds generally resistant to biological degradation
Acrylic acid	Some ethers
Aliphatic	Ethylene chlorohydrin
Aliphatic alcohols (normal, iso, secondary)	Isoprene, methyl vinyl ketone
Aliphatic aldehydes	Morpholine
Aliphatic esters	Oil
Alkyl benzene sulphonates with exception of propylene-based benzaldehyde	Polymeric compounds, polypropylene benzene sulfonates
Aromatic amines	Selected hydrocarbons
Dichlorophenols	Aliphatics–branched
Ethanolamines	Aromatics–substitute
Glycols	Alkyl–aryl groups
Ketones	Tertiary aliphatic alcohols
Methacrylic	Tertiary aliphatic sulfonates
Methyl methacrylate	Tertiary aliphatic sulfonates
Monochlorophenols	Trichlorophenols
Nitriles	
Phenols	
Primary aliphatic amines	
Styrene	
Vinyl acetate	

[a] Some compounds can be degraded biologically only after extended periods of seed acclimation.
[From Eckenfelder, W. W., Jr. (1989). "Industrial Water Pollution Control." McGraw-Hill, New York.]

fluents reaches several thousand milligrams of oxygen per liter of effluent. Such an amount of oxygen must be provided from the gas to the liquid phase by aeration systems. An adequate rate of oxygen transfer from gas to liquid is essential, hence, proper process equipment design is required for good performance of the process. Rate of oxygen transfer can be calculated from the following equation:

$$dC/dt = K_La(C_{WS} - C)$$

where, dC/dt = mass of oxygen per unit time per unit volume, K_La = overall oxygen transfer coefficient (1/time unit), and $(C_{WS} - C)$ = the difference between dissolved oxygen saturation in liquid and actual oxygen concentration in the bulk of the liquid phase (mass of oxygen/volume unit).

For a continuous aerobic treatment process, which is designed for removal of carbonaceous organic matter, the steady state operating dissolved oxygen level range is between 0.5 and 1.5 mg/liter.

During biological oxidation, several mineral elements are essential for the metabolism of organic matter. The nitrogen requirement is about 4.3 kg N/100 kg BOD$_{removed}$ and the phosphorous requirement is 0.6 kg P/100 kg BOD$_{removed}$. Usually a concentration of trace elements in a carrier water is sufficient for microbial growth at aerobic conditions.

2. Performance of Aerobic Process
a. Activated Sludge Processes

The activated sludge process is performed in a tank (i.e., reactor) supplied with liquid feed for microorganisms (wastewaters + nutrients required) and air. During aeration, intensive mixing must be provided. When mixing and aeration are stopped, microorganisms settle and can be separated from the clear liquid. Separation may be performed in the same reactor (then the process is performed in a batch mode) or in another tank, called a settler (then the process may be performed in a continuous mode and concentrated microorganisms, called sludge, may be recycled), dependent on reactor design and operation.

TABLE XVII
Reaction Rate Coefficients for Organic Wastewaters

Wastewater	K (1/day)	Temperature (°C)
Potato processing	36.0	20
Peptone	4.03	22
Sulfite paper mill	5.0	18
Vinyl acetate monomer	5.3	20
Polyester fiber	14.0	21
Formaldehyde, propanol, methanol	19.0	20
Cellulose acetate	2.6	20
AZO dyes, epoxy, optical brighteners	2.2	18
Petroleum refinery	9.1	20
Vegetable tannery	1.2	20
Organic phosphates	5.0	21
High nitrogen organics	22.2	22
Organic intermediates	20.6	26
	5.8	8
Viscose, rayon and nylon	8.2	19
	6.7	11
Soluble fraction of domestic sewage	8.0	20

[From Eckenfelder, W. W., Jr. (1989). "Industrial Water Pollution Control." McGraw-Hill, New York.]

A continuous system for wastewater treatment is shown in Fig. 1. A definition of symbols from this figure is presented in Table XVIII.

Activated sludge processes are usually fast and highly effective but require a large amount of energy for mixing and aeration. A large amount of sludge (70–120 kg dry solids/1000 m³) is produced, depending on the process and origin of the wastewaters. This sludge is generally difficult to filter and its disposal is cumbersome.

b. Aerated Lagoons

Aerated lagoons are basins having a depth varying from 4 to 12 ft (1.2 to 3.6 m), in which oxygenation of wastewaters is accomplished by aeration units. The difference between aerated lagoons and the activated sludge system is that lagoons are flow-through devices and no recycle of sludge is provided.

Solid concentration in the lagoon is a function of wastewaters' characteristics and retention time. It is usually between 80 and 200 mg TSS/liter, i.e., much lower than that for conventional activated sludge units (2000–3000 mg TSS/liter). Because of this difference, lagoons require much more area for wastewater oxidation than do activated sludge pro-

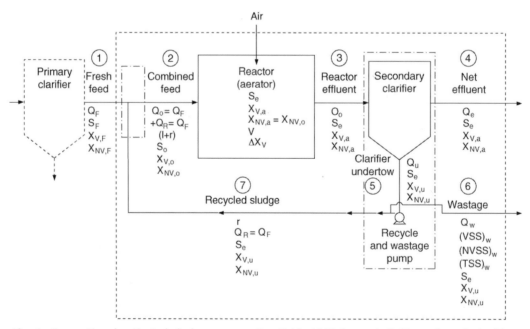

Fig. 1. Conventional activated sludge process. See Table XVIII for a definition of symbols. [From Ramalho, R. S. (1983). "Introduction to Wastewater Treatment Process." Academic Press, New York.]

TABLE XVIII
Definition of Symbols Used in Figure 1

Symbols

1. **Flow rates**

    Q_F, fresh feed; m^3

    Q_R, recycle; m^3

    r, recycle ratio; dimensionless ($r = Q_R/Q_F$)

    Q_o, Combined feed; m^3; $Q_o = .Q_F + Q_R = Q_F (1 + r)$ (stream 2)

    (volume of combined feed = volume of reactor effluent, i.e., Q_o (stream 2) = Q_o (stream 3)

    Q_e, net effluent; m^3 (stream 4)

    Q_w, wasteage; m^3 (stream 6) (notice that ; $Q_F = Q_e + Q_w$)

    Q_u, clarifier underflow; m^3 $Q_u = Q_w, + Q_R = Q_w, + Q_F$ (stream 5)

2. **Concentration (mg/liter) of soluble BOD**

    S_F, soluble BOD of fresh feed

    S_o, soluble BOD of combined feed

    S_e, soluble BOD of effluent

3. **Concentrations (mg/liter) of volatile suspended solids (VSS)**

    $X_{V,F}$, VSS in fresh feed

    $X_{V,o}$, VSS in combined feed

    $X_{V,a}$, VSS in rector; this also is equal to concentration of VSS in reactor effluent (complete mix reactor at steady state)

    $X_{v,u}$, VSS in secondary clarifier underflow

    $X_{V,e}$, VSS in net effluent

4. **Concentrations (mg/liter) of nonvolatile suspended solids (NVSS)**

    $X_{NV,F}$, NVSS in fresh feed

    $X_{NV,o}$, NVSS in combined feed

    $X_{NV,a}$, NVSS in reactor ($X_{NV,a} = X_{NV,o}$); this also equals concentration of NVSS in reactor effluent (complete mix reactor at steady state)

    $X_{NV,u}$, NVSS in secondary clarifier underflow

    $X_{NV,e}$, NVSS in net effluent

5. **Wastage**

    $(VVS)_w$, kg/day of VSS in wasteage

    $(NVVS)_w$, kg/day of NVSS in wasteage

    $(TVS)_w$, kg/day of TSS in wasteage

6. **Reactor volume**

    V, reactor volume, m^3

7. **Sludge production**

    X_V (kg/day)

Key

For suspended solids, double subscripts are utilized (e.g., $X_{V,i}$, $X_{NV,i}$.) The first subscript (V or NV) designates volatile and nonvolatile suspended solids, respectively. The second subscript (i) refers to the specific stream in question:

F, fresh feed (stream 1); o, combined feed (stream 2); a, reactor effluent (stream 3); e, net effluent (stream 4); u, underflow from secondary clarifier (stream 5).

[From Ramalho, R. S. (1983). "Introduction to Wastewater Treatment Process." Academic Press, New York.]

TABLE XIX
Performance of Lagoon Systems: Summary of Average Data from
Aerobic and Facultative Ponds

Industry	Area/1000 (m^2)	Depth (m)	Detention ($days$)	Loading [$kg/(m\_day)$]	BOD removal (%)
Meat and poultry	5.3	3.0	0.9	0.0080	80
Canning	27.9	5.8	1.8	0.0157	98
Chemical	125.4	5.0	1.5	0.0175	87
Paper	340	5.0	1.5	0.0116	80
Petroleum	62.7	5.0	1.5	0.0031	76
Wine	28.3	1.5	0.5	0.0245	
Dairy	30.4	5.0	1.5	0.0024	95
Textile	12.5	4.0	1.2	0.0183	45
Sugar	80.9	1.5	0.5	0.0096	67
Rendering	8.9	4.2	1.3	0.0040	76
Hog feeding	2.4	3.0	0.9	0.0396	
Laundry	8.0	3.0	0.9	0.0058	
Miscellaneous	60.7	4.0	1.2	0.0062	95
Potato	10.2	5.0	1.5	0.0123	

[From Eckenfelder, W. W., Jr. (1989). "Industrial Water Pollution Control." McGraw-Hill, New York.]

cesses. Performance of lagoon systems for waste-waters from several industries is presented in Table XIX.

C. Anaerobic Treatment of Industrial Effluents

1. Characteristics of Anaerobic Processes

Organic materials, in the absence of exogenous electron acceptors, such as oxygen, nitrate, and sulfate, can be converted into methane and carbon dioxide through a complex series of microbial interaction. Anaerobic degradation of organic compounds is shown in Fig. 2. In this process, most of the chemical energy in the starting substrate is released as methane and may be recovered.

The aerobic conversion of 1 kg COD requires 2 kW of electricity (for mixing and oxygen supply) and produces 0.5 kg of biomass (dry weight). Anaerobically, 1 kg COD gives rise to 0.35 m^3 biogas (equivalent to about 0.4 liter of liquid fuel) and 0.1 kg of biomass, which can be dewatered if required.

Traditionally, anaerobic digestion was utilized almost exclusively for the stabilization of sewage sludge. The process received little application in the

treatment of organic industrial wastes due to several limitations, including the low achievable rates of performance, the inability to withstand hydraulic and organic shockloads, and poor process control. These problems, all inherent to conventional digesters,

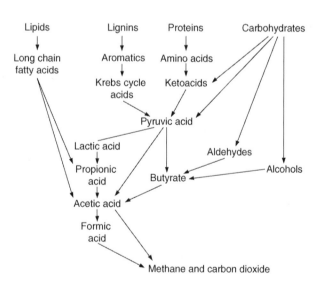

Fig. 2. Biological sequences for breakdown of solids to methane and carbon dioxide. [From Kosaric, N., and Blaszczyk, R. (1990). *Adv. Biochem. Eng. Biotechnol.* **42**, 27–62, by courtesy of Mercel Dekker, Inc.]

were associated with difficulties in retaining biomass within the digester and with a very long retention time (up to 50 days). New reactor designs and new methods of bacterial bed preparation allowed this time to be shortened considerably, down to a few hours.

Anaerobic treatment can accommodate very strong industrial wastewaters, but success is not controlled by BOD concentrations but rather by a number of other very important design criteria. Anaerobic volumetric loading rates are often 10 times higher than possible aerobically. Due to higher loading rates, anaerobic biotechnology applied to industrial wastewaters utilizes a much smaller reactor and produces a greatly reduced solids waste biomass than would be necessary if an aerobic process were applied to the same effluent. The associated significant financial savings and reduction of environmental impact warrant serious consideration, usually more than compensating for prolonged start-up time requirements to achieve design biomass concentrations.

When a comparison of anaerobic and aerobic technologies is conducted, certain points can be underlined:

- volumetric organic loading rates 5–10 times higher than for aerobic processes
- biomass synthesis rates of only 5–20% of those for aerobic processes
- nutrient requirements of only 5–20% of those for aerobic processes
- anaerobic biomass preserved for months or years without serious deterioration in activity
- no aeration energy requirements for anaerobic processes vs 500–2000 kWh/1000 kg COD for aerobic processes
- methane production of 12,000,000 BTU/1000 kg COD destroyed.

Acetate is the most important compound quantitatively produced in the fermentation of organic substrate by the bacterial population, with propionate production of secondary consequence. A number of microorganisms can convert soluble organic matter into acetic acid, some of these are *Lactobacillus, Escherichia, Staphylococcus, Micrococcus, Bacillus, Pseudomonas, Desulfovibrio, Selenomonas, Veillonella,*

Sarcinia, Streptococcus, Desulfobacterium, and *Desulfomonas.*

Methanogenic bacteria convert acetate and H_2/CO_2 into methane. These bacteria are very sensitive to oxygen and are obligate anaerobes. Some of the notable species that have been classified are *Methanobacterium formicicum, M. bryanti,* and *M. thermoautotrophicum, Methanobrevibacter ruminantium, M. arboriphilus,* and *M. smithii; Methanococcus vannielli* and *M. votae; Methanomicrobium mobile, Methanogenium cariaci,* and *M. marinsnigri; Methanospirillum hungatei,* and *Methanosarcina barkei.*

2. Anaerobic Reactors

a. Anaerobic Contact Reactor

In the anaerobic contact reactor system, anaerobic microorganisms are recycled and added to wastewaters and mixed. Because of relatively small biomass accumulation during anaerobic digestion, microorganisms must be separated by settling, centrifugation, or flotation. Separation is difficult and takes a long time. A schematic of an anaerobic contact reactor (ACR) is shown in Fig. 3.

b. Anaerobic Filter

In the anaerobic filter (AF), anaerobic micro organisms attach to a solid packing material within the reactor and remain there for a long time (solids retention time (SRT) > 100 d relative to hydraulic retention time (HRT) of 0.5–2 d). The process is very effective. The main limitation is plugging of the reactor by accumulated bacteria. Downflow and

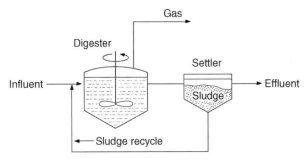

Fig. 3. Schematic of an anaerobic contact reactor. [From Kosaric, N., and Blaszczyk, R. (1990). *Adv. Biochem. Eng. Biotechnol.* **42,** 27–62, by courtesy of Marcel Dekker, Inc.]

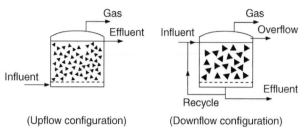

Fig. 4. Schematic of anaerobic filters.

TABLE XX
Comparison of Anaerobic Digestion Systems

Reactor type	Typical loading rate (kg COD/m³ day)	Hydraulic retention time (days)
CSTR	0.25–3.0	10–60
Contact	0.25–4.0	12–15
UASBR	10–30	0.5–7.0
CASBER	4.0–5.0	0.5–12
RBC	0.005–0.02	0.4–1.0
Anaerobic filter	1.0–40	0.2–3.0
AAFEB	1.0–50	0.2–5.0
AFB	1.0–100	0.2–5.0

Note. CSTR, continuous stirrer tank reactor; UASBR, upflow anaerobic sludge blanket; CASBER, carrier assisted sludge bed reactor; RBC, rotation biological contactor; AAFEB, anaerobic attached film expanded bed; AFB, anaerobic fluidized bed. [From Stronach, S. M. et al. (1986). "Biotechnology Monograph." Springer-Verlag, Berlin.]

upflow configuration is recognized. A schematic of upflow and downflow AF is shown in Fig. 4.

c. Upflow Anaerobic Sludge Blanket Reactor

In the upflow anaerobic sludge blanket reactor (UASBR) system, anaerobic microorganisms attach to each other and develop spherical aggregates with diameters of 1–3 mm. No support for microorganisms is necessary. These aggregates settle very well. Their settling velocity is in the range of 20–90 m/h, which results in a low retention time of even a few hours. A schematic of a UASBR is shown in Fig. 5.

d. Fluidized or Expanded Bed Reactor

In a fluidized or expanded bed reactor, microorganisms attach to the small particles of the carrier, which may be nonporous, such as sand, sepiolite, carbon, and different types of plastics, or porous, such polyurethane foam and pumice. Settling veloc-

ity of particles covered with microorganisms is up to 50 m/h. A proper superficial liquid rate expands the bed. When the bed expands up to 30%, the reactor is named the expanded bed reactor. When the expansion is between 30 and 100%, the reactor is named the fluidized bed reactor. A schematic of the fluidized/expanded bed reactor is shown in Fig. 6.

Ranges of process parameters for the preceding types of reactors are presented in Table XX. Performance of the anaerobic process for selected industries is shown in Table XXI.

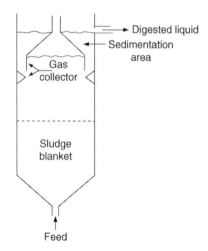

Fig. 5. Schematic of an upflow anaerobic sludge blanket reactor.

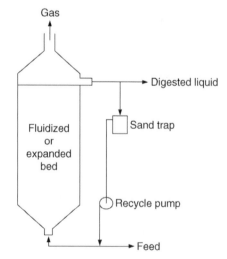

Fig. 6. Schematic of a fluidized (expanded) bed reactor.

TABLE XXI
Performance of Anaerobic Processes

Wastewater	Process	Loading [kg/(m³·day)]	HRT (h)	Temperature (°C)	Removal (%)
Meat packing	Anaerobic contact	3.2 (BOD)	12	30	95
Meat packing		2.5 (BOD)	13.3	35	95
Keiring		0.085 (BOD)	62.4	30	59
Slaughterhouse		3.5 (BOD)	12.7	35	95.7
Citrus		3.4 (BOD)	32	34	87
Synthetic	Upflow filter	1.0 (COD)	–	25	90
Pharmaceutical		3.5 (COD)	48	35	98
Pharmaceutical		0.56 (COD)	36	35	80
Guar gum		7.4 (COD)	24	37	60
Rendering		2.0 (COD)	36	35	70
Landfill leachate		7.0 (COD)	–	25	89
Paper mill foul condensate		10–15 (COD)	24	35	77
Synthetic	Expanded bed	0.8–4.0 (COD)	0.33–6.0	10–3.0	80
Paper mill foul condensate		35–48 (COD)	8.4	35	88
Skimmed milk	USAB	71 (COD)	5.3	30	90
Sauerkraut		8.0–9.0 (COD)	–	–	90
Potato		25–45 (COD)	4.0	35	93
Sugar		22.5 (COD)	6.0	30	94
Champagne		15 (COD)	6.8	30	91
Sugar beet		10 (COD)	4.0	35	80
Brewery		95 (COD)	–	–	83
Potato		10 (COD)	–	–	90
Paper mill foul		4.0–5.0 (COD)	70	35	87

[From Eckenfelder, W. W., Jr. (1989). "Industrial Water Pollution Control." McGraw-Hill, New York.]

e. Anaerobic Ponds

The biological process of anaerobic ponds is the same as that occuring in anaerobic digestion reactors but lasts much longer, due to limited circulation and diffusion. A large process area is needed. Open ponds can not be used near populated areas because of the unpleasant odor. Anaerobic ponds are particularly suitable for treatment of high-strength organic wastewater that also contains a high concentration of solids. Anaerobic ponds have a depth up to 6 m. The BOD removal efficiency of this system is between 50 and 70%. The partially clarified effluent is usually discharged to another process for further treatment.

f. Facultative Ponds

The facultative pond is divided by loading stream and thermal stratification into an aerobic surface and an anaerobic bottom. The aerobic surface layer has a diurnal variation, increasing in oxygen content during the daylight hours, due to algal photosynthesis, and decreasing during the night. Sludge deposited on the bottom undergoes anaerobic decomposition. Odors will be produced if an aerobic layer is not maintained. Between the aerobic and anaerobic zones is an intermediate zone that is partly aerobic and partly anaerobic.

Performance of anaerobic and facultative ponds for some industries is presented in Table XXII.

IV. SLUDGE TREATMENT

Under either aerobic or anaerobic conditions, only 25–40% of the resulting biomass which is synthesized may be further biodegraded; the remaining 60–75% of the residual biomass is so refractory that it

TABLE XXII
Performance of Anaerobic and Facultative Ponds

Industry	Area/1000 m^3	Depth (m)	Detention (days)	Loading (kg/m^3day)	BOD removal (%)
Summary of average data from anaerobic ponds					
Canning	10	1.8	15	0.0435	51
Meat and poultry	4	2.2	16	0.1401	80
Chemical	0.6	1.1	65	0.0060	89
Paper	287	1.8	18.4	0.0386	50
Textile	8.9	1.8	3.5	0.1593	44
Sugar	142	2.1	50	0.0267	61
Wine	15	1.2	8.8		
Rendering	4	1.8	245	0.0178	37
Leather	10.5	1.3	6.2	0.3336	68
Potato	40	1.2	3.9		
Summary of average data from combined aerobic–anaerobic ponds					
Canning	22	1.5	22	0.0686	91
Meat and poultry	3.2	1.2	43	0.0296	94
Paper	10200	1.7	136	0.0031	94
Leather	18.6	1.2	152	0.0056	92
Miscellaneous industrial wastes	570	1.2	66	0.0142	

[From Eckenfelder, W. W., Jr. (1989). "Industrial Water Pollution Control." McGraw-Hill, New York.]

cannot practically be destroyed by any natural means, except maybe by incineration or chemical hydrolysis. The sludge separated during industrial effluent treatment contains most of the undesired compounds removed from wastewaters. Formerly, aerobic environmental control process held a virtual monopoly on the industrial wastewater market. Biomass disposal problems caused by large volumes of refractory biomass and, in some cases, even the possible long-term contamination of secondary and tertiary lands by disposal of waste-activated sludge biomass, were largely ignored. Under anaerobic conditions, ordinarily, more than 90% of the wastewater COD is converted to methane gas as an end-product. This energy equivalent is not available for biomass synthesis. Aerobic treatment normally produces about 10 times more biomass than its anaerobic counterpart. Therefore, both the financial and waste biomass disposal site requirements are reduced. A special treatment procedure must be involved before disposal of the remaining sludge. Sludge lagoons and on-site landfill areas storing waste aerobic biomass may in-

creasingly be restricted in the future. A list of possible procedures is presented in Table XXIII.

Waste sludge may be disposed of on land or in a landfill. New environmental regulations have restricted the application of incineration for sludge disposal. This reflects the concern of producing gases which may contain toxic matter.

The waste sludge contains 2 to 5% nitrogen, 0.5 to 2% phosphorus, and 0.1 to 0.5% potassium. These elements are important plant nutrients and are slowly released when sludge, is applied on land. The organic carbon content of sludge increases the humus content of soil. In general, land application of sludge improves soil permeability and provides essential nutrients for plant growth.

In addition to nutrients and organic carbon, waste sludge may contain heavy metals and pathogenic microorganisms. Heavy metals originate from industrial wastewaters discharged to the treatment plant. High concentrations of heavy metals may be toxic to microorganisms, plants, or humans consuming sludge-treated plants. The amount of heavy metals

TABLE XXIII
Sludge Processing and Disposal Methods

Processing or disposal function	Unit operation, unit process, or treatment method
Preliminary operations	Sludge pumping
	Sludge grinding
	Sludge blending and storage
	Sludge degritting
Thickening	Gravity thickening
	Flotation thickening
	Centrifugation
	Gravity belt thickening
	Rotary drum thickening
Stabilization	Lime stabilization
	Heat treatment
	Aerobic digestion
	Anaerobic digestion
	Composting
Conditioning	Chemical conditioning
	Heat treatment
Disinfection	Pasteurization
	Long-term storage
Dewatering	Vacuum filter
	Centrifuge
	Belt press filter
	Filter press
	Sludge drying beds
	Lagoons
Heat drying	Dryer variations
	Multiple effect evaporator
Thermal reduction	Multiple heat incineration
	Fluidized bed incineration
	Co-incineration with solid wastes
	Wet air oxidation
	Vertical deep well reactor
Ultimate disposal	Land application
	Distribution and marketing
	Landfill
	Lagooning
	Chemical fixation

[From Metcalf & Eddy, Inc. (1972). "Wastewater Engineering: Treatment, Disposal, Reuse." McGraw-Hill, New York.]

accumulated in the sludge may affect the performance of a sludge treatment process and restrict land application of the waste sludge.

Pathogenic microorganisms are present in large numbers in both primary and waste-activated sludge.

Strict regulations have been implemented for sludge handling and disposal to prevent the contamination of man, animals, and plants by pathogenic microorganisms (USEPA, 1979).

A. Biological Sludge Treatment

The objectives of sludge treatment are the following:

- Stabilize sludge
- Reduce/eliminate pathogens
- Reduce the volume of sludge and
- Make treated sludge more marketable for land application.

The organic matter, or volatile solids content, of sludge is reduced to 30 to 50% during stabilization. This low volatile solids content slows down the decomposition process to such an extent that no odor is produced during disposal.

A sludge free of pathogens is essential when sludge is applied on crop land or used by the public for gardening. The protection of treatment plant personnel also requires that the pathogen content of sludge be reduced or eliminated.

The raw primary and biological waste sludge contains 96–99% water and the solids phase contains 60–80% volatile matter.

The cost of disposal is a function of volume of waste sludge. To operate a cost effective sludge disposal program, the volume of sludge has to be reduced. This can be achieved by sludge thickening and dewatering. A minimum of 35% percent solids content is required for landfill disposal.

A reduction in the volatile solids content reduces the total solids content of sludge and, in most cases, improves the dewaterability of sludge.

The treated sludge, in addition to the above technical constrains, should also have an appearance which is acceptable for the final user. This includes texture, odor, appearance, and plant nutrient supplement. Sludge separated after biological treatment of industrial effluent is mainly composed of bacterial cells, which contains a large part of initial BOD from the industrial effluents. The reduction of BOD in the sludge may be performed aerobically or anaerobically.

1. Aerobic Digestion of Sludge

In this process, a mixture of primary digestible sludge from primary treatment and activated sludge from aerobic biological treatment is aerated for an extended period of time. This results in cellular destruction with a decrease of volatile suspended solids and, consequently, reduction of the amount of sludge that is to be disposed. The reduction results from conversion by oxidation of a substantial part of the sludge into volatile products, namely, CO_2, NH_3, and H_2. This oxidation occurs when the substrate in an aerobic system is insufficient for energy maintenance and synthesis (endogenous decay).

During aerobic digestion, the organic content of sludge is reduced by 50 to 60%. The digested sludge is biologically stable.

The aerobic digestion process is very sensitive to temperature and for all practical purposes, stops at or below 8°C. The digestion rate increases by a factor of three from 10° to 20°C. The aerobic digestion process is performed in an open and unheated tank. Consequently, the operating parameters for the digestion process are selected, such that it would allow the digestion to proceed throughout the year. This can be achieved by using a long sludge age and low volumetric organic load rate.

The design parameters for aerobic digestion are summarized below:

- Volatile solids load: 1.6 kg VSS/m³d

- Sludge age: 45 days. This includes the sludge age of the activated sludge system

- Oxygen requirement: 1.8 to 2.1 kg O_2/kg VSS

- Power requirement: 13 to 110 W/m³

The advantages of the aerobic digestion process can be summarized in the following points:

- Capital costs are generally lower than for anaerobic systems for plants under 5 million gallons per day (220 L/s)

- Easy to operate
- Does not generate nuisance odors
- Supernatant low in BOD, suspended solids, and ammonia
- Reduces the number of pathogens to a low level
- Reduces volatile solids content of sludge by 50 to 60%
- It can be used for sludges which contain high levels of heavy metals or toxic organic compounds.

The following are the disadvantages of the aerobic digestion process:

- Digested sludge has very poor dewatering characteristics. This may require larger dewatering equipment than that for the dewatering of the same volume of anaerobically digested sludge.
- High power costs. Power is required for aeration.
- Performance is affected by the temperature.

2. Anaerobic Digestion of Sludge

Anaerobic sludge digestion is among the oldest forms of biological wastewater treatment. It is performed as the following: (a) Standard-rate digestion in a single-stage process, in which the function of digestion, sludge thickening, and supernatant formation are carried out simultaneously; (b) Single-stage high-rate digestion, in which the sludge is mixed intimately by gas recirculation, mechanical mixers, pumping, or draft tube mixers. When a two-stage, high-rate digestion process is employed, the second stage is used for storage and concentration of digested sludge and for the formation of relatively clear supernatant.

During anaerobic digestion of sludge, a gas (65–70% CH_4, and a small amount of N_2, H_2, and H_2S) are produced. Typical biogas values vary from 0.75 to 1.1 m³/kg of volatile solids destroyed.

The optimum operating temperature for sludge digestion is between 33° and 38°C. Most aerobic digesters are operated at 35°C. At this temperature, the conversion of volatile solids to methane is completed in 10 to 15 days. Consequently, anaerobic digesters are with a minimum of 15 days hydraulic retention time

If the digester is loaded with more volatile solids than is allowed, the digester content may turn acidic and the conversion of the organic matter to methane stops.

High-rate digesters receive high food loads. This high load rate may be sufficient to produce local acidic islands at the feed port. To avoid the local build-up of volatile acids, the digester content has to be mixed. Mixing is also necessary to maintain a constant and uniform temperature in the digester. A sudden change of more than $\pm 2°C$ can upset the methanogenic bacteria and result in the build-up of volatile acids. The digester content can be mixed with mechanical or diffused gas mixers. Mechanical mixers are typically located at the top of the digester. Gas-mixing is performed with the biogas produced during digestion. Gas may be introduced through diffusers of spargers.

The suspended solids content of the digester's feed is usually increased to 4 to 10%. This is essential for cost-effective digestion operation. A high suspended solids content in the feed is necessary to maintain a high biomass inventory in the digester. This is the prerequisite to operate the digester with a high atile solids load. The volume of biogas produced increases with increased volumetric load to the digester. Under these high load conditions, significantly more biogas is produced than is required to heat the digester content.

The following operating parameters are recommended for anaerobic digestion:

- Volatile solids load 1.6 to 6 kg/m³ d Suspended solids content of feed 4 to 10%
- Minimum hydraulic retention time: 15 days
- Mixing:
  - Gas mixing: 5 to 8 W/m³
  - Mechanical: 6 to 7 W/m³
- Temperature: 33 and 38°C (35°C)
- pH between 6.5 and 7.5
- Alkalinity > 2000 mg/l.

The advantages of the digestion process can be summarized in the following points:

- Biogas production: The biogas contains 60 to 75% methane and the rest is carbon dioxide. At 35°C, the digester produces 760 l biogas for every kg of volatile solids digested.
- Reduces the volatile solids content of sludge by 40 to 55%.
- Digested sludge is easy to dewater.
- Digested sludge is suitable for use as a soil conditioner.
- Digestion process inactivates pathogenic microorganisms.

The following are the disadvantages of the anaerobic digestion process:

- High capital cost. The process requires a complex system, consisting of reactor tanks, gas collection and utilization system, boiler, heat exchanger, pumps, mixing system, process monitoring equipment, and safety devices.
- The process is susceptible to upsets, which may be caused by a sudden temperature change, increased organic load, heavy metal build-up, and toxic organic compounds.
- Poor quality supernatant, which increases the BOD_5 and suspended solids load of the activated sludge process.
- Slow growth rate of biomass in digester. The doubling time for methanogens is about 4 to 6 days. This time is required to produce a new generation of methanogens and replace the old biomass. Consequently, at least the same length of time is required for the digester to respond to any change in the operating conditions.

3. Composting

The composting process involves complex destruction of organic material, coupled with the production of humic matter, to produce a stabilized end-product. The microorganisms involved fall into three major categories: bacteria, actinomycetes, and fungi. Approximately 20–30% of the volatile solids are converted to CO_2 and H_2O. During the process, the compost heats to temperatures in the pasteurization range and enteric pathogenic organisms are destroyed. Composting may by accomplished under anaerobic and aerobic conditions. Aerobic composting accelerates material decomposition and results in a higher rise in the temperature, necessary for pathogenic destruction, and minimizes the potential for the nuisance odors.

Environmental factors influence the activities of the bacteria, fungi, and actinomycetes in this oxidation decomposition process and affect the speed and course of composting cycles. The composting process is considered complete when the product can be stored without giving rise to nuisances such as odors, and when pathogenic organisms have been reduced to a level such that the material can be handled with minimum risk.

Compost can provide a portion of the nutrient requirements for growth of crops. It can improve the quality of soils containing excessive amounts of sand or clay. The following operating parameters are maintained for successful composting:

- Moisture content between 50 and 60%. The compost can not be heated up to the optimum operating temperature if the raw material mixture has a higher moisture content. At low moisture content, the temperatures rise above the optimum and the microorganisms are killed or inhibited. This slows down or stops the composting process.
- The raw material must be a porous and structurally stable mixture. This is essential for good aeration of large-sized piles.
- The composted material must reach a temperature of 55 to 65°C. The composted material has to be exposed to this temperature for at least 15 days. This is essential to kill pathogens.
- Optimum carbon to nitrogen ratio: 30.
- Air supply between 5 and 15% by volume.

The above optimum conditions can be established by co-composting raw sludge with sawdust, wood chips, or municipal refuse. Raw sludge is typically dewatered to 20–30% before composting.

V. NEW TRENDS IN WASTEWATER TREATMENT

A combination of higher discharge standards, tighter restrictions on odor control, ground water contamination, sludge disposal site criteria, and energy efficiencies have led researchers and end-users to seek alternative methods of wastewater treatment. In the future, the most appropriate solutions will be novel technologies with lower operating costs and/ or capital costs.

The following can be cited as desirable design criteria for future processes:

- High volumetric loading rate
- Versatile biomass inventory for relative ease of biodegradation of the pollutants
- Inherent alkalinity-generation potential
- Targeted effluent quality
- More compact reactor configuration
- Biomass immobilization to achieve low reactor volume/biomass ratios
- Thermophilic activity
- Low sludge production
- Low energy requirements
- Low maintenance and operator costs
- Integrated processes where a high-added-value product is produced in parallel to waste treatment.

The microbial biomass increases in proportion to the energy available from the metabolism of the organic pollutant. If the microbes are intentionally concentrated within the system (biomass immobilization) to enhance the rate of removal of the organic contaminant, the required reactor volume will be reduced, increasing process stability.

Relative ease of biodegradation of the pollutants is a key consideration in the design decision because of its governing the minimum hydraulic retention time. The amount of pollutant removed is the result of the ease of biodegradation times the amount of biomass times the contact time. Most simple molecules like organic acids, alcohols, and sugars can be metabolized within a matter of minutes to hours if an active biomass is provided, thus allowing short retention times and, therefore, high loading rates, even though the BOD concentration is low. More complex pollutants, such as starches and proteins, may require relatively long hydrolysis times with associated longer retention times, necessitating greater anaerobic reactor volume.

TABLE XXIV
Summary of Industrial Waste: Its Origin, Characteristics, and Treatment

Industries producing waste	Origin of major wastes	Major characteristics	Major treatment and disposal methods
Apparel			
Textiles	Cooking of fibers; desizing of fabric	Highly alkaline, colored, high BOD and temperature, high suspended solids	Neutralization, chemical precipitation, biological treatment, aeration and/or trickling filtration
Leather goods	Unhairing, soaking, deliming, and bating of hides	High total solids, hardness, salt, sulfides, chromium, pH, precipitated lime, and BOD	Equalization, sedimentation, and biological treatment
Laundry trades	Washing of fabrics	High turbidity, alkalinity, and organic solids	Screening, chemical precipitation, floatation, and adsorption
Food and drugs			
Canned goods	Trimming, culling, juicing, and blanching of fruits and vegetables	High in suspended solids, colloidal and dissolved organic matter	Screening, lagooning, soil absorption, or spray irrigation
Dairy products	Dilutions of whole milk, separated milk, buttermilk, and whey	High in dissolved organic matter, mainly protein, fat, and lactose	Biological treatment, aeration, trickling filtration, activated sludge
Brewed and distilled beverages	Steeping and pressing of grain, residue from distillation of alcohol, condensate from stillage evaporation	High in dissolved organic solids, containing nitrogen and fermented starches for their products	Recovery, concentration by centrifugation and evaporation, trickling filtration: using feeds; digestion of slops
Meat and poultry products	Stockyards: slaughtering of animals; rendering of bones and fats; residues in condensates; grease and fresh water: picking of chickens	High in dissolved and organic matter, blood, other proteins, and fats	Screening, settling, and or floatation, trickling filtration
Animal feedlots	Excreta from animals	High in organic suspended soilds and BOD	Land disposal and anaerobic lagoons
Beet sugar	Transfer screening and juicing waters; drainings from lime sludge; condensates after evaporation; juice and extracted sugar	High in dissolved and suspended organic matter containing sugar and protein	Reuse of wastes, coagulation, and lagooning
Pharmaceutical products	Mycelium, spent filtrate, and wash waters	High in suspended and dissolved organic matter, including vitamins	Evaporation and drying: feeds
Yeast	Residue from yeast filtration	High in solids (mainly organic) and BOD	Anaerobic digestion, trickling, filtration
Pickles	Lime water; brine, alum and turmeric, syrup, seeds and pieces of cucumber	Variable pH, high suspended solids, color and organic matter	Good housekeeping, screening, equalization
Coffee	Pulping and fermenting of coffee bean	High BOD and suspended solids	Screening, settling and trickling filtration
Fish	Rejects from centrifuge; pressed fish; evaporator and other washwater waste	Very high BOD, total organic solids, and odor	Evaporation of total waste; barge remainder to sea
Rice	Soaking, cooking, and washing of rice	High BOD total and suspended solids (mainly starches)	Lime coagulation, digestion

continues

Continued

Industries producing waste	Origin of major wastes	Major characteristics	Major treatment and disposal methods
Soft drinks	Bottle washing; floor and equipment cleaning; syrup-storage tank drains	High pH, suspended solids and BOD	Screening, plus discharge to municipal sewer
Bakeries	Washing and greasing of pans; floor washings	High BOD, grease, floor washings, sugars, flour, detergents	Amenable to biological oxidation
Water production	Filter backwash; lime–soda sludge; brine; alum sludge	Minerals and suspended solids	Direct discharge to streams or indirectly through holding lagoons
Materials			
Pulp and paper	Cooking, refining, washing of fibers, screening of paper pulp	High or low pH, color high suspended, colloidal, and dissolved solids, inorganic fillers	Settling, lagooning, biological treatment, aeration, recovery of by-products
Photographic products	Spent solutions of developer and fixer	Alkaline, containing various organic and inorganic reducing agent	Recovery of silver; discharge of wastes into municipal sewer
Steel	Cooking of coal, washing of blast-furnace flue gases and pickling of steel	Low pH, acids, cyanogen, phenol, ore, coke, limestone, alkali, oils, mill scale, and fine suspended solids	Neutralization, recovery and reuse, chemical coagulation
Metal-plated products	Stripping of oxides, cleaning and plating of metals	Acid, metals, toxic, low volume, mainly mineral water	Alkaline chlorination of cyanide; reduction and precipitation of chromium; lime precipitation of other metals
Iron-foundry products	Wasting of used sand by hydraulic discharge	High suspended solids, mainly sand: some clay and coal	Selective screening, drying of reclaim sand
Oil fields and refineries	Drilling mud, salt, oil, and some natural gas: acid sludges and miscellaneous oils from refining	High dissolved salts from field; high BOD, odor, phenol and sulfur compounds from refinery	Diversion, recovery, injection of salts; acidification and burning of alkaline sludges
Fuel oil use	Spills from fuel-tank filling waste; auto crankcase oils	High in emulsified and dissolved oils	Leak and spill prevention, flotation
Rubber	Washing of latex, coagulated rubber, exuded impurities from crude rubber	High BOD and odor, high suspended solids, variable pH, high chlorides	Aeration, chlorination, sulfonation, biological treatment
Glass	Polishing and cleaning of glass	Red color, alkaline nonsettleable suspended solids	Calcium–chloride precipitation
Naval stores	Washing of stumps, drop solution, solvent recovery, and oil-recovery water	Acid, high BOD	By-product recovery, equalization, recirculation and reuse, trickling filtration
Glue manufacturing	Lime wash, acid washes, extraction of nonspecific proteins	High COD, BOD, pH, chromium, periodic strong mineral acids	Amenable to aerobic biological treatment, flotation, chemical precipitation
Wood preserving	Steam condensates	High in COD, BOD, solids, phenols	Chemical coagulation; oxidation pond and other
Candle manufacturing	Wax spills, stearic acid condensates	Organic (fatty) acids	Anaerobic digestion
Plywood manufacturing	Glue washings	High BOD, pH, phenols, potential toxicity	Settling ponds, incineration

continues

Continued

Industries producing waste	Origin of major wastes	Major characteristics	Major treatment and disposal methods
Chemicals			
Acids	Dilute wash waters; many varied diluted acids	Low pH, low organic content	Upflow or straight neutralization, burning when some organic matter is present
Detergents	Washing and purifying soaps and detergents	High in BOD and saponified soaps	Flotation and skimming, precipitation with $CaCl_2$
Cornstarch	Evaporator condensate or bottoms when not reused or recovered, syrup final washes, waste from "bottling-up process"	High BOD and dissolved organic matter; mainly starch and related material	Equalization, biological filtration, anaerobic digestion
Explosives	Washing TNT and guncotton for purification, washing and pickling of cartridges	TNT, colored, acid, odors, and contains organic acids and alcohol from powder and cotton, metals, acid, oils, and soaps	Flotation, chemical precipitation, biological treatment, aeration, chlorination of TNT
Pesticides	Washing and purification products such as 2.4-D and DDT	High organic matter, benzene ring-structure, toxic to bacteria and fish, acid	Dilution, storage, activated-carbon adsorption, alkaline chlorination
Phosphate and phosphorus	Washing, screening, floating rock, condenser bleed-off from phosphate reduction plant	Clays, slimes, and tall oils, low pH, high suspended solids, phosphorus, silica and fluoride	Lagooning, mechanical clarification, coagulation and settling of refined waste
Formaldehyde	Residues from manufacturing synthetic resins and from dyeing synthetic fibers	Normally high BOD and HCHO, toxic to bacteria in high concentrations	Trickling filtration, adsorption on activated charcoal
Plastics and resins	Unit operations from polymer preparation and use; spills and equipment washdowns	Acid, caustic, dissolved organic matter, such as phenols, formaldehyde, etc.	Discharge to municipal sewer, reuse, controlled-discharge
Energy			
Steam power	Cooling water, boiler blowdown, coal drainage	Hot, high volume, high inorganic and dissolved solids	Cooling by aeration, storage of ashes, neutralization of excess acid wastes
Coal processing	Cleaning and classification of coal, leaching of sulfur strata with water	High suspended solids, mainly coal; low pH, high H_2SO_4 and $FeSO_4$	Settling, froth flotation, drainage control, and sealing of mines
Nuclear power and radioactive materials	Processing ores; laundering of contaminated clothes; research lab wastes; processing of fuel; power plant cooling waters	Radioactive elements, can be very acid and "hot"	Concentration and containing or dilution and dispersion

[From Nemerow, N. L. (1992). "Liquid Waste and Industry," Addisson-Wesley Publishing Co, New York.]

The concentration of the substrate and the acclimation level of the biomass also affect the pollutant biodegradation rate.

In a large anaerobic treatment facility, the methane may be advantageously utilized for heating or electricity generation. Warm or cool *concentrated* wastewaters may be treated by anaerobic processes, since the heat transferred by burning the methane produced is sufficient to raise the influent temperature 3.3°C for each 1000 mg/L of COD converted into methane.

Table XXIV summarizes major group of industrial wastes, their characteristics, and existing treatment methods.

Traditionally, the usual aim of waste management has been minimum treatment and disposal. Recently, a new concept has been introduced to waste management: recycling and reutilization. This new concept regards waste materials as valuable raw materials from which high-value-added products can be produced. There is a growing interest currently in the development of policies and practices that will minimize the wasting of material resources.

Recycling is applicable to all types of wastes and the basic concept is that wastes should not be thrown away, but should be processed to return them back to society as useful products. This has led to novel approaches evoking the implementation of new policies for wise management of natural resources.

However, a successful waste reutilization program can not be attained unless certain points are taken into consideration:

- Directing the recovered products into versatile areas
- Investigating the marketability and profitability

- Economic and social acceptability
- Identification of appropriate reprocessing technologies.

See Also the Following Articles

Acetic Acid Production • Meat and Meat Products • Methanogenesis • Oil Pollution • Pulp and Paper

Bibliography

Eckenfelder, W. W., Jr. (1989). "Industrial Water Pollution Control. " McGraw-Hill, New York.

Ganoulis, J. G. (1994). "Engineering Risk Analysis of Water Pollution: Probabilities and Fuzzy Sets. " Weinheim, New York.

Kosaric, N., and Blaszczyk, R. (1990). Microbial aggregates in aerobic wastewater treatment. *In* "Advances in Biochemical Engineering/Biotechnology." Vol. 42, pp. 27–62. Springer Verlag, Berlin/Heidelberg, Germany.

Masters, G. M. (1991). "Introduction to Environmental Engineering and Science." Prentice-Hall, Englewood Cliffs, NJ.

Metcalf & Eddy, Inc. (1991). "Wastewater Engineering: Treatment, Disposal, Reuse." McGraw-Hill, New York.

Nemerow, N. L. (1972). "Liquid Waste of Industry." Addison-Wesley Publishing Co., New York.

Ramalho, R. S. (1983). "Introduction to Wastewater Traetment Process." Academic Press, New York.

Speece, R. E. (1996). "Anaerobic Biotechnology for Industrial Wastewater."

Stronach, S. M., Rudd, T., and Lester, J. N. (1986). Anaerobic digestion process in industrial wastewater treatment. *In* "Biotechnology Monographs," Vol. 2. (S. Aiba, L. T. Fan, A. Fiechter, and K. Schügerl, eds.). Springer-Verlag, Berlin, Germany.

Vriens, L., van Soest, H., and Verachtert, H. (1990). *Critical Rev. Biotech.* **10**, 1–46. United States Environmental Protection Agency. (1979). "Process Design Manual for Sludge Treatment and Disposal." Washington, DC.

Industrial Fermentation Processes

Thomas M. Anderson

Archer Daniels Midland BioProducts Plant, Decatur, IL

GLOSSARY

aseptic   In medicine, referring to the absence of infectious organisms. In fermentation, referring to the absence of contaminating microorganisms.

auxotrophy   The inability of a culture to produce an essential component or growth requirement.

bacteriophage   Viruses that attack bacteria and destroy a fermentation.

bioreactor   A closed vessel used for fermentation or enzyme reaction.

BOD   Biological oxygen demand; the amount of oxygen required by microorganisms to carry out oxidative metabolism in water containing organic matter, such as sewage.

mass transfer   The movement of a given amount of fluid or the material carried by the fluid. In bioreactors, it refers to the dispersal and solubilization of nutrients and gases in the fermentation medium.

scale up   To take a process from the laboratory to full scale production, usually through intermediate steps.

sterilization   The removal of all microorganisms from a material or space. Standard conditions for steam sterilization are 121°C, 15 psig for 15 minutes.

MANY PRODUCTS   are made by large scale fermentation today, including amino acids, enzymes, organic acids, vitamins, antibiotics, solvents and fuels (Table I).

There are several advantages to making products by fermentation:

- Complex molecules that occur naturally, such as antibiotics, enzymes, and vitamins, are impossible to produce chemically.
- Optically active compounds, such as amino acids and organic acids, are difficult and costly to prepare chemically.
- "Natural" products that can be economically derived by chemical processes, but, for food purposes, are better produced by fermentation, such as beverage ethanol and vinegar (acetic acid).
- Fermentation usually uses renewable feedstocks instead of petrochemicals.
- Reaction conditions are mild, in aqueous media, and most reaction steps occur in one vessel.
- The by-products of fermentation are usually environmentally benign compared to the organic chemicals and reaction by-products of chemical manufacturing. Often, the cell mass and other major by-products are highly nutritious and can be used in animal feeds.

There are, of course, drawbacks to fermentation processes:

- The products are made in complex solutions at low concentrations compared to chemically derived compounds.
- It is difficult and costly to purify the product.
- Microbial processes are much slower than chemical processes, increasing the fixed costs of the process.
- Microbial processes are subject to contamination by competing microbes, requiring the steril-

TABLE I
Fermentation Products

Category	Examples	Function
Organic acids	Lactic, citric, acetic, formic, acrylic	Food acidulants, textiles, tanning, cleaners, intermediate for plastics and organic chemicals
Organic chemicals	Ethanol, glycerol, acetone, butanediol, propylene glycol	Solvents, intermediates for plastics, rubber and chemicals, antifreeze, cosmetics, explosives
Amino acids	MSG, L-Lysine, L-tryptophane, L-phenylalanine	Animal feeds, flavors, sweetners
Enzymes	Amylases, cellulases, glucose isomerase, proteases, lipases	Grain processing, cheesemaking, tanning, juice processing, high fructose corn syrup
Antibiotics	Penicillins, streptomycin, tetracycline, chloramphenicol	Human and animal health care
Bioplymers	Xanthan, dextran, poly-β-hydroxybutyrate	Stabilizers and thickeners in foods, oil well drilling, plastics
Vitamins	Vitamin B_{12}, biotin, riboflavin	Nutritional supplements, animal feed additives
Cell mass	Yeast, lactic acid bacteria	Single cell protein, baker's and brewer's yeast, starter cultures

ization of the raw materials and the containment of the process to avoid contamination.

- Most microorganisms do not tolerate wide variations in temperature and pH and are also sensitive to upsets in the oxygen and nutrient levels. Such upsets not only slow the process, but are often fatal to the microorganisms. Thus, careful control of pH, nutrients, air, and agitation requires close monitoring and control.
- Although nontoxic, the waste products are high in BOD, requiring extensive sewage treatment.

Other processes considered large-scale microbial processes are industrial or municipal sewage treatment, bioremediation of contaminated soil and water, and bioleaching of metal ores. Among the largest fermentation facilities are sewage treatment plants, consisting of anaerobic primary treatment and aerobic secondary treatment facilities. Bioremediation of environmental contamination is an emerging technology. Initially, native consortia of microorganisms were used to degrade contaminants. Now, cultures are selected, developed, and grown to more rapidly rehabilitate contaminated environmental sites.

The vast majority of large-scale fermentations use bacteria, yeast and fungi, but some processes use algae, plant, and animal cells. Several cellular activities contribute to fermentation products:

- Primary metabolites: ethanol, lactic acid, and acetic acid.
- Energy storage compounds: glycerol, polymers, and polysaccharides.
- Proteins: extracellular and intracellular enzymes, single-cell proteins and foreign proteins.
- Intermediary metabolites: amino acids, citric acid, vitamins, and malic acid.
- Secondary metabolites: antibiotics.
- Whole-cell products; single-cell protein, baker's yeast and brewer's yeast, bioinsecticides.

Fermentation products can be growth associated or non-growth associated. Primary metabolites, such as ethanol and lactic acid, are generally growth associated, as are cell mass products, while secondary metabolites, energy storage compounds, and polymers are non-growth associated. Other products, such as proteins, depend on the cellular or metabolic function.

I. HISTORY

Humans have used fermentation from the beginning of recorded history to provide products for everyday use. For many centuries, most microbial processing was to preserve or alter food products for human consumption. Fermentation of grains or fruit produced bread, beer, and wine that retained much of the nutrition of the raw materials, while keeping the product from spoiling. The natural yeasts that caused the fermentation added some vitamins and other nutrients to the bread or beverage. Lactic acid bacteria fermented milk to yogurt and cheeses, extending the life of milk products. Other food products were preserved or enhanced in flavor by fermentation, such as pickled vegetables and the fermentation of tea leaves and coffee beans.

Fermentation was an art until the second half of the 19th century. A batch was begun with either a "starter," a small portion of the previous batch, or with the cultures residing in the product or vessel. The idea that microbes were responsible for fermentations was not introduced until 1857, when Louis Pasteur published a paper describing the cause of failed industrial alcohol fermentations. He also quantitatively described microbial growth and metabolism for the first time and suggested heat treatment (pasteurization) to improve the storage quality of wines. This was the first step toward sterilization of a fermentation medium to control fermentation conditions. In 1883, Emil Christian Hansen began using pure yeast cultures for beer production in Denmark. Beer and wine were produced at relatively large scale starting in the eighteenth century to satisfy the demands of growing urban populations. In the mid-nineteenth century, the introduction of denaturation freed ethanol from the heavy beverage tax burden so that it could be used as an industrial solvent and fuel.

The first aseptic fermentation on a large scale was the acetone–butanol fermentation, which both Britain and Germany pursued in the years preceding World War I. The initial objective was to provide butanol as a precursor for butanediol, for use in synthetic rubber. After the beginning of the war, the focus of the process in Britain became acetone, which was used for munitions manufacture. Britain had previously been importing acetone from Germany. One of the people instrumental in the development of the fermentation was Chaim Weizmann, who later became the first prime minister of Israel. As the process was developed and scaled up, it was found that the producing culture, *Clostridium acetobutylicum*,

would become overwhelmed by competing bacteria introduced from the raw materials. Thus, the culture medium had to be sterilized and the process run under aseptic conditions. All penetrations on the reaction vessel were steam sealed to prevent contamination. Production eventually took place in Canada and the United States, due to the availability of cheap raw materials.

In the 1920s and 1930s, the emphasis in fermentation shifted to organic acids, primarily, lactic acid and citric acid. In the United States, where prohibition had outlawed alcoholic beverages, facilities and raw materials formerly used for alcoholic beverage production became available. Lactic acid is currently used as an acidulant in foods, a biodegradable solvent in the electronic industry (as ethyl lactate), and a precursor for biodegradable plastics. Citric acid is used in soft drinks, as an acidulant in foods, and as a replacement for phosphates in detergents. As part of an effort to find uses for agricultural products during the depression, the USDA Northern Regional Research Laboratory (NRRL) pioneered the use of surplus corn products, such as corn steep liquor, and the use of submerged fungal cultures in fermentation. Previously, fungi had been grown on solid media or in the surface of liquid media. This set the stage for the large-scale production of penicillin, which was discovered in 1929 in the Britain, developed in the 1930s, and commercialized in 1942 in the United States. It was the first "miracle" drug, routinely curing bacterial infections that had previously caused serious illness or death. The demand was very high during World War II and the years following. Penicillin was initially produced as a surface culture in one-quart milk bottles. The cost, availability, and handling of bottles severely limited the expansion of production. Scientists at the NRRL discovered a new production culture on a moldy cantaloupe and developed a submerged culture fermentation. This led to significant increases in productivity per unit volume and the ability to greatly increase the scale of production by using stirred tank bioreactors. The success of penicillin inspired pharmaceutical companies to launch massive efforts to discover and develop many other antibiotics in the 1940s and 1950s. Most of these fermentations were highly aerobic, requiring high aeration and agitation. As the scale of production increased, it was found that mass transfer became limiting. The field of biochemical engineering emerged as a distinct field at this time, to study mass transfer problems in fermentation and to design large-scale fermentors capable of high transfer rates.

In the 1960s, amino acid fermentations were developed in Japan. Initially, L-glutamic acid, as monosodium glutamate, was produced as a flavor enhancer, to supplant MSG extracted from natural sources. Using cultures derived from glutamic acid bacteria, production of other amino acids followed. Amino acids are used in foods as nutrients, sweeteners, and flavor enhancers, and in animal feeds to increase the efficiency of low protein feeds. Commercial production of enzymes for use in industrial processes began on a large scale in the 1970s as well. Microbial enzymes account for 80% of all enzymes in commercial use, including grain processing, sugar production, juice and wine clarification, detergents, and high fructose corn syrup. The discovery of the tools of genetic engineering expanded the possibilities for products made by fermentation. Insulin was the first genetically engineered fermentation commercialized, developed in 1977. Since then, many genetically engineered products have been produced on a large scale.

II. CULTURE SELECTION AND DEVELOPMENT

A. Selection

Microbial processes begin with the culture used for production. Once the desired product or microbial activity is defined, the selection process begins. A culture should produce, or have the metabolic potential to produce, the desired product. Other attributes are important, such as substrate specificity, growth factor requirements, growth characteristics (pH, temperature, aeration, shear sensitivity), fermentation by-products, effect of the organism on downstream processing, and environmental and health effects. Literature searches can narrow the range of cultures to be screened, saving valuable time in culture selection. Culture collections often have cultures with some or all of the desired characteristics, but it is sometimes necessary to screen cultures from the environment. Determining the type of environment in which organisms with the desired characteristics live and how to separate them from the other cultures present takes considerable care. A thorough understanding of the physiology of the desired microorganism is necessary to design a successful isolation and selection strategy. Successful isolation of a useful culture often requires a combination of several enrichment or selective methods. Also, a system of storing and cataloging potentially useful isolates is very important so that commercially viable cultures are not lost. After an initial screening, there are usually many potentially useful isolates and secondary screening is necessary to eliminate false positives and evaluate the potential of the remaining candidates. The secondary screening is usually semi-quantitative or quantitative. The list of candidates is narrowed, as much as possible, using mass screening methods, such as agar plates with selective growth inhibitors or metabolic indicators. Large-scale screening of individual isolates in shake flask culture is very time consuming and expensive. Once a few isolates have been selected, culture and process development usually begin in parallel to condense the timeline to production.

B. Development

It is very rare that an isolate from the environment produces the desired product cost-effectively. Often, cultures make only minute quantities of the product. Increasing the productivity of the initial isolates requires a program of genetic improvement. Classical mutation and selection methods are used with most cultures selected from the environment because little is known about their genetics and whether the cultures possess cloning vectors such as plasmids, transposons, or temperate bacteriophage. Typically, a mutagen, such as ultraviolet light, ionizing radiation, or a chemical mutagen, is applied and the culture is grown in the presence of a selective growth inhibitor or toxin. The survivors are isolated and tested. This is usually an iterative process; mutant strains are screened, remutagenized, and reselected several times, often using higher concentrations or different selective agents, until a culture with commercial potential is obtained. Even after a process is successfully brought to production, culture improvement is ongoing to improve profitability and maintain a competitive advantage.

In the past two decades, the use of genetic engineering has supplemented, and sometimes replaced, classical genetic techniques. Insertion of genes into plasmids has improved the ease with which genes from one culture are transferred to another and has allowed the production of human and other mammalian proteins, such as insulin and interferons, in microbial fermentations. Plasmids with high copy numbers and strong transcriptional promoters have dramatically improved production of many proteins and enzymes. Knowledge of enzyme structure and function has led to site-specific mutagenesis, increasing the efficiency of mutagenesis programs and reducing the deleterious effects of "shotgun" mutagenesis, used in classical genetics. The introduction of polymerase chain reaction (PCR) technology has led to the ability to genetically sample environments where isolation of cultures is extremely difficult or impossible, such as cold benthic environments barely above freezing, deep sea thermal vents where microorganisms live in temperatures well above 100°C, or acidic hot springs, such as those in Yellowstone National Park. DNA libraries from these environments can be screened for enzymes that perform under extreme conditions without having to culture the microorganisms.

III. PROCESS DEVELOPMENT AND SCALE-UP

A. Development

Process development usually overlaps culture development. The purpose of process development is the formulation of media, optimization of culture conditions, and determination of the biochemical engineering parameters used to design the full-scale bioreactors. The early stages of process development are usually performed in shake flasks, where the nutritional requirements of the culture are determined and potential media components are screened. Initial growth and production studies can be performed in shake flasks as well. The limitation of shake flask culture is that pH, oxygen content, and other environmental factors cannot be easily monitored and controlled and mass transfer studies are difficult. The next stage in process development, performed in laboratory scale fermentors, is determination of fermentation characteristics, such as pH optimum, oxygen uptake rate, growth and production rates, sensitivity to nutrients and by-products, broth viscosity, heat generation, and shear sensitivity. This information is used to determine what mode of fermentation will be used and to develop the fermentation parameters, as well as to determine the mass transfer characteristics of the fermentation used in the design of the bioreactor. In addition, the medium is developed and optimized at this stage.

A variety of biochemical engineering methods for scale-up of bioreactors have been applied over the years, including constant oxygen transfer rate, constant agitation power per volume, constant impeller tip speed, equal mixing times, or similar momentum factors or feedback control to try to maintain important environmental factors as constant as possible. Each has limitations in predicting the effect of scale-up on the process. It is more difficult to predict how a biological process will react at the commercial scale, based on laboratory and pilot plant studies, than a chemical process. This is due to the complexity of reactions and interactions that occur in a bioreactor. There are often unforseen consequences of changing the scale of operation due to the effect of heat and mass transfer on microorganisms. For example, mixing times in a laboratory or pilot scale fermentor are a few seconds, while in large-scale fermentors, mixing times can be two minutes or more. The average conditions are the same as in a smaller fermentor, but an individual microorganism encounters a variety of suboptimal conditions for a significant period of time. Also, as the fermentor size increases, the heat generation increases proportional to the volume, while the cooling capacity increases proportional to the surface area. Therefore, larger vessels require internal cooling coils to supplement the water jacket. This can aggravate mixing problems further. The properties of the fermentation broth (viscosity, osmotic pressure, substrate, and product and waste product concentrations) and gas/liquid interactions

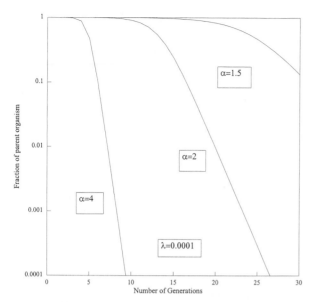

Fig. 1. Effect of growth rate differences between parent and variant strain on survival of parent strain. [Reprinted from Asenjo, J. A., and Merchuk, J. C., eds. (1995). ''Bioreactor System Design,'' Marcel Dekker, Inc., New York. Reprinted by courtesy of Marcel Dekker, Inc.]

(gas lineal velocity, surface tension, pressure gradients) are also scale dependent. Thus, fermentation scale-up is often highly empirical and based on the experiences and training of the scientists and engineers involved. The cost and risk of scale-up are higher for biological systems than for other chemical systems, due to the intermediate scale pilot plant steps required to successfully predict the outcome of full-scale operations. Some of this cost can be reduced by using existing facilities or rented facilities to test the process or by using seed fermentors (typically, 5–10% as large as production fermentors) for the final pilot plant scale.

A important microbiological factor that affects the scale-up biological systems is the increase in the number of generations required for full-scale operation. Nonproducing variant strains often arise from the parent population. Most commercially used microorganisms are mutated and selected for increased product yield and rates of production. This often decreases the growth rate and hardiness of the culture. Therefore, a variant that either reverts to a previous condition or that short-circuits the selected pathways by additional mutations will have a com-

petitive advantage. The percent of variants in the population in a given generation depends on the rate at which variants appear and the relative growth rate of the variant to the parent population, as expressed by the equation

$$\frac{X_n}{X_n + X_m} = \frac{\alpha + \lambda + 1}{(\alpha - 1) + \lambda^{N_g(\alpha + \lambda + 1)}},$$

where X_n is the number of parent culture cells, X_m is the number of variant culture cells, α is the ratio of the specific growth rate of the variant to the parent, λ is the rate of appearance of variants per genome per generation, and N_g is the number of generations. The effect of specific growth rate and the effect of appearance of variants on the parent strain population are shown graphically in Fig. 1 and Fig. 2, respectively. An unstable culture can cause serious disruptions to production in a large-scale fermentation plant. Variant strains are usually less sensitive to adverse conditions (extremes in temperature or pH, nutrient quality, anaerobic conditions) than parent strains, as well. A production strain that appears stable for the required number of generations under

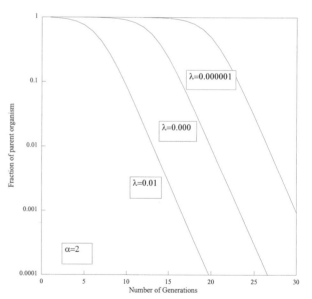

Fig. 2. Effect of variant formation rate on survival of parent strain. [Reprinted from Asenjo, J. A., and Merchuk, J. C., eds. (1995). ''Bioreactor System Design,'' Marcel Dekker, Inc., New York. Reprinted by courtesy of Marcel Dekker, Inc.]

laboratory conditions may exhibit instability when introduced to a full-scale plant. It is important to test the stability of potential production strains under conditions that are as similar as possible to the conditions expected in the production environment. It is also important to establish a system of culture storage and management that minimizes the number of generations required to reach full-scale production and maintains the seed stock under stable conditions. Genetically engineered cultures have additional stability problems, which can be expressed in a similar mathematical expression as variant formation. Often, the genes for producing the desired product are located on plasmids that also include selectible markers, usually antibiotic resistance. In the laboratory, these cultures are maintained in media that contain the selective agent. It is not practical to use antibiotics in a large-scale plant; therefore, production cultures must have highly stable plasmids. The rate of plasmid loss can be much higher than variant formation in production cultures. Even if they aren't completely lost, reduction of plasmid number can drastically reduce productivity per cell. Plasmid maintenance usually requires a compromise between the level expression of the cloned genes and culture growth rate.

C. Media Development and Optimization

The development of a suitable, economical medium is a balance between the nutritional requirements of the microorganism and the cost and availability of the medium components. The chemical constituents of the medium are determined by the composition of the cell mass and product, the stoichiometry of cell and product formation, and from yield coefficients, which can be estimated from shake flask or lab scale fermentor experiments. On a dry weight basis, 90–95% of microbial biomass consists of carbon, oxygen, hydrogen, nitrogen, sulfur, magnesium, and potassium. The remaining 5–10% is microelements (required in small amounts), primarily calcium, manganese, iron, copper, and zinc. The stoichiometry can be determined from the general equation:

$$CH_aO_b(\text{Carbohydrate}) + A(O_2) + B(NH_3)$$
$$\rightarrow Y_{x/s}(CH_aO_\beta N_\gamma)\,(\text{Biomass})$$
$$+ Y_{p/s}(CH_{\alpha 1}O_{\beta 1}N_{\gamma 1})(\text{product}) + Y_{CO_{2/s}}(CO_2) + D(H_2O)$$

The terms $Y_{x/s}$, $Y_{p/s}$, and $Y_{CO_{2/s}}$ are the yield coefficients for cell mass, product and CO_2, respectively. In most large-scale fermentations, the carbon usually comes from carbohydrates, such as sugars or starches. The nitrogen is from a variety of organic and inorganic nitrogen sources and from ammonia added to control the pH. Phosphate, sulfur, and magnesium are added as salts or in complex nutrients. The micronutrients are often derived from the water or other raw materials or added as mineral salts. Some commonly used fermentation substrates are shown in Table II.

Many microbes are auxotrophs for specific growth factors, such as vitamins or amino acids, and will be growth limited without adequate quantities of the compound in the medium. Other cultures may not have an absolute requirement for a growth factor,

TABLE II
Fermentation Raw Materials

Function	Raw material
Carbon	Glucose
	Sucrose
	Lactose
	Corn syrup
	Starch
	Ethanol
	Paraffins
	Vegetable oils
Carbon, vitamins, micronutrients	Beet molasses
	Cane molasses
Simple nitrogen	Ammonia
	Urea
Simple nitrogen, sulfur	Ammonium sulfate
Amino nitrogen	Cottonseed meal
	Casein
	Soy flour
Amino nitrogen, vitamins, micronutrients	Brewer's yeast, yeast extract
	Corn steep liquor
	Distiller's dried solubles
Carbon, amino nitrogen, vitamins, micronutrients	Whey

but grow slowly in the absence of the specific factor. In many cases, the exact requirements are not known, but complex substrates are required for optimal growth. It is often necessary to screen a variety of potential nutrients and combinations of nutrients to satisfy the nutritional requirements and minimize the raw materials costs. Another problem encountered with some components of culture medium is seasonal or hidden auxotrophy. This is variation in growth and production encountered due to either seasonal changes in processing or with new crops of vegetable sources of raw materials. Often, this is not seen in the laboratory, but at the large scale, due to poorer mixing in large-scale fermentors. Other considerations for selection of nutrients are regulatory approval and kosher certification for some foods.

IV. LARGE-SCALE OPERATION

A. Inoculum Production

The starting place for large-scale fermentations is in the inoculum laboratory. The fermentation is doomed to failure without pure, active inoculum. This starts with the storage of culture. It is important to store the culture under conditions that retain both genetic stability and viability. A variety of methods have been used, all with advantages and disadvantages. Storing cultures in agar stabs or on agar slants is one of the oldest methods. It is simple and cultures can be stored at room temperature or under refrigeration, but viability and stability are limited and cultures must be transferred frequently. Lyophilization (freeze drying) requires no refrigeration or freezing and cultures can be stored almost indefinitely at room temperature and can serve as a long-term method of safely storing culture in the event of catastrophe. Lyophilization takes more skill and equipment than slant preparation and often has significant viability loss. Cultures are not easily revived, making it unsuitable for daily use. Cultures stored in aqueous glycerol solutions at very low temperatures, $-80°C$ or lower, generally have high viability and stability, are easy to produce and revive, but require cryogenic freezers or liquid nitrogen, and are, thus, susceptible to power outages and equipment failure.

Once the cultures are safely stored, they must be revived, grown in the inoculum laboratory, and transferred to growth or "seed" fermentors that feed the production fermentors. Cultures are usually grown in shake flask culture in the laboratory and transferred to a suitable container for transfer to the plant. In the laboratory, culture transfers are exposed to the air, so precautions are taken to prevent incidental contamination, such as performing transfers in rooms or hoods with HEPA-filtered air and the wearing of sterile coveralls, gloves, and masks by technicians. Once out in the plant, all transfers are made through steam-sterilized piping or hoses using differential pressure.

B. Types of Bioreactors

There are a wide variety of of bioreactor designs (Fig. 3 and Fig. 4). Selection of a reactor design for a particular process depends on a variety of factors, including mass transfer considerations, mixing, shear sensitivity, broth viscosity, oxygen demand, reliability of operation, sterilization considerations, and the cost of construction and operation. Due to the complex nature of fermentation scale-up, a few basic reactor designs are used for most applications. The two most common are the stirred tank and the air lift reactors.

Stirred tank reactors use sparged air and submerged impellers to aerate and mix the broth. They are versatile and are especially adapted to highly aerobic cultures and highly viscous fermentations. The drawbacks are high energy input and the use of rotating seals on the agitator shaft, which are a contamination risk. Even within this category, there are many variations in design, such as the style, number, and placement of impellers, the height to diameter ratio, the number and placement of coils or baffles, that affect the mixing characteristics of the vessel. Due to high risk of scale-up and the high capital cost of building large-scale fermentation facilities, most plants install sirred-tank reactors.

Airlift fermentors mix the broth with air from the sparger. Some designs have an internal draft tube to direct the flow of fluid. Most airlift designs have a much greater height-to-diameter ratio than stirred tank vessels to improve oxygen transfer. The mixing

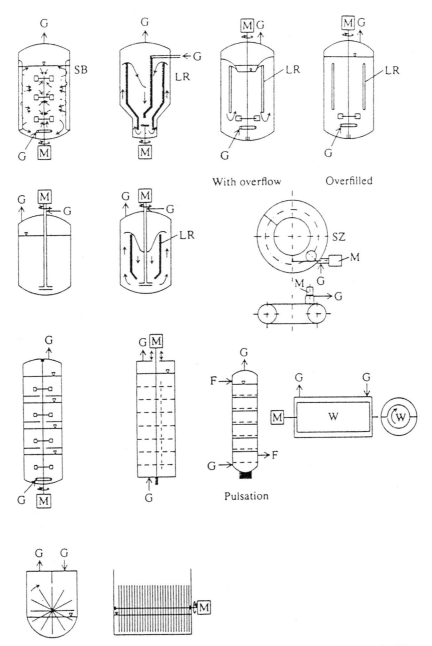

Fig. 3. Bioreactors with mechanical stirring. M, motor; g, gas (air); SB, baffles; LR, conduit tube; W, roller; F, liquid. [Reprinted from Schugerl, K. (1982). New bioreactors for aerobic processes, *Int. Chem. Eng.* **22**, 591.]

is not as good as in a stirred tank but the energy input and shear forces are much lower, thus useful for shear sensitive cultures or in processes where the energy cost of agitation is a significant factor. Also at very large scale, the heat added to a stirred-tank vessel by the agitator becomes increasingly difficult

to remove, adding to the cooling costs. Most large-scale airlift fermentors are used for plant effluent treatment, production of baker's yeast, or for fungal fermentations where the size of the mycelial pellets is controlled by shear forces.

In addition to the mass transfer characteristics of

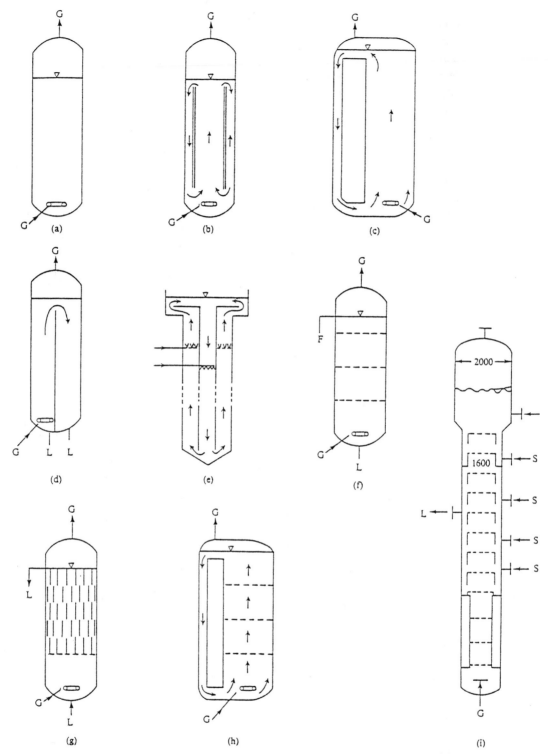

Fig. 4. Bioreactors with pneumatically stirring. ID, injection nozzle; SK, float; HFP, helical flow promoter. See Fig. 3 for other abbreviations. [Reprinted from Schugerl, K. (1982). New bioreactors for aerobic processes, *Int. Chem. Eng.* **22**, 591.]

a fermentor, other factors are important in the operation of a bioreactor, such as the ability to clean the vessel, sterile integrity of the vessel, and maintenance costs. Some reactor designs that have excellent characteristics in the pilot plant are not practical choices for large-scale operation due to mechanical complexity that causes sterility and maintenance problems on scale-up.

C. Modes of Operation

There are three basic modes of operation: batch, fed batch, and continuous, with variations of these three basic modes (Fig. 5). In batch mode, all the

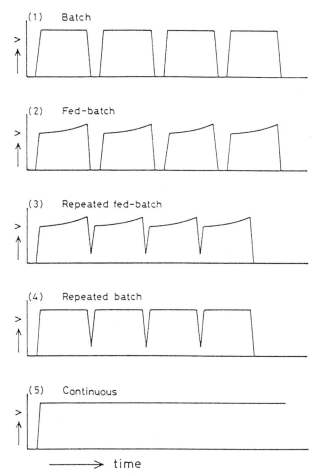

Fig. 5. Changes in volume of culture broth with elapsed time for various modes of operation in bioreactors. [Reprinted from Asenjo, J. A., and Merchuk, J. C., eds. (1995). "Bioreactor System Design," Marcel Dekker, Inc., New York. Reprinted by courtesy of Marcel Dekker, Inc.]

ingredients required for fermentation except pH control chemicals, usually ammonia, are added to the fermentor prior to inoculation. The fermentation is run until the nutrients are exhausted, then the broth is harvested. The advantage is simplicity of operation and reduced risk of contamination. It is useful in fermentations with high yield-per-unit substrate and with cultures that can tolerate high initial substrate concentrations.

Fed batch mode starts with some of the nutrients in the fermentor before inoculation. Concentrated nutrients are added as the fermentation progresses. The advantages are the ability to add large quantities of nutrients to the fermentor by adding them gradually and the ability to control the rate of nutrient addition. This allows for high product concentrations without subjecting the culture to inhibition by high levels of nutrients. It also allows for control of culture growth rate, which is required in some fermentations to maximize productivity and yield. Overall, the use of the fermentor time is better in fed batch than straight batch fermentation, reducing fixed costs. The disadvantages are increased risk of contamination, due to the addition of nutrients through a continuous sterilizer, and increased equipment costs for continuous sterilization and flow control equipment for feed streams.

Both batch and fed batch can be run in repeated mode, with a small portion of the previous batch left in the fermentor for inoculum. The medium is then added through a continuous sterilizer. Use of the fermentor is increased by eliminating turnaround time, but the risks of contamination and genetic degradation of the culture are increased. In any case, repeated batch mode cannot be repeated indefinitely, due to maintenance and cleaning needs. Usually, repeated fermentations are run for two or three batches.

In continuous mode, the starting medium and inoculum are added to the fermentor. After the culture is grown, the fermentor is fed nutrients and broth is withdrawn at the same rate, maintaining a constant volume of broth in the fermentor. In continuous mode with cell recycle, the cell mass is returned to the fermentor using microfiltration with bacteria or screens with fungal mycelia. Continuous mode maximizes the use of the vessel and is especially good for fermentations that take a long time to reach high productivity. The disadvantages are increased risk of contamination, especially since it is difficult to keep contamination from growing through the continuous harvest line. As with repeated batch mode, continuous fermentations cannot be run indefinitely, but fermentations of several hundred hours' duration are possible.

D. Monitoring and Control

Ideally, a fermentation is best controlled using online, real-time measurements of pH, oxygen, cell mass, substrate and product concentrations, metabolic state, and nutrient flow rates and applying the data to a precise model of how the culture will respond to changes in any of the parameters. Only a few of the parameters are actually measurable in real time and the models of culture behavior are often imprecise extrapolations of experimental data, although technological advances are moving in the direction of better monitoring and control.

Most physical parameters—temperature, pressure, power input, impeller speed, gas and liquid flow—can be measured accurately without invasive instrumentation. Reliable online measurement of the chemical environment has been limited to pH, dissolved oxygen, and off gas analysis. Advances in infrared analysis have recently made it possible to measure several parameters of broth composition, such as substrate, product, and ammonia concentration, simultaneously on-line. Various optical density probes, to measure cell mass, and enzyme probes, to measure nutrient or product concentration, are available, but are often subject to fouling or cannot be sterilized. Measurement of physiological characteristics, such as intracellular ATP, DNA, and RNA content, are not currently feasible and must be inferred from the physical and chemical data.

At a basic level, most fermentations are run under temperature and pH control. In other fermentations, the dissolved oxygen is controlled, usually by changing the agitator speed. In a batch fermentation, not much more is usually needed. In fed batch and continuous fermentations, it is often necessary to control the nutrient concentration at a limiting concentration in order to control the growth rate of the culture,

maximize product yield, or avoid anaerobic conditions.

Substrate feed can be controlled using feedback control from pH, dissolved oxygen, or off gas analysis, either directly or by calculating physiological parameters, such as substrate utilization rate or oxygen uptake rate. In a pH-controlled feed system, the consumption of carbohydrate drives the pH down, and base (usually ammonia) is added to maintain a constant pH. When the carbohydrate becomes limiting, the pH is driven up instead. When this occurs, the feed is increased, driving the pH downward again. Sophisticated control schemes can be devised by measuring the rate of base consumption and the rate of pH increase when substrate is limiting. Without proper control loops, a pH controlled feed can bog down or overfeed, as the measurement can only tell whether there is not enough feed, not when there is too much. A dissolved oxygen (DO) controlled fermentation is maintained between two setpoints, usually between 20 and 40% dissolved oxygen, by increasing the feed when the DO rises and decreasing the feed when it drops. Without some damping of the response in DO-controlled fermentations, there is a tendency for large amplitude swings of DO and feed rate, due to delayed response and long mixing times.

E. Sterilization and Contamination Control

Sterilization is the process of eliminating all viable organisms from equipment and materials. For a fermentation plant, this requires:

- Presterilization of equipment (fermentor vessels, inoculation and feed piping, and air filters);
- Sterilization of feedstocks, either in the fermentor vessel or through a continuous sterilizer;
- Maintenance of aseptic conditions throughout the entire fermentation process.

Most fermentation processes require aseptic conditions for optimal productivity and yield. Some fermentations are relatively insensitive to contamination: yeast and fungal fermentations performed at low pH, fermentations with short cycle times (high growth rates), or fermentations that produce inhibitory or toxic products (alcohols, antibiotics, organic acids) tend to be resistant to contamination, but rampant, out-of-control contamination can destroy productivity altogether. Other processes are highly sensitive to contamination: repeated batch or continuous fermentations and slow-growing bacterial fermentations are more susceptible to contamination and, therefore, require more stringent conditions to maintain aseptic conditions. In addition, some contaminants can be pathogenic or produce toxins that can render products unusable or be a health risk to plant workers or consumers. Therefore, complete asepsis is the operating philosophy of most large-scale plants. The philosophy of aseptic operation must be an integral part of every aspect of the operation, from design and construction to operating procedures and personnel training, for a plant to operate aseptically.

Most process equipment and medium components are sterilized by steam under pressure. The rate of cell death depends on the time and temperature of exposure to steam. The spores of the thermophile *Bacillus stearothermophilius* are the usual benchmark for determining sterilization conditions. Standard conditions for steam sterilization are 121°C for 15 minutes, but the time required for sterilization is less at higher temperatures. Most fermentation broths and continuous feeds are sterilized by steam sterilization either *in situ* or in a continuous sterilization system, unless the medium contains heat-labile constituents.

Sterilization of the fermentor broth *in situ* requires a significant amount of time in a large vessel for heating and cooling the broth, which reduces the productive time of the fermentor. Large amounts of cooling water are required as well, and the long time required for sterilization can cause degradation of some medium components. *In situ* sterilization, however, has the advantage of being simple, thus reducing the risk of upset in the process resulting in contamination.

Continuous sterilization is sometimes used for fermentor broths but more often for nutrients added to the fermentor during the fermentation, especially for fed-batch or continuous fermentations. This is usually accomplished by either direct steam injection into the nutrient stream or the transfer of heat through a heat exchanger with no direct steam contact. After heating, a holding loop maintains the temperature for a specified time, followed by a second heat exchanger to reduce the heat of the medium before it is added to the fermentor. Continuous sterilization can be carried out at increased temperature and reduced time, to minimize the heat damage to the medium and reduce the length of holding tube required or increase the flow of fluid through the system. As a practical matter, heat from fluid returning from the holding loop is used, through a heat exchanger, to preheat broth going to the sterilizer, thus reducing the requirements for both steam to heat the incoming broth and cooling water to cool the sterilized broth. Duplicate systems are required for systems to be cleaned and maintained properly.

Filter sterilization is used for liquids with heat-sensitive components or nonaqueous liquids with low boiling points, which are added to the fermentor after sterilization and cooldown. Air and other gases are also sterilized by filtration. Two basic types of filters have been used. Depth filters consist of layers of glass wool and were mostly used to sterilize air, but can be difficult to sterilize and dry. Depth filters have been mostly replaced by absolute membrane filters, which are thin membranes with pores no larger than their rated size. Most bacteria and spores are retained by filters with 0.2 μm absolute pore size. Hydrophobic filters are used to sterilize air, usually preceded by prefilters to remove solids and water from the compressed air. Bacteriophage, which are retained on dry filters by electrostatic forces, are small enough to pass through if the air is wet, making it very important to keep the process air as dry as possible. While microbial contamination usually just reduces productivity and yield, bacteriophage can destory the production culture within a few hours of introduction into the fermentor. Fermentation plant managers live in fear of bacteriophage.

During the fermentation, the contents of the reactor must be isolated from potential contaminating sources. Besides filtering the air, all penetrations into the fermentor must be sealed from the outside. Agitators are sealed with double mechanical seals that have steam between the rotating seals. Dissolved oxygen and pH electrodes are sealed with O-rings or gaskets. Piping that penetrates the vessel, such as media fill lines, harvest lines, and inoculation lines, are isolated with steam seals, which are sections of piping between two valves that connect the fermentor to the outside "septic" environment. There is a steam source and a relief valve to vent steam condensate and maintain steam pressure above 15 psig when the line is not in use. Although valves are routinely tested for leaks, there are no valves with fine enough tolerances to prevent contamination from growing across the contact surfaces.

Of all the problems encountered in a fermentation plant, contamination is the most persistent. Contamination control is a daily, ongoing activity. The culture is monitored for contamination by microscopic examination and plating on nutrient agar, from the first shake flask through the production fermentor. This helps to prevent contaminated culture from being used in production and to trace the source of contamination whenever possible. In addition to testing the culture for contamination, other potential sources of contamination, such as sterile feed streams into the fermentor, are monitored. Unfortunately, the size of sample that is practical to test for contamination is very minute in comparison to the size of the process being tested. Contamination is usually not detected until it is a serious problem. Therefore, preventive maintenance is very important to control contamination. Fermentors, piping, valves, and filters are routinely tested for integrity. Leaky valves and cracks in cooling jackets or coils are common sites for microbial contamination to enter the process, and cooling water leaks are a common source of bacteriophage contamination. Proper procedures for sterilization, culture transfer, and fermentor operation are critical to maintaining an aseptic process. This involves careful evaluation of the effect of procedures on process integrity, writing procedures clearly, training fermentor operators properly, and monitoring the effectiveness of training.

F. Utilities

Although fermentation processes usually are performed at near ambient temperatures, the energy consumption is high due to the need for sterilization, cooling, agitation, and aeration.

1. Steam

Steam is required for sterilization of the medium, either in the fermentor or through continuous sterilizers. During the fermentation itself, a certain amount of steam is required to maintain the microbiological integrity of the fermentor vessel by the use of steam seals. Fermentation plants often require large quantities of steam for evaporation and drying in downstream processing, due to the low concentration of product in the broth.

2. Cooling Water

After sterilization, fermentors need to be cooled before they can be inoculated; during the fermentation, cultures generate great quantities of metabolic heat, and agitators add some heat to the broth. This heat must be removed to maintain the proper incubation temperature. For downstream processing, cooling water is needed for certain types of crystallizers and for condensers on evaporators. Although cooling water is recycled, there is high energy input for cooling towers and chillers and evaporative water loss in cooling towers.

3. Process Water

Much of the fermentor broth is water. The quality of the water used depends on the sensitivity of the culture to minerals in the water, effect of water quality on downstream processing, and regulatory requirements. In some cases, water must be purified by reverse osmosis for use in fermentation. Other fermentations can use water recycled from process condensers. Water is also used for some downstream process steps and for cleaning and rinsing the fermentors.

4. Electricity

Agitation, compressed air, and water chillers require a great amount of electricity. Where economies of scale are adequate, cogeneration of steam and electricity are very cost effective. Boilers using natural gas, oil, or coal produce steam used to feed generators for electricity. The "waste" steam produced in the generation process, which is at high enough pressure for use in fermentation, is captured and used as the process steam for the plant.

5. Sewage Treatment

Fermentation generates waste with high BOD, mainly in the form of spent cell mass. While much of it can be used for animal feeds, some low solids streams are generated and must be treated as sewage. A lot of waste water is also generated, making the throughput for a treatment plant high. In many cases, on-site primary and secondary treatment is required to avoid paying high municipal sewage charges and fines.

G. Downstream Processing

Since fermentation products are often produced in dilute aqueous solution that also includes the cell mass, metabolic by-products, and various salts, downstream processing is a major part of producing a fermentation product. The degree of purification of the product depends on the type of product and the end use. Industrial enzymes, for example, are often separated from the cell mass and concentrated by ultrafiltration to make a crude extract. On the other extreme, antibiotics for human use undergo many steps of purification. Common methods include microfiltration and ultrafiltration to remove cell mass and other debris or retain larger molecular weight proteins; concentration by evaporation or reverse osmosis; and crystallization by chilling or evaporation, pH or solvent precipitation, centrifugation, and chromatography.

See Also the Following Articles

AMINO ACID PRODUCTION • BIOREACTORS • FERMENTATION • MUTAGENESIS

Bibliography

Asenjo, J. A., and Merchuk, J. C., eds. (1995). "Bioreactor System Design." Marcel Dekker Inc., New York.

Atkinson, B., and Mavituna, F. (1991). "Biochemical Engineering and Biotechnology Handbook." Stockton Press, New York.

Bud, R. (1993). "The Uses of Life: A History of Biotechnology." Cambridge Univ. Press, Cambridge.

Humphrey, A. (1998). Shake flask to fermentor: What have we learned." *Biotechnology Progress* **14**:1, 3–7.

Pons, M.-N. ed. (1991). "Bioprocess Monitoring and Control." Hanseer Publishers, Munich, Germany.

Wang, D. I. C., Cooney, C. L., Demain, A. L., Dunhill, P., Humphrey, A. E., and Lilly, M. D. (1979). "Fermentation and Enzyme Technology." John Wiley & Sons, New York.

Infectious Waste Management

Gerald A. Denys

Clarian Health Methodist Hospital

GLOSSARY

disposal Dilution and dispersal into air or into sewers or containment in a landfill.

guidelines Recommendations issued by government agencies or professional organizations.

hazardous waste Material or substance that poses a significant threat to human health or environment, requiring special handling, processing, or disposal; hazardous waste often refers to chemical, radioactive, or mixed (multihazardous) wastes.

infectious waste Material capable of producing an infectious disease; a subset of medical waste.

medical waste Material generated as a result of patient care, diagnosis, treatment, or immunization.

multihazardous waste Waste that is infectious, chemically hazardous, and/or radioactive.

regulations Requirements developed by government at the federal, state, and local levels that are mandatory and enforceable by law.

sharps Hypodermic needles, syringes, disposable pipettes, capillary tubes, microscope slides, coverslips, and broken glass.

standard Performance of activities or quality of products established by professional organizations; standards have no force of law.

standard (universal) precautions All blood and body fluids are considered potentially infectious, and appropriate procedures and barrier precautions (to skin and mucus membranes) should be used to prevent personnel exposure.

treatment A process that reduces or eliminates the hazardous properties or reduces the amount of waste.

SINCE 1988, the subject of infectious medical waste has gained considerable public and governmental attention. This has, in part, been due to the public alarm over identifiable medical waste washed up on our beaches and fears created by the Acquired Immune Deficiency Syndrome (AIDS) epidemic.

Today's health care industry is faced with new laws and regulations to control infectious medical waste. As a result, infectious waste management programs have expanded, along with specialized commercial services for disposal of infectious waste. The cost of these management programs and waste treatment and disposal systems has increased considerably. For effective management of infectious waste, a comprehensive waste management plan is essential. Such a plan includes all aspects of waste management: definitions of waste types, source separation, containment, handling, storage, transport, treatment, disposal, quality control procedures, training, and waste minimization.

I. IDENTIFICATION OF INFECTIOUS WASTE

A. Definitions

Medical waste refers to waste generated by health care institutions as the result of the care, diagnosis, or treatment of a patient, as well as during medical and pharmaceutical research. Infectious waste is usually considered a subset of medical waste and also includes waste from veterinary clinics and industrial laboratories; however, there is no consensus on an exact definition. The definition of infectious waste and items included in these definitions varies according to federal, state, and local authorities. A number of terms for infectious waste have accompanied the development of infectious waste regulations with no consistent terminology. These terms include infective, biomedical, biological, microbiological, pathological, biohazardous, medical, hospital, and red bag waste. The Environmental Protection Agency (EPA) defines infectious waste as "waste capable of producing an infectious disease." The Centers for Disease Control and Prevention (CDC) definition includes "microbiological waste (e.g., cultures, stocks), blood and blood products, pathological waste, and sharps." The landfill operator often regulates waste by his or her own criteria. If red bags, or items resembling medical waste, are found in a load of trash and appears to be infectious, even if previously treated, he or she will not accept it.

Identifying waste as infectious is difficult because there is no epidemiological evidence that medical waste poses a risk to public health. Decisions as to what is infectious must be made by determining the potential risk for infection or injury to individuals who segregate, handle, store, treat, or dispose of medical waste. Because infectious waste requires special treatment, definitions by regulatory agencies can have serious economic implications. Without a national standard, a range of 3 to 90% of hospital waste may be interpreted as infectious.

B. Sources

Regulated medical waste represents only about 1% of all municipal solid waste created in the United States each year. Within a medical facility, a variety of wastes are generated, consisting of infectious, chemical, radioactive, and multihazardous waste. Other wastestreams include waste water, food wastes, and general trash. Infectious waste is only a small segment of this total. The primary generation points of infectious waste in hospitals are the surgical and autopsy suites, isolation wards, laboratories, and dialysis units. The quantity of waste produced per patient is dependent on how infectious waste is classified. A decrease trend in length of patient hospital stay has resulted in the growth of home health care. Medical waste generated at home may be regulated

by some states. Other generators of infectious waste are varied and are included in Table I. The volume of infectious waste will most likely increase as an aging population requires more surgical intervention and long-term health care.

C. Types

Although federal agencies, such as the CDC and EPA, have published general guidelines for categories of infectious waste, some state and local regulators may be more explicit. The CDC designates infectious waste relevant to patients with AIDS and preventing the transmission of Human Immunodeficiency Virus (HIV) and Hepatitis B Virus (HBV) to health care workers. The EPA also considers waste types relevant to highly communicable diseases. Isolation waste from patients in isolation precaution are classified as "regulated medical waste" under the EPA medical tracking regulations (Table II).

All cultures and stocks of infectious agents from medical, research, and industrial laboratories should be managed as infectious waste. These waste types contain large numbers of microorganisms in high concentrations and present a great risk to exposure if not treated. Discarded biologicals and waste from vaccine production by pharmaceutical companies for human and veterinary use should also be managed as infectious waste because pathogens may be present. Pathological waste includes tissues, organs, body parts, and body fluids removed during surgery or

---

TABLE I
Generators of Infectious Waste

Healthcare and related facilities[a]
Home care settings
Academic and industrial research laboratories
Pharmaceutical industry
Veterinary hospitals and offices
Funeral homes
Food, drug, and cosmetic industry

[a] Includes hospitals, outpatient clinics, ambulatory surgery centers, medical and diagnostic labs, blood centers, dialysis centers, nursing homes and hospices, physician and dental offices, diet or health care clinics, emergency medical health providers, and home health agencies.

TABLE II

Categories of Waste Designated as Infectious by Federal Agencies

Category	CDC[a]	OSHA[b]	EPA[c]
Microbiological[d]	Yes	Yes	Yes
Blood and blood products[e]	Yes	Yes	Yes
Pathological[f]	Yes	Yes	Yes
Sharps, contaminated[g]	Yes	Yes	Yes
Sharps, uncontaminated	No	No	No
Isolation waste[h]	No	No	Yes
Contaminated animal carcasses[i]	No	Maybe[j]	Yes
Surgery and autopsy waste[k]	No	Maybe[l]	Maybe[l]
Contaminated lab waste[m]	No	Maybe[l]	Maybe[l]
Hemodialysis waste[n]	No	Yes	Maybe[l]
Contaminated equipment	No	Maybe[l]	No

[a] Centers for Disease Control and Prevention's guidelines for prevention and transmission of human immunodeficiency virus (HIV) and hepatitis B virus (HBV).

[b] U.S. Department of Labor, Occupational Safety and Health Administration's exposure to blood-borne pathogens rule.

[c] U.S. Environmental Protection Agency's standards for the tracking and management of waste.

[d] Includes cultures and stocks of infectious agents.

[e] Includes all human blood, serum, plasma, blood products, and items saturated or soaked with blood.

[f] Includes tissues, organs, body parts, and body fluids.

[g] Includes contaminated and unused needles, syringes, scalpel blades, disposable pipettes, capillary tubes, microscope slides, coverslips, and broken glass.

[h] Includes waste from patients with contagious, highly communicable diseases (e.g., Lassa, Marburg, and Ebola viruses).

[i] Includes body parts and bedding.

[j] If waste is from a facility that works with HIV or HBV.

[k] Includes soiled dressing, sponges, drapes, and gloves.

[l] If waste contains liquid, semiliquid, or dried blood or other potentially infectious material.

[m] Includes specimen containers, gloves, and lab coats.

[n] Includes tubing, filters, sheets, and gloves.

autopsy. Special handling of this category of waste is indicated because of the infectious potential and for aesthetic reasons. The Occupational Safety and Health Administration (OSHA) addresses infectious waste as an aspect of occupational exposure to blood-borne pathogens. As recommended by CDC's Standard (Universal) Precautions, human blood and blood products should be managed as infectious. Waste in this category includes all human blood, serum, plasma, and blood products (Table II). A major concern is the risk of acquiring HIV and HBV. Contaminated and unused sharps have been known to be an occupational hazard to all handlers of medical waste and require special containers to protect against injury and disease. Sharps include hypodermic needles, syringes, scalpel blades, disposable pipettes, capillary tubes, microscope slides, coverslips, and broken glass. Animal carcasses, body parts, and bedding exposed to an infectious agent and used for medical research are capable of transmitting an infectious disease. These waste types are similar to pathological and blood categories. Other categories of waste which are not designated as infectious but may be considered potentially infectious include wastes from surgery and autopsy that have come in contact with pathological waste (e.g., soiled dressings, sponges, drapes, and gloves). Contaminated laboratory waste (e.g., specimen containers, gloves, and lab coats), hemodialysis waste (e.g., tubing, filters, sheets, and gloves), and contaminated equipment and parts which may have come in contact with an infectious agent also are included in this category. All these waste categories should be managed as infectious if grossly contaminated with blood or body fluids.

D. Source Separation

One problem many medical facilities experience is the high volume of noninfectious waste, such as flowers, newspapers, paper towels, and cups, which end up in the infectious waste red bags. As much as 50 to 60% of red bag waste is noninfectious material, which could be handled as general trash. A clear definition of infectious waste needs to be established, to assure that the waste is separated at the point of generation. Reducing the amount of waste requiring treatment can also reduce waste-handling costs. Infectious waste should also be segregated from other wastestreams until final treatment and disposal. Waste containers should be used and labeled "biohazard," as required by OSHA regulation. Source segregation and waste segregation, however, requires educating employees, providing proper containers and labels, and frequent monitoring.

II. PUBLIC HEALTH RISKS

The probability of developing an infection from medical waste is extremely low. Four conditions must be met for infection to occur. First, a sufficient number of viable pathogens must be present in the waste. Second, the pathogen must exhibit a virulence factor. Infection does not necessarily result in disease. Third, the pathogen must gain entry into the host. Transmission of microorganisms may occur by four routes: breaks in the skin by cuts, scrapes, or puncture wounds; contact with mucus membranes by splashing of eyes, nose, or mouth; inhalation; and ingestion. The last requirement for infection is a susceptible host for the pathogen to infect. If any of these conditions are not met, disease transmission cannot occur and the risk of infection is eliminated. The only medical waste that has been associated with the transmission of an infectious disease are contaminated sharps. The risk of infection occurs when sharps are mishandled and are not properly discarded or contained.

Medical waste-related HIV and HBV infections are a public health concern for selected health care associated occupations, such as janitorial and laundry workers, nurses, emergency medical personnel, and medical waste handlers. There is virtually no infectious hazard of medical waste to the public or the environment. The general public may come into contact with medical waste from home health care and intravenous drug users' related waste. Studies have shown that general household waste contains more potentially pathogenic microorganisms than does medical waste. There is also no epidemiologic evidence that medical waste treated by chemical, physical, or biologic means have caused disease in the community. In addition, untreated medical waste can be disposed of in properly operated sanitary landfills, if workers follow proper procedures to prevent contact with waste during handling and disposal operations. The public impacts of medical waste disposal are confined to environmental issues, such as improperly contained disposal or poorly operated incinerators or other medical waste treatment technologies.

III. TREATMENT AND DISPOSAL OPTIONS

There are many options for the treatment and disposal of infectious medical waste. An overview of available methods for on-site treatment is presented in Table III. A brief description of many of these treatment methods is provided to help in the understanding of the treatment processes.

A. Traditional Methods

Incineration is very effective in treating most medical waste. In this process, combustible material is transformed into noncombustible ash. Of the incinerators currently available, the controlled-air type is most widely used. The principle of a controlled-air incinerator involves two sequential combustion processes which occur in two separate chambers (Fig. 1). Waste is fed into a primary chamber, where the combustion process begins in an oxygen-starved atmosphere. The combustion temperature (1600–1800°F) is regulated by air to volatilize and oxidize the fixed carbon in the waste. The combustion process is completed in the upper chamber where excess air is introduced and the volatile gas–air mixture is burned at a higher temperature (1800+°F). For efficient decontamination and burning of combustible material, the proper temperature must be maintained for an appropriate period of time and mixed with a sufficient amount of oxygen. Although incineration is most appropriate for treatment of chemotherapy drug waste and disposal of human and animal tissue, it has its drawbacks. Extreme variations in waste content or an improperly operated incinerator can result in the release of viable microorganisms to the environment via stack emissions, ash residue, or wastewater. Toxic metals can concentrate in the ash and some units cannot handle large quantities of glass or bulk fluids. Regulatory requirements for air emissions and disposal of liquid and solid residue require expensive pollution control measures. Community awareness and more stringent permitting and licensing requirements, especially in populous areas, have shut down or limited operations.

Approximately one-third of hospitals in the United States treat their microbiological waste using a steam sterilizer or autoclave. An autoclave is an insulated pressure chamber in which saturated steam is used to obtain elevated temperatures. Air is removed from the chamber by either gravity displacement or using a prevacuum cycle. A gravity displacement autoclave is most commonly used for medical waste (Fig. 2).

TABLE III
Comparison of On-Site Medical Waste Treatment Technologies[a]

| | Traditional methods | | Alternative treatment methods | | |
| | Incineration | Steam/Autoclave | Steam/ Mechanical[b] | Steam/ Compaction[c] | Steam/Heat/ Alkali[d] |
Factor					
Type of waste[e]	RMW/P/C	RMW	RMW/P	RMW	RMW/P/C
Equipment operation	Complex	Easy	Easy/automated	Easy/automated	Easy/automated
Operator requirement	Trained and certified	Trained	Trained	Trained	Trained
Load standardization	Needed[f]	Needed[f]	Needed[f]	Needed[f]	Needed
Capacity[g]	1200 lb/h	200 lb/h	200–1370 lb/h	25–2000 lb/h	10–10,000 lb/h
Effect of waste treatment	Burned	Unchanged	Shredded/ ground	Compacted	Digested
Volume reduction	85–95%	30%	85–90%	60–80%	98%
Potential side benefits	Energy recovery	None	Yes[h]	Yes[h]	Yes[h]
Disposal of residue					
Liquids	Treated[i]	Sanitary sewer[j]	Sanitary sewer[j]	Sanitary sewer[j]	Sanitary sewer[j]
Solids	Hazardous ash[k]	Sanitary landfill[l]	Sanitary landfill	Sanitary landfill	Sanitary landfill

Alternative Treatment Methods

Factor	Chlorine dioxide/ Mechanical[m]	Peracetic acid/ Mechanical[n]	Electrothermal/ Mechanical[o]	Microwave/ Mechanical[p]	Pyrolysis/ Oxidation[q]
Type of waste[e]	RMW/P	RMW	RMW	RMW[t]	RMW/P/C
Equipment operation	Easy	Easy	Complex	Easy	Easy/automated
Operator requirement	Trained	Trained	Trained	Trained	Trained
Load standardization	Needed[f]	Needed[f]	Needed[f,s]	Needed[f]	Needed
Capacity[g]	600 lb/h	20 lb/h	n/a	550–990 lb/h	100–2500 lb/h
Effect of waste treatment	Shredded	Pulverized	Shredded	Shredded	Vaporized
Volume reduction	85–95%	Up to 85%	Up to 85%	Up to 85%	99%
Potential side benefit	Yes[h]	Yes[h]	Yes[h]	Yes[h]	Energy recovery[h]
Disposal of residue					
Liquids	Sanitary sewer[j,s]	Sanitary sewer[j]	None[t]	None[t]	Sanitary sewer[j]
Solids	Sanitary landfill	Sanitary landfill	Sanitary landfill	Sanitary landfill	Sanitary landfill

[a] Information provided by manufacturers.
[b] Rotoclave by Tempico, Inc., Madisonville, LA.
[c] San-I-Pak, High Vacuum Autoclaves, Tracy, CA.
[d] Waste Reduction by Waste Reduction, Inc., Indianapolis, IN.
[e] Infectious waste types; RMW, regulated medical waste; P, pathological waste; C, chemotherapeutic waste. Waste segregation may be needed to eliminate pathological, chemotherapeutic, and/or nontreatable waste from the wastestream.
[f] To eliminate bulk fluids, such as dialysis fluid, or to ensure that sufficient solid waste is present in the load together with the bulk fluid.
[g] Includes a range of models.
[h] No harmful by-products; lower volume to commercial wastestreams.
[i] If there is scrubber water present, it must be treated at POTW or on-site.
[j] Low volume or intermittent drain to sanitary sewer.
[k] RCRA-permitted landfill.
[l] Potential problem with recognizable red bag waste.
[m] World Environmental Services Co., Ramona, CA.
[n] EcoCycle 10 by STERIS Corporation, Mentor, OH.
[o] Stericycle, Inc., Deerfield, IL.
[p] Sanitec, Inc., West Caldwell, NJ.
[q] Bio-Oxidation Services, Inc., Harrisburg, PA.
[r] Point of generation treatment.
[s] Potential formation of carcinogenic compounds in chlorine-based systems.
[t] Moisture retained in solids or held within the unit.
[Adapted from Marsik, F. J., and Denys, G. A. (1994). Sterilization, decontamination, and disinfection procedures for the microbiology laboratory, *Manual of Clinical Microbiology* (6th ed.). (T. R. Murray, E. J. Baron, M. A. Pfaller, F. C. Tenover, and R. H. Yolken, eds.). American Society for Microbiology, Washington, DC.]

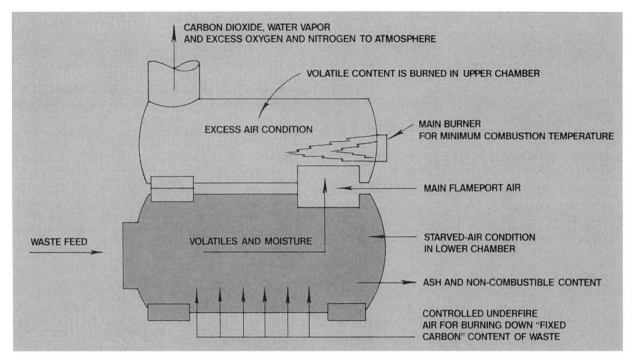

Fig. 1. Controlled air incinerator involves two sequential combustion processes taking place in a dual chamber. Combustion process begins in an oxygen-starved atmosphere (lower chamber) and is completed in an air/excess-air condition at a higher temperature (upper chamber). (Courtesy of Joy Energy Systems, Inc., Charlotte, NC.)

Lighter steam is fed into the chamber to displace heavier air. Commercial autoclaves usually operate at a temperature of 132°C (270°F), pressurized to about 60 to 75 psi, and at an operational cycle time of 1 hour. For steam sterilization to be effective, time, temperature, and direct steam contact with the infectious agents are critical. Factors that can influence treatment effectiveness include waste density, physical state and size, and organic content. Because waste composition is so variable, sterilization may not be achieved. The treated waste is recognizable and can be offensive. Waste that should not be autoclaved include antineoplastic agents, toxic chemicals, and radioisotopes, which may not be destroyed, or volatile chemicals, which could be vaporized and disseminated by heat. Steam sterilizers range in size from small tabletop models to large commercial units.

The sanitary landfill can be used to dispose of untreated medical waste. Studies have shown that pathogenic microorganisms are significantly reduced in a properly operated landfill. Many states, however, have restricted the use of landfills for disposal of

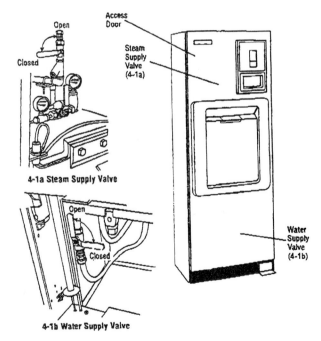

Fig. 2. Steam sterilizer or autoclave operates as an insulated pressure chamber in which saturated steam is used to elevate temperature. (Courtesy of STERIS Corporation, Mentor, OH.)

regulated medical waste. Landfills are currently used for the final disposal of treated and reduced waste. Hazardous incinerator ash must be disposed of in a Resource Conservation and Recovery Act (RCRA) permitted landfill.

Many health-care facilities dispose of their wastewater through a sanitary sewer system. Unless prohibited by local authorities, bulk blood, body fluids, and other liquids may be poured down a designated sink. This type of waste stream comprises a small portion of sanitary sewer discharges and is diluted by large amounts of residential sewage. Secondary water treatment methods are very effective in treating biological wastes.

Other methods described for decontaminating medical waste include chemical treatment, such as sodium hypochlorite and ethylene oxide gas, and radiation. These methods are less popular and are considered an adjunct to previously described technologies.

B. Alternative Technologies

Several alternative technologies for medical waste treatment and disposal have been developed. These include chemical, heat, steam, electrothermal, radiation, microwave, and pyrolysis. Most treatment systems also grind or shred the waste to improve decontamination and reduce waste volume. The efficacy of each method is dependent on the contact time, number of microorganisms in the waste to be treated, organic content, volume, and physical state of the waste.

Steam/mechanical treatment (Rotoclave, Tempico, Inc., Madisonville, LA) utilizes a pressure vessel with a rotating internal drum and agitating device (Fig. 3). Infectious wastes in bags or containers are automatically loaded into the drum and closed tightly. Pressurized steam is introduced into the drum where the waste is subjected to agitation to assure thorough steam penetration. The unit operates at a temperature of 292°F, pressurized to 48 psi, and at an operational cycle time of 65 min. After the process is complete, all material passes through a grinder to render it unrecognizable. Exhaust steam is discharged through a condensing cooler and charcoal filter. Un-

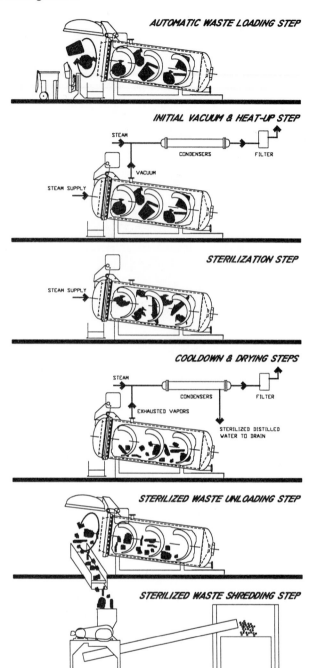

Fig. 3. Steam/mechanical treatment process. Tempico, Inc., Rotoclave process cycle steps. Infectious waste is steam-treated in a pressure vessel with a rotating internal drum for constant agitation of materials. After sterilization, waste is uploaded into a shredder for final destruction. (Courtesy of Tempico, Inc., Madisonville, LA.)

recognizable solid material is reduced up to 85% and considered general refuse.

Steam/compaction waste handling systems (San-I-Pak, Tracy, CA) employ a high-vacuum autoclave combined with a compactor (Fig. 4). Infectious waste is first placed into the sterilizer chamber and steam treated. After loading, the system is self-operating. After a 55-min cycle time, the chamber liner begins moving treated waste out of the sterilizer chamber. Treated waste is then dropped into the compaction chamber. A piston compacts the waste into a roll-off container. The compacted waste is reduced 50% but is not physically destroyed. Waste is then sent to a landfill or municipal incinerator for final disposal. This system is also designed for the compaction and disposal of general refuse.

Microwave/mechanical treatment (Sanitec, Inc., West Caldwell, NJ) is a process which shreds wastes and then heats it by microwaving (Fig. 5). Infectious waste is automatically fed into a hopper for shredding and sprayed with steam. The waste moves up a screw conveyer where it is heated at 203°F for 30 min, with multiple microwaves. The waste then enters a temperature holding section before being discharged into a waste container. Air is discharged through HEPA filters. To ensure proper processing, the waste should contain less than 10% liquid by weight and less than 1% metallic content. The volume reduction of waste is up to 85% before final disposal into a landfill or municipal incinerator.

Chemical/mechanical treatment systems incorporate grinding or shredding of waste, either before or after chemical treatment. In large units (World Environmental Services Co., Ramona, CA), waste in collection bags is fed into a cutting chamber where the waste is reduced and then treated with disinfec-

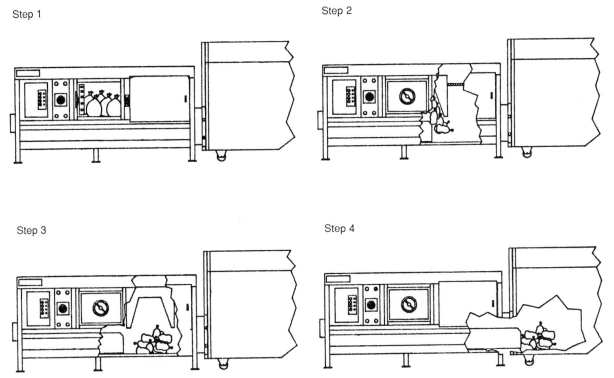

Fig. 4. Order of operation for infectious waste in the steam/compaction treatment process. Infectious waste is steam-treated in a sterilizer chamber and then compacted before final disposal. Step 1, Infectious waste is placed into the sterilizer chamber, door is closed, and "start" button depressed; Step 2, Upon completion of the sterilization cycle, the liner moves the waste out of the sterilizer chamber; Step 3, The sterilized waste is discharged into the compaction chamber; Step 4, The waste is then compacted into the roll-off container. (Courtesy of San-I-Pak, Inc., Tracy, CA.)

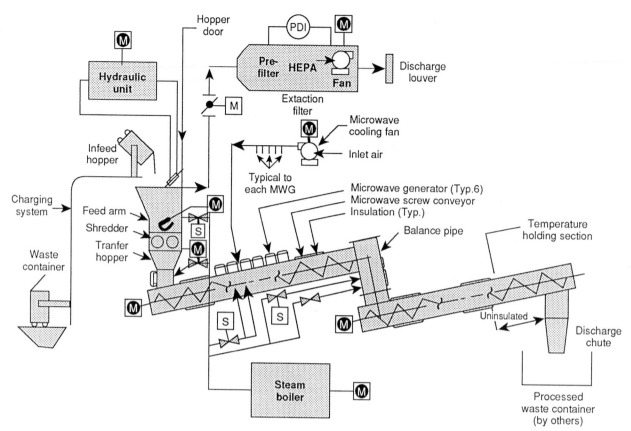

Fig. 5. Microwave/mechanical treatment process. Infectious waste is shredded and heated by multiple microwaves before final disposal. (Courtesy of Sanitec, Inc., West Caldwell, NJ.)

tant (chlorine dioxide). Fresh disinfectant must be added to the treatment solution to keep it active. Solid waste is removed through a continuous-action screw auger into a collection cart. Liquid effluent is discharged into the sanitary sewer. Air released is vented through HEPA filters and carbon filter systems. Smaller units (STERIS Corporation, Mentor, OH) have been designed for ease of use and point-of-generation treatment (Fig. 6). Infectious waste and sterilant concentrate (peracetic acid) are placed into a grinding chamber that simultaneously pulverizes and decontaminates. After a 10-min process cycle, the treated waste is mechanically dewatered and deposited into the normal trash. Alkaline Hydrolysis Tissue Digestion (Waste Reduction by Waste Reduction, Inc., Indianapolis, IN) is an alternative to incineration for the treatment and disposal of infectious animal waste. Animal carcasses and tissues are loaded into a digestor, which consists of a pressure vessel lined with a basket. The lid is closed and sealed and water is fed into the unit with a measured amount of alkali (sodium or potassium hydroxide). The digestor is operated at 110 to 120°C, at 15 psi, for 18 hours to complete digestion of organic matter. After completion of the cycle, the vessel contents are cooled and liquids drained into the sewer. Any remaining residue (bones) is removed from the basket for crushing and/or disposal.

The Stericycle electrothermal deactivation system (Stericycle, Inc., Deerfield, IL) utilizes low-frequency radio waves. Waste is shredded and exposed to an oscillating, high-intensity electrical field that generates temperature high enough to inactivate microorganisms. Since this technology does not destroy or change material, shredding of waste before treatment is necessary. The system is maintained under negative pressure, and air is released through a HEPA filter system.

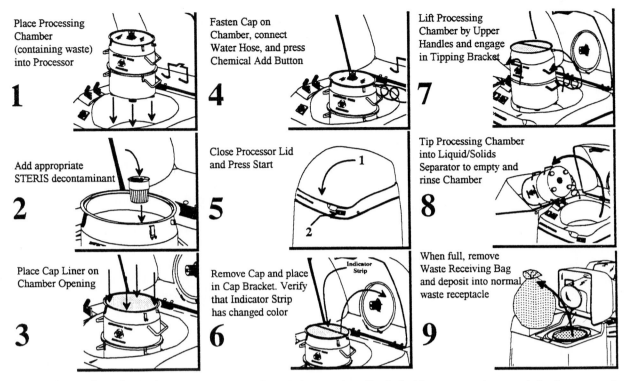

1 Place Processing Chamber (containing waste) into Processor

2 Add appropriate STERIS decontaminant

3 Place Cap Liner on Chamber Opening

4 Fasten Cap on Chamber, connect Water Hose, and press Chemical Add Button

5 Close Processor Lid and Press Start

6 Remove Cap and place in Cap Bracket. Verify that Indicator Strip has changed color

7 Lift Processing Chamber by Upper Handles and engage in Tipping Bracket

8 Tip Processing Chamber into Liquid/Solids Separator to empty and rinse Chamber

9 When full, remove Waste Receiving Bag and deposit into normal waste receptacle

Fig. 6. Chemical/mechanical treatment process. STERIS Corporation, EcoCycle 10 process cycle steps. Infectious waste is destroyed and decontaminated at point of generation. A premeasured container of decontaminant is added to the processing chamber containing waste. The processing chamber is sealed and processor lid closed. Water is automatically added to the processing chamber when the start button is depressed and waste is ground and pulverized. After decontamination, the processing chamber is emptied and automatically rinsed in the liquid/solids separator. Treated materials are captured in a special bag where the waste is rinsed and spin-dried for deposit into the normal trash receptacle. Liquids are filtered and pumped to the sewer. (Courtesy of STERIS Corporation, Mentor, OH.)

The Bio-Oxidizer system (Bio-Oxidation Services, Inc., Harrisburg, PA) uses pyrolysis and oxidation to treat a wide range of infectious waste. The waste is automatically loaded into the unit and heated to a temperature between 200 and 1000°F inside the pyrolysis chamber. Vaporized solid and liquid waste are drawn by induced draft into a two-stage oxidation chamber operating at 1800–2000°F. Waste is converted to an inert, sterile residue with negligible air emissions.

Other approaches to the treatment of infectious waste are currently being developed and evaluated, while some are no longer available in the United States. These technologies include chemical, enzyme, plasma, radiation, and thermal processes. Some are now only used at large off-site commercial facilities.

C. Selection Considerations

A number of factors should be considered when selecting an infectious waste treatment method. No single technology is applicable for all types of infectious waste and for all institutions. More than one treatment method may be used for different infectious wastestreams. For example, the sanitary sewer may be suitable for the disposal of liquid waste and incineration is preferred for pathological waste. The advantages and disadvantages of onsite versus offsite treatment options should also be considered. Onsite treatment allows one to control costs and reduces the potential for liability of the waste leaving a facility. Sharing a treatment facility may be a attractive alternative. Federal, state, and local regulations may have impact on the treatment selection. In addition to regulatory requirements, cost, availability to generator, occupational risks, and impact on the environment are also very important when selecting treatment method. Once a treatment method has been selected, it is important to establish a quality assurance program to ensure that it is functioning well. Important elements of a quality assurance plan should include indicator tests of equipment operations and treatment effectiveness (e.g., biological indicator tests of steam treatment, test of incinerator stack gas, and HEPA filter tests of mechanical treatment), criteria to measure effectiveness (e.g., operation and maintenance schedule), and records of data and review for compliance (e.g., temperature and calibration charts).

IV. WASTE MINIMIZATION

Waste minimization is an important component of infectious waste management. After establishing that a waste type is infectious, the next step is to properly handle, store, and dispose of it. This ensures that special wastes are segregated from normal trash and recyclables. Smaller quantities of infectious waste generated will result in less handling and treatment at a cost savings.

Other methods for reducing waste generation include product substitution, reusable supplies, repackaging, and recycling materials. Product substitution is important in minimizing infectious, as well as other, hazardous wastes. Examples include needleless intravenous systems that reduce sharps waste and new technologies which use small quantities of nonhazardous reagents. In addition to substitution, reprocessing and reusing items intended to be used once have been implemented in some institutions. Other noninfectious materials that lend themselves to recycling include solvents, paper, packaging material, and aluminum cans. Waste treatment technologies, such as incineration, compaction, shredding, and grinding, are also methods of reducing the volume of solid waste.

A successful waste minimization program will help reduce the cost of regulatory compliance. The program should include an employee awareness compo-

nent, employee training, purchasing strategies, and inventory control. Waste minimization takes effort and management support to facilitate every step from the initial purchase of a product through proper storage and disposal.

V. INFECTIOUS WASTE MANAGEMENT PLAN

A. Regulatory Compliance

Regulations, guidelines, and standards can influence how infectious waste is managed. Compliance with all regulatory requirements is mandatory. At the federal level, the EPA, OSHA, and the U.S. Department of Transportation (DOT) have issued specific regulations governing the management of infectious waste. The EPA and the Agency for Toxic Substances and Disease Registry under RCRA was directed by Congress to implement a 2-year demonstration project to track medical waste from start to finish for data collection and analysis (Medical Waste Tracking Act of 1988). To date, the EPA's Final Report to Congress is still in preparation. The Clean Air Act of 1990 also directed the EPA to regulate medical/infectious waste incinerators more rigorously. Other federal regulations relevant to infectious waste management include rulings by OSHA on hazard communication/right-to-know and the prevention of occupational exposure to blood-borne pathogens. Employers are required to inform workers who handle infectious waste about the risks of exposure to these wastes. The DOT regulates the containment of regulated medical waste for transport. Although regulatory requirements may differ, many states also regulate the management of infectious waste. Local ordinances also affect infectious waste treatment and disposal methods.

Rather than issuing regulations, various federal, state, and professional organizations have published guidelines on infectious waste management. Guidelines have been published by federal agencies, such as the EPA, CDC, OSHA, and the National Institute for Occupational Safety and Health (NIOSH). The National Committee for Clinical Laboratory Standards has also prepared guidelines for protecting laboratory workers from infectious diseases transmitted by blood and tissue.

Standards relevant to infectious waste management have also been established. The Joint Commission on Accreditation of Healthcare Organizations (JCAHO), which accredits hospitals, has published Standard #PL.1.10 that pertains to hazardous materials and wastes. The American Society for Testing and Materials (ASTM) develops standards for testing the strength of plastic used to manufacture red bags and defining puncture resistance in sharps containers.

B. Components

A comprehensive infectious waste management plan should be established at each institution. Table IV lists the components which should be included in such a plan. Figure 7 illustrates the complex waste management planning and implementation process. The waste should be managed at every step of the process: purchase of materials, waste generation, collection, handling, storage, transport, treatment, and final disposal.

A waste management system is needed for the management of all wastes. It should be based on regulatory requirements and institutional policy.

TABLE IV
Components of the Infectious Waste Management Plan

Waste management system
Identification of infectious waste
Waste discard
Waste handling and collection
Waste storage
Waste treatment
Disposal of treated waste
Off-site transport for treatment
Contingency planning
Emergency planning
Training
Recordkeeping
Quality assurance/quality control
Incident/accident analysis
Review procedures and practices

Written procedures and standard operating practices should be established and personnel responsible for implementing the system identified. Policy and procedures should be presented, training provided, and implementation required.

Waste identified as infectious should be discarded in designated types of containers or red bags, to protect waste handlers and minimize waste handling. All sharps should be placed directly into impervious, rigid, and puncture-resistant containers. Glass and liquids should be placed in disposable cardboard containers and leakproof boxes or containers with secured lids. Mixed waste containing infectious and radioactive or infectious and toxic chemical waste must be segregated and directed to the appropriate treatment procedure, based on the severity of the hazardous materials. Waste handling and collection procedures should be established to minimize the potential for exposure. Collection carts should be covered and disinfected after use. When waste is transferred offsite for treatment and disposal, plastic bags should be packed in rigid containers, such as plastic barrels or heavy cartons. The waste should be transported in closed and leakproof dumpsters or trucks. All infectious waste packages and containers should be marked with the universal biohazard symbol (Fig. 8). Treatment of infectious waste within the same day of collection may not be possible, therefore, plans for storage must be made. Factors to consider when storing infectious waste include the integrity of the packaging, storage temperature, storage time, and storage area. The storage area must be kept locked, with limited access to authorize personnel, and be kept free from rodents and vermin. Various treatment technologies for infectious waste are available. Even after treatment, certain types of waste, such as sharps and pathological waste, may require additional treatment according to state and local requirements to render it unrecognizable. Special treatment considerations are needed for mixed infectious and radioactive and toxic chemical wastes. After the waste is treated, solid residues are usually buried in a landfill or incinerated. Treated liquid wastes are discharged to the sanitary sewer system.

A contingency plan for the treatment of infectious waste should be established for when equipment fails or personnel problems arise. Alternatives are waste

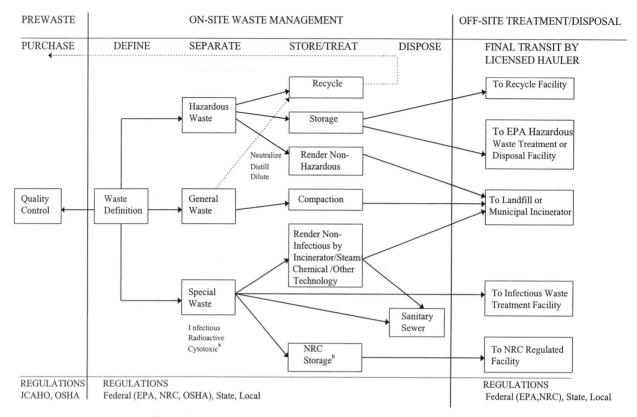

Fig. 7. Waste management planning and implementation process.

Fig. 8. Infectious waste biohazard symbol.

storage, an exchange agreement with another institution, and offsite treatment by a licensed contractor. Emergency planning should be established in the event of an accidental spill or loss of containment. This plan should include cleanup procedures, use of personal protective equipment, and disposal of spill residue.

Training of all personnel who generate or handle infectious waste is necessary for implementation of the system. Training programs should be based on standard (universal) precautions to minimize occupational exposure to potentially infectious material within the health care setting. Home health care providers should also be encouraged to educate patients and clients on how to safely dispose of sharps in their home. According to OSHA's hazardous communication/right-to-know regulations and blood-borne pathogens rule, education and training of personnel are required. Educational programs need to be developed and implemented when the infectious waste management plan is first developed, new employees are hired, and whenever infectious management practices are changed.

Recordkeeping is also essential to conform with regulatory requirements and document compliance. Quality assurance and quality control procedures are needed to ensure that the infectious waste management system is working. Examples of quality assurance activities may include spot checks for proper waste collection, transport and storage; documentation of treatment effectiveness; and analysis of accident and incident records. Review of accident or incident reports also help identify occupational issues and deficiencies in the waste management system or practices.

C. Implementation

Implementation is the most difficult part of any infectious waste management plan. Planning requires developing policy and procedures, training, and assigning responsibility. Policy should be determined by administrative operations personnel, generating departments, and service departments and should be based on the user group and legal considerations. Once a draft policy is determined to be operational, it should be presented to appropriate administrative authorities, such as infection control committee, safety committee, or nursing, laboratory, and hospital administration for input.

Development of effective educational materials and training programs for all employees is also important. Small group education and training should be given to managers and supervisors with followup inservice given to all employees. Waste haulers and treatment personnel should also be educated on policy. The responsibility of infectious and medical waste management has traditionally been assigned to service personnel. Shared responsibility among many departments should be considered. This will require administrative support, coordination of assignments, and communication. Formation of an interdisciplinary waste management committee should also be considered to develop waste management strategies and long-range planning.

There are several advantages to implementing a comprehensive infectious waste management plan. An effective plan provides compliance with all regulatory requirements and meets standards required for certification. Other benefits includes minimizing hazard and occupational exposures, minimizing risks and liability, achieving quality assurance, and reducing costs.

See Also the Following Articles

BIOMONITORS OF ENVIRONMENTAL CONTAMINATION • WASTEWATER TREATMENT, MUNICIPAL

Bibliography

Agency for Toxic Substances and Disease Registry. (1990). "The Public Health Implications of Medical Waste: A Report to Congress." U. S. Department of Health and Human Services (PB91-100271), Atlanta, GA.

Gordon, J. G., and Denys, G. A. (1996). Minimization of waste generation in medical laboratories. *In* "Pollution Prevention and Waste Minimization in Laboratories" (P. A. Reinhardt, K. L. Leonard, and P. C. Ashbrook, eds.), CRC/Lewis Publishers, Boca Raton, FL.

Gordon, J. G., Reinhardt, P. A., Denys, G. A., and Alvarado, C. J. (in press). Medical waste management. *In* "Hospital Epidemiology and Infection Control" (2nd ed.) (C. Glen Mayhall, ed.). Williams and Wilkins, Baltimore, MD.

Marsik, F. J., and Denys, G. A. Sterilization, decontamination, and disinfection procedures for the microbiology laboratory. *In* "Manual of Clinical Microbiology" (6th ed.) (T. R. Murray, E. J. Baron, M. A. Pfaller, F. C. Tenover, and R. H. Yolkens, eds.). American Society for Microbiology, Washington, DC.

Reinhardt, P. A., and Gordon, J. G. (1991). "Infectious and Medical Waste Management." Lewis Publishers, Chelsea, MI.

Rutala, W. A., Odette, R. L., and Samsa, G. P. (1989). Management of infectious waste by U.S. hospitals. *JAMA* **262**, 1635–1640.

Rutala, W. A., and Weber, D. J. (1991). Infectious waste—Mismatch between science and policy. *N. Engl. J. Med.* **325**, 578–582.

Thomas, C. S. (1997). Management of infectious waste in the home care setting." *J. Intravenous Nursing* **20** (4), 188–192.

U. S. Department of Health and Human Services, Centers for Disease Control (1989). "Guidelines for prevention and transmission of human immunodeficiency virus and hepatitis B virus to health care and public safety workers." *Morbid. Mortal. Weekly Rep.* **38** (suppl 5–6), 1–37.

U. S. Department of Health and Human Services, Centers for Disease Control. (1990). "Public health service statement on management of occupational exposure to human immunodeficiency virus, including considerations regarding Zidovudine postexposure use." *Morbid. Mortal. Weekly Rep.* **39** (no. RR-1).

U. S. Department of Labor, Occupational Safety and Health Administration. (1991). Occupational exposure to bloodborne pathogens; Final rule. *Fed. Regist.* **56**, 64003–64182.

U. S. Environmental Protection Agency. (1986). "Guide for Infectious Waste Management." Washington DC. (EPA/530-SW-86-014).

U. S. Environmental Protection Agency. (1989). Standards for the tracking and management of medical waste. *Fed. Regist.* **54**, 12326–12395.

U. S. Environmental Protection Agency. (1997). Standards for performance for new stationary sources and emission guidelines for existing sources. Hospital/medical/infectious waste incinerators; Final rule. *Fed. Regist.* **62**, 48347.

Influenza Viruses

Christopher F. Basler and Peter Palese

Mount Sinai School of Medicine

I. Viral Structure and Classification
II. Viral Genes and Proteins
III. The Influenza Virus Replication Cycle
IV. Pathogenesis and Epidemiology
V. Vaccines and Antivirals
VI. Genetic Manipulation of Influenza Viruses

GLOSSARY

antigenic drift The gradual accumulation of amino acid changes in the surface glycoproteins of influenza A and B viruses.

antigenic shift The sudden appearance of antigenically novel surface glycoproteins in influenza A viruses.

pandemic A worldwide epidemic.

reassortant A virus containing segments derived from two different influenza viruses which resulted from the co-infection of a single cell.

reverse genetics Techniques which allow the introduction of specific mutations into a gene of an RNA virus.

transfectant virus A virus carrying a segment introduced by reverse genetics techniques.

INFLUENZA A, B, AND C VIRUSES each occupy a separate genus of the family Orthomyxoviridae. The common characteristics of these viruses include a plasma membrane-derived envelope studded with viral glycoproteins and a segmented, negative-sense RNA genome. Influenza viruses can be significant pathogens for humans and animals. Regular epidemics of influenza A and B viruses continue to occur, causing significant illness and death, while influenza C viruses may cause nonepidemic childhood disease.

Pandemics caused by influenza A viruses appear to have occurred for centuries at irregular intervals of one to several decades. These pandemics have been the cause of great morbidity and mortality, with the "Spanish flu" of 1918 probably having been the most severe, resulting in the deaths of an estimated 20 million people worldwide. In addition, influenza A viruses are associated with periodic outbreaks of high pathogenicity in poultry, horses, pigs, and other species. Such epizootics can have devastating effects on these animal populations.

Because of their importance as human pathogens, the molecular biology of influenza viruses has been extensively studied. Furthermore, influenza viruses serve as a paradigm by which the interplay between host immune responses and viral evolution can be explored. In particular, infection with an influenza virus induces expression of antiviral antibodies. However, the error-prone replication of the viral RNA genome results in the gradual accumulation of changes within a particular viral strain. Viral mutants which can escape host antibody response are thus selected for by the immune system. The resulting gradual accumulation of mutations within the surface proteins of viral strains is known as "antigenic drift." Influenza viruses which have undergone antigenic drift are able to evade host immune responses of previously infected individuals. This characteristic correlates with the ability of influenza viruses to cause annual epidemics. Periodically, a new viral strain emerges through the "mating" of a human and an animal strain. The resulting "reassortant" virus may have antigenic properties which differ dramatically from those of the previously circulating strains. When this antigenic shift occurs, most people have no preexisting protective immunity to the new reassortant and an influenza virus pandemic may begin.

I. VIRAL STRUCTURE AND CLASSIFICATION

A. Structure

Influenza virions are spherical or pleomorphic enveloped particles with a diameter of 80–120 nm and an overall molecular weight of approximately 2.5×10^8 (Fig. 1). The viral envelope is derived from the host cell plasma membrane and makes up 18–37% of the virion weight. Embedded within the viral envelope are viral glycoproteins (predominantly, the hemagglutinin (HA) and neuraminidase (NA), for influenza A and B viruses) which comprise approximately 50% of the virion mass, with 5% of the total virus mass consisting of associated carbohydrates. Underlying the viral membrane is the viral matrix, which consists of a single protein (M1). This matrix is associated with the lipids of the viral envelope and may interact with the cytoplasmic tails of one or more of the viral glycoproteins. Within the viral matrix are

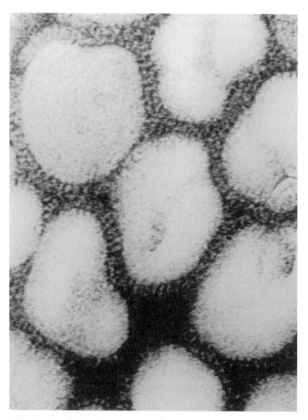

Fig. 1. Electron micrograph of influenza A virus particles. Provided by M.-T. Hsu and Peter Palese.

ribonucleoprotein particles (also, RNPs or nucleocapsids), consisting of the viral genomic RNA segments (vRNAs) coated with viral nucleoprotein (NP) and associated with the heterotrimeric viral polymerase. The viral RNA genome is segmented (8 segments in influenza A and B viruses and 7 segments in influenza C viruses), and each viral RNA segment (vRNA) is of negative sense. That is, rather than directly encoding a viral protein, the vRNA acts as template for complementary mRNA synthesis. It is these positive-sense mRNAs which direct the synthesis of viral proteins. Because the genome is negative-sense, the heterotrimeric viral polymerase is present in the virion, with at least one polymerase trimer bound to each RNA segment. Thus, when the viral RNPs enter the nucleus, the polymerase is available to initiate viral gene expression. In addition to these, small amounts of other viral proteins are present in influenza virus particles. For example, the M2 protein of influenza A viruses and the NB protein of influenza B viruses are also found in the viral envelope. Small amounts of the influenza A virus NEP (NS2) protein are also found associated with viral particles. In influenza A virus infected cells, one nonstructural protein, NS1, is produced (Fig. 2).

B. Taxonomic Classification and Nomenclature

Influenza viruses are members of the family Orthomyxoviridae, which has been divided into four genuses: the influenza A viruses, the influenza B viruses, the influenza C viruses, and the Thogoto viruses. Common features of these viruses include the host cell-derived viral envelope, the presence of virus encoded membrane bound glycoproteins, a segmented, negative-sense RNA genome which replicates in the nucleus of infected cells, and the use of endonuclease cleaved host cell mRNAs as primers for viral transcription. The viruses differ in the number of proteins encoded by the genome, the types of glycoproteins found on the cell surface, and the number of virion RNA segments. For instance, influenza A and B viruses have eight genome segments (Fig. 3), the influenza C viruses possess 7 RNAs, and the Thogoto viruses possess only 6 RNAs.

Influenza viruses are typically named according to

the following convention: influenza serotype (A, B, or C)/site of origin/isolate number/strain designation/year of origin. Influenza A viruses are further classified based on the antigenic properties of their major surface glycoproteins, the hemagglutinin (H) and neuraminidase (N), and the host from which the virus was first isolated (omitted if human host). An example would be: influenza A/duck/Ukraine/1/63 (H3N8) virus.

For influenza A viruses, 15 distinct HA types (H1 through H15) and 9 distinct NA types (N1 through N9) have been identified. Members of each HA type or NA type show antigenic similarity, but different isolates from the same viral subtypes possess nucleotide and amino acid differences within their surface antigens. Although there are 15 HA types and 9 NA types, not all possible combinations of HA and NA types are seen in humans. This probably reflects the coevolution of HA and NA function. For example, widespread human respiratory disease in this century has only been caused by H1N1, H2N2, and H3N2 viruses.

In contrast to influenza A viruses, type B viruses do not possess different HA and NA types. These

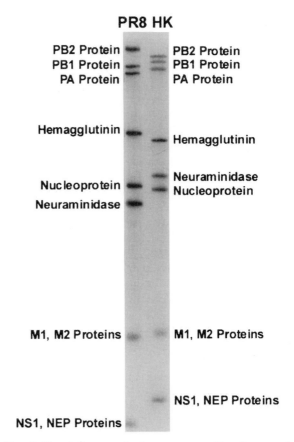

Fig. 3. The influenza A virus genome. The 8 genomic (vRNA) segments of two different influenza A virus strains are shown after their electrophoretic separation on a polyacrylamide gel. RNAs from influenza A PR/8/34 (H1N1) virus and influenza A Hong Kong/8/68 (H3N2) virus are shown. The protein products of each RNA segment are shown beside the corresponding bands.

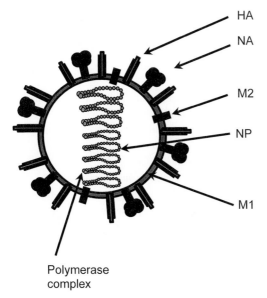

Fig. 2. Diagram of an influenza virus particle. HA, hemagglutinin; NA, neuraminidase; M2, M2 ion channel; NP, nucleoprotein covering the single-strand RNA segments; M1, matrix protein; polymerase complex, a heterotrimer composed of the proteins PA, PB1, and PB2.

viruses, while undergoing antigenic drift, are not subject to major antigenic shifts and are limited in their host range to humans. The influenza C viruses, which possess the HEF (hemagglutinin, esterase, and fusion) protein rather than separate HA and NA proteins, are also not normally subclassified according to their surface antigens and, like the influenza B viruses, are isolated only from humans.

II. VIRAL GENES AND PROTEINS

The influenza A virus vRNAs range in size from greater than 2300 to less than 900 nucleotides in length. Each single-stranded viral RNA segment contains conserved 5' and 3' ends of 13 and 12 nucleotides, respectively (5' AGUAGAAACAAGG and 3' UCG(U/C)UUUCGUCC). These conserved noncoding sequences have been shown to act as promoters for both the transcription and replication of viral genes. The vRNAs contain additional noncoding sequences unique to particular segments, as well as polyadenylation signals used in mRNA synthesis. The vRNAs serve as templates for the synthesis of two forms of RNA: in transcription, they direct mRNA synthesis and, in replication, they direct cRNA synthesis. As templates for protein encoding mRNAs, the vRNAs encode either one or two viral proteins apiece. By this strategy, influenza A viruses encode 10 proteins on 8 genomic RNA segments (Table I).

Segments 1, 2, and 3, the three largest vRNAs in the viral genome, each encode one of the viral polymerase proteins (PA, PB1, or PB2). As noted above, a heterotrimeric complex of PA, PB1, and PB2 serves as the viral polymerase.

Segment 4 encodes the HA protein. The mature influenza HA is an integral membrane, homo-trimeric glycoprotein, found on the surface of virions and at the surface of infected cells. The HA serves several important functions in the replication of influenza viruses. First, the HA binds to the cell surface receptor (sialic acid), thus mediating viral attachment to, and subsequent endocytosis into, the host cell. Second, the HA is the viral fusion protein, mediating fusion between viral and host cell membranes following virion endocytosis (see Section III.A). This ability of HA to mediate membrane fusion requires its cleavage into two disulfide-linked subunits, HA1 and HA2. Because this cleavage is dependent on host cell proteases, the nature of the HA cleavage site can affect viral host range and pathogenicity. Finally, the HA is the primary target of neutralizing antibodies generated by the infected host (see Section IV). Because of its importance, the HA has been extensively studied and detailed x-ray crystal structures have been solved for the H3 type.

Rather than possessing two separate glycoproteins which have receptor binding, fusion and receptor destroying activities (see following), influenza C virus possesses a single glycoprotein which performs these functions. This protein, the hemagglutinin, esterase, fusion protein (HEF), binds to 9-O-acetyl-N-

TABLE I
Influenza A Virus[a] Genes and Their Encoded Proteins

Segment	Length in nucleotides	Encoded proteins	Predicted molecular weight[b]	Molecules per virion	Function
1	2341	PB2	85,700	30–60	Polymerase subunit, binds host cell mRNA cap, endonuclease
2	2341	PB1	86,500	30–60	Polymerase subunit, RNA elongation
3	2233	PA	82,400	30–60	Polymerase subunit
4	1778	HA	61,468[c]	500	Binds cell surface receptor, fusion protein, possesses neutralizing epitopes
5	1565	NP	56,101	1000	RNA binding protein, function in replication
6	1413	NA	50,087[d]	100	Neuraminidase, promotes virus release
7	1027	M1	27,801	3000	Forms viral matrix, regulates RNP trafficking in and out of nucleus
		M2	11,010	20–60	Ion channel
8	890	NS1	26,815	0	Interferon antagonist, possible role in viral gene expression
		NEP (NS2)	14,216	130–200	Nuclear export factor

[a] Influenza A/PR/8/34 virus.

[b] Predicted molecular weight of unmodified, monomeric proteins.

[c] HA is a glycoprotein which forms trimers. A precursor form, HA0, is cleaved into two disulfide-linked subunits, HA1 and HA2.

[d] NA is a glycoprotein which forms tetramers.

acytlneuraminic acid, has a low-pH triggered fusion activity, which requires proteolytic cleavage into two disulfide-joined subunits, and has receptor destroying neuraminate-O-acetyl esterase activity.

Segment 6 of influenza A and B viruses encodes the NA (neuraminidase) protein, the second major glycoprotein and second important antigenic determinant found on the surface of virions (Fig. 2). The neuraminidase (acylneuraminyl hydrolase, EC 3.2.1.18) is an enzyme which cleaves the α-ketosidic linkage connecting terminal sialic acid residues to adjacent D-galactose or D-galactosamine residues. Thus, the NA possesses a "receptor-destroying" activity and removes from glycoconjugates the sialic acid residues which might otherwise bind HA. This function is important in viral release. Antibodies against the NA protein inhibit release of virus from infected cells and, in this way, may facilitate recovery from influenza virus infections. The NA protein has been studied extensively and detailed x-ray crystal structures are available for the N2, N9, and B types. Effective anti-influenza virus drugs have been developed which inhibit NA enzymatic activity.

In the case of influenza B virus, a second glycoprotein of unknown function, NB, is also encoded on segment 6. The NB protein is translated from the same mRNA as the influenza B virus NA, but the two proteins begin at different AUGs and are translated in different reading-frames. Because influenza C virus encodes only one protein, the HEF (on segment 4), to carry out its receptor binding, fusion, and receptor destroying activities, there is no NA segment in influenza C virus.

Segment 7 of influenza A viruses encodes two proteins, the M1 or matrix protein and the M2 protein. The M1 protein is encoded by an unspliced mRNA, while M2 is encoded by a spliced mRNA. The M1 protein is a 28 kDa phosphoprotein, which can associate with lipids and with RNA. M1 is an abundant structural protein, but also helps regulate movement of RNPs into and out of the nucleus and may terminate viral RNA synthesis late in infection. M2 is an ion channel protein which facilitates viral entry and helps the HA protein maintain its proper configuration during its processing in the trans-Golgi network.

The influenza B virus segment 7 also encodes two proteins, M1 and BM2. In this case, the two proteins are encoded by a single transcript. In the case of influenza C virus, the matrix protein is produced from a spliced mRNA, while a small integral membrane protein, CM2, is encoded by the unspliced segment 6 mRNA.

Segment 8 of influenza A and B viruses also encodes 2 proteins, the NS1 and the NEP (NS2). The nonstructural NS1 protein is a phosphoprotein which is expressed at high levels in infected cells. Pleotropic NS1 functions have been described; however, the major function of NS1 appears to be that of an interferon antagonist. In fact, a genetically engineered influenza virus lacking the NS1 gene only replicates efficiently in interferon-deficient systems. The NEP (nuclear export protein or, previously, NS2) exhibits nuclear export activity, interacts with the M1 protein, and may facilitate the export of viral RNPs from the nucleus late in the viral replication cycle. Like the M1 and M2 proteins, the influenza A virus NS1 and NEP are produced from unspliced and spliced mRNAs, respectively. The splicing occurs in such a way that NS1 and NEP have the same 9 amino-terminal amino acids, but the remainder of the two proteins are unique. Influenza B and C viruses also appear to encode NS1 and NS2 proteins from unspliced and spliced mRNAs, respectively.

III. THE INFLUENZA VIRUS REPLICATION CYCLE

A. Entry

Influenza viruses initiate entry into host cells via the binding of the HA protein to sialic acid residues on the cell surface. These sugars (sialic acids/neuraminic acids) are found on glycoconjugates covalently joined to host cell proteins and/or lipids. One sialic acid binding site is found on each HA monomer at a site distal from the membrane. Each binding site consists of a pocket containing several conserved amino acid residues. Depending on the presence of particular amino acids within the sialic acid binding pocket, different HAs show preferential binding to sialic acid residues with either $\alpha2,3$ or $\alpha2,6$ linkages, and these differences may influence viral host range.

Once the virus has bound its receptor, it is internalized in a clathrin-coated vesicle. The vesicle then fuses with an endosome. Due to the action of proton pumps in the endosomal membrane, the interior of the endosome is acidified. This drop in pH triggers dramatic changes in the conformation of the HA. As a result of these drastic conformational changes, a hydrophobic "fusion peptide," which was previously buried within the HA structure, becomes exposed, intercalates into the endosomal membrane, and mediates membrane fusion. This fusion permits the release of viral RNPs into the cytoplasm (Fig. 4).

The RNPs must then migrate to the nucleus of the infected cell in order to initiate viral gene expression. However, the virion RNPs are associated with M1, and studies suggest that RNPs bound by M1 will not enter the nucleus. However, the presence of the M2 ion channel in the viral envelope permits an acid-activated flow of ions into the virus. This drop in pH appears to dissociate M1 from the viral RNPs. Once the RNPs are deposited into the cytoplasm, they migrate to the nucleus and initiate viral gene expression. The anti-influenza virus drugs amantadine and rimantadine appear to act primarily by inhibiting the role of M2 in uncoating.

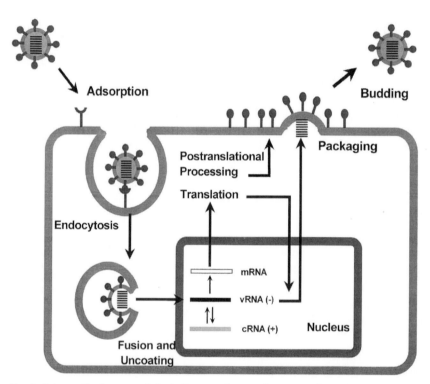

Fig. 4. Schematic diagram of the influenza virus replication cycle. Adsorption occurs through the interaction of the viral HA protein and sialic acid residues on cell surface glycoproteins or glycolipids. During endocytosis, the interior of the endocytic vesicle becomes acidified. This drop in pH triggers conformational changes in the HA protein, which result in fusion between viral and endosomal membranes and release of viral particles into the cytoplasm. Also, the pH drop dissociates the M1 protein from the RNP complexes, freeing these complexes to enter the nucleus. Nuclear import of RNPs occurs through nuclear pores and is required for viral gene expression. In the nucleus, viral RNAs are transcribed into protein encoding mRNAs, and viral RNAs are replicated. Late in infection, new RNPs, assembled in the nucleus, are exported to the cytoplasm with the aid of the M1 and NEP (NS2) proteins. Viral assembly occurs at the plasma membrane and includes the packaging of viral genomes, in the form of RNPs, into budding virus particles.

B. Nuclear Import of Viral RNPs

The RNPs must migrate to the cell nucleus for viral gene expression and genome replication to take place. Movement of the RNPs into the nucleus occurs by transport through the nuclear pore complex. It has been shown that a nuclear localization signal (NLS) at the amino-terminus of the viral NP can interact with an NLS-receptor, either NPI-1 or NPI-3 (also termed karyopherin $\alpha1$ and karyopherin $\alpha2$, respectively). NPI-1 or -3, associated with NP, then interacts with other components of the cellular nuclear import machinery. The RNP complex is then transported across the nuclear pore.

C. Viral RNA Synthesis

Once the viral RNAs, NP, and polymerase proteins enter the nucleus, viral transcription and genome replication can begin. Viral RNA synthesis requires cis-acting sequences, including promoters which are found at the 5′ and 3′ ends of the viral RNAs, and the viral PA, PB1, PB2, and NP proteins. Also required is the continued production of host cell mRNAs by RNA polymerase II. Three forms of viral-specific RNA are produced in infected cells (Fig. 5). In viral transcription, the negative-sense vRNA serves as template for the synthesis of positive-sense mRNAs, which direct synthesis of viral proteins. In viral replication, the vRNA directs synthesis of plus-sense cRNA, and cRNA, in turn, serves as template for new vRNAs, some of which will be the genomic segments in newly formed virions.

1. Transcription

During transcription, the synthesis of viral encoded mRNAs, the negative-sense genomic vRNA serves as template for the generation of 5′ capped, 3′ polyadenylated transcripts. Transcription is initiated using a 5′ capped oligonucleotide (10–13 nucleotides long), which is cleaved from a host cell mRNA and serves as a primer for viral mRNA synthesis. The PB2 polymerase subunit binds the cap and is required for the endonuclease activity which cleaves the cellular mRNA. Once initiated, transcription proceeds until a polyadenylation signal, consisting of 5–7 uridine residues 15–22 nucleotides from the 5′ end of the vRNA, is reached. When the polymerase reaches the polyU stretch near the vRNA 5′ end, a reiterative addition of A residues occurs. This results in the production of an mRNA which terminates with a polyA tail.

2. Replication

During viral RNA replication, RNAs of both positive- and negative-sense are generated. In contrast to transcription, replication results in the production of RNAs which are uncapped and not polyadenylated. First, full-length, positive-sense complements of the genomic vRNA, known as cRNA, are made. These cRNAs are, in turn, used as templates for the synthesis of new full-length, negative-sense vRNAs. The mechanism which permits uncapped initiation and the bypass of polyadenylation signals is not well understood. However, *in vitro* RNA synthesis experiments suggest that the presence of a soluble form of NP (not associated with RNP) is required for bypass of the polyadenylation site, a process known as anti-termination.

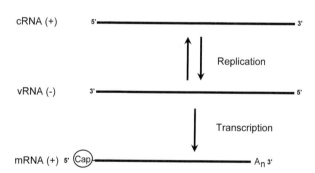

Fig. 5. Forms of RNA generated during influenza virus infection. The negative-sense vRNA serves as template for the synthesis of cRNA and mRNA. cRNA is a full-length complement of vRNA and is uncapped and not polyadenylated. Viral mRNAs possess 5′ capped oligonucleotides which are cleaved from host cell mRNAs, and polyA tails, which are added at a site 15–22 nucleotides upstream of the 5′ end of the template vRNA. Therefore, mRNA transcripts are not full-length complements of vRNA.

D. Assembly and Release

Assembly of the new virus particles takes place at the host cell plasma membrane. This requires the transport of viral glycoproteins (HA, NA, M2) to the plasma membrane, movement of newly synthesized RNPs out of the nucleus, and accumulation of M1 protein at assembly sites.

Viral glycoproteins reach the plasma membrane using the cellular secretory pathway. As the proteins migrate from the endoplasmic reticulum, through the Golgi apparatus and to the plasma membrane, they acquire their native conformation, receive post-translational modifications, and form multimers. For some influenza virus strains, the M2 ion channel functions to help maintain proper HA folding during its transport to the cell surface. The HA of these strains, when expressed in the absence of M2 function, prematurely assumes its fusion mediating conformation during transit through the acidic trans Golgi network (TGN). Coexpression of M2 keeps this nonfunctional version of HA from forming.

The mechanism by which viral RNPs reach the plasma membrane is poorly understood. Late in infection, viral RNPs must be exported from the nucleus into the cytoplasm, prior to transport to the plasma membrane. RNP nuclear export appears to require the presence of the M1 protein, which may associate with RNPs in the nucleus and facilitate their export. Once in the cytoplasm, RNPs bound by M1 do not reenter the nucleus. The viral NEP (nuclear export protein, previously NS2) may cooperate with M1 in mediating RNP export. The NEP has been shown to possess a functional nuclear export signal and is able to interact with M1, suggesting that NEP provides the signal which mediates RNP export.

At the plasma membrane, viral components are assembled, and new viruses bud from the host cell. Efficient release of these newly formed particles requires the receptor destroying activity of the viral neuraminidase. By removing sialic acid residues from glycoproteins at the cell surface, the NA promotes efficient release of influenza viruses from the infected cell. Thus, when viruses carrying temperature-sensitive NA proteins are grown at nonpermissive temperature, newly formed virions aggregate near the cell surface. This aggregation is caused by the binding of HA to sialic acid containing glycoproteins.

IV. PATHOGENESIS AND EPIDEMIOLOGY

A. Influenza Illness

Influenza virus transmission between humans is thought to most often occur via aerosolized droplets produced upon sneezing or coughing. Infection can initiate in the upper respiratory tract but may also involve the lower respiratory tract, where replication appears to be more efficient. The incubation period is 1 to 4 days, and detectable virus is shed from day 1 to day 3–5, with a peak of virus shedding 72 hours postinfection. Influenza virus infections can be asymptomatic but can also result in pneumonia and death. Typical symptoms (Table II) of influenza A and B virus infection are similar and include a sudden onset, fever, chills, headache, myalgia, cough, and malaise. Fever usually peaks 24 hours after onset and persists for 1 to 5 days. Recovery from uncomplicated influenza virus infection occurs over one to two weeks. Such symptoms are similar to those seen upon infection with other upper respiratory pathogens, although the presence of high fever, severe symptoms, and knowledge of an ongoing epidemic may suggest influenza virus illness. Therefore, confirmation of influenza virus infection requires laboratory documentation.

Complications of influenza virus infection include secondary bacterial infections (including secondary bacterial pneumonia), primary viral pneumonia, Reye's syndrome, and, possibly, rare neurological

TABLE II
Influenza

Symptoms and signs	Complications
malaise	primary viral pneumonia
myalgia	secondary bacterial pneumonia
fever	otitis media
chills	Reye's syndrome
headache	encephalitis (Parkinson's
respiratory symptoms	Disease)
cough	
substernal burning	
gastrointestinal symptoms (rare)	

complications. Those most at risk for complications related to influenza virus infections are the elderly, young children, and individuals with underlying medical conditions, such as chronic cardiac disease, chronic pulmonary disease, and diabetes.

Secondary bacterial infections are the most frequent cause of influenza virus-related pneumonia. Such secondary infections usually appear 5 to 10 days after the onset of illness and are caused most frequently by *Streptococcuus pneumoniae, Staphylococcus aureus,* and *Hemophilus influenzae.* Although potentially fatal, these infections usually respond to appropriate antibiotic treatment.

Primary viral pneumonia is a less common but more often fatal complication. It occurs more frequently after influenza A virus infection than influenza B virus infection. This pneumonia usually appears 3 to 5 days after onset of illness and often progresses rapidly, is seen predominantly in those predisposed to influenza-associated complications, and is relatively rare. When it does occur, primary viral pneumonia is characterized by diffuse pulmonary infiltrates, acute respiratory failure, and high mortality.

Influenza virus infection, particularly influenza B virus infection, has also been associated with the development of Reye's syndrome in children and adolescents. Symptoms of Reye's syndrome include sudden onset of noninflammatory encephalopathy and hepatic disfunction. This syndrome, which has been associated with other viral agents as well, has been reported to have a fatality rate of 22 to 44%, although milder forms of the syndrome may exist. The role of viral infection in its etiology is not understood, but its occurrence has been associated with the use of salicylates, such as aspirin.

It has also been suggested that some influenza virus infections result in other neurologic complications. Encephalopathy and encephalitis occurring during or after influenza virus infection have been described. Encephalopathy during viral infection occurs at the time of maximal symptoms. Postinfluenza encephalitis is difficult to document since virus has rarely been recovered from brain. The 1918 influenza pandemic was associated with "encephalitis lethargica" followed by postencephalitic Parkinson's disease approximately 10 years after the appearance of the first

syndrome. However, a causal relationship between viral infection and this disorder has not been firmly established.

B. Immune Response to Influenza Virus Infection

Infection with influenza viruses induces interferon α, the production of anti-influenza virus antibodies, and the production of anti-influenza virus T cell responses. Interferon can be detected in serum and respiratory tract secretions during influenza virus illness. The production of interferon is thought to contribute to viral clearance, because influenza viruses have been shown to be sensitive to the antiviral effects of interferons. Transgenic mice defective in signal transduction pathways required for interferon induction are more susceptible to influenza virus infection than are mice with an intact interferon system. The mechanisms by which interferons inhibit influenza viral replication have not been fully defined, but it is known that interferon α induces Mx protein expression. The Mx proteins are GTPases, which are able to inhibit primary transcription by influenza viruses. The influenza A virus NS1 protein appears to be an important factor in counteracting the effects of interferon α on viral replication; viruses lacking the NS1 induce high levels of interferon in the host and only replicate efficiently in interferon-deficient systems.

Antibodies are generated against a variety of viral proteins during infection and can be detected both in serum and in respiratory secretions. Antibody found in the serum is predominantly of the IgG-type and that in secretions of the IgA-type. Of greatest importance are neutralizing antibodies generated against the HA. In particular, five epitopes (A–E) on HA have been identified, and their sequences vary from HA type of HA type. These are also the sites which are altered during antigenic drift. In addition, antibodies against the NA protein are able to decrease the severity of illness and to speed recovery. Presumably these effects are due to inhibition of viral spread rather than actual neutralization of viral infection. In general, the humoral immune response is thought to provide protection from future infections by strains of similar antigenicity.

Influenza virus infection also induces the production of virus-specific CD4$^+$ and CD8$^+$ T lymphocytes. The CD8$^+$ cells are cytotoxic T cells (CTL) which lyse infected target cells and, thus, are thought to be important in clearance of viral infection. This function is supported by experiments using mouse models. CD4$^+$ cells (helper T cells) appear to facilitate anti-influenza virus antibody production, and perhaps viral clearance, by "helping" in the expansion of influenza virus specific B cells.

C. Epidemics and Pandemics

Influenza A and B virus epidemics appear suddenly and rapidly, with many individuals becoming ill within a short time span. Epidemics are usually associated with increases in absenteeism from schools and work, an increase in hospital admissions, and an increase in "influenza and pneumonia related deaths." Epidemics arise when there is a sufficient number of individuals susceptible to the circulating virus, such that rapid and widespread (epidemic) transmission occurs. Therefore, the ability of viruses with certain HA and NA types (e.g., H3N2) to repeatedly cause epidemics is related to the antigenic drift, which allows the virus to evade the immune response of previously exposed individuals. Epidemics usually peak in 2 to 3 weeks and wane in 5 to 8 weeks.

Influenza A virus pandemics have occurred periodically for at least several centuries. Some of these, particularly the 1918 pandemic, have had profound effects on human populations (Fig. 6). Pandemic viruses arise given two conditions. First, most individuals must be susceptible to infection because they lack preexisting immunity to the circulating virus. Second, the virus must replicate efficiently in humans. When these two criteria are met, the new influenza virus may spread rapidly throughout the world. Typically, the emergent strain largely replaces the previously circulating strain(s), although the reasons for this are unclear. For example, the 1918 pandemic was caused by an H1N1 virus, and H1N1 viruses circulated until 1957, when an H2N2 strain appeared. The appearance of the H2N2 viruses coincided with the disappearance of H1N1 viruses. H2N2 viruses circulated until 1968 when the H3N2 pandemic virus arose. Since 1968, H3N2 viruses have

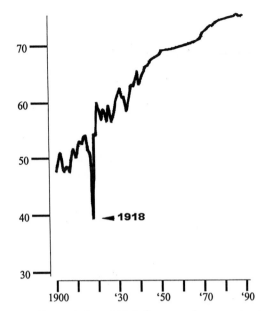

Fig. 6. Effect of the 1918 influenza virus pandemic on average life span in the United States. The precipitous drop in life expectancy (years of age; *y*-axis) of United States citizens in 1918–1919 was due primarily to the 1918 influenza virus pandemic caused by an H1N1 virus. Modified from data provided by J. Lederberg.

been the major cause of influenza illness. However, in 1977, H1N1 viruses reemerged and did not replace the H3N2 viruses. Rather, the H3N2 and H1N1 viruses have been cocirculating since 1977 (Fig. 7).

D. Emergence of Pandemic Strains

Because new influenza virus pandemics carry a heavy toll in morbidity and mortality, it is of great interest to study the mechanism by which pandemic strains introduce themselves into the human population. It is known that avian species, waterfowl in particular, harbor viruses with all possible HA and NA types, and that influenza viruses are enzootic in these birds. Based on phylogenetic analysis of the sequences from different influenza virus isolates, it appears that the current influenza viruses evolved from ancient, avianlike viruses. Typically, the avian viruses cannot efficiently replicate in humans. This inability of influenza viruses to readily cross species barriers is probably a multigenic phenomenon. In addition, the receptor specificity of the HA may con-

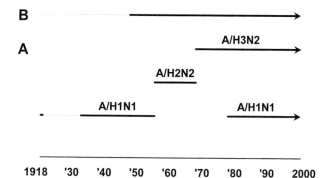

B

A

A/H3N2

A/H2N2

A/H1N1   A/H1N1

1918 '30 '40 '50 '60 '70 '80 '90 2000

Fig. 7. Epidemiology of influenza A and B viruses in the twentieth century. The prevalent human influenza virus strains of the twentieth century are indicated. The appearance of the new influenza A virus subtypes in 1957, 1968, 1977, and probably 1918 resulted in pandemics. The thin dotted lines indicate supposed prevailing strains (before the first isolation of influenza A and B viruses). Sequence analysis of viral RNA obtained from formalin-fixed tissue samples indicates that the 1918 pandemic strain was indeed an H1N1 virus.

tribute to species specificity. The HA of avian species prefers to bind sialic acid residues with $\alpha2,3$-linkages, whereas human virus HAs prefer sialic acid residues with $\alpha2,6$-linkages. It has been shown that the gastrointestinal tract of birds is rich in sialic acid residues with $\alpha2,3$ linkages, while the human trachea contains predominantly sialic acid residues with $\alpha2,6$ linkages.

Despite such barriers, transmission of avian viruses directly to mammals has been documented. In 1989, for example, there appears to have been the direct transmission of avian H3N8 virus to horses. Also, avian viruses appear to have caused significant amounts of disease and death in seals and in farm-raised minks. Occasional direct transmission of avian viruses to humans has also been documented. Most notably, in 1997 an H5N1 virus, highly pathogenic in chickens, was transmitted to humans. These transmissions resulted in 18 documented infections and 6 deaths in Hong Kong. However, there is no evidence of avian viruses becoming established in human populations.

Despite the fact that avian viruses are rarely a direct cause of human disease, phylogenetic evidence suggests that avian influenza viruses were the source of the novel genes found in the 1957 and 1968 pan-

demic strains. The 1957 pandemic H2N2 strain most likely derived its HA, NA, and PB1 genes derived from an avian virus and the remainder from the previously circulating human virus. Likewise, the 1968 pandemic H3N2 strain is thought to have derived its HA and PB1 genes from an avian virus and the remaining six genes from the previously circulating human H2N2 virus. New HA (and NA) gene(s) provided the pandemic strain with the ability to evade preexisting anti-influenza antibodies.

Both avian and human viruses are able to replicate efficiently in pigs, perhaps because cells lining the pig trachea possess surface sialic acid residues with both $\alpha2,3$ and $\alpha2,6$ linkages. It has, therefore, been proposed that pigs may serve as "mixing vessels" in which avian and mammalian viruses can reassort and adapt to growth in a mammalian species. This process is thought to occur most frequently in China, which appears to be a hot spot for the development of novel influenza viruses and was the likely source of the 1957 and 1968 pandemic strains. Generation of avian–mammalian virus reassortants in China is likely due to agricultural practices in which birds, pigs, and humans are frequently found in close contact. However, evidence for interspecies transmission and mixing in pigs has been found elsewhere, including Europe, suggesting that China may not be the only location where new pandemic strains can arise.

In addition to the potential role of avian viruses as the contributing source of new genes for pandemic viruses, the presence of influenza viruses in other species, such as pigs, horses, and sea mammals, suggests that these species might also contribute novel genes to a pandemic strain. Also, the process of reassortment may occur in animals other than pigs, including humans, suggesting possible mixing vessels other than pigs.

V. VACCINES AND ANTIVIRALS

A. Vaccines

Effective influenza virus vaccines are available and are recommended for the elderly and for individuals with underlying medical conditions which put them at risk for influenza-related medical complications.

These vaccines contain a mixture of inactivated influenza A and B viruses and are protective because they induce production of neutralizing anti-HA antibodies. Production of anti-NA antibodies may provide some further protection from illness as well but appears not to be critical for vaccine efficacy. Since the vaccine-induced protective immune response is relatively short lived, and since influenza viruses undergo frequent changes in antigenicity, annual reimmunization is recommended.

Currently, influenza virus vaccines are generated from virus grown in embryonated chicken eggs. Two general types of vaccine are currently available. "Whole virus vaccine" consists of purified, formalin-inactivated virus. "Split virus vaccine" can consist of either purified, detergent-disrupted virus or purified surface antigens. Typically, whole virus vaccine is administered to adults, while split virus vaccine is administered to children, since this vaccine induces fewer side effects in the young. These inactivated vaccines have been shown to be 70–90% effective in reducing the severity of influenza virus illness when vaccine antigenicity closely matches the circulating viral strains. However, the vaccines are less effective in children and the elderly than in healthy adults. Vaccine failure occurs when either the vaccine preparation does not elicit strong immune responses or when the vaccine antigenicity differs significantly from that of the circulating viral strains.

The propensity of influenza viruses to undergo changes in antigenicity requires that the influenza virus vaccine be reevaluated every year. Surveillance for changes in the antigenicity of circulating influenza viruses is carried out at various laboratories throughout the world. Based on this surveillance, recommendations for the composition of the next influenza season's vaccine are made. Recent vaccines have included three components, an H3N2 virus, an H1N1 virus, and a type B virus, to provide protection against the known circulating strains. Once viruses with suitable antigenic properties are identified, these must be adapted to growth in embryonated chicken eggs or reassorted with other strains to generate a virus that has the desired surface antigens and that grows well in eggs.

Recently, live, attenuated influenza virus vaccines have been developed. It is hoped that such vaccines will offer advantages over the inactivated virus vaccines. These advantages might include generation of a CTL response against conserved, internal viral antigens. It is also hoped that a live vaccine inoculated intranasally will generate an increased mucosal immune response, which may provide better protection from a respiratory virus. Live, cold-adapted influenza A and B virus vaccines, which were attenuated by repeated passage at low temperatures, have proven to be efficacious in clinical trials in the United States. The development of "reverse-genetics" systems for influenza viruses, which allow the introduction of specific changes into influenza virus genes, offers the ability to design specific attenuating mutations or to introduce additional (foreign) epitopes into influenza viruses. These techniques offer the possibility of generating even more effective influenza virus vaccines in the future.

B. Anti-influenza Virus Drugs

Viruses often possess unique functions which are attractive targets for antiviral drugs. Theoretically, such novel functions can be inhibited without damaging the cells of the infected host. Several drugs with anti-influenza virus activity have been described. These include amantadine and the related compound, rimantadine, which are in clinical use today. Ribavirin is another drug which has anti-influenza virus activity but which is not commonly used to treat influenza virus infections. Neuraminidase inhibitors, which were first described in the early 1970s, have recently been developed into effective antiviral agents that should become available for widespread clinical use in the near future. Additionally, hemagglutinin-based inhibitors are presently being developed.

1. Amantadine and Rimantadine

The related anti-influenza virus drugs amantadine and rimantadine were discovered in the early 1960s by screening for compounds which inhibit viral replication in tissue culture. These drugs are specific and effective inhibitors of the influenza A virus M2 ion channel. By inhibiting M2 function, two distinct steps are targeted in viral replication, but these effects do not extend to influenza B virus. Amantadine appears to interact with the M2 transmembrane domains which, in the M2 tetramer, form the pore through which ions pass. As has been described, the M2 ion channel protein facilitates, within the endosome, the acidification of the viral core and the dissociation of the M1 protein from RNP. Once the RNPs are free of M1, they are able to enter the nucleus and initiate viral replication. When the ion channel activity is blocked by amantadine, RNP uncoating is blocked and replication is inhibited. Additionally, some influenza A virus strains possess HAs which are particularly acid sensitive. For these viruses, passage through the acidic trans Golgi network can trigger conformational changes in the HA unless the M2 ion channel is also expressed. Thus, when M2 activity is inhibited by amantadine or rimantadine, these strains produce an inactive HA. The result is a decrease in viral infectivity.

Amantadine and rimantadine are clinically effective in reducing symptoms of influenza A virus infections and in blocking infection of individuals who receive the drug prophylactically. Both drugs have been associated with side effects, including neurological reactions, but rimantadine seems to be more readily tolerated than amantadine. Use of both amantadine and rimantadine does result in the rapid appearance of drug resistant mutant viruses, both *in vitro* and in humans. Such resistant viruses can even be transmitted from one individual to another. However, to this point, resistant viruses do not appear to have increased pathogenicity as compared to sensitive strains and resistant viruses are not widespread in nature.

2. Ribavirin

Ribavirin is a nucleoside analog, which was first synthesized in 1972. This drug inhibits replication of a variety of viruses, including influenza viruses. The mechanism by which ribavirin exerts its antiviral effects is not clear but specific mechanisms probably differ for different viruses. In the case of influenza viruses, ribavirin may inhibit function of the viral polymerase. Clinical trials with aerosolized ribavirin demonstrated that it can shorten influenza A and B virus clinical illness, but ribavirin is currently approved in the United States only for treatment of respiratory syncytial virus infections.

3. Neuraminidase Inhibitors

The influenza virus neuraminidase was the first virus-associated enzyme discovered. It has long seemed a desirable target for antiviral therapy. In the early 1970s, analogs of the NA substrate, sialic acid, were developed and found to inhibit viral NA activity *in vitro* and to inhibit viral replication in tissue culture. For example, addition of the sialic acid analog FANA (2-deoxy-2,3-dehydro-N-trifluoroacetylneuraminic acid) to influenza virus infected cells resulted in a decease in viral plaque size and an inhibition of multicycle viral replication. Subsequent studies showed that infected cells treated with FANA had much more sialic acid at their cell surface than did infected, untreated cells. Additionally, in the presence of the inhibitor, viruses formed large aggregates at the cell surface rather than being efficiently released. Apparently, without NA function, the presence of sialic acid on the cell surface and on viral glycoproteins resulted in this aggregation. However, when these early NA inhibitors were used to treat influenza virus infection in animals, no effect was seen.

More recently, new inhibitors of NA have been developed based on the crystal structure of NA complexed with its substrate. The specific substrate-interacting amino acids in the NA active site were identified, as were additional conserved residues in the vicinity of the active site. Based on the active site structure, compounds were synthesized which could bind with high affinity in the active site. Two such compounds, 4-guanidino-Neu5Ac2en (zanamivir, also GG167) and GS4104 (ester form of GS4071) are efficient inhibitors of influenza A and B virus NAs and are presently being used in humans. Zanamivir must be administered intranasally or by inhalation for it to show antiviral activity *in vivo*, but has been shown to be effective in reducing influenza virus symptoms in infected individuals and in preventing infection when the drug was administered prior to infection. The compound GS4104 is orally bioavailable and, therefore, may be taken as a pill.

As for amantadine, growth of virus in the presence of NA inhibitors may result in the selection of drug resistant mutants. Two classes of NA-inhibitor resistant mutants have been identified. The less frequently appearing but perhaps more obvious class of resistant mutants possesses changes in the NA active site. Such resistant mutants may have a decreased affinity for the inhibitors. The more frequently isolated drug-resistant mutants possess changes in the HA protein. These HA mutations decrease the affinity of the HA for sialic acid. Viruses with these mutant HAs are, therefore, able to elute from the infected cell surface when the NA receptor destroying activity is low. Interestingly, some HA mutant viruses exhibit drug dependence, that is, the viruses grow more efficiently in the presence than in the absence of NA inhibitor. Presumably, these viruses have an HA with sufficiently low receptor affinity that infection is only efficient when the receptor-destroying activity of NA is decreased. Despite the appearance of NA inhibitor resistant mutants *in vitro*, such mutants have not been isolated *in vivo*. Perhaps resistant mutants are too impaired to be isolated *in vivo*. Also promising is the observation that viruses isolated from the wild have not shown significant resistance to the NA inhibitors. It remains to be seen, however, whether drug resistant viruses may appear in nature following widespread use of these drugs.

4. HA Inhibitors

Inhibitors of HA protein function are also under development. HA inhibitors target one of two functions, either HA binding to sialic acid or HA fusion activity. The first step in viral infection is binding of the HA protein to its receptor, sialic acid. Therefore, sialic acid analogs have also been tested as HA inhibitors. However, it has been more difficult to identify analogs which inhibit HA-sialic acid binding than to find analogs which are effective NA inhibitors.

Since HA-mediated membrane fusion is required for influenza virus infection, this function is also an attractive target for antiviral development. A number of fusion inhibitors have been described. These compounds are thought to bind HA in such a way as to interfere with the conformational changes of HA required for membrane fusion. Because the structures bound by fusion inhibitors may be less well conserved than the sialic acid binding site, it remains

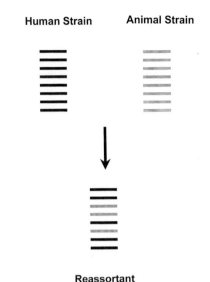

Fig. 8. Reassortment. Reassortment is the exchange of genome segments by two viruses of the same type (i.e., A, B, or C) upon co-infection of the same cell. Since there are 8 genomic RNA segments in influenza A viruses, reassortment between two influenza A virus strains can theoretically result in the appearance of 256 vRNA combinations. However, not all gene combinations are readily obtained by reassortment because some viral gene products coevolve.

to be seen whether fusion inhibitors can be developed with activity against all HA types.

VI. GENETIC MANIPULATION OF INFLUENZA VIRUSES

A. Reassortment

Influenza viruses can exchange genetic information with one another. This exchange occurs in a process called "reassortment," in which chimeric viruses derive segments from two "parental" viruses (Fig. 8). Reassortment occurs when two viruses of the same type infect the same cell, however, it does not occur across type (i.e., influenza A viruses only exchange segments with other A viruses and B or C viruses only with other B or C viruses, respectively).

Reassortment is a useful technique for defining which segment encodes a particular viral phenotype. Such analyses have been useful for determining the functions of various viral proteins, defining attenuating mutations and identifying segments encoding resistance to antiviral drugs. For example, an interesting mutation responsible for a particular viral phenotype might arise spontaneously, following experimental mutagenesis, or in response to growth of virus under selective conditions. The genetic lesions responsible for the phenotype can be mapped using reassortment. Once the phenotype is mapped, the mutant may be used to identify the function of a viral protein. For example, a ts lesion for the influenza A/WSN/33 virus *ts3* strain was mapped to the NA gene. With this knowledge, it was shown that, at nonpermissive temperature, *ts3* did not efficiently bud from infected cells, thus demonstrating the importance of the NA in viral release. Reassortment has also been successfully used to transfer high growth or attenuation characteristics to viruses with novel HA and NA proteins. Reassortants deriving 6 genes from a high yield parent and 6 genes from a cold-adapted parent show high growth in embryonated eggs and attenuation characteristics in humans, respectively. The high yielding influenza virus reassortants have been extraordinarily helpful in the manufacture of killed influenza virus vaccines. Reassortants with a cold-adapted background (6:2 reassortants) are presently being developed as live virus vaccine candidates.

B. Reverse Genetics

Despite the usefulness of mutagenesis and reassortment, these techniques do not permit the introduction of site-specific mutations into the viral genome, a technique which has proved powerful for the analysis of DNA viruses, retroviruses, and positive-strand RNA viruses. Such techniques were relatively straightforward for these viruses because an infectious particle can be generated following transfection of cells with either the cloned genomic DNA or with a cDNA derived from genomic viral RNA. Transfection of such constructs results in the production of viral proteins which can replicate the transfected genome. Because the RNA of negative-strand RNA viruses is not infectious by itself, techniques were needed to introduce both viral RNA and viral polymerase into cells such that a recombinant virus could be recovered.

A reverse genetics system was developed by which a cloned cDNA was transcribed *in vitro* in the presence of purified viral polymerase to generate a polymerase-associated vRNA replica. The resulting RNP complex was transfected into cells and, following transfection, the cells were infected with a "helper virus." The virus yield emerging from the transfection/infection is a mixture of the helper virus and the "transfectant," virus which has acquired the transfected segment. In order to obtain a virus in which one gene has been replaced, a selection system is used to prevent growth of the helper virus while permitting the growth of the transfectant viruses (Fig. 9). For example, the NA gene of influenza A/WSN/33 confers upon that virus the ability to grow on MDBK cells in the absence of exogenous trypsin. A reassortant virus, WSN-HK, which has seven seg-

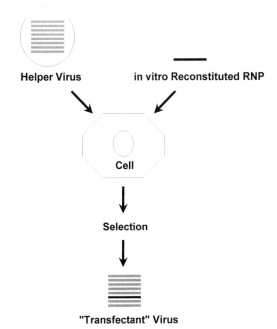

Fig. 9. Method of RNP transfection used to introduce new genetic information into influenza viruses. A vRNA is synthesized *in vitro* in the presence of viral polymerase to generate the reconstituted RNP. The reconstituted RNP is transfected into cells, and the cells are subsequently infected with a helper virus. The resulting virus which emerges from the transfection/infection is a mixture of the helper virus and new "transfectant viruses" carrying the mutant gene segment. A selection procedure is then performed to separate out the transfectant virus.

ments from "WSN" but the sixth (NA) segment from influenza A/Hong Kong/8/68 is unable to grow on MDBK cells in the absence of trypsin. Therefore, the WSN-HK can be used as a helper virus to rescue altered forms of the "WSN" NA. After transfection, the mixture of transfectant and helper virus is grown on MDBK cells in the absence of trypsin. Only viruses that have acquired the new NA segment from the "WSN" virus are able to grow.

Selection systems that permit the rescue of transfectant viruses have been described for all of the influenza A virus segments with the exception of the PA and PB1 segments. Rescue of mutant influenza B virus HA and NA genes has also been described.

These techniques have been useful in defining many aspects of influenza virus biology. For example, much of what we know about viral promoter function in the context of viral replication was obtained using reverse genetics technology. Also, structure/function analyses of the HA, NA, and NS proteins have been significantly aided by these techniques.

Recently, entirely plasmid-based systems which allow the production of transfectant influenza A viruses have been described (Neumann *et al.*, 1999; Fodor *et al.*, 1999). In these systems, cells are transfected with twelve plasmids, four plasmids encoding the viral proteins (PA, PB1, PB2, and NP) required for replication of viral RNAs, plus eight plasmids which produce replicas of the negative-sense vRNAs. Following transfection, virus is obtained without the need for helper virus. Because selection against a helper virus is no longer needed to rescue a desired mutation into virus, genetic manipulation of any segment of the influenza virus genome should now be possible.

As noted above, reverse genetics techniques have also been applied to influenza virus vaccine development. Introduction of specific, stable, attenuating mutations into virus should facilitate the development of better live, attenuated vaccines. Additionally, reverse genetics has been used to engineer influenza viruses that express foreign genes. These viruses are able to elicit strong immune responses against the expressed foreign epitopes. For example, such constructs have been used to elicit protective immune responses in mice against a mouse malaria parasite and against a tumor cell line. Thus, these new techniques promise not only to further our understanding of influenza virus replication but should facilitate the development of new vaccines against influenza viruses, other pathogens, and tumor antigens.

See Also the Following Articles

ANTIGENIC VARIATION • ANTIVIRAL AGENTS • VIRUS INFECTION • ZOONOSES

Bibliography

Colman, P. M. (1994). Influenza virus neuraminidase: Structure, antibodies, and inhibitors. *Protein Science* 3, 1687–1696.

Doherty, P. C. *et al.* (1997). Effector CD4+ and CD8+ T-cell mechanisms in the control of respiratory virus infections. *Immunological Reviews* 159, 105–117.

García-Sastre, A., and Palese, P. (1993). Genetic manipulation of negative-strand RNA virus genomes. *Ann. Rev. Microbiol.* 47, 765–790.

Gerhard, W. *et al.* (1997). Role of B-cell response in recovery from primary influenza virus infection. *Immunological Reviews* 159, 95–103.

Hayden, F. G., and Palese, P. (1997). Influenza virus. *In* "Clinical Virology" (D. D. Richman *et al.*). Churchill Livingstone, New York.

Lamb, R. A., and Krug, R. M. (1996). Orthomyxoviridae: The viruses and their replication. *In* "Fields Virology" (3rd ed.) (Fields *et al.*, eds.). Lippincott-Raven Publishers, Philadelphia.

Murphy, B. R., and Webster, R. G. (1996). Orthomyxoviruses. *In* "Fields Virology" (3rd ed.) (Fields *et al.*, eds.). Lippincott-Raven Publishers, Philadelphia.

Palese, P. *et al.* (1997). Development of novel influenza virus vaccines and vectors. *J. Infectious Disease* 176, Suppl. 1, S45–49.

Smith, C. B. (1998). Influenza viruses. *In* "Infectious Diseases" (2nd ed.) (Gorbach *et al.*, eds.). W. B. Saunders Company, Philadelphia.

"Textbook of Influenza." (1998). (K. G. Nicholson, *et al.*, eds.). Blackwell Science Inc., Malden, MA.

Webster *et al.* (1992). Evolution and ecology of influenza viruses. *Microbio. Rev.* 56, 152–179.

Insecticides, Microbial

Allan A. Yousten
Virginia Polytechnic Institute and State University

Brian Federici
University of California, Riverside

Donald Roberts
Utah State University

I. Bacteria
II. Viruses
III. Fungi

GLOSSARY

biological control Use of natural enemies, including predators, parasites, and diseases, to control insect pests of agricultural, medical, or veterinary importance.

cry genes Genes found in *Bacillus thuringiensis* and a few other bacteria that encode proteins toxic to insects; the genes are frequently located on large plasmids that may be transmissible by conjugation between bacteria.

delta endotoxin A protein toxin produced by the bacterium *Bacillus thuringiensis* that is lethal when ingested by insects of a specific order.

granulosis virus Type of enveloped, double-stranded DNA virus reported only from insects of the order Lepidoptera (moths, butterflies) that replicate initially in the nucleus and later in the cytoplasm of infected cells; virions occluded individually in small occlusion bodies called granules.

integrated pest management Practice for managing insect pest populations whereby chemical, biological, and cultural techniques are integrated to achieve effective, environmentally safe pest control; emphasis on the reduction or elimination of synthetic chemical insecticides.

nuclear polyhedrosis virus Type of enveloped, double-stranded DNA virus reported from insects and other invertebrate organisms; replicates in nuclei of infected cells and occluded in large protein crystals known as polyhedra.

parasporal inclusion body Aggregate of one or more proteins formed inside a bacterial cell at the time of sporulation; sometimes called crystals; may be toxic to insects that ingest them.

recombinant viral insecticide Insect virus that has been genetically engineered, using recombinant DNA technology, to improve insecticidal properties.

MICROBIAL INSECTICIDES are composed of microorganisms that produce disease and, ultimately, death in insects. The microorganisms are introduced into the environment where they may persist and limit insect populations for extended periods of time. More commonly, however, the microorganisms alone or the microorganisms with an associated toxin are delivered into the insect-infested area, where they will produce disease only for limited periods. Even though these microorganisms may be detectable in the area for a long time, they are incapable of maintaining themselves in numbers sufficient to provide effective insect control. When used in this way, they are similar to synthetic organic chemical insecticides; however, they differ in important ways.

A significant advantage of microbial insecticides is safety, which results from the high degree of specificity for the target insect. Nontarget animals, beneficial insects, and plants are not affected. Another advantage is that relatively little resistance has been found among insects exposed to certain microbial insecticides for many years. This has allowed their use in situations where synthetic chemicals have become ineffective. The intelligent use of microorganisms in integrated pest management programs may extend their effectiveness for a long time. In addition to these advantages, microbial insecticides may often be brought onto the market more rapidly and at a lower cost than synthetic chemicals.

Although specificity is an advantage when considered in terms of safety, it is also a disadvantage in some agricultural situations where several different insect pests are present simultaneously. Another disadvantage is that some microbial insecticides are more affected by environmental factors such as humidity and sunlight than are synthetic chemicals. Knowledge of insect ecology and of the environment are important to the successful use of microbial insecticides. The entomopathogenic microorganisms that are likely to make the most important contributions to agriculture and to public health in future years are bacteria, viruses, and fungi.

I. BACTERIA

A. Introduction

A large number of bacterial species have been isolated from insects, and some of these have proven to be pathogenic; however, considerations of safety and effectiveness have resulted in only a few being developed as insecticides. The following species have been proven effective and have been produced commercially.

B. *Bacillus thuringiensis*

Bacillus thuringiensis is the most successful and most widely used microbial insecticide. This is a gram-positive, spore-forming, facultative bacterium. Comparisons of *B. thuringiensis* and *B. cereus*, using a large number of phenotypic tests, were unable to distinguish between the two species; however, *B. thuringiensis* has been retained as a distinct species because it produces several related proteins that are toxic when ingested by insects. Within the species, the strains have been grouped into 70 serotypes, based on flagellar H antigens. In nature, these bacteria have been isolated from diseased insects, from the soil, from granary dust, and from the leaf surface of many plants. This bacterium is easily grown, sporulates well under laboratory conditions, and uses a variety of carbohydrates as carbon and energy sources.

The pathogenicity of *B. thuringiensis* is caused by the toxic proteins that it produces. These toxins, referred to as "delta endotoxins," are synthesized beginning at about stage III of sporulation. They aggregate within cells as parasporal inclusion bodies, or "crystals." The inclusions are readily visible by phase contrast microscopy, and their presence adjacent to an elliptical spore in an unswollen sporangium allows a tentative identification of the species. The parasporal inclusions may be bypyramidal (Fig. 1), elliptical, spherical, rhomboidal, or irregular in shape. These inclusions contain one or more proteins toxic to either Lepidoptera, Diptera, Coleoptera, or, occasionally, to two of these insect orders. Many isolates have been recovered that are not toxic to any insect tested. Some isolates produce a cytolytic/hemolytic toxic protein that is also found in the parasporal inclusion. Although this toxin is less effective against insects than the delta endotoxin, it appears to act synergistically with the delta endotoxins. In addition to these protein toxins, some isolates produce a soluble β-exotoxin, an adenine–glucose–allaric acid compound that functions as an RNA synthesis inhibitor.

The nucleotide sequence of many (176 at this writing) of the genes (cry genes) encoding toxic proteins has been determined and these have been classified

Fig. 1. Carbon replica of *B. thuringiensis* parasporal inclusions (crystals) containing lepidopteran-active toxin.

based on sequence similarities (see *http://www.biols.-susx.ac.uk/Home/Neil_Crickmore/Bt/index.html*). This system has replaced an earlier classification based on insect host specificity. The cytolytic proteins found in some strains are not related to the insecticidal delta endotoxins. The delta endotoxins are synthesized in a variety of sizes, ranging from about 65 kDa to 135 kDa. The larger proteins (protoxins) are cleaved by insect gut proteases to a smaller size, which constitutes the active toxin. The active fragment is found in the N-terminal end of the protoxin and often contains five conserved sequence blocks. X-ray crystallography of delta endotoxin has revealed a protein composed of three domains, each of which is belived to have a role in toxin binding to a target receptor, insertion into the host cell membrane, or some combination of these roles. Membrane insertion by the toxin results in formation of a pore which upsets ion and osmotic balance causing gut cell lysis and larval death.

Any *B. thuringiensis* isolate may possess one or more of the cry genes. Because the toxins encoded by these genes differ in their toxicity for any one insect, different isolates possess different levels of toxicity for insects. This information has opened up the possibility of combining different toxin genes in a single strain to increase the host range of the bacteria included in an insecticide. It may also be possible to use site-directed mutagenesis to change the amino acid sequence of a toxin to increase the toxicity or to vary the spectrum of activity of the toxin.

Following ingestion by larvae, inclusions dissolve in the alkaline gut. For many of the proteins, this is followed by proteolytic cleavage, converting a protoxin into an active toxin.

Bacillus thuringiensis is produced on a large scale by industrial fermentation techniques. At the completion of growth, sporulation, and associated toxin production, the cells are recovered by centrifugation or filtration. Subsequent formulation depends on the target insect (i.e., whether the toxin is directed at agricultural pests on plants, forest insects, or mosquito or blackfly larvae in water). The bacteria must be sprayed repeatedly, much like a chemical insecticide. This is because the bacteria lack the ability to reproduce in numbers adequate to provide effective insect control and the toxins are degraded by sunlight and are washed off plants by rain. Despite worldwide use of *B. thuringiensis* in agriculture for about 40 years and heavy use of a dipteran-active strain in an onchocerciasis control program in Africa during the 1980s and 1990s, there have been only a few isolated reports of the development of resistance to these toxins in the field. Insertion of individual cry genes into plants such as corn, cotton, and potato has given rise to the fear that long-term exposure to single toxins may increase the likelihood of selection for insect resistance. Schemes to avoid this have been proposed but are of unproven effectiveness.

C. *Bacillus sphaericus*

These bacteria are aerobic, gram-positive bacilli that form round spores in a swollen sporangium. They are common inhabitants of soil and are relatively inert in their metabolism, failing to metabolize glucose or several other sugars because of missing enzymes for sugar transport and catabolism. They are readily grown in media containing proteinaceous substrates. The majority of strains found in culture collections have no pathogenicity for insects and the first pathogenic isolate was described in 1964. Strains that are pathogenic to mosquito larvae are found in a single DNA similarity group within the species and are not readily distinguished from nonpathogenic strains by common phenotypic tests. Pathogenicity is limited to mosquito larvae and is caused by production of a protein toxin that accumulates within the cell as a parasporal inclusion at the time of sporulation (Fig. 2). Although similar to the delta endotoxins of *B. thuringiensis* in cell location, the toxins of the two species are not related. Unlike the dipteran-active *B. thuringiensis* toxins, the *B. sphaericus* toxin has no activity against blackflies. The *B. sphaericus* toxin is unusual in that it is found in a binary form; proteins of 41.9 kDa and 51.4 kDa are both found in the parasporal body and both are required to kill larvae. Upon release from the parasporal inclusion in the alkaline gut, the toxins are cleaved by gut proteases to smaller forms that have increased toxicity to cultured mosquito cells. The genes for both of the proteins have been cloned and only small differences in amino acid sequence have been found among toxins from different strains. However, these

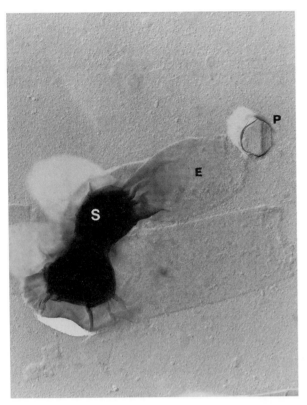

Fig. 2. Carbon replica of *B. sphaericus* spore (S), exosporium (E), and parasporal inclusion (P), the latter containing proteins toxic to mosquito larvae.

differences do result in somewhat different levels of toxicity for different mosquito species. The mode of action of this toxin is unknown. In addition to the binary toxin, *B. sphaericus* produces three additional toxins (Mtx, Mtx2, Mtx3) of 100, 31.8, and 35.8 kDa, respectively. These do not appear to play a significant role in pathogenicity. This pathogen appears to persist somewhat longer in the aquatic environment after application than does *B. thuringiensis*. There have been a few reports of mosquito resistance to the binary toxin.

D. *Paenibacillus* (formerly, *Bacillus*) *popilliae* and *Paenibacillus* (formerly, *Bacillus*) *lentimorbus*

Paenibacillus popilliae and *P. lentimorbus* are the etiological agents of milky disease in larvae of the family Scarabaeidae (Coleoptera). In North America,

the best known of these insects is the Japanese beetle. Similar beetle larvae are major agricultural pests in other parts of the world. These two closely related bacterial species are facultative, nutritionally fastidious, spore forming, and present a gram-positive cell wall profile, although the gram stain is reported to be negative during vegetative growth. Some have suggested there should be a single species (*P. popilliae*) and that *P. lentimorbus* should be a subspecies within that species. However, DNA similarity studies have shown that, although closely related, the two species are distinct. The species were originally separated by the production of a parasporal inclusion by *P. popilliae* and the absence of this body in *P. lentimorbus*. However, the DNA similarity study demonstrated that some isolates (although not the type strain) of *P. lentimorbus* also produce paraspores. *P. popilliae*, the subject of most metabolic studies, appears to lack a complete tricarboxylic acid cycle and is catalase negative. It has been suggested that these enzyme deficiencies may be related to the inability of these bacteria to sporulate *in vitro*.

There is no direct evidence that the protein composing the parasporal body plays any role in the course of the disease process. The gene encoding the parasporal protein has been cloned and sequenced from a European isolate and was shown to have significant similarity to genes encoding the parasporal proteins in *B. thuringiensis*.

The course of infection in the Japanese beetle larva is initiated following ingestion of spores. Over the period of 2 to 4 weeks, spores germinate in the larval gut, vegetative cells penetrate the epithelium, proliferate in the hemolymph, and finally, sporulate in the hemolymph. The dead larva displays a turbid or "milky" hemolymph containing up to 5×10^{10} spores/ml. With the disintegration of the cadaver, the spores are released into the soil to be consumed by additional larvae as they feed on plant roots. Since the bacteria persist in the soil, they do not need to be reapplied, as do *B. thuringiensis* and *B. sphaericus*.

Insecticides containing spores of *P. popilliae* are produced by collecting infected larvae from beneath sod or by collecting uninfected larvae and infecting them in the laboratory. The spores from infected larvae are blended with inert carriers and sold as a powder, designed to be introduced into the soil. A U.S. government-sponsored program using powder produced in this way distributed spores widely in the northeast United States in the 1940s and 1950s. Following its registration in 1948, the powder has been produced commercially on a small scale. Although the bacteria grow on bacteriological media, only a small percentage sporulate. This has frustrated attempts to produce spores by large-scale fermentation. Although a patent issued in 1989 described a process for obtaining *P. popilliae* spores *in vitro,* it now appears that this process is unsuccessful.

E. *Serratia entomophila*

Serratia entomophila causes amber disease in the New Zealand grass grub. Bacteria are ingested with food particles and adhere to the foregut membrane. The insect ceases feeding, takes on an amber coloration, and, after about one month, the bacteria invade the hemolymph, producing a fatal septicemia. Genes essential to the disease process are located on a large plasmid. First isolated in 1981, the bacteria are produced commercially in New Zealand and seeded as vegetative cells into pasture soils.

F. Future Developments

Screening programs continually uncover new bacterial insect pathogens, as well as strains of known pathogens having unique toxins. The anaerobic mosquito pathogen, *Clostridium bifermentans,* has promise as a biocontrol agent but has not yet been produced commercially. The vast numbers of anaerobic bacteria have been largely neglected as potential entomopathogens and may yield more useful strains. The description of a high molecular weight toxic protein complex, produced by the nematode symbiont *Photorhabdus luminescens,* may lead to the insertion of this toxin into plants. Genetic manipulation of *B. thuringiensis* toxins is already producing strains having better insecticidal performance. Insertion of the genes for these toxins into plants as an alternative to traditional formulation techniques will undoubtedly continue, despite reservations about the effect of this approach on resistance development.

II. VIRUSES

A. Introduction

Viruses are obligate intracellular parasites that are known from all types of organisms. Biochemically, they consist of a DNA or RNA genome surrounded by a layer of protein subunits, collectively known as the capsid. Some of the more complex types of viruses also contain lipid in the form of an envelope that, typically, is derived from membranes of the host cell.

As they do in other types of organisms, viruses cause important diseases in insects, and it has been known for more than 50 years that some types of viruses attacking insects periodically cause spectacular declines in insect populations. The lethal effects that these viruses have on insect populations suggested that viruses could be used as "natural enemies" to control insect pests and, over the past few decades, there has been a considerable effort to develop insect viruses as either classical biological control agents or viral insecticides. A classical biological control agent is a living organism or virus that, when introduced into a pest population, results in permanent establishment of the agent, accompanied by reduction of the population on a long-term basis, to a point where it no longer is a significant economic pest. A viral insecticide, on the other hand, does not lead to long-term control, but rather must be applied periodically, much like a synthetic chemical insecticide, though not as frequently.

Though the use of insect viruses as classical biological control agents provides an ideal solution to pest control problems, this tactic, unfortunately, is not effective in most cases. A noteworthy exception is control of the European spruce sawfly by a nuclear polyhedrosis virus in Canada. This pest was introduced into Canada during the early part of the century and was a major forest pest by the 1930s. Around 1935, a sawfly nuclear polyhedrosis virus was introduced along, with parasitic wasps, to control this pest and, within 5 years, the sawfly was effectively controlled by these natural enemies. Subsequent studies showed that the virus was the principal factor responsible for this control. In most cases, however, viruses are used as insecticides.

The viruses used as insecticides have a very narrow host spectrum and, typically, are only active against the target insect pest and a few closely related species, making them much safer environmentally than most of the broad-spectrum chemical insecticides. Though this is certainly an advantage, viruses also have disadvantages, a key one being that, as obligate parasites, mass production for commercial use requires that they be grown in living insect hosts. In this section, the properties of the major types of viruses that attack insects will be reviewed, with the discussion focusing on the nuclear polyhedrosis viruses, the virus type that has been most widely used as microbial insecticide and that shows the most promise for more widespread use in the future.

B. Major Types of Viruses

1. Occluded and Nonoccluded Viruses

The viruses that attack insects are divided into two broad nontaxonomic categories, the occluded and nonoccluded viruses. The occluded viruses are so named because, after formation in infected cells, the mature virus particles (virions) are occluded within a protein matrix, forming protective paracrystalline bodies that are generically referred to as either inclusion or occlusion bodies. In the nonoccluded viruses, the virions occur freely or occasionally form paracrystalline arrays of virions that are also known as inclusion bodies; these, however, have no occlusion body protein interspersed among the virions.

The occluded viruses of insects include the cytoplasmic polyhedrosis viruses (cytoplasmic RNA viruses), entomopoxviruses (cytoplasmic DNA viruses), nuclear polyhedrosis viruses (nuclear DNA viruses), and granulosis viruses (nuclear/cytoplasmic DNA viruses), but the overwhelming majority of those developed or under development as microbial insecticides belong to the latter two groups, both of which belong to the insect virus family Baculoviridae. The baculoviruses are large DNA viruses that form enveloped rod-shaped virus particles that are packaged by the virus into the occlusion bodies already noted and illustrated in Figs. 3a and 3b. Literally thousands of nuclear polyhedrosis viruses (NPVs) and granulosis viruses (GVs) have been dis-

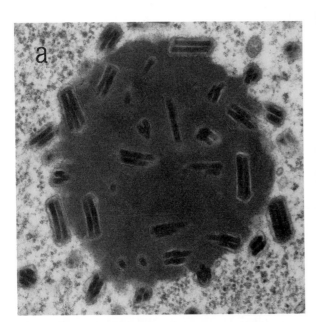
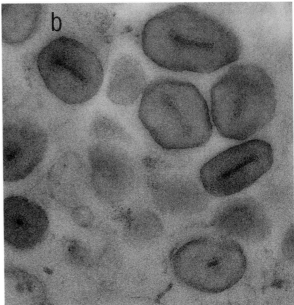

Fig. 3. Electron micrographs illustrating the internal structure of occlusion bodies of a nuclear polyhedrosis virus (a) and a granulosis virus (b). In the nuclear polyhedrosis viruses, many virions are occluded in each occlusion body, whereas in the granulosis viruses, there is only one virion peer occlusion body. The nuclear polyhedrosis viruses and granulosis viruses are the most common types of insect viruses used as insecticides. Magnification is approximately 52,000×.

covered, but only those that attack important lepidopteran (caterpillar) or hymenopteran (sawfly) pests are used or considered for use as insecticides.

2. General and Molecular Biology

The NPVs and GVs cause infectious diseases of insects and gain entry to their hosts primarily by being eaten. In the case of lepidopterans, insect larvae ingest occlusion bodies while feeding, and the virus invades and, initially, colonizes the host's stomach. Once established, the virus spreads to most other tissues, where it infects and grows within the cell nuclei (NPVs) or in nuclei and the cytoplasm (GVs), eventually killing the host. The time between infection and death varies but depends mostly on the amount of virus consumed and the stage of larval development at the time of infection. Young larvae that consume several polyhedra, for example, can be killed within 48–72 hr. but older larvae that feed on the same amount of virus may not die for 1 week or longer. The initial transmission and infection process is similar for the NPVs in sawfly larvae, except that the infections are restricted to stomach cells.

Most NPVs and GVs are very specific in their host range, attacking only one or several closely related species. A few, such as the NPV isolated from the alfalfa looper, *Autographa californica,* have a broad host spectrum that extends to over 40 lepidopteran species. In comparison to chemicals, this is still a very narrow spectrum of activity and is one of the reasons that these viruses are so safe for nontarget organisms.

With respect to their molecular biology, the genome of the NPVs and GVs is a large, circular, double-stranded DNA molecule, with a size of approximately 130 kilobase pairs. It is estimated that the genome of each virus is capable of encoding about 50 proteins, making these viruses among the largest and most complex known. During viral replication, the synthesis of these proteins is divided into four major sequential phases. Virions are assembled and occluded in occlusion bodies during the last phase, which is also the time period during which the occlusion body protein, a very highly expressed protein of 29 kDa, is also synthesized. In the NPVs, numerous virions are occluded within each occlusion body, and, as these are typically polyhedral in shape,

they are often referred to as polyhedra. In the GVs, each virion is occluded in a single small occlusion body, and these, being granular in shape, are referred to as granules. In the case of NPVs, hundreds of polyhedra form within each infected nucleus and, in the GVs, thousands of granules form per cell.

C. Conventional Viral Insecticides

The major insect pests, be they of agricultural or medical importance, fall into two broad groups with respect to their feeding—they are either chewing or sucking insects. Sucking insects feed directly in host tissues and are not good targets for viral insecticides because, unlike fungi, they cannot penetrate through insect cuticle. Thus, the NPVs and GVs, which must be eaten to be effective, can only be used from a practical standpoint against insects with chewing mouthparts. Fortunately, many of the most important agricultural insect pests, caterpillars, for example, are chewing insects. In addition, many NPVs and GVs have been isolated from these important insect pests.

1. Target Pests: Key Examples

The primary targets for viral insecticides are the numerous caterpillar (cutworms, loopers, armyworms, and bollworms) and sawfly pests that attack crops, such as vegetables, cotton, maize, sorghum, rice, fruit crops, and forests throughout the world. Some particularly important target pests include species of *Heliothis,* such as *H. zea, H. virescens,* and *H. armigera,* that are major pests of grain crops and cotton in many regions of the world; species of *Spodoptera,* such as *S. littoralis* and *S. litura,* major pests of cotton in the Middle East and Asia, and *S. exigua* and *S. frugiperda,* major pests of vegetable crops in many areas of the world; and species of *Mamestra* that attack vegetable crops in Europe. With respect to forests, the major targets include sawfly larvae of the genus *Neodiprion* in North America and Europe and caterpillar pests, such as larvae of the gypsy moth and Douglas fir tussock moth, in the United States.

For many of the above pests, viral insecticides serve as useful methods of control when used either alone or in conjunction with reduced levels of chemical insecticides in integrated pest management programs. As in the case of chemical insecticides, the use of an insect virus as a control agent, in many countries, requires that it be registered as an insecticide. The following are examples of viral insecticides registered by the U.S. Environmental Protection Agency: the NPV of *Heliothis* spp., developed for use in cotton (commercial name, Elcar); the NPV of the Douglas fir tussock moth, for use in forests (TM-Biocontrol 1); the NPV of the gypsy moth, for use in hardwood forests (Gypcheck), and the NPV of *Neodiprion sertifer,* for use against the pine sawfly. The GV of the codling moth, an important pest of apples, is currently in the process of being registered. A similar range of viruses have been registered in Europe.

Though viruses are registered and used in industrialized countries, their use in less developed nations has actually been much more extensive. For example, a variety of different NPVs and GVs, especially those infective for *Spodoptera* and *Heliothis* spp., are widely used for control of vegetable and field crop pests in many parts of the world. The reasons for this are that registration is relatively simple or not required, labor to produce the virus is inexpensive, and the viruses are cheaper than the chemical insecticides available locally. The viruses in most of these countries are produced by farmers' cooperatives or "cottage industries." Usage appears to be increasing, which testifies to the efficacy of these viruses, at least by the standards used within these countries.

2. Production and Use

As obligate intracellular parasites, viruses can only be grown in living cells. Thus, NPVs and GVs must be produced in their larval hosts. To do this, large cultures of the target pest are reared, usually in large environmental chambers, either on an artificial diet or on a natural food source. When the larvae are slightly beyond the middle phase of their growth, they are fed an amount of virus sufficient to kill them before they pupate. This ensures maximum yield of virus. Just prior to or shortly after death, the virus-infected larvae are collected and formulated into an insecticidal powder or liquid concentrate. These insecticide powders or liquids are then applied to crops or forests, using equipment and methods similar to those used to apply chemical insecticides. Another possibility for production is to grow these viruses *in vitro* in large cell cultures, but cell culture technology has not yet made this commercially feasible.

3. Limitations

Due largely to the slow speed-of-kill, narrow host spectrum, and necessity of using larvae for mass production, relatively few viruses have been registered as insecticides in the United States or Europe. One reason for this is that viruses have had to compete with available chemical insecticides, which, in general, have been cheaper; however, the increasing costs of chemical insecticides and the development of resistance to these in many insect populations have resulted in increased interest in developing more viruses as insecticides. In addition, the development of recombinant DNA techniques offers the potential for increasing the host range of insect viruses in a controlled manner and the possibility for increasing the speed-of-kill.

D. Improved Recombinant Viral Insecticides

As has been noted, two commercial limitations of viral insecticides are that they have a very narrow host range and, if not used against young larvae, which is often not practical, they can take from 5–10 days to kill the target pest, a period that can result in significant economic loss. To overcome these limitations, techniques of recombinant DNA technology are being used to genetically engineer more efficacious viral insecticides. The basic objective here is to find genes which can be deleted or added to improve efficacy. One gene for deletion, egt—encoding ecdysone glucosyl transferase—has been identified. When deleted, it reduces the time from infection until insect death by 25%. Other genes are for insecticidal proteins that can be introduced into the viral genome under the control of a strong promoter and that will result in an expanded host range, but that are still limited to important insect pests and an increased speed-of-kill. Several candidate genes have been identified, including those for insect peptide hormones, regulatory enzymes, and insecticidal proteins from mites, spiders, and scorpions. Several recombinant viruses that express these proteins have been constructed, and it has been shown that the speed-of-kill has been reduced to a period of several days in late instar larvae, as opposed to a week or more for the wild-type virus. These results indicate that the goal of producing a genetically engineered viral insecticide that is environmentally safe and commercially feasible should be a reality within the next few years.

E. Conclusions

The successful use of viruses in developed countries in the future will continue to depend on the regulatory environment, their efficacy, and their economic competitiveness. As long as other cheaper materials, such as chemicals and *Bacillus thuringiensis*, are available and effective, they will be used. The viruses likely to be used in the future will be conventional viruses, targeted against major pests for which chemicals and bacterial insecticides are not available, and recombinant viruses that kill target insects more quickly and have a broader host range.

III. FUNGI

A. Introduction

The entomopathogenic fungi fill an important niche in microbial control of insect pests. Virtually all insect orders are susceptible to fungal diseases. Fungi are particularly important in the control of pests that feed by sucking plant juices, because these insects have no means of ingesting pathogens, and they are important for Coleoptera control because viral and bacterial diseases are unknown for many of the coleopteran pests. There are approximately 700 species of entomopathogenic fungi in almost 100 genera. Accordingly, there is the potential for developing microbial control programs with fungi for virtually all pest insect species. Only a very small percentage of possible fungus/insect combinations, however, have been tested for their potential as microbial control systems.

From the 1880s through the early 1900s, the spec-

tacular epizootics caused by entomopathogenic fungi led to studies of their potential use for pest control. Interest in fungi as pest control agents waned, however, as synthetic chemical insecticides were used more frequently. More recently, because of the myriad difficulties that have been gradually encountered in the use of chemical insecticides, the field of biological control has been undergoing a renaissance. In particular, the knowledge of entomopathogenic fungi is increasing rapidly.

B. Diversity of Entomopathogenic Fungi

Insect-infecting fungi are found in virtually all taxonomic groups except the higher Basidiomycetes (mushrooms) and the dematiaceous Hyphomycetes (melanized imperfect fungi). Entomogenous fungal species are also diverse in their degree of virulence; they range from obligate pathogens, through facultative pathogens attacking only weakened hosts, to commensal or symbiotic fungi. Research aimed at pest control usually targets facultative and obligate pathogens of economically important pests. Taxonomically more primitive fungi with flagellated motile spores have been one focus for mosquito control efforts in aquatic habitats; however, the majority of entomogenous fungi with control potential are in the Order Entomophthorales (Class Zygomycetes) or the Class Hyphomycetes. The Entomophthorales are generally characterized by species with heightened host specificity and great epizootic potential. Many species in the Hyphomycetes have broader host ranges and are generally easier to grow *in vitro*. As a result, almost all species that have been registered for use are Hyphomycetes.

In addition to having a wide spectrum of hosts, entomopathogenic fungi often have wide geographic ranges. Considerable genetic diversity can be found among various isolates of single species from different hosts and localities. In addition to naturally occurring fungal isolates, there is the potential to modify characteristics of entomopathogenic fungi through genetic manipulation techniques. Three of the most common hyphomycetous entomopathogenic fungi have been transformed in this way.

C. Pathogenicity

In virtually all cases, the infective unit for an entomopathogenic fungal species is a spore, usually, a conidium. The course of disease development is initiated by the spore adhering to the cuticle of an insect. The spore germination and the resulting germ tube either penetrates the cuticle directly or produces an appressorium, from which an infection peg grows into the insect cuticle. An understanding of the mechanisms for triggers, receptors, and the biochemical machinery for appressorium initiation and their relationship to the expression of virulence by insect pathogens has been slow to form; however, the development of an *in vitro* system, whereby differentiation of *Metarhizium anisopliae* is induced when conidia germinate on a flat hydrophobic surface in a sparse nutritional environment, has facilitated meaningful examinations of the biological significance and function of appressoria. A model for appressorium formation has been proposed in which a localized induction signal disrupts the apical Ca^{2+} gradient required for localized exocytosis and maintenance of polar germ-tube growth. This initiates appressorium formation by turgor pressure against an expanded area of the cell wall, produced by randomly dispersed exocytosis of new wall material over the entire cell surface. The fungus penetrates to the hemocoel, where it normally grows in a yeastlike phase, called hyphal bodies or blastospores. A fungus may produce toxins in the hemocoel that aid in overcoming the immune response of the host or in causing other disruptions of host physiology. The fungus eventually invades virtually all internal organs of the insect after which it penetrates to the outside of the cuticle and produces new conidia. In dry conditions, the fungus may lie dormant within the dead host for long periods of time, rather than emerging to the exterior and producing conidia. Some Entomophthorales produce heavy-walled resting spores that remain within cadavers and serve as the overwintering stage of these fungi.

A full understanding of the factors affecting the ability of the fungus to enter, kill, and subsequently sporulate on its host is paramount in the development of fungi as control agents. Factors that interfere in any of the critical steps could render the pathogen less efficient, if not ineffective. Only recently have the complex interactions occurring between the host integument and fungal pathogen been explored in detail. Features affecting these interactions include molting, host defense mechanisms, fungal strain, presence of nutrients on host integument, temperature, and humidity. Behavior of the host insect can be important. Fungus-infected grasshoppers and houseflies seek warm sites or bask in sunlight to raise their blood temperature to well above that which allows fungal growth, thereby escaping or delaying death from the fungus.

A possible indirect use of entomopathogenic fungi for insect control would be the detection, isolation, characterization, and, finally, commercial development of toxins produced in culture by these fungi. The secondary metabolites of entomopathogenic fungi include a number of compounds toxic to insects. The known toxic compounds are quite diverse chemically but a number of them (e.g., destruxins and bassianolide) are depsipeptides. A second indirect use, and one currently being researched, is to transfer to plants fungal genes for products (enzymes and toxins) debilitating for insects—and thereby protect the plants—or to other insect pathogens, e.g., viruses—to enhance their effectiveness in pest control.

D. Current Approaches to Field Use

1. Permanent Introduction

Frequently called classical biological control, permanent introduction entails establishment of a fungal species in an area with host populations where the pathogen does not occur. This method for control is clearly the least labor-intensive and least costly over the long term, because it involves a limited number of releases of relatively small amounts of fungal propagules. Its results are aimed at self-sustaining, long-term control.

The pathogen, *Entomophaga maimaiga* has been used successfully in both classical biological control and inoculative release efforts against the gypsy moth. *Entomophaga maimaiga* was introduced from Japan to northeastern North America in 1910–1911 to control gypsy moth populations that had been introduced from France four decades earlier. This fungus was not believed to have become established until, in 1989 and 1990, it caused extensive mortality in gypsy moth populations in 10 northeastern states. Current thinking is that this outbreak was due to an unintentional, rather recent, introduction of a highly virulent *E. maimaiga* strain. *Entomophaga maimaiga* is now distributed in those areas where the gypsy moth has been established for some time. Therefore, at present, efforts are under way to introduce this fungus to the leading edge of the ever-increasing gypsy moth distribution.

Although classical biological control introductions are frequently attempted with parasitic insects, colonization attempts of exotic fungi to control exotic pests have seldom been reported in the literature. Obviously, this tactic is underemployed in biological control programs.

2. Inoculative Augmentation

Inoculative augmentation involves releasing a pathogen in the field, with the expectation that it will cycle in the host population to provide effective control. Frequently, inoculative releases are repeated during a season (mycoinsecticide), and it is not expected that effective populations of the pathogen will carry through to the next year. This method has been used with fungi, most frequently, for pest control on an annual basis.

The coleopterous family Scarabaeidae contains many species whose larval stages are devastating to roots and whose adults may or may not cause foliar damage to agricultural and horticultural plants. The group includes Japanese beetles, European cockchafers, and rhinoceros beetles. The larval stages, which may last 1 to 3 years, usually remain within the root zone. This makes application of fungi, as well as other types of control, difficult. A novel, long-term approach to control of the cockchafer is under development in Switzerland where the insect has a 3-year life cycle and the population is synchronous. Adults are present only once every 3 years. Adults aggregate at the edges of pastures, where they previously developed through the immature stages, and feed and mate in the surrounding trees. *Beauveria brongniartii* blastospores are sprayed on these border trees. The adult female, after feeding, returns to the pasture where she burrows several centimeters into the

ground to oviposit. Many fungus-exposed adults die in the field, either underground or on the surface. The fungus sporulates on the cadavers, thereby introducing foci of large numbers of conidia throughout the field. It has been noted that, after the second generation following application, the fungus maintains the cockchafer population at nondamaging levels.

Metarhizium flavoviride isolates from West Africa and northeastern Brazil are effective pathogens of grasshoppers and locusts; and developing of mass production of conidia, formulation, and field-application techniques are under way. Spittle bugs are serious pests in sugar cane and improved pasture lands in Brazil. Conidia of *Metarhizium anisopliae* have been used for control of these insects for more than 10 years. Production of the fungus is by small companies or grower cooperatives. The spittle bug populations in sugar cane fields normally are reduced by about 40% following fungus application. Another attractive feature of fungus use, in contrast to chemical insecticides, is that hymenopterous parasites used for control of a lepidopterous borer of cane are not affected. Since 1986, approximately 100,000 hectares of cane have been treated annually. Perhaps the largest program entailing fungi for insect control is that of the People's Republic of China to treat pine forests with *Beauveria bassiana* conidia for control of pine moth larvae. At least 1 million hectares are involved and applications are usually needed at 3-year intervals.

3. Conservation or Environmental Manipulation

This involves manipulation of the host environment to enhance activity of the fungal pathogen. Such manipulations have proven very successful in systems that are well understood, and use of this approach is clearly a goal of integrated pest management. This approach would include such tactics as selection of chemical pesticides or timing of chemical pesticide application to cause minimal damage to the entomopathogenic fungi. Because of moisture requirements of most fungi infecting insects, environmental manipulations generally center around maintaining the pathogen in a moist environment where pests are also abundant. An elegant example of environmental manipulation of fungal entomopathogens is a system developed for control of the alfalfa weevil. This is based on cutting alfalfa early and leaving it in windrows for several days. The adult weevils aggregate in the windrows, which provide a moist and warm microclimate for transmission and development of *Erynia* spp. Simulation modeling of this system, with the addition of early season sampling to regulate insecticide application, has led to altered alfalfa weevil control recommendations and projected increased net profits to growers.

4. Compatibility in Integrated Pest Management

The restricted host ranges of fungi offer control with limited threat to nontarget organisms, including compatibility with parasites and parasitoids and other pathogens. The most successful examples of integrated use are found in greenhouse pest control where *Verticillium lecanii* and *Aschersonia aleyrodis* can be used with predatory mites and beneficial wasps for effective control of aphids and whiteflies, respectively. Also, *Metarhizium* is used in Brazil for spittle bug control in cane fields, simultaneously with release of *Apanteles* wasps parasitic on sugar cane borers.

E. Mass Production and Formulation

Although some potentially useful entomopathogenic fungi are difficult to mass produce, many grow profusely on very simple, inexpensive media. The most common procedure for large-scale production is to grow the fungus on plant materials, such as bran, cracked barley, rice, or even peat soil, until sporulation. The resulting conidia, with or without plant substrate, are introduced into the field, usually by means of standard pesticide application equipment. A recent breakthrough has been the development of a technique for producing dry, viable mycelium that can be stored under refrigeration. When introduced into fields, particles of dry mycelium rapidly produce infectious conidia following rehydration from dew, rain, or soil moisture. This method can be used with many species of fungi and has been successfully field tested. Formulation is frequently very simple or nonexistent. Nevertheless, development of proper formulations is crucially needed. These would be designed to improve shelf life, persistence in the field, the range of conditions under which the fungus is effective, spread in the insect habitat, tolerance to solar irradiation, and adherence to substrates where insects reside.

F. Biotic and Abiotic Limitations

As with other biological control agents, fungi are limited by an array of biotic and abiotic factors. Biotic limitations are poorly understood. These include microbial antagonists on the host integument, leaf surface, or in the soil; host feeding behavior, physiological condition, including humoral and cellular immune reactions, and age; and fungal strain.

Abiotic factors include inactivation by sunlight, desiccation, certain insecticides or fungicides, and temperature and humidity thresholds for germination and growth. Traditionally, because of a germination requirement for free water or a relative humidity of over 90%, relative macrohumidity has been considered as the most serious constraint on the use of fungi for insect control; however, studies indicate that fungal infections can occur at relatively low macrohumidities. Microhumidities at the surface of the host integument or on the foliage may be sufficient for spore germination and host penetration. Incompatibility with other components of pest control such as insecticides and fungicides can also be a serious limitation.

G. Future Research Approaches with Fungi

Although entomopathogenic fungi have obvious high potential for pest control, this potential has seldom been properly exploited. With the increased interest in use of fungi in other biotechnologies, major improvements in mass cultivation, storage, and formulation technology can be expected. These advances, coupled with increased understanding of pathogenesis with emphasis on epizootiology, host specificity, virulence, and use of recombinant DNA technology, fungi should play an increasingly important role in integrated pest management.

See Also the Following Articles

Biopesticides, Microbial • Pesticide Biodegradation • Recombinant DNA, Basic Procedures

Bibliography

Burges, H. D. (ed.) (1981). "Microbial Control of Pests and Plant Diseases 1970–1980." Academic Press, London, UK.

Federici, B. (1999). A perspective on pathogens as biological control agents for insect pests. *In* "Handbook of Biological Control" (T. S. Bellows and T. W. Fisher, eds.), pp. 517–548, Academic Press, San Diego.

Granados, R., and Federici, B. (1986). "The Biology of Baculoviruses. Vols. I and II." CRC Press, Boca Raton, FL.

Hajek, A. E. (1997). Ecology of terrestrial fungal entomopathogens. *Advances in Microbial Ecology* **15**, 193–249.

Laird, M., Lacey, L., and Davidson, E. (ed.) (1989). "Safety of Microbial Insecticide." CRC Press, Boca Raton, FL.

Porter, A. G., Davidson, E., and Liu, J. (1993). Mosquitocidal toxins of bacilli and their genetic manipulation for effective biological control of mosquitoes. *Microbiol. Rev.* **57**, 838–861.

Prieto-Samsonov, D. L., Vazquez-Padron, R., Ayra-Pardo, C., Gonzalez-Cabrera, J., and de la Riva, G. (1997). *Bacillus thuringiensis:* from biodiversity to biotechnology. *J. Indus. Microbiol. Biotechnol.* **19**, 202–219.

Roberts, D. W., Fuxa, J., Gaugler, R., Goettel, M., Jaques, R., and Maddox, J. (1991). Use of pathogens in insect control. *In* "Handbook of Pest Management in Agriculture" (2nd ed.) (D. Pimentel, ed.), pp. 243–278. CRC Press, Boca Raton, FL.

Roberts, D. W., and Hajek, A. E. (1992). Entomopathogenic fungi as bioinsecticides. *In* "Frontiers in Industrial Mycology" (G. F. Leatham, ed.), pp. 144–159. Chapman and Hall, New York.

Samson, R., Evans, H., and Latge, J. P. (1988). "Atlas of Entomopathogenic Fungi." Springer-Verlag, The Netherlands.

Schnepf, E., Crickmore, N., VanRie, J., Lereclus, D., Baum, J., Feitelson, J., Zeigler, D., and Dean, D. 1998. *Bacillus thuringiensis* and its pesticidal crystal proteins. *Microbiol. Mol. Biol. Rev.* **62**, 775–806.

Wraight, S. J., and Carruthers, R. E. (1998). Production, delivery and use of mycoinsecticides for control of insect pests of field crops. *In* "Biopesticides: Use and Delivery" (F. Hall and J. Menn, eds.) (In press) Humana Press, Totowa, NJ.

Interferons

Bryan R. G. Williams

Cleveland Clinic Foundation

GLOSSARY

antiviral state Intracellular resistance to virus infection.

apoptosis Genetically programmed cell death.

cytokine Intercellular signaling protein.

DNA methylation Addition of methyl ($-CH_3$) groups to DNA.

natural killer cells Large granular white blood cells that mediate innate, nonspecific immunity.

RNA polymerase II Enzyme which catalyzes the synthesis of RNA complementary to a DNA template. RNA polymerase II transcribes all genes except those encoding small RNAs such as transfer RNAs or ribosomal RNAs.

transcription factors Regulatory proteins that bind short, specific DNA sequences which control the transcription of genes to which they are linked.

THE INTERFERONS (IFNs) are antiviral cytokines synthesized and secreted by vertebrate cells in response to virus infections and other stimuli. Cells exposed to IFNs respond by acquiring resistance to subsequent virus infection. IFNs are also able to inhibit the proliferation of cancer cells, and because of these combined properties of inducing antiviral resistance and regulating cell growth IFNs are used as therapeutic agents, for the treatment of both viral infections and cancer. IFNs are also able to modulate the immune response and this has led to their clinical use in diseases with underlying immunological etiologies, including forms of multiple sclerosis. IFNs were discovered and named for their capability of inducing resistance to virus infection and were the first described members of the large family of biologically important regulatory proteins termed cytokines.

Studies on the mechanism of action of IFN have provided important information relevant to the mechanism of action of other cytokines, including the nature of receptors and mechanisms of signal transduction and transcriptional regulation of inducible genes. In a physiological context, IFNs act in concert with a complex network of other cytokines.

I. DISCOVERY AND DEFINITION

The interferons (IFNs) were discovered in 1957 as antiviral agents synthesized in influenza virus-infected chick embryo cells. IFNs are produced by cells in response to viral infection and confer an antiviral state on cells exposed to them (Fig. 1). In addition to avian, IFNs are also present in fish, reptiles, and mammals. The cloning and characterization of IFNs revealed the existence of two types, I and II (Table I). The type I IFN family is predominately IFN-α and -β, but it also includes two other subfamily members, IFN-ω and IFN-τ. Type II comprises only IFN-γ.

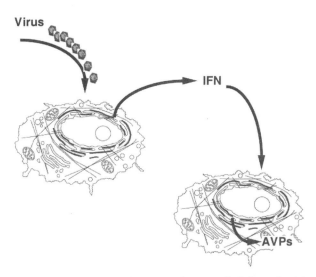

Fig. 1. Production and action of IFN. Cells infected with viruses produce IFNs which act on adjacent cells to induce an antiviral state mediated by the activity of proteins induced by IFNs.

A. Type I IFNs

The genes and corresponding proteins of the type I IFN superfamily are structurally related and the genes in human are clustered within 400 kb on the

TABLE I
Human IFN Genes and Proteins

Gene	Protein
IFNA1	IFN-α_1, IFN-αD
IFNA2	IFN-α_2 (IFN-α_{2b}), IFN-αA (IFN-α_{2a}), IFN-α_{2c}
IFNA4	IFN-α_{4a} (IFN-α76), IFN-α_{4b}
IFNA5	IFN-α_5, IFNα-G, IFN-α61
IFNA6	IFN-α_6, IFN-αK, IFN-α54
IFNA7	IFN-α_7, IFN-αJ, IFNαJ1
IFNA8	IFN-α_8, IFN-αB2, IFN-αB
IFNA10	IFN-αC, ψIFN-α10, ψIFN-αL, IFN-α6L
IFNA13	IFN-α13 (sequence identical to that of IFN-α_1)
IFNA14	IFN-α14, IFN-αH, IFN-αH1
IFNA16	IFN-α16, IFN-αWA, IFN-αO
IFNA17	IFN-α17, IFN-αI, IFN-α88
IFNA21	IFN-α21, IFN-αF
IFNA22	ψIFN-αE
IFNB1	IFN-β
IFNW1	IFN-ω
IFNG	IFN-γ

short arm of chromosome 9. Fourteen human genes comprise the IFN-α family and 12 IFN-α proteins are produced from the 14 genes (2 of the genes are pseudogenes) (Table I). In primates and rodents, there is only a single IFN-β gene. In contrast, ungulates have several IFN-β genes. In humans, the IFN-β gene is located at the telomeric end of the a cluster. IFN-ω genes are represented by six loci in humans but only one is functional. Several genes also exist in cattle, sheep, pig, horse, and rabbit. No IFN-ω has been described in mice. The IFN-ω has a characteristic 6-amino acid extension at the carboxyl termini compared to IFN-α and the gene maps to the end of the IFN-α cluster between IFN-α2 and -β. IFN-τ (trophoblast) is unique to cattle and sheep, in which it is essential for the development and maintenance of pregnancy. The type I IFN-α proteins have 165 or 166 amino acid residues. IFN-β has 166 amino acid residues, and both IFN-ω and -τ have 172 residues. There are four conserved cysteines in IFN-α proteins which form two intramolecular disulfide bonds (Cys1–Cys98 and Cys29–Cys138). Human IFN-α proteins do not contain N-glycosylation sites and are not N-glycosylated. In contrast, IFN-β has an N-glycosylation site (at position 80) and is glycosylated as a mature protein. IFN-τ, like IFN-ω, has a 6-amino acid carboxyl-terminal extension when aligned with IFN-α. Because of the solution of the crystal structures of both IFN-α2 and IFN-β, these are placed in a class of helical cytokines that are defined by five short helices labeled A–E. Two surfaces on opposite sides of the molecule are sensitive to amino acid substitutes supporting a model for single-ligand interaction with two receptor chains analagous to growth hormone receptor interactions.

B. Type II IFN (IFN-γ)

Type II IFN is acid labile (in contrast to type I IFNs which are acid stable) and is commonly referred to as IFN-γ. Whereas IFN-α and IFN-ω are products of leukocytes and -β; is produced by fibroblasts, IFN-γ is produced by activated T cells and natural killer (NK) cells. IFN-γ has no structural similarity with type I IFNs but shares their antiviral properties. The single human IFN-γ gene maps to chromosome 12 and has three introns, in contrast to the type I

IFNs which are encoded by intronless genes. The 166-amino acid, mature IFN-γ protein has two N-glycosylation sites and is glycosylated as a mature protein. However, unlike the type I IFNs, there are no cysteine residues in the mature IFN-γ molecule. X-ray crystallographic studies indicate that functionally active IFN-γ is a homodimer.

II. INDUCTION OF IFN SYNTHESIS

Type I IFNs can be induced by different infectious agents in almost all cell types, although virus infection is the most common cause of IFN production. Bacteria, mycoplasma, and protozoa can all induce IFN in different cells and low-molecular-weight inducers of IFNs in animals have been described. Bacterial lipopolysaccharide is an efficient inducer of type I IFNs in mononuclear phagocytes. With the exception of viruses, the mechanisms of induction by these other agents have not been investigated. Viral induction of IFN synthesis is mediated by double-stranded RNA (dsRNA), although in some cases in constituents of viruses such as hemagglutinin (influenza virus) may act as IFN inducers in animals and humans. Viral dsRNA, which is produced during the replication of both DNA and RNA viruses, is a strong inducer of type I IFNs in cell culture. The type I IFN genes are usually transcriptionally silent unless activated through a signal transduction pathway mediated by dsRNA. Activators of T cells or NK cells induce the synthesis of type II IFN-γ. These activators include interleukin-12 or -18, antigens, cytokines, or nonspecific T cell activators such as phytohemagglutinin or concanavalin A. The regulation of type I IFN synthesis occurs at both the transcriptional and the posttranscriptional levels. The transcriptional activation of type I IFN genes involves the formation of active transcriptional complexes on these genes as a result of posttranscriptional activation of preexisting transcription factors as well as displacement of repressors which, in the absence of dsRNA-mediated signals, silence the type I IFN genes. Type I IFN messenger RNAs are rapidly degraded through the recognition of destabilization sequences in the 3' untranslated regions of their transcripts. IFN induction can be enhanced by exposure of cells to small amounts of type I IFN which primes the cells for maximum response to inducers such as dsRNA or viruses. Inhibitors of protein synthesis can result in the superinduction of IFN messenger RNA through the inhibition of synthesis of negative regulatory factors. The molecular mechanism involved in the transcriptional induction of type I IFN synthesis is complex and involves an interplay between *cis*-acting negative and positive regulatory DNA elements and interaction with their cognate transcription factors.

A. Induction of IFN-β Synthesis

The induction of type I IFNs has been most thoroughly studied for the IFN-β gene. The IFN-β promoter consists of overlapping regulatory elements, termed positive regulatory domains (PRDI-IV), and a negative regulatory element (NRE) (Fig. 2).

The transcriptional activation of the human IFN-β gene by viruses requires the assembly of a higher order transcription enhancer complex, termed the enhancersome. The assembly of the enhancersome requires several transcription factors and is coordinated through specific interactions with the high-mobility group protein, HMG1(Y). HMG1(Y) promotes the cooperative binding of the transcription factors and is necessary for them to act synergistically *in vivo*. PRDI overlaps with PRDIII and binds members of IFN-regulatory factor 1 (IRF-1) and other IRF family members. PRDII is recognized by nuclear factor kappa B (NF-κB) and PRDIV by heterodimer consisting of activating transcription factor 2 (ATF-2) and c-*jun*. The transcriptional activator proteins, IRF3 and IRF7, as well as the transcriptional coactivators, P300 and CBP, are required for induction of IFN-β synthesis. The enhancersome complex also includes transcription factors ATF-2/C-jun and NF-κB. Several DNA-binding proteins have been described that have no activation potential but can still bind to the enhancer region; these include IRF-2, PRDI-binding factor-1, and homodimers of the NF-κB component, p50. These proteins may repress the IFN promoter in the uninduced state. Following virus infection, HMG1(Y) acts as an essential component to assemble cooperatively the different transcriptional activators. Once assembled, the activation

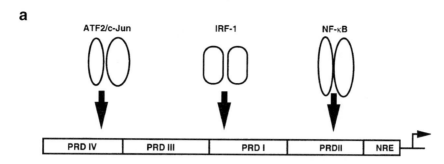

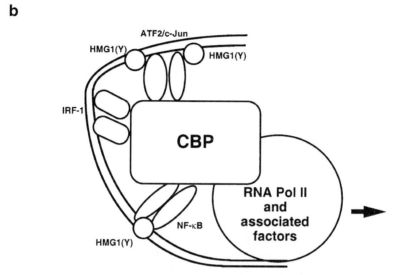

Fig. 2. Transcriptional activation of the IFN-β promoter. (a) Different transcription factors are activated as a result of virus infection and bind cognate DNA elements arranged in a unique pattern in the promoter region of the IFN-β gene. (b) The high-mobility group protein 1 (Y) is required for assembly of the complex, and the transcriptional coactivator CBP integrates the action of several transcription factors.

domains of these factors form a high-affinity binding site for the recruitment of the Creb-binding protein (CBP). CBP interacts with the RNA polymerase II holoenzyme via its carboxyl-terminal domain, whereas the activation domains of the other transcriptional activators contact components of the basal transcriptional apparatus. CBP, which contains an intrinsic histone acetyltransferase, subsequently acetylates HMG1(Y), destabilizing the enhancersome and thereby terminating IFN-β gene transcription. Thus, the coordinate activation of many transcriptional activated proteins and their cooperative assembly into a transcriptional enhancer complex are required to drive transcription of the IFN-β gene as a result of virus infection or dsRNA treatment of cells (Fig. 2).

B. Induction of IFN-γ Synthesis

The control of IFN-γ gene expression and synthesis is organized through a combination of positive and negative *cis*-regulating elements, located both in the 5′ flanking region and in the first intron of the gene. In addition to specific antigen stimulation and production during virus infection, IFN-γ production is increased by interleukin-1 and -2, growth factors, and estrogens. Glutocorticoids, transforming growth factor-β, and interleukin-10 are negative regulators of IFN-γ production. In nonstimulated cells, suppression of IFN-γ gene transcription is mediated by DNA methylation at CpG dinucleotide islands in the promoter region. Unlike type I IFNs, which are regulated at both the transcriptional and posttranscriptional levels, the production of IFN-γ is regulated primarily at the transcriptional level. These CpG islands in the promoter are conserved in IFN-γ genes of different species. In general, in cells not known to express IFN-γ, the CpG island is methylated, whereas in cells that express IFN-γ in response to different inducers the CpG island is unmethylated. Both negative and positive *cis*-elements are involved in induction of IFN-γ synthesis. The nuclear factor YY1 interacts with a silencer region in the IFN-γ promoter, whereas the transcription factor AP2 binds an adjacent site and is required for stimulus-driven expression of the IFN-γ gene. Members of the NF-κB family bind to a positive element present in the first intron of the IFN-γ gene and other positive elements which confer inducibility have been identified in the 5′ flanking region of the gene. These positive regulatory elements bind transcription factors GATA and AP1, members of the Creb/ATF family, and the transcription factor C-jun. A combination of complex transcription factor requirement and differential methylation of the promoter restricts the expression of the IFN-γ gene to specific lymphoid populations, including T, NK, and dendritic cells.

III. IFN RECEPTORS

IFNs mediate their effects by binding to high specific-affinity receptors on cellular membranes. This is a necessary first step to generate the signals required for transcriptional induction of IFN-responsive genes encoding the specific effectors for the different activities of IFNs. IFN-α and -β bind to the same type I receptor, whereas IFN-γ binds to a different type II receptor. IFNs exhibit restricted host range in their activities and this is mediated at the level of receptor binding. In general, human IFNs act on human cells and show only limited activity on cells from other species, although there are exceptions to this among the several IFN-α subtypes.

A. The Type I IFN Receptor

The type I IFN receptor is present at only a few thousand binding sites per cell and exhibits an affinity for IFN in the subnanomolar range. To elicit a cellular response, the type I IFN-α/β receptor requires two subunits, ifnar1 and ifnar2, which belong to the class II family of helical cytokine (hcrII) receptors (Fig. 3). The ifnar2 chain of the receptor encoded by a gene on human chromosome 21 exists in three forms: ifnar-2a, a soluble form of the extracellular domain; ifnar-2b, an alternatively spliced variant with a short cytoplasmic domain; and ifnar-2c. Only ifnar-2c is capable of transducing a signal upon ligand binding to the receptor complex. Neither ifnar1 nor ifnar2 alone bind to IFN-α/β with high affinity. Both ifnar1 and ifnar2 are required to constitute a high-affinity α/β receptor. IFN-β, which shares only 35% identity with IFN-α and possibly some of the IFN-α subtypes, engages the α/β receptor in different ways. This can result in the activation of selective subsets of genes reflecting distinct structural differences that are transmitted through the membrane to the intercytoplasmic domains of the receptors which then mediate a differential response.

The ifnar1 subunit of the IFN-α/β receptor is a 530-amino acid polypeptide encoded by a gene on human chromosome 21. The domains of the ifnar1 subunit include a short transmembrane-spanning region of 22 amino acids and a cytoplasmic tail of 99 amino acids. The solution structure of other members of the hcrII family (IFNGR1 and tissue factor) allows the modeling of the extracellular domains of ifnar1 and -2. Thus, the characteristic external domain is divided into two subdomains of approximately 100 amino acids (sd100 A and B from the N terminus) which each adopt an immunoglobulin-like fold with seven β strands (S1–S7) in two β sheets. The ifnar1 extracellular domain (317 amino acids) has a duplication of these subdomains. The structural integrity and formation of an active IFN-binding site not only require ifnar1 and -2 but also associated intracellular Jaks. The cytoplasmic tail contains a tyrosine residue at 466 which when phosphorylated constitutes a binding site for the transcription factor Stat 2. There is also a binding site for the signal-transducing Janus (JAK) kinase, Tyk 2 (Fig. 3), on

Class II cytokine receptor

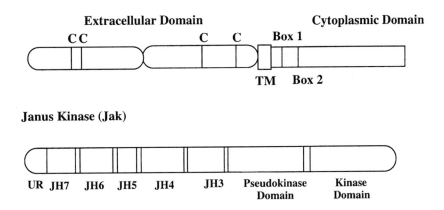

Janus Kinase (Jak)

Signal Transducer and Activator of Transcription (Stat)

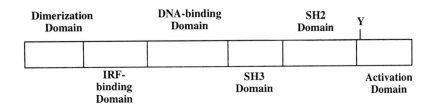

Interferon regulatory factor (IRF)

Fig. 3. Essential components of IFN signaling. In the class II cytokine receptor, C denotes conserved cysteine residues, and TM denotes the transmembrane domain. In the Janus kinase, UR denotes a unique region, and JH denotes Jak homology domain. In the Stat proteins, Y denotes a regulatory tyrosine residue. In IRF, W2 denotes conserved tryptophan residues.

the cytoplasmic tail of the ifnar1 receptor subunit. The ifnar2-c subunit is 315 amino acids. The first 217 amino acids constitute the ligand-binding domain followed by a transmembrane-spanning region of 22 amino acid and a cytoplasmic tail of 76 amino acids. There is a site on the cytoplasmic tail of ifnar2-c that binds Stat2 in unstimulated cells. There is also a binding site for another member of the JAK family, JAK1. The high-affinity binding and signaling receptor for IFN-α/β requires both ifnar1 and ifnar2-c and their associated kinases.

B. The Type II IFN-γ Receptor

IFN-γ receptors are expressed on most cells and exhibit species specificity in ligand binding. They consist of two polypeptide chains—IFNGR1, a 472-amino acid (90-kDa) polypeptide encoded on human chromosome 6, and IFNGR2, a 315-amino acid (62-kDa) polypeptide encoded by a gene on human chromosome 21. IFNGR1 is the major ligand-binding chain of the IFN-γ receptor, with the ligand-binding extracellular domain of 228 amino acids organized into the conserved domains characteristic of hcrII cytokine receptors. The transmembrane domain of IFNGR1 is 24 amino acids and the cytoplasmic tail of this receptor consists of 221 amino acid residues. Human IFN-γ is a dimer in solution and binds two IFNGR1 extracellular domains, inducing receptor dimerization. The intracellular domain of IFNGR1 has been functionally defined in terms of its role in dictating the IFN-γ cellular response. Residues 256–301, designated region 1, are required for both ligand internalization and biological responses. The tyrosine residue 440 becomes phosphorylated upon ligand engagement with the receptor and is involved in recruiting Stat1α into the receptor complex to activate the signal transduction pathway. A 4-residue sequence, 266 LPKS 296 in the membrane proximal region of the IFNGR1 intracellular domain, binds to JAK1. JAK2 binds to a 12-residue proline-rich sequence 263 PPSIPLQIEEYL 274 in the membrane-proximal region of the intracellular domain of IFNGR2. Unlike the type I IFN receptor, the structural integrity of the ligand binding ability of the type 2 IFN receptor is not influenced by Jaks.

IV. CELLULAR RESPONSE TO IFNs

IFNs are expressed ubiquitously and most cells have a common response to signals initiated by the binding of IFNs to cellular receptors. Common components included in this response are the receptor subunits, JAKs and Stats (Fig. 4). These components are activated by phosphorylation of specific tyrosines following ligand receptor interactions. Stats form homo- or heterodimers through mutual phosphotyrosine src homology region 2 (SH-2) interactions. Stat1 homodimers bind to DNA elements termed gamma-activated sites (GAS elements) which drive the expression of target genes. Stat1 and -2 heterodimers bind to p48, a member of the IRF family, and the resulting complex termed IFN-stimulated gene factor 3 (ISGF3) binds to IFN-stimulated regulatory ele-

ments (ISREs) upstream of different IFN responsive genes.

A. Type I IFN Signal Transduction

IFN-α/β signaling can be divided into five steps:

1. IFN-driven dimerization of the receptor outside the cell
2. Initiation of a tyrosine phosphorylation cascade inside the cell
3. Dimerization of the phosphorylated Stats
4. Transport of the Stats into the nucleus
5. Binding to specific DNA sequences and stimulation of transcription

Stat1 and Stat2 are preassociated in latent form with the ifnar2-c subunit of the IFN-α/β receptor. Ifnar1 is associated with Tyk2 and ifnar2 with Jak1. Following binding of IFN-α/β to ifnar2-c, the JAK kinases are brought together and cross-phosphorylated on tyrosine and activated. The tyrosine at residue 466 on ifnar1 is phosphorylated and latent Stat2 is then transferred to this docking site on ifnar1 through an interaction requiring its specific SH-2 domain and the phosphotyrosine residue at 466 on ifnar1. Stat2 then becomes tyrosine phosphorylated and provides a docking site for recruitment of Stat1 to ifnar1, which is then tyrosine phosphorylated and activated. The arrangement of the IFN-α/β preferentially results in the formation of heterodimers of STAT2 and STAT1. This requires the activity of cytosolic phospholipase A2 which is associated with Jak1. The heterodimers then translocate to the nucleus and form a high-affinity DNA-binding complex for the ISRE with p48.

B. Type II IFN Signal Transduction

Following IFN-γ-induced dimerization of the IFN-γ receptor chains, there is cross-phosphorylation of Jak1 and Jak2. The activated Jaks phosphorylate the IFNGR1 chain on a tyrosine residue and Stat1α is recruited to this docking site through an SH-2 domain interaction with the phosphorylated tyrosine residue. The bound STAT1α is phosphorylated on tyrosine residue 701 and disengages from the receptor forming a homodimeric complex with tyrosine phosphorylated STAT1α molecules through their mutual SH-2 domain phosphotyrosine interactions. STAT1 homodimers are then translocated to the nucleus, where they interact with GAS sites.

V. COMPONENTS OF THE IFN SIGNAL TRANSDUCTION CASCADE

A. JAK Kinases

JAKs are a family of cytoplasmic protein tyrosine kinases that are essential for IFN action. Three members of the JAK kinase family, Jak1, Jak2, and Tyk2, are involved in IFN signaling. The JAK proteins are 100–140 kDa and each possesses a tyrosine kinase activity mediated through a conserved catalytic domain. The JAKs have seven highly conserved regions of shared homology designated JAK homology (JH) domains 1–7 (Fig. 3). The C-terminal, JH1 domain has the conserved residues consistent with protein tyrosine kinase activity. N terminal to this is the JH2 kinase-like domain. This contains several amino acid substitutions which render it inert for kinase activity; however, in the case of the JH2 domain of Tyk2, it is essential for IFN-α signaling. The functions of the other conserved domains remain to be established. Following interaction of IFN with its receptor, the JAK kinases are phosphorylated on tyrosine and activated. The phosphotyrosine residue on Tyk2 activated as a result of IFN-α/β binding to ifnar-2c is in the JH1 kinase domain. The JAKs are preassociated with receptors prior to ligand binding. In the case of the α/β receptor, Tyk2 is associated with ifnar1 and Jak1 with ifnarR2. For the IFN-γ receptor, Jak1 is associated with IFNGR1, whereas Jak2 is associated with IFNGR2. Ligand-dependent receptor aggregation results in the activation of JAK activity and subsequent downstream signaling events. The specificity of the JAKs is determined by their association with the particular IFN receptor subunits and not by a high degree of substrate specificity.

B. Signal Transducers and Activators of Transcription

IFN-induced receptor aggregation and activation of the JAK kinases result in the phosphorylation and assembly of signal transducers and activators of transcription (Stat) proteins. Phosphorylation of Stats on tyrosine allows the dimerization by reciprocal SH-2 phosphotyrosine interaction and entry into the nucleus to regulate the transcription of the many different IFN-responsive genes. The Stat proteins most closely associated with the IFN response include Stat1 α/β and Stat2. Stat1 α/β are encoded by a single gene and are generated by alternative pre-mRNA splicing. Stat1 α/β are identical through the first 712 amino acids but differ in that there are 38 additional C-terminal residues present in Stat1 α polypeptide. The transcription factor complex ISGF3 is composed of Stat1α/Stat2, or Stat1β/Stat2 heterodimers along with the IRF1 family member p48. Single tyrosine residues (701 on Stat1 and 699 on Stat2) are phosphorylated following the activation of the JAK kinases. These tyrosine residues are situated in a phosphotyrosine-binding pocket of SH-2 domains and are essential for recruitment of the Stat proteins to the cytoplasmic tail of the activated receptor and for forming dimers, either between Stat1 and Stat2, in the case of the formation of the complex ISGF3, or Stat1 homodimers which result from engagement of the IFN-γ receptor. In each case, dimerization occurs via association of the corresponding phosphotyrosine residues in the C terminus of each partner with the SH-2 domain of the other member of the complex. Signal specificity in the IFN system is governed by recruitment of specific Stats to the cytoplasmic tails of the corresponding receptors and this is mediated by the SH-2 domains of the individual Stat molecules. The C-terminal domains of Stats are involved in transcriptional activation and in their ligand-dependent activation. As described previously, the conserved tyrosine residue (701 on Stat1α) is responsible for the ability of Stat proteins to homodimerize, and phosphorylation of this residue is the hallmark of Stat activation. The transcriptional activation domains of Stats are located in the C terminus in Stat1α, in which the 38 C-terminal amino acids form an essential transcriptional activation domain. Stat1β lacks these residues and accordingly also lacks transcriptional activation function. Serine residue 727 at the C terminus of Stat1α is required for full transcrip-

a

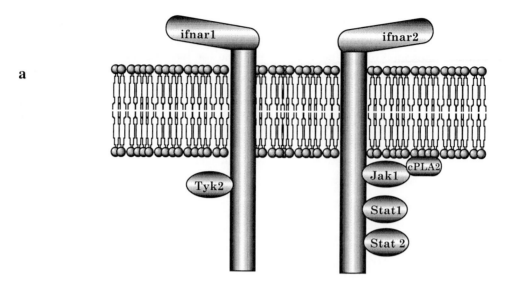

b

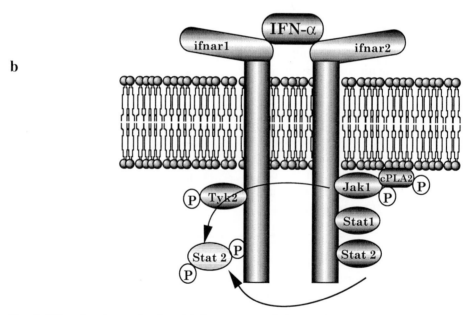

Fig. 4. IFN-α signal transduction. (a) Components of the signal transduction pathway are preassociated with the unliganded receptor. (b) IFN-α binding activates Jak kinases resulting in phosphorylation of different components of the pathway. Stat2 moves to its docking site on ifnar1. (c) Phosphorylated Stat1 interacts via SH-2 domains with Stat2, and the heterodimers are released from the receptor and interact with p48 to form ISGF3. (d) ISGF3 migrates to the nucleus.

c

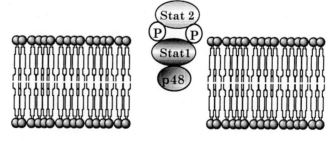

Cytoplasm

d

Nucleus

Fig. 4. *Continued*

tional activation and is phosphorylated by an unknown serine threonine kinase. Stat2 does not have a C-terminal serine phosphorylation site but has a highly acidic transactivation domain downstream from the tyrosine phosphorylation site at residue 699. The terminal 181 amino acids of Stat2 are important for binding and association with the basal transcription machinery. The SH-2 of Stat proteins are the determinants of signaling specificity. Stat2 interacts specifically with the cytoplasmic domain of ifnar2 through an SH-2 phosphotyrosine action. The SH-2 domain of Stat1 mediates an interaction with the phosphotyrosine 699 on Stat2, transmitting the signal through the IFN-α/β receptor. In the case of IFN-γ signal transduction, the Stat1 SH-2 domain interacts with the activated IFN-γ receptor.

These domains are also essential for homodimerization and heterodimerization among Stat1 and Stat2, although Stat1 homodimers predominate following IFN-γ receptor activation. Other functional domains on Stats include the NH2 domain, which is involved in binding protein tyrosine phosphatases and also in Stat dimer/dimer formation. The N-terminal domain with Stat2 is involved in binding to p48 via lysine residue 161 and the N-terminal of Stat1 interacts with CBP-p300. Finally, the interaction of Stat2 and Stat1 is also mediated by the N-terminal domain of Stat2.

STAT1 is involved in signaling through both the IFN-γ and IFN-α/β receptors. In the case of IFN-α/β receptor activation, Stat1 is an essential component of ISGF3 along with Stat2 and p48. Following IFN-γ receptor activation, Stat1 is activated to form a dimer which recognizes GAS sites upstream of IFN-γ responsive genes. Stat1 and Stat2 heterodimers do not bind DNA strongly unless they are complexed with p48. Stat2 homodimers bind DNA weakly. Dimers of Stat1, however, are competent to bind DNA and drive the activation of transcription through GAS sites.

C. p48

p48 (also known as ISGF3-γ) is a member of a family of transcription factors, the IFN regulatory factors, which can function as either activators or repressors of transcription. p48 is a 398-amino acid protein which shares significant homology in its N-terminal, 115-amino acid DNA-binding domain with other members of the IRF family. The signature feature of this domain is a cluster of five tryptophan residues, three of which are essential for contacting DNA. The C-terminal region of p48 contains an IRF association domain which mediates interaction between p48 and Stat1 and Stat2 heterodimers. p48 recognizes and binds to IFN stimulated response elements and serves as an essential DNA-binding domain subunit of ISGF3. The association of p48 with Stat1 and Stat2 increases p48 DNA binding activity by 25-fold. Although the DNA-binding domain of p48 is sufficient for DNA binding, the first N-terminal 9 amino acids of p48 are essential for DNA target specificity.

VI. IFN-INDUCED PROTEINS

Treatment of cells with either type I or type II IFNs induces the synthesis of many cellular RNAs and their corresponding proteins. In both type I and type II IFNs, the numbers of these proteins exceed 200. However, only a small number of these proteins have been assigned specific IFN-regulated functions. IFNs-α and -β induce the same set of proteins, but IFN-β also induces unique β-specific proteins. IFN-γ induces a different set of proteins, although some of these overlap with those induced by the type I IFNs. The spectrum of proteins induced by IFNs depends on cell type, but most type I IFN-inducible mRNAs are induced as a primary response which does not require protein synthesis but makes use of the STAT proteins that preexist in the cell. The induction of protein synthesis by IFN is transient, peaking after several hours and then declining to the uninduced level. The element in the promoters of IFN-α/β-responsive genes required for the response to IFN-α/β is the ISRE. GAS elements are also able to mediate a response to IFN-α but to a lower level. IFN-γ-inducible genes respond through the binding of STAT1 homodimers to the GAS element. Several IFN-γ-inducible genes are not induced as a primary response but require protein synthesis for their induction.

VII. THE IFN-INDUCED ANTIVIRAL PATHWAYS

Although IFN induces the synthesis and accumulation of many cellular RNAs and proteins that may differ in type and amount with different cells, only three IFN-regulated pathways have been definitely assigned an antiviral function. However, IFNs are known to inhibit virus replication at different stages of infection, including entry into the cell, uncoating of the virus particle, DNA and RNA protein synthesis, virus assembly, and release from the cell. IFN-induced proteins which have yet to be ascribed a function could operate at any of these steps.

The three pathways which have been characterized in detail and which mediate the antiviral activity of IFNs are the Mx, PKR, and the 2-5A pathways (Fig. 5). The latter two were discovered as a result of an investigation of the inhibitory effects of dsRNA on protein synthesis performed in extracts from IFN-treated cells. The key enzyme intermediates in both pathways require dsRNA for activation. During virus infection, dsRNA can be derived from either the incoming viral RNA genome or viral replication intermediates.

A. The Mx Proteins

The Mx proteins are IFN-inducible, 70- to 80-kDa GTPases which belong to the Dynamin superfamily. Mx proteins, like other Dynamins, self-assemble into horseshoe- and ring-shaped helices and other helical structures. The Dynamins play important roles in transport processes such as endocytosis and intracellular vesicle transport, and in plants they are important for cell formation. The human MxA protein is induced by type I IFNs and accumulates in the cytoplasm of cells where it forms tight oligomeric complexes. MxA interferes with the multiplication and spread of orthomyxoviruses, rhabdoviruses, paramyxoviruses, and bunyoviruses. Synthesis of MxA is induced during acute viral infections and impairs the growth of influenza (an orthomyxovirus) and other viruses at the level of transcription. MxA works in the cytoplasm to bind viral nucleocapsidis and block their normal movement into the nucleus. The murine Mx-1 protein inhibits the primary transcription of viruses in the nucleus. The induction of Mx genes by IFNs requires the transcription factor complex ISGF3, which binds to its cognate ISRE in the 5′ flanking regions of the Mx genes.

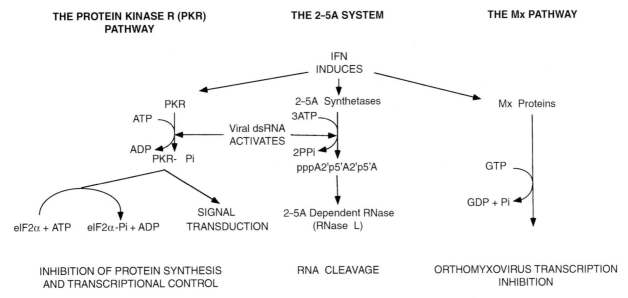

Fig. 5. Three known pathways mediating the antiviral activities of IFNs.

B. dsRNA-Dependent Protein Kinase

The IFN-inducible dsRNA-dependent kinase (PKR) plays roles in different cellular processes, including growth regulation, antiviral protection, signal transduction, and differentiation. The synthesis of PKR is induced by IFNs through an ISRE, but most cells also express constitutive levels of the protein. PKR is normally inactive, but on binding dsRNA it undergoes autophosphorylation and then dsRNA-independent phosphorylation of substrates. The binding of dsRNA requires the presence of two conserved dsRNA-binding motifs present in the N-terminal regulatory domain of PKR. The first motif is the most critical for binding dsRNA and is highly conserved within a large family of dsRNA-binding proteins. A single dsRNA-binding motif can recognize as few as 11 base pairs of dsRNA, but there is no sequence specificity in this recognition. Since dsRNA is a by-product of virus infection, there are opportunities for most viruses to activate PKR during the infection cycle. dsRNA also occurs as a structural feature of cellular RNAs and these may also activate PKR as part of a cellular regulatory control mechanism. Exposure of cells to stress such as heat shock can also activate the transcription of cellular Alu RNAs which can regulate PKR activity.

The binding of dsRNA to PKR causes a conformational change in the protein exposing the catalytic domain for activation. The antiviral activity of PKR is mediated by its phosphorylation of the α subunit of initiation factor eIF2 (Fig. 5). Phosphorylation of eIF2 prevents recycling of eIF2:GDP2 to eIF2:GTP, trapping the recycling factor eIF2α and resulting in rapid inhibition of translation. PKR also mediates programmed cell death (apoptosis) induced by dsRNA. Both inhibition of protein synthesis and induction of apoptosis restrict viral replication and spread.

PKR is also a signal-transducing kinase for the transcription factor NF-κB. PKR activation by dsRNA leads to the phosphorylation of IκB, the inhibitor of transcription NF-κB causing its release from the inactive NF-κB/IκB complex and activation of NF-κB, allowing it to regulate NF-κB-dependent genes.

The basal levels of PKR can be directly activated by IFNs independent of the need for further protein synthesis. In these circumstances, the activation of PKR leads to the phosphorylation of downstream activators of transcription resulting in the transcriptional activation of different genes involved in the cellular immune response, including the synthesis of chemokines, class I MHC, and molecules involved in apoptosis. The loss of PKR function results in a decreased antiviral activity of IFNs. Many viruses have employed different mechanisms to inhibit PKR and thereby overcome the antiviral effects of IFN. These mechanisms include the synthesis of inhibitory dsRNAs, the synthesis of proteins which can bind and sequester dsRNA activators of PKR, the synthesis of protein inhibitors of PKR or proteases that cleave PKR, or the activation of cellular inhibitors of PKR that are normally sequestered but are only activated in response to virus infection.

C. The 2-5A Synthetase/RNase L System

The 2-5 A synthetase/RNase L system is a multienzyme cellular RNA degradation pathway activated by IFN treatment of cells. 2-5A synthetase genes induced by IFN via ISREs are activated by dsRNAs to polymerize short $2',5'$ oligoadenylates (2-5A) that activate the 2-5A-dependent ribonuclease RNase L. Thus, the enzymes in this pathway are inactive or latent and require first dsRNA as a cofactor in the case of the 2-5A synthetase enzymes and then 2-5A as the activator for the latent ribonuclease, RNase L. This ribonuclease is present at a basal level in most cells, but its synthesis can also be stimulated by IFNs.

Latent RNase L exists as a monomer and 2-5A induces the formation of a homodimeric active enzyme. The activation of RNase L is reversible because 2-5A is rapidly degraded by phosphodiesterases and phosphatases. This results in deactivation of the 2-5A pathway. The 2-5A synthetases are encoded by multiple genes which map to chromosome 12q21 in humans. The individual enzymes reside in different subcellular locations. The different forms of 2-5A synthetase have differing activities. The largest isoform, 100 kDa, is active as a monomer and preferentially synthesizes dimers of 2-5A. The 69- and 71-kDa isoforms are myristoylated and glycosylated and exist as membrane-bound dimers. The 71-kDa isoform is the result of alternative splicing. The 2-5A synthetases are present in all cellular compartments. The small human 2-5A synthetase isoforms result from alternative splicing of the same gene yielding a 364-residue, 40 kDa protein encoded for by a 1.6-kB mRNA and a 400-residue, 46-kDa protein encoded for by a 1.8-kB RNA. Although the amino-terminal 346 residues of both proteins are identical, the 54 C-terminal unique residues of the 49-kDa isoform are highly hydrophobic and this form of the isozyme is membrane bound, whereas the 46-kDa form which lacks this hydrophobic C-terminal sequence is cytoplasmic. The first 346 amino acids of 40- and 46-kDa isoforms constitute a domain that is duplicated in the 68- and 71-kDa isoforms and occurs three times in the 100-kDa isoform. Specific antiviral functions for the different 2-5A synthetases have not been described. The dsRNA-binding domains of the 2-5A synthetase isoforms bear no structural similarity to the dsRBD of PKR and require a higher concentration of dsRNA for activation than does PKR.

RNase L is a single isozyme and is encoded by a unique gene mapping to human chromosome 1p31. The human RNase L protein has 741 residues, the N-terminal half of which contains a repeated P loop motif and nine ankryin repeats, both involved in 2-5A binding function to repress RNase activity. The C-terminal half of the protein contains a region of protein kinase homology, a cysteine-rich domain, and the ribonuclease domain. The isolated C-terminal half of RNase L cleaves RNA in the absence of 2-5A. RNase L binds 2-5A with high affinity and the active ribonuclease activity is sequence nonspecific but cleaves U–U and U–A bonds preferentially. Inhibition of RNase L activity by expression of dominant-negative derivatives or deletion of the gene from mice result in a defective anti-EMCV effect of IFN-α.

D. Other Antiviral Pathways Regulated by IFNs

The ability of IFNs to confer an antiviral state is a fundamental property and allows their classification and individual activity measurement. Different stages of virus replication may be inhibited by IFNs, including entry and/or uncoating, transcription initiation of translation, maturation, assembly, and release. Since the pathways described previously are primarily directed toward translational control mechanisms, it is obvious that other IFN-regulated antiviral proteins exist to interfere with different stages of virus replication. However, despite identification of a large number of IFN-regulated genes and proteins, the assignment of specific antiviral activities to these remains to be achieved.

VIII. INHIBITION OF CELL GROWTH BY IFNs

The cell growth inhibitory properties of IFNs are assumed to contribute to their utility as anticancer agents. However, IFNs also regulate cellular responses to inducers of apoptosis and modulate the immune response, and these effects underlie antitumor activity. The prolonged activation of Stat1 and the induction of synthesis of several proapoptotic genes by IFNs may sensitize cells to apoptosis inducers, including chemotherapeutic agents. Thus, the modulatory effects of IFNs influence both the development of resistance to infection and antitumor activities. Direct inhibition of cell growth by IFN is displayed against a variety of cultured cells *in vitro*. This is due in part to the promotion of differentiation which is particularly apparent when used in combination with other agents, such as retinoids. Significant IFN-induced gene products have not been linked directly to antiproliferative activity. However, IFN-α does regulate properties of different components of the cell cycle, including c-Myc, pRB, cyclin D3, and cdc25A. The mechanisms by which the cell cycle is suppressed may differ in different cell types. Thus, CDK inhibitors, such as p21, p27, and p57^{KIP}, have been invoked in some cells, whereas in others these inhibitors are not involved.

IX. EFFECTS OF IFNs ON THE IMMUNE SYSTEM

IFNs affect all phases of innate and adaptive immune responses. Type I and II IFNs have overlapping effects on the immune system. IFN-γ is predominantly an immunomodulatory factor produced by T cells and NK cells in response to immune and/or inflammatory stimuli. IFN-γ functions to stimulate the development and actions of immune effector cells. Type I IFNs are largely responsible for increasing the effectiveness of the adaptive immune response to resist viral infection, including the stimulation of CD44 memory T cells. Both type I and type II IFNs enhance the expression of MHC class I proteins to promote the development of CD8+ T cell responses. MHC class I expression is dependent on the transcription factor IRF-1, and mutations in genes linked to the IRF-1 signal transduction pathway result in failure to actively upregulate of MHC class I proteins on cell surfaces in response to IFN. IFN-γ induces the expression of MHC class II proteins on cells promoting an enhanced CD4+ T cell response. This response depends on the transactivating factor CIITA, which is regulated at a posttranscriptional level by IFN-γ. Type I and type II IFNs are able to upregulate the expression of the different protein components which constitute the proteosome protein processing pathway, which is responsible for generating antigenic peptides expressed by MHC Class II. Accordingly, IFNs can enhance immunogenicity by increasing the quantity and repertoire of peptides displayed in association with MHC class I proteins which are dependent on an active proteosome-mediated pathway.

X. CLINICAL USE OF IFNs

Since their discovery, IFNs have held the promise of treatment of viral diseases. This was not realized until the production and testing of recombinant IFNs which are now approved for the treatment of hepatitis B and C. IFN-α is the only effective treatment for patients with chronic hepatitis C. However, it was approval of IFN for treatment of a rare form of leukemia (hairy cell) that led to increased testing and approval of IFNs (type 1) for a range of malignant, viral, and autoimmune diseases (Table II). In addition to hairy cell leukemia, both natural and recombinant IFNs are used for treating chronic myelogenous leukemia, myeloma, melanoma, renal cell carcinoma, lymphoma, and Karposi's sarcoma. IFNs are either used alone or as adjuvant therapy, and new methods

TABLE II
Approved Clinical Uses for Interferons

Disease category	Type
Cancer	Hairy cell leukemia
	Chronic myelogenous leukemia
	Myeloma
	Renal cell carcinoma
	Kaposi's sarcoma
	Melanoma
Viral	Hepatitis B
	Hepatitis C
	Hepatitis D
	Papillomas
Immune dysfunction	Multiple sclerosis
	Chronic granulomatous disease

for stabilizing IFN increasing its bioavailability are impacting on the clinical utility. In addition to viral hepatitis, IFNs are also used for treatment of herpetic keratitis and laryngeal and genital papillomas. Although the use of IFN-γ in multiple sclerosis was quickly terminated because of adverse outcomes, it is approved for treating chronic granulomatous disease. Recombinant IFN-β produced either as an unglycosylated form in *Escherichia coli* or as a glycosylated protein in mammalian cell cultures is used extensively in the treatment of relapsing/remitting multiple sclerosis and is an effective treatment in terms of relapse rate, defined disability, and all magnetic resonance imaging outcomes.

TABLE III
Side Effects Associated with IFN Therapy

Initial injections
  Chills/rigors
  Fever
  Myalgias
  Mild neutropenia
Chronic administration
  Fatigue
  Anorexia
  Mild neutropenia
  Transaminase elevations
  Weight loss
  Depression

IFNs are increasingly being used in combination with other agents. For example, in patients with chronic hepatitis C who relapse after treatment with IFN, combination with the oral antiviral agent ribavirin results in higher rates of sustained virologic, biochemical, and histologic response. Although clinical experience with therapy of solid tumors has not shown the success seen with some leukemias, combination therapy may lead to better outcomes. Thus, the regimen of 5-fluorouracil and IFN-β is active in patients with advanced colorectal carcinoma, and survival with this regimen is comparable to or better than that with other modulating regimens. Importantly, the rapid protein clearance and systemic toxicities seen with unmodified IFNs can be modulated by conjugation with polyethylene glycol (PEG). PEG conjugation of proteins makes them tolerogenic, inhibiting antibody responses.

Neutralizing antibody has been reported in patients receiving IFN-α therapy. IFN therapy at therapeutic doses causes many dose-related side effects (Table III). All these are reversible on termination of therapy and can be tolerated with effective patient management. The most common adverse effects are flu-like syndrome, fatigue, anorexia, and depression. The acute flu-like symptoms are dose related and tolerance generally follows several days of therapy. Fatigue, anorexia, and depression are more likely to be chronic and can be dose limiting. The mechanisms of IFN-induced toxicity are not well understood but since IFNs can regulate the expression of more than 300 genes, including many proinflammatory cytokines, secondary activities likely account for many of the side effects of therapy. In addition to the combination of systemic IFN administration with cytotoxic or differentiation-inducing agents, gene therapeutic approaches which deliver high concentrations of IFN into the tumor bed are under investigation for treatment of solid tumors. Low-dose oral IFN has also shown efficacy in animal models and may be appropriate for treatment of diseases for which high-dose systemic administration and its associated toxicity are inappropriate.

See Also the Following Articles

Antiviral Agents • Cellular Immunity • Methylation of Nucleic Acids and Proteins • Protein Biosynthesis • T Lymphocytes

Bibliography

Bach, E. A., Aguet, M., and Schreiber, R. D. (1997). The IFN gamma receptor: A paradigm for cytokine receptor signaling. *Annu. Rev. Immunol.* **15**, 563–591.

Borden, E. C., and Parkinson, D. (1998). A perspective on the clinical effectiveness and tolerance of interferon-alpha. *Sem. Oncol.* **25**(Suppl. 1), 3–8.

Darnell, J. E., Jr., Kerr, I. M., and Stark, G. R. (1994). Jak-STAT pathways and transcriptional activation in response to IFNs and other extracellular signaling proteins. *Science* **264**(5164), 1415–1421.

Pfeffer, L. M., Dinarello, C. A., Herberman, R. B., Williams, B. R., Borden, E. C., Bordens, R., Walter, M. R., Nagabhushan, T. L., Trotta, P. P., and Pestka, S. (1998). Biological properties of recombinant alpha-interferons: 40th anniversary of the discovery of interferons. *Cancer Res.* **58**(12), 2489–2499.

Stark, G. R., Kerr, I. M., Williams, B. R., Silverman, R. H., and Schreiber, R. D. (1998). How cells respond to interferons. *Annu. Rev. Biochem.* **67**, 227–264.

International Law and Infectious Disease

David P. Fidler

Indiana University School of Law–Bloomington

GLOSSARY

customary international law A main source of international legal rules; customary rules of international law form when States exhibit a general and consistent pattern of behavior and follow the pattern out of a sense of legal obligation.

emerging infectious diseases Infectious diseases whose incidence in humans, animals, and plants has increased in the past 20 years or threatens to increase in the future.

globalization of public health Phenomenon under which the processes of globalization (e.g., trade, travel) affect the ability of the sovereign state to protect the health of its public.

harmonization The process of rendering national legal rules identical or substantially similar to serve an international objective.

legal regime A set of legal rules pertaining to a distinct area or problem that regulates behavior in that specific context; in international law, international organizations often administer such legal regimes.

xenotransplantation The transplantation of animal tissues into the human body.

INTERNATIONAL LAW has long played an important role in international efforts to control and prevent infectious diseases. States have used international law as an instrument of infectious disease control by establishing specific rules of behavior and creating international organizations charged with controlling and preventing the international spread of infectious diseases. In addition, infectious diseases arise as issues in many different areas of international law.

The global crisis in emerging infectious diseases has raised the profile of international law in infectious disease control by stimulating new attention on international law by the World Health Organization (WHO) and international lawyers. This renewed interest in international law and infectious diseases includes concern about emerging infectious disease problems that may require international legal activity.

I. NEED FOR INTERNATIONAL LAW IN INFECTIOUS DISEASE CONTROL

A. The System of International Law

International law is the body of rules that has arisen within the decentralized, anarchic environment of relations between independent sovereign States. Although international organizations and individuals are also subjects of international law, it contains, preponderantly, rules directed towards States. For States to deal with common problems or pursue common interests, they almost always interact through the procedures and according to the rules established in international law. The structure and dynamics of the international political and economic system impel States to use and further develop international law. International law embodies what States believe are common interests and values, contains the rules States establish to achieve those common objectives, and helps States create common institutions to bolster the pursuit of mutual ends.

B. The Globalization of Public Health

Since at least the mid-nineteenth century, States have not been able to control infectious diseases without international cooperation. Infectious disease control has long been a global problem. The globalization of public health forces States to work with each other because a single State or small group of States cannot effectively deal with global infectious disease threats. The globalization of public health places infectious disease control on the diplomatic and legal agendas in international relations, making it part of the political dynamics of the international system.

C. Sources of International Law on Infectious Disease Control

The main sources of international law are treaties, customary international law, and those general principles of domestic law recognized widely in the international system. Judicial decisions and the writings of international lawyers are subsidiary means for interpreting rules of international law formed through one of the primary sources. The treaty has been, and will continue to be, the most important source of international law relating to infectious diseases.

II. HISTORY OF INTERNATIONAL LAW AND INFECTIOUS DISEASES

The importance of international law to infectious disease control became clear at the dawn of international health cooperation in the mid-nineteenth century. Beginning with the first International Sanitary Conference in 1851, States concerned about the international spread of cholera, plague, and yellow fever attempted to fashion international legal rules on infectious disease control. From 1851 to 1892, these efforts were not successful, despite repeated attempts to negotiate and adopt treaties on control of infectious diseases in humans. In 1878, some European States concluded a treaty on combating the spread of a plant pest, *phylloxera vastatrix*. After 1892, States adopted numerous treaties addressing the control of infectious diseases of humans, animals, and plants (see Table I). These extensive international legal activities in the latter half of the nineteenth century and first half of the twentieth century focused at different times on four objectives, outlined here.

A. To Protect Europe and North America from Foreign Diseases

A powerful political impetus for the international diplomatic and legal activity on cholera, plague, and yellow fever in the late nineteenth and early twentieth centuries was the desire of European States, and later the United States, to protect themselves against the importation of diseases from the Middle East, Near East, and tropical regions. Diseases endemic in the northern hemisphere did not become the subject of international legal rules until well into the twentieth century. The objective was to protect the European and North American peoples from "Asiatic" and tropical diseases.

B. To Regulate National Quarantine Measures

The most powerful political catalyst for international legal activity on infectious diseases in the latter half of the nineteenth century was the desire to reduce the commercial and economic burdens that national quarantine systems were imposing on international trade. As the speed and volume of international trade increased in the nineteenth century, national quarantine systems became an increasingly painful thorn in the side of merchants and their government supporters, particularly within major maritime powers, such as Great Britain. Much of the

TABLE I
List of Selected Treaties Relevant to Infectious Disease Control, 1878–1945

Year	Treaty	Description
1878	Convention Respecting Measures to be Taken Against *Phylloxera Vastatrix*	Established international regime to combat spread of plant pest
1892	International Sanitary Convention	Focused on spread of cholera through Suez Canal and related to Mecca pilgrimages
1893	International Sanitary Convention	Focused on cholera control
1894	International Sanitary Convention	Focused on disease threats relating to Mecca pilgrimages and maritime traffic in Persian Gulf
1897	International Sanitary Convention	Focused on control of plague
1902	International Conference of American States	Established the Pan American Sanitary Bureau
1903	International Sanitary Convention	Consolidated the 1892, 1893, 1894, and 1897 conventions
1907	Rome Agreement Establishing the Office International de l'Hygiéne Publique	Established international regime for cooperation on infectious diseases in humans
1912	International Sanitary Convention	Revised the 1903 convention
1924	Scheme for Permanent Health Organization of the League of Nations	Established the Health Office of the League of Nations
1924	International Agreement for the Creation at Paris of the Office International des Epizooties	Established international regime for combating animal diseases
1924	Agreement Respecting Facilities to be Given to Merchant Seamen for the Treatment of Venereal Disease	Established obligations to provide venereal disease treatment to all merchant seamen regardless of nationality
1926	International Sanitary Convention	Revised the 1912 convention
1929	International Convention for the Protection of Plants	Established international regime to combat plant diseases and pests
1930	Convention Concerning Anti-Diptheritic Serum	Established international quality-control regime for antidiptheritic serum
1933	International Sanitary Convention on Aerial Navigation	Created international legal rules for new challenges posed by air travel and transport
1934	International Convention for Mutual Protection Against Dengue Fever	Created international regime to combat the spread of dengue fever
1935	International Convention on the Campaign Against Infectious Diseases in Animals	Created international regime to combat the appearance and spread of infectious diseases of animals
1938	International Sanitary Convention	Revised the 1926 convention

international legal activity, from 1851 to the establishment of WHO, concentrated on the harmonization of national quarantine systems to reduce the burdens they imposed on international commerce.

Quarantine harmonization efforts did not progress until scientists, such as Louis Pasteur and Robert Koch, began to provide more accurate scientific understandings about pathogenic microbes. As "germ theory" prevailed, States found a common scientific foundation on which to base international legal rules on quarantine harmonization.

C. To Establish Surveillance Systems

As the scientific understanding of infectious diseases developed, States began to realize that disease surveillance was an important element of any international infectious disease control strategy. Thus, the creation of international surveillance systems became an objective of international diplomatic and legal activity. International sanitary conventions imposed on the States duties to report to the others outbreaks of specified diseases. Later, the duty to report channeled surveillance information through international health organizations.

D. To Create Permanent International Health Organizations

The frustrating and inefficient process of convening ad hoc international sanitary conferences, combined with the growing importance of surveillance, led States to believe that effective international infectious disease control required the presence of a permanent international health organization. Four permanent international health organizations were established in the first 25 years of the twentieth century: the Pan American Sanitary Bureau (1902), the Office International de l'Hygiéne Publique (1907), the Health Organization of the League of Nations (1924), and the Office International des Epizooties (1924). These international health organizations not only played an important function in the international legal regimes on infectious disease but they also helped to develop international law in this area by overseeing revision of existing treaties and fostering the adoption of treaties in new areas.

III. INTERNATIONAL LEGAL REGIMES SPECIFIC TO INFECTIOUS DISEASE CONTROL

Three distinct international legal regimes currently exist that specifically address infectious disease control (see Table II).

A. Human Diseases

WHO's International Health Regulations (IHR) constitute the main international legal rules addressing the control of human infectious diseases. In 1951, WHO adopted the International Sanitary Regulations pursuant to Article 21 of the WHO constitution. These Regulations replaced the main international sanitary conventions adopted between 1892 and 1944. WHO changed the name of these Regulations to the IHR in 1969.

The purpose of the IHR is to provide maximum protection against the international spread of disease with minimum interference with world traffic. To achieve this objective, the IHR establish a surveillance system for three diseases (cholera, plague, and yellow fever), establish the maximum measures

TABLE II
Three Major International Legal Regimes on Infectious Diseases

Disease area	Main agreements	International organization	Purpose
Human diseases	WHO Constitution and International Health Regulations	World Health Organization	Maximum protection against the international spread of disease with minimum interference with world traffic
Animal diseases	1924 International Agreement and International Animal Health Code	Office International des Epizooties	(1) To inform governments about animal diseases and their control; (2) to harmonize regulations for trade in animals; and (3) to coordinate research on surveillance and control of animal diseases
Plant diseases	1951 International Plant Protection Convention	Food and Agriculture Organization	To secure common and effective action against the spread of pests and diseases of plants and plant products

WHO Member States can take to prevent disease importation, and impose requirements on modes and locations of transport to prevent the spread of disease vectors, such as rats and mosquitoes.

The IHR have, however, not been successful at achieving either maximum protection against the international spread of disease or minimum interference with world traffic. WHO Member States have routinely failed to report disease outbreaks and have frequently applied excessive measures against the travel and trade of States suffering infectious disease outbreaks. These problems contributed to WHO's decision in 1995 to revise the IHR (see Section V).

B. Animal Diseases

The international legal regime concerning animal diseases is centered on the Office International des Epizooties (OIE), located in Paris, France. In 1998, the OIE had approximately 150 Member States. The animal disease regime contains three basic objectives: (1) collecting and disseminating information about animal disease outbreaks; (2) harmonizing regulations for trade in animals and animal products to prevent the international spread of animal diseases and the application of unjustified trade restrictions; and (3) promoting scientific research on animal diseases and the dissemination of such research internationally. These core features of the international legal regime on animal diseases echo central features of the IHR, namely, surveillance and regulatory harmonization. The International Animal Health Code constitutes the main body of rules for OIE Member States in connection with surveillance and regulatory harmonization. Like WHO in connection with human diseases, the OIE has an important role in scientific research on animal diseases. OIE standards also play an important role in the Agreement on the Application of Sanitary and Phytosanitary Measures (SPS Agreement) of the World Trade Organization (WTO) (see Section IV.A).

C. Plant Diseases

The International Plant Protection Convention (IPPC) constitutes the central instrument in the in-

ternational legal regime on plant diseases. The IPPC was adopted pursuant to Article XIV of the Constitution of the United Nations Food and Agriculture Organization (FAO). In 1998, the IPPC had 106 Member States. Similar to the IHR and the International Animal Health Code, the IPPC seeks to prevent the international spread of disease through harmonization of national plant quarantine measures. The IPPC also establishes a global reporting service on plant diseases and pests, which disseminates reports on outbreaks and effective disease and pest control techniques. In 1997, the FAO Council submitted to IPPC States parties amendments to the IPPC that would bring the IPCC into closer alignment with the WTO's SPS Agreement, which provides that the IPPC is the relevant international organization for the setting of plant health standards (see Section IV.A).

IV. OTHER INTERNATIONAL LEGAL REGIMES RELEVANT TO INFECTIOUS DISEASES

Infectious diseases affect many areas of international law beyond the international legal regimes specifically designed to control them. Section IV briefly identifies some important areas of international law that are relevant to global infectious disease control (see Table III).

A. International Trade Law

As noted, infectious diseases and international trade have a long history of interaction. Much of the international health law already sketched had trade objectives in mind, as well as health objectives, as illustrated by the IHR's objective of minimum interference with world traffic. International trade law, likewise, recognizes health as a value worth protecting. The growth of food-borne infectious diseases fostered by the liberalization of international trade constitutes a key threat in the globalization of public health and directly implicates international trade law.

International trade law recognizes that the protection of the life and health of humans, animals, and

TABLE III
Areas of International Law Important to Infectious Disease Control

International trade law	International trade is a channel for the spread of infectious diseases. International trade law recognizes a State's right to restrict trade for public health but imposes scientific and trade-related disciplines on the exercise of this right.
International human rights law	Establishes criteria for government handling of persons suffering from infectious diseases and contains the right to health
International humanitarian law	Imposes obligations on belligerents in both international and civil armed conflicts not to create conditions conducive to the spread of infectious diseases and makes violations of these duties war crimes.
International law on arms control	Prohibits the development, production, stockpiling, and use of biological weapons
International environmental law	Environmental change is a factor in the emergence and re-emergence of infectious diseases, making international environmental law important in addressing infectious disease threats.

plants is a legitimate reason for restricting trade flows. The sovereign right to restrict trade for health purposes is, however, subject to scientific disciplines and trade-related disciplines. As provided in the WTO's SPS Agreement, a trade-restricting health measure has to be based on a risk assessment and be supported by sufficient scientific evidence. Scientifically justified measures must then satisfy the trade-related disciplines, which require that the measure in question not be an arbitrary or disguised restriction on trade and be the least trade-restrictive measure possible. The scientific and trade-related disciplines of the SPS Agreement have been applied in WTO dispute settlement cases to measures that restricted trade for infectious disease control purposes.

The SPS Agreement also provides that any trade-restricting health measures that conform to or are based on scientific standards established by the relevant international organization satisfy the science-based disciplines of the SPS Agreement. Thus, the SPS Agreement encourages WTO Member States to harmonize their sanitary and phytosanitary measures using internationally established scientific standards. The SPS Agreement specifically mentions the Codex Alimentarius Commission (Codex) as the relevant international organization for food safety standards,

the OIE as the relevant standard-setting body for animal diseases, and the IPPC as the relevant organization for standards on plant diseases. Although WHO is not specifically mentioned in the SPS Agreement, it clearly constitutes a relevant international organization that sets health-related standards. Although most of the standards developed by these international organizations are initially developed as nonbinding recommendations, the SPS Agreement gives these standards much more legal relevance. For example, Codex, OIE, and IPPC standards have played roles in WTO cases, settling disputes arising under the SPS Agreement.

The basic approach toward trade-restricting health measures can also be found in regional international trade agreements, such as the North American Free Trade Agreement and the European Community (EC). In the EC context, the European Commission also has power to restrict trade for health reasons, as illustrated in the EC's worldwide ban on British beef imposed in connection to the possible transmission of bovine spongiform encephalopathy (BSE) to humans via prions. The United Kingdom challenged this ban before the European Court of Justice (ECJ), but the ECJ upheld the EC's actions as justified in the face of an uncertain, but potentially deadly, possibility of BSE transmission to humans.

Another important aspect of the international trade regime for infectious disease control is the protection of intellectual property rights found in the WTO's Agreement on the Trade-Related Aspects of Intellectual Property Rights (TRIPS). Through TRIPS, patent protection for pharmaceuticals is being harmonized along standards common in developed but not developing countries. Western pharmaceutical companies argue that such increased patent protection is necessary to provide adequate private-sector incentives for the development of new antimicrobial products. Developing countries and many nongovernmental organizations fear that TRIPS will decrease access to essential drugs in the developing world by increasing the cost of drugs. Some developing countries are considering compulsory licensing of pharmaceutical patents that, while allowed by TRIPS under certain conditions, is fiercely opposed by Western pharmaceutical companies and their governments. The controversy over compulsory licensing was coming to a boil in 1999 in connection with HIV/AIDS therapies and efforts by countries, such as South Africa, to improve access to such therapies through compulsory licensing.

B. International Human Rights Law

Infectious diseases create concerns in connection with the protection of human rights under international law. The widespread mistreatment of persons with HIV/AIDS by governments all over the world led public health experts and human rights lawyers and activists to make common cause against these human rights violations. International human rights law recognizes that a government may restrict civil and political rights for public health reasons, but such restrictions have to satisfy four criteria: the restrictions must be (1) prescribed by law, (2) applied in a nondiscriminatory manner, (3) applied for a compelling public interest, and (4) the least restrictive measure possible to achieve the compelling public interest. The criteria exist to balance the individual's enjoyment of rights with the community's interest in being protected from serious health threats.

The human right to health also relates to infectious diseases. Although the exact scope and content of the right to health remains controversial, many people believe that a government's failures to provide safe drinking water, basic sanitation, and immunizations against infectious diseases violate the right to health, because these failures lead to infectious disease morbidity and mortality. A complicating aspect of the right to health is the principle of progressive realization, which holds that a government's duty to fulfill the right to health depends on its available economic resources. The principle of progressive realization makes it difficult to establish clearly that a government's failures to provide safe water, basic sanitation, and immunizations violate the right to health.

It is sometimes argued that developed states have an international legal duty to provide financial and technological assistance to developing states in connection with health development. The argument variously relies on the human rights to health, to development, and to the benefits of scientific progress. The substance and parameters of none of these human rights are, however, clear in international law, which weakens assertions that international law requires developed states to assist developing states to improve their public health and health care systems.

C. International Humanitarian Law

The laws of war impose many obligations on belligerents in international and civil armed conflicts to protect combatants, prisoners of war (POWs), and civilians from infectious diseases. Some of the protections are indirect, such as the prohibition on attacking civilian targets or resources critical to the survival of the civilian population. Other protections are more direct, such as the detailed requirements for belligerents to provide POWs and civilian detainees with adequate food, safe water, sanitary housing, and medical care. Failures to fulfill the duties not to attack civilian targets and to treat POWs and civilian detainees humanely expose people to infectious diseases and constitute war crimes under international humanitarian law.

D. International Law on Arms Control

International law prohibits the use of pathogenic microbes as weapons of war. The 1925 Geneva Protocol established this prohibition, which has since entered into customary international law. The Geneva Protocol did not, however, ban the development and production of biological weapons; and many countries, including the United States and the Soviet Union, engaged in offensive and defensive biological weapons programs for many decades.

The United States' unilateral renunciation of its offensive biological weapons program in the late 1960s paved the way for the adoption of the 1972 Biological Weapons Convention (BWC), which prohibited the development, production, and stockpiling of biological weapons. The proliferation of biological weapons capabilities in many countries since 1972 and the growing threat of biological terrorism have suggested, however, that the BWC's prohibitions were not entirely successful. Negotiations are currently under way to create a Protocol to the BWC that will establish a compliance regime to strengthen the BWC's prohibitions.

E. International Environmental Law

Public health experts often argue that environmental change and degradation contribute to the emergence and reemergence of infectious diseases. Fears that the depletion of the ozone layer will cause adverse human health problems, including weakening of the human immune system, produced the diplomatic effort that created the international legal regime on ozone depletion. The creation of the United Nations Framework Convention on Climate Change and the Kyoto Protocol has been accompanied by concerns that global warming will expand the geographic habitat of disease vectors, such as the mosquito. Efforts to protect biodiversity through international law have as one aim the preservation of plant species that might have potential antibiotic or other pharmaceutical uses against diseases. The international legal regime on biodiversity suffers, however, from tension over developing countries' accusations of "biopiracy" against Western pharmaceutical companies that seek to exploit biological resources in developing countries. Some developing countries have taken national steps to protect their biodiversity from exploitation by Western drug companies. Although international environmental law has developed in some areas of concern in connection with infectious diseases, in other relevant areas, such as local air and water pollution, marine pollution, and deforestation, international environmental law is not well developed.

V. EMERGING INFECTIOUS DISEASES AND THE REVISION OF THE IHR

The global crisis in emerging infectious diseases forced WHO to reevaluate its international legal approach to infectious diseases. The IHR had not been a successful regime, as has been noted. The emergence and reemergence of infectious diseases also made the IHR's focus on only three diseases anachronistic. In 1995, WHO decided the time had come to revise the IHR to meet the challenges posed by emerging infectious diseases and the globalization of public health.

While the IHR revision process will not be finished for a number of years, the revisions proposed to date radically change the nature of the IHR. For example, WHO proposes to require syndrome reporting rather than reporting of specific diseases. Syndrome reporting would expand the surveillance system operated under the IHR. For example, tuberculosis (TB) is not a disease subject to the current IHR; but, under the proposed move to syndrome reporting, outbreaks or epidemics of TB that meet specified criteria would be reportable under the category of "acute respiratory syndrome." Other proposed revisions suggest WHO wants to expand the flow of epidemiological information to and from it and to monitor compliance with the new IHR more vigorously than it has done in the past. WHO also hopes that the global scope of the media and the revolution in information technologies (i.e., the Internet and electronic mail) will help make States less reluctant to report disease outbreaks because they can no longer hide such public health events from the world. Technology could, thus, aid compliance with the international legal rules.

VI. GROWING INTEREST IN THE ROLE OF INTERNATIONAL LAW IN GLOBAL PUBLIC HEALTH

As a historical matter, WHO proved uninterested in international law as an instrument of global public health. WHO has preferred to issue nonbinding recommendations under Article 23 of the WHO Constitution rather than develop international legal rules under Articles 19 and 21 of the Constitution. To the extent they had any legal relevance, WHO's activities resemble a "soft law" approach to global public health. The lack of legal interest partly explains why WHO let the IHR atrophy as an international legal regime. WHO's historical antipathy toward international law may, however, be changing, as suggested by the IHR revision and other international legal developments, such as the proposed framework convention on tobacco control. WHO may finally be showing some appreciation that international law plays a role in global public health in many ways,

including in connection with infectious diseases (see Table IV).

This perceived change in WHO's attitude toward international law may mean some emerging issues in infectious diseases stimulate international legal activity. Experts have identified, for example, the safety of blood and other biological products moving in international commerce, antimicrobial resistance, and xenotransplantation as potential areas where international law could be fruitfully used for infectious disease control purposes. These possibilities, combined with the continued relevance of many different international legal regimes, underscore the important place of international law in global infectious disease strategies.

See Also the Following Articles

BIOLOGICAL WARFARE • EMERGING INFECTIONS • GLOBAL BURDEN OF INFECTIOUS DISEASES • PATENTING OF LIVING ORGANISMS AND NATURAL PRODUCTS • SURVEILLANCE OF INFECTIOUS DISEASES

Bibliography

Fidler, D. P. (1998). The future of the World Health Organization: What role for international law? *Vanderbilt Journal of Transnational Law* **31**, 1079–1126.

Fidler, D. P. (1996). Globalization, international law, and emerging infectious diseases. *Emerging Infectious Diseases* **2**, 77–84.

Fidler, D. P. (1997). The globalization of public health: Emerging infectious diseases and international relations. *Indiana Journal of Global Legal Studies* **5**, 11–51.

Fidler, D. P. (1999). "International Law and Infectious Diseases." Clarendon Press, Oxford, UK.

Fidler, D. P. (1999). Legal challenges posed by the use of antimicrobials in food animal production. *Microbes and Infection* **1**, 29–38.

Fidler, D. P. (1998). Legal issues associated with antimicrobial drug resistance. *Emerging Infectious Diseases* **4**, 169–177.

Fidler, D. P. (1997). The role of international law in the control of emerging infectious diseases. *Bulletin Institut Pasteur* **95**, 57–72.

Fidler, D. P. (1997). Trade and health: The global spread of diseases and international trade. *German Yearbook of International Law* **40**, 300–355.

Fluss, S. S. (1997). International public health law: An overview. *In* "Oxford Textbook of Public Health" (R. Detels *et al.*, eds.) (3rd ed.), pp. 371–390, Oxford Univ. Press, Oxford, UK.

Gostin, L. O., and Lazzarini, Z. (1997). "Human Rights and

TABLE IV
WHO Awareness of the Importance of International Law in Global Public Health

1995	WHO begins revision of the International Health Regulations
1996	WHO begins effort to draft framework convention on global tobacco control
1997	International Conference on Global Health Law in New Delhi, India, cosponsored by WHO, at which the participants in the New Delhi Declaration on Global Health Law called on WHO to play a more active legal role
1998	New Health for All in the Twenty-First Century policy emphasizes the need to protect and promote human rights and states the need to use international law in its mission, and new WHO Director-General Gro Harlem Brundtland emphasizes importance of international rules, standards, and norms

Public Health in the AIDS Pandemic." Oxford Univ. Press, Oxford, UK.

L'hirondel, A., and Yach, D. (1998). Develop and strengthen public health law. *World Health Statistics Quarterly* **51**, 79–87.

Plotkin, B. J., and Kimball, A.-M. (1997). Designing the international policy and legal framework for emerging infectious diseases: First steps. *Emerging Infectious Diseases* **3**, 1–9.

Taylor, A. L. (1997). Controlling the global spread of infectious diseases: Toward a reinforced role for the International Health Regulations. *Houston Law Review* **33**, 1327–1362.

Tomasevski, K. (1995). Health. *In* "United Nations Legal Order, Vol. 2" (O. Schachter and C. Joyner, eds.), pp. 859–906. Cambridge Univ. Press, Cambridge, UK.

Yach, D., and Bettcher, D. (1998). The globalization of public health, I: Threats and opportunities. *American Journal of Public Health* **88**, 735–738.

Intestinal Protozoan Infections in Humans

Adolfo Martínez-Palomo and Martha Espinosa-Cantellano

Center for Research and Advanced Studies, Mexico

I. Amebiasis
II. Giardiasis
III. Cryptosporidiosis

GLOSSARY

asymptomatic carrier A person harboring an infectious microbe, without symptoms of the disease, but capable of transmitting it to other persons.

immunocompetence The ability or capacity to develop an immune response (i.e., antibody production and/or cell-mediated immunity) following antigenic challenge.

immunocompromised patient A person that has lost immunocompetence, either because of a disease (e.g., AIDS) or immunosuppresive therapy (e.g., cancer treatment).

parasite A plant or animal that lives upon or within another organism, at whose expense it obtains some advantage.

protista One of the four main groups of eukaryotes, the other three being animals, plants, and fungi. Protists are single-celled organisms with a nucleus and include protozoa, algae, and slime-molds, among others.

protozoa A taxonomic group of nonphotosynthetic, single-celled organisms with a nucleus. The name "protozoa" literally means "first animals." Although some protozoa are relatively simple, others have a complex life cycle.

zoonosis A disease of animals that may be transmitted to man.

A NUMBER OF PATHOGENIC PROTOZOAN PARASITES are capable of producing intestinal infections in humans. These include *Entamoeba histolytica, Giardia lamblia, Cryptosporidium parvum, Blastocystis hominis, Dientamoeba fragilis, Balantidium coli, Isospora belli,* and several genera of microsporidia. In terms of clinical impact, however, the first three are the most important protozoan infections of the gastrointestinal tract, and this article will, therefore, be devoted to them. Humans are the only hosts for *E. histolytica* and *G. lamblia,* in whom they produce amebiasis and giardiasis, respectively. Except for *D. fragilis,* a trichomonad infrequently associated with gastrointestinal symptoms, the other parasites produce mainly zoonotic infections that can eventually infect humans. In recent years, however, increasing attention has been paid to these parasites because of their association with diarrheal syndromes in AIDS or other immunocompromised patients. *Cryptosporidium* has been most frequently identified as the causative agent of these syndromes; brief mention will be given to this parasite.

I. AMEBIASIS

Amebiasis is the infection of the human gastrointestinal tract by *Entamoeba histolytica,* a parasite capable of invading the intestinal mucosa that may spread to other organs, mainly the liver, giving rise to intestinal and extraintestinal lesions. Invasive amebiasis is still a major health problem in certain areas of Africa, Asia, and Latin America. In most industrialized countries, however, the number of severe cases of amebiasis is much lower. Nevertheless, knowledge of the disease in these regions is also important, since failure to identify an amebic infection may result in a lethal outcome, e.g., intestinal amebiasis may be treated as chronic ulcerative colitis. In addition, high infection rates can exist among certain immigrant groups and epidemic outbreaks can occur in institutions, such as schools or mental hospitals.

Encyclopedia of Microbiology, Volume 2
SECOND EDITION

852

In areas of high prevalence, invasive amebiasis characteristically occurs in endemic form. In 1984, it was estimated that 40 million people developed disabling colitis or liver abscesses. At least 40,000 deaths that year were attributable to amebiasis, mostly as a consequence of liver abscesses. Therefore, on a global scale, amebiasis comes third among all parasitic causes of death, behind only malaria and schistosomiasis.

A. The Life Cycle of *E. histolytica*

The motile form of the parasite, the trophozoite, is a highly dynamic pleomorphic cell that measures 20–60 μm in diameter (Fig. 1). It lives in the lumen of the large intestine, where it multiplies and differentiates into a cyst, a tetranucleated round or slightly oval body 8–20 μm in diameter, surrounded by a thick refractile wall. Cysts are the resistant forms responsible for the transmission of the infection. Cysts are excreted in stools and may be ingested by a new host in contaminated food or water. The cyst wall is dissolved in the upper gastrointestinal tract and the parasite excysts in the terminal ileum, each excysted cell giving rise to 8 uninucleated trophozo-

ites. Trophozoites may invade the colonic mucosa and produce dysentery and, through blood-borne spreading, give rise to extraintestinal lesions, mainly, liver abscesses. The course of dysentery is usually self-limited, but amebic liver abscess is potentially fatal, unless it is diagnosed promptly and treated appropriately.

B. Clinical Features

Depending on the affected organ, the clinical manifestations of amebiasis are intestinal or extraintestinal, the former being the more common. Both localizations can occur at the same time, but they are usually manifested separately.

1. Intestinal Amebiasis

The high rates of asymptomatic amebic infections (~90%) puzzled clinicians, researchers, and epidemiologists for over a century. It is now known that most asymptomatic intestinal infections formerly attributed to "nonpathogenic" strains of *E. histolytica* are, in fact, due to a different species of ameba, *E. dispar,* morphologically similar to

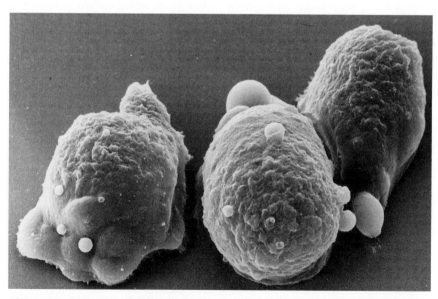

Fig. 1. Trophozoites of *Entamoeba histolytica.* Trophozoites live in the lumen of the large intestine. They are characteristically pleomorphic, with a rough membrane surface and several pseudopodia, extending from different parts of the cell. The parasite moves through the extension of these pseudopods. Scanning electron micrograph.

E. histolytica but of a noninvasive nature. Carrier states, however, can be attributed to *E. histolytica.*

The acceptance of the existance of *E. dispar* as a separate species from *E. histolytica* has deep epidemiological and therapeutical implications. Until diagnostic methods that distinguish infection of *E. histolytica* from *E. dispar* are available for the clinical laboratories, the real incidence of amebiasis (i.e., the infection by *E. histolytica,* regardless of the presence of symptoms) can not be established. Furthermore, these diagnostic tools will also eliminate unnecessary treatment currently administered to carriers of *E. dispar.* Several tests have been developed, but they are still not easily available for diagnostic laboratories, especially in those countries where amebiasis is most prevalent.

When *E. histolytica* invades the colonic mucosa, the clinical spectrum includes acute colitis with dysentery or bloody diarrhea, fulminating colitis, amebic appendicitis, and ameboma of the colon. The first manifestation may follow a relatively benign course, but the other three are severe forms of the disease, requiring prompt medical treatment.

Dysenteric and diarrheic syndromes account for 90% of cases of invasive intestinal amebiasis. The onset is gradual, with intense abdominal pain. Initially, there are loose watery stools that rapidly become blood-stained with mucus. Watery diarrhea or loose stools without blood may be present for few days, particularly when distal regions of the colon are involved. Fever and systemic manifestations are generally absent. The clinical course is moderate and symptoms disappear rapidly with treatment; spontaneous remissions may be observed.

2. *Amebic Liver Abscess*

This is the most common extraintestinal form of invasive amebiasis. It results from the migration of *E. histolytica* trophozoits from the colon to the liver via the portal circulation. Amebic abscesses may be found in all age groups, but are 10 times more frequent in adults than in children, and show a higher frequency in males than in females. Lesions are usually single and localized to the right lobe of the liver.

The onset is usually abrupt, with pain in the upper abdomen and high fever, with rigors and profuse sweating. There is anorexia and rapid weight loss;

nausea and vomiting may occur and, in some cases, diarrhea or dysentery may be present. Physical examination generally reveals a pale, wasted patient with an enlargd, tender liver. Digital pressure in the right lower intercostal spaces will produce intense pain and there is often marked tenderness on percussion over the right lower ribs in the posterior region. Movement of the right side of the chest and diaphragm is greatly restricted, as well as the intensity of respiratory sounds.

C. Pathogenesis and Pathology

The lytic and invasive characteristics of *E. histolytica* are related to multifactorial mechanisms that include a striking motility and phagocytic capacity of the trophozoites, release of membrane pore-forming peptides, and proteases that produce contact-dependent lysis of target cells, and degradation of extracellular matrix components.

Invasion of the colonic and cecal mucosa by *E. histolytica* begins in the interglandular epithelium. Cell infiltration around invading amebas leads to rapid lysis of inflammatory cells and tissue necrosis; thus, acute inflammatory cells are seldom found in biopsy samples or in scrapings of rectal mucosal lesions. Ulcerations may deepen and progress under the mucosa to form typical "flask ulcers," which extend into the submucosa, producing abundant microhemorrhages. This explains the finding of hematophagous amebas—amebas containing erythrocytes in the cytoplasm—in stool specimens or in rectal scrapings, still the best indication of the amebic nature of a case of dysentery or bloody diarrhea. Macroscopically, the ulcers are initially superficial, with hyperemic borders, a necrotic base, and normal mucosa between the sites of invasion. Further progression of the lesions may produce loss of the mucosa and submucosa covering the muscle layers and eventually leads to the rupturing of the serosa.

Complications of intestinal amebiasis include perforation, direct extension to the skin, and dissemination, mainly to the liver. The early stages of hepatic amebic invasion have not been studied in humans. In experimental animals, inoculation of *E. histolytica* trophozoites into the portal vein produces multiple foci of neutrophil accumulation around the parasites, followed by focal necrosis and granulomatous infiltration. As the lesions extend in size, the granulomas are gradually substituted by necrosis, until the lesions coalesce and necrotic tissue occupies progressively larger portions of the liver. Human liver abscesses consist of areas in which the parenchyma has been completely substituted by a semisolid or liquid material, composed of necrotic matter and a few cells. Neutrophils are generally absent, and amebas tend to be located at the periphery of the abscess.

If properly treated, invasive amebic lesions in humans, whether localized in the large intestine, liver, or skin, almost invariably heal without the formation of scar tissue.

D. Diagnosis

1. *Intestinal Amebiasis*

Trophozoites can be identified by microscopic examination of recently evacuated stools and are most likely present in the bloody mucus. The presence of motile, hematophagous trophozoites of *E. histolytica* establishes the diagnosis of amebiasis, but the microscopic examination of amebas has several drawbacks, including the requirement for a skilled technician. Rectosigmoidoscopy and colonoscopy of benign cases show small ulcerations with linear or oval contours, 3 to 5 mm in diameter, covered by a yellowish exudate containing many trophozoites. As has been mentioned, specialized laboratory methods to distinguish nonhematophagous trophozoites of *E. histolytica* from those of the noninvasive *E. dispar* in asymptomatic carriers are available. Cyst detection usually requires concentration methods, including flotation or sedimentation procedures. The development of diagnostic tests to distinguish cysts of *E. histolytica* from those of *E. dispar* is still in the research phase.

2. *Amebic Liver Abscess*

This condition should be suspected particularly in endemic areas, or when there is a history of travel to those countries, in patients who present with spiking fever, weight loss, and abdominal pain in the upper right quadrant, or epigastrium, with tenderness in the liver area. Other signs include leucocytosis, elevated alkaline phosphatase, and an elevated right diaphragm in chest films. Liver imaging with sonography or computed tomography will demonstrate a space-occupying lesion in 75–95% of the cases, according to the procedure and course of the illness. This should be followed by the detection of antiamebic antibodies, which are elevated in more than 90% of cases. The tests currently used are indirect hemagglutination, counterimmunoelectrophoresis, and enzyme immunoassays.

E. Management

Metronidazole and related nitroimidazole compounds have contributed greatly to decreasing the morbidity and mortality of amebiasis; they are reasonably well tolerated and, in spite of their reported carcinogenic effect in rodents and their mutagenic potential in bacteria, no such effects have been reported in humans. Other amebicides for invasive intestinal amebiasis are diiodohydroxyquin, diloxanide furoate, and paromomycin. Emetine hydrochloride, dehydroemetine, and chloroquine are seldomly used, at present.

Amebic liver abscess should be treated with chemotherapy; surgery is rarely indicated. Percutaneous drainage of an amebic liver abscess is performed when rupture of a large abscess is imminent, as a complementary therapy to shorten the course of the disease when response to chemotherapy has been slow, or when pyogenic or mixed infection is suspected. Drainage should be done under ultrasound or CT guidance; catheters should not be left for drainage and should be rapidly removed to avoid contamination of the track and skin. Indications for surgical drainage include imminent rupture of inaccessible liver abscess, especially of the left lobe, risk of peritoneal leakage of necrotic fluid after aspiration, or rupture of a liver abscess.

F. Prevention and Control

The main reservoir of *E. histolytica* is humans. The cysts may remain viable and infective for a few days in feces. Since they are killed by desiccation, cyst-laden dust is not infective. They are also killed by temperatures higher than 68°C, so that boiled water is safe. The amount of chlorine used to purify ordinary water is insufficient to kill cysts; higher levels of chlorine are effective, but the water must be de-chlorinated before use.

The control of invasive amebiasis could be achieved through improvement of living standards and the establishment of adequate sanitary conditions in countries where the disease is prevalent. Community strategies include improvement of environmental sanitation of water supply, food safety, and health education to prevent fecal–oral transmission; individual actions should aim at early detection and treatment of cases of infection and/or disease. Mass chemotherapy of high-risk populations has been attempted, with only partially successful results. Individual or collective chemoprophylaxis is not indicated.

II. GIARDIASIS

Giardiasis is the infection of the human small intestine by the protozoan parasite *Giardia lamblia,* also known as *G. intestinalis* or *G. duodenalis*. The parasite has a worldwide distribution; it is the most commonly diagnosed flagellate in the intestinal tract, being more prevalent in children than in adults.

A. The Life Cycle of *Giardia lamblia*

As with *E. histolytica,* the life cycle of *G. lamblia* involves a motile trophozoite stage and an infective cyst. Trophozoites are pear-shaped cells with two nuclei, four pairs of flagella used for locomotion, and a ventral sucking disc that allows the parasite attachment to the intestinal epithelium (Fig. 2). These small protozoa (10–20 μm in length and 5–15 μm in width) dwell in the proximal portion of the small intestine, the duodenum, where they multiply by binary fission and, under conditions still not understood, transform into four nucleated cysts, the infective stage of the parasite. Cysts are excreted in the feces and ingested by another host in contaminated food or water. Cysts are round or oval cells, measuring 11–14 μm in length and 7–10 μm in width, surrounded by a thick wall. They excyst in the duodenum, giving rise to two trophozoites.

B. Clinical Features

Most individuals are asymptomatic, but an unspecified percentage of those infected may develop acute or chronic symptoms. The former include the sudden onset of explosive, watery, foul diarrhea with flatulence, cramps, and abdominal distention, and virtual absence of blood or cellular exudate in stools. Subacute or chronic infections may be accompanied by recurrent, brief episodes of loose foul stools, flatulence, cramps, abdominal distention, and belching. Between episodes of mushy stools, the patient may be constipated. Several malabsorption syndromes have been associated with giardiasis; these include steator-

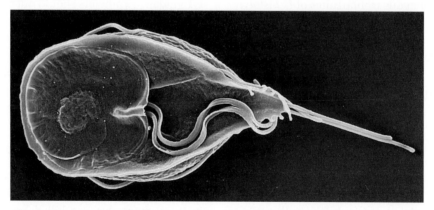

Fig. 2. Trophozoites of *Giardia lamblia*. The trophozoite stage of the parasite dwells in the proximal portion of the small intestine, the duodenum. Trophozoites are pear-shaped cells with four pairs of flagella, used for locomotion, and a ventral sucking disc that allows the parasite attachment to the intestinal epithelium. Scanning electron micrograph.

rhea, disaccharidase deficiency, vitamin B$_{12}$ malabsorption, hypocarotenemia, low serum folate and protein-losing enteropathy. In addition, lactose intolerance may be present. Spontaneous resolution of the infection seems to be common, although in some patients, particularly children, symptoms may last for months.

C. Pathogenesis and Pathology

There is still no explanation for the high rates of asymptomatic giardia infections, and the mechanisms by which the parasite causes alterations in small bowel function in some patients are largely unknown. Trophozoites adhere to the epithelium through the ventral disk; although a few reports have identified some organisms inside the intestinal mucosa, most authors agree on the noninvasive nature of the parasite. Although mechanical interference from the epithelial lining of large numbers of trophozoites has been held responsible for the various malabsorption syndromes seen in some patients, this hypothesis has not been proven and does not explain why uptake of some substances absorbed at other intestinal levels is prevented. Thus, the malabsorption syndromes remain unexplained. Most infections show an unaltered intestinal morphology, but in a few, usually chronically infected, symptomatic individuals, flattened villi and varying inflammation of the lamina propria have been identified in jejunal mucosa biopsy samples. The pathogenesis of diarrhea in giardiasis is unknown.

D. Diagnosis

Diagnosis is carried out by the finding of trophozoites or cysts in the microscopic examination of stools or duodenal aspirate. Because of the cyclical shedding of parasites, repeated stool examinations are recommended. Tests to detect parasite antigen in stool are now commercially available, but are relatively expensive (Fig. 3).

E. Management

Treatment is usually effective with metronidazole, quinacrine, or furazolidone. Most patients respond

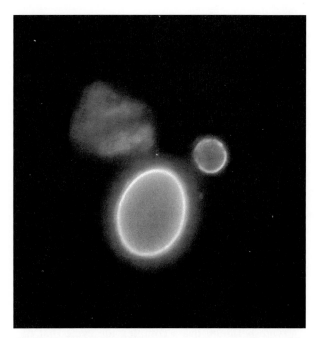

Fig. 3. A *Cryptosporidium* oocyst and a *Giardia lamblia* cyst, found in the same specimen. Monoclonal antibodies have been used to identify these parasites. The small *Cryptosporidium parvum* oocysts (mean size 5.0 × 4.5 μm) can be clearly differentiated from the much larger *Giardia lamblia* cysts (11–14 μm in length and 7–10 μm in width). Immunofluorescence microscopy.

well to therapy; if infection persists after repeated treatment, reinfection from family members, close personal contacts, and environmental sources should be considered.

F. Prevention and Control

Although giardiasis is extremely infectious, disease can be prevented by improved personal hygiene, routine hand-washing, better preparation and storage of food and water, and control of mammals and other sources that may contaminate food and water. Cysts are not killed by routine chlorination methods and remain viable for months in cold fresh water, but boiling and filtration eliminates them effectively.

III. CRYPTOSPORIDIOSIS

Cryptosporidium, an intestinal coccidian parasite of domestic and wild animals, has only recently been found to be a leading cause of protozoal diarrhea in humans, especially in children less than 2 years of age. The first cases of cryptosporidiosis were reported in the mid-1970s, followed by a few other reports, until it was recognized that *C. parvum* could produce a short-term diarrheal disease in immunocompetent persons and a prolonged, life-threatening, choleralike illness in immunocompromised patients. Higher prevalence rates (3–20%) have been reported for the general population in Asia, Australia, Africa, and Central and South America than in Europe (1–2%) or North America (0.6–4.3%). In AIDS patients, *Cryptosporidium* has been identified as the cause of chronic diarrhea in 11–21% cases in industrialized countries and in 12–85% in developing countries.

A. The Life Cycle of *Cryptosporidium*

Cryptosporidium has a complex life cycle that includes both asexual and sexual stages and an autoinfection mechanism that maintains the parasite in the host without repeated oral exposure to resistant forms. Infection is initiated with the consumption of oocysts, which excyst in the lumen of the intestine to liberate sporozoites, that penetrate the intestinal epithelial cells. Sporozoites develop into trophozoites, which divide asexually to give rise to merozoites in type I meronts. After being released, the invasive merozoites may enter adjacent host cells, to form additional type I meronts, or transform into type II meronts. These enter the host cell to form the sexual stages, micro and macrogamonts, that fertilize to form zygotes. About 80% of the zygotes develop into environmentally resistant, thick-walled oocysts that undergo sporogony to form sporulated oocysts containing four sporozoites; oocysts are responsible for the transmission of infection from one host to another. The rest of the zygotes (~20%) transform into thin-walled sporulated oocystes that release the sporozoites in the intestinal lumen and, thus, represent the autoinfective life cycle forms of the parasite.

B. Clinical Features

In immunocompetent patients, the infection is usually self-limited and symptoms include watery, nonbloody diarrhea, abdominal pain, nausea, vomiting, and low-grade fever. In contrast, in immunocompromised patients, particularly in AIDS cases, symptoms may last for several months and produce profound weight loss. In these patients, *C. parvum* is not always confined to the gastrointestinal tract, but may give rise to extraintestinal infections responsible for a variety of respiratory problems, cholecystitis, hepatitis, and pancreatitis. In AIDS patients, cryptosporidiosis may be a life-threatening illness.

C. Pathogenesis and Pathology

The pathophysiological mechanisms of *Cryptosporidium*-induced diarrhea are poorly understood. *In vitro* studies suggest that there may be a toxin-mediated hypersecretion into the gut, but no toxin has yet been isolated from the organism. Malabsorption and impaired digestion have been reported during infection; the severity seems to be related to the number of organisms present in the intestine.

Histologically, different life-cycle forms can be observed in a low columnar or cuboidal epithelium, and microvilli may be absent at the site of parasite attachment. Intestinal lesions include mild to severe villous atrophy and inflammatory infiltrate into the subepithelial lamina propria. Although the parasite can be found at all levels of the intestinal tract, the jejunum is usually the most heavily infected.

D. Diagnosis

The diagnosis is based on the microscopical finding of *Cryptosporidium* oocysts in stool samples, using concentration methods and acid-fast stains to differentiate the parasite from yeasts. Fixation of stool and other body fluid specimens, such as bile, is recommended because of potential biohazard considerations. Newer methods include the use of monoclonal antibodies-based reagents in fluorescence or enzyme immunoassays (Fig. 3).

E. Management

An effective drug for cryptosporidiosis in humans is not available at present, although paromomycin has given promising results. Paromomycin treatment decreases stool frequency and oocyst excretion in some patients, but biliary disease progresses despite long-term therapy. Surgical treatment with cholecystectomy and sphincterotomy has achieved variable therapeutic success. Management of the patient includes supportive care with fluid and electrolyte replacement and administration of antidiarrheal agents.

F. Prevention and Control

Because effective treatment for *Cryptosporidium* infection is not available, prevention is the most realistic method of control. Although zoonotic transmission is important for rural populations, person-to-person transmission through direct or indirect contact with stool material is common in urban environments. Pets, especially when less than 6 months old or with diarrhea, should be monitored.

Most recommendations concerning cryptosporidiosis are directly related to water consumption, since water-borne outbreaks are well documented. Water sources and water treatment plants should be carefully monitored. High-volume filters and immunofluorescent detection methods have identified *Cryptosporidium* oocysts in surface and drinking waters and in sewage effluents from many geographic regions, but there is still no method available to asess their viability.

Oocysts are not killed by routine chlorination methods, but boiling and filtration are effective. Because of the small size of the oocysts, it is important to select filters that remove particles 0.1–1 μm in size. Drinking high quality bottled water when traveling abroad reduces the risk of *Cryptosporidium*-induced traveler's diarrhea.

See Also the Following Articles

Diagnostic Microbiology • Food-borne Illnesses • Water, Drinking

Bibliography

Fayer, R. (ed.) (1997). *"Cryptosporidium* and Cryptosporidiosis." CRC Press, Boca Raton, FL.

Hill, D. R. (1995). *Giardia lamblia. In* "Principles and Practice of Infectious Diseases." (G. L. Mandell, J. E. Bennett, and R. Dolin, eds.), (4th ed.) Chapter 259, pp. 2487–2493. Churchill Livingstone, New York.

Marshall, M. M., Naumovitz, D., Ortega, Y., and Sterling, C. R. (1997). Waterborne protozoan pathogens. *Clin. Microbiol. Rev.* **10**, 67–85.

Martínez-Palomo, A., and Espinosa-Cantellano, M. (1998). Intestinal amoebae. *In* "Topley & Wilson's Microbiology and Microbial Infections." (Leslie Collier, ed.-in-chief), (A. Balows and M. Sussman, general eds.), (9th ed.), Vol. 5, Ch. 14, pp. 157–177. Arnold, London, UK.

Iron Metabolism

Charles F. Earhart
The University of Texas at Austin

GLOSSARY

bacterioferritin (BFR) A subclass of prokaryotic ferritin that contains heme.

ferritin (FTN) A large intracellular iron storage protein, present in both prokaryotic and eukaryotic cells.

haptoglobin (Hp) A serum protein that scavenges hemoglobin.

heme (Hm) A metal ion, customarily, iron in the FeII state, chelated in a tetrapyrrole ring.

hemopexin A serum protein that strongly binds heme.

hemophore (Hbp) A secreted bacterial protein that shuttles environmental heme to the bacterial surface.

lactoferrin (LF) A glycoprotein present in the mucosal secretions and phagocytic cells of vertebrates that binds and transports iron.

siderophore A small molecule secreted by bacteria and fungi that chelates environmental FeIII and, as the ferrisiderophore, then binds to the microbial surface.

transferrin (TF) Iron-binding and transport protein found in the serum and lymph of vertebrates.

A KNOWLEDGE OF IRON METABOLISM is essential for a complete understanding of microbial growth and survival. In microbes, iron is necessary for processes such as electron transport, nitrogen fixation, removal of toxic forms of oxygen, synthesis of DNA precursors, tRNA modifications, and syntheses of certain amino acids and tricarboxylic acid cycle intermediates; some bacteria can even oxidize iron to obtain energy.

Iron is such a valuable and versatile nutrilite that, of all the organisms on earth, only certain lactic acid bacteria can manage without it. The utility of iron stems primarily from the fact that its redox couple (FeII/FeIII) can have a range of potentials of -300 to $+700$ mV, depending on the nature of the ligands and the environment surrounding the coordinated iron ions.

A major consideration in iron metabolism is that, although iron is abundant on the surface of the earth, it is relatively unavailable. At neutral pH and in an oxidizing environment, which includes most common microbial habitats, iron exists in the $+3$ valence state and, as such, is extremely insoluble. For a microbe inhabiting an animal host, iron availability is restricted by the presence of iron storage and transport proteins, such as ferritin, lactoferrin, and transferrin. Thus, although relatively low concentrations of iron (5 μM) are generally sufficient for maximum growth yields, bacteria frequently find themselves in iron-deficient environments and must devote a significant amount of their resources to obtaining this metal. Also important is the fact that it is possible for bacteria to suffer from iron overload and iron assimilation must, therefore, be precisely controlled.

This article covers prokaryotic iron metabolism only. Descriptions of fungal iron assimilation are included in the reviews by Guerinot (1994) and Leong and Winkelmann (1998).

I. IRON UPTAKE

Bacteria can obtain iron from a variety of sources but, regardless of its origin, iron must be transported through the several microbial surface layers to reach the cytoplasm. For gram-negative bacteria, these layers minimally include an outer membrane, a monolayer of peptidoglycan, and an innermost cytoplasmic membrane. The peptidoglycan cell wall is located in the periplasm, the space between the outer and inner membranes. Gram-positive cells, in contrast, contain only a thick, highly cross-linked peptidoglycan cell wall external to the cytoplasmic membrane.

A great variety of iron transport systems can be distinguished on the basis of the iron source and the form in which iron is mobilized, but, in general, they follow a pattern (Fig. 1). Thus, for iron complexed to a carrier, first, passage through the outer membrane requires an outer membrane receptor protein whose synthesis is iron-regulated. A receptor protein has specificity for a given iron-carrier complex, binds that complex only, and is sometimes synthesized in abundance only when that particular iron complex is available. In the two best-studied cases, receptor proteins have been shown to be gated pores. Second, to permit entry of complexed or free iron into the periplasm, cytoplasmic membrane proteins TonB,

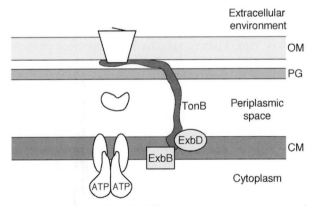

Fig. 1. Generalized high affinity iron transport system of gram-negative bacteria. The three basic components are shown; they include (i) an outer membrane receptor protein, (ii) a TonB system for energizing the receptor protein, and (iii) a periplasmic binding protein-dependent ABC transporter, located in the cytoplasmic membrane. OM, outer membrane; PG, peptidoglycan; CM, cytoplasmic membrane. See text for details.

TABLE I
Consensus Sequences Related to Iron

TonB Box[a]	AspThr Hyd Val Val Thr Ala Glu
Fur binding site (Fur box, iron box)	GATAATGATAATCATTATC
DtxR binding site	TAAGGTAAGCTAACATA T G T

[a] Positions two and five are the most conserved. Position three is the most variable; Hyd indicates hydrophobic amino acid (Ile, Leu, Val, Met, Phe).

ExbB, and ExbD are required. These proteins function as a group to utilize the electrochemical potential of the cytoplasmic membrane to open the gated receptor pores, permitting iron and iron-chelates to pass into the periplasm. Generally, each bacterium has just one set of TonB/Exb proteins, capable of interacting with multiple receptors, but *Vibrio cholerae* was recently shown to have two distinct TonB systems. TonB is anchored in the cytoplasmic membrane but spans the periplasmic space, so as to be able to physically contact all receptors that require it. ExbB has three transmembrane domains and is present primarily in the cytoplasm, while ExbD has only one transmembrane segment and extends into the periplasm. Determination of the details by which TonB and its accessory proteins act is an area of active research. A straightforward mechanism would have ExbB sense the proton motive force and, with ExbD, use it to induce a conformational change in TonB, that, in turn, could bring about, by physical contact, allosteric changes in outer membrane receptors. TonB-dependent receptors have a heptapeptide (TonB box) near their amino termini that is thought to serve as a recognition sequence (Table I). Third, transport across the cytoplasmic membrane employs members of the ABC supertransporter family. In this case, transport components consist of a peripheral cytoplasmic membrane ATPase, present in two copies and possessing a defining ATP binding site motif, and two hydrophobic cytoplasmic membrane proteins. These latter proteins can be similar but distinct polypeptides, one large fusion protein, or a homodimer. In addition, iron transporters have a component exterior to the cytoplasmic membrane. For gram-negative cells, the component is a periplasmic binding protein, which recognizes the iron substrate and presents it to the cytoplasmic membrane complex. In gram-positive cells, the periplasmic protein is, in some cases, replaced by a lipoprotein tethered in the cytoplasmic membrane but protruding into the peptidoglycan region.

In summary, entry of iron into the cytoplasm of gram-negative bacteria requires an outer membrane receptor protein, a TonB system, and an ABC transporter. Passage through the outer and cytoplasmic membranes depends on the proton motive force and ATP, respectively. Outer membrane receptor proteins bind just one specific iron complex, TonB systems have broad specificity, and ABC transporters often recognize several iron-complexes, provided these are structurally related.

A. Siderophore-Mediated Systems

A major mechanism by which bacteria obtain iron is through the production and secretion of small (600–1000 Da), iron-chelating molecules termed siderophores. Many different siderophores have been isolated and characterized but they can generally be classified as being either of the catecholate or hydroxamate type (Fig. 2). Siderophores are synthesized in iron-deficient environments; after release from cells, siderophores bind iron and the siderophore-iron complex is subsequently internalized using the type of transport systems that will be described.

Siderophore biosynthesis is an interesting and growing area of research. Siderophores, generally, are built up of amino acids and hydroxy acids and, although they contain amide bonds, their synthesis does not involve ribosomes. Instead, a thiotemplate process with strong similarities to that employed for synthesis of certain peptide antibiotics is used. Non-ribosomal peptide synthetases link activated precursors to the enzyme-bound cofactor 4′-phosphopantetheine (P-pant). The thioesterified precursors are then covalently linked by amide bonds in the sequence determined by their position on the synthetase. The synthetase enzymes can be huge; in some cases, molecular weights of greater than 350,000 have been reported. Also, studies on siderophore biosynthesis resulted in the discovery of a new class of enzymes, the phosphopanthetheinyl transferases, which donate the P-pant to the peptide synthetases. This essential posttranslational modification of synthetases permits them to bind activated precursors and then join the precursors together to yield the final product.

Other molecules, which do not meet the rigorous definition of siderophores, can also provide iron to cells by means of specific outer membrane receptor proteins acting in concert with binding protein-dependent ABC transporters. Citrate, when present in environmental concentrations of greater than 0.1 mM, is one such molecule, as are dihydroxybenzoic acid, a precursor of the common siderophore enterobactin, and dihydroxybenzoylserine, a breakdown product of enterobactin (Fig. 2).

Release of iron from siderophores is not well understood. Ferrisiderophores enter the cytoplasm and free siderophores, or modified versions, are released back into the medium. Because siderophores (i) bind FeII less avidly than FeIII and (ii) the cytoplasm is a reducing environment, enzymatic reduction of iron is thought to be the release mechanism. *In vitro* experiments have demonstrated the presence of a variety of ferrisiderophore reductase activities that employ reduced flavins to convert iron to the ferrous state. These enzymes have broad specificity and their abundance is not affected by iron availability. The Fes protein of *E. coli* (ferrienterobactin esterase) is the one case where a specific iron-regulated protein is necessary for release of siderophore-bound iron. This enzyme hydrolyzes Ent but this esterase activity may be secondary to its primary reductase role.

Release of newly synthesized siderophores is similarly unclear. In only one case (*Mycobacterium smegmatis*) has a mutant been isolated that appears to be defective in siderophore secretion. The defective protein has no homologs in gene banks. There is indirect evidence that siderophore biosynthetic enzymes have an unusual intracellular cytoplasmic location. Osmotic shock, a procedure that releases periplasmic proteins, has been shown to release several enterobactin biosynthetic enzymes. This result could indicate a peripheral association with the cytoplasmic membrane, which has long been postulated for these enzymes.

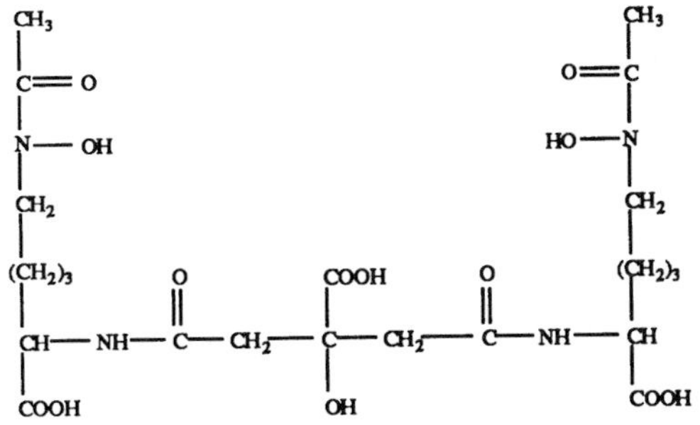

Aerobactin

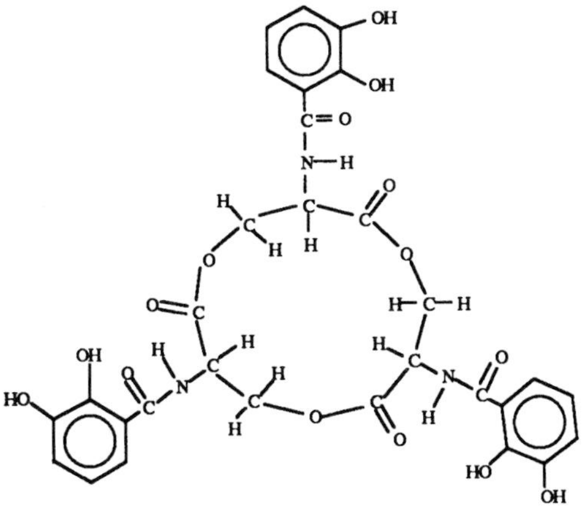

Enterobactin

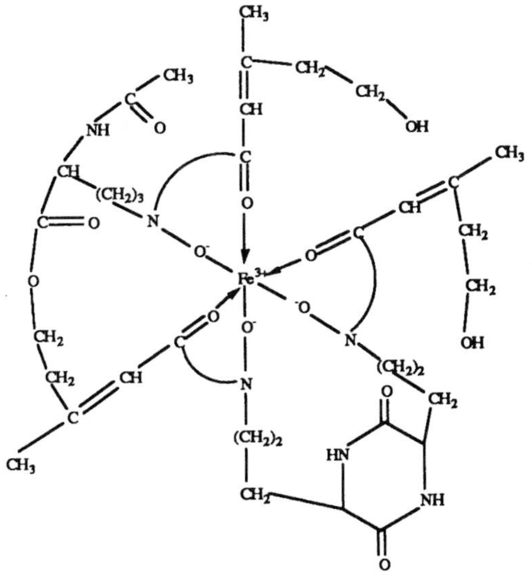

Coprogen

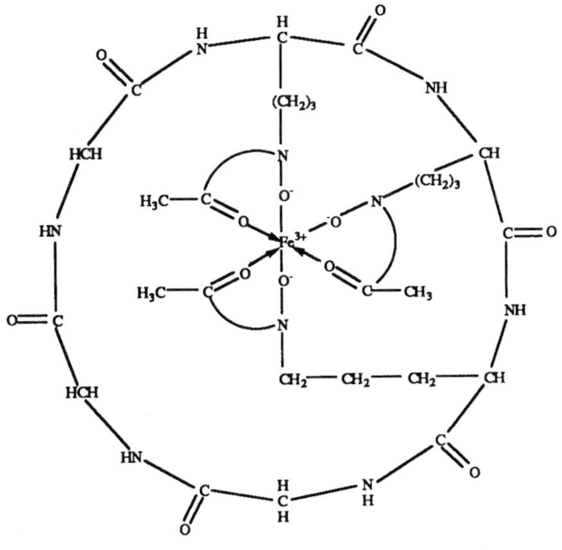

Ferrichrome

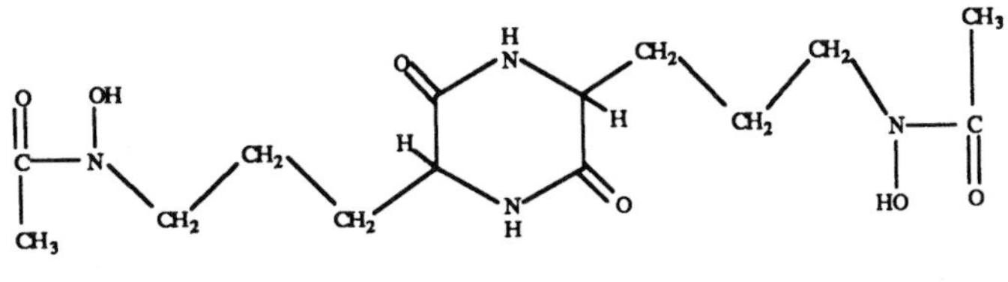

Rhodotorulic acid

Fig. 2. Some representative siderophores [modified from Earhart (1996) with permission]. Aerobactin and enterobactin are bacterial products; rhodotorulic acid and the nonferrated forms of ferrichrome and coprogen are synthesized by fungi. Enterobactin is a catechol-type siderophore and the other four are hydroxamate-type siderophores. Remarkably, *Escherichia coli* can utilize all of these.

B. Uptake of Ferrous and Ferric Ions

Ferrous iron transport is useful to microbes capable of inhabiting oxygen-restricted environments, such as swamps, intestines, and marshes, or acidic locales, where reduced iron is stable and soluble. An *E. coli* ferrous iron transport system (*feo*) has been identified; two proteins, FeoA and FeoB, participate in FeII uptake. FeoB is a large cytoplasmic membrane protein with a nucleotide binding site, suggesting that ATP hydrolysis is the energy source for transport. The FeoA protein is small (less than 10,000 Da) and of unknown function. *Salmonella typhimurium*, a facultative anaerobe like *E. coli*, has *feo* genes and *Methanococcus jannischii* has a homolog of *feoB*. The latter finding is of interest in that *M. jannischii* is classified in the domain Archaea, not in the domain Bacteria. Lastly, certain aerotolerant bacteria, such as the gram-positive organism *Streptococcus mutans*, obtain iron by using a reductase exposed on the cell exterior to convert surface-bound FeIII to FeII. A ferrous ion transporter then delivers the iron to the cytoplasm.

A ferric iron acquisition system of the ABC type is present in a number of gram-negative genera including *Serratia*, where it was discovered and termed Sfu type transport, *Haemophilus*, *Yersinia*, *Actinobacillus*, and *Neisseria*. A periplasmic binding protein accepts FeIII and presents it to the cytoplasmic components of the transporter, which internalize the iron. Ferric iron transporters can also function in concert with uptake systems that have outer membrane components. In these cases, they become the terminal portion of the assimilation path. Thus, iron is removed from transferrin and lactoferrin at the outer membrane (see following), a process which requires a receptor and active TonB system, and then translocated into the cytoplasm by the ferric transporter. Also, iron can be transferred from citrate to a periplasmic binding protein for entry into the cytoplasm.

C. Uptake of Iron from Heme

The vast majority of iron in animals is present intracellularly as heme (Hm). Heme, in turn, serves as a prosthetic group of proteins, primarily hemoglobin (Hb), but also myoglobin and Hm-containing proteins such as cytochromes. Cell lysis is necessary for release of these proteins. Freed Hb, such as that present following hemolysis of erythrocytes, is bound by the serum glycoprotein haptoglobin (Hp). Hb not bound by Hp can oxidize, in the process releasing Hm from globin. This extracellular Hm is bound by the plasma protein hemopexin and, less specifically, by serum albumin. Therefore, extracellular Hm, available as a possible source of iron (and Hm), occurs infrequently by itself as it is usually bound to Hb, Hb–Hp complexes, and hemopexin. In no case has a siderophore been able to scavenge iron from Hm. However, as will be described, each Hm source is capable of being utilized by one or another group of bacteria. These uptake systems vary greatly. An additional complication is that there is such variability among strains of a given species that it is often not possible to list the systems present in a given species. For instance, *Escherichia coli* lab strains cannot use either Hm or Hb as iron sources but pathogenic *E. coli* strains can often use both.

Iron assimilation pathways that recognize free Hm resemble those for iron-siderophore complexes; they require (i) a single, ligand-specific outer membrane receptor protein that is TonB-dependent, and (ii) for cytoplasmic membrane passage, an ABC transporter. For several of these systems, including those of *Yersinia enterocolitica*, *Vibrio cholerae*, and *Shigella dysenteriae*, it has been deduced that the entire heme molecule enters the cytoplasm. (In these cases, heme supported porphyrin growth requirements, as well as providing iron.) Little is known regarding the intracellular release of iron from Hm. HemS of *Y. enterocolitica* may provide this function, and HmnO of *Corynebacterium diphtheriae*, one of only a few gram-positive species able to assimilate Hm, may be a Hm oxygenase activity.

Several genera can remove Hm from Hb. *Neisseria* and *Haemophilus* spp. have TonB-dependent Hb-binding proteins in their outer membranes. Remarkably, both of these genera have additional TonB-dependent receptors that function in iron acquisition from Hb–Hp complexes. Each of these Hb–Hp receptors may consist of two different proteins. A different mechanism for the initial steps in obtaining iron from Hm or Hb is found in *Serratia marcescens*. This organism uses an ABC transporter to secrete a small protein (HasA) that functions as a hemophore (Hbp). That is, HasA is an extracellular Hm binding protein that is necessary for uptake of Hm, either free or bound to Hb. It shuttles its bound Hm to an outer membrane receptor (HasR). *E. coli* strains harboring the virulence plasmid pColV-K30 also secrete a Hbp. Unlike HasA, Hbp is autotransported out of the cell and is bifunctional. It has protease activity, degrading Hb as well as binding and transporting the released Hm. A chromosomally encoded outer membrane receptor protein (ChuA), required for the well-known *E. coli* pathogen O157:H7 to use Hb or Hm, may be the Hbp–Hm receptor.

Only *Haemophilus* strains are known to utilize Hm associated with hemopexin. The system is not well characterized but three genes are required and one appears to encode a large secreted Hbp (HxuA). HxuA binds Hm–hemopexin, removes the Hm, and carries Hm to an outer membrane receptor. The other two genes encode proteins concerned with the secretion of HxuA. Like all *Haemophilus* Hm transport systems (free Hm, Hb:Hp, Hm:albumin), the Hm–hemopexin system requires a functional TonB protein.

D. Acquisition of Iron from Transferrin and Lactoferrin

Transferrin (TF) and lactoferrin (LF) are extracellular iron transport molecules present in the fluids of many vertebrates; each of these related glycoproteins can bind two ferric ions. Many siderophores are capable of removing iron from these glycoproteins. *Neisseria* and *Haemophilus* produce no siderophores, however, and, instead, they have specific TonB-dependent outer membrane receptors that bind these transport proteins. An ABC transporter necessary for all nonheme iron uptake pathways (TF, LF, or iron chelates) is present in *Neisseria*; iron from these sources passes through the outer membrane and is bound by periplasmic protein FbpA, prior to entry into the cytoplasm. Outer membrane TF and LF receptors are unusual in that they appear to be bipartite; one protein has the characteristics of a typical TonB-dependent receptor, while the second is a surface-exposed lipoprotein. These receptors in *Neisse-*ria and *Haemophilus* show high specificity for glycoproteins of their normal hosts, such as humans.

E. Low Affinity Iron Transport

The high affinity iron transport systems described immediately above are generally not expressed in environments containing more than 5 to 10 μM iron. Remarkably, the means by which bacteria assimilate iron in such iron-replete environments, oxic or anoxic, is not known. This so-called low affinity iron uptake may first require that iron be in the FeII form. Ascorbic acid, which reduces FeIII, stimulates low affinity iron uptake. FeII could be assimilated either by (i) the repressed levels of the *feo* system proteins or (ii) a cytoplasmic membrane transporter with broad specificity for divalent cations, such as the CorA protein of *E. coli* and *Salmonella typhimurium*. Also, several types of small molecules (monocatecholates, α-keto acids, and α-hydroxy acids) can, in certain genera, function as siderophores; they could mobilize extracellular ferric iron and, using a variety of transporters, provide iron.

II. IRON-DEPENDENT REGULATION

Genes encoding proteins necessary for high affinity iron uptake are regulated by iron availability. The key regulatory protein in most bacteria, gram-positive and gram-negative, is Fur. Fur, a small, histidine-rich polypeptide, is an aporepressor; in the presence of its corepressor FeII, it binds DNA. The holorepressor binds operators termed iron boxes, which are AT rich 19 bp sequences with dyad symmetry (Table I). Iron boxes are located in promoters of iron-regulated genes, such that steric hindrance prevents the repressor and RNA polymerase from binding simultaneously. Adequate intracellular iron supplies thus prevent transcription of genes for transport proteins and siderophore biosynthetic enzymes. In *E. coli*, there are approximately 50 such Fur-regulated genes.

Negative control of genes by Fur does not completely explain the regulatory effects of iron. Although most iron-regulated genes are repressed by Fur in iron-replete conditions, some are positively controlled by Fur and some are induced by iron only in the absence of Fur. This complexity arises because Fur can influence the expression of certain regulatory systems and because some iron-regulated genes are also subject to other global control molecules. This latter observation becomes understandable when the key role of iron in metabolism is considered. For instance, iron requirements should depend in part on whether the fueling reactions being utilized require an electron transport system (respiration), with its requisite Hm and iron–sulfur proteins and propensity to generate toxic oxygen species, or not (fermentation).

Gram-positive organisms whose DNA has a high GC content, like *Corynebacterium* and *Streptomyces,* have no Fur but, instead, accomplish iron-regulated transcriptional control with DtxR proteins. The protein sequence of DtxR is unlike that of Fur but DtxR, nonetheless, functions in a Fur-like manner. FeII–DtxR binding sites are 19 bp palindromes (Table I) and are positioned in the promoters of genes they regulate.

Some iron transport systems are produced only in iron-deficient environments that contain their cognate ferrisiderophores. That is, the systems are synthesized only if the intracellular iron concentration, as sensed by Fur, is low and if their specific ferrisiderophore is bound to the cell. As established in the ferridicitrate system of *E. coli,* the specific outer membrane receptor must be present, as must a functional TonB system. Binding of ferrisiderophore to the receptor transmits a signal to a cytoplasmic membrane protein that has both periplasmic and cytoplasmic domains. The signal is then passed to a cytoplasmic sigma-factorlike protein that, upon activation, associates with RNA polymerase and stimulates transcription of the necessary transport genes. Unique regulatory elements for these systems include an outer membrane receptor capable of interacting with the cytoplasmic membrane protein, the cytoplasmic membrane protein, and its partner sigmalike factor. Regulation initiated at the cell surface is common among Pseudomonads, which have a large number of quite distinct transport systems.

Regulation of iron assimilation not only conserves bacterial carbon and energy resources but also prevents overaccumulation of iron. Iron overload is a legitimate concern in bacteria as excess free iron can promote the formation of the toxic superoxide (O_2^-) and hydroxyl (OH ·) radicals. The significance of DNA damage brought about by iron-generated hydroxyl radicals is demonstrated by the finding that *E. coli* Fur⁻ mutants, which assimilate too much iron, that are also defective in DNA repair, cannot survive in oxic environments.

The intracellular oxidation state of iron is also used for regulation. Several proteins that control genes whose functions are related to oxygen levels contain iron–sulfur clusters; the iron present in these clusters is used as a sensing device. The *E. coli* Fnr protein, a transcriptional activator needed for anaerobic growth, is active only when its Fe–S center is reduced. In contrast, the sensor protein for superoxide stress, SoxR, is a positive regulator that is active with an oxidized Fe–S cluster.

III. INTRACELLULAR "FREE" AND STORED IRON

Two types of iron storage proteins, ferritin (FTN) and bacterioferritin (BFR), are found in bacteria. BFRs are a subfamily of FTNs, distinguished by the fact that they contain heme units. Like eukaryotic ferritins, the bacterial proteins are large (ca. 500,000 Da) and composed of 24 subunits.

The actual role of these proteins is uncertain. Unlike eukaryotic FTNs, which can accommodate over 4000 iron atoms, maximum iron contents for FTN and BFR are much less. BFR is synthesized primarily during periods of slow or no growth and may serve as an iron donor upon resumption of growth. FTN, on the other hand, is present at a constant low level throughout the growth cycle and, for at least *Escherichia coli* and *Campylobacter jejuni,* has been shown to protect against iron-catalyzed oxidative damage.

A major uncertainty regarding bacterial iron metabolism is the distribution and form of much of the intracellular iron. Even the quantity of iron in a bacterium is unclear; for the well-studied organism *E. coli,* estimates of the number of iron atoms/cell range from 100,000 to 750,000. In a variety of bacteria, Hm-containing proteins account for less than 10% of the total iron, in contrast to the situation in animals. Iron–sulfur proteins also contain 10% or less of the bacterial iron, as do FTN and BFR when cells grown under iron-deficient conditions are examined. The remainder of the bacterial iron, the majority, exists in a mobile pool about which little is known. It is this mobile iron that presumably is responsible for iron regulation and for the toxic effects of iron overload. Whether or not iron passes through this pool before being placed by ferrochelatase into protoporphyrin *IX* to form Hm, before incorporation into Fe–S clusters of proteins by unknown means, or before being converted to FeIII by the ferroxidase activity of BFR and FTN and deposited in the core of these molecules is unknown.

IV. IRON IN PRIMARY FUELING REACTIONS

Some bacteria can use the oxidation of iron compounds as their primary energy source. Bacteria capable of using inorganic, rather than organic, molecules for their fueling reactions are termed chemolithotrophs and iron-oxidizing bacteria are a major group in this nutritional category. Iron-oxidizing bacteria typically live in acidic, aerobic environments rich in both reduced iron and sulfur compounds; they grow poorly at pH values greater than 4. Among other things, low pH is critical in keeping FeII from being spontaneously oxidized to FeIII.

Thiobacillus ferrooxidans, the best studied microbe in this group, oxidizes iron using proteins in the cell envelope. An outer membrane complex oxidizes iron to the ferric form, and a periplasmic protein transfers the electrons to cytochromes in the cytoplasmic membrane; these, in turn, pass the electrons to oxygen, the ultimate electron acceptor, in the cytoplasm.

T. ferrooxidans is a major cause of water pollution, specifically, acid mine drainage. Ferrous sulfide is common in many coal and ore sites and its chemical and bacterial oxidation leads to both acidification and addition of dissolved metals to the water. Downstream, as the pH of the mine water becomes less acidic, insoluble ferric precipitates form.

In contrast to the relatively few bacteria which can oxidize iron as the initial step in generating energy, in the absence of oxygen, many bacteria can use FeIII as the ultimate electron acceptor in fueling reactions.

This is an example of anaerobic respiration and, for iron, has the geochemical significance of solubilizing iron.

V. IRON AND PATHOGENICITY

A major stimulus to current studies on microbial iron metabolism is their medical significance. A brief review of the genera studied in Section I will emphasize this fact.

Bacterial pathogens find themselves in an iron-deficient environment when they invade a eukaryotic host; conditions facilitating their acquisition of iron would be expected to increase virulence. Several general observations support this idea. Humans with higher than normal iron levels show enhanced susceptibility to bacterial infections. Similarly, at least 20 bacterial pathogens are more virulent when their animal host is injected with iron compounds prior to infection. Because excess iron can have a number of effects, including some that are detrimental to host immune defenses, these data are somewhat ambiguous. They are bolstered, however, by *in vitro* experiments showing that the antibacterial effects of body fluids can be uniquely reversed by iron supplementation.

Bacteria must synthesize an appropriate iron uptake system to overcome the bacteriostatic conditions in their hosts. Pathogens which multiply extracellularly, such as in blood or on mucosal surfaces, and which do not lyse host cells must be able to obtain iron from TF or LF. This can be accomplished by siderophores or by TF- or LF-specific receptor proteins. Septicemic bacteria with defective siderophore systems are, in fact, less virulent. On the other hand, the major source of iron for intracellular pathogens is Hm and siderophore-deficient mutants of these organisms are still virulent. The required iron is taken from Hm-compounds, in this case.

Remarkably, it has been difficult to identify specific iron uptake systems that are essential for virulence. The multiplicity of means for assimilating iron in every species studied complicates such attempts. However, when all high affinity systems are inactivated by use of mutations in *tonB, Salmonella typhimurium, Vibrio cholerae,* and *Haemophilus influenzae* are avirulent. That a specific iron assimilation pathway can be a virulence factor has been demonstrated most clearly in certain strains of *Vibrio anguillarum*. This marine microbe is a pathogen of salmonid fish, such as salmon; approximately 10 bacteria per fish are sufficient to cause vibriosis, a fatal disease characterized by hemorrhagic septicemia. A plasmid-mediated iron-uptake system that uses the siderophore anguibactin was implicated in virulence. An outer membrane receptor protein (FatA) and an ABC transporter (FatB-D) for anguibactin are encoded by the plasmid, as are genes which positively control synthesis of the Fat transport proteins and siderophore biosynthesis. Plasmid mutations which abolish transport protein production or anguibactin synthesis greatly reduce virulence, as does loss of the plasmid from the cells.

Last, many pathogens utilize the low iron concentrations in the host as an environmental signal to synthesize virulence factors. Fur and DtxR-like proteins control not only iron acquisition systems but synthesis of toxins and hemolysins in these organisms.

See Also the Following Articles

ABC TRANSPORT • ORE LEACHING BY MICROBES

Bibliography

Braun, V., Hantke, K., and Köster, W. (1998). Bacterial iron transport: Mechanisms, genetics, and regulation. *In* "Metal Ions in Biological Systems," Vol. 35, "Iron Transport and Storage in Microorganisms, Plants, and Animals" (A. Sigel and H. Sigel, eds.), pp. 67–145. Marcel Dekker, Inc., New York.

Byers, B. R., and Arceneaux, J. E. L. (1998). Microbial iron transport: Iron acquisition by pathogenic microorganisms. *In* "Metal Ions in Biological Systems," Vol. 35, "Iron Transport and Storage in Microorganisms, Plants, and Animals" (A. Sigel and H. Sigel, eds.), pp. 37–66. Marcel Dekker, Inc., New York.

Crosa, J. H. (1997). Signal transduction and transcriptional and posttranscriptional control of iron-regulated genes in bacteria. *Microbiol. Mol. Biol. Rev.* **61**, 319–336.

Earhart, C. F. (1996). Uptake and metabolism of iron and molybdenum. *In* "*Escherichia coli* and *Salmonella*: Cellular and molecular biology," Vol. 1 (2nd ed.) (F. C. Neidhardt, R. Curtiss, III, J. L. Ingraham, E. C. C. Lin, K. B. Low, B. Magasanik, W. S. Reznikoff, M. Riley, M. Schaechter, and H. E. Umbarger, eds.), pp. 1075–1090. ASM Press, Washington, DC.

Expert, D., Enard, C., and Masclaux, C. (1996). The role of iron in plant host–pathogen interactions. *Trends Microbiol.* **4**, 232–237.

Guerinot, M. L. (1994). Microbial iron transport. *Annu. Rev. Microbiol.* **48**, 743–772.

Leong, S. A., and Winkelmann, G. (1998). Molecular biology of iron transport in fungi. *In* "Metal Ions in Biological Systems," Vol. 35, "Iron Transport and Storage in Microorganisms, Plants, and Animals" (A. Sigel and H. Sigel, eds.), pp. 147–186. Marcel Dekker, Inc., New York.

Neilands, J. B. (1995). Siderophores: Structure and function of microbial iron transport compounds. *J. Biol. Chem.* **270**, 26723–26726.

Otto, B. R., Verweij-van Voght, A. M. J. J., and MacLaren, D. M. (1992). Transferrins and heme-compounds as iron sources for pathogenic bacteria. *Crit. Revs. Microbiol.* **18**, 217–233.

Payne, S. M. (1988). Iron and virulence in the family enterobacteriaceae. *Crit. Revs. Microbiol.* **16**, 81–111.

Winkelmann, G., van der Helm, D., and Neilands, J. B. (eds.). (1987). "Iron Transport in Microbes, Plants and Animals." VCH Press, Weinheim, Germany.

ISBN 0-12-226802-4

90038